Quarks and Nuclear Physics
QNP 2006

Società Italiana
di Fisica

Springer

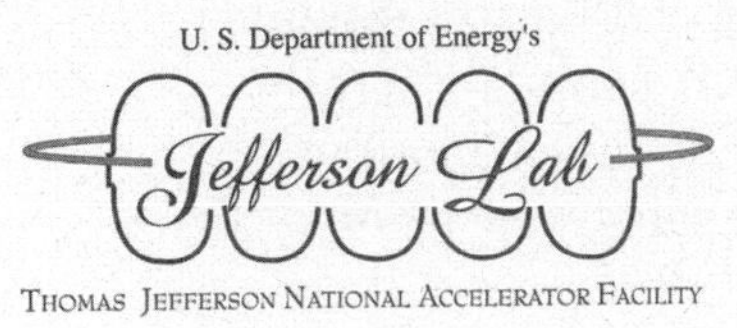

The IVth International Conference on Quarks and Nuclear Physics

QNP 2006

June 5–10, 2006

Madrid, Spain

edited by

A. Dobado, F.J. Llanes-Estrada and V. Vento

Prof. Antonio Dobado Gonzalez
Prof. Felipe J. Llanes-Estrada
Departamento de Física Teórica I
Universidad Complutense de Madrid
28040 Madrid, Spain

Prof. Vicente Vento
Departamento de Física Teórica
Universidad de Valencia
46100 Burjassot (Valencia), Spain

A part of the articles in this book originally appeared in the journal
The European Physical Journal A − Hadrons and Nuclei
Volume 31, Number 4
ISSN 1434-6001
© SIF and Springer-Verlag Berlin Heidelberg 2007

ISBN-10 3-540-72515-6 Springer Berlin Heidelberg New York

ISBN-13 978-3-540-72515-2 Springer Berlin Heidelberg New York

Library of Congress Control Number: 2007926261

Springer is a part of Springer Science+Business Media
springer.com

The use of general descriptive names, registered names, trademarks, etc. in this publication does not imply, even in the absence of a specific statement, that such names are exempt from the relevant protective laws and regulations and therefore free for general use.

Typesetting and Cover design: SIF Production Office, Bologna, Italy
Printing and Binding: Tipografia Compositori, Bologna, Italy

Printed on acid-free paper   SPIN: 11891543 − 5 4 3 2 1 0

QNP 2006 Organization

International **Advisory** **Committee**	Peter Barnes	Los Alamos
	Nora Brambilla	Milano
	Stanley J. Brodsky	SLAC, Stanford
	Stephen R. Cotanch	NCSU, Raleigh
	Sergei Denisov	IHEP, Serpukhov
	Dmitri Diakonov	NORDITA, Copenhagen
	John Domingo	Jefferson Lab, Newport News
	Alex Dzierba	Indiana U.
	Torleif Ericson	CERN, Genève and Uppsala
	Amand Faessler	Tübingen
	Lidia Ferreira	IST, Lisbon
	Ettore Fiorini	Milano
	Avraham Gal	Hebrew U.
	Michel Garçon	Saclay
	Don Geesaman	Argonne
	Klaus Goeke	Bochum
	Roy Holt	Argonne
	Harold Jackson	Argonne
	Robert Jaffe	MIT, Cambridge, USA
	Alan D. Krisch	Michigan U.
	Ulf G. Meißner	U. Bonn-FZ Jülich
	Eduardo de Rafael	Marseille
	Jose Emilio Ribeiro	IST, Lisbon
	Mark Strikman	Pennsylvania U.
	Toru Sugitate	Hiroshima
	Adam P. Szczepaniak	Indiana U.
	Anthony W. Thomas	Jefferson Lab, Newport News
	Hiroshi Toki	Osaka
	Jochen Wambach	Darmstadt
Local **Advisory** **Committee**	Ramon Fernández Alvarez-Estrada	U. Complutense, Madrid
	Jose María Gómez Gómez	U. Complutense, Madrid
	Elvira Moya de Guerra	U. Complutense, Madrid
	Antonio González Arroyo	U. Autonoma, Madrid
	Juan Terrón	U. Autonoma, Madrid
	Domenec Espríu	Barcelona
	Cristina Manuel	Barcelona
	Francisco Fernández	Salamanca
	Eulogio Oset	Valencia
	Antonio Pich	Valencia
	Vicente Vento	Valencia
	Carlos Pajares	Santiago
Scientific **Convenors**	Pedro Bicudo	IST, Lisbon
	Elena Ferreiro	Santiago
	Angel Gomez Nicola	U. Complutense, Madrid
	Jose Ramón Peláez	U. Complutense, Madrid
	Jose M. Udías Moinelo	U. Complutense, Madrid
	Antonio Pineda	Barcelona
	Stefan Sint	U. Autonoma, Madrid

List of participants

Guests from foreign institutions

Ricardo Alarcon
Arizona State U.

Konrad Aniol
California State U.

Alejandro Arrizabalaga
NIKHEF, Amsterdam

Oliver Baer
Humboldt U., Berlin

Ian Balitsky
Jefferson Lab, Newport News

Peter Barnes
Los Alamos National Laboratory

Marco Battaglieri
INFN, Genova

Cesare Bini
U. "La Sapienza", Roma and
INFN, Roma

Ignazio Bombaci
U. Pisa

Tomas Brauner
Charles U., Prague

Stanley J. Brodsky
SLAC, Stanford

Fabien Buisseret
U. Mons-Hainault

Daniel Cabrera
Texas A&M

Tim Van Cauteren
U. Gent

Gianluigi Cibinetto
INFN, Ferrara

Ian Cloet
Jefferson Lab, Newport News

Gilberto Colangelo
U. Bern

Alberto Correa dos Reis
Centro Brasileiro de Pesquisas Fisicas
Rio de Janeiro

Stephen R. Cotanch
NCSU, Raleigh

Catalina Oana Curceanu
INFN, Frascati

Annalisa d'Angelo
U. "Tor Vergata", Roma

Donal Day
Virginia U.

Wouter Deconinck
U. Michigan, Ann Arbor

T. William Donnelly
MIT, Cambridge, USA

David d'Enterria
CERN, Genève

Wolfgang Eyrich
U. Erlangen

Amand Faessler
U. Tübingen

Christian S. Fischer
GSI, Darmstadt

Frederic Fleuret
LLR-Ecole Polytechnique, Palaiseau

Justin Foley
Swansea U.

Phillippe de Forcrand
ETH, Zürich

Justin Frantz
Stony Brook U.

Jan Friedrich
T.U. München

Ingo Froehlich
U. Frankfurt

Nicolas Garron
DESY, Zeuthen

Paola Gianotti
INFN, Frascati

Albrecht Gillitzer
FZ, Jülich

Sergio Giudici
U. Pisa

Daniel Gomez Dumm
U. Nacional de la Plata

Thomas Gutsche
U. Tübingen

Simon Hands
Swansea U.

Christoph Hanhart
FZ, Jülich

Urs M. Heller
American Physical Society
Ridge, New York and
Brookhaven National Laboratory

Harold Jackson
Argonne National Laboratory

Karl Jansen
NIC, DESY, Zeuthen

Peter Janssens
U. Gent

Xiaobin Ji
BES, Beijing

Sonja Kabana
U. Nantes

Gabriel Karl
U. Guelph

Roman Kezerashvili
NY City College of Technology

Daniil Kirilov
FZ, Jülich

Yoshiaki Koma
DESY, Hamburg

Mikhail Kopytin
DESY, Zeuthen

Alexander Korzenev
U. Mainz

Alexander Kozlov
Weizmann Institute of Science

Boris Krippa
U. Manchester

Alan D. Krisch
Michigan U.

Bernd Krusche
U. Basel

Yoshinobu Kuramashi
U. Tsukuba

Olga Lakhina
U. Pittsburgh

Michael Leitch
Los Alamos National Laboratory

Randy Lewis
Regina U.

Hongbo Liao
BES, Beijing

Javier Lopez Albacete
Ohio State U.

Valery V. Lyuboshitz
JINR, Dubna

Aneesh Manohar
U. California, San Diego

Vincent Mathieu
U. Mons-Hainault

Takayuki Matsuki
U. Tokyo

Erin McDermott
Kettering U., Michigan

Larry McLerran
Brookhaven National Laboratory

Ulf G. Meißner
U. Bonn

Zein-Eddine Meziani
Temple U.

Chris Michael
Liverpool U.

Gerald Miller
U. Washington

Stanislaw Mrowczinsky
Institute of Nuclear Studies, Warsaw

Ulrich Mueller
U. Mainz

Jeffrey Mitchell
Brookhaven National Laboratory

Takashi Nakano
RCNP, U. Osaka

Ilia Narodetskiy
ITEP, Moscow

Silvia Niccolai
IPN, Orsay

Kostas Orginos
Jefferson Lab, Newport News

Simone Pacetti
INFN, Frascati

Carlos Pena Ruano
CERN, Genève

Owe Philipsen
U. Münster

Klaus Rabbertz
U. Karlsruhe

Brian Raue
Florida International U.

Paul Reimer
Argonne National Laboratory

Avraham Rinat
Weizmann Institute of Science

Craig Roberts
Argonne National Laboratory

Sinead Ryan
Trinity College, Dublin

Jan van Ryckebusch
U. Gent

Marius Sadzikowski
Jagellonian U., Krakow

Bijan Saghai
CEA-Saclay

Koichi Saito
Tokyo U. of Science

Juan Jose Sanz Cillero
IPN, Orsay

David Saxon
U. Glasgow

Shinya Sawada
KEK, Tsukuba

Susan Schadmand
IKP, Jülich

Thomas Schaefer
NCSU, Raleigh

Carlo Schaerf
U. "Tor Vergata", Roma

Carlos Schat
U. Buenos Aires

Gunar Schnell
U. Gent

Wolfgang Schroeder
U. Erlangen-Nueremberg

Wolfram Schroers
DESY, Zeuthen

Norberto Scoccola
CNEA, Buenos Aires

Olga Shekhovtsova
INFN, Frascati

Fernando Silveira Navarra
U. Sao Paulo

Bernard Silvestre-Brac
LPSC, Grenoble

Silvano Simula
U. "Roma III", Roma

Jonivar Skullerud
Trinity College, Dublin

Rainer Sommer
DESY, Zeuthen

Feliciano de Soto
LPSC, Grenoble

Brijesh Srivastava
Purdue U., Indiana

Paul Stoler
Rennselaer Polytechnic Institute

Eric S. Swanson
U. Pittsburgh

Adam P. Szczepaniak
Indiana U.

Chung i Tan
Brown U.

Kunihiko Terasaki
Kanazawa U.

Andreas Thomas
U. Mainz

Anthony W. Thomas
Jefferson Lab, Newport News

Peter Tandy
Kent State U.

Laura Tolos
GSI, Darmstadt

Silvano Tosi
INFN, Genova

Michael Tytgat
U. Gent

Antonio Vairo
U. Milano

Raffaella de Vita
INFN, Genova

Michele Viviani
U. Pisa

Daniel Watts
Edinburgh U.

Dennis Weygand
Jefferson Lab, Newport News

Herbert Weigel
U. Siegen

Christian Weiss
Jefferson Lab, Newport News

Michael Wood
U. Massachussets, Amherst

Weilin Yu
U. Giessen

Adriano Zallo
INFN, Frascati

James Zanotti
U. Edinburgh

Participants from institutions in the Iberian peninsula (Spain and Portugal)

Bernardo Adeva
U. Santiago

Deborah Aguilera
U. Alicante

Conrado Albertus Torres
U. Granada

Nestor Armesto
U. Santiago Compostela

Eef van Beveren
U. Coimbra

Pedro Bicudo de Almeida
IST, Lisbon

Francisco Cao
U. Complutense, Madrid

Pedro Costa
U. Coimbra

Aurore Courtoy
U. Valencia

Leticia Cunqueiro
U. Santiago

Antonio Dobado Gonzalez
U. Complutense, Madrid

Michael Doering
U. Valencia

Rafel Escribano
U. Autonoma, Barcelona

Francisco Fernandez
U. Salamanca

Ramon Fernandez Alvarez-Estrada
U. Complutense, Madrid

Daniel Fernandez Fraile
U. Complutense, Madrid

Cesar Fernandez Ramirez
CSIC and U. Complutense, Madrid

Carmen Garcia Recio
U. Granada

Ruben Garcia Marin
U. Complutense, Madrid

Jose Maria Gomez Gomez
U. Complutense, Madrid

Angel Gomez Nicola
U. Complutense, Madrid

Pedro Gonzalez
U. Valencia

Murat Kaskulov
U. Valencia

Felipe J. Llanes-Estrada
U. Complutense, Madrid

Volodymyr Magas
U. Barcelona

Cristina Manuel
CSIC, Barcelona

Eugenio Megias
U. Granada

Rita Alexandra Monteiro
U. Coimbra

Andre Luiz Mota
U. Granada/U. S. Joao del Rei

Elvira Moya de Guerra
U. Complutense and CSIC, Madrid

Juan Nieves
U. Granada

Santiago Noguera
U. Valencia

Orlando Oliveira
U. Coimbra

Jose A. Oller
U. Murcia

Eulogio Oset
U. Valencia

Carlos Pajares
U. Santiago

Joannis Papavassiliou
U. Valencia

Jose R. Pelaez Sagredo
U. Complutense, Madrid

Antonio Pineda
U. Barcelona

Angels Ramos
U. Barcelona

Luis Roca
U. Murcia

David Rodriguez Entem
U. Salamanca

Jose Rodriguez Quintero
U. Huelva

Maria Ruivo
U. Coimbra

Enrique Ruiz Arriola
U. Granada

George Rupp
IST, Lisbon

Kenji Sasaki
U. Valencia

Stefan Sint
U. Autonoma, Madrid

Joan Soto
U. Barcelona

Kazuo Tsushima
U. Salamanca

Jose M. Udias Moinelo
U. Complutense, Madrid

Lourdes del Valle Tabares
U. Complutense, Madrid

Jose Maria Verde Velasco
U. Salamanca

Vicente Vento
U. Valencia

Manuel Vicente Vacas
U. Valencia

Javier Vijande
U. Salamanca

Francisco Yndurain
U. Autonoma, Madrid

Contents

Contributions marked with an asterisk have appeared in Eur. Phys. J. A, Vol. **31**, no. 4 (2007).

■ Baryon Physics

■ Chiral Dynamics

■ Matter under Extreme Conditions

Preface

The *IVth International Conference on Quarks and Nuclear Physics* convened in Madrid between June 5th and 10th, 2006. The meeting was attended by more than two hundred participants and over hundred presentations took place. These proceedings are an attempt to capture for the posterity the intense scientific exchange that took place during those days. The presentations were selected by the meeting convenors in collaboration with the advisory committees. A selection of refereed contributions of these proceedings appeared in Eur. Phys. J. A, Vol. **31**, issue no. 4 (2007).

The scientific content of the conference was broad, covering topics in Nuclear Structure and Interactions, Electron Scattering, Nuclear Astrophysics, Field Theory and Lattice, Quark Models, Chiral Lagrangians, etc, and of particular great interest were the presentations by most of the experimental collaborations in the field. Precisely this was the richness of the meeting, the possibility to attend talks in the same field of physics but with diverse approaches.

The meeting also covered topics relatively new to Nuclear Physics, such as Heavy Quark Physics, that is to play an important role in the future FAIR experimental program, and had a session dedicated to the 50th anniversary of the discovery of the Proton Form Factor in 1956.

We have gathered in this volume over a hundred contributions that we hope will convey to you, the practicing physicist, the excitement that meeting attendees felt. The page limits were 3 pages for research contributions and 6 pages for reviews. We have been flexible with this rule as the referees sometimes required clarification or expansion of the ideas presented.

We have classified the contributions according to a few main lines which, in some way, respects the classification of the various sessions of the conference as approved by the International Advisory Committee.

We would like to thank our endorsing and sponsoring organizations, the American Physical Society (through its topical group in Hadron Physics), and the Real Sociedad Española de Física, the Thomas Jefferson National Accelerator Facility, the Ministerio de Educación y Ciencia of Spain through grants FPA 2004-02602, 2005-02327 and FPA 2005-23849-E, and the host institution, the Universidad Complutense de Madrid, through a grant from Vicerrectorado de Investigacion, and financing the seminar Matter under Extreme Conditions through the Vicerrectorado de Relaciones Internacionales.

We are indebted to many people for their enthusiastic support and collaboration, especially the conference convenors, the members of the Advisory Committees, the Dean of Physics at the Universidad Complutense, José María Gómez, Anthony Thomas from Jefferson Lab, and Craig Roberts from APS. We would also like to thank our support staff Chon, Alvaro, Carlos, Guillermo, Javier, Juan, Lourdes, Rubén, for their help.

We wish this volume becomes useful for the future readers, both graduate students and practicing physicists. We hope, for those who attended the meeting, that it will bring back some good memories of a very successful encounter and the days spent in the beautiful and lively city of Madrid. Finally, we would like to thank all participants for making *QNP 2006* a reference in Nuclear and Particle Physics.

Vicente Vento

Editor of EPJ A

Antonio Dobado
Felipe J. Llanes-Estrada

The Editors

Madrid and Valencia, January 2007

Eur. Phys. J. A **31**, 397–402 (2007)

DOI 10.1140/epja/i2006-10170-1

THE EUROPEAN
PHYSICAL JOURNAL A

Special Article – QNP 2006

Modern theory of nuclear forces A.D. 2006

U.-G. Meißner[a]

HISKP (Th), Universität Bonn, D-53115 Bonn, Germany and
IKP, FZ Jülich, D-52425 Jülich, Germany

Received: 24 September 2006
Published online: 16 February 2007 – © Società Italiana di Fisica / Springer-Verlag 2007

Abstract. I present and discuss recent results on nuclear forces and few-nucleon systems obtained in the framework of chiral effective nuclear field theory.

PACS. 13.75.Gx Pion-baryon interactions – 25.30.Rw Electroproduction reactions – 12.39.Fe Chiral Lagrangians

1 Introduction

One of the most challenging problems of strong QCD is the derivation of nuclear forces. The underlying QCD fields, quarks and gluons, are confined within hadrons, therefore nuclear forces are the residual forces between colorless objects, much like the van der Waals forces in molecular physics. Furthermore, typical energy scales in nuclear physics correspond to a low-resolution microscope, *e.g.* producing a neutral pion at threshold by a real photon requires a photon laboratory energy of about 150 MeV. Stated differently, nuclei are made of protons and neutrons plus virtual mesons —their QCD substructure is effectively masked. Since the nuclear binding energies are much smaller than the nuclear masses, we essentially have to deal with a non-relativistic problem. It appears thus appropriate to analyze the nuclear A-body problem by solving the Schrödinger equation

$$H\Psi_A = E_A\Psi_A, \qquad H = T + V = \sum_A \frac{p_A^2}{2m_{\mathrm{N}}} + V, \quad (1)$$

where the potential V is a string of terms, $V = V_{\mathrm{NN}} + V_{3\mathrm{N}} + V_{4\mathrm{N}} + \dots$. Traditionally, the two-nucleon potential V_{NN} is reconstructed from the large body of pp and np scattering data to a high accuracy. However, the two-nucleon forces alone do not give the proper nuclear binding energies and level schemes —a small three-nucleon force (TNF) is needed to cure this problem. Making an ansatz for such a TNF with a few adjustable parameters, the pattern of binding energies and excited states for nuclei up to $A \simeq 12$ based on *ab initio* Monte Carlo simulations is amazingly well described (for a recent status report, see *e.g.* [1]). However, there are important open problems: 1) Why is there this hierarchy $V_{2\mathrm{N}} \gg V_{3\mathrm{N}} \gg V_{4\mathrm{N}}$? 2) Gauge and chiral symmetries are difficult to include

and 3) what is the connection to QCD? As will be discussed in what follows, chiral effective field theory (EFT) offers an approach that a) is linked to QCD via its symmetries; b) allows for systematic calculations with a controlled theoretical error; c) explains the observed hierarchy of the nuclear forces and gives consistent two-, three-, and four-body forces; d) matches nucleon structure to nuclear dynamics; e) offers the possibility of a consistent inclusion of strange quarks (hyper-nuclear physics); f) allows for a lattice formulation/chiral extrapolations; and g) puts nuclear physics on a sound basis.

2 Effective field theory for nuclear forces

In this section, I briefly discuss the formulation of the EFT for few-baryon interactions (NN, NNN, YN, ...). As discussed before, the underlying fields are ground-state baryons and the octet of Goldstone bosons of QCD. The starting point is the chiral effective Lagrangian (for the moment, I restrict myself to the two-flavor case of pions and nucleons),

$$\mathcal{L}_{\mathrm{EFF}} = \mathcal{L}_{\pi\pi} + \mathcal{L}_{\pi\mathrm{N}} + \mathcal{L}_{\mathrm{NN}} + \dots, \quad (2)$$

which allows for a systematic expansion in powers of Q/Λ_χ and M_π/Λ_χ, where Q denotes any external soft scale, M_π is the pion mass related to explicit chiral symmetry breaking and $\Lambda_\chi \simeq 1\,\mathrm{GeV}$ is the hard scale related to spontaneous chiral symmetry breaking. The pion and pion-nucleon sectors are perturbative in Q, the corresponding EFT is the chiral perturbation theory (CHPT). The parameters in $\mathcal{L}_{\pi\pi}$ and $\mathcal{L}_{\pi\mathrm{N}}$ are known from CHPT studies, these are the so-called low-energy constants (LECs). Matters are more difficult/interesting for systems with two or more nucleons. The small nuclear binding energies (or large S-wave scattering lengths) require a non-perturbative resummation to have a useful approach up

[a] e-mail: `meissner@itkp.uni-bonn.de`

to the pion production threshold. Following Weinberg [2], the organization of the theory takes place on the level of the effective potential V_{eff} which is then injected into a regularized Lippmann-Schwinger equation to generate the bound and scattering states. The various terms in the effective potential are ordered according to

$$V_{\text{eff}} \equiv V_{\text{eff}}(Q, g, \mu) = \sum_{\nu} Q^{\nu} \, \mathcal{V}_{\nu}(Q/\mu, g) \,, \qquad (3)$$

where Q is the soft scale (either a baryon three-momentum, a Goldstone boson four-momentum or a Goldstone boson mass), g is a generic symbol for the pertinent low-energy constants, μ a regularization scale, $\mathcal{V}_{\nu}$ is a function of order one (naturalness), and $\nu \geq 0$ is the chiral power. It can be expressed as (for connected diagrams)

$$\nu = 2 - B + 2L + \sum_{i} v_i \, \Delta_i \,, \quad \Delta_i = d_i + \frac{1}{2} b_i - 2 \,, \qquad (4)$$

with B the number of incoming (outgoing) baryon fields, L counts the number of Goldstone boson loops, and v_i is the number of vertices with dimension Δ_i. The vertex dimension is expressed in terms of derivatives (or Goldstone boson masses) d_i and the number of internal baryon fields b_i at the vertex under consideration. The leading-order (LO) potential is given by $\nu = 0$, with $B = 2$, $L = 0$ and $\Delta_i = 0$. Using eq. (4) it is easy to see that this latter condition is fulfilled for two types of interactions: a) non-derivative four-baryon contact terms with $b_i = 4$ and $d_i = 0$ and b) one-meson exchange diagrams with the leading meson-baryon derivative vertices allowed by chiral symmetry ($b_i = 2, d_i = 1$). At next-to-leading order (NLO), one encounters the first contribution from two-pion exchange and so on. Also, three- (four-) nucleon forces first appear at NN(N)LO. This parametrical suppression explains naturally the observed hierarchy of nuclear forces. The potential requires further regularization when injected into the Lippmann-Schwinger equation, here the so-called spectral function regularization has become the method of choice to further separate the short- and the long-distance physics (for details, see [3]). For further details on nuclear EFT, I refer to the recent and comprehensive review by Epelbaum [4].

3 The forces between two nucleons

The two-nucleon forces have been analyzed to a high precision, more precisely to N^{3}LO [5,6]. The one- and two-pion exchanges are given in terms of LECs from the pion-nucleon Lagrangian, these have been previously determined in studies of pion-nucleon scattering and pion production in pion-nucleon collisions (for an update on the dimension two LECs, see [7] and some of the third-order LECs were determined in refs. [8,9]). Three-pion exchange also appears at N^{3}LO, but it has been shown to contribute negligibly [10]. This information from πN scattering constitutes the direct link to QCD via its symmetries and their realizations. The four-nucleon couplings must

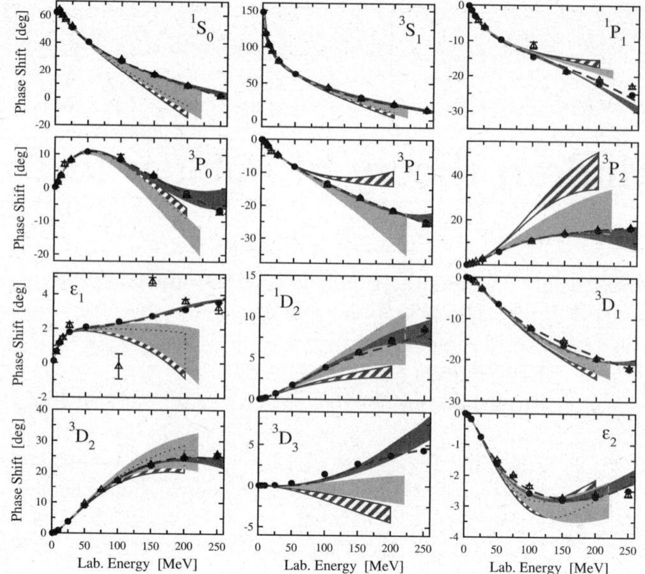

Fig. 1. Low np phase shifts as a function of the nucleon laboratory energy. The hatched, light-shaded, and dark-shaded bands denote the NLO, NNLO and N^{3}LO results, respectively. The N^{3}LO results from ref. [5] are shown by the dashed lines. Also shown are the results from the Nijmegen [11] (filled circles) and the Virginia Tech [12] (open triangles) partial-wave analyses.

be determined from a fit to the low NN phases (obtained, *e.g.*, from the Nijmegen partial-wave analysis). At LO, N(N)LO and N^{3}LO, there are 2, 7, and 15 four-nucleon independent couplings (note that at NNLO one has no new contact terms due to parity). In addition, one has to account for the effects of electromagnetism and other strong isospin breaking effects in the pp, np and nn channels. This machinery has also been developed in the past years, see *e.g.* the review [4]. The resulting description of the S-, P- and D-waves of np scattering at NLO, NNLO and N^{3}LO is shown in fig. 1. As expected for a converging EFT, the theoretical uncertainty decreases with increasing order. The description of the low phases is excellent and the resulting S-wave scattering lengths and effective range parameters are in good agreement with the ones obtained from the Nijmegen PWA. Also, the deuteron properties are well reproduced. For any practical application, the EFT description of the two-nucleon system now matches the accuracy of the so-called high-precision (semi)phenomenological potentials —quite a milestone for the EFT program for nuclear physics. What remains to be done is the consistent construction of the electroweak currents —work along these lines is in progress.

4 Three-nucleon forces

One of the most appealing features of the EFT approach is the consistent derivation of two- and three-nucleon forces —this was simply not possible in the conventional approach based on meson-exchanges and alike. The leading

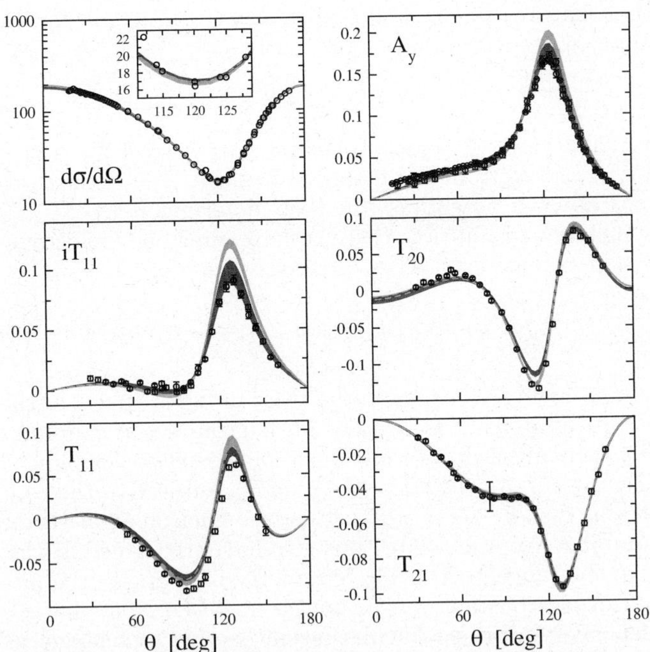

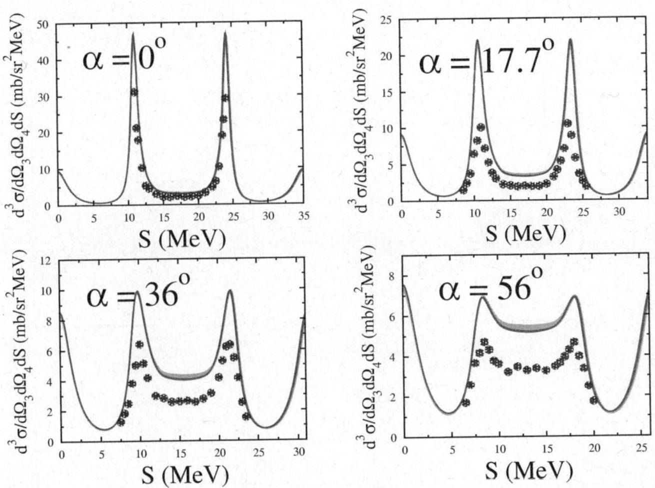

Fig. 3. Differential cross-section for pd break-up in the symmetric constant relative energy configuration at $E_d = 19\,\mathrm{MeV}$ for $\alpha = 0°, 17.7°, 36°$ and $56°$ compared to the NNLO calculation.

Fig. 2. nd elastic scattering observables (differential cross-section, vector and tensor analyzing powers) at 10 MeV at NLO (light-shaded bands) and at NNLO (dark-shaded bands). Note that at NLO, the 3NF is still absent. The data can be traced back from ref. [14].

3NF appears at NNLO and is given in terms of three topologies. These are the tree level two-pion exchange, the one-pion exchange (between a 4N contact term and the third nucleon) and a genuine 6N contact interaction. While the LECs related to the TPE topology are known from πN scattering, the other two contributions contain one unknown LEC, respectively. These LECs are called c_D and c_E, respectively. As already discussed some time ago, the second topology also features in pion production in pp collisions [13], this again shows the strength of EFT connecting many different processes and reactions. To pin down the LECs, one has to use two low-energy input data. In ref. [14], the triton binding energy and the doublet nd scattering length $^2a_{nd}$ were used to determine c_E and c_D. An update is given in the review [4] and a study of the 3NF in ^{7}Li in the framework of the no-core shell model was presented in [15]. Already at this order in the chiral expansion, one obtains an excellent description of many pd and nd scattering and break-up data, see, e.g., fig. 2.

However, there are various good reasons to work out the 3NF at N^{3}LO. First, of course, one wants to achieve consistency with the two-nucleon force which is already available at N^{3}LO (see above). Second if one considers the same observables as in fig. 2 for higher energies, e.g. for $E_n = 65\,\mathrm{MeV}$, the theoretical uncertainty is uncomfortably large. Third, at very low energies there are still discrepancies in the description of the vector analyzing power A_y and also, recent measurements at Cologne [16] of pd break-up in the "symmetric relative constant energy" configuration at $E_d = 19\,\mathrm{MeV}$ show significant deviations from the NNLO prediction with increasing angle α (the angle between the space-star configuration of the three outgoing nucleons and the incoming deuteron), see fig. 3. Note that the also measured analyzing powers A_{yy} agree much better with the theoretical predictions, for details see [16]. Thus, it is mandatory to calculate the 3NF at N^{3}LO. This is quite a formidable task, but work in progress by Bernard, Epelbaum and others looks very promising. Furthermore, recent progress in the inclusion of Coulomb effects in three-nucleon systems [17] has to be built into the EFT. It is also worth mentioning that the 4NF, that first appears at N^{4}LO, has recently been presented by Epelbaum [18]. For a first estimate of the effects of this force in ^{4}He, see [19].

5 Hyperon-nucleon forces

Strange quark effects in nuclei are investigated in the field of hyper-nuclear physics. To address such issues requires the knowledge of the fundamental hyperon-nucleon (YN) interactions. Before discussing these in the framework of a chiral EFT, let me stress the differences to the NN case. First, there exist not many data for low and moderate energies and also, their precision is mostly limited. A partial-wave analysis is therefore not available. However, the ambitious hyper-nuclear programs at KEK, CEBAF, MAMI, DAΦNE and JPARC will provide further specific information on the YN interaction by mapping out precisely the level schemes of a large variety of hyper-nuclei. Second, one has to consider channel coupling in the physical basis due to the Λ-Σ^0 mixing. Third, due to the larger strange quark mass, bigger explicit chiral symmetry-breaking effects are expected.

In ref. [20], we have considered the YN interaction in chiral effective field theory to leading order in the power counting, cf. eq. (4) (for an earlier analysis treating the

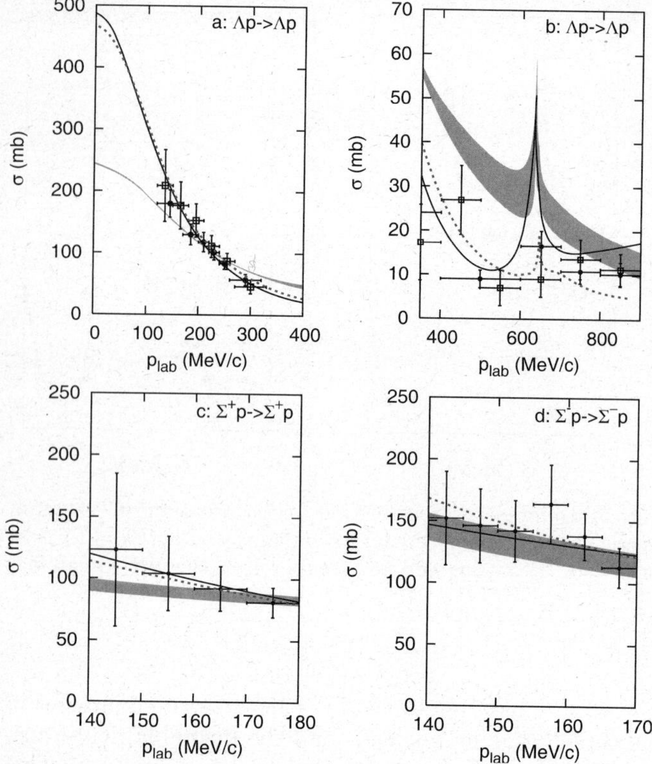

Fig. 4. "Total" cross-section σ (as defined in eq. (9)) as a function of p_{lab}. The shaded band is the chiral EFT potential for $\Lambda = 550, \ldots, 700\,\text{MeV}$, the dashed curve is the Jülich '04 model [23], and the solid curve is the Nijmegen NSC97f model [24].

boson exchanges perturbatively, see [21]). Such an exploratory study is motivated by the fact that a) in the YN system the S-wave scattering lengths (that are not precisely known) are not as unnaturally large as in the NN sector and b) the three-flavor theory is much richer at leading order. More precisely, we have to consider the contact interactions of the baryon octet,

$$B = \begin{pmatrix} \frac{\Sigma^0}{\sqrt{2}} + \frac{\Lambda}{\sqrt{6}} & \Sigma^+ & p \\ \Sigma^- & \frac{-\Sigma^0}{\sqrt{2}} + \frac{\Lambda}{\sqrt{6}} & n \\ -\Xi^- & \Xi^0 & -\frac{2\Lambda}{\sqrt{6}} \end{pmatrix}. \quad (5)$$

Here, we identify the physical η with the octet particle —this is correct modulo NLO corrections. It is shown in ref. [20] that there are 6 independent four-baryon contact terms without derivatives, from which 5 combinations appear in the YN system (the sixth combination only contributes to the $\Lambda\Lambda$ and ΞN channels). Thus, one has to determine the 5 corresponding LECs. This is done best in the partial-wave basis:

$$
\begin{aligned}
V_{1S0}^{\Lambda\Lambda} &= C_{1S0}^{\Lambda\Lambda}, & V_{3S1}^{\Lambda\Lambda} &= C_{3S1}^{\Lambda\Lambda}, \\
V_{1S0}^{\Sigma\Sigma} &= C_{1S0}^{\Sigma\Sigma}, & V_{3S1}^{\Sigma\Sigma} &= C_{3S1}^{\Sigma\Sigma}, \\
\widetilde{V}_{1S0}^{\Sigma\Sigma} &= 9C_{1S0}^{\Lambda\Lambda} - 8C_{1S0}^{\Sigma\Sigma}, & \widetilde{V}_{3S1}^{\Sigma\Sigma} &= C_{3S1}^{\Lambda\Lambda}, \\
V_{1S0}^{\Lambda\Sigma} &= 3\left(C_{1S0}^{\Lambda\Lambda} - C_{1S0}^{\Sigma\Sigma}\right), & V_{3S1}^{\Lambda\Sigma} &= C_{3S1}^{\Lambda\Sigma}, \quad (6)
\end{aligned}
$$

that features singlet and triplet waves (supplemented by appropriate isospin factors),

$$V^{(0)} = C_S^{BB} + C_T^{BB}\, \boldsymbol{\sigma}_1 \cdot \boldsymbol{\sigma}_2. \quad (7)$$

In [20], we have chosen to search for $C_{1S0}^{\Lambda\Lambda}$, $C_{3S1}^{\Lambda\Lambda}$, $C_{1S0}^{\Sigma\Sigma}$, $C_{1S0}^{\Sigma\Sigma}$, and $C_{3S1}^{\Lambda\Sigma}$ in the fitting procedure. The other three partial-wave potentials are then determined by $SU(3)$-symmetry. In addition, there is the leading one-Goldstone-boson exchange from the pseudoscalar octet (π, K, η)

$$\mathcal{L} = \left\langle \frac{D}{2}\bar{B}\gamma^\mu\gamma_5 \{u_\mu, B\} + \frac{F}{2}\bar{B}\gamma^\mu\gamma_5 [u_\mu, B] \right\rangle, \quad (8)$$

where the brackets denote the trace in flavor space. In the $SU(3)$ limit (that is to LO), all baryon-meson couplings can be expressed in terms of the pion-nucleon coupling f and the $SU(3)$ ratio $\alpha = D/(D+F)$, subject to the constraint that $F + D = g_A$, with g_A the nucleon axial-vector coupling. We use $\alpha = 0.4$ but also have performed fits for α in the range $[0.36, 0.44]$. Symmetry breaking in the meson decay constants only appears at NLO. Note that we also have performed calculations neglecting η exchange, as it is often done in meson-exchange models. The fits are not very sensitive to this, but it is remarkable that the consistent inclusion of the eta leads to a somewhat improved plateau (that is a smaller variation of the χ^2/dof as the cut-off in the LS equation is varied). We also include the leading Coulomb effects using the Vincent-Phatak procedure properly formulated for the EFT, see, e.g., ref. [22]. Altogether, we fit to 34 total cross-section data points and the $\Sigma^- p$ capture ratio at rest. The total cross-sections are found by simply integrating the differential cross-sections, except for the $\Sigma^+ p \to \Sigma^+ p$ and $\Sigma^- p \to \Sigma^- p$ channels. For those channels the experimental total cross-sections were obtained via

$$\sigma = \frac{2}{\cos\theta_{\max} - \cos\theta_{\min}} \int_{\cos\theta_{\min}}^{\cos\theta_{\max}} \frac{d\sigma(\theta)}{d\cos\theta} d\cos\theta, \quad (9)$$

for various values of $\cos\theta_{\min}$ and $\cos\theta_{\max}$. Following [24], we use $\cos\theta_{\min} = -0.5$ and $\cos\theta_{\max} = 0.5$ in our calculations for the $\Sigma^+ p \to \Sigma^+ p$ and $\Sigma^- p \to \Sigma^- p$ cross-sections, in order to stay as close as possible to the experimental procedure. We also impose the constraint that the fits should produce a bound hyper-triton with roughly the correct binding energy. For a reasonable cut-off variation, the four-baryon LECs come out of natural size,

$$
\begin{aligned}
4\pi\, m_B^2\, C_{1S0}^{\Lambda\Lambda} &= -0.6, \ldots, -0.4, \\
4\pi\, m_B^2\, C_{3S1}^{\Lambda\Lambda} &= -0.3, \ldots, -0.03, \\
4\pi\, m_B^2\, C_{1S0}^{\Sigma\Sigma} &= -1.1, \ldots, -1.0, \\
4\pi\, m_B^2\, C_{3S1}^{\Sigma\Sigma} &= 3.2, \ldots, 3.5, \\
4\pi\, m_B^2\, C_{3S1}^{\Lambda\Sigma} &= -0.1, \ldots, 0.05, \quad (10)
\end{aligned}
$$

with $m_B \simeq 1.1\,\text{GeV}$ the average octet baryon mass, and the corresponding description of the total cross-sections for some of the channels is shown in fig. 4. The shaded band is obtained by varying the cut-off in the Lippmann-Schwinger equation between 550 and 700 MeV, the corresponding total χ^2 varies from 29.6 to 34.6. The resulting

Table 1. The YN singlet and triplet scattering lengths (a) and effective ranges (r) (in fm) and the hyper-triton binding energy, E_B (in MeV) as a function of the cut-off Λ (in MeV). The binding energies for the hyper-triton (last row) are calculated using the Idaho-N^{3}LO NN potential [5]. The experimental value of the hyper-triton binding energy is $-2.354(50)$ MeV.

Λ	550	600	650	700
$a_S^{\Lambda p}$	-1.90	-1.91	-1.91	-1.91
$r_S^{\Lambda p}$	1.40	1.40	1.36	1.35
$a_T^{\Lambda p}$	-1.22	-1.23	-1.23	-1.23
$r_T^{\Lambda p}$	2.05	2.13	2.20	2.27
$a_S^{\Sigma^+ p}$	-2.24	-2.32	-2.36	-2.39
$r_S^{\Sigma^+ p}$	3.74	3.60	3.53	3.63
$a_T^{\Sigma^+ p}$	0.70	0.65	0.60	0.56
$r_T^{\Sigma^+ p}$	-2.14	-2.78	-3.55	-4.36
E_B	-2.35	-2.34	-2.34	-2.36

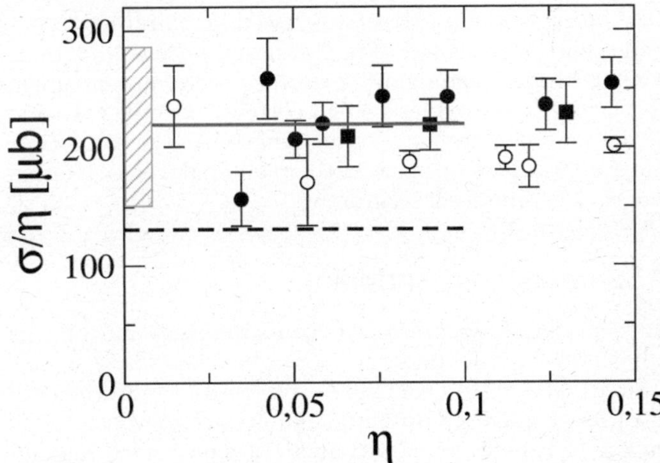

Fig. 5. Comparison of the NLO results to experimental data for $pp \to d\pi^+$. The dashed and the solid curves show the LO and the NLO result, respectively. The grey hatched area depicts the theoretical uncertainty at NLO. The data can be traced back from ref. [30].

singlet and triplet scattering lengths, the effective ranges and the corresponding hyper-triton binding energy are collected in table 1. Note that a Λp singlet scattering lengths of about -1.9 fm leads to the correct hyper-triton binding energy (within 0.5% of the empirical value). This value for $a_S^{\Lambda p}$ differs considerably from the one obtained in meson-exchange models [23,25].

Our findings show that the chiral effective field theory scheme, applied earlier to the NN interaction, also works well for the YN interaction. In the future it will be interesting to study the convergence of the chiral EFT for the YN interaction by doing NLO and NNLO calculations. In view of hyper-nucleus calculations, three-baryon forces that naturally arise in chiral EFT, should be investigated too (for a study of the hyper-triton in an EFT with contact interactions, see [26]). Furthermore, a combined NN and YN study in chiral EFT, starting with a NLO calculation, needs to be performed. Work in this direction is in progress.

6 Pion production in proton-proton collisions

As noted before, (threshold) pion production in proton-proton collisions encodes further complementary information on the structure of the few-nucleon forces. Also, there exist a waste amount of precise data for the reactions $pp \to pp\pi^0, pp \to d\pi^+, \dots$ from COSY at Jülich, IUCF at Bloomington, TRIUMF at Vancouver and from TSL at Uppsala, for a recent and comprehensive review see [27]. These reactions are characterized by a fairly large momentum transfer squared already at threshold:

$$t \simeq -m_{\mathrm{N}} M_\pi = -(360\,\mathrm{MeV})^2 . \qquad (11)$$

Consequently, the power counting has to be adjusted correspondingly, the appropriate expansion parameter is $\chi = \sqrt{M_\pi/m_{\mathrm{N}}} \simeq 0.4$. It was already demonstrated in

ref. [28] that such a power counting leads to the correct ordering of the various contributions at threshold (which had earlier been evaluated in the standard Weinberg counting with some rather strange findings, see *e.g.* [29]). Furthermore, in contrast to the construction of 2N and 3N forces, in pion production one must include the $\Delta(1232)$-resonance explicitly in the EFT. Another important issue, which was only understood recently [30], is the way one has to consistently include the initial-state and the final-state interactions together with the interaction kernel —otherwise one is left with amplitudes that cannot be controlled by analytic contributions. In [30] we have performed a NLO calculation of the threshold cross-section σ_{thr} for $pp \to d\pi^+$ incorporating all these ingredients. In the threshold region, the cross-section can be written as

$$\sigma_{\mathrm{thr}} = \alpha\eta + \beta\eta^3 + \mathcal{O}(\eta^5), \quad \eta = p_\pi/M_\pi . \qquad (12)$$

The leading-order calculation leads to a too small cross-section by about a factor of $2/3$, $\alpha^{\mathrm{LO}} = 131\,\mu$b, see the dashed line in fig. 5. At NLO, this factor is effectively recovered, without any adjustable parameter one finds (for details, see [30])

$$\alpha^{\mathrm{NLO}} = 220\,\mu\mathrm{b}, \qquad (13)$$

as shown by the solid line in fig. 5. However, at this order one still has a sizeable uncertainty, which is depicted by the grey hatched area in the figure. Needless to say that much work needs to be done to sharpen these conclusions, in particular, the NNLO corrections, which contain for the first time counter terms, have to be worked out. Also, the intricate reaction $pp \to pp\pi^0$, that is sensitive to the small isoscalar pion-nucleon scattering amplitude, can now be addressed systematically.

Let me point out that the NNπ intermediate state plays an important role in the calculation of the dispersive and absorptive corrections to the complex-valued pion-deuteron scattering length. In ref. [31] a parameter-free calculation of these corrections based on chiral perturbation theory is presented. It is shown that once *all* diagrams contributing to leading order to this process are included, their net effect provides a small correction to the real part of the pion-deuteron scattering length. At the same time the sizable imaginary part of the pion-deuteron scattering length is reproduced accurately.

7 Summary and outlook

In this talk, I have presented the foundations and various applications of the modern theory of nuclear forces. It has a direct relation to QCD via its symmetries (and their realizations) and is a systematic and precise approach based on the chiral effective Lagrangian of pions, nucleons and external sources. In this framework, it is possible to derive three-nucleon forces (see sect. 4) and external electroweak currents that are consistent with the dominant NN forces. Furthermore, utilizing again the effective chiral Lagrangian, one can derive consistently nucleon and nuclear properties, which is of particular importance for the model-independent extraction of neutron properties from light nuclear targets such as deuterium or ^{3}He. As discussed in sect. 5, the extension of this scheme to include strange quarks, more precisely the exploratory study for the hyperon-nucleon interactions, look quite promising. Chiral effective-field theory also allows for a variation of the fundamental QCD parameters, like *e.g.* the quark masses, and thus provides chiral extrapolation functions for the analysis of lattice QCD sector (for a status report on these activities, see [32]).

Clearly, there is lots of work ahead to make further progress:

a) the three-nucleon forces should be worked out to N^{3}LO,

b) the electroweak current operators have to be constructed to N^{3}LO,

c) further systematic studies of pion production in proton-proton collisions at NNLO are to be carried out,

d) a combined analysis of NN and YN interactions at (N)NLO would certainly shed further light on the details of the YN interactions, and

e) more work should also be devoted to extend this scheme to medium and heavy nuclei (halos, cluster structures, no-core shell model, ...), which will be of importance for the nuclear-structure program at the future FAIR facility and other radioactive-beam facilities world-wide.

In summary, let me say that a new era of nuclear physics has just begun.

I am grateful to all my collaborators from Bonn, Jülich, Strasbourg, ... for very pleasant collaborations on the topics reported here. I also thank Evgeny Epelbaum for a careful reading of the manuscript. This work is supported in parts by the EU Integrated Infrastructure Initiative Hadron Physics Project under contract number RII3-CT-2004-506078 and by DFG (SFB/TR 16, "Subnuclear Structure of Matter").

References

1. S. Pieper, Nucl. Phys. A **751**, 516 (2005).
2. S. Weinberg, Nucl. Phys. B **363**, 3 (1991).
3. E. Epelbaum, W. Glöckle, U.-G. Meißner, Eur. Phys. J. A **19**, 125 [nucl-th/0304037]; 401 [nucl-th/0308010] (2004).
4. E. Epelbaum, Prog. Part. Nucl. Phys. **57**, 654 (2006) [arXiv:nucl-th/0509032].
5. D.R. Entem, R. Machleidt, Phys. Rev. C **68**, 041001 (2003) [arXiv:nucl-th/0304018].
6. E. Epelbaum, W. Glöckle, U.-G. Meißner, Nucl. Phys. A **747**, 362 (2005) [arXiv:nucl-th/0405048].
7. U.-G. Meißner, PoS (LAT2005) 009 [arXiv:hep-lat/0509029].
8. N. Fettes, U.-G. Meißner, Nucl. Phys. A **676**, 311 (2000) [arXiv:hep-ph/0002162].
9. N. Fettes, U.-G. Meißner, Nucl. Phys. A **693**, 693 (2001) [arXiv:hep-ph/0101030].
10. N. Kaiser, Phys. Rev. C **61**, 014003 (2000) [arXiv:nucl-th/9910044]; **62**, 024001 (2000) [arXiv:nucl-th/9912054].
11. V.G.J. Stoks *et al.*, Phys. Rev. C **48**, 792 (1993).
12. SAID on-line program, http://gwdac.phys.gw.edu.
13. C. Hanhart, U. van Kolck, G.A. Miller, Phys. Rev. Lett. **85**, 2905 (2000) [arXiv:nucl-th/0004033].
14. E. Epelbaum, A. Nogga, W. Glöckle, H. Kamada, U.-G. Meißner, H. Witala, Phys. Rev. C **66**, 064001 (2002) [arXiv:nucl-th/0208023].
15. A. Nogga, P. Navratil, B.R. Barrett, J.P. Vary, Phys. Rev. C **73**, 064002 (2006) [arXiv:nucl-th/0511082].
16. J. Ley *et al.*, Phys. Rev. C **73**, 064001 (2006).
17. A. Deltuva, A.C. Fonseca, P.U. Sauer, Phys. Rev. C **71**, 054005 (2005) [arXiv:nucl-th/0503012].
18. E. Epelbaum, Phys. Lett. B **639**, 456 (2006) [arXiv:nucl-th/0511025].
19. D. Rozpedzik *et al.*, arXiv:nucl-th/0606017.
20. H. Polinder, J. Haidenbauer, U.-G. Meißner, Nucl. Phys. A **779**, 244 (2006) [arXiv:nucl-th/0605050].
21. C.L. Korpa, A.E.L. Dieperink, R.G.E. Timmermans, Phys. Rev. C **65**, 015208 (2002) [arXiv:nucl-th/0109072].
22. M. Walzl, U.-G. Meißner, E. Epelbaum, Nucl. Phys. A **693**, 663 (2001) [arXiv:nucl-th/0010019].
23. J. Haidenbauer, U.-G. Meißner, Phys. Rev. C **72**, 044005 (2005) [arXiv:nucl-th/0506019].
24. Th.A. Rijken, V.G.J. Stoks, Y. Yamamoto, Phys. Rev. C **59**, 21 (1999).
25. T.A. Rijken, Y. Yamamoto, Phys. Rev. C **73**, 044008 (2006) [nucl-th/0603042].
26. H.W. Hammer, Nucl. Phys. A **705**, 173 (2002) [arXiv:nucl-th/0110031].
27. C. Hanhart, Phys. Rep. **397**, 155 (2004) [arXiv:hep-ph/0311341].
28. C. Hanhart, N. Kaiser, Phys. Rev. C **66**, 054005 (2002) [nucl-th/0208050].
29. B.Y. Park, F. Myhrer, J.R. Morones, T. Meissner, K. Kubodera, Phys. Rev. C **53**, 1519 (1996) [arXiv:nucl-th/9512023].
30. V. Lensky, V. Baru, J. Haidenbauer, C. Hanhart, A.E. Kudryavtsev, U.-G. Meißner, Eur. Phys. J. A **27**, 37 (2006) [arXiv:nucl-th/0511054].
31. V. Lensky, V. Baru, J. Haidenbauer, C. Hanhart, A. Kudryavtsev, U.-G. Meißner, arXiv:nucl-th/0608042; to be published in Phys. Lett. B (2007).
32. M.J. Savage, arXiv:nucl-th/0601001.

Eur. Phys. J. A **31**, 403–408 (2007)
DOI 10.1140/epja/i2006-10261-y

Special Article – QNP 2006

Effective-field theory and the nuclear many-body problem

T. Schäfer[a]

Department of Physics, North Carolina State University, Raleigh, NC 27695, USA

Received: 23 November 2006
Published online: 12 March 2007 – © Società Italiana di Fisica / Springer-Verlag 2007

Abstract. We review many-body calculations of the equation of state of dilute neutron matter in the context of effective-field theories of the nucleon-nucleon interaction.

PACS. 21.65.+f Nuclear matter – 24.85.+p Quarks, gluons, and QCD in nuclei and nuclear processes

1 Introduction

One of the central problems of nuclear physics is to calculate the properties of nuclear matter starting from the two-body scattering data and the binding energies of few-body bound states [1,2]. The nuclear-matter problem is notoriously difficult. Some of the problems that are often mentioned are

- the large short-range repulsive core in the nucleon-nucleon interaction,
- the large scattering length in the $^1S_0$ channel, the small binding energy of the deuteron, and the small saturation density,
- the need to include three (and possibly four) body forces,
- the need to include non-nucleonic degrees of freedoms, such as isobars, mesons, quarks, etc.

Ever since the discovery of QCD the classic nuclear-matter problem has evolved into the broader question of how the properties of nuclear matter are related to the parameters of the QCD, the QCD scale parameter and the masses of the light quarks.

Over the last couple of year much progress has been made in understanding these kinds of questions in the case of nuclear two- and three-body bound states [3]. Using effective-field theory methods it was shown that

- the short-range behavior of the nuclear force is not observable. Using the renormalization group the short-distance behavior can be modified without changing low-energy scattering data and binding energies [4,5];
- effective field theories can accommodate the large scattering lengths in the nucleon-nucleon system [6,7]. The scattering lengths depend sensitively on the quark masses, see fig. 1, and the large value of $a(^1S_0)$ observed in nature appears to be accidental [8,9];

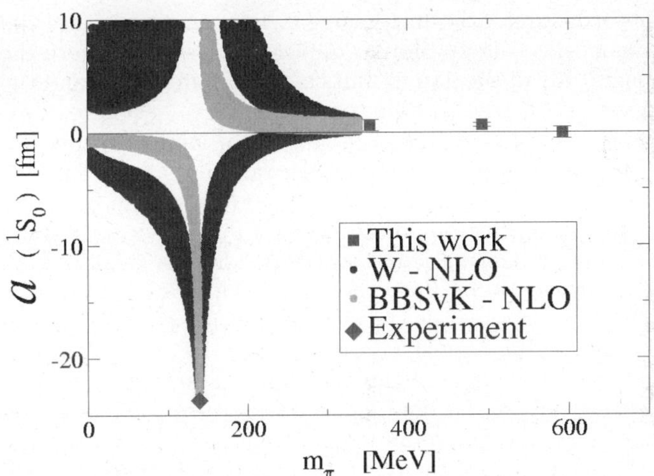

Fig. 1. (Color online) Quark mass dependence of the scattering length in the $^1S_0$ channel. The plot shows a combination of unquenched lattice QCD results (red points, from Beane *et al.* [13]) and chiral extrapolations. The experimental point is shown in purple. Two different power-counting schemes were employed to constrain the quark mass dependence, the BBSvK scheme [6,14] and the W (Weinberg) scheme [3].

- a local three-body force is necessary to renormalize the two-body force already at leading order[1] [10]. As a consequence, one cannot predict three-body binding energies based on two-body scattering data alone;
- non-nucleonic degrees of freedom, quark effects, relativistic effects etc. can be absorbed in local operators.

Effective-field theories have also achieved remarkable quantitative success in describing the available nucleon-nucleon scattering data below the pion production threshold [11,12]. The long-term goal is to achieve a similar

---

[1] An exception is the original Weinberg scheme in which three- and four-body forces are considered higher-order corrections.

[a] e-mail: `thomas_schaefer@ncsu.edu`

qualitative and quantitative understanding of the nuclear many-body problem.

In this contribution we shall study a simple limiting case of the nuclear-matter problem. We shall concentrate on pure neutron matter at densities significantly below the nuclear-matter saturation density. The neutron-neutron scattering length is $a_{nn} = -18\,\mathrm{fm}$ and the effective range is $r_{nn} = 2.8\,\mathrm{fm}$. This means that there is a range of densities for which the inter-particle spacing is large compared to the effective range but small compared to the scattering length. Neutron matter in this regime exhibits interesting universal properties. We are interested in the limit $(k_F a_{nn}) \to \infty$ and $(k_F r_{nn}) \to 0$, where k_F is the Fermi momentum. From a dimensional analysis it is clear that the energy per particle at zero temperature has to be proportional to the energy per particle of a free Fermi gas at the same density:

$$\frac{E}{A} = \xi \left(\frac{E}{A}\right)_0 = \xi \frac{3}{5}\left(\frac{k_F^2}{2m}\right). \tag{1}$$

The constant ξ is universal, *i.e.* independent of the details of the system. Similar universal constants govern the magnitude of the gap in units of the Fermi energy and the equation of state at finite temperature.

Universality also implies that the properties of this system can be studied using atoms rather than nuclei. The scattering length of certain fermionic atoms can be tuned using Feshbach resonances, see [15] for a review. A small negative scattering length corresponds to a weak attractive interaction between the atoms. This case is known as the BCS limit. As the strength of the interaction increases the scattering length becomes larger. It diverges at the point where a bound state is formed. The point $a = \infty$ is called the unitarity limit, since the scattering cross-section saturates the s-wave unitarity bound $\sigma = 4\pi/k^2$. On the other side of the resonance the scattering length is positive. In the BEC limit the interaction is strongly attractive and the fermions form deeply bound molecules.

2 Numerical calculations

The calculation of the dimensionless quantity ξ is a nonperturbative problem. In this section we shall describe an approach based on lattice field theory methods. The physics of the unitarity limit is captured by an effective Lagrangian of point-like fermions interacting via a short-range interaction. The Lagrangian is

$$\mathcal{L} = \psi^\dagger \left(i\partial_0 + \frac{\boldsymbol{\nabla}^2}{2m}\right)\psi - \frac{C_0}{2}\left(\psi^\dagger\psi\right)^2. \tag{2}$$

The standard strategy for dealing with the four-fermion interaction is to use a Hubbard-Stratonovich transformation. The partition function can be written as [16]

$$Z = \int DsDcDc^* \exp\left[-S\right], \tag{3}$$

where s is the Hubbard-Stratonovich field and c is a Grassmann field. S is a discretized Euclidean action:

$$\begin{aligned}
S = \sum_{n,i} &\left[e^{-\hat{\mu}\alpha_t} c_i^*(\boldsymbol{n})c_i(\boldsymbol{n}+\hat{0})\right.\\
&\left. -e^{\sqrt{-C_0\alpha_t}s(n)+\frac{C_0\alpha_t}{2}}(1-6h)c_i^*(\boldsymbol{n})c_i(\boldsymbol{n})\right]\\
&-h\sum_{n,l_s,i}\left[c_i^*(\boldsymbol{n})c_i(\boldsymbol{n}+\hat{l}_s)+c_i^*(\boldsymbol{n})c_i(\boldsymbol{n}-\hat{l}_s)\right]\\
&+\frac{1}{2}\sum_{n}s^2(\boldsymbol{n}). \tag{4}
\end{aligned}$$

Here i labels spin and $\boldsymbol{n}$ labels lattice sites. Spatial and temporal unit vectors are denoted by $\hat{l}_s$ and $\hat{0}$, respectively. The temporal and spatial lattice spacings are b_τ and b. The dimensionless chemical potential is given by $\hat{\mu} = \mu b_\tau$. We define α_t as the ratio of the temporal and spatial lattice spacings and $h = \alpha_t/(2\hat{m})$. Note that for $C_0 < 0$ the action is real and standard Monte Carlo simulations are possible.

The four-fermion coupling is fixed by computing the sum of all two-particle bubbles on the lattice. Schematically,

$$\frac{m}{4\pi a} = \frac{1}{C_0} + \frac{1}{2}\sum_p \frac{1}{E_{\boldsymbol{p}}}, \tag{5}$$

where the sum runs over discrete momenta on the lattice and $E_{\boldsymbol{p}}$ is the lattice dispersion relation. A detailed discussion of the lattice regularized scattering amplitude can be found in [17,18,16]. For a given scattering length a the four-fermion coupling is a function of the lattice spacing. The continuum limit correspond to taking the temporal and spatial lattice spacings b_τ, b to zero

$$b_\tau\mu \to 0, \qquad bn^{1/3} \to 0, \tag{6}$$

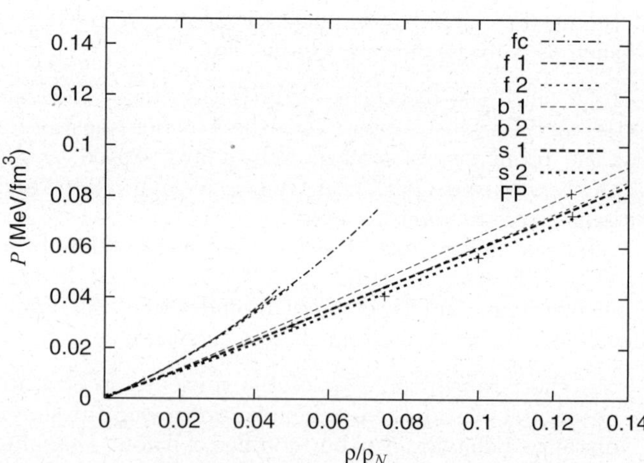

Fig. 2. Equation of state of pure neutron matter at $T = 4\,\mathrm{MeV}$ from lattice simulations of an effective-field theory. Figure taken from Lee and Schäfer [16]. The curves labeled fc, f1, f2 show results for a free gas on the lattice and in the continuum, the curves labeled b1, b2 show ladder sums, and s1, s2 are numerical results on different lattices. We also compare to the variational results of Friedman and Pandharipande (FP).

where μ is the chemical potential, n is the density and $an^{1/3}$ is fixed. We performed numerical simulations at non-zero temperature and concluded that $\xi = (0.09\text{–}0.42)$. Lee studied canonical $T = 0$ simulations and obtained $\xi = 0.25$ [19]. Green Function Monte Carlo calculations give $\xi = 0.44$ [20], and finite-temperature lattice simulations have been extrapolated to $T = 0$ to yield similar results [21,22].

Lattice results for the equation of state of dilute neutron matter at $T = 4\,\text{MeV}$ are shown in fig. 2. For comparison, we show variational results obtained by Friedman and Pandharipande using a phenomenological potential [23]. We observe that the lattice calculations agree very well with the variational result. The pressure is very similar to that of non-interacting neutrons scaled by a factor $\sim 1/2$. The lattice calculation can be extended to higher densities by including explicit pionic degrees of freedom in the effective Lagrangian [24]. In this case a mild sign problem returns, but at $T \neq 0$ this sign problem can be handled with standard methods.

3 Analytical approaches: large-N expansion

It is clearly desirable to find a systematic analytical approach to the dilute Fermi liquid in the unitarity limit. Various possibilities have been considered, such as an expansion in the number of fermion species [25,26] or the number of spatial dimensions [27–30].

We begin with a brief description of the large-N approach. The physics of the large-N limit depends on the symmetries of the interaction. One possibility is a $SU(N)$ symmetric interaction [25]

$$\mathcal{L} = \frac{C_0}{2} \left(\psi_f^\dagger \psi_f \right)^2 , \tag{7}$$

where $f = 1, \ldots, N$ is a flavor label. A smooth large-N limit is achieved by keeping $C_0 N = c_0$ constant as $N \to \infty$. The large-N limit is most easily studied by introducing a Hubbard-Stratonovich field σ coupled to the density $\psi_f^\dagger \psi_f$. The leading contribution to the free energy comes from the free-fermion term and the mean-field contribution, both of which scale as N. Subleading $1/N$ corrections arise from particle-hole ring diagrams. The problem is that at any fixed order in the large-N expansion the free energy diverges as the scattering length is taken to infinity.

This problem is related to the fact that particle ladders need to be summed in the unitarity limit. This can be achieved by considering a $Sp(2N)$ symmetric interaction of the form [26]

$$\mathcal{L} = \frac{G_0}{2}(\psi_f \mathcal{J}^{fg} \psi_g)(\psi_h^\dagger \mathcal{J}^{hi} \psi_i^\dagger), \tag{8}$$

where $\mathcal{J} = (\sigma_2) \otimes \ldots \otimes (\sigma_2)$ and $G_0 N = g_0$ constant as $N \to \infty$. This interaction can be bosonized using a difermion field $\Phi = (\psi_f \mathcal{J}^{fg} \psi_g)/N$. At large-$N$ the leading

contribution corresponds to the mean-field BCS approximation. The thermodynamic potential in the unitarity limit is

$$\Omega = -N \int \frac{\mathrm{d}^3 p}{(2\pi)^3} \left\{ \sqrt{\epsilon_p^2 + \Phi^2} - \epsilon_p - \frac{m\Phi^2}{p^2} \right\} \cdot \tag{9}$$

with $\epsilon_p = E_p - \mu$. This function can be minimized numerically. We find $\Phi_0 = 1.16\mu$ and $\xi = 0.59$. Nikolic and Sachdev studied $1/N$ corrections near T_c [26]. These effects are not small. They find, for example, $(\mu/T)(T_c) = 1.50 + 2.79/N + O(1/N^2)$.

4 Large-d expansion

Steele suggested that the many-body problem of non-relativistic fermions near the unitarity limit can be studied using an expansion in $1/d$, where d is the number of spatial dimensions [27]. The main idea is that phase space factors associated with hole lines are suppressed as $d \to \infty$ so that the leading-order contribution comes from 2-particle ladders, and higher-order corrections correspond to the hole line expansion of Bethe and Brueckner.

Consider the effective Lagrangian in eq. (2). We first study perturbative corrections to the ground-state energy in d spatial dimensions. The leading-order correction to the energy per particle is

$$\frac{E_1}{A} = \frac{1}{d} \left[\frac{\Omega_d C_0 k_F^{d-2} M}{(2\pi)^d} \right] \left(\frac{k_F^2}{2M} \right). \tag{10}$$

This expression indicates that the large-d limit should be taken in such a way that

$$\lambda \equiv \left[\frac{\Omega_d C_0 k_F^{d-2} M}{d(2\pi)^d} \right] \xrightarrow{d \to \infty} \text{const.} \tag{11}$$

In the following we wish to study whether this limit is smooth even if the theory is non-perturbative. Consider the in medium two-particle scattering amplitude in d spatial dimensions. The result is

$$\int \frac{\mathrm{d}^d q}{(2\pi)^d} \frac{\theta_q^+}{k^2 - q^2 + i\epsilon} = f_{vac}(k) + \frac{k_F^{d-2}\Omega_d}{2(2\pi)^d} f_{PP}^{(d)}(\kappa, s). \tag{12}$$

The theta-function $\theta_q^+ \equiv \theta(k_1 - k_F)\theta(k_2 - k_F)$ with $k_{1,2} = P/2 \pm k$ requires both fermion momenta to be above the Fermi surface. The first term on the RHS is the vacuum contribution. In dimensional regularization the vacuum term is purely imaginary and does not contribute to the ground-state energy. The second term is the medium contribution which depends on the scaled relative momentum $\kappa = k/k_F$ and center-of-mass momentum $s = P/(2k_F)$. In the large-d limit we find

$$f_{PP}^{(d)}(s, \kappa) = \frac{1}{d} f_{PP}^{(0)}(s, \kappa) \left(1 + O\left(\frac{1}{d}\right) \right), \tag{13}$$

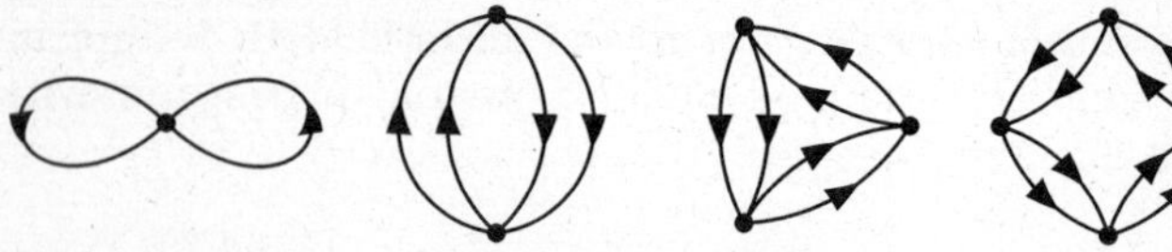

Fig. 3. The particle ladder diagrams shown in this figure give the leading-order contribution to the ground-state energy in the large-d limit.

which implies that all two-particle ladder diagrams are of the same order, see fig. 3. The sum of all ladder diagrams can be calculated by noting that, except for the logarithmic (BCS) singularity at $s = 0, \kappa = 1$, the particle-particle bubble is a smooth function of the kinematic variables s and κ. Hole-hole phase space, on the other hand, is strongly peaked at $\bar{s} = \bar{\kappa} = 1/\sqrt{2}$ in the large-d limit. We find that $f_{PP}^{(d)}(\bar{s},\bar{\kappa}) = 4/d \cdot (1 + O(1/d))$. The ladder sum is a simple geometric series and [28]

$$\frac{E}{A} = \left\{ 1 + \frac{\lambda}{1 - 2\lambda} + O\left(\frac{1}{d}\right) \right\} \left(\frac{k_F^2}{2M}\right), \qquad (14)$$

where λ is the coupling constant defined in eq. (11). We observe that if the strong-coupling limit $\lambda \to \infty$ is taken after the limit $d \to \infty$ the universal parameter ξ is given by $1/2$. We have also studied the role of pairing in the large-d limit. The pairing gap is

$$\Delta = \frac{2e^{-\gamma} E_F}{d} \exp\left(-\frac{1}{d\lambda}\right) \left(1 + O\left(\frac{1}{d}\right)\right). \qquad (15)$$

There is no exponential suppression in $d \to \infty$ limit, but Δ/E_F is down by a power of $1/d$. As a consequence the pairing energy is sub-leading compared to the result in eq. (14).

5 Epsilon expansion near four dimensions

Nussinov and Nussinov observed that the fermion many-body system in the unitarity limit reduces to a free Fermi gas near $d = 2$ spatial dimensions, and to a free Bose gas near $d = 4$ [29]. Their argument was based on the behavior of the two-body wave function as the binding energy goes to zero. For $d = 2$ it is well known that the limit of zero binding energy corresponds to an arbitrarily weak potential. In $d = 4$ the two-body wave function at $a = \infty$ has a $1/r^2$ behavior and the normalization is concentrated near the origin. This suggests the many-body system is equivalent to a gas of non-interacting bosons.

A systematic expansion based on the observation of Nussinov and Nussinov was studied by Nishida and Son [30,31]. In this section we shall explain their approach. We begin by restating the argument of Nussinov and Nussinov in the effective-field theory language. In dimensional regularization $a \to \infty$ corresponds to $C_0 \to \infty$. The fermion-fermion scattering amplitude is given by

$$\mathcal{A}(p_0, \boldsymbol{p}) = \frac{\left(\frac{4\pi}{m}\right)^{\frac{d}{2}}}{\Gamma\left(1 - \frac{d}{2}\right)} \frac{i}{(-p_0 + E_p/2 - i\delta)^{\frac{d}{2}-1}}, \qquad (16)$$

where $\delta \to 0+$. As a function of d the Gamma-function has poles at $d = 2, 4, \ldots$ and the scattering amplitude vanishes at these points. Near $d = 2$ the scattering amplitude is energy and momentum independent. For $d = 4 - \epsilon$ we find

$$\mathcal{A}(p_0, \boldsymbol{p}) = \frac{8\pi^2 \epsilon}{m^2} \frac{i}{p_0 - E_p/2 + i\delta} + O(\epsilon^2). \qquad (17)$$

We observe that at leading order in ϵ the scattering amplitude looks like the propagator of a boson with mass $2m$. The boson-fermion coupling is $g^2 = (8\pi^2\epsilon)/m^2$ and vanishes as $\epsilon \to 0$. This suggests that we can set up a perturbative expansion involving fermions of mass m weakly coupled to bosons of mass $2m$. In the unitarity limit the Hubbard-Stratonovich transformed Lagrangian reads

$$\mathcal{L} = \Psi^\dagger \left[i\partial_0 + \sigma_3 \frac{\boldsymbol{\nabla}^2}{2m} \right] \Psi + \mu \Psi^\dagger \sigma_3 \Psi + \left(\Psi^\dagger \sigma_+ \Psi \phi + \text{h.c.}\right), \qquad (18)$$

where $\Psi = (\psi_\uparrow, \psi_\downarrow^\dagger)^T$ is a two-component Nambu-Gorkov field, σ_i are Pauli matrices acting in the Nambu-Gorkov space and $\sigma_\pm = (\sigma_1 \pm i\sigma_2)/2$. In the superfluid phase ϕ acquires an expectation value. We write

$$\phi = \phi_0 + g\varphi, \qquad g = \frac{\sqrt{8\pi^2\epsilon}}{m} \left(\frac{m\phi_0}{2\pi}\right)^{\epsilon/4}, \qquad (19)$$

where $\phi_0 = \langle\phi\rangle$. The scale $M^2 = m\phi_0/(2\pi)$ was introduced in order to have a correctly normalized boson field. The scale parameter is arbitrary, but this particular choice simplifies some of the loop integrals. In order to get a well-defined perturbative expansion we add and subtract a kinetic term for the boson field to the Lagrangian. We include the kinetic term in the free part of the Lagrangian:

$$\mathcal{L}_0 = \Psi^\dagger \left[i\partial_0 + \sigma_3 \frac{\boldsymbol{\nabla}^2}{2m} + \phi_0(\sigma_+ + \sigma_-) \right] \Psi$$
$$+ \varphi^\dagger \left(i\partial_0 + \frac{\boldsymbol{\nabla}^2}{4m} \right) \varphi. \qquad (20)$$

The interacting part is

$$\mathcal{L}_I = g\left(\Psi^\dagger \sigma_+ \Psi \varphi + \text{h.c}\right) + \mu \Psi^\dagger \sigma_3 \Psi$$
$$- \varphi^\dagger \left(i\partial_0 + \frac{\boldsymbol{\nabla}^2}{4m} \right) \varphi. \qquad (21)$$

Note that the interacting part generates self-energy corrections to the boson propagator which, by virtue of eq. (17), cancel against the kinetic term of the boson field. We have also included the chemical potential term in $\mathcal{L}_I$. This is motivated by the fact that near $d = 4$ the system reduces to a non-interacting Bose gas and $\mu \to 0$. We will count μ as a quantity of $O(\epsilon)$.

The Feynman rules are quite simple. The fermion and boson propagators follow from eq. (20) and the fermion-boson vertices are $ig\sigma^\pm$. Insertions of the chemical potential are $i\mu\sigma_3$. Both g^2 and μ are corrections of order ϵ. In order to verify that the ϵ expansion is well defined we have to check that higher-order diagrams do not generate powers of $1/\epsilon$. Studying the superficial degree of divergence of

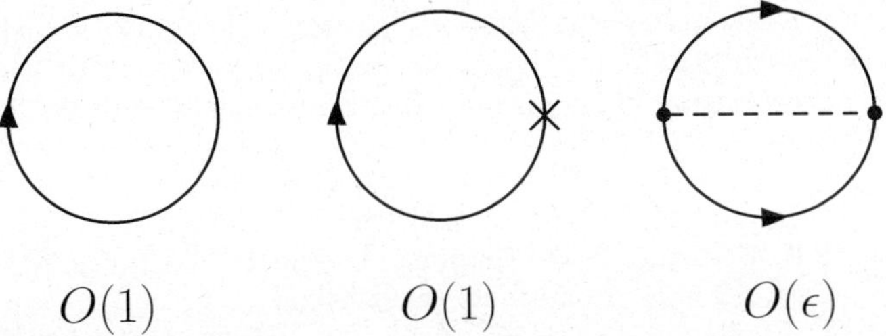

Fig. 4. Leading-order contributions to the ground-state energy in the $\epsilon = 4 - d$ expansion. Solid lines are fermion propagators, dashed lines are boson propagators, and the cross is an insertion of the chemical potential.

diagrams one can show that there are only a finite number of one-loop diagrams that generate $1/\epsilon$ terms.

The leading-order diagrams that contribute to the effective potential are shown in fig. 4. The first diagram is the free-fermion loop which is $O(1)$. The second diagram is the μ insertion which is $O(1)$ because the loop diagram is divergent in $d = 4$. The sum of these two diagrams is

$$V_1 = -\int \frac{\mathrm{d}^d p}{(2\pi)^d} \left\{ \sqrt{E_{\boldsymbol{p}}^2 + \phi_0^2} - \frac{\mu E_{\boldsymbol{p}}}{\sqrt{E_{\boldsymbol{p}}^2 + \phi_0^2}} \right\}. \quad (22)$$

The integral can be computed analytically. Expanding to first order in $\epsilon = 4 - d$, we get

$$V_0 = \left\{ \frac{\phi_0}{3} \left[1 + \frac{7 - 3(\gamma + \log(2))}{6} \epsilon \right] \right.$$
$$\left. - \frac{\mu}{\epsilon} \left[1 + \frac{1 - 2(\gamma - \log(2))}{4} \epsilon \right] \right\} \left(\frac{m\phi_0}{2\pi} \right)^{d/2}. \quad (23)$$

Nishida and Son also computed the two-loop contribution shown in fig. 4. The result is

$$V_2 = -C\epsilon \left(\frac{m\phi_0}{2\pi} \right)^{d/2}, \quad (24)$$

where $C \simeq 0.14424$. We can now determine the minimum of the effective potential. We find

$$\phi_0 = \frac{2\mu}{\epsilon} \left[1 + (3C - 1 + \log(2)) \epsilon + O(\epsilon^2) \right]. \quad (25)$$

The value of $V = V_1 + V_2$ at ϕ_0 determines the pressure and $n = \partial P/\partial \mu$ gives the density. We find

$$n = \frac{1}{\epsilon} \left[1 - \frac{1}{4} (2\gamma - 1 - \log(2)) \epsilon \right] \left(\frac{m\phi_0}{2\pi} \right)^{d/2}. \quad (26)$$

We can compare this result with the density of a free Fermi gas in d dimensions. This equation determines the relation between $\epsilon_F \equiv k_F^2/(2m)$ and the density. We get

$$\epsilon_F = \frac{2\pi}{m} \left[\frac{n}{2} \Gamma \left(\frac{d}{2} + 1 \right) \right]^{2/d}. \quad (27)$$

We determine ϵ_F for the interacting gas by inserting n from eq. (26) into eq. (27). The universal parameter is $\xi = \mu/\epsilon_F$. We find

$$\xi = \frac{1}{2} \epsilon^{3/2} + \frac{1}{16} \epsilon^{5/2} \log(\epsilon) - 0.025 \epsilon^{5/2} + \ldots = 0.475, \quad (28)$$

where we have set $\epsilon = 1$. The calculation has been extended to $O(\epsilon^{7/2})$ by Arnold *et al.* [32]. Unfortunately, the next term is very large and it appears necessary to combine the expansion in $4 - \epsilon$ dimensions with a $2 + \epsilon$ expansion in order to extract useful results.

6 Epsilon expansion near two dimensions

Near two spatial dimensions the scattering amplitude in the unitarity limit vanishes linearly in $\bar{\epsilon} = d - 2$

$$\mathcal{A}(p_0, p) = i \frac{2\pi}{m} \bar{\epsilon} + O(\bar{\epsilon}^2). \quad (29)$$

The coefficient of $\bar{\epsilon}$ is momentum end energy independent. This means that we can set up a perturbative expansion with an effective four-fermion coupling $g^2 = 2\pi\epsilon/m$. This expansion is very similar to the perturbative $(k_F a)$ expansion studied by Huang, Lee and Yang [33,34], see fig. 5, but it is not restricted to the weak-coupling limit. To $O(\bar{\epsilon})$ the effective potential is given by

$$V_0 + V_1 = -P_{free} - \frac{g^2}{4} \rho^2 + O(\bar{\epsilon}^2), \quad (30)$$

where P_{free} is the pressure of free fermions expanded to $O(\bar{\epsilon})$ and the density is given by

$$\rho = 2 \int \frac{\mathrm{d}^2 p}{(2\pi)^2} \Theta(\mu - E_p). \quad (31)$$

From the total pressure we can compute the density and Fermi energy as in the previous section. The universal parameter ξ is given by

$$\xi = 1 - \bar{\epsilon} + O(\bar{\epsilon}^2) = 0 \quad (\bar{\epsilon} = 1). \quad (32)$$

Similar to the perturbative expansion pairing is exponentially suppressed in the $\bar{\epsilon}$ expansion. The pairing gap is [31]

$$\phi_0 = \frac{2\mu}{e} \exp \left(-\frac{1}{\bar{\epsilon}} \right) (1 + O(\bar{\epsilon})), \quad (33)$$

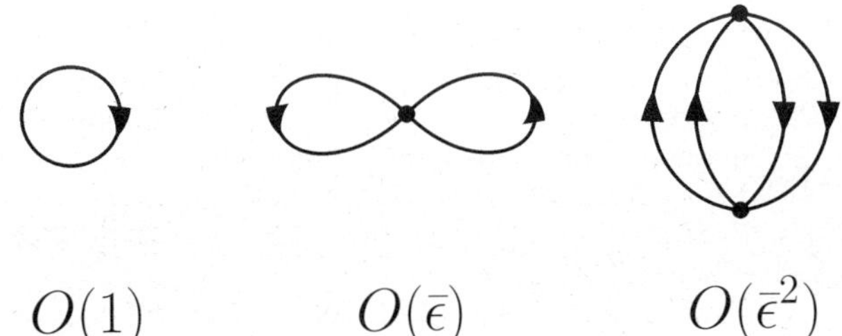

Fig. 5. Leading-order contributions to the ground-state energy in the $\bar{\epsilon} = d - 2$ expansion.

which corresponds to the perturbative result of Gorkov and Melik-Barkhudarov [35]. Equation (32) shows that the $\bar{\epsilon}$ expansion is poorly convergent. However, the $\bar{\epsilon}$ expansion is useful in improving the convergence of the $\epsilon = d - 4$ expansion, and in connecting the perturbative $(k_F a)$ expansion with the physics of the unitarity limit.

7 Outlook

In this contribution we focused on an idealized systems of neutrons at very low density. The obvious question is to what extent these methods can be extended to nuclear systems near saturation density.

The lattice simulations can easily be extended to include finite-range effects, explicit pions, and isospin. Some of these refinements will cause a sign problem in the simulation, but in most cases the sign problem can be handled with standard methods. A significant amount of work will be required in order to reduce discretization errors to the point where the interactions are quantitatively reliable all the way up to momenta on the order of the Fermi momentum in nuclear matter.

The large-N, large-d, or epsilon expansions are easily extended to interactions with a finite scattering length and a finite effective range [26,36,37]. There is also no obvious obstacle to including explicit pion degrees of freedom. It will be interesting to extend these methods to systems of protons and neutrons. In this case three-body forces have to be included. The central question is whether saturation can be achieved, and whether effective-field theories provide a qualitative understanding of the Coester line [38].

We should also note that the many-body physics that governs the equation of state of nuclear matter near saturation density may well be simpler than the physics of the unitarity limit. Effective-range corrections suppress the two-body scattering amplitude and nuclear matter is more perturbative than dilute neutron matter. As a consequence, perturbative calculations using soft potentials or effective interactions adjusted to nuclear-matter properties may well be reliable [39,40].

The work described in this contribution was done in collaboration with S. Cotanch, C.-W. Kao, A. Kryjevski, D. Lee and G. Rupak. This work is supported in part by the US Department of Energy grant DE-FG-88ER40388.

References

1. H.A. Bethe, Annu. Rev. Nucl. Part. Sci. **21**, 93 (1971).
2. A.D. Jackson, Annu. Rev. Nucl. Part. Sci. **33**, 105 (1983).
3. S. Weinberg, Phys. Lett. B **251**, 288 (1990).
4. G.P. Lepage, preprint, nucl-th/9706029.
5. S.K. Bogner, T.T.S. Kuo, A. Schwenk, Phys. Rep. **386**, 1 (2003) [nucl-th/0305035].
6. D.B. Kaplan, M.J. Savage, M.B. Wise, Nucl. Phys. B **534**, 329 (1998) [nucl-th/9802075].
7. U. van Kolck, Nucl. Phys. A **645**, 273 (1999) [nucl-th/9808007].
8. S.R. Beane, M.J. Savage, Nucl. Phys. A **717**, 91 (2003) [nucl-th/0208021].
9. E. Epelbaum, U.G. Meissner, W. Gloeckle, Nucl. Phys. A **714**, 535 (2003) [nucl-th/0207089].
10. P.F. Bedaque, H.W. Hammer, U. van Kolck, Phys. Rev. Lett. **82**, 463 (1999) [nucl-th/9809025].
11. E. Epelbaum, Prog. Part. Nucl. Phys. **57**, 654 (2006) [nucl-th/0509032].
12. D.R. Entem, R. Machleidt, Phys. Rev. C **68**, 041001 (2003) [nucl-th/0304018].
13. S.R. Beane, P.F. Bedaque, K. Orginos, M.J. Savage, Phys. Rev. Lett. **97**, 012001 (2006) [hep-lat/0602010].
14. S.R. Beane, P.F. Bedaque, M.J. Savage, U. van Kolck, Nucl. Phys. A **700**, 377 (2002) [nucl-th/0104030].
15. C. Regal, PhD Thesis, University of Colorado (2005) cond-mat/0601054.
16. D. Lee, T. Schäfer, Phys. Rev. C **72**, 024006 (2005) [nucl-th/0412002]; **73**, 015201 (2006) [nucl-th/0509017]; **73**, 015202 (2006) [nucl-th/0509018].
17. J.W. Chen, D.B. Kaplan, Phys. Rev. Lett. **92**, 257002 (2004) [hep-lat/0308016].
18. S.R. Beane, P.F. Bedaque, A. Parreno, M.J. Savage, Phys. Lett. B **585**, 106 (2004) [hep-lat/0312004].
19. D. Lee, Phys. Rev. B **73**, 115112 (2006) [cond-mat/0511332].
20. J. Carlson, J.J. Morales, V.R. Pandharipande, D.G. Ravenhall, Phys. Rev. C **68**, 025802 (2003) [nucl-th/0302041].
21. A. Bulgac, J.E. Drut, P. Magierski, Phys. Rev. Lett. **96**, 090404 (2006) [cond-mat/0505374].
22. E. Burovski, N. Prokof'ev, B. Svistunov, M. Troyer, Phys. Rev. Lett. **96**, 160402 (2006) [cond-mat/0602224].
23. B. Friedman, V.R. Pandharipande, Nucl. Phys. A **361**, 502 (1981).
24. D. Lee, B. Borasoy, T. Schäfer, Phys. Rev. C **70**, 014007 (2004) [nucl-th/0402072].
25. R.J. Furnstahl, H.W. Hammer, Ann. Phys. (N.Y.) **302**, 206 (2002) [nucl-th/0208058].
26. P. Nikolic, S. Sachdev, preprint cond-mat/0609106.
27. J.V. Steele, preprint nucl-th/0010066.
28. T. Schäfer, C.W. Kao, S.R. Cotanch, Nucl. Phys. A **762**, 82 (2005) [nucl-th/0504088].
29. Z. Nussinov, S. Nussinov, preprint cond-mat/0410597.
30. Y. Nishida, D.T. Son, preprint cond-mat/0604500.
31. Y. Nishida, D.T. Son, preprint cond-mat/0607835.
32. P. Arnold, J.E. Drut, D.T. Son, preprint cond-mat/0608477.
33. T.D. Lee, C.N. Yang, Phys. Rev. **105**, 1119 (1957).
34. K. Huang, C.N. Yang, Phys. Rev. **105**, 767 (1957).
35. L.P. Gorkov, T.K. Melik-Barkhudarov, Sov. Phys. JETP **13**, 1018 (1961).
36. G. Rupak, preprint nucl-th/0605074.
37. J.W. Chen, E. Nakano, preprint cond-mat/0610011.
38. F. Coester, S. Cohen, B. Day, C.M. Vincent, Phys. Rev. C **1**, 769 (1970).
39. S.K. Bogner, A. Schwenk, R.J. Furnstahl, A. Nogga, Nucl. Phys. A **763**, 59 (2005) [nucl-th/0504043].
40. N. Kaiser, S. Fritsch, W. Weise, Nucl. Phys. A **697**, 255 (2002) [nucl-th/0105057].

Eur. Phys. J. A **31**, 409–414 (2007)
DOI 10.1140/epja/i2006-10163-0

THE EUROPEAN
PHYSICAL JOURNAL A

Special Article – QNP 2006

Neutrino-nucleus interactions

T.W. Donnelly[a]

Center for Theoretical Physics, Laboratory for Nuclear Science and Department of Physics, Massachusetts Institute of Technology, Cambridge, MA 02139, USA

Received: 24 September 2006
Published online: 16 February 2007 – © Società Italiana di Fisica / Springer-Verlag 2007

Abstract. Inclusive electron scattering cross-sections in the quasielastic and resonance regions for few GeV electrons are well represented in terms of scaling functions and scaling variables, the so-called superscaling analysis (SuSA). The concepts of scaling of the first and second kinds and superscaling are discussed, as are several mechanisms which are known to yield scaling violations. Given the high quality of scaling for cross-sections at appropriate kinematics, it is shown how the ideas can be turned around to provide predictions for both charge-changing and neutral current neutrino reactions with nuclei at comparable kinematics.

PACS. 23.40.Bw Weak-interaction and lepton (including neutrino) aspects – 24.10.Jv Relativistic models – 25.30.-c Lepton-induced reactions – 25.30.Fj Inelastic electron scattering to continuum

1 Scaling of the first kind

A basic contention in the present discussions is the following: While it may not be sufficient, it is necessary that in modeling electroweak interactions with nuclei at few GeV energies a good understanding of existing inclusive electron scattering data must be reached before one can have much confidence in predictions of neutrino reactions with nuclei. Presently, modeling is not completely able to provide good enough understanding, and consequently other approaches must also be pursued. In particular, the approach followed here uses concepts of scaling [1–14] (0th, 1st and 2nd kinds and superscaling: the superscaling analysis, SuSA).

For inclusive semi-leptonic electroweak processes, in addition to the lepton scattering angle, one has energy transfer ω and 3-momentum transfer q (or, equivalently, Q^2 and ν). First, one replaces ω with the scaling variable $y = y(q, \omega)$, given by the lowest value of the missing momentum at the lowest missing energy kinematically allowed for semi-inclusive knockout of nucleons from the nucleus. While an exact formula can be written [1], the following is a reasonable approximation [2,3] for the y-scaling variable:

$$y = y(q, \omega) \cong \sqrt{\tilde{\omega}(2m_N + \tilde{\omega})} - q \,,$$

where $\tilde{\omega} \equiv \omega - E_s$ with E_s the separation energy and m_N the nucleon mass. This choice is motivated by the understanding that the main contributions to quasielastic

(QE) inclusive electroweak cross-sections arise from the above kinematic region, making y play a special role.

Second, one defines the function

$$F(q, y) \equiv \frac{\mathrm{d}^2\sigma/\mathrm{d}\Omega_e\mathrm{d}\omega}{A\Sigma_{eN}^{eff}} \,,$$

where the numerator is the double-differential inclusive electron scattering cross-section and the denominator

$$A\Sigma_{eN}^{eff} = Z\overline{\sigma}_{ep}^{elastic} + N\overline{\sigma}_{en}^{elastic}$$

is proportional to the effective (*i.e.*, incorporating relativistic effects; see [2,3]) single-nucleon cross-section with proton and neutron numbers as weighting factors. Typical results are shown in fig. 1 (and more may be found in [1]). As the momentum transfer becomes very large one sees that the results become independent of q and therefore $F(q, y)$ tends towards a universal function which depends only on the scaling variable

$$F(q, y) \to F(y) \equiv F(\infty, y);$$

one calls this behavior *scaling of the 1st kind*. Note that in the region above the QE peak ($y \cong 0$) where resonances, meson production and the start of Deep Inelastic Scattering (DIS) enter, 1st-kind scaling is, not unexpectedly, badly violated.

[a] e-mail: `donnelly@mit.edu`

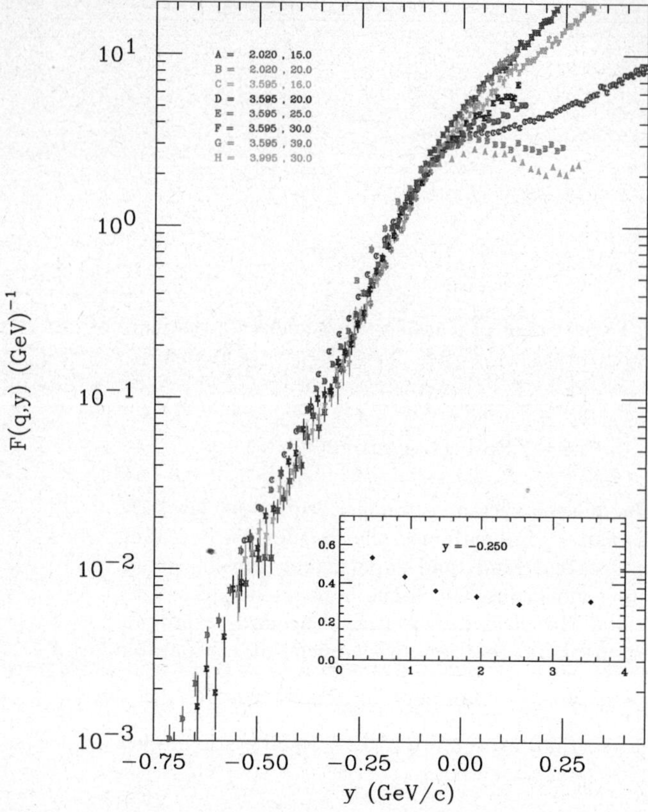

Fig. 1. Scaling of the 1st kind in ^{56}Fe. The various data sets correspond to different beam energies and scattering angles, and therefore to different values of momentum transfer. The inset shows the approach to scaling as a function of q at $y = -0.25$. See [1] for references to the data.

2 Scaling of the second kind

Next, one introduces a characteristic momentum scale for a given nuclear species

$$k_A = \sqrt{\langle k^2 \rangle_A}$$

and uses this to define a dimensionless function

$$f(q, y) \equiv k_A \cdot F(q, y).$$

Correspondingly, one wishes to introduce a dimensionless scaling variable ψ and then to plot $f(q, \psi)$ *versus* ψ for fixed kinematics but different nuclear species. The Relativistic Fermi Gas (RFG) model [4] is used to motivate the choice of the scaling variable. In the RFG one has

$$[k_A]^{RFG} = k_F$$

and the dimensionless RFG scaling variable is given by

$$\psi = \frac{1}{\sqrt{\xi_F}} \frac{\lambda - \tau}{\sqrt{(1 + \lambda)\tau + \kappa\sqrt{\tau(1 + \tau)}}}$$

$$\cong \frac{1}{\eta_F} \left[\lambda\sqrt{1 + 1/\tau} - \kappa \right]$$

$$\cong y/k_A \,,$$

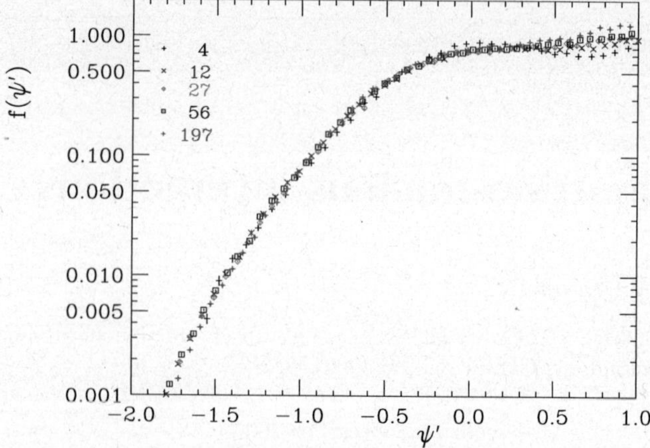

Fig. 2. Scaling of the 2nd kind at 3.6 GeV beam energy and a scattering angle of 16 degrees for nuclei ranging from ^{4}He to ^{197}Au. See [2,3] for references to the data.

where $\kappa \equiv q/2m_N$, $\lambda \equiv \omega/2m_N$, $\tau \equiv |Q^2|/4m_N^2 = \kappa^2 - \lambda^2 > 0$, $\eta_F \equiv k_F/m_N \sim 0.25$ and $\xi_F \equiv \sqrt{1 + \eta_F^2} - 1 \cong \eta_F^2/2 \sim 0.03$. In performing detailed analyses of data, the scaling variable actually used is denoted ψ' and has a small empirical shift in energy loss, E_{shift}, in going from ω to $\omega - E_{shift}$ in the above expressions.

The results are displayed in fig. 2. In the scaling region ($\psi' < 0$) a universal behavior is seen, with very little dependence on the nuclear species; that is, one observes *scaling of the 2nd kind*. Again, in the region above $\psi' = 0$ where resonances, meson production and the start of DIS enter the 2nd-kind scaling is not as good (see the discussions in sect. 4, however).

3 Superscaling

Although the amount of data separated into longitudinal (L) and transverse (T) responses is small, one can attempt a scaling analysis with what does exist. The inclusive cross-section may be written

$$\frac{d^2\sigma}{d\Omega_e d\omega} = \sigma_M \left[v_L R_L(q, \omega) + v_T R_T(q, \omega) \right],$$

where $v_L = |Q^2/q^2|^2$ and $v_T = \frac{1}{2}|Q^2/q^2| + \tan^2\theta_e/2$ are the usual Rosenbluth factors for inclusive electron scattering and σ_M is the Mott cross-section. From the individual response functions one can define corresponding scaling functions:

$$F_{L,T}(q, y) \equiv \frac{R_{L,T}(q, \omega)}{\left[A\Sigma_{eN}^{eff} \right]_{L,T} / \sigma_M v_{L,T}},$$

$$f_{L,T}(q, y) \equiv k_A \cdot F_{L,T}(q, y).$$

Let us focus on the longitudinal scaling function. In contrast to the transverse sector, here one does not expect large contributions from either meson-exchange currents (MEC) and their associated correlations [6–8] or inelastic

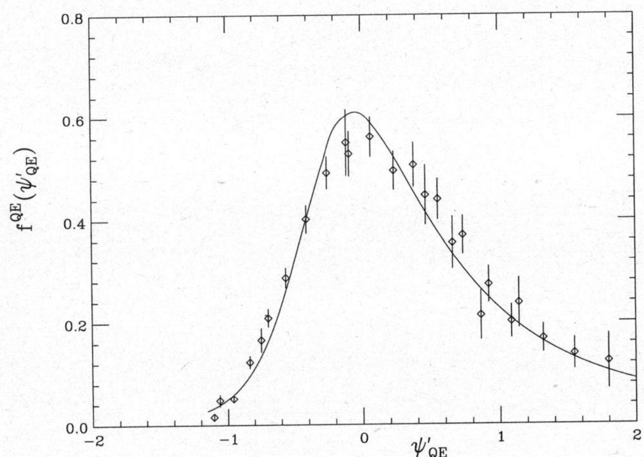

Fig. 3. Longitudinal scaling function from the analysis in [15], together with a parametrized fit from [5].

contributions from pion production and electroproduction of resonances such as the Δ (for example, see [9] for further discussions). Accordingly, the longitudinal scaling function provides a unique window into the roles played by the nuclear dynamics, both from initial-state energy and momentum distributions and from final-state interaction effects.

What results for the longitudinal scaling function is shown in fig. 3 where data together with a parametrization are shown plotted *versus* the scaling variable (which is now denoted ψ'_{QE} to emphasize the fact that it builds in the kinematics of elastic eN scattering, but no inelasticity as in the following section). This is seen to be both independent of q (scaling of the 1st kind) and also independent of nuclear species (scaling of the 2nd kind); that is, one has *superscaling*. These phenomenological results should be compared with the RFG model where the scaling function is given by a parabola which lies in the range $-1 < \psi'_{QE} < +1$ and peaks at the value 0.75. Clearly the RFG, while roughly correct, is not so in detail: the peak value is about 25% too high and the scaling function is symmetrical, whereas the experimental one has a pronounced asymmetry with a tail extending to high-energy loss (positive scaling variable). In recent modeling this behavior has also been observed in some cases [10,11] and appears to occur only when relatively strong final-state interactions are present.

Furthermore, in the RFG one has

$$[f_L]^{RFG} = [f_T]^{RFG} = [f]^{RFG}$$

which has been called *scaling of the 0th kind*. Indeed, if it were not for contributions from resonances, meson production and DIS (which should not scale, since they involve different elementary cross-sections, not elastic eN scattering, and since the scaling variables constructed above are appropriate only for QE scattering; see the discussions to follow), and for effects from MEC and their associated correlations one would expect scaling of the 0th kind to be found. In recent modeling [10–14] this behavior has been verified.

4 Scaling in the resonance region

In recent work [12] the resonance region has been studied as well as the QE region. First, the QE scaling variable introduced above must be replaced by a new one when, instead of eN elastic scattering, one excites the nucleon to a resonance of mass m_*:

$$\psi \to \psi_* = \frac{1}{\sqrt{\xi_F}} \frac{\lambda - \tau\rho_*}{\sqrt{(1+\lambda\rho_*)\tau + \kappa\sqrt{\tau(1+\tau\rho_*^2)}}},$$

where $\rho_* \equiv 1 + (m_*^2 - m_N^2)/|Q^2| \geq 1$ (and $\rho_* = 1$ for elastic, *i.e.*, QE, scattering). As above, ψ'_* is defined using the same expression, but with the energy shift E_{shift}.

Second, the fact that the elementary cross-section is not that of elastic eN scattering has to be taken into account. One begins by writing the inclusive cross-section in the form

$$\frac{\mathrm{d}^2\sigma}{\mathrm{d}\Omega_e\mathrm{d}\omega} \equiv \Sigma^{QE} + \Sigma',$$

namely, into a piece that is called quasielastic and is assumed to obey scaling of the 0th kind, plus a piece that contains the remaining contributions. The QE piece is thus assumed to be

$$\Sigma^{QE} \equiv \frac{A\Sigma_{eN}^{eff}}{k_A} \cdot f_L(\psi'_{QE}),$$

i.e., it employs the *longitudinal* scaling function for both L and T contributions to the cross-section. The remainder, Σ', should then contain (at least) the contributions from inelastic eN scattering and MEC+correlation effects.

As a first approximation in [12] it was assumed that the dominant effect in the 1 GeV regime is from electroexcitation of the Δ, and that MEC+correlation effects, while not absent, are corrections to this. One then proceeds as follows: first, the QE contribution (called $[\Sigma^{QE}]_{expt}$, since the experimental scaling function was used) is subtracted from the total experimental cross-section to isolate the remainder:

$$[\Sigma']_{expt} = \left[\frac{\mathrm{d}^2\sigma}{\mathrm{d}\Omega_e\mathrm{d}\omega}\right]_{expt} - [\Sigma^{QE}]_{expt}.$$

Second, following the approach pursued in [12], one makes the assumption that the next most important contribution in this range of kinematics arises from electroproduction of the Δ. Of course this is only a rough approximation, since non-resonant pion production, the tails of other resonances or, alternatively, contributions from DIS are not totally absent; likewise MEC effects with their associated correlations can also play a role. However, the Δ is certainly important and, accordingly, the analysis proceeded by invoking dominance of this mode, implying that the remainder is to be divided by the effective $N \to \Delta$ electron scattering cross-section to define a new function

$$F^\Delta(q,y) \equiv \frac{[\Sigma']_{expt}}{A\Sigma_{N\Delta}^{eff}},$$

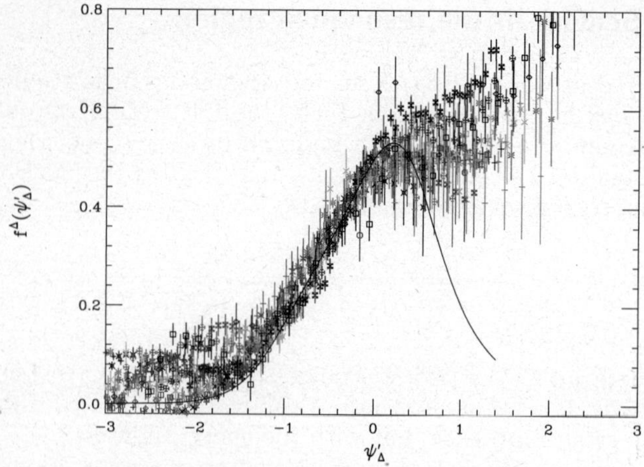

Fig. 4. Scaling in the resonance region (see text for discussion).

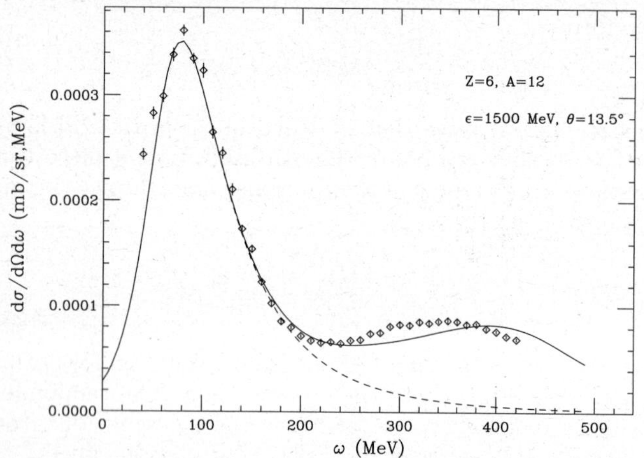

Fig. 5. Inclusive electron scattering from ^{12}C at beam energy 1.5 GeV and scattering angle 13.5 degrees, together with the results of using the phenomenological scaling functions discussed in the text.

where

$$A\Sigma_{N\Delta}^{eff} = Z\overline{\sigma}_{p\to\Delta^+}^{inelastic} + N\overline{\sigma}_{n\to\Delta^0}^{inelastic}\,.$$

As before, a dimensionless scaling function may also be defined:

$$f^\Delta(q,y) \equiv k_A \cdot F^\Delta(q,y)\,.$$

The results are then plotted in fig. 4 as a function of an inelastic scaling variable which in this initial analysis has been constructed using the centroid mass of the Δ, also an approximation:

$$\psi_\Delta' \equiv \psi_*'(m_* \to m_\Delta)\,.$$

Rather good scaling behavior is observed in the region below the Δ peak where $\psi_\Delta' = 0$, although, as expected, not at higher-energy loss where higher-lying resonances, DIS, etc., take over. In the region below the Δ peak there is some residual which is presumably at least partially due to effects from MEC and their associated correlations.

This means that one should have a reasonable representation of the total inclusive electron scattering cross-section for the 1 GeV energy regime for energies ranging

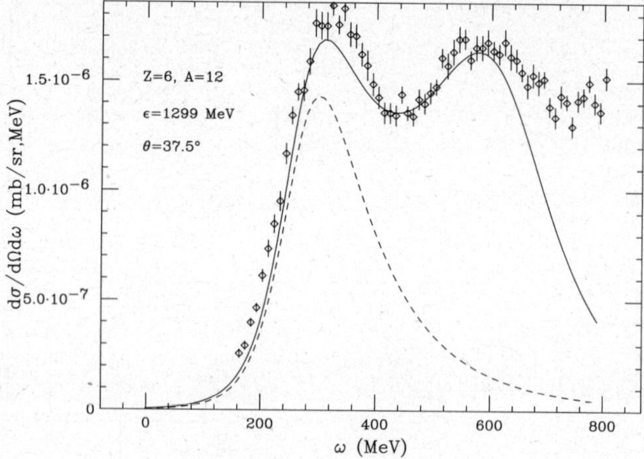

Fig. 6. As for the previous figure, but now at beam energy 1.3 GeV and scattering angle 37.5 degrees.

from below the QE peak (typically around $\psi_\Delta' \sim -1$ in fig. 4) up through the peak of the Δ. To test this one can reassemble the complete inclusive cross-section using the QE and Δ scaling functions with their attendant single-baryon cross-sections. Typical results are shown in figs. 5 and 6. Clearly, one has a rather successful understanding of the EM inclusive cross-section for this range of kinematics, leaving a residual typically of perhaps 10–15% to be accounted for by effects likely from MEC and their associated correlations.

5 Neutrino-nuclear cross-sections

Just as for the electron scattering reactions in the QE and Δ regions, the scaling functions determined above are employed, but now multiplied by the corresponding charge-changing (CC) neutrino reaction cross-sections for the Z protons and N neutrons in the nucleus. The CCν cross-section may be written:

$$\left[\frac{\mathrm{d}^2\sigma}{\mathrm{d}\Omega_{kk'}\,\mathrm{d}k'}\right]_\chi = \sigma_0 R_\chi\,,$$

where

$$\sigma_0 \equiv \frac{(G\cos\theta_c)^2}{2\pi^2}\left[k'\cos\widetilde{\theta}_{kk'}\right]^2$$

is the elementary cross-section (the analog of the Mott cross-section in sect. 3) containing the Cabibbo angle θ_c, the charged lepton momentum k' and an effective scattering angle $\theta_{kk'}$ defined via

$$\tan^2\widetilde{\theta}_{kk'} \equiv \frac{|Q^2|}{4\epsilon\epsilon' - |Q^2|}\,,$$

where ϵ and ϵ' are the incident neutrino and final-state charged lepton energies, respectively. The subscript $\chi = +(-)$ labels the specific case, *i.e.*, neutrinos (antineutrinos). The response function above may be decomposed into individual angle-independent responses, just as when

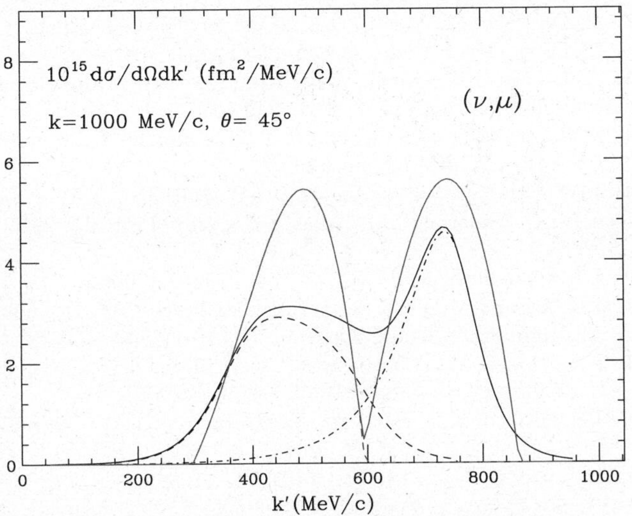

Fig. 7. Charge-changing muon neutrino cross-sections at neutrino energy $k = 1\,\text{GeV}$ and scattering angle 45 degrees. The right-hand peaks are from QE contributions while the left-hand ones are from the Δ. The upper curves are obtained using the RFG, while the lower ones are using the phenomenological scaling analysis.

discussing inclusive electron scattering. This time, since there is a vector (V)/axial-vector (A) parity-violating interference and since the axial-vector current is not conserved, one has more terms:

$$R_\chi = \left[\widehat{V}_{CC}R_{CC} + 2\widehat{V}_{CL}R_{CL} + \widehat{V}_{LL}R_{LL} + \widehat{V}_T R_T\right]$$
$$+ \chi\left[\widehat{V}_{T'}R_{T'}\right],$$

where $R_K = R_K^{VV} + R_K^{AA}$ for $K = CC,\ CL,\ LL,\ T$ and R_K^{VA} for $K = T'$. The various contributions are labeled with C for charge, L for longitudinal and T or T' for the two types of transverse responses which can enter. The kinematic factors $\widehat{V}_K$ are generalizations of the familiar Rosenbluth factors given in sect. 3 which account for the finite mass of the charged lepton (see [12] for specifics). For charge-changing muon neutrino processes in the QE region one has the elementary reactions $\nu_\mu + n \rightarrow p + \mu^-$ and $\overline{\nu}_\mu + p \rightarrow n + \mu^+$, while in the Δ region one has $\nu_\mu + p \rightarrow \Delta^{++} + \mu^-$, $\nu_\mu + n \rightarrow \Delta^+ + \mu^-$, $\overline{\nu}_\mu + p \rightarrow \Delta^0 + \mu^+$ and $\overline{\nu}_\mu + n \rightarrow \Delta^- + \mu^+$.

Typical results are shown in fig. 7. The two peaks are from QE (right) and Δ (left) regions, with the higher-lying curves for the RFG and the lower-lying from the phenomenological scaling analyses. Clearly, there are significant differences and, while providing a rough measure of the electroweak responses, the RFG results are higher and more localized than the SuSA results. Although not shown here, it should be noted that non-relativistic approximations to the kinematics and electroweak currents yield cross-sections that are even further from the SuSA curves in fig. 7.

6 Conclusions and discussion

Several conclusions emerge from the SuSA approach presented here:

- Scaling of the 1st and 2nd kinds for inclusive electron scattering appears to be well satisfied at the QE peak and below (the scaling region); the region above the QE peak clearly has contributions which do not scale, at least in the same way as the QE cross-sections do. These scaling violations are at least partially due to inelastic processes such as electroexcitation of the Δ.
- For the longitudinal response superscaling is observed, as expected, since this response sector has only relatively small contributions from Δ production and from MEC effects.
- In recent work [12] these ideas have been extended into the Δ region. In that approach the QE contributions were subtracted and the remainder analyzed in terms of a Δ-dominated scaling function and scaling variable; again good scaling is seen.
- The SuSA approach has been used to make predictions for charge-changing neutrino reactions in the few GeV energy region. Significant differences are observed with respect to conventional RFG modeling.

In on-going studies the scaling ideas are being extended in several ways:

- A relativized shell model study of neutrino reactions was undertaken [13], yielding successful scaling behavior.
- A relativistic impulse approximation study of the superscaling function was undertaken in [10,11]. What was found is that the universal superscaling function arises naturally from relativistic mean-field modeling, although not from the relativistic plane-wave impulsive approximation (RPWIA) or from modeling which employs the real parts of conventional optical potentials.
- 1p-1h MEC/correlation effects have been explored in detail in [6], while 2p-2h MEC/correlation effects are presently being incorporated in relativistic modeling of EW processes [7,8]. Although the latter study has not yet been fully executed, the expectation is that 10–15% of the transverse EM response in the region between the QE peak and the Δ peak could be due to such effects. These contributions enter as scaling violating effects. One should also note an asymmetric feature of the MEC effects: in the transverse contributions, which are dominant for neutrino reactions in the 1 GeV energy regime, the MEC contribute to the vector current, but not in leading order to the axial-vector current. Hence, it is necessary to understand such effects when attempting to model the neutrino reaction cross-sections to better than 10–15%.
- Extensions to neutral-current (NC) neutrino scattering in the QE region (u-channel inclusive *versus* t-channel inclusive) have been undertaken [14]. Further extensions to include NC processes in the Δ region are being contemplated, as are both CCν and NCν reactions at higher inelasticity than the Δ region.

This work is supported by the U.S. Department of Energy (DOE) under cooperative agreement No. DE-FC02-94ER40818.

References

1. D.B. Day, J.S. McCarthy, T.W. Donnelly, I. Sick, Annu. Rev. Nucl. Part. Sci. **40**, 357 (1990).
2. T.W. Donnelly, I. Sick, Phys. Rev. Lett. **82**, 3212 (1999).
3. T.W. Donnelly, I. Sick, Phys. Rev. C **60**, 065502 (1999).
4. W.M. Alberico, A. Molinari, T.W. Donnelly, E. Kronenberg, J.W. Van Orden, Phys. Rev. C **38**, 1801 (1988).
5. C. Maieron, T.W. Donnelly, I. Sick, Phys. Rev. C **65**, 025502 (2002).
6. J.E. Amaro, M.B. Barbaro, J.A. Caballero, T.W. Donnelly, A. Molinari, Phys. Rep. **368**, 317 (2002) and references therein.
7. A. De Pace, M. Nardi, W.M. Alberico, T.W. Donnelly, A. Molinari, Nucl. Phys. A **726**, 303 (2003).
8. A. De Pace, M. Nardi, W.M. Alberico, T.W. Donnelly, A. Molinari, Nucl. Phys. A **741**, 249 (2004).
9. M.B. Barbaro, J.A. Caballero, T.W. Donnelly, C. Maieron, Phys. Rev. C **69**, 035502 (2004).
10. J.A. Caballero, J.E. Amaro, M.B. Barbaro, T.W. Donnelly, C. Maieron, J.M. Udias, Phys. Rev. Lett. **95**, 252502 (2005).
11. J.A. Caballero, Phys. Rev. C **74**, 015502 (2006).
12. J.E. Amaro, M.B. Barbaro, J.A. Caballero, T.W. Donnelly, A. Molinari, I. Sick, Phys. Rev. C **71**, 015501 (2005).
13. J.E. Amaro, M.B. Barbaro, J.A. Caballero, T.W. Donnelly, C. Maieron, Phys. Rev. C **71**, 065501 (2005).
14. J.E. Amaro, M.B. Barbaro, J.A. Caballero, T.W. Donnelly, Phys. Rev. C **73**, 035503 (2006).
15. J. Jourdan, Nucl. Phys. A **603**, 117 (1996).

Eur. Phys. J. A **31**, 415–416 (2007)
DOI 10.1140/epja/i2006-10293-3

**THE EUROPEAN
PHYSICAL JOURNAL A**

Special Article – QNP 2006

Extraction of G_M^n from inclusive electron scattering on D, ^{4}He

A.S. Rinat[1], M.F. Taragin[1], and M. Viviani[2,a]

[1] Weizmann Institute of Science, Department of Particle Physics, Rehovot 76100, Israel
[2] INFN, Sezione di Pisa and Physics Department, University of Pisa, I-56100, Italy

Received: 22 December 2006
Published online: 22 March 2007 – © Società Italiana di Fisica / Springer-Verlag 2007

Abstract. We use the relation between Structure Functions (SFs) of nuclei A and nucleons N in order to fomulate a criterion which isolates the QE part out of the total inclusive cross-section. From data points around the QEP we extract the reduced neutron magnetic form factor $\langle \alpha_n = G_M^n/\mu_n G_d \rangle$. The latter shows an unexpected decrease up to $Q^2 = 10 \,\mathrm{GeV}^2$, the largest measured.

PACS. 14.20.Dh Protons and neutrons – 13.60.Hb Total and inclusive cross-sections (including deep-inelastic processes) – 25.30.Fj Inelastic electron scattering to continuum

1 Extraction of G_M^n

We report on the extraction of the neutron form factor (FF) $G_M^n(Q^2) = \mu_n G_d(Q^2)\alpha_n(Q^2)$ from reduced total cross-sections $K^A(E,\theta,\nu)$ of unpolarized electrons from a target A, defined as

$$K^A(E,\theta,\nu) = \frac{\mathrm{d}^2\sigma^A(E;\theta,\nu)}{\mathrm{d}\Omega\,\mathrm{d}\nu} \Big/ \sigma_M(E;\theta,\nu)$$

$$= \left[\frac{2xM}{Q^2} F_2^A(x,Q^2) + \frac{2}{M} F_1^A(x,Q^2)\mathrm{tg}^2(\theta/2) \right], \quad (1)$$

E,θ,ν are beam energy, scattering angle and energy loss; M, Q^2 and $x = Q^2/2M\nu$ are the nucleon mass, squared 4-momentum and the Bjorken variable; σ_M is the Mott cross-section.

Nuclear SFs in (1) are related to those for p,n by [1]

$$F_k^A(x,Q^2) = \int_x^A \frac{\mathrm{d}z}{z^{2-k}} f^{PN,A}(z,Q^2)\left[F_k^N\left(\frac{x}{z},Q^2\right) \right],$$

$$F_k^N = [F_k^p + F_k^n]/2, \quad (2)$$

where $f^{PN,A}$ is a calulable SF for a fictituous nucleus, composed of point-nucleons.

SFs of nucleons N may be decomposed in elastic and inelastic parts and eq. (2) defines the same for composite nuclei. The elastic N parts $NE^N = \sigma^{N;NE}$ are proportional to $\delta(1-x)$ and contain combinations of 4 FFs. $NI^p = \sigma^{NI;p}$ for a p has been measured, but there is no direct parallel information on the n. Indirect methods for F^n have to be devised [2], yielding in the end NI^N.

Given the above information on the averaged N, the nuclear counterparts $K^{A;NE,NI}$ follow from eq. (2). In view of the x-dependence of NE^N, $K^{A;NE}(x,Q^2)$ contains in addition to the FFs in $K^{N;NE}$ a factor $f^{PN,A}(x,Q^2)$. NI^A has to be computed from (1), using NI^N and $f^{PN,A}$.

For nearly the entire x-range $NI^A \gg NE^A$, but in the QE region both parts compete. An extraction of FFs in general, and of α_n in particular, requires an accurate isolation of $K^{A,NI}$ from $K^{A,data}$, eq. (1). The criterion is the approximate x-independence of the following LHS:

$$[K^{A,data}(x,Q^2) - K^{A,NI}(x,Q^2)]/f^{PN,A}(x,Q^2)$$
$$\Longleftrightarrow K^{A;NE}(x,Q^2)/f^{PN,A}(x,Q^2). \quad (3)$$

When fullfilled, one identifies the LHS of eq. (3) with $K^{A;NE}$. α_n can then be extracted provided the remaining 3 FFs are known. For G_E^p we choose the parametrizations in ref. [3] and for G_E^n the Galster one, as up-dated in ref. [4]. We tested two conflicting versions for the proton E/M ratio G_E^p/G_M^p: I) from a polarization transfer measurements [5], II) from a Rosenbluth separation.

The above identification leads to α_n, which from eq. (3) could be x- and A-dependent, but both dependences are very weak. One thus selects a continuous x-range around $x \approx 1$ (the QEP), the region which is maximally sensitive to changes in α_n. An average $\langle \alpha_n \rangle$ can then be determined over the selected range for each set. By construction, $NE^A(\langle \alpha_n \rangle) + NI(\mathrm{comp})$ fits the data in that region on the average.

2 Results and discussion

In figs. 1, 2 we display two out of the 30 analyzed data sets, namely for the D with $E = 4.045 \,\mathrm{GeV}$, $\theta = 55°$ [6], respectively $E = 18.476 \,\mathrm{GeV}$, $\theta = 10°$ [7]. For the former only the part up to $\nu \approx 2.75 \,\mathrm{GeV}$ is shown. Filled circles

a e-mail: michele.viviani@pi.infn.it

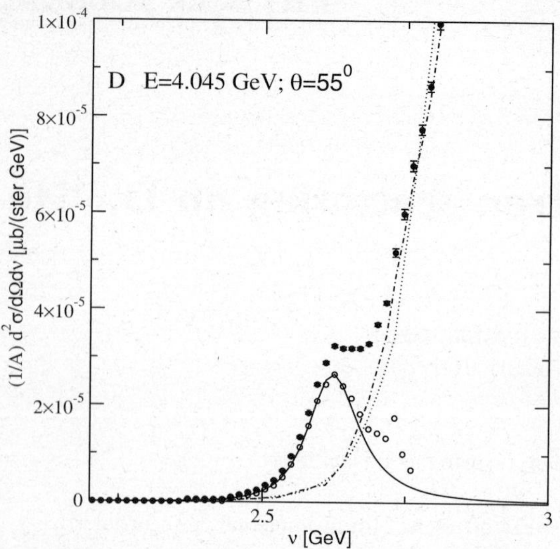

Fig. 1. For D: $E = 4.045\,\text{GeV}$, $\theta = 55°$; $\bar{Q}^2 = 4.900\,\text{GeV}^2$ [6]. NE curve for $\alpha_n = 0.958$.

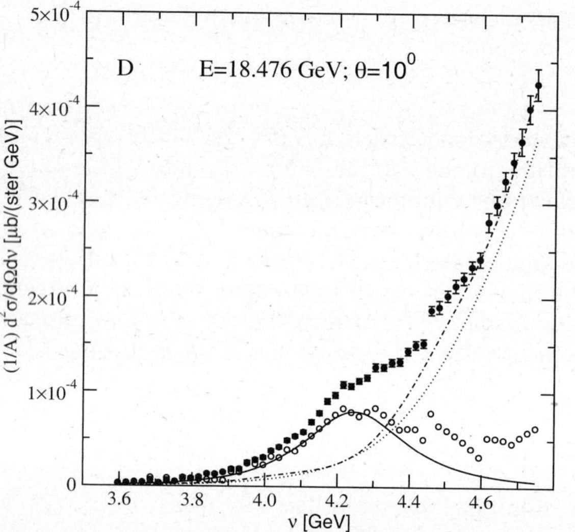

Fig. 2. For D: $E = 18.476\,\text{GeV}$, $\theta = 10°$; $\bar{Q}^2 = 8.03\,\text{GeV}^2$ [7]. NE curve for $\alpha_n = 0.879$.

are data with error bars, dotted lines represent $NI^D(\text{calc})$. Solid lines are $NE(\langle\alpha_n\rangle)$, while empty circles correspond to $NE^D(\text{extr}) = \text{data}^D - NI^D(\text{calc})$.

For a perfect theory and data $NE^D(\langle\alpha_n\rangle) + NI(\text{comp})$ should fit the data over the entire (x,ν)-range, but on the inelastic side of the QEP, about starting where $NI^D \approx NE^D$, the above-determined $NE^D(\text{extr})$ is not smooth and $< 15\%$ larger than $NE(\langle\alpha_n\rangle)$. It reaches a peak in cross-sections around the first maximum, which (if noticible) reflects inclusive N-Δ excitation on a bound nucleon, beyond which the discrepancy rapidly vanishes. Part of the observation may be due to noise in the data, but the near universality of the relative size and shape of the discrepancy makes one believe that the input F^p is at fault. That SF has been represented by the inclusive excitation of 5 resonances, and a slight increase of the N-Δ strength will only locally affect the small tail of F^p, and thus of

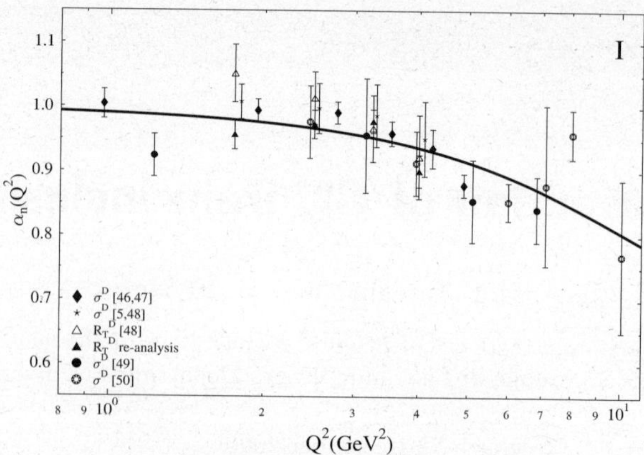

Fig. 3. $\alpha_n(\bar{Q}^2)$, extracted from D data.

F^A on the elastic tail side. It is not possible to change a single parameter out of the 20 odd parameters of the used representation without spoiling the fit [8]. In the figures we mark the local effect of an empirical, dot-dashed lines $NI^D(\text{emp})$, which causes $NE^D(\text{extr})$ and $NE^D(\langle\alpha_n\rangle)$ to coincide over the entire $x(\nu)$ interval of the data set. Incidentally, beyond the above-mentioned peak region, the unmodfified $F^{p,A}$ produce good, and frequently excellent agreement with the data.

Since different data sets occasionally overlap in Q^2, a consistency test requires the same $\alpha_n(Q^2)$ for those Q^2. That requirement appears well obeyed.

A mayor result of the analysis is the behavior of $\alpha_n(\bar{Q}^2)$, as a function of $\bar{Q}^2$, the representative value at the QEP. We display only the results for option I from D data (fig. 3), which show an unexpected, continuously decreasing α_n down to the largest measured $Q^2 \approx 10\,\text{GeV}^2$. For option II the decrease is even steeper. Preliminary CLAS data appear to be flatter [9]. The only available scarce and older He data draw on low-Q^2 data. For those, the underlying theory is less reliable than for higher values. For overlapping Q^2 one finds approximately the same α.

A much more extended account can be found in refs. [10].

References

1. R.L. Jaffe, *Proceedings of Los Alamos Summer School (1983)* (Wiley, New York, 1984) p. 360; S.A. Gurvitz, A.S. Rinat, Phys. Rev. C **65**, 024310 (2002); A.S. Rinat, M.F. Taragin, Phys. Rev. C **73**, 045201 (2006).
2. A.S. Rinat, M.F. Taragin, Phys. Lett. B **551**, 284 (2003).
3. H. Budd, A. Bodek, J. Arrington, arXiv:hep-ex/0308005.
4. A.F. Krutov, V.E. Troitsky, Eur. Phys. J. A **16**, 285 (2003).
5. M. Jones *et al.*, Phys. Rev. Lett. **84**, 1398 (2000); O. Gayou *et al.*, Phys. Rev. Lett. **88**, 092301 (2002).
6. I. Niculescu *et al.*, Phys. Rev. Lett. **85**, 1182 (2000).
7. S. Rock, *et al.*, Phys. Rev. Lett. **49**, 1139 (1982); Phys. Rev. D **46**, 24 (1992).
8. Y. Liang *et al.*, arXiv:hep-ph/0403058.
9. W. Brooks, private communication; Nucl. Phys. A **750**, 261 (2005).
10. A.S. Rinat, M.F. Taragin, M. Viviani, Phys. Rev. C **70**, 014003 (2004); Nucl. Phys. A **784**, 25 (2007) .

Eur. Phys. J. A **31**, 417–423 (2007)
DOI 10.1140/epja/i2006-10232-4

THE EUROPEAN
PHYSICAL JOURNAL A

Special Article – QNP 2006

Hard collisions of spinning protons: Past, present and future

A.D. Krisch[a]

Spin Physics Center, University of Michigan, Ann Arbor, MI 48109-1120, USA

Received: 8 November 2006
Published online: 21 March 2007 – © Società Italiana di Fisica / Springer-Verlag 2007

Abstract. There will be a review of the history of polarized proton beams, and a discussion of the unexpected and still unexplained large transverse spin effects found in several high-energy proton-proton spin experiments at the ZGS, AGS and Fermilab. Next, there will be a discussion of present and possible future experiments on the violent elastic collisions of polarized protons at IHEP-Protvino's 70 GeV U-70 accelerator in Russia and the new high-intensity 50 GeV J-PARC facility being built at Tokai in Japan.

PACS. 13.88.+e Polarization in interactions and scattering – 13.85.Dz Elastic scattering – 41.75.Ak Positive-ion beams

I will first discuss the violent elastic collisions of unpolarized protons. Figure 1 shows the cross-section for proton-proton elastic scattering plotted against a scaled P_t^2-variable that was proposed in 1963 [1] and 1967 [2]. This plot is from updates by Peter Hansen and me [3,4]. Notice that at small P_t^2 the cross-section drops off with a slope of about $10 \, (\mathrm{GeV}/c)^{-2}$. Fourier-transforming this slope gives the size and shape of the proton-proton interaction in the diffraction peak; it is a Gaussian with a radius of about 1 fermi. At medium P_t^2 there is a component with a slope of about $3 \, (\mathrm{GeV}/c)^{-2}$; however, this component disappears rapidly with increasing energy; at lab energies of a few TeV, it has totally disappeared. Thus, one can see a sharp destructive interference between the small-P_t^2 diffraction peak and the large-P_t^2 hard-scattering component. Since the diffraction peak is mostly *diffractive*, its amplitude must be mostly imaginary, as has been experimentally verified. Thus, the sharp destructive interference implies that the large-P_t^2 component is also mostly imaginary; thus, it is probably mostly *diffractive*. This large-P_t^2 component is probably the elastic *diffractive* scattering due to the *direct* interactions of the proton's constituents; its slope of about $1.5 \, (\mathrm{GeV}/c)^{-2}$ implies that these *direct* interactions occur within a Gaussian-shaped region of radius about 0.3 fermi.

Since the medium-P_t^2 component disappears at high energy, it is probably the *direct* elastic scattering of the two protons. This view is supported by the experimental fact that proton-proton elastic scattering is the only exclusive process that still can be precisely measured at TeV energies. To understand this, note that *direct* elastic scattering and all other exclusive processes must compete with each other for the total p-p cross-section, which is less than 100 millibarns. At TeV energies, there are certainly more than 10^5 exclusive channels in this competition; thus, each channel has an average cross-section of less than 1 microbarn. Moreover, since the medium-P_t^2 elastic component does not interfere strongly with either the large-P_t^2 or small-P_t^2 components, its amplitude is probably real. Also note that the large-P_t^2 component intersects the cross-section axis at about 10^{-5} below the small-P_t^2 *diffractive* component.

An earlier version of fig. 1 got me started in the spin business. In 1966, we carefully measured p-p elastic scattering at the ZGS at exactly 90°_{cm} from 5 to 12 GeV [5]; the sharp slope-change, shown by the stars, was apparently the first direct evidence for constituents in the proton. Dividing these 90°_{cm} p-p elastic cross-sections by 4 (due to the protons' particle identity) made all then existing proton-proton elastic data, above a few GeV, fit on a single curve [2]. During a 1968 visit to Ann Arbor, Robert Serber informed me that, in dividing the 90°_{cm} points by 4, I had made an assumption about the ratio of the spin singlet and triplet p-p elastic-scattering amplitudes. I recall being astounded and saying that I knew nothing about spin and certainly had not measured the spin of either proton. He said with a smile that both statements might be true; nevertheless, my nice fit required this assumption. Professor Serber, as usual, spoke quietly; however, as a student, I had learned that he was almost always right. Thus, I looked for data on proton-proton elastic scattering, above a few GeV, in the singlet and triplet spin states. I found that none existed and decided to try to polarize the protons in the ZGS.

At the 1969 New York APS Meeting, I learned that EG&G was the representative for a new polarized proton ion source made by ANAC in New Zealand. I dis-

[a] e-mail: krisch@umich.edu

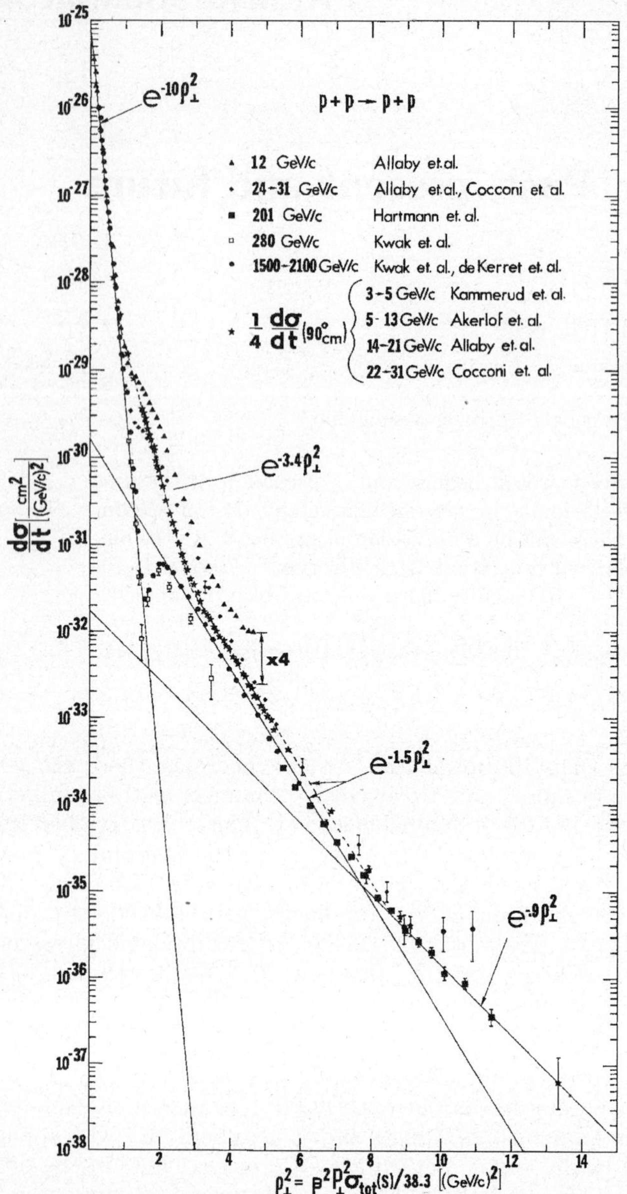

Fig. 1. Proton-proton elastic cross-sections plotted against the scaled P_t^2-variable. The 12 GeV/c points of Allaby *et al.* were not corrected for 90_{cm}° particle identity effects.

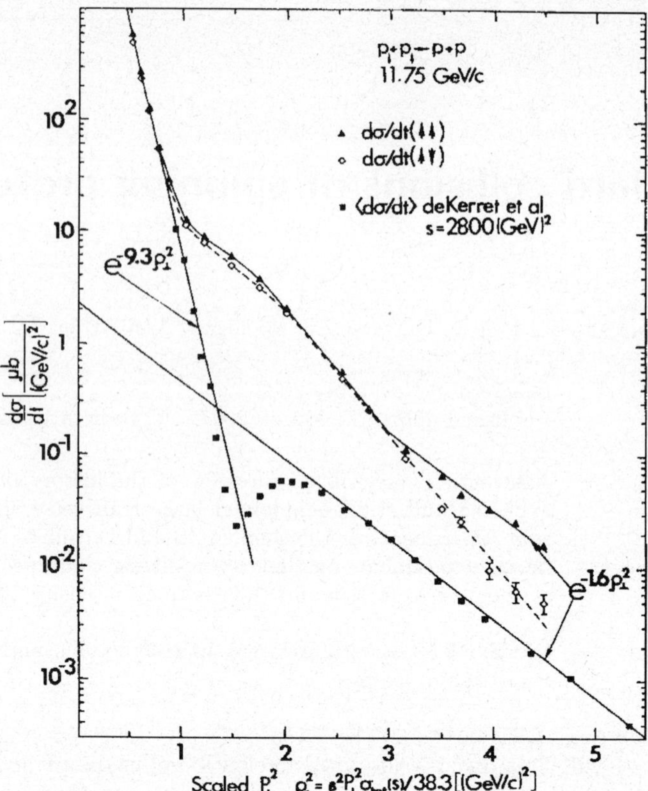

Fig. 2. The proton-proton elastic cross-section near 12 GeV in pure initial spin states is plotted against the scaled P_t^2-variable.

ized protons at a strong-focusing accelerator, such as the AGS, we probably would have failed and abandoned the polarized proton beam business. Fortunately, it worked at the weak-focusing ZGS, and experiments [6] soon showed that the p-p total cross-section had significant spin dependence; this surprised many people, including me.

Figure 2 shows our perhaps most important result [7] from the ZGS polarized proton beam. The 12 GeV proton-proton elastic cross-section in pure initial spin states is plotted against the scaled P_t^2-variable; in the diffraction peak the spin-parallel and spin-antiparallel cross-sections are essentially equal to each other and to the unpolarized data from the CERN ISR at $s = 2800\,\text{GeV}^2$; thus, in small-angle *diffractive* scattering, the protons in different spin states (and at different energies) all have about the same cross-section. The medium-P_t^2 component, which still exists near 12 GeV, has only a small spin dependence; again note that it has totally disappeared at $2800\,\text{GeV}^2$. However, the behavior of the large-P_t^2 hard-scattering component was a great surprise. When the protons' spins are parallel, they seem to have exactly the same behavior as the much higher-energy unpolarized ISR data; however, when their spins are antiparallel their cross-section drops with the medium-P_t^2 component's steeper slope. When this data first appeared in 1977 and 1978, people were totally astounded; most had thought that spin effects would disappear at high energies. In the following years, many theoretical papers tried to explain this unexpected behav-

cussed this with my long-time colleague, Larry Ratner, and then with Bruce Cork, Argonne's Associate Director, and Robert Duffield, Argonne's Director. They apparently decided it was a good idea; Duffield soon hired me as a consultant to Argonne at $100 per month. In 1973, after a lot of hard work by many people, the ZGS accelerated the world's first high-energy polarized proton beam.

One needed some hardware to overcome both intrinsic and imperfection depolarizing resonances. Fortunately, both types of resonances were fairly weak at the 12 GeV ZGS, which was the highest-energy weak-focusing accelerator ever built. All higher-energy accelerators wisely use strong focusing, which makes the depolarizing resonances much stronger. If we had first tried to accelerate polar-

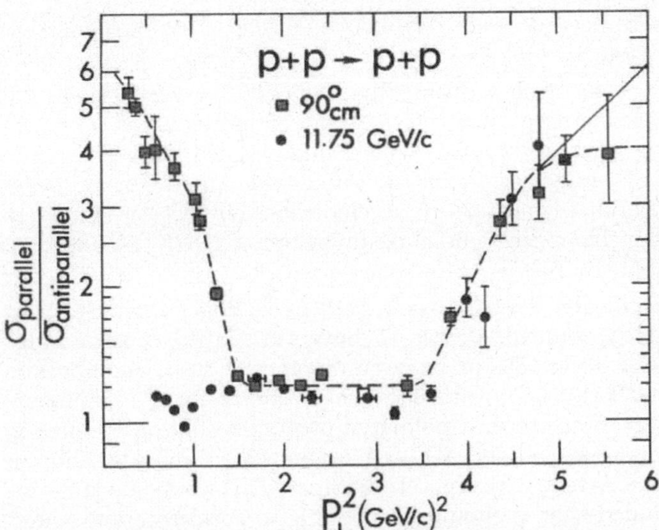

Fig. 3. The measured spins-parallel/spins-antiparallel cross-section ratio ($\sigma_{\uparrow\uparrow}/\sigma_{\uparrow\downarrow}$) is plotted against P_t^2.

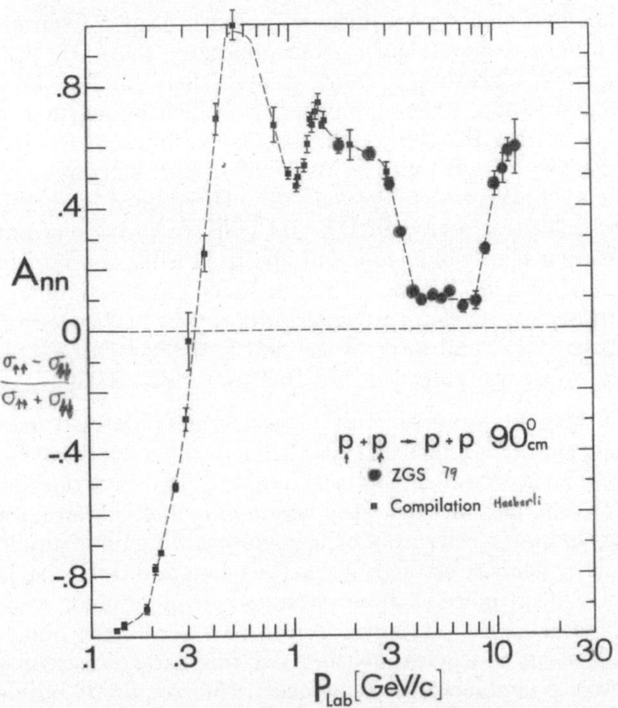

Fig. 4. $A_{nn} \equiv (\sigma_{\uparrow\uparrow} - \sigma_{\uparrow\downarrow})/(\sigma_{\uparrow\uparrow} + \sigma_{\uparrow\downarrow})$ is plotted against P_{Lab}.

ior; none were fully successful. In particular, the theory that is now called QCD, has been unable to deal with this data; Glashow once called this experiment "the thorn in the side of QCD". In his summary talk at Blois 2005, Stan Brodsky called this result "one of the unsolved mysteries of Hadronic Physics".

I learned something important from questions during two seminars about this result. Two distinguished physicists, Professor Weisskopf at CERN and then Professor Bethe at Copenhagen a week later, asked the same question, apparently independently. Each said that our big spin effect at large P_t^2 was quite interesting; but at 12 GeV, the spins-parallel/spins-antiparallel ratio was only big near 90°_{cm}, where particle identity was important for p-p scattering. They asked: how could one be sure that our large spin effect was due to hard-scattering at large P_t^2, rather than particle identity near 90°_{cm}? One would be foolish to ignore the comments of two such distinguished theorists, which were similar to Professor Serber's comment 10 years earlier.

However, it seemed that their question could not be answered theoretically; thus, we tried to answer it experimentally with a second ZGS experiment, which varied P_t^2 by holding the p-p scattering angle fixed at exactly 90°_{cm}, while varying the energy of the proton beam. This 90°_{cm} p-p elastic fixed-angle data [8] is plotted against P_t^2 in fig. 3, along with the fixed-energy data of fig. 2. There are large differences at small P_t^2, where the 90°_{cm} data are at very low energy; however, above P_t^2 of about $1.5\,(\mathrm{GeV}/c)^2$, the two sets of data fall right on top of each other. The point at $P_t^2 = 2.5\,(\mathrm{GeV}/c)^2$, where the ratio is near 1, is just as much at 90°_{cm}, as the $5\,(\mathrm{GeV}/c)^2$ point, where the ratio is 4. This data apparently convinced Professors Bethe and Weisskopf that the large spin effect was not due to 90°_{cm} particle identity and was a large-P_t^2 hard-scattering effect.

Figure 4 shows A_{nn} for p-p elastic scattering [4] plotted against the lab momentum, P_{Lab}; it includes the ZGS data from fig. 3 [8] plus some lower-energy data obtained from Willy Haeberli who is an expert on low-energy p-p spin experiments. At the lowest momentum (near $T = 10\,\mathrm{MeV}$) A_{nn} is very close to -1; thus, two protons with parallel spins can never scatter at 90°_{cm}. Next A_{nn} climbs rapidly to $+1$; then protons with antiparallel spins can never scatter at 90°_{cm}. Then at medium energy, there are some oscillations that were once thought to be due to dibaryon resonances, but are probably due to the onset of N^*-resonance production. In the ZGS region, A_{nn} first drops rapidly; it is next small and constant over a large range; it then rises rapidly to 0.6. These huge and sharp oscillations of A_{nn} are quite impressive.

I now turn to money and politics. In 1972 the AEC had agreed to shut down the ZGS in 1975 to get funding for PEP at SLAC. When the unique ZGS polarized beam started operating in 1973, the wisdom of this decision was questioned; AEC then set up a committee which extended ZGS operations through 1977. A second committee was set up in 1976; it extended operations of the unique ZGS polarized beam until 1979 [9]. Henry Bohm, the President of AUA (which operated Argonne), asked ERDA (was AEC) to set up a third committee to again extend ZGS running. But OMB objected, so there was no third committee; however, this effort had some benefit. When James Kane, of ERDA, responded negatively to Bohm, his justification for this was that it might now be possible to accelerate polarized protons in a strong-focusing accelerator, such as the AGS; he copied me on this letter.

We had also started interacting with Ernest Courant and others at Brookhaven about polarizing the AGS: first at a 1977 Workshop in Ann Arbor [10] and then at a 1978 Workshop at Brookhaven [11]. When he learned of Kane's letter, Brookhaven's Associate Director, Ronald Rau, asked for a copy of it. With this letter, he convinced William Wallenmeyer, the long-time Director of High Energy Physics at AEC, ERDA and DoE, to provide about $8 Million to Brookhaven, and about $2 Million split between Michigan, Argonne, Rice and Yale, for the challenging project of accelerating polarized protons in the strong-focusing AGS, and later in the 400 GeV ISABELLE collider, which was canceled, but then reborn as RHIC.

It was far more difficult to accelerate polarized protons in the strong-focusing AGS than in the weak-focusing ZGS. The strong-focusing principal, invented by Courant, Livingston and Snyder [12], made possible all modern large circular accelerators by using alternating quadrupole magnetic fields to strongly focus the beam and thus keep it small. Unfortunately, these strong quadrupole fields were very good at depolarizing protons. To accelerate polarized protons to 22 GeV at the AGS, one had to overcome 45 strong depolarizing resonances. This required: building lots of challenging hardware; significantly upgrading the AGS controls; and spending lots of time individually overcoming the 45 depolarizing resonances. Michigan built the 12 ferrite quadrupole magnets that were installed in the AGS to overcome its 6 intrinsic resonances by rapidly jumping the AGS's vertical betatron tune through each resonance. Brookhaven was building their 12 power supplies; each power supply had to provide 1500 amps at 15000 volts (about 22 MW) during each quadrupole's $1.6\,\mu s$ rise time. Overcoming the many imperfection depolarizing resonances (occurring every 520 MeV) required programming the AGS's 96 small correction dipole magnets to form a horizontal B-field wave of 4 oscillations at the instant when the proton energy passed through $G\gamma = 4$; then, about 20 ms later in the AGS cycle, when $G\gamma$ was 5, the 96 magnets had to form a horizontal B-field wave with 5 oscillations, etc. ($G = 1.79285$ is the proton's anomalous magnetic moment, while $\gamma = E/m$.).

After all this hardware was installed, an even larger problem was tuning the AGS. In 1988, when we accelerated polarized protons to 22 GeV, we needed 7 weeks of exclusive use of the AGS; this was difficult and expensive. Once a week, Nicholas Samios, Brookhaven's Director, would visit the AGS Control Room to politely ask me how long the tuning would continue and to remind us that it was costing $1 million a week. Moreover, it was soon clear that, except for Larry Ratner (then at Brookhaven) and me, no one could tune through these many resonances; thus, for some weeks, Larry and I worked 12-hour shifts 7 days each week. Larry was older than me; after 5 weeks he collapsed. While I was younger than Larry, I thought it unwise to try to work 24-hour shifts every day. Thus, I asked our Postdoc, Thomas Roser, who until then had worked mostly on polarized targets and scattering experiments, if he wanted to learn accelerator physics in a hands-on way for 12 hours every day. Thomas apparently learned

well; he now leads Brookhaven's Accelerator-Collider Division.

One benefit from this difficult 7-week period [13] was learning that our method of individually overcoming each resonance, which had worked so well at the ZGS, might work at the AGS, but would not be practical at higher-energy accelerators. This lesson helped to launch our Siberian-snake programs at IUCF [14] and then SSC [15,16].

In the 1980s, a new proton collider, the SSC, was being planned; it was to have two 20 TeV proton rings each about 80 km in circumference. Owen Chamberlain and Ernest Courant encouraged me to form a collaboration to insure that polarized protons would be possible in this new Collider. We first organized a 1985 workshop in Ann Arbor, with Kent Terwilliger. This workshop [15] concluded that it should be possible to accelerate and maintain the polarization of 20 TeV protons in the SSC, but only if the new Siberian-snake concept of Derbenev and Kondratenko [17] really worked; otherwise, it would be totally impractical. Recall that it took 49 days to correct the 45 depolarizing resonances at the AGS —about one a day. Each 20 TeV SSC ring would have about 36000 depolarizing resonances to correct. These higher-energy resonances would be much stronger and harder to correct; but even at one per day, this would require about 100 years of tuning for each ring. The workshop also concluded that one must prove experimentally that the *too-good-to-be-true* Siberian snakes really worked; otherwise, there would be no approval to install the 26 Siberian snakes needed in each SSC ring.

Fortunately, Indiana's IUCF was then building a new 500 MeV synchrotron Cooler Ring. Some of us workshop participants then collaborated with Robert Pollock and others at IUCF to build and test the world's first Siberian snake in the Cooler Ring. We brought experience with synchrotrons and high-energy polarized beams, while the IUCF people brought experience with low-energy polarized beams and the CE-01 detector, which was our polarimeter. In 1989, we demonstrated that a Siberian snake could easily overcome a strong imperfection depolarizing resonance [14]. For 13 years we continued these experiments and learned many things about spin-manipulating polarized beams; after the Cooler Ring shut down in 2002, this program was continued at the 3 GeV COSY in Juelich.

In 1990 we formed the SPIN Collaboration and submitted Expression of Interest EOI001 to SSC: to accelerate and store polarized protons at 20 TeV, and to study spin effects in 20 TeV p-p collisions. It was submitted a week before the deadline, which made it SSC EOI001 [16]. Thus, we made the first presentation to the SSC PAC before a huge audience that included many newspaper reporters and TV cameras. Perhaps partly because of this publicity, we were soon *partly* approved by SSC Director Roy Schwitters. By *partly* I mean that he decided to add 26 empty spaces for Siberian snakes in each SSC Ring; each space was about 20 m long, which added about 0.5 km in each Ring. Unfortunately, the SSC was canceled around 1993, before it was finished, but after $2.5 billion was spent. Nevertheless, our detailed studies of the behavior and spin-manipulation of polarized protons at IUCF and COSY helped in developing polarized beams around the world: Brookhaven now has 250 GeV polarized protons in each RHIC ring [18]; perhaps someday CERN's 7 TeV LHC might have polarized protons.

We eventually accelerated polarized protons to 22 GeV in the AGS [13] and obtained some A_{nn} data [19] well above the ZGS energy of 12 GeV; but we never had enough beam time to get precise A_{nn} data at high-P_t^2. However, during tune-up runs for the A_{nn} experiment, we used the unpolarized AGS proton beam to test our polarized proton target and double-arm magnetic spectrometer by measuring A_n in 28 GeV proton-proton elastic scattering; this data resulted in an interesting surprise. Despite QCD's inability to explain the big A_{nn} from the ZGS, our QCD friends had made a firm prediction that the one-spin A_n must go to 0; moreover, this prediction would become more firm at higher energies and in more violent collisions. But above $P_t^2 = 3\,(\mathrm{GeV}/c)^2$, A_n instead began to deviate from 0 and was quite large at $P_t^2 = 6\,(\mathrm{GeV}/c)^2$. This led to more controversy [20]; some QCD supporters said that our A_n data must be wrong.

Experimenters take such accusations seriously. Thus, we started preparing an experiment that could study A_n at high-P_t^2 with better precision. Our spectrometer worked well, but we could only use about 0.1% of the AGS beam intensity, because a higher-intensity beam would heat our Polarized Proton Target (PPT) and depolarize it. Thus, we started building a new PPT that could operate with 20 times more beam intensity; this required ^{4}He evaporation cooling at 1 K, which has much more cooling power than our earlier ^{3}He evaporation PPT at 0.5 K. However, to maintain a target polarization near 50% at 1 K, the PPT model required increasing the B-field from 2.5 to 5 tesla. Thus, we ordered a high-quality 5 T superconducting magnet from Oxford Instruments, with a B-field uniformity of a few 10^{-5} over the PPT's 3 cm diameter volume. We also obtained a Varian 20 W at 140 GHz Extended Interaction Oscillator; it was apparently the highest-power 140 GHz microwave source available. As the PPT assembly started in 1989, we worried that, if the PPT model was wrong, the polarization might be only 10%; instead, we were very lucky; it was 96% [21].

Moreover, the target polarization averaged 85% for a 3-month-long run with high-intensity AGS beam in early 1990. As shown in fig. 5, this let us precisely measure A_n at even larger P_t^2. When these precise new data were published [22], some theorists seemed quite unhappy; they still believed the QCD prediction that A_n must go to 0, but they now refused to state at what P_t^2 or energy this prediction would become valid. They also now said that QCD might not work for elastic scattering, which they now considered less fundamental than inelastic scattering, where they said QCD should work. Thus, one result of our experiments was to make both elastic-scattering experiments and spin experiments unpopular in some circles.

Other experimenters started doing inclusive polarization experiments, especially at Fermilab. Figure 6 shows

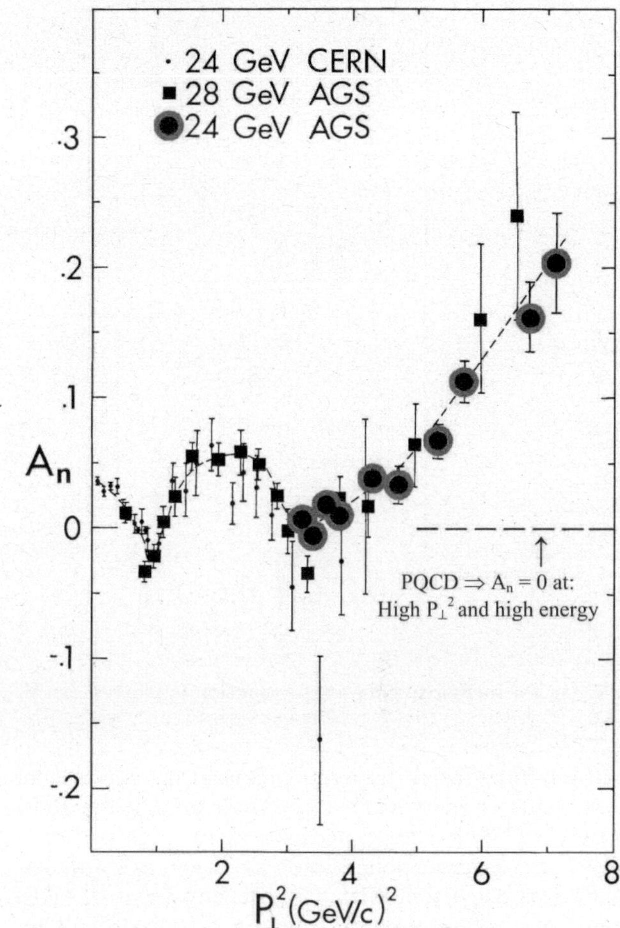

Fig. 5. $A_n \equiv (\sigma_\uparrow - \sigma_\downarrow)/(\sigma_\uparrow + \sigma_\downarrow)$ is plotted against P_t^2 for p-p elastic scattering.

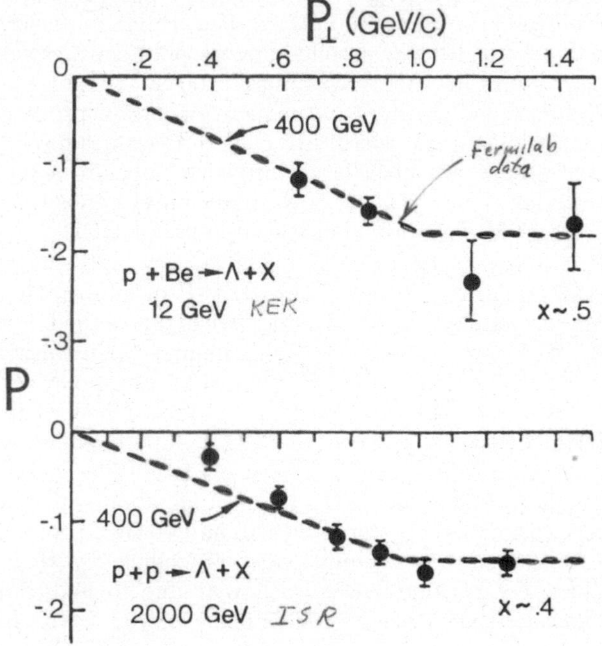

Fig. 6. The inclusive Λ polarization is plotted against P_t.

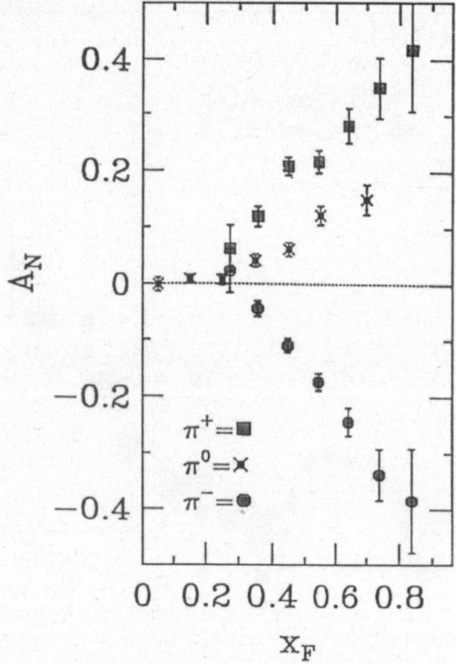

Fig. 7. A_n for inclusive π-meson production is plotted *vs.* X_F.

the 400 GeV inclusive hyperon polarization experiments from the 1970s and 1980s, led by Pondrum, Devlin, Heller and Bunce [23]; it clearly shows a small polarization at small P_t and a larger polarization at larger P_t. Moreover, their data is consistent with 12 GeV data from the KEK PS and with 2000 GeV data from the CERN ISR. These data do not support the QCD prediction that inelastic spin effects disappear at high energy or high P_t^2.

Another group at Fermilab, led by Yokosawa, developed a polarized secondary beam using the polarized protons from polarized hyperon decay. The beam's intensity was only about 10^5 per second, but its polarization was about 50% and its energy was about 200 GeV. They obtained some nice A_n data on inclusive π-meson production [24], which are shown in fig. 7. The A_n values for π^+ and π^- mesons are both large but with opposite signs, while A_n for the π^0 data is 50% smaller and is positive. These 200 GeV data provide little support for QCD.

We tried to measure spin effects in very-high-energy p-p scattering at UNK, which IHEP-Protvino started building around 1986; IHEP and Michigan signed the NEPTUN-A Agreement in 1989. Michigan's main contribution was a 12 tesla at 0.16 K Ultra-cold Spin-polarized Jet. UNK's circumference would be 21 km with 3 rings: a 400 GeV warm ring and two 3 TeV superconducting rings; its injector was IHEP's existing 70 GeV accelerator, U-70. By 1998 the UNK tunnel and about 80% of its 2200 warm magnets were finished, and 70 GeV protons were transferred into its tunnel with 99% efficiency. However, progress became slower each year due to financial problems; in 1998 Russia's MINATOM placed UNK on long-term standby.

IHEP Director, A.A. Logunov, had earlier suggested moving our experiment to IHEP's existing 70 GeV U-70

accelerator. By March 2002 the resulting SPIN@U-70 Experiment on 70 GeV p-p elastic scattering at high P_t^2 was fully installed, except for our detectors and Polarized Proton Target (PPT). However, just before our 4 tons of detectors, electronics and computers were to be shipped, the US Government suspended the US-Russian Peaceful Use of Atomic Energy Agreement started by President Eisenhower in 1953. Nevertheless, DoE asked us to send the shipment, since under the terms of the PUOAE Agreement, the experiment could be done exactly as planned. DoE faxed us a copy of the Agreement; thus, we sent the shipment; it arrived at Moscow airport on March 11, 2002. However, Russian Customs impounded it for 8 months before returning it to Michigan.

Despite this problem, we remain friends with our IHEP colleagues and there have been four SPIN@U-70 test runs using Russian detectors and an unpolarized target; we participated in the November 2001 and April 2002 runs. We hope that the US-Russian PUOAE Agreement is soon restarted so that the SPIN@U-70 experiment can continue and measure A_n at P_t^2 near $12\,(\mathrm{GeV}/c)^{-2}$.

However, over 4 years have now passed. If this international problem, involving the Iranian reactor, continues much longer, we may try to do a similar experiment at Japan's new very-high-intensity 50 GeV proton accelerator, J-PARC, which should run in 2008. If J-PARC can accelerate polarized protons to 50 GeV, then one could study the large and mysterious elastic spin effects in both A_n and A_{nn} for the first time in decades [25].

References

1. A.D. Krisch, Phys. Rev. Lett. **11**, 217 (1963); Phys. Rev. **135**, B1456 (1964).
2. A.D. Krisch, Phys. Rev. Lett. **19**, 1149 (1967); Phys. Lett. **44**, 71 (1973).
3. P.H. Hansen, A.D. Krisch, Phys. Rev. D **15**, 3287 (1977).
4. A.D. Krisch, Z. Phys. C **46**, S113 (1990).
5. C.W. Akerlof *et al.*, Phys. Rev. Lett. **17**, 1105 (1966); Phys. Rev. **159**, 1138 (1967).
6. E.F. Parker *et al.*, Phys. Rev. Lett. **31**, 783 (1973); W. deBoer *et al.*, Phys. Rev. Lett. **34**, 558 (1975).
7. J.R. O'Fallon *et al.*, Phys. Rev. Lett. **39**, 733 (1977); D.G. Crabb *et al.*, Phys. Rev. Lett. **41**, 1257 (1978).
8. A.M.T. Lin *et al.*, Phys. Lett. B **74**, 273 (1978); E.A. Crosbie *et al.*, Phys. Rev. D **23**, 600 (1981).
9. T. Khoe *et al.*, Part. Accel. **6**, 213 (1975); R.L. Walker, R.E. Diebold, G. Fox, J.D. Jackson, A.D. Krisch, T.A. O'Halloran, D.D. Reeder, N.P. Samios, H.K. Ticho, Report: *AEC Review Panel on the ZGS* (1976).
10. A.D. Krisch, A.J. Salthouse (Editors), AIP Conf. Proc. **42** (1978).
11. B. Cork, E.D. Courant, D.G. Crabb, A. Feltman, A.D. Krisch, E.F. Parker, L.G. Ratner, R.D. Ruth, K.M. Terwilliger, *Accelerating Polarized Protons in the AGS* (1978).
12. E.D. Courant, M.S. Livingston, H.S. Snyder, Phys. Rev. **88**, 1190 (1952).
13. F.Z. Khiari *et al.*, Phys. Rev. D **39**, 45 (1989).
14. A.D. Krisch *et al.*, Phys. Rev. Lett. **63**, 1137 (1989).

15. A.D. Krisch, A.M.T. Lin, O. Chamberlain (Editors), AIP Conf. Proc. **145** (1985).
16. SSC-EOI001 SPIN: Michigan, Indiana, Protvino, Dubna, Moscow, KEK, Kyoto (1990).
17. Ya.S. Derbenev, A.M. Kondratenko, AIP Conf. Proc. **51**, 292 (1979).
18. M. Bai *et al.*, Phys. Rev. Lett. **96**, 174801 (2006); *SPIN 2006 Symposium, Kyoto*, to be published in AIP Conf. Proc. (2007).
19. D.G. Crabb *et al.*, Phys. Rev. Lett. **60**, 2351 (1988).
20. A.D. Krisch, *Collisions of spinning protons*, Sci. Am. **257**, 42 (August 1987).
21. D.G. Crabb *et al.*, Phys. Rev. Lett. **64**, 2627 (1990).
22. D.G. Crabb *et al.*, Phys. Rev. Lett. **65**, 3241 (1990).
23. K. Heller *et al.*, Phys. Rev. Lett. **51**, 2025 (1983).
24. D.L. Adams *et al.*, Z. Phys. C **56**, 181 (1992).
25. I gave a somewhat similar talk at *Blois2005, Elastic and Diffractive Scattering*, edited by M. Haguenauer, B. Nicolescu, J.T.T. Van (Thê Gioi Publishers, Vietnam, 2006) p. 69.

Eur. Phys. J. A **31**, 424–428 (2007)
DOI 10.1140/epja/i2006-10189-2

THE EUROPEAN
PHYSICAL JOURNAL A

Special Article – QNP 2006

The neutrinoless double-beta decay: A test for new physics

A. Faessler[a]

Institute for Theoretical Physics, University of Tuebingen, Germany

Received: 8 October 2006
Published online: 22 February 2007 – © Società Italiana di Fisica / Springer-Verlag 2007

Abstract. The neutrinoless double-beta decay is not allowed in the Standard Model (SM) but it is allowed in most Grand Unified Theories (GUTs). The neutrino must be a Majorana particle (identical with its antiparticle) and must have a mass to allow the neutrinoless double-beta decay. Apart of one claim that the neutrinoless double-beta decay in ^{76}Ge is measured, one has only upper limits for this transition probability. But even the upper limits allow to give upper limits for the electron Majorana neutrino mass and upper limits for parameters of GUTs and the minimal R-parity violating supersymmetric model. One further can give lower limits for the vector boson mediating mainly the right-handed weak interaction and the heavy mainly right-handed Majorana neutrino in left-right symmetric GUTs. For that, one has to assume that the specific mechanism is the leading one for the neutrinoless double-beta decay and one has to be able to calculate reliably the corresponding nuclear matrix elements. In the present contribution, one discusses the accuracy of the present status of calculating the nuclear matrix elements and the corresponding limits of GUTs and supersymmetric parameters.

PACS. 23.40.-s β decay; double β decay; electron and muon capture – 21.60.-n Nuclear structure models and methods – 12.60.-i Models beyond the standard model

1 Physics beyond the standard model and the double-beta decay

In the standard model the neutrinoless double-beta decay is forbidden but it is allowed in most Grand Unified Theories (GUTs) where the neutrino is a Majorana particle (identical with its antiparticle) and where the neutrinos have a mass.

In GUTs, each beta decay vertex in fig. 1 can occur in eight different ways (see fig. 2).

The hadronic current changing a neutron into a proton can be left or right handed, the vector boson exchanged can be the light one W_1 or the heavy orthogonal combination W_2 mediating mainly a right-handed weak interaction and two different leptonic currents changing a neutrino into an electron. So the simple vertex for the beta decay can occur in eight different ways.

With two such vertices and the exchange of a light or a heavy mainly right-handed Majorana neutrino one already has 128 different matrix elements describing the neutrinoless double-beta decay [1] (see fig. 3).

At the R-parity violating vertex (see fig. 4) of the up- and the down-quarks and the SUSY electron $\tilde{e}$, one has the R-parity violating coupling constant λ'_{111} which is new compared to the standard model [2]. The formation of a pion by a down- and an up-quark produces a long-range

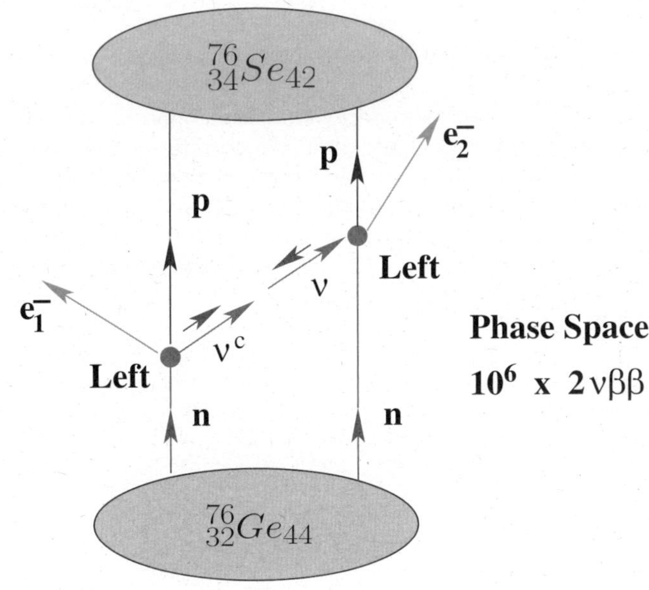

Fig. 1. Neutrinoless double-beta decay of ^{76}Ge through ^{76}As to the final nucleus ^{76}Se. The neutrino must be a Majorana particle that means identical with its antiparticle and must have a mass to allow this decay.

[a] e-mail: amand.faessler@uni-tuebingen.de

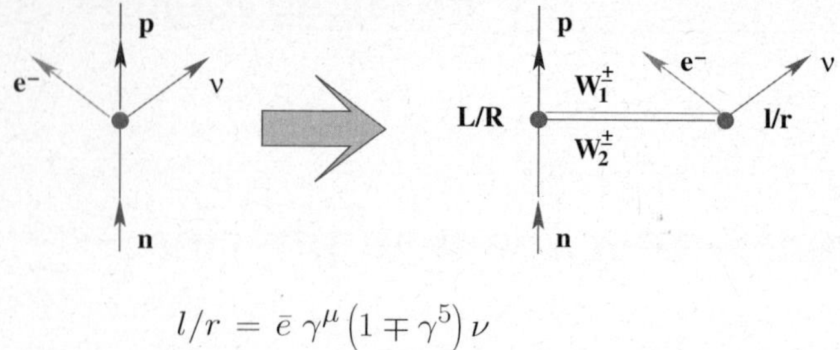

$$l/r = \bar{e}\,\gamma^\mu\left(1 \mp \gamma^5\right)\nu$$

$$L/R = p^+\left(g_V \mp g_A\,\vec{\sigma}\right)n$$

$$g_V = 1 \;;\; g_A = 1.25$$

Fig. 2. The single-beta decay changes a neutron into a proton by the emission of an electron and (in the standard model) an antineutrino. The microscopic version is shown on the right side of the figure.

$$\hat{H}^{\text{GUT}}_{\text{Weak}} = \frac{G_F \cos\vartheta_c}{\sqrt{2}}\left[1\cdot l\cdot L + tg\,\vartheta\, r\cdot L + tg\,\vartheta\, R\cdot l + \frac{M_1^2}{M_2^2}\, r\cdot R\right]$$

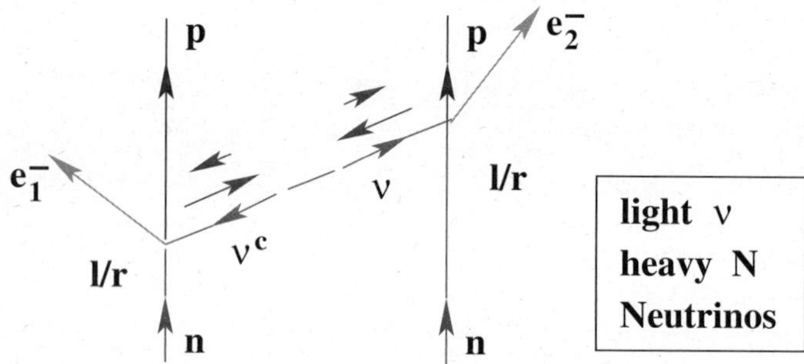

Fig. 3. Diagram of the double-beta decay.

neutrinoless double-beta decay transition operator. Due to the short-range Brueckner repulsive correlations between two nucleons, this is increasing the transition probability by a factor 10000. The upper limit derived from upper limits of the neutrinoless transition probability for λ'_{111} is therefore more stringent and one can derive from the neutrinoless transition probability in ^{76}Ge an upper limit for $|\lambda'_{111}| < 10^{-4}$.

To calculate the neutrinoless double-beta decay transition probability, we use Fermi's Golden Rule in second order:

$$T = \sum_k \frac{\langle f|\hat{H}_W|k\rangle\langle k|\hat{H}_W|i\rangle}{E_i - E_k},$$

$$E_i = E_{0+}(^{76}\text{Ge}),$$

$$E_k = E_k(e^-) + E_k(\nu) + E_{\text{nucleus}}(^{76}\text{As};k),$$

$$T = M_m\langle m_\nu\rangle + M_\theta\langle tg\,\vartheta\rangle$$

$$+ M_{WR}\left\langle\left(\frac{M_2}{M_R}\right)^2\right\rangle$$

$$+ M_{SUSY}\lambda'^2_{111} + M_{VR}\left\langle\frac{m_p}{M_{VR}}\right\rangle,$$

$$w = \frac{2\pi}{\hbar}|T|^2 p_f \le 4.4\cdot10^{-33}\ [\text{s}^{-1}].\tag{1}$$

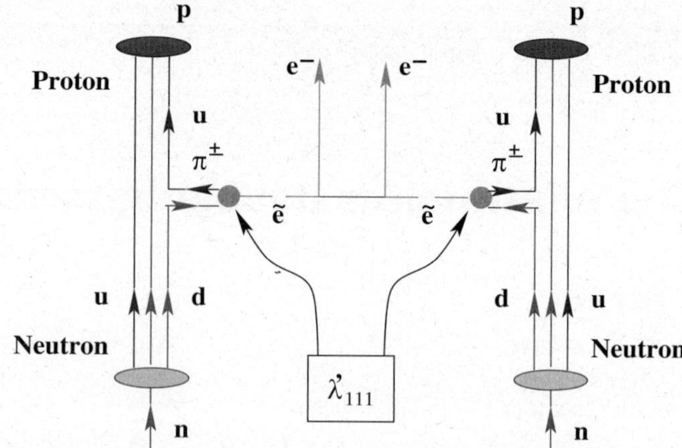

Fig. 4. Matrix element for the neutrinoless double-beta decay within R-parity violating supersymmetry.

Here $|k\rangle$ are the intermediate nuclear states in ^{76}As with a Majorana neutrino and one electron. The sum over k includes also an integration over neutrino energies. E_{0+} (^{76}Ge) is the ground-state energy of ^{76}Ge.

To calculate the nuclear matrix elements, the most reliable method has turned out to be the Quasiparticle Random Phase Approximation (QRPA) to calculate in the example of ^{76}Ge the wave function of the initial ^{76}Ge, the excited states of the intermediate nucleus ^{76}As and the ground state of the final nucleus ^{76}Se [3].

The QRPA approach describes the intermediate excited nuclear states $|m\rangle$ for example in ^{76}As as a coherent superposition of two quasiparticle excitations and two quasiparticle annihilations relative to the initial ground state (in our example) of ^{76}Ge:

$$a_i^+ = u_i c_i^+ - v_i c_{\bar{i}},$$

$$A_\alpha^+ = [a_i^+ a_k^+]_{J\alpha},$$

$$Q_m^+ = \sum_\alpha \left[X_\alpha^m A_\alpha^+ - Y_\alpha^m A_\alpha\right],$$

$$|m\rangle = Q_m^+|g\rangle.\tag{2}$$

The QRPA equation for the determination of the intermediate states by X_α^m and Y_α^m is derived from the many-body Schroedinger equation by using quasiboson commutation relations for the quasiparticle pairs $A_\alpha^\dagger$. One uses therefore for deriving the QRPA equations that two quasiparticle states behave like bosons. This is a usual approximation one often is using in physics. For example, one describes a pair of quarks and antiquarks as a meson and treats it as a boson, although it consists out of a fermion pair.

To test the quality of this approximation for the two neutrino and the neutrinoless double-beta decay, we include the exact commutation relations of the fermion pairs at least as ground-state expectation values. This is called renormalized-QRPA (R-QRPA) applied in [4] to the two neutrino double-beta decay and in [5] for the first time to the interesting neutrinoless double-beta decay.

The R-QRPA includes the Pauli principle into the QRPA and reduces by that the number of quasiparticles

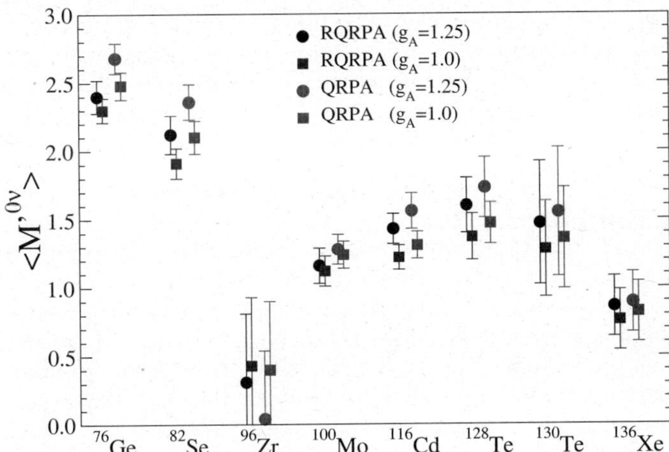

Fig. 5. Neutrinoless double-beta decay matrix elements at the Majorana neutrino mass calculated for different initial double-beta decay nuclei (for more details see text).

Table 1. This table gives limits for the lifetime and some parameters derived from data of ^{76}Ge. A lower limit of the experimental lifetimes for the 0ν double-beta decay is indicated in the second row with the experimental reference in the third row. The fourth row indicates the upper limit of the Majorana neutrino mass derived with the help of our QRPA matrix element in an intermediate basis (three oscillator shells). The fifth row gives the ratio of the proton mass over heavy Majorana neutrino mass for ^{76}Ge. This ratio is smaller than 0.79 which means that the averaged heavy Majorana neutrino, which is mainly right-handed, should have a mass larger than 1.2 GeV. The last row gives the upper limit of the R-parity violating coupling constant λ'_{111}. The upper limit of about 10^{-4} derived from ^{76}Ge is by one order of magnitude more stringent than other published values.

Nucleus	^{76}Ge
$\tau^{1/2}$ (exp) [y.]	$> 1.9\ 10^{25}$
Klapdor *et al.*	hep-ph/0512263
$\langle m \rangle$ [eV]	< 0.47
$m(p)/M(\nu)$	< 0.79
$\lambda'_{111}[10^{-4}]$	< 1.1

in the ground state. R-QRPA stabilizes the intermediate nuclear wave functions compared to the QRPA approach. One is relatively close to a phase transition to stable proton-neutron Gamow-Teller correlations. Since they are not included in the basis, the QRPA solution is collapsing if the 1^+ proton-neutron nucleon-nucleon two-body matrix elements are increased. One usually studies this by multiplying all nucleon-nucleon particle-particle matrix elements of the Brueckner reaction matrix of the Bonn (or the Nijmegen or Argonne) potential with a factor g_{pp} (typical: $g_{pp} \approx 0.9$). With an increasing g_{pp} the QRPA solution collapses like a spherical solution is collapsing if a nucleus gets deformed by increasing the quadrupole force. The inclusion of the Pauli principle in the R-QRPA approach moves this instability to much larger nucleon-nucleon matrix elements (larger factors g_{pp}) and the agreement for the two neutrino double-beta decay with the experimental value is more in the range of $g_{pp} = 1$.

Figure 5 shows the neutrinoless double-beta decay matrix elements calculated in 36 different ways for each initial nucleus indicated in the figure [6]. The approach includes three different forces (Bonn, Nijmegen, Argonne) and three different basis sets (about two oscillator shells, about three oscillator shells and about five to six oscillator shells) and four different approaches QRPA and the renormalized QRPA (RQRPA) with two different axial vector coupling constants $g_A = 1.25$ and with the quenched value $g_A = 1.0$. In each of the 36 calculations for each nucleus, the factor g_{pp} in front of the two nucleon-nucleon matrix element is adjusted to reproduce the experimental value of the two neutrino double-beta decay. The error bars given in the figure include the 1σ error of the theory plus the experimental error of the two neutrino double-beta decay transition probability. In several nuclei, like ^{76}Ge and ^{100}Mo the errors are as small as $\pm15\%$. In most cases the error is less than $\pm30\%$. But for ^{96}Zr one has an error of the neutrinoless double-beta decay matrix element of $\pm100\%$. The main part of the error indicated in fig. 5 are the experimental error of the two neutrino double-

beta decay probability. So when this measurements are improved, one is also automatically improving the prediction of the neutrinoless double-beta decay matrix elements. Results beyond the Standard Model using the most restrictive limit for the neutrinoless double-beta decay (hep-ph/0512263) with our matrix elements are given in table 1.

The present calculation includes the higher-order currents according to Towner and Hardy [7] and short-range correlations in the neutrinoless double-beta decay transition probability with a Jastrow factor.

The results of other groups seem partially not to agree with our results given above:

Muto's [8] results agree with our results if we include into his calculations the higher-order currents. The induced pseudoscalar current required by PCAC reduces his results by 30%. In addition we have a nuclear radius $R = 1.1 \cdot A^{1/3}$ [fm] of the Woods-Saxon potential compared to $R = 1.2 \cdot A^{1/3}$ [fm] of Muto. This difference reduces his values further by 10%. If we reduce his values for ^{76}Ge by 40% his values $M^{0\nu} = 4.59$ (QRPA) and 3.88 (R-QRPA) are reduced to 2.76 (QRPA) and 2.34 (R-QRPA), which agree nicely with our result of 2.68 (QRPA) and 2.41 (R-QRPA).

Civitarese and Suhonen have consistently higher values.

The calculation of Civitarese and Suhonen [9] are different from ours in several respects (see table 2):

1. They have fitted the factor g_{pp} not to the two-neutrino double-beta decay but to three known single-beta decay transition probabilities from a 1^+ ground state in intermediate nuclei to the final nuclear ground state.
2. They have included higher-order terms of the weak nucleon current according to Suhonen, Kadhkikar and Faessler [10] which is based on the simple quark model with a harmonic-oscillator potential for the quarks

Table 2. Comparison of the results of Civitarese and Suhonen (CS) with ours.

Nucleus	CS (QRPA, 1.254)	Our (QRPA, 1.25)
^{76}Ge	3.33	2.68 ± 0.12
^{100}Mo	2.97	1.30 ± 0.10
^{130}Te	3.49	1.56 ± 0.47
^{136}Xe	4.64	0.90 ± 0.20

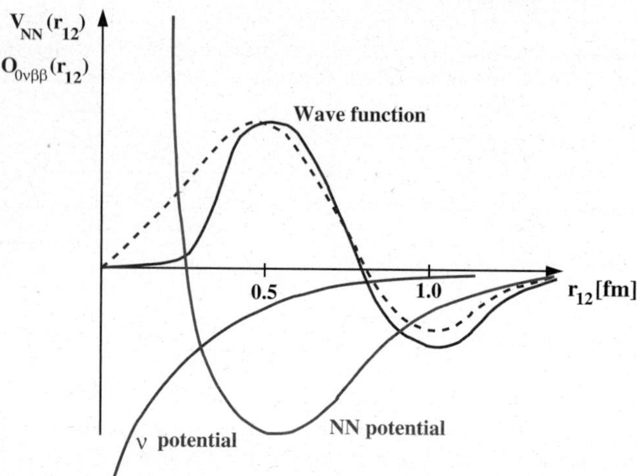

Fig. 6. Qualitative dependence on relative distance r_{12} of the two nucleons involved in the $0\nu\beta\beta$ of the nucleon-nucleon potential, the uncorrelated wave function (dashed line), the correlated wave function and the transition operator (ν potential).

leading to a Gaussian form factor for the weak currents and yielding weak magnetism as the main higher contribution. In our approach we included the higher-order currents according to Towner and Hardy [7] who fitted the higher-order currents to allowed beta decays, to first forbidden beta decays, to beta-gamma angular correlations, to beta-alpha angular correlations and to muon capture in nuclei. Towner and Hardy use dipole form factors and the induced pseudoscalar term and not weak magnetism yields the largest effect.

3. Short-range Brueckner correlations are not included in the calculation of Civitarese and Suhonen [9]. In our calculation, we include short-range Brueckner repulsion between the nucleons for the neutrinoless transition operator in form of a Jastrow factor. The inclusion of the short-range repulsion in the work of Civitarese and Suhonen would reduce the matrix element for the neutrinoless transition probability and could bring it into agreement with our value if also a more accurate expression for the higher-order weak currents and the different adjustment of g_{pp} are included.

Figure 6 shows qualitatively the nucleon-nucleon interaction with the short-range repulsion. In addition the relative wave function of the two nucleons involved in the double-beta decay is given with and without short-range repulsion considered (dashed line). The relative radial dependence of the transition operator for the $0\nu\beta\beta$ decay is also given qualitativeley. The difference in the $0\nu\beta\beta$ decay matrix element with and without short-range correlation is roughly given by the square of the difference between the correlated and uncorrelated wave function weighted by the transition potential for the neutrino (ν potential).

2 Summary

In this contribution we investigated the accuracy of the matrix elements for the neutrinoless double-beta decay. This accuracy determines the accuracy of the Majorana neutrino mass extracted from the neutrinoless double-beta decay. We used the Quasiparticle Random Phase Approach (QRPA) and the renormalized-QRPA (RQRPA) approximation which includes the Pauli principle and reduces the ground-state correlations and stabilises the wave function of the excited states of the intermediate nuclei. To get a feeling for the uncertainty we calculated the neutrinoless double-beta decay for three different nucleon-nucleon forces (Bonn, Nijmegen, Argonne) and three different basis sets (small with about two oscillator shells, intermediate with about three oscillator shells and large with about five oscillator shells). In addition we performed the calculations with QRPA, with R-QRPA (including the Pauli principle) and for two different axial vector coupling constants $g_A = 1.25$ and $g_A = 1.00$. In each of the 36 calculations for every nucleus we adjust a factor g_{pp} multiplying the nucleon-nucleon two-body matrix elements of the order between 0.85 and 1.1 to fit the experimental two-neutrino transition probabilities. We give the theoretical error of the neutrinoless transition probability as the 1σ error of the nine calculations for each method plus the experimental error of the two-neutrino transition probability. In many cases the theoretical error of the neutrinoless nuclear matrix element is less than $\pm 20\%$ but in one case, ^{96}Zr, it is $\pm 100\%$. But the main part of the error as indicated in fig. 5 comes from the experimental error of the two-neutrino transition probability. An improvement of the measurement of the two-neutrino transition probability would even further reduce the uncertainty and by that further reduce the error of the Majorana neutrino mass which can be determined if the neutrinoless double-beta decay is measured.

2.1 Problems to be solved

One open problem is connected with the fact that two-quasi particle excitations of the initial and the final nucleus to the intermediate states yield a slightly different Hilbert space and thus also slightly different intermediate states. This is treated by including an overlap matrix between these states. We are working on a better treatment and methods to see which error is induced by this effect.

A second problem is connected with the proton and neutron number non-conservation in the BCS treatment of pairing. Particle number projection before variation and the use of the Lipkin-Nogami method did show that this is only a problem if one of the nuclei involved has a closed shell [11].

A third problem are nuclear deformations for example for ^{150}Nd [12].

I would like to thank Dr. F. Simkovic, Dr. V. Rodin and Prof. P. Vogel for important contributions to the results presented here.

References

1. T. Tomoda, A. Faessler, Phys. Lett. B **199**, 475 (1987).
2. A. Faessler, S. Kovalenko, F. Simkovic, J. Schwieger, Phys. Rev. Lett. **78**, 183 (1997).
3. O. Civitarese, A. Faessler, T. Tomoda, Phys. Lett. B **194**, 11 (1987).
4. J. Toivanen, J. Suhonen, Phys. Lett. **75**, 410 (1995) and Phys. Rev. C **55**, 2314 (1997).
5. J. Schwieger, F. Simkovic, A. Faessler, Nucl. Phys. A **600**, 179 (1996); F. Simkovic, J. Schwieger, M. Veselsky, G. Pantis, A. Faessler, Phys. Lett. B **393**, 267 (1997).
6. V.A. Rodin, A. Faessler, F. Simkovic, P. Vogel, Phys. Rev. C **68**, 044302 (2003); Nucl. Phys. A **766**, 107 (2006) nucl-th/0503063.
7. I.S. Towner, J.C. Hardy, preprint nucl-th/9504015 v1 (12 Apr 1995).
8. K. Muto, Phys. Lett. B **391**, 243 (1997).
9. O. Civitarese, J. Suhonen, Nucl. Phys. A **729**, 867 (2003); **752**, 53 (2005).
10. J. Suhonen, S.B. Khadkikar, A. Faessler, Nucl. Phys. A **529**, 727 (1991).
11. Ch.C. Moustakidis, T.S. Kosmas, F. Simkovic, Amand Faessler, nucl-th/0511013.
12. R. Alvarez-Rodriguez, P. Sarriguren, E. Moya de Guerra, L. Pacearescu, Amand Faessler, F. Simkovic, Phys. Rev. C **70**, 064309 (2004) nucl-th/0411039.

Eur. Phys. J. A **31**, 429–434 (2007)

DOI 10.1140/epja/i2006-10263-9

THE EUROPEAN
PHYSICAL JOURNAL A

Special Article – QNP 2006

Few- and many-body methods in nuclear physics

M. Viviani[a]

Istituto Nazionale di Fisica Nucleare, Sezione di Pisa, Largo Pontecorvo, 3, I-56127 Pisa, Italy

Received: 23 November 2006
Published online: 15 March 2007 – © Società Italiana di Fisica / Springer-Verlag 2007

Abstract. In this contribution a brief overview of the status and perspectives of the theoretical methods for studying light and heavy nuclear systems is presented.

PACS. 21.60.-n Nuclear structure models and methods – 24.10.Cn Many-body theory – 21.45.+v Few-body systems

1 Introduction

Current nuclear-physics research includes a very rich variety of phenomena. The structure of hadrons and nuclei, the properties of the different phases of nuclear matter, the origin of the elements through primordial and stellar nucleo-sysnthesis, the test of fundamental symmetries in nuclei represent very broadly the main themes and questions in nuclear science today [1,2]. In this contribution, we shall concentrate on the primary goal of nuclear physics, namely the study of the structure and dynamics of nuclear systems. Some of the main questions in this case are the following: 1) How are complex nuclei built from their constituents? 2) How well can nuclear reaction rates be computed? Etc.

The "standard" starting point in trying to answer these questions is to consider the nucleus as composed of nucleons interacting via non-relativistic potentials mediated by pions and other mesons. In this model there are also several issues still not well established, in spite of decades of studies. For example, the nature of the nucleon-nucleon (NN) and of the three-nucleon (3N) interactions, the importance of many-body (correlation) effects with respect to single-particle properties of the nuclear medium and the limits of the nuclear stability, namely the so-called neutron and proton drip lines, etc., are still open problems.

Moreover, in recent years new and relevant developments have taken place. In fact, whereas only the stable nuclei could be studied until a few years ago, nowadays it is possible to observe and perform measurements of the specific properties (still with some limitations) of a large number of unstable nuclei of short lifetime. Most of the properties of these new nuclei are somewhat different from those already known. For example, they have a large "halo", as seen from their size which has been found to be larger than that given by the standard rule $R = r_0 A^{1/3}$.

Also the shell structure of the nuclei around the drip lines has offered some surprises.

The remarkable advances in the treatment of the nuclear many-body problem have greatly improved our understanding of various nuclear systems (as well as the technical developments of accelerators, detectors and data analysis techniques). In this contribution a brief overview of the status and perspectives of the theoretical methods used to study light and heavy nuclear systems is presented. Clearly, due to the limited space, we have selected a few, illustrative topics. First of all, a detailed discussion of our current understanding of the nuclear interaction is reported in sect. 2. Some important recent advances in the study of the medium-heavy and light systems are examined in sects. 3 and 4, respectively. Finally, the conclusions are given in sect. 5.

2 The nuclear interaction

The starting point of "standard" nuclear physics is the interaction between nucleons. It is a rather complicated problem and a well-founded theory has not yet been achieved. The NN interaction is known very well only for large interparticle separations where the one-pion exchange potential (OPEP) is the dominant term. At shorter distances the situation is rather complicated. Usually one resorts to NN scattering experiments to obtain valid parameterizations in the inner regions. In the next three subsections, the different approaches used to parameterize the NN interaction are summarized. Finally, in the last subsection, a brief discussion of the current understanding of the 3N force is reported.

2.1 Phenomenological models

About two decades ago, the Nijmegen group started a program to improve the phase shift analysis (PSA) of NN

a e-mail: michele.viviani@pi.infn.it

scattering data below 350 MeV [3]. In their final result, the Nijmegen PSA, they consider all the NN data taken from 1955 to 1992 and reject the points with improbably high χ^2 (about 30% of the data was disregarded). The remaining data could be fitted with an energy-dependent PSA [3] having a χ^2/datum of 0.99.

Based on this "pruned" database, in the period 1994–2001 a new class of charge-dependent NN potentials was constructed. The new potentials constructed were: 1) the Nijmegen potentials (Nijm-I, Nijm-II and Reid93) [4]; 2) the Argonne potential (AV18) [5]; 3) the Bonn potential (CD Bonn) [6]; and 4) the non-local potentials of Doleschall and collaborators [7]. All these potentials have in common the use of approximately 45 parameters and fit the Nijmegen data set with χ^2/datum ≈ 1. These modern potentials use different short- and intermediate-range parameterizations (they have all in common the long-range OPEP). Moreover, they differ in other properties, for example the CD Bonn potential is non-local in coordinate space while AV18 is local.

2.2 Nuclear forces from effective field theories

Alternatively, one can derive NN forces from an effective field theory (EFT) starting from a chiral Lagrangian. In this approach, which was pioneered by Weinberg [8], one starts from a Lagrangian describing pions and nucleons which takes into account the (approximate) chiral symmetry. Such Lagrangians reproduce correctly the physics of πN scattering at low energies [9]. The central point of this approach is the definition of a power counting scheme which permits a systematic truncation of the possible contributions to the interaction (chiral perturbation theory).

The most serious attempts to construct an NN interaction from an EFT have been performed in refs. [10] and [11], where potentials have been developed to the next-to-next-to-next-to-leading-order (N3LO) in the chiral expansion. In their current form, these potentials contains 27 parameters (coming from the unknown coupling constants of the contact terms in the chiral Lagrangian). These parameters have been fitted to the NN database, which nowadays has been noticeably enriched with respect to the Nijmegen database, with a χ^2 very close to 1 (the best result has been achieved for the so-called N3LO potential of ref. [11]). Therefore, these chiral potentials have reached an accuracy comparable to those achieved for the phenomenological potentials discussed in sect. 2.1.

2.3 Effective forces

The standard description of the NN interaction emerging from the approaches described above is a short-range repulsion and a longer-range attraction. In recent years however, it has been argued that, if one is interested in low-energy observables, a low-momentum interaction could be used [12,13]. This potential V_{low-k} can be derived from one of the "bare" NN interactions discussed previously by "eliminating" the high-momentum part for

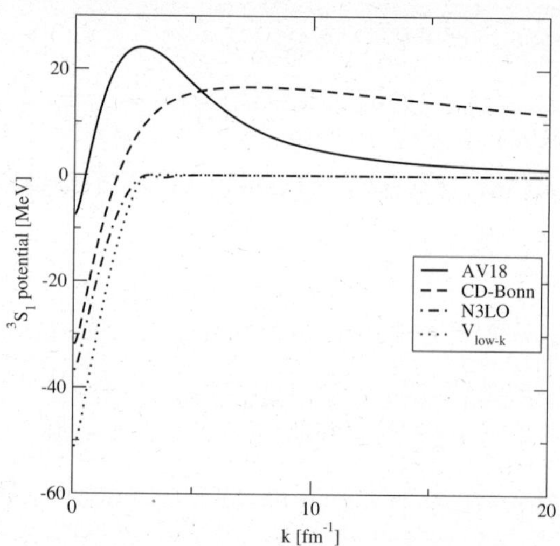

Fig. 1. Diagonal part of the $^3S_1$ potential in momentum space for the AV18 [5]. CD Bonn [6], N3LO [11] and one version of the V_{low-k} [13] interaction models.

$k > \Lambda \approx 2\,\mathrm{fm}^{-1}$ using renormalization group techniques or with a unitary transformation. This approach has a number of attractive features. First of all, for $k < \Lambda$, V_{low-k} reproduces all the NN observables (including the deuteron properties). Secondly, it is universal, in the sense that starting from different bare interactions an essentially unique V_{low-k} is derived. Finally, it can allow for shell model calculations in heavy nuclei (see sect. 3.2). Similar approaches have been proposed in refs. [14] and [15].

To give an idea of the behavior of these interactions, the on-shell part of some of these potentials in momentum space is plotted in fig. 1. As can be seen from the figure, the "phenomenological" AV18 and CD Bonn potentials have large high-momentum tails, coming from the strong repulsion at short interparticle distances. The N3LO potential is much softer (it vanishes for $k > 5\,\mathrm{fm}^{-1}$). Finally, the V_{low-k} potential is even softer (it does not contain any repulsion in coordinate spaced).

2.4 The 3N forces

In the study of systems with $A \geq 3$, one should also add appropriate many-nucleon forces. It is currently believed that only the 3N forces should play a relevant role in light systems (for a first quantitative estimate of the magnitude of a four-nucleon force, see ref. [16]). Several models of 3N forces have been proposed, mainly based on the exchange of pions among the three nucleons. The best known of these models are the Tucson-Melbourne [17] (TM) and the Brazil [18] (BR) 3N forces. Another model, the Urbana [19] (UR) 3N force is based on the Fujita-Miyazawa mechanism (exchange of two pions with the excitation of a Δ). Other models are discussed, for example, in ref. [20].

New mechanisms for the 3N force have also been proposed using the chiral approach [21]. The development of a 3N force at N^3LO is currently under way [22]. In this approach, it is possible to prove that the 3N forces compare at a large order (N^2LO) in the chiral power counting; their effects are thus suppressed compared to NN interactions [23].

The Urbana-Argonne group has in recent years developed [19,24] a formidable technique, the Green Function Monte Carlo (GFMC) method (see below), which can be successfully applied to calculate the spectra of nuclei up to $A = 12$. They have shown that using the AV18 NN potential + UR 3N force (version IX) one does not obtain a good description of the spectra of such nuclei. For this reason a new model of 3N interaction has been proposed [25], the Illinois 3N force, incorporating three pion exchange "box" diagrams. Using this model together with the NN AV18 potential, a good description of the full spectrum of the light nuclei with $A \leq 12$ has been reached. However, this potential model contains some terms which do not respect the chiral counting order, and for this reason it is not equivalent with the models developed using the chiral approach. This issue is still under debate.

As we have seen, several new models of NN and 3N interactions have recently been developed. It is therefore of interest to perform tests in order to understand which of these different models should be more appropriate for describing the very complex nature of the various nuclear systems.

3 Many-body methods

To study the structure of medium to heavy nuclei a large variety of techniques has been used. Some of them are discussed briefly in the following subsections.

3.1 Microscopic approaches

One of the traditional approaches for studying the nuclear structure is to try to derive the properties of the nuclei from the "bare" interaction between nucleons, as derived from the PSA of NN data (microscopic approach), by taking the strong repulsion at short interparticle distances fully into account. To be successful, one has to take into account a proper 3N force in these approaches.

The best-known solution to this problem lies in the Brückner theory [26,27]. Nowadays, nuclear and neutron matter calculations have been performed up to three hole lines diagrams and extended up to rather high densities, of the order of 6 times the saturation density. The calculated equation of state (EOS), namely the binding energy per nucleon in function of the density, for high density is a critical quantity necessary for understanding the properties of compact stars. Spin- and isospin-polarized nuclear-matter calculations have also been performed [28].

One of the key points of a calculation of the EOS is the reproduction of the empirical saturation point of nuclear matter (namely, the binding energy per particle B/A of nuclear matter should have a maximum of 15 MeV at the empirical density $\rho_0 = 0.16\,\mathrm{fm}^{-3}$). It is well known that for obtaining that value with a local potential like the AV18 potential, it is essential to include a strong repulsive core and a proper 3N force [27]. Note that for reproducing the empirical saturation point, one has to change the parameters of the 3N force, whose values therefore differ from those derived in the $A = 3, 4$ systems [29].

Recently, the effect of using non-local potentials has been investigated. The cases of CD Bonn and Doleschall potentials have been studied in refs. [30,31] and ref. [32], respectively. In both cases, the value of B/A is found to be rather larger than the empirical one. This indicates that to get the correct saturation point, one needs an amount of 3N force larger than usual.

One of the key problems is to know the uncertainties of the computed EOS. From this point of view, the comparison with the EOS calculated using other techniques is very helpful. Another approach used extensively for studying nuclear matter and medium-heavy nuclei starting directly from the bare NN interaction is the Correlated Basis Function (CBF) theory [33]. In this approach, one starts with a basis of states where the correlations are already built-in. It is therefore a method very well adapted to treating cases where the interaction has a strong repulsion at short range.

A detailed comparison between the Brückner and CBF results has been performed recently in ref. [34]. The results are found to be in agreement up to densities $\sim 1.5\rho_0$. For larger densities the differences may become rather sizeable especially in the presence of a 3N force. Further studies will be necessary to clarify this discrepancy.

The Brückner and CBF theory have also been used for making calculations in finite systems [35–38] with good results. An application where the Brückner theory is used to calculate the rate of the double beta decay is reported in ref. [39].

As already mentioned, another powerful approach for studying the structure of ($A \leq 12$) nuclei is the GFMC technique developed by the Urbana group [19,24]. It would be interesting to apply this model also to the study of the medium-heavy nuclei and nuclear matter. Some calculations in this direction have been performed in ref. [40]. However, the application of the GFMC technique becomes more and more time consuming for $A > 13$ due to the large number of spin-isospin states.

A way of overcoming this problem has been proposed in ref. [41]. This method, the so-called auxiliary field diffusion Monte Carlo (AFDMC), is still based on a Monte Carlo technique. The summation over the spin-isospin degrees of freedom is simplified and this allows for the extensions to larger systems. Applications has been presented for nuclear and neutron matter, small neutron droplets and medium-heavy nuclei [41,42].

3.2 Effective approaches

As discussed in sect. 2.2, a number of "effective" interactions have recently been devised, where the repulsive part has been eliminated. Thus, these "effective" potentials can be used in shell model calculations without the need to use the Brückner resummation.

A study of nuclear matter using the $V_{low\text{-}k}$ potential has been performed in ref. [43]. In this case, also a TM-like 3N force has been included in the calculation (note that the parameters of the 3N force were chosen to be the same as determined from the ^{3}H and ^{4}He binding energies [44]). The EOS, computed in the Hartree-Fock approximation plus dominant second-order corrections of the standard perturbation theory, is found to be rather realistic. The computed saturation point is close to the empirical value. A more complete calculation is currently underway.

Several shell model calculations for finite nuclei have also been performed. For example, with the $V_{low\text{-}k}$ potential, the binding energies and radii of the closed-shell nuclei ^{4}He, ^{16}O, and ^{40}Ca have been found to be in agreement with the experimental data for less than 1%, and that the theoretical findings depend only slightly on the bare NN potential used as starting point [45,46]. Other shell model calculations have been performed, for instance for ^{100}Sn and ^{132}Sn. These, respectively, proton- and neutron-rich nuclei are unstable and, as mentioned in the introduction, understanding the shell structure far from stability is one of the new frontiers of nuclear physics. In general, the result of shell model calculations are able to reproduce well the experimental data for the low-lying levels [47].

Analogously, the potential derived in ref. [14] using the unitary correlation operator method (UCOM) has been used for studying a large number of nuclei. A remarkable agreement of the calculated ground-state energies has been found over the whole mass range from ^{4}He to ^{208}Pb [48].

3.3 Other approaches

For reasons of space, a number of other approaches have to be ignored. Amongst them, we can mention the use of simpler effective forces (like the Skyrme and Gogny forces), the in-medium chiral perturbation theory approach, the various algebraic methods (such as the interacting boson model), the energy functional methods, and many others. For a general review, see, for example, ref. [49].

4 Few-body methods

Historically, the study of the nuclear interaction has been the central goal in few-nucleon systems. In fact, it is possible to achieve numerically accurate solutions of the associated quantum-mechanical problem, allowing for unambiguous comparisons between theory and experiments [50]. Also in this case, the lack of space forces us to select a few topics. This section is organized as follows: in sect. 4.1, a brief description of the theoretical methods which are applied most to studying few-body systems are briefly recalled. The results obtained for the $A = 3$ bound states are reported in sect. 4.2. In sect. 4.3, a discussion of the $N - d$ A_y "puzzle" is presented. Finally, a brief discussion of the reaction $d + d \rightarrow {}^4\text{He} + \pi^0$ is reported in sect. 4.4.

4.1 Theoretical methods

A great deal of effort has been dedicated to the development of techniques for the numerical solution of the non-relativistic Schrödinger equation,

$$H\Psi(1, 2, \ldots, A) = E\Psi(1, 2, \ldots, A), \tag{1}$$

where H is a nuclear Hamiltonian. In this section, we will limit ourselves to studying $A = 3, 4$ systems. In the Faddeev equation (FE) approach [51–54], eq. (1) is transformed to a set of coupled equations for the Faddeev amplitudes, which are then solved directly (in momentum or coordinate space) after a partial wave expansion.

In the GFMC method already mentioned, one computes $\exp(-\tau H)\Phi(1, 2, \ldots, A)$, where $\Phi(1, 2, \ldots, A)$ is a trial wave function, using a stochastic procedure to obtain, in the limit of large τ, the exact ground-state wave function Ψ [19,24].

The stochastic variational method (SVM) [55,56] and the coupled rearrangement channel Gaussian method (CRCG) [57,58] provide a variational solution of eq. (1) by expanding the (radial part of the) wave function in Gaussians. Another widely applied variational method is the expansion of the wave function using the hyperspherical harmonic (HH) basis [59,60].

Very recently two other new techniques have been proposed. In the no-core shell model (NCSM) method [61, 62] the calculations are performed using a (translationally invariant) harmonic-oscillator (HO) finite basis P and introducing an effective P-dependent Hamiltonian H_P to replace H in eq. (1). The operator H_P is constructed so that the solution of the equation $H_P\Psi(P) = E_P\Psi(P)$ provides eigenvalues which quickly converge to the exact ones as P is enlarged. The effective interaction hyperspherical harmonic (EIHH) method [63,64] is based on a similar idea, but the finite basis P is constructed in terms of the HH functions.

4.2 Bound states of the A = 3 system

All of the methods discussed above provide very accurate results for the $A = 3, 4$ bound states [65]. For example, results for a few representative potential models are reported in table 1 for the ^{3}H binding energy and other properties. The NN potentials considered are the CD Bonn, N3LO and AV18 potential models.

There is good agreement, at the level of 0.1%, between the results obtained with different techniques for all the quantities considered in the tables. A similar precision is nowadays also achieved for the $A = 4$ nucleus.

4.3 The puzzle of the N − d vector analyzing powers

The $N - d$ elastic scattering process has been studied extensively in recent years. Most of the calculations have been performed using the "phenomenological" potentials described in sect. 2.1, with the eventual inclusion of the

Table 1. The triton binding energies B (MeV), the expectation values of the kinetic energy operator $\langle T \rangle$ (MeV), the mean square radii $\sqrt{\langle r^2 \rangle}$ (fm), and the P and D probabilities (all in %), calculated with the CD Bonn, N3LO, and AV18 potentials. The results obtained with various techniques have been reported.

Interaction	Method	B	$\langle T \rangle$	$\sqrt{\langle r^2 \rangle}$	P_P	P_D
CD Bonn	HH [66]	7.998	37.630	1.721	0.047	7.02
	FE [67]	7.997	37.620	–	0.047	7.02
	FE [68]	7.998	37.627	–	0.047	7.02
	NCSM [69]	7.99(1)	–	–	–	–
N3LO	HH [66]	7.854	34.555	1.758	0.037	6.31
	FE [67]	7.854	34.546	–	0.037	6.32
	FE [68]	7.854	34.547	–	0.037	6.32
	NCSM [69]	7.85(1)	–	–	–	–
AV18	HH [70]	7.618	46.707	1.770	0.066	8.511
	FE [67]	7.621	46.73	–	0.066	8.510

TM, BR or UR 3N force. The agreement between these calculations and the large amount of experimental data is in general impressive, except for the vector analyzing power observables. This is a fairly old problem, already reported about 20 years ago [71,72] in the case of $n - d$ and later confirmed also in the $p-d$ case [73]. It consists of the fact that all the theoretical calculations based on phenomenological NN potentials (even including TM, BR and UR 3N forces) underestimate the measured nucleon vector analyzing power A_y by about 20–30%. The same problem also occurs for the vector analyzing power of the deuteron iT_{11} [73], while the deuteron tensor analyzing powers T_{20}, T_{21} and T_{22} are reasonably well described [51].

In most of the calculations, the electromagnetic (EM) interaction is usually approximated with the point Coulomb potential (sometime also this term is disregarded). Given the extreme sensitivity of the observables A_y and iT_{11} to the spin-orbit interaction [74,75], it has been recently investigated [76,77] whether this problem could be related to the absence of the magnetic-moment (MM) interaction between the nucleons in the theoretical studies. The results obtained in ref. [76] for A_y at $E_{lab} = 1$ and 3 MeV for $p - d$ elastic scattering have been reported in fig. 2. As can be seen, the effect of the MM interaction is quite sizable, but insufficient to solve the disagreement with the data.

The calculation of A_y with the chiral potentials is still under way. Preliminary results have shown that the situation is unclear as this observable depends rather strongly on the the cutoff Λ of the theory [80]. Calculation with the effective interactions are still to be performed.

The attention has also been focused on exotic 3N force terms not contemplated so far [81,82]. Calculations are under way to see the effect of the Illinois 3N force and the chiral 3N force at the N^3LO level. In the next few years, there is hope of solving this puzzle.

Similar discrepancies have also been observed in the four-nucleon system [83]. As an example, the experimental total $n-{}^3H$ cross-sections for $E_n = 3$–5 MeV have been found at variance with all the theoretical calculations performed so far [84].

4.4 The reaction $d + d \to \alpha + \pi^0$

The good accuracy which can nowadays be obtained in the numerical solution of four-nucleon problems allows for interesting studies and applications. Among them, the study of charge symmetry breaking (CSB) terms in the NN interaction by means of the $d + d \to \alpha + \pi^0$ reaction is rather appealing. The CSB terms in the nuclear interaction come from both the u- and d-quark mass difference (a fundamental quantity poorly known) and from EM effects. The recent observation of the $d + d \to \alpha + \pi^0$ reaction at IUCF [85] is a direct probe of CSB. The theoretical study of this reaction is currently under way [86].

5 Conclusions

In recent years, there have been important developments in the field of many- and few-nucleon systems. Here, we have concentrated our attention mainly on studying the infinite nuclear matter, heavy nuclei and few-nucleon systems. The link between these fields lies in the use of the same interaction models to describe their properties. The hope is to provide a "unified" picture of all the nuclear systems starting from the microscopic NN + 3N interaction.

An important point to be stressed is that nuclear physics is embedded into the context of other sciences. A large part of problems in nuclear and atomic/molecular physics are nowadays tackled using the same methods. The few-body techniques described here are also currently used for studying the structure of hadrons in terms of constituent quarks. Studies in nuclear physics have also a big impact in particle physics, cosmology and astrophysics. In fact, the linkage between these fields is fundamental for our understanding of the base structure of our universe. The increase of the capability to provide model-independent predictions of nuclear structure and reactions will contribute strongly to the knowledge of our world.

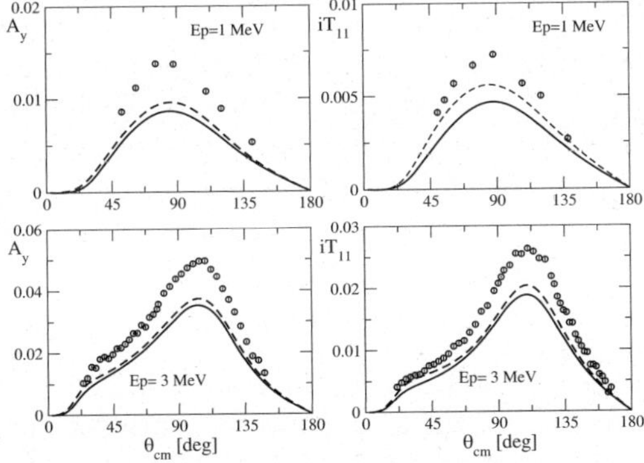

Fig. 2. The $p - d$ A_y and iT_{11} calculated using AV18 (solid lines) and AV18 + MM (dashed lines). Experimental points are from ref. [78] (1 MeV) and ref. [79] (3 MeV).

References

1. A. Richter, Nucl. Phys. A **751**, 3c (2005).
2. B.K. Jennings, A. Schwenk, nucl-th/0512013.
3. V.G.J. Stoks *et al.*, Phys. Rev. C **48**, 792 (1993).
4. V.G.J. Stoks *et al.*, Phys. Rev. C **49**, 1950 (1994).
5. R.B. Wiringa, V.G.J. Stoks, R. Schiavilla, Phys. Rev. C **51**, 38 (1995).
6. R. Machleidt, Phys. Rev. C **63**, 024001 (2001).
7. P. Doleschall, I. Borbély, Phys. Rev. C **62**, 054004 (2000); P. Doleschall *et al.*, Phys. Rev. C **67**, 064005 (2003).
8. S. Weinberg, Physica A **96**, 327 (1979); Phys. Lett. B **251**, 288 (1990).
9. See, for example, J.D. Walecka, *Theoretical Nuclear and Subnuclear Physics* (Oxford University Press, New York, 1995).
10. E. Epelbaum, W. Gloeckle, Ulf-G. Meissner, Nucl. Phys. A **747**, 362 (2005).
11. D.R. Entem, R. Machleidt, Phys. Rev. C **68**, 041001 (2005) (2003).
12. S.K. Bogner, T.T.S. Kuo, A. Schwenk, Phys. Rep. **386**, 1 (2003).
13. L. Coraggio *et al.*, Phys. Rev. C **71**, 014307.
14. R. Roth *et al.*, Nucl. Phys. A **745**, 3 (2004); Phys. Rev. C **72**, 034002 (2005).
15. A.M. Shirokov *et al.*, Phys. Rev. C **70**, 044005 (2004).
16. D. Rozpedzik *et al.*, nucl-th/0606017.
17. S.A. Coon *et al.*, Nucl. Phys. A **317**, 242 (1979).
18. M.R. Robilotta, H.T. Coelho, Nucl. Phys. A **460**, 645 (1986).
19. B.S. Pudliner *et al.*, Phys. Rev. C **56**, 1720 (1997).
20. J.L. Friar, D. Hüber, U. van Kolk, Phys. Rev. C **59**, 53 (1999).
21. E. Epelbaum *et al.*, Phys. Rev. C **66**, 064001 (2002).
22. Ulf-G. Meissner, private communication.
23. U. van Kolck, Phys. Rev. C **49**, 2932 (1999).
24. R.B. Wiringa *et al.*, Phys. Rev. C **62**, 074001 (2000).
25. S.C. Pieper *et al.*, Phys. Rev. C **64**, 014001 (2001).
26. See, for example, M. Baldo (Editor), *Nuclear Methods and the Equation of State* (World Scientific, Singapore, 1999).
27. M. Baldo *et al.*, Phys. Rev. C **65**, 017303 (2002).
28. W. Zuo, Caiwan Shen, U. Lombardo, Phys. Rev. C **67**, 037301 (2003).
29. M. Baldo, private communication.
30. P. Bozek, P. Czerski, Acta Phys. Pol. B **34**, 2759 (2003).
31. Kh. Gad, Eur. Phys, J. A **22**, 405 (2004).
32. M. Baldo, C. Maieron, Phys. Rev. C **72**, 034005 (2005).
33. S. Fantoni, A. Fabrocini, *Microscopic Quantum Many-Body Theories and Their Applications*, Lect. Notes Phys., Vol. **510** (Springer-Verlag, Berlin, 1998).
34. I. Bombaci *et al.*, Phys. Lett. B **609**, 232 (2005).
35. K.W. Schmid, H. Müther, R. Machleidt, Nucl. Phys. A **530**, 12 (1991).
36. W.H. Dickoff, C. Barbieri, Prog. Part. Nucl. Phys. **52**, 377 (2004).
37. G. Co' *et al.*, Nucl. Phys. A **549**, 2 (1992).
38. C. Bisconti *et al.*, Phys. Rev. C **73**, 054304 (2006).
39. A. Faessler, these proceedings.
40. S. Fantoni, A. Sarsa, K.E. Schmmidt, Phys. Rev. Lett. **87**, 181101 (2001).
41. F. Pederiva *et al.*, Nucl. Phys. A **742**, 255 (2004).
42. S. Gandolfi *et al.*, Phys. Rev. C **73**, 044704 (2006).
43. S.K. Bogner *et al.*, Nucl. Phys. A **763**, 59 (2005).
44. A. Nogga, S.K. Bogner, A. Schwenk, Phys. Rev. C **70**, 061002(R) (2004).
45. L. Coraggio *et al.*, Phys. Rev. C **68**, 034320 (2003).
46. L. Coraggio *et al.*, Phys. Rev. C **73**, 014304 (2006).
47. P. Guazzoni *et al.*, Phys. Rev. C **69**, 024619 (2004).
48. R. Roth *et al.*, Phys. Rev. C **73**, 044312 (2006).
49. See, for example, *the Proceedings of the International Nuclear Physics Conference 2004, Goteborg (Sweden)*, in Nucl. Phys. A **751** (2005).
50. H. Arenhoevel *et al.*, nucl-th/0412039.
51. W. Glöckle *et al.*, Phys. Rep. **274**, 107 (1996).
52. N.W. Schellingerhout, J.J. Schut, L.P. Kok, Phys. Rev. C **46**, 1192 (1992).
53. A. Nogga *et al.*, Phys. Rev. C **65**, 054003 (2002).
54. R. Lazauskas, J. Carbonell, Phys. Rev. C **70**, 044002 (2004).
55. K. Varga, Y. Suzuki, Phys. Rev. C **52**, 2885 (1995).
56. Y. Suzuki, K. Varga, *Stochastic Variational Approach to Quantum Mechanical Few-Body Problems* (Springer-Verlag, 1998).
57. H. Kameyama, M. Kamimura, Y. Fukushima, Phys. Rev. C **40**, 974 (1989).
58. M. Kamimura, H. Kameyama, Nucl. Phys. A **508**, 17c (1990).
59. A. Kievsky, S. Rosati, M. Viviani, Nucl. Phys. A **577**, 511 (1994).
60. M. Viviani, A. Kievsky, S. Rosati, Phys. Rev. C **71**, 024006 (2005).
61. P. Navrátil, J.P. Vary, B.R. Barrett, Phys. Rev. Lett. **84**, 5728 (2000); Phys. Rev. C **62**, 054311 (2000).
62. P. Navrátil, W.E. Ormand, Phys. Rev. Lett. **88**, 152502 (2002).
63. N. Barnea, W. Leidemann, G. Orlandini, Phys. Rev. C **61**, 054001 (2000).
64. N. Barnea, W. Leidemann, G. Orlandini, Phys. Rev. C **67**, 054003 (2003).
65. H. Kamada *et al.*, Phys. Rev. C **64**, 044001 (2001).
66. M. Viviani *et al.*, Few-Body Syst. **39**, 159 (2006).
67. A. Nogga *et al.*, Phys. Rev. C **67**, 034004 (2003).
68. A. Deltuva, R. Machleidt, P.U. Sauer, Phys. Rev. C **68**, 024005 (2003).
69. P. Navrátil, E. Caurier, Phys. Rev. C **69**, 014311 (2004).
70. A. Kievsky *et al.*, Few-Body Syst. **22**, 1 (1997).
71. Y. Koike, J. Haidenbauer, Nucl. Phys. A **463**, 365c (1987).
72. H. Witala, W. Glöckle, T. Cornelius, Nucl. Phys. A **491**, 157 (1988).
73. A. Kievsky *et al.*, Nucl. Phys. A **607**, 402 (1996).
74. H. Witala, W. Glöckle, Nucl. Phys. A **528**, 48 (1991).
75. T. Takemiya, Prog. Theor. Phys. **86**, 975 (1991).
76. A. Kievsky, M. Viviani, L.E. Marcucci, Phys. Rev. C **69**, 014002 (2004).
77. H. Witala *et al.*, Phys. Rev. C **67**, 064002 (2003).
78. M.H. Wood *et al.*, Phys. Rev. C **65**, 034002 (2002).
79. K. Sagara *et al.*, Phys. Rev. C **50**, 576 (1994).
80. A. Nogga, private communications.
81. A. Kievsky, Phys. Rev. C **60**, 096003 (1999).
82. L. Canton, W. Schadow, Phys. Rev. C **64**, 031001(R) (2001).
83. B.M. Fisher *et al.*, Phys. Rev. C **74**, 034001 (2006).
84. R. Lazauskas *et al.*, Phys. Rev. C **71**, 034004 (2005).
85. E.J. Stephenson *et al.*, Phys. Rev. Lett. **91**, 142302 (2003).
86. A. Nogga *et al.*, Phys. Lett. B **639**, 465 (2006).

Eur. Phys. J. A **31**, 435–440 (2007)
DOI 10.1140/epja/i2006-10262-x

Special Article – QNP 2006

Overview of COSY

W. Eyrich[a]

University of Erlangen-Nuremberg, Germany

Received: 23 November 2006
Published online: 27 March 2007 – © Società Italiana di Fisica / Springer-Verlag 2007

Abstract. The COSY facility is sketched together with its experiments. The hadron physics program is discussed highlighting recent results with particular emphasis on nucleon-nucleon interaction, precision experiments, and meson and strangeness production.

PACS. 29.40.-n Radiation detectors – 13.75.-n Hadron-induced low- and intermediate-energy reactions and scattering (energy $\leq 10\,\mathrm{GeV}$) – 14.40.-n Mesons – 14.20.Jn Hyperons

1 Accelerator and experiments

COSY is a synchrotron at the Research Centre Juelich. It delivers unpolarized as well polarized proton and deuteron beams in the momentum range from $300\,\mathrm{MeV}/c$ up to about $3650\,\mathrm{MeV}/c$. Electronic cooling is used at the injection energy whereas stochastic cooling is provided for higher energies. COSY can be used as a storage ring for internal experiments and delivers external beams by stochastic extraction with spill times between a few seconds and several minutes.

The excellent beam quality of the extracted beam with an admittance of about 0.5 mm mrad allows precise tracking close to the target. This is of special importance for measurements close to reaction thresholds. In fig. 1 a scheme of the COSY facility is shown together with the internal experiments ANKE, COSY-11, EDDA and PISA and the external experiments BIG KARL, COSY-TOF and JESSICA. Both PISA and JESSICA, which have been used for the investigation of spallation processes, are not discussed in this talk. Moreover the place of the WASA detector, which is in the commissioning, is indicated. Information about the COSY-facility and the various experiments can be found at the COSY homepage [1].

ANKE and COSY-11 are magnetic spectrometers with a wide momentum acceptance for charged particles; both well suited for studying meson production processes close to threshold with particle emission into forward direction. Whereas COSY-11 employs one accelerator dipole, ANKE is a chicane consisting of three dipoles with the middle one as analysing part. COSY-11, a scheme of which is shown in fig. 2, has been upgraded during the last years for registering neutrons, deuterons, spectator protons and to extend the acceptance for kaons.

[a] e-mail: eyrich@physik.uni-erlangen.de

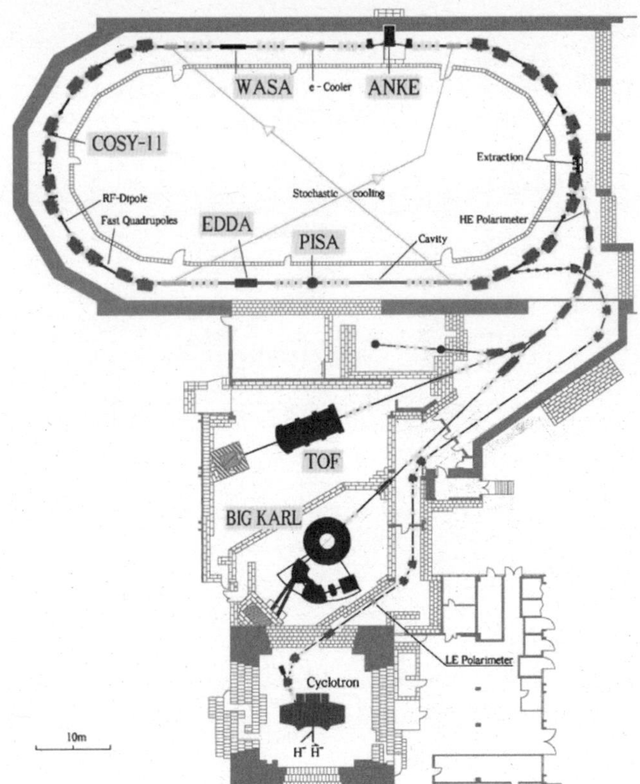

Fig. 1. The COSY facility with internal and external experiments.

As can be seen in fig. 3, the detection system of ANKE comprises tracking detectors and scintillator hodoscopes in forward direction for positive and negative reaction products. In addition, dedicated range hodoscopes and Cherenkov detectors allow the identification of charged

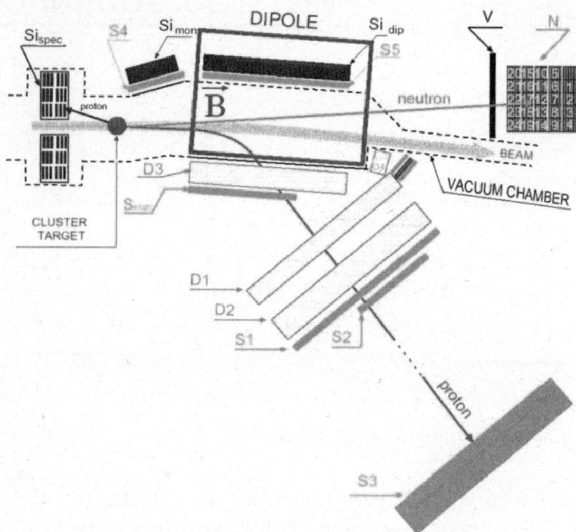

Fig. 2. Scheme of the COSY-11 detection system. D_i denotes drift chambers, S_i and V scintillation, C Cherenkov, N neutron, and Si silicon detectors, whereas the last ones are used for spectator protons and negatively charged particles.

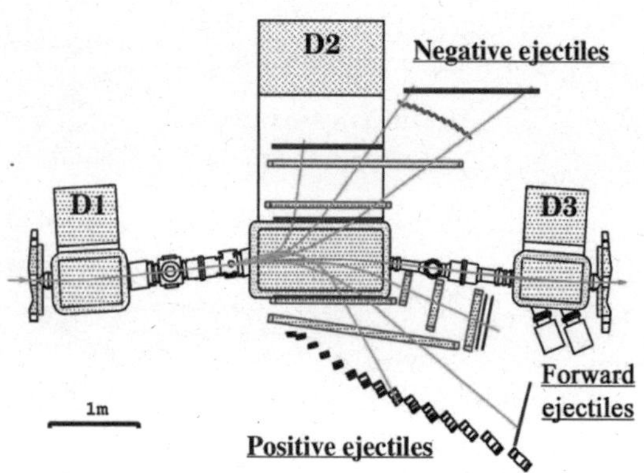

Fig. 3. Scheme of the ANKE detection system.

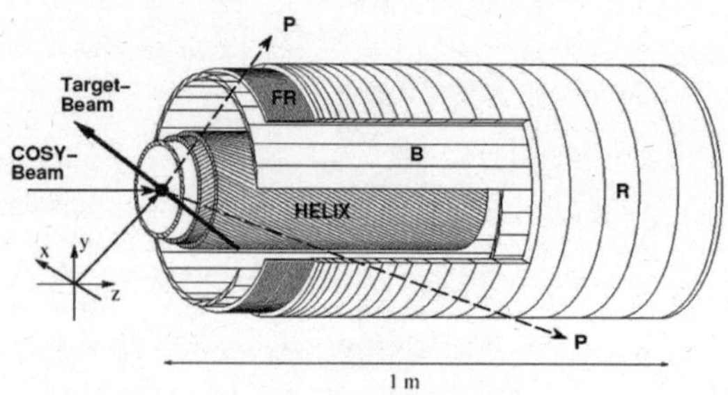

Fig. 4. Scheme of the EDDA detector.

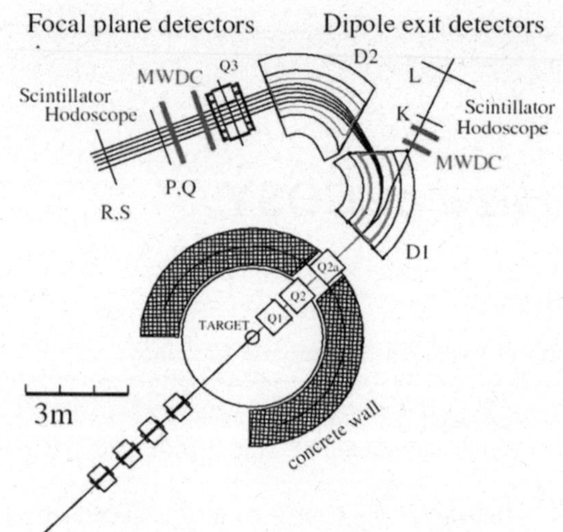

Fig. 5. Scheme of the BIG KARL spectrometer.

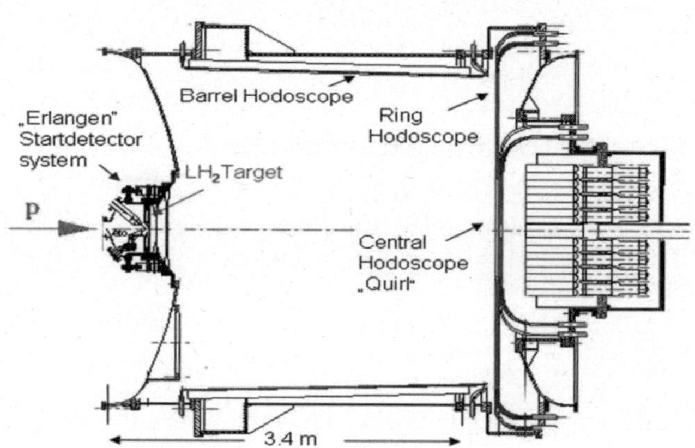

Fig. 6. Scheme of the COSY-TOF detector.

preparation of polarized proton and deuteron beams. All three internal experiments discussed here use gas (cluster) jet targets or, alternatively, thin foils or fibres. In addition, EDDA and ANKE possess polarized atomic beam targets, moreover ANKE has the option to use a storage cell.

BIG KARL, a setup of which is shown in fig. 5, is a high-resolution magnetic spectrometer of 3Q2D type with a momentum resolution of $\Delta p/p < 5 \cdot 10^{-5}$ and a momentum acceptance of $\pm 4.5\%$. The ejectiles are detected by multiwire drift chambers. Particle identification is done in combination with time-of-flight measurements using scintillator hodoscopes. In the target region the spectrometer is supplemented at large emission angles by a stack of four highly granulated annular germanium detectors, and a three-layer scintillator detector (ENSTAR).

The non-magnetic COSY-TOF apparatus (see fig. 6) is a huge vacuum vessel covered by several layers of scintillators which are used for tracking and together with the detectors in the target region for time-of-flight measurement. The target area detectors, consisting of a highly segmented annular silicon microstrip detector and two fibre hodoscopes, are suited for tracking and especially the reconstruction of delayed decays. Moreover, the setup comprises

kaons. Around the target silicon detectors allow to detect slow particles.

The non-magnetic EDDA detector, schematically shown in fig. 4, consists of two cylindrical detector shells of overlapping scintillator bars (outer) and helix wounded scintillating fibres (inner). The solid angle coverage is 30° to 150° in $\Theta_{c.m.}$ for elastic proton-proton scattering. It is now used as a very efficient beam polarimeter for the a central calorimeter in the end cap of the vessel. The goal is the examination of particle production in proton-proton and proton-deuteron collisions. Due to the kinematical completeness of the measurement the full set of variables can be determined, including total and differential cross-sections, angular distributions, and Dalitz plots.

So far the detectors at COSY are photon-blind. This will change with the WASA detector which was transferred form CELSIUS at Uppsala to COSY in 2005 and will start operation in the second half of 2006. This facility can detect both neutral and charged particles in a large angular and momentum range. Dedicated information on the measurements planned with WASA can be found in [1].

2 Physics at COSY

A broad spectrum of physics is covered by the different experiments at COSY. The excellent cooled beams are especially well suited for precise particle production measurements in the threshold regions. In the following a few recent results from experiments covering different topics are briefly discussed.

2.1 Nucleon-nucleon interaction

Elastic pp scattering investigations at EDDA resulted in accurate data for excitation functions [2], analysing power [3] and spin correlation observables [4]. Examples are shown in fig. 7.

These data extend drastically the base for phase shift analysis and are important for an improved understanding of the elementary pp interaction. EDDA also looked intensively for dibaryon resonances with a negative result.

ANKE is continuing the study of nucleon-nucleon reactions especially investigating pn-observables using the polarized deuteron beam and the polarized target which is now available.

2.2 Precision experiments

The precise COSY beam makes different types of precision experiments feasible. At the BIG KARL spectrometer a precision measurement of the η-mass, which still is poorly known, has been performed. Minimal systematic errors could by achieved by a self-calibrating measurement where the brilliance of the COSY beam plays an important role. The three reactions $p + d \rightarrow t + \pi^+$, $p + d \rightarrow \pi^+ + t$ and $p + d \rightarrow {}^3\text{He} + \eta$ were measured simultaneously with one setting of the spectrometer, where always the first particle produced was detected. In fig. 8 the reconstructed missing-mass spectrum of the η-particle is shown.

The final value of

$$m(\eta) = 547.311 \pm 0.028(\text{stat.}) \pm 0.032(\text{syst.})\,\text{MeV}/c^2$$

(see ref. [5]) is in agreement with older data but in disagreement with a recent result from NA48, which dominates the PDG average. Details of the BIG KARL experiment are discussed in the talks of Daniel Kirilov; very

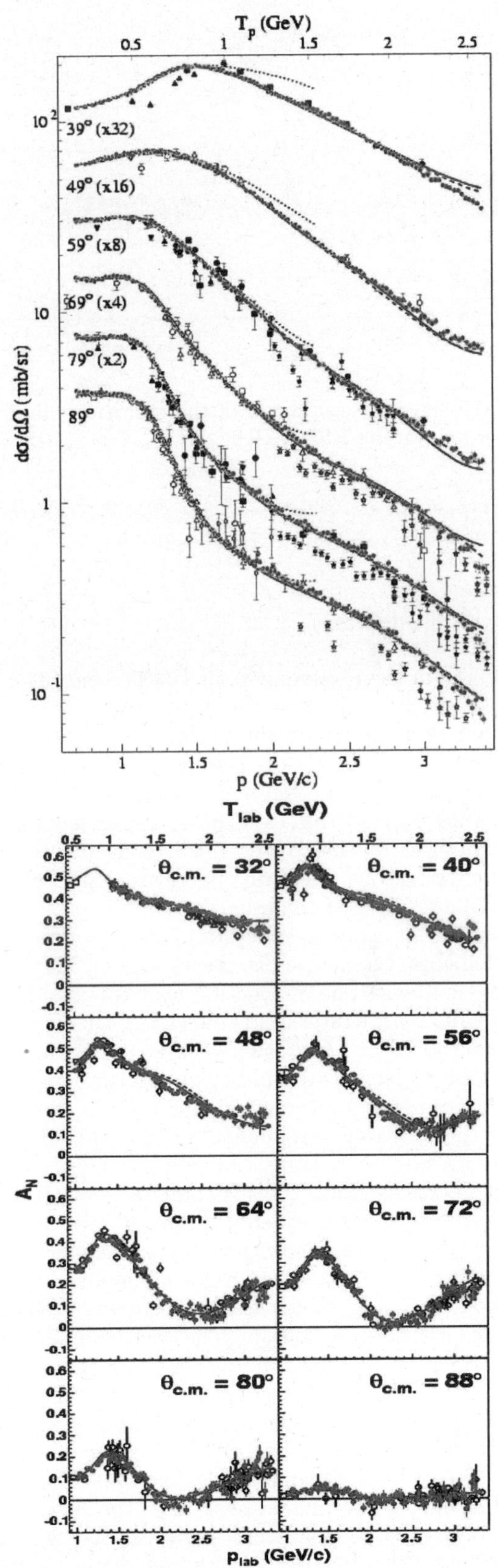

Fig. 7. Excitation functions of pp differential cross-sections and analyzing power for several scattering angles; EDDA data (in grey) together with data from former experiments.

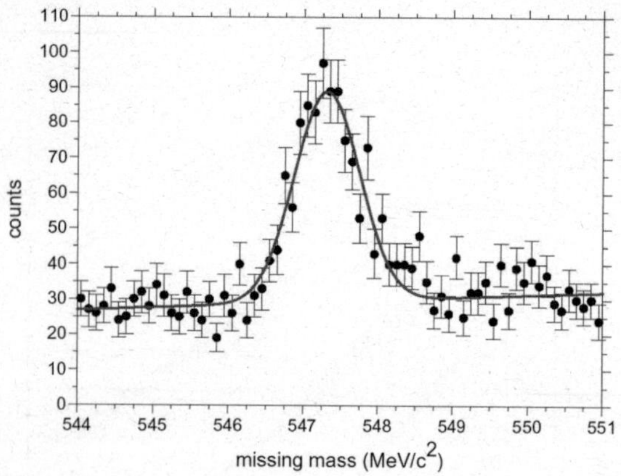

Fig. 8. Missing-mass spectrum of the η-particle obtained by the experiment using BIG KARL.

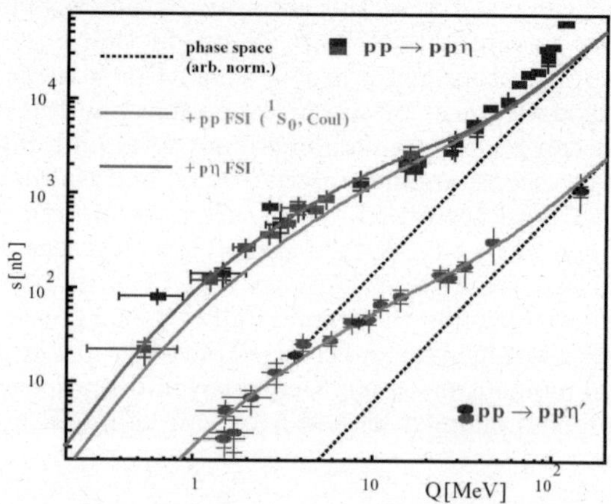

Fig. 9. Excitation functions of the reactions $pp \to pp\eta'$ and $pp \to pp\eta$. The grey points are the COSY-11 results.

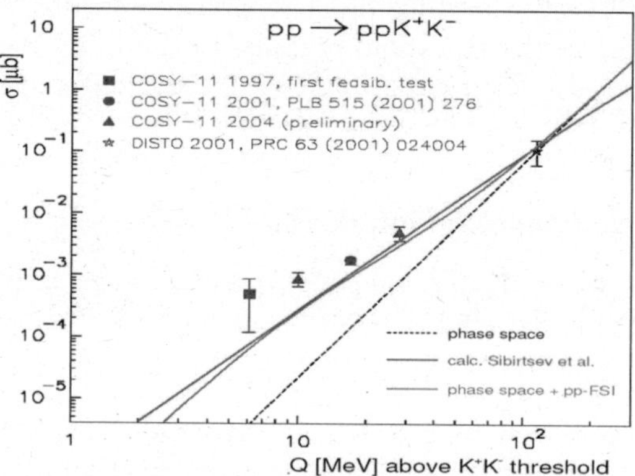

Fig. 10. Excitation functions of the reaction $pp \to ppK^+K^-$ very close to threshold.

recent data from the KLOE experiment on this topic are presented by Cesare Bini at this conference.

2.3 Meson production

A large part of the experiments at COSY concentrates on physics of the near-threshold meson production in different flavour and spin channels. Meson production is studied in nucleon-nucleon reactions as well as in nucleon-nuclei reactions. The measurements performed by the various internal and external experiments span from pion- to phi-production. In this paper I will concentrate on the elementary reactions. Medium effects are discussed by Albrecht Gillitzer at this conference.

Of special interest to understand the reaction dynamics in elementary reactions is the role of N^*-resonances, and connected with the production near threshold, the influence of final-state interaction effects. As an example the results of COSY-11 comparing η- and η'-production are sketched here. Near-threshold η-production differs considerably from pion production due to the strong coupling of the $S_{11}(1535)$-resonance to the η-nucleon decay channel, leading to a threshold enhancement of the η-production cross-section. This is compatible with a rather large η-nucleon scattering length. The η'-proton interaction near threshold appears to be much weaker than the η-proton interaction. Using the stochastically cooled proton beam COSY-11 performed measurements for both systems very close to threshold [6]. The energy dependence of the cross-sections are shown in fig. 9. Comparing the data to phase space (dashed line) reveals that for both reactions the final-state interactions enhance the total cross-sections by more than one order of magnitude at small excess energies. While the η'-production data are well described by modulation of the phase space with the pp FSI (solid curve), for the η one this is not sufficient. The remaining discrepancies close to threshold can be explained by the influence of the attractive η-proton interaction. The upmost curve in fig. 9 corresponds to a simple model where an incoherent pairwise interaction between the outgoing particles is used.

COSY-11 also investigated the channel $pp \to ppK^+K^-$ very close to threshold [7]. The extracted excitation function is shown in fig. 10 together with a data point at somewhat higher energy from DISTO. Again there is a strong influence of the pp final-state interaction which leads to a significant enhancement very close to threshold.

Vector meson production is studied at COSY in the reaction $pp \to pp\phi$ by ANKE and in the channel $pp \to pp\omega$ by COSY-TOF. One of the interesting topics in this field is the value of the cross-section ratio of both reactions, which is discussed in the context of the violation of the OZI rule. In fig. 11 the ϕ cross-section is shown, obtained by ANKE for very low excess energies together with a data point from DISTO at somewhat higher energy. The curves correspond to phase space (full) and a calculation including pp FSI (dashed), which is normalized to the data points at higher energies.

COSY-TOF measured total and differential cross-sections of the reaction $pp \to pp\omega$ for various excess energies. Results on the differential cross-sections are shown in fig. 12 together with data on $pp \to pp\phi$ from DISTO. The angular distributions behave similarly for both channels. Combining the data from ANKE and COSY-TOF

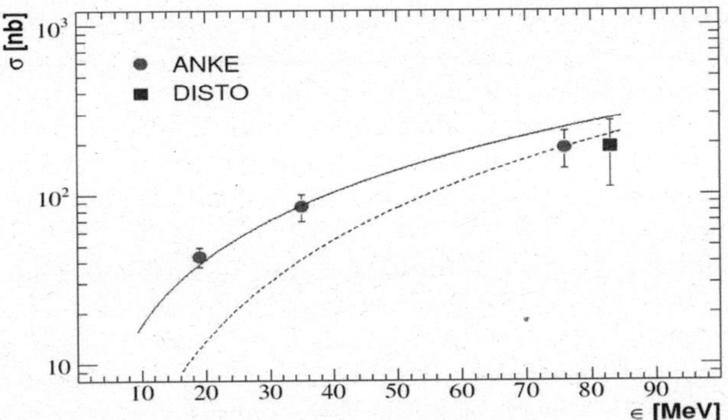

Fig. 11. Cross-section of $pp \to pp\phi$ obtained by ANKE together with a data point from DISTO.

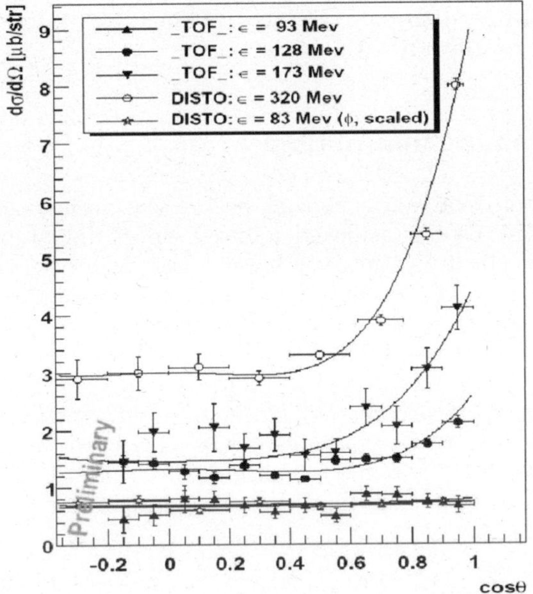

Fig. 12. Angular distributions of the reaction $pp \to pp\omega$ from COSY-TOF together with DISTO data on the reaction $pp \to pp\phi$.

the obtained ratio ϕ to ω is about 7 times larger than the ratio obtained from the OZI rule. This value of about 7 close to threshold is larger than the value of about three obtained in average from experiments at higher energies. For final conclusion this has to be discussed by investigating the reaction mechanism.

2.4 Strangeness production

An essential part of the program at COSY-TOF aims at the production of Λ-, Σ^0-, and Σ^+-hyperons in proton-proton collisions and in a further stage also at proton-neutron reactions. The main interest in the investigation of the associated strangeness production in elementary reactions close to threshold is to get insight into the dynamics of the $s\bar{s}$-production. The questions especially concern the role of N^*-resonances and the hyperon-nucleon

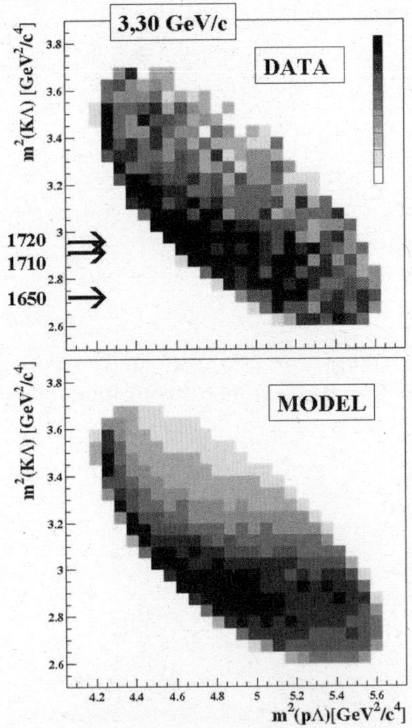

Fig. 13. Dalitz plots of the reaction $pp \to K^+\Lambda p$ at a beam momentum of $3.30\,\mathrm{GeV}/c$. Top: data from COSY-TOF. Bottom: model of Sibirtsev with adjusted parameters (see text). The arrows indicate the $N^*(1650, 1710, 1720)$-resonances.

final-state interaction which is known to be of special importance close to threshold. A further topic is also the search for exotic resonances as the pentaquark Θ^+. To come to conclusive results, precise measurements of the observables are needed, concentrating on exclusive data, covering the full phase space.

The external experiment COSY-TOF is a wide-angle, non-magnetic spectrometer with an inner detector system, which is optimized for strangeness production measurements. This inner system with small beam holes as well as the outer detector system covers the full angular range of the reaction products. It allows the complete reconstruction of the $pp \to K^+\Lambda p$ and $pp \to K^0\Sigma^+ p$ events, including a precise measurement of the delayed decay of the Λ-hyperon and the K^0-meson, respectively. A more detailed discussion of the strangeness production program at COSY-TOF is given in a talk by Wolfgang Schroeder at this conference.

The reaction $pp \to K^+\Lambda p$ has been investigated at momenta from threshold up to nearly the COSY limit for the external beam ([8,9]). For reactions with more than two particles in the final state, as in the case of the reaction $pp \to K^+\Lambda p$, Dalitz plots are a powerful tool to extract information about the reaction mechanism. Whereas pure phase space leads to a homogeneous strength distribution, in particular resonances should lead to significant deviations from this. Figure 13 (top) shows as an example the Dalitz plot extracted from the experiment at the beam momenta of $3.30\,\mathrm{GeV}/c$. It obviously shows strong deviations from phase space. Using a parametrization of

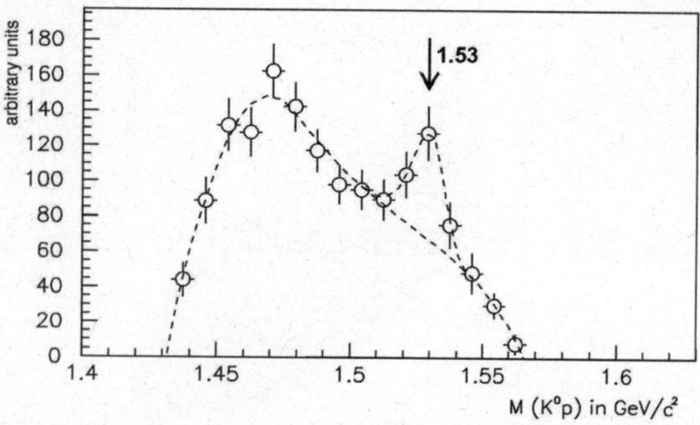

Fig. 14. Invariant-mass spectrum of the K^0p system from the reaction $pp \to K^0\Sigma^+p$ at a beam momentum of $p_{beam} = 2.95\,\mathrm{GeV}/c$.

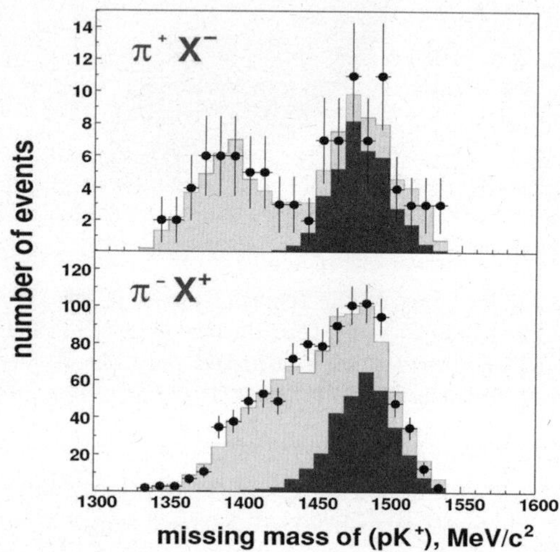

Fig. 15. Missing-mass $MM(pK^+)$ spectra for the reactions $pp \to pK^+\pi^+X^-$ (top) and $pp \to pK^+\pi^-X^+$ (bottom). The contribution of the newly found Y^{0*}-resonance is shown as a solid histogram (dark).

A. Sibirtsev [10], which includes the $N^*(1650, 1710, 1720)$-resonances, a non-resonant term and the $p\Lambda$ final-state interaction on the amplitude base, the strengths of the various contributions were adjusted individually to achieve a best fit for the Dalitz plots. The results are shown on the lower plot of fig. 13. The data are well described by the model fits. Measurements between $2.85\,\mathrm{GeV}/c$ and $3.30\,\mathrm{GeV}/c$ show a pronounced energy dependence of the contributions of the different resonances (see talk of W. Schroeder).

Apart from the Λ-production also the Σ-production is exclusively studied in the COSY-TOF experiment. This includes the search for the exotic pentaquark Θ^+ using the reaction $pp \to K^0\Sigma^+p$. In the K^0p system evidence was found for a narrow resonance at a mass of $1530\pm5\,\mathrm{MeV}/c^2$ with strangeness $S = 1$ [11]. The published spectrum is shown in fig. 14.

To come to a final decision on the existence of the Θ^+ in the investigated reaction very recently COSY-TOF performed a measurement which will give results with strongly improved statistical accuracy. More details on this topic are presented at this conference.

Also at ANKE strangeness production is a very important part of the experimental program. Besides the study of K-production in proton-nucleus reaction strangeness production is also intensively investigated in pp and pn reactions.

One of the most interesting results in this field is a possible new excited hyperon resonance for which evidence was found in the reaction $pp \to pK^+Y$ using a beam momentum of $3.65\,\mathrm{GeV}/c$ ([12]). Analysing the missing-mass $MM(pK^+)$ spectra for the two reaction chains $pp \to pK^+\pi^+X^-$ and $pp \to pK^+\pi^-X^+$, which are shown in fig. 15, and comparing with MC simulations including all known hyperon resonances, the necessity for an additional hyperon resonance Y^{0*} was found. Mass and width have been deduced to be $M(Y^{0*}) = (1480 \pm 15)\,\mathrm{MeV}/c^2$ and $\Gamma(Y^{0*}) = (60 \pm 15)\,\mathrm{MeV}/c^2$.

3 Summary and outlook

Using internal and external beams the various experiments at COSY achieved a broad spectrum of precise results which significantly extend the data base in different fields of hadron physics in the corresponding energy regime. In the future hadron physics at COSY will be performed especially by the three major experiments ANKE, COSY-TOF, WASA. The main topics will be spin physics, spectroscopy and symmetries. Using the precise COSY beam these experiments will contribute on the way to understand how hadrons are made in nature and will help to better understand the strong interaction.

References

1. http://www.fz-juelich.de/ikp/en/experiments.shtml.
2. EDDA Collaboration (D. Albers *et al.*), Phys. Rev. Lett. **78**, 1652 (1997).
3. EDDA Collaboration (M. Altmeier *et al.*), Phys. Rev. Lett. **85**, 1819 (2000).
4. EDDA Collaboration (F. Bauer *et al.*), Phys. Rev. Lett. **90**, 142301 (2003).
5. M. Abdel-Bary *et al.*, Phys. Lett. B **619**, 281 (2005).
6. COSY-11 Collaboration (P. Moskal *et al.*), Phys. Rev. C **69**, 025203 (2004); COSY-11 Collaboration (A. Khoukaz *et al.*), Eur. Phys. J. A **20**, 235 (2004).
7. COSY-11 Collaboration (P. Winter *et al.*), Phys. Lett. B **635**, 23 (2006).
8. COSY-TOF Collaboration (R. Bilger *et al.*), Phys. Lett. B **420**, 217 (1998).
9. COSY-TOF Collaboration (S. Abd El-Samad *et al.*), Phys. Lett. B **632**, 27 (2006).
10. A. Sibirtsev, private communication, 2002, 2005.
11. COSY-TOF Collaboration (M. Abdel-Bary *et al.*), Phys. Lett. B **595**, 127 (2004).
12. ANKE Collaboration (I. Zychor *et al.*), Phys. Rev. Lett. **96**, 0123002 (2006).

The kaon optical potential modified by Θ^+ Pentaquark excitation

D. Cabrera[1], L. Tolós[2], A. Ramos[3], and A. Polls[3]

[1] Cyclotron Institute and Physics Department, Texas A&M University, College Station, Texas 77843-3366, USA
[2] Gesellschaft für Schwerionenforschung, Planckstrasse 1, D-64291 Darmstadt, Germany
[3] Departament d'Estructura i Constituents de la Matèria, Universitat de Barcelona, Diagonal 647, 08028 Barcelona, Spain

Abstract. We study the kaon nuclear optical potential including the effect of the Θ^+ pentaquark in a selfconsistent approach, starting from an extension of the Jülich meson-exchange potential as bare kaon-nucleon interaction. Significant differences between a fully self-consistent calculation and the low-density $T\rho$ approximation are observed. Whereas the influence of the $KN \to \Theta^+$ absorption process is found to be negligible, due to the small $\Theta^+ KN$ coupling, the two-nucleon mechanism ($KNN \to \Theta^+ N$), estimated from the two-meson coupling of the pentaquark, provides the necessary additional kaon absorption to reconcile with data the systematically low theoretical K^+-nucleus reaction cross sections.

PACS. 13.75.Jz – 25.75.-q – 21.30.Fe – 21.65.+f – 12.38.Lg – 14.40.Ev – 25.80.Nv

1 Introduction

The possible existence of the Θ^+ pentaquark with positive strangeness [1,2] and a very small width has attracted renovated interest on the properties of the KN interaction [3,4]. The medium properties of kaons are of particular interest as they are considered suited probes of the hot and dense matter created in heavy ion collisions and of partial restoration of chiral symmetry in dense matter. The in-medium KN interaction, believed to be smooth in the absence of baryonic resonances with $S = +1$, has been usually implemented in a $T\rho$ approximation, leading to a repulsive single-particle kaon potential of around 30 MeV at normal nuclear matter density [5,6] (the more recent self-consistent approach in [7] showing a kaon mass shift of 36 MeV at $\rho = \rho_0$). However, calculations of the kaon optical potential based on the $T\rho$ approximation share a common lack of success in reproducing K^+-nucleus total and reaction cross sections (see [8] and references therein), underestimating the data by about 10-15%, a problem which remains unsolved although a wide variety of mechanisms has been explored [9–12].

In this work we start by comparing the $T\rho$ approximation of the kaon optical potential to a fully self-consistent calculation of the in-medium KN effective interaction. The latter, based on the Jülich meson-exchange model including the coupling of the Θ^+ to the KN system [3], allows us to investigate the changes on the kaon optical potential due to the one-nucleon induced excitation of the Θ^+ in the medium. In addition, following the suggestion of [13], in which substantially improved fits to the data were obtained by incorporating K^+ absorption by nucleon pairs

$(KNN \to \Theta^+ N)$, we evaluate a microscopic calculation of this mechanism according to the model of Ref. [14] for the Θ^+ interaction with a $K\pi$ cloud, where the pion couples to particle-hole (ph) and Delta-hole (Δh) excitations. We study the effect of this mechanism in the kaon optical potential and obtain improved K^+ nuclear cross sections in close agreement with the experimental data [15].

2 In-medium KN interaction

We evaluate the in-medium KN interaction in a $G-$matrix approach, including Pauli blocking on the nucleonic intermediate states and the dressing of the K meson and nucleon. Schematically,

$$\langle KN \mid G \mid KN \rangle = \langle KN \mid V \mid KN \rangle$$
$$+ \langle KN \mid V \mid KN \rangle \frac{Q_{KN}}{\Omega - E_K - E_N + i\eta} \langle KN \mid G \mid KN \rangle .$$

$$(1)$$

The bare interaction, V, is obtained from the meson-exchange Jülich model for KN scattering with a Θ^+ pole term [3]. The Θ^+ bare mass and coupling to KN are chosen as to reproduce the physical Θ^+ mass and a width of 5 MeV in free space (widths higher than a few MeV have been excluded by recent analysis of the K^+N and K^+d data [16–19]). The kaon single-particle energy in Eq. (1) is obtained self-consistently as $E_K(\mathbf{q}; \rho) = \sqrt{m_K^2 + \mathbf{q}^2} + \mathrm{Re}\, U_K(E_K, \mathbf{q}; \rho)$, with U_K the complex single-particle potential which, in the Brueckner-Hartree-Fock approach, is

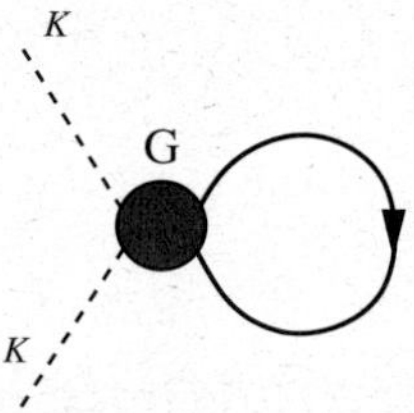

Fig. 1. One-nucleon contribution to the kaon self-energy from the in-medium $G-$matrix.

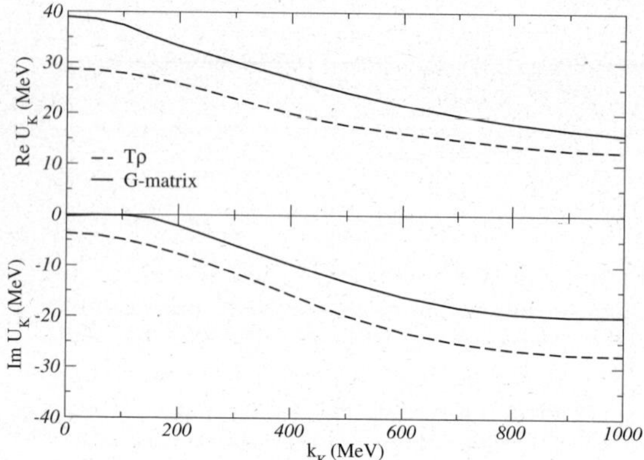

Fig. 2. Kaon optical potential as a function of the kaon momentum at normal nuclear matter density for the $T\rho$ approximation (dashed lines) and the G-matrix calculation (solid lines).

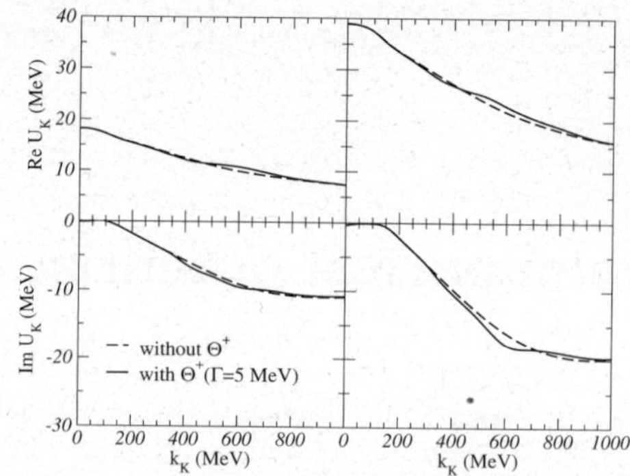

Fig. 3. Kaon optical potential for 0.5 ρ_0 (left panels) and ρ_0 (right panels), without Θ^+ (dashed lines) and including a Θ^+ resonance with a width of 5 MeV (solid lines).

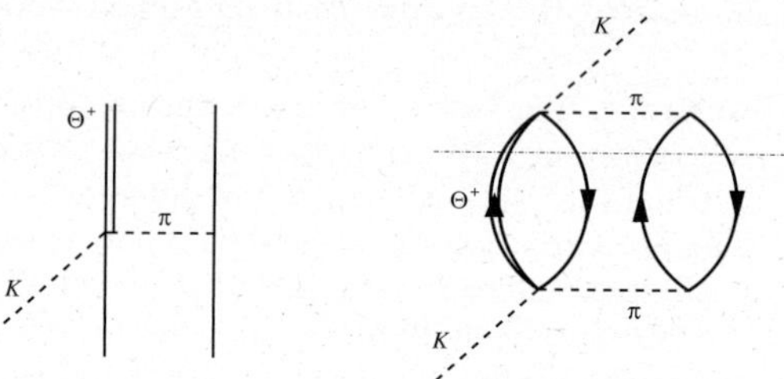

Fig. 4. Two-nucleon kaon absorption mechanism by excitation of Θ^+ (left) and corresponding manybody diagram (right).

given by

$$U_K(E_K, \mathbf{q}; \rho) =$$
$$\sum_{N \leq F} \langle KN \mid G_{KN \to KN}(\Omega = E_N + E_K) \mid KN \rangle , \qquad (2)$$

as diagrammatically represented in Fig. 1. The nucleon single-particle energies are taken from a relativistic $\sigma - \omega$ model with density-dependent scalar and vector coupling constants [20], which provides an attraction of -80 MeV at saturation density.

We show in Fig. 2 the kaon optical potential at $\rho = \rho_0$ in absence of the coupling to Θ^+, for the self-consistent $G-$matrix calculation and a $T\rho$ approximation (replacing G in Eq. (2) by the free amplitude T). The latter, which ignores in-medium effects on the KN interaction as well as on the K, N energies, leads to a kaon optical potential (at zero momentum) 10 MeV less repulsive than the $G-$matrix result, which amounts to 39 MeV. The imaginary part of the potential indicates kaon widths of about 18 MeV at about 400 MeV/c momentum. The selfconsistent implementation of medium effects generates relevant features of the KN interaction which reflect on the kaon properties in the nuclear medium in comparison with the $T\rho$ approximation.

3 Kaon absorption from Θ^+ excitation

Turning on the $KN\Theta^+$ coupling incorporates the effect of one-body Θ^+ excitation in the effective KN interac-

tion. Interestingly, the $G-$matrix exhibits the expected structure associated to the Θ^+ with practically unchanged properties (mass and width) up to $\rho = \rho_0$ as compared to the free $T-$matrix amplitude, except for the in-medium modified KN threshold. This result is highly non-trivial since the properties of the Θ^+ in this approach result from multiple rescattering in the $T-$matrix equation and interferences between the polar and non-polar terms of the KN interaction. The selfconsistent kaon optical potential (Fig. 3) exhibits rather small changes (starting above 300 MeV/c momentum) from the $KN \to \Theta^+$ mechanism, which is related to a small $KN\Theta^+$ coupling (as implied by the small Θ^+ width).

The kaon optical potential may also receive contributions from the two-nucleon process $KNN \to \Theta^+ N$ which can be realized from the Θ^+ coupling to $K\pi$ with subsequent absorption of the pion by a ph or Δh excitation (Fig. 4). The $\Theta^+ K\pi N$ coupling was evaluated in a study of the two-meson cloud effects on the baryon antidecuplet binding in vacuum [14] and latter applied in a calculation of the Θ^+ self-energy in nuclear matter [21]. Its contribution to the kaon optical potential can be obtained from the kaon self-energy diagram in Fig. 4 (right),

$$\Pi_K^{2N}(q^0, \mathbf{q}; \rho) = i \int \frac{d^4 k}{(2\pi)^4} \left[D_\pi^{(0)}(k) \right]^2 \Pi_\pi(k; \rho)$$
$$\times (-9) \sum_{j=S,V} \mid t^{(j)}(k,q) \mid^2 U_\Theta(q - k; \rho) , \qquad (3)$$

where $D_\pi^{(0)}(k)$ is the free pion propagator, $\Pi_\pi(k; \rho)$ stands for the $ph+\Delta h$ contribution to the pion self-energy including short range correlations, $U_\Theta(q - k; \rho)$ represents the

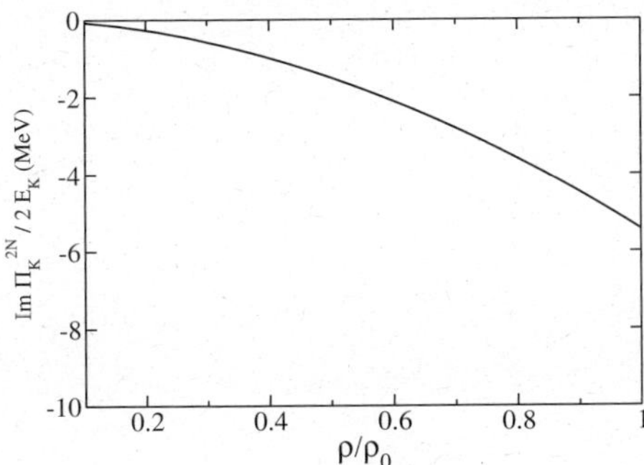

Fig. 5. Imaginary part of the two-nucleon contribution to the kaon optical potential as a function of the density for a kaon momentum of 500 MeV/c.

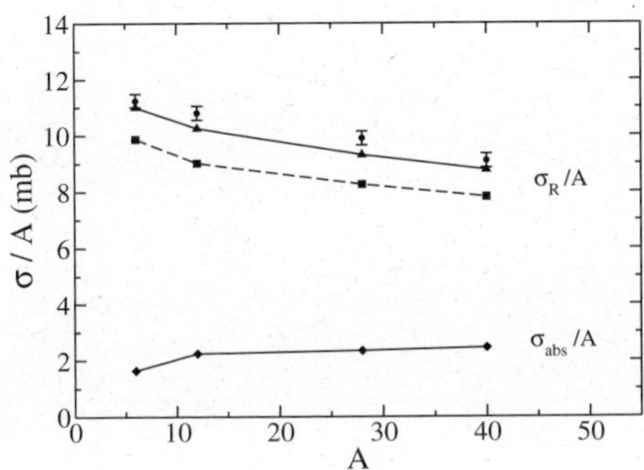

Fig. 6. Absorption and reaction cross sections per nucleon for a kaon laboratory momentum of 488 MeV/c, from the $G-$matrix kaon optical potential (dashed line) and including, in addition, the $2N-$absorption mechanism (solid line). Data for σ_R/A are taken from the analysis of [8].

pentaquark-hole Lindhard function and $t^{(j)}$ denotes the scalar (S) and vector (V) $\Theta^+ K\pi N$ amplitudes [15]. The contribution of this mechanism to the imaginary part of the kaon optical potential is shown in Fig. 5 as a function of the density for a kaon momentum of 500 MeV/c, exhibiting the expected ρ^2 dependence of a two-nucleon process and a significant size, comparable to the one-nucleon mechanisms depicted in Fig. 2.

4 K-nucleus cross section

We calculate kaon nuclear cross sections in the eikonal formalism according to

$$\sigma = \int d^2b \left[1 - \exp\left(-\int_{-\infty}^{\infty} -\frac{1}{q} \mathrm{Im}\, \Pi(q; \rho(\mathbf{b}, z)) dz \right) \right], \tag{4}$$

where Π is given by the two-nucleon component of the kaon self-energy (Π_K^{2N}) for the absorption cross section (σ_{abs}), or by the total self-energy including as well the contribution of the one-nucleon mechanisms for the reaction cross section (σ_R), respectively. In Fig. 6 the calculated cross sections per nucleon in ^{6}Li, ^{12}C, ^{28}Si and ^{40}Ca are compared to experimental data. The results for σ_R/A

from the $G-$matrix calculation of the kaon optical potential including the one-nucleon Θ^+ excitation mechanism underestimate the data by about 15% (dashed lines). On the other hand, the absorption cross sections per nucleon obtained from the $2N$ mechanism (Figs. 4 and 5) are about 2-3 mb, right below the upper bound of 3.5 mb established in Ref. [13]. The reaction cross sections per nucleon obtained with the complete imaginary part of the kaon self-energy, including both the one- and two-nucleon processes, lie very close to the experimental data (upper solid line). Our model for the two-nucleon kaon absorption mechanism provides the required strength to bring the reaction cross sections in agreement with experiment, thereby giving a possible answer to a long-standing anomaly in the physics of kaons in nuclei.

We thank J. Haidenbauer for providing us with the extended Jülich code, and L. Roca and M.J. Vicente-Vacas for fruitful discussions. L.T. acknowledges support from the Alexander von Humboldt Foundation and D.C. from Ministerio de Educación y Ciencia (Spain). This work is partly supported by DG-ICYT contract BFM2002-01868, the Generalitat de Catalunya contract SGR2001-64, and the E.U. EURIDICE network contract HPRN-CT-2002-00311. This research is part of the EU Integrated Infrastructure Initiative Hadron Physics Project under contract number RII3-CT-2004-506078.

References

1. T. Nakano et al. [LEPS Collaboration], Phys. Rev. Lett. **91**, 012002 (2003).
2. www.rcnp.osaka-u.ac.jp/~hyodo/research/Thetapub.html
3. J. Haidenbauer and G. Krein, Phys. Rev. **C 68**, 052201 (2003).
4. A. Sibirtsev et al., Phys. Lett. **B 599**, 230 (2004); *ibid* Eur. Phys. J. **A23**, 491 (2005).
5. N. Kaiser et al., Nucl. Phys. **A 594**, 325 (1995); N. Kaiser et al., Nucl. Phys. **A 612**, 297 (1997).
6. E. Oset and A. Ramos, Nucl. Phys. **A 635**, 99 (1998).
7. C. L. Korpa and M. F. M. Lutz, Acta Phys. Hung. A **22**, 21 (2005).
8. E. Friedman et al., Phys. Rev. **C 55**, 1304 (1997); E. Friedman et al., Nucl. Phys. **A625**, 272 (1997).
9. P.B. Siegel et al., Phys. Rev. **C 30**, 1256 (1984); P.B. Siegel et al., Phys. Rev. **C 31**, 2184 (1985).
10. G.E. Brown et al., Phys. Rev. Lett. **60**, 2723 (1988).
11. M.F. Jiang and D.S. Koltun. Phys. Rev. **C 46**, 2462 (1992).
12. C. Garcia-Recio et al., Phys. Rev. **C 51**, 237 (1995).
13. A. Gal and E. Friedman, Phys. Rev. Lett. **94**, 072301 (2005); *ibid* Phys. Rev. C **73** (2006) 015208.
14. A. Hosaka et al., Phys. Rev. C **71**, 045205 (2005).
15. L. Tolos et al., Phys. Lett. B **632** (2006) 219.
16. D. Diakonov et al., Z. Phys. **A 359**, 305 (1997).
17. S. Nussinov, arXiv: hep-ph/0307357.
18. R. A. Arndt et al., Phys. Rev. **C 68**, 042201 (2003).
19. W. R. Gibbs, Phys. Rev. **C 70**, 045208 (2004).
20. R. Machleidt, Adv. Nucl. Phys. **19**, 189 (1989).
21. D. Cabrera et al., Phys. Lett. B **608** (2005) 231.

A critical analysis on deeply bound kaonic states in nuclei and the KEK experiment

E. Oset[1] and H. Toki[2]

[1] Departamento de Física Teórica and IFIC, Centro Mixto Universidad de Valencia-CSIC, Institutos de Investigación de Paterna, Aptd. 22085, 46071 Valencia, Spain
[2] Research Center for Nuclear Physics, Osaka University, Ibaraki, Osaka 567-0047, Japan

© Società Italiana di Fisica / Springer-Verlag 2007

Abstract. We critically discuss the theoretical developments that led to predictions of very deeply bound kaon states and then revise the data of the KEK experiment from where claims for evidence of deeply bound kaon atoms in nuclei were made. We conclude that the peaks seen in the experiment correspond to the absorption of kaons by a pair of nucleons leading to Σp and Λp and leaving the daughter nucleus as a spectator. These conclusions have been reconfirmed by a recent experiment at FINUDA.

PACS. 25.80 – 13.75.Jz – 36.10.-k

1 Introduction

The success of the theoretical predictions and posterior finding of deeply bound pionic atoms lead to a search for other possible deeply bound states in nuclei. The case of deeply bound kaons was a good candidate since from data on kaon atoms it was known that the strong interaction of negative kaons with nuclei was attractive. This was in spite of the fact that the low density theorem $\Pi = t\rho$ gives repulsion since the isoscalar $K^- N$ amplitude at threshold is repulsive. The consideration of Pauli blocking in intermediate states was shown to be responsible for this change of sign in the interaction [1] although further considerations of selconsistency in the calculations led to a much weaker attraction than that produced by Pauli blocking alone [2–4]. These latter works were performed within the context of chiral unitary theory and all them led to moderate potentials of the order of 50 MeV attraction at normal nuclear matter density. By means of these, deeply bound K^- states could be accommodated in medium and heavy nuclei with a binding of about 30 MeV, but unfortunately a very large width of the order of 100 MeV, which would make the spectroscopy of such states a hopeless task. It was then a surprise that a theoretical paper appeared predicting a much larger kaon attraction in nuclei with three and four nucleons with 108 MeV binding in ^3He with $I = 0$ [5]. Inspired by this prediction an experiment was done at KEK with K^- absorption at rest leading to proton emission. A peak in the proton momentum spectrum was found, but if it were interpreted as corresponding to a bound kaon atom the binding energy would have been 195 MeV and the isospin would be $I = 1$. In view of the clear discrepancy from the predictions a paper was published claiming the discovery of a strange tribaryon [6]. This experimental finding stimulated further work and the authors of [5] made some corrections to their earlier work from where they found a stronger potential of 618 MeV attraction in the center of the nucleus due among other rough approximations to allowing the nucleus get compressed to **ten times** the nuclear matter density [7]. From that moment on the KEK work was presented in Conferences as an evidence for deeply bound kaons in nuclei. In a posterior paper [8], the authors of the present work made a critical review of the theoretical approach of [5] and found a natural intepretation for the peaks seen in [6], the essence of which we present here, together with further developments spurred by that publication.

2 Criticism of the theoretical potential leading to deeply bound kaon atoms

In [8] we make a thorough discussion of the calculations involved in the chiral models [2–4] which serves as a basis of discussion for the differences and defficiencies of the work of [5]. Limitation of space only allows here a brief description of these defficiencies:

1) In the chiral models a coupled channel approach is made which contains the channels $\bar{K}N, \pi\Sigma, \pi\Lambda, \eta\Sigma, \eta\Lambda, K\Xi$. In the work of [5] only $\bar{K}N, \pi\Lambda, \pi\Sigma$ channels are considered but the couplings of $\pi\Lambda$ and $\pi\Sigma$ to themselves is neglected in spite of their large strength obtained from chiral Lagrangians. This restriction prevents the authors to get two $\Lambda(1405)$ states which are strongly supported experimentally now [9].

2) The authors of [5] assume the $\Lambda(1405)$ to be a bound state of $\bar{K}N$, while in chiral theories the two poles are a complicated mixture of coupled channles. These assumptions lead to a $\bar{K}N$ amplitude below threshold that goes from a factor two about 50 % larger than chiral models depending on the model.

3) One of the most serious problem in the approach of [5] is the lack of selfconsistency. In the chiral calculations the $\bar{K}$ potential is obtained and inserted in the $\bar{K}$ propagators of the loops and the procedure is itererated till convergence is found. This is a must in the calculations because of the presence of the $\Lambda(1405)$ resonance close to threshold. Indeed changes in the masses of the particles move the resonance up and down and one goes easily from attraction to repulsion of the $\bar{K}$. Needless to say that with a potential of 618 MeV attraction as in [7] the selfconsistent consideration its absolutely necessary, but it is not done in [7]. This lack is sufficient too make the results unreliable.

4) Since not enough attraction is obtained to get deep and narrow states the nuclear matter is compressed to **ten times** nuclear matter density in [5]. We consider this utterly exagerated.

3 Discussion on the KEK experiment

On the basis of the discussion in the former section it is quite clear that the KEK experiment [6] could not be interpreted in terms of the creation of deeply bound kaonic states. From this perspective we would like to interpret the meaning of the peak seen there. The experiment is

$$\text{stopped } K^- + {}^4He \rightarrow S + p \qquad (1)$$

where S has a mass of about 3115 MeV, and has the quantum numbers of YNN with zero charge where Y is a S=-1 hyperon. The state S would have $I_3 = -1$ and hence cannot be $I = 0$ as predicted originally in Ref. [5], which was already noted in Ref. [6]. The peak is also quite narrow, around or smaller than 20 MeV. In [8] a discussion is made of possible mechanisms producing the peak and all them are eliminated for inconsistencies with other data. Only one possible reaction passed all the experimental tests and we describe it below.

In view of the previous unsuccessful trials it was concluded in [8] that the reaction which is most likely to happen is K^- absorption by two nucleons in ${}^4He$ leaving the other two nucleons as spectators. This kaon absorption process should happen from some K^- atomic orbits which overlap with the tail of the nuclear density and hence the Fermi motion of the nucleons is small. Then we would have $K^-NN \rightarrow \Lambda N$, and ΣN, and the two baryons are emitted back to back with the momentum for the proton of 562, and 488 MeV/c respectively. These results are very interesting: the peak of the proton momentum in Ref. [6], before proton energy loss corrections, appears at 475 MeV/c (see Fig. 5 of this reference). This matches well with the 488 MeV/c proton momentum from a $K^-NN \rightarrow \Sigma N$ event, and the proton would lose about 13 MeV/c when

crossing the thick target. This energy loss is compatible with the estimate of about 30 MeV/c in Ref. [6], particularly taking the width of the peak also into account.

This suggestion sounds good, but then one could ask oneself: what about $K^-NN \rightarrow \Lambda N$? Should not there be another peak around 550 MeV/c, counting also the energy loss? The logical answer is yes, and curiously one sees a second peak around 545 MeV/c in the experiment. The peak is clearly visible although less pronounced than the one at 475 MeV/c and it appears in the region of fast decline of the cross section.

There are other arguments supporting our suggestion. Indeed, as mentioned above, the pion momenta from Λ decay are smaller than those from Σ decay. As a consequence of this we should expect the peak associated with $p\Lambda$ emission to appear in the low momentum side of the pion (the range of pion momenta is from 61 MeV/c to 146 MeV/c from phase space considerations). Actually, this is the case in the experiment of Ref. [6] as one can see in Fig. 5d of this reference, corresponding to the spectrum when the low pion cut is applied, where the peak of higher momentum stands out more clearly. On the other hand, by working out the phase space for Σ decay, the pion momenta range from 162 Mev/c to 217 MeV/c and the pion could be seen in the two regions of pion momenta of Ref. [6], as it is indeed the case (see figs. 5c, 5d of Ref. [6]).

One can even argue about the size of the peaks and their relative strength. For this the information of Ref. [10] is very useful. There we find the following results for events per stopped K^-:

$$\Sigma^- p\, d \qquad 1.6\ \% \qquad\qquad (2)$$
$$\Sigma^- ppn \qquad 2.0\ \% \qquad\qquad (3)$$
$$\Lambda(\Sigma^0)pnn \qquad 11.7\ \% \qquad\qquad (4)$$

with errors of 30-40 %.

From these data and further anaylsis is was concluded in [8] that one could understand the larger strength of the peak coming from Σp versus that from Λp. It was also suggested that because the $\Sigma^- p\, d$ reaction leaves indeed the daugher nucleus (d) untouched, this channel is the best candidate to explain the peak caused by the Σp emission, leading to a momentum of the proton of 482 MeV/c.

The hypothesis advanced should have other consequences. Indeed, this peak should not be exclusive of the small nuclei. This should happen for other nuclei. Actually in other nuclei, let us say $K^-\ {}^7Li$, the signals that we are searching for should appear when a proton as well as a Σ or a Λ would be emitted back to back and a residual nuclear system remains as a spectator and stays nearly in its ground state. We would thus expect two new features: first, the two peaks should be there. However, since now the spectator nuclear systems remain nearly in their ground states, only about the binding energy of the two participant nucleons will have to be taken from the kaon mass, instead of the 28 MeV in ${}^4He$ for a full break up, as a consequence of which the proton momenta should be a little bigger. We make easy estimates of 502 MeV/c for the proton momentum in the case of $p\Sigma$ emission and 574

MeV/c for the case of $p\Lambda$ emission. Curiously the FINUDA data [11] exhibits two peaks in ^7Li around 505 MeV/c and 570 MeV/c (see comments at the end about a recent publication).

There is one more prediction we can make. The process discussed has to leave the remnant nucleus in nearly its ground state. This means that one has to ensure that the nucleus is not broken, or excited largely, when the energetic protons go out of the nuclear system. Theoretically one devises this in terms of a distortion factor that removes events when some collision of the particles with the nucleons takes place. Obviously this distortion factor would reduce the cross sections more for heavier nuclei and, hence, we should expect the signals to fade away gradually as the nuclear mass number increases. This is indeed a feature of the FINUDA data [11].

Similarly, we can also argue that the spectator nucleus, with a momentum equal to that of the combined pair on which K^- absorption occurs, will have smaller energies for heavier nuclei since their mass is larger. Hence, the spreading of the energy of the emitted proton should become narrower for heavier nuclei. This is indeed a feature of the FINUDA data [11].

This sequence of predictions of our hypothesis, confirmed by the data of [6] and [11], provides a strong support for the mechanism suggested of K^- absorption by a pair of nucleons leaving the rest of the nucleons as spectators. Certainly, further tests to support this idea, or eventually refute it, should be welcome. An obvious test is to search for Σ or Λ in coincidence and correlated back to back with the protons of the peak.

in a system of three particles are out of scale, we looked for a plausible explanation of present experiments which could be interpreted differently than creating these deeply bound K^- states. After using information and the analysis of Ref. [6] discarding potential alternatives with chain reactions, we were led by elimination to a source of explanation deceivingly simple, and which, however, passes the present experimental tests: the association of the observed proton peaks to K^- absorption by a nucleon pair leaving the rest of the nucleons as spectators. From the emission of $p\Sigma$ we explained the peak found in Ref. [6] and from the emission of $p\Lambda$ we predicted another peak which is indeed present in the experiment of Ref. [6]. The hypothesis made led us to conclude that these peaks should also be visible in other nuclei at slightly larger proton momenta, should be narrower and their strength should diminish with increasing mass number of the nucleus. Fortunately, all these predictions could be tested with present data from FINUDA [11] and all the predictions were confirmed by these data.

The FINUDA data quoted in this work as [11] have been recently published [12] and there they reconfirm our claims for the interpretation of these peaks, that are seen in other light nuclei and tend to disappear in heavy ones, as we predicted.

This work is partly supported by DGICYT contract number BFM2003-00856, the Generalitat Valenciana, the CSIC-JSPS collaboration agreement and the E.U. EURIDICE network contract no. HPRN-CT-2002-00311. This research is part of the EU Integrated Infrastructure Initiative Hadron Physics Project under contract number RII3-CT-2004-506078.

4 Conclusions

In this paper we have made a thorough review of the theoretical developments that led to predictions of deeply bound kaonic atoms in light nuclei. We could show that there were many approximations done, which produced unreliably deep potentials. Three main reasons made the approximations fail dramatically: the problem of the coupled channels to produce the two $\Lambda(1405)$ states was reduced to only one channel, the $\bar{K}N$, and only one $\Lambda(1405)$, which was assumed to be a bound state of the $\bar{K}N$ potential was considered. The second serious problem was the lack of selfconsistency in the intermediate states, which makes the results absolutely unreliable when one is in the vicinity of a resonance, as is the case here. The last serious problem is allowing the matter to be compressed to **ten time** nuclear matter density. Altogether one obtains a potential as large as ten times what one gets without making these approximations. We also state that the width becomes zero in the I=0 channel due to the binding energy being lower than the pion-Σ threshold. However, the selfconsistency consideration automatically produces the two body kaon absorption processes, which provides large widths.

With the weakness of the theoretical basis exposed and the realization that binding energies of 200 MeV for a kaon

References

1. V. Koch, Phys. Lett. B **337** (1994) 7 [arXiv:nucl-th/9406030].
2. M. Lutz, Phys. Lett. B **426** (1998) 12 [arXiv:nucl-th/9709073].
3. A. Ramos and E. Oset, Nucl. Phys. A **671** (2000) 481 [arXiv:nucl-th/9906016].
4. J. Schaffner-Bielich, V. Koch and M. Effenberger, Nucl. Phys. A **669** (2000) 153 [arXiv:nucl-th/9907095].
5. Y. Akaishi and T. Yamazaki, Phys. Rev. C **65** (2002) 044005.
6. T. Suzuki *et al.*, Phys. Lett. B **597** (2004) 263.
7. Y. Akaishi, A. Dote and T. Yamazaki, Phys. Lett. B **613** (2005) 140 [arXiv:nucl-th/0501040].
8. E. Oset and H. Toki, Phys. Rev. C **74** (2006) 015207 [arXiv:nucl-th/0509048].
9. V. K. Magas, E. Oset and A. Ramos, Phys. Rev. Lett. **95** (2005) 052301 [arXiv:hep-ph/0503043].
10. P. A. Katz, K. Bunnell, M. Derrick, T. Fields, L. G. Hyman and G. Keyes, Phys. Rev. D **1** (1970) 1267.
11. N. Grion and S. Piano, private communication. See preliminary results in the talk of T. Yamazaki at the chiral05 Conference, Tokyo, 2005, http://chiral05.riken.jp/
12. M. Agnello at al. (FINUDA Collaboration), Nucl. Phys. A775 (2006) 35

Chiral approach to antikaons in dense matter

Laura Tolós[1], Angels Ramos[2], and Eulogio Oset[3]

[1] Gesellschaft für Schwerionenforschung, Planckstrasse 1, D-64291 Darmstadt, Germany
[2] Departament d'Estructura i Constituents de la Matèria, Universitat de Barcelona, Diagonal 647, 08028 Barcelona, Spain
[3] Departamento de Física Teórica and IFIC, Centro Mixto Universidad de Valencia-CSIC, Institutos de Investigación de Paterna, Ap. Correos 22085, E-46071 Valencia, Spain

Abstract. Antikaons in dense nuclear matter are studied using a chiral unitary approach which incorporates the s- and p-waves of the $\bar{K}N$ interaction. We include, in a self-consistent way, Pauli blocking effects, meson self-energies modified by nuclear short-range correlations and baryon binding potentials. We show that the on-shell factorization cannot be applied to evaluate the in-medium corrections to p-wave amplitudes. We also obtain an attractive shift for the Λ and Σ masses of -30 MeV at saturation density while the Σ^* width gets sensibly increased to about 80 MeV. The moderate attraction developed by the antikaon does not support the existence of very deep and narrow bound states.

PACS. 13.75.-n – 13.75.Jz – 14.20.Jn – 14.40.Aq – 21.65.+f – 25.80.Nv

Understanding the interaction of $\bar{K}$ with nucleons and nuclei has recently been a matter of interest [1]. The $\bar{K}N$ interaction below threshold is governed by the presence of the $\Lambda(1405)$ resonance. Phenomenology of kaonic atoms indicate that the $\bar{K}$ feels an attractive potential at low densities. Theoretically, this attraction can be traced back to the modified $\Lambda(1405)$ in the medium due to Pauli blocking effects [2] combined with the self-consistent consideration of the $\bar{K}$ self-energy [3,4] and the inclusion of self-energies of the mesons and baryons in the intermediate states [5]. Moderate attractions of the order of -50 MeV at normal nuclear matter density ρ_0 are found by different approaches, in particular using chiral theories and the unitarization extensions in coupled-channels [5].

Further theoretical advances have been achieved by looking at the p-wave contribution to the $\bar{K}N$ optical potential. While the p-wave contributions for atoms are negligible [6], heavy-ion collisions provide an excellent scenario for testing high-momenta kaons and, therefore, further partial-wave contributions. In fact, some work has been done in that direction by studying the $\bar{K}$ at finite momenta from a self-consistent calculation using the Jülich meson exchange interaction [7]. An evaluation of antikaons in matter based on chiral dynamics including s-, p-, d-waves has been also performed [8].

This work investigates the $\bar{K}$ in the nuclear medium following a chiral unitary approach using s- and p-wave contributions to $\bar{K}N$ and incorporating the corresponding many-body corrections. We will show that the on-shell factorization cannot be applied for p-waves in the medium as well as the importance of including the self-energy of mesons modified by nuclear short-range correlations and the binding energy of baryons. We will obtain the $\bar{K}$ self-energy in nuclear matter and, as a byproduct, the self-energy of $\Lambda(1115)$, $\Lambda(1405)$, $\Sigma(1195)$ and $\Sigma^*(1385)$ [9].

1 In-medium $\bar{K}N$ interaction

The properties of the $\bar{K}$ in the nuclear medium are obtained by incorporating medium modifications to the effective $\bar{K}N$ interaction. The effective $\bar{K}N$ interaction is built solving the coupled-channel Bethe-Salpeter equation using tree level contributions as the kernel of the equation. The kernel for the s-wave amplitude is derived from the lowest order chiral lagrangian that couples the octet of pseudoscalar mesons to the octet of $1/2^+$ baryons [10]. The main contribution to the p-wave amplitudes comes from the Λ, Σ and Σ^* pole terms [11]. For meson-baryon scattering the kernel can be factorized on the mass shell in the loop functions [10,12]. The loop function is then regularized by means of a cutoff or dimensional regularization. The formal result is schematically given by

$$T = [1 - VG]^{-1}V , \tag{1}$$

where V is the kernel and G the loop function.

The medium modifications arise from the inclusion of Pauli blocking effects on the nucleons and the dressing of mesons and baryons in the intermediate loops. The binding effects for baryons are taken within the mean-field approach [9]. For $\bar{K}$ and pions the medium modifications are included via the corresponding self-energy [5].

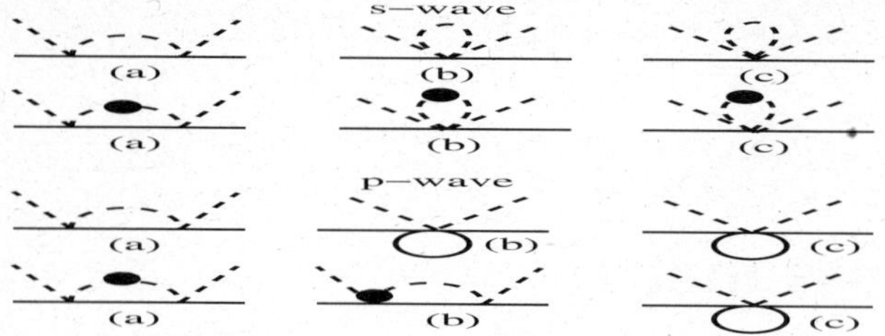

Fig. 1. On-shell (a), off-shell (b) and tadpole (c) contributions for s- (upper two rows) and p-wave (lower two rows) in free space and with self-energy insertions.

In order to calculate the in-medium amplitudes we will perform a similar unitarization procedure as in free space. The evaluation of the free space amplitudes rely on the factorization of the on-shell interaction kernel out of the loop function. In this section we show that the on-shell factorization is still valid for in-medium s-wave amplitudes while not for p-wave contributions.

In the s-wave amplitude of Fig. 1, the off-shell dependence of the two vertices in the loop function eliminates a baryon propagator. Hence, this off-shell term ((b) in row 1) is cancelled by the presence of a tadpole term ((c) in row 1), in a suitable renormalization scheme. When we make self-energy insertions in the meson line, the cancellation between the off-shell part ((b) in row 2) and the tadpole ((c) in row 2) still holds. Then, the in-medium s-wave amplitudes are obtained by solving Eq. (1) with the loop functions modified by meson self-energies and baryon binding corrections.

The situation is different for the p-wave amplitudes due to the different $\mathbf{q}^2$ dependence (see Fig. 1). The off-shell part ((b) in row 3) eliminates a meson propagator (see Ref. [13]) and, as a consequence, it can be cancelled by the tadpole term ((c) in row 3). However, when the meson propagator is dressed, the off-shell term cancels only one of the two intermediate meson propagators. Hence, in the medium, we do not find the cancellation between the off-shell part ((b) in row 4) and tadpole term ((c) in row 4) that we found before for the s-wave amplitude. The use of the previous on-shell factorization could then lead to a violation of causality. This is solved by adding to the free loop the medium corrections calculated using the full off-shell $\mathbf{q}^2$ contribution of the vertices

$$G_l^p(s) \to G_l^p(s) + \frac{1}{\mathbf{q}_{on}^2}[I_{\mathrm{med}}(s) - I_{\mathrm{free}}(s)] \,,$$

$$I_{\mathrm{med}}(s) = i \int \frac{d^4q}{(2\pi)^4} \mathbf{q}^2 D_M(q) G_B(P-q)$$

$$I_{\mathrm{free}}(s) = i \int \frac{d^4q}{(2\pi)^4} \mathbf{q}^2 D_M^0(q) G_B^0(P-q) \,. \qquad (2)$$

with D_M and G_B being the meson and baryon propagators. The total four-momentum in the lab frame is $P = q + p$, where q and p are the meson and baryon four-momentum in this frame. Another ingredient to be considered when dealing with p-wave amplitudes in the medium is the inclusion of nuclear short-range correlations. The $\pi(\bar{K})$ propagators should account for the fact

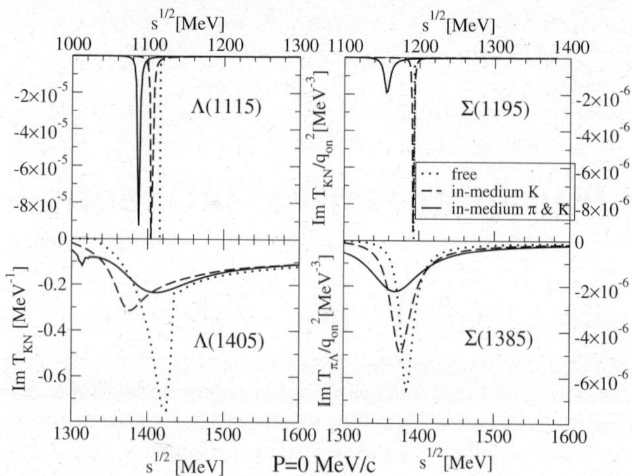

Fig. 2. $\Lambda(1115)$, $\Lambda(1405)$, $\Sigma(1195)$ and $\Sigma^*(1385)$ resonances.

that the nucleon-nucleon (hyperon-nucleon) interaction is not only driven by one-pion (one-kaon) exchange (see details in Ref. [9]).

The in-medium p-wave amplitudes are finally obtained from Eq. (1) using the in-medium meson-baryon propagators of Eq. (2), which incorporate the right $\mathbf{q}^2$ dependence, as well as Pauli blocking effects, dressing of mesons, baryon binding potentials and short-range correlations.

The $\bar{K}$ self-energy in nuclear matter is obtained self-consistently summing the in-medium $\bar{K}N$ for s- and p-waves over the Fermi sea of nucleons $n(\mathbf{p})$ according to

$$\Pi_{\bar{K}}(q^0, \mathbf{q}, \rho) = 4 \int \frac{d^3p}{(2\pi)^3} n(\mathbf{p}) T_{\bar{K}N}(P^0, \mathbf{P}, \rho) \,. \qquad (3)$$

Then, the $\bar{K}$ spectral function reads

$$S_{\bar{K}}(q^0, \mathbf{q}, \rho) = -\frac{1}{\pi} \frac{\mathrm{Im}\,\Pi_{\bar{K}}(q^0, \mathbf{q}, \rho)}{\mid (q^0)^2 - \mathbf{q}^2 - m_{\bar{K}}^2 - \Pi_{\bar{K}}(q^0, \mathbf{q}, \rho) \mid^2} \,. \qquad (4)$$

2 Hyperons and $\bar{K}$ self-energy

In Fig. 2 we display the results for the $\Lambda(1115)$, $\Lambda(1405)$, $\Sigma(1195)$ and $\Sigma^*(1385)$ by showing the imaginary part of the in-medium effective amplitude as function of c.m. energy $\sqrt{s}$ for a total momentum $P = 0$. The free amplitudes (dotted lines) are compared to the in-medium ones at ρ_0 dressing the antikaons self-consistently (dashed lines) and also considering the in-medium effects on pions (solid lines). The $\Lambda(1115)$ acquires an attractive shift of -10 MeV for the first approach while the shift is increased to -28 MeV when pions are dressed due to the appearance of new decay channels, namely $\Lambda N \to \Sigma N$. This is in accordance to hypernuclear spectroscopy data [14]. The $\Lambda(1405)$ is generated dynamically close to the free position and gets strongly diluted when the in-medium properties of pions are considered as a consequence of new channels ($\Lambda N N^{-1}$, $\Sigma N N^{-1}$). The cusp seen for 1320 MeV corresponds to the opening of the $\pi\Sigma$ channel. With regards to the $I = 1$ resonances, the $\Sigma(1195)$ shows an attraction

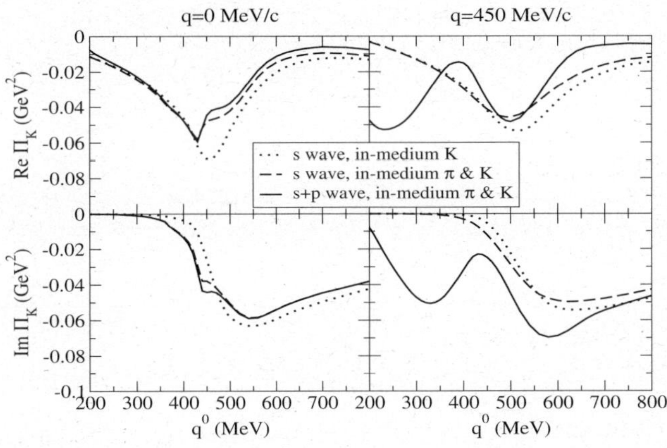
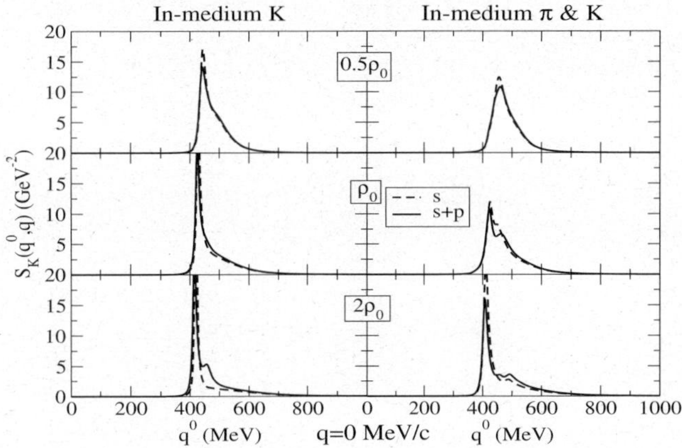

Fig. 3. $\bar{K}$ self-energy as function of $\bar{K}$ energy for two different kaon momenta (left). $\bar{K}$ spectral function as function of $\bar{K}$ energy for zero momentum for the two approaches considered in the text (right).

of -35 MeV when pions are dressed in line with [15] and in contrast with -10 MeV [8] or even the repulsion in [16]. The only reliable evidence, obtained from kaonic atoms, is that the $\Sigma(1195)$ requires attraction at small densities. The $\Sigma^*(1385)$ stays close to the free position for both approaches, in contrast to [8], and increases the width from 30 MeV in free space to 80 MeV when pions are dressed.

The self-energy of $\bar{K}$ at ρ_0 as function of energy is shown in Fig. 3 (left) for two momenta. We show the s-wave component only dressing kaons (dotted lines), dressing also pions (dashed lines) and the p-wave contributions for the latest approach (solid lines). We observe a small imaginary part at subthreshold for zero momentum coming from $\bar{K}NN \to \Sigma N, \Lambda N$. The small p-wave strength is due the Fermi motion of nucleons which produces a slight repulsion in the real part of the self-energy since the energies that come into play are above the Λ, Σ and Σ^* excitations. At finite momentum of 450 MeV/c, the imaginary part shows the ΣN^{-1} component at 300 MeV and the $\Sigma^* N^{-1}$ one around 550 MeV. In this case, the p-wave components determine the imaginary part. The optical potential calculated from the full self-energy as $\Pi_k(q, \varepsilon(q))/(2\varepsilon(q))$, where $\varepsilon(q)$ is the quasiparticle energy $\varepsilon(q)^2 = m_{\bar{K}}^2 + q^2 + \mathrm{Re}\Pi(q, \varepsilon(q))$, changes from -40 MeV at $\rho_0/2$ to -70 MeV at $2\rho_0$ for $\mathbf{q} = 0$ when only $\bar{K}$ are dressed self-consistently. A similar shift from -30 MeV to -80 MeV in the range of $\rho_0/2$ to $2\rho_0$ is obtained when pions are also dressed. Moreover, the imaginary part is sizeable for both approaches, in agreement with the self-energy.

Finally, the antikaon spectral function at zero momentum for different densities and the two approaches is shown in Fig. 3 (right). The spectral function does not show a Breit-Wigner behaviour. The slow fall off on the right-hand side of the quasiparticle peak is due to $\Lambda(1405)N^{-1}$ excitation and the p-wave components are the result of the Fermi motion of nucleons. With increasing density, the quasiparticle peak gains attraction and the spectral function is diluted. The small peak observed on the right-hand side of the quasiparticle peak at $2\rho_0$ is due to $\Sigma^*(1385)N^{-1}$.

In summary, we have investigated the properties of the s- and p-wave $\bar{K}$ self-energy in nuclear matter within the context of a chiral unitary approach. We have shown that the on-shell factorization of the amplitudes cannot be applied for the in-medium corrections to the p-wave amplitudes. Furthermore, we have studied the properties of the Λ, Σ and Σ^* hyperons in nuclear matter. While Λ and Σ feel an attractive potential of -30 MeV at ρ_0, the Σ^* barely changes its mass but develops a width of 80 MeV. The contribution of the p-wave components to the $\bar{K}$ self-energy and, hence, to the spectral function are small for low-momenta but considerable at subthreshold energies for finite momenta due to $\bar{K}N \to \Sigma$ conversion. The antikaon potential obtained from the self-energy can only give $\bar{K}$ states bound by no more than 50 MeV but with a sizeable width of 100 MeV. Therefore, deep and bound $\bar{K}$ states are not expected, in agreement with previous self-consistent calculations.

References

1. J. A. Oller, E. Oset and A. Ramos, Prog. Part. Nucl. Phys. **45**, (2000) 157.
2. V. Koch, Phys. Lett. B **337**, (1994) 7.
3. M. Lutz, Phys. Lett. B **426**, (1998) 12.
4. J. Schaffner-Bielich, V. Koch and M. Effenberger, Nucl. Phys. A **669**, (2000) 153.
5. A. Ramos and E. Oset, Nucl. Phys. A **671**, (2000) 481 .
6. C. Garcia-Recio, E. Oset, A. Ramos and J. Nieves, Nucl. Phys. A **703**, (2002) 271.
7. L. Tolos, A. Ramos, A. Polls and T. T. S. Kuo, Nucl. Phys. A **690**, (2001) 547.
8. M. F. M. Lutz and C. L. Korpa, Nucl. Phys. A **700**, (2002) 309.
9. L. Tolos, A. Ramos and E. Oset, Phys. Rev. C **74**, (2006) 015203
10. E. Oset and A. Ramos, Nucl. Phys. A **635**, (1998) 99.
11. D. Jido, E. Oset and A. Ramos, Phys. Rev. C **66**, (2002) 055203.
12. J. A. Oller and U. G. Meissner, Phys. Lett. B **500**, (2001) 263.

13. D. Cabrera and M. J. Vicente Vacas, Phys. Rev. C **67**, (2003) 045203.
14. T. Hasegawa et al., Phys. Rev. C **53**, (1996) 1210; P.H. Pile et al., Phys. Rev. Lett. **66**, (1991) 2585; D.H. Davis, J. Pniewski, Contemp. Phys. **27**, (1986) 91; M. May et al., Phys. Rev. Lett. **78** (1997) 4343.
15. C. J. Batty et al., Phys. Lett. B **74**, (1978) 27.
16. N. Kaiser, Phys. Rev. C **71**, (2005) 068201.

Eur. Phys. J. A **31**, 441–445 (2007)
DOI 10.1140/epja/i2006-10272-8

THE EUROPEAN
PHYSICAL JOURNAL A

Special Article – QNP 2006

Meson photoproduction on the nucleon with polarized photons

The GRAAL Collaboration

A. D'Angelo[3,10,a], O. Bartalini[3,10], V. Bellini[1,6], J.P. Bocquet[13], P. Calvat[13], M. Capogni[3,10,b], L. Casano[10], M. Castoldi[8], J.-P. Didelez[15], R. Di Salvo[10], A. Fantini[3,10], D. Franco[3,10], G. Gervino[4,11], F. Ghio[9,12], G. Giardina[2,7], B. Girolami[9,12], A. Giusa[1,7], M. Guidal[15], E. Hourani[15], V. Kouznetsov[14], R. Kunne[15], A. Lapik[14], P. Levi Sandri[5], A. Lleres[13], F. Mammoliti[1,7], G. Mandaglio[2,7], D. Moricciani[10], A.N. Mushkarenkov[14], V. Nedorezov[14], L. Nicoletti[3,10,13], C. Randieri[1,5], D. Rebreyend[13], F. Renard[13], N. Rudnev[14], T. Russew[13], G. Russo[1,7], C. Schaerf[3,10], M.-L. Sperduto[1,7], M.-C. Sutera[7], A. Turinge[14], and V. Vegna[3,10]

[1] Dipartimento di Fisica ed Astronomia, Università di Catania, Via Santa Sofia 64, I-95123 Catania, Italy
[2] Dipartimento di Fisica, Università di Messina, Salita Sperone 31, I-98166 Messina, Italy
[3] Dipartimento di Fisica, Università di Roma "Tor Vergata", Via della Ricerca Scientifica 1, I-00133 Roma, Italy
[4] Dipartimento di Fisica Sperimentale, Università di Torino, Via P. Giuria, I-00125 Torino, Italy
[5] INFN - Laboratori Nazionali di Frascati, Via E. Fermi 40, I-00044 Frascati, Italy
[6] INFN - Laboratori Nazionali del Sud, Via di Santa Sofia 44, I-95123 Catania, Italy
[7] INFN - Sezione di Catania, Via Santa Sofia 44, I-95123 Catania, Italy
[8] INFN - Sezione di Genova, Via Dodecanneso 33, I-16146 Genova, Italy
[9] INFN - Sezione di Roma, Piazzale Aldo Moro 2, I-00185 Roma, Italy
[10] INFN - Sezione di Roma "Tor Vergata", Via della Ricerca Scientifica 1, I-00133 Roma, Italy
[11] INFN - Sezione di Torino, I-10125 Torino, Italy
[12] Istituto Superiore di Sanità, Viale Regina Elena 299, I-00161 Roma, Italy
[13] IN2P3, Laboratoire de Physique Subatomique et de Cosmologie, 38026 Grenoble, France
[14] Institute for Nuclear Research, 117312 Moscow, Russia
[15] IN2P3, Institut de Physique Nucléaire, 91406 Orsay, France

Received: 8 December 2006
Published online: 20 March 2007 – © Società Italiana di Fisica / Springer-Verlag 2007

Abstract. Meson photoproduction with polarized photons has proved to be a powerful tool to identify contributions of baryon resonances that are not evident in the differential cross-sections. It provides information that are complementary to those extracted using pion-nucleon scattering data. Extensive results have been produced in the past on beam asymmetries by the Graal collaboration for η and π^0 on the proton. New results are now available for the same reactions on the quasi-free neutron and for the K^+ photoproduction on the proton. Contributions from hitherto undetected baryon resonances may be important to understand the results.

PACS. 13.60.Le Meson production – 13.88.+e Polarization in interactions and scattering – 25.20.Lj Photoproduction reactions

1 Introduction

The determination of baryon resonances properties, such as masses, widths, helicity amplitudes and partial decay widths in nucleon-meson channels, is fundamental to improve our understanding of the internal structure of the nucleons. Most of our knowledge of resonance parameters come from πN scattering data, however recent precise data on photonuclear reactions have provided comple-

mentary information such as resonance photocouplings [1]. Quark model calculations are presently able to reproduce most of the properties of the excited states of the nucleon [2–4], but several issues still remain open. The number of observed states is smaller than the predicted one: the hitherto undetected resonances are called *missing resonances* and the search of experimental evidence of their existence is presently of great interest. One possible explanation is that the *missing resonances* weakly couple to the πN channels and their contribution to πN scattering and π photoproduction reactions is negligible. The study of different reactions such as K, η and ω photoproduction,

---

[a] e-mail: `annalisa.dangelo@roma2.infn.it`
[b] *Present affiliation*: ENEA - C.R. Casaccia, Via Anguillarese 301, I-00060 Roma, Italy.

where the resonant state couples to different meson-N decay channel, may provide new information on the subject.

The access to polarization observables, which are sensitive to the interference between different multipoles, has the potential to reveal small resonance contributions which remain hidden in the differential cross-section under the dominant terms.

The Graal Collaboration has extensively studied η photoproduction on the proton using polarized photons, providing new important constraints to models that extract resonance properties from multipole analysis of the reaction amplitudes. The helicity amplitude $A^p_{1/2}$ of the $S_{11}(1535)$-resonance, which dominates the reaction at lower energies, has been extracted from η photoproduction data and the result obtained ((100 ± 3) $10^{-3}\,\mathrm{GeV}^{-1/2}$) is higher than the one obtained from πN scattering data ((78 ± 4)$10^{-3}\,\mathrm{GeV}^{-1/2}$). (See [5] for a review.) More experimental information is necessary to clarify the situation, since quark models predict a much higher value for the $A^p_{1/2}$ parameter than the experimental ones.

Complementary information may be provided by the neutron helicity amplitudes, which are related to those of the proton by isospin decomposition relations: $A^p = A^{IS} + A^{IV}$ and $A^n = A^{IS} - A^{IV}$, where A^{IS} and A^{IV} are the isoscalar and the isovector components of the electromagnetic excitation of the resonance. However, experimental results on the neutron are scarce and must be extracted from bound nucleons in light nuclei.

We present samples of the first measurements obtained for the beam asymmetry Σ on quasi-free protons and quasi-free neutrons for the η and π^0 photoproduction reactions on a deuteron target. Comparison of results of the same channels from free protons are also presented to evaluate the contribution from the nucleon binding effects. Some recent results on the $K^+\Lambda$ and $K^+\Sigma^0$ photoproduction beam asymmetries have also been obtained on a proton target, being the first world result on the subject.

2 The GRAAL experimental setup

Present results are obtained at the GRAAL facility, which consists of a tagged and polarized Compton backscattering photon beam, located at the European Synchrotron Radiation Facility in Grenoble, complemented by a large-solid-angle detector (Laγrange). The tagged photon energy spectrum extends from $600\,\mathrm{MeV}$ to $1500\,\mathrm{MeV}$, and the beam may be linearly polarized to a degree higher than 75% on the whole tagged range. The central part of the Laγrange detector (see [6] for details) consists of a BGO calorimeter, complemented by a plastic scintillator barrel and an internal tracker made of two cylindrical multi-wire proportional chambers (MWPC), covering all azimuthal angles corresponding to polar angles in the interval $25°$–$155°$; forward polar angles, lower than $25°$, are covered by two couples of tracking planar MWPCs, a double wall of plastic scintillators and a shower wall, consisting of four layers of lead and plastic scintillators.

The BGO calorimeter has excellent energy resolution for electrons and photons detection (3% FWHM at $1\,\mathrm{GeV}$

energy), and good response to protons for energies up to $300\,\mathrm{MeV}$. Charged particles are tracked with high spatial resolution by the MWPCs (from $1.5°$ to $3.5°$ FWHM) and their energy is well determined by time-of-flight (TOF) measurements by the forward walls. Neutrons may be detected either in the BGO calorimeter, with no information on their energy, or in the forward shower wall, the energy being given by TOF measurements.

The whole apparatus is optimized for the detection of mesons that decay into photons, but it is also able to detect with reasonable performances final states having multiple charged particles.

A cryogenic target of $6\,\mathrm{cm}$ length may be filled either with liquid hydrogen or liquid deuterium, so that reactions on the free proton or bound protons and neutrons may be studied, respectively.

During data taking the beam polarization direction is periodically rotated to be either horizontal or vertical. Data analysis selects photonuclear events, in both polarization conditions. By constructing the ratio

$$\frac{N_{||}/F_{||}}{N_{||}/F_{||} + N_\perp/F_\perp} = \frac{1}{2}(1 - P\Sigma\cos(2\varphi)), \qquad (1)$$

where $N_{||}$ ($N_\perp$) is the number of reaction events and $F_{||}$ ($F_\perp$) is the corresponding photon flux for horizontal (vertical) beam polarization, the Σ beam polarization observable may be extracted from the fit of the azimuthal dependence of the ratio, P being the known degree of beam polarization. We observe that the value of Σ may be determined without knowing the absolute efficiency of the event selection procedure.

3 η photoproduction from quasi-free neutrons

η-meson photoproduction events are selected requiring that two photons, having an invariant mass in the interval 0.35–$0.7\,\mathrm{GeV}$, are detected by the BGO calorimeter. When a deuterium target is used, data are analyzed to select quasi-free kinematic conditions on the bound nucleons: $\gamma d \to N\eta(spectator)$. The η-meson is produced by the photon interaction with one of the nucleons whose momentum is given by the Fermi distribution, while the other nucleon acts as a spectator and most of the times does not escape the target. By detecting a proton or a neutron in the BGO calorimeter or in the forward walls, in coincidence with the η-meson and satisfying two-body kinematics constraints, slightly smeared by Fermi motion effects, it is possible to select an η photoproduction event on the quasi-free proton or neutron, respectively.

The Σ beam polarization observable has been determined from eq. (1) for the η photoproduction reaction on both bound nucleons, for eleven incoming beam energy bins from reaction threshold to $1.5\,\mathrm{GeV}$, and nine angular bins from $30°$ to $170°$ for the meson polar angle in the centre-of-mass reference.

Applying the same selection criteria on data taken using a proton target, the Σ beam asymmetry have been

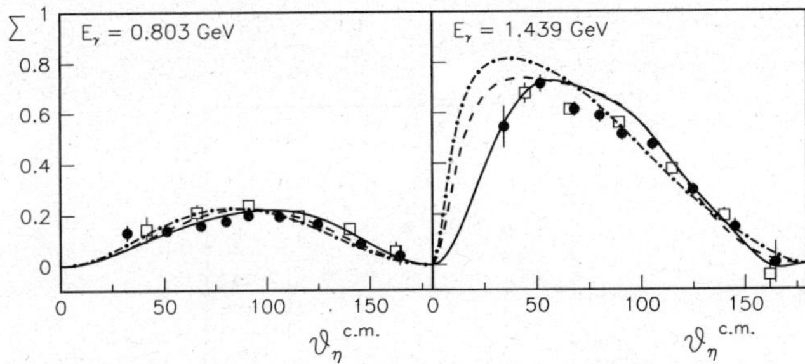

Fig. 1. Comparison of Σ beam asymmetry for η photoproduction on free (open squares) and bound (full circles) protons. Predictions from SAID multipole analysis are plotted as solid curves. MAID(2001) and MAID(2003) isobar models are shown as dashed and dot-dashed curves, respectively.

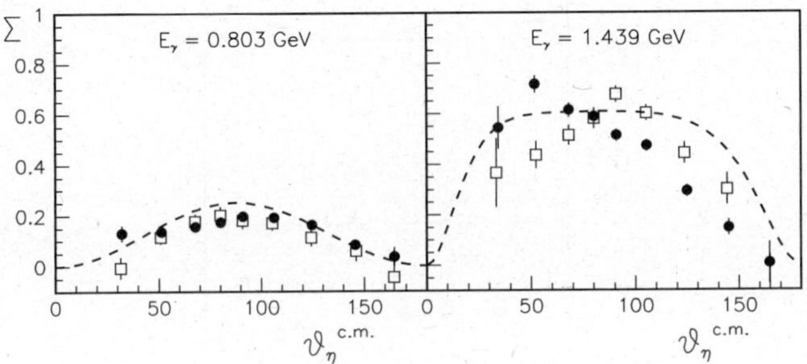

Fig. 2. Comparison of Σ beam asymmetry results for η photoproduction from quasi-free neutrons (open squares) and quasi-free protons (full circles). Predictions from MAID(2001) isobar model are shown as dashed curves.

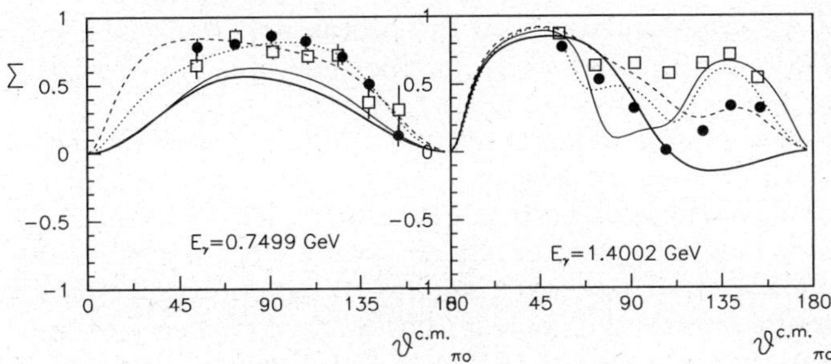

Fig. 3. Comparison of Σ beam asymmetry for π^0 photoproduction from quasi-free neutrons (open squares) and quasi-free protons (full circles). Predictions from SAID multipole analysis are plotted as dotted and thin solid curves for π^0 production on the proton and the neutron, respectively. MAID(2003) isobar models are shown as dashed and thick solid curves for π^0 production on the proton and the neutron, respectively.

obtained also for the free proton. Figure 1 shows the comparison of the results obtained on the free and the bound proton for two energy bin values, close to threshold and in the higher-energy range, respectively. We find that the behaviour is very similar for the free and the bound nucleon for almost all energies and angles and we deduce that the binding effects in deuteron are almost negligible when the beam asymmetry is considered.

Data are also compared with existing predictions from SAID partial wave analysis [7] (solid curve), η-MAID(2001) [8] isobar model (dashed curve) and the new reggized version of the same model η-MAID(2003) [9] (dot-dashed curve). A general good agreement is obtained between both models and our results, the SAID prediction being closer to our data for the higher-energy values.

Results for the Σ beam asymmetry for the η photoproduction from quasi-free neutrons (open squares) are shown in fig. 2, compared with results from quasi-free protons (full circles) and MAID(2001) predictions from the free neutron (dashed curve), which is presently the only available model for the neutron target. We note that the behaviour of the beam asymmetry is quite similar for the two nucleons, the neutron presenting a more backward-forward symmetrical peak at higher energies. Comparison with MAID model is generally not satisfactory, being too high at energies close to 1 GeV, and showing an angular distribution which is flat compared to data. The present

data, being the first ones from quasi-free neutrons are expected to be useful for a better understanding of the isospin dependence of the η-nucleon photoproduction.

4 π^0 photoproduction from quasi-free neutrons

Extensive results have been obtained by the GRAAL collaboration also on the π^0 photoproduction on the proton [6]. The main difference with respect to the η photoproduction channel is due to the fact that pions have isospin $I = 1$ while η-mesons are iso-singlets. This implies that both Δ and N^* resonances may be excited in pion photoproduction reactions, while only N^* baryon resonances are involved in the η photoproduction channel. This makes the multipole analysis of pion photoproduction more complicated, but due to the higher cross-section the existing experimental database is much wider and very precise data are available.

Starting from the same data set to which we refer in sect. 3 and using the same events selection criteria described therein, imposing that the invariant mass of the two detected photons falls into the interval 90–175 MeV we obtained results on the Σ beam asymmetry for the π^0 photoproduction from quasi-free protons and quasi-free neutrons, using the deuteron target and from the free proton using the hydrogen target. The behaviour of the beam asymmetry for the free and the quasi-free proton is identical within the experimental errors showing also for the π^0 channel that binding and/or rescattering effects are negligible.

Results for π^0 photoproduction beam asymmetry off the quasi-free neutron are shown in fig. 3 (open squares) together with those off the quasi-free proton (closed circles), for two energy bins. While the trend for the two nucleons is similar at lower energies, the Σ beam asymmetry is sensibly higher for the quasi-free neutron than for the proton. Comparison with MAID(2003) [10] isobar model and SAID partial-wave amplitude predictions are also plotted as dashed and dotted curves, respectively, for

π^0 photoproduction from the proton and thick and thin solid curves, respectively, for the same reaction from the neutron.

We observe that prediction for the π^0p channel are not in agreement with our experimental results and revised versions of both SAID and MAID solutions have been created including Graal data on the proton in the experimental database. Moreover predictions for the π^0 production off the neutron show more complicated structure than our results. After several decades of results on the pion photoproduction channels, constraints from polarization observables are still able to add new information on the reaction mechanism.

5 K^+ photoproduction from the proton

Using the hydrogen target and taking advantage of the high resolution for angular information from the tracking detectors, it has been possible to analyze events from the K^+ photoproduction on the proton; both $\gamma p \to K^+ \Lambda$ and $\gamma p \to K^+ \Sigma^0 \to K^+ \Lambda\gamma$ reactions could be studied, considering the charged decay channel of the Λ ($\Lambda \to p\pi^-$).

Events with three charged particles in the final state have been analyzed; particle identification and momentum calculation was performed using two different kinematics procedures, the former starting from a three-body kinematics approach, the latter treating the reactions as a cascade of two-body reactions ($K\Lambda$ production followed by the Λ decay in $p\pi^-$). The detection of the photon from the Σ^0 decay, which has 76.96 MeV energy in the reference frame where Σ^0 is at rest, was used to discriminate among the two reactions.

The Λ decay, being governed by weak interactions, allows for the measurement of the Λ degree of polarization, so that both the recoil polarization P and the Σ beam polarization observables could be measured.

Results for the first observables have been found in agreement with data from SAPHIR [11] and CLAS [12]. Those for the Σ beam polarization are the first available in the energy range from reaction threshold to 1.5 GeV incoming photon energies.

Figure 4 shows the angular distribution of the beam asymmetry for three incoming photon energies, compared with presently available model predictions: kaon-MAID(2000) isobar model [13] (dotted line), Saclay-Argonne-Pittsburgh dynamical coupled-channel model [14] (dashed line), Ghent isobar and Regge plus Resonance models [15] (dot-dashed lines) and Bonn coupled-channel partial-wave analysis [16] (solid line).

The kaon-MAID(2000) model includes a *missing* resonance: the $D_{13}(1900)$. It may well reproduce the angular distribution of the unpolarized differential cross-section from SAPHIR, but it over estimates the present measurements of the Σ beam asymmetry.

The Saclay-Argonne-Pittsburgh dynamical coupled-channel model includes resonances in the frame of a chiral constituent-quark model and finds the necessity to include three new resonances: $S_{11}(1806)$, $P_{13}(1893)$ and $D_{13}(1954)$, the last one being the most evident. Its ability

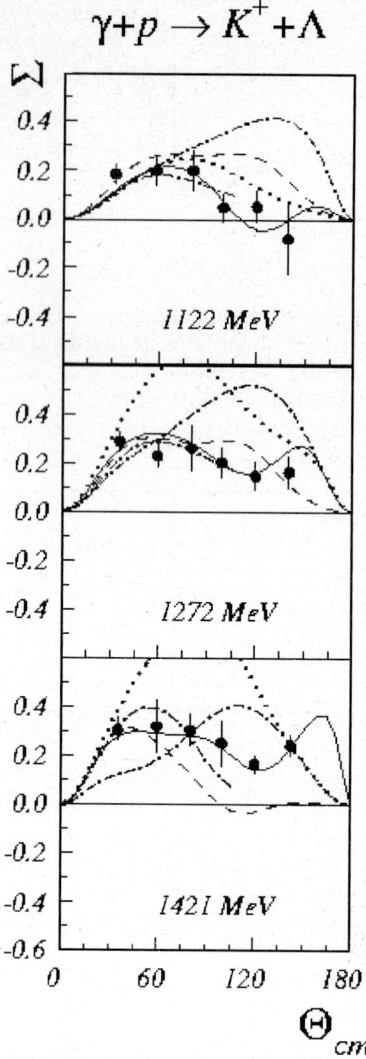

Fig. 4. Angular distributions for the beam asymmetry for the $\gamma p \to K^+\Lambda$ reaction. Data are compared with kaon-MAID(2000) isobar model (dotted line), Saclay-Argonne-Pittsburgh dynamical coupled-channel model (dashed line), Ghent isobar and Regge plus Resonance models (dot-dashed lines) and Bonn coupled-channel partial-wave analysis (solid line).

to reproduce the new beam asymmetry data is scarce at backward angles.

The Ghent isobar model includes a *missing* $D_{13}(1900)$ state, in addition to standard Born, background and resonance terms; its reggized version, plotted only for forward angles, better agrees with present beam asymmetry results.

Finally the Bonn coupled-channel partial-wave model finds the necessity of including several new N^*-resonances above 1800 MeV and it strongly demands the presence of a new $D_{13}(1875)$ state. Its ability to reproduce the Graal results on the beam asymmetry is remarkable.

We conclude that most models confirm the need of including a *missing* D_{13}-resonance of mass around 1900 MeV, the Bonn model being able to well reproduce

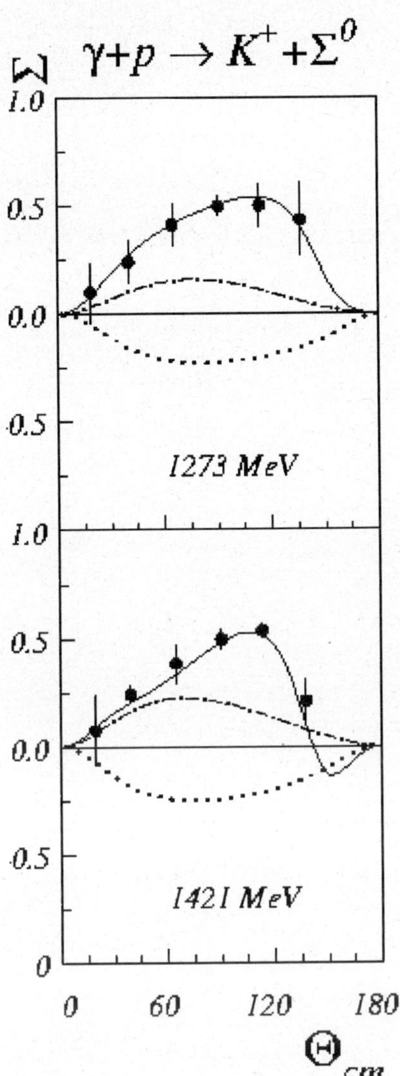

Fig. 5. Angular distributions for the beam asymmetry for the $\gamma p \rightarrow K^+ \Sigma^0$ reaction. Data are compared with kaon-MAID(2000) isobar model (dotted line), Ghent isobar (dot-dashed line) and Bonn coupled-channel partial-wave analysis (solid line).

all the new Graal results on the Σ beam asymmetry for the $\gamma p \rightarrow K^+ \Lambda$ channel.

Similar results for the $\gamma p \rightarrow K^+ \Sigma^0$ channel are shown in fig. 5 for two energy bins. Also these data are the first available for the beam polarization asymmetry at low energies. They are compared with kaon-MAID(2000) isobar model (dotted line), Ghent isobar model (dot-dashed line) and the Bonn coupled-channel partial-wave analysis (solid line). Again the Bonn model has the impressing ability to fully reproduce our results.

6 Summary

First results on the Σ beam asymmetry observable have been obtained for incoming photon energies up to 1500 MeV for the η-meson and the π^0 photoproduction from the quasi-free neutron. The data are not reproduced by the presently available model predictions and may contribute to a better understanding of the isospin dependence of the reaction mechanisms. First new data are also available for the K^+ photoproduction from the proton for both $K^+ \Lambda$ and $K^+ \Sigma^0$ channels. Comparison with existing predictions show a very good agreement with the Bonn coupled-channel partial-wave analysis, which supports the need for the introduction of a new $D_{13}(1900)$-resonance.

References

1. Particle Data Group (K. Hagiwara *et al.*), Phys. Rev. D **66**, 0100001 (2002).
2. S. Capstick, W. Roberts, Phys. Rev. D **49**, 4570 (1994).
3. L.Y. Glozman *et al.*, Phys. Rev. D **58**, 094030 (1998).
4. U. Loering *et al.*, Eur. Phys. J. A **10**, 395 (2001).
5. B. Krusche, S. Schadmand, Prog. Part. Nucl. Phys. **51**, 399 (2003).
6. O. Bartalini, V. Bellini, J.P. Bocquet, M. Capogni, L. Casano, M. Castoldi, P. Calvat, A. D'Angelo, R. Di Salvo, A. Fantini, C. Gaulard, G. Gervino, F. Ghio, B. Girolami, A. Giusa, V. Kouznetsov, A. Lapik, P. Levi Sandri, A. Lleres, D. Moricciani, A.N. Mushkarenkov, V. Nedorezov, L. Nicoletti, C. Perrin, D. Rebreyend, F. Renard, N. Rudnev, T. Russew, C. Schaerf, M.-L. Sperduto, M.-C. Sutera, A. Turinge, Eur. Phys. J. A **26**, 399 (2005).
7. http://gwdac.phys.gwu.edu/.
8. G. Knchlein, D. Drechsel, L. Tiator, Z. Phys. A **352**, 327 (1995); W.-T. Chiang, S.-N. Yang, L. Tiator, D. Drechsel, Nucl. Phys. A **700**, 429 (2002); *eta-maid* at www.kph.uni-mainz/MAID.
9. W.-T. Chiang, S.N. Yang, L. Tiator, M. Vanderhaeghen, D. Drechsel, Phys. Rev. C **68**, 045202 (2003); *eta-maid* at www.kph.uni-mainz/MAID.
10. D. Drechsel, S.S. Kamalov, L. Tiator, Nucl. Phys. A **645**, 145 (1999); *maid 2003* at www.kph.uni-mainz/MAID.
11. K.-H. Glander *et al.*, Eur. Phys. J. A **19**, 251 (2004).
12. R. Bradford *et al.*, Phys. Rev. C **73**, 035202 (2006).
13. T. Mart, C. Bennhold, H. Haberzettl, L. Tiator, *Kaon-MAID 2000* at www.kph.uni-mainz/MAID.
14. B. Juliá-Díaz, B. Saghai, T.-S.H. Lee, F. Tabakin, Phys. Rev. C **73**, 055204 (2006).
15. D.G. Ireland, S. Janssen, J. Ryckebusch, Nucl. Phys A **740**, 147 (2004); T. Corthals, J. Ryckebusch, T. Van Cauteren, Phys. Rev. C **73**, 045207 (2006).
16. A.V. Anisovich *et al.*, Eur. Phys. J. A **25**, 427 (2005); A.V. Sarantsev *et al.*, Eur. Phys. J. A **25**, 441 (2005).

Eur. Phys. J. A **31**, 446–450 (2007)

DOI 10.1140/epja/i2006-10183-8

Recent KLOE results on hadron physics

C. Bini[a]

Università "La Sapienza" and INFN Roma, Roma, Italy

Received: 25 October 2006

Published online: 19 February 2007 – © Società Italiana di Fisica / Springer-Verlag 2007

Abstract. The KLOE experiment at the Frascati e^+e^- collider DAFNE has completed this year its data taking. An integrated luminosity of $2.7\,\mathrm{fb}^{-1}$ has been collected mostly at the ϕ-resonance peak. A wide experimental program is in progress. The detection of ϕ radiative decays allows to study the properties of the lowest-mass scalar and pseudoscalar mesons and to obtain information on their structure. The main results are reviewed together with the prospects for low-energy e^+e^- physics at Frascati.

PACS. 13.66.Bc Hadron production in e^-e^+ interactions – 14.40.-n Mesons

1 The KLOE experiment at DAFNE

The Frascati e^+e^- collider DAFNE has been working since 1999 at a center-of-mass energy $\sqrt{s} \sim m_\phi = 1019.4\,\mathrm{MeV}$. In this last year it has reached an instantaneous luminosity of $1.5 \times 10^{32}\,\mathrm{cm}^{-2}\mathrm{s}^{-1}$ that is the highest luminosity ever reached by any e^+e^- collider in this energy region. DAFNE has two experimental regions: one is occupied by the KLOE experiment; in the other one the FINUDA and DEAR experiments have been run in different times.

The KLOE experiment has collected in 5 years an integrated luminosity of $2.7\,\mathrm{fb}^{-1}$: $2.5\,\mathrm{fb}^{-1}$ at the ϕ peak (corresponding to 6×10^9 ϕ decays) and $0.2\,\mathrm{fb}^{-1}$ at a slightly lower center-of-mass energy, $\sqrt{s} = 1\,\mathrm{GeV}$, out of the ϕ-resonance region. The detector consists of a very large cylindrical drift chamber [1] around the interaction region, a hermetic calorimeter [2] done with scintillating fibers

[a] e-mail: cesare.bini@roma1.infn.it; On behalf of the KLOE Collaboration: F. Ambrosino, A. Antonelli, M. Antonelli, C. Bacci, P. Beltrame, G. Bencivenni, S. Bertolucci, C. Bini, C. Bloise, V. Bocci, F. Bossi, D. Bowring, P. Branchini, S.A. Bulychjov, R. Caloi, P. Campana, G. Capon, T. Capussela, F. Ceradini, S. Chi, G. Chiefari, P. Ciambrone, S. Conetti, E. De Lucia, A. De Santis, P. De Simone, G. De Zorzi, S. Dell'Agnello, A. Denig, A. Di Domenico, C. Di Donato, S. Di Falco, B. Di Micco, A. Doria, M. Dreucci, G. Felici, A. Ferrari, M.L. Ferrer, G. Finocchiaro, C. Forti, P. Franzini, C. Gatti, P. Gauzzi, S. Giovannella, E. Gorini, E. Graziani, M. Incagli, W. Kluge, V. Kulikov, F. Lacava, G. Lanfranchi, J. Lee-Franzini, D. Leone, M. Martini, P. Massarotti, W. Mei, S. Meola, S. Miscetti, M. Moulson, S. Müller, F. Murtas, M. Napolitano, F. Nguyen, M. Palutan, E. Pasqualucci, A. Passeri, V. Patera, F. Perfetto, L. Pontecorvo, M. Primavera, P. Santangelo, E. Santovetti, G. Saracino, B. Sciascia, A. Sciubba, F. Scuri, I. Sfiligoi, T. Spadaro, M. Testa, L. Tortora, P. Valente, B. Valeriani, G. Venanzoni, S. Veneziano, A. Ventura, R.Versaci, G. Xu.

Table 1. Main ϕ decay channels together with their branching ratios.

Channel	Branching ratio
K^+K^-	49.2%
$K^0\overline{K^0}$	34.0%
$\rho\pi + \pi^+\pi^-\pi^0$	15.3%
$\eta\gamma$	1.3%
$\pi^0\gamma$	0.125%
$\eta'\gamma$	6.2×10^{-5}
$\pi^0\pi^0\gamma$	1.1×10^{-4}
$\eta\pi^0\gamma$	8.3×10^{-5}

and lead surrounding the chamber and a superconducting solenoid providing a 0.6 T magnetic field. The whole detector is about $4\,\mathrm{m}$ long and has a radius of about $3\,\mathrm{m}$. The drift chamber allows the tracking of charged particles and the measurements of their momenta. The calorimeter measures energy, position and time of neutral particles (photons, K_L) and, due to its excellent timing capabilities, provides also a good particle identification of charged particles (electrons *vs.* muons *vs.* pions).

The main decay channels of the ϕ-resonance are summarised in table 1. It can be seen that a ϕ-factory is mostly a kaon factory but, even if at a lower rate, it is also a factory of η ($\sim 1\%$), η' (few $\times 10^{-5}$) and, through the $\pi\pi\gamma$ and $\eta\pi\gamma$ radiative decays, a factory of scalar mesons, $f_0(980)$, $f_0(600) \to \pi\pi$ and $a_0(980) \to \eta\pi$. A large amount of $\pi^+\pi^-\gamma$ events are also observed. The large majority of them are due to the initial-state radiation that allows to study the $e^+e^- \to \pi^+\pi^-$ cross-section from the two-pion threshold up to the ϕ mass. However, a small fraction of these events is due to the decay chain $\phi \to f_0\gamma \to \pi^+\pi^-\gamma$. In the following the results obtained by KLOE concerning the ϕ radiative decays to scalars and pseudoscalar mesons are presented and discussed. A review of the kaon physics done at KLOE can be found in ref. [3], and the first KLOE results on the $\pi^+\pi^-$ cross-section below the ϕ peak obtained using the initial-state radiation can be found in ref. [4].

2 Results on scalar mesons

Standing at the ϕ-resonance energy, the lowest mass $I = 0$ (the $f_0(980)$ and the controversial $f_0(600)$ or σ) and $I = 1$ (the $a_0(980)$) scalar mesons are accessible through ϕ radiative decays. In particular the $\pi^+\pi^-\gamma$ and $\pi^0\pi^0\gamma$ final states are sensitive to the $I = 0$ sector and the $\eta\pi^0\gamma$ to the $I = 1$ sector. The nature of these scalar mesons is still today controversial. In fact it is hard to describe their spectrum and their properties as standard $q\bar{q}$ states. Several hints suggest a 4-quark, $qq\bar{q}\bar{q}$ structure [5,6]. For what concerns the two almost mass degenerate states $f_0(980)$ and $a_0(980)$, whose mass is very close to twice the kaon mass, the possibility that they are $K\overline{K}$ molecules has been also considered [7,8].

The KLOE data on ϕ radiative decays are interesting essentially for two reasons:

1. the possibility to estimate the coupling of the scalar mesons to the ϕ that is an almost pure $s\bar{s}$ quark state, either directly [9] or through a kaon loop [10];
2. the possibility to look for the presence of the σ in the mass spectra.

The extraction of these information from the data is not straightforward since a huge unreducible background is present at least for the $\pi\pi$ channels. A fit of the mass spectra has to be done in any case using a parametrisation for the signal and for the background and including in a proper way the interference between the different amplitudes. In the end one has to take into account several unknown parameters and it is not easy to obtain model-independent conclusions.

The results of the KLOE analysis of the $\pi^+\pi^-\gamma$ channel are shown in figs. 1 and 2. The signal due to the $f_0(980)$ is clearly seen in the $\pi^+\pi^-$ invariant-mass spectrum as a peak in the f_0 mass region (fig. 1) and in the profile of the forward-backward asymmetry (fig. 2) [11]. In the case of the $\pi^0\pi^0\gamma$ channel [12] the $f_0(980)$ signal can be observed as a population in the right corner of the Dalitz plot shown in fig. 3.

In the $\pi^+\pi^-\gamma$ channel the main background is provided by the initial-state radiation with the typical structure due to the ρ^0 peak and to the ρ^0-ω interference (the main structure in the spectrum of fig. 1(a)). In the case of the $\pi^0\pi^0\gamma$ channel the main background is provided by the reaction $e^+e^- \to \omega\pi^0$ with $\omega \to \pi^0\gamma$ (see fig. 3).

A good description of these data is provided by the kaon-loop model [10] but also by the model described in ref. [9]. In the first case the ϕ is assumed to couple to a kaon pair and these in turn, after the irradiation of the photon, annihilate producing the scalar meson. The amplitude is parametrised in terms of the f_0 mass and of the

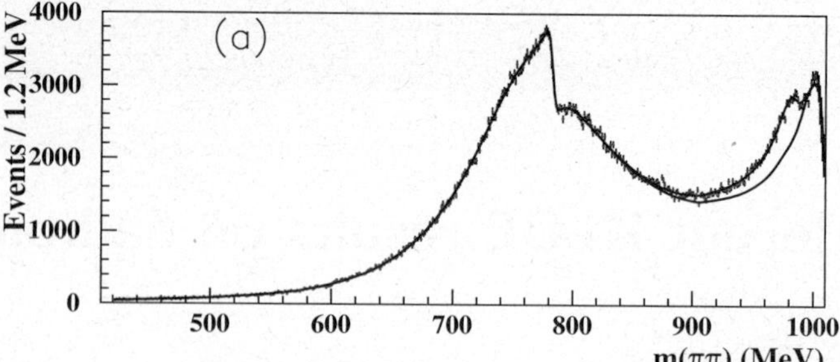

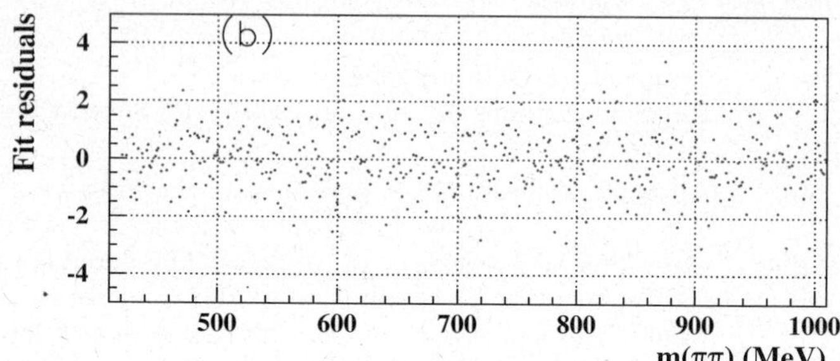

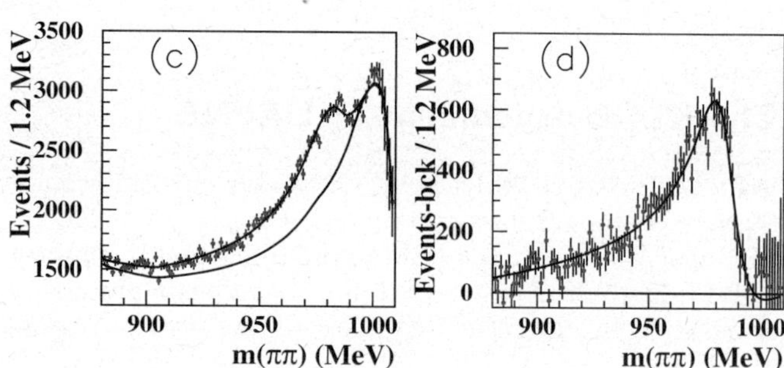

Fig. 1. Fit of the $\pi^+\pi^-$ invariant-mass spectrum for the $\pi^+\pi^-\gamma$ sample. (a) Experimental data compared to the fit function and to the background evaluated by the fit itself; (b) residuals of the fit; (c) expanded view of the fit in the $f_0(980)$ region and (d) experimental spectrum after background subtraction in the same mass region.

Table 2. Summary of the KLOE results on the most significant $f_0(980)$ parameters. The mass and the coupling to the ϕ are obtained using the fits done according to ref. [9] while the ratio R is obtained using the kaon-loop fit; the $B.R.$ is obtained by integrating the square of the resulting scalar amplitude.

Parameter	$\pi^+\pi^-\gamma$ [11]	$\pi^0\pi^0\gamma$ [12]
m_{f_0} (MeV)	973–981	981–987
$g_{\phi f_0\gamma}$ (GeV^{-1})	1.2–2.0	2.5–2.9
$R = g^2_{f_0 K^+ K^-}/g^2_{f_0\pi^+\pi^-}$	2.2–2.8	3.0–7.2
$B.R.(\times 10^{-4})$	2.1–2.4	1.0–1.2

couplings $g_{f_0 K^+ K^-}$ and $g_{f_0\pi^+\pi^-}$. In the second case the ϕ is directly coupled to the scalar mesons through the emission of the photon. The main parameter is in this case the coupling $g_{\phi f_0\gamma}$.

The couplings of the f_0 to KK to $\pi\pi$ and to the ϕ-meson obtained by the fits are shown in table 2. An

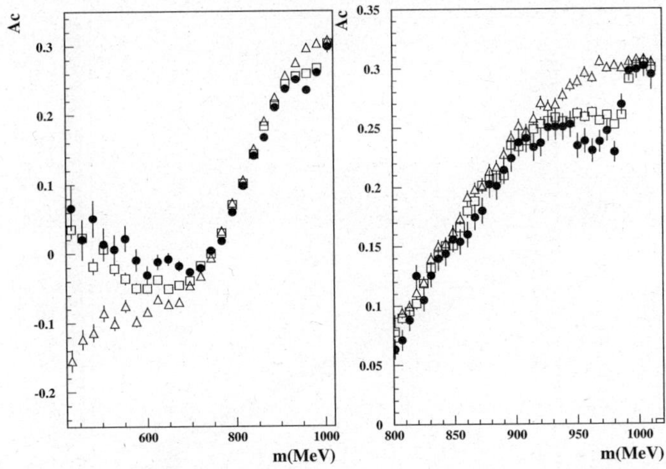

Fig. 2. The forward-backward asymmetry of $\pi^+\pi^-\gamma$ events. Data (full circles) are compared to the Monte Carlo expectations with no f_0 (open triangles), and with f_0 (open squares). The right plot shows the detail of the comparison in the f_0 region.

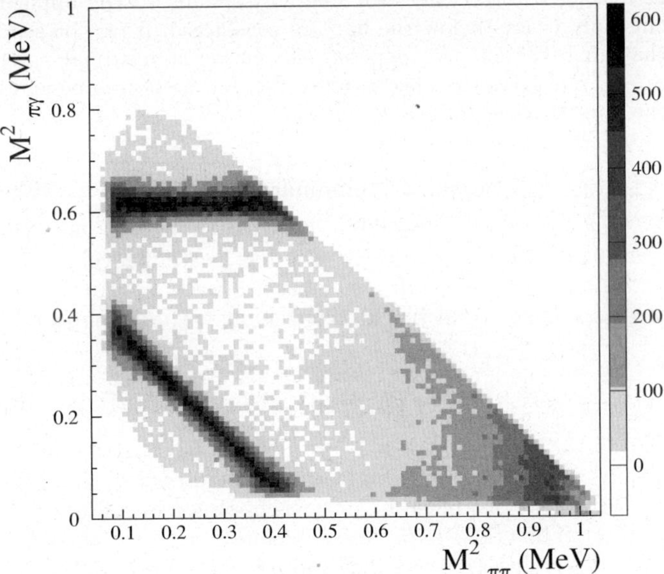

Fig. 3. Dalitz plot of the $\pi^0\pi^0\gamma$ sample. The two narrow bands are due to events $e^+e^- \to \omega\pi^0$ with $\omega \to \pi^0\gamma$, while the excess of the events in the right corner is due to the radiative decay $\phi \to f_0(980)\gamma$ with $f_0 \to \pi^0\pi^0$.

estimate of the branching ratio $B.R.(\phi \to f_0(980)\gamma \to \pi\pi\gamma)$ obtained by integrating the resulting scalar amplitudes, is also obtained and presented in table 2.

Looking at the results of the fit we make the following considerations:

- the mass of the $f_0(980)$ is in good agreement with the range reported by the Particle Data Group $m_{f_0} = 980 \pm 10\,\mathrm{MeV}$ [13];
- the coupling to the ϕ is larger then $1\,\mathrm{GeV}^{-1}$ in both cases; for comparison notice that the same couplings for the lowest-mass pseudoscalar mesons are all well

below $1\,\mathrm{GeV}^{-1}$ (between ~ 0.1 for the π^0 and $\sim 0.7\,\mathrm{GeV}^{-1}$ for the η and η');
- the ratio R is also larger than 1, showing that the f_0 has a larger coupling to kaons than to pions;
- the branching ratios of the two channels differ by a factor 2 as it should be based on isospin arguments.

For what concerns the σ the results are somehow contradictory. In fact the kaon-loop fit requires the presence of the σ in the case of the $\pi^0\pi^0\gamma$ analysis to account for the interference pattern with the $\omega\pi^0$ amplitude, but it does not require it in the $\pi^+\pi^-\gamma$ analysis. Moreover, the direct coupling fit describes the low-energy part of the amplitude as a polynomial background, so that even in this case the σ is not required. The conclusion is that it is not possible to obtain a model-independent indication on the σ.

Finally, the KLOE analysis of $\phi \to a_0\gamma$ [14] gives a lower branching ratio, $B.R.(\phi \to a_0\gamma \to \eta\pi^0\gamma) = (7.4 \pm 0.7) \times 10^{-5}$ and lower couplings to kaons, the ratio R between the coupling to kaons and to $\eta\pi^0$ being $R = g^2_{a_0 K^+ K^-}/g^2_{a_0\eta\pi^0} = 0.74 \pm 0.05$. A new analysis with a larger statistics is in progress and will be published soon.

3 Results on pseudoscalar mesons

Large samples of pseudoscalar mesons are also accessible at KLOE through the ϕ radiative decays (see table 1). In particular about 8×10^7 monochromatic η-mesons with $p = 360\,\mathrm{MeV}/c$ and about 4×10^5 η'-mesons with $p = 60\,\mathrm{MeV}/c$ are present in the full KLOE data sample.

A measurement of the η-η' mixing angle has already been published [15] based on the detection of the final state $\pi^+\pi^-3\gamma$ in the first $20\,\mathrm{pb}^{-1}$ of data collected in the year 2000. Improved upper limits on the branching ratios of the forbidden decays $\eta \to \gamma\gamma\gamma$ [16] and $\eta \to \pi^+\pi^-$ [17] have been also obtained and the dynamic of the $\eta \to 3\pi$ decays has been studied by means of a Dalitz-plot analysis [18].

We present here two recent results: the first is an improved measurement of the η-η' mixing angle done using the $\pi^+\pi^-7\gamma$ and 7γ final states, the second is a new measurement of the η mass based on the 3γ final states $\phi \to \eta\gamma$ with $\eta \to \gamma\gamma$.

The pseudoscalar mixing angle in the flavour basis φ_P is obtained by the ratio

$$R = \frac{B.R.(\phi \to \eta'\gamma)}{B.R.(\phi \to \eta\gamma)} \tag{1}$$

according to ref. [19]. To get the $B.R.(\phi \to \eta'\gamma)$ we use the $\pi^+\pi^-7\gamma$ final states that are due either to the decay chain $\eta' \to \eta\pi^+\pi^-$ with $\eta \to 3\pi^0$ or to $\eta' \to \eta\pi^0\pi^0$ with $\eta \to \pi^+\pi^-\pi^0$. For the $B.R.(\phi \to \eta\gamma)$ we use the 7γ final states due to the decay chain $\eta \to 3\pi^0$. Out of a $427\,\mathrm{pb}^{-1}$ data sample we select 3750 η' events and 1.7×10^6 η events. The estimated background is about 345 events (less than 10%) for the η' sample and is negligible for the η sample. The result for the ratio R is:

$$R = (4.79 \pm 0.09_{stat} \pm 0.20_{syst}) \times 10^{-3}, \tag{2}$$

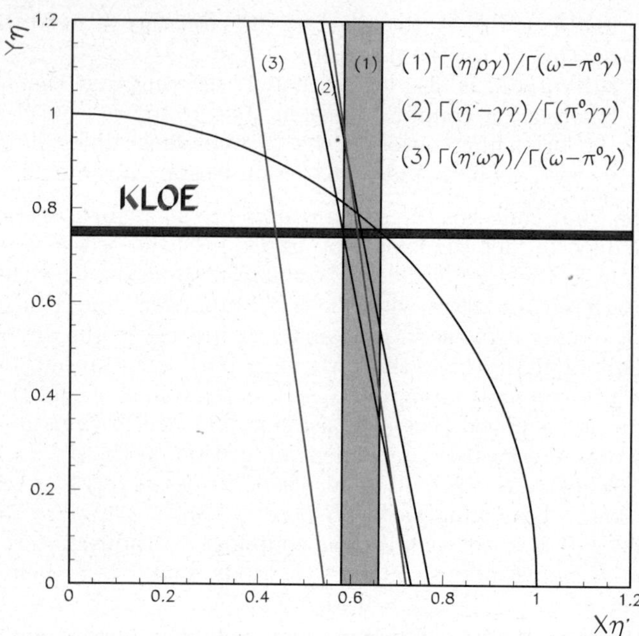

Fig. 4. Bounds in the $(Y_{\eta'}, X_{\eta'})$-plane coming from KLOE (horizontal band) and from the other experimental quantities indicated in the insert.

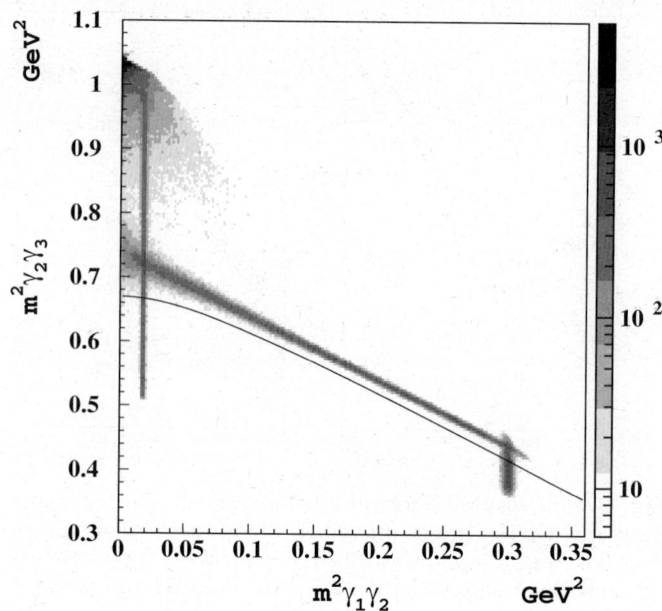

Fig. 5. Dalitz plot of 3γ events with the photons ordered in increasing energy. The solid line corresponds to the applied cut: only events below the line are considered. It can be seen that after the cut two populations survive at fixed values of $m^2_{\gamma_1\gamma_2}$: one gives the mass peak of the π^0, the other one gives the mass peak of the η.

where the systematic uncertainty is dominated by the knowledge of the intermediate branching ratios of the η' that is not better than 3%. The mixing angle in the flavour basis is

$$\varphi_P = (41.4 \pm 0.3_{stat} \pm 0.7_{syst} \pm 0.6_{th})^\circ, \qquad (3)$$

where we add a further theoretical uncertainty due to the knowledge of the parameters entering in the angle determination [19]. By applying the relation $\theta_P = \varphi_P - \mathrm{arctg}\sqrt{2}$ we get the mixing angle in the octet-singlet basis to be

$$\theta_P = (-13.3 \pm 0.3_{stat} \pm 0.7_{syst} \pm 0.6_{th})^\circ \qquad (4)$$

in agreement with the recent BES results $\theta_P = (-15.9 \pm 1.2)^\circ$ [20] obtained using J/ψ radiative decays.

The KLOE result is used to check the amount of possible gluonium content in the η' wave function. We define the three coefficients $X_{\eta'}$, $Y_{\eta'}$ and $Z_{\eta'}$ as the coefficients of the quark and gluon component in the η' wave function:

$$|\eta'\rangle = X_{\eta'} \frac{1}{\sqrt{2}} |u\bar{u} + d\bar{d}\rangle + Y_{\eta'} |s\bar{s}\rangle + Z_{\eta'} |\text{gluonium}\rangle. \quad (5)$$

In case of no gluonium content ($Z_{\eta'} = 0$) the mixing angle φ_P is simply related to $Y_{\eta'}$ through the

$$|Y_{\eta'}| = \cos\varphi_P \qquad (6)$$

so that the KLOE result corresponds to a horizontal band in the $(Y_{\eta'}, X_{\eta'})$-plane shown in fig. 4 where other bands based on different measurements are included. A cross-check of the $Z_{\eta'} = 0$ hypothesis is obtained by checking that the bands meet in a region compatible with the

$X^2_{\eta'} + Y^2_{\eta'} = 1$ curve. A preliminary result of this analysis gives $X^2_{\eta'} + Y^2_{\eta'} = 0.93 \pm 0.06$ that constrains the gluonium content of the η' to be below the few percent level.

The two most recent and most accurate measurements of the η mass are in bad disagreement with each other:

$$m_\eta = (547.843 \pm 0.030 \pm 0.041) \text{ MeV} \quad (\text{NA48}), \qquad (7)$$
$$m_\eta = (547.311 \pm 0.028 \pm 0.032) \text{ MeV} \quad (\text{GEM}). \qquad (8)$$

The NA48 experiment at CERN has looked at the decay $\eta \to \pi^0\pi^0\pi^0$ from a η sample of very high energy (in average 110 GeV) produced by protons hitting a beryllium target [21]. The GEM experiment has measured the missing-mass distribution of the reaction $p + d \to {}^3\text{He} + \eta$ [22]. All the previous experiments quoted in ref. [13] have larger uncertainties but central values somehow more consistent with the GEM result.

KLOE performs the η mass measurement by exploiting the large samples of 3γ events mostly due to the decay chains: $\phi \to \eta\gamma$ with $\eta \to \gamma\gamma$ and $\phi \to \pi^0\gamma$ with $\pi^0 \to \gamma\gamma$. The kinematic fit of the 3γ events uses all the KLOE calorimeter information, in particular the very good timing and position resolutions, resulting in a very good resolution on the invariant masses of any photon pair. From the Dalitz plot shown in fig. 5 one can easily separate the events due to the η from those due to the π^0 and select a region of the plot from which to obtain two very narrow mass peaks, one for each meson. The preliminary results are

$$m_{\pi^0} = (134.990 \pm 0.006 \pm 0.030) \text{ MeV}, \qquad (9)$$
$$m_\eta = (547.822 \pm 0.005 \pm 0.069) \text{ MeV}. \qquad (10)$$

The statistical uncertainties are negligible, the systematic ones are mostly due to the knowledge of the effective center-of-mass energy and interaction region position. The value of the π^0 mass gives an absolute calibration of the method. In fact it is in agreement with the more precise value reported by the Particle Data Group $m_{\pi^0} = 134.9766 \pm 0.0006$ [13] within the systematic uncertainty.

The KLOE result on the η mass confirms the NA48 value and is in disagreement with the GEM value. At the moment there is no explanation for this inconsistency between different kinds of measurements of the same quantity.

4 Prospects for e^+e^- at Frascati

Here, the prospects of e^+e^- physics at Frascati are briefly discussed. More detailed information on accelerator developments and on experimental proposals can be found in ref. [23].

The e^+e^- program of the Frascati laboratory is well defined for the next 2 years. The FINUDA run is now in progress at DAFNE, aiming to collect an overall luminosity 5 times the one collected up to now. The main goal of the new FINUDA run is to confirm or disprove the evidence for a deeply bound hypernuclear state obtained in the first run [24]. After the FINUDA run there will be the run of the SIDDHARTA experiment [25] aiming to study the formation of exotic atoms with an improved version of the DEAR experiment.

The laboratory has not yet taken any decision concerning the program after these runs. However, the possibility to continue the e^+e^- program of DAFNE with two main improvements is strongly considered: an increase of luminosity at the ϕ peak up to $\sim 10^{33}$ cm^{-2}s^{-1} and an increase of center-of-mass energy up to about 2.5 GeV.

Based on this accelerator improvement scheme, 3 experimental proposals have been submitted, briefly described in the following:

- The KLOE2 proposal aims to continue the kaon and η/η' physics program at the ϕ peak, but also to take data at center-of-mass energies up to 2.5 GeV in order to perform a precision measurement of the multi-hadronic and $\gamma\gamma$ cross-sections.
- The AMADEUS proposal aims to refine the study of the deeply bound hypernuclear states using a large rate of monochromatic charged kaons at the ϕ peak and possibly exploiting a more general-purpose detector.
- The DANTE proposal wants to measure the time-like baryon form factors, in particular neutron and proton, by detecting the reactions $e^+e^- \to n\bar{n}$ and $e^+e^- \to p\bar{p}$

possibly measuring the polarisations of the nucleons in the final state.

A discussion of most of the physics items included in these proposals can be found in ref. [26]

References

1. M. Adinolfi *et al.*, Nucl. Instrum. Methods A **488**, 1 (2002).
2. M. Adinolfi *et al.*, Nucl. Instrum. Methods A **482**, 364 (2002).
3. KLOE Collaboration (A. Aloisio *et al.*), in *Le Rencontres de la Vallee d'Aoste - Results and Perspectives in Particle Physics, February 2005*, edited by M. Greco, Frascati Phys. Ser., Vol. **XXXIX**, p. 233.
4. KLOE Collaboration (A. Aloisio *et al.*), Phys. Lett. B **606**, 12 (2005).
5. R.L. Jaffe, Phys. Rev. D **15**, 15 (1977).
6. L. Maiani *et al.*, Phys. Rev. Lett. **93**, 212002 (2004).
7. J. Weinstein, N. Isgur, Phys. Rev. Lett. **48**, 10 (1982).
8. V. Baru *et al.*, Phys. Lett. B **586**, 53 (2004).
9. G. Isidori *et al.*, JHEP **0605**, 049 (2006).
10. N.N. Achasov, V.N. Ivanchenko, Nucl. Phys. B **315**, 465 (1989).
11. KLOE Collaboration (F. Ambrosino *et al.*), Phys. Lett. B **634**, 148 (2006).
12. KLOE Collaboration (F. Ambrosino *et al.*), Eur. Phys. J. C **49**, 473 (2007).
13. Particle Data Group (W.M. Yao *et al.*), J. Phys. G **33**, 1 (2006).
14. KLOE Collaboration (A. Aloisio *et al.*), Phys. Lett. B **536**, 209 (2002).
15. KLOE Collaboration (A. Aloisio *et al.*), Phys. Lett. B **541**, 45 (2002).
16. KLOE Collaboration (A. Aloisio *et al.*), Phys. Lett. B **591**, 49 (2004).
17. KLOE Collaboration (F. Ambrosino *et al.*), Phys. Lett. B **606**, 276 (2005).
18. KLOE Collaboration (F. Ambrosino *et al.*), in AIP Conf. Proc. **814**, 463 (2006).
19. A. Bramon *et al.*, Eur. Phys. J. C **7**, 24 (1999).
20. BES Collaboration (M. Ablikim *et al.*), Phys. Rev. D **73**, 052008 (2006).
21. NA48 Collaboration (A. Lai *et al.*), Phys. Lett. B **533**, 196 (2002).
22. GEM Collaboration (M. Abdel-Bary *et al.*), Phys. Lett. B **619**, 281 (2005).
23. See the web site: `http://www.lnf.infn.it/lnfadmin/direzione/roadmap`.
24. FINUDA Collaboration (M. Agnello *et al.*), Phys. Rev. Lett. **94**, 212303 (2005).
25. C. Curceanu, these proceedings.
26. F. Ambrosino *et al.*, submitted to Eur. Phys. J. C, hep-ex/0603056.

Eur. Phys. J. A **31**, 451–453 (2007)
DOI 10.1140/epja/i2006-10195-4

THE EUROPEAN
PHYSICAL JOURNAL A

Special Article – QNP 2006

Exclusive meson production at HERMES

M. Tytgat[a]

On behalf of the HERMES Collaboration
University of Gent, Department of Subatomic and Radiation Physics, Proeftuinstraat 86, 9000 Gent, Belgium

Received: 8 October 2006
Published online: 21 February 2007 – © Società Italiana di Fisica / Springer-Verlag 2007

Abstract. Hard exclusive production of mesons in deep-inelastic scattering allows one to probe the so-far unknown Generalized Parton Distributions (GPDs) of the nucleon. The HERMES experiment has measured several different observables in exclusive meson production by scattering the 27.6 GeV HERA lepton beam off an internal fixed gaseous target. Recent results on exclusive π^+, ρ^0 and pion pair production will be presented.

PACS. 13.60.-r Photon and charged-lepton interactions with hadrons – 13.60.Le Meson production – 13.88.+e Polarization in interactions and scattering – 14.20.Dh Protons and neutrons

1 Introduction

Exclusive reactions in electron-nucleon scattering have recently gained interest within the framework of the Generalized Parton Distributions (GPDs) [1]. In the QCD factorization theorem for hard exclusive electroproduction of mesons with longitudinal photons [2], GPDs appear in the parametrization of the target nucleon. They extend our description of the nucleon spin structure beyond the form factors and standard parton distributions which we know from (semi)-inclusive deep-inelastic scattering. GPDs are sensitive to correlations between partons with different momenta and can probe the nucleon transverse and longitudinal momentum structure in a coherent manner. The second moment of these GPDs is in fact related to the total angular-momentum contribution of partons to the nucleon spin. In leading twist the nucleon can be parametrized using two unpolarized GPDs, H and E, and two polarized ones, $\tilde{H}$ and $\tilde{E}$, where the quantum numbers of the produced final state determine the sensitivity to different GPDs.

The data presented here were collected from 1996 to 2004 with the HERMES spectrometer [3], using hydrogen and deuterium gas targets internal to the 27.6 GeV HERA lepton storage ring at DESY.

2 Exclusive π^+ production

Exclusive π^+ production on the nucleon is sensitive to the polarized GPDs $\tilde{E}$ and $\tilde{H}$ without the need for beam or target polarization and is studied at HERMES by requiring the missing mass M_X for the reaction $ep \to eX\pi^+$

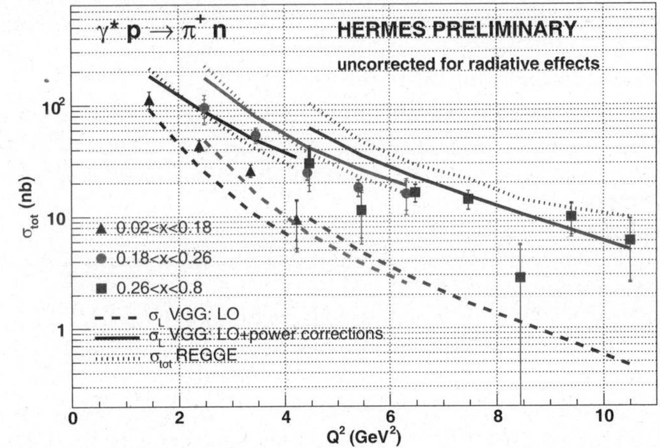

Fig. 1. The exclusive π^+ production cross-section as a function of Q^2 for several x_B ranges compared to GPD [4] and Regge [6] model calculations.

to be near the nucleon mass. Due to the limited resolution in M_X, the exclusive channel $ep \to en\pi^+$ cannot be separated easily from the neighboring channels and non-exclusive background and hence, as the process $ep \to en\pi^-$ is forbidden, a normalized sample of $ep \to eX\pi^-$ events was subtracted from the data to obtain a pure exclusive π^+ signal.

For the determination of the absolute production cross-section, the detection probability was derived using two Monte Carlo simulations for exclusive π^+ production based on different GPD parametrizations [4,5]. The measured Q^2-dependence of the total photoproduction cross-section is shown in fig. 1. Radiative effects have not been accounted for in this result and are estimated to

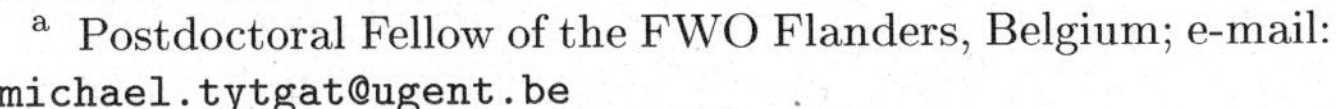

---

[a] Postdoctoral Fellow of the FWO Flanders, Belgium; e-mail: michael.tytgat@ugent.be

be about 20% with little x_B or Q^2-dependence. A separation of the cross-section into its transverse and longitudinal components is not feasible at HERMES. However, with a virtual photon polarization parameter ϵ ranging from 0.80 to 0.95 at HERMES and the predicted power suppression of the transverse part [2], the total cross-section $\sigma_{tot} = \sigma_T + \epsilon\sigma_L$ is expected to be dominated by the longitudinal component at large Q^2. The data in fig. 1 are compared to GPD model calculations [4] for the longitudinal cross-section. While the Q^2 behavior is in agreement with the data, the leading-order calculations underestimate the cross-section, whereas including power corrections overestimates the data. The Regge model calculations [6] for the total cross-section shown in the figure are seen to overestimate the data. In the latter model the transverse part of the cross-section is suppressed with respect to the longitudinal contribution by at least a factor four. The observed Q^2 behavior of the measured cross-section is in agreement with a $1/Q^6$-dependence at fixed x_B as predicted by the factorization theorem for the longitudinal cross-section [2].

3 Exclusive ρ^0 production

Hard exclusive ρ^0 production provides a probe for the unpolarized GPDs H and E. At HERMES this process can be cleanly identified by reconstructing the ρ^0 in its 2-pion decay channel and by requiring $\Delta E = (M_X^2 - M^2)/2M$, which is a measure of the missing energy in the $ep \to eX\rho^0$ reaction, to be close to zero. Non-exclusive background in the ρ^0 event sample is corrected for using Monte Carlo simulations.

In hard exclusive ρ^0 production the transverse target single-spin asymmetry (TTSSA) is sensitive to the interference between GPD E and H [7]. Whereas in the cross-section GPD E is kinematically suppressed with respect to GPD H, the asymmetry has a linear dependence on GPD E and is therefore expected to provide a good handle on the latter. Furthermore, the TTSSA also appears sensitive to the total u-quark angular momentum [8]. The asymmetry is calculated as

$$A_{UT}(\phi, \phi_S) = \frac{1}{|S_\perp|} \frac{N^\uparrow(\phi, \phi_S) - N^\downarrow(\phi, \phi_S)}{N^\uparrow(\phi, \phi_S) + N^\downarrow(\phi, \phi_S)}$$
$$= A_{UT}^{\sin(\phi - \phi_S)} \sin(\phi - \phi_S). \tag{1}$$

$|S_\perp|$ is the average target polarization and $N^{\uparrow(\downarrow)}$ represent the luminosity normalized yields for the two polarized target helicity states. ϕ and ϕ_S are the azimuthal angles of the ρ^0 production plane and the transverse component of the target spin with respect to the lepton scattering plane, respectively.

Figure 2 displays the preliminary HERMES results for the $\sin(\phi - \phi_S)$ moment, $A_{UT}^{\sin(\phi - \phi_S)}$, of the asymmetry as a function of x_B. The data were taken during 2002–2004 on a transversely polarized hydrogen target with $|S_\perp| = 0.76 \pm 0.05$. The moment is corrected for non-exclusive and non-resonant background contributions, however no longitudinal/transverse separation has

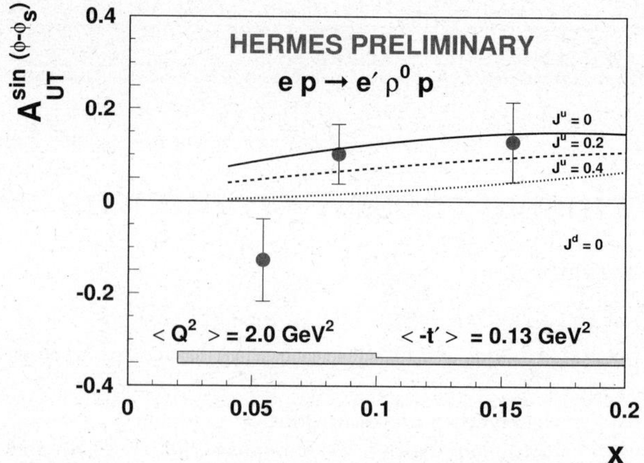

Fig. 2. The $\sin(\phi - \phi_S)$ moment of the exclusive ρ^0 TTSSA as a function of x_B, compared to GPD-based calculations [8].

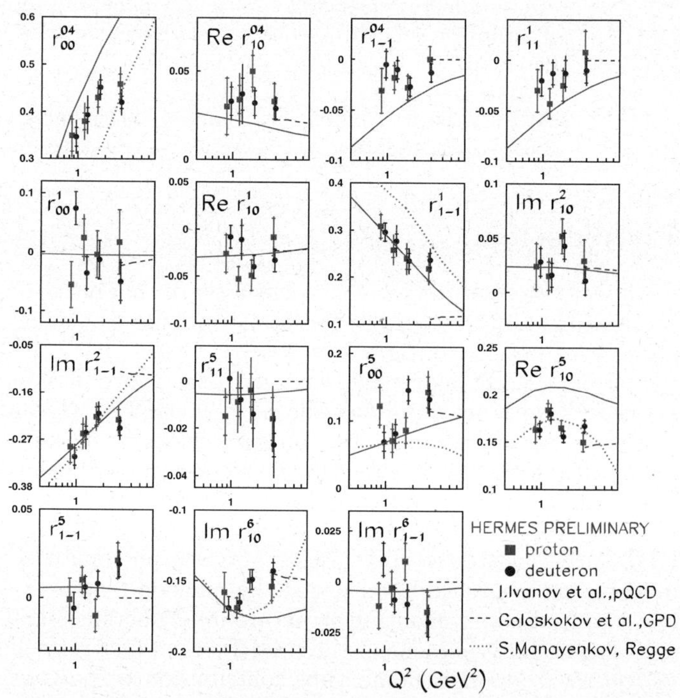

Fig. 3. The Q^2-dependence of the 15 unpolarized SDMEs for the proton and deuteron compared to theoretical calculations [10–12].

been performed yet. The observed x_B-dependence agrees with the behavior predicted by GPD models [7]. In the figure the data are compared to GPD-based calculations for different values of the quark total angular-momentum contribution [8].

Another set of observables in exclusive ρ^0 production are the Spin Density Matrix Elements (SDMEs), which describe the helicity transfer from the virtual photon to the vector meson and can be accessed through measurements of angular distributions of the meson and its decay pions [9]. With the polarized HERA lepton beam, and for both unpolarized hydrogen and deuterium targets, the full set of 15 unpolarized plus 8 beam polarization-dependent

SDME was determined via a maximum likelihood fit of Monte Carlo events to the measured data.

Figure 3 displays the unpolarized SDMEs for the proton and deuteron as a function of Q^2. A clear violation of the s-channel helicity conservation hypothesis is observed through, $e.g.$, r_{00}^5 and Re r_{10}^{04} being significantly non-zero. The data are compared to calculations based on pQCD k_T factorization [10], Regge phenomenology [11] and on a GPD-based model [12]. Although a reasonable description of the data is obtained, these models are primarily based on Pomeron exchange at high energies and do not include quark exchange, which at HERMES kinematics is believed to be the dominant production mechanism [13].

4 Hard exclusive pion pair production

Hard exclusive electroproduction of $\pi^+\pi^-$ pairs off nucleons represents another probe for the GPD H and E. The process can proceed via $q\bar{q}$ or two-gluon exchange. In the former case the pair is produced with either ρ-meson (strong isospin $I = 1$, total angular momentum $J = 1, 3 \ldots$ and C-parity $C = -1$) or f-meson ($I = 0$, $J = 0, 2 \ldots$, $C = +1$) quantum numbers. The $C = +1$ ($C = -1$) $q\bar{q}$ exchange is described by flavour singlet (non-singlet) parton combinations and due to C-parity conservation the pion pairs have $C = -1$ ($C = +1$) [14]. In the two-gluon exchange channel only ρ-meson–like pion pairs are created. Since this process allows one to distinguish two-gluon and $q\bar{q}$ exchange, new information can be gained on both the nucleon quark and gluon GPDs [15]. Furthermore, the interference between the two isospin channels may reveal information on the amplitude level of the weak isoscalar channel. To study the interference between the P-wave ($I = 1$) and S, D-wave ($I = 0$) production, the Legendre moments $\langle P_1(\cos\theta)\rangle$ and $\langle P_3(\cos\theta)\rangle$ were investigated [16]. These are defined as

$$\langle P_n(\cos\theta)\rangle^{\pi^+\pi^-} = \frac{\int_{-1}^1 \mathrm{d}\cos\theta\, P_n(\cos\theta) \frac{\mathrm{d}\sigma^{\pi^+\pi^-}}{\mathrm{d}\cos\theta}}{\int_{-1}^1 \mathrm{d}\cos\theta \frac{\mathrm{d}\sigma^{\pi^+\pi^-}}{\mathrm{d}\cos\theta}}, \quad (2)$$

where θ is the polar angle of the π^+ with respect to the direction of the $\pi^+\pi^-$ pair in the center-of-momentum frame of the virtual photon and target nucleon. The production cross-section $\mathrm{d}\sigma^{\pi^+\pi^-}/\mathrm{d}\cos\theta$ can be expressed in terms of the spin density matrix $\rho_{\lambda\lambda'}^{JJ'}$ of the pion pair, where the diagonal elements $\rho_{\lambda\lambda}^{JJ}$ give the probability to produce the pion pair with angular momentum J and projection λ and the off-diagonal terms describe the interference terms. Contributions with $J > 2$ are expected to be negligible in the range of the pion pair invariant mass $M_{\pi\pi}$ covered by HERMES. The two Legendre moments then become $\langle P_1\rangle = \frac{1}{\sqrt{15}}[4\sqrt{3}\rho_{11}^{21} + 4\rho_{00}^{21} + 2\sqrt{5}\rho_{00}^{10}]$ and $\langle P_3\rangle = \frac{1}{7\sqrt{5}}[-12\rho_{11}^{21} + 6\sqrt{3}\rho_{00}^{21}]$, $i.e.$ $\langle P_1\rangle$ is sensitive to the P-wave interference with S- and D-waves, whereas $\langle P_3\rangle$ is sensitive to the P-wave interference with a D-wave only.

Figure 4 shows the measured $\langle P_1\rangle$ moment for a hydrogen target. At $M_{\pi\pi}$ values around the ρ^0 mass, the

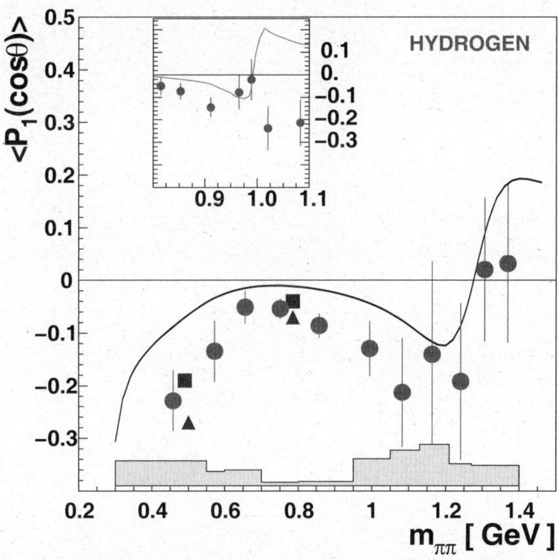

Fig. 4. The $M_{\pi\pi}$-dependence of the $\langle P_1\rangle$ moment for hydrogen data compared to leading-twist predictions including gluon exchange (solid curve) and without gluon exchange (squares at $x_B = 0.1$ and triangles at $x_B = 0.2$) [15]. The insert displays the $f_0(980)$ region with finer bins together with calculations including f_0 contributions [17].

absolute value of the moment shows a minimum, indicating the dominant role of ρ^0 production in the denominator of the moment. At higher and lower masses one can observe the interference of the ρ^0 tails with the non-resonant S-wave amplitude. At $M_{\pi\pi} \approx 1\,\mathrm{GeV}$, one sees a hint of the interference of the ρ^0 tail and the S-wave $\pi^+\pi^-$ production from the narrow $f_0(980)$-resonance. In the $f_2(1270)$ meson region a sign change is suggested due to the interference between the ρ^0 upper tail and the f_2 D-wave. A fair agreement is found between the data and the GPD-based model predictions [15].

References

1. D. Müller $et\ al.$, Fortschr. Phys. **42**, 101 (1994); A.V. Radyushkin, Phys. Lett. B **380**, 417 (1996); X. Ji, Phys. Rev. Lett. **78**, 610 (1997).
2. J. Collins $et\ al.$, Phys. Rev. D **56**, 2982 (1997).
3. K. Ackerstaff $et\ al.$, Nucl. Instrum. Methods A **417**, 230 (1998).
4. M. Vanderhaeghen $et\ al.$, Phys. Rev. D **60**, 094017 (1999).
5. L. Mankiewicz $et\ al.$, Eur. Phys. J. C **10**, 307 (1999).
6. J.M. Laget, Phys. Rev. D **70**, 054023 (2004).
7. K. Goecke $et\ al.$, Prog. Part. Nucl. Phys. **47**, 401 (2001).
8. F. Ellinghaus $et\ al.$, hep-ph/0506264.
9. K. Schilling, G. Wolf, Nucl. Phys. B **61**, 381 (1973).
10. I.P. Ivanov $et\ al.$, hep-ph/0501034, DESY-04-243.
11. S.I. Manayenkov, Eur. Phys. J. C **33**, 397 (2004).
12. S.V. Goloskokov, P. Kroll, Eur. Phys. J. C **42**, 281 (2005).
13. M. Diehl, A.V. Vinnikov, Phys. Lett. B **609**, 286 (2005).
14. M. Diehl, Phys. Rep. **388**, 41 (2003).
15. B. Lehmann-Dronke $et\ al.$, Phys. Rev. D **63**, 114001 (2001); Phys. Lett. B **475**, 147 (2000).
16. A. Airapetian $et\ al.$, Phys. Lett. B **599**, 212 (2004).
17. Ph. Hägler $et\ al.$, Eur. Phys. J. A **18**, 389 (2003).

Eur. Phys. J. A **31**, 454–457 (2007)

DOI 10.1140/epja/i2006-10291-5

Special Article – QNP 2006

The $\Phi \to K^0 \bar{K}^0 \gamma$ decay

R. Escribano[a]

Grup de Física Teòrica and IFAE, Universitat Autònoma de Barcelona, E-08193 Bellaterra (Barcelona), Spain

Received: 22 December 2006
Published online: 20 March 2007 – © Società Italiana di Fisica / Springer-Verlag 2007

Abstract. The scalar contributions to the radiative decay $\phi \to K^0 \bar{K}^0 \gamma$ are studied within the framework of the Linear Sigma Model (LσM). Theoretical predictions for the associated subprocesses $\phi \to f_0 \gamma$ and $\phi \to a_0 \gamma$ as well as the ratio $\phi \to f_0 \gamma / a_0 \gamma$ are also given.

PACS. 12.39.-x Phenomenological quark models – 11.30.Rd Chiral symmetries – 13.40.Hq Electromagnetic decays – 13.75.Lb Meson-meson interactions

1 Introduction

The radiative decays of the light vector mesons ($V = \rho, \omega, \phi$) into a pair of neutral pseudoscalars ($P = \pi^0, K^0, \eta$), $V \to P^0 P^0 \gamma$, are an excellent laboratory for investigating the nature and extracting the properties of the light scalar meson resonances ($S = \sigma, a_0, f_0$) [1]. In addition, their study also complements other analysis based on central production, D and J/ψ decays, etc. [2]. Renewed interest on these radiative decays from both the theoretical and experimental sides are found in refs. [3–5] and ref. [6], respectively. Particularly interesting is the $\phi \to K^0 \bar{K}^0 \gamma$ decay, which, as we will see, can provide us with valuable information on the properties of the $f_0(980)$ and $a_0(980)$ resonances. The ratio $\phi \to f_0 \gamma / a_0 \gamma$ can also be used to extract relevant information on the scalar mixing angle. At present, there is not yet experimental data available for the $\phi \to K^0 \bar{K}^0 \gamma$ process, while for the ratio $\phi \to f_0 \gamma / a_0 \gamma$, the experimental value measured by the KLOE Collaboration is $R(\phi \to f_0 \gamma / a_0 \gamma) = 6.1 \pm 0.6$ [7].

The purpose of this contribution is to compute the $\phi \to K^0 \bar{K}^0 \gamma$ decay where the scalar effects are known to be dominant. This process is interesting to study, on one side, because it allows for a direct measurement of the $K \bar{K}$ couplings to the f_0 and a_0 mesons thus avoiding a model-dependent extraction and, on the other side, since it could pose a background problem for testing CP-violation at DAΦNE. The direct measurement of the couplings seems to be feasible in the near future with the higher luminosity expected at DAFNE-2. Having $50 \, \mathrm{fb}^{-1}$, the number of the expected $K^0 \bar{K}^0 \gamma$ final state is in the range 2–8×10^3 [6]. The analysis of CP-violating decays in $\phi \to K^0 \bar{K}^0$ has been proposed as a way to measure the ratio ϵ'/ϵ [8], but because this means looking for a very small

effect, a $B(\phi \to K^0 \bar{K}^0 \gamma) \gtrsim 10^{-6}$ will limit the precision of such a measurement. Related to the $\phi \to K^0 \bar{K}^0 \gamma$ decay, there are also the processes $\phi \to (f_0, a_0)\gamma$ which are the main contributions to the former through the decay chain $\phi \to (f_0 + a_0)\gamma \to K^0 \bar{K}^0 \gamma$. An accurate measurement of the production branching ratio and of the mass spectra for $\phi \to f_0(980)/a_0(980)\gamma$ decays can clarify the controversial nature of these well-established scalar mesons.

In sect. 2, we present a short review of the approaches used in the literature to study these processes emphasizing the different treatments of the scalar contribution. The $\phi \to K^0 \bar{K}^0 \gamma$ decay is discussed in sect. 3. The subprocesses $\phi \to f_0 \gamma$ and $\phi \to a_0 \gamma$ as well as the ratio $\phi \to f_0 \gamma / a_0 \gamma$ are discussed in sect. 4. Concluding remarks are presented in sect. 5.

2 Theoretical framework

An early attempt to explain the $V \to P^0 P^0 \gamma$ decays was done in ref. [9] using the vector meson dominance (VMD) model. In this framework, the $V \to P^0 P^0 \gamma$ decays proceed through the decay chain $V \to V P^0 \to P^0 P^0 \gamma$, where the intermediate vectors exchanged are $V = \bar{K}^{*0}$ and $V' = K^{*0}$ for $\phi \to K^0 \bar{K}^0 \gamma$. The calculated branching ratio is found to be $B^{\mathrm{VMD}}_{\phi \to K^0 \bar{K}^0 \gamma} = 2.7 \times 10^{-12}$ [9]. Later on, the $V \to P^0 P^0 \gamma$ decays were studied in a Chiral Perturbation Theory (ChPT) context enlarged to include on-shell vector mesons [10]. In this formalism, $B^{\chi}_{\phi \to K^0 \bar{K}^0 \gamma} = 7.6 \times 10^{-9}$. Taking into account both chiral and VMD contributions, one finally obtains $B^{\mathrm{VMD}+\chi}_{\phi \to K^0 \bar{K}^0 \gamma} = 7.6 \times 10^{-9}$ [10]. Additional contributions are those coming from the exchange of scalar resonances. A first model including the scalar resonances explicitly is the *no structure model*, where the $V \to P^0 P^0 \gamma$ decays proceed through the decay chain

---

[a] e-mail: `Rafel.Escribano@ifae.es`

$V \to S\gamma \to P^0 P^0 \gamma$ and the coupling $VS\gamma$ is considered as pointlike. A second model is the *kaon loop model* [11], where the initial vector decays into a pair of charged kaons that, after the emission of a photon, rescatter into a pair of neutral pseudoscalars through the exchange of scalar resonances. In ref. [11], it was shown for the first time the convenience of studying the $\phi \to K^0 \bar{K}^0 \gamma$ process to investigate the nature of the f_0 and a_0 scalar mesons. In this pioneer work, $B(\phi \to (f_0 + a_0)\gamma \to K^0 \bar{K}^0 \gamma) = 1.3 \times 10^{-8}$, for a four-quark structure of the f_0 and a_0, and 2.0×10^{-9}, for a two-quark structure. The previous two models include the scalar resonances *ad hoc*, and the pseudoscalar rescattering amplitudes are not chiral invariant. This problem is solved in the next two models which are based not only on the *kaon loop model* but also on chiral symmetry. The first one is a chiral unitary approach (Uχ) where the scalar resonances are generated dynamically by unitarizing the one-loop pseudoscalar amplitudes. In this approach, $B^{U\chi}_{\phi \to K^0 \bar{K}^0 \gamma} = 5 \times 10^{-8}$ [12]. The second model is the Linear Sigma Model (LσM), a well-defined $U(3) \times U(3)$ chiral model which incorporates *ab initio* the pseudoscalar and scalar mesons nonets. The advantage of the LσM is to incorporate explicitly the effects of scalar meson poles while keeping the correct behaviour at low invariant masses expected from ChPT.

In the next two sections, we discuss the scalar contributions to the $\phi \to K^0 \bar{K}^0 \gamma$ decay and the ratio $\phi \to f_0\gamma/a_0\gamma$ in the framework of the LσM.

3 $\phi \to K^0 \bar{K}^0 \gamma$

As stated in the introduction, this process is the only radiative decay where the f_0 and a_0 scalar mesons contribute simultaneously. This will allow, once this decay is measured presumably at DAFNE-2, to extract relevant information on the nature and couplings of both mesons, and to compare it with the one obtained from the already experimentally measured $\phi \to \pi^0 \pi^0 \gamma$ and $\phi \to \pi^0 \eta \gamma$ decays. Therefore, a theoretical prediction for the different contributions to the branching ratio and mass spectrum of this process is welcome and useful. In addition, our analysis will serve to certify that the decay $\phi \to K^0 \bar{K}^0 \gamma$ cannot pose a background problem for testing CP-violation at DAΦNE since the calculated branching ratio is well below 10^{-6} (see below).

The scalar contribution to this process is driven by the decay chain $\phi \to K^+ K^- (\gamma) \to K^0 \bar{K}^0 \gamma$. The contribution from pion loops is known to be negligible due to G-parity and the Zweig rule. The amplitude for $\phi(q^*, \epsilon^*) \to K^0(p)\bar{K}^0(p')\gamma(q, \epsilon)$ is given by [1]

$$\mathcal{A}^{L\sigma M}_{\phi \to K^0 \bar{K}^0 \gamma} = \frac{eg_s}{2\pi^2 m^2_{K^+}} \{a\} \, L(s) \times \mathcal{A}^{L\sigma M}_{K^+ K^- \to K^0 \bar{K}^0}, \quad (1)$$

where $\{a\} = (\epsilon^* \cdot \epsilon)(q^* \cdot q) - (\epsilon^* \cdot q)(\epsilon \cdot q^*)$, $m^2_{K^0 \bar{K}^0} \equiv s$ is the dikaon invariant mass and $L(m^2_{K^0 \bar{K}^0})$ is a loop integral function. The $\phi K \bar{K}$ coupling constant g_s takes the value $|g_s| \simeq 4.5$ to agree with $\Gamma^{\text{exp}}_{\phi \to K^+ K^-} = 2.09 \, \text{MeV}$ [13]. The

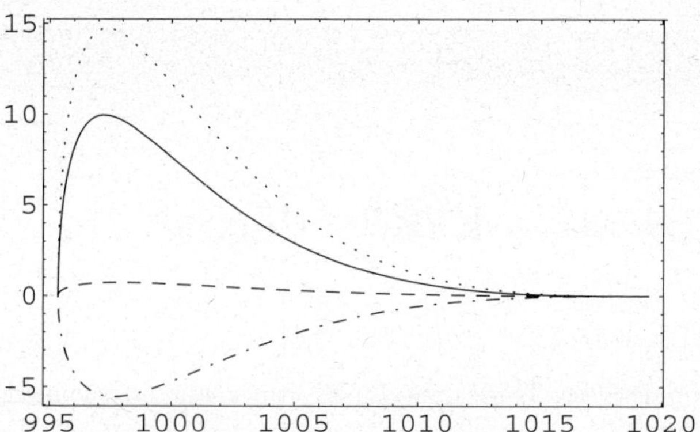

Fig. 1. $\mathrm{d}B(\phi \to K^0 \bar{K}^0 \gamma)/\mathrm{d}m_{K^0 \bar{K}^0} \times 10^9$ (in units of MeV^{-1}) as a function of the dikaon invariant mass $m_{K^0 \bar{K}^0}$ (in MeV). The dotted, dashed and dot-dashed lines correspond to the separate contributions from f_0, a_0 and their interference, respectively. The solid line is the total result.

$K^+ K^- \to K^0 \bar{K}^0$ amplitude in eq. (1) is calculated from the LσM at tree level and turns out to be[1]

$$\mathcal{A}^{L\sigma M}_{K^+ K^- \to K^0 \bar{K}^0} = \frac{s - m^2_K}{4f^2_K} \left[\frac{m^2_K - m^2_\sigma}{D_\sigma(s)} (c\phi_S - \sqrt{2}s\phi_S)^2 \right.$$

$$\left. + \frac{m^2_K - m^2_{f_0}}{D_{f_0}(s)} (s\phi_S + \sqrt{2}c\phi_S)^2 - \frac{m^2_K - m^2_{a_0}}{D_{a_0}(s)} \right] \quad (2)$$

$$+ \frac{m^2_K - s/2}{2f^2_K},$$

where $D_S(s)$ are the $S = \sigma, f_0, a_0$ propagators, ϕ_S is the scalar mixing angle in the quark-flavour basis and $(c\phi_S, s\phi_S) \equiv (\cos \phi_S, \sin \phi_S)$. A Breit-Wigner propagator is used for the σ, while for the f_0 and a_0 a complete one-loop propagator taking into account finite-width corrections is preferable (see ref. [1] for details). A few remarks on the four-pseudoscalar amplitude in eq. (2) are of interest. First, for $m_S \to \infty$ ($S = \sigma, f_0, a_0$), the LσM amplitude (2) reduces to the corresponding chiral-loop amplitude $s/4f^2_K$ [10], thus satisfying the chiral constraints and making the whole analysis quite reliable. Hence, the amplitude (2) has to be considered as an improved expression of its chiral-loop counterpart mentioned before taking into account the (pole-dominated) s-channel scalar dynamics. Second, the large widths of the scalar resonances break chiral symmetry if they are naively introduced in amplitudes, an effect already noticed in ref. [14]. Accordingly, we introduce the widths in the scalar meson propagators only *after* chiral cancellation of constant terms in the original LσM amplitude. In this way the pseudo-Goldstone nature of kaons is preserved.

The final results for $\mathcal{A}(\phi \to K^0 \bar{K}^0 \gamma)$ are then the sum of the LσM contribution in eq. (1) plus the VMD contribution that can be found in ref. [1]. The $K^0 \bar{K}^0$ invariant-mass distribution, with the separate contributions from f_0, a_0 and their interference, as well as the total result,

---

[1] For a detailed derivation of eq. (2) starting from the $U(3) \times U(3)$ LσM, see ref. [1].

are shown in fig. 1. These mass spectra are computed assuming masses for the scalar resonances of $m_{f_0} = 985\,\mathrm{MeV}$ and $m_{a_0} = 984.7\,\mathrm{MeV}$, and a pseudoscalar (scalar) mixing angle of $\phi_P = 41.8°$ ($\phi_S = -8°$). The values of the f_0 mass and the scalar mixing angle are obtained from the $\phi \to \pi^0\pi^0\gamma$ analysis in ref. [1] whereas the a_0 mass is taken from ref. [13] and ϕ_P from the ratio $\phi \to \eta'\gamma/\eta\gamma$ [15]. As seen, the f_0 contributes more strongly than the a_0 due to a smaller imaginary part of the propagator and a larger coupling to kaons. The interference is negative since isospin invariance implies $g_{f_0 K^+ K^-} = g_{f_0 K^0 \bar{K}^0}$ and $g_{a_0 K^+ K^-} = -g_{a_0 K^0 \bar{K}^0}$. Integrating the $K^0 \bar{K}^0$ invariant-mass spectrum one obtains for the scalar contribution $B(\phi \to K^0 \bar{K}^0 \gamma)_{\mathrm{L}\sigma\mathrm{M}} = 7.5 \times 10^{-8}$. It is worth mentioning that the value obtained is very sensitive to the scalar mixing angle, a change of $1°$ modifies the branching ratio around 20%. The whole scalar contribution includes not only the f_0 and a_0 effects but also the σ ones. However, the latter are negligible due to the suppression of the $\sigma K \bar{K}$ coupling if[2] $m_\sigma \simeq m_K$ and for kinematical reasons. Numerically, they amount to less than 1% of the total result. Finally, the exchange of intermediate K^{*0} and $\bar{K}^{*0}$ vector mesons is calculated to be $B(\phi \to K^0 \bar{K}^0 \gamma)_{\mathrm{VMD}} = 2.0 \times 10^{-12}$ and therefore negligible.

4 $\phi \to (f_0, a_0)\gamma$ and $\phi \to f_0\gamma/a_0\gamma$

In the kaon loop model, these two processes are driven by the decay chain $\phi \to K^+ K^-(\gamma) \to f_0\gamma$ and $a_0\gamma$. The amplitudes are given by

$$A = \frac{eg_s}{2\pi^2 m_{K^+}^2} \{a\}\, L(m_{f_0(a_0)}^2) \times g_{f_0(a_0)K^+K^-} \,, \qquad (3)$$

where the scalar coupling constants are fixed within the LσM to

$$g_{f_0 K^+ K^-} = \frac{m_K^2 - m_{f_0}^2}{2f_K}(s\phi_S + \sqrt{2}c\phi_S),$$

$$g_{a_0 K^+ K^-} = \frac{m_K^2 - m_{a_0}^2}{2f_K}. \qquad (4)$$

The ratio of the two branching ratios is thus

$$R_{\phi \to f_0\gamma/a_0\gamma}^{\mathrm{L}\sigma\mathrm{M}} = \frac{|L(m_{f_0}^2)|^2}{|L(m_{a_0}^2)|^2} \frac{\left(1 - m_{f_0}^2/m_\phi^2\right)^3}{\left(1 - m_{a_0}^2/m_\phi^2\right)^3} \times \frac{g_{f_0 K^+ K^-}^2}{g_{a_0 K^+ K^-}^2}$$

$$\simeq (s\phi_S + \sqrt{2}c\phi_S)^2 \,, \qquad (5)$$

where the approximation is valid for $m_{f_0} \simeq m_{a_0}$. Integrating the corresponding amplitude in eq. (3) one obtains the branching ratios $B(\phi \to f_0\gamma) = 2.6 \times 10^{-4}$ and

$B(\phi \to a_0\gamma) = 1.6 \times 10^{-4}$, respectively. For the ratio (5) one gets $R_{\phi \to f_0\gamma/a_0\gamma}^{\mathrm{L}\sigma\mathrm{M}} \simeq 1.6$ for $\phi_S = -8°$. These predictions should be compared with the experimental measurements $B(\phi \to f_0\gamma) = (4.40 \pm 0.21) \times 10^{-4}$, $B(\phi \to a_0\gamma) = (7.6 \pm 0.6) \times 10^{-5}$ [13] and $R_{\phi \to f_0\gamma/a_0\gamma}^{\mathrm{KLOE}} = 6.1 \pm 0.6$ [7]. As seen, the discord is clear. However, the first value, which is based on the KLOE study of the $\phi \to \pi^0\pi^0\gamma$ decay [17], is obtained from a large destructive interference between the $f_0\gamma$ and $\sigma\gamma$ contributions to $\phi \to \pi^0\pi^0\gamma$, in disagreement with other experiments [18]. Consequently, the experimental value for the ratio (5) could be overestimated. In addition, the approximate expression for this ratio is only valid when the f_0 mass is below the charged kaon threshold ($2m_{K^+} \simeq 987\,\mathrm{MeV}$). If not, the steep behaviour of the loop function after threshold makes the approximation meaningless and the exact expression has to be taken. In this case, $R_{\phi \to f_0\gamma/a_0\gamma}^{\mathrm{L}\sigma\mathrm{M}}$ can be much smaller than the given prediction (independently of the mixing angle value). Conversely, if the a_0 mass, which we have kept fixed to $m_{a_0} = 984.7\,\mathrm{MeV}$, is moved to a value above threshold, then the ratio can be even larger than the experimental result. In this case, our prediction for $B(\phi \to a_0\gamma)$ decreases and agrees with the experimental result for $m_{a_0} \simeq 989\,\mathrm{MeV}$. Therefore, a confirmation of the $\phi \to f_0\gamma$ and $\phi \to a_0\gamma$ measurements from the analysis of the $\phi \to \pi^0\pi^0\gamma$ and $\pi^0\eta\gamma$ decays together with a precise fit of the f_0 and a_0 masses using these processes is mandatory before drawing definite conclusions on the validity of our predictions. One should keep in mind, however, that the experimental data on $\phi \to \pi^0\pi^0\gamma$ and $\pi^0\eta\gamma$ are satisfactorily accommodated in our framework (see ref. [1] for comparison).

5 Conclusions

The radiative decay $\phi \to K^0 \bar{K}^0 \gamma$ has been shown to be very useful to extract relevant information on the properties of the $f_0(980)$ and $a_0(980)$ scalar mesons. Our predicted branching ratio including these scalar resonances explicitly is $B(\phi \to K^0 \bar{K}^0 \gamma) = 7.5 \times 10^{-8}$. This value is in agreement with previous phenomenological estimates [12, 19, 20] (see also ref. [19] for a review of earlier predictions). Notice that the branching ratio obtained here is one order of magnitude larger than the chiral-loop prediction $B(\phi \to K^0 \bar{K}^0 \gamma) = 4.1 \times 10^{-9}$. However, it is still one order of magnitude smaller than the limit, $\mathcal{O}(10^{-6})$, in order to pose a background problem for testing CP-violating decays at DAΦNE. A measurement of this process at DAFNE-2 would be welcome and can serve as an additional test of the whole approach. We have also shown that the ratio $\phi \to f_0\gamma/a_0\gamma$ can be used to obtain valuable information on the scalar mixing angle and on the nature of the $f_0(980)$ and $a_0(980)$ scalar states.

This work was supported in part by the Ramon y Cajal program, the Ministerio de Educación y Ciencia under grant FPA2005-02211, the EU Contract No. MRTN-CT-2006-035482, "FLAVIAnet", and the Generalitat de Catalunya under grant 2005-SGR-00994.

---

[2] A value of $m_\sigma = 478\,\mathrm{MeV}$ is taken from the analysis of $D^+ \to \pi^-\pi^+\pi^+$ performed by the E791 Collaboration [16] at Fermilab.

References

1. R. Escribano, arXiv:hep-ph/0606314.
2. F.E. Close, N.A. Tornqvist, J. Phys. G **28**, R249 (2002) [arXiv:hep-ph/0204205].
3. G. Isidori, L. Maiani, M. Nicolaci, S. Pacetti, JHEP **0605**, 049 (2006) [arXiv:hep-ph/0603241].
4. D. Black, M. Harada, J. Schechter, Phys. Rev. D **73**, 054017 (2006) [arXiv:hep-ph/0601052].
5. N.N. Achasov, A.V. Kiselev, Phys. Rev. D **73**, 054029 (2006) [arXiv:hep-ph/0512047].
6. F. Ambrosino *et al.*, arXiv:hep-ex/0603056.
7. KLOE Collaboration (A. Aloisio *et al.*), Phys. Lett. B **536**, 209 (2002) [arXiv:hep-ex/0204012].
8. I. Dunietz, J. Hauser, J.L. Rosner, Phys. Rev. D **35**, 2166 (1987).
9. A. Bramon, A. Grau, G. Pancheri, Phys. Lett. B **283**, 416 (1992).
10. A. Bramon, A. Grau, G. Pancheri, Phys. Lett. B **289**, 97 (1992).
11. N.N. Achasov, V.N. Ivanchenko, Nucl. Phys. B **315**, 465 (1989).
12. J.A. Oller, Phys. Lett. B **426**, 7 (1998) [arXiv:hep-ph/9803214].
13. Particle Data Group (S. Eidelman *et al.*), Phys. Lett. B **592**, 1 (2004).
14. N.N. Achasov, G.N. Shestakov, Phys. Rev. D **49**, 5779 (1994).
15. KLOE Collaboration (A. Aloisio *et al.*), Phys. Lett. B **541**, 45 (2002) [arXiv:hep-ex/0206010].
16. E791 Collaboration (E.M. Aitala *et al.*), Phys. Rev. Lett. **86**, 770 (2001) [arXiv:hep-ex/0007028].
17. KLOE Collaboration (A. Aloisio *et al.*), Phys. Lett. B **537**, 21 (2002) [arXiv:hep-ex/0204013].
18. M.N. Achasov *et al.*, Phys. Lett. B **485**, 349 (2000) [arXiv:hep-ex/0005017].
19. F.E. Close, N. Isgur, S. Kumano, Nucl. Phys. B **389**, 513 (1993) [arXiv:hep-ph/9301253]; N. Brown, F.E. Close, in *The DAΦNE Physics Handbook*, edited by L. Maiani, G. Pancheri, N. Paver (INFN-LNF, 1992) p. 447.
20. N.N. Achasov, V.V. Gubin, Phys. Rev. D **64**, 094016 (2001) [arXiv:hep-ph/0106321].

Eur. Phys. J. A **31**, 458–460 (2007)

DOI 10.1140/epja/i2006-10220-8

THE EUROPEAN
PHYSICAL JOURNAL A

Special Article – QNP 2006

Tests of the final-state radiation model at DAFNE near $\pi^+\pi^-$ threshold

G. Pancheri[1], O. Shekhovtsova[1,2,a], and G. Venanzoni[1]

[1] INFN Laboratori Nazionale di Frascati, Frascati (RM) 00044, Italy
[2] NSC "Kharkov Institute for Physics and Technology", Institute for Theoretical Physics, Kharkov 61108, Ukraine

Received: 8 November 2006
Published online: 6 March 2007 – © Società Italiana di Fisica / Springer-Verlag 2007

Abstract. Effects due to the non-pointlike behaviour of pions in the process $e^+e^- \to \pi^+\pi^-\gamma$ can arise for hard photons in the final state. By means of a Monte Carlo event generator, which also includes the contribution of the direct decay $\phi \to \pi^+\pi^-\gamma$, we estimate these effects in the framework of the resonance perturbation theory. We consider angular cuts used in the KLOE analysis of the pion form factor at threshold. A method to reveal the effects of the non-pointlike behaviour of pions in a model-independent way is proposed.

PACS. 13.25.Jx Decays of other mesons – 12.39.Fe Chiral Lagrangians – 13.40.Gp Electromagnetic form factors

Final-state radiation (FSR) is the main irreducible background in radiative return measurements of the hadronic cross-section [1] which is important for the anomalous magnetic moment of the muon [2]. Besides being of interest as an important background source, this process could be of interest in itself, because a detailed experimental study of FSR allows us to get information about the pion-photon interaction at low energies. Differently from ISR, whose accuracy is limited by the numerical precision on the evaluation of high-order QED diagrams (see, for example, [3,4] and discussion there), the FSR evaluation relies on specific models for the coupling of hadrons to photons. Usually, the FSR amplitude in the process $e^+e^- \to \pi^+\pi^-\gamma$ is evaluated in the scalar QED (sQED) model, where the pions are treated as point-like particles and the total FSR amplitude is multiplied by the pion form factor computed in the VMD model [4,5]. While this assumption is generally valid for relatively soft photons, it can fail for low values of the invariant mass of the hadronic system, *i.e.* when the intermediate hadrons are far off shell.

In general case the cross-section of the reaction

$$e^+(p_1) + e^-(p_2) \to \pi^+(p_+)\pi^-(p_-)\gamma(k)$$

with the photon emitted in the final state, can be written as

$$d\sigma = \frac{1}{2s(2\pi)^5} \int \delta^4(P - p_+ - p_- - k)\frac{d^3p_+ d^3p_- d^3k}{8E_+E_-\omega}|M|^2, \tag{1}$$

where $P = p_1 + p_2$, $s = P^2$,

$$M = \frac{e}{s}M^{\mu\nu}\bar{u}(-p_1)\gamma_\mu u(p_2)\epsilon_\nu^\star \tag{2}$$

and the tensor $M^{\mu\nu}$ describes the process

$$\gamma^*(P) \to \pi^+(p_+)\pi^-(p_-)\gamma(k).$$

Based on charge conjugation symmetry, photon crossing symmetry and gauge invariance, $M^{\mu\nu}$ can be expressed by three gauge invariant tensors

$$M^{\mu\nu}(P,k,l) \equiv -ie^2 M_F^{\mu\nu}(P,k,l) =$$
$$-ie^2(\tau_1^{\mu\nu}f_1 + \tau_2^{\mu\nu}f_2 + \tau_3^{\mu\nu}f_3), \quad l = p_+ - p_-,$$
$$\tau_1^{\mu\nu} = k^\mu P^\nu - g^{\mu\nu}k \cdot P,$$
$$\tau_2^{\mu\nu} = k \cdot l(l^\mu P^\nu - g^{\mu\nu}k \cdot l) + l^\nu(k^\mu k \cdot l - l^\mu k \cdot P),$$
$$\tau_3^{\mu\nu} = P^2(g^{\mu\nu}k \cdot l - k^\mu l^\nu) + P^\mu(l^\nu k \cdot P - P^\nu k \cdot l).$$

While the last decomposition is general and does not depend on the FSR mechanism, the exact value of the scalar functions f_i (form factors), each depending in terms of three independent variables, is determined by the specific FSR models. In the paper [6] the prediction for f_i in the framework of the resonance perturbation theory (RPT) was considered. We would like to remind that RPT is a model based on chiral perturbation theory (χPT) with the explicit inclusion of the vector and axial-vector mesons, $\rho_0(770)$ and $a_1(1260)$ [7]. Whereas χPT gives correct predictions for the pion form factor at very low energy, RPT is the appropriate framework to describe the pion form factor at intermediate energies ($E \sim m_\rho$) [7].

a e-mail: olga.shekhovtsova@lnf.infn.it

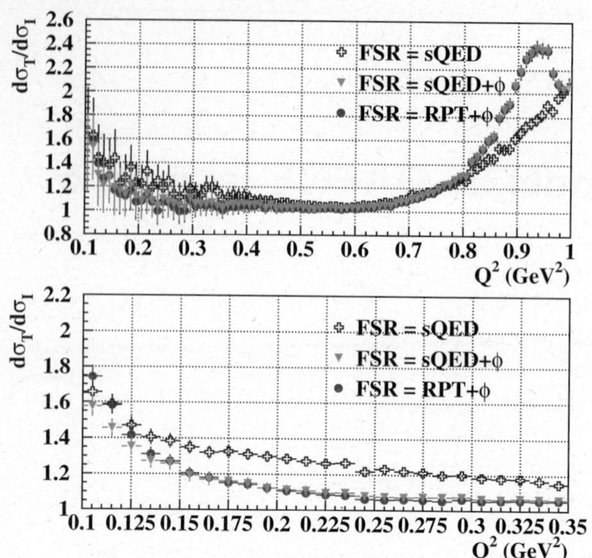

Fig. 1. The ratio $d\sigma_T/d\sigma_I$ as a function of the invariant mass of the two pions, in the region $50° \leq \theta_\gamma \leq 130°$, $50° \leq \theta_\pi \leq 130°$, for different models of FSR at $s = m_\phi^2$.

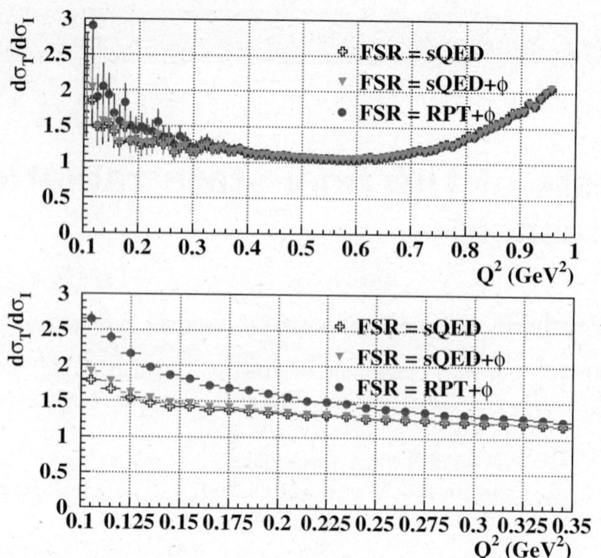

Fig. 2. The ratio $d\sigma_T/d\sigma_I$ as a function of the invariant mass of the two pions, in the region $50° \leq \theta_\gamma \leq 130°$, $50° \leq \theta_\pi \leq 130°$, for different models of FSR at $s = 1\,\mathrm{GeV}^2$.

At $s = m_\phi^2$, DAFNE energies, an additional complication arises: the presence of the direct decay $\phi \to \pi^+\pi^-\gamma$. Its contribution should be included in a realistic Monte Carlo generator (MC).

It is convenient to write down the differential cross-section for the reaction $e^+e^- \to \pi^+\pi^-\gamma$, where the FSR amplitude (M_{FSR}) receives contributions both from RPT (M_{RPT}) and the $\phi \to \pi^+\pi^-\gamma$ decay (M_ϕ), as

$$d\sigma_T \sim |M_{ISR} + M_{FSR}|^2 = d\sigma_I + d\sigma_F + d\sigma_{IF},$$
$$d\sigma_I \sim |M_{ISR}|^2, \quad d\sigma_{IF} \sim 2\,\mathrm{Re}\{M_{ISR} \cdot (M_{RPT} + M_\phi)^*\},$$
$$d\sigma_F \sim |M_{RPT}|^2 + |M_\phi|^2 + 2\,\mathrm{Re}\{M_{RPT} \cdot M_\phi^*\}. \tag{3}$$

The ϕ direct decay is described by Achasov four-quark model with parameters extracted from the fit to $\phi \to \pi^0\pi^0\gamma$ [8]. The interference term $d\sigma_{IF}$ is equal to zero for symmetric cuts on the polar angle of the pions. We consider only the case of destructive interference between the two amplitudes ($\mathrm{Re}(M_{RPT} \cdot M_\phi^*) < 0$). Published data from the KLOE experiment [1] are in favour of this assumption, which we will use in the following.

In figs. 1 and 2 we show the values of $d\sigma_T/d\sigma_I$ for the angular cuts of the KLOE large-angle analysis $50° \leq \theta_\gamma \leq 130°$, $50° \leq \theta_\pi \leq 130°$, with and without contributions from RPT and ϕ direct decay, for a hard-photon radiation with energies $E_\gamma > 20\,\mathrm{MeV}$ for $s = m_\phi^2$ and $s = 1\,\mathrm{GeV}^2$.

For $s = m_\phi^2$ three distinctive features can be noted: 1) the peak at about $1\,\mathrm{GeV}^2$ corresponds to the f_0 intermediate state for the $\phi \to \pi\pi\gamma$ amplitude; 2) the presence of RPT terms in the FSR is relevant at low Q^2, where they give an additional contribution up to 40% to the ratio $d\sigma_{RPT+\phi}/d\sigma_{sQED+\phi}$; 3) the destructive interference with the ϕ direct decay amplitude reduces $d\sigma_F$ at low Q^2 (see fig. 1, down). Also the dependence of the cross-section

on the FSR model is decreased due to the destructive interference.

In the case of $s = 1\,\mathrm{GeV}^2$ the ϕ resonant contribution is suppressed ($d\sigma_T$ with and without the ϕ direct decay almost coincide, see fig. 2. Therefore the main contribution beyond sQED to the FSR cross-section comes from RPT. As expected the presence of RPT gives relevant effects at the low-Q^2 region, while the presence of a bump at high Q^2 is due to the ϕ direct decay. For a more detailed discussion and results for forward-backward asymmetry see [9,10].

Contributions to FSR beyond sQED, as in the case of RPT, can lead to sizeable effects on the cross-section and asymmetry at threshold, as shown in figs. 1 and 2. A precise measurement of the pion form factor in this region needs to control the FSR cross-section at an accuracy better than 1% [11]. This looks like a rather difficult task, if one thinks that effects beyond sQED, as well as the contribution from $\phi \to \pi^+\pi^-\gamma$, are model dependent. In ref. [10] a method for a model-independent analysis of the FSR contribution beyond sQED was proposed. The main idea of this method is the following: we propose to consider a quantity that can be related to experimental spectra and that has a very well-described behaviour in sQED. In our opinion, this quantity can be determined as

$$\Delta Y(Q^2) = Y_{s_1}(Q^2) - Y_{s_2}(Q^2),$$
$$Y_s(Q^2) = \frac{\left(\frac{d\sigma_T}{dQ^2}\right)_s - \left(\frac{d\sigma_{sQED+\phi}}{dQ^2}\right)_s}{H_s(Q^2)},$$
$$= |F_\pi(Q^2)|^2 + \Delta F_s(Q^2), \tag{4}$$

where s_1 and s_2 are two different c.m. energy of e^+e^-, for KLOE setup $s_1 = 1\,\mathrm{GeV}^2$ and $s_2 = m_\phi^2$, $\frac{d\sigma_T}{dQ^2}$ is the differential cross-section of the process $e^+e^- \to \pi^+\pi^-\gamma$ (photon

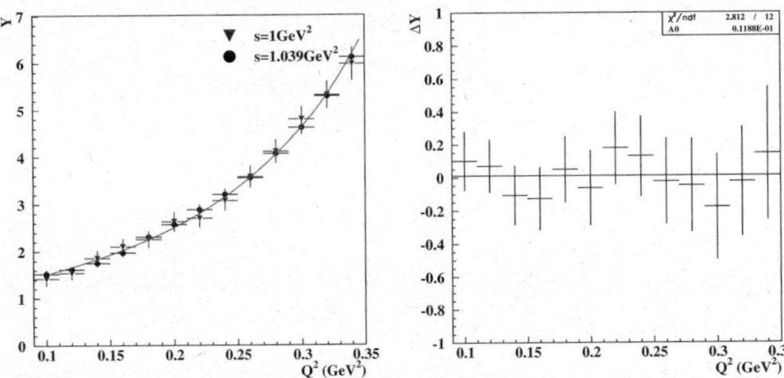

Fig. 3. Left: $Y_s(Q^2)$ at $s = 1\,\mathrm{GeV}^2$ (triangles), and at $s = m_\phi^2$ (circles), when FSR includes only sQED and ϕ contribution. The pion form factor $|F_\pi(Q^2)|^2$ is shown by the solid line. Right: the difference $\Delta Y(Q^2)$.

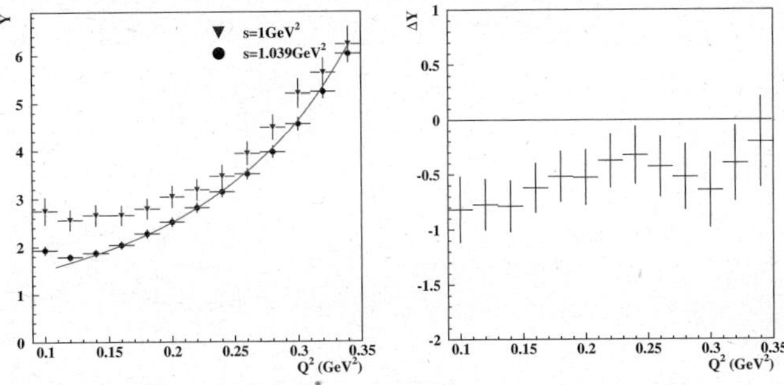

Fig. 4. Left: $Y_s(Q^2)$ at $s = 1\,\mathrm{GeV}^2$ (triangles), and at $s = m_\phi^2$ (circles), when FSR includes RPT and ϕ contribution. The pion form factor $|F_\pi(Q^2)|^2$ is shown by the solid line. Right: the difference $\Delta Y(Q^2)$.

can be radiated both from final state and by leptons) and $\dfrac{\mathrm{d}\sigma_{sQED+\phi}}{\mathrm{d}Q^2}$ is the differential cross-section only for FSR in the framework of sQED and due to ϕ direct decay. In the framework of sQED the value of $Y(Q^2)$ coincides with the square of the pion form factor and does not depend on the initial energy. In other words for sQED $\Delta Y(Q^2) = 0$. That means that any deviation from zero will be in favour of some contribution beyond sQED. In experimental conditions this theoretical idea can be affected by the statistical uncertainty. To check if the number of events collected by KLOE ($200\,\mathrm{pb}^{-1}$ at $1\,\mathrm{GeV}^2$ and $2.5\,\mathrm{fb}^{-1}$ at m_ϕ^2) is enough to show a possible deviation from sQED, we applied the results of our MC when FSR is calculated by sQED or RPT. Figure 3 shows the quantity $Y_s(Q^2)$ at $s_1 = 1\,\mathrm{GeV}^2$ and at $s_2 = m_\phi^2$ and the value $\Delta Y(Q^2)$, when FSR is described by sQED. As expected, each of the quantities Y_{s_1} and Y_{s_2} coincides with the square of the pion form factor $|F_\pi(Q^2)|^2$, shown by solid line. The value of ΔY is consistent with zero. A combined fit of Y_{s_1} and Y_{s_2} to the pion form factor is also possible:

$$F_\pi(Q^2) \simeq 1 + p_1 * Q^2 + p_2 * q^4. \tag{5}$$

It gives the following values: $p_1 = 1.4 \pm 0.186\,\mathrm{GeV}^{-2}$, $p_2 = 8.8 \pm 0.73\,\mathrm{GeV}^{-4}$, $\chi^2/\nu = 0.25$.

A different situation appears if the FSR emission from pions is modeled by RPT. In this case, as shown in fig. 4, the difference $\Delta Y(Q^2) \neq 0$ and the quantities $Y_s(Q^2)$ cannot be anymore identified with $|F_\pi(Q^2)|^2$. A combined fit of Y_{s_1} and Y_{s_2} is no longer possible.

At the end we would like to remind the main points of this work:

- Test of FSR at threshold in the process $e^+e^- \to \pi^+\pi^-\gamma$ is an important issue to get information about pion-photon interaction when the intermediate hadrons are far off shell.
- At $s = m_\phi^2$ an additional complication arises: the presence of the direct decay $\phi \to \pi^+\pi^-\gamma$ whose amplitude and relative phase can be described according to some model.
- We show that the low-Q^2 region is sensitive to the inclusion of additional terms beyond sQED in the FSR cross-section.
- A method to study FSR by model-independent way based on the experimental spectra is proposed.

Work is in progress to include the $f_0 + \sigma$ parametrization for the ϕ direct decay (instead of only the f_0 one, as it has been done in this paper) and consider the $\phi \to \rho\pi \to \pi\pi\gamma$ decay.

References

1. KLOE Collaboration (A. Aloisio *et al.*), Phys. Lett. B **606**, 12 (2005) [arXiv:hep-ex/0407048].
2. S. Eidelman, F. Jegerlehner, Z. Phys. C **67**, 585 (1995).
3. A. Hoefer, J. Gluza, F. Jegerlehner, Eur. Phys. J. C **24**, 51 (2002) [arXiv:hep-ph/0107154]; C. Glosser, S. Jadach, B.F.L. Ward, S.A. Yost, Phys. Lett. B **605**, 123 (2005) [arXiv:hep-ph/0406298].
4. S. Binner, J.H. Kühn, K. Melnikov, Phys. Lett. B **459**, 279 (1999) [arXiv:hep-ph/9902399]; H. Czyż, A. Grzelinska, J.H. Kühn, G. Rodrigo, Eur. Phys. J. C **27**, 563 (2003) [arXiv:hep-ph/0212225].
5. J. Schwinger, in *Particles, Sources and Fields*, Vol. **3** (Addison-Wesley, Redwood City, USA, 1989) p. 99.
6. S. Dubinsky, A. Korchin, N. Merenkov, G. Pancheri, O. Shekhovtsova, Eur. Phys. J. C **40**, 41 (2005) [arXiv:hep-ph/0411113].
7. G. Ecker, J. Gasser, A. Pich, E. de Rafael, Nucl. Phys. B **321**, 311 (1989); G. Ecker, J. Gasser, H. Leutwyler, A. Pich, E. de Rafael, Phys. Lett. B **223**, 425 (1989).
8. KLOE Collaboration (A. Aloisio *et al.*), Phys. Lett. B **537**, 21 (2002) [arXiv:hep-ph/0204013].
9. G. Pancheri, O. Shekhovtsova, G. Venanzoni, arXiv:hep-ph/0506332.
10. G. Pancheri, O. Shekhovtsova, G. Venanzoni, arXiv:hep-ph/0605244.
11. K. Melnikov, Int. J. Mod. Phys. A **16**, 4591 (2001) [arXiv:hep-ph/0105267].

Eur. Phys. J. A **31**, 461–464 (2007)

DOI 10.1140/epja/i2006-10205-7

Special Article – QNP 2006

The study of light scalar mesons at BESII

Liao Hongbo[a]

For the BES Collaboration
Institute of High Energy Physics, CAS, Beijing 100049, China

Received: 25 October 2006
Published online: 28 February 2007 – © Società Italiana di Fisica / Springer-Verlag 2007

Abstract. The production of σ and κ in J/ψ decays is presented using 58 million J/ψ events collected at BES II detector. We also report the study of the light scalar mesons $f_0(980)$, $f_0(1370)$, $f_0(1500)$ and $f_0(1710)$ etc. in J/ψ decays.

PACS. 12.39.Mk Glueball and nonstandard multi-quark/gluon states – 13.25.Gv Decays of J/ψ, Υ, and other quarkonia – 14.40.-n Mesons

1 Introduction

Since the discovery of the J/ψ at Brookhaven [1] and SLAC [2] in 1974, more than one hundred exclusive decay modes of the J/ψ have been reported. The J/ψ decays provide an excellent source of events to study light-hadron spectroscopy and search for glueballs, hybrids, and exotic states. Recently, 5.8×10^7 J/ψ events have been obtained with the upgraded Beijing Spectrometer (BESII) [3]. Many important results on study of the light scalar mesons and the measurements of J/ψ decays are performed based on this data sample.

2 Study of scalar mesons

2.1 σ in $J/\psi \to \omega\pi^+\pi^-$

In $J/\psi \to \omega\pi^+\pi^-$, there are conspicuous $\omega f_2(1270)$ and $b_1(1235)\pi$ signals. At low $\pi\pi$ mass, a large and broad peak due to the σ is observed.

Figure 1 shows the $\pi^+\pi^-$ invariant mass distribution from $J/\psi \to \omega\pi^+\pi^-$. Partial-wave analyses have been performed on this channel using two methods and different parametrization of σ are applied [4].

Different analysis methods and different parameterizations of the σ amplitude give consistent results for the σ pole. The average pole position is determined to be $(541 \pm 39) - i(252 \pm 42)\,\mathrm{MeV}/c^2$.

2.2 κ in $J/\psi \to K^+K^-\pi^+\pi^-$

We find evidence for the κ in the process $J/\psi \to K^*(890)\kappa$, $\kappa \to (K\pi)_S$. We select a $K^+\pi^-$ pair in the

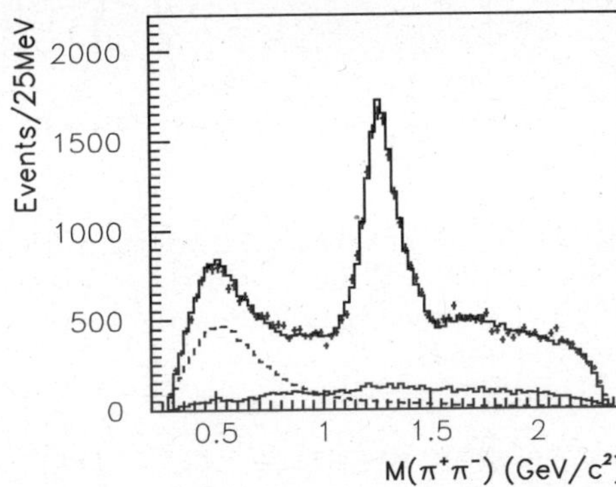

Fig. 1. The $\pi^+\pi^-$ invariant-mass distribution from $J/\psi \to \omega\pi^+\pi^-$ (crosses). The upper full histogram shows the maximum-likelihood fit, the lower full histogram corresponds to the background estimated from ω sidebands, and the dashed histogram shows the σ contribution.

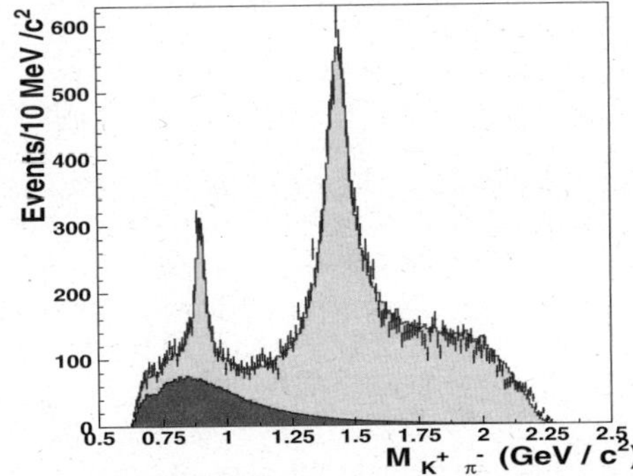

Fig. 2. $K^+\pi^-$ combinations are selected in the mass range $892 \pm 100\,\mathrm{MeV}/c^2$ from $J/\psi \to K^+K^-\pi^+\pi^-$ data. The figure shows the invariant-mass distribution of accompanying $K^-\pi^+$ pairs (crosses). The upper full histogram shows the maximum-likelihood fit, the lower one shows the κ contribution.

[a] e-mail: liaohb@mail.ihep.ac.cn

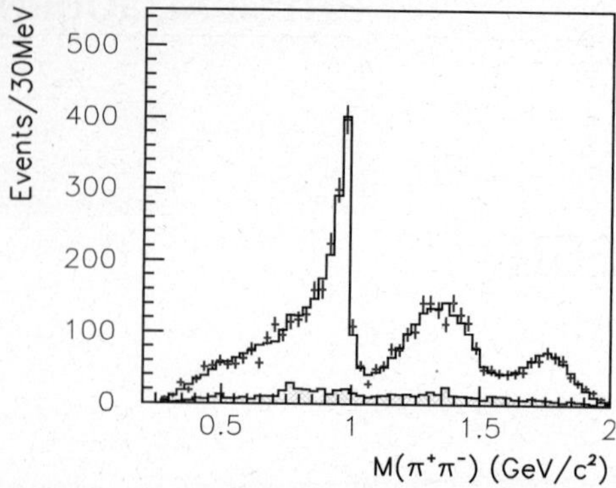

Fig. 3. The $\pi^+\pi^-$ invariant-mass distribution from $J/\psi \to \phi\pi^+\pi^-$ (crosses). The full histogram shows the maximum-likelihood fit and the shaded histogram the background estimated from ϕ sidebands.

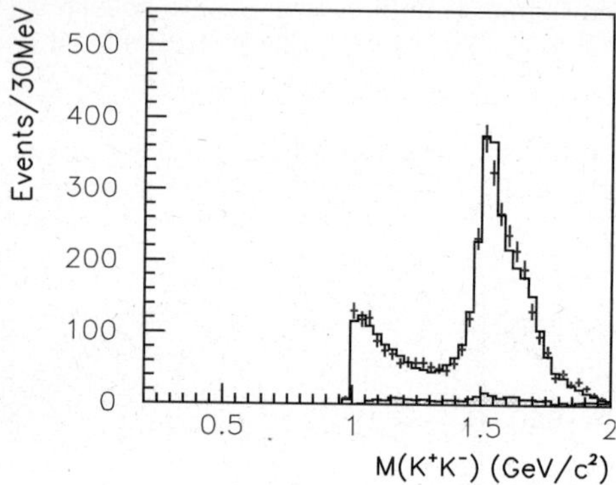

Fig. 4. The K^+K^- invariant-mass distribution from $J/\psi \to \phi K^+K^-$ (crosses). The full histogram shows the maximum-likelihood fit and the shaded histogram the background estimated from ϕ sidebands.

K^* mass range $892 \pm 100\,\mathrm{MeV}/c^2$; fig. 2 then shows the projection of the mass of the other $K^-\pi^+$ pair.

Two independent PWA analyses have been performed. Both favor strongly the fact that the low mass enhancement of the $K^+\pi^-$ system is a resonance. The 0^+ resonances κ is highly necessary in both fits. The average values of pole position for κ is determined to be $(841 \pm 78^{+81}_{-73}) - i(309 \pm 91^{+48}_{-72})\,\mathrm{MeV}/c^2$ [5].

2.3 Study of $J/\psi \to \phi\pi^+\pi^-$ and $J/\psi \to \phi K^+K^-$

Figures 3 and 4 show the $\pi^+\pi^-$ and K^+K^- invariant mass distribution from $J/\psi \to \phi\pi^+\pi^-$ and $J/\psi \to \phi K^+K^-$, respectively. In figs. 3 and 4, the shaded histogram corresponds to the background estimated from ϕ sidebands.

The $\phi\pi^+\pi^-$ and ϕK^+K^- data are fitted simultaneously by using partial-wave analysis [6], constraining resonance masses and widths to be the same in both sets of data. The full histogram in figs. 3 and 4 show the maximum-likelihood fit.

The $f_0(980)$ is observed clearly in both sets of data. The Flatté form

$$f = \frac{1}{M^2 - s - i(g_1\rho_{\pi\pi} + g_2\rho_{K\bar{K}})} \tag{1}$$

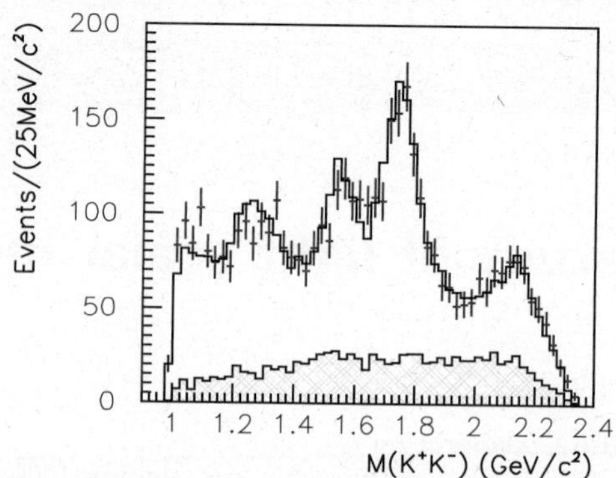

Fig. 5. The K^+K^- invariant-mass distribution from $J/\psi \to \omega K^+K^-$ (crosses). The full histogram shows the maximum-likelihood fit and the shaded histogram the background estimated from ω sidebands.

has been used to fit the $f_0(980)$ amplitude. Here ρ is the Lorentz invariant phase space, $2k/\sqrt{s}$, where k refers to π or K momentum in the rest frame of the resonance. The present data offer the opportunity to determine the parameters of $f_0(980)$ accurately: $M = 0.965 \pm 0.008(\mathrm{stat})\pm 0.006(\mathrm{sys})\,\mathrm{GeV}/c^2$, $g_1 = 0.165\pm0.010(\mathrm{stat})\pm 0.015(\mathrm{sys})\,\mathrm{GeV}/c^2$, $g_2/g_1 = 4.21 \pm 0.25(\mathrm{stat}) \pm 0.21(\mathrm{sys})$.

The $\phi\pi\pi$ data also exhibit a strong peak centred at $M = 1335\,\mathrm{MeV}/c^2$. It may be fitted with $f_2(1270)$ and a dominant 0^+ signal made from $f_0(1370)$ interfering with a smaller $f_0(1500)$ component. There is definite evidence that the $f_0(1370)$ signal is resonant, from interference with $f_2(1270)$. The mass and width of $f_0(1370)$ are determined to be: $M = 1350 \pm 50\,\mathrm{MeV}/c^2$ and $\Gamma = 265 \pm 40\,\mathrm{MeV}/c^2$.

There is also a definite signal from $f_0(1790) \to \pi^+\pi^-$ with $M = 1790^{+40}_{-30}\,\mathrm{MeV}/c^2$, $\Gamma = 270^{+60}_{-30}\,\mathrm{MeV}/c^2$. It cannot arise from $f_0(1710)$, since the branching fraction ratio $K\bar{K}/\pi\pi$ for $f_0(1790)$ is a factor 14 lower than that reported in ref. [4] for $f_0(1710)$. The large discrepancy in branching fractions implies the existence of two distinct states $f_0(1790)$ and $f_0(1710)$, the $f_0(1790)$ decaying dominantly to $\pi\pi$ and the $f_0(1710)$ dominantly to $K\bar{K}$. The $f_0(1790)$ is a natural candidate for the radial excitation of $f_0(1370)$ and behaves like $f_0(1370)$.

For ϕK^+K^- data, there is a conspicuous peak due to $f_2'(1525)$, but there is a shoulder on its upper side. This shoulder is fitted mostly by $f_0(1710)$ interfering with $f_0(1500)$; there is also a possible small contribution from $f_0(1790)$ interfering with $f_0(1500)$.

2.4 Study of $J/\psi \to \omega K^+K^-$

Figure 5 shows the K^+K^- invariant-mass distribution from $J/\psi \to \omega K^+K^-$. The shaded area indicates background events from the sideband estimation. A partial-wave analysis has been performed [7], the full histogram in fig. 5 shows the maximum-likelihood fit.

A dominant feature of $J/\psi \to \omega K^+K^-$ is $f_0(1710)$, the present data are consistent with earlier studies which identify $J = 0$. The fitted $f_0(1710)$ optimises at $M = 1738 \pm 30\,\mathrm{MeV}/c^2$, $\Gamma = 125 \pm 20\,\mathrm{MeV}/c^2$.

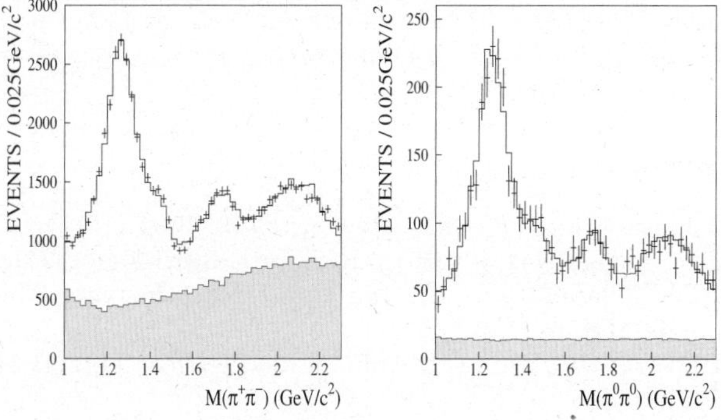

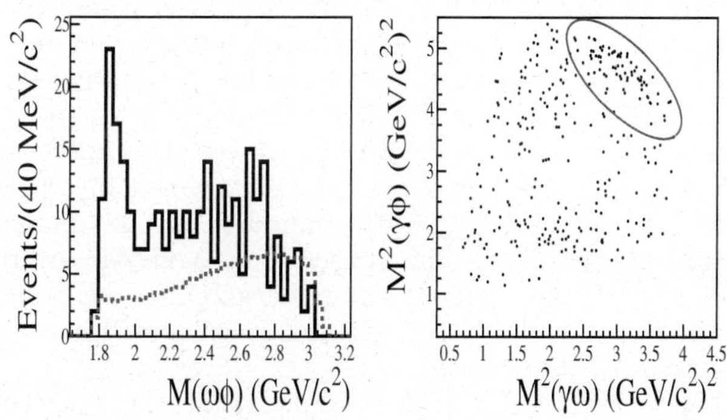

Fig. 6. Left: the $\pi^+\pi^-$ invariant-mass distribution from $J/\psi \to \gamma\pi^+\pi^-$ (crosses). The full histogram shows the maximum-likelihood fit and the shaded histogram corresponds to the $\pi^+\pi^-\pi^0$ background. Right: the $\pi^0\pi^0$ invariant-mass distribution from $J/\psi \to \gamma\pi^0\pi^0$. The crosses are data, the full histogram shows the maximum-likelihood fit, and the shaded histogram corresponds to the background.

In $J/\psi \to \omega\pi^+\pi^-$ [4], there is no definite evidence for the presence of $f_0(1710)$; if its mass is scanned, there is no optimum around $1710\,\mathrm{MeV}/c^2$, and the fitted $f_0(1710)$ is only $0.43 \pm 0.21\%$ of $\omega\pi^+\pi^-$. In the ωK^+K^- data presented here, the $f_0(1710)$ intensity is $(38\pm6)\%$ of the data within the same acceptance as for $\omega\pi^+\pi^-$. The branching fraction for $J/\psi \to \omega f_0(1710)$, $f_0(1710) \to K^+K^-$ is $(6.6 \pm 1.3) \times 10^{-4}$. We find at the 95% confidence level

$$\frac{BR(f_0(1710) \to \pi\pi)}{BR(f_0(1710) \to K\bar{K})} < 0.11, \qquad (2)$$

where all charge states for decay are taken into account.

2.5 Study of $J/\psi \to \gamma\pi\pi$

Figure 6 shows the $\pi^+\pi^-$ and $\pi^0\pi^0$ invariant-mass distribution from $J/\psi \to \gamma\pi^+\pi^-$ and $J/\psi \to \gamma\pi^0\pi^0$. A partial-wave analysis is carried out in the 1.0–$2.3\,\mathrm{MeV}/c^2$ $\pi\pi$ mass range [8]. There are two 0^{++} states at ~ 1.45 and $1.75\,\mathrm{GeV}/c^2$, respectively. One 0^{++} state peaks at a mass of $1466 \pm 6 \pm 20\,\mathrm{MeV}/c^2$ with a width of $108^{+14}_{-11} \pm 25\,\mathrm{MeV}/c^2$, which is approximately consistent with $f_0(1500)$. However, due to the large interference between S-wave states, a possibility contribution from $f_0(1370)$ cannot be excluded. The mass and width of another 0^{++} state is $1765^{+4}_{-3} \pm 13\,\mathrm{MeV}/c^2$ and $145 \pm 8 \pm 69\,\mathrm{MeV}/c^2$, respectively.

A strong production of the $f_0(1710)$ signal was observed in the PWA of $J/\psi \to \gamma K\bar{K}$ [9], with a mass of $1765 \pm 4^{+10}_{-25}\,\mathrm{MeV}/c^2$ and a width of $166^{+5+15}_{-8-10}\,\mathrm{MeV}/c^2$. If the 0^{++} state at $\sim 1.75\,\mathrm{GeV}/c^2$ observed here is interpreted as coming from $f_0(1710)$, we obtain the $\pi\pi$ to $K\bar{K}$ branching ratio as $\frac{\Gamma(f_0(1710)\to\pi\pi)}{\Gamma(f_0(1710)\to K\bar{K})} = 0.41^{+0.11}_{-0.17}$. This value is slightly higher than in $\omega\pi^+\pi^-$ and ωK^+K^- [7,10]. Hence, an alternative interpretation for this 0^{++} state is the $f_0(1790)$. This 0^{++} state may also be a superposition of $f_0(1710)$ and $f_0(1790)$.

Fig. 7. Left: the $K^+K^-\pi^+\pi^-\pi^0$ invariant-mass distribution for the $J/\psi \to \gamma\omega\phi$ candidate events. The dashed curve indicates the acceptance varying with the $\omega\phi$ invariant mass. Right: Dalitz plot.

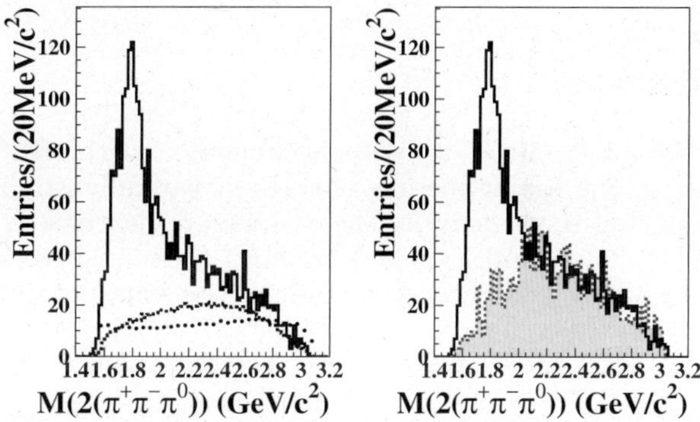

Fig. 8. Left: the $2(\pi^+\pi^-\pi^0)$ invariant-mass distribution for candidate events. The dashed curve is the phase space invariant mass distribution, and the dotted curve shows the acceptance *versus* the $\omega\omega$ invariant mass. Right: the $2(\pi^+\pi^-\pi^0)$ invariant mass of the inclusive Monte Carlo sample (shaded histogram).

2.6 0^{++} states in $J/\psi \to \gamma\omega\phi$ and $J/\psi \to \gamma\omega\omega$

The left panel of fig. 7 shows the $K^+K^-\pi^+\pi^-\pi^0$ invariant-mass distribution for the $J/\psi \to \gamma\omega\phi$ candidate events. The dashed curve indicates the acceptance varying with the $\omega\phi$ invariant mass. The right panel is the Dalitz plot. The detailed analysis of $J/\psi \to \gamma\omega\phi$ is described in ref. [11]. The decay modes of $\omega \to \pi^+\pi^-\pi^0$ and $\phi \to K^+K^-$ are used in the analysis.

The significance of the $\omega\phi$ threshold enhancement is more than 10σ. From a partial-wave analysis with covariant helicity coupling amplitudes, the spin-parity of the $X = 0^{++}$ with an S-wave $\omega\phi$ system is favored. The mass and width of the enhancement are determined to be $M = 1812^{+19}_{-26}$ (stat) ± 18 (syst) MeV/c^2 and $\Gamma = 105 \pm 20$ (stat) ± 28 (syst) MeV/c^2, and the product branching fraction is $\mathcal{B}(J/\psi \to \gamma X) \cdot \mathcal{B}(X \to \omega\phi) = (2.61 \pm 0.27$ (stat) ± 0.65 (syst)$) \times 10^{-4}$. The mass and width of this state are not compatible with any known scalars listed in the Particle Data Group (PDG) [10]. It could be an unconventional state [12–16]. However, more statistics and further studies are needed to clarify this.

In the left of fig. 8, it is the $2(\pi^+\pi^-\pi^0)$ invariant-mass distribution for $J/\psi \rightarrow \gamma\omega\omega$ ($\omega \rightarrow \pi^+\pi^-\pi^0$) candidate events. The dashed curve is the phase space invariant-mass distribution, and the dotted curve shows the acceptance *versus* the $\omega\omega$ invariant mass. The right one is $2(\pi^+\pi^-\pi^0)$ invariant mass too, the shaded histogram is the inclusive Monte Carlo sample. In our analysis of $J/\psi \rightarrow \gamma\omega\omega$ [17], the presence of a signal around $1.76\,\mathrm{GeV}/c^2$ and its pseudoscalar character are confirmed, and the mass, width, and branching fraction are measured by partial-wave analysis.

The partial-wave analysis shows that the structure around $1.76\,\mathrm{GeV}/c^2$ in the $\omega\omega$ invariant-mass spectrum is predominantly pseudoscalar. But one 0^{++} component is needed in the fit around $1.8\,\mathrm{GeV}/c^2$. If we use the mass and width of which was determined in $J/\psi \rightarrow \gamma\omega\phi$, the statistical significance of the 0^{++} component is 6.5σ.

3 Summary

Based on 5.8×10^7 J/ψ events accumulated at the BESII detector, the scalar mesons have been carefully studied with partial-wave analysis. The σ and κ are clearly seen in $J/\psi \rightarrow \omega\pi^+\pi^-$ and $J/\psi \rightarrow K^*K\pi, K^+K^-\pi^+\pi^-$, respectively and the corresponding pole positions are obtained. The light scalar mesons $f_0(980)$, $f_0(1370)$, $f_0(1500)$ and $f_0(1710)$ etc., have been studied in some J/ψ decay channels. The 0^{++} $\omega\phi$ threshold enhancement in $J/\psi \rightarrow \gamma\omega\phi$ is observed. And in $J/\psi \rightarrow \gamma\omega\omega$, one 0^{++} state is needed around $1.8\,\mathrm{GeV}/c^2$ in the fit of $\omega\omega$ invariant mass spectrum.

These works of BES Collaborations are done by many people inside and outside BES Collaboration. Thank all people who contributed to these works.

References

1. J.J. Aubert *et al.*, Phys. Rev. Lett. **33**, 1404 (1974).
2. J.E. Augustin *et al.*, Phys. Rev. Lett. **33**, 1406 (1974).
3. BES Collaboration (J.Z. Bai *et al.*), Nucl. Instrum. Methods A **458**, 627 (2001).
4. BES Collaboration (M. Ablikim *et al.*), Phys. Lett. B **598**, 149 (2004).
5. BES Collaboration (M. Ablikim *et al.*), Phys. Lett. B **633**, 681 (2006).
6. BES Collaboration (M. Ablikim *et al.*), Phys. Lett. B **607**, 243 (2005).
7. BES Collaboration (M. Ablikim *et al.*), Phys. Lett. B **603**, 138 (2004).
8. BES Collaboration (M. Ablikim *et al.*), Phys. Lett. B **642**, 441 (2006).
9. BES Collaboration (J.Z. Bai *et al.*), Phys. Rev. D **68**, 052003 (2003).
10. Particle Data Group (W.-M. Yao *et al.*), J. Phys. G **33**, 1 (2006).
11. BES Collaboration (M. Ablikim *et al.*), Phys. Rev. Lett. **96**, 162002 (2006).
12. Bing An Li, hep-ph/0602072.
13. Xiao-Gang He, Xue-Qian Li *et al.*, Phys. Rev. D **73**, 051502 (2006).
14. Pedro Bicudo *et al.*, hep-ph/0602172.
15. Kuang-Ta Chao, hep-ph/0602190.
16. D.V. Bugg, hep-ph/0603018.
17. BES Collaboration (M. Ablikim *et al.*), Phys. Rev. D **73**, 112007 (2006).

Eur. Phys. J. A **31**, 465–467 (2007)

DOI 10.1140/epja/i2007-10004-8

THE EUROPEAN
PHYSICAL JOURNAL A

Special Article – QNP 2006

On the coherent inelastic processes in the interaction of hadrons and γ-quanta with nuclei at ultrarelativistic energies

V.L. Lyuboshitz and V.V. Lyuboshitz[a]

Joint Institute for Nuclear Research, 141980 Dubna, Moscow Region, Russia

Received: 8 January 2007

Published online: 23 March 2007 – © Società Italiana di Fisica / Springer-Verlag 2007

Abstract. The coherent inelastic processes of the type $a \to b$, which may take place in the collisions of hadrons and γ-quanta with nuclei at very high energies (the nucleus remains the same), are theoretically investigated. The influence of matter inside the nucleus is taken into account by using the optical model based on the concept of refraction index. Analytical formulas for the effective cross-section $\sigma_{\mathrm{coh}}(a \to b)$ are obtained, taking into account that at ultrarelativistic energies the main contribution into $\sigma_{\mathrm{coh}}(a \to b)$ is provided by very small transferred momenta in the vicinity of the minimum longitudinal momentum transferred to the nucleus.

PACS. 13.85.-t Hadron-induced high- and super-high-energy interactions (energy > 10 GeV) – 13.85.Fb Inelastic scattering: two-particle final states

1 Momentum transfer at ultrarelativistic energies and coherent reactions on nuclei

In the present work we will investigate theoretically the processes of inelastic coherent scattering at collisions of particles with nuclei at very high energies. It is essential that at ultrarelativistic energies the minimum longitudinal momentum transferred to a nucleus tends to zero, and in connection with this the role of coherent processes increases.

Let
$$f_{a+N \to b+N}(\mathbf{q}) = [Z f_{a+p \to b+p}(\mathbf{q}) + (A - Z) f_{a+n \to b+n}(\mathbf{q})]/A$$
be the average amplitude of an inelastic process $a + N \to b + N$ on a separate nucleon in the rest frame of the nucleus (laboratory frame). Here Z is the number of protons in the target nucleus, $(A - Z)$ is the number of neutrons in the target nucleus, $\mathbf{q} = \mathbf{k}_b - \mathbf{k}_a$ is the momentum transferred to the nucleon, $\mathbf{k}_a$ and $\mathbf{k}_b$ are the momenta of the particles a and b, respectively. In the framework of the impulse approximation [1], taking into account the interference phase shifts at the inelastic scattering of a particle a on the system of nucleons, the expression for the effective cross-section of the coherent inelastic process $a \to b$ on a nucleus can be presented in the following form:

$$\sigma_{\mathrm{coh}}(a \to b) = \int |f_{a+N \to b+N}(\mathbf{q})|^2 P(\mathbf{q}) \mathrm{d}\Omega_b, \quad (1)$$

where $\mathrm{d}\Omega_b$ is the element of the solid angle of flight of the particle b in the laboratory frame, and the magnitude

$P(\mathbf{q})$ has the meaning of the probability of the event that at the collision with the particle a all the nucleons will remain in the nucleus and the quantum state of the nucleus will not change. Let us introduce the nucleon density $n(\mathbf{r})$ normalized by the total number of nucleons in the nucleus: $\int_V n(\mathbf{r})\mathrm{d}^3\mathbf{r} = A$, where the integration is performed over the volume of the nucleus. Then

$$P(\mathbf{q}) = \left| \int_V n(\boldsymbol{\rho}, z) \exp(-i\mathbf{q}_\perp \boldsymbol{\rho}) \exp(-iq_\parallel z) \mathrm{d}^2\boldsymbol{\rho}\,\mathrm{d}z \right|^2. \quad (2)$$

Here the axis z is parallel to the initial momentum $\mathbf{k}_a$, $\mathbf{q}_\perp$ and $q_\parallel$ are the transverse and longitudinal components of the transferred momentum, respectively.

It is easy to see that the momenta $|\mathbf{q}| \lesssim 1/R$, transferred to a nucleon (R is the radius of a nucleus), give the main contribution to the effective cross-section of the coherent inelastic process $a \to b$ on the nucleus. At ultrarelativistic energies, when $E_a \gg 1/R$, $E_b \gg 1/R$, the recoil energy of the nucleon $E_{\mathrm{rec}} \approx |\mathbf{q}|^2/m_N \lesssim (m_N R^2)^{-1}$ and the much smaller recoil energy of the nucleus can be neglected. In doing so, the effective flight angles for the particle b are very small: $\theta \lesssim 1/kR \ll 1$, where $k = E_a \approx E_b$. Then it is possible to assume in eqs. (1) and (2) that the transverse and longitudinal transferred momenta are as follows:

$$|\mathbf{q}_\perp| = k\theta, \qquad q_\parallel = q_{\min} = \frac{m_a^2 - m_b^2}{2k}, \quad (3)$$

where m_a and m_b are the masses of the particles a and b, respectively. Here $q_{\min}$ is the *minimum* transferred momentum corresponding to the "forward" direction.

---

[a] e-mail: Valery.Lyuboshitz@jinr.ru

In most cases the characteristic momentum transferred to the nucleus at the inelastic coherent scattering ($|\mathbf{q}| \sim 1/R$) is small as compared with the characteristic momentum transferred to the nucleon in the process $a + N \to b + N$. In connection with this, the amplitude $f_{a+N\to b+N}(\mathbf{q})$ in eq. (1) can be replaced by its value $f_{a+N\to b+N}(0)$ corresponding to the flight of the particle b in the "forward" direction. Taking into account that at small angles θ the solid angle in eq. (1) is $\mathrm{d}\Omega_b = \sin\theta\mathrm{d}\theta\mathrm{d}\phi \approx \mathrm{d}^2\mathbf{q}_\perp/k^2$ and using the properties of the two-dimensional δ-function, we obtain, as a result of integrating the expression (1) over the transverse transferred momenta and over the volume of the nucleus, the following equation:

$$\sigma_{\mathrm{coh}}(a \to b) = \frac{4\pi^2}{k^2}|f_{a+N\to b+N}(0)|^2$$
$$\times \int \left(\left|\int_{-\infty}^{\infty} n(\boldsymbol{\rho}, z)\exp(-iq_{\min}z)\,\mathrm{d}z\right|^2\right)\mathrm{d}^2\boldsymbol{\rho}, \qquad (4)$$

where $q_{\min}$ is determined by eq. (3).

In the case of a spherical nucleus with the radius R and the constant density of nucleons $n_0 = 3A/4\pi R^3$, eq. (4) gives at sufficiently high energies, when $q_{\min}R \ll 1$,

$$\sigma_{\mathrm{coh}}(a \to b) = \frac{8\pi^3}{k^2}n_0^2|f_{a+N\to b+N}(0)|^2 R^4 =$$
$$\frac{9\pi}{2k^2R^2}A^2|f_{a+N\to b+N}(0)|^2. \qquad (5)$$

2 Effect of matter inside the nucleus on coherent processes

In the relations obtained above the multiple scattering of the initial and final particles on nucleons of the nucleus was neglected. This is possible when the mean free paths of particles a and b inside the nucleus are much greater than the nuclear radius R. Actually, the role of matter inside the nucleus may be essential, especially in the case of medium and heavy nuclei. For the analysis of the effects of matter inside the nucleus we will apply the optical model of the nucleus at high energy based on the concept of refraction index [1,2].

Taking into account the refraction indices of the particles a and b, the influence of matter inside the nucleus on the coherent inelastic processes implies the introduction of the additional complex phase shift into eq. (4): the exponential factor $\exp(-iq_{\min}z)$ is replaced by $Q = \exp[-iq_{\min}z + i\delta(\boldsymbol{\rho}, z)]$. In the case of the spherical nucleus with the constant density $n(\rho, z) = n_0$ inside the interval $0 \le z \le \sqrt{R^2 - \rho^2}$ ($\rho = |\boldsymbol{\rho}|$) and $n(\rho, z) = 0$ outside this interval, the additional phase inside the considered interval is described by the equation

$$\delta(\rho, z) = (\chi_a - \chi_b)z + 2\chi_b\sqrt{R^2 - \rho^2}, \qquad (6)$$

where

$$\chi_a = \frac{2\pi n_0}{k}f_{a+N\to a+N}(0),$$

$$\chi_b = \frac{2\pi n_0}{k}f_{b+N\to b+N}(0). \qquad (7)$$

Here $k = E_a$ is the initial energy in the rest frame of the nucleus (laboratory frame); $f_{a+N\to a+N}(0)$ and $f_{b+N\to b+N}(0)$ are the average amplitudes of the elastic scattering of the particles a and b on a nucleon at the zero angle in the laboratory frame; the complex magnitudes χ_a and χ_b describe the phase shifts and the absorption of the particles a and b at their passage through the matter inside the nucleus, connected with the difference of the refraction indices from unity. The relations (7) hold at $|\chi_a|/k \ll 1$, $|\chi_b|/k \ll 1$. Let us note that the quantity $\mathrm{Re}(\chi_b - \chi_a)$ determines the additional longitudinal transferred momentum connected with the presence of the matter.

Using the optical theorem [3], we can rewrite the relations (7) in the form

$$\chi_a = i n_0(1 - i\alpha_a)\sigma_{aN}/2, \quad \chi_b = i n_0(1 - i\alpha_b)\sigma_{bN}/2,$$

where σ_{an} and σ_{bn} are the total interaction cross-sections of the particles a and b with nucleons, averaged over the protons and neutrons of the nucleus, α_a and α_b are the ratios of the real parts of the amplitudes $f_{a+N\to a+N}(0)$ and $f_{b+N\to b+N}(0)$, respectively, to their imaginary parts.

Taking into account eq. (6), after the replacement $q_{\min}z \to q_{\min}z - \delta(\boldsymbol{\rho}, z)$ in eq. (4) and the integration over z, we obtain the following expression for the cross-section of the coherent reaction $a \to b$ on a nucleus:

$$\sigma_{\mathrm{coh}}(a \to b) = \frac{8\pi^3}{k^2}n_0^2\frac{|f_{a+N\to b+N}(0)|^2}{|q_{\min} + \Delta\chi|^2}$$
$$\times \int_0^R \left|\exp\left[-2i(q_{\min} - \chi_a)\sqrt{R^2 - \rho^2}\right]\right.$$
$$\left. - \exp\left[2i\chi_b\sqrt{R^2 - \rho^2}\right]\right|^2 \rho\,\mathrm{d}\rho, \qquad (8)$$

where $\Delta\chi = \chi_a - \chi_b$.

3 Dependence of the cross-sections of the inelastic coherent processes on the nuclear radius

The results of the sect. 1 are valid when all effects connected with the rescattering of particles in the matter inside the nucleus are practically absent. In this situation the probabilities of absorption of the particles a and b and the additional phase shifts at their passage through the nucleus are close to zero. In the case of a spherical nucleus with the constant density of nucleons, this leads to the restriction $|\chi_a|R \ll 1$, $|\chi_b|R \ll 1$ or $L_a \gg R$, $L_b \gg R$, where $L_a = (n_0\sigma_{aN})^{-1}$ and $L_b = (n_0\sigma_{bN})^{-1}$ are the mean free paths inside the nucleus.

In the case of medium and heavy nuclei the radius of the nucleus $R \approx 1.1 \cdot 10^{-13}A^{1/3}$ cm; then the density of nucleons, incorporated in eq. (8), amounts to $n_0 \approx 0.28 \cdot 10^{39}$ cm^{-3}.

It follows from eq. (8) that when both the mean free paths are small as compared with the nuclear radius ($L_a \ll R$, $L_b \ll R$), the coherent processes are conditioned only by the peripheral collisions of the initial particle a with the nucleons located in the surface layer of the nucleus. In the considered case, neglecting in eq. (8) the particle masses ($|q_{\min}| \ll |\Delta\chi|$), we obtain at $f_{b+N \to b+N}(0) \neq f_{a+N \to a+N}(0)$:

$$\sigma_{\mathrm{coh}}(a \to b) =$$

$$\pi \frac{|f_{a+N \to b+N}(0)|^2}{|f_{b+N \to b+N}(0) - f_{a+N \to a+N}(0)|^2} \left[\frac{L_a^2}{2} + \frac{L_b^2}{2} \right.$$

$$\left. + 4 L_a^2 L_b^2 \, \mathrm{Re} \left(\frac{1}{L_a + L_b + i(L_a \alpha_b - L_b \alpha_a)} \right)^2 \right]. \quad (9)$$

Let us consider now the situation when the total cross-section of the interaction of the initial particle a with nucleons is small, so that $\sigma_{aN} \ll \sigma_{bN}$, $L_a \gg R$, $L_b \lesssim R$; in doing so, the relation $|f_{a+N \to b+N}(0)| \ll |f_{b+N \to b+N}(0)|$ should hold. In particular, we can deal with the coherent production of vector mesons ρ^0, ω, ϕ at the interaction of very high-energy photons with nuclei.

In the considered case eq. (8) (without the terms, depending on the masses m_a and m_b) gives

$$\sigma_{\mathrm{coh}}(a \to b) = \pi R^2 \left| \frac{f_{a+N \to b+N}(0)}{f_{b+N \to b+N}(0)} \right|^2$$

$$\times \left\{ 1 + \frac{1}{x^2} \left[\frac{1}{2}(1 - e^{-2x}) - 4 \frac{1 - \alpha^2}{(1 + \alpha^2)^2} (1 - e^{-x} \cos \alpha x) \right. \right.$$

$$\left. - \frac{8\alpha}{(1 + \alpha^2)^2} e^{-x} \sin \alpha x \right] + \frac{1}{x} \left[\frac{4}{1 + \alpha^2} e^{-x} \cos \alpha x \right.$$

$$\left. \left. - \frac{4\alpha}{1 + \alpha^2} e^{-x} \sin \alpha x - e^{-2x} \right] \right\}, \quad (10)$$

where $\alpha \equiv \alpha_b$, $x = n_0 \sigma_{bN} R = R/L_b$. At $x \gg 1$ (large cross-sections σ_{bN}, heavy nuclei) we obtain the simple expression

$$\sigma_{\mathrm{coh}}(a \to b) = \pi R^2 \left| \frac{f_{a+N \to b+N}(0)}{f_{b+N \to b+N}(0)} \right|^2. \quad (11)$$

Let us emphasize that, according to eq. (11), the effective cross-section of the coherent process $a \to b$ on a nucleus at very high energies has the *same* dependence on the number of nucleons (proportional to $A^{2/3}$) as the scattering cross-section of the final particle b on the "*black*" nucleus, despite the smallness of the interaction cross-section of the initial particle a (for example, γ-quantum) with a separate nucleon (in connection with this, see [4,5]).

For the coherent process $\gamma \to \rho^0$ on the lead nucleus ($R = 1.1 \cdot 10^{-13} A^{1/3} \, \mathrm{cm} \approx 6.5 \, \mathrm{Fm}$, $L_\rho \sim 1.5 \, \mathrm{Fm}$, $|f_{\gamma+N \to \rho+N}(0)/f_{\rho+N \to \rho+N}(0)|^2 \sim 10^{-3}$), the formula (11) is applicable at the energies of γ-quanta above several tens of GeV in the nucleus rest frame ($k \gg m_\rho^2 L_\rho \sim 4.5 \, \mathrm{GeV}$). In doing so, $\sigma_{\mathrm{coh}}(\gamma + \mathrm{Pb} \to \rho^0 + \mathrm{Pb}) \sim 1.3 \, \mathrm{mbn}$.

When, on the contrary, $\sigma_{aN} \gg \sigma_{bN}$, $L_b \gg R$, $L_a \sim R$, $|f_{a+N \to b+N}(0)| \ll |f_{a+N \to a+N}(0)|$, then the effective

cross-section of the coherent production of the particle b is described by the same formulae (10), (11), in which one should take $x = R/L_a$, $\alpha \equiv \alpha_a$ and replace the amplitude $f_{b+N \to b+N}(0)$ by $f_{a+N \to a+N}(0)$.

Taking into account that

$$f_{b+N \to b+N}(0) = i \, k \, \sigma_{bN}(1 - i\alpha_b)/4\pi,$$

it is easy to verify that the expansion of the expression (10) into the power series over the parameter x leads at $x \ll 1$ to the relation (5), just as one would expect at the conditions $L_a \gg R, L_b \gg R$. In this limit $\sigma_{\mathrm{coh}}(a \to b)$ is proportional to R^4 (or to $A^{4/3}$).

4 Summary

In the present work the coherent processes at the interaction of ultrarelativistic particles with atomic nuclei are investigated. The role of these processes essentially increases at very high energies due to the fact that the minimum momentum, transferred to a nucleon, tends to zero with increasing energy. For the purpose of the analysis of the influence of matter inside the nucleus on coherent reactions, the concept of refraction index is used. The relations, describing the dependence of the effective cross-sections of the inelastic processes on the nuclear radius and the mean free paths of the initial and final particles in the matter inside the nucleus, are obtained.

We did not consider the reverse transitions at the propagation of final particles in the matter inside the nucleus. In principle, the contribution of these transitions could be studied in the framework of the theory taking into account the distinction of the stationary states in the matter from the stationary states in the vacuum due to the mixing of the vacuum states. One may expect that really, with existing sizes of nuclei, the corresponding effects are relatively small.

This work is supported by Russian Foundation of Basic Research (Grant No. 05-02-16674)

References

1. M. Goldberger, K. Watson, *Collision Theory* (New York-London-Sydney, 1964) Chapt. 11.
2. R. Jastrov, Phys. Rev. **82**, 261 (1951).
3. L.D. Landau, E.M. Lifshitz, *Quantum Mechanics. Nonrelativistic Theory* (Moscow, Nauka, 1989) §125 (translation: Pergamon Press, New York, 1977).
4. V.N. Gribov, Zh. Eksp. Teor. Fiz. **57**, 1307 (1969).
5. L. Frankfurt, V. Guzey, M. McDermott, M. Strikman, Phys. Rev. Lett. **87**, 192301 (2001).

Eur. Phys. J. A **31**, 468–473 (2007)
DOI 10.1140/epja/i2006-10186-5

THE EUROPEAN
PHYSICAL JOURNAL A

Special Article – QNP 2006

Scalar and axial-vector mesons

E. van Beveren[1,a] and G. Rupp[2]

[1] Centro de Física Teórica, Departamento de Física, Universidade de Coimbra, P-3004-516 Coimbra, Portugal
[2] Centro de Física das Interacções Fundamentais, Instituto Superior Técnico, Edifício Ciência, P-1049-001 Lisboa, Portugal

Received: 25 October 2006
Published online: 22 February 2007 – © Società Italiana di Fisica / Springer-Verlag 2007

Abstract. Almost thirty years ago, Penny G. Estabrooks asked "Where and what are the scalar mesons?" (P. Estabrooks, Phys. Rev. D **19**, 2678 (1979)). The first part of her question can now be confidently responded (E. van Beveren *et al.*, Z. Phys. C **30**, 615 (1986)). However, with respect to the "What" many puzzles remain unanswered. Scalar and axial-vector mesons form part of a large family of mesons. Consequently, though it is useful to pay them some extra attention, there is no point in discussing them as isolated phenomena. The particularity of structures in the scattering of —basically— pions and kaons with zero angular momentum is the absence of the centrifugal barrier, which allows us to "see" strong interactions at short distances. Experimentally observed differences and similarities between scalar and axial-vector mesons on the one hand, and other mesons on the other hand, are very instructive for further studies. Nowadays, there exists an abundance of theoretical approaches towards the mesonic spectrum, ranging from confinement models of all kinds, *i.e.*, glueballs, and quark-antiquark, multiquark and hybrid configurations, to models in which only mesonic degrees of freedom are taken into account. Nature seems to come out somewhere in the middle, neither preferring pure bound states, nor effective meson-meson physics with only coupling constants and possibly form factors. As a matter of fact, apart from a few exceptions, like pions and kaons, Nature does not allow us to study mesonic bound states of any kind, which is equivalent to saying that such states do not really exist. Hence, instead of extrapolating from pions and kaons to the remainder of the meson family, it is more democratic to consider pions and kaons mesonic resonances that happen to come out below the lowest threshold for strong decay. Nevertheless, confinement is an important ingredient for understanding the many regularities observed in mesonic spectra. Therefore, excluding quark degrees of freedom is also not the most obvious way of describing mesons in general, and scalars and axial-vectors in particular.

PACS. 14.40.-n Mesons – 14.40.Cs Other mesons with $S = C = 0$, mass $< 2.5\,\text{GeV}$ – 14.40.Ev Other strange mesons – 14.40.Lb Charmed mesons

Introduction

Since all known mesonic resonances have quantum numbers (spin J, parity P, charge-conjugation parity C, flavour/isospin) which agree with the quantum numbers of a confined pair of a quark and an antiquark, it is reasonable to set out a description in terms of a confined two-particle system, like the harmonic oscillator (HO) [1]. However, the experimental mass spectrum of mesonic resonances does not agree with the spectrum of the ordinary HO [2]: masses from the HO and from experiment differ substantially, while, furthermore, many HO states are not seen in experiment.

Recent discoveries [3,4] seem to indicate that the present experimental spectrum for mesonic resonances is still very incomplete, but masses are usually well determined. Hence, the HO classification of states may work well, but it fails to reproduce the masses accurately. However, reproducing the data is not only reproducing resonance positions and possibly widths. One must also reproduce the full scattering data of experiment, even at energies where no resonances show up, and in all possible scattering channels. Consequently, it is not at all clear that the HO does not work well for confinement, as long as the effects of all possible meson loops have not yet been accounted for [5]. Moreover, upon exploring the Weyl-conformal-invariance property of QED/QCD, one obtains Anti-DeSitter (AdS) confinement [6] with flavour-independent HO-like level spacings of spectra. The linear-plus-Coulomb type of confinement is in the AdS approach obtained from one-gluon exchange at short distances [7]. Furthermore, in lattice-based results, one also seems to observe a HO-like spectrum [8], but the common practice of fitting lattice parameters to the ground states of the spectrum leads to too large level splittings at higher energies [9].

[a] e-mail: `eef@teor.fis.uc.pt`

Table 1. $c\bar{c}$ states expected in the mass region 4.0 ± 0.2 GeV.

J^{PC}	0^{-+}	1^{--}	1^{+-}	0^{++}	1^{++}
Degeneracy	1	2	1	1	1
J^{PC}	2^{++}	2^{-+}	2^{--}	3^{--}	3^{+-}
Degeneracy	2	1	1	2	1
J^{PC}	3^{++}	4^{++}	4^{-+}	4^{--}	5^{--}
Degeneracy	1	1	1	1	1

Unquenching the confinement spectrum has been studied by various groups, and for a variety of different confinement mechanisms [5,10–13]. The procedure usually amounts to the inclusion of meson loops in a $q\bar{q}$ description, or, equivalently, the inclusion of quark loops in a model for meson-meson scattering, resulting in resonance widths, central masses that do not coincide with the pure confinement spectrum, mass shifts of bound states, resonance lineshapes that are very different from the usual Breit-Wigner ones, threshold effects and cusps. In particular, it should be mentioned that mass shifts are large and negative for the ground states of the various flavour configurations [14]. Unquenching the lattice is still in its infancy, at least for the light scalars, as we conclude from ref. [15]. However, its effects should not be underestimated. Hence, ground-state levels of quenched approximations for $q\bar{q}$ configurations in relative S-waves must be expected to come out far above the experimental masses.

From pure 2-body harmonic-oscillator confinement we expect to find 18 $c\bar{c}$ states in the mass region of 4.0 ± 0.2 GeV, according to the quantum numbers given in table 1. Nevertheless, we encounter only three states with established quantum numbers in ref. [16], *i.e.*, $\psi(3770)$ at 3.77 GeV, $\psi(4040)$ at 4.04 GeV and $\psi(4160)$ at 4.16 GeV. The discovery of possibly four more $c\bar{c}$ states in this mass region [17] starts filling the many gaps in the experimental $c\bar{c}$ spectrum. But for unquenching HO confinement, one also needs some detailed experimental information on charmed decay modes. However, the only experiment which addresses this issue [18] dates from 1977, and reports results that are at odds with naive expectations [19].

In this context it is opportune to quote the remark of E. Swanson in his excellent review on the newly discovered states [20]: *"It is worth noting that attempts to unquench the quark model are fraught with technical difficulty and a great deal of effort is required before we can be confident in the results of any model."*

Coupled channels

Using coupled-channel techniques, one can simulate unquenching. In refs. [5,21], quark-pair creation was modelled by coupling $q\bar{q}$ confinement and meson-meson scattering channels. Yu. Kalashnikova [22] applied this method to describe the new $X(3872)$ state [17] by coupling $c\bar{c}$ with $J^P = 1^+$ to D-meson pairs ($D\bar{D}$, $D\bar{D}^*$, $D^*\bar{D}^*$, $D_s\bar{D}_s$, $D_s\bar{D}_s^*$ and $D_s^*\bar{D}_s^*$). In refs. [5,21], the same technique was

applied to bound states below the lowest threshold for strong decay. Hence, in this philosophy, such states contain components of virtual meson pairs. C. Albertus described in his talk a method to actually observe the virtual $B^*\pi$ component of the B-meson in semileptonic $B \to \pi\ell\nu$ decay [23].

In other approaches, the composition of confinement channels is not *a priori* known, since their effects are replaced by resonance-pole exchanges [24–26]. These methods have the advantage that the difficult issue of confinement does not have to be addressed when analysing scattering data. In particular, in the scalar-meson sector it has shown to be a convincing strategy.

$D^*_{s0}(2317)$ and its first radial excitation

Confinement dictates the quantum numbers of the states, and indicates the mass region where one may expect to observe them. Fine structure follows from additional interactions, which may even generate "dynamically" extra resonances.

When in the model of ref. [14] one determines the mass of the lowest $c\bar{s}$ state in a relative P-wave, one obtains 2.545 GeV for pure HO confinement, so even larger than the mass of 2.48 GeV predicted by S. Godfrey and N. Isgur [27]. But under the creation of a non-strange quark pair this system couples to $D(c\bar{n}) + K(n\bar{s})$, which has its threshold at 2.363 GeV. Unquenching the $c\bar{s}$ state, by allowing it to couple to DK [28], brings the ground-state mass of the full system down to 2.32 GeV, exactly where it has been found in experiment [29,30]. Similar results have been obtained by D.S. Hwang and D.-W. Kim [31], and by Yu. Simonov and J.A. Tjon [32].

There exist many alternative explanations for the mass of the $D^*_{s0}(2317)$. Exploiting the full Dirac structure of confined $q\bar{q}$ states, T. Matsuki and collaborators got 2.446 GeV in ref. [33] and, after refining their parameters, 2.330 GeV in ref. [34]. Applying the resonating-group method to a chiral-symmetric quark model, P. Bicudo obtained short-range meson-meson attraction and so DK molecules [35]. This interesting result partly confirmed the description in ref. [28]. In the latter work, the $D^*_{s0}(2317)$ was considered a two-component object: 1) a pure $c\bar{s}$ state in a relative P-wave; 2) an S-wave DK state. Hence, it certainly contains a virtual DK molecular-like component.

Th. Mehen and R. Springer [36] concluded that including counterterms is critical for fitting current data of scalar and axial-vector charmed mesons with one-loop chiral corrections. By imposing simultaneously the constraints from chiral symmetry and heavy-quark spin symmetry on effective theories of heavy-light hadrons, M. Nowak and J. Wasiluk [37] advocated that $D^*_{s0}(2317)$ and $D_{s1}(2460)$ must be viewed as the chiral doublers of D_s and D_s^*, which was contested by P. Bicudo [38]. A. Zhang [39] found that the slopes of Regge trajectories decrease with increasing quark mass, and predicted 2.35 GeV for the mass of the missing $J^P = 1^+$ D meson. We certainly endorse his recommendation that *"predicted states should be searched for and more (strong) decay modes should be detected."*

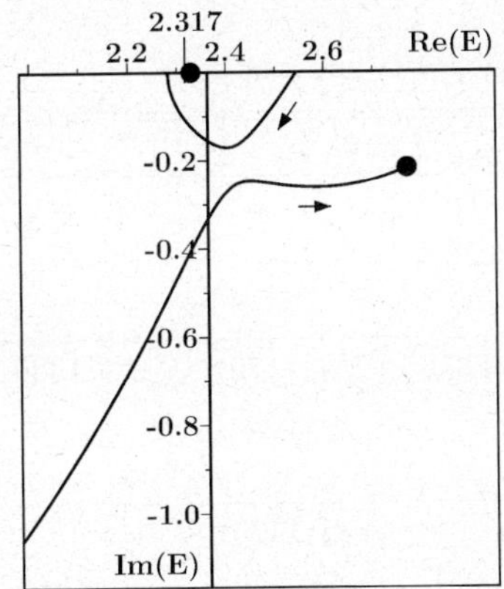

Fig. 1. Trajectories of poles in the DK S-wave scattering amplitude for the model of ref. [42], as a function of the amount of unquenching. In the quenched approximation, the dynamically generated pole has negative infinite imaginary part, whereas the confinement ground state comes out at $\sqrt{s} = 2.454\,\mathrm{GeV}$ on the real axis. The arrows indicate how the poles move when unquenching increases. The model's physical values are indicated by dots. The imaginary axis is drawn at the DK threshold.

Since the mass of the $D^*_{s0}(2317)$ ends up below the threshold of the lowest OZI-allowed [40] decay mode (*i.e.* DK), it represents a bound state in this specific selection of decay channels, which we consider the most important. Consequently, the $D^*_{s0}(2317)$ may be represented by a pole on the real energy axis. The pole representing the first radial excitation of the $c\bar{s}$ system in a relative P-wave comes out well above the DK threshold in the model of ref. [41]. Consequently, this pole has a negative imaginary part [42, 43], which gives rise to the width of the radial excitation of the $D^*_{s0}(2317)$. In ref. [42], two poles were found, one at $2.32\,\mathrm{GeV}$ and a second at $(2.85 - i0.024)\,\mathrm{GeV}$, representing the ground state and the first radial excitation of the $J^P = 0^+$ $c\bar{s}$ system, respectively. Experiment [4] reported a $c\bar{s}$ structure at $2.86\,\mathrm{GeV}$, with precisely the same lineshape as the theoretical prediction [43], and being compatible with $J^P = 0^+$ quantum numbers. However, an alternative explanation exists [44].

But in ref. [42] an additional pole showed up in the scattering amplitude of the model of ref. [41]. Its theoretical position was reported at $(2.78 - i0.23)\,\mathrm{GeV}$. The appearance of such *dynamically generated* poles is no surprise, since similar (at that time "unexpected") poles had been observed and reported [45] for the model of ref. [14], explaining well the existence of the $J^P = 0^+$ phenomena below $1.0\,\mathrm{GeV}$, as the natural consequence of combining confinement and quark-pair creation.

In fig. 1 we show the trajectories of the two lowest-lying poles in the scattering matrix for increasing $c\bar{s}$-DK coupling. The BABAR Collaboration reported in ref. [4] on the possible existence of a broad $c\bar{s}$ resonance, which might correspond to the dynamically generated pole.

E. Kolomeitsev and M. Lutz [46] obtained a bound state at $2303\,\mathrm{MeV}$ for a coupled DK-$D_s\eta$ system from their non-linear chiral $SU(3)$ Lagrangian.

Four-quark configurations have been proposed to explain the mass of the $D^*_{s0}(2317)$, namely by L. Maiani *et al.* [47], by H. Cheng and W. Hou [48], and by K. Terasaki [49]. The latter author argued in his talk [50] why the doubly charged partners have not been seen by BABAR [30]. In ref. [51], it was concluded that the $q\bar{q}$ picture is not adequate for charmed scalar mesons, based on the conflict between theory and experiment for chiral-loop corrections to the mass differences between the scalar and pseudoscalar heavy-light mesons: *"the unitarised meson model of ref. [45] or the 4-quark picture for the scalar mesons [48,49] may be a useful remedy in explaining the scalar states containing one heavy quark."*

Positive-parity mesons

In table 2, we show the experimental spectrum of light positive-parity mesons. Only the f_2 states allow for comparison with HO confinement. We therefore assume that the states in the first two f_2 columns contain mostly non-strange $q\bar{q}$ pairs, and in the next two predominantly $s\bar{s}$ pairs. Furthermore, the quark pair may come in a relative P-wave (1st and 3rd column), or in an F-wave (2nd and 4th column). From the values for the central mass positions as given in ref. [16], we collect in table 3 mass differences for a selected set of f_2 states. For HO confinement one has a level spacing of $0.38\,\mathrm{GeV}$ [14], which agrees well with the splittings in table 3. The degeneracy lifting of P-wave and F-wave is some $40 \pm 10\,\mathrm{MeV}$ for non-strange (1st and 2nd f_2 column in table 2), and about $165 \pm 20\,\mathrm{MeV}$ for strange (3rd and 4th column). But the non-degenerate ground states $f_2(1270)$ and $f_2(1430)$ do seem too high in mass with respect to the splittings in table 3. In order to understand that, we have to return to fig. 1.

The two trajectories shown in fig. 1 come close to each other for certain values of the $c\bar{s}$-DK coupling. Upon a variation of one other model parameter, this becomes a saddle point. Depending on the value of this parameter, the trajectories may interchange. In that case the end points are connected differently, making the $D^*_{s0}(2317)$ the dynamically generated state, whereas the other pole then seems to stem from the confinement ground state. This is actually what appears to happen for the light positive-parity ground-state mesons and makes them move up in energy when unquenching is turned on. For the scalar mesons, those states correspond to the $f_0(1370)$, $f_0(1500)$, $K^*_0(1430)$ and $a_0(1450)$. The dynamically generated poles correspond to the lower-lying scalar mesons [45].

The light scalar mesons

The scalar mesons have been a source of inspiration and controversy since the 1960s. Particularly, the $\epsilon(600)$ and $\kappa(900)$ [52], nowadays called $f_0(600)$ (or simply σ) and $K^*_0(800)$ [16], respectively, disappeared from the "Particle Listings" in the late 1970s [53]. Their nature is still not settled, nor for their nonet partners $f_0(980)$ and $a_0(980)$.

Table 2. The experimentally observed light positive-parity mesons.

	$I = 1$		$I = \frac{1}{2}$	$I = 0$			
0^+	$a_0(1450)$		$K_0(1430)$	$f_0(1370)$	$f_0(1500)$		
			$K_0(1980)$	$f_0(1710)$	$f_0(2020)$		
				$f_0(2200)$			
1^+	$a_1(1260)$	$b_1(1235)$	$K_1(1270)$	$f_1(1285)$	$f_1(1420)$	$h_1(1170)$	$h_1(1380)$
	$a_1(1640)$		$K_1(1400)$	$f_1(1510)$		$h_1(1595)$	
			$K_1(1650)$				
2^+	$a_2(1320)$		$K_2(1340)$	$f_2(1270)$		$f_2(1430)$	
	$a_2(1700)$			$f_2(1525)$	$f_2(1565)$	$f_2(1640)$	$f_2(1810)$
			$K_2(1980)$	$f_2(1910)$	$f_2(1950)$	$f_2(2010)$	$f_2(2150)$
				$f_2(2300)$	$f_2(2340)$		

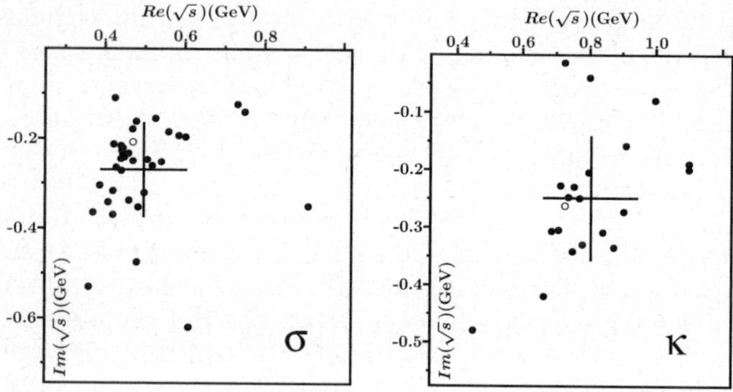

Fig. 2. Pole positions found in the literature, for the two most disputed scalar mesons: $f_0(600)$ and $K_0^*(800)$. From the above collection of real and imaginary parts, we find on average for the $f_0(600)$ pole position $(497 \pm 109) - i(269 \pm 106)$ MeV, and for the $K_0^*(800)$ $(805 \pm 143) - i(249 \pm 109)$ MeV. The figures are taken from `http://cft.fis.uc.pt/eef/sigkap.htm` where, moreover, all references are given. The open circles correspond to the scattering poles in ref. [45].

Table 3. The experimentally [16] observed mass differences for a selected set of light positive-parity mesons.

States	Mass difference
$m(f_2(1910)) - m(f_2(1525))$	0.39 ± 0.01 GeV
$m(f_2(2300)) - m(f_2(1910))$	0.38 ± 0.03 GeV
$m(f_2(1950)) - m(f_2(1565))$	0.40 ± 0.02 GeV
$m(f_2(2340)) - m(f_2(1950))$	0.39 ± 0.04 GeV
$m(f_2(2010)) - m(f_2(1640))$	0.38 ± 0.05 GeV
$m(f_2(2150)) - m(f_2(1810))$	0.34 ± 0.02 GeV

But it seems that we are converging towards the idea that low-energy scattering data are best described by a nonet of scalar S-matrix poles [54]. In fig. 2 we collect some of the pole positions encountered in the literature for the σ and the κ.

As early as in 1968, P. Roy [55], using current-algebra sum rules, estimated that a κ-meson might exist, with a central mass at 725 MeV and a width of $\geq 28 \pm 2$ MeV.

Later, J. Basdevant and B. Lee [56] determined, for various values of f_π, the σ mass and width within the σ model. For $f_\pi = 95$ MeV, they found $M_\sigma = 425$ MeV and $\Gamma_\sigma = 220$ MeV. In ref. [57], a Lagrangian for vector-meson exchange was employed to determine masses of 460 MeV and 665 MeV, and widths of 675 MeV and 840 MeV, for the σ and the κ, respectively. In 1982, M.D. Scadron [58] extended to the entire scalar nonet the dynamical spontaneous breakdown of chiral symmetry for the QCD quark theory, elaborated with R. Delbourgo in ref. [59], resulting in $\sigma(750)$, $\kappa(800)$, $f_0(980)$ and $a_0(985)$, with respective widths of 280 MeV, 80 MeV, 24 MeV and 58 MeV.

In the meantime, based on the phenomenology of tetraquarks ($qq\bar{q}\bar{q}$), R.L. Jaffe [60] also had come to the conclusion that "*there should be a very broad kaon-pion enhancement at roughly 900* MeV." The reason for considering multi-quark configurations was based on the observation that naive confinement models, taylormade for $c\bar{c}$ and $b\bar{b}$, produce spectra for positive-parity mesons comparable to the spectrum shown in table 2, with no sign of the light scalar mesons, and, furthermore, in order to explain the degeneracy of $f_0(980)$ and $a_0(980)$. In ref. [45] (1986), however, it was shown that the coupled-channel approach, representing quark-pair creation, is capable of describing the spectra of heavy and light quarks with one set of parameters, and that, as a bonus, the full light scalar nonet pops up without even being anticipated, with $f_0(980)$ and $a_0(980)$ almost degenerate, and having widths comparable to experiment (the σ and κ results for ref. [45] are shown in fig. 2). The effects of S-wave thresholds have been studied in refs. [7,61,62].

Nevertheless, many years later, the Jülich group did not find a κ pole in their meson-exchange model [63], nor did N. Törnqvist and M. Roos [64], nor the K-matrix analysis of A. Anisovich and A. Sarantsev [65]. Moreover, in order to explain part of the scalar-meson nonet with the lowest-lying 0^{++} glueball, P. Minkowski and W. Ochs argued that rather a distorted nonet containing the $f_0(1500)$ as a partner [66] is consistent with what can be expected theoretically. The issue culminated in a paper by S. Cherry and M. Pennington [67], which claimed in the title "*There is no $\kappa(900)$*", followed by ref. [68], in which M. Boglione and M. Pennington even argued that the κ pole might be found on the real axis, below the $K\pi$ threshold.

Many more works have appeared on the matter of the light scalar mesons and, in particular, on the existence and position of the κ pole. The Ishidas, with K. Takamatsu and T. Tsuru, applying the method of interfering Breit-Wigner amplitudes to a reanalysis of the $K\pi$ S-wave phase shifts, found evidence for the existence of a $\kappa(900)$ [69]. D. Black, A. Fariborz, F. Sannino and J. Schechter, studying meson-meson scattering with the use of an effective chiral Lagrangian, found a $\kappa(900)$, together with $\sigma(560)$, $a_0(980)$ and $f_0(980)$ [70]. But they also remarked that fitting the light scalars into a nonet pattern suggests that the underlying structure is closer to diquark-antidiquark than to quark-antiquark. M. Volkov and V. Yudichev proposed that the $f_0(1500)$ should be composed mostly of the scalar glueball [71]. Y. Dai and Y. Wu [72] concluded, using a non-linear effective chiral Lagrangian for meson fields, obtained from integrating out the quark fields by using a finite regularisation method, that the lightest nonet of scalar mesons, which appear as the chiral partners of the nonet of pseudoscalar mesons, should be composite Higgs bosons with masses below the chiral-symmetry-breaking scale $\Lambda_\chi \sim 1.2\,\text{GeV}$. H.Q. Zheng and collaborators [73] showed that the κ-resonance exists, if the scattering length parameter in the $I = 1/2$ and $J = 0$ channel does not deviate much from its value predicted by chiral perturbation theory. T. Kunihiro *et al.* [74] reported on very heavy κ-mesons in unquenched lattice calculations. Y. Oh and H. Kim [75] proposed that the scalar $\kappa(800)$-meson may play an important role in K^* photoproduction, and in particular that the parity asymmetry can separate the κ-meson contribution in K^* photoproduction.

A breakthrough came from the E791 experiment, with a clear $\kappa(800)$ signal [76]. But the analysis of production data is far from trivial, and seems to be still under study [77]. However, in ref. [78] J. Oller showed that scattering data for $\pi\pi$ and $K\pi$ [79] are in perfect agreement with the more recent production data [76,80], using the unitarised chiral perturbation method, earlier developed with E. Oset and J. Peláez [81]. Recently, also the BES Collaboration found evidence for κ-meson production, *i.e.*, in $J/\psi \to \bar{K}^{*0}(892)K^+\pi^-$ [82]. In his analysis of the *combined* LASS, E791 and BES data, D.V. Bugg concluded that $K\pi$ is fitted well with a κ pole at $(750 \pm 30) - i(342 \pm 60)\,\text{MeV}$ and the usual $K_0(1430)$-resonance [83].

In ref. [84], a scalar nonet $\kappa(1045)$, $\sigma(600)$, $a_0(873)$ and $f_0(980)$ was obtained, using the extended three-flavour NJL model, which has no quark confinement, and including 4- and 6-quark interactions. In ref. [85], it was shown that one needs to include at least also 8-quark interactions in the NJL model in order to obtain globally stable vacuum solutions. This has, moreover, a considerable effect on the mass of the σ-resonance [86]. Finally, ref. [87] demonstrated how, in the model of ref. [28], by just varying the flavour-mass parameters (and at the same time the related threshold masses), the poles of the universal S-matrix transform into one another, thus relating *e.g.* $\kappa(800)$ to $D_{s0}^*(2317)$ via a continuous process.

Conclusions

In the past decade, great progress has been made in the theoretical understanding of the mesonic sector of hadronic resonances, in particular of the light-scalar nonet. The efforts in setting up and carefully analysing new experiments have greatly contributed to this achievement. Nevertheless, it should be recognised that still a lot of additional data are needed and must be unravelled, in particular on resonance positions and strong-decay branching ratios, before we can hope for a more complete understanding of strong interactions.

We thank the organisers of the conference for their hospitality and their efforts to bring together many specialists in a great variety of areas. It has been a very inspiring and fruitful week. This work was partly supported by the *Fundação para a Ciência e a Tecnologia* of the *Ministério da Ciência e do Ensino Superior* of Portugal, under contract POCI/FP/63437/2005.

References

1. E. van Beveren, G. Rupp, at http://cft.fis.uc.pt/eef/QNP06_Beveren/qnp06talk/spectrum/ccbarho.htm.
2. E. van Beveren, G. Rupp, at http://cft.fis.uc.pt/eef/QNP06_Beveren/qnp06talk/spectrum/ccbar.htm; bbbar.htm; nsbar.htm; nnbart.htm and nnbars.htm.
3. BELLE Collaboration (S.K. Choi *et al.*), Phys. Rev. Lett. **89**, 102001 (2002); **89**, 129901, (2002)(E); CLEO Collaboration (J.L. Rosner *et al.*), Phys. Rev. Lett. **95**, 102003 (2005); CDF Collaboration (F. Abe *et al.*), Phys. Rev. Lett. **81**, 2432 (1998); CLEO Collaboration (G. Bonvicini *et al.*), Phys. Rev. D **70**, 032001 (2004); SELEX Collaboration (A.V. Evdokimov), Phys. Rev. Lett. **93**, 242001 (2004).
4. BABAR Collaboration (B. Aubert), Phys. Rev. Lett. **97**, 222001 (2006).
5. E. van Beveren, C. Dullemond, G. Rupp, Phys. Rev. D **21**, 772 (1980); **22**, 787 (1980)(E).
6. C.J. Isham, A. Salam, J.A. Strathdee, Nature Phys. Sci. **244**, 82 (1973); H.B. Nielsen, A. Patkos, Nucl. Phys. B **195**, 137 (1982); E. van Beveren, C. Dullemond, T.A. Rijken, Phys. Rev. D **30**, 1103 (1984); S.J. Brodsky, G.F. de Teramond, Phys. Lett. B **582**, 211 (2004).
7. E. van Beveren, T.A. Rijken, C. Dullemond, G. Rupp, Lect. Notes Phys. **211**, 331 (1984).
8. O. Oliveira, R.A. Coimbra, arXiv:hep-ph/0603046.
9. K.J. Juge, A. O'Cais, M.B. Oktay, M.J. Peardon, S.M. Ryan, PoS (LAT2005) 029 (2006).
10. N.A. Törnqvist, Ann. Phys. (N.Y.) **123**, 1 (1979).
11. E. Eichten, in *Cargese 1975, Part A* (Plenum Press, New York, 1976) pp. 305-328.
12. E.E. Kolomeitsev, M.F.M. Lutz, AIP Conf. Proc. **717**, 665 (2004).
13. P. Geiger, E.S. Swanson, Phys. Rev. D **50**, 6855 (1994).
14. E. van Beveren, G. Rupp, T.A. Rijken, C. Dullemond, Phys. Rev. D **27**, 1527 (1983).
15. E.B. Gregory *et al.*, PoS (LAT2005) 027 (2006).

16. Particle Data Group Collaboration (S. Eidelman *et al.*), Phys. Lett. B **592**, 1 (2004).

17. BELLE Collaboration (S.K. Choi *et al.*), Phys. Rev. Lett. **91**, 262001 (2003); CDF II Collaboration (D. Acosta *et al.*), Phys. Rev. Lett. **93**, 072001 (2004); BELLE Collaboration (K. Abe *et al.*), Phys. Rev. Lett. **94**, 182002 (2005); arXiv:hep-ex/0507019; arXiv:hep-ex/0507033.

18. G. Goldhaber *et al.*, Phys. Rev. Lett. **37**, 255 (1976); Phys. Lett. B **69**, 503 (1977).

19. A. de Rújula, H. Georgi, S.L. Glashow, Phys. Rev. Lett. **37**, 398 (1976).

20. E.S. Swanson, Phys. Rep. **429**, 243 (2006).

21. E. Eichten, K. Gottfried, T. Kinoshita, K.D. Lane, T.M. Yan, Phys. Rev. D **17**, 3090 (1978); **21**, 313 (1980)(E).

22. Yu.S. Kalashnikova, Phys. Rev. D **72**, 034010 (2005).

23. C. Albertus, arXiv:hep-ph/0610062, these proceedings.

24. N.A. Törnqvist, Z. Phys. C **68**, 647 (1995).

25. M. Harada, F. Sannino, J. Schechter, Phys. Rev. D **54**, 1991 (1996).

26. M.F.M. Lutz, E.E. Kolomeitsev, Nucl. Phys. A **730**, 392 (2004).

27. S. Godfrey, N. Isgur, Phys. Rev. D **32**, 189 (1985).

28. E. van Beveren, G. Rupp, Phys. Rev. Lett. **91**, 012003 (2003); Mod. Phys. Lett. A **19**, 1949 (2004).

29. BABAR Collaboration (B. Aubert *et al.*), Phys. Rev. Lett. **90**, 242001 (2003).

30. S. Tosi, these proceedings.

31. D.S. Hwang, D.-W. Kim, Phys. Lett. B **601**, 137 (2004).

32. Yu.A. Simonov, J.A. Tjon, Phys. Rev. D **70**, 114013 (2004).

33. T. Matsuki, T. Morii, Phys. Rev. D **56**, 5646 (1997) (Aust. J. Phys. **50**, 163 (1997)).

34. T. Matsuki, T. Morii, K. Sudoh, arXiv:hep-ph/0605019.

35. P. Bicudo, Nucl. Phys. A **748**, 537 (2005).

36. T. Mehen, R.P. Springer, Phys. Rev. D **72**, 034006 (2005).

37. M.A. Nowak, J. Wasiluk, Acta Phys. Pol. B **35**, 3021 (2004).

38. P. Bicudo, Phys. Rev. D **74**, 036008 (2006).

39. A. Zhang, Phys. Rev. D **72**, 017902 (2005).

40. S. Okubo, Phys. Lett. **5**, 165 (1963); G. Zweig, CERN Reports TH-401 and TH-412, see also *Developments in the Quark Theory of Hadrons*, edited by D.B. Lichtenberg, S.P. Rosen, Vol. **1** (Hadronic Press, 1981) pp. 22-101; J. Iizuka, K. Okada, O. Shito, Prog. Theor. Phys. **35**, 1061 (1966).

41. E. van Beveren, G. Rupp, Int. J. Theor. Phys. Group Theor. Nonlin. Opt. **11**, 179 (2006).

42. E. van Beveren, G. Rupp, Phys. Rev. Lett. **97**, 202001 (2006).

43. G. Rupp, arXiv:hep-ph/0610188, these proceedings.

44. P. Colangelo, F. De Fazio, S. Nicotri, Phys. Lett. B **642**, 48 (2006).

45. E. van Beveren, *et al.*, Z. Phys. C **30**, 615 (1986).

46. E.E. Kolomeitsev, M.F.M. Lutz, Phys. Lett. B **582**, 39 (2004).

47. L. Maiani, F. Piccinini, A.D. Polosa, V. Riquer, PoS (HEP2005) 105 (2006).

48. H.Y. Cheng, W.S. Hou, Phys. Lett. B **566**, 193 (2003).

49. K. Terasaki, arXiv:hep-ph/0309119.

50. K. Terasaki, arXiv:hep-ph/0609223, these proceedings.

51. D. Bećirević, S. Fajfer, S. Prelovšek, Phys. Lett. B **599**, 55 (2004).

52. Particle Data Group (T.A. Lasinski *et al.*), Rev. Mod. Phys. **45**, S1 (1973).

53. Particle Data Group (C. Bricman *et al.*), Rev. Mod. Phys. **52**, S1 (1980).

54. F.E. Close, N.A. Törnqvist, J. Phys. G **28**, R249 (2002).

55. P. Roy, Phys. Rev. **168**, 1708 (1968); **172**, 1850 (1968)(E).

56. J.L. Basdevant, B.W. Lee, Phys. Rev. D **2**, 1680 (1970).

57. D. Iagolnitzer, J. Zinn-Justin, J.B. Zuber, Nucl. Phys. B **60**, 233 (1973).

58. M.D. Scadron, Phys. Rev. D **26**, 239 (1982).

59. R. Delbourgo, M.D. Scadron, Phys. Rev. Lett. **48**, 379 (1982).

60. R.L. Jaffe, Phys. Rev. D **15**, 267 (1977).

61. E. van Beveren, G. Rupp, arXiv:hep-ph/0207022.

62. J.L. Rosner, Phys. Rev. D **74**, 076006 (2006).

63. G. Janssen, B.C. Pearce, K. Holinde, J. Speth, Phys. Rev. D **52**, 2690 (1995).

64. N.A. Törnqvist, M. Roos, Phys. Rev. Lett. **76**, 1575 (1996).

65. A.V. Anisovich, A.V. Sarantsev, Phys. Lett. B **413**, 137 (1997).

66. P. Minkowski, W. Ochs, Eur. Phys. J. C **9**, 283 (1999).

67. S.N. Cherry, M.R. Pennington, Nucl. Phys. A **688**, 823 (2001).

68. M. Boglione, M.R. Pennington, Phys. Rev. D **65**, 114010 (2002).

69. S. Ishida, M. Ishida, Taku Ishida, Kunio Takamatsu, Tsuneaki Tsuru, Prog. Theor. Phys. **98**, 621 (1997); M. Ishida, AIP Conf. Proc. **688**, 18 (2004).

70. D. Black, A.H. Fariborz, F. Sannino, J. Schechter, Phys. Rev. D **59**, 074026 (1999); D. Black, A.H. Fariborz, J. Schechter, in *Kyoto 2000*, KEK-Proceedings-2000-4, 105 (2000).

71. M.K. Volkov, V.L. Yudichev, Eur. Phys. J. C **10**, 223 (2001).

72. Y.B. Dai, Y.L. Wu, Eur. Phys. J. C **39**, S1 (2005).

73. H.Q. Zheng, Z.Y. Zhou, G.Y. Qin, Z.G. Xiao, J.J. Wang, N. Wu, Nucl. Phys. A **733**, 235 (2004).

74. SCALAR Collaboration (T. Kunihiro, S. Muroya, A. Nakamura, C. Nonaka, M. Sekiguchi, H. Wada), Nucl. Phys. Proc. Suppl. **129**, 242 (2004).

75. Y. Oh, H. Kim, Phys. Rev. C **74**, 015208 (2006).

76. E791 Collaboration (E.M. Aitala *et al.*), Phys. Rev. Lett. **89**, 121801 (2002).

77. A. Reis, these proceedings.

78. J.A. Oller, Phys. Rev. D **71**, 054030 (2005).

79. P. Estabrooks, R.K. Carnegie, A.D. Martin, W.M. Dunwoodie, T.A. Lasinski, D.W. Leith, Nucl. Phys. B **133**, 490 (1978); LASS Collaboration (D. Aston *et al.*), Nucl. Phys. B **296**, 493 (1988).

80. E791 Collaboration (E.M. Aitala *et al.*), Phys. Rev. Lett. **86**, 770 (2001).

81. J.A. Oller, E. Oset, J.R. Peláez, Phys. Rev. D **59**, 074001 (1999); **60**, 099906 (1999)(E); J.A. Oller, E. Oset, Phys. Rev. D **60**, 074023 (1999).

82. BES Collaboration (M. Ablikim *et al.*), Phys. Lett. B **633**, 681 (2006).

83. D.V. Bugg, Phys. Lett. B **632**, 471 (2006).

84. P. Costa, M.C. Ruivo, C.A. de Sousa, Yu.L. Kalinovsky, Phys. Rev. D **71**, 116002 (2005).

85. A.A. Osipov, B. Hiller, J. da Providência, Phys. Lett. B **634**, 48 (2006).

86. A.A. Osipov, B. Hiller, A.H. Blin, J. da Providência, to be published in Ann. Phys. (N.Y.) doi:10.1016/j.aop.2006.08.004, arXiv:hep-ph/0607066.

87. E. van Beveren, J.E.G. Costa, F. Kleefeld, G. Rupp, Phys. Rev. D **74**, 037501 (2006).

Eur. Phys. J. A **31**, 474–478 (2007)
DOI 10.1140/epja/i2006-10249-7

Special Article – QNP 2006

The Kπ and ππ S-wave amplitude from D-meson decays

A.C. dos Reis[a]

Centro Brasileiro de Pesquisas Físicas, Rio de Janeiro, RJ 22290-180, Brazil

Received: 23 November 2006
Published online: 9 March 2007 – © Società Italiana di Fisica / Springer-Verlag 2007

Abstract. The issue of $K\pi$ and $\pi\pi$ S-wave amplitude is addressed using decays of D-mesons. Model-independent measurements of the phases of the $\pi^+\pi^+$ and $K^-\pi^+$ S-wave amplitude from $D^+ \to \pi^-\pi^+\pi^+$ and $D^+ \to K^-\pi^+\pi^+$ decays are discussed. The result indicates a deviation from the phase of the $K^-\pi^+$ S-wave amplitude obtained by scattering experiments. This could be interpreted as an indication of the presence of 3-body final-state interaction, or in other words, that the phases from production and scattering process cannot be directly compared.

PACS. 11.80.Et Partial-wave analysis – 11.80.La Multiple scattering – 13.30.Eg Hadronic decays – 13.30.Ce Leptonic, semileptonic, and radiative decays

1 Introduction

Hadronic decays of heavy flavor are a key tool for the studies of a number of phenomena, from hadron dynamics to CP violation. Specifically, decays of D-mesons to light hadrons is a unique source of information on the dynamics of the strong interaction at low energy.

In hadronic multi-body D decays resonances are abundantly produced. The bulk of the hadronic decay rate can be described by combining simple tree level quark diagrams and regular $q\bar{q}$ resonances. The memory of the quark content of the initial state is preserved: in the case intermediate channels having a vector, axial-vector or tensor states, there is a strong correlation between the final state quarks of the D decay and the quark content of the produced resonances. Decays of D-meson could, thus, be seen as an anti-glueball filter. Extending to the scalar sector the above-mentioned correlation, one could infer the quark content of the scalar states. For instance, the large fraction of the $\sigma\pi$ component in $D^+ \to \pi^-\pi^+\pi^+$ and the absence of such channel in the $D_s^+ \to \pi^-\pi^+\pi^+$ suggests a small $\bar{s}s$ and a strong $\bar{n}n$ component the σ wave function.

Another advantage of D decays is that one can continuously cover the $K\pi$ and $\pi\pi$ spectra from threshold up to $1.7\,\mathrm{GeV}$ ($M_D - m_\pi$), filling the gaps on the existing data on $\pi\pi$ and $K\pi$. Data from K_{e4} decays go from $\pi\pi$ threshold up to $380\,\mathrm{MeV}$, while the CERN-Munich data starts only at $580\,\mathrm{MeV}$. In the case of the $K\pi$, the LASS data starts only at $825\,\mathrm{MeV}$.

An important feature of D decays is the absence of the Adler zeroes, which suppress the production of the

lowest-mass scalar resonances σ and κ in $\pi\pi$ and $K\pi$ scattering. Moreover, the scalar component is very large in the hadronic decays of the D-meson to three-body final states with a pair of identical particles. These features make the $D^+, D_s^+ \to \pi^-\pi^+\pi^+$ and $D^+ \to K^-\pi^+\pi^+$ the golden modes for studies of the $\pi\pi$ and $K\pi$ S-wave amplitude.

2 The ππ S-wave amplitude

A few years ago the E791 Collaboration published a series of Dalitz plot analyzes showing evidence for low mass, broad scalar states, identified as the σ [1] and κ [2]. Soon after, CLEO [3], BaBaR [4], Belle [5], using D decays, and BES [6], using J/ψ decays, showed further evidence for these states in different reactions. Nowadays, the existence of these broad scalars is well established.

In all of the above analyses the S-wave component of the decay amplitude included a Breit-Wigner for the σ and for the κ. It is well known that the Breit-Wigner is not a correct representation for such states [7] but, nevertheless, it yields very good fits to the data in all cases. One should look at the Breit-Wigner as an effective representation of the amplitude for real values of energy, where we have the experimental data. Even though there is a remarkable agreement between the σ Breit-Wigner parameters obtained from different analyzes of D decays (BES results using J/ψ decays yield a somewhat different set of parameters), one should not use them to extract the σ pole position.

There has been an intense debate on how to represent the S-wave amplitude. The basic point here is that the phase of the S-wave amplitude from D decays does not

[a] e-mail: alberto@cbpf.br

match that of $\pi\pi$ scattering. One possible reason for this discrepancy would be the fact that the experimentalists adopted a wrong model in their analysis. There has been attempts to reconcile the S-wave phases. An example is the work done by J. Oller [8]. The argument is as follows. If one writes the amplitude as $N(s)/D(s)$, where $N(s)$ is a real function accounting for the production of the $\pi\pi$ pair and $D(s)$ a complex function describing the $\pi\pi$ final-state interaction, then different process like $\pi\pi$ scattering and D decays would have different $N(s)$, but should share the same function $D(s)$. According to this point of view $D(s)$ would be a universal function for all process involving a $\pi\pi$ pair. After all, if a resonance is a real particle, it should have the same mass and width in whichever process it appears. A good fit to E791 data is obtained in [8], but at the expense of very large interference terms. In practice, it is a bit difficult to distinguish between the non-resonant contribution and broad structures with no distinct angular distribution in the Dalitz plot. The magnitudes and phases of the non-resonant term, and the mass and width of broad states are correlated, so can have similar distributions constructed from different combinations of these parameters.

There has been also attempts to use the K-matrix approach, which is also based on the assumption of an isolated $\pi\pi$ system. An example is the analysis of Fermilab FOCUS experiment [9]. Again, a good fit is obtained, but the same data can be also fitted with a σ Breit-Wigner, yielding mass and width which is very similar to the E791 parameters. There are several problems with the K-matrix approach: the physical interpretation is a bit obscure, and there is enough freedom in the model for the $\pi\pi$ production to accommodate the σ peak without considering it explicitly as one of the K-matrix poles.

For an isolated $\pi\pi$ system, the matching of the phases would be a consequence of the Watson theorem [10]. In spite of being often used, this assumption is far from trivial, since the $\pi\pi$ is embedded in a multi-body, strongly interacting final state. Actually, there is no experimental evidence to support this assumption.

From the experimental side, a high statistics, model-independent measurement of the S-wave phase from D decays is in order. It would be very interesting also to have a comparative study of the semi-leptonic decays $D^+ \to \pi^-\pi^+l^+\nu$, where the $\pi\pi$ system is really isolated.

Recently, a model-independent measurement of the S-wave phase was performed on the E791 data by Miranda and Bediaga [11], based on the so-called amplitude difference method. This method uses the interference between the S- and the D-waves (essentially the σ and $f_2(1270)$) in the decay $D^+ \to \pi^-\pi^+\pi^+$ to extract the S-wave phase with no assumption. The result confirmed the resonant behavior at low $\pi\pi$ mass, with a phase motion that is different from that of $\pi\pi$ scattering. However, the statistics is still limited, and this measurement should be repeated when larger samples become available. There is another way to access the S-wave phase without any assumption about its content. This method has been applied to the $K\pi$ system and will be described in the next section.

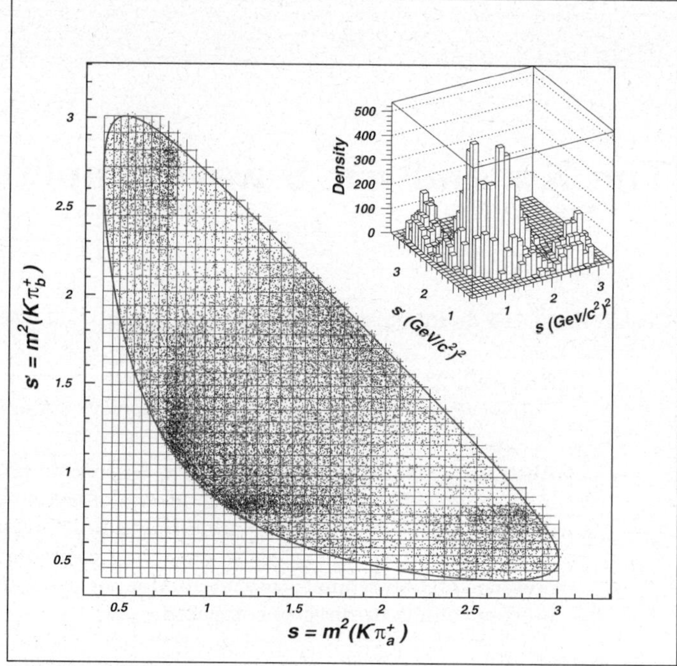

Fig. 1. Dalitz plot for the $D^+ \to K^-\pi^+\pi^+$. The plot is symmetrized, each event appearing twice.

3 The Kπ S-wave amplitude

3.1 The D$^+ \to$ K$^-\pi^+\pi^+$ decay

A model-independent partial-wave analysis (MIPWA) of the $D^+ \to K^-\pi^+\pi^+$ decay was recently performed by E791 [12]. The Dalitz plot is shown in fig. 1. The bands corresponding to the $K^*(982)\pi$ mode is clearly seen. One of the most remarkable features of this Dalitz plot is the asymmetry in each of the $K^*(982)$ bands. This is most readily explained by the interference with the $K\pi$ S-wave component. One can use this interference effect to infer the structure of the S-wave, provided the other components are correctly modeled.

In this work the S-wave amplitude was parameterized by a generic complex function, $\mathcal{A} = f(s)e^{i\phi(s)}$, with s being the $K\pi$ invariant mass squared. The P- and D-waves were parameterized in the framework of the isobar model, as in [2]. The $K\pi$ mass spectrum was divided in bins and, for each bin, the magnitude and the phase of $\mathcal{A}(s_k) = a_k e^{i\phi_k}$ were determined by the fit.

Considering the simplest, tree level diagrams for the $D^+ \to K^-\pi^+\pi^+$ decay, the $K^-\pi^+$ system should be in isospin $I = 1/2$ state. The decay amplitude is written as a sum of $K^-\pi^+$ partial waves labeled by the orbital angular momentum L,

$$A(s_a, s_b) = \sum_L (-2pq)^L P_L(\cos\theta) f_D^L(q, r_D) C_L(s_a), \quad (1)$$

where s_a, s_b are the two independent $K^-\pi^+$ mass squared, $\mathbf{p}$ and $\mathbf{q}$ are the momenta for the K^- and the bachelor pion π_b^+, respectively, in the $K^-\pi_a^+$ rest frame, θ

is the helicity angle, $\cos\theta = \hat{p}\cdot\hat{q}$, and f_D^L is a form factor for the D^+ vertex. The amplitude must be Bose symmetrized with respect to the identical pions π_a^+ and π_b^+.

$$\mathcal{A}(s_a, s_b) = A(s_a, s_b) + A(s_b, s_a). \tag{2}$$

The main goal is to measure $C_0(s)$,

$$C_0(s) = a_0(s_k)e^{i\phi_0(s_k)} = c_k e^{i\gamma_k}. \tag{3}$$

The reference waves are

$$C_1(s) = [BW_{K^*(892)}(s) + b_1 BW_{K_1^*(1680)}(s)]f_R^1(p, r_R) \tag{4}$$

and

$$C_2(s) = b_2 BW_{K_2^*(1430)}(s)f_R^2(p, r_R). \tag{5}$$

A maximum likelihood fit, including representation of the background, effects of the finite detector resolution and correction for the non-uniform acceptance, was performed in order to determine the values of c_k and γ_k. The likelihood function is

$$\mathcal{L} = \prod_{\text{events}} [P_S(M; s_a, s_b) + P_B(M; s_a, s_b)], \tag{6}$$

where the signal and background probability distribution functions are

$$P_S(M; s_a, s_b) = \frac{g(M)\varepsilon(s_a, s_b)\,|\mathcal{A}(s_a, s_b)|^2}{\int ds_a ds_b dM g(M)\varepsilon(s_a, s_b)\,|\mathcal{A}(s_a, s_b)|^2} \tag{7}$$

and

$$P_B(M; s_a, s_b) = \sum b_k(M)f_k \frac{\mathcal{B}_k(s_a, s_b)}{N_k}, \tag{8}$$

In the above equations, $g(M)$ and $b_k(M)$ are functions describing the $K^-\pi^+\pi^+$ invariant-mass distribution for the signal and background, and $\varepsilon(s_a, s_b)$ is the acceptance.

The $K\pi$ elastic cross-section measurement is largely dominated by the LASS experiment [13]. LASS measured the $K\pi$ S-wave amplitude from the reaction $K + N \to K\pi + N$. The one-pion exchange is assumed to be the dominant mechanism for this reaction. What LASS actually measured was the scattering amplitude of a real kaon and a virtual pion, off the mass shell, if one neglects other contributions like ρ exchange. The $I = 1/2$ S-wave amplitude, $\mathcal{T} = a(s)e^{i\delta(s)}$, is predominantly elastic up to the $K\eta'$ threshold. If the bachelor pion, π_b^+ in the $D^+ \to K^-\pi_a^+\pi_b^+$ decay acts like a mere spectator, then Watson theorem should hold and the dynamics of the final state would be entirely determined by the $K\pi_a$ system. In this case, the relation between $C_L(s)$ and the corresponding amplitudes $T_L(s)$, measured in scattering experiments, is

$$C_L(s) = \frac{\sqrt{s}}{p}\frac{\mathcal{P}(s)T_L(s)}{p^L f_D^L}, \tag{9}$$

where $\mathcal{P}(s)$ is an unknown function describing the $K\pi$ production in D decays, with no s-dependent phase. As a consequence, the phases $\phi_L(s)$ and $\delta_L(s)$ should match,

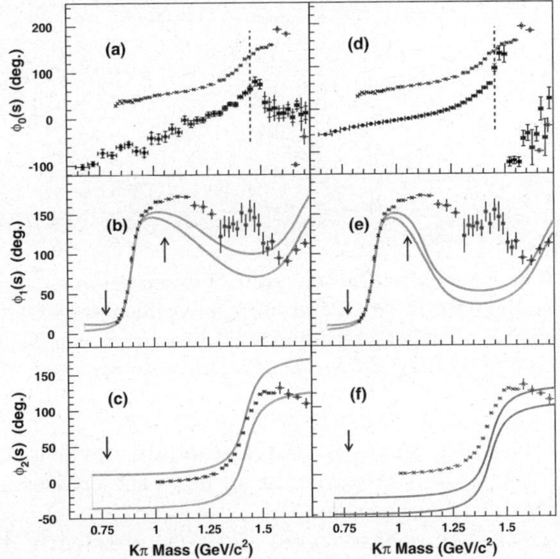

Fig. 2. Results of the best fit to E791 data (panels (a) to (c)), compared to the LASS results (crosses). In (a) the E791 phases are shown as solid circles, whereas in (b) and (c) the solid curves are the E791 P- and D-wave phases (with errors). Panels (d) to (f) show the same quantities from a fit where the S-wave amplitude was fixed to that of the LASS experiment.

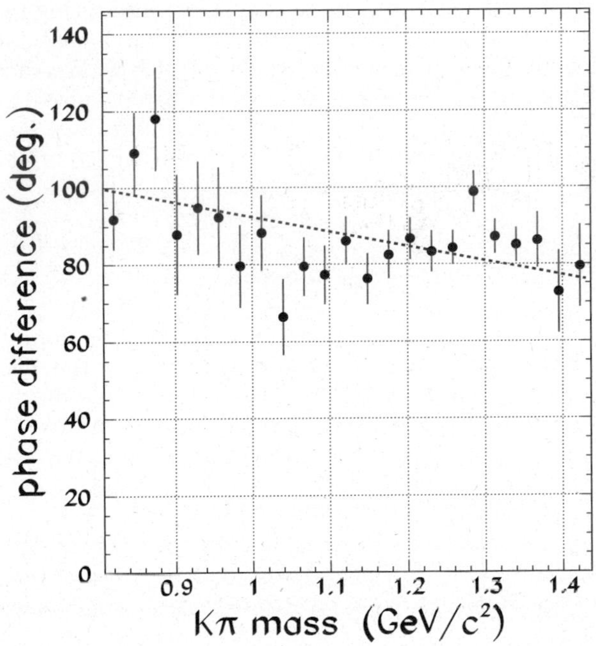

Fig. 3. Difference between S-wave phases from LASS and E791 as a function of the $K\pi$ mass. Error bars are largely dominated by E791 data.

up to a constant, for all waves. The validity of the Watson theorem relies on the absence of FSI between $K\pi_a$ and π_b. It has been often assumed to hold, but it has never been tested.

In figs. 2(a)-(c) the result of the MIPWA is shown. In fig. 2(a) the solid circles are the phase $\phi_0(s)$ determined by the MIPWA, whereas in figs. 2(b) and (c) the solid

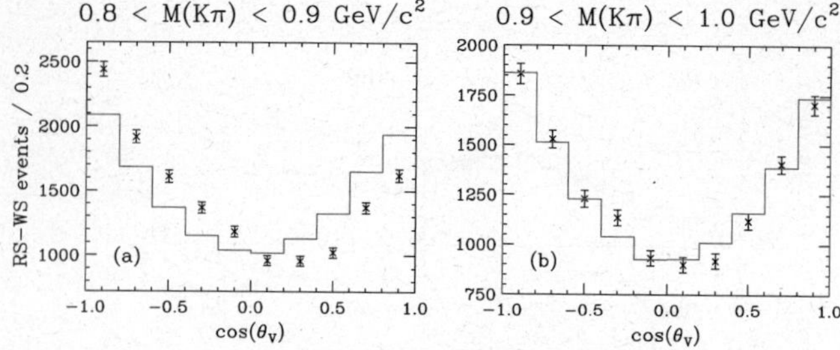

Fig. 4. The $\cos\theta_V$ distribution, split between samples above and below the $0.9\,\mathrm{GeV}/c^2$. Points with error bars are FOCUS data (background subtracted) and solid histogram is a Monte Carlo simulation of the $D^+ \to \bar{K}^*(892)^0\mu^+\nu$ decay.

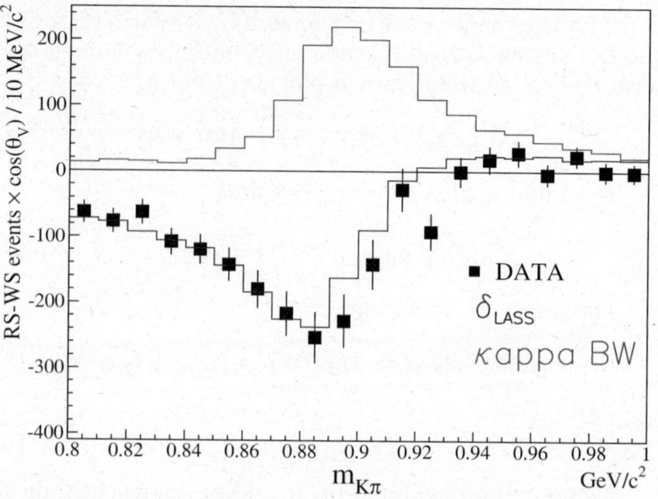

Fig. 5. The background-subtracted $m_{K\pi}$ distribution weighted by $\cos\theta_V$. Data show good agreement with the LASS non-resonant parameterization (solid histogram), but not with the κ.

curves enclose the zones within 1 standard deviation of the MIPWA P- and D-waves. In all plots the crosses are the LASS $I = 1/2$ phases $\delta_L(s)$.

We see that the phases $\phi_0(s)$ and $\delta_0(s)$ are clearly different, not only by an overall shift. They show a different s-dependence, which can be see more clearly in fig. 3, where the two phases have been subtracted. Errors (essentially from E791) are still large, but they will be reduced by a factor of two when the result of the FOCUS experiment MIPWA analysis is released.

A possible origin of the observed discrepancy could be a wrong modeling of the P- and D-waves, which would reflect on the measurement of $\phi_0(s)$. It is interesting to fit the data constraining the shape of the S-wave phase $\phi_0(s)$ to follow that of LASS. The results of such fit are shown in figs. 2(d)-(e). The large offset of the S-wave phase persists, while both P- and D-waves now show even larger differences than before. The phase variation required by Watson theorem is not observed.

3.2 The $D^+ \to K^-\pi^+\mu^+\nu$ decay

Recently, the Fermilab FOCUS Collaboration found a small $K^-\pi^+$ ($\simeq 5\%$) S-wave component in the decay $D^+ \to K^-\pi^+\mu^+\nu$ [14]. The kinematics of this decay is described by five variables. Usually, these variables are the $K^-\pi^+$ and $\mu^+\nu$ invariant mass squared, the angle between the planes defined by the $K^-\pi^+$ and $\mu^+\nu$, the angle between the μ^+ and the D line of flight, in the $\mu^+\nu$ frame, and the angle between the K^- and the D line of flight, in the $K^-\pi^+$ frame, $\cos\theta_V$. The presence of the S-wave is manifested by an asymmetry in the $\cos\theta_V$ distribution, shown in fig. 4. If the $K^-\pi^+$ resulted only from the $K^*(892)$, the $\cos\theta_V$ would be symmetric. A full angular analysis was performed in order to disentangle the S-wave component. It was found that this component was empirically described by a constant amplitude, $\mathcal{A}_0 = 0.36e^{i\pi/2}$.

It is very interesting to study the $K^-\pi^+$ amplitude in semi-leptonic decays. Since in this case there is no strong FSI between the $K^-\pi^+$ and the lepton pair, the Watson theorem should strictly hold. In a second study of this channel [15], the content of the S-wave was investigated. The line shape of the $K^-\pi^+$ peak is rather insensitive to the form of the S-wave, so one has to look

at the $\cos\theta_V$ distribution in order to distinguish between models. This distribution is well represented if instead of the $\mathcal{A}_0 = 0.36e^{i\pi/2}$ S-wave amplitude one uses the LASS effective-range formula,

$$\mathcal{A}_0 = \frac{p^*}{\sqrt{s}} \sin\delta_0(s)e^{i\delta_0(s)}, \qquad (10)$$

with s being the $K^-\pi^+$ mass squared, p^* being the kaon momentum in the $K\pi$ frame and $\delta_0(s)$ given by

$$\cot\delta_0(s) = \frac{1}{ap} + \frac{bp}{2}. \qquad (11)$$

This equation represents a non-resonant contribution defined by a scattering length a and an effective range b. A Breit-Wigner representing a possible contribution from the κ-meson would also work, but in this case we should add a $\pi/2$ overall phase.

4 Conclusion

Based on the analysis of high-statistics data on D decays, one can conclude that the $K^-\pi^+$ S-wave phase variation is consistent with the $I = 1/2$ S-wave phase from $K^-\pi^+$ scattering in semi-leptonic decays, but differs substantially in hadronic decays.

There are several possible reasons for that. First, the $K^-\pi^+$ phase extracted by LASS relies on the assumption of a particular reaction mechanism, the one-pion exchange. Other mechanisms could also contribute. In this case the LASS $K^-\pi^+$ phase would not correspond to a pure $K^-\pi^+$ scattering.

In another scenarium, the $K^-\pi^+$ system from the $D^+ \to K^-\pi^+\pi^+$ decay would be a mixture of $I = 1/2$ and $I = 3/2$ states, whereas in the $D^+ \to K^-\pi^+\mu^+\nu$ decay the $K^-\pi^+$ system would be produced only in a

$I = 1/2$ state, assuming the decay mechanism to be the tree level W-radiation amplitude. The $I = 1/2/I = 3/2$ mixture would be different in $D^+ \to K^-\pi^+\pi^+$ decay and $K^-\pi^+$ scattering [16]. Compatibility between LASS and E791 would be achieved comparing the E791 phase to a sum of $I = 1/2$ and $I = 3/2$ phases from LASS. One problem of such hypotesis is that the required amount of the $I = 3/2$ amplitude is large, which implies a large non-resonant component (recall that the $I = 3/2$ amplitude is entirely non-resonant). Such a large non-resonant component would be unique to this particular channel.

The difference between LASS $I = 1/2$ and E791 phases could also be explained by the final-state interactions, present in hadronic final states, but absent in semi-leptonic decays [17]. The effect of the FSI would be to distort the phase variation of the pure $K^-\pi^+$ scattering. Recent calculations [18] show that, once FSI is taken into account, the phases extracted from production and scattering match nicely.

In any case, one cannot extract the κ pole position directly from the S-wave phase measured by the MIPWA.

For the $\pi^-\pi^+$ S-wave amplitude, the method developed by Bediaga and Miranda [11] should be applied for higher statistics samples whenever available. Also, the MIPWA technique used by E791 on the $D^+ \to K^-\pi^+\pi^+$ decay should be performed on the $D^+ \to \pi^-\pi^+\pi^+$, as an alternative method to extract the S-wave phase. In addition, the semi-leptonic decay $D^+ \to \pi^-\pi^+\mu^+\nu$ should be analyzed looking for an S-wave component, in which the phase of the $\pi^-\pi^+$ amplitude could be measured in the absence of FSI. Here, the S-wave $\pi^+\pi^-$ system can be only in $I = 0$ state. If the FSI play no role, the S-wave phase from $D^+ \to \pi^-\pi^+\pi^+$ and from the $I = 0$ CERN-Munich and K_{l4} data should match.

On the theoretical side, the FSI correction to the pure $\pi^-\pi^+/K^-\pi^+$ scattering amplitude should be computed. If this is the origin of the observed discrepancies, then the content of the $\pi^-\pi^+$ and $K^-\pi^+$ S-waves could be extracted from D-meson decays and, in particular, that would allow the determination of the position of the σ and κ poles.

I would like to thank Felipe Llanes Estrada and Antonio Dobado Gonzalez for the excellent, perfectly organized Conference, and for the warm and kind hospitality. It was a real pleasure to participate in *QNP 2006*.

References

1. E791 Collaboration (E.M. Aitala *et al.*), Phys. Rev. Lett. **86**, 770 (2001).
2. E791 Collaboration (E.M. Aitala *et al.*), Phys. Rev. Lett. **89**, 121801 (2002).
3. CLEO Collaboration (H. Muramatsu *et al.*), Phys. Rev. Lett. **89**, 251802 (2002).
4. BaBar Collaboration (B. Aubert *et al.*), hep-ex/0408088.
5. Belle Collaboration (K. Abe *et al.*), hep-ex/0308043.
6. BES Collaboration (M. Ablikim *et al.*), Phys. Lett. B **633**, 681 (2006) hep-ex/0406038.
7. S. Gardner, U.G. Meissner, Phys. Rev. D **65**, 094004 (2002).
8. J.A. Oller, Phys. Rev. D **71**, 054030 (2005).
9. FOCUS Collaboration (J.M. Link *et al.*), Phys. Lett. B **585**, 200 (2004).
10. K.M. Watson, Phys. Rev. **88**, 1163 (1952).
11. I. Bediaga, J. Miranda, Phys. Lett. B **633**, 167 (2006).
12. E791 Collaboration (E.M. Aitala *et al.*), Phys. Rev. D **73**, 032004 (2006).
13. LASS Collaboration (D. Aston *et al.*), Nucl. Phys. B **296**, 493 (1988).
14. FOCUS Collaboration (J.M. Link *et al.*), Phys. Lett. B **535**, 43 (2002).
15. FOCUS Collaboration (J.M. Link *et al.*), Phys. Lett. B **621**, 72 (2005).
16. L. Edera, M.R. Pennington, Phys. Lett. B **623**, 55 (2005).
17. I. Caprini, hep-ex/0603250.
18. M.R. Pennington, hep-ex/0608016.

Eur. Phys. J. A **31**, 479–484 (2007)

DOI 10.1140/epja/i2007-10006-6

THE EUROPEAN
PHYSICAL JOURNAL A

Special Article – QNP 2006

Forward dispersion relations and Roy equations in $\pi\pi$ scattering

R. Kamiński[1], J.R. Peláez[2,a], and F.J. Ynduráin[3]

[1] Department of Theoretical Physics Henryk Niewodniczański Institute of Nuclear Physics, Polish Academy of Sciences, 31-242, Kraków, Poland

[2] Departamento de Física Teórica, II (Métodos Matemáticos), Facultad de Ciencias Físicas, Universidad Complutense de Madrid, E-28040, Madrid, Spain

[3] Departamento de Física Teórica, C-XI Universidad Autónoma de Madrid, Canto Blanco, E-28049, Madrid, Spain

Received: 8 January 2007
Published online: 27 March 2007 – © Società Italiana di Fisica / Springer-Verlag 2007

Abstract. We first review the results of an analysis of $\pi\pi$ interactions in S, P and D waves for the two-pion effective mass from threshold to about 1.4 GeV. In particular, we show a recent improvement of this analysis above the $K\bar{K}$ threshold using more data for phase shifts and including the $S0$-wave inelasticity from $\pi\pi \to K\bar{K}$. In addition, we have improved the fit to the $f_2(1270)$-resonance and used a more flexible P-wave parametrization above the $K\bar{K}$ threshold and included an estimation of the $D2$-wave inelasticity. The better accuracy thus achieved also required a refinement of the Regge analysis above 1.42 GeV. Finally, in this work we check that the $\pi\pi$ scattering amplitudes obtained in this approach satisfy remarkably well forward dispersion relations and Roy's equations.

PACS. 13.75.Lb Meson-meson interactions

1 Introduction

In a previous analysis [1] a set of fits to different data sets on $\pi\pi$ scattering was presented together with a detailed description of the mathematical methods used in calculations. Forward dispersion relations (FDRs) were then used in order to test the correctness of the amplitudes thus constructed. Remarkably, it was found that some of the very frequently used sets of phase shifts do not satisfy FDRs below 1 GeV. Thus, FDRs were shown to give strong constraints to fits which, when used later as a constraint, lead to an improved and precise representation of $\pi\pi$ scattering amplitudes below, roughly 1 GeV. In the regions from about 1 GeV to 1.4 GeV there was still some mismatch between the real part of the amplitudes and the results of dispersive evaluations in [1] (especially for $\pi^0\pi^0$ scattering).

In a subsequent article [2], in order to improve the agreement with the constraints given by FDRs, we have reanalysed the parametrizations of the $S0$ above $K\bar{K}$ threshold, the $D0$-wave and, to a lesser extent, the P and $D2$ waves. In the $S0$-wave we took into account systematically the elasticity data from the $\pi\pi \to K\bar{K}$ reaction [3–7], included in the fit more data on phase shifts above the $K\bar{K}$ threshold [3–6] and used a more flexible parametrization from 0.932 GeV to 1.4 GeV. In the $D0$-wave we have used experimental data from [3,6,8], information on low-energy parameters (the scattering length and slope) and included in the fit the width and mass of the $f_2(1270)$-resonance as

given by the PDG [9]. The result is that, for both $S0$ and $D0$ waves, we have obtained more accurate parametrizations with smaller errors compared to those in the previous approach [1]. In the P-wave we have exploited a more flexible parametrization between the $K\bar{K}$ threshold and 1.42 GeV and in the $D2$-wave we have included its estimated inelasticity above 1 GeV.

This more accurate determination of the $\pi\pi$ amplitudes below 1.42 GeV allowed us to refine the Regge analysis that had been used in [1,10]. This has been done by removing the degeneracy condition $\alpha_\rho(0) = \alpha_{P'}(0)$ which thus modifies slightly the central values of the intercepts $\alpha_\rho(0)$ (by $\sim 11\%$) and $\alpha_{P'}(0)$ (by $\sim 4\%$), but yields smaller errors than those in [1].

We have found that the $\pi\pi$ amplitudes with the new parametrizations of phase shifts and inelasticities in the S, P and D waves together with the just discussed small changes in $\alpha_\rho(0)$ and $\alpha_{P'}(0)$ allow for a much better fulfilment of FDRs than in [1]. The biggest improvement in χ^2 (about 66%) is achieved for the forward $\pi^0\pi^0$ dispersion relation and a smaller one (about 15%) for $\pi^0\pi^+$. In the case of the forward dispersion relation for isopin 1 in the t-channel a very tiny deterioration has been found (χ^2 increased by about 26%), which is still acceptable, since, considering all FDRs together, there is a considerable overall improvement in their fulfilment. It is worth noting that this has been achieved despite the improved data fits have smaller errors than in [1].

Finally, in the present work, and using those improved parametrizations, we will test the Roy equations, which,

---

[a] e-mail: jrpelaez@fis.ucm.es

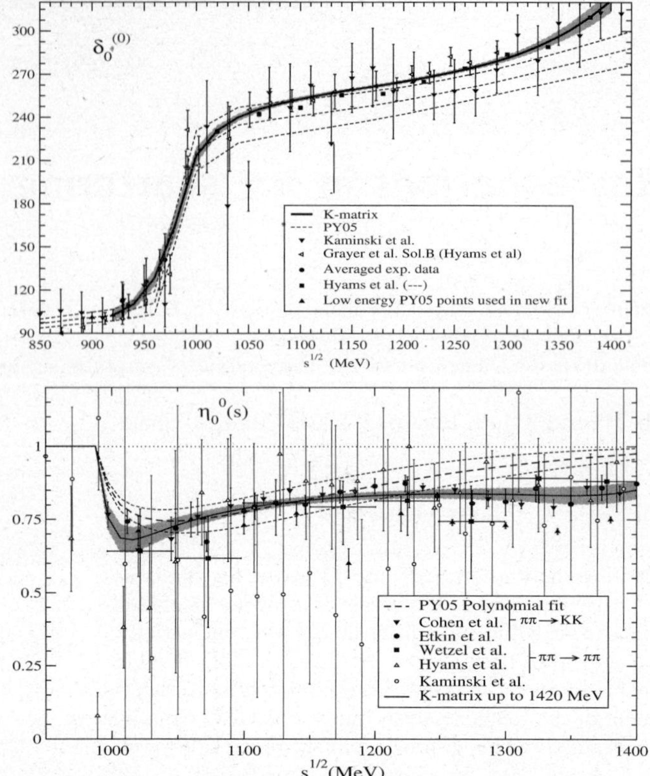

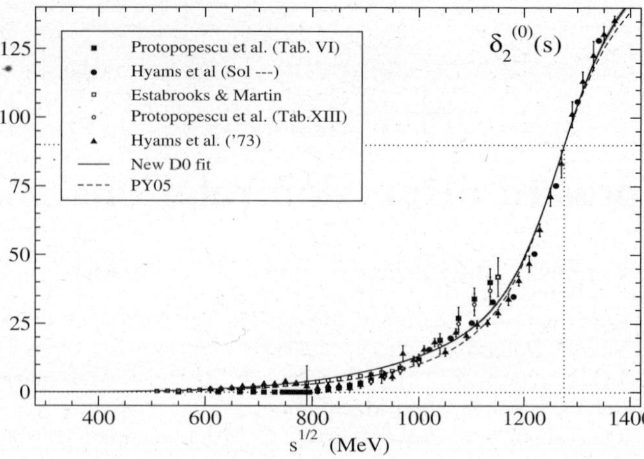

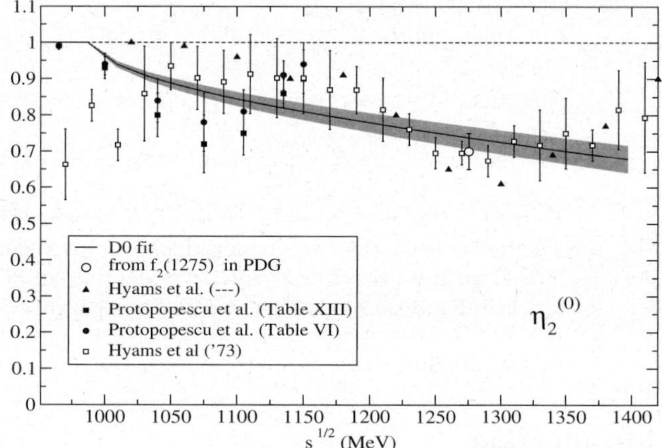

Fig. 1. Phase shifts and inelasticities of the $S0$-wave fitted using the **K**-matrix approach of [2] (solid lines). The dotted lines represent the results of [1].

Fig. 2. The $D0$-wave phase shifts and inelasticities determined in [2] (solid lines) and in [1] (dotted line —only for phase shifts). Dark areas denote the errors, which for the phase shifts have just the thickness of the line.

contrary to FDRs, incorporate s-t crossing, by calculating the difference between the real parts of the input amplitudes and those obtained from Roy's equations. We have found that, on average, and up to almost the KK threshold, the deviation from zero is smaller that 1.05 times the value of the errors for the $S0$-wave, smaller that 1.2 for the $S2$-wave and smaller than 0.65 for the P-wave.

2 S, P and D waves at higher energies but below 1.42 GeV

In this section the main features of the new parameterizations of S, P and D waves between roughly the $K\bar{K}$ threshold and 1.42 GeV are presented. Details of each parametrization can be found in [2]. Since the description of the $S2$-wave was not changed in [2], any information on this wave is available in [1].

2.1 The S0-wave

In the present approach we obtain both the tangent of the phase shifts $\tan \delta_0^0$ and the inelasticity η_0^0 above 0.932 GeV as functions of **K**-matrix elements

$$K_{ij}(s) = \frac{\mu \alpha_i \alpha_j}{M_1^2 - s} + \frac{\mu \beta_i \beta_j}{M_2^2 - s} + \frac{1}{\mu}\gamma_{ij}, \qquad (1)$$

where $i, j = 1, 2$ denote π or K, respectively, and we set the mass scale $\mu = 1\,\text{GeV}$. All α_i, β_i and γ_i are determined from the fit. Note that $M_1 = 0.9105 \pm 0.0070\,\text{GeV}$ simulates the left-hand cut of the **K**-matrix located at $2\sqrt{M_K^2 - m_\pi^2} = 0.952\,\text{GeV}$ and the pole at $M_2 = 1.324 \pm 0.006\,\text{GeV}$ is connected with δ_0^0 passing through 270°. The parametrizations of [1] (below 0.932 GeV) and of [2] (above 0.932 GeV) are matched at 0.932 MeV. In the fit, all data on phase shifts below and above the $K\bar{K}$ threshold [3–6] have been used simultaneously. For η_0^0, data from $\pi\pi \to K\bar{K}$ have been used together with data on $\pi\pi \to \pi\pi$ [3–7]. The resulting fit yields $\chi^2/\text{d.o.f.} = 0.6$ and can be seen in fig. 1.

2.2 The D0-wave

For this wave we proceeded by fitting simultaneously below and above the $K\bar{K}$ a parametrization

$$\cot \delta_2^{(0)} = \frac{s^{1/2}}{2k^5}(M_{f_2}^2 - s)m_\pi^2 (B_0 + B_1 w(s)) \qquad (2)$$

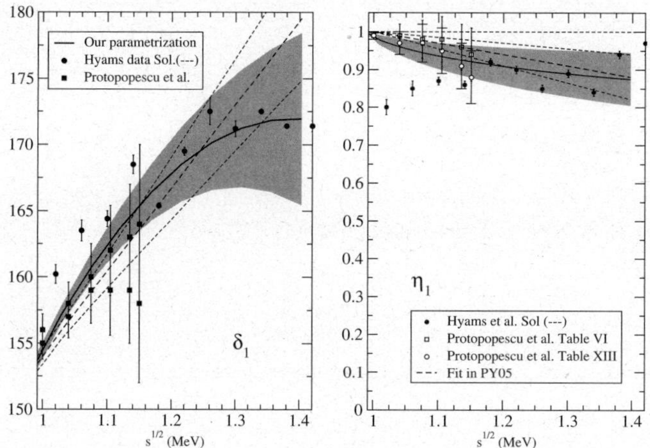

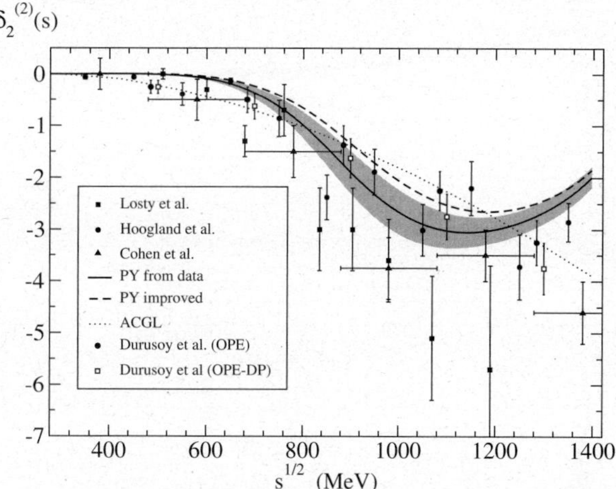

Fig. 3. Fits to the P-wave phase shifts and inelasticity (solid lines). Dark areas show the errors of our results. The dotted lines represent results obtained in [1].

Fig. 4. Results of the fit to the $D2$-wave. The solid line denotes the fit to the experimental data only and the dashed one the fit to the data and FDR [1]. For the data enclosed in the figure see references in [2].

with $w(s) = \dfrac{\sqrt{s}-\sqrt{s_0-s}}{\sqrt{s}+\sqrt{s_0-s}}$, but using different B_i and s_0 parameters above and below KK threshold. We also required both parametrizations to match at $\sqrt{s} = 2m_K$, thus eliminating one parameter. In the present approach, the mass of the $f_2(1270)$-resonance M_{f_2} was fixed to the PDG value [9]. The B_i parameters have been obtained for those two energy regions from fits to experimental data points [3,6,8] together with three other constraints: the width of the $f_2(1270)$-resonance from [9], plus the scattering length and the slope parameter calculated from the Froissart-Gribov representation. The resulting $\chi^2/\mathrm{d.o.f.} = 0.65$.

The inelasticity is parametrized in the same way as in [1] and fitted to the experimental data of refs. [3,6,8]

$$\eta_2^{(0)}(s) = 1 - \epsilon\,\frac{k_2(s)}{k_2(M_{f_2}^2)}. \tag{3}$$

Results of the fits for phase shifts and inelasticities are presented in fig. 2.

2.3 The P-wave

In the P-wave, above the $K\bar{K}$ threshold, we have used a more flexible parametrization than in [1]:

$$\delta_1(s) = \lambda_0 + \sum_{i=1}^{2} \lambda_i \left(\sqrt{s/4m_K^2} - 1\right)^i, \tag{4}$$

$$\eta_1(s) = 1 - \sum_{i=1}^{2} \epsilon_i \left(\sqrt{1 - 4m_K^2/s}\right)^i, \tag{5}$$

where λ_0 is fixed by the phase shift at $2m_K$ which is obtained from the fit to the pion form factor [11]. We have then fitted data from [6,8] obtaining $\chi^2/\mathrm{d.o.f.} = 0.6$ and $\chi^2/\mathrm{d.o.f.} = 1.1$ for the phase shifts and inelasticity, respectively. The results are presented in fig. 3.

2.4 The D2-wave

In the $D2$-wave we have used one single parametrization up to $1.42\,\mathrm{GeV}$ with four free parameters

$$\cot\delta_2^{(2)} = \frac{s^{1/2}}{2k^5}\left(B_0 + B_1 w(s) + B_1 w(s)^2\right)\frac{m_\pi^4 s}{4(m_\pi^2 + \Delta^2) - s}, \tag{6}$$

where Δ fixes zero of the phase shift near the $\pi\pi$ threshold. Since the data on this wave are not accurate we have added one more constrain using the scattering length calculated from the Froissart-Gribov representation [1]. As a result we have obtained the fit presented in fig. 4.

The lack of experimental data on inelasticity led us to estimate it from a model (see ref. [2]) writing

$$\eta_2^{(2)} = 1 - \epsilon(1 - \hat{s}/s)^3, \tag{7}$$

with $\sqrt{\hat{s}} = 1.05\,\mathrm{GeV}$ and $\epsilon = 0.2 \pm 0.2$. The inelasticity is very small and even negligible below $1.25\,\mathrm{GeV}$.

3 Regge parametrization

In the analysis of [1,10] the fits were made with the assumption of "exact degeneracy" of the intercept parameters $\alpha_\rho = \alpha_{P'}$ for ρ and f_2 exchange. In our new approach this degeneracy has been lifted. As a consequence, there was a very small change in the high-energy behaviour of scattering amplitudes (especially a little for higher energies) but, as can be seen in next section, even such a small change could be significant given the level of precision achieved in our FDR calculation. The energy dependence of the new scattering amplitudes after eliminating the degeneracy is seen in fig. 5.

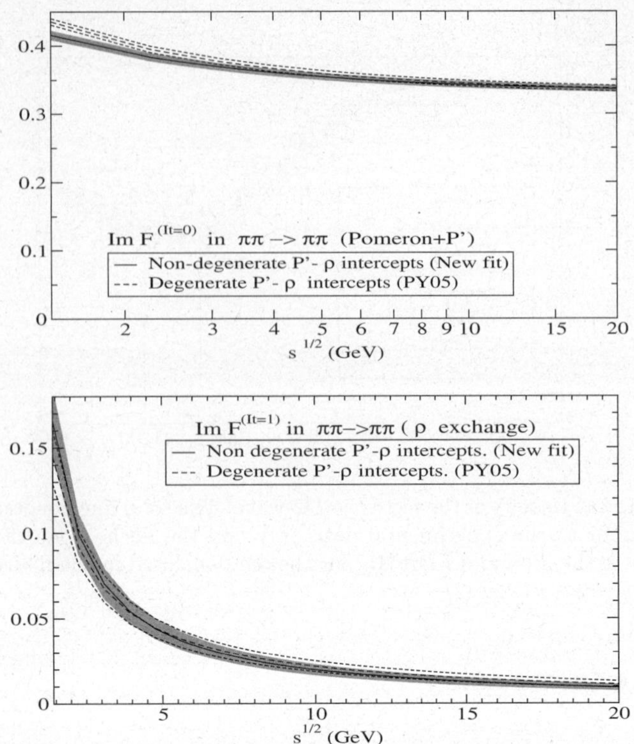

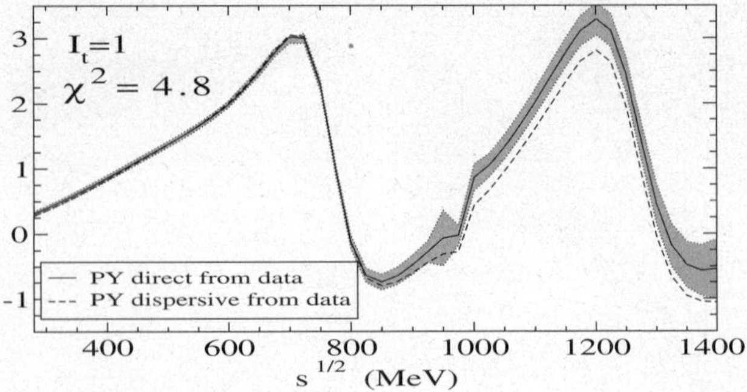

Fig. 5. The scattering amplitudes calculated with "exact degeneracy" (broken lines) and without this condition (solid lines). Dark bands stand for uncertainties.

4 Implementation of forward dispersion relations

The S, P and D waves presented in sect. 2 together with the improved Regge description in the previous section have been examined in the same way as in [1], by checking the FDRs, but without imposing them as constraints. Thus, in fig. 6 we present the results from the amplitudes in [1] obtained from fits to data. In contrast, in fig. 7 we show results using the improved fits given in [2], that we are reviewing here before checking that they also satisfy the Roy equations. The F_{00}, F_{0+} and $I_t = 1$ names used in figs. 6 and 7 correspond to the FDRs for the $\pi^0\pi^0$, $\pi^0\pi^+$ and t-channel isospin-1 scattering amplitudes, whose full mathematical expressions can be found in [1] and [2]. The word "dispersive" denotes results obtained from the integrals in the FDRs whereas "direct" means the real parts evaluated directly from parametrizations.

We provide in table 1 the FDRs averaged χ^2 obtained over the range from threshold up to 930 MeV or 1420 MeV. Note that the modifications in the S and P waves above $K\bar{K}$ threshold, as well as of the D-wave, lead to a significant improvement of the accuracy in the FDRs for the $\pi^0\pi^0$ scattering amplitudes when compared with the previous results in [1]. The final decrease of the χ^2 for $\pi^0\pi^+$ is also due to the influence of the new Regge amplitude. Note that in the $I_t = 1$ case there is a tiny deterioration despite a significant χ^2 decrease due to the Regge part.

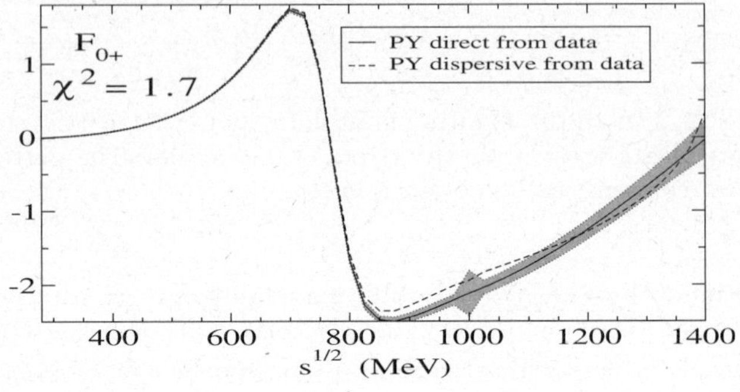

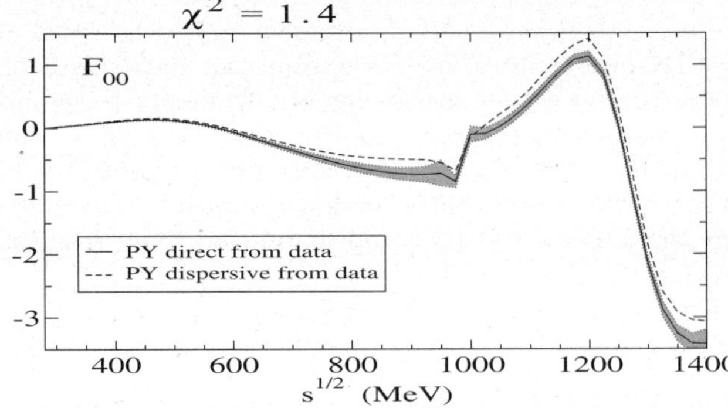

Fig. 6. The $\pi^0\pi^0$, $\pi^0\pi^+$ and t-channel 1 forward dispersion relations described in previous analysis [1]. The values of the χ^2 denote averaged values over all 25 points chosen in the energy range from the $\pi\pi$ threshold to 1.42 GeV.

Table 1. Comparison of averaged χ^2 for different FDRs obtained in a previous analysis [1] and in the presented one (new δ, η and new Regge) in two energy ranges. Numbers correspond to fits to experimental data only (without constraints from FDRs).

Results of [1]	New δ, η	New Regge	Energy range
For $\pi^0\pi^0$ dispersion relations			
3.8	1.52	1.41	$s^{1/2} < 930\,\mathrm{MeV}$
4.8	1.76	1.63	$s^{1/2} < 1420\,\mathrm{MeV}$
For $\pi^0\pi^+$ dispersion relations			
1.7	1.75	1.60	$s^{1/2} < 930\,\mathrm{MeV}$
1.7	1.60	1.44	$s^{1/2} < 1420\,\mathrm{MeV}$
For $I_t = 1$ scattering amplitudes			
0.2	0.57	0.32	$s^{1/2} < 930\,\mathrm{MeV}$
1.4	2.32	1.76	$s^{1/2} < 1420\,\mathrm{MeV}$

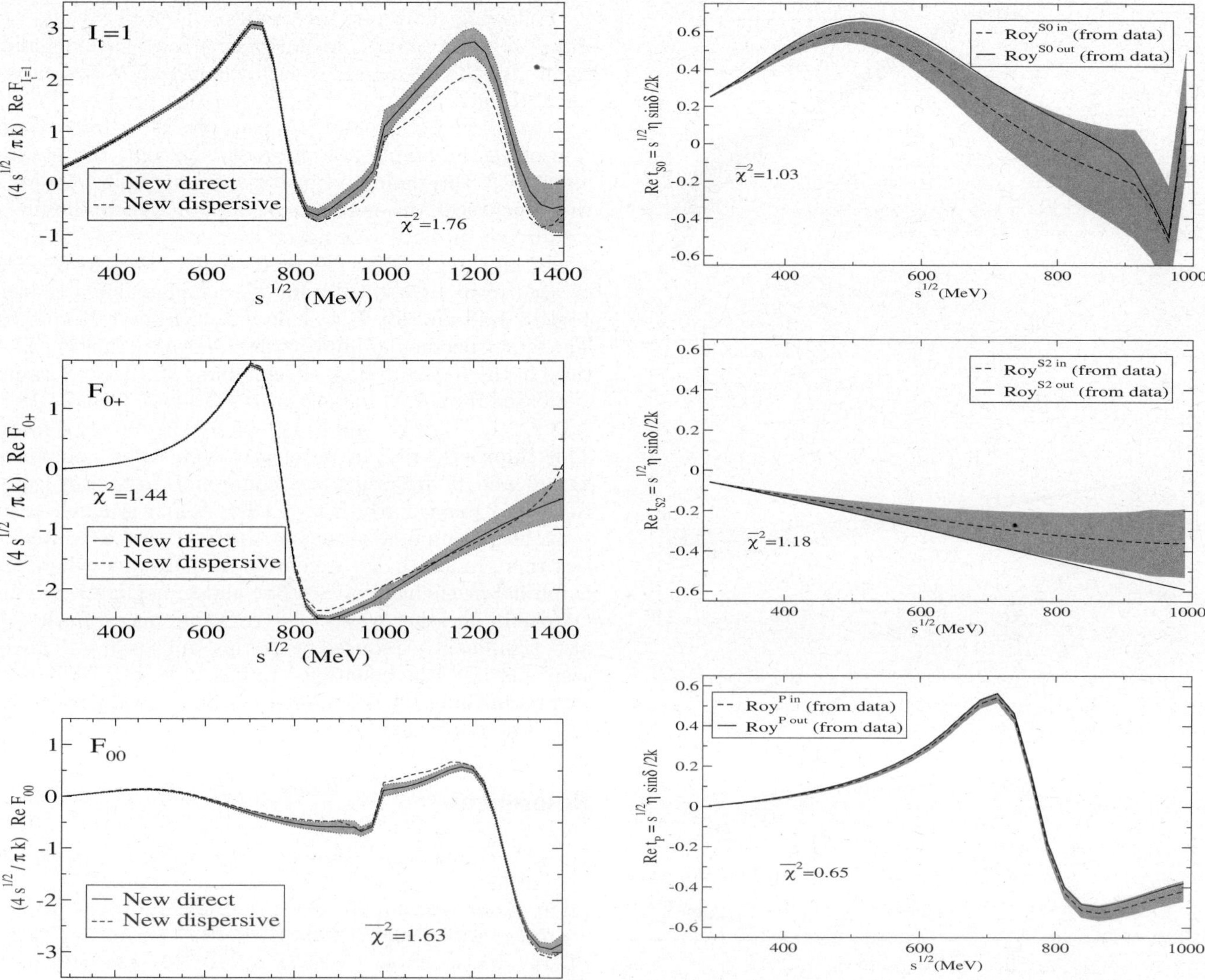

Fig. 7. Dispersion relations for new S, P and D waves described in this paper. The χ^2 definition as in fig. 6.

Fig. 8. Differences between real parts of amplitudes calculated directly from amplitudes and those from the integral representation of Roy's equations. The notation is explained in the sect. 5.

5 Tests using Roy's equations

We here present an advance of our ongoing analysis where we test our new $\pi\pi$ scattering amplitudes using Roy's equations [12–14]:

$$\mathrm{Re}\, f_\ell^I(s) = a_0^0 \delta_{I0}\delta_{\ell 0} + a_0^2 \delta_{I2}\delta_{\ell 0}$$

$$+(2a_0^0 - 5a_0^2)\left(\delta_{I0}\delta_{\ell 0} + \frac{1}{6}\delta_{I1}\delta_{\ell 1} - \frac{1}{2}\,\delta_{I2}\delta_{\ell 0}\right)\frac{s-4\mu^2}{12\mu^2}$$

$$+\sum_{I'=0}^{2}\sum_{\ell'=0}^{1}\int_{4\mu^2}^{s_{max}} \mathrm{d}s'\, K_{\ell\ell'}^{II'}(s,s')\,\mathrm{Im}\, f_{\ell'}^{I'}(s')$$

$$+\mathrm{d}_\ell^I(s,s_{max}), \tag{8}$$

where

$$f_\ell^I(s) = \sqrt{\frac{s}{s-4\mu^2}}\frac{1}{2i}\left(\eta_\ell^I e^{2i\delta_\ell^I} - 1\right), \tag{9}$$

with a_0^0 and a_0^2 being the $S0$ and $S2$ scattering lengths, $K_{\ell\ell'}^{II'}(s,s')$ known kernels and $d_\ell^I(s,s_{max})$ the so-called driving terms. In our calculations we have chosen $s_{max} = 103m_\pi^2$.

In fig. 8 we show the real part of the $S0$, $S2$ and P partial waves obtained from eq. (8) (continuous line, called Roy^{out}) *versus* the real part obtained directly from our parametrizations (dashed line, called Roy^{in}).

The agreement is remarkable, taking into account the uncertainties (the dark areas in fig. 8). Furthermore, the agreement is even more impressive, taking into account that we have not imposed any constraints from FDRs or Roy's equations themselves and that the amplitudes come just from fits to data (that is why they are labeled "from data" in the figure). Moreover, we use the new S, P and D waves described in sect. 2 and the Regge model with different intercepts $\alpha_\rho(0)$ and $\alpha_{P'}(0)$, and all of them have experimental errors even smaller than those of [1].

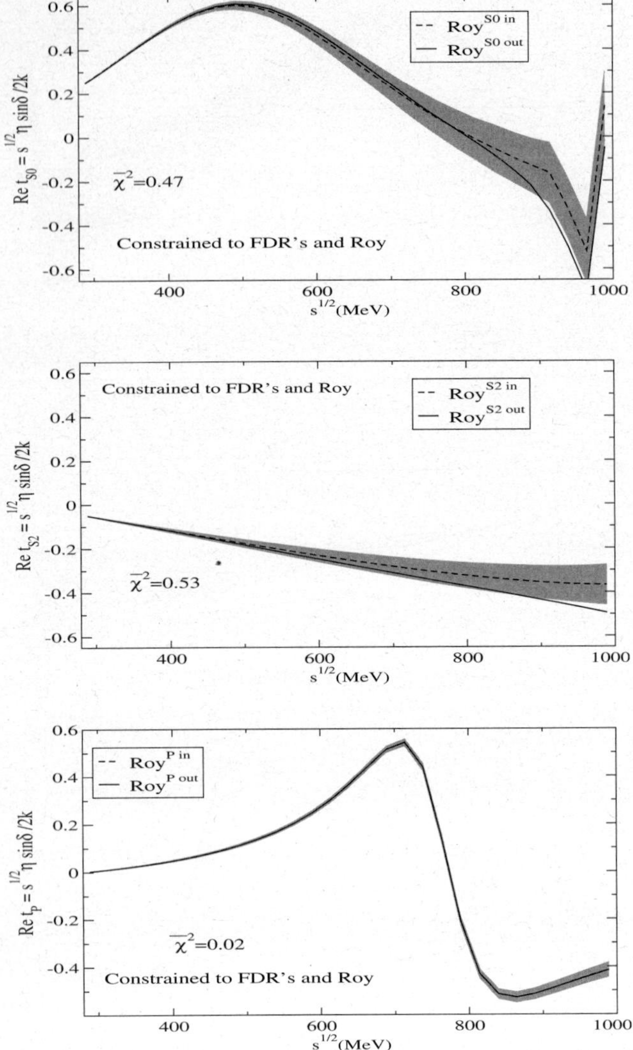

Fig. 9. As in fig. 8 but for data fits constrained by FDRs and Roy's equations.

6 Conclusions

The results reviewed here indicate that the improvement in the fits to data in the S and P waves above $K\bar{K}$ threshold and the D-wave described in sect. 2 together with a slight improvement in Regge trajectories, allowing for non $\rho - f$ degeneracy, also improves the fulfillment of forward dispersion relations. Despite the smaller errors of those amplitudes, the averaged χ^2 is indeed lower than in previous analysis [1].

But also here, as the main novelty, we have shown that those amplitudes fulfill also quite well Roy's equations, and therefore, crossing symmetry, up to roughly 1 GeV.

Following, however, the analysis done in [1] one can think about a wider implementation of FDR including them into the fits together with the already fitted experimental data. We report briefly on our progress in this approach, where, for the moment, we allow for a variation of all the amplitude parameters except, the P-wave above $K\bar{K}$ threshold and just the α and β parameters in the **K**-matrix. Although our results are just preliminary, we already noticed significant decreases of the averaged χ^2 for all three FDRs. The preliminary values for F_{00} decreased from 1.63 to 0.42, for F_{0+} changed slightly from 1.44 to 1.48 and for $I_t = 1$ decreased from 1.76 to 0.89. The more spectacular improvement, however, has been noticed in the Roy equations. Preliminary averaged χ^2 values decreased from 1.03 to 0.47 for the $S0$-wave, from 1.18 to 0.53 for the $S2$-wave and from 0.65 to 0.02 for the P-wave. This improvement can be clearly seen when comparing figs. 8 and 9. Imposing the constrains from Roy equations and particularly from FDRs, which is much stringent, leads to modifications in the S, P and D waves by less than 1σ (with the exception for $D2$-wave where the empirical fit changed by $\sim 1.3\sigma$) and to negligible modifications in all other waves. The resulting uncertainties are also significantly reduced with this approach as can be seen, just for Roy equations, in figs. 8 and 9. At present we are finishing the determination of the final parameters and their uncertainties.

References

1. J.R. Peláez, F.J. Ynduráin, Phys. Rev. D **71**, 074016 (2005).
2. R. Kamiński, J.R. Peláez, F.J. Ynduráin, Phys. Rev. D **74**, 014001 (2006); 079903 (2006)(E).
3. B. Hyams *et al.*, Nucl. Phys. B **64**, 134 (1973); P. Estabrooks, A.D. Martin, Nucl. Phys. B **79**, 301 (1974).
4. G. Grayer *et al.*, Nucl. Phys. B **75**, 189 (1974).
5. R. Kamiński, L. Leśniak, K. Rybicki, Z. Phys. C **74**, 79 (1997); EPJdirect C **4**, 4 (2002).
6. B. Hyams *et al.*, Nucl. Phys. B **100**, 205 (1975).
7. W. Wetzel *et al.*, Nucl. Phys. B **115**, 208 (1976); D. Cohen *et al.*, Phys. Rev. D **22**, 2595 (1980); E. Etkin *et al.*, Phys. Rev. D **25**, 1786 (1982).
8. S.D. Protopopescu *et al.*, Phys. Rev. D **7**, 1279 (1973).
9. S. Eidelman *et al.*, Phys. Lett. B **592**, 1 (2004).
10. J.R. Pelaez, F.J. Yndurain, Phys. Rev. D **69**, 114001 (2004).
11. J.F. de Trocóniz, F.J. Ynduráin, Phys. Rev. D **65**, 093001 (2002); **71**, 073008 (2005).
12. S.M. Roy, Phys. Lett. B **36**, 353 (1971).
13. B. Ananthanarayan, G. Colangelo, J. Gasser, H. Leutwyler, Phys. Rep. **353**, 207 (2001).
14. R. Kamiński, L. Leśniak, B. Loiseau, Phys. Lett. B **551**, 241 (2003).

Strangeness conservation and pair correlations of neutral kaons with low relative momenta produced in inclusive multiparticle processes

V.L. Lyuboshitz and V.V. Lyuboshitz[a]

Joint Institute for Nuclear Research, 141980 Dubna, Moscow Region, Russia

Abstract. The phenomenological structure of inclusive cross-sections of the production of two neutral K-mesons in hadron-hadron, hadron-nucleus and nucleus-nucleus collisions is investigated taking into account the strangeness conservation in strong and electromagnetic interactions. The relations describing the dependence of the correlations of two short-lived and two long-lived neutral kaons $K_S^0 K_S^0$, $K_L^0 K_L^0$ and the correlations of "mixed" pairs $K_S^0 K_L^0$ at small relative momenta upon the space-time parameters of the generation region of K^0 and $\bar{K}^0$- mesons, which involve the contributions of Bose-statistics and S-wave strong final-state interaction, have been obtained. It is shown that under the strangeness conservation the correlation functions of the pairs $K_S^0 K_S^0$ and $K_L^0 K_L^0$, produced in the same inclusive process, coincide, and the difference between the correlation functions of the pairs $K_S^0 K_S^0$ and $K_S^0 K_L^0$ is conditioned by the production of the pairs of non-identical neutral kaons $K^0 \bar{K}^0$.

PACS. 11.30.Hv $-$ 13.85.Ni $-$ 14.40.Aq $-$ 25.75.Gz

1 Consequences of the strangeness conservation

In the work [1] the properties of the density matrix of two neutral K-mesons, following from the strangeness conservation in strong and electromagnetic interactions, have been investigated. By definition, the diagonal elements of the non-normalized two-particle density matrix coincide with the two-particle structure functions, which are proportional to the double inclusive cross-sections.

Strangeness is the additive quantum number. Taking into account the strangeness conservation, the pairs of neutral kaons $K^0 K^0$ (strangeness $S = +2$), $\bar{K}^0 \bar{K}^0$ (strangeness $S = -2$) and $K^0 \bar{K}^0$ (strangeness $S = 0$) are produced incoherently. This means that in the K^0-$\bar{K}^0$- representation the non-diagonal elements of the density matrix between the states $K^0 K^0$ and $\bar{K}^0 \bar{K}^0$, $K^0 K^0$ and $K^0 \bar{K}^0$, $\bar{K}^0 \bar{K}^0$ and $K^0 \bar{K}^0$ are equal to zero. However, the non-diagonal elements of the two-kaon density matrix between the two states $|K^0\rangle^{(\mathbf{P}_1)}|\bar{K}^0\rangle^{(\mathbf{P}_2)}$ and $|\bar{K}^0\rangle^{(\mathbf{P}_1)}|K^0\rangle^{(\mathbf{P}_2)}$ with the zero strangeness are not equal to zero, in general. Here $\mathbf{p}_1$ and $\mathbf{p}_2$ are the momenta of the first and second kaons.

The internal states of K^0 -meson ($S = 1$) and $\bar{K}^0$ -meson ($S = -1$) are the superpositions of the states $|K_S^0\rangle$ and $|K_L^0\rangle$, where K_S^0 is the short-lived neutral kaon and

K_L^0 is the long-lived one. Neglecting the small effect of CP non-invariance, the CP-parity of the state K_S^0 is equal to $(+1)$, and the CP-parity of the state K_L^0 is equal to (-1); in doing so,

$$|K^0\rangle = \frac{1}{\sqrt{2}}\Big(|K_S^0\rangle + |K_L^0\rangle\Big), \quad |\bar{K}^0\rangle = \frac{1}{\sqrt{2}}\Big(|K_S^0\rangle - |K_L^0\rangle\Big).$$

It is clear that both the quasistationary states of the neutral kaon have no definite strangeness.

It is easy to show that

$$|K^0\rangle^{(\mathbf{P}_1)} \otimes |K^0\rangle^{(\mathbf{P}_2)} =$$

$$= \frac{1}{2}\Big(|K_S^0\rangle^{(\mathbf{P}_1)} \otimes |K_S^0\rangle^{(\mathbf{P}_2)} + |K_L^0\rangle^{(\mathbf{P}_1)} \otimes |K_L^0\rangle^{(\mathbf{P}_2)} +$$

$$+ |K_S^0\rangle^{(\mathbf{P}_1)} \otimes |K_L^0\rangle^{(\mathbf{P}_2)} + |K_L^0\rangle^{(\mathbf{P}_1)} \otimes |K_S^0\rangle^{(\mathbf{P}_2)}|\Big), \quad (1)$$

$$|\bar{K}^0\rangle^{(\mathbf{P}_1)} \otimes |\bar{K}^0\rangle^{(\mathbf{P}_2)} =$$

$$= \frac{1}{2}\Big(|K_S^0\rangle^{(\mathbf{P}_1)} \otimes |K_S^0\rangle^{(\mathbf{P}_2)} + |K_L^0\rangle^{(\mathbf{P}_1)} \otimes |K_L^0\rangle^{(\mathbf{P}_2)} -$$

$$- |K_S^0\rangle^{(\mathbf{P}_1)} \otimes |K_L^0\rangle^{(\mathbf{P}_2)} - |K_L^0\rangle^{(\mathbf{P}_1)} \otimes |K_S^0\rangle^{(\mathbf{P}_2)}\Big). \quad (2)$$

It follows from the Bose-symmetry of the wave function of two neutral kaons with respect to the total permutation of internal states and momenta that the CP-parity of the system $K^0 \bar{K}^0$ is always positive [2] (the C-parity is $(-1)^L$,

[a] e-mail: `Valery.Lyuboshitz@jinr.ru`

the space parity is $P = (-1)^L$, where L is the orbital momentum).

The system of two non-identical neutral kaons $K^0\bar{K}^0$ in the symmetric internal state, corresponding to even orbital momenta, is decomposed into the schemes $|K_S^0\rangle|K_S^0\rangle$ and $|K_L^0\rangle|K_L^0\rangle$ [2]:

$$|\psi^+\rangle = \frac{1}{\sqrt{2}}\left(|K^0\rangle^{(\mathbf{p_1})} \otimes |\bar{K}^0\rangle^{(\mathbf{p_2})} + |\bar{K}^0\rangle^{(\mathbf{p_1})} \otimes |K^0\rangle^{(\mathbf{p_2})}\right) =$$

$$= \frac{1}{\sqrt{2}}\left(|K_S^0\rangle^{(\mathbf{p_1})} \otimes |K_S^0\rangle^{(\mathbf{p_2})} - |K_L^0\rangle^{(\mathbf{p_1})} \otimes |K_L^0\rangle^{(\mathbf{p_2})}\right); \quad (3)$$

meantime, the system $K^0\bar{K}^0$ in the antisymmetric internal state, corresponding to odd orbital momenta, is decomposed into the scheme $|K_S^0\rangle|K_L^0\rangle$ [2]:

$$|\psi^-\rangle = \frac{1}{\sqrt{2}}\left(|K^0\rangle^{(\mathbf{p_1})} \otimes |\bar{K}^0\rangle^{(\mathbf{p_2})} - |\bar{K}^0\rangle^{(\mathbf{p_1})} \otimes |K^0\rangle^{(\mathbf{p_2})}\right) =$$

$$= \frac{1}{\sqrt{2}}\left(|K_S^0\rangle^{(\mathbf{p_1})} \otimes |K_L^0\rangle^{(\mathbf{p_2})} - |K_L^0\rangle^{(\mathbf{p_1})} \otimes |K_S^0\rangle^{(\mathbf{p_2})}\right). \quad (4)$$

The strangeness conservation leads to the fact that all the double inclusive cross-sections of production of pairs $K_S^0 K_S^0$, $K_L^0 K_L^0$ and $K_S^0 K_L^0$ (two-particle structure functions) prove to be symmetric with respect to the permutation of momenta $\mathbf{p}_1$ and $\mathbf{p}_2$:

$$f_{SS}(\mathbf{p}_1, \mathbf{p}_2) = f_{SS}(\mathbf{p}_2, \mathbf{p}_1); \quad f_{LL}(\mathbf{p}_1, \mathbf{p}_2) = f_{LL}(\mathbf{p}_2, \mathbf{p}_1);$$

$$f_{SL}(\mathbf{p}_1, \mathbf{p}_2) = f_{SL}(\mathbf{p}_2, \mathbf{p}_1). \quad (5)$$

Besides, due to the strangeness conservation, the structure functions of neutral K-mesons produced in inclusive processes are invariant with respect to the replacement of the short-lived state K_S^0 by the long-lived state K_L^0, and *vice versa* [1]:

$$f_{SS}(\mathbf{p}_1, \mathbf{p}_2) = f_{LL}(\mathbf{p}_1, \mathbf{p}_2) =$$

$$= \frac{1}{4}\left[f_{K^0 K^0}(\mathbf{p}_1, \mathbf{p}_2) + f_{\bar{K}^0 \bar{K}^0}(\mathbf{p}_1, \mathbf{p}_2) + f_{K^0 \bar{K}^0}(\mathbf{p}_1, \mathbf{p}_2) +\right.$$

$$\left. + f_{\bar{K}^0 K^0}(\mathbf{p}_1, \mathbf{p}_2)\right] + \frac{1}{2}\operatorname{Re}\rho_{K^0\bar{K}^0 \to \bar{K}^0 K^0}(\mathbf{p}_1, \mathbf{p}_2), \quad (6)$$

$$f_{SL}(\mathbf{p}_1, \mathbf{p}_2) = f_{LS}(\mathbf{p}_1, \mathbf{p}_2) =$$

$$= \frac{1}{4}\left[f_{K^0 K^0}(\mathbf{p}_1, \mathbf{p}_2) + f_{\bar{K}^0 \bar{K}^0}(\mathbf{p}_1, \mathbf{p}_2) + f_{K^0 \bar{K}^0}(\mathbf{p}_1, \mathbf{p}_2) +\right.$$

$$\left. + f_{\bar{K}^0 K^0}(\mathbf{p}_1, \mathbf{p}_2)\right] - \frac{1}{2}\operatorname{Re}\rho_{K^0\bar{K}^0 \to \bar{K}^0 K^0}(\mathbf{p}_1, \mathbf{p}_2), \quad (7)$$

where $\rho_{K^0\bar{K}^0 \to \bar{K}^0 K^0}(\mathbf{p}_1, \mathbf{p}_2) = (\rho_{\bar{K}^0 K^0 \to K^0 \bar{K}^0}(\mathbf{p}_1, \mathbf{p}_2))^*$ are the non-diagonal elements of the two-kaon density matrix. The difference between the two-particle structure functions f_{SS} and f_{SL} is connected just with the contribution of these non-diagonal elements.

It is evident that the one-particle structure functions for the production of K_S^0 and K_L^0 are equal to each other. After integrating the relations (6) over the momentum distribution of neutral kaons one can obtain the mutual equality of the average multiplicities of the K_S^0 and K_L^0-states, as well as the mutual equality of the average squares of multiplicities:

$$\langle n_S \rangle = \langle n_L \rangle, \qquad \langle n_S^2 \rangle = \langle n_L^2 \rangle. \quad (8)$$

2 Structure of pair correlations of identical and non-identical neutral kaons with close momenta

Now let us consider, within the model of one-particle sources [2-7], the correlations of pairs of neutral K-mesons with close momenta (see also [8]). In the case of the identical states $K_S^0 K_S^0$ and $K_L^0 K_L^0$ we obtain the following expressions for the correlation functions R_{SS}, R_{LL} (proportional to the structure functions), normalized to unity at large relative momenta:

$$R_{SS}(\mathbf{k}) = R_{LL}(\mathbf{k}) = \lambda_{K^0 K^0}\left[1 + F_{K^0}(2\mathbf{k}) + 2\,b_{\mathrm{int}}(\mathbf{k})\right] +$$

$$+ \lambda_{\bar{K}^0 \bar{K}^0}\left[1 + F_{\bar{K}^0}(2\mathbf{k}) + 2\,\tilde{b}_{\mathrm{int}}(\mathbf{k})\right] +$$

$$+ \lambda_{K^0 \bar{K}^0}\left[1 + F_{K^0 \bar{K}^0}(2\mathbf{k}) + 2\,B_{\mathrm{int}}(\mathbf{k})\right]. \quad (9)$$

Here $\mathbf{k}$ is the momentum of one of the kaons in the c.m. frame of the pair, and the quantities $\lambda_{K^0 K^0}$, $\lambda_{\bar{K}^0 \bar{K}^0}$ and $\lambda_{K^0 \bar{K}^0}$ are the relative fractions of the average numbers of produced pairs $K^0 K^0$, $\bar{K}^0 \bar{K}^0$ and $K^0 \bar{K}^0$, respectively ($\lambda_{K^0 K^0} + \lambda_{\bar{K}^0 \bar{K}^0} + \lambda_{K^0 \bar{K}^0} = 1$) . The "formfactors" $F_{K^0}(2\mathbf{k})$, $F_{\bar{K}^0}(2\mathbf{k})$ and $F_{K^0 \bar{K}^0}(2\mathbf{k})$ appear due to the contribution of Bose-statistics:

$$F_{K^0}(2\mathbf{k}) = \int W_{K^0}(\mathbf{r})\,\cos(2\mathbf{kr})\,d^3\mathbf{r},$$

$$F_{\bar{K}^0}(2\mathbf{k}) = \int W_{\bar{K}^0}(\mathbf{r})\,\cos(2\mathbf{kr})\,d^3\mathbf{r},$$

$$F_{K^0 \bar{K}^0}(2\mathbf{k}) = \int W_{K^0 \bar{K}^0}(\mathbf{r})\,\cos(2\mathbf{kr})\,d^3\mathbf{r}, \quad (10)$$

where $W_{K^0}(\mathbf{r})$, $W_{\bar{K}^0}(\mathbf{r})$ and $W_{K^0 \bar{K}^0}(\mathbf{r})$ are the probability distributions of distances between the sources of emission of two K^0-mesons, between the sources of emission of two $\bar{K}^0$-mesons and between the sources of emission of the K^0-meson and $\bar{K}^0$-meson, respectively, in the c.m. frame of the kaon pair. Meantime, the quantity $b_{\mathrm{int}}(\mathbf{k})$ describes the contribution of the S-wave interaction of two K^0-mesons, the quantity $\tilde{b}_{\mathrm{int}}(\mathbf{k})$ describes the contribution of the S-wave interaction of two $\bar{K}^0$-mesons and the quantity $B_{\mathrm{int}}(\mathbf{k})$ describes the contribution of the S-wave interaction of the K^0-meson with the $\bar{K}^0$-meson. Due to the CP-invariance, the quantities $b_{\mathrm{int}}(\mathbf{k})$ and $\tilde{b}_{\mathrm{int}}(\mathbf{k})$ can be expressed by means of averaging the same function $b(\mathbf{k}, \mathbf{r})$ over the different distributions:

$$b_{\mathrm{int}}(\mathbf{k}) = \int W_{K^0}(\mathbf{r})b(\mathbf{k}, \mathbf{r})d^3\mathbf{r},$$

$$\tilde{b}_{\mathrm{int}}(\mathbf{k}) = \int W_{\bar{K}^0}(\mathbf{r})b(\mathbf{k}, \mathbf{r})d^3\mathbf{r}.$$

The quantity $B_{\mathrm{int}}(\mathbf{k})$ has the structure

$$B_{\mathrm{int}}(\mathbf{k}) = \int W_{K^0 \bar{K}^0}(\mathbf{r})B(\mathbf{k}, \mathbf{r})d^3\mathbf{r},$$

where $B(\mathbf{k}, \mathbf{r}) \neq b(\mathbf{k}, \mathbf{r})$.

Let us emphasize that when the pair of non-identical neutral kaons $K^0 \bar{K}^0$ is produced but the pair of identical quasistationary states $K_S^0 K_S^0$ (or $K_L^0 K_L^0$) is registered over decays, the two-particle correlations at small relative momenta have the same character as in the case of usual identical bosons with zero spin [2].

For the pairs of non-identical kaon states $K_S^0 K_L^0$ the correlation functions at small relative momenta have the form:

$$R_{SL}(\mathbf{k}) = R_{LS}(\mathbf{k}) = \lambda_{K^0 K^0}\left[1 + F_{K^0}(2\mathbf{k}) + 2\,b_{\text{int}}(\mathbf{k})\right] +$$

$$+ \lambda_{\bar{K}^0 \bar{K}^0}\left[1 + F_{\bar{K}^0}(2\mathbf{k}) + 2\,\tilde{b}_{\text{int}}(\mathbf{k})\right] +$$

$$+ \lambda_{K^0 \bar{K}^0}\left[1 - F_{K^0 \bar{K}^0}(2\mathbf{k})\right]. \tag{11}$$

In accordance with Eq.(11), at the production of the pair of non-identical neutral kaons $K^0 \bar{K}^0$ and the registration of the two-particle state $K_S^0 K_L^0$ over decays the pair correlations are analogous to the correlations of two identical fermions with the same spin projections. This is connected with the fact that in the considered case the pair $K_S^0 K_L^0$ has odd orbital momenta [2].

It follows from Eqs.(9) and (11) that the correlation functions of pairs of neutral K-mesons with close momenta, which are created in inclusive processes, satisfy the relation

$$R_{SS}(\mathbf{k}) + R_{LL}(\mathbf{k}) - R_{SL}(\mathbf{k}) - R_{LS}(\mathbf{k}) =$$

$$= 2\left[R_{SS}(\mathbf{k}) - R_{SL}(\mathbf{k})\right] =$$

$$= 4\lambda_{K^0 \bar{K}^0}\left[F_{K^0 \bar{K}^0}(2\mathbf{k}) + B_{\text{int}}(\mathbf{k})\right]. \tag{12}$$

We see that the difference between the correlation functions of the pairs of identical neutral kaons $K_S^0 K_S^0$ and pairs of non-identical neutral kaons $K_S^0 K_L^0$ is conditioned exclusively by the generation of $K^0 \bar{K}^0$-pairs.

The relations connecting the contribution of the S-wave strong interaction into the pair correlations of particles at small relative momenta with the parameters of low-energy scattering were obtained earlier in the papers [4-7]. It is essential that the "formfactors" (10) and the functions $b_{\text{int}}(\mathbf{k})$, $\tilde{b}_{\text{int}}(\mathbf{k})$ and $B_{\text{int}}(\mathbf{k})$ depend on the space-time parameters of the generation region of neutral kaons and tend to zero at high values of the relative momentum $q = 2|\mathbf{k}|$ of two neutral kaons. [1]

---

[1] Let us note that, in principle, the P-wave resonance (ϕ-meson, $M = 1021$ MeV/c^2, $\Gamma = 4$ MeV) may influence the $K_S^0 K_L^0$ - correlations. But this influence manifests itself only in the narrow region of comparatively large relative momenta $2k \approx 220$ MeV/c, and it is strongly suppressed at small relative momenta which are discussed here.

3 Contribution of the S-wave $K^0\bar{K}^0$-interaction

The function $B(\mathbf{k}, \mathbf{r})$, describing the contribution of the final-state interaction between K^0 and $\bar{K}^0$ mesons into the $K_S^0 K_S^0$- correlations and into the difference of the correlation functions $(R_{SS}(\mathbf{k}) - R_{SL}(\mathbf{k}))$, may be calculated analytically using the approximation of the superposition of the plane and spherical waves, if characteristic distances r_0 between sources of K^0- and $\bar{K}^0$-mesons are: $r_0 \gg d_0$, where d_0 is the radius of action of short-range forces between the K^0-meson and $\bar{K}^0$- meson (really, already at $r_0 > d_0$) [4]. In so doing

$$B(\mathbf{k}, \mathbf{r}) = |A_{K^0 \bar{K}^0}(k)|^2 \frac{1}{r^2} +$$

$$+ 2\,\text{Re}\left(A_{K_0 \bar{K}_0}(k)\frac{\exp(ikr)\cos \mathbf{kr}}{r}\right), \tag{13}$$

where $A_{K^0 \bar{K}^0}(k) \equiv A_{K^0 \bar{K}^0 \to K^0 \bar{K}^0}(k)$ is the amplitude of the S-wave $K^0 \bar{K}^0$-scattering, $k = |\mathbf{k}|$, $r = |\mathbf{r}|$.

Now let us take into account the effect of the possible transition $K^+ K^- \to K^0 \bar{K}^0$ between the pairs of oppositely charged and neutral kaons on the pair correlations of two identical kaons with small relative momenta. We can use here the theory of the S-wave multichannel scattering [7]. Assuming that pairs $K^+ K^-$ and $K^0 \bar{K}^0$ are emitted with equal probabilities by the same pairs of isotopically unpolarized sources, the function $B(\mathbf{k}, \mathbf{r})$ should be replaced by $\widetilde{B}(\mathbf{k}, \mathbf{r}) = B(\mathbf{k}, \mathbf{r}) + \Delta B(\mathbf{k}, \mathbf{r})$, where [7]

$$\Delta B(\mathbf{k}, \mathbf{r}) =$$

$$= \left|A_{K^+ K^- \to K^0 \bar{K}^0}^{(c)}(k)\right|^2 \frac{\cos^2 \tilde{k}r + C(\tilde{k}a_c)\sin^2 \tilde{k}r}{r^2}. \tag{14}$$

Here $\tilde{k} = \sqrt{k^2 + 2M_K \Delta M_K}$ and k are the moduli of the momentum of each of the charged kaons and of each of the neutral kaons, respectively, in the c.m. frame of the pair $K^0 \bar{K}^0$,
$M_K = (M_{K^0} + M_{K^+})/2 \approx 495.6$ MeV/c^2,
$\Delta M_K = M_{K^0} - M_{K^+} \approx 4$ MeV/c^2,
$a_c = 2\hbar^2/(M_{K^+}e^2) \approx 108.5$ Fm is the Bohr radius of the $K^+ K^-$-system,

$$C(\tilde{k}a_c) = \frac{2\pi/\tilde{k}a_c}{1 - \exp(-2\pi/\tilde{k}a_c)}$$

is the Coulomb factor corresponding to the attraction of the oppositely charged kaons, $A_{K^+ K^- \to K^0 \bar{K}^0}^{(c)}(k)$ is the effective amplitude of the reaction $K^+ K^- \to K^0 \bar{K}^0$, renormalized by the Coulomb interaction. [2]

---

[2] The S-wave amplitude, determining directly the effective cross-section of the process $K^+ K^- \to K^0 \bar{K}^0$ according to the standard formula

$$\sigma_{K^+ K^- \to K^0 \bar{K}^0}(k) = 4\pi \left|A_{K^+ K^- \to K^0 \bar{K}^0}(k)\right|^2 \frac{k}{\tilde{k}},$$

is $A_{K^+ K^- \to K^0 \bar{K}^0}(k) = \sqrt{C(\tilde{k}a_c)}\,A_{K^+ K^- \to K^0 \bar{K}^0}^{(c)}(k)$.

At $k = 0$ the modulus of the momentum of the K^+ (K^-)-meson in the c.m. frame of the pair $K^0\bar{K}^0$ is equal to $\tilde{k}_0 = \sqrt{2M_K\Delta M_K} \approx 62.8$ MeV/c. As a result, the Coulomb factor incorporated in Eq.(14) is close to unity:

$$1 < C(\tilde{k}a_c) \leq C(\tilde{k}_0 a_c) \approx 1.0934.$$

Thus, with the precision of the order of 10% we obtain

$$\Delta B(\mathbf{k}, \mathbf{r}) = |A_{K^+K^-\to K^0\bar{K}^0}(k)|^2 \frac{1}{r^2}. \tag{15}$$

It is known that the amplitudes $A_{K^0\bar{K}^0\to K^0\bar{K}^0}(k)$ and $A_{K^+K^-\to K^0\bar{K}^0}(k)$ are determined by the contribution of the sub-threshold S-wave resonances $f_0(980)$ (isotopic spin $T = 0$) and $a_0(980)$ (isotopic spin $T = 1$) [4,7]:

$$A_{K^0\bar{K}^0\to K^0\bar{K}^0}(k) = \frac{1}{2}\left[A^{(T=0)}(k) + A^{(T=1)}(k)\right],$$

$$A_{K^+K^-\to K^0\bar{K}^0}(k) = \frac{1}{2}\left[A^{(T=0)}(k) - A^{(T=1)}(k)\right]. \tag{16}$$

According to Eqs. (13), (15) and (16), we come to the following approximate relation:

$$\tilde{B}(\mathbf{k}, \mathbf{r}) = B(\mathbf{k}, \mathbf{r}) + \Delta B(\mathbf{k}, \mathbf{r}) =$$

$$= \left[|A^{(T=0)}(k)|^2 + |A^{(T=1)}(k)|^2\right]\frac{1}{2r^2} +$$

$$+ \mathrm{Re}\left(\left[A^{(T=0)}(k) + A^{(T=1)}(k)\right]\frac{\exp(ikr)\cos(\mathbf{kr})}{r}\right). \tag{17}$$

Recently the first statistically meaningful experimental results on the Bose-correlations of two K_S^0-mesons, produced in collisions of relativistic heavy ions, have been presented by the international STAR Collaboration [9].

4 Summary

1. It is shown that, taking into account the strangeness conservation, the double inclusive cross-sections of the production of two short-lived neutral K-mesons and two long-lived neutral K-mesons are equal to each other. This result is the direct consequence of the strangeness conservation.

2. Within the model of one-particle sources the formulas for the correlation functions $R_{SS} = R_{LL}$ and $R_{SL} = R_{LS}$ are obtained, which involve the contributions of Bose-statistics, of the S-wave final-state interaction of two K^0 ($\bar{K}^0$)-mesons as well as of a K^0-meson with a $\bar{K}^0$-meson, and also the contribution of transitions $K^+K^- \to K^0\bar{K}^0$, and depend upon the relative fractions of produced pairs K^0K^0, $\bar{K}^0\bar{K}^0$ and $K^0\bar{K}^0$.

3. It is shown that the production of $K^0\bar{K}^0$-pairs with the zero strangeness leads to the difference between the correlation functions R_{SS} and R_{SL} of two neutral kaons.

This work is supported by Russian Foundation of Basic Research (Grant No. 05-02-16674).

References

1. V.L.Lyuboshitz. Yad. Fiz. **23** (1976) 1266 [Sov. J. Nucl. Phys. **23** (1976) 673].
2. V.L.Lyuboshitz and M.I.Podgoretsky. Yad. Fiz. **30** (1979) 789 [Sov. J. Nucl. Phys. **30** (1979) 407].
3. M.I.Podgoretsky. Fiz. Elem. Chast. At. Yadra **20** (1989) 628 [Sov. J. Part. Nucl. **20** (1989) 266].
4. R.Lednicky and V.L.Lyuboshitz. Yad. Fiz. **35** (1982) 1316 [Sov. J. Nucl. Phys. **35** (1982) 770].
5. V.L.Lyuboshitz. Yad. Fiz. **41** (1985) 820 [Sov. J. Nucl. Phys. **41** (1985) 523].
6. V.L.Lyuboshitz. Yad. Fiz. **48** (1988) 1501 [Sov. J. Nucl. Phys. **48** (1988) 956].
7. R.Lednicky, V.V.Lyuboshitz and V.L.Lyuboshitz. Yad. Fiz. **61** (1998) 2161 [Phys. At. Nucl. **61** (1998) 2050].
8. V.L.Lyuboshitz and V.V.Lyuboshitz, *Proceedings of XVII International Baldin Seminar on High Energy Physics Problems (ISHEPP XVII)*, JINR **E1,2-2005-103**, vol. I, Dubna, 2005, p.361.
9. B.I.Abelev et al. (STAR Collaboration), nucl-ex/0608012 (2006).

Eur. Phys. J. A **31**, 485–490 (2007)

DOI 10.1140/epja/i2006-10225-3

Special Article – QNP 2006

Photoexcitation of baryon resonances on nucleons and nuclei

An example: recent results from photoproduction of η-mesons

B. Krusche[a], I. Jaegle, and T. Mertens

For the CBELSA/TAPS Collaboration
Department of Physics and Astronomy, University of Basel, Ch-4056 Basel, Switzerland

Received: 8 November 2006
Published online: 1 March 2007 – © Società Italiana di Fisica / Springer-Verlag 2007

Abstract. Photoproduction of η-mesons off the proton, the deuteron and other light nuclei, and off heavy nuclei has been studied in detail during the last decade at the electron accelerators MAMI in Mainz, ELSA in Bonn, ERSF in Grenoble, CEBAF in Newport News, at KEK, and at Tohoku. The physics topics range from the detailed investigation of the properties of known nucleon resonances (in particular $S_{11}(1535)$), over the search of new states, to the in-medium properties of baryon resonances, and the possible formation of η-nucleus (quasi)bound states (η-mesic nuclei). It is thus an excellent example for the different aspects of the photoexcitation of nucleon resonances. Here we report new preliminary results from the CBELSA/TAPS experiment for this reaction at higher incident photon energies for the deuteron and heavy nuclei.

PACS. 13.60.Le Meson production – 25.20.Lj Photoproduction reactions

1 Introduction

The photoexcitation of baryon resonances is connected to two important aspects of non-perturbative, low-energy QCD. The excitation spectrum of the free nucleon itself is far from being understood. In particular, it is still not known if the problem of "missing resonances" is caused by experimental bias or if baryon models use inapt effective degrees of freedom. This problem is sketched in fig. 1 where the experimentally observed excitation spectrum [1] is compared to the predictions from typical quark model calculations [2]. Two problems are evident: the ordering of some of the lowest-lying states is not reproduced. In particular, the lowest-lying $N^\star$, the $P_{11}(1440)$ ("Roper") resonance and the lowest lying excited Δ, the $P_{33}(1600)$, which in the quark model belong to the $N = 2$ oscillator shell, appear below the states from the $N - 1$ shell. This is a severe problem for almost all quark models. Furthermore, many more states are predicted than have been observed, and the observed states are mostly clustered into a few narrow bands, a pattern which is not apparent for the quark model results. Most excited nucleon states have been first observed with hadron-induced reactions, in particular elastic scattering of charged pions. It is thus possible that the data base is biased for states that couple only weakly to πN. However, the large progress in accelerator and detector technology made during the last two decades, now allows to study the electromagnetic ex-

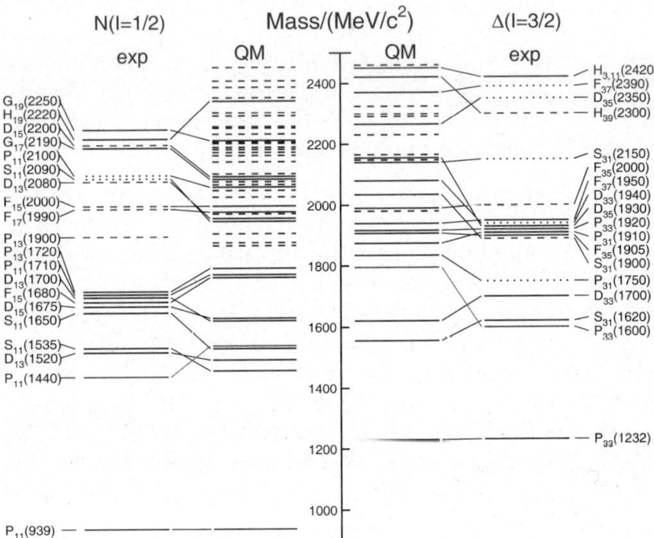

Fig. 1. Excitation scheme of the nucleon. Left-hand side: isospin $I = 1/2$ N-states, right-hand side: isospin $I = 3/2$ Δ-states. States labeled "exp": experiment (Review of Particle Properties [1]), full lines: three or four stars, dashed: two, dotted: one star. States labeled "QM": quark model [2], all states $N = 1, 2$ bands, low lying states $N = 3, 4, 5$. Full lines: (tentative) assignment to observed states, dashed lines: no observed counterparts.

---

[a] e-mail: Bernd.Krusche@unibas.ch

citation of resonances via photon-induced reactions with comparable precision, although the cross-sections are typically smaller by two orders of magnitude. These experiments have developed into two directions: measurements of photon-induced meson production reactions up to high excitation energies and for many different nucleon-meson final states with the aim to identify at least some of the "missing" states (see, *e.g.*, [3]) and precise investigations of the properties of known low-lying nucleon resonances (see, *e.g.*, [4]).

The other QCD aspect are the in-medium properties of hadrons, which are hotly discussed. Unlike any other composite systems, hadrons are objects, which are build out of constituents with masses (5–15 MeV for u, d quarks) that are negligible compared to the total mass. Most of the mass is generated by dynamical effects from the interaction of the quarks and an important role is played by the spontaneous breaking of chiral symmetry, the fundamental symmetry of QCD. The symmetry breaking, which is connected to a non-zero expectation value of scalar $q\bar{q}$ pairs in the vacuum, the chiral condensate, is reflected in the hadron spectrum. Without it, hadrons would appear as mass degenerate parity doublets, which is neither true for baryons nor for mesons. However, model calculations (see, *e.g.*, ref. [5]) indicate a temperature and density dependence of the condensate, which is connected to a partial restoration of chiral symmetry. Although there is no direct relation between the quark condensate and the in-medium properties of hadrons (masses, widths etc.), there is an indirect one via QCD sum rules, which connect the QCD picture with the hadron picture. In the latter, the in-medium modifications arise from the coupling of mesons to resonance-hole states and the coupling of the modified mesons to resonances. Recently, Post, Leupold, and Mosel [6] have calculated the hadron in-medium spectral functions for π-, η-, and ρ-mesons and baryon resonances in a self-consistent coupled channel approach.

Experimentally, one of the clearest, although still not fully explained, in-medium effects has been observed in the excitation function of the total photoabsorption reaction (TPA). This is shown in fig. 2, where the TPA off the proton [7] and the deuteron [8] are compared to the average of the nuclear cross-sections [9], all normalized to the mass numbers A. The bump in the elementary cross-sections around 700 MeV incident photon energy, which corresponds to the excitation of the $P_{11}(1440)$, $D_{13}(1520)$, and $S_{11}(1535)$ resonances, is not seen in the nuclear data. Many different effects have been discussed in the literature. They include trivial ones like nuclear Fermi motion, which certainly contributes to the broadening of the structure, but cannot explain its complete disappearance. Collisional broadening of the resonances due to additional decay channels like $NN^{\star} \to NN$ has been studied in detail in the framework of transport models of the Boltzmann-Uehling-Uhlenbeck (BUU) type (see, *e.g.*, [10]), but can also not fully explain the data. The situation is complicated since already on the free nucleon the so-called second resonance bump consists of a superposition of reaction channels with different energy dependen-

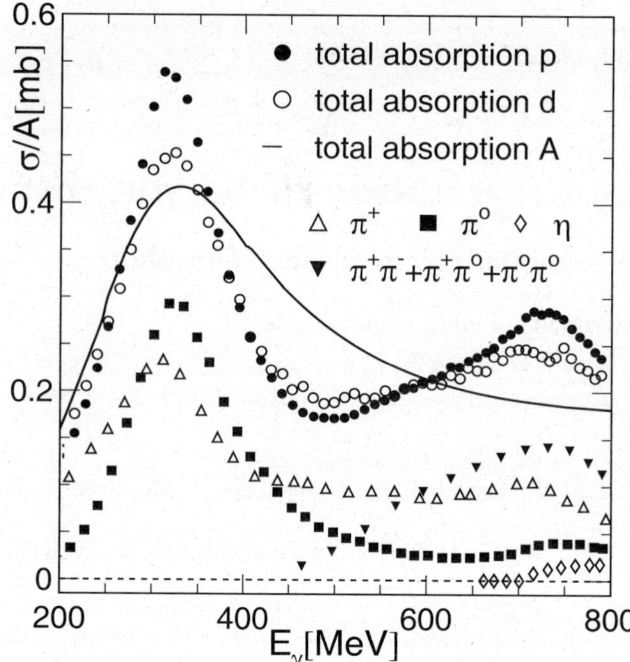

Fig. 2. Excitation functions for total photoabsorption off the proton [7], the deuteron [8] and the average for nuclei [9]. Also shown are the partial meson production cross-sections for the proton [11–18].

cies (see fig. 2). Inclusive reactions like TPA alone do not allow to study in-medium properties of individual nucleon resonances. A study of the partial reactions channels is desirable, but their experimental identification is more involved, and final-state interaction effects (FSI) [19] as well as experimental bias due to the averaging over the nuclear density [20] must be accounted for (see ref. [21] for a recent summary). Of special interest are meson production reactions which are dominated in the energy region of interest by one of the three resonances. Single- and double-pion production reactions have been employed for the study of the D_{13}-resonance in the nuclear medium [22,23], although up to now without conclusive results.

In this paper we will discuss new results for the photoproduction of η-mesons, which can address interesting aspects of both, the study off excited states of the free nucleon and the in-medium properties of nucleon resonances.

The excitation function of η photoproduction of the free proton is shown in fig. 3. From threshold throughout the second resonance region it is completely dominated by the $S_{11}(1535)$-resonance [21,24] ($N\eta$ decay branching ratio $\approx 50\%$ [1]), there is some influence of an interference between the excitation of the $S_{11}(1535)$ and $S_{11}(1650)$ resonances and only a very weak contribution from the $D_{13}(1520)$ ($N\eta$ decay branching ratio $\approx 0.23\%$ [1]). An analysis of η photo- and electroproduction data in the framework of an isobar model [25] revealed weak contributions of four further resonances up to excitation energies of 1700 MeV and a recent combined analysis of η- and π-photoproduction reports contributions of another four states between $W = 2000$–2200 MeV [26].

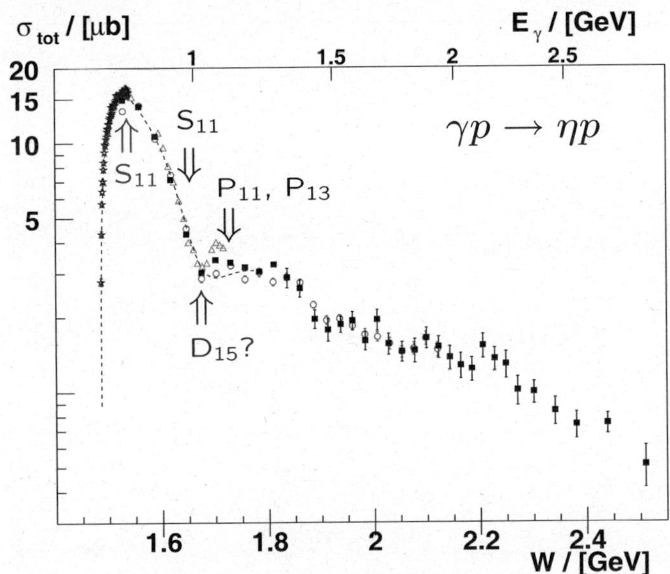

Fig. 3. Excitation function for photoproduction of η-mesons off the proton [13,27,28,26].

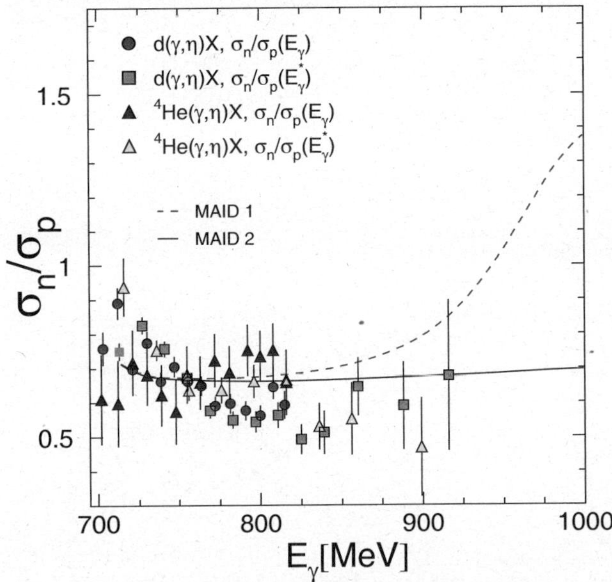

Fig. 4. Experimental results for the ratio for proton and neutron cross-section measured with deuteron [34] and ^{4}He [31] targets compared to the prediction of ref. [25] (curve labeled MAID1) and from the same model if only $S_{11}(1535)$ contributes (curve labeled MAID2).

2 Photoproduction of η-mesons off the neutron

The study of η photoproduction off the neutron aims at the investigation of the isospin structure of the electromagnetic excitation of resonances. Due to the non-availability of free neutron targets, photoproduction off light nuclei, in particular the deuteron, must be explored. The investigation of quasifree and coherent η photoproduction off ^{2}H and ^{3,4}He [29–35] has clarified the isospin structure of the $S_{11}(1535)$ electromagnetic excitation, which was found to be dominantly isovector (see [4] for a summary) and in addition provided interesting insights into the η-nucleus interaction. Typical results for the neutron/proton cross-section ratio are summarized in fig. 4. In the region of the $S_{11}(1535)$ it is almost constant around a value of 2/3. The slight rise to very low energies results from a breakdown of the participant-spectator picture at energies below the production threshold on the free nucleon. A quantitative analysis of all measurements results in a value of $|A_{1/2}^n|/|A_{1/2}^p| = 0.82\pm0.02$ for the ratio of the electomagnetic helicity couplings of the $S_{11}(1535)$ [34,23].

At energies above the $S_{11}(1535)$ models predict a significant rise of the ratio due to the contribution of higher-lying resonances (see fig. 4 [25]). In the work of Chiang *et al.* [25] ("Eta-MAID"), the largest contributions comes from the $D_{15}(1675)$-resonance, which is known to have a stronger electromagnetic coupling to the neutron than to the proton [1]. However, the model uses a decay branching ratio of this resonance to $N\eta$ of 17%, while the Particle Data Review quotes $(0\pm1)\%$ [1]. On the other hand, also in the framework of the chiral soliton model [36] a state is predicted in this energy range, with has much stronger photon couplings to the neutron than to the proton and a large decay branching ratio into $N\eta$. This state is the nucleon-like member of the predicted anti-decuplet of pen-

taquarks, which would be a P_{11} state. Therefore it would be very interesting to extend the data into the region of incident photon energies around 1 GeV.

In the following we will discuss preliminary results from a measurement of quasifree η photoproduction off neutrons and protons bound in the deuteron. The experiment was done at the Bonn ELSA accelerator with the combined Crystal Barrel [37] and TAPS [38,39] detectors. The η-mesons were detected via the $\eta \rightarrow 3\pi^0 \rightarrow 6\gamma$ decay chain. First the invariant masses of the three pions and then the overall invariant mass of the η were used for the identification. The recoil protons and neutrons were identified with charged-particle identification detectors and a time-of-flight *versus* energy analysis.

At incident photon energies above the $\eta\pi$ production threshold (930 MeV for the free nucleon, effectively lowered for the deuteron due to Fermi smearing) significant background from this channel appears. Part of it can be suppressed with the almost 4π solid angle coverage of the detector by the condition that not more than six photons and one recoil nucleon have been detected. However, there is still residual background for example from the final state $NN\eta\pi^\pm$ where the charged pion goes down the beam pipe or is misidentified as a proton, while a recoil neutron has escaped detection. This background was removed with a missing mass analysis based on the four-vectors of the six photons. For quasifree reactions off the deuteron, the missing mass peaks are only moderately broadened by Fermi smearing compared to the free nucleon case [16]. A comparison of the inclusive cross-sections for the final states $np\eta$ and $NN\eta X$ is shown in fig. 5.

The results for the excitation functions of η photoproduction off the deuteron are summarized in fig. 6. The

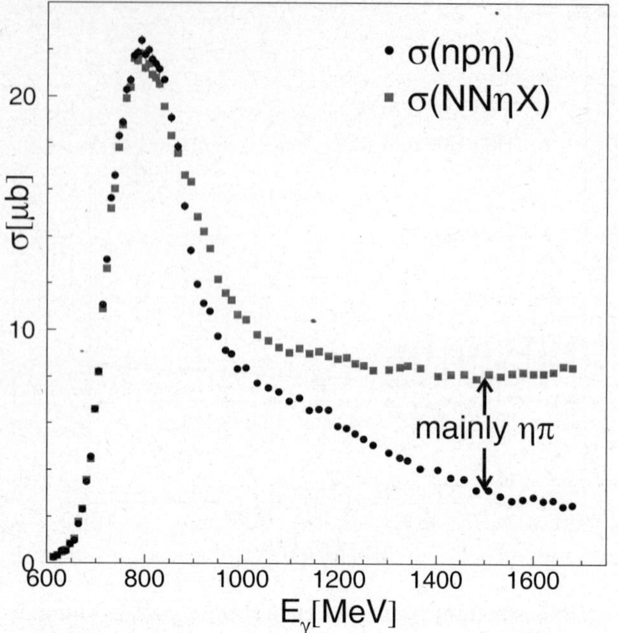

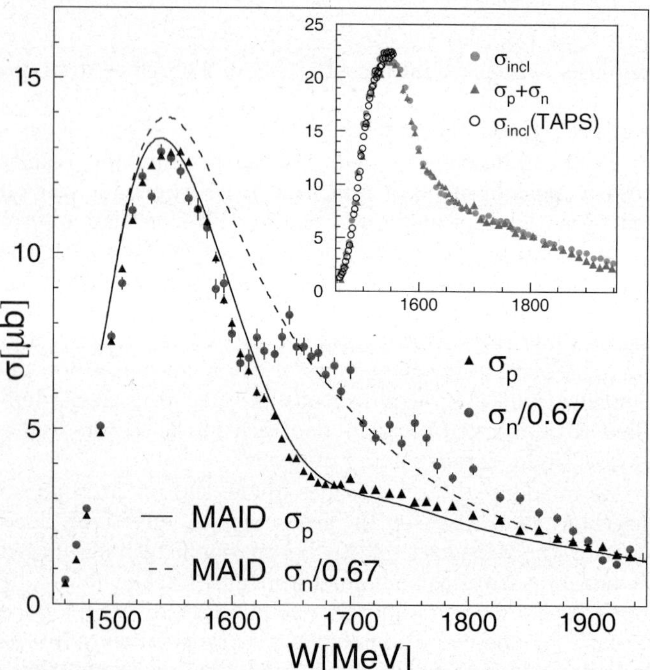

Fig. 5. Excitation functions for single η photoproduction ($\gamma d \to np\eta$) and the inclusive reaction $\gamma d \to NNX\eta$.

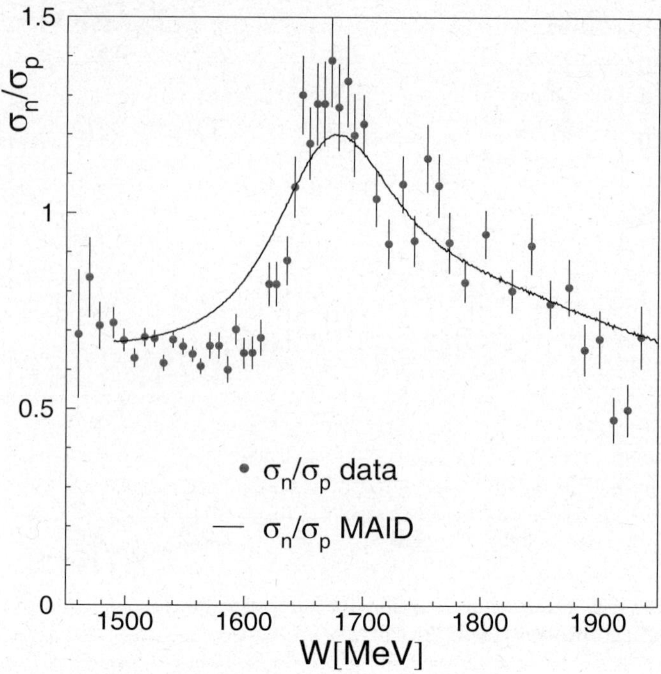

Fig. 7. Ratio of quasifree neutron and proton cross-sections compared to the prediction of the Eta-MAID model [25] (folded with momentum distribution) as a function of invariant mass W.

Fig. 6. Total cross-sections for the final states ηp (σ_p, participant proton) and ηn (σ_n, participant neutron) normalized by $3/2$ compared to the results of the MAID model [25] folded with the momentum distribution of the bound nucleons. Insert: total inclusive cross-section for the final state ηnp. Open symbols: results from ref. [34], dots: present result, triangles: sum of quasifree exclusive cross-sections $\sigma_p + \sigma_n$.

cross-section σ_{incl} corresponds to single quasifree η photoproduction off the deuteron, *i.e.* all events with an η identified by invariant mass and production of further mesons

excluded by missing mass were accepted. For the exclusive cross-sections σ_p and σ_n coincident detection of a participant recoil proton, respectively recoil neutron, was required. The absolute normalization of the cross-sections was obtained from the target thickness, the incident photon flux, the decay branching ratios ($\eta \to 3\pi^0$, $\pi^0 \to \gamma\gamma$), the detection efficiency of the η-mesons and for the exclusive cross-sections also the detection efficiencies of the recoil nucleons. The insert of fig. 6 shows that the inclusive cross-section agrees with previous results [29,34] available for incident photon energies below 800 MeV. Furthermore, the sum of the two exclusive cross-sections agrees well with the total inclusive cross-section, which is an additional test for the correctness of the efficiency correction applied for the recoil nucleon detection. The excitation functions for quasifree production off the proton and off the neutron shown in the main frame of the figure are in reasonable agreement with the predictions of the Eta-MAID model [25] folded with the nuclear momentum distribution. The excitation function off the neutron shows indeed a resonance structure around invariant masses of $W \approx 1675\,\mathrm{MeV}$, which is not seen for the proton. This is particularly clear in the ratio of the two cross-sections shown in fig. 7. It is close to 2/3 in the $S_{11}(1535)$ range, but then it peaks at an invariant mass around 1675, indicating the contribution of a resonance which couples much stronger to the neutron than to the proton. A similar structure has been reported from the GRAAL experiment [40] The analysis of the angular distributions, which may carry information about the quantum numbers of this resonance, is still under way.

3 Photoproduction of η-mesons off nuclei

Photoproduction of η-mesons in the second resonance region is an excellent tool for the study of the $S_{11}(1535)$-resonance, which completely dominates this reaction [13, 24]. A first search for possible in-medium effects on the S_{11} spectral function was done with the TAPS experiment at MAMI [41]. However, the experiment covered only incident photon energies up to 800 MeV, *i.e.* approximately up to the peak position of the resonance. The experimental results were in good agreement with BUU-model calculations (see, *e.g.*, [10]). Furthermore, an almost perfect scaling of the cross-sections with $A^{2/3}$ (A = nuclear mass number) was found, which indicates strong final state interaction effects. Subsequently, measurements at KEK [42] and Tohoku [43] extended the energy range up to 1.1 GeV. The KEK experiment reported some collisional broadening of the S_{11}-resonance, however possible background from $\eta\pi$ final states, which can fake such a result, was neglected. The Tohoku experiment pointed to a significant contribution of a higher-lying resonance to the $\gamma n \to n\eta$ reaction. However, none of these experiments covered the full line shape of the S_{11}.

Preliminary results for η photoproduction off ^{2}H and off heavier nuclei from the Crystal Barrel/TAPS experiment at the Bonn ELSA accelerator are summarized in fig. 8. The data agree with the previous results [41] at incident photon energies below 800 MeV, where they scale with $A^{2/3}$ (respectively A for the deuteron). This scaling changes at higher incident photon energies, where the cross-sections are approximately proportional to A. It is clear from the discussion in sect. 2 that at this energies $\eta\pi$ final states contribute. However, there is no good reason why this contributions, which are also subject to strong FSI, should modify the scaling. Indeed, for the low-energy

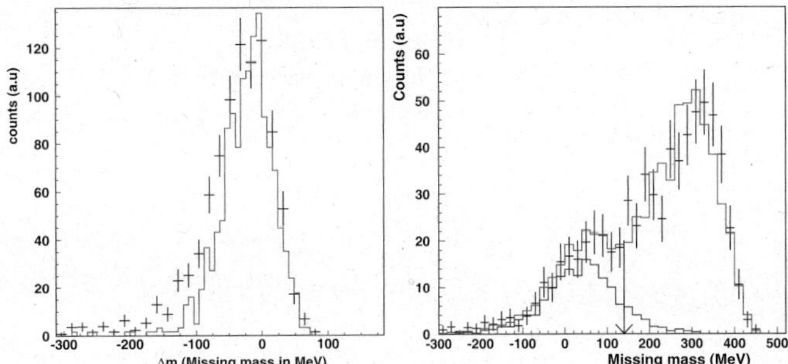

Fig. 9. Missing-mass spectra for the measurement off lead nuclei for incident photon energies around 800 MeV (left-hand side) and 1500 MeV (right-hand side). The histograms are results of a BUU calculation [44] folded with the detector resolution.

regime it was shown in [23] that due to FSI all investigated reactions (η, single π, double π) obey approximately an $A^{2/3}$ scaling law. Another possibility would be that the η mean free path increases as function of η momentum, so that FSI for η-mesons gets effectively weaker for higher incident photon energies. However, a detailed analysis of the scaling of the momentum-differential cross-sections, which will be discussed elsewhere, points to another explanation. The η mean free paths seems to be fairly constant over a wide range of η momentum. However, at higher incident photon energies secondary processes, *e.g.*, of the type $\gamma N \to N\pi$, $\pi N \to \eta N$ become more and more important and these contribution increase of course as function of the mass number.

Both effects, the contribution of $\eta\pi$ final states and the secondary η production processes, obscure the line shape of the S_{11} at photon energies above 800 MeV. A suppression of these contributions is again possible via cuts on the reaction kinematic, which are however much less selective than for the deuteron due to the stronger Fermi smearing. Typical missing mass spectra calculated under the assumption of quasifree single η photoproduction are shown in fig. 9. The missing mass ΔM is given by

$$\Delta M = \sqrt{\left(E_b + m_N - \sum_{i=1}^{6} E_{\gamma_i}\right)^2 - \left(\boldsymbol{P}_b - \sum_{i=1}^{6} \boldsymbol{P}_{\gamma_i}\right)^2} - m_N, \tag{1}$$

where E_{γ_i} and $\boldsymbol{P}_{\gamma_i}$ are energies and momenta of the decay photons and m_N is the nucleon mass. At low incident photon energies a clear peak centered around zero is visible, at higher energies a large contribution from $\eta\pi$ and secondary processes appears at positive missing energies. The distributions are in good agreement with results from the BUU model [44]. The indicated cut removes most of the unwanted contributions. Excitation functions for single, quasifree η photoproduction after the missing mass cut are shown in figs. 10, 11 for carbon and lead. They are compared to the average nuclear cross-section, obtained from the deuteron data, folded with the momentum distributions of the nucleons bound in the heavy nuclei. Already these preliminary results show, that the in-medium spec-

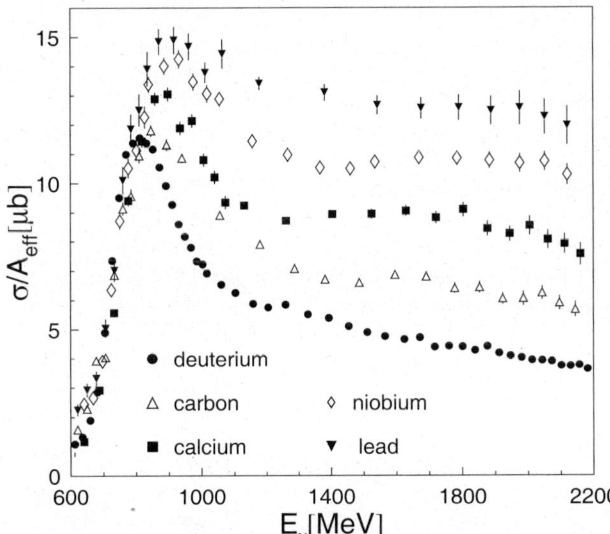

Fig. 8. Total inclusive (final state ηX, where X can also include pions) η photoproduction off the deutron and off nuclei from the present experiment, scaled by $A_{eff} = 2$ for the deuteron and $A_{eff} = A^{2/3}$ for all other nuclei.

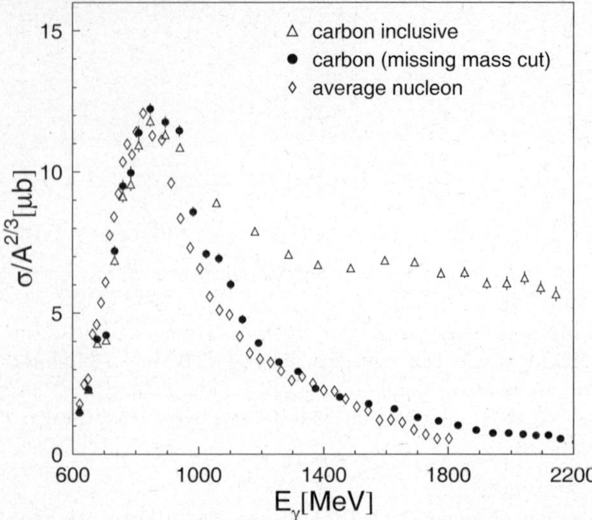

Fig. 10. Total cross-section for η production off carbon. Inclusive excitation function for final states with at least one η and after a cut on reaction kinematics for single, quasifree η production compared to the Fermi smeared average neutron cross-section.

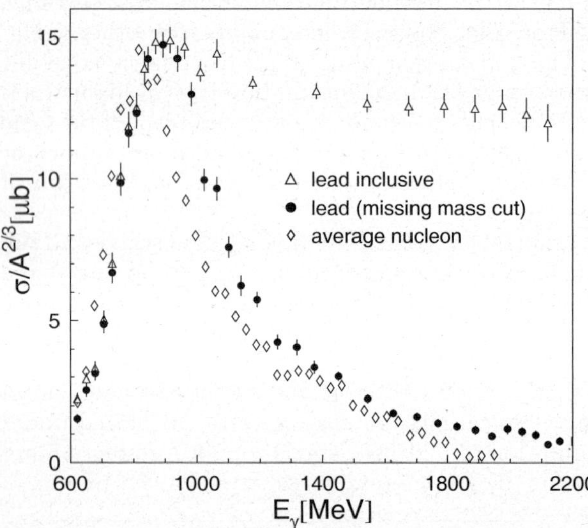

Fig. 11. Same as fig. 10 for lead nuclei.

tral function of the S_{11}-resonance is not strongly modified. In case of carbon, the data agree almost perfectly with the Fermi-smeared average nucleon data. There is no need for a significant collisional broadening, which was reported in [42] on the basis of data without kinematical cuts. The agreement for the lead data is less good, it would be consistent with some upward shift of the resonance position together with a broadening. However, in particular for lead the suppression of $\eta\pi$ final states and secondary processes by the missing mass cut is not perfect. A more detailed analysis of the data including a comparison to BUU predictions is still under way.

The experimental results have been obtained by the CB-ELSA/TAPS Collaboration and are part of the PhD theses of I. Jaegle and T. Mertens. We gratefully acknowledge many stimulating discussions with U. Mosel and P. Mühlich (BUU results), and with L. Tiator (Eta-MAID). This work was supported by Schweizerischer Nationalfonds and Deutsche Forschungsgemeinschaft.

References

1. W.-M. Yao *et al.*, J. Phys. G **33**, 1 (2006).
2. S. Capstick, W. Roberts, Phys. Rev. D **49**, 4570 (1994); **57**, 4301 (1998); **58**, 074011 (1998).
3. V.D. Burkert, T.-S. Lee, Int. J. Mod. Phys. E **13**, 1035 (2004).
4. B. Krusche, S. Schadmand, Prog. Part. Nucl. Phys. **51**, 399 (2003).
5. M. Lutz, S. Klimt, W. Weise, Nucl. Phys. A **542**, 521 (1992).
6. M. Post, S. Leupold, U. Mosel, Nucl. Phys. A **741**, 81 (2004).
7. T.A. Armstrong *et al.*, Phys. Rev. D **5**, 1640 (1972).
8. T.A. Armstrong *et al.*, Nucl. Phys. B **41**, 445 (1972).
9. N. Bianchi *et al.*, Phys. Lett. B **325**, 333 (1994).
10. J. Lehr, M. Effenberger, U. Mosel, Nucl. Phys. A **671**, 503 (2000).
11. K. Büchler *et al.*, Nucl. Phys. A **570**, 580 (1994).
12. A. Braghieri *et al.*, Phys. Lett. B **363**, 46 (1995).
13. B. Krusche *et al.*, Phys. Rev. Lett. **74**, 3736 (1995).
14. M. MacCormick *et al.*, Phys. Rev. C **53**, 41 (1996).
15. F. Härter *et al.*, Phys. Lett. B **401**, 229 (1997).
16. B. Krusche *et al.*, Eur. Phys. J. A **6**, 309 (1999).
17. M. Wolf *et al.*, Eur. Phys. J. A **9**, 5 (2000).
18. W. Langgärtner *et al.*, Phys. Rev. Lett. **87**, 052001 (2001).
19. B. Krusche *et al.*, Eur. Phys. J. A **22**, 347 (2004).
20. J. Lehr, U. Mosel, Phys. Rev. C **64**, 042202 (2001).
21. B. Krusche, Prog. Part. Nucl. Phys. **55**, 46 (2005).
22. B. Krusche *et al.*, Phys. Rev. Lett. **86**, 4764 (2001).
23. B. Krusche *et al.*, Eur. Phys. J. A **22**, 277 (2004).
24. B. Krusche *et al.*, Phys. Lett. B **397**, 171 (1997).
25. W.-T. Chiang *et al.*, Nucl. Phys. A **700**, 429 (2002).
26. V. Crede *et al.*, Phys. Rev. Lett. **94**, 012004 (2005).
27. F. Renard *et al.*, Phys. Lett. B **528**, 215 (2002).
28. M. Dugger *et al.*, Phys. Rev. Lett. **89**, 222002 (2002).
29. B. Krusche *et al.*, Phys. Lett. B **358**, 40 (1995).
30. P. Hoffmann-Rothe *et al.*, Phys. Rev. Lett. **78**, 4697 (1997).
31. V. Hejny *et al.*, Eur. Phys. J. A **6**, 83 (1999).
32. J. Weiss *et al.*, Eur. Phys. J. A **11**, 371 (2001).
33. V. Hejny *et al.*, Eur. Phys. J. A **13**, 493 (2002).
34. J. Weiss *et al.*, Eur. Phys. J. A **16**, 275 (2003).
35. M. Pfeiffer *et al.*, Phys. Rev. Lett. **92**, 252001 (2004).
36. R.A. Arndt *et al.*, Phys. Rev. C **69**, 035208 (2004).
37. E. Aker *et al.*, Nucl. Instrum. Methods A **321**, 69 (1992).
38. R. Novotny, IEEE Trans. Nucl. Sci. **38**, 379 (1991).
39. A.R. Gabler *et al.*, Nucl. Instrum. Methods A **346**, 168 (1994).
40. V. Kuznetsov for the GRAAL Collaboration, hep-ex/0409032 (2004).
41. M. Röbig-Landau *et al.*, Phys. Lett. B **373**, 45 (1996).
42. T. Yorita *et al.*, Phys. Lett. B **476**, 226 (2000).
43. T. Kinoshita *et al.*, Phys. Lett. B **639**, 429 (2006).
44. P. Mühlich, University of Giessen, private communication (2006).

Eur. Phys. J. A **31**, 491–494 (2007)
DOI 10.1140/epja/i2006-10273-7

Special Article – QNP 2006

Large-N Weinberg-Tomozawa interaction and spin-flavor symmetry

C. García-Recio[a], J. Nieves, and L.L. Salcedo

Departamento de Física Atómica, Molecular y Nuclear, Universidad de Granada, E-18071 Granada, Spain

Received: 8 December 2006
Published online: 20 March 2007 – © Società Italiana di Fisica / Springer-Verlag 2007

Abstract. The construction of an extended version of the Weinberg-Tomozawa Lagrangian, in which baryons and mesons form spin-flavor multiplets, is reviewed and some of its properties discussed, for an arbitrary number of colors and flavors. The coefficient tables of spin-flavor irreducible representations related by crossing between the s-, t- and u-channels are explicitly constructed.

PACS. 14.20.Gk Baryon resonances with $S = 0$ – 11.15.Pg Expansions for large numbers of components ($e.g.$, $1/N_c$ expansions) – 11.10.St Bound and unstable states; Bethe-Salpeter equations – 11.30.Rd Chiral symmetries

1 Introduction

Spin-flavor symmetry, the symmetry by which $SU(N_f)$ is promoted to $SU(2N_f)$, was fashionable in the early days of the quark model [1] due to its predictive power and successes in the description of hadronic properties, but was largely left aside in favor of QCD when the latter theory was developed. In addition symmetries including together Poincaré and internal groups were shown to be of necessity of approximated type [2]. As it turned out, large-N QCD (one of the few techniques available to attack QCD in the non perturbative regime) was shown to display spin-flavor symmetry for the lightest baryon of the theory [3].

The success of chiral unitarization schemes in the study of meson-baryon resonances within a Bethe-Salpeter approach [4], indicated the opportunity of considering their large-N versions. In particular, in the s-wave 0^- meson–$1/2^+$ baryon sector, appropriate to investigate negative-parity baryons, a large simplification of the formalism was obtained if the introduction of spin-flavor multiplets for baryon was complemented with the similar approach in the meson sector, thereby including vector mesons in the game [5]. This naturally leads to an extended version of the s-wave Weinberg-Tomozawa (WT) Lagrangian [6] with spin-flavor symmetry, the construction of which is reviewed below, in the first part of the paper. This interaction turns out to have a very constrained form which follows entirely from chiral symmetry considerations. Chiral symmetry consistent with spin-flavor has been advocated by Caldi and Pagels [7]. In the second part of the paper we present new and detailed results concerning crossing coefficients. They describe how interactions driven by specific $SU(2N_f)$ irreducible representations in the t- or u-channels are decomposed in the s-channel.

2 WT interaction and spin-flavor symmetry

In what follows N denotes the number of colors, N_f the number of flavors and $n = 2N_f$. In the large-N limit, the spin-flavor $SU(n)$ symmetry becomes exact for the QCD baryon [3]. The baryon is described by a field $\mathcal{B}^{i_1\cdots i_N}(x)$. The label i contains spin and flavor and runs from 1 to n. This field is a fully symmetric $SU(n)$ tensor, $i.e.$, falls in the representation with Young tableau $[N]$. (The fully antisymmetric color factor of the wave function is implicit.) For $N = N_f = 3$ this is the irrep **56** of $SU(6)$, so for general N and N_f it is sometimes denoted by "**56**". Assuming spin-flavor symmetry in the mesonic sector (not a direct consequence of large N QCD), the lightest mesons fall in the adjoint representation of $SU(n)$, "**35**" with tableau $[2, 1^{n-2}]$, with linear field $\Phi^i{}_j(x)$ and non-linear field $\mathcal{U} = \exp(2i\Phi/f)$.

The baryon and linear meson fields transform and are normalized as $Q^{i_1} \cdots Q^{i_N}$ and $Q^i\overline{Q}_j - (1/n)Q^k\overline{Q}_k\delta^i{}_j$, respectively, where Q^i and $\overline{Q}_i$ are quark and the antiquark fields,

$$Q^i \to U^i{}_j Q^j, \ \overline{Q}_i \to (U^i{}_j)^*\overline{Q}_j = \overline{Q}_j(U^{-1})^j{}_i, \ U \in SU(n). \tag{1}$$

As in the $SU(3)$ case, the WT Lagrangian is obtained from the free baryonic Lagrangian by replacing the derivative by a covariant derivative which includes the

[a] e-mail: g_recio@ugr.es

mesons

$$\mathcal{L} = \frac{1}{N!}\mathcal{B}^\dagger_{i_1\cdots i_N}\left(i\nabla_0 - M + \frac{1}{2M}\boldsymbol{\nabla}^2\right)\mathcal{B}^{i_1\cdots i_N}. \qquad (2)$$

The covariant derivative $\nabla_\mu = \partial_\mu + A_\mu$ has connection

$$A_\mu(x) = \frac{1}{2}(u^\dagger\partial_\mu u + u\partial_\mu u^\dagger)$$
$$= \frac{1}{2f^2}[\Phi,\partial_\mu\Phi] + \mathcal{O}(\Phi^3) \qquad (3)$$

(where $u^2 = \mathcal{U}$). This yields a spin-flavor extended WT interaction

$$\mathcal{L}^{\mathrm{sf}}_{\mathrm{WT}} = \frac{iN}{2f^2}[\Phi,\partial_\mu\Phi]^j{}_k\frac{1}{N!}\mathcal{B}^\dagger_{ji_2\cdots i_N}\mathcal{B}^{ki_2\cdots i_N}. \qquad (4)$$

As shown in [5] this extended Lagrangian includes that of $SU(3)$ in the case $N = N_f = 3$, for the 0^- meson–$1/2^+$ baryon sector.

Because the covariant derivative acts on each quark index in $\mathcal{B}^{i_1\cdots i_N}$, there is an extra factor of N beyond that coming from the meson weak-decay constant. This would imply a large-N dependence of $\mathcal{O}(N^0)$ for the WT amplitude, of the same order as that of the p-wave amplitude [3]. This holds for generic baryons. The situation is more subtle if only baryons with finite spin and flavor in the large-N limit are retained [5]. In this case the spin-flavor extended WT amplitude is $\mathcal{O}(N^0)$ only if vector (and/or flavored) mesons are involved, but $\mathcal{O}(N^{-1})$ in the pseudoscalar (or flavorless) meson sector. This is the standard result. Indeed, the commutator of the absorbed and emitted meson fields in $\mathcal{L}^{\mathrm{sf}}_{\mathrm{WT}}$ just produces a spin-flavor generator to be measured by the baryonic matrix element. If the mesons carry only flavor but not spin, the generator will also be purely flavor-like, and flavor was finite by assumption. This gives a power of N less than for a generic baryon.

Note the group structure [5]

$$\mathcal{L}^{\mathrm{sf}}_{\mathrm{WT}} \sim ((M^\dagger \otimes M)_{\text{``}\mathbf{35}_A\text{''}} \otimes (B^\dagger \otimes B)_{\text{``}\mathbf{35}\text{''}})_{\mathbf{1}} \qquad (5)$$

(M, $M^\dagger$, B and $B^\dagger$ representing annihilation and creation operators for mesons and baryons) which indicates a purely antisymmetric adjoint coupling in the t-channel. As we will see below (cf. (7)), in the s-channel (or u-channel) there are four possible $SU(n)$ irreps, and so four independent couplings for a generic $SU(n)$ invariant meson-baryon Hamiltonian[1]. In the t-channel one has instead

$$\text{``}\mathbf{35}\text{''} \otimes \text{``}\mathbf{35}\text{''} = \mathbf{1} \oplus \text{``}\mathbf{35}_S\text{''} \oplus \text{``}\mathbf{35}_A\text{''} \oplus \text{``}\mathbf{405}\text{''} \oplus \cdots,$$
$$\text{``}\mathbf{56}\text{''} \otimes \text{``}\mathbf{56}^*\text{''} = \mathbf{1} \oplus \text{``}\mathbf{35}\text{''} \oplus \text{``}\mathbf{405}\text{''} \oplus \cdots, \qquad (6)$$

where all the irreps not explicited differ from mesons to baryons. Thus there are again four independent coupling constants in the t-channels, as it should be. "$\mathbf{405}$" is the representation with tableau $[4,2^{n-2}]$. The detailed group structure of WT comes as a consequence of the $SU(n)$ chiral-symmetry requirement [7].

Phenomenological consequences of the extended WT interaction in the negative-parity baryon sector are analyzed in [5].

---

[1] This is the generic case. Some irreps may not exits for particular low values of n or N, see (8).

3 Crossed projectors

The coupling of baryons and mesons gives four $SU(n)$ irreps

$$\text{``}\mathbf{56}\text{''} \otimes \text{``}\mathbf{35}\text{''} = \text{``}\mathbf{56}\text{''} \oplus \text{``}\mathbf{70}\text{''} \oplus \text{``}\mathbf{700}\text{''} \oplus \text{``}\mathbf{1134}\text{''} \qquad (7)$$

with tableaus $[N]$, $[N-1,1]$, $[N+2,1^{n-2}]$ and $[N+1,2,1^{n-3}]$, respectively. In what follows they will be called simply irreps 1, 2, 3, and 4. They have dimensions

$$d_1 = \binom{n+N-1}{N}, \qquad d_2 = \frac{(N-1)(n-1)}{n+N-1}d_1,$$
$$d_3 = \frac{(n+N-1)(n-1)}{N+1}d_1,$$
$$d_4 = \frac{nN(n+N)(n-2)}{(n+N-1)(N+1)}d_1. \qquad (8)$$

For convenience we introduce a vector notation so that $\boldsymbol{d} = [d_1,d_2,d_3,d_4]$.

An $SU(n)$ invariant Hamiltonian takes therefore the form (again using a vector/tensor notation in irrep space)

$$H = \sum_{\alpha=1}^{4} h_\alpha P_\alpha = \boldsymbol{h}\cdot\boldsymbol{P}, \qquad (9)$$

where P_α denotes the projector of meson-baryon states on the irrep α. For the extended WT interaction, a direct calculation using Wick contractions [5] gives

$$\boldsymbol{h}_{\mathrm{WT}} = -\boldsymbol{\lambda}_{\mathrm{WT}}(\sqrt{s}-M)/f^2,$$
$$\boldsymbol{\lambda}_{\mathrm{WT}} = [n, N+n, -N, 1]. \qquad (10)$$

The WT amplitude is therefore $\mathcal{O}(N^0)$ for the irreps 2 and 3 and $\mathcal{O}(N^{-1})$ for 1 and 4. It is noteworthy the relation

$$\boldsymbol{\lambda}_{\mathrm{WT}}\cdot\boldsymbol{d} = 0, \qquad (11)$$

which embodies the property $\mathrm{tr}(H_{\mathrm{WT}}) = 0$, in turn coming from the vanishing of the trace of the meson operator living in the adjoint representation (obtained from the commutator of the meson fields).

As said the WT amplitude is purely "$\mathbf{35}$" in the t-channel and $\boldsymbol{\lambda}_{\mathrm{WT}}$ reflects this group structure as seen in the s-channel. In what follows we investigate how the various irrep in the t- and u-channels are seen in the s-channel. The u-channel corresponds to crossing of the meson legs. For instance a direct baryon-pole (s-channel) mechanism would give a pure "$\mathbf{56}$" coupling[2], i.e., $\boldsymbol{h}_{\mathrm{dbp}} \propto [1,0,0,0]$, while a crossed baryon-pole (u-channel) mechanism

$$((M \otimes B^\dagger)_{\text{``}\mathbf{56}\text{''}_*} \otimes (M^\dagger \otimes B)_{\text{``}\mathbf{56}\text{''}})_{\mathbf{1}} \qquad (12)$$

gives instead (see Appendix D of [5a])]

$$\boldsymbol{\lambda}_{\mathrm{cbp}} = \left[N+n-1-\frac{N}{n}-\frac{n}{N}, -1-\frac{n}{N}, 1, -\frac{1}{N}\right] \qquad (13)$$

---

[2] We assume here an hypothetical $SU(n)$ invariant meson-baryon-baryon coupling. The physical coupling is p-wave, hence the orbital momentum would have to be included.

Table 1. Matrix a.

$$a = \begin{bmatrix} \dfrac{n(n+N)(N-1)-N^2}{N(n-1)(n+N)} & -\dfrac{n}{N(n-1)} & \dfrac{n}{(n-1)(n+N)} & -\dfrac{n}{N(n-1)(n+N)} \\[2ex] -\dfrac{n(N-1)}{N(n+N-1)} & \dfrac{n}{N(n+N-1)} & \dfrac{N-1}{(n+N-1)} & -\dfrac{N-1}{N(n+N-1)} \\[2ex] \dfrac{n(n+N+1)}{(N+1)(n+N)} & \dfrac{n+N+1}{(N+1)} & \dfrac{n}{(n+N)(N+1)} & \dfrac{n+N+1}{(n+N)(N+1)} \\[2ex] -\dfrac{n^2(n-2)}{(n-1)(N+1)(n+N-1)} & -\dfrac{n(n-2)(n+N)}{(n-1)(N+1)(n+N-1)} & \dfrac{nN(n-2)}{(n-1)(N+1)(n+N-1)} & \dfrac{N(n-1)(n+N)+1}{(n-1)(N+1)(n+N-1)} \end{bmatrix}$$

(the normalization is arbitrary). While the effect of crossing for other irreps could be computed as in [5], by using Wick contractions, we present here a shortcut. To this end we introduce the crossed projectors

$$\langle m_1, b_1 | P_\alpha^c | m_2, b_2 \rangle = \langle \overline{m_2}, b_1 | P_\alpha | \overline{m_1}, b_2 \rangle, \tag{14}$$

where $|\overline{m}\rangle = C|m\rangle$ indicates the crossed meson state.

Since the P_α^c themselves are invariant operators, we have

$$P_\alpha^c = \sum_\beta a_{\alpha\beta} P_\beta, \quad \text{or} \quad \boldsymbol{P}^c = a \cdot \boldsymbol{P}, \tag{15}$$

where the (real) crossing coefficients $a_{\alpha\beta}$ are those to be determined. They describe the u-channel irreps as seen in the s-channel.

For the group $SU(2)$ the (antilinear) operator C is just $|j,m\rangle = (-1)^{j-m}|j,-m\rangle$, however, in the general case, the only property needed of C is to be a unitary conjugation, *i.e.*, $CC^\dagger = C^2 = 1$. Together with general projector properties

$$P_\alpha = P_\alpha^\dagger, \qquad P_\alpha P_\beta = \delta_{\alpha\beta} P_\alpha,$$
$$\operatorname{tr} P_\alpha = d_\alpha, \qquad \sum_\alpha P_\alpha = 1, \tag{16}$$

one easily establishes the relations

$$\sum_\gamma a_{\alpha\gamma} a_{\gamma\beta} = \delta_{\alpha\beta}, \qquad \sum_\gamma a_{\alpha\gamma} a_{\beta\gamma} d_\gamma = d_\alpha \delta_{\alpha\beta},$$
$$\sum_\alpha a_{\alpha\beta} = 1, \qquad \sum_\beta a_{\alpha\beta} d_\beta = d_\alpha. \tag{17}$$

The first two relations imply that the matrix

$$\hat{a}_{\alpha\beta} = d_\alpha^{-1/2} a_{\alpha\beta} d_\beta^{1/2} \tag{18}$$

satisfies $\hat{a} = \hat{a}^T = \hat{a}^{-1}$, *i.e.*, it has four orthonormal eigenvectors $\hat{e}_\lambda$ with eigenvalues ± 1. As a consequence $\hat{a}$ can be written as $P_{+1} - P_{-1} = 1 - 2P_{-1}$ ($P_{\pm 1}$ being the orthogonal projectors on the ± 1 subspaces of the four dimensional irrep space) and a is just $\hat{a}$ expressed in a non-orthonormal basis. The third and fourth relations in (17) identify $[1,1,1,1]$ as one such left eigenvector of a, with eigenvalue $+1$, and d as its right eigenvector version.

The left eigenvector $\boldsymbol{\lambda_1} = [1,1,1,1]$ can also be identified as the s-channel view of the t-channel coupling $((M^\dagger \otimes M)_\mathbf{1} \otimes (B^\dagger \otimes B)_\mathbf{1})_\mathbf{1}$ (this operator just counts

the number of mesons and baryons), and the eigenvalue $+1$ indicates that $\mathbf{1}$ comes from a symmetric coupling of "**35**" $\otimes$ "**35**".

On the other hand, the WT interaction (irrep "**35**$_A$" in the t-channel) is antisymmetric under meson crossing and so $\boldsymbol{\lambda}_{\mathrm{WT}}$ is a left eigenvector of a with eigenvalue -1 (correspondingly, it is orthogonal to $\boldsymbol{\lambda_1}$, see (11).) The two other left eigenvectors correspond to "**35**$_S$" and "**405**" in the t-channel. Since they are both symmetric under crossing, they carry eigenvalue $+1$.

In view of the fact that $\boldsymbol{\lambda}_{\text{``}\mathbf{35}_A\text{''}} = \boldsymbol{\lambda}_{\mathrm{WT}}$ is the unique eigenvector with eigenvalue -1, we can directly reconstruct a as follows[3]

$$a_{\alpha\beta} = \delta_{\alpha\beta} - 2 \frac{\lambda_\alpha \lambda_\beta d_\alpha}{\sum_\gamma \lambda_\gamma^2 d_\gamma}. \tag{19}$$

The matrix is displayed in table 1. Note that $\boldsymbol{\lambda}_{\mathrm{cbp}}$ is correctly reproduced (up to a factor) by the first row of a. This matrix contains all u-channel results. For the "**35**$_S$" t-channel, a direct Wick computation gives

$$\boldsymbol{\lambda}_{\text{``}\mathbf{35}_S\text{''}} = \left[\left(\frac{2}{n} + \frac{1}{N}\right)(n-2), \left(\frac{1}{n} + \frac{1}{N}\right)(n-2), \frac{n-2}{n}, -\frac{1}{N} - \frac{2}{n}\right]. \tag{20}$$

As can be verified it is a $+1$ left eigenvector of a and is orthogonal to $\boldsymbol{\lambda_1}$ and $\boldsymbol{\lambda}_{\text{``}\mathbf{35}''_A}$. Finally, by orthogonality we obtain

$$\boldsymbol{\lambda}_{\text{``}\mathbf{405}\text{''}} = \left[n, -\frac{n+N}{N-1}, -\frac{N}{n+N+1}, \frac{1}{n}\right]. \tag{21}$$

Supported by DGI and FEDER (FIS2005-00810), Junta de Andalucía (FQM-225), EURIDICE (HPRN-CT-2002-00311), and Hadron Physics Project (RII3-CT-2004-506078).

References

1. F. Gursey, L.A. Radicati, Phys. Rev. Lett. **13**, 173 (1964); A. Pais, Phys. Rev. Lett. **13**, 175 (1964); B. Sakita, Phys. Rev. **136**, B1756 (1964).
2. S.R. Coleman, J. Mandula, Phys. Rev. **159**, 1251 (1967).

---

[3] Since $\hat{a} = 1 - 2P_{-1} = 1 - 2\hat{e}_{-1} \otimes \hat{e}_{-1}$.

3. R. Dashen, E. Jenkins, A.V. Manohar, Phys. Rev. D **49**, 4713 (1994).
4. N. Kaiser, P.B. Siegel, W. Weise, Nucl. Phys. A **594**, 325 (1995); Phys. Lett. B **362**, 23 (1995); E. Oset, A. Ramos, Nucl. Phys. A **635**, 99 (1998); M.F.M. Lutz, E.E. Kolomeitsev, Nucl. Phys. A **700**, 193 (2002); J. Oller, U. Meißner, Phys. Lett. B **500**, 263 (2001); J. Nieves, E. Ruiz Arriola, Phys. Rev. D **64**, 116008 (2001); C. García-Recio, J. Nieves, E. Ruiz Arriola, M.J. Vicente-Vacas, Phys. Rev. D **67**, 076009 (2003).
5. a) C. García-Recio, J. Nieves, L.L. Salcedo, Phys. Rev. D **74**, 036004; b) 034025 (2006).
6. S. Weinberg, Phys. Rev. Lett. **17**, 616 (1966); Y. Tomozawa, Nuovo Cimento A **46**, 707 (1966).
7. D.G. Caldi, H. Pagels, Phys. Rev. D **14**, 809 (1976); **15**, 2668 (1977).

Eur. Phys. J. A **31**, 495–498 (2007)
DOI 10.1140/epja/i2006-10191-8

Special Article – QNP 2006

Collective resonances in the soliton model approach to meson-baryon scattering

H. Weigel[a]

Fachbereich Physik, Siegen University, D-57068 Siegen, Germany

Received: 8 October 2006
Published online: 22 February 2007 – © Società Italiana di Fisica / Springer-Verlag 2007

Abstract. The proper description of hadronic decays of baryon resonances has been a long-standing problem in soliton models for baryons. In this paper I present a solution to this problem in the three-flavor Skyrme model that is consistent with large-N_C consistency conditions. As an application I discuss hadronic pentaquark decays and show that predictions based on axial current matrix elements are erroneous.

PACS. 11.15.Pg Expansions for large numbers of components (*e.g.*, $1/N_c$ expansions) – 12.39.Dc Skyrmions – 13.75.Jz Kaon-baryon interactions

1 Introduction

Commonly hadronic decays of baryon resonances are described by a Yukawa interaction of the generic structure

$$\mathcal{L}_{\text{int}} = g\,\bar{\psi}_{B'}\,\phi\,\psi_B\,, \tag{1}$$

where B' is the resonance that decays into baryon B and meson ϕ and g is a coupling constant. It is crucial that this interaction Lagrangian is *linear* in the meson field.

The situation is quite different in soliton models that are based on action functionals of only meson degrees of freedom, $\Gamma = \Gamma[\Phi]$. These action functionals contain classical (static) soliton solutions, Φ_{cl}, that are identified as baryons. The interaction of these baryons with mesons is described by the (small) meson fluctuations about the soliton: $\Phi = \Phi_{\text{cl}} + \phi$. By pure definition, we have

$$\left.\frac{\delta\Gamma[\Phi]}{\delta\Phi}\right|_{\Phi=\Phi_{\text{cl}}} = 0\,. \tag{2}$$

Thus, there is no term linear in ϕ to be associated with the Yukawa interaction, eq. (1). This puzzle has become famous as the Yukawa problem in soliton models. However, this does not mean that hadronic decays of resonances cannot be described in soliton models. Rather they have to be extracted from meson baryon scattering amplitudes, just as in experiment. In soliton models two-meson processes acquire contributions from the second-order term

$$\Gamma^{(2)} = \frac{1}{2}\,\phi\,\left.\frac{\delta^2\Gamma[\Phi]}{\delta^2\Phi}\right|_{\Phi=\Phi_{\text{cl}}}\phi\,. \tag{3}$$

This expansion simultaneously represents an expansion in N_C, the number of color degrees of freedom: $\Gamma = \mathcal{O}(N_C)$,

$\Gamma^{(2)} = \mathcal{O}(N_C^0)$ while terms $\mathcal{O}(\phi^3)$ vanish in the limit $N_C \to \infty$. This implies that $\Gamma^{(2)}$ contains all large-N_C information about hadronic decays of resonances. We may reverse that statement to argue about computations of hadronic decay widths in soliton models: Their results and those obtained from $\Gamma^{(2)}$ *must* be identical in the limit $N_C \to \infty$. Unfortunately, the most prominent baryon resonance, the Δ, becomes degenerate with the nucleon as $N_C \to \infty$. It is stable in that limit and its decay is not subject to the above-described litmus test. The situation is more interesting when extending soliton models to flavor $SU(3)$. In the so-called rigid-rotator approach (RRA) that generates baryon states as (flavor) rotational excitations of the soliton, resonances emerge that dwell in the anti-decuplet representation of flavor $SU(3)$. The most discussed (and disputed) such state is the Θ^+ pentaquark with zero isospin and strangeness $S = +1$. In the limit $N_C \to \infty$ the anti-decuplet states maintain a non-zero mass difference with respect to the nucleon. Therefore, the decay properties of Θ^+ as predicted in any soliton model must also be seen in the S-matrix for koan-nucleon scattering as computed from $\Gamma^{(2)}$. In the $S = -1$ sector the resulting equations of motion for ϕ yield a P-wave bound state whose occupation serves to describe the ordinary hyperons, Λ, Σ, Σ^*, etc. Therefore this treatment of hyperon states is called the bound-state approach (BSA). The above-discussed litmus test requires that the BSA and RRA give identical results for the Θ^+ properties as $N_C \to \infty$. This did not seem to be true and it was argued that the prediction of pentaquarks would be a mere artifact of the RRA [1]. Here we will show that this conclusion is premature and that pentaquark states do indeed emerge in both approaches. Furthermore, the comparison between the BSA and RRA provides an unambiguous computation

[a] e-mail: `weigel@physik.uni-siegen.de`

of pentaquark widths: It differs substantially from previous approaches based on assuming pertinent transition operators for $\Theta^+ \to KN$ [2,3]. Details of these studies are contained in ref. [4] and ref. [5] may be consulted for a review on $SU(3)$ soliton models.

2 Constrained fluctuations and Θ^+ width

When restricted to modes spanned by the soliton's rigid rotations, the P-wave fluctuations in strangeness direction have two bound states with eigenenergies

$$\omega_\pm = \frac{1}{2}\left[\sqrt{\omega_0^2 + \frac{3\Gamma}{2\Theta_K}} \pm \omega_0\right], \qquad (4)$$

where Θ_K is the moment of inertia for the rotation of the soliton into strangeness direction and Γ is the functional of the soliton that measures flavor symmetry breaking. The latter would be zero if the masses of the strange and non-strange quarks were equal. Both functionals are $\mathcal{O}(N_C)$. The subscript on $\omega_\pm$ refers to the strangeness quantum number of the bound state. Hence $\omega_0 = N_C/(4\Theta_K)$ removes the degeneracy between $S = \pm 1$ baryons. This contribution stems from the Wess-Zumino term in $\Gamma[\Phi]$. In accordance with the above discussion $\omega_\pm = \mathcal{O}(N_C^0)$. While ω_- is the energy of the above-mentioned bound state describing ordinary hyperons, ω_+ is eventually utilized to construct pentaquark states.

In the RRA the collective coordinates $A(t) \in SU(3)$ that parameterize the flavor orientation of the soliton are canonically quantized. The resulting Hamiltonian is (numerically) exactly diagonalized for arbitrary N_C [4] and symmetry breaking [6]. The so-computed mass difference between the states that for $N_C = 3$ correspond to the Λ (Θ^+) and the nucleon approaches ω_- (ω_+) as $N_C \to \infty$. This suggests that indeed the RRA and BSA are identical in that limit. This identity has a caveat when the restriction that BSA modes are spanned by the rigid rotation is removed. Though $\omega_- < m_K$ still corresponds to a true bound state, ω_+ is a continuum state. Thus, a pronounced resonance structure would be expected in the BSA phase shift around $\omega = \omega_+$. Unfortunately, that is not the case, as seen from fig. 1. The BSA phase shift hardly reaches $\pi/2$ rather than quickly passing through this value. The ultimate comparison requires to generalize the RRA to the rotation-vibration approach (RVA)

$$U(\boldsymbol{x},t) = A(t)\xi_0(\boldsymbol{x})\exp\left[\frac{i}{f_\pi}\sum_{\alpha=4}^{7}\lambda_\alpha\widetilde{\eta}_\alpha(\boldsymbol{x},t)\right]\xi_0(\boldsymbol{x})\,A^\dagger(t),$$

$$(5)$$

where $\xi_0(\boldsymbol{x}) = \exp[i\hat{\boldsymbol{x}}\cdot\boldsymbol{\tau}F(|\boldsymbol{x}|)/2]$ is the chiral field representation of the soliton ($\Phi_{\rm cl}$) and $A(t) \in SU(3)$ parameterizes the collective rotations. Modes that correspond to the collective rotations must be excluded from the fluctuations $\widetilde{\eta}$, i.e. the fluctuations must be orthogonal to the zero mode $z \sim \sin(F/2)$. Imposing the corresponding constraints for these fluctuations (and their conjugate momenta) yields integro-differential equations listed in

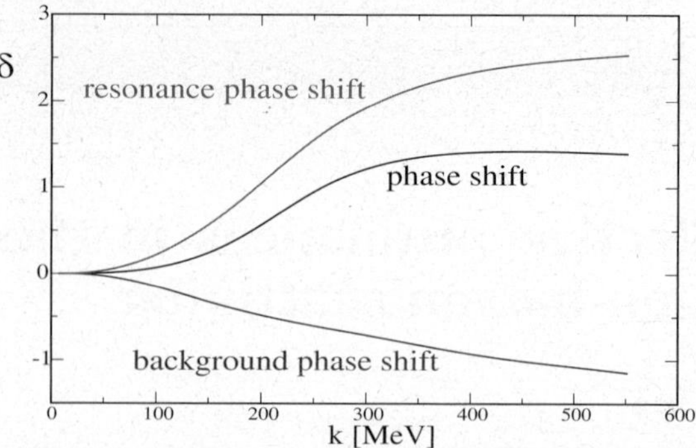

Fig. 1. (Colour on-line) Phase shift computed in the BSA (middle, black line) and the resonance phase (top, red line) shift after removal of the background (bottom, blue line) contribution in the RVA.

ref. [4]. In the BSA $A(t)$ is restricted to $SU(2)$ but there is no constraint on the fluctuations η. The above-discussed litmus test requires that the scattering data computed from $\widetilde{\eta}$ and η be identical when N_C is sent to infinity.

For the moment let us omit the coupling between $\widetilde{\eta}$ and the collective soliton excitations. This truncation defines the background wave function $\overline{\eta}$ (also orthogonal to the zero mode). Treating $\overline{\eta}$ as an harmonic fluctuation provides the background phase shift shown as the blue on-line curve in fig. 1. Remarkably, the difference between the phase shifts of $\overline{\eta}$ and η clearly exhibits a distinct resonance structure. This is the resonance phase shift to be associated with the Θ^+ pentaquark in the limit $N_C \to \infty$!

The parameterization, eq. (5) is not a solution to the classical equation of motion, The strategy rather is to solve them order by order in the N_C expansion. Hence the arguments deduced from eq. (2) do not apply and an interaction Hamiltonian that is linear in the fluctuations indeed emerges. This generates Yukawa couplings between the collective soliton excitations and the fluctuations $\widetilde{\eta}$. In ref. [4] we have derived this Hamiltonian keeping all contributions that survive as $N_C \to \infty$. The corresponding Yukawa exchanges extend the integro-differential equations for $\overline{\eta}$ by a separable potential V_Y, therewith providing the equations of motion for $\widetilde{\eta}$. For $N_C \to \infty$ the equation of motion for $\widetilde{\eta}$ is solved by $\widetilde{\eta} = \eta - \langle z|\eta\rangle z$. The phase shifts extracted from η and $\widetilde{\eta}$ are identical because $z(|\boldsymbol{x}|)$ is localized in space. Thus, the BSA and RVA yield the same spectrum and are indeed equivalent in the large-N_C limit. But, the RVA provides a distinction between resonance and background contributions to the scattering amplitude. Applying the R-matrix formalism on top of the constrained fluctuations $\overline{\eta}$ shows that V_Y *exactly* contributes the resonance phase shift shown in fig. 1 when the Yukawa coupling is computed for $N_C \to \infty$. This identifies the exchange of a state predicted in the RRA which thus is no artifact. In contrast, pentaquarks are also predicted by the BSA; just well hidden. Nevertheless, collective coordinates are mandatory to obtain finite-N_C corrections to the BSA for the properties of Θ^+. Though not all $\mathcal{O}(1/N_C)$ operators were included in ref. [4], subleading effects are substantial. For example, in the case $m_K = m_\pi$ the mass

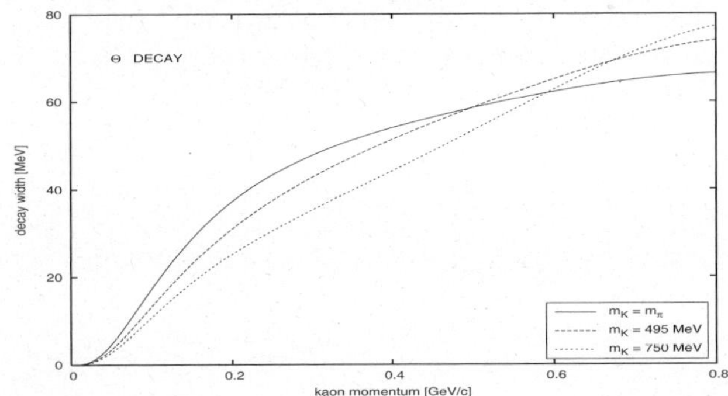

Fig. 2. Skyrme model prediction for the decay width, $\Gamma(\omega)$ of Θ^+ for $N_C = 3$ as a function of the kaon momentum $k = \sqrt{\omega^2 - m_K^2}$, cf. eq. (6). Top curve: $m_K = m_\pi$, bottom curves: $m_K \neq m_\pi$.

difference with respect to the nucleon increases by a factor two from ω_0 to $(N_C + 3)/4\Theta_K$ for $N_C = 3$.

The separable potential V_Y also yields the general expression for the width as a function of the kaon energy $\omega_k = \sqrt{k^2 + m_K^2}$ from the R-matrix formalism [4]:

$$\Gamma(\omega_k) = 2k\omega_0 \left| X_\Theta \int_0^\infty r^2 dr\, z(r) 2\lambda(r)\overline{\eta}_{\omega_k}(r) \right.$$

$$\left. + \frac{Y_\Theta}{\omega_0}\left(m_K^2 - m_\pi^2\right) \int_0^\infty r^2 dr\, z(r)\overline{\eta}_{\omega_k}(r) \right|^2 . \quad (6)$$

Here $\overline{\eta}_{\omega_k}(|\boldsymbol{x}|)$ is the P-wave projection of the background wave function $\overline{\eta}$ for a prescribed energy ω_k and $\lambda(|\boldsymbol{x}|)$ is a radial function that stems from the Wess-Zumino term. The matrix elements (X_Θ and Y_Θ) of the collective coordinate operators that enter eq. (6) are computed from the eigenstates of the collective coordinate Hamiltonian. The resulting width is shown for $N_C = 3$ in fig. 2 for both the flavor symmetric case and the physical kaon-pion mass difference. As a function of momentum, there are only minor differences between these two cases. Assuming the observed resonance to be the (disputed) $\Theta^+(1540)$ a width of roughly 40 MeV is read off from fig. 2 [4]. However, fig. 1 suggests that the resonance should be about 200 MeV above threshold which corresponds to $M_{\Theta^+} \approx 1.65$ GeV. Hence it seems very unlikely that chiral soliton models predict a light and very narrow pentaquark, though the numerical results for masses and widths of pentaquarks are model dependent.

3 Conclusion

In this paper I have presented a thorough comparison [4] between the bound state (BSA) and rigid-rotator approaches (RRA) to chiral soliton models in flavor $SU(3)$. Though I have only considered the simplest such model, the actual analysis merely concerns the treatment of kaon degrees of freedom. Therefore the qualitative results are valid for *any* chiral soliton model.

A sensible comparison with the BSA requires the consideration of harmonic oscillations in the RRA as well.

They are incorporated via the rotation-vibration approach (RVA), however constraints must be implemented to ensure that the introduction of such fluctuations does not double-count any degrees of freedom. The RVA clearly shows that the prediction of pentaquarks is not an artifact of the RRA, pentaquarks are genuine within chiral soliton models. Only within the RVA chiral soliton models generate interactions for hadronic decays. Technically, the derivation of this Hamiltonian is quite involved, however, the result is as simple as convincing: In the limit $N_C \to \infty$, in which the BSA is undoubtedly correct, the RVA and BSA yield identical results for the baryon spectrum and the kaon-nucleon S-matrix. This identity also holds when flavor symmetry breaking is included. This demonstrates that collective coordinate quantization may be successfully applied regardless of whether or not the respective modes are zero modes.

In the flavor symmetric case the interaction Hamiltonian contains only a *single* structure (X_Θ in eq. (6)) of $SU(3)$ matrix elements for the $\Theta^+ \to KN$ transition. Any additional $SU(3)$ structure only enters via flavor symmetry breaking. This proves earlier approaches [2,3] incorrect that adopted any possible structure that would contribute in the large-N_C limit and fitted coefficients from a variety of hadronic decays under the assumption of $SU(3)$ relations. The study presented in this talk thus suggests that it is not worthwhile to bother about the obvious arithmetic error in ref. [2] that was discovered earlier [7,8] because the conceptual deficiencies in such width calculations are more severe. Assuming $SU(3)$ relations among hadronic decays is not a valid procedure in chiral soliton models. The embedding of the classical soliton breaks $SU(3)$ and thus yields different structures for different hadronic transitions.

Even in case pentaquarks turn out not to be what some recent experiments have suggested, they have definitely been very beneficial in combining the bound state and rigid-rotator approaches and solving the Yukawa problem in the kaon sector; both long-standing puzzles in chiral soliton models.

I am very appreciative to Hans Walliser for the very fruitful collaboration on which this talk is based. I am grateful to the organizers for this pleasant conference. Attendance of the conference has been made possible by the DFG under contract We 1254/12-1.

References

1. N. Itzhaki *et al.*, Nucl. Phys. B **684**, 264 (2004); T.D. Cohen, Phys. Lett. B **581**, 175 (2004); Phys. Rev. D **70**, 014011 (2004); A. Cherman *et al.*, Phys. Rev. D **72**, 094015 (2005).

2. D. Diakonov, V. Petrov, M. Polyakov, Z. Phys. A **359**, 305 (1997).

3. J.R. Ellis, M. Karliner, M. Praszałowicz, JHEP **05**, 002 (2004); M. Praszałowicz, Acta Phys. Pol. B **35**, 1625 (2004); M. Praszałowicz, K. Goeke, Acta Phys. Pol. B **36**, 2255 (2005).

4. H. Walliser, H. Weigel, Eur. Phys. J. A **26**, 361 (2005); H. Weigel, arXiv:hep-ph/0510111.
5. H. Weigel, Int. J. Mod. Phys. A **11**, 2419 (1996).
6. H. Yabu, K. Ando, Nucl. Phys. B **301**, 601 (1988).
7. H. Weigel, Eur. Phys. J. A **2**, 391 (1998);
8. R.L. Jaffe, Eur. Phys. J. C **35**, 221 (2004).

Eur. Phys. J. A **31**, 499–502 (2007)

DOI 10.1140/epja/i2006-10203-9

Special Article – QNP 2006

Resonances and the Weinberg-Tomozawa 56-baryon–35-meson interaction

C. García-Recio, J. Nieves[a], and L.L. Salcedo

Departamento de Física Atómica, Molecular y Nuclear, Universidad de Granada, E-18071 Granada, Spain

Received: 25 October 2006
Published online: 22 February 2007 – © Società Italiana di Fisica / Springer-Verlag 2007

Abstract. Vector meson degrees of freedom are incorporated into the Weinberg-Tomozawa (WT) meson-baryon chiral Lagrangian by using a scheme which relies on spin-flavor $SU(6)$ symmetry. The corresponding Bethe-Salpeter approximation successfully reproduces previous $SU(3)$-flavor WT results for the lowest-lying s-wave negative-parity baryon resonances, and it also provides some information on the dynamics of the heavier ones. Moreover, it also predicts the existence of an isoscalar spin-parity $\frac{3}{2}^-$ K^*N bound state (strangeness +1) with a mass around 1.7–1.8 GeV, unstable through K^* decay. Neglecting d-wave KN decays, this state turns out to be quite narrow ($\Gamma \leq 15\,\text{MeV}$) and it might provide clear signals in reactions like $\gamma p \to \bar{K}^0 p K^+ \pi^-$ by looking at the three-body $pK^+\pi^-$ invariant mass.

PACS. 11.30.Hv Flavor symmetries – 11.10.St Bound and unstable states; Bethe-Salpeter equations – 11.30.Rd Chiral symmetries – 11.80.Gw Multichannel scattering

1 Introduction

We present results obtained from a scheme where it is assumed that the light-quark–light-quark interaction is approximately spin independent as well as $SU(3)$ independent. This corresponds to treating the six states of a light quark (u, d or s with spin up, $\uparrow$, or down, $\downarrow$) as equivalent, and leads us to the invariance group $SU(6)$. Despite the fact that the no-go Coleman-Mandula theorem [1] forbids this hybrid symmetry (mixing the compact, purely internal flavor symmetry, with the noncompact Poincare symmetry of spin angular momentum) to be exact, there exist several $SU(6)$ predictions (relative closeness of baryon octet and decuplet masses, the axial current coefficient ratio $F/D = 2/3$, the magnetic moment ratio $\mu_p/\mu_n = -3/2$) which are remarkably well satisfied in nature [2]. This suggests that $SU(6)$ could be a good approximate symmetry. Though in general the spin-flavor symmetry is not exact for excited baryons even in the large N_c limit (being N_c the number of colors)[1], in the real world ($N_c = 3$), the zeroth order spin-flavor symmetry breaking turns out to be similar in magnitude to $\mathcal{O}(N_c^{-1})$ breaking effects [5]. Spin-flavor symmetry in the meson sector is not a direct consequence of large-N_c QCD either. However, vector mesons ($K^*, \rho, \omega, \bar{K}^*, \phi$) do exist and they are known to play a relevant role in hadronic

physics. Inescapably, they will couple to baryons and will presumably influence the properties of the baryonic resonances. Lacking better theoretically founded models to include vector mesons, we regard the spin-flavor symmetric scenario as reasonable first step. The large-N_c consequences of this scheme have been pursued in [6].

We will consider the s-wave interaction between the $SU(6)$ lowest-lying meson multiplet (**35**) and the lowest-lying baryons (**56**-plet) at low energies. The meson multiplet contains the octet of pseudoscalar ($K, \pi, \eta, \bar{K}$) and the nonet of vector ($K^*, \rho, \omega, \bar{K}^*, \phi$) mesons, while the baryon one is constructed from the (N, Σ, Λ, Ξ) octet of spin-1/2 baryons and the ($\Delta, \Sigma^*, \Xi^*, \Omega$) decuplet of spin-3/2 baryons. Assuming that the s-wave effective meson-baryon Hamiltonian is $SU(6)$ invariant, and since the $SU(6)$ decomposition of the product of the **35** (meson) and **56** (baryon) representations yields

$$\mathbf{35} \otimes \mathbf{56} = \mathbf{56} \oplus \mathbf{70} \oplus \mathbf{700} \oplus \mathbf{1134}, \tag{1}$$

there are only four, Wigner-Eckart irreducible matrix elements (WEIMEs), free functions of the meson-baryon Mandelstam variable s. Similar ideas were already explored in the late sixties, within the effective-range approximation [7]. In this work, based on the findings of ref. [8], two major improvements have been introduced: i) The use of the underlying Chiral Symmetry (CS), which would allow to determine the value of the $SU(6)$ irreducible matrix elements from the Weinberg-Tomozawa (WT) interaction (leading term of the chiral Lagrangian involving Goldstone bosons and the octet of spin-1/2

---

[a] e-mail: jmnieves@ugr.es

[1] In the large-N_c limit [3,4], there exists an exact spin-flavor symmetry for ground-state baryons.

baryons), ii) the use of Bethe-Salpeter Equations (BSE) [9, 10], in coupled channels and with an appropriated Renormalization Scheme (RS) [11,12], to determine the scattering amplitudes, going thus beyond the effective-range approximation.

2 $SU(6)$ extension of the meson-baryon WT interaction

We will work with well-defined total isospin (I), angular momentum (J) and hypercharge (Y) meson-baryon states constructed out of the $SU(6)$ **35** (mesons) and **56** (baryon) multiplets. By imposing that the effective s-wave meson-baryon *potential* (V) is a $SU(6)$ invariant operator, we find [8]

$$\langle \mathcal{M}'\mathcal{B}'; JIY|V|\mathcal{M}\mathcal{B}; JIY \rangle = \sum_{\phi} V_{\phi}(s)\mathcal{P}^{\phi,JIY}_{\mathcal{M}\mathcal{B},\mathcal{M}'\mathcal{B}'}, \quad (2)$$

where $\mathcal{M} \equiv [(\mu_M)_{2J_M+1}, I_M, Y_M]$ stands for meson states and similarly $\mathcal{B}$ for baryon ones and the labels μ and ϕ denote $SU(3)$ and $SU(6)$ representations, respectively. In the above equation ϕ runs over the **56**, **70**, **700** and **1134** irreducible representations (irreps), as inferred from eq. (1). The projectors are given in terms of the $SU(3)$ isoscalar [13], and the $SU(6)$-multiplet coupling [14,15] factors,

$$\mathcal{P}^{\phi,JIY}_{\mathcal{M}\mathcal{B},\mathcal{M}'\mathcal{B}'} = \sum_{\mu,\alpha} \begin{pmatrix} \mathbf{35} & \mathbf{56} \\ \mu_M J_M & \mu_B J_B \end{pmatrix} \begin{vmatrix} \phi \\ \mu J \alpha \end{pmatrix}$$

$$\times \begin{pmatrix} \mu_M & \mu_B \\ I_M Y_M & I_B Y_B \end{pmatrix} \begin{vmatrix} \mu \\ IY \end{pmatrix} \begin{pmatrix} \mu'_{M'} & \mu'_{B'} \\ I'_{M'} Y'_{M'} & I'_{B'} Y'_{B'} \end{pmatrix} \begin{vmatrix} \mu \\ IY \end{pmatrix}$$

$$\times \begin{pmatrix} \mathbf{35} & \mathbf{56} \\ \mu'_{M'} J'_{M'} & \mu'_{B'} J'_{B'} \end{pmatrix} \begin{vmatrix} \phi \\ \mu J \alpha \end{pmatrix}, \quad (3)$$

where α accounts for the multiplicity of each of the μ_{2J+1} $SU(3)$ multiplets of spin J (for $L = 0$, J is given by the total spin of the meson-baryon system) entering in the representation ϕ. The $SU(6)$ WEIMEs, $V_{\phi}(s)$, might be constrained by demanding that the above interaction, when restricted to the Goldstone pseudoscalar meson and the lowest $J^P = \frac{1}{2}^+$ baryon octet space, will reduce to that deduced from $SU(3)$ chiral symmetry. At leading order in the chiral expansion, this latter one is obtained from the WT Lagrangian, which besides hadron masses only depends on the $f \simeq 93\,\mathrm{MeV}$, the pion weak decay constant, and it can be exactly recovered from eq. (2) by setting the WEIMEs as follows [8]:

$$V_{\phi}(s) = \bar{\lambda}_{\phi} \frac{\sqrt{s} - M}{2f^2}, \quad (4)$$

with $\bar{\lambda}_{\mathbf{56}} = -12$, $\bar{\lambda}_{\mathbf{70}} = -18$, $\bar{\lambda}_{\mathbf{700}} = 6$ and $\bar{\lambda}_{\mathbf{1134}} = -2$ and M the common octet and decuplet baryon mass^2. This

---

^{2} The $SU(6)$ extension thus obtained (eqs. (2) and (4)) also leads to the *potentials* used in refs. [16,17] to study the $(\mathbf{10_4})$baryon-$(\mathbf{8_1})$meson sector.

is not a trivial fact and it is intimately linked to the group structure of the WT term. Indeed, the underlying reason for this is CS, since the WT Lagrangian is not just $SU(3)$ symmetric but also chiral ($SU_L(3)\otimes SU_R(3)$) invariant [8].

3 Meson-baryon scattering matrix

We solve the coupled-channel BSE with an interaction kernel determined by eqs. (2) and (4). In a given JIY sector, the solution for the coupled-channel s-wave scattering amplitude, $T^J_{IY}(\sqrt{s})$ (normalized as the t-matrix defined in eq. (33) of [18]), in the *on-shell* scheme [9,10,19,11,20] reads,

$$T^J_{IY}(\sqrt{s}) = \frac{1}{1 - V^J_{IY}(\sqrt{s})\, J^J_{IY}(\sqrt{s})} V^J_{IY}(\sqrt{s}) \quad (5)$$

with $V^J_{IY}(\sqrt{s}) = \langle \mathcal{M}'\mathcal{B}'; JIY|V|\mathcal{M}\mathcal{B}; JIY \rangle$, and $J^J_{IY}(\sqrt{s})$ a diagonal matrix of loop functions [18,21]. Those are logarithmically divergent and hence to make them finite an ultraviolet (UV) cutoff or a subtraction point μ^{JIY}_i, such that

$$\left[J^J_{IY}(\sqrt{s} = \mu^{JIY}_i) \right]_{ii} = 0 \quad (6)$$

with the index i running in the coupled-channel space, is needed.

By setting $\mu^{JIY}_i = \sqrt{M^2_i + m^2_i}$, with M_i and m_i the masses of the baryon and meson entering in the channel i of the sector JIY, we recover previous results, deduced from the WT $SU(3)$ chiral Lagrangian [11,12, 18–21], and make new predictions. For instance, in the $I = 0, S = -1$ ($Y = 0$) sector, looking for poles in the second Riemann sheet:

- For $J^P = \frac{1}{2}^-$, we obtain the $\Lambda(1390)$, $\Lambda(1405)$, $\Lambda(1670)$ resonances, but we also find strength around

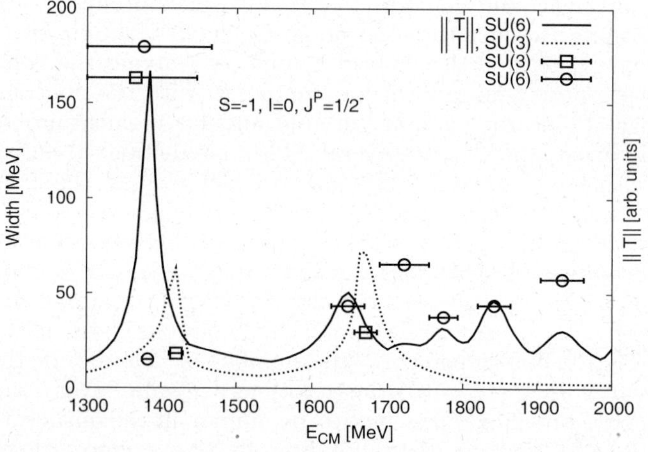

Fig. 1. Spin-parity $J^P = \frac{1}{2}^-$ resonance properties in the $I = 0, S = -1$ ($Y = 0$) sector. Solid and dotted lines stand for the $SU(6)$ and $SU(3)$ T-matrix norms, respectively. The norm is calculated as $\|T\| = \mathrm{Max}(\sum_{j=1}^{n}|T_{ij}|, i = 1,\ldots,n)$ where i,j run over the number of coupled channels. Finally, the points with error bars are defined from the masses and widths of the found resonances as $(M_R \pm \Gamma_R/2, \Gamma_R)$.

1800 MeV, which might correspond to the three star $\Lambda(1800)$, which has a sizeable $N\bar{K}^*$ coupling (see fig. 1).

- For $J^P = \frac{3}{2}^-$, we get signals for some d-wave resonances (not accessible with the $SU(3)$ WT chiral Lagrangian): the four star $\Lambda(1520)$ and $\Lambda(1690)$ states with large couplings to the $\Sigma^*\pi$ channel, and the $\Lambda(2325)$ resonance which couples to $\Lambda\omega$ channel.

Similar results are found for the other I, Y sectors. $SU(6)$ symmetry breaking effects such that the use of a common weak decay constant (f) for all channels lead to changes of the order of 30% in resonance widths and excitation energies. These uncertainties are similar to those stemming from the use of different UV cutoffs or subtraction points to renormalize the amplitudes.

4 Exotic states

The $SU(6)$ meson-baryon interaction constructed in sect. 2 turns out to be attractive in the **1134**-irrep space ($\bar{\lambda}_{\mathbf{1134}} = -2$), which might lead to the existence of some exotic states. For instance, in the $Y = -3, I = J = 1/2$ sector, we have $|\bar{K}^*\Omega\rangle = |\mathbf{1134}; \mathbf{35_2}\rangle$ or for $Y = +2, I = 0, J = 3/2\,|K^*N\rangle = -|\mathbf{1134}; \mathbf{10_4^*}\rangle$, hence one finds attractive $\bar{K}^*\Omega$ or K^*N interactions with these exotic quantum numbers. Whether these interactions are strong enough or not to bind the meson-baryon system will depend on the RS employed to make finite the BSE and on the nature of the further s- and d-wave contributions to the interaction matrix.

Neglecting any further correction to the *potential*, we present here results for the K^*N system in the $I = 0, J = 3/2$ sector (see ref. [8] for some more details). We find a pole in the first Riemann sheet corresponding to a K^*N bound state which we call Θ^{*+}. This state is unstable since the K^* decays into $K\pi$. To estimate the Θ^{*+} width, we model the Θ^*NK^* coupling as

$$\mathcal{L}_{\Theta^*NK^*} = -\frac{g}{\sqrt{2}}\overline{\Theta}^\mu \left(K_\mu^{*0}p - K_\mu^{*+}n \right) + \text{h.c.}, \quad (7)$$

where Θ^μ is a Rarita-Schwinger field, p and n the nucleon fields, K_μ^{*0} and K_μ^{*+} the Proca fields which annhilate and create neutral and charged K^* and $\bar{K}^*$ mesons. The subsequent K^* decay is described following ref. [22]. The coupling g is determined by the residue at the pole of $T_{02}^{\frac{3}{2}}$ (*i.e.*, $T_{02}^{\frac{3}{2}} \approx g^2 \times 2M_{\Theta^*}/(s - M_{\Theta^*}^2)$).

Resonance mass, residue and width depend on the RS employed. We have used an UV cutoff (Λ) to evaluate the loop function $J(\sqrt{s})$, which is equivalent to choose a scale $\bar{\mu}$ such that $J(\sqrt{s} = \bar{\mu}) = 0$. Results are shown in fig. 2. For $\bar{\mu}$ ranging from 0.05 GeV ($\Lambda \approx 1.08$ GeV) to 1.7 GeV ($\Lambda \approx 0.46$ GeV) the resonance mass (width) varies from 1.688 GeV (0.3 MeV), close to the ($M_N + m_\pi + m_K$) threshold, to 1.831 GeV (9 MeV, but the width does not grow monotonously, see figure), close to the ($M_N + m_{K^*}$) threshold. Other mechanisms for K^*N scattering (d-wave KN,

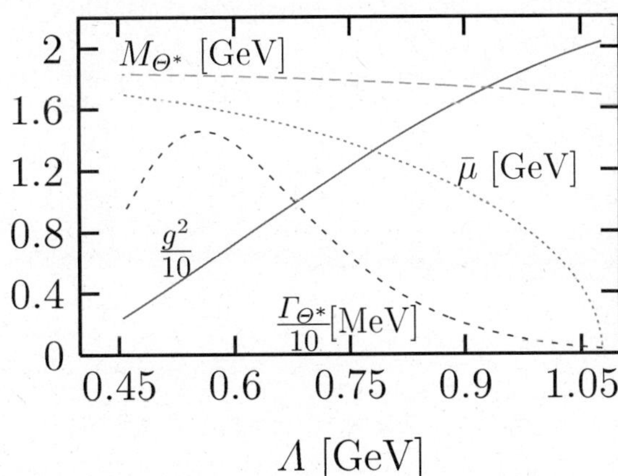

Fig. 2. Resonance Θ^{*+} properties as a function of the UV cutoff Λ or the subtraction scale $\bar{\mu}$.

K^*N contributions, u-channel pole graph, single pion exchange between K^* and N, sequential exchange of two pions with an intermediate K meson, corresponding to a box graph $K^*N \to KN \to K^*N, \ldots$) might quantitatively modify these results. However, we do not expect such corrections to be large enough to affect the existence of the Θ^* pentaquark [8]. Possible production and identification mechanisms for this resonance could be found in reactions like $\gamma p \to \bar{K}^0 pK^+\pi^-$ by measuring the three-body $pK^+\pi^-$ invariant mass.

This work was supported by DGI, FEDER, UE and Junta de Andalucía funds (FIS2005-00810, HPRN-CT-2002-00311, FQM225).

References

1. S.R. Coleman, J. Mandula, Phys. Rev. **159**, 1251 (1967).
2. R.F. Lebed, Phys. Rev. D **51**, 5039 (1995).
3. G.'t Hooft, Nucl. Phys. B **72**, 461 (1974).
4. E. Witten, Nucl. Phys. B **160**, 57 (1979).
5. J.L. Goity, C.L. Schat, N.N. Scoccola, Phys. Rev. D **66**, 114014 (2002).
6. C. García-Recio, J. Nieves, L.L. Salcedo, Phys. Rev. D **74**, 036004 (2006).
7. D.C. Carey, Phys. Rev. **169**, 1368 (1968).
8. C. García-Recio, J. Nieves, L.L. Salcedo, Phys. Rev. D **74**, 034025 (2006).
9. J. Nieves, E. Ruiz Arriola, Phys. Lett. B **455**, 30 (1999).
10. J. Nieves, E. Ruiz Arriola, Nucl. Phys. A **679**, 57 (2000).
11. M.F.M. Lutz, E.E. Kolomeitsev, Nucl. Phys. A **730**, 392 (2004).
12. C. García-Recio, M.F.M. Lutz, J. Nieves, Phys. Lett. B **582**, 49 (2004).
13. J.J. de Swart, Rev. Mod. Phys. **35**, 916 (1963).
14. J.C. Carter, J.J. Coyne, S. Meshkov, Phys. Rev. Lett. **14**, 523 (1965).
15. J.C. Carter, J.J. Coyne, S. Meshkov, Phys. Rev. Lett. **14**, 850 (1965).
16. E.E. Kolomeitsev, M.F.M. Lutz, Phys. Lett. B **585**, 243 (2004).

17. S. Sarkar, E. Oset, M.J. Vicente Vacas, Nucl. Phys. A **750**, 294 (2005).
18. J. Nieves, E. Ruiz Arriola, Phys. Rev. D **64**, 116008 (2001).
19. E. Oset, A. Ramos, Nucl. Phys. A **635**, 99 (1998).
20. D. Jido, J.A. Oller, E. Oset, A. Ramos, U.G. Meissner, Nucl. Phys. A **725**, 181 (2003).
21. C. García-Recio, J. Nieves, E. Ruiz Arriola, M.J. Vicente Vacas, Phys. Rev. D **67**, 076009 (2003).
22. G. Ecker, J. Gasser, H. Leutwyler, A. Pich, E. de Rafael, Phys. Lett. B **223**, 425 (1989).

Eur. Phys. J. A **31**, 503–505 (2007)
DOI 10.1140/epja/i2006-10276-4

Special Article – QNP 2006

Hyperon production at COSY-TOF

W. Schroeder[a]

For the COSY-TOF Collaboration
University of Erlangen-Nuremberg, Germany

Received: 23 November 2006
Published online: 23 March 2007 – © Società Italiana di Fisica / Springer-Verlag 2007

Abstract. The strangeness production program at the COSY-TOF experiment is discussed. The apparatus is shown emphasizing the technique to measure delayed decays. Results obtained for the reactions $pp \to K^+ \Lambda p$ and $pp \to K^0 \Sigma^+ p$ are discussed.

PACS. 13.75.-n Hadron-induced low- and intermediate-energy reactions and scattering (energy $\leq$ 10 GeV) – 14.20.Gk Baryon resonances with $S = 0$ – 14.20.Jn Hyperons

1 Introduction

The main interest in the investigation of the associated strangeness production in elementary reactions close to threshold is to get insight into the dynamics of the production. The questions especially concern the role of N^*-resonances and the hyperon-nucleon final-state interaction which is known to be of special importance close to threshold. To get to conclusive results precise observables are needed, concentrating on exclusive data, covering the full phase space.

2 Experimental setup

The external experiment COSY-TOF [1] is a wide-angle, non-magnetic spectrometer. A cross-section is shown in fig. 1. A few millimeters behind the very small liquid-hydrogen target the start (inner) detector system, which is optimized for strangeness production measurements, is installed. The stop detector with a length of about 3 m consists of various scintillator detectors (quirl, ring and barrel).

Except for small beam holes, the inner detector system, as well as the outer detector system covers the full angular range of the reaction products for the channels $pp \to K^+ \Lambda p$ and $pp \to K^0 \Sigma^+ p$. This allows a complete reconstruction of the events, including a precise measurement of the delayed decay of the Λ-hyperon and the K^0, respectively. A schematic view of the start detector system, consisting of the start torte (two layers of wedge-shaped scintillators), a double-sided microstrip detector and two scintillating fibre hodoscopes is shown together with events of the type $pp \to K^+ \Lambda p$ (fig. 2) and $pp \to K^0 \Sigma^+ p$ (fig. 3), respectively.

[a] e-mail: schroeder@physik.uni-erlangen.de

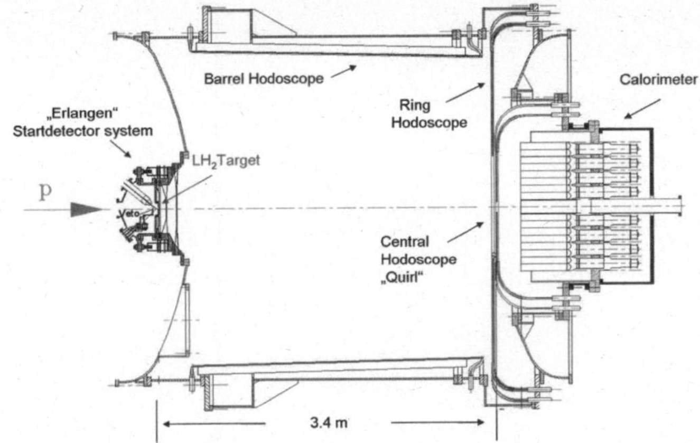

Fig. 1. Cross-section of the COSY-TOF spectrometer.

3 Results

A particular focus in the investigation of the reaction $pp \to K^0 \Sigma^+ p$ was laid on the search for the possible exotic pentaquark resonance Θ^+ for which evidence was reported for the first time by the LEPS Collaboration [2]. Among other experiments COSY-TOF found evidence for a peak at $1.53\,\text{GeV}/c^2$ with a statistical significance of about 4σ in the $K^0 p$ invariant-mass spectrum at a beam momentum of $2.95\,\text{GeV}/c$ [3], which is shown in fig. 4. To get to a final decision on the existence of the Θ^+ in the investigated channel recently COSY-TOF performed a measurement that will give results with a significantly improved statistical accuracy.

The reaction $pp \to K^0 \Lambda^+ p$ has been investigated at several beam momenta between $2.5\,\text{GeV}/c$ and $3.3\,\text{GeV}/c$ [4,5], that means from close to threshold up to nearly the COSY limit for the external beam. For reactions with more than two particles in the final state, as

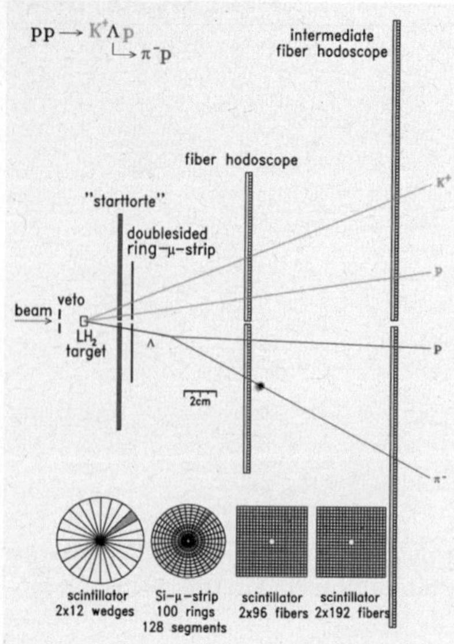

Fig. 2. Schematic view of the start detector with an event of type $pp \to K^+ \Lambda p$.

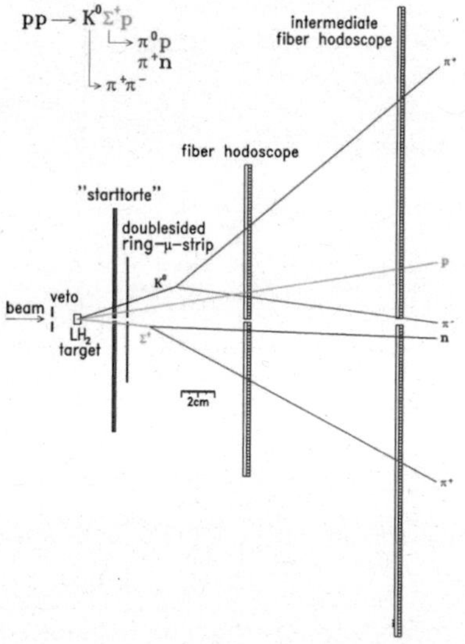

Fig. 3. Schematic view of the start detector with an event of type $pp \to K^0 \Sigma^+ p$.

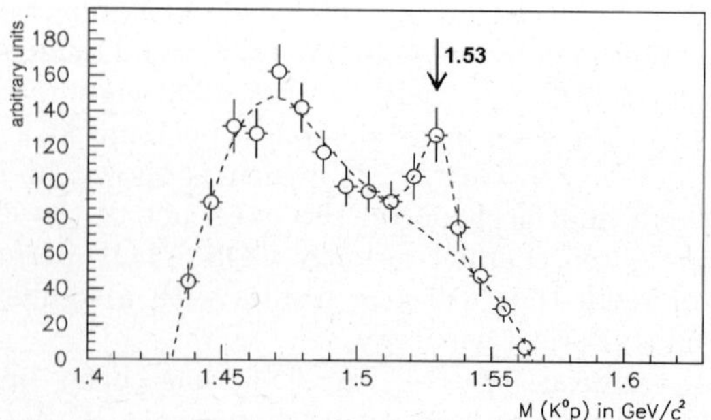

Fig. 4. $K^0 p$ invariant-mass spectrum of the reaction $pp \to K^0 \Sigma^+ p$ at a beam momentum of $2.95\,\mathrm{GeV}/c$.

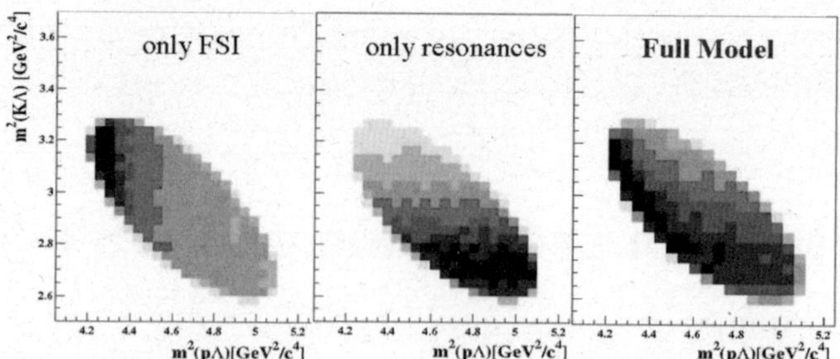

Fig. 5. Model calculations using parametrization of Sibirtsev [7] without resonances (left), without FSI (middle) and full calculation (right) for $2.95\,\mathrm{GeV}/c$.

in the case of the reaction $pp \to K^+ \Lambda p$, Dalitz plots are a powerful tool to extract information about the reaction mechanism. Whereas pure phase-space leads to a homogeneous distribution of the strength, in particular resonances should lead to significant deviations from this.

In fig. 6 (upper part) Dalitz plots for experimental data at beam momenta of $2.95\,\mathrm{GeV}/c$, $3.20\,\mathrm{GeV}/c$ and $3.30\,\mathrm{GeV}/c$, respectively are shown. They obviously show strong deviations from phase space. Monte Carlo simulations show that higher partial waves influence the Dalitz plot distributions only in a minor way. The strength coefficients of these partial waves have been extracted from the experimental angular distributions, which show some anisotropy for all three ejectiles. From theoretical work and our previous investigations [4], it is most likely that the observed anisotropy has its origin in the influence of the $p\Lambda$ final-state interaction and/or N^*-resonances. To obtain more insight into the various contributions, the data were compared with a model parametrization prepared by A. Sibirtsev [6], which includes the $N^*(1650, 1710, 1720)$-resonances, a non-resonant term and the $p\Lambda$ final-state interaction on the amplitude base (see eq. (1)):

$$\frac{\mathrm{d}^2\sigma}{\mathrm{d}m_{K\Lambda}^2 \mathrm{d}m_{p\Lambda}^2} =$$

$$fl\Phi \left| \left(\sum_R (C_R \cdot A_R) + C_N \right) \cdot (1 + C_{FSI} \cdot A_{FSI}) \right|^2 . \quad (1)$$

The quantity fl gives the normalization to the total cross-section, Φ is a phase-space factor. The third factor gives the deviation from an equally distributed Dalitz plot. A_R are the amplitudes of the Breit-Wigner-shapes of the three considered N^*-resonances. A_{FSI} denotes the amplitude of the $p\Lambda$ final-state interaction as given by the Juelich YN-model [7]. The strength of the individual resonances C_R, of the non-resonant contribution C_N (which includes the kaon exchange) and of the $p\Lambda$ final state C_{FSI} can be adjusted individually. For illustration the results of the model, obtained without resonances ("only FSI"), without FSI ("only resonances"), and the full calculation ("full model") are shown in fig. 5. Obviously only the full model can reproduce the data (see also fig. 6).

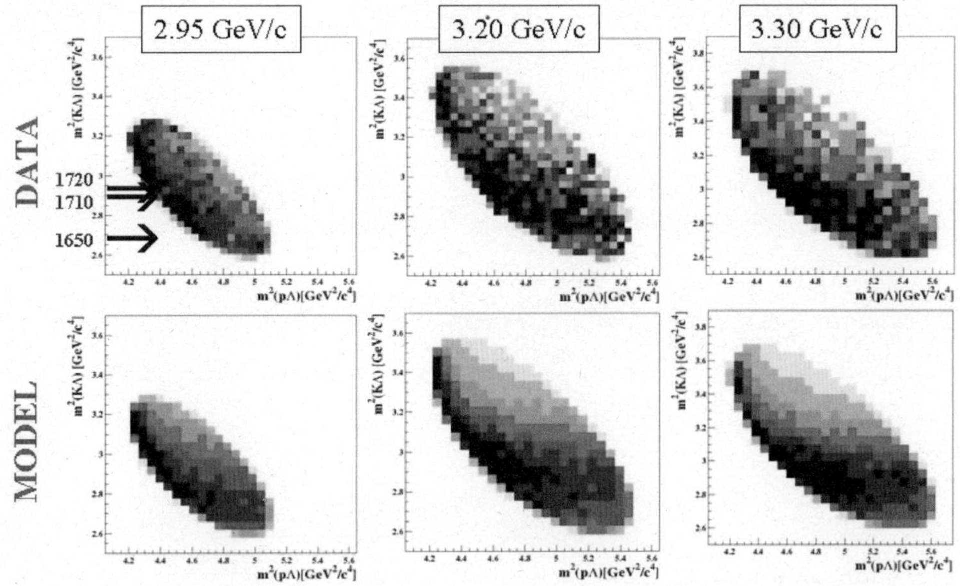

Fig. 6. Dalitz plots of the reaction $pp \to K^+ \Lambda p$; data (upper) compared to model fits (lower).

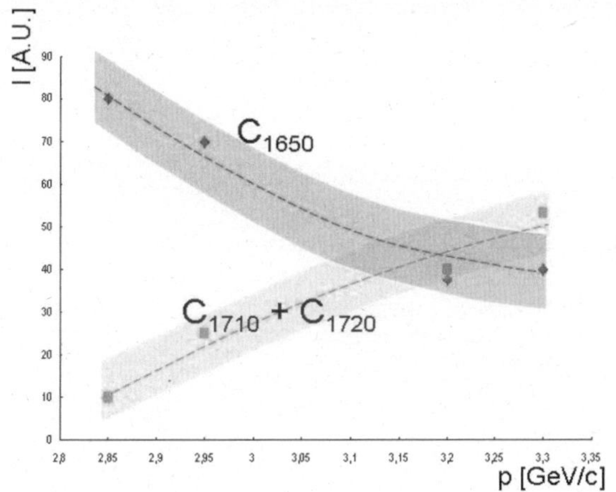

Fig. 7. Contribution of $N^*(1650)$ compared to the sum of $N^*(1710) + N^*(1720)$ as a function of the beam momentum.

The strengths of the various contributions were adjusted individually to achieve a best fit for the various Dalitz plots. The results are shown in fig. 6 (lower part). The data are well described by the model fits. The obtained reduced χ^2-values are between 1.4 and 2.0. The most interesting result is that the strength of the contribution of the $N^*(1650)$-resonance compared to the sum of the contributions of the $N^*(1710)$ and $N^*(1720)$ changes dramatically with the beam momentum. This is shown in fig. 7. The given bands correspond to a 3σ-error interval of the extracted strengths of the resonances. The amplitude of the non-resonant contribution is smaller by a factor of about ten compared to the sum of the three contributing resonances. The influence of the $p\Lambda$ final-state interaction is significant even for the highest momentum; within errors is the corresponding amplitude independent of the beam momentum. From these results it has to be concluded that there is a dominant exchange of non-strange mesons. Only these are able to contribute to the observed leading mechanism via N^*-resonances.

4 Summary and outlook

The COSY-TOF experiment is well suited for hyperon production experiments.

A final decision on the existence of the Θ^+ should be possible by using the extended data set of the reaction $pp \to K^0 \Sigma^+ p$, which is analyzed at the moment.

For the reaction channel $pp \to K^+ \Lambda p$ a strong influence of N^*-resonances was observed.

Data with much larger event samples, which are under investigation, will allow to study the resonance parameters in detail and to search for unknown resonances. In this context the inclusion of other reaction channels accessible through np-reactions by the use of a deuterium target, and the use of a polarized beam, which have already been started (tested), will be of special importance.

References

1. http://www.fz-juelich.de/ikp/COSY-TOF/index_e.html.
2. LEPS Collaboration (T. Nakano *et al.*), Phys. Rev. Lett. **91**, 012002 (2003).
3. COSY-TOF Collaboration, Phys. Lett. B **595**, 127 (2004).
4. COSY-TOF Collaboration, Phys. Lett. B **420**, 217 (1998).
5. COSY-TOF Collaboration, Phys. Lett. B **632**, 27 (2006).
6. A. Sibirtsev, private communications, 2002, 2005.
7. A. Reuber, K. Holinde, J. Speth, Nucl. Phys. A **570**, 543 (1994).

Eur. Phys. J. A **31**, 506–508 (2007)

DOI 10.1140/epja/i2006-10180-y

Special Article – QNP 2006

Decays of excited baryons in the large-N_c expansion of QCD

J.L. Goity[1,2] and N.N. Scoccola[3,4,5,a]

[1] Department of Physics, Hampton University, Hampton, VA 23668, USA
[2] TJNAF, Newport News, VA 23606, USA
[3] Physics Department, CNEA, (1429) Buenos Aires, Argentina
[4] Universidad Favaloro, Solís 453, (1078) Buenos Aires, Argentina
[5] CONICET, Rivadavia 1917, (1033) Buenos Aires, Argentina

Received: 8 October 2006
Published online: 16 February 2007 – © Società Italiana di Fisica / Springer-Verlag 2007

Abstract. We present the analysis of the decay widths of excited baryons in the framework of the $1/N_c$ expansion of QCD. These studies are performed up to order $1/N_c$ and include both positive- and negative-parity excited baryons.

PACS. 11.15.Pg Expansions for large numbers of components (*e.g.*, $1/N_c$ expansions) – 13.30.-a Decays of baryons – 14.20.Gk Baryon resonances with $S = 0$

1 Introduction

The $1/N_c$ expansion of QCD [1] has proven to be a useful tool for analyzing the baryon spectrum [2]. This success is mostly a consequence of the emergent contracted spin-flavor symmetry in the large-N_c limit [3,4]. Although spin-flavor is not an exact symmetry in the excited baryon sector [5], its N_c^0 breaking turns out to be small as several analyses of the baryon spectrum have shown [6–8]. We apply here the $1/N_c$ expansion of QCD to the study of the decays of the non-strange excited baryons belonging to the **70**-plet of negative parity and the **56**-plets of positive parity.

2 Strong decays

The ℓ_P partial-wave strong decay width of a resonance with total angular momentum, spin and isospin J^*, S^*, I^*, respectively, into a ground-state baryon with quantum numbers J, I and a pseudo-scalar meson with isospin I_P can be expressed as

$$\Gamma^{[\ell_P, I_P]} = \frac{k_P}{8\pi^2} \frac{M_B}{M_B^*} \frac{|B(\ell_P, I_P, J, I, J^*, I^*, S^*)|^2}{(2J^* + 1)(2I^* + 1)}, \quad (1)$$

where $B(\ell_P, I_P, J, I, J^*, I^*, S^*)$ are the reduced matrix elements of the baryonic operator. For each meson angular-momentum projection μ and isospin projection α, such

Table 1. Basis operators for the strong decays of the non-strange **70**-plet baryons. The operators for eta D-wave decays are not listed since no empirical information is available for those decays. nB indicates the n-bodyness of each operator.

		Name	Operator	Order
	1B	$O_1^{[0,1]}$	$\left(\xi^{(1)}\, g\right)^{[0,1]}$	1
Pion		$O_2^{[0,1]}$	$\frac{1}{N_c}\left(\xi^{(1)}\,(s\,T_c)^{[1,1]}\right)^{[0,1]}$	$1/N_c$
S-wave	2B	$O_3^{[0,1]}$	$\frac{1}{N_c}\left(\xi^{(1)}\,(t\,S_c)^{[1,1]}\right)^{[0,1]}$	$1/N_c$
		$O_4^{[0,1]}$	$\frac{1}{N_c}\left(\xi^{(1)}\,(g\,S_c)^{[1,1]}\right)^{[0,1]}$	$1/N_c$
	1B	$O_1^{[2,1]}$	$\left(\xi^{(1)}\, g\right)^{[2,1]}_{[i,a]}$	1
		$O_2^{[2,1]}$	$\frac{1}{N_c}\left(\xi^{(1)}\,(s\,T_c)^{[1,1]}\right)^{[2,1]}$	$1/N_c$
Pion		$O_3^{[2,1]}$	$\frac{1}{N_c}\left(\xi^{(1)}\,(t\,S_c)^{[1,1]}\right)^{[2,1]}$	$1/N_c$
D-wave	2B	$O_4^{[2,1]}$	$\frac{1}{N_c}\left(\xi^{(1)}\,(g\,S_c)^{[1,1]}\right)^{[2,1]}$	$1/N_c$
		$O_5^{[2,1]}$	$\frac{1}{N_c}\left(\xi^{(1)}\,(g\,S_c)^{[2,1]}\right)^{[2,1]}$	$1/N_c$
		$O_6^{[2,1]}$	$\frac{1}{N_c}\left(\xi^{(1)}\,(s\,G_c)^{[2,1]}\right)^{[2,1]}$	1
	3B	$O_7^{[2,1]}$	$\left(\frac{\xi^{(1)}}{N_c^2}\left(s\,(\{S_c, G_c\})^{[2,1]}\right)^{[2,1]}\right)^{[2,1]}$	$1/N_c$
		$O_8^{[2,1]}$	$\left(\frac{\xi^{(1)}}{N_c^2}\left(s\,(\{S_c, G_c\})^{[2,1]}\right)^{[3,1]}\right)^{[2,1]}$	$1/N_c$
eta	1B	$O_1^{[0,0]}$	$\left(\xi^{(1)}\, s\right)^{[0,0]}$	1
S-wave	2B	$O_2^{[0,0]}$	$\frac{1}{N_c}\left(\xi^{(1)}\,(s\,S_c)^{[1,0]}\right)^{[0,0]}$	$1/N_c$

[a] e-mail: scoccola@tandar.cnea.gov.ar

Table 2. Fit parameters. LO: there is two-fold ambiguity for the angle θ_1. For θ_3 there is an *almost* two-fold ambiguity given by the two values indicated in parenthesis and which only differ in the two slightly different values of $C_6^{[2,1]}$. NLO: due to lack of empirical data, 3B operators and the subleading operator for η emission are not included. No degeneracy in θ_1 but *almost* two-fold ambiguity in θ_3 given by the two values indicated in parenthesis. Values of coefficients which differ in the corresponding fits are indicated in parenthesis.

	LO	NLO
$C_1^{[0,1]}$	31 ± 3	23 ± 3
$C_2^{[0,1]}$	–	$(7.4, 32.5) \pm (27, 41)$
$C_3^{[0,1]}$	–	$(20.7, 26.8) \pm (12, 14)$
$C_4^{[0,1]}$	–	$(-26.3, -66.8) \pm (39, 65)$
$C_1^{[2,1]}$	4.6 ± 0.5	3.4 ± 0.3
$C_2^{[2,1]}$	–	-4.5 ± 2.4
$C_3^{[2,1]}$	–	$(-0.01, 0.08) \pm 2$
$C_4^{[2,1]}$	–	5.7 ± 4.0
$C_5^{[2,1]}$	–	3.0 ± 2.2
$C_6^{[2,1]}$	$(-1.86, -2.25) \pm 0.4$	-1.73 ± 0.26
$C_1^{[0,0]}$	11 ± 4	17 ± 4
θ_1	1.56 ± 0.15 0.35 ± 0.14	0.39 ± 0.11
θ_3	$(3.00, 2.44) \pm 0.07$	$(2.82, 2.38) \pm 0.11$
χ^2_{dof}	1.5	0.9
dof	10	3

Table 3. Basis operators for the decays of the low-lying positive-parity baryons. Note that in the case of the $\ell = 0$ (Roper) baryons the operator $O_3^{[\ell_P,1]}$ is absent.

	Name	Operator	Order
1B	$O_1^{[\ell_P,1]}$	$\frac{1}{N_c} \left(\xi^{(\ell)}\, G \right)^{[\ell_P,1]}$	1
2B	$O_2^{[\ell_P,1]}$	$\frac{1}{N_c^2} \left(\xi^{(\ell)}\, ([S\,,\,G])^{[1,1]} \right)^{[\ell_P,1]}$	$1/N_c$
	$O_3^{[\ell_P,1]}$	$\frac{1}{N_c^2} \left(\xi^{(\ell)}\, (\{S\,,\,G\})^{[2,1]} \right)^{[\ell_P,1]}$	$1/N_c$

Table 4. Fit parameters corresponding to the pion P-wave decays of $\ell = 0$ excited baryons.

Pion P-waves	LO	NLO
χ^2_{dof}	4.05	0.1
dof	3	2
$C_1^{[1,1]}$	18.7 ± 2.4	17.0 ± 1.6
$C_2^{[1,1]}$	–	24.4 ± 6.3

operator admits an expansion in $1/N_c$ of the form [9,10]

$$B_{[\mu,\alpha]}^{[\ell_P,I_P]} = \left(\frac{k_P}{\Lambda} \right)^{\ell_P} \sum_{q,j} C_q^{[\ell_P,I_P]} \left(\xi^{(\ell)}\, \mathcal{G}_q^{[j,I_P]} \right)_{[\mu,\alpha]}^{[\ell_P,I_P]}. \quad (2)$$

The factor $\left(\frac{k_P}{\Lambda} \right)^{\ell_P}$ is included to account for the chief meson momentum dependence of the partial wave, where we set the scale Λ to $200\,\mathrm{MeV}$. The operator $\xi^{(\ell)}$ drives the transition from the $(2\ell + 1)$-plet to the singlet $O(3)$ state. The operators $\mathcal{G}$ give transitions within the spin-flavor representations in which the excited and GS baryons reside. They can be written as products of the generators of the spin-flavor algebra acting on the excited quark state $\lambda = s_i, t_a, g_{ia}$ and on the core $\Lambda_c = (S_c)_i, (T_c)_a, (G_c)_{ia}$. The relevant operators for the non-strange members of the **70**-plet are given in table 1.

The dynamics of the decays is encoded in the effective dimensionless coefficients $C_q^{[\ell_P,I_P]}$. Here, they are obtained by fitting the available empirical decay widths [11]. In the case of the negative-parity baryons we have, in addition, two extra unknown parameters: the mixing angles θ_{2J} between the two sets of excited nucleon states N_J^*, where $J = 1/2, 3/2$. The results of the corresponding fits are shown in table 2. From those results we observe that, in general, the LO operators already provide a reasonable description of the decay widths. The NLO operators play some role in improving the results, although their coefficients are not well determined due to the large error bars in the empirical widths.

The basis operators for the decays of the positive-parity baryons are given in table 3. Note that, since the corresponding spin-flavor wave functions are completely symmetric, only the total generators $\Lambda = \lambda + \Lambda_c$ are required.

The results of the fits corresponding to the pion P-wave decays of $\ell = 0$ excited (Roper multiplet) baryons are given in table 4. We observe that in this case the LO fit leads to rather poor results. In fact, the NLO operator is essential to obtain a good description of these decays. It should be noted, however, that the coefficient of the LO order operator remains stable, and that the one of the NLO operator is of natural size.

The results for the decays of the $\ell = 2$ baryons are given in table 5. For the P-wave decays, the LO analysis already provides an excellent fit. In fact, the NLO analysis implies that the coefficients of the 2B operators are compatible with zero. In the case of the F-wave decays, the LO analysis gives a reasonable fit provided the empirical errors are taken to be 30% or more, which is the expected accuracy of such an analysis. If in the NLO fit the errors are taken at its empirical estimates the χ^2_{dof} turns out to be rather large. The main contribution to this quantity comes from the $\pi\Delta$ decay of the $N(1680)$ for which only an upper bound is given in ref. [11]. As shown in table 5, if such a decay is removed from the analysis a quite good fit is obtained.

Another interesting observation that points to the consistency of the framework based on the approximate $O(3) \times SU(2N_f)$ symmetry is that the predicted

Table 5. Fit parameters corresponding to the pion P- and F-wave decays of $\ell = 2$ excited baryons. In the case of the LO fit of the F-wave decays those empirical errors that are less than 30% have been increased up to that value. In the corresponding NLO fit the empirical value of the decay $N(1680) \to \pi\,\Delta$ has not been considered.

Pion P-waves	LO	NLO
χ^2_{dof}	0.15	0.44
dof	3	1
$C_1^{[1,1]}$	6.83 ± 0.77	4.09 ± 0.47
$C_2^{[1,1]}$	–	0.11 ± 2.41
$C_3^{[1,1]}$	–	0.43 ± 6.09
Pion F-waves	LO	NLO
χ^2_{dof}	1.73	0.38
dof	4	1
$C_1^{[3,1]}$	1.01 ± 0.09	0.84 ± 0.04
$C_2^{[3,1]}$	–	-1.30 ± 0.21
$C_3^{[3,1]}$	–	-0.26 ± 0.47

suppression of the η channels in the decays of positive-parity baryons is clearly displayed by the observed decays. This represents a strong experimental confirmation that those baryons do belong primarily to a symmetric representation of $SU(2N_f)$. In the case of the negative-parity baryons the η channel is not suppressed and is indeed very important. This implies that these states belong primarily to the mixed-symmetric representation of $SU(2N_f)$. Thus, η channels serve as a selector of the spin-flavor structure of excited baryons.

3 Radiative decays

The situation concerning the radiative decays is in a more preliminary stage. In ref. [12] a LO analysis of the radiative decays of the $(\mathbf{70}, 1^-)$-plet states has been performed, and a reasonable fit obtained. Perhaps the most important conclusion of that study is that, although there are already 2B operators contributing to this order, the amplitudes are dominated by 1B operators. Of course, to have a clearer understanding a NLO analysis is required. Concerning the radiative decays of the positive-parity resonances belonging to the $\mathbf{56}$-plet, a study which includes up to NLO corrections is currently under way [13]. Preliminary results indicate that, as in the case of the strong decays, the LO analysis seem to be problematic for Roper baryons. On the other hand, at least some LO model independent relations seems to work well for the $(\mathbf{56}, 2^+)$-plet states. For example, for the decays of the $\Delta(1950)$ one obtains the model-independent relation $A_{3/2}^N / A_{1/2}^N = 1.29$ to be compared with the empirical value 1.27 ± 0.24.

4 Conclusions

To conclude, the $1/N_c$ expansion provides a systematic approach to the properties of the excited baryons. The analysis of the masses shows that the N_c^0 breaking of the spin-flavor symmetry is small. For strong decays one finds, in general, a dominance of the one-body LO operators. In some cases, as *e.g.* the D-wave decays of the negative-parity excited baryons, the $1/N_c$ corrections are not well established due to the rather large uncertainties of the empirical data. In the particular case of the Roper baryons, two-body NLO operators seem to be required to obtain a good description of the empirical decay widths. In the case of the radiative decays, there is work in progress. Preliminary results indicate that the situation is similar to that of strong decays. Finally, we should notice that, here, we have followed the so-called operator approach to excited baryons in large-N_c QCD. Since the decay widths of such baryons are of order N_c^0, there is also an alternative large-N_c description of such states as resonances in meson-baryon scattering [14].

This work was partially supported by the National Science Foundation (USA) through grant # PHY-9733343 (JLG) and by the CONICET and ANPCYT (Argentina) through grants PIP 02368 and PICT 03-08580 (NNS).

References

1. G.'t Hooft, Nucl. Phys. B **72**, 461 (1974); E. Witten, Nucl. Phys. B **160**, 57 (1979).
2. For reviews see: A.V. Manohar, hep-ph/9802419; E. Jenkins, hep-ph/0111338; R.F. Lebed, Czech. J. Phys. **49**, 1273 (1999). For more recent developments, J.L. Goity *et al.* (Editors), *Large Nc QCD 2004: Proceedings of the Workshop, Trento, Italy, 5-11 July 2004* (World Scientific, Singapore, 2005).
3. J.L. Gervais, B. Sakita, Phys. Rev. Lett. **52**, 87 (1984); Phys. Rev. D **30**, 1795 (1984).
4. R. Dashen, A.V. Manohar, Phys. Lett. B **315**, 425; 438 (1993); E. Jenkins, Phys. Lett. B **315**, 441 (1993).
5. J.L. Goity, Phys. Lett. B **414**, 140 (1997).
6. C.E. Carlson, C.D. Carone, J.L. Goity, R.F. Lebed, Phys. Lett. B **438**, 327 (1998); Phys. Rev. D **59**, 114008 (1999).
7. C.L. Schat, J.L. Goity, N.N. Scoccola, Phys. Rev. Lett. **88**, 102002 (2002); J.L. Goity, C.L. Schat, N.N. Scoccola, Phys. Rev. D **66**, 114014 (2002); J.L. Goity, C. Schat, N.N. Scoccola, Phys. Lett. B **564**, 83 (2003).
8. N. Matagne, F. Stancu, Phys. Rev. D **71**, 014010 (2005); Phys. Lett. B **631**, 7 (2005); Phys. Rev. D **73**, 114025 (2006).
9. J.L. Goity, C.L. Schat, N.N. Scoccola, Phys. Rev. D **71**, 034016 (2005).
10. J.L. Goity, N.N. Scoccola, Phys. Rev. D **72**, 034024 (2005).
11. Particle Data Group (S. Eidelman *et al.*), Phys. Lett. B **592**, 1 (2004).
12. C.E. Carlson, C.D. Carone, Phys. Rev. D **58**, 053005 (1998).
13. J.L. Goity, N.N. Scoccola, in preparation.
14. For short reviews on this approach, see T.D. Cohen, hep-ph/0501090; R.F. Lebed, hep-ph/0501021.

Eur. Phys. J. A **31**, 509–511 (2007)
DOI 10.1140/epja/i2006-10248-8

Special Article – QNP 2006

The HERMES Recoil Detector

W. Yu[a]

On behalf of the HERMES Collaboration
II.Physikalisches Institut, University of Giessen, 35392 Giessen, Germany

Received: 23 November 2006
Published online: 21 March 2007 – © Società Italiana di Fisica / Springer-Verlag 2007

Abstract. The HERMES Collaboration installed a new Recoil Detector to upgrade the existing spectrometer to study hard exclusive processes which provide access to generalised parton distributions (GPDs) and hence to the orbital angular momentum of quarks. The HERMES Recoil Detector mainly consists of three components: a silicon detector surrounding the target cell inside the beam vacuum, a scintillating fibre tracker and a photon detector with three layers of tungsten and scintillator bars in three different orientations. All three detectors are located inside a solenoidal magnet which provides a 1 T longitudinal magnetic field. The Recoil Detector was installed in January 2006 and data taking will last until July of 2007.

PACS. 29.30.Aj Charged-particle spectrometers: electric and magnetic – 14.20.Dh Protons and neutrons – 29.40.Wk Solid-state detectors – 29.40.Mc Scintillation detectors

1 Introduction

HERMES is one of the three experiments at HERA, and uses the 27.5 GeV longitudinal polarised electron/positron beam and a polarised or unpolarised gas target internal to the storage ring to explore and disentangle the different contributions to the spin of the nucleon. The HERMES spectrometer [1] is a forward-angle instrument with a dipole magnet providing an integrated field of 1.3 Tm. The spectrometer consists of two identical halves located above and below the electron/positron beam pipe, and has an angular acceptance of ± 170 mrad horizontally, and $\pm(40\text{–}140)$ mrad vertically. Three drift chambers in the front region, three multi-wire-proportional chambers (MCs) inside the magnet and four drift chambers in the backward region compose the tracking system of the spectrometer. The track reconstruction is based on a pattern-matching algorithm and momentum look-up method [1]. The average angular resolution is better than 0.6 mrad and the average momentum resolution is better than 2%. These resolutions degraded somewhat after the insertion of the ring imaging Cherenkov detector (RICH) in 1998 with more materials in the path of the particles. Particle identification (PID) is provided by a lead-glass calorimeter, a pre-shower detector, a transition radiation detector, and a RICH. The PID system provides electron/positron identification with an average efficiency of 98–99% and a hadron contamination of less than 1%. The trigger for the scattered electron/positron is formed by requiring hits in three scintillator hodoscopes together with energy deposited in two adjacent columns of the calorimeter.

Due to the limited acceptance and energy and momentum resolutions of the HERMES spectrometer, the HERMES Collaboration installed a new Recoil Detector in January 2006, whose main purpose is to study hard exclusive processes by detecting the recoiling particles. These hard exclusive processes provide access to the generalised parton distributions (GPDs) [2]. GPDs offer a possibility to determine the orbital angular momentum of quarks, and can be accessed by studying deeply virtual Compton scattering (DVCS) [3]. In the reaction $ep \to e'\gamma p$, the DVCS process interferes with the Bethe-Heitler process where the incoming or outgoing electron/positron radiates a hard real photon in the Coulomb field of the nucleon. This interference leads to measurable beam spin and beam charge asymmetries around the virtual photon direction.

2 The Recoil Detector

The energy resolution for the electrons/positrons and photons of the existing HERMES spetrometer is not sufficient to select a clean sample of the exclusive DVCS process, and it has no acceptance to detect recoiling particles under large angles. The Recoil Detector [4] has been designed to upgrade the HERMES spectrometer to mainly study this exclusive DVCS process. Its objectives are positive identification of recoiling protons, measuring the momenta of these particles to improve the resolution of the transverse momentum and rejecting non-exclusive background events.

---

[a] e-mail: `yuwl@mail.desy.de`

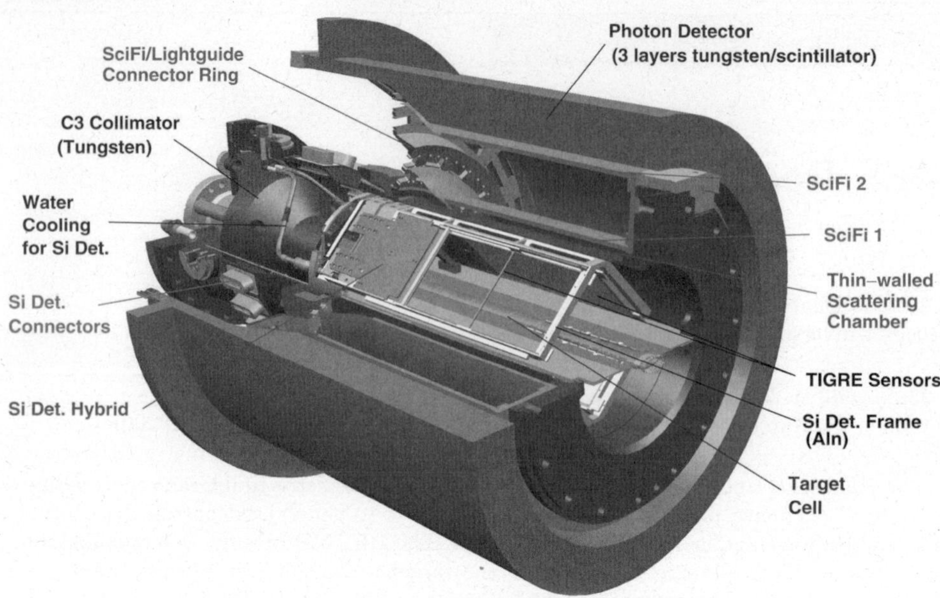

Fig. 1. Schematic overview of the HERMES Recoil Detector.

The Recoil Detector shown in fig. 1 consists of three main components: a silicon strip detector surrounding the target cell inside the beam vacuum, a scintillating fibre tracker and a photon detector consisting of three layers of tungsten and scintillator bars. All three detectors are placed in a longitudinal magnetic field of 1 T generated by a solenoidal magnet. An additional collimator (C3) was installed in the front part of the target chamber to suppress the beam-related background in the Recoil Detector. The fibre tracker and the silicon detectors are optimized to detect recoiling protons from hard exclusive processes. They cover a momentum range around $0.1\,\mathrm{GeV}/c$ to $1.4\,\mathrm{GeV}/c$ and provide the particle identification properties to discriminate protons from pions. The photon detector is used for particle identification of high-momentum particles and for background rejection of neutral particles. All three detectors provide a measurement of the deposited energy of the traversing particles. Momentum determination for low-momenta particles stopping inside the silicon detector is performed using the total energy deposition in both layers of silicon in combination with the reconstructed track direction. The momentum of fast particles is determined by their bending in the longitudinal magnetic field.

2.1 The silicon strip detector

The silicon strip detector (SSD) measures the momentum of recoiling protons in the range $135\text{–}400\,\mathrm{MeV}/c$ via their energy deposition and also provides space points for the tracking of minimum ionising particles (MIPs). It consists of 16 double-sided silicon sensors based on the TTT [5] design of Micron Semiconductors, with 128 strips per side and a strip width of $758\,\mu\mathrm{m}$. The sensors measure $99 \times 99\,\mathrm{mm}^2$ with a thickness of $300\,\mu\mathrm{m}$ and are arranged around the target cell in two layers of four modules, each

containing two sensors. The necessary high dynamic range of up to 70 MIPs is established by a charge division readout where the signal is split into a high-gain and a low-gain path. Both signals are read out by Helix 3.0 readout chips. A signal-to-noise ratio of 6.5 for a minimum ionising particle could be reached at test beams with an efficiency of close to 99%. The sensors have been calibrated with low-energy protons at the Erlangen tandem accelerator to better than 2%.

2.2 The scintillating-fibre tracker

The scintillating-fibre tracker (SFT) measures the particles of higher momenta above $250\,\mathrm{MeV}/c$ using two barrels consisting of 4 layers of 1 mm Kuraray SCSF-78 scintillating fibres each. A stereo angle of 10° between each two adjacent layers of each barrel allows space point reconstruction. Each layer was assembled in modules such that single modules with defects could have been discarded. The diameters of the barrels are 220 mm and 370 mm, respectively. The signals of the fibres are read out via 3.5 m long light guides of Kuraray clear fibres coupled to Hamamatsu H-7548 64-channel PMTs. The readout of the SFT is based on front-end cards using GASSIPLEX chips which are hold-and-sample chips. Due to this fact, the readout integrates over 6 adjacent HERA bunches, which would increase the random background by the same factor. In order to avoid this, the dynode 12 signals of the PMTs are used to extract additional fast-timing information to discriminate between different HERA bunches. The cylinders are built from 72 SciFi strips as self-supporting structures to minimize the material transversed by the particles. A gain monitoring system (GMS) was built to monitor changes in the response of the PMTs. The momenta of particles are measured through the deflection of the tracks in the solenoidal 1 T longitudinal magnetic field. From a GSI test experiment with a mixed proton and pion beam at GSI/Darmstadt, the detection efficiency of the SFT modules for a minimum ionising particle using a threshold of 1 photon electron has been determined to be $\approx 99\%$, efficiencies for protons even exceed this value. In addition to the tracking information the energy deposition in the fibres will be used to obtain pion/proton particle identification up to about $800\,\mathrm{MeV}/c$.

2.3 The photon detector

The photon detector consists of 3 layers of scintillator bars with tungsten as a converter. The scintillators are 1.1 cm thick and are read out using wavelength shifting fibres. The innermost scintillators are parallel to the beam, the two further layers have stereo angles of $\pm 45°$, respectively so that charged-particle tracks can be reconstructed in space. The detection of photons, especially those from the Δ^+ decays, will allow the further reduction of non-exclusive background events. The photon detector also provides a cosmic-ray trigger for test and alignment measurements, and provides additional pion/proton particle identification at higher momenta.

3 The commissioning of the Recoil Detector

Before the installation of the Recoil Detector in January 2006, the completely assembled detector was tested in 2005 using cosmic particles with the magnetic field turned on at a test site in the HERMES experimental hall. This test not only provided the chance to commission the whole hardware and the data acquisition system, but also gave the opportunity to accumulate cosmic data, to develop the required software, and to take data for the internal alignment of the detector.

After a rough alignment, the residuals from cosmic tracks were measured to be 0.27 mm for the silicon detector and 0.35 mm for the SFT, about 20% higher than what would be expected for a perfectly aligned detector. Efficiencies for the SFT and the photon detector were in agreement with the test experiments, but the MIP detection efficiency for the silicon strip detector was with 80–90% significantly lower due to high common mode noise present in the cosmic test. The electrical set-up of the detector was revised and the noise level was reduced to the level known from the previous test measurements.

The Recoil Detector was installed in January 2006 and started data taking in February. Data taking using the HERA electron beam has been successfully carried out for the scintillating-fibre tracker and the photon detector. The target cell was damaged by accidental beam loss in March, which caused problems for the silicon detector due to radio frequency noise introduced by the beam bunches. At the end of June 2006 the HERA beam polarity was changed from electrons to positrons. The data taking was continued with the fully installed Recoil Detector after the repair of the target in July 2006.

References

1. HERMES Collaboration (K. Ackerstaff *et al.*), Nucl. Instrum. Methods A **417**, 230 (1998).
2. X. Ji, Phys. Rev. Lett. **78**, 610 (1997); Phys. Rev. D **55**, 7114 (1997).
3. V.A. Korotkov, W.-D. Nowak, Nucl. Phys. A **711**, 175 (2002).
4. R. Kaiser *et al.*, HERMES Internal Note 02-033 (2002).
5. http://www.micronsemiconductor.co.uk/pdf/ttt.pdf.

Eur. Phys. J. A **31**, 512–514 (2007)
DOI 10.1140/epja/i2006-10226-2

Special Article – QNP 2006

Search for missing baryon resonances via associated strangeness photoproduction

B. Saghai[1,a], J.-C. David[1], B. Juliá-Díaz[2], and T.-S.H. Lee[3]

[1] Laboratoire de Recherche sur les Lois Fondamentales de l'Univers, DAPNIA/SPhN, CEA/Saclay, Gif-sur-Yvette, France
[2] ECM, Facultat de Fisica, Universitat de Barcelona, E-08028 Barcelona, Spain
[3] Physics Division, Argonne National Laboratory, Argonne, IL 60439, USA

Received: 8 November 2006
Published online: 1 March 2007 – © Società Italiana di Fisica / Springer-Verlag 2007

Abstract. Differential cross-section and single polarization observables in the process $\gamma p \to K^+ \Lambda$ are investigated within a constituent-quark model and a dynamical coupled-channel formalism. The effects of two new nucleon resonances and of the $K^*(892)$- and $K1(1270)$-exchanges are briefly presented.

PACS. 11.80.-m Relativistic scattering theory – 13.60.Le Meson production – 14.20.Gk Baryon resonances with $S = 0$ – 24.10.Eq Coupled-channel and distorted-wave models

1 Introduction

Photoproduction of mesons appears to be a promising area to investigate issues related to the missing-baryon resonances, predicted by different QCD-inspired approaches [1]. Recent works, summarized, *e.g.*, in ref. [2], show indications on few of them. In that latter publication, we have reported on significant contributions from new S_{11} and D_{13} resonances to the process $\gamma p \to K^+ \Lambda$ studied in the total center-of-mass energy range $W \equiv \sqrt{s} \approx$ 1.6 GeV to 2.6 GeV, which corresponds to the baryon resonances mass region.

In this short report, we focus on the interplay between s- and t-channel contributions and the duality hypothesis.

2 Theoretical frame and t-channel issues

In order to study the kaon photoproduction on the proton, we have developed [2,3] multistep coupled-channel formalisms for the reactions $\pi N \to \pi N$, $\pi N \to KY$, $KY \to KY$, and $\gamma p \to KY$. The $\pi N \to \pi N$ potential comes from an advanced version of effective Lagrangians approaches [4] using a unitary transformation method. The same method is also used to derive from effective Lagrangians the basic non-resonant $\pi N \to KY$ and $KY \to KY$ transition potentials. The direct channel $\gamma p \to K^+ \Lambda$ is handled within a chiral constituent quark model [2,5] based on the $SU(6) \otimes O(3)$ broken symmetry,

with the starting point being the low-energy QCD Lagrangian [6]. The four components for the photoproduction of pseudoscalar mesons based on the QCD Lagrangian are

$$\mathcal{M}_{fi} = \mathcal{M}_{seagull} + \mathcal{M}_s + \mathcal{M}_u + \mathcal{M}_t. \tag{1}$$

The first term in eq. (1) is a seagull term. It is generated by the gauge transformation of the axial vector A_μ in the QCD Lagrangian. The second and the third terms correspond to the s- and u-channels, respectively. The last term is the t-channel contribution and contains two parts: i) K^+-exchange; ii) K^*- and $K1$-exchanges. In our previous investigations [2,7], the dynamics of our models were partially based on the duality hypothesis, according to which, at any given energy one could use *either* the s-channel resonance prescription *or* the t-channel Regge pole description, provided that we sum over an *infinite* number of terms. An approximation to this idea and widely discussed in the literature [8], is to express the physical scattering as a series of s-channel resonances plus a general background. In those works, we had adopted this approximation, given that our approach allows us to take into account individual contributions from all known nucleon resonances in the first and second resonance regions, and treat as degenerate higher-mass resonances. However, with the advent of data at high energies ($W \geq 2.4$ GeV), it is desirable to find out how well higher-mass resonances are handled and if there is need to include t-channel resonances. If so, we ought to study the drawback of such treatment on the missing resonances issues.

a e-mail: bijan.saghai@cea.fr

3 Results and discussion

We have extended the formalism presented in ref. [2] to embody the t-channel K^*- and $K1$-exchanges [9].

The fitted data base contains 1029 data points released recently: differential cross-sections from SAPHIR [10], recoil-Λ polarization from JLab [11] and GRAAL [12], as well as polarized beam asymmetry from GRAAL [12]. Those data span the following angular and energy ranges in the phase space: $18° \leq \theta_K^{c.m.} \leq 162°$, $0.912\,\mathrm{GeV} \leq E_\gamma^{lab} \leq 2.575\,\mathrm{GeV}$, corresponding to the center-of-mass energy range $1.6\,\mathrm{GeV} \leq W \leq 2.4\,\mathrm{GeV}$. At this stage of our study, differential cross-section data from JLab [13] and LEPS [14] have not been included in the fitted data base in order to avoid possible confusion between t-channel effects and consequences of inconsistencies among different data base, as, for example, discussed in refs. [2,15].

In order to make clear the respective roles played by t-channel contributions and the new resonances, we have performed minimizations for three configurations with respect to the reaction mechanism, namely, a) full model: it includes all known nucleon and hyperon resonances, t-channel contributions from the exchange of $K^*(892)$ and $K1(1270)$, as well as contributions from new S_{11}- and D_{13}-resonances. In the subsequent two configurations, the following contributions have been *swictched-off*: b) K^*- and $K1$-exchanges; c) new resonances.

In figs. 1–3 the results of those three configurations are compared with the data around $E_\gamma^{lab} = 1.075\,\mathrm{GeV}$ and $1.375\,\mathrm{GeV}$, which correspond to the total center-of-mass energy $W \approx 1.7$ and $1.9\,\mathrm{GeV}$, respectively. Because of lack of space, here we single out results at only two energies,

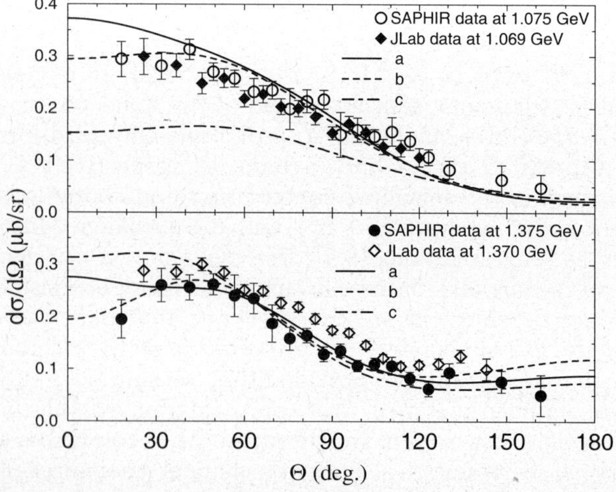

Fig. 1. Differential cross-section for the reaction $\gamma p \to K^+ \Lambda$ as a function of the outgoing kaon angle in the center-of-mass frame. The curves are: a) full model (full curves), b) full model but with t-channel K^* and $K1$ contributions switched-off (dashed curves), c) full model, but with contributions from the new 3rd S_{11}- and D_{13}-resonances swictched-off (dot-dashed curves). The curves in the upper box are for $E_\gamma^{lab} = 1.075\,\mathrm{GeV}$ and in the lower one for $E_\gamma^{lab} = 1.375\,\mathrm{GeV}$. Data are from SAPHIR [10] and JLab [13].

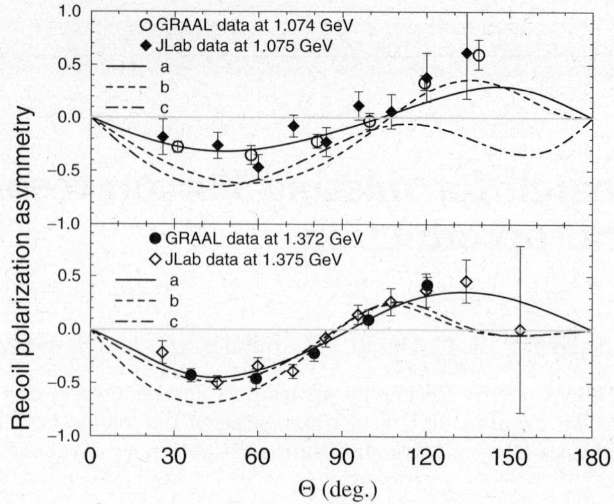

Fig. 2. Same as fig. 1, but for recoil Λ polarization asymmetry. Data are from GRAAL [12] and JLab [11].

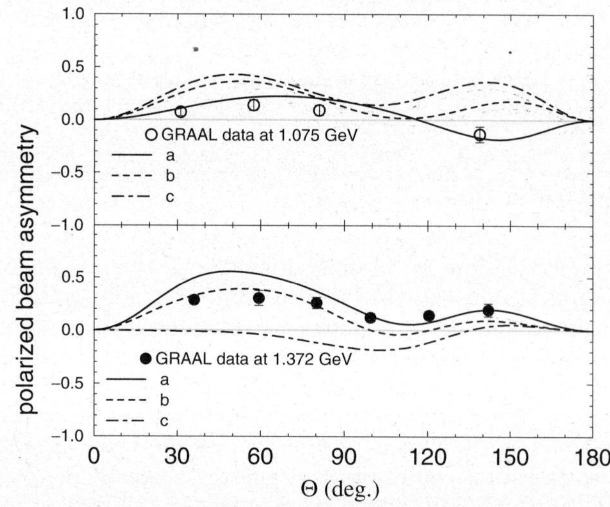

Fig. 3. Same as fig. 1, but for polarized beam asymmetry. Data are from GRAAL [12].

Table 1. Reduced χ^2s for the reaction mechanism configurations, as explained in the text.

Configuration	a	b	c
$\chi^2_{d.o.f}$	1.83	3.48	5.25

which are in the mass region of the new resonances. This choice is also related to the fact that the GRAAL [12] data are limited to $E_\gamma^{lab} \leq 1.5\,\mathrm{GeV}$.

Results for the differential cross-section of the reaction $\gamma p \to K^+ \Lambda$ are depicted in fig. 1, where the JLab data [13] (not fitted) are also shown. Figures 2 and 3 embody results for single polarization asymmetries for recoil-hyperon polarization ($\gamma p \to K^+ \vec{\Lambda}$) and polarized beam ($\vec{\gamma} p \to K^+ \Lambda$).

Table 2. Sensitivity of the investigated observables to the contributions from the t-channel (TC) K^*- and $K1$-exchanges, and from the two new resonances (NR), corresponding to the configurations b) and c), as explained in the text. The most significant angular regions per energy are given in columns 3 to 6.

Observable	Switched-off	$E_\gamma^{lab} = 1.075$ GeV		$E_\gamma^{lab} = 1.375$ GeV	
		Forward hemisphere	Backward hemisphere	Forward hemisphere	Backward hemisphere
a) Cross-section	TC	$\leq 40°$	–	$\leq 30°$	–
	NR	$30°$ to $90°$	–	$\leq 40°$	–
b) Recoil asymmetry	TC	$30°$ to $90°$	–	$20°$ to $50°$	–
	NR	$30°$ to $70°$	$\geq 130°$	–	$\geq 130°$
c) Beam asymmetry	TC	$30°$ to $60°$	$130°$ to $150°$	–	–
	NR	$30°$ to $60°$	$130°$ to $150°$	$30°$ to $90°$	$90°$ to $120°$

The full model (full curves) leads to a $\chi^2_{d.o.f}$ of 1.83 (table 1), and the extracted mass and width for the two new resonances are: $S_{11}(M = 1.820\,\text{GeV},\ \Gamma = 240\,\text{MeV})$ and $D_{13}(M = 1.920\,\text{GeV},\ \Gamma = 160\,\text{MeV})$. The full model gives a reasonable account of the whole fitted data base.

Removing the t-channel resonances (dashed curves) and refitting the data, increases the $\chi^2_{d.o.f}$ by almost a factor of 2 (table 1). In fig. 1, the extreme forward angle data are better reproduced in the absence of t-channel contributions. This is not the case for data at higher energies ($W \geq 2\,\text{GeV}$), as expected from the duality hypothesis. Finally, minimization within the configuration c), which embody no new resonances, leads to a deterioration of the $\chi^2_{d.o.f}$ by about a factor of 3 (table 1). Switching-off the t-channel resonances or the new nucleon resonances produces significant effects on the investigated observables in various angular regions, depending on the incident photon energy. Those features appearing in figs. 1–3 are summarized in table 2.

4 Conclusions

Among the three observables reported here, t-channel contributions show the largest effects in the recoil-Λ polarization. The new resonances introduced here produce noticeable features on all three observables in a rather broad range of phase space. Those features are not mimicked by the t-channel resonances studied here, showing that the s-channel resonances embodied in our approach satisfy to a large extent the duality requirements. The extracted Mass and width values for those resonances are $S_{11}(M = 1.820\,\text{GeV},\ \Gamma = 240\,\text{MeV})$ and $D_{13}(M = 1.920\,\text{GeV},\ \Gamma = 160\,\text{MeV})$. Those values are compatible with findings from other works, namely, for S_{11} refs. [2,5,16] and for D_{13} refs. [2,15,17].

We are indebted to F. Tabakin for his contribution to this work at earlier stages and for enlightening discussions. We wish to thank the GRAAL Collaboration and especially, Annick Lleres and Dominique Rebreyend for having provided us with their data [12] prior to publication.

References

1. See, *e.g.*, S. Capstick, W. Roberts, Prog. Part. Nucl. Phys. **45**, 5241 (2000) and references therein.
2. B. Juliá-Díaz, B. Saghai, T.-S.H. Lee, F. Tabakin, Phys. Rev. C **73**, 055204 (2006).
3. W.-T. Chiang, B. Saghai, F. Tabakin, T.-S.H. Lee, Phys. Rev. C **69**, 065208 (2004).
4. T. Sato, T.-S.H. Lee, Phys. Rev. C **54**, 2660 (1996); **63**, 055201 (2001); A. Matsuyama, T. Sato, T.-S.H. Lee, arXiv: nucl-th/0608015.
5. B. Saghai, Z. Li, Eur. Phys. J. A **11**, 217 (2001).
6. A. Manohar, H. Georgi, Nucl. Phys. B **234**, 189 (1984); Z. Li, H. Ye, M. Lu, Phys. Rev. C **56**, 1099 (1997).
7. W.-T. Chiang, B. Saghai, F. Tabakin, T.-S.H. Lee, Phys. Lett. B **517**, 101 (2001).
8. See, *e.g.*, P. Collins, *An Introduction to Regge Theory and High Energy Physics* (Cambridge University Press, Cambridge, 1977).
9. J.C. David, C. Fayard, G.H. Lamot, B. Saghai, Phys. Rev. C **53**, 2613 (1996).
10. The SAPHIR Collaboration (K.H. Glander *et al.*), Eur. Phys. J. A **19**, 251 (2004).
11. The CLAS Collaboration (J.W.C. McNabb *et al.*), Phys. Rev. C **69**, 042201 (2004).
12. The GRAAL Collaboration (A. Lleres *et al.*), Eur. Phys. J. A **31**, 79 (2007); D. Rebreyend, private communication (2006).
13. The CLAS Collaboration (R. Bradford *et al.*), Phys. Rev. C **73**, 035202 (2006).
14. The LEPS Collaboration (R.G.T. Zegers *et al.*), Phys. Rev. Lett. **91**, 092001 (2003); The LEPS Collaboration (M. Sumihama *et al.*), Phys. Rev. C **73**, 035214 (2006).
15. T. Mart, A. Sulaksono, arXiv: nucl-th/0609077.
16. Z. Li, R. Workman, Phys. Rev. C **53**, R549 (1996); A. Švarc, S. Ceci, arXiv: nucl-th/0009024; G.-Y Chen *et al.*, Nucl. Phys. A **723**, 447 (2003); B. Saghai, Z. Li, *Proceedings of NSTAR 2002 Workshop on the Physics of Excited Nucleons, Pittsburgh, PA (USA), 2002*, edited by S.A. Dytman, E.S. Swanson (World Scientific, New Jersey, 2003) arXiv: nucl-th/0305004.
17. N.G. Kelkar, M. Nowakowski, K.P. Khemchandani, S.R. Jain, Nucl. Phys. A **730**, 121 (2004); A.V. Anisovich *et al.*, Eur. Phys. J. A **25**, 427 (2005); A.V. Sarantsev *et al.*, Eur. Phys. J. A **25**, 441 (2005).

Eur. Phys. J. A **31**, 515–518 (2007)

DOI 10.1140/epja/i2006-10190-9

THE EUROPEAN
PHYSICAL JOURNAL A

Special Article – QNP 2006

A $SU(4) \otimes O(3)$ scheme for nonstrange baryons

P. González[1,a], J. Vijande[1], A. Valcarce[2], and H. Garcilazo[3]

[1] Departament de Física Teórica and IFIC, Universidad de Valencia - CSIC, E-46100 Burjassot, Valencia, Spain
[2] Grupo de Física Nuclear and IUFFyM, Universidad de Salamanca, E-37008 Salamanca, Spain
[3] Escuela Superior de Física y Matemáticas, Instituto Politécnico Nacional, Edificio 9, 07738 Mexico D.F., Mexico

Received: 8 October 2006
Published online: 26 February 2007 – © Società Italiana di Fisica / Springer-Verlag 2007

Abstract. We show that nonstrange baryon resonances can be classified according to multiplets of $SU(4) \otimes O(3)$. We identify spectral regularities and degeneracies that allow us to predict the high-spin spectrum from 2 to 3 GeV.

PACS. 12.39.Jh Nonrelativistic quark model – 14.20.-c Baryons (including antiparticles) – 14.20.Gk Baryon resonances with $S = 0$

1 Introduction

The theoretical study of the high-energy part of the baryonic spectrum has been a subject of interest in the last decade, the aim being to get a better understanding of the dynamics (in particular the confinement mechanism of quarks in the baryon and the decay hadronization process), or, at least, the symmetries involved. In particular, the idea of a parity multiplet classification scheme at high excitation energies as due to chiral symmetry was suggested some years ago [1] and put in question later on [2]. The lack of precise and complete data prevents, at the current moment, to extract any definitive conclusion.

From the point of view of dynamics quark model refined potentials with linear confinement have provided a reasonably accurate description of the whole nonstrange light baryon spectrum once the coupling to πN formation channels is taken into account [3]. The low probability obtained for many resonances to be formed would explain their no experimental detection providing a solution to the so-called missing state problem (the difference between the number of predicted states above 1 GeV excitation energy —infinite with a linear potential— and the number of known resonances).

Alternatively, the use of a quark-quark screened potential [4–6], motivated by recent unquenched QCD lattice calculations showing string breaking in the static potential between two quarks [7,8], allows to obviate the missing state problem (up to the limit of applicability of the model). In refs. [4] and [6] a correct prediction of the number and ordering of the known N and Δ resonances, up to 2.4 GeV mass or 1.5 GeV excitation energy, is obtained.

a e-mail: pedro.gonzalez@uv.es

Above this limit the 3-free quark state is energetically favored pointing out the need to implement the coupling to the continuum. Nonetheless the unambiguous assignment of quantum numbers to experimental states in the region of applicability translates, as we shall show, into a well-defined symmetry pattern.

In this article we identify the symmetry pattern as the one corresponding to $SU(4) \otimes O(3)$, $SU(4)$ containing $SU(2)_{\text{spin}} \otimes SU(2)_{\text{isospin}}$ and $O(3)$ standing for the orbital symmetry, and we analyze spectral regularities and degeneracies according to it. The extension of this pattern to energies above the applicability limit of the model allows us to predict the spectrum in the range 2–3 GeV where only incomplete and non-precise data exist.

2 $SU(4) \otimes O(3)$ Pattern

In ref. [6] a quark model including confinement and minimal one gluon exchange (Coulomb + hyperfine) interactions has been developed. Screening is imposed by requiring that the interaction potential saturates (*i.e.*, becomes constant) at a certain distance to be fixed phenomenologically. Though one cannot obtain a precise fit to the spectrum with such a simplistic model it is amazing that one can make an unambiguous assignment of quantum numbers to the dominant configuration of any J^P ground and first non-radial states up to $J = 11/2$. This assignment agrees completely with the ones available in the literature (only up to $J = 7/2$) with much more refined theoretical models [9] or purely phenomenological analysis [10].

In tables 1 and 2 we group experimental resonances according to their dominant configuration (the symmetry

Table 1. Positive-parity N and Δ states (masses in MeV) for different dominant spatial-spin configurations up to $\simeq 3\,$GeV. Experimental data are from PDG [11]. Stars have been omitted for four-star resonances. States denoted by a question mark correspond to predicted resonances that do not appear in the PDG (their predicted masses appear in table 3).

$(K, L, Symmetry)$	$S = 1/2$	$S = 3/2$
$(0, 0, [3])$	$N(1/2^+)(940)$	
		$\Delta(3/2^+)(1232)$
$(2, 2, [3])$	$N(5/2^+)(1680),\ N(3/2^+)(1720)$	
		$\Delta(7/2^+)(1950)$
$(4, 4, [3])$	$N(9/2^+)(2220)$	
		$\Delta(11/2^+)(2420)$
$(6, 6, [3])$	$N(13/2^+)(**)(2700)$	
		$\Delta(15/2^+)(**)(2950)$
$(2, 0, [21])$	$N(1/2^+)(***)(1710)$	
	$\Delta(1/2^+)(1750)$	
$(2, 2, [21])$	$N(5/2^+)(**)(2000)$	
		$N(7/2^+)(**)(1990)$
	$\Delta(5/2^+)(1905)$	
$(4, 4, [21])$	$N(9/2^+)(2220)$	
		$N(11/2^+)(?)$
	$\Delta(9/2^+)(**)(2300)$	
$(6, 6, [21])$	$N(13/2^+)(2700)$	
		$N(15/2^+)(?)$
	$\Delta(13/2^+)(?)$	

Table 2. Negative-parity N and Δ states (masses in MeV) for different dominant spatial-spin configurations up to $\simeq 3\,$GeV. Experimental data are from PDG [11]. Stars have been omitted for four-star resonances. States denoted by a question mark correspond to predicted resonances that do not appear in the PDG (their predicted masses appear in table 3).

$(K, L, Symmetry)$	$S = 1/2$	$S = 3/2$
$(1, 1, [21])$	$N(3/2^-)(1520),\ N(1/2^-)(1535)$	
		$N(5/2^-)(1675)$
	$\Delta(3/2^-)(1700),\ \Delta(1/2^-)(1620)$	
$(3, 3, [21])$	$N(7/2^-)(2190)$	
		$N(9/2^-)(2250)$
	$\Delta(7/2^-)(*)(2200)$	
$(5, 5, [21])$	$N(11/2^-)(***)(2600)$	
		$N(13/2^-)(?)$
	$\Delta(11/2^-)(?)$	
$(3, 3, [3])$	$N(7/2^-)(?)$	
		$\Delta(9/2^-)(**)(2400)$
$(5, 5, [3])$	$N(11/2^-)(?)$	
		$\Delta(13/2^-)(**)(2750)$

pattern obtained has been extended up to $2\,$GeV excitation energy). To express the spatial part we use the hyperspherical harmonic notation, *i.e.*, the quantum numbers $(K, L, Symmetry)$. The so-called great orbital, K, defines the parity of the state, $P = (-)^K$, and its centrifugal barrier energy, $\frac{\mathcal{L}(\mathcal{L}+1)}{2m\langle\rho^2\rangle}$ ($\mathcal{L} = K + \frac{3}{2}$, ρ : hyperradius). L is the total orbital angular momentum. *Symmetry* specifies the spatial symmetry ([3] : symmetric, [21] : mixed,

[111] : antisymmetric) which combines to the spin, S, and isospin, T, symmetries ($S, T = 3/2$: symmetric; $S, T = 1/2$: mixed) to have a symmetric wave function (the color part is antisymmetric). More precisely, $T = 1/2$ for N and $T = 3/2$ for Δ, hence the spatial-spin wave function must be mixed for N and symmetric for Δ.

A look at the tables makes manifest the underlying symmetry.

Table 3. Predicted N and Δ states in the interval $[2.2, 3.0]$ MeV. We denote by a black dot the first non-radial excitation.

	N		Δ	
$J = 7/2$	$N(7/2^+)^\bullet(2220)$	$N(7/2^-)^\bullet(2250)$		$\Delta(7/2^-)^\bullet(2400)$
$J = 9/2$	$N(9/2^+)^\bullet(2450)$	$N(9/2^-)^\bullet(2600)$	$\Delta(9/2^+)^\bullet(2420)$	$\Delta(9/2^-)^\bullet(2650)$
$J = 11/2$	$N(11/2^+)(2450)$			$\Delta(11/2^-)(2650)$
	$N(11/2^+)^\bullet(2700)$	$N(11/2^-)^\bullet(2650)$	$\Delta(11/2^+)^\bullet(2850)$	$\Delta(11/2^-)^\bullet(2750)$
$J = 13/2$		$N(13/2^-)(2650)$	$\Delta(13/2^+)(2850)$	
	$N(13/2^+)^\bullet(2900)$		$\Delta(13/2^+)^\bullet(2950)$	
$J = 15/2$	$N(15/2^+)(2900)$			

For positive parity each box in the upper part of table 1 groups four $N(J^P)$ ground states —two isospin and two spin projections— and sixteen $\Delta(J^P)$ ground states —four isospin and four spin projections— corresponding to an orbitally symmetric configuration. It is worth to mention that the N states have also some probability of mixed orbital symmetry with the same values of K and L. This mixing explains the appearance of an additional $N(3/2^+)$, which is the symmetry partner of $\Delta(1/2^+)$, in the second upper box as a consequence of the bigger hyperfine attraction for orbitally symmetric $S = 1/2$ states. Another consequence of the mixing is the presence of corresponding N excitations with reverse probabilities that appear in the boxes of the lower part of table 1. Any of these excitations adds four states —two isospin and two spin projections— to the eight N's and eight Δ's ground states present in each box.

For negative parity, table 2, the same box pattern repeats with different combinations of N's and Δ's.

This 20-member box picture where all the members of the same box have the same parity given by $P = (-)^L$, is a reflection of an underlying $SU(4) \otimes O(3)$ symmetry providing a $(20, L^P)$ classification scheme, the 20plet structure coming out naturally from the product of irreducible quark representations: $4 \otimes 4 \otimes 4 = 20_S \oplus 20_M \oplus 20_M \oplus \overline{4}$. The only difference in content between a box and the corresponding 20plet refers to mixed N resonances being a linear combination of N members of the 20plets with well-defined orbital symmetry.

It is worth to emphasize that $SU(4)$ goes beyond a factorization $SU(2) \otimes SU(2)$ as can be checked through the N-Δ degeneracies appearing within the same box when the $SU(4)$ breaking spin-spin interaction plays a minor role.

3 Spectral regularities and degeneracies

From the spectral pattern represented by tables 1 and 2 experimental regularities and degeneracies for $J \geq 5/2$ ground states come out:

i) $E_{N,\Delta}(J + 2) - E_{N,\Delta}(J) \approx 400\text{--}500 \text{ MeV}$,

ii) $N(J^\pm) \approx \Delta(J^\pm)$ for $J = \frac{4n+3}{2}$, $n = 1, 2 \ldots$,

iii) $N(J^+) \approx N(J^-)$ for $J = \frac{4n+1}{2}$, $n = 1, 2$.

These rules can also be obtained theoretically by refitting the quark model to reproduce precisely the $J \geq 5/2$ states. Rule i) expresses the increasing of the centrifugal barrier between states with the same orbital symmetry and the slowly varying spin-spin contribution for the same S when increasing L for $L \geq 2$. Rule ii) for positive parity reflects the small spin-spin contribution for $S = 3/2$ when $L \geq 2$, and for negative parity reflects the $SU(4) \otimes O(3)$ degeneracy for N's and Δ's in the same multiplet once the centrifugal barrier suppresses greatly the hyperfine splitting. Rule iii) for N parity doublets comes from the balance between a bigger repulsion (due to bigger K and L) and a bigger hyperfine attraction (due to lower S) for $N(J^+)$ against $N(J^-)$.

For excited states the absence of spin-orbit and tensor forces in our dynamical model suggests a new rule for $J \geq 5/2$:

$$\text{iv) } (N(J), \Delta(J))^\bullet \approx (N(J + 1), \Delta(J + 1)),$$

say the first non-radial excitation of $N(J)$ and the ground state of $N(J+1)$ are almost degenerate (the same for Δ). This rule is well satisfied by experimental data.

Taking into account rules i)-iv) and the developed symmetry pattern we can make predictions for, until now, unknown states from 2 to 3 GeV, table 3. Though some of our predicted states might be masked by experimental uncertainties and others could not be easily detected (small coupling to formation channels) we hope the results in table 3 may be of some help to guide future experimental searches.

This work has been partially funded by MCyT under Contract No. FPA2004-05616, by JCyL under Contract No. SA104/04, and by GV under Contract No. GV05/276.

References

1. D. Jido, T. Hatsuda, T. Kunihiro, Phys. Rev. Lett. **84**, 3252 (2000); D. Jido, M. Oka, A. Hosaka, Prog. Theor. Phys. **106**, 873 (2001); T.D. Cohen, L.Ya. Glozman, Phys. Rev. D **65**, 016006 (2002).
2. R.L. Jaffe, D. Pirjol, A. Scardicchio, Phys. Rev. Lett. **96**, 121601 (2006); hep-ph/0602010.
3. S. Capstick, W. Roberts, Prog. Part. Nucl. Phys. **45**, S241 (2000) and references therein.

4. J. Vijande, P. González, H. Garcilazo, A. Valcarce, Phys. Rev. D **69**, 074019 (2004).
5. Z. Zhang, Y. Yu, P. Shen, X. Shen, Y. Dong, Nucl. Phys. A **561**, 595 (1993).
6. P. González, J. Vijande, A. Valcarce, H. Garcilazo, Eur. Phys. J. A **29**, 235 (2006).
7. G.S. Bali, Phys. Rep. **343**, 1 (2001).
8. SESAM Collaboration (G.S. Bali, H. Neff, T. Düssel, T. Lippert, K. Schilling), Phys. Rev. D **71**, 114513 (2005).
9. S. Capstick, W. Roberts, Phys. Rev. D **47**, 1994 (1993).
10. E. Klempt, Phys. Rev. C **66**, 058201 (2002).
11. S. Eidelman *et al.*, Phys. Lett. B **592**, 1 (2004).

Eur. Phys. J. A **31**, 519–521 (2007)

DOI 10.1140/epja/i2006-10282-6

THE EUROPEAN
PHYSICAL JOURNAL A

Special Article – QNP 2006

Strange-baryon production asymmetry in K$^{\pm}$N interactions

G.H. Arakelyan[1], C. Merino[2,a], and Yu.M. Shabelski[3]

[1] YerPhI, Armenia
[2] Departamento de Física de Partículas, Facultade de Física, and IGAE, Universidade de Santiago de Compostela, Galicia, Spain
[3] SPNPI, Gatchina, St. Petersburg, Russia

Received: 18 December 2006
Published online: 21 March 2007 – © Società Italiana di Fisica / Springer-Verlag 2007

Abstract. The asymmetry of strange-baryon production in Kp interactions at high energies is considered in the framework of the Quark-Gluon String Model. The contribution of the string-junction mechanism to the strange baryon production is analysed.

PACS. 12.40.-y Other models for strong interactions – 13.60.Rj Baryon production – 13.75.Jz Kaon-baryon interactions

The Quark-Gluon String Model (QGSM) is based on the Dual Topological Unitarization (DTU) and it describes quite reasonably and in a theoretically consistent way many features of high-energy production processes, including the inclusive spectra of different secondary hadrons, their multiplicities, etc., both in hadron-nucleon and hadron-nucleus collisions [1–3]. High-energy interactions are considered as taking place via the exchange of one or several Pomerons, and all elastic and inelastic processes result from cutting through or between those exchanged Pomerons [4,5]. The possibility of different numbers of Pomerons to be exchanged introduces absorptive corrections to the cross-sections which are in agreement with the experimental data on production of hadrons consisting of light quarks. Inclusive spectra of hadrons are related to the corresponding fragmentation functions of quarks and diquarks, which are constructed in terms of the intercepts of well-known Regge trajectories [6].

In the string models baryons are considered as configurations consisting of three strings attached to three valence quarks and connected in one point, which is called string junction (SJ) [7]. Thus the SJ mechanism has a nonperturbative origin in QCD.

It is very important to understand the role of the SJ mechanism in the dynamics of high-energy hadronic interactions, in particular in processes implying baryon number transfer. Significant results on this question were obtained in [8–11], where the SJ mechanism was used to analyse the strange-baryon production in πp and pp interactions. The

detailed analysis of SJ contribution into strange-baryon production in KN interaction was presented in [12].

The present report is devoted to the calculation of asymmetry of strange-baryon production in the case of kaon beams. We analyse the existing data on asymmetry of Λ and $\bar{\Lambda}$ production on K-beams [13], and we compare the experimental data with the result of our calculations for a value of the SJ intercept $\alpha_{SJ} = 0.9$.

The $\bar{\Lambda}/\Lambda$ asymmetry is defined as

$$A(\bar{\Lambda}/\Lambda) = \frac{N_{\Lambda} - N_{\bar{\Lambda}}}{N_{\Lambda} + N_{\bar{\Lambda}}} \tag{1}$$

for each x_F bin.

The formula describing the inclusive spectrum (*i.e.*, Feynman-x, x_F, distribution) of a secondary hadron h in KN scattering in QGSM is given by the following expression [1]:

$$\frac{x_E}{\sigma_{inel}} \cdot \frac{\mathrm{d}\sigma}{\mathrm{d}x_F} = \sum_{n=1}^{\infty} w_n \cdot \varphi_n^h(x_F), \tag{2}$$

where $x_E = E/E_{max}$, and

$$w_n = \sigma_n \bigg/ \sum_{n=1}^{\infty} \sigma_n \tag{3}$$

is the weight of the diagram with n cut Pomerons. The n cut Pomeron cross-sections σ_n are calculated using the quasi-eikonal approximation with a supercritical Pomeron [5]:

$$\sigma_n = \frac{\sigma_P}{n \cdot z} \left(1 - e^{-z} \sum_{k=0}^{\infty} \frac{z^k}{k!}\right), \ n \geq 1, \tag{4}$$

a e-mail: Merino@fpaxp1.usc.es

$$z = \frac{2C_P\gamma_P}{R_P^2 + \alpha_P'\ln(s/s_0)} \cdot \left(\frac{s}{s_0}\right)^\Delta , \tag{5}$$

$$\sigma_P = 8\pi\gamma_P \cdot \left(\frac{s}{s_0}\right)^\Delta , \tag{6}$$

where s is the cms energy, σ_P is the Pomeron contribution to the total cross-section, $\Delta = \alpha_P(0) - 1$ is the excess of the Pomeron intercept over 1 (supercritical Pomeron), and parameters γ_P, R_P^2, and C_P take the values for the case of Kp interactions presented in [14]. The normalization parameter $s_0 = 1$. and Pomeron slope $\alpha_P' = 0.21$ are well-known parameters (see, for instance [1]).

The function $\varphi_n^h(x_F)$ in eq. (2) determines the contribution of the diagram, in which n Pomerons are cut. In the case of Kp collisions this function has the form [3]:

$$\varphi_n^{Kp\to h}(x_F) = f_{\bar{q}}^h(x_+,n) \cdot f_q^h(x_-,n)$$
$$+ f_q^h(x_+,n) \cdot f_{qq}^h(x_-,n)$$
$$+ 2(n-1)f_s^h(x_+,n) \cdot f_s^h(x_-,n) , \tag{7}$$

with

$$x_\pm = \frac{1}{2}\left[\left(\frac{4m_\perp^2}{s} + x_F^2\right)^{\frac{1}{2}} \pm x_F\right] . \tag{8}$$

The quantities f_{qq}, f_q, $f_{\bar{q}}$, and f_s in eq. (7) correspond to the contributions of the diquark, the valence quark and antiquark, and the sea quarks, while the contributions of the incident particle and the target proton depend on the variables x_+ and x_-, respectively.

The values of $f_{qq}^h(x_\pm,n)$, $f_q^h(x_\pm,n)$, $f_{\bar{q}}^h(x_\pm,n)$, and $f_s^h(x_\pm,n)$ can be obtained through the convolution of the corresponding momentum distribution of the diquarks, valence quarks, and sea quarks in the colliding hadrons, $u(x)$, and the fragmentation function $G^h(z)$ of either diquarks or quarks into secondary hadrons:

$$f_i^h(x_\pm,n) = \int_{x_\pm}^1 u_i(x_1,n) \cdot G_i^h(x_\pm/x_1)\mathrm{d}x_1, \quad i=q,\bar{q}, \tag{9}$$

$$f_{qq}^h(x_-,n) = \frac{2}{3}\int_{x_-}^1 u_{ud}(x_1,n) \cdot G_{ud}^h(x_-/x_1)\mathrm{d}x_1$$
$$+ \frac{1}{3}\int_{x_-}^1 u_{uu}(x_1,n) \cdot G_{uu}^h(x_-/x_1)\mathrm{d}x_1, \tag{10}$$

$$f_s^h(x_\pm,n) = \frac{1}{2+\delta}$$
$$\cdot \left[\int_{x_\pm}^1 u_{\bar{u}}(x_1,n)\frac{G_{\bar{u}}^h(x_\pm/x_1) + G_u^h(x_\pm/x_1)}{2}\mathrm{d}x_1\right.$$
$$+ \int_{x_\pm}^1 u_{\bar{d}}(x_1,n)\frac{G_{\bar{d}}^h(x_\pm/x_1) + G_d^h(x_\pm/x_1)}{2}\mathrm{d}x_1$$
$$\left.+ \delta \cdot \int_{x_\pm}^1 u_{\bar{s}}(x_1,n)\frac{G_{\bar{s}}^h(x_\pm/x_1) + G_s^h(x_\pm/x_1)}{2}\mathrm{d}x_1\right] . \tag{11}$$

The parameter $\delta \sim 0.2$–0.32 determines here the relative suppression of strange quarks in the sea. As shown

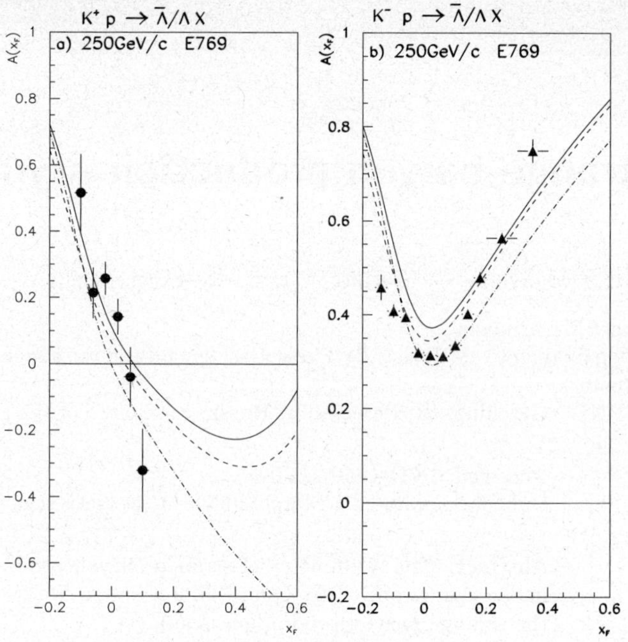

Fig. 1. The $\bar{\Lambda}/\Lambda$ asymmetry in (a) K^+p and (b) K^-p collisions. Experimental data at $250\,\mathrm{GeV}/c$ [13] and the corresponding QGSM description. The solid curve corresponds to $\alpha_{SJ} = 0.9$, $\varepsilon = 0.024$, and $\delta = 0.32$, the dashed curve to $\delta = 0.2$, and the dashed-dotted curve to $\varepsilon = 0$.

in [15] (see also discussion in [12]), the better agreement of QGSM with data on strange-baryon production on nucleus was obtained with $\delta = 0.32$, instead of the previous value $\delta = 0.2$. In principle one cannot exclude the possibility that the value of δ would be different for secondary baryons and for mesons (*i.e.*, for Λ-baryon and for kaon).

The detailed description of high-energy hadron-nucleon cross-sections on the base of the Reggeon calculus has been presented in many papers (see, for instance [1,3]).

The diquark and quark distribution functions and the fragmentation functions are determined by the Regge intercepts [6].

The complete set of distribution and fragmentation functions used in this paper is presented in the appendix of ref. [12].

The SJ mechanism has a nonperturbative origin and since it is at present not possible to determine the value of α_{SJ} in QCD from first principles. Thus we treat α_{SJ} and ε (the weight of the diagramm describing the SJ contribution, see [12]) as phenomenological parameters which should be determined from experimental data. In the present calculation, we use the values $\alpha_{SJ} = 0.9$ and $\varepsilon = 0.024$, as done in [9,12,15].

The fragmentation functions into $\bar{\Lambda}$ do not depend on the SJ mechanism, so the $\bar{\Lambda}$ spectra obtained for different values of α_{SJ} are the same, and they have a very small dependence on the strange quark suppression factor δ (see [12]).

In fig. 1 we show the comparison of the QGSM calculations with the data on the $\bar{\Lambda}/\Lambda$ asymmetry $A(\bar{\Lambda}/\Lambda)$,

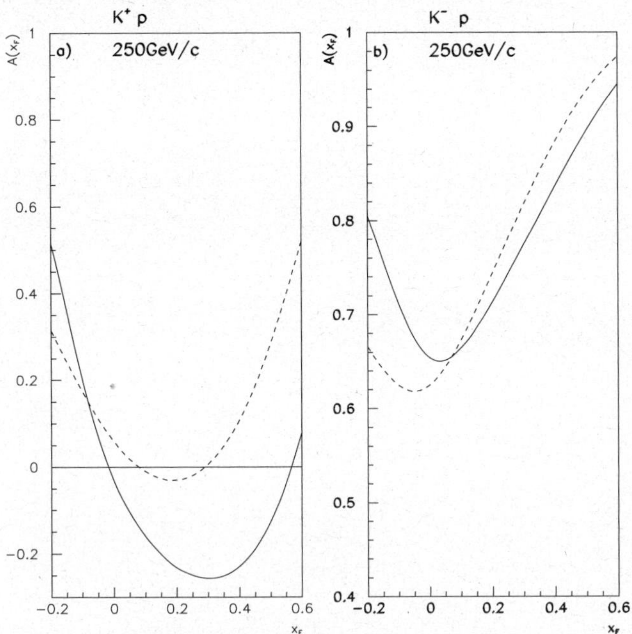

Fig. 2. QGSM prediction for the x_F-dependence of the asymmetry of heavy strange baryons in (a) K^+p and (b) K^-p collisions at 250 GeV/c. Solid curves are for Ξ^-, dashed curves for Ω^-.

produced in K^+p (fig. 1a) and K^-p (fig. 1b) interactions at 250 GeV/c [13].

The asymmetry data are rather interesting. In the proton fragmentation region the values of $A(\bar{\Lambda}/\Lambda)$ are close to unity, and that is natural since a proton fragments into Λ with significantly larger probability than into $\bar{\Lambda}$. In the kaon fragmentation region $A(\bar{\Lambda}/\Lambda)$ becomes negative and decreases very fast in the case of K^+ beam at the x_F values where experimental data exist. In the case of K^- beam $A(\bar{\Lambda}/\Lambda)$ increases very fast with x_F and the variation among calculations with different values of the parameters is rather small. Both these behaviors are also natural since the K^+ contains a $\bar{s}$ valence quark which preferably fragments into $\bar{\Lambda}$, while the valence s quark in the K^- fragments rather often into Λ. However, in both cases the $A(\bar{\Lambda}/\Lambda)$ experimental x_F-dependences are much steeper than the theoretical predictions. This is a probable indication that the fragmentation functions $s \to \Lambda$ and $\bar{s} \to \bar{\Lambda}$ should be further enhanced.

In the K^+ fragmentation region (fig. 1a) the predicted values of $A(\bar{\Lambda}/\Lambda)$ at $x_F > 0.4$ show a change of behavior since they start increasing. In this region the contribution of the direct fragmentation of $\bar{s} \to \bar{\Lambda}$ which makes $A(\bar{\Lambda}/\Lambda)$ to decrease becomes smaller than the effect of SJ diffusion which increases the multiplicity of Λ. The measurement of the asymmetry $A(\bar{\Lambda}/\Lambda)$ in the region $x_F \geq 0.4$ in K^+p collisions could make the situation clearer.

The predictions for the asymmetry in Ξ and Ω baryon production in K^+p and K^-p interactions are presented in fig. 2. Here the general situation is similar to that of the case of asymmetry in Λ production shown in fig. 1.

In fig. 2 we present the predictions for asymmetries in Ξ^- and Ω production in K^+p and K^-p collisions at 250 GeV/c. In the central region of K^+p collisions the yields of Ξ^- and $\bar{\Xi}^+$, as well as those of Ω^- and $\bar{\Omega}^+$ are predicted to be practically the same. The smaller fragmentation function of valence $\bar{s}$-quark into strange baryon is compensated by the larger fragmentation function of the target diquark. In the case of K^- beam (fig. 2b) the asymmetry for both Ξ and Ω productions increases in the whole region of positive x_F.

The situation for K^- beam seems worse than in the K^+ case since all curves are, as a rule, below the experimental data, but both the number and the quality of experimental data on K^- beam are not not very high.

The QGSM predicts a weak energy dependence of the Λ and $\bar{\Lambda}$ production cross-section in Kp collisions at the considered energies.

The experimental data on high-energy Λ production are not in contradiction with the possibility of baryon charge transfer over large rapidity distances, and the $\bar{\Lambda}/\Lambda$ asymmetry is provided by SJ diffusion through baryon charge transfer.

The presence of baryon asymmetry in the projectile hemisphere for Kp collisions provides good evidence for such a mechanism.

This work was partially financed by CICYT of Spain through contract FPA2002-01161, and by Xunta de Galicia (Spain) through contract PGIDIT03PXIC20612PN. G.H.A. and C.M. were also supported by NATO grant CLG.980335 and Yu.M.S. by grants NATO PDD (CP) PST.CLG 980287 and RCGSS-1124.2003.2. G.H.A. thanks Xunta de Galicia (Spain) for financial support.

References

1. A.B. Kaidalov, K.A. Ter-Martirosyan, Yad. Fiz. **39**, 1545 (1984); **40**, 211 (1984).
2. A. Capella, U. Sukhatme, C.I. Tan, J. Tran Thanh Van, Phys. Rep. **236**, 225 (1994).
3. Yu.M. Shabelski, Yad. Fiz. **49**, 1081 (1989).
4. V.A. Abramovski, V.N. Gribov, O.V. Kancheli, Yad. Fiz. **18**, 595 (1973).
5. K.A. Ter-Martirosyan, Phys. Lett. B **44**, 377 (1973).
6. A.B. Kaidalov, Sov. J. Nucl. Phys. **45**, 902 (1987); Yad. Fiz. **43**, 1282 (1986).
7. X. Artru, Nucl. Phys. B **85**, 442 (1975).
8. G.H. Arakelyan, A. Capella, A.B. Kaidalov, Yu.M. Shabelski, Eur. Phys. J. C **26**, 81 (2002).
9. F. Bopp, Yu.M. Shabelski, Yad. Fiz. **68**, 2155 (2005); hep-ph/0406158 (2004).
10. G.H. Arakelyan, C. Merino, Yu.M. Shabelski, Yad. Fiz. **69**, 884 (2006); hep-ph/0505100.
11. O.I. Piskounova, to be published in Yad. Fiz. **70**, N. 6 (2007); hep-ph/0604157.
12. G.H. Arakelyan, C. Merino, Yu.M. Shabelski, to be published in Yad. Fiz. **70**, N. 6 (2007); hep-ph/0604103.
13. E769 Collaboration (G.A. Alves *et al.*), Phys. Lett. B **559**, 179 (2003); hep-ex/0303027.
14. P.E. Volkovitsky, A.M. Lapidus, V.I. Lisin, K.A. Ter-Martirosyan, Yad. Fiz. **24**, 1237 (1976).
15. F. Bopp, Yu.M. Shabelski, hep-ph/0603193 (2006).

Quark Models for Excited Baryons

Gabriel Karl

Department of Physics, University of Guelph, Guelph, Canada, Perimeter Institute, Waterloo, Canada

Abstract. I briefly review Quark Models for excited baryons and stress the importance of measurements of the electric quadrupole of the omega minus hyperon, as well as the appearance of contact terms in quadrupole-quadrupole interactions.

PACS. 12.39.Pn

1 Introduction and acknowledgements

In this talk I stress the importance of experimental determination of the electric quadrupole moment of the omega minus hyperon. The written version is distorted by the avoidance of equations, due to my unfamiliarity with the latex medium. So everything is written out in words. I start with a little history and acknowledgements to my teachers and collaborators, on excited baryons. Excited nucleons were first observed about 1950, after the second world war.The quark model was proposed in 1964 by Gell-Mann and Zweig, in part to classify the hadrons then known.I became interested in the subject as a result of lectures given in Toronto by Bill Sharp, on group theory and SU3, and in November 1965 went to an APS meeting in Chicago where Richard(Dick)H.Dalitz(1925-2006) gave an invited talk on Quark Models for excited baryons. So the subject is about 40 years old. Dick died in January 2006, and is of course well known for many other contributions in Physics such as his Dalitz plot, Dalitz pairs,etc.He was my teacher in Quark Models.His work was very important in starting this field although many other theorists contributed significantly-such as O.W.Greenberg, H.Lipkin, G.Morpurgo,Y.Nambu and many others. I spent the period 1966-69 as a postdoc in Dick's research group at Oxford and collaborated with friends on quark models: Edward Obryk (from Poland), Les Copley (from Canada) and Frank Close(from England). We had a good time working on models and fitting what little data was known at the time. Our most notable results were in two papers on the photo-excitation of higher nucleonic and baryonic states as understood in quark models. The model we used required light quarks and the features of the data arose from the interference of electric and magnetic interactions of quarks. Our model was very successful and a few years later Feynman, Kisslinger and Ravndal adopted it, (with some changes, and references to earlier work) and managed to syphon citations to this work. Later on, after the discovery of J/psi and psi-prime I had the good fortune to

work with Nathan Isgur(1947-2001)on what are now called QCD inspired quark models, and we proposed a specific quark model for baryons, which in turn syphoned citations from earlier work like that of Dalitz, Greenberg,etc.Such is the fate of research in Physics. I will describe some of this work below. More recently, I have been lucky to collaborate with Victor Novikov(from ITEP,Moscow)and will describe some of our work also below.I learnt from this experience that it is most beneficial to collaborate with people who are smarter than me.

2 QCD "inspired" quark models for excited baryons

These models were stimulated by the discovery and interpretation of charmonium states and essentially assumed that the light quarks can be treated similarly to charmed quarks.The first paper along these lines was due to deRujula,Georgi and Glashow for ground state hadrons, but very soon Isgur et al extended the model to excited baryons. The main ingredient was the use of "gluon" exchange between quarks, so nowadays this model is called OGE- for One Gluon Exchange.But the name is a little pretentious since this is in not the gluon of perturbative QCD. Eventually other models of similar type were proposed: Boson Exchange(Glozman et al), Instanton Exchange(Metsch et al) which have very similar features- three quarks confined, with an "residual" exchange interaction to provide splittings and mixing of states. While I do not intend to review in detail these models here, it is fair to say that they all fit the data equally well (or badly).As in the comment on OGE, all these models have somewhat pretentious names- they are phenomenological models, somewhat similar the shell model of Nuclear Physics. None are derivable from first principles(unfortunately). The main differences are in mixings induced by these residual interactions.These show up in issues like the quadrupole moment of the omega minus hyperon,quadrupole contributions to the photo excitation of the P33 resonance, mixings of quartet and doublet states in the negative parity nucleons. These are more delicate issues, where the models differ substantially in their predictions. This is a point which the late Nathan Isgur also emphasized shortly before his death, and I support his point of view. Nathan also felt that the OGE model was more fundamental, but I am less sure about this point. In any case the quadrupole moment of the omega minus is not known experimentally and is a good quantity to try to predict. I suspect that within a decade we will have some data available.

3 Electric quadrupole moment of Ω^-

Among the predictions of Quark Models for ground state baryons,one of the most interesting is the Electric Quadrupole Moment of the omega minus hyperon. This is rather sensitive to the so- called tensor forces between quarks. The omega minus has spin 3/2 and therefore is allowed to have a quadrupole moment,and also a higher moment. Different quark models make different predictions for this model due to the different angular behaviour

of the residual forces. To measure this moment it will be necessary to capture the omega minus by a nucleus and observe the X-rays which are emitted in transitions between the bound states. This procedure was proposed by M.Goldhaber and Sternheimer some time ago. These authors have proposed to study the fine structure of the bound states to disentangle the quadrupole moment of the hyperon. Victor Novikov and the speaker have recently proposed (Phys.Rev.C, 2006) to use the hyperfine structure of the system, using a nucleus which also has a quadrupole moment. The interaction of the two quadrupoles has a contact term which is measurable in P-states of the system. This kind of contact term has never been seen, but it is analogous to the Fermi contact interaction which however occurs only in s-states. Thus the observation of these contact interactions is interesting in itself, quite apart from the measurement of the hyperon's quadrupole. The LHC which will produce beams of hadrons of high energy offers the promise of creation of large numbers of omega minus hyperons, which make the above spectroscopic measurements feasible.

The speaker is indebted to the organizers of the meeting for the invitation to speak and also for their help with transparencies, and support.

Eur. Phys. J. A **31**, 522–526 (2007)
DOI 10.1140/epja/i2006-10230-6

THE EUROPEAN
PHYSICAL JOURNAL A

Special Article – QNP 2006

Measurement of pionium lifetime with the Dirac spectrometer

B. Adeva[a], A. Romero Vidal, and O. Vázquez Doce

IGFAE, Universidade de Santiago de Compostela, 15706 Santiago de Compostela, Spain

Received: 8 November 2006
Published online: 5 March 2007 – © Società Italiana di Fisica / Springer-Verlag 2007

Abstract. Pionium ($\pi^+\pi^-$ bound state) lifetime is measured with improved precision with respect to earlier work, and the $\pi\pi$ s-wave scattering length difference between $I = 0$ and $I = 2$ amplitudes $|a_0 - a_2|$ is determined to 5% precision.

PACS. 13.75.Lb Meson-meson interactions – 12.39.Fe Chiral Lagrangians

1 Introduction

Pionium is a Coulomb $\pi^+\pi^-$ bound state, with Bohr radius $r_B = 387$ fm. Its ground state ($1s$) lifetime τ_{1s} is dominated by the short-range reaction $\pi^+\pi^- \to \pi^0\pi^0$, which largely exceeds the $\gamma\gamma$ decay, neglected in the present analysis:

$$\Gamma = \frac{1}{\tau} = \Gamma_{2\pi^0} + \Gamma_{\gamma\gamma} \qquad (1)$$

with $\frac{\Gamma_{\gamma\gamma}}{\Gamma_{2\pi^0}} \sim 4 \times 10^{-3}$. At lowest order in QCD and QED, the total width can be expressed as a function of the s-wave $I = 0$ (a_0) and $I = 2$ (a_2) $\pi\pi$ scattering lengths and the next-to-leading order has been calculated

$$\Gamma_{1s} = \frac{1}{\tau_{1s}} = \frac{2}{9}\alpha^3 p |a_0 - a_2|^2 (1 + \delta)\, M_{\pi+}, \qquad (2)$$

where $p = \sqrt{M_{\pi+}^2 - M_{\pi^0}^2 - \frac{1}{4}\alpha^2 M_\pi^2}$. A significant correction $\delta = (5.8 \pm 1.2) \times 10^{-2}$ arises with respect to lowest order, once a non-singular relativistic amplitude at threshold is built [1]. Therefore a 5% precision can be achieved in the measurement of $|a_0 - a_2|$ provided a 10% lifetime error is reached. Note should be taken that this method implies access to the physical reaction threshold (Bohr momentum $P_B \sim 0.5$ MeV/c). The $\pi\pi$ scattering lengths have been calculated in the framework of chiral perturbation theory with small errors [2]. Independently, a self-consistent representation of these amplitudes has been recently evaluated [3]. The former implies a lifetime prediction $\tau_{1s} = 2.9 \pm 0.1$ fs. There exists an ample and detailed literature about the chiral expansion of $\pi\pi$ amplitudes, including error estimates from experimental parameter uncertainies [4].

Pionium, with 4-momentum $p_A = (E_A, \boldsymbol{p}_A)$, is produced by Coulomb final-state interaction in ns states ac-

cording to the expression

$$\frac{\mathrm{d}\sigma}{\mathrm{d}\boldsymbol{p}_A} = (2\pi)^3 \frac{E_A}{2M_\pi} |\psi_n(0)|^2 \left(\frac{\mathrm{d}\sigma_s^0}{\mathrm{d}\boldsymbol{p}_1 \mathrm{d}\boldsymbol{p}_2}\right)_{\boldsymbol{p}_1 = \boldsymbol{p}_2 = \boldsymbol{p}_A/2}, \qquad (3)$$

where σ_s^0 denotes the Coulomb-uncorrected semi-inclusive $\pi^+\pi^-$ cross-section [5], which is enhanced by a high-energy and high-intensity proton beam colliding on a nuclear target foil.

Produced atoms propagate inside the target foil (decay length is only a few microns) before they are ionized (broken up) into $\pi^+\pi^-$ pairs in the continuum spectrum, which are subsequently triggered and detected by the DIRAC spectrometer. Due to the very small momentum transfer induced by the electric field near the target nuclei, the signal is detected as an excess with respect to the $\pi^+\pi^-$ Coulomb-correlated spectrum at very low center-of-mass momentum $Q(\sim 0.5\,\mathrm{MeV}/c)$.

A first measurement of pionium lifetime was published by DIRAC with extremely conservative error assesment [6]. Considerable work has taken place since then, and the results presented here arise from a better knowledge of the main error sources, as well as from a complete use of the spectrometer detectors.

2 Spectrometer setup

DIRAC is a double-arm spectrometer [7] where $\pi^+\pi^-$ pairs are collected into a narrow window of $10 \times 10\,\mathrm{cm}^2$ aperture at 2.5 m from the target foil, and then split by a 1.65 T dipole magnet. It is installed at the East Hall 24 GeV/c proton beam line of CERN PS. A top view is depicted in fig. 1. The spectrometer acceptance is elevated by 5.7° with respect to the beam line, in order to avoid backgrounds from the induced radiation. The beam was normally operated at 10^{11} protons per spill.

---

[a] e-mail: adevab@usc.es

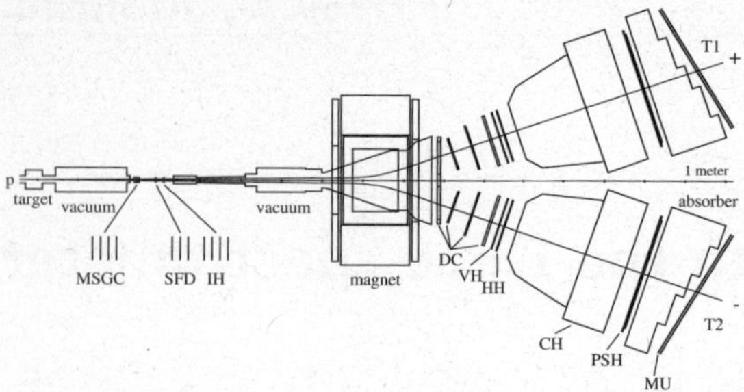

Fig. 1. Top view of the DIRAC spectrometer.

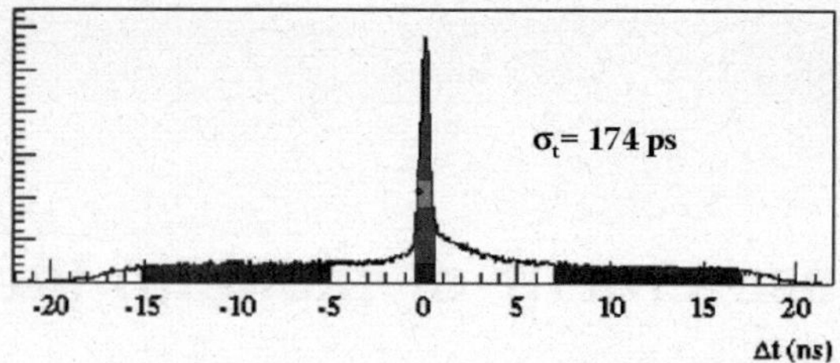

Fig. 2. Time difference between the negative and positive arms measured by the vertical hodoscope. The central peak corresponds to prompt events and the dark region to accidental pairs.

In addition to the double arm downstream the magnet, an upstream arm has been built with the following purpose:

– Improve the Q_T and Q_L resolution by measuring the two-pion opening angle, thus reducing background and pion decay path.
– Perform identification of very close pairs by means of pulse-height analysis of double ionization.

The upstream arm is composed by 4 MSGC/GEM planes, 2 Scintillator Fibre Detectors (SFD) and 4 Ionization Hodoscope (IH) planes. A third SFD was introduced in 2002 and 2003 runs.

The downstream two-arm spectrometer is made of 5 fast Drift Chambers (DC) as main tracking detector, two Cherenkov counters (CH) to provide efficient electron veto, a high-resolution Vertical Hodoscope (VH) system as Time-of-Flight (TOF) detector, Horizontal Hodoscopes (HH) to trigger horizontal splitting, Pre-Shower detectors (PSH) for further electron rejection and Muon Counters (MU) to veto pion decays.

The whole setup provides Q_T and Q_L resolutions of $0.1\,\text{MeV}/c$ and $0.5\,\text{MeV}/c$ respectively. Due to the small pionium rate compared to the Coulomb interacting background, low-Q selection ($Q < 30\,\text{MeV}/c$) must be achieved at trigger level with uniform acceptance, which requires a sophisticated multi-level structure. Wide TOF trigger cuts are applied, in order to accept a sizeable fraction of accidental pairs. In addition, asymmetric triggers are used to select $\Lambda \to p\pi^-$ decay for calibration purposes (Λ mass resolution is $\sigma_\Lambda = 0.395\,\text{MeV}/c^2$).

3 Reconstruction method

Pions are reconstructed in DIRAC after performing independent tracking in the upstream and downstream arms. The opening angle is determined with high precision (only limited by multiple scattering inside the target) making use of 4 MSGC/GEM detectors, in conjunction with TDC information from SFD.

Upstream track pairs are matched in space and time with DC tracks with uniform matching efficiency. When only a single unresolved track can be matched, a double-ionization pulse-height signal is required in IH.

A high-precision TOF measurement ($\sim 170\,\text{ps}$) is provided by the VH, which allows to perform clean separa-

tion between time-correlated (prompt) pairs and accidental pairs coming from different proton-nucleus interactions (see fig. 2). Under the prompt peak in the Δt spectrum we find both the real Coulomb-correlated pairs and the corresponding pionium signal fraction. It is, however, inevitable (despite the excellent time resolution) that a fraction of non-Coulomb pairs are also selected, arising from inclusive pion "long lifetime" decays in the subpicosecond range. Of course, the accidental pair fraction can be determined in addition from fig. 2.

Prompt experimental pairs are finally selected for the analysis with relative momentum cuts $|Q_L| < 20\,\text{MeV}/c$ and $Q_T < 5\,\text{MeV}/c$. Protons are removed with a $P_+ > 4\,\text{GeV}/c$ cut over the positive arm.

4 Analysis of Q-spectrum

The prompt two-pion spectrum in (Q_T, Q_L)-plane has been χ^2-analysed by comparison with the following input spectra:

– Monte Carlo describing the *Coulomb* final-state interaction (CC) by means of the Sakharov-Gamow factor $A_C(Q) = 2\pi(p_B/Q)/(1 - \exp(-2\pi p_B/Q))$, p_B being the pion Bohr momentum.
– Monte Carlo describing *accidental* coincidences taken by the spectrometer (AC), simulated isotropically in their center-of-mass frame.
– Monte Carlo describing *non-Coulomb* $\pi^+\pi^-$ pairs (NC). It simulates the additional fraction of events from decay of long-lifetime resonances which are still detected as time correlated. In practice, it differs from the previous one very slightly, only for the different lab frame pion momentum spectrum.
– Pionium *atom* Monte Carlo model (AA) which is used to cross-check and fit the observed deviation with respect to the continuum background constructed from the previous Monte Carlo input.

The laboratory frame pair momentum spectrum has been generated according to the one really observed with spectrometer data.

A two-dimensional analysis of the $\pi^+\pi^-$ spectrum in the center-of-mass frame has been carried out, choosing the transverse $Q_T = \sqrt{Q_X^2 + Q_Y^2}$ and longitudinal $Q_L =$

$|Q_Z|$ components (with respect to the pair direction of flight Z) as independent variables. The analysis has been done independently at ten individual $600\,\mathrm{MeV}/c$ bins of the lab frame momentum p of the pair.

The total number of events used for the above Monte Carlo samples are denoted by N_{CC}, N_{AC}, N_{NC} and N_{AA} respectively, whereas N_p represents the total number of prompt events in the analysis, under the reference cuts $Q_T < 5\,\mathrm{MeV}/c$ and $Q_L < 20\,\mathrm{MeV}/c$. Index k runs over all (i,j) bins of the (Q_T, Q_L) histograms, and we denote by N_{CC}^k the number of Coulomb events observed in each particular bin (i,j). Similarly for the other input spectra, namely N_{AC}^k, N_{NC}^k and N_{AA}^k. Normalised spectra are used to fit the data, and we denote them by small letters, $n_{CC}^k = N_{CC}^k/N_{CC}$ and likewise for the rest. The ratios $x_{CC} = N_{CC}/N_p$, $x_{AC} = N_{AC}/N_p$, $x_{NC} = N_{NC}/N_p$ and $x_{AA} = N_{AA}/N_p$ help define the statistical errors. The χ^2 analysis is based upon the expression

$$\chi^2 =$$
$$\sum_k \frac{\left(N_p^k - \beta\alpha_1 n_{CC}^k - \beta\alpha_2 n_{AC}^k - \beta\alpha_3 n_{NC}^k - \beta\gamma n_{AA}^k\right)^2}{\beta\left(n_p^k + n_{CC}^k\left(\frac{\alpha_1^2}{x_{CC}}\right) + n_{AC}^k\left(\frac{\alpha_2^2}{x_{AC}}\right) + n_{NC}^k\left(\frac{\alpha_3^2}{x_{NC}}\right) + n_{AA}^k\left(\frac{\gamma^2}{x_{AA}}\right)\right)},$$
$$(4)$$

where α_1 represents the fraction of $\pi^+\pi^-$ Coulomb-interacting pairs, the sum $\alpha_2 + \alpha_3$ that of non-Coulomb pairs, and γ the atom pair fraction. It should be noted that α_2 represents specifically the fraction of accidental pairs, which was fixed to the experimentally observed values determined from the spectrum in fig. 2. Minimization was carried out in $(0.5\times0.5)\,(\mathrm{MeV}/c)^2$ bins over the entire (Q_T, Q_L)-plane, under the constraint $\alpha_1 + \alpha_2 + \alpha_3 + \gamma = 1$, with α_3 and γ as free parameters. The β parameter, which represents the overall Monte Carlo normalisation, is actually determined by the number of prompt events in the domain under fit, and it does not need to be varied. The $\alpha_3 = 1 - \alpha_1 - \alpha_2 - \gamma$ fraction then measures the long-lifetime component.

We define the atom signal as the difference between the prompt spectrum and the Monte Carlo with the pionium component (AA) removed. This 2D signal, which reveals the excess with respect to the calculated Coulomb interaction enhancement is compared with the Monte Carlo prediction for atom production. The difference between the two is shown in fig. 3 under the form of a lego plot, where a signed transverse component Q_{xy} has been defined by projecting the measured value of Q_T over a randomly selected azimuth ϕ ($Q_{xy} = Q_T \cos\phi$).

In fig. 4 the p-integrated Q_T spectrum is shown, together with the Monte Carlo and the atom signal extracted from their difference. The longitudinal spectrum is displayed in fig. 5 as well as that of the relative momentum magnitude Q. The observed atom signal is confined to the $Q_L < 2\,\mathrm{MeV}/c$ region as expected from Monte Carlo, whereas the Q_T distribution is wider, due to the larger effect of multiple scattering in the transverse projection.

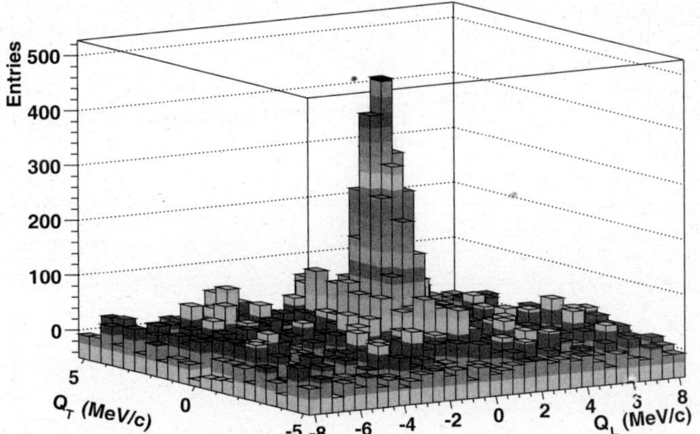

Fig. 3. Lego plot showing pionium break-up in the $(Q_T, Q_L = |Q_Z|)$-plane.

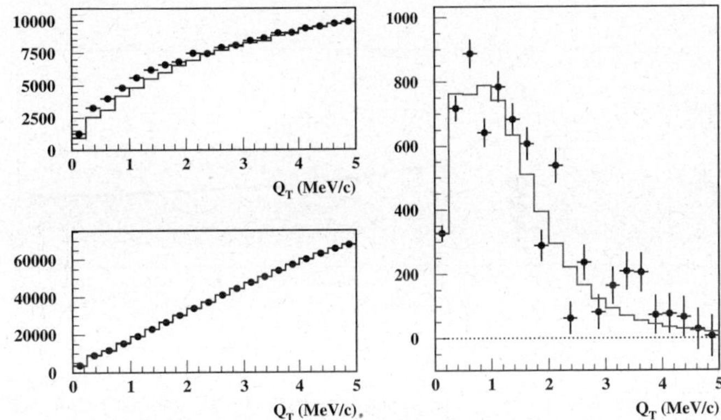

Fig. 4. (Colour on-line) Two-dimensional fit projection onto Q_T. The data are shown separately for $Q_L < 2\,\mathrm{MeV}/c$ (top left) and $Q_L > 2\,\mathrm{MeV}/c$ (bottom left). The difference between prompt data (dots) and Monte Carlo (continuous blue line) is amplified and compared with the pionium Monte Carlo production as red line (right).

5 Breakup probability and pionium lifetime

Once the number of observed atom pairs n_A, and the number of background Coulomb pairs N_C have been determined from the fit in the same kinematical region (by means of the α_1 parameter), pionium breakup probabilities P_{Br} can be determined by using the concept of K-factors. The P_{Br} has been defined as the ratio of n_A over the number of pionium pairs originally produced by the final-state interaction N_A, $P_{Br} = n_A/N_A$. The number of atoms N_A produced in a given phase-space volume is analytically calculated in quantum mechanics and can be related to the number N_C of produced Coulomb pairs by means of the theoretical K-factor:

$$K^{th} = \frac{N_A}{N_C} = 8\pi^2 Q_0^2 \frac{\sum_1^\infty \frac{1}{n^3}}{\int A_C(Q)\mathrm{d}^3 Q}, \qquad (5)$$

where $Q_0 = \alpha M_\pi$ is two times the atom Bohr momentum p_B. However, the direct measurement obtained in DIRAC for n_A and N_C has been influenced by several reconstruction biases, and that is why we define the experimental

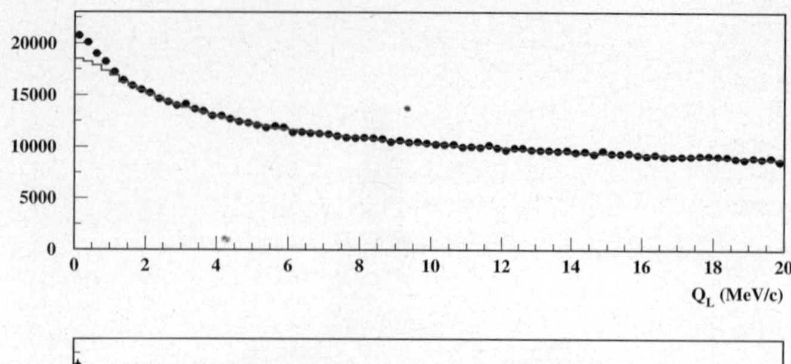

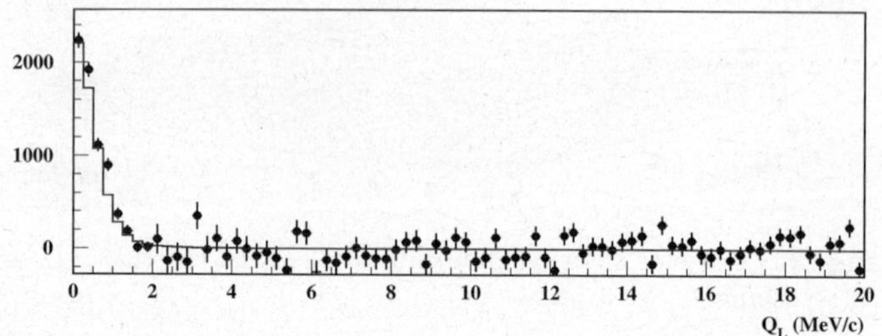

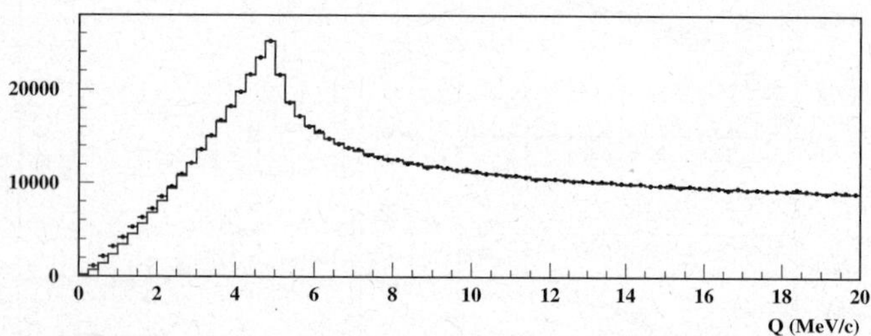

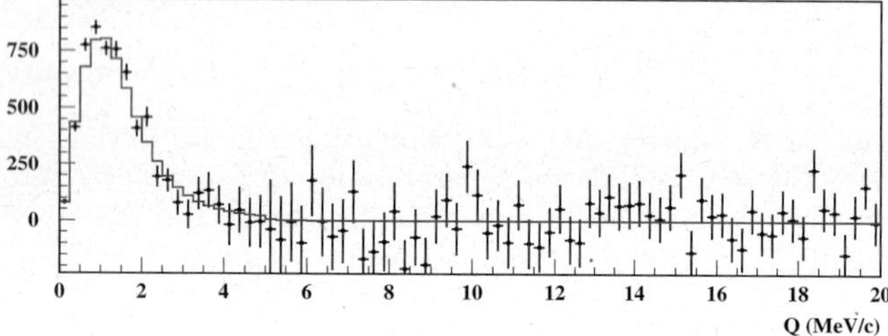

Fig. 5. Two-dimensional fit projection onto Q_L (top panels). The difference between prompt data (dots) and Monte Carlo (line) is amplified at the bottom, where the signal is compared with the pionium atom production Monte Carlo. Projection onto Q is displayed in the bottom panels

K-factor as $K^{exp}(\Omega) = K^{th}(\Omega)\epsilon_A(\Omega)/\epsilon_C(\Omega)$, where ϵ_A and ϵ_C represent the efficiency of the reconstruction chain for atoms and Coulomb pairs, respectively, in a given kinematical region Ω. The P_{Br} is then determined as

$$P_{Br} = \frac{n_A}{N_C K^{exp}(\Omega)}. \tag{6}$$

The chosen domain $Q_T < 5\,\mathrm{MeV}/c$ and $Q_L < 2\,\mathrm{MeV}/c$ ensures that the pionium signal is entirely contained.

Figure 6 shows the measured breakup probability as a function of the atom momentum. P_{Br} values are compatible with a smooth increase with increasing atom momentum, as predicted by Monte Carlo tracking inside the target foil [8]. The $1s$ pionium lifetime (τ_{1s}) and statistical error can then be determined by χ^2 minimization with respect to the latter prediction, having τ_{1s} as the only

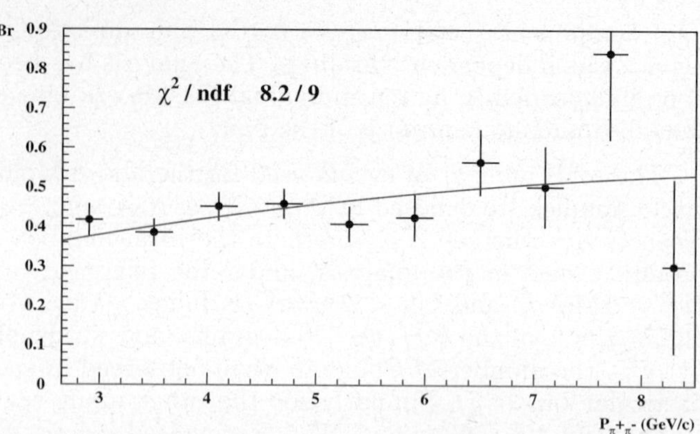

Fig. 6. Pionium break-up probabilities as a function of atom momentum, as compared to best fit Monte Carlo prediction with average Ni foil thickness ($\tau_{1s} = 2.58\,\mathrm{fs}$).

free parameter. Alternatively, the individual P_{Br} measurements at each momentum bin can be combined with independent statistical errors.

6 Systematic error

We have studied the magnitude of possible systematic errors in the measurement of breakup probability, which are summarized in table 1. Generally, they are associated with imperfections of the Monte Carlo simulations of the performance of the apparatus. Given the fact that these are tuned with actual spectrometer data in all cases, the uncertainty ultimately comes from the quality and consistency of a certain number of reference distributions. Concerning the atom propagation inside the target foil, the description of breakup probability is achieved with 1% precision [8]. ω, η' and finite-size nuclear effects on the Coulomb interaction spectrum [9] have been taken into account as a small correction to the CC Monte Carlo spectrum ($\Delta P_{Br} = -0.01$ on average). Its related uncertainty has been estimated by changing the ω fraction by $\pm 25\%$. The material budget of the upstream arm is known to 1.5% precision [10] and furthermore the utilization of the first planes of the MSGC/GEM detector significantly reduces its influence in the measurement. Small biases in Q_L trigger acceptance have been corrected by means of accidental pairs, the corresponding precision being lim-

Table 1. Estimated contributions to systematic error on P_{Br} from the most relevant sources.

Source	σ
Multiple scattering angle	± 0.003
Q_L trigger acceptance	± 0.007
Simulation of MSGC background	± 0.006
Double-track resolution simulation	± 0.003
Atom signal shape	± 0.003
Finite-size effects and η'/ω contamination	± 0.003
Total	0.009

ited by statistics (specially at large momentum). The error estimates in table 1 correspond to maximum reasonable variations, until contradiction with specific reference distributions is encountered. They have been statistically combined, by convoluting step functions defined within the indicated error limits, to provide an overall systematic error in the P_{Br} of ± 0.009, which can be used as a 1σ estimate. This error has been converted into an asymmetric lifetime error $\Delta\tau_{1s} = {}^{+0.15}_{-0.14}$ fs using the Monte Carlo propagation code.

A contamination of K^+K^- pairs in the 2001 $\pi^+\pi^-$ data sample has been studied and it appears to be $(2.38 \pm 0.35) \times 10^{-3}$ at $p = 2.9\,\mathrm{GeV}/c$. The Coulomb spectrum of K^+K^- is known and this originates a small correction to τ_{1s}, with negative sign. Another small correction (with positive sign) is also necessary, due to a slight lower-Z contamination in the target foil. Both topics are being finalized at the moment of writing these proceedings, and the overall systematic uncertainty is not expected to increase significantly, as a result of these studies.

7 Summary

Following the analysis of previous sections, the $1s$ lifetime of pionium atom has been determined to be $\tau_{1s} = 2.58^{+0.26}_{-0.22}(stat)^{+0.15}_{-0.14}(syst)$ fs. A quadrature of both sources of error yields the combined result

$$\tau_{1S} = 2.58^{+0.30}_{-0.26}\,\text{fs},$$

which can be converted into a measurement of the s-wave amplitude difference $|a_1 - a_0| = 0.280 {}^{+\,0.016}_{-\,0.014}\,M_\pi^{-1} = (0.280 \pm 0.015)\,M_\pi^{-1}$.

Minor corrections to the previous measurement are still being investigated before a completely final result will be issued.

We thank the DIRAC Collaboration as a whole for the use of these data. We would like to acknowledge the funding received from the spanish Ministery of Education and Science (MEC), under projects AEN99-0488 and FPA2005-06441, and from the PGIDT of Xunta de Galicia under project PXI20602PR. This publication would not have been possible without the strong computing support received from Centro de Supercomputación de Galicia (CESGA). We thank in particular Carlos Fernández Sánchez, in charge of the SVGD cluster at CESGA, and Andrés Gómez Tato. Our special gratitude to J.J. Saborido Silva and M. Sánchez Garcia, for helping the implementation of our GRID computing strategy. We are also endebted to Cibrán Santamarina Rios for advice concerning his pionium propagation code, as well as to Valery Yazkov for clarification of momentum resolution aspects in DIRAC.

References

1. J. Gasser, V.E. Lyubovitskij, A. Rusetsky, A. Gall, Phys. Rev. D **64**, 016008 (2001).
2. G. Colangelo, J. Gasser, H. Leutwyler, Nucl. Phys. B **603**, 125 (2001).
3. J.R. Peláez, F.J. Ynduráin, Phys. Rev. D **71**, 074016 (2005).
4. J. Nieves, E. Ruiz-Arriola, Eur. Phys. J. A **8**, 377 (2000).
5. B. Adeva *et al.*, DIRAC proposal, CERN/SPSLC 95-1, SPSLC/P 284 (1994).
6. DIRAC Collaboration (B. Adeva *et al.*), Phys. Lett. B **619**, 50 (2005).
7. DIRAC Collaboration (B. Adeva *et al.*), Nucl. Instrum. Methods A **515**, 467 (2003).
8. C. Santamarina *et al.*, J. Phys. B **36**, 4273 (2003).
9. R. Lednicky, nucl-th/0501065.
10. B. Adeva, A. Romero Vidal, O. Vázquez Doce, *Study of Multiple Scattering in Upstream Detectors in DIRAC*, DIRAC Note 05-16, http://dirac.web.cern.ch/DIRAC/i_notes.html.

Eur. Phys. J. A **31**, 527–533 (2007)
DOI 10.1140/epja/i2006-10185-6

Special Article – QNP 2006

Aspects of strangeness -1 meson-baryon scattering

J.A. Oller[1,a], J. Prades[2], and M. Verbeni[1]

[1] Departamento de Física, Universidad de Murcia, E-30071 Murcia, Spain
[2] CAFPE and Departamento de Física Teórica y del Cosmos, Universidad de Granada, Campus de Fuente Nueva,
E-18002 Granada, Spain

Received: 25 October 2006
Published online: 22 February 2007 – © Società Italiana di Fisica / Springer-Verlag 2007

Abstract. We consider meson-baryon interactions in S-wave with strangeness -1. This is a sector populated by plenty of resonances interacting in several two-body coupled channels. We consider a large set of experimental data, where the recent experiments are remarkably accurate. This requires a sound theoretical description to account for all the data and we employ Unitary Chiral Perturbation Theory up to and including $\mathcal{O}(p^2)$. The spectroscopy of our solutions is studied within this approach, discussing the rise of the two $\Lambda(1405)$-resonances and of the $\Lambda(1670)$, $\Lambda(1800)$, $\Sigma(1480)$, $\Sigma(1620)$ and $\Sigma(1750)$. We finally argue about our preferred fit.

PACS. 13.75.Jz Kaon-baryon interactions – 12.39.Fe Chiral Lagrangians – 11.80.-m Relativistic scattering theory – 11.80.Gw Multichannel scattering

1 Introduction

The study of strangeness -1 meson-baryon dynamics comprising the $\bar{K}N$ plus coupled channels, has been renewed both from theoretical and experimental sides. Experimentally, we have new exciting data like the increasing improvement in precision of the measurement of the α line of kaonic hydrogen accomplished recently by DEAR [1], and its foreseen better determination, with an expected error of a few eV, by the DEAR/SIDDHARTA Collaboration [2]. This has established a challenge to theory in order to match such precision. In this respect, ref. [3] provides an improvement over the traditional Deser formula for relating scattering at threshold with the spectroscopy of hadronic atoms [4]. In addition, one needs a good scattering amplitude to be implemented in this equation. The study of strangeness -1 meson-baryon scattering has a long history [5–12] within K-matrix models, dispersion relations, meson-exchange models, quark models, cloudy-bag models or large-N_c QCD, just to quote a few. However, in more recent years it has received a lot of attention from the application of $SU(3)$ baryon Chiral Perturbation Theory (CHPT) to this sector together with a unitarization procedure, see, e.g., [13–22]. Recently, ref. [3] pointed out the possible inconsistency of the DEAR measurement on kaonic hydrogen and K^-p scattering, since the unitarized CHPT results, able to reproduce the scattering data, were not in agreement with DEAR. Later on, ref. [20] insisted on this fact based on its own fits, although they only included partially the $\mathcal{O}(p^2)$ CHPT amplitudes [21]. However, the situation changed with ref. [22], as it was shown

that one can obtain fits in Unitary CHPT (UCHPT), including full $\mathcal{O}(p^2)$ CHPT amplitudes, which are compatible both with DEAR and with K^-p scattering data. In addition, ref. [23] extended the work of ref. [22] by including additional experimental data, recently measured with remarkable precision by the Crystal Ball Collaboration, for the reactions $K^-p \to \eta\Lambda$ [24] and $\pi^0\pi^0\Sigma^0$ [25]. The importance of including the latter data in any analysis of K^-p interactions has been singled out in ref. [26]. We will report here about the series of works [22,21,23] on meson-baryon CHPT. Recently, we also presented the first full and minimal $SU(3)$ CHPT meson-baryon Lagrangians to $\mathcal{O}(p^3)$ in ref. [27].

The study of K^-p plus coupled channel interactions offers, from the theoretical point of view, a very challenging test ground for chiral effective field theories of QCD since one has there plenty of experimental data, Goldstone bosons dynamics and large and explicit $SU(3)$ breaking. In addition, this sector shows a very rich spectroscopy with many $I = 0, 1$ S-wave resonances that will be the object of our study as well. Apart from that, these interactions are interrelated with many other interesting areas, e.g., possible kaon condensation in neutron-proton stars, large yields of K^- in heavy-ions collisions, kaonic atoms or non-zero–strangeness content of the proton.

2 Formalism and results

A general meson-baryon partial wave in coupled channels can be written in matrix notation as [17]

$$T = [1 + \mathcal{K}g]^{-1}\mathcal{K}, \tag{1}$$

a e-mail: oller@um.es

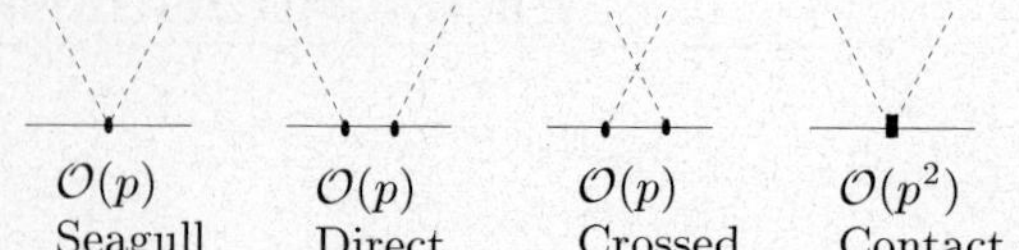

Fig. 1. Diagrams for the calculation of the baryon CHPT scattering amplitudes up to and including $\mathcal{O}(p^2)$. The first three diagrams are of $\mathcal{O}(p)$ while the last one is of $\mathcal{O}(p^2)$.

Table 1. A-type fits that agree with the DEAR data, eq. (5). The $\sigma_{\pi N}$ value enforced in the fits is given in the first row. For precise definitions of the parameters f, b_0, b_D, b_F, b_i and a_i see ref. [23]. Three among the parameters b_0, b_D, b_F and b_i are fixed.

Units	$\sigma_{\pi N}$	20^*	30^*	40^*
	(MeV)			
MeV	f	75.2	71.8	67.8
GeV^{-1}	b_0	-0.615	-0.750	-0.884
GeV^{-1}	b_D	$+0.818$	$+0.848$	$+0.873$
GeV^{-1}	b_F	-0.114	-0.130	-0.138
GeV^{-1}	b_1	$+0.660$	$+0.670$	$+0.676$
GeV^{-1}	b_2	$+1.144$	$+1.169$	$+1.189$
GeV^{-1}	b_3	-0.297	-0.316	-0.315
GeV^{-1}	b_4	-1.048	-1.181	-1.307
	a_1	-1.786	-1.591	-1.413
	a_2	-0.519	-0.454	-0.386
	a_5	-1.185	-1.170	-1.156
	a_7	-5.251	-5.209	-5.123
	a_8	-1.316	-1.310	-1.308
	a_9	-1.186	-1.132	-1.050

where g is a diagonal matrix that comprises the unitarity bubble for every channel and $\mathcal{K}$ is the interaction kernel that is determined from meson-baryon CHPT. This is accomplished by performing a power expansion of T calculated from CHPT and then matched, order by order, with the chiral expansion of eq. (1),

$$T_1+T_2+T_3+\mathcal{O}(p^4) = \mathcal{K}_1+\mathcal{K}_2+\mathcal{K}_3-\mathcal{K}_1\cdot g\cdot\mathcal{K}_1+\mathcal{O}(p^4), \quad (2)$$

taking into account that g is of chiral order one. We calculate $\mathcal{K}$ up to $\mathcal{O}(p^2)$, $\mathcal{K}_1 = T_1$ and $\mathcal{K}_2 = T_2$. The lowest-order result, T_1, contains the seagull, direct and crossed exchange diagrams, while the next–to–leading-order amplitudes, T_2, come from pure contact interactions. This is shown diagramatically in fig. 1. Once the kernel $\mathcal{K} = T_1 + T_2$ has been calculated, we insert it in eq. (1) and evaluate the S-wave amplitudes.

In ref. [22] a large set of meson-baryon scattering data was fitted which was enlarged in ref. [23] including new precise ones from the Crystal Ball Collaboration. Namely, ref. [22] took into account the $\sigma(K^-p \to K^-p)$ elastic cross-section [28–31], the $\sigma(K^-p \to \bar{K}^0 n)$ charge exchange one [28,29,31–33], and several hyperon production reactions, $\sigma(K^-p \to \pi^+\Sigma^-)$ [28–30], $\sigma(K^-p \to \pi^-\Sigma^+)$ [29–31], $\sigma(K^-p \to \pi^0\Sigma^0)$ [29] and $\sigma(K^-p \to \pi^0\Lambda)$ [29]. In our normalization the corresponding cross-

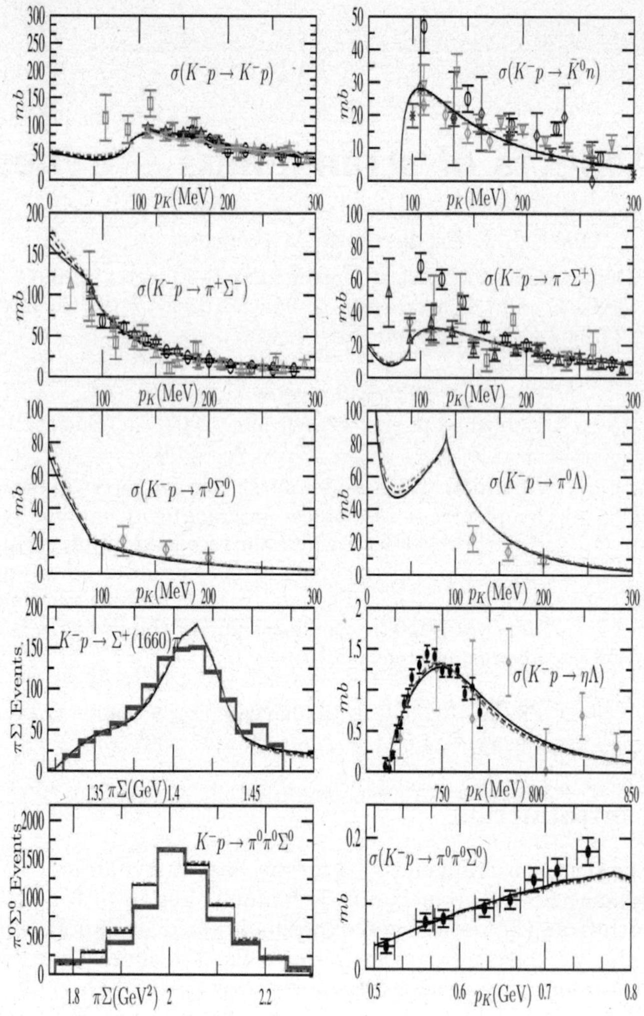

Fig. 2. The solid lines correspond to the $\sigma = 40^*$ MeV fit, the dashed lines to the 30^* MeV fit, and the dash-dotted curves to the 20^* MeV one of table 1. The different lines can be barely distinguished. For experimental references see ref. [23].

section, keeping only the S-wave, is given by

$$\sigma(K^-p \to MB) = \frac{1}{16\pi s}\frac{p'}{p}|T_{K^-p\to MB}|^2, \quad (3)$$

where MB denotes the final meson-baryon system, p' the final CM three-momentum and p the initial one.

In addition, we also fit the precisely measured ratios at the K^-p threshold [34,35]:

$$\gamma = \frac{\sigma(K^-p \to \pi^+\Sigma^-)}{\sigma(K^-p \to \pi^-\Sigma^+)} = 2.36 \pm 0.04, \quad (4)$$

$$R_c = \frac{\sigma(K^-p \to \text{charged particles})}{\sigma(K^-p \to \text{all})} = 0.664 \pm 0.011,$$

$$R_n = \frac{\sigma(K^-p \to \pi^0\Lambda)}{\sigma(K^-p \to \text{all neutral states})} = 0.189 \pm 0.015.$$

The first two ratios, which are Coulomb corrected, are measured with 1.7% precision, which is of the same order

Table 2. Table of results of the A-type fits, given in table 1. The $\sigma_{\pi N}$ value enforced in the fits is given in the first row.

$\sigma_{\pi N}$	20^*	30^*	40^*
γ	2.36	2.36	2.37
R_c	0.629	0.628	0.628
R_n	0.168	0.171	0.173
ΔE (eV)	194	192	192
Γ (eV)	324	302	270
ΔE_D (eV)	204	204	207
Γ_D (eV)	361	338	305
$a_{K^- p}$ (fm)	$-0.49 + i\,0.44$	$-0.49 + i\,0.41$	$-0.50 + i\,0.37$
a_0 (fm)	$-1.07 + i\,0.53$	$-1.04 + i\,0.50$	$-1.02 + i\,0.45$
a_1 (fm)	$0.44 + i\,0.15$	$0.40 + i\,0.15$	$0.33 + i\,0.14$
$\delta_{\pi\Lambda}(\Xi)$ (°)	3.4	4.5	5.7
m_0 (GeV)	1.2	1.1	1.0
a_{0+}^+ $(10^{-2} \cdot M_\pi^{-1})$	-2.0	-2.2	-2.2

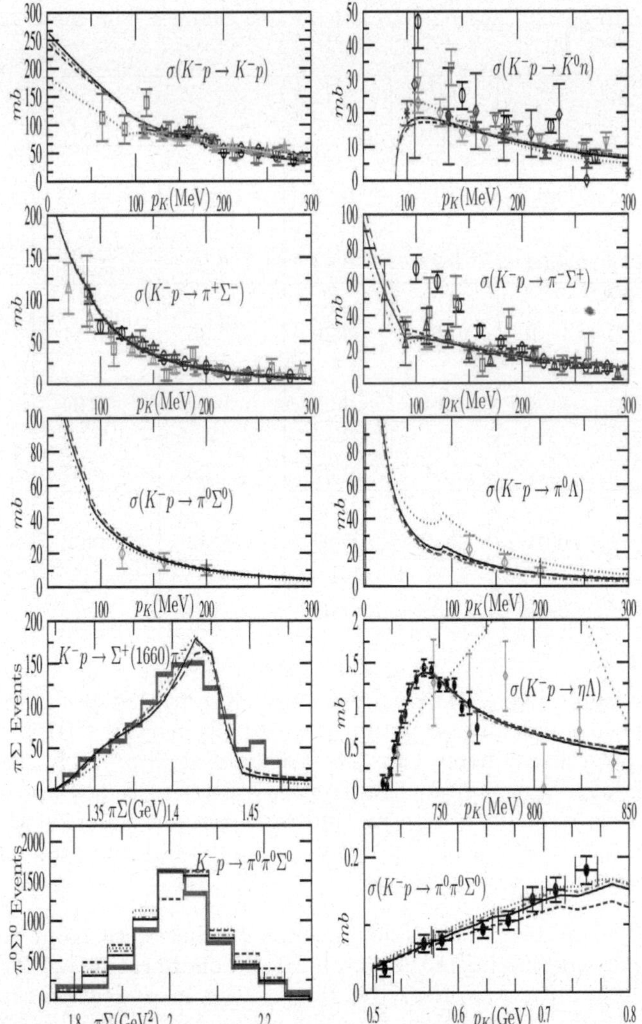

Fig. 3. The solid lines correspond to the $\sigma_{\pi N} = 40^*$ MeV fit, the dashed lines to the 30^* MeV fit, the dash-dotted curves to the 20^* MeV one and the dotted lines to the $\mathcal{O}(p)$ fit of table 3.

Table 3. B-type fits that do not agree with the DEAR data, eq. (5). The enforced $\sigma_{\pi N}$ value in the fit is shown in the first line. For precise definitions of the parameters f, b_0, b_D, b_F, b_i and a_i see ref. [23]. Three among the parameters b_0, b_D, b_F and b_i are fixed.

Units	$\sigma_{\pi N}$ (MeV)	20^*	30^*	40^*	$\mathcal{O}(p)$
MeV	f	95.8	113.2	100.0	93.9
GeV^{-1}	b_0	-0.201	-0.159	-0.487	0^*
GeV^{-1}	b_D	-0.005	-0.297	0.127	0^*
GeV^{-1}	b_F	-0.133	-0.157	-0.188	0^*
GeV^{-1}	b_1	$+0.122$	$+0.016$	$+0.135$	0^*
GeV^{-1}	b_2	-0.080	-0.151	-0.037	0^*
GeV^{-1}	b_3	-0.533	-0.281	-0.494	0^*
GeV^{-1}	b_4	$+0.028$	-0.291	-0.173	0^*
	a_1	$+4.037$	$+4.188$	$+2.930$	-2.958
	a_2	-2.063	-3.129	-2.400	-1.479
	a_5	-1.131	-1.214	-1.225	-1.330
	a_7	-3.488	-3.000	-2.795	-1.805
	a_8	-0.347	$+0.642$	$+2.906$	-0.655
	a_9	-1.767	-2.109	-1.913	-1.918

as the expected isospin violations. Indeed, all the other observables we fit have uncertainties larger than 5%.

Since we are just considering the S-wave amplitudes, we only include in the fits those data points for the several K^-p cross-sections with laboratory frame K^- three-momentum $p_K \leq 0.2\,\mathrm{GeV}$. This also enhances the sensitivity to the lowest-energy region in which we are particularly interested. We also include in the fits the $\pi^\pm\Sigma^\mp$ event distributions from the chain of reactions $K^-p \to \Sigma^+(1660)\pi^-$, $\Sigma^+(1660) \to \pi^+\Sigma\pi$ [36].

The number of data points included in each fit, without the data for the energy shift and width of kaonic hydrogen, is 97. Unless the opposite is stated, we also include in the

Table 4. Table of results for the B-type fits, given in table 3.

$\sigma_{\pi N}$	20^*	30^*	40^*	$\mathcal{O}(p)$
γ	2.34	2.35	2.34	2.32
R_c	0.643	0.643	0.644	0.637
R_n	0.160	0.163	0.176	0.193
ΔE (eV)	436	409	450	348
Γ (eV)	614	681	591	611
ΔE_D (eV)	418	385	436	325
Γ_D (eV)	848	880	844	775
$a_{K^- p}$ (fm)	$-1.01 + i\,1.03$	$-0.93 + i\,1.07$	$-1.06 + i\,1.02$	$-0.79 + i\,0.94$
a_0 (fm)	$-1.75 + i\,1.15$	$-1.65 + i\,1.30$	$-1.79 + i\,1.10$	$-1.50 + i\,1.00$
a_1 (fm)	$-0.13 + i\,0.39$	$-0.14 + i\,0.36$	$-0.12 + i\,0.46$	$0.32 + i\,0.46$
$\delta_{\pi\Lambda}(\Xi)$ (°)	-1.4	1.7	-1.2	-1.4
m_0 (GeV)	0.8	0.6	0.7	–
a_{0+}^+ ($10^{-2} \cdot M_\pi^{-1}$)	-0.5	-1.4	$+0.3$	–

Table 5. Fit I, $I = 0$ poles. The pole positions are given in MeV and the couplings in GeV. The symbol $|\gamma_i|_I$ means the coupling of the corresponding pole to the state with definite isospin I made up by the charged states of the i-th channel. The couplings to the $I = 1, 2$ channels are always close to zero.

Re(Pole) $|\gamma_{\pi\Lambda}|$	$-$Im(Pole) $|\gamma_{\pi\Sigma}|_0$	Sheet $|\gamma_{\pi\Sigma}|_1$	$|\gamma_{\pi\Sigma}|_2$	$|\gamma_{\bar{K}N}|_0$	$|\gamma_{\bar{K}N}|_1$	$|\gamma_{\eta\Lambda}|$	$|\gamma_{\eta\Sigma}|$	$|\gamma_{K\Xi}|_0$	$|\gamma_{K\Xi}|_1$
1301	13	1RS							
0.03	1.12	0.02	0.01	5.83	0.05	0.41	0.04	2.11	0.03
1309	13	2RS							
0.02	3.66	0.02	0.02	4.46	0.04	0.21	0.04	3.05	0.03
1414	23	2RS							
0.14	4.24	0.13	0.01	4.87	0.39	0.85	0.20	9.35	0.11
1388	17	3RS							
0.02	3.81	0.02	0.02	1.33	0.04	0.42	0.04	9.55	0.04
1676	10	3RS							
0.01	1.28	0.03	0.00	1.67	0.01	2.19	0.07	5.29	0.07
1673	18	4RS							
0.01	1.26	0.02	0.00	1.82	0.01	2.13	0.06	5.32	0.06
1825	49	5RS							
0.02	2.29	0.02	0.00	2.10	0.02	0.89	0.03	7.43	0.09

fits the DEAR measurement of the shift and width of the $1s$ kaonic hydrogen energy level [1],

$$\Delta E = 193 \pm 37(\text{stat}) \pm 6(\text{syst.}) \text{ eV},$$
$$\Gamma = 249 \pm 111(\text{stat.}) \pm 39(\text{syst.}) \text{ eV}, \tag{5}$$

which is around a factor of two more precise than the KEK previous measurement [37], $\Delta E = 323 \pm 63 \pm 11$ eV and $\Gamma = 407 \pm 208 \pm 100$ eV. To calculate the shift and width of the $1s$ kaonic hydrogen state we use the results of [3] incorporating isospin-breaking corrections up to and including $\mathcal{O}(\alpha^4, (m_d - m_u)\alpha^3)$. We further constrain our fits by computing at $\mathcal{O}(p^2)$ in baryon $SU(3)$ CHPT several πN observables with the values of the low-energy constants involved in the fit. Unitarity corrections in the πN sector are not as large as in the $S = -1$ sector, $e.g.$, there is not anything like a $\Lambda(1405)$-resonance close to threshold, and hence a calculation within pure $SU(3)$ baryon CHPT is more reliable here. Thus, we calculate at $\mathcal{O}(p^2)$, a_{0+}^+, the isospin-even pion-nucleon S-wave scattering length, $\sigma_{\pi N}$, the pion-nucleon σ-term, and m_0 from the value of the proton mass m_p. In this way we fix three of our free parameters.

In addition, ref. [23] included further data from recent experiments of the Crystal Ball Collaboration [24,25]. These data comprises the $K^- p \to \eta\Lambda$ cross-section and $\Sigma\pi$ event distributions from the reaction $K^- p \to \pi^0\pi^0\Sigma^0$. As noted in ref. [23] these data cannot be reproduced from the fits given in ref. [22] and then new fits were considered in the former reference that from the beginning included the data from [24,25]. As in ref. [22] two types of fits were

Table 6. Fit I, $I = 1$ poles. The pole positions are given in MeV and the couplings in GeV. The couplings to the $I = 0, 2$ channels are always close to zero. The notation is like in table 5.

Re(Pole) $\|\gamma_{\pi\Lambda}\|$	$-$Im(Pole) $\|\gamma_{\pi\Sigma}\|_0$	Sheet $\|\gamma_{\pi\Sigma}\|_1$	$\|\gamma_{\pi\Sigma}\|_2$	$\|\gamma_{\bar{K}N}\|_0$	$\|\gamma_{\bar{K}N}\|_1$	$\|\gamma_{\eta\Lambda}\|$	$\|\gamma_{\eta\Sigma}\|$	$\|\gamma_{K\Xi}\|_0$	$\|\gamma_{K\Xi}\|_1$
1425	6.5	2RS							
1.35	0.24	1.66	0.01	0.35	3.92	0.05	4.23	0.49	2.98
1468	13	2RS							
2.80	0.16	5.96	0.02	0.23	8.74	0.04	10.66	0.19	2.48
1433	3.7	3RS							
0.65	0.08	0.80	0.00	0.12	1.58	0.02	5.82	0.20	2.14
1720	18	4RS							
1.82	0.02	1.21	0.00	0.02	0.95	0.02	6.78	0.05	5.31
1769	96	6RS							
2.65	0.00	0.61	0.00	0.00	2.48	0.00	3.32	0.01	4.22
1340	143	3-4RS							
1.33	0.14	5.50	0.02	0.02	1.58	0.00	3.28	0.03	1.20
1395	311	3-4RS							
2.08	0.01	1.49	0.01	0.00	1.24	0.00	7.63	0.01	3.97

Table 7. Fit II, $I = 0$ poles. The pole positions are given in MeV and the couplings in GeV. The notation is like in table 5.

Re(Pole) $\|\gamma_{\pi\Lambda}\|$	$-$Im(Pole) $\|\gamma_{\pi\Sigma}\|_0$	Sheet $\|\gamma_{\pi\Sigma}\|_1$	$\|\gamma_{\pi\Sigma}\|_2$	$\|\gamma_{\bar{K}N}\|_0$	$\|\gamma_{\bar{K}N}\|_1$	$\|\gamma_{\eta\Lambda}\|$	$\|\gamma_{\eta\Sigma}\|$	$\|\gamma_{K\Xi}\|_0$	$\|\gamma_{K\Xi}\|_1$
1347	36	2RS							
0.02	6.48	0.12	0.02	2.60	0.10	1.42	0.01	0.32	0.07
1427	18	2RS							
0.12	3.87	0.23	0.01	6.99	0.23	3.49	0.05	1.64	0.32
1340	41	3RS							
0.07	5.92	0.08	0.01	0.62	0.08	2.33	0.01	0.75	0.04
1667	8	4RS							
0.03	0.77	0.05	0.00	0.59	0.01	3.32	0.02	12.17	0.08
1667	8	5RS							
0.03	0.77	0.05	0.00	0.59	0.01	3.32	0.03	12.17	0.06

Table 8. Fit II, $I = 1$ poles. The pole positions are given in MeV and the couplings in GeV. The notation is like in table 5.

Re(Pole) $\|\gamma_{\pi\Lambda}\|$	$-$Im(Pole) $\|\gamma_{\pi\Sigma}\|_0$	Sheet $\|\gamma_{\pi\Sigma}\|_1$	$\|\gamma_{\pi\Sigma}\|_2$	$\|\gamma_{\bar{K}N}\|_0$	$\|\gamma_{\bar{K}N}\|_1$	$\|\gamma_{\eta\Lambda}\|$	$\|\gamma_{\eta\Sigma}\|$	$\|\gamma_{K\Xi}\|_0$	$\|\gamma_{K\Xi}\|_1$
1399	41	2RS							
1.49	0.09	5.58	0.01	0.13	4.92	0.08	0.73	0.03	4.99
1424	3.6	2RS							
0.54	0.14	1.58	0.00	0.20	1.17	0.10	0.61	0.04	3.76
1311	122	3-4RS							
2.63	0.05	4.61	0.01	0.02	3.44	0.02	0.60	0.03	3.60
1426	3	3RS							
0.56	0.04	1.18	0.00	0.07	0.77	0.04	0.61	0.02	3.74

found. The so-called A-type fits, table 1, that together with scattering data also reproduce the DEAR measurement on kaonic hydrogen, and the B-type fits, table 3, that reproduce the former but not the latter. In fig. 2 and table 2 we show the reproduction of the data by the A-type fits and in fig. 3 and table 4 the same is done for the B-type fits. In the last column of table 3 we include the lowest-order fit, only with $\mathcal{K}_1$, also shown in fig. 3.

It is worth stressing the good reproduction of the data accomplished by the A-type fits, including the most recent measurement of the kaon hydrogen lowest-energy level and width. One also observes a factor of 2 of difference between the K^-p scattering lengths of the A- and B-type fits. So, if finally the DEAR measurement [1] is confirmed by the results of the DEAR/SIDDHARTA Collaboration [2], it would give rise to an important step forward in the knowledge of kaon-nucleon interactions. This difference in the scattering lengths makes also that only the A-type fits have a pattern of isospin violation in the calculations of the shift and width of kaonic hydrogen of expected size [3]. For the B-type fits the isospin violations turn out to be rather large, 30%, while for the A-type ones only 14%. In addition, we also observe from tables 2 and 4 that the values of the scattering lengths are rather independent of the values given to the sigma terms.

3 Spectroscopy

We show in tables 5 and 6 the $I = 0$ and 1 poles, respectively, corresponding to the so-called fit I of ref. [23], given in the fifth column of table 1.

By passing continuously from one Riemann sheet to the other some of the poles in the tables with the same isospin are connected and represent the same resonance. One observes poles corresponding to the $\Lambda(1405)$, $\Lambda(1670)$ and $\Lambda(1800)$ in good agreement with the mass and width given to those resonances in the PDG [38]. In addition, there is a lighter resonance around 1310 MeV, not quoted in the PDG, and this has to do with the so-called two $\Lambda(1405)$-resonances, although for fit I it appears much lighter than in ref. [18]. For $I = 1$ one has the $\Sigma(1750)$-resonance. Fit-I amplitudes also show in $I = 1$ a broad bump at around 1.6 GeV corresponding to the $\Sigma(1620)$. There are also other poles around the $\bar{K}N$ threshold which mix up giving rise to a clear bump structure from 1.4 to 1.45 GeV. These poles could be related to the $\Sigma(1480)$ [39]. Finally, we also observe an $I = 2$ pole for fit I at $1722 - i\,181$ MeV.

In tables 7 and 8 the $I = 0$, 1 poles positions for the fit II of ref. [23], given in the fifth column of table 4, are shown. There are also poles corresponding to the $\Lambda(1405)$, $\Lambda(1670)$ but not for the $\Lambda(1800)$. There is no $\Sigma(1750)$-resonance either and the bumps for the $\Sigma(1620)$ have disappeared in several amplitudes.

In summary we have reviewed the works of refs. [22,21, 23]. We have shown two types of fits that agree with scattering experimental data but only one type agrees with the DEAR measurement of kaonic hydrogen [1]. These latter fits are also those that offer a remarkable agreement with

spectroscopic information [38]. Hence, taken into consideration the present experimental information, the so-called fits A, table 1, are preferred over the fits B, table 3.

This work has been supported in part by the MEC (Spain) and FEDER (EC) Grants Nos. FPA2003-09298-C02-01 (J.P.), FPA2004-03470 (J.A.O. and M.V.), the Fundación Séneca grant Ref. 02975/PI/05 (J.A.O. and M.V.), the European Commission (EC) RTN Network EURIDICE under Contract No. HPRN-CT2002-00311 and the HadronPhysics I3 Project (EC) Contract No. RII3-CT-2004-506078 (J.A.O.) and by Junta de Andalucía Grants Nos. FQM-101 (J.P. and M.V.) and FQM-347 (J.P.).

References

1. DEAR Collaboration (G. Beer *et al.*), Phys. Rev. Lett. **94**, 212302 (2005).
2. DEAR/SIDDHARTA Collaboration (D.L. Sirghi, F. Sirghi), *The physics of kaonic atoms at DAFNE*, http://www.lnf.infn.it/esperimenti/dear/DEAR_RPR.pdf.
3. U.-G. Meißner, U. Raha, A. Rusetsky, Eur. Phys. J. C **35**, 349 (2004).
4. S. Deser *et al.*, Phys. Rev. **96**, 774 (1954); T.L. Trueman, Nucl. Phys. **26**, 57 (1961).
5. R.H. Dalitz, S.F. Tuan, Phys. Rev. Lett. **2**, 425 (1959); Ann. Phys. (N.Y.) **8**, 100 (1959).
6. A.D. Martin, N.M. Queen, G. Violini, Nucl. Phys. B **10**, 481 (1969); P.M. Gensini, R. Hurtado, G. Violini, PiN Newslett. **13**, 291 (1997); B. Di Claudio, A.M. Rodriguez-Vargas, G. Violini, Z. Phys. C **3**, 75 (1979).
7. A.D. Martin, Nucl. Phys. B **179**, 33 (1979).
8. R. Buttgen, K. Holinde, J. Speth, Phys. Lett. B **163**, 305 (1985); R. Buettgen, K. Holinde, A. Mueller-Groeling, J. Speth, P. Wyborny, Nucl. Phys. A **506**, 586 (1990); A. Mueller-Groeling, K. Holinde, J. Speth, Nucl. Phys. A **513**, 557 (1990).
9. T. Hamaie, M. Arima, K. Masutani, Nucl. Phys. A **591**, 675 (1995).
10. P.J. Fink, G. He, R.H. Landau, J.W. Schnick, Phys. Rev. C **41**, 2720 (1990).
11. E.A. Veit, B.K. Jennings, R.C. Barrett, A.W. Thomas, Phys. Lett. B **137**, 415 (1984).
12. J.L. Goity, C.L. Schat, N.N. Scoccola, Phys. Rev. D **66**, 114014 (2002); Phys. Rev. Lett. **88**, 102002 (2002).
13. N. Kaiser, P.B. Siegel, W. Weise, Nucl. Phys. A **594**, 325 (1995).
14. J.A. Oller, E. Oset, Nucl. Phys. A **620**, 438 (1997); **652**, 407 (1999)(E).
15. E. Oset, A. Ramos, Nucl. Phys. A **635**, 99 (1998).
16. J.A. Oller, E. Oset, A. Ramos, Prog. Part. Nucl. Phys. **45**, 157 (2000).
17. J.A. Oller, U.-G. Meißner, Phys. Lett. B **500**, 263 (2000).
18. D. Jido, J.A. Oller, E. Oset, A. Ramos, U.-G. Meißner, Nucl. Phys. A **725**, 181 (2003).
19. C. Garcia-Recio, M.F.M. Lutz, J. Nieves, Phys. Lett. B **582**, 49 (2004).
20. B. Borasoy, R. Nissler, W. Weise, Phys. Rev. Lett. **94**, 213401 (2005); **96**, 199201 (2006); Eur. Phys. J. A **25**, 79 (2005); Phys. Rev. C **74**, 055201 (2006).

21. J.A. Oller, J. Prades, M. Verbeni, Phys. Rev. Lett. **96**, 199202 (2006).
22. J.A. Oller, J. Prades, M. Verbeni, Phys. Rev. Lett. **95**, 172502 (2005).
23. J.A. Oller, Eur. Phys. J. A **28**, 63 (2006).
24. Crystal Ball Collaboration (A. Starostin *et al.*), Phys. Rev. C **64**, 055205 (2001).
25. Crystal Ball Collaboration (S. Prakhov *et al.*), Phys. Rev. C **70**, 034605 (2004).
26. V.K. Magas, E. Oset, A. Ramos, Phys. Rev. Lett. **95**, 052301 (2005).
27. J.A. Oller, M. Verbeni, J. Prades, JHEP **0609**, 079 (2006) and hcp-ph/0701096 (Erratum); M. Frink, U.-G. Meißner, Eur. Phys. J. A **29**, 255 (2006).
28. W.E. Humphrey, R.R. Ross, Phys. Rev. **127**, 1305 (1962).
29. J.K. Kim, Phys. Rev. Lett. **14**, 29 (1965).
30. M. Sakitt *et al.*, Phys. Rev. B **139**, 719 (1965).
31. J. Ciborowski *et al.*, J. Phys. G **8**, 13 (1982).
32. W. Kittel, G. Otter, I. Wacek, Phys. Lett. **21**, 349 (1966).
33. D. Evans *et al.*, J. Phys. G **9**, 885 (1983).
34. R.J. Nowak *et al.*, Nucl. Phys. B **139**, 61 (1978).
35. D. Tovee *et al.*, Nucl. Phys. B **33**, 493 (1971).
36. R.J. Hemmingway, Nucl. Phys. B **253**, 742 (1985).
37. M. Iwasaki *et al.*, Phys. Rev. Lett. **78**, 3067 (1997); Phys. Rev. C **58**, 2366 (1998).
38. Particle Data Group (W.M. Yao *et al.*), J. Phys. G **33**, 1 (2006).
39. I. Zychor *et al.*, Phys. Rev. Lett. **96**, 012002 (2006).

Eur. Phys. J. A **31**, 534–536 (2007)

DOI 10.1140/epja/i2006-10292-4

THE EUROPEAN
PHYSICAL JOURNAL A

Special Article – QNP 2006

Pseudoscalar meson masses in unitarized chiral perturbation theory

J.A. Oller and L. Roca[a]

Departamento de Física, Universidad de Murcia, E-30071 Murcia, Spain

Received: 22 December 2006

Published online: 22 March 2007 – © Società Italiana di Fisica / Springer-Verlag 2007

Abstract. This contribution summarizes the work explained in arXiv:hep-ph/0608290 where we perform a non-perturbative chiral study of the masses of the lightest pseudoscalar mesons. The pseudoscalar self-energies are calculated by the evaluation of the scalar self-energy loops with full S-wave meson-meson amplitudes taken from Unitary Chiral Perturbation Theory (UCHPT). These amplitudes, among other features, contain the lightest nonet of scalar resonances σ, $f_0(980)$, $a_0(980)$ and κ. The self-energy loops are regularized by a proper subtraction of the infinities within a dispersion relation formulation of the amplitudes. Values for the bare masses of pions and kaons and the η_8 mass are obtained. We then match to the self-energies from standard Chiral Perturbation Theory (CHPT) to $\mathcal{O}(p^4)$ and resum higher orders from our calculated scalar self-energies. The dependence of the self-energies on the quark masses allows a determination of the ratio of the strange-quark mass over the mean of the lightest-quark masses, $m_s/\hat{m}$, in terms of the $\mathcal{O}(p^4)$ CHPT low-energy constant combinations $2L_8^r - L_5^r$ and $2L_6^r - L_4^r$. In this way, we give a range for the values of these low-energy counterterms and for $3L_7 + L_8^r$, once the η-meson mass is invoked. The low-energy constants are further constraint by performing a fit to the recent MILC lattice data on the pseudoscalar masses, and $m_s/\hat{m} = 25.6 \pm 2.5$ results. This value is consistent with 24.4 ± 1.5 from CHPT and phenomenology and more marginally with the value 27.4 ± 0.5 obtained from pure perturbative chiral extrapolations of the MILC lattice data to physical values of the lightest-quark masses.

PACS. 14.65.Bt Light quarks – 12.40.Yx Hadron mass models and calculations – 14.40.Aq π, K, and η mesons – 12.39.Fe Chiral Lagrangians

Three-flavour CHPT has already reached a very sophisticated stage with present calculations at the level of next-to-next-to-leading order (NNLO). In ref. [1] the masses and decay constants of the lightest octet of pseudoscalar mesons are worked out at NNLO. One striking fact concerns the much larger $\mathcal{O}(p^6)$ two-loop contribution than the full $\mathcal{O}(p^4)$ one to the self-energies of the pseudoscalars, by around one order of magnitude. As stated in ref. [1] the $\mathcal{O}(p^6)$ counterterms "seem severely overestimated" due to the complicated nature of the scalar sector. Hence, this study rises the interesting question of how one can improve the calculation of the contributions from the scalar sector. This is an object of study of the investigation done in ref. [2], which is the work we summarize in the present contribution. Concerning the importance of the S-wave dynamics, it is well known that the scalar sector is the source of large higher-order corrections to the CHPT series, because of the enhancement of unitarity diagrams, which definitely slow down the convergence of the perturbative chiral series. A novel example, where these large corrections play a role, as we show in this work, concerns the determinations of the light-quark masses from present lattice calculations of the pseudoscalar masses [3, 4] with three dynamical fermions. In these evaluations one confronts with the problematic S-wave dynamics plagued of non-perturbative effects. The MILC Collaboration reports a value for the ratio $m_s/\hat{m} = 27.4 \pm 0.5$, with $\hat{m} = (m_u + m_d)/2$, inconsistent with its determination from CHPT and phenomenology [5], $m_s/\hat{m} = 24.4 \pm 1.5$. Because Unitary CHPT (UCHPT) has proved very successful in dealing, precisely, with the strong meson-meson S-wave interactions, it is worth employing this scheme as a parameterization to bring down the results from lattice QCD to the physical lightest-quark masses. Here, we will employ $\mathcal{O}(p^4)$ CHPT and supply it with the corrections from UCHPT of $\mathcal{O}(p^6)$ and higher.

In the following, we briefly summarize the formalism and results of ref. [2]. The starting point in ref. [2] is the S-wave meson-meson partial waves both for resonant isospins $(I) = 0, 1, 1/2$ as well as for the much smaller and non-resonant ones, $I = 3/2$ and 2. For the former set we take the amplitudes of ref. [6]. It is worth stressing that this is just a subset of all possible S-wave two-body rescattering diagrams. Nonetheless, these have been

------

[a] e-mail: `luisroca@um.es`

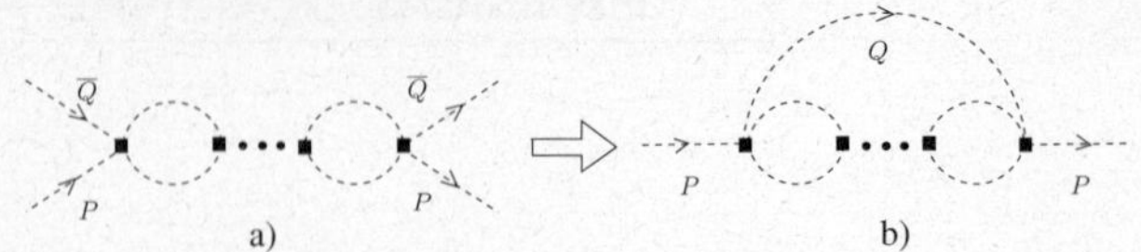

Fig. 1. a) represents the S-wave amplitude $P\overline{Q} \to P\overline{Q}$ and b) is the diagram for the calculation of the self-energy of the pseudoscalar P due to the intermediate pseudoscalar Q.

shown to be the dominant contributions and enough to explain a wide range of phenomenology like, *e.g.*, meson-meson scattering [7,6,8], $\gamma\gamma \to$ meson-meson [9], meson form factors [10–12], ϕ [13], J/Ψ [10,14], η [15], B and D [16] decays, determinations of light-quark masses and V_{us} [17], etc.

The interaction kernel employed in ref. [6] comprises the lowest-order CHPT amplitudes together with the exchange of s-channel scalar resonances in a chiral symmetric invariant way. The low-lying scalar resonances are generated even when no explicit resonances at tree level are included [7,6]. The basic point in UCHPT is to resum the right-hand or unitarity cut to all orders, which is the source of the large corrections produced by the S-wave meson-meson interactions, and perform a chiral expansion of the rest, the so-called interaction kernel, calculating it perturbatively from CHPT [18,19,6].

The set of diagrams enhanced in S-wave meson-meson scattering are represented in fig. 1a, for the process $P\overline{Q} \to P\overline{Q}$. Our expression for the self-energy of the pseudoscalar P due to the loops schematically represented in fig. 1 can be expressed by the sum,

$$\Sigma_P^U = -\sum_Q \int \frac{\mathrm{d}^3 k}{(2\pi)^3 2E_Q(\mathbf{k})} T_{P\overline{Q} \to P\overline{Q}}(s_1), \qquad (1)$$

with $Q = \{\pi^+, \pi^-, \pi^0, K^+, K^-, K^0, \overline{K}^0, \eta\}$ and $s_1 = (M_P - E_Q(\mathbf{k}))^2 - \mathbf{k}^2$ for $p = (M_P, \mathbf{0})$. In addition one has to add a tadpole contribution coming from the t-crossed process $P\overline{P} \to Q_i\overline{Q}_i$ which has a similar expression. The loop function of eq. (1) is quadratically divergent. By performing a once subtracted dispersion relation for the scattering amplitudes, it is easy to identify the origin of the infinites and get rid of them up to logarithmic pieces. (See ref. [2] for details.)

The mass equation one has to solve is

$$M_P^2 = \overset{\circ}{M}{}_P^{\,2} + \Sigma_P(\overset{\circ}{M}{}_Q^{\,2}; M_Q^2). \qquad (2)$$

The dependence on the bare masses $\overset{\circ}{M}{}_Q^{\,2}$ originates from the dependence on the explicit inclusion of the quark mass matrix in CHPT from where the interaction kernel $\mathcal{K}$ is calculated [6]. By solving exactly eq. (2) we find $\overset{\circ}{M}{}_\pi = 126 \pm 4\,\mathrm{MeV}$, $\overset{\circ}{M}{}_K = 476 \pm 10\,\mathrm{MeV}$, $M_{\eta_8} = 635 \pm 15\,\mathrm{MeV}$. Regarding the sizes of the self-energies, we have, $\Sigma_P(\overset{\circ}{M}{}_Q^{\,2}; M_Q^2)/M_P^2 = 0.11 \pm 0.06$, 0.08 ± 0.04, 0.26 ± 0.06, for pions, kaons and etas, respectively. We observe that most of the physical masses of the pseudoscalars

is due to the bare mass. Taking into account the expression for the bare masses in terms of the quark masses and the values of the former ones given above, one get,

$$r_m = \frac{m_s}{\hat{m}} = 2\frac{\overset{\circ}{M}{}_K^{\,2}}{\overset{\circ}{M}{}_\pi^{\,2}} - 1 = 27.1 \pm 2.5.$$

On the other hand, one can split the self-energy into the $\mathcal{O}(p^4)$ contribution and higher orders:

$$M_P^2 = \overset{\circ}{M}{}_P^{\,2} + \Sigma_P^{4\chi}(\overset{\circ}{M}{}_Q^{\,2}; L_i^r) + \Sigma_P^H(\overset{\circ}{M}{}_Q^{\,2}; M_Q^2), \qquad (3)$$

where $\Sigma_P^H(\overset{\circ}{M}{}_Q^{\,2}; M_Q^2)$ is the self-energy calculated removing explicitly the $\mathcal{O}(p^4)$ contribution and $\Sigma_P^{4\chi}(\overset{\circ}{M}{}_Q^{\,2}; L_i^r)$ is the $\mathcal{O}(p^4)$ CHPT contribution from [20]. Equation (3) incorporates the combinations of the low-energy constants $L_{(5,8)} \equiv 2L_8^r - L_5^r$, $L_{(4,6)} \equiv 2L_6^r - L_4^r$ and $L_{(7,8)} \equiv 3L_7 + L_8^r$. As a function of $L_{(4,6)}$ and $L_{(5,8)}$, we solve first for the $\overset{\circ}{M}{}_\pi^{\,2}$ and $\overset{\circ}{M}{}_K^{\,2}$ bare masses in eqs. (3). Once we calculate $\overset{\circ}{M}{}_\pi^{\,2}$ and $\overset{\circ}{M}{}_K^{\,2}$, we fix $\overset{\circ}{M}{}_\eta^{\,2}$ using the Gell-Mann-Okubo relation and from the η equation, we then solve for $L_{(7,8)}$. In fig. 2 we show by contour plot lines the calculated values for $r_m = m_s/\hat{m}$ as a function of $L_{(4,6)}$ and $L_{(5,8)}$. (All the values of the LECs displayed are given in units of 10^{-3}.) If one takes into account the value of $m_s/\hat{m}$ from CHPT [5], 24.4 ± 1.5, the preferred region of $2L_6^r - L_4^r$ and $2L_8^r - L_5^r$ would be about in between the 23 and 26 contour lines. The shadowed areas below and above these lines represent the uncertainties of our calculation in these lines due to the variation in the renormalization scale, $\mu \sim [0.5, 1.2]\,\mathrm{GeV}$, and in the input parameters for the S-waves. Hence, they delimit the allowed bounded region of $2L_6^r - L_4^r$ and $2L_8^r - L_5^r$ that is permitted if the result $m_s/\hat{m} = 24.4 \pm 1.5$ [5] is taken.

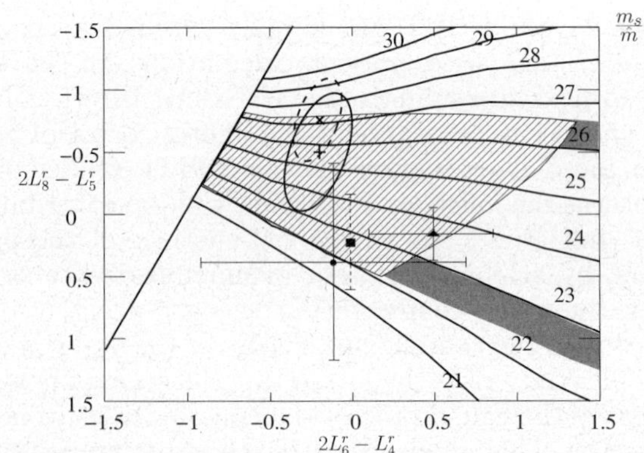

Fig. 2. Contour plot for $r_m = m_s/\hat{m}$ as a function of $L_{(4,6)}$ and $L_{(5,8)}$ showing also our point from the result of the fit to lattice data together with our theoretical uncertainty represented by the solid ellipse. The dashed ellipse represents the value for the LECs, within uncertainty, needed to reproduce the bare masses obtained with the full dynamical model. On the other hand, the triangle point corresponds to the lattice extrapolation from Staggered CHPT [21]. The circle and square points correspond, respectively, to the $\mathcal{O}(p^4)$ and $\mathcal{O}(p^6)$ CHPT results from refs. [22,23].

In order to obtain a sharper determination for the self-energies and quark mass ratio r_m, we employ our eq. (3) to reproduce the quark mass dependence of the recent lattice determinations of pseudoscalar masses from the MILC Collaboration [3] with three dynamical quarks, u, d and s and two lattice spacings $a_{coarse} = 0.12\,\text{fm}$ and $a_{fine} = 0.09\,\text{fm}$. Our aim here in ref. [2] was just to have more points with which to compare, including those for unphysical values of quark masses and not only the physical ones, so that we can obtain a more constraint determination for $L_{(5,8)}$ and $L_{(4,6)}$. The MILC Collaboration [3] provides pseudoscalar meson masses as a function of the quark ones, measured in terms of the lattice spacing a, which is known with an error of about 1.2%. Current quark masses and those of lattice are proportional [4], so we write the bare meson masses in terms of the lattice quark masses as, $\overset{\circ}{M}{}^2_\pi = 2B_0C(a)a\hat{m}$, $\overset{\circ}{M}{}^2_K = B_0C(a)a(\hat{m} + m_s)$, $\overset{\circ}{M}{}^2_\eta = \frac{2}{3}B_0C(a)a(\hat{m} + 2m_s)$, and treat $B_0C(a)$ as a free parameter in our fits to the three flavour runs of ref. [3]. We notice that $C(a_{fine}) = 1.49C(a_{coarse})$ [2]. We obtain a very good fit to the lattice data. From the fit we get the values, $2L_8^r - L_5^r = -0.52 \pm 0.43$, $2L_6^r - L_4^r = -0.20 \pm 0.17$. These values can be compared with the result from the MILC Collaboration [21] $2L_8^r - L_5^r = 0.16 \pm 0.2$, $2L_6^r - L_4^r = 0.4 \pm 0.4$, given by the triangle point in fig. 2. The corresponding ratio of quark masses is then,

$$m_s/\hat{m} = 25.6 \pm 2.5. \tag{4}$$

This value is in perfect agreement with 24.4 ± 1.5 from ref. [5]. The previous result is also in agreement, at the level of one sigma, with the value 27.4 ± 0.5 determined in ref. [4]. However, the difference between the central values for r_m obtained and the one of ref. [4] is rather large, 2 units. At this point, it is worth stressing that we have been able to give, with a non-perturbative chiral parameterization, a remarkably good reproduction of the lattice data while giving rise to a value for r_m significantly smaller than that of ref. [4] and in the range of values previously predicted from CHPT and phenomenology [5]. Thus, we certainly can conclude that it is not a *necessary* feature of the lattice runs of pseudoscalar masses by the MILC Collaboration [4] having a significantly larger $m_s/\hat{m}$ value than that previously determined in ref. [5]. With respect to the bare masses one has, $\overset{\circ}{M}_\pi = 125.0 \pm 4.3\,\text{MeV}$, $\overset{\circ}{M}_K = 456.2 \pm 20.8\,\text{MeV}$. Regarding $L_{(7,8)}$ we get $L_{(7,8)} = -0.6 \pm 0.6$, in good agreement with typical results in the literature, although with large errors. The dashed ellipse in fig. 2 represents the value for the LECs needed to reproduce the results for the bare masses in the full dynamical case. From there one can conclude that the dynamical scalar resonance saturation of $L_{(4,6)}$ is very good but poorer for $L_{(5,8)}$, although still compatible within errors.

We also explicitly show in table 1 the relative sizes of $\Sigma_P^{4\chi}$ and Σ_P^H. We also get that the ratio between the $\mathcal{O}(p^4)$ and the $\mathcal{O}(p^6)$ and higher-order contributions is quite sensitive to the LECs considered. The sizes of our self-energies from Σ_P^H are rather similar to those determined in ref. [23] to $\mathcal{O}(p^6)$. Reference [1] obtains:

Table 1. Relative sizes of the $\mathcal{O}(p^4)$, $\mathcal{O}(p^6)$ and higher-order contributions, $\Sigma_P^{4\chi}$ and Σ_P^H, respectively, to the self-energies.

	π	K	η
$\Sigma_P^{4\chi}/M_P^2$	-0.0789 ± 0.060	-0.221 ± 0.087	-0.410 ± 0.193
Σ_P^H/M_P^2	0.222 ± 0.04	0.375 ± 0.07	0.506 ± 0.09
$\Sigma_P^{4\chi}/\Sigma_P^H$	-0.355 ± 0.287	-0.589 ± 0.236	-0.810 ± 0.384

$\Sigma_\pi^{6\chi}/M_\pi^2 = 0.132\text{--}0.355$, $\Sigma_K^{6\chi}/M_K^2 = 0.194\text{--}0.423$ and $\Sigma_\eta^{6\chi}/M_\eta^2 = 0.234\text{--}0.521$. These values are well inside our bulk of results for $\Sigma_\pi^H/M_\pi^2 = 0.222 \pm 0.04$, $\Sigma_K^H/M_K^2 = 0.375 \pm 0.07$, $\Sigma_\eta^H/M_\eta^2 = 0.506 \pm 0.09$. Thus, the calculations of ref. [23] to $\mathcal{O}(p^6)$, although showing that this order is much larger than the $\mathcal{O}(p^4)$, does not imply necessarily the lack of convergence of the chiral series since our calculation, estimating higher-orders corrections by incorporating physical S-waves which include both resonant and non-resonant physics, gives us values of similar size to those of this reference up to $\mathcal{O}(p^6)$.

References

1. G. Amoros, J. Bijnens, P. Talavera, Nucl. Phys. B **568**, 319 (2000).
2. J.A. Oller, L. Roca, arXiv:hep-ph/0608290.
3. C. Aubin *et al.*, Phys. Rev. D **70**, 094505 (2004).
4. C. Aubin *et al.*, Phys. Rev. D **70**, 031504 (2004).
5. H. Leutwyler, Phys. Lett. B **378**, 313 (1996).
6. J.A. Oller, E. Oset, Phys. Rev. D **60**, 074023 (1999).
7. J.A. Oller, E. Oset, Nucl. Phys. A **620**, 438 (1997); **652**, 407 (1999)(E).
8. J.A. Oller, E. Oset, J.R. Pelaez, Phys. Rev. Lett. **80**, 3452 (1998); Phys. Rev. D **59**, 074001 (1999); **60**, 099906 (1999)(E).
9. J.A. Oller, E. Oset, Nucl. Phys. A **629**, 739 (1998).
10. U.G. Meissner, J.A. Oller, Nucl. Phys. A **679**, 671 (2001); T.A. Lahde, U.G. Meissner, arXiv:hep-ph/0606133.
11. J.A. Oller, E. Oset, J.E. Palomar, Phys. Rev. D **63**, 114009 (2001).
12. M. Jamin, J.A. Oller, A. Pich, Nucl. Phys. B **622**, 279 (2002).
13. J.A. Oller, Phys. Lett. B **426**, 7 (1998); Nucl. Phys. A **714**, 161 (2003); J.E. Palomar, L. Roca, E. Oset, M.J. Vicente Vacas, Nucl. Phys. A **729**, 743 (2003).
14. L. Roca, J.E. Palomar, E. Oset, H.C. Chiang, Nucl. Phys. A **744**, 127 (2004).
15. E. Oset, J.R. Pelaez, L. Roca, Phys. Rev. D **67**, 073013 (2003).
16. J.A. Oller, Phys. Rev. D **71**, 054030 (2005).
17. M. Jamin, J.A. Oller, A. Pich, JHEP **0402**, 047 (2004); Eur. Phys. J. C **24**, 237 (2002); arXiv:hep-ph/0605095.
18. J.A. Oller, U.G. Meissner, Phys. Lett. B **500**, 263 (2001).
19. J.A. Oller, Phys. Lett. B **477**, 187 (2000).
20. J. Gasser, H. Leutwyler, Nucl. Phys. B **250**, 465 (1985).
21. MILC Collaboration (C. Aubin *et al.*), Phys. Rev. D **70**, 114501 (2004).
22. J. Bijnens, G. Colangelo, J. Gasser, Nucl. Phys. B **427**, 427 (1994).
23. G. Amoros, J. Bijnens, P. Talavera, Nucl. Phys. B **585**, 293 (2000); **598**, 665 (2001)(E).

Eur. Phys. J. A **31**, 537–539 (2007)
DOI 10.1140/epja/i2006-10197-2

THE EUROPEAN
PHYSICAL JOURNAL A

Special Article – QNP 2006

Precision measurements of kaonic atoms at DAΦNE and future perspectives

C. Curceanu (Petrascu)[1,a], M. Bazzi[1], G. Beer[2], L. Bombelli[3], A.M. Bragadireanu[1,4], M. Cargnelli[5], M. Catitti[1], C. Fiorini[3], T. Frizzi[3], F. Ghio[6], B. Girolami[6], C. Guaraldo[1], M. Iliescu[1], T. Ishiwatari[5], P. Kienle[5,7], P. Lechner[8], P. Levi Sandri[1], A. Longoni[3], V. Lucherini[1], J. Marton[5], D. Pietreanu[1], T. Ponta[4], D.L. Sirghi[1,4], F. Sirghi[1], H. Soltau[8], L. Struder[9], E. Widmann[5], and J. Zmeskal[5]

[1] INFN, Laboratori Nazionali di Frascati, C.P. 13, Via E. Fermi 40, I-00044 Frascati (Roma), Italy
[2] Department of Physics and Astronomy, University of Victoria, P.O. Box 3055, Victoria B.C., Canada V8W3P6
[3] Politecnico di Milano, Sezione di Elettronica, Via Golgi 40, I-20133 Milano, Italy
[4] IFIN-HH, P.O. Box MG-6, R-76900 Magurele, Bucharest, Romania
[5] Stefan Meyer Institut für subatomare Physik, Boltzmanngasse 3, A-1090, Vienna, Austria
[6] INFN Sezione di Roma I and Istituto Superiore di Sanità I-00161, Roma, Italy
[7] Technische Universität München, Physik Department, James-Franck-Strasse, D-85748 Garching, Germany
[8] PNSensors GmbH, Rönnerstr. 28, D-80803 München, Germany
[9] MPI for Extraterrestrial Physics, Giessenbachstr. D-85740 Garching, Germany

Received: 8 October 2006
Published online: 22 February 2007 – © Società Italiana di Fisica / Springer-Verlag 2007

Abstract. The DAΦNE electron-positron collider at the Frascati National Laboratories has made available a unique "beam" of negative kaons providing unprecedented conditions for the study of the low-energy kaon-nucleon interaction, a field still largely unexplored. The DEAR (DAΦNE Exotic Atom Research) experiment at DAΦNE and its successor SIDDHARTA (SIlicon Drift Detector for Hadronic Atom Research by Timing Application) aim at a precision measurement of the strong-interaction shift and width of the fundamental $1s$ level, via the measurement of the X-ray transitions to this level, for kaonic hydrogen and kaonic deuterium. The aim is to extract the isospin-dependent antikaon-nucleon scattering lengths and to contribute to the understanding of aspects of chiral symmetry breaking in the strangeness sector.

PACS. 36.10.-k Exotic atoms and molecules (containing mesons, muons, and other unusual particles) – 13.75.Jz Kaon-baryon interactions – 29.30.Kv X- and γ-ray spectroscopy

1 The SIDDHARTA scientific case

The precision measurements of kaonic atoms at the DAΦNE accelerator [1] of the LNF-INFN Laboratories are going to be performed in the framework of the SIDDHARTA international Collaboration [2]. The SIDDHARTA experiment will continue, deepen and enlarge the successful scientific line, initiated by the DEAR experiment [3], in performing precision measurements of X-ray transitions in exotic (kaonic) atoms at DAΦNE.

The aim of the experiment is a precise determination of the isospin-dependent antikaon-nucleon scattering lengths, through an eV measurement of the K_α line shift and width in kaonic hydrogen, and a similar, first time, measurement of kaonic deuterium. SIDDHARTA measures the X-ray transitions occurring in the cascade processes of kaonic atoms. A kaonic atom is formed when a negative kaon (from the decays of ϕ's, produced at DAΦNE) enters a target, loses its kinetic energy through the ionization and excitation of the atoms and molecules of the medium, and is eventually captured, replacing the electron, in an excited orbit. Via different cascade processes (Auger effect, Coulomb de-excitation, scattering, electromagnetic transitions) the kaonic atom de-excites to lower states. When a low-n state with small angular momentum is reached, the strong interaction with the nucleus comes into play. This strong interaction is the reason for a shift in energy of the lowest-lying level from the purely electromagnetic value and for a finite lifetime of the state, due to nuclear absorption of the kaon.

For kaonic hydrogen and deuterium the K-series transitions are of primary experimental interest since they are the only ones affected by the strong interaction. The K_α lines are clearly separated from the higher-K transitions. The shift ϵ and the width Γ of the $1s$ state of kaonic hydrogen are related in a fairly model-independent way to the

---

[a] e-mail: `petrascu@lnf.infn.it`

real and imaginary part of the complex s-wave scattering length, a_{K^-p}:

$$\epsilon + i\Gamma/2 = 412 a_{K^-p} \,\mathrm{eV\,fm^{-1}}. \tag{1}$$

This expression in known as the Deser-Trueman formula [4]. A similar relation applies to the case of kaonic deuterium and to its corresponding scattering length, a_{K^-d}.

The measured scattering lengths are then related to the isospin-dependent scattering lengths, a_0 and a_1:

$$a_{K^-p} = 1/2(a_0 + a_1); \quad a_{K^-n} = a_1. \tag{2}$$

The extraction of a_{K^-n} from a_{K^-d} requires a more complicated analysis than the impulse approximation (K^- scattering from each free nucleon): higher-order contributions associated with the K^-d three-body interaction have to be taken into account. This requires solving the three-body Faddeev equations by the use of potentials, taking into account the coupling among the multichannel interactions.

An accurate determination of the K^-N isospin-dependent scattering lengths will place strong constraints on the low-energy K-N dynamics, which in turn constrains the $SU(3)$ description of chiral symmetry breaking [5].

In 2002, the DEAR experiment performed the most precise measurement to date of kaonic hydrogen X-ray transitions to the $1s$ level [6]:

$$\epsilon = -193 \pm 37(\text{stat.}) \pm 6(\text{syst.})\,\text{eV}, \tag{3}$$

$$\Gamma = 249 \pm 111(\text{stat.}) \pm 30(\text{syst.})\,\text{eV}. \tag{4}$$

This measurement has triggered new interest from the theoretical groups working in the low-energy kaon-nucleon interaction field, and as well it is related to non-perturbative QCD tests [7–9].

The new experiment, SIDDHARTA, aims to improve the precision obtained by DEAR by an order of magnitude and to perform the first measurement ever of kaonic deuterium.

Other measurements (kaonic helium, sigmonic atoms, precise determination of the charged kaon mass) are also considered in the scientific program.

2 The SIDDHARTA setup

SIDDHARTA represents a new phase in the study of kaonic atoms at DAΦNE. The DEAR precision was limited by a signal/background ratio of about 1/70. To significantly improve this ratio, a breakthrough is necessary. An accurate study of the background sources present at DAΦNE was redone. The background includes two main sources:

– synchronous background: coming together with the kaons —related to K^- interactions in the setup materials and also to the ϕ-decay processes; it can be defined as hadronic background;

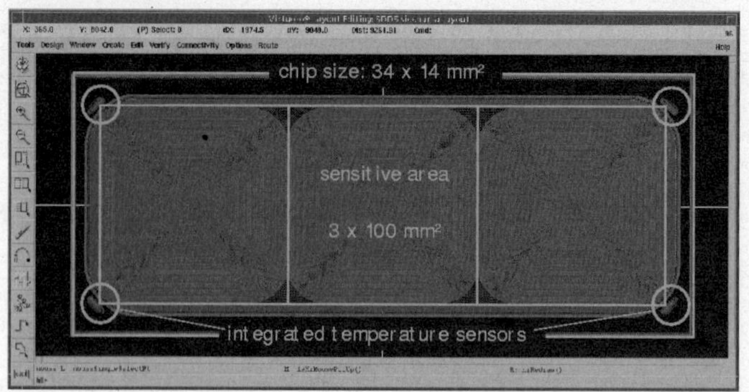

Fig. 1. SDD layout on the readout side: 3 SDD cells, read independently, each of $1\,\mathrm{cm}^2$ area, monolithically integrated on one chip.

– asynchronous background: final products of electromagnetic showers in the machine pipe and in the setup materials originating from particles lost from primary circulating beams either due to the interaction of particles in the same bunch (Touschek effect) or due to the interaction with the residual gas. Accurate studies performed by DEAR showed that the main background source in DAΦNE is of the second type, which shows the way to reduce it. A fast trigger correlated to a kaon entering into the target would cut the main part of the asynchronous background.

X-rays were detected by DEAR using CCDs (Charge-Coupled Devices) [10], which are excellent X-ray detectors, with very good energy resolution (about $140\,\mathrm{eV}$ FWHM at $6\,\mathrm{keV}$), but having the drawback of being non-triggerable devices (since the read-out time per device is at the level of $10\,\mathrm{s}$). A recently developed device, which preserves all good features of CCDs (energy resolution, stability and linearity), but additionally is triggerable — $i.e.$ fast (at the level of $1\,\mu\mathrm{s}$), was implemented. This new detector is a large-area Silicon Drift Detector (SDD), specially designed for spectroscopic application. The development of the new $1\,\mathrm{cm}^2$ SDD device is partially performed under the Joint Research Activity JRA10 of the I3 project "Study of strongly interacting matter (HadronPhysics)" within FP6 of the EU.

The trigger in SIDDHARTA will be given by a system of scintillators which will recognize a kaon entering the target making use of the back-to-back production mechanism of the charged kaons at DAΦNE from ϕ-decay.

Successful tests of SDD prototypes were performed in 2003 and 2004 at the Beam Test Facility of Frascati (BTF), in realistic ($i.e.$ DEAR-like) conditions. The results of these tests were very encouraging: a trigger rejection factor of 5×10^{-5} was measured. Extrapolated to SIDDHARTA conditions, this number translates for the kaonic hydrogen measurement into a S/B ratio in the region of interest of about 20/1. By triggering the SDDs, the asynchronous e.m. background (mainly due to the Touschek effect) can therefore be eliminated. Taking into account the synchronous background contribution, we can estimate a total S/B ratio of about 4/1.

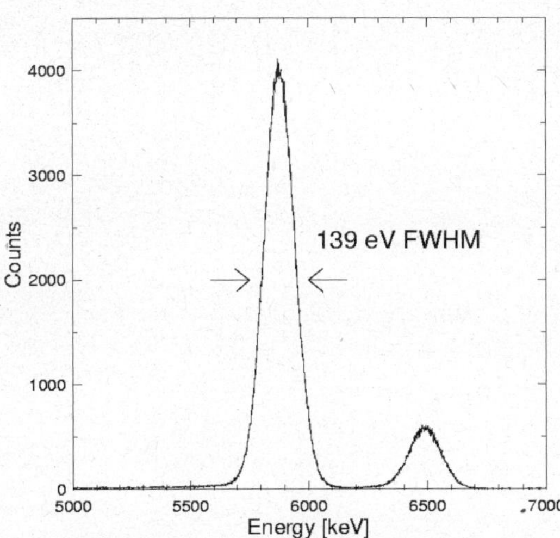

Fig. 2. The X-ray spectrum from an iron source as measured in the laboratory with an SDD chip prototype. Te experimental resolution, FWHM (Full-Width at Half Maximum) at 5.9 keV is 139 eV.

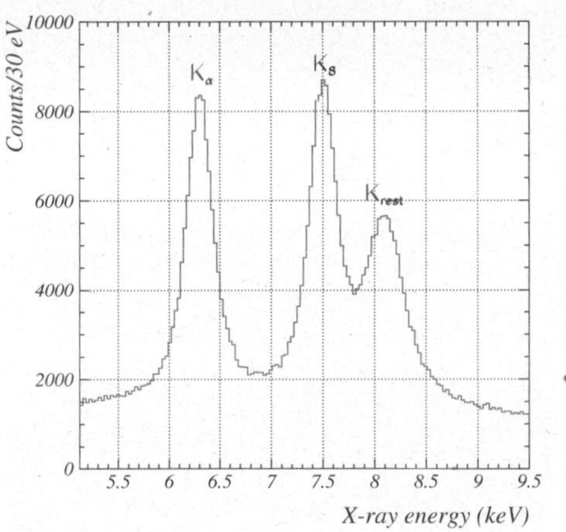

Fig. 3. The kaonic hydrogen Monte Carlo simulated spectrum for about $300\,\mathrm{pb}^{-1}$ of integrated luminosity in SIDDHARTA and a signal/background ratio equal to 4 : 1.

The SDD production, with 3 SDD cells, each of $1\,\mathrm{cm}^2$ area, monolithically integrated on one chip, is finished, in the configuration shown in fig. 1. Presently, the SDDs are under test. The first results show a very good experimental resolution, fig. 2, and a stability of the order of 2-3 eV at 6 keV (by using a 1 mV stabilized power supply developed in the framework of SIDDHARTA).

The SIDDHARTA setup will contain about 200 SDD chips of $1\,\mathrm{cm}^2$ each, placed around a cylindrical target, containing high-density gaseous hydrogen (deuterium). The setup will be installed above the beam pipe. The various elements of the SIDDHARTA setup are under production and testing, such as to be ready to install at DAΦNE to start taking data in autumn 2007.

The kaonic hydrogen simulated spectrum obtainable from about $300\,\mathrm{pb}^{-1}$ of integrated luminosity in SIDDHARTA, with a signal/background ratio of about 4/1 is shown in fig. 3.

With such a spectrum, a precision at the level of an eV for kaonic hydrogen is reachable.

3 Conclusions

DAΦNE has unique features as a kaon source which is intrinsically clean and of low momentum —a situation unattainable with fixed target machines— especially suitable for kaonic atom research.

The DEAR/SIDDHARTA experiments combine the newly available techniques with the good kaon beam quality to initiate a renaissance in the investigation of the low-energy kaon-nucleon interaction.

DEAR has performed the most precise measurement of kaonic hydrogen; the eV precision measurement of the strong-interaction shift and width of the fundamental level in kaonic hydrogen will be performed by SIDDHARTA. The first measurement of kaonic deuterium is also planned. These results will open new windows in the study of the kaon-nucleon interaction, in particular chiral symmetry breaking in the strangeness sector, via the determination of the kaon nucleon sigma terms.

The measurement of kaonic helium, feasible in SIDDHARTA, allows study of the behaviour of the sub-threshold resonance $\Lambda(1405)$ in nuclei. Other light kaonic atoms can be studied in SIDDHARTA as well.

DAΦNE proves to be a real and ideal "kaonic atom" factory.

Part of this work was supported by "Transnational access to Research Infrastructure" TARI - INFN, Laboratori Nazionali di Frascati, HadronPhysics I3, Contract No. RII3-CT-2004-506078.

References

1. G. Vignola, *Proceedings of the 5th European Particle Accelerator Conference, Sitges (Barcelona)*, edited by S. Myres *et al.* (Institute of Physics Publishing, Bristol and Philadelphia, 1996) p. 22.
2. M. Cargnelli on behalf of the SIDDHARTA Collaboration, Acta Phys. Slovaca **55**, 65 (2005).
3. S. Bianco *et al.*, Riv. Nuovo Cimento **22**, No. 11, 1 (1999).
4. S. Deser *et al.*, Phys. Rev. **96**, 774 (1954); T.L. Truemann, Nucl. Phys. **26**, 57 (1961); A. Deloff, Phys. Rev. C **13**, 730 (1976).
5. C. Guaraldo, *Proceedings of the DAFNE'99 Workshop, November 16-19, 1999, Frascati, Italy*, edited by S. Bianco, F. Bossi, G. Capon, F.L. Fabbri, P. Gianotti, G. Isidori, F. Murtas, Frascati Physics Series, Vol. **XVI** (INFN, LNS, 1999) p. 643.
6. G. Beer *et al.*, Phys. Rev. Lett. **94**, 212302 (2005).
7. B. Borasoy, R. Nissler, W. Weise, Phys. Rev. Lett. **94**, 213401 (2005); **96**, 199201 (2006).
8. J.A. Oller, J. Prades, M. Verbeni, Phys. Rev. Lett. **95**, 172502 (2005); **96**, 199202 (2006).
9. J. Mares, E. Friedman, A. Gal, Nucl. Phys. A **770** (2006); Phys. Lett. B **606**, 295 (2005).
10. T. Ishiwatari *et al.*, Nucl. Instrum. Methods A **556**, 509 (2006).

Eur. Phys. J. A **31**, 540–542 (2007)

DOI 10.1140/epja/i2006-10241-3

Special Article – QNP 2006

Meson-baryon s-wave resonances with strangeness -3

C. García-Recio[a], J. Nieves, and L.L. Salcedo

Departamento de Física Atómica, Molecular y Nuclear, Universidad de Granada, E-18071 Granada, Spain

Received: 8 November 2006

Published online: 14 March 2007 – © Società Italiana di Fisica / Springer-Verlag 2007

Abstract. Starting from a consistent $SU(6)$ extension of the Weinberg-Tomozawa (WT) meson-baryon chiral Lagrangian (Phys. Rev. D **74**, 034025 (2006)), we study the s-wave meson-baryon resonances in the strangeness $S = -3$ and negative-parity sectors. Those resonances are generated by solving the Bethe-Salpeter equation with the WT interaction used as kernel. The considered mesons are those of the **35**-$SU(6)$-plet, which includes the pseudoscalar (PS) octet of pions and the vector (V) nonet of the rho-meson. For baryons we consider the **56**-$SU(6)$-plet, made of the $1/2^+$ octet of the nucleon and the $3/2^+$ decuplet of the Delta. Quantum numbers $I(J^P) = 0(3/2^-)$ are suggested for the experimental resonances $\Omega^*(2250)^-$ and $\Omega^*(2380)^-$. Among other, resonances with $I = 1$ are found, which minimal quark content is $sss\bar{l}l'$, being s the *strange* quark and l, l' any of the the light *up* or *down* quarks. A clear signal for such a pentaquark would be a baryonic resonance with strangeness -3 and electric charge -2 or 0, in proton charge units. We suggest looking for $K^-\Xi^-$ resonances with masses around 2100 and 2240 MeV in the sector $1(1/2^-)$, and for $\pi^\pm\Omega^-$ and $K^-\Xi^{*-}$ resonances with masses around 2260 MeV in the sector $1(3/2^-)$.

PACS. 11.30.Hv Flavor symmetries – 11.30.Ly Other internal and higher symmetries – 11.10.St Bound and unstable states; Bethe-Salpeter equations – 11.30.Rd Chiral symmetries

1 Introduction

Using a spin-flavor-$SU(6)$ extended Weinberg-Tomozawa (WT) meson-baryon interaction [1][1], we study the s-wave resonances with strangeness $S = -3$, isospin $I = 0, 1$ and spin-parity $J^P = 1/2^-, 3/2^-, 5/2^-$. The resonances are generated by solving the Bethe-Salpeter equation with the extended WT meson-baryon interaction used as a kernel. In this model, the involved mesons are those of the **35**-$SU(6)$-plet $= 8_1 \oplus 8_3 \oplus 1_3$, which includes the PS meson octet of the pions, $(\pi, \eta, K, \bar{K}) \in 8_1$ and the V nonet of the rho-meson, $(\rho, \omega, \phi, K^*, \bar{K}^*) \in 8_3 \oplus 1_3$. We approximate the η-meson as the isospin singlet state of the PS $SU(3)$-octet. For the ω- and ϕ-vector mesons, we assume ideal mixing among the V singlet and octet mesons. The baryons are those of the **56**-$SU(6)$-plet $= 8_2 \oplus 10_4$, which contains the $1/2^+$ octet of the N $(N, \Lambda, \Sigma, \Xi)$ and the $3/2^+$ decuplet of the Δ $(\Delta, \Sigma^*, \Xi^*, \Omega)$. Masses, widths and couplings of the resonances found are calculated and, when possible, comparison with experimental ones is attempted.

Unitary extensions of chiral perturbation theory to study meson-baryon interactions using a coupled channel scheme were introduced some time ago [3]. They have

been successfully applied in the theoretical microscopical description of meson-baryon scattering and of well-known lowest-lying baryon resonances, which were shown to be dynamically generated. Thus different $J^P = 1/2^-$ s-wave resonances ($\subset 8_\pi \times 8_N$, made of PS mesons of the pion octet and of the baryons of the nucleon octet) like $N^*(1535)$, $\Lambda(1405)$, $\Lambda(1670)$, $\Sigma(1620)$ and $\Xi(1620)$ [4] and, more recently, the $J^P = 3/2^-$ d-wave resonances ($\subset 8_\pi \times 10_\Delta$, made of PS mesons of the pion octet and of the baryons of the delta decuplet) like $\Lambda(1520)$, $\Sigma(1670)$ and $\Xi(1820)$ [5] have been found and their properties studied. Those previous chiral Bethe-Salpeter coupled channels unitary approaches using the WT kernel have included hadron multiplets belonging to the flavor $SU(3)$ irreducible representations. In the case of mesons, the only ingredient has been the octet of PS mesons.

2 Model

Motivations for extending the previous $SU(3)$-based models to a $SU(6)$ extended model are the following. First, in the large-N_c limit, 8_N and 10_Δ are degenerated and form a 56-multiplet of spin-flavor $SU(6)$. Second, vector mesons do exist, interact and couple to baryons. Third, there are baryonic resonances decaying to a PS meson and a baryon, but also to a V meson and a baryon, for instance, the strangeness -3 resonance $\Omega^*(2380)^-$ decays to $K^-\Xi^{*0}$

---

[a] e-mail: g_recio@ugr.es

[1] This WT interaction has also been extended to arbitrary number of colors and flavors in ref. [2].

and to $\bar{K}^{*0}\Xi^-$ with similar strengths and of the same order as the other known decay mode ($\bar{K}^-\pi^0\Xi^-$) [6]. All of these call for a "$SU(6)$" model which deals all together with the 56-baryons and 35-mesons like that of ref. [1]. We consider this spin-flavor symmetric scenario as a reasonable first approach. In the "$SU(6)$" approach the interacting kernel V^{IYJ} for a sector with hypercharge Y (strangeness plus one for baryons), isospin I and spin J is given by

$$\langle m_i, B_i | V^{IYJ}(s) | m_j, B_j \rangle = D_{ij}^{IYJ} \frac{2\sqrt{s} - M_{B_i} - M_{B_j}}{4 f^2},$$

$$m_i, m_j \in \mathbf{35}, B_i, B_j \in \mathbf{56}, \quad D^{IYJ} = \sum_\nu \bar{\lambda}_\nu \, \hat{P}_\nu^{IYJ},$$

$$\bar{\lambda}_{\mathbf{56}} = -12, \quad \bar{\lambda}_{\mathbf{70}} = -18, \quad \bar{\lambda}_{\mathbf{700}} = 6, \quad \bar{\lambda}_{\mathbf{1134}} = -2,$$

$\hat{P}_\nu^{IYJ}$ is the projector of the meson-baryon into the ν-$SU(6)$-representation, and ν runs over $\mathbf{56}, \mathbf{70}, \mathbf{700}, \mathbf{1134}$ because $\mathbf{35} \otimes \mathbf{56} = \mathbf{56} \oplus \mathbf{70} \oplus \mathbf{700} \oplus \mathbf{1134}$.

The $\bar{\lambda}_\nu$ positive (negative) means that in channel ν the meson-baryon interaction is repulsive (attractive). When this kernel is restricted to m_i, m_j being only PS mesons, it coincides with the "$SU(3)$" lowest order WT kernels previously used in refs. [4,5].

Note that D is $SU(6)$ invariant, this symmetry being explicitly broken by the different masses of baryons and mesons. In addition we replace the f^2 of the interaction by $f_{m_i} f_{m_j}$ for the ij matrix element. We use $f_\pi = 92.4\,\text{MeV}$, $f_K = 1.15 f_\pi$, $f_\eta = 1.2 f_\pi$ [7], and $f_{K^*} = f_K$, $f_\rho = f_\pi$, $f_\omega = f_\phi = f_\eta$. This interaction is used to solve the Bethe-Salpeter coupled channel equation for the meson-baryon T-matrix

$$T^{-1}(s) = V^{-1}(s) - J(s), \tag{1}$$

where $J(s)$ is the diagonal matrix of the meson-baryon loop functions [4,5]. For each channel (m_i, B_i) it is ultraviolet regularized by subtracting a constant so that $J(s = m_{m_i}^2 + M_{B_i}^2) = 0$.

We test that the inclusion of the new "$SU(6)$" channels (those involving V mesons not included in the "$SU(3)$" calculations) does not spoil previous results which were successful. See ref. [8] for comparison in the sector ($S = -1$, $I = 0$, $J^P = 1/2^-$).

3 Results

We solve the coupled-channel Bethe-Salpeter equation and look for the T-matrix poles in the second Riemann sheet. Close to a pole the T-matrix behaves as

$$T_{ij} \sim \frac{g_i g_j}{\sqrt{s} - (M_R - i\Gamma_R/2)} \tag{2}$$

and the position of the pole and its residue define the mass M_R, width Γ_R and complex coupling constants g_i to different i channels of the found resonance.

All the calculations are done neglecting the widths of the baryons of the decuplet and of the V mesons. When the orbital angular momentum of the meson-baryon system is zero, the odd-parity $S = -3$ resonances formed by

Table 1. Masses, widths and absolute values of coupling constants for each channel, in the sector $I = 0$, $J^P = 1/2^-$ and $S = -3$. The underlining indicates open channels. "$SU(6)$" stands for the full $\mathbf{35} \times \mathbf{56}$ result. "$SU(3)$" for the $\mathbf{8} \times \mathbf{8}$ result.

$I = 0, J^P = 1/2^-$, "$SU(6)$"						
M_R	Γ_R			$|g_i|$		
[MeV]		$\bar{K}\Xi$	$\bar{K}^*\Xi$	$\bar{K}^*\Xi^*$	$\omega\Omega$	$\phi\Omega$
1309	0	1.22	0.61	3.27	0.17	5.78
1871	70	<u>1.57</u>	3.63	4.17	1.39	3.23
2201	2.8	<u>0.22</u>	1.63	2.36	0.48	1.38
2334	44	<u>0.76</u>	<u>0.37</u>	0.66	3.44	1.08
2454	33	<u>0.38</u>	<u>0.22</u>	<u>0.97</u>	0.18	4.32
–	–	–				"$SU(3)$"

Table 2. Same as table 1 for the sector $I = 0$, $J^P = 3/2^-$.

$I = 0, J^P = 3/2^-$, "$SU(6)$"							
M_R	Γ_R				$|g_i|$		
[MeV]		$\bar{K}\Xi^*$	$\bar{K}^*\Xi$	$\eta\Omega$	$\bar{K}^*\Xi^*$	$\omega\Omega$	$\phi\Omega$
1969	0	1.78	2.43	2.37	1.44	0.00	2.51
2265	82	<u>1.19</u>	<u>0.34</u>	<u>0.28</u>	3.71	1.21	0.55
2343	36	<u>0.42</u>	<u>0.84</u>	<u>0.04</u>	1.01	3.31	0.16
2437	80	<u>0.07</u>	<u>0.09</u>	1.19	<u>0.12</u>	0.00	4.38
2051	86	<u>1.97</u>		3.34			"$SU(3)$"

Table 3. Same as table 1 for the sector $I = 0$, $J^P = 5/2^-$.

$I = 0, J^P = 5/2^-$, "$SU(6)$"				
M_R	Γ_R		$|g_i|$	
[MeV]		$\bar{K}^*\Xi^*$	$\omega\Omega$	$\phi\Omega$
2376	0	1.41	2.78	0.00

Table 4. Same as table 1 for the sector $I = 1$, $J^P = 1/2^-$.

$I = 1, J^P = 1/2^-$, "$SU(6)$"					
M_R	Γ_R			$|g_i|$	
[MeV]		$\bar{K}\Xi$	$\bar{K}^*\Xi$	$\bar{K}^*\Xi^*$	$\rho\Omega$
2100	118	<u>1.47</u>	3.39	1.1	2.1
2241	63	<u>0.88</u>	<u>0.87</u>	1.67	3.87
–	–	–			"$SU(3)$"

coupling $\mathbf{35}$-mesons to $\mathbf{56}$-baryons can have the following $I(J^P)$ quantum numbers: $0(1/2^-)$, $0(3/2^-)$, $0(5/2^-)$, $1(1/2^-)$, $1(3/2^-)$ and $1(5/2^-)$. The mass, width and absolute values of the coupling constants of the resonances found for each of those sectors are shown in tables 1, 2, 3, 4, 5 and 6, respectively. Resonances with width below $125\,\text{MeV}$ are displayed in fig. 1.

Several ($S = -3$, $I = 1$) resonances have been predicted (tables 4, 5 and 6 and full symbols in fig. 1). These resonances have, at least, three s quarks to provide strangeness $S = -3$ and a pair of light (u or d)

Table 5. Same as table 1 for the sector $I = 1$, $J^P = 3/2^-$.

M_R [MeV]	Γ_R	$\pi\Omega$	$\bar{K}\Xi^*$	$\bar{K}^*\Xi$	$\bar{K}^*\Xi^*$	$\rho\Omega$
				$\|g_i\|$		
2018	267	2.19	1.64	2.14	1.75	0.25
2258	101	0.54	1.23	0.13	3.19	2.33
2288	32	0.33	0.17	0.93	2.29	3.11
2146	359	2.30	2.47			"$SU(3)$"

Header for the table: $I = 1, J^P = 3/2^-$, "$SU(6)$"

Table 6. Same as table 1 for the sector $I = 1$, $J^P = 5/2^-$.

$I = 1$, $J^P = 5/2^-$, "$SU(6)$"

M_R [MeV]	Γ_R	$\bar{K}^*\Xi^*$	$\rho\Omega$
		$\|g_i\|$	
2324	0	1.79	3.09

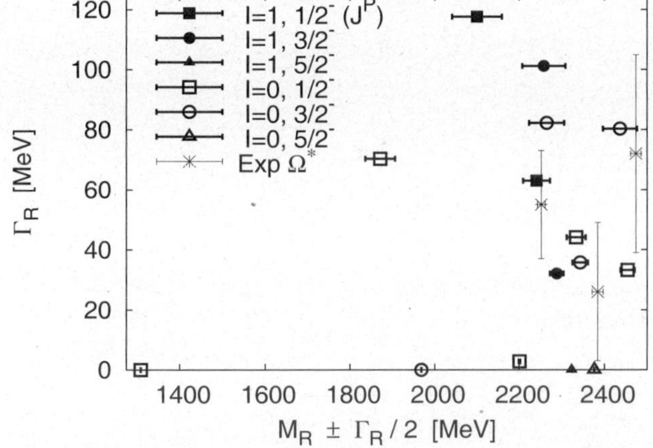

Fig. 1. Spin-parity $J^P = \frac{1}{2}^-$, $\frac{3}{2}^-$ and $\frac{5}{2}^-$ resonance properties in the $I = 0$ (empty symbols) and $I = 1$ (filled symbols), $S = -3$ ($Y = -2$) sectors. The points with error-bars are defined from the masses (M_R) and widths (Γ_R) of the found resonances as ($M_R \pm \Gamma_R/2, \Gamma_R$). The experimental Ω^* resonance masses and widths with their error bars are from refs. [7,6,9,10].

quark-antiquark to achieve isospin $I = 1$ quantum number. Hence, those dynamically generated ($S = -3$, $I = 1$) resonances are pentaquarks and in principle have a clear experimental signature. A clear signal of them would be a resonance with strangeness $S = -3$ and electric charge $Q = -2$ (this is $I_z = -1$) with minimal quark content $sssd\bar{u}$. Another clear signature would be a $S = -3$ and $Q = 0$ resonance with a minimal quark content of $sssu\bar{d}$. From the tables, the best ways for observing pentaquarks would be, in the sector $I(J^P) = 1(1/2^-)$ looking for $K^-\Xi^-$ resonances with masses around 2100 and 2240 MeV, and in the sector $I(J^P) = 1(3/2^-)$ looking for $\pi^\pm\Omega^-$ and $K^-\Xi^{*-}$ resonances with masses around 2260 MeV.

All the experimentally known Ω^* resonances are listed in table 7. Tentatively, we identify the experimental isoscalar $\Omega^*(2250)^-$ to the theoretical $\Omega^*(2265)^-$ of sector $0(3/2^-)$, because of the relative closeness of

Table 7. Experimentally known Ω^* resonances [7,6,9,10]. The ** and *** refer to the status rating used by the PDG. The branching ratios are relative.

Resonance $I(J)^P$	Mass Width [MeV]		Decay modes	Branching ratios
$\Omega^*(2250)^-$ $0(?^?)$ ***	2252 ± 9	55 ± 18	$\Xi^-\pi^+K^-$	1
			$\Xi^{*0}K^-$	0.7 ± 0.2
$\Omega^*(2380)^-$ $?(?^?)$ **	2384 ± 13	26 ± 23	$\Xi^-\pi^+K^-$	1
			$\Xi^{*0}K^-$	< 0.4
			$\Xi^-\bar{K}^{*0}$	0.5 ± 0.3
$\Omega^*(2470)^-$ $?(?^?)$ **	2474 ± 12	72 ± 33	$\Omega^-\pi^+\pi^-$	

masses and widths and, also, of observed decay channels. Likewise, the experimental $\Omega^*(2380)^-$ could be assigned to the found $\Omega^*(2343)^-$ of sector $0(3/2^-)$. We do not find a clear assignment for the experimentally observed $\Omega^*(2470)$; the decay to $\Omega^-\pi^+\pi^-$ through $\Omega\rho$ in our model takes place only for smaller resonance masses. Presumably, in this decay other mechanisms, for instance involving p-wave, could be at work.

More details will be given elsewhere.

We acknowledge discussions with Drs Angels Ramos and Manuel José Vicente-Vacas. This work was supported by DGI, FEDER, UE and Junta de Andalucía funds (FIS2005-00810, HPRN-CT-2002-00311, FQM225).

References

1. C. Garcia-Recio, J. Nieves, L.L. Salcedo, Phys. Rev. D **74**, 034025 (2006).
2. C. Garcia-Recio, J. Nieves, L.L. Salcedo, Phys. Rev. D **74**, 036004 (2006); hep-ph/0610204, these proceedings (DOI: 10.1140/epja/i2006-10203-9).
3. N. Kaiser, P.B. Siegel, W. Weise, Nucl. Phys. A **594**, 325 (1995); N. Kaiser, T. Waas, W. Weise, Nucl. Phys. A **612**, 297 (1997); E. Oset, A. Ramos, Nucl. Phys. A **635**, 99 (1998); J.A. Oller, U.G. Meissner, Phys. Lett. B **500**, 263 (2001).
4. J. Nieves, E. Ruiz Arriola, Phys. Rev. D **64**, 116008 (2001); T. Inoue, E. Oset, M.J. Vicente-Vacas, Phys. Rev. C **65**, 035204 (2002); D. Jido et al., Nucl. Phys. A **725**, 181 (2003); C. Garcia-Recio et al., Phys. Rev. D **67**, 076009 (2003); C. Garcia-Recio, M.F.M. Lutz, J. Nieves, Phys. Lett. B **582**, 49 (2004); J.A. Oller, arXiv:hep-ph/0603134.
5. E.E. Kolomeitsev, M.F.M. Lutz, Phys. Lett. B **585**, 243 (2004); S. Sarkar, E. Oset, M.J. Vicente Vacas, Nucl. Phys. A **750**, 294 (2005); Eur. Phys. J. A **24**, 287 (2005); L. Roca, S. Sarkar, V.K. Magas, E. Oset, Phys. Rev. C **73**, 045208 (2006).
6. S.F. Biagi et al., Z. Phys. C **31**, 33 (1986).
7. Particle Data Group (W.-M. Yao et al.), J. Phys. G **33**, 1 (2006).
8. C. Garcia-Recio, J. Nieves, L.L. Salcedo, arXiv:hep-ph/0610127, these proceedings.
9. D. Aston et al., Phys. Lett. B **194**, 579 (1987).
10. D. Aston et al., Phys. Lett. B **215**, 799 (1988).

Eur. Phys. J. A **31**, 543–548 (2007)
DOI 10.1140/epja/i2006-10193-6

THE EUROPEAN
PHYSICAL JOURNAL A

Special Article – QNP 2006

Towards an understanding of the light scalar mesons

C. Hanhart[a]

Institut für Kernphysik, Forschungszentrum Jülich, D-52425 Jülich, Germany

Received: 8 October 2006
Published online: 23 February 2007 – © Società Italiana di Fisica / Springer-Verlag 2007

Abstract. Although studied for many years the nature of the light scalar mesons remains controversial. Here we shall present a method, applicable for s-wave states located close to a threshold, that allows one to quantify the molecular part of a given state. When applied to the $f_0(980)$ a dominance of the molecular component is found. In the second part, we show that requirements of field-theoretic consistency and chiral symmetry, when applied to the scattering of light pseudo-scalars, naturally lead to the appearance of dynamical poles in the scalar sector. A program is proposed on how to further investigate experimentally the mixing between these dynamical states and possible genuine quark states.

PACS. 13.60.Le Meson production – 13.75.-n Hadron-induced low- and intermediate-energy reactions and scattering (energy ≤ 10 GeV) – 14.40.Cs Other mesons with $S = C = 0$, mass < 2.5 GeV

1 Introduction

At first glance, it comes as a surprise that the lowest scalar excitations of QCD are still not fully understood. First, there has been a long debate on how many scalar states there are below 1 GeV. Besides the well-established isovector $a_0(980)$ and the isoscalar $f_0(980)$ there was experimental as well as theoretical evidence for two more states, namely the $f_0(600)$ —often called the σ-meson— and the isodoublet κ. Together these states could fill the lowest scalar nonet.

Although recent efforts in dispersion theory in combination with chiral perturbation theory unambiguously determined the existence as well as the position of the σ-pole [1], the discussion of the very nature of this state and its relatives is far from settled. Analyses can be found in the literature that identify these structures with conventional $q\bar{q}$ states (see, *e.g.*, refs. [2]) —in some analyses with a sizable admixture from the continuum [3,4]— compact $qq\,\bar{q}\bar{q}$ states [5,6] or loosely bound $\bar{K}K$ molecules [7,8].

What is clearly called for is a study that identifies those cases when the nature of a particular state can be read off from an experimental observable. Based on an old proposal by Weinberg [9], a first step in this direction was taken in ref. [10] —specifically, the original argument was extended to also allow for the presence of inelasticities. The conditions where this method can be applied were found to be

- the state must be an s-wave with respect to the continuum states[1];

- the binding energy ϵ must be *much* smaller than any intrinsic scale of the problem;
- any inelastic threshold must be "far away" (in units of the binding energy) from the elastic threshold of interest.

Once the above conditions are met, it was shown in ref. [10] that also for inelastic interactions the value of the effective coupling of a resonance to the continuum state of interest[2] is a direct measure of the molecular component; especially, its value gets maximum (up to higher-order corrections) in the case of a pure molecule. The derivation of this result is sketched below. More details can be found in ref. [10]. We would also like to stress that a model-independent analysis is possible only if the above conditions are met. Thus, as a matter of principle, our analysis cannot be applied to, *e.g.*, the σ-meson. For a recent attempt to pin down the nature of this resonance we refer to ref. [11].

In this brief note we will focus more on a discussion of why this works and how to further exploit this insight. Thus, in the next section, the role of the effective coupling will be discussed and the scheme will be applied to the $f_0(980)$. In sect. 3 we argue that a prominent molecular structure of the light scalar mesons emerges quite naturally from the properties of the meson-meson scattering amplitude near threshold controlled by chiral symmetry. In sect. 4 we shall briefly comment on possible further experiments to investigate the mixing of the light scalar mesons with the (heavier) quark states.

---

[a] e-mail: `c.hanhart@fz-juelich.de`

[1] Not to be obscured with the quantum numbers with respect to the quark constituents.

[2] To be more specific: what is meant is the corresponding residuum at the resonance pole.

2 How does this work?

For simplicity, let us focus on a situation where two spin-zero mesons of mass m couple to a single, isolated resonance state. The possible presence of inelasticity will be discussed later. The two-point function $g(s)$ for the resonance state may then be written as

$$g(s) = \frac{1}{s - M_0^2 - i\Sigma(s)} + \text{regular terms}$$
$$= \frac{Z}{s - M^2} + \text{regular terms}, \qquad (1)$$

where M_0 (M^2), Σ, Z, denote the bare (physical) mass, the self-energy, and the wave function renormalization of the state of interest. It is straightforward to show that (for a non-relativistic system) the Z-factor measures the bare-state admixture of the physical state [9,10]. Thus, $Z = 0$ ($Z = 1$) corresponds to a pure molecule (compact state).

By assumption the binding energy is much smaller than any intrinsic scale of the problem —assumed especially to be smaller than the inverse of the range of forces. In this situation the vertex that couples the state of interest to the continuum can be safely assumed to be point-like and the self-energy is just the standard scalar loop function times a strength parameter that we will call G. One then easily derives

$$\frac{1}{Z} - 1 = \frac{G}{4}\sqrt{\frac{m}{\epsilon}} + \mathcal{O}(\epsilon R), \qquad (2)$$

where R denotes the range of forces. Weinberg applied this to the deuteron, where $R \sim 1/m_\pi$. For the scalar mesons, on the contrary, we look at the scattering of two pseudo-scalar mesons. Thus the lightest particle that can be exchanged in the t-channel is the ρ-meson from which we get $R \sim 0.25$ fm, which is of the order of the extension of conventional mesons. Under the conditions assumed the $1/\sqrt{\epsilon}$ term should dominate the right-hand side of eq. (2) thus allowing one to express the effective coupling G in terms of Z, which "measures" the nature of the state. On the other hand, the effective coupling ZG is —in principle— a measurable quantity: it is the residue of the scattering matrix at the resonance pole and it can be related to the scattering length and the effective range for the meson-meson scattering. This follows directly from matching the expression for the meson-meson scattering matrix —$Gg(s)$— to the corresponding effective range expansion:

$$Gg(s) = \frac{G}{s - M^2 + iMG\sqrt{s - 4m^2}/2}$$
$$= \frac{1}{2m}\left(\frac{G/2}{\epsilon + k^2/m + G/2(ik + \sqrt{m\epsilon})}\right) + \mathcal{O}\left(\frac{k^2}{m^2}\right)$$
$$= -\frac{1}{2m}\left(\frac{1}{1/a + r/2\,k^2 - ik}\right) + \mathcal{O}\left(\frac{k^2}{m^2}\right), \qquad (3)$$

which leads (up to higher orders) to

$$-\frac{1}{a} = \frac{2\epsilon}{G} + \sqrt{m\epsilon}, \qquad -\frac{1}{2}r = \frac{2}{Gm}. \qquad (4)$$

Using eq. (2) one may equivalently express a and r directly in terms of Z.

Note that the coupling G controls the relative importance of the term non-analytic in s to that analytic in s. Since the only source of a non-analyticity are the unitarity cuts, this is exactly what controls the amount of molecular admixture. For a recent discussion of s-wave thresholds on amplitudes we refer to ref. [12] and references therein. We shall come back to this below.

For later use it is convenient to introduce g_{eff}, defined as

$$\frac{g_{eff}^2}{4\pi} = Z8m^2 G$$
$$= 32(1 - Z)m\sqrt{\epsilon m} \leq 32m\sqrt{\epsilon m}. \qquad (5)$$

It is this effective coupling that controls the resonance coupling in inelastic reactions like $\phi \to \gamma\pi\pi$. This reaction will be discussed in more detail below. It is important to observe that the formalism sketched here gives a result that is intuitively clear: the larger the effective coupling of the physical resonance to the continuum the larger the probability to find the continuum state in the physical state or, stated differently, the larger its molecular component. Equation (5), however, makes an even stronger statement: the maximum coupling is constrained from above and the value of this maximal coupling is controlled solely by the binding energy. Below we will call the model with $g_{eff}^2/4\pi = 32m\sqrt{\epsilon m}$ the *naive molecular model*.

What changes now if we introduce an inelastic channel? By assumption this new channel is not allowed to introduce any new small scale into the problem. Thus we may assume that the inelastic threshold is, when measured in units of the binding energy, very far away. Then its leading effect is to introduce a constant (or at most weakly energy dependent) imaginary part $i\Gamma$ to the denominator of $g(s)$, or —equivalently— to the self energy. We may therefore write

$$g(s) = \frac{1}{s - M^2 + iMG\sqrt{s - 4m^2}/2 + i\Gamma}. \qquad (6)$$

Thus, the appearance of Γ does not change the relative importance of the $s - M^2$ piece and the $iMg\sqrt{s - 4m^2}/2$ piece, and consequently G still measures the amount of molecular admixture with respect to the elastic channel [10]. A discussion on the subleading corrections is provided in ref. [13].

Equation (6) is nothing but the standard Flatté form used for the parameterization of resonance signals near thresholds [14]. As stated above, for a state with a negligible continuum admixture, the term linear in G of eq. (6) can be neglected and the resulting distribution for the resonance is that of a standard Breit-Wigner. This is sketched in fig. 1a. On the other hand, if the state is predominantly a molecule, it is the G-term that controls the dynamics near the meson-meson threshold and the variation of s in the first term may be neglected. The resonance signal then produces a very pronounced cusp structure, for

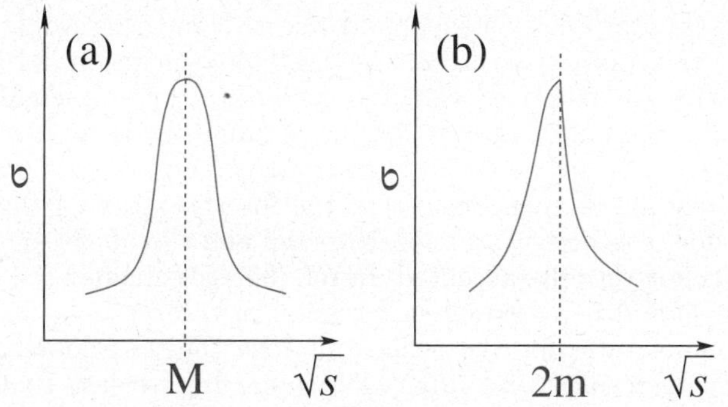

Fig. 1. Typical resonance signals for a genuine state (a) and a dominant molecular state (b).

which a typical case is shown in fig. 1b. (Many different shapes are possible; see the discussion in ref. [14].) It therefore appears a straightforward task to use measured Flatté distributions to extract G and deduce from this the molecular admixture. Unfortunately, due to a scale invariance of the amplitude, an extraction of the absolute value of G directly from the mass distributions is very difficult. (See ref. [15] for a detailed discussion.) As we shall see, inelastic processes might be more useful here.

Thus, qualitatively one comes to the conclusion that the more distorted a Breit-Wigner distribution gets through the opening of a production threshold the more molecular component there is in the corresponding resonance. In ref. [10] we tried to put this on more quantitative grounds. Thus, the strongly asymmetric mass distribution seen for the $f_0(980)$ by the BES Collaboration [16] in the reaction $J/\Psi \to \phi\pi\pi$ already provides strong evidence for a predominantly molecular composition of the f_0.

3 The physics of light scalar mesons

It is surprising that the lowest scalar excitations in QCD seem to be of quite complicated structure —even more complicated than the higher excited states, where we assume to find the $\bar{q}q$ states. In this section we will argue that the appearance of low-lying dynamical states is a natural consequence of chiral symmetry. Note, here "natural" should not be confused with necessary. In this context see discussion in ref. [17].

The observation essential for our argument is the energy dependence of the $\pi\pi$ scattering amplitude near its threshold. It reads in the scalar-isoscalar channel [18]

$$V_{\pi\pi} = (s - m_\pi^2/2)/f_\pi^2. \tag{7}$$

Corrections to this expression can be calculated using chiral perturbation theory in a controlled way [19].

One may ask if such an energy dependence near threshold can emerge solely from an s-channel resonance. The answer is "no" for two reasons and may be read off the corresponding scattering matrix directly. The scattering potential for scattering through a resonance has the form

$$V_R = G_R^2/(s - m^2), \tag{8}$$

thus, matching to $V_{\pi\pi}$ would give first of all a relation between the masses of scalar resonances and f_π that looks very counterintuitive; but, even more importantly, for V_R to employ the energy dependence of $V_{\pi\pi}$ calls for

$$G_R \propto \sqrt{s - m_\pi^2/2}. \tag{9}$$

In order to give the correct energy dependence near threshold the amplitude must contain an unphysical branch point. Clearly, in elastic $\pi\pi$ scattering this will not show up, because only G_R^2 appears. Production amplitudes, however, are linear in G_R and thus one would start to notice the unphysical cut in calculations based on eq. (8). Although not allowed by analyticity, several works use the vertex function of eq. (9) (see, $e.g.$, ref. [20]) or similar ones (see, $e.g.$, ref. [21]).

To cure this problem one option would be to use the linear sigma model. There a four-pion contact term is present in addition to the σ pole diagrams and a threshold amplitude of the form of eq. (7) emerges naturally. Then, one finds that the contact term as well as the t- and u-channel exchanges play a prominent role in the dynamics below $1\,\text{GeV}$ [22]. An alternative and theoretically more appealing approach (see discussion in ref. [23]) is chiral perturbation theory. There only pions (and in the $SU(3)$ extension also kaons and eta mesons) appear as dynamical fields that interact through contact interactions constructed consistent with chiral symmetry. The corresponding leading-order Lagrangian automatically gives a potential of the form of eq. (7) —see ref. [19] and references therein. In this scheme there is a strongly energy-dependent, non-resonant π-π interaction to be included in the theory, whose s-dependence is so strong that the tree level potential of eq. (7) hits the unitarity bound already at quite low energies. Or, stated differently, a properly unitarized amplitude will naturally employ a pole in the complex plane to prevent the amplitude from growing beyond what is allowed by unitarity. This pole should be identified with the $f_0(600)$ —the σ-meson. This picture was introduced in refs. [23,6] and was further supported by studies within unitarized chiral perturbation theory [24]. For a recent review see ref. [25].

Here one comment is necessary. Although it seems obvious that a pole, produced just by unitarizing a non-pole meson-meson interaction, is generated dynamically, this is not necessarily the case. However, large-N_c arguments, with N_c the number of colours, can be used to show that the intuitive picture is indeed correct for the scalar mesons [26].

So far the argument was given for $\pi\pi$ scattering only, however studies within both unitarized chiral perturbation theory discussed above as well as phenomenological models [27] show that the same holds for the scattering of all the pseudo-Goldstone bosons with each other. As a consequence, in addition to the lowest pole in the $\pi\pi$ channel, coupling to the $\bar{K}K$ system leads to a pole close to the $\bar{K}K$ threshold, identified with the $f_0(980)$. Also in the $\pi\eta$-$\bar{K}K$ coupled system a pole appears close to the $\bar{K}K$ threshold, interpreted as $a_0(980)$ and in the πK channel the κ pole appears.

To summarize this part, we find as a natural consequence of the analytic properties of the scattering amplitude for the scattering of the ground-state pseudo-scalar mesons with each other that dynamical poles with scalar quantum numbers are produced. This insight is fully in line with the experimental evidence that the $f_0(980)$ is predominantly of dynamical origin. What remains to be seen is the mixing pattern of those molecular states with the non-molecular states. In the rest of this presentation we shall argue that exploiting the matrix element for scalars coupling to a photon and a vector meson in various kinematic regimes is what should provide important information in this direction.

4 $\phi \to \gamma\pi\pi$ and possible further experiments

In recent years a lot of experimental as well as theoretical effort went into studies of the reactions $\phi \to \gamma\pi\pi$ and $\phi \to \gamma\pi\eta$, both believed to shed light on the nature of the light scalar mesons f_0 and a_0, respectively. Following the reasoning of ref. [28] and that of the previous sections, in this section we shall argue that in these reactions the molecular component of the scalar mesons was measured. In addition, we shall show that looking at the same matrix elements for different masses of the vector meson will allow one to study the mixing of this molecular component with possible compact states.

There is strong experimental evidence that the reaction $\phi \to \gamma\pi\pi$ in the upper part of the available phase space runs predominantly through a kaon loop followed by a resonance formation into the $f_0(980)$. The kaon loop dominance was observed by Achasov and Kiselev (see ref. [29] and references therein). The reason why one can in the reaction $\phi \to \gamma \to \pi$ almost "see" the kaon loops in the spectrum is the following: gauge invariance demands the transition amplitude for $\phi \to \gamma\pi\pi$ to vanish for vanishing values of ω —the energy of the outgoing photon. As a consequence, the ϕ decay rate has to scale as ω^3 at the upper end of phase space. On the other hand, the experimental spectrum of refs. [30,31] is (almost) identical to that expected from an undistorted f_0 spectral function. What looks contradictory at first glance is quite natural, since the $\bar{K}K$ threshold is very close to the mass of both the ϕ and the f_0. Consequently, there is a pronounced cusp structure in the kaon loop that effectively compensates the mentioned ω^3 suppression at the upper end of phase space.

In the previous section we argued that the effective coupling of the scalar mesons to kaons is a measure for the importance of the molecular component of the $f_0(980)$. We now see that the kaon loop dominates the transition rate $\phi \to \gamma\pi\pi$. We therefore have to conclude that the radiative decay of the ϕ into a pion pair measures the molecular component of the scalar meson. Even more, we may use the naive molecular model introduced above to estimate the rate for $\phi \to \gamma\pi\pi$ based on eq. (5) with $Z = 0$ and a typical value of $\epsilon = 10\,\mathrm{MeV}$. Then, we get [28] $\Gamma \sim 0.6\,\mathrm{keV}$ to be compared to the experimental value of $0.4\,\mathrm{keV}$ [31].

This we interpret as strong evidence in favor of a prominent molecular structure of the f_0. This conclusion is in line with the results of calculations within the unitarized chiral perturbation theory for the ϕ radiative decay [32–34].

It should be mentioned that the interpretation of the data for $\phi \to \gamma\pi\pi$ is still controversial and the above picture is not yet fully accepted. In ref. [35] it is claimed that there should be a strong suppression of the $\phi \to \gamma f_0/a_0$ branching ratio for the scalars in case they are loosely bound molecules as compared to point-like scalars that correspond to compact quark states, (10^{-5} *vs.* 10^{-4}). A study by Achasov *et al.* [36], where the finite width of scalars was taken into account, arrived at the same conclusion. Thus, the authors of [35] and [36] stress that data for this branching ratio should allow to prove or rule out the molecular model of the scalars. However, this conclusion is based on a confusion between the notion of a wave function and that of a vertex function. Indeed, the relevant scale in the spatial extension of a wave function of a molecule is indeed the binding energy: the smaller the binding the larger the extension. What enters, however, in the loops depicted in fig. 2 is not the wave function but the vertex function, whose scale is set by the range of forces —for a molecule made of two pseudo-scalars this is of the order of the mass inverse of the ρ-meson, much smaller than what derives from the binding energy. The wave function, on the other hand, is proportional to the vertex function times the two-meson propagator (see eq. (22) in ref. [37]). This very propagator, however, is already included explicitly in the integral for the kaon loop[3].

Recently, an attempt was made to set up an analysis for the data on $\phi \to \gamma\pi\pi$ that avoids the use of the kaon loops [40]. The authors parameterize the $\phi \to \gamma\pi\pi$ amplitude as a polynomial in s and succeeded in fitting the ϕ-decay spectra once a sufficient number of terms was included in the polynomial. One might interpret this as an indication that the kaon loops are not of as high relevance as indicated in the previous paragraphs. However, a polynomial in s will not have the non-analytic pieces that are provided by the kaon loops. A truly model-independent analysis would therefore include both the mentioned polynomial as well as the kaon loop. A fit then needs to decide how much loop is really necessary. Given the comments above, it is likely that such a fit will call for a quite small contribution from the terms analytic in s.

Thus far, we have argued that the ϕ radiative decay measures the molecular component of the scalar mesons. The corresponding matrix element may be written as

$$i\mathcal{M}(p'^{\,2}{=}m_s^2,\, p^2{=}m_\phi^2)_\mu = \langle s(p')|A_\mu|\phi(p)\rangle,$$

where A^μ denotes the photon field operator and p' (p) are the four-momenta of the scalar (vector) meson. What remains unanswered so far is the amount of mixing of molecular states with genuine quark states. As argued in ref. [28], further insight into this issue can be gained from

---

[3] For more details on this discussion see the comment on ref. [28] in ref. [38] and our reply to this in ref. [39].

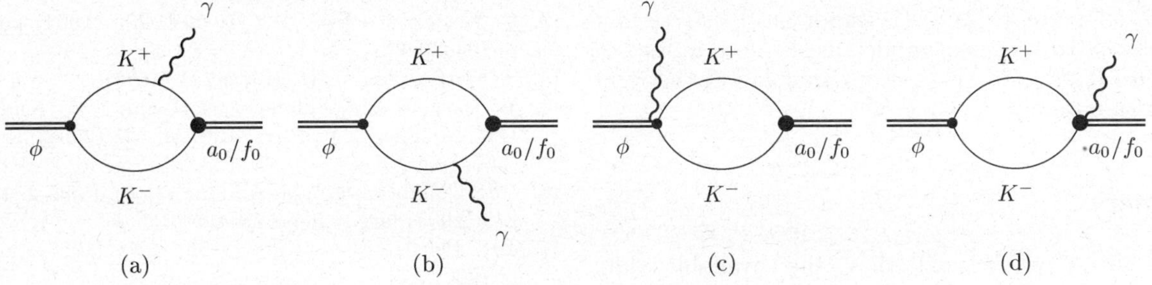

Fig. 2. Kaon loop contributions to the radiative decay amplitude.

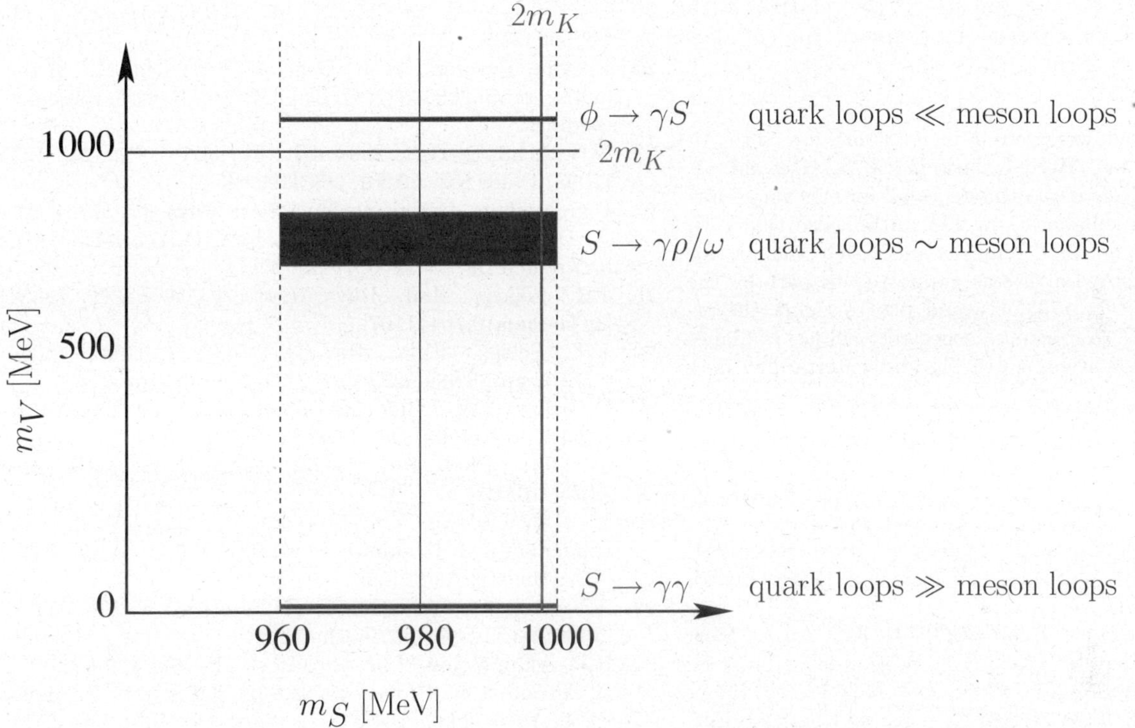

Fig. 3. Illustration of various kinematical regimes probed by radiative decays involving scalar and vector mesons. Here m_S (m_V) denotes the mass of the scalar (vector) meson involved in the decay. The extention of the boxes refers to the mass ranges probes by the corresponding reactions. E.g., the hight of the box that centers around 770 MeV in m_V resembles the width of the ρ-meson.

a systematic study of radiative decays of scalar mesons into vectors, corresponding to the matrix element

$$i\mathcal{M}(m_s^2, m_V^2)_\mu^\dagger = \langle V|A_\mu|s\rangle.$$

The physics behind this is quite easy to understand: the reason why the ϕ radiative decays have large kaon loop contributions is the proximity of the $\bar{K}K$ threshold to both the mass of the ϕ as well as of the scalars of interest here. Looking at the radiative decays of the scalars allows us to study the same matrix element in different kinematics: we can now change the mass of the vector meson away from that of the ϕ to that of the ω and ρ, or even to $m = 0$ for the decay $s \to \gamma\gamma$. As explained above, the unitarity cut related to the $\bar{K}K$ threshold introduces a significant energy dependence to the kaon loop. As a consquence, the absolute value of the corresponding transition matrix el-

ement gets smaller the further we move away from the threshold. On the other hand, quark loops do not feel the kaon threshold and therefore a much weaker energy dependence is expected for these. This means that, as we move from the $\phi s\gamma$ vertex to the $(\rho\omega)s\gamma$, and eventually the $\gamma s\gamma$ vertex, it should be straightforward to disentangle the kaon loop part of the matrix element from the quark lines. This logic is sketched in fig. 3, where the size of the boxes corresponds to the widths of the particles involved in the decay[4].

In oder to be more quantitative, particular models need to be employed for the quark loop contributions. For

---

[4] Whenever quark loops and meson loops are considered simultaneously there is the possibility of double counting. However, here this is not the case since the relevant meson loops are finite.

details we refer to ref. [28]. It is important to stress that already the naive molecular model, as sketched in sect. 2, gives values for the decays $s \to \gamma\gamma$ that are of the right order of magnitude.

5 Summary

To summarize, we argued that the available data —especially that on the ϕ radiative decays— are compatible with a model that assumes the f_0 to be predominantly of molecular nature. In order to determine the mixing scheme of this molecule and its $SU(3)$ relatives with possible quark states a series of experiments that studies the radiative decays of the scalars was proposed.

The results presented were found in collaboration with M. Evers, J. Haidenbauer, Yu.S. Kalashnikova, S. Krewald, A. Kudryavtsev, A.V. Nefediev —thanks to all for this very fruitful and enlightening collaboration. The author also thanks J. Durso for numerous editorial remarks and U.-G. Meißner for critical reading. This research was supported in part by the grant DFG-436 RUS 113/733. I would like to thank the organizers of the *QNP conference*, especially Felipe J. Llanes Estrada, for an informative, educating and entertaining conference.

References

1. I. Caprini, G. Colangelo, H. Leutwyler, Phys. Rev. Lett. **96**, 132001 (2006) [arXiv:hep-ph/0512364]; G. Colangelo, these proceedings.
2. D. Morgan, Phys. Lett. B **51**, 71 (1974); K.L. Au, D. Morgan, M.R. Pennington, Phys. Rev. D **35**, 1633 (1987); D. Morgan, M.R. Pennington, Phys. Lett. B **258**, 444 (1991); Phys. Rev. D **48**, 1185 (1993); A.V. Anisovich *et al.*, Eur. Phys. J. A **12**, 103 (2001); S. Narison, hep-ph/0012235.
3. E. Van Beveren, T.A. Rijken, K. Metzger, C. Dullemond, G. Rupp, J.E. Ribeiro, Z. Phys. C **30**, 615 (1986).
4. N.A. Tornqvist, M. Roos, Phys. Rev. Lett. **76**, 1575 (1996) [arXiv:hep-ph/9511210].
5. R. Jaffe, Phys. Rev. D **15**, 267 (1977); N.N. Achasov, A.V. Kiselev, Phys. Rev. D **68**, 014006 (2003) [arXiv:hep-ph/0212153].
6. D. Black, A.H. Fariborz, F. Sannino, J. Schechter, Phys. Rev. D **59**, 074026 (1999) [arXiv:hep-ph/9808415].
7. J. Weinstein, N. Isgur, Phys. Rev. Lett. **48**, 659 (1982); Phys. Rev. D **27**, 588 (1983); **41**, 2236 (1990); G. Janssen *et al.*, Phys. Rev. D **52**, 2690 (1995); J.A. Oller, E. Oset, Nucl. Phys. A **620**, 438 (1997); [**652**, 407 (1999)(E)].
8. J.A. Oller, E. Oset, J.R. Pelaez, Phys. Rev. Lett. **80**, 3452 (1998) [arXiv:hep-ph/9803242].
9. S. Weinberg, Phys. Rev. **130**, 776 (1963); **131**, 440 (1963); **137** B672 (1965).
10. V. Baru *et al.*, Phys. Lett. B **586**, 53 (2004).
11. M.R. Pennington, Phys. Rev. Lett. **97**, 011601 (2006).
12. J.L. Rosner, arXiv:hep-ph/0608102.
13. B. Kerbikov, Phys. Lett. B **596**, 200 (2004) [arXiv:hep-ph/0402022].
14. S. Flatté, Phys. Lett. B **63**, 224 (1976).
15. V. Baru, J. Haidenbauer, C. Hanhart, A. Kudryavtsev, U.G. Meißner, Eur. Phys. J. A **23**, 523 (2005) [arXiv:nucl-th/0410099].
16. BES Collaboration (M. Ablikim *et al.*), Phys. Lett. B **607**, 243 (2005) [arXiv:hep-ex/0411001].
17. J.R. Pelaez, Phys. Rev. D **55**, 4193 (1997) [arXiv:hep-ph/9609427].
18. S. Weinberg, Phys. Rev. Lett. **17**, 616 (1966).
19. G. Colangelo, J. Gasser, H. Leutwyler, Nucl. Phys. B **603**, 125 (2001) [arXiv:hep-ph/0103088].
20. N.A. Tornqvist, Z. Phys. C **68**, 647 (1995) [arXiv:hep-ph/9504372].
21. E. van Beveren, D.V. Bugg, F. Kleefeld, G. Rupp, arXiv:hep-ph/0606022; G. Rupp, E. van Bevern, these proceedings.
22. N.N. Achasov, G.N. Shestakov, Phys. At. Nucl. **56**, 1270 (1993) [Yad. Fiz. **56N9**, 206 (1993)].
23. U.G. Meißner, Comments Nucl. Part. Phys. **20**, 119 (1991).
24. A. Dobado, J.R. Pelaez, Phys. Rev. D **47**, 4883 (1993) [arXiv:hep-ph/9301276].
25. J.R. Pelaez, Mod. Phys. Lett. A **19**, 2879 (2004) [arXiv:hep-ph/0411107].
26. J.R. Pelaez, Phys. Rev. Lett. **92**, 102001 (2004) [arXiv:hep-ph/0309292].
27. D. Lohse, J.W. Durso, K. Holinde, J. Speth, Nucl. Phys. A **516**, 513 (1990); G. Janssen, B.C. Pearce, K. Holinde, J. Speth, Phys. Rev. D **52**, 2690 (1995) [arXiv:nucl-th/9411021].
28. Y. Kalashnikova, A. Kudryavtsev, A.V. Nefediev, J. Haidenbauer, C. Hanhart, Phys. Rev. C **73**, 045203 (2006) [arXiv:nucl-th/0512028].
29. N.N. Achasov, A.V. Kiselev, Phys. Rev. D **73**, 054029 (2006) [arXiv:hep-ph/0512047].
30. R.R. Akhmetshin *et al.*, Phys. Lett. B **462**, 480 (1999).
31. A. Aloisio *et al.*, Phys. Lett. B **536**, 209 (2002); A. Aloisio *et al.*, Phys. Lett. B **537**, 21 (2002); C. Bini, these proceedings.
32. E. Marko, S. Hirenzaki, E. Oset, H. Toki, Phys. Lett. B **470**, 20 (1999); J.E. Palomar, S. Hirenzaki, E. Oset, Nucl. Phys. A **707**, 161 (2002); J.E. Palomar, L. Roca, E. Oset, M.J. Vicente Vacas, Nucl. Phys. A **729**, 743 (2003).
33. V.E. Markushin, Eur. Phys. J. A **8**, 389 (2000).
34. J.A. Oller, Phys. Lett. B **426**, 7 (1998); J.A. Oller, Nucl. Phys. A **714**, 161 (2003).
35. F.E. Close, N. Isgur, S. Kumano, Nucl. Phys. B **389**, 513 (1993).
36. N.N. Achasov, V.V. Gubin, V.I. Shevchenko, Phys. Rev. D **56**, 203 (1997).
37. Y.S. Kalashnikova, A.E. Kudryavtsev, A.V. Nefediev, C. Hanhart, J. Haidenbauer, Eur. Phys. J. A **24**, 437 (2005) [arXiv:hep-ph/0412340].
38. N.N. Achasov, A.V. Kiselev, arXiv:hep-ph/0606268.
39. Yu.S. Kalashnikova, A.E. Kudryavtsev, A.V. Nefediev, J. Haidenbauer, C. Hanhart, arXiv:hep-ph/0608191.
40. G. Isidori, L. Maiani, M. Nicolaci, S. Pacetti, JHEP **0605**, 049 (2006) [arXiv:hep-ph/0603241]; KLOE Collaboration (F. Ambrosino), Phys. Lett. B **634**, 148 (2006); KLOE Collaboration (F. Nguyen), talk at *ICHEP06*.

Eur. Phys. J. A **31**, 549–552 (2007)
DOI 10.1140/epja/i2006-10294-2

THE EUROPEAN
PHYSICAL JOURNAL A

Special Article – QNP 2006

Deuteron radial moments for renormalized chiral potentials

E. Ruiz Arriola[a] and M. Pavon Valderrama

Departamento de Física Atómica, Molecular y Nuclear, Universidad de Granada, E-18071 Granada, Spain

Received: 22 December 2006
Published online: 20 March 2007 – © Società Italiana di Fisica / Springer-Verlag 2007

Abstract. We calculate deuteron positive and negative radial moments involving any bilinear function of the deuteron S and D wave functions for renormalized OPE and TPE chiral potentials. The role played by the strong singularities of the potentials at the origin and the short-distance insensitivity of the results when the potentials are fully iterated is emphasized as compared to realistic potentials.

PACS. 21.30.Fe Forces in hadronic systems and effective interactions – 11.10.Gh Renormalization – 13.75.Cs Nucleon-nucleon interactions – 21.45.+v Few-body systems

Chiral dynamics has played an important role in the theoretical description of low-energy hadronic reactions [1] and so far is the only known vestige of the underlying fundamental QCD theory of strong interactions in nuclear physics. There is a number of low-energy theorems based on chiral symmetry which provide a quantitative and model-independent insight into low-energy processes involving pions and nucleons, due to the clear scale separation between nuclear physics and QCD. For compound systems which at low energies disclose their composite nature the theoretical description necessarily becomes very involved and probably dependent on arbitrary assumptions. On the contrary, for weakly bound systems such as the deuterium nucleus one expects important simplifications leading to a more scheme-independent and possibly systematic description of these systems. This possibility motivated the introduction of Effective Field Theory (EFT) approaches [2] for nuclear physics based on the chiral symmetry of QCD, and the derivation of low-energy theorems, as, for example, pion-deuteron scattering [3] (for comprehensive reviews see, e.g., refs. [4–6]). In many cases most of the information needed for reactions involving the deuteron can be encoded by simple deuteron matrix elements.

Guided by earlier work [7,8], we have proposed [9–13] to renormalize the NN interaction in a non-perturbative way, highlighting model-independent long-distance correlations among physical observables. In our approach the long-distance chiral NN One-Pion Exchange (OPE) and Two-Pion Exchange (TPE) potentials, computed within perturbation theory in refs. [14–16], are iterated to all orders in the Schrödinger equation very much in the spirit of the original Weinberg approach [2]. However, some subtleties are found [10,12,17], which impose strong con-

straints on the admissible short-distance physics based on orthogonality, uniqueness and finiteness of the results.

In the $^3S_1 - {}^3D_1$ channel, the relative proton-neutron state for negative energy is described by the coupled equations

$$\begin{pmatrix} -\frac{\mathrm{d}^2}{\mathrm{d}r^2} + U_s(r) & U_{sd}(r) \\ U_{sd}(r) & -\frac{\mathrm{d}^2}{\mathrm{d}r^2} + \frac{6}{r^2} + U_d(r) \end{pmatrix} \begin{pmatrix} u \\ w \end{pmatrix} = \\ -\gamma^2 \begin{pmatrix} u \\ w \end{pmatrix} . \tag{1}$$

Here $\gamma = \sqrt{MB}$, with B the deuteron binding energy and M the nucleon mass, $U(r) = MV(r)$ are the reduced potentials and $u(r)$ and $w(r)$ are S- and D-wave deuteron reduced wave functions, respectively. At long distances they satisfy,

$$\begin{pmatrix} u \\ w \end{pmatrix} \to A_S\, e^{-\gamma r} \begin{pmatrix} 1 \\ \eta \left[1 + \frac{3}{\gamma r} + \frac{3}{(\gamma r)^2} \right] \end{pmatrix} , \tag{2}$$

where η is the asymptotic D/S ratio parameter and A_S is the asymptotic normalization factor, which is such that the deuteron wave functions are normalized to unity. For conventions and numerical values of parameters we use refs. [11,12] throughout.

In this work we report on the radial moments

$$\langle r^n \rangle_u = \int_0^\infty r^n u(r)^2 \mathrm{d}r , \tag{3}$$

$$\langle r^n \rangle_w = \int_0^\infty r^n w(r)^2 \mathrm{d}r , \tag{4}$$

$$\langle r^n \rangle_{uw} = \int_0^\infty r^n u(r)w(r)\mathrm{d}r , \tag{5}$$

for $-3 \leq n \leq 2$ which appear in many situations of interest, such as the calculation of the matter radius, the

---

[a] Spokesperson; e-mail: `earriola@ugr.es`

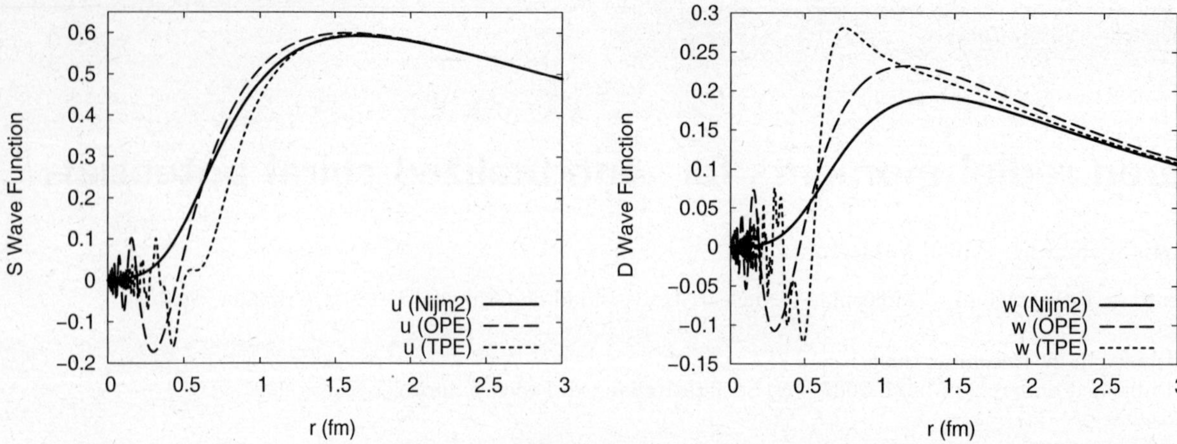

Fig. 1. The OPE and TPE deuteron wave functions, u (left) and w (right), as a function of the distance (in fm) compared to the Nijmegen-II wave functions [22]. The asymptotic normalization $u \to e^{-\gamma r}$ has been adopted and the asymptotic D/S ratio is taken $\eta = 0.0256(4)$ in the TPE case (for OPE $\eta = 0.026333$). We use the set IV of chiral couplings (see ref. [12]).

deuteron quadrupole moment and deuteron magnetic moment for the positive powers, as well as πd and Kd elastic scattering and neutral pion photoproduction, $\gamma d \to \pi^0 d$, in the case of the negative powers.

An important issue is the finiteness of the negative radial moments, a topic which has been recently discussed for OPE [11,18,19] and TPE [20]. The remarkable finding is that chiral potentials [14–16], when fully iterated, have an increasing number of finite inverse radial moments due to the near the origin singularities of the potential. They smoothen the short-distance behaviour of the wave functions and hence improve the convergence of the inverse radial moments. This is in sharp contrast with perturbative approaches [21], for which the perturbative wave functions diverge at the origin [11,12][1], or conventional (regular) phenomenological potentials [22] where the S- and D-wave short-distance behaviour of the wave functions, $u \sim r$ and $w \sim r^5$ respectively, is enough to render $\langle 1/r \rangle_u$ and $\langle 1/r^2 \rangle_u$ finite, but produce divergent higher inverse moments. We illustrate the situation below.

At distances much shorter than the pion Compton wavelength, the OPE potential behaves as

$$\begin{pmatrix} U_s^{\mathrm{OPE}}(r) & U_{sd}^{\mathrm{OPE}}(r) \\ U_{sd}^{\mathrm{OPE}}(r) & U_d^{\mathrm{OPE}}(r) \end{pmatrix} \to \frac{1}{r^3} \begin{pmatrix} R_s & R_{sd} \\ R_{sd} & R_d \end{pmatrix} \tag{6}$$

with $R_s = 0$, $R_{sd} = 2\sqrt{2}R$, $R_d = 4R$ and $R = 3g_A^2 M/32\pi f^2$ ($= 1.07764\,\mathrm{fm}$). This behaviour of the potential is strong enough to overcome the centrifugal barrier at short distances, thus modifying the usual short-distance behaviour of the wave functions, which can schematically be written as

$$u(r) \sim w(r) \sim \left(\frac{r}{R}\right)^{3/4} f\left(\frac{r}{R}\right), \tag{7}$$

where $f(r/R)$ represents some linear combination of $\sin\left(4\sqrt{R/r}\right)$, $\cos\left(4\sqrt{R/r}\right)$ and $\exp\left(-4\sqrt{2}\sqrt{R/r}\right)$ (for a

complete analysis, see ref. [11]). The elimination of the diverging exponential fixes $\eta_{\mathrm{OPE}} = 0.0263$. From this short-distance behaviour of the wave functions, one finds that the $\langle 1/r \rangle_u$ and $\langle 1/r^2 \rangle_u$ moments are finite for the OPE potential, while $\langle 1/r^3 \rangle_u$ and higher moments diverge, as it would happen for a regular potential.

The short-distance behaviour of the TPE (NNLO) has been exploited in ref. [12]. The potential at short distances behaves as [14–16]

$$\begin{pmatrix} U_s^{\mathrm{TPE}}(r) & U_{sd}^{\mathrm{TPE}}(r) \\ U_{sd}^{\mathrm{TPE}}(r) & U_d^{\mathrm{TPE}}(r) \end{pmatrix} \to \frac{1}{r^6} \begin{pmatrix} R_s^4 & R_{sd}^4 \\ R_{sd}^4 & R_d^4 \end{pmatrix}, \tag{8}$$

where

$$(R_s)^4 = \frac{3g_A^2}{128 f^4 \pi^2}(4 - 3g_A^2 + 24\bar{c}_3 - 8\bar{c}_4),$$

$$(R_{sd})^4 = -\frac{3\sqrt{2}g_A^2}{128 f^4 \pi^2}(-4 + 3g_A^2 - 16\bar{c}_4),$$

$$(R_d)^4 = \frac{9g_A^2}{32 f^4 \pi^2}(-1 + 2g_A^2 + 2\bar{c}_3 - 2\bar{c}_4), \tag{9}$$

and $\bar{c}_i = Mc_i$ are the low-energy chiral couplings appearing in πN scattering. As in the OPE case, this potential is strong enough at short distances to modify the short-distance behaviour of the wave function, which now reads

$$u(r) \sim w(r) \sim C_+ \, (r/R_+)^{3/2} \, f_+ \, (r/R_+) \\ + C_- \, (r/R_-)^{3/2} \, f_- \, (r/R_-), \tag{10}$$

where R_+^4 and R_-^4 are the eigenvalues of the matrix in eq. (8), and $f_\pm(r/R_\pm)$ represents a linear combination of $\sin\left(R_\pm^2/2r^2\right)$ and $\cos\left(R_\pm^2/2r^2\right)$ leaving η_{TPE} as a free parameter [12,20]. From this short-distance behaviour, the $\langle 1/r \rangle_u$, $\langle 1/r^2 \rangle_u$ and $\langle 1/r^3 \rangle_u$ radial moments are finite for the TPE potential, while higher moments diverge (although they would become finite for higher-order potentials).

---

[1] This divergency holds for perturbations both on boundary conditions or on distorted (fully iterated) OPE waves.

Table 1. Deuteron radial moments (in units of powers of fm). We consider the OPE and TPE potentials; in the case of the OPE potential we have taken $g_{\pi NN} = 13.08$ (*i.e.*, $g_A = 1.29$, OPE) and $g_A = 1.26$ (OPE*), while in the TPE case we show the results corresponding to the four set of chiral couplings considered along our previous works [12, 20]. In the OPE case the error is estimated by varying the semiclassical matching radius [11, 20] in the 0.1–0.2 fm range, while in the TPE case the error comes from the experimental uncertainty of the D/S ratio, $\eta = 0.0256(4)$. TPE Sets I, II, II and IV refer to the chiral parameters, c_1, c_3 and c_4 of refs. [23, 16, 24] and [25], respectively. NijmII and Reid93 are calculated from ref. [22] or taken from ref. [26].

	Short	OPE	OPE*	TPE-SetI	TPE-SetII	TPE-SetIII	TPE-SetIV	NijmII	Reid93
$\gamma\ (\mathrm{fm}^{-1})$	Input	Input	Input	Input	Input	Input	Input	0.231605	0.231605
η	0.0	0.026333	0.025547	Input	Input	Input	Input	0.02521	0.02514
$\langle r^2 \rangle_u$	9.3213	14.582(6)	14.424(6)	15.60(9)	15.61(11)	15.3(3)	15.09(13)	15.129	15.147
$\langle r^2 \rangle_w$	0.0	0.3849(2)	0.36371(15)	0.37(3)	0.38(2)	0.37(2)	0.38(2)	0.3438	0.3129
$\langle r^2 \rangle_{uw}$	0.0	2.0883(9)	2.0144(8)	2.14(4)	2.15(3)	2.11(3)	2.09(3)	2.035	2.032
$\langle r^1 \rangle_u$	2.1589	3.0400(14)	3.0185(13)	3.21(2)	3.20(2)	3.16(2)	3.12(3)	3.138	3.139
$\langle r^1 \rangle_w$	0.0	0.14287(6)	0.13691(6)	0.134(12)	0.139(12)	0.136(12)	0.146(12)	0.1204	0.1206
$\langle r^1 \rangle_{uw}$	0.0	0.5898(3)	0.5739(2)	0.586(14)	0.589(14)	0.581(15)	0.584(13)	0.5594	0.5590
$\langle r^0 \rangle_u$	1.0	0.9270(4)	0.9287(4)	0.935(9)	0.930(9)	0.931(10)	0.918(10)	0.9436	0.9430
$\langle r^0 \rangle_w$	0.0	0.07312(3)	0.07146(2)	0.065(9)	0.070(9)	0.069(10)	0.081(10)	0.05635	0.05699
$\langle r^0 \rangle_{uw}$	0.0	0.23989(11)	0.23691(10)	0.222(7)	0.225(7)	0.225(8)	0.233(7)	0.2166	0.2172
$\langle r^{-1} \rangle_u$	∞	0.4259(3)	0.4336(2)	0.382(5)	0.377(5)	0.388(6)	0.384(5)	0.4160	0.4163
$\langle r^{-1} \rangle_w$	0.0	0.052498(5)	0.05256(3)	0.042(8)	0.048(7)	0.048(10)	0.063(10)	0.03419	0.03520
$\langle r^{-1} \rangle_{uw}$	0.0	0.14120(7)	0.14239(3)	0.112(5)	0.115(5)	0.117(6)	0.128(6)	0.1153	0.1166
$\langle r^{-2} \rangle_u$	∞	0.3464(8)	0.3582(3)	0.210(4)	0.205(4)	0.220(5)	0.221(3)	0.2607	0.2646
$\langle r^{-2} \rangle_w$	0.0	0.0771(2)	0.0783(3)	0.038(8)	0.044(7)	0.045(12)	0.064(11)	0.02613	0.02780
$\langle r^{-2} \rangle_{uw}$	0.0	0.1551(4)	0.1589(3)	0.072(4)	0.075(4)	0.079(4)	0.093(6)	0.08122	0.08413
$\langle r^{-3} \rangle_u$	∞	∞	∞	0.159(3)	0.155(3)	0.173(4)	0.1851(8)	∞	∞
$\langle r^{-3} \rangle_w$	0.0	∞	∞	0.053(10)	0.059(9)	0.066(14)	0.091(14)	0.02465	0.02783
$\langle r^{-3} \rangle_{uw}$	0.0	∞	∞	0.0626(14)	0.068(2)	0.071(2)	0.097(6)	0.07342	0.08064

The wave functions for OPE and TPE as compared to the NijmII ones have been depicted in fig. 1. The radial moments are tabulated in table 1. As we see, and despite the very different behaviour at short distances between the deuteron wave functions corresponding to renormalized chiral potentials and to phenomenological potentials, the convergent moments are fairly similar (the $u + w$ combination works better) despite that NijmII and Reid93 contain no explicit TPE components. For inverse moments the trend improves clearly when going from OPE to TPE, which we interpret as a correct implementation of model-independent long-distance correlations generated by chiral symmetry and renormalization constraints.

This work is supported by Spanish DGI and FEDER funds with grant no. BFM2002-03218, Junta de Andalucía grant No. FQM-225, and EU RTN Contract CT2002-0311 (EURIDICE).

References

1. T.E.O. Ericson, W. Weise, *Pions and Nuclei* (Clarendon, Oxford, 1988).
2. S. Weinberg, Phys. Lett. B **251**, 288 (1990).
3. S. Weinberg, Phys. Lett. B **295**, 114 (1992).
4. P.F. Bedaque, U. van Kolck, Annu. Rev. Nucl. Part. Sci. **52**, 339 (2002).
5. D.R. Phillips, Czech. J. Phys. **52**, B49 (2002).
6. D.R. Phillips, J. Phys. G **31**, S1263 (2005).
7. D.W.L. Sprung, W. van Dijk, E. Wang, D.C. Zheng, P. Sarriguren, J. Martorell, Phys. Rev. C **49**, 2942 (1994).
8. S.R. Beane, P.F. Bedaque, M.J. Savage, U. van Kolck, Nucl. Phys. A **700**, 377 (2002) nucl-th/0104030.
9. M. Pavon Valderrama, E. Ruiz Arriola, Phys. Lett. B **580**, 149 (2004) nucl-th/0306069.
10. M. Pavon Valderrama, E. Ruiz Arriola, Phys. Rev. C **70**, 044006 (2004) nucl-th/0405057.
11. M. Pavon Valderrama, E. Ruiz Arriola, Phys. Rev. C **72**, 054002 (2005) nucl-th/0504067.
12. M.P. Valderrama, E.R. Arriola, Phys. Rev. C **74**, 054001 (2006).
13. M. Pavon Valderrama, E. Ruiz Arriola, Phys. Rev. C **74**, 064004 (2006).
14. N. Kaiser, R. Brockmann, W. Weise, Nucl. Phys. A **625**, 758 (1997).
15. J.L. Friar, Phys. Rev. C **60**, 034002 (1999).
16. M.C.M. Rentmeester, R.G.E. Timmermans, J.L. Friar, J.J. de Swart, Phys. Rev. Lett. **82**, 4992 (1999).
17. A. Nogga, R.G.E. Timmermans, U. van Kolck, Phys. Rev. C **72**, 054006 (2005) nucl-th/0506005.
18. A. Nogga, C. Hanhart, Phys. Lett. B **634**, 210 (2006).

19. L. Platter, D.R. Phillips, Phys. Lett. B **641**, 164 (2006).
20. M.P. Valderrama, E.R. Arriola, `nucl-th/0605078`.
21. B. Borasoy, H.W. Griesshammer, Int. J. Mod. Phys. E **12**, 65 (2003).
22. V.G.J. Stoks, R.A.M. Klomp, C.P.F. Terheggen, J.J. de Swart, Phys. Rev. C **49**, 2950 (1994).
23. P. Buettiker, U.G. Meissner, Nucl. Phys. A **668**, 97 (2000) `hep-ph/9908247`.
24. D.R. Entem, R. Machleidt, Phys. Rev. C **66**, 014002 (2002) `nucl-th/0202039`.
25. D.R. Entem, R. Machleidt, Phys. Rev. C **68**, 041001 (2003) `nucl-th/0304018`.
26. J.J. de Swart, C.P.F. Terheggen, V.G.J. Stoks, `nucl-th/9509032`.

Eur. Phys. J. A **31**, 553–556 (2007)

DOI 10.1140/epja/i2006-10218-2

THE EUROPEAN
PHYSICAL JOURNAL A

Special Article – QNP 2006

Dimension-2 condensates and Polyakov Chiral Quark Models

E. Megías[a], E. Ruiz Arriola, and L.L. Salcedo

Departamento de Física Atómica, Molecular y Nuclear, Universidad de Granada, E-18071 Granada, Spain

Received: 8 November 2006

Published online: 1 March 2007 – © Società Italiana di Fisica / Springer-Verlag 2007

Abstract. We address a possible relation between the expectation value of the Polyakov loop in pure gluodynamics and full QCD based on Polyakov Chiral Quark Models where constituent quarks and the Polyakov loop are coupled in a minimal way. To this end, we use a center symmetry-breaking Gaussian model for the Polyakov loop distribution which accurately reproduces gluodynamics data above the phase transition in terms of dimension-2 gluon condensate. The role played by the quantum and local nature of the Polyakov loop is emphasized.

PACS. 12.39.Fe Chiral Lagrangians – 11.10.Wx Finite-temperature field theory – 12.38.Lg Other nonperturbative calculations

The phase transition of QCD matter at finite temperature from hadrons to a quark-gluon plasma was established long ago [1–3] (for a review see, *e.g.*, [4]). The precise definition of the phase transition requires a proper identification of the relevant order parameters. In the heavy-quark limit, the Polyakov loop vacuum expectation value signals the breaking of the center symmetry corresponding to the deconfinement phase transition when changing from zero to one [5]. In the limit of light quarks the chiral condensate determines the restoration of chiral symmetry when the chiral condensate vanishes above the critical temperature. In the real QCD case with dynamical massive quarks both chiral and center symmetries are explicitly broken, and neither the Polyakov loop nor the chiral condensate are truly order parameters, although a rather sharp crossover is expected across the phase transition for these quantities [4].

Polyakov-Chiral Quark Models allow to study the interplay between chiral symmetry restoration and center symmetry breaking [6] in a quantitative manner [7–13]. At zero temperature the constituent quark mass, M, is dynamically generated via the spontaneous breaking of chiral symmetry, inducing a exponentially small, $\sim e^{-M/T}$, breaking of the center symmetry at low temperatures [9–11]. This provides the rationale for keeping the Polyakov loop as an order parameter also in the unquenched case. However, although the coupling of the Polyakov loop to quarks is rather unique, details regarding the postulated purely gluonic action differ [7–13]. This additional information is beyond the chiral quark model capabilities and must always be postulated. In this regard, it seems natural to constrain the gluonic action to reproduce pure gluody-

namics results. In the present work we address a possible relation between the expectation value of the Polyakov loop in pure gluodynamics and full QCD based on the coupling to quarks generated by the fermion determinant within a Polyakov NJL model (PNJL).

One of the problems one must also face when comparing models with lattice data has to do with the difficult but necessary renormalization of the Polyakov loop. Indeed, after renormalization the Polyakov loop falls outside the unitary group [14,15]. The spontaneous breaking of the center symmetry above some critical temperature occurs already at the level of pure gluodynamics and is interpreted as the signal of deconfinement [16]. Full dynamical QCD lattice simulations account, in addition, for an explicit center symmetry breaking at low temperatures due to the presence of fermions [17]. In a recent paper [18] we have shown that the behaviour of the Polyakov loop above the deconfinement phase transition can be naturally accommodated by a dimension-2 condensate, instead of the more conventional perturbative calculations. The quality of our fits leaves little doubt on the veracity of the description of the lattice data. To our knowledge this is the first time that power corrections in temperature have been advocated to describe QCD right above the deconfinement phase transition. This is in sharp contrast to the common believe that logarithmic thermal corrections should be considered instead yielding to rather unnatural, in fact inaccurate, fits to the existing lattice results.

Following our study on the role of dimension-2 condensate in the center symmetry-breaking as a guideline [18] we make here a Gaussian ansatz for the purely gluonic contribution to the Polyakov loop distribution which reproduces remarkably well the lattice data for pure glu-

---

[a] Spokesperson; e-mail: `emegias@ugr.es`

odynamics [16]. Then, using the PNJL model [7–13] we may *predict* the value of the Polyakov loop in the *unquenched* case and compare with full dynamical calculations. In our view this is the best possible scenario for a Polyakov-Chiral quark model to work.

Let us explain how the dimension-2 gluon condensate can generate a powerlike behaviour of the Polyakov loop at temperatures slightly above the deconfinement phase transition in gluodynamics. In the Polyakov gauge the (expectation value of the) Polyakov loop is defined as

$$L = \left\langle \frac{1}{N_c} \operatorname{tr}_c e^{ig A_0(\boldsymbol{x})/T} \right\rangle , \tag{1}$$

where $\langle\ \rangle$ denotes vacuum expectation value and tr_c is the (fundamental) color trace. A_0 is the gluon field in the (Euclidean) time direction, $A_0 = \sum T_a A_{0,a}$, T_a being the Hermitian generators of the $SU(N_c)$ Lie algebra in the fundamental representation, with the standard normalization $\operatorname{tr}_c(T_a T_b) = \delta_{ab}/2$. Note that in this gauge we have the invariance under the large gauge transformation $g A_0 \to g A_0 + 2\pi T$.

If we expand the Polyakov loop at high temperatures we get the cumulant expansion

$$L = \sum_{k=0}^{\infty} \frac{1}{(2k)!} \left(\frac{-g^2}{T^2} \right)^k \langle A_{0,a_1} \dots A_{0,a_{2k}} \rangle$$
$$\times \frac{1}{N_c} \operatorname{tr}_c(T_{a_1} \dots T_{a_{2k}}), \tag{2}$$

where odd terms have been discarded on the assumption that global colour symmetry breaking does not occur. Averaging over the $N_c^2 - 1$ colour gluonic degrees of freedom we get

$$\langle A_{0,a_1} \dots A_{0,a_{2k}} \rangle = \frac{\langle (A_{0,a})^{2k} \rangle (N_c^2 - 3)!!}{(N_c^2 + 2k - 3)!!} \delta_{a_1,\dots,a_{2k}} \tag{3}$$

where $\delta_{a_1,\dots,a_{2k}}$ is the fully symmetrized product of the Kronecker delta symbols δ_{ab}. The $SU(N_c)$ Casimir invariant is $C_2 = \sum_a T_a T_a = (N_c^2 - 1)/(2N_c)\mathbf{1}_{N_c}$. In the large N_c limit only the contractions of adjacent indices (including cyclic permutations) survive. For instance, if we twist two adjacently contracted generators we get $T_a T_b T_a = -T_b/2N_c$ so that $(T_a T_b)^2$ is suppressed by two powers of N_c as compared to $T_a^2 T_b^2$. Then, neglecting these terms we obtain

$$L = \sum_{k=0}^{\infty} \frac{1}{(2k)!} \left(\frac{-g^2}{T^2} \right)^k \frac{\langle (A_{0,a})^{2k} \rangle}{(2N_c)^k} . \tag{4}$$

Now, we expect $\langle (A_{0,a})^{2k} \rangle$ to scale as N_c^{2k}, in which case and for large T the limit $N_c \to \infty$ with $g^2 N_c$ fixed is finite and non-trivial, at least in the absence of radiative corrections. Finally, if we assume vacuum saturation of condensates we can reduce further the expression since

$$\langle A_0^{2k} \rangle = (2k - 1)!! \langle A_0^2 \rangle^k + \text{n.v.c.}, \tag{5}$$

where n.v.c. stands for non-vacuum–connected terms. After summing up the series we get the result

$$L = \exp\left[-\frac{g^2 \langle A_{0,a}^2 \rangle}{4 N_c T^2} \right] . \tag{6}$$

The exponentiation means that temperature is not necessarily very high since we summed up all the series and, actually, for small T yields $L \to 0$. Of course, as one approaches the true critical temperature one expects non-analytical behaviour not encoded in the high-temperature expansion[1]. On top of eq. (6) one must take radiative corrections into account yielding, in fact, $L > 1$ [14,15] because of renormalization effects at high temperatures. This is equivalent to multiply eq. (6) by a slowly varying temperature-dependent factor [18].

The previous behaviour suggests a Gaussian ansatz for the Polyakov loop distribution, *i.e.* the probability of having a configuration with a given value of Ω. Unfortunately, not much is known about this object in QCD (see however [11]). In the Polyakov gauge, where A_0 is both static and diagonal we can write the piece of the vacuum wave functional depending on the temporal gluon fields as

$$\rho_0(A_3, A_8) = C e^{-(A_3^2 + A_8^2)/a^2}, \tag{7}$$

where we assume that the gluon variables take any real values on the real axis. The normalization constant C and the width of the distribution a can be determined from the definition of the dimension-2 condensate

$$\langle A_3^2 + A_8^2 \rangle_0 = a^2. \tag{8}$$

Although the previous Gaussian reproduces the result of the cumulant expansion of the Polyakov loop, it does not incorporate the known gauge invariance which corresponds to a periodicity condition on the gluon field distribution and the integration measure.

Once the Polyakov loop distribution has been determined for pure gluodynamics we set out to describe its expectation value when dynamical quarks are included. To do so, we use the PNJL model. For a constant value of the Polyakov loop, Ω, and the scalar field $S = M$ (which we identify with the constituent quark mass) the quark

---

[1] Nonetheless, although eq. (6) does not predict a phase transition one could still make an estimate of the critical temperature related to the point where L presents no curvature, *i.e.* the inflexion point which yields, $T_c^2 = g^2 \langle A_0^2 \rangle / 6 N_c$. On the other hand, we expect the condensate to scale with N_c and the number of Euclidean dimensions as $g^2 \langle A_\mu^2 \rangle_\mu \sim 4\pi\alpha(\mu)(N_c^2 - 1)D\Lambda_{\text{QCD}}^2$. This shows that the large-N_c limit of L is well defined for *fixed* T. Taking $\Lambda_{\text{QCD}} = 240\,\text{MeV}$ and the scale $\mu = 2\,\text{GeV}$ with the conversion factor $\mu = 2\pi T$ and $N_c = 3$ we get $g^2 \langle A_\mu^2 \rangle = (2.2\,\text{GeV})^2$ in rough agreement with the determinations from lattice data. Then $T_c^2 \sim 4\pi\alpha(2\pi T_c)(N_c^2 - 1)\Lambda_{\text{QCD}}^2 / 6 N_c$ yielding the estimate $T_c = 1.1\Lambda_{\text{QCD}}$.

effective action is given by

$$\frac{T}{V}\Gamma_Q = \frac{1}{4G_S}\operatorname{Tr}_f(M - \hat{M}_0)^2 - 2N_f \int \frac{\mathrm{d}^3 k}{(2\pi)^3}$$

$$\times \left(N_c \epsilon_k + T \operatorname{tr}_c \log\left[1 + e^{-\epsilon_k/T}\,\Omega\right] + \text{c.c.}\right), \quad (9)$$

where we have only retained the vacuum contribution, so there is no contribution of meson fields. V is three-dimensional volume and $\epsilon_k = +\sqrt{k^2 + M^2}$ is the energy of a constituent quark with mass M. We define the Polyakov-loop–averaged action

$$e^{-\Gamma_Q(M,T)} = \int \mathrm{d}\Omega\,\rho_0(A_3, A_8)\,e^{-\Gamma_Q(M,\Omega,T)}. \quad (10)$$

The value of M is determined by minimization of $\Gamma_Q(M,T)$ with respect to M, $\partial\Gamma_Q(M,T)/\partial M = 0$, which is the gap equation and determines M at a given temperature T, denoted as $M^* = M(T)$. This procedure allows to compute the integration in meson fields at the mean-field level. In addition, the relation between the (single flavour) chiral quark condensate, $\langle\bar{q}q\rangle$, and the constituent quark mass, reads

$$2G_S\langle\bar{q}q\rangle^* = -(M^* - m_q). \quad (11)$$

Any observable is obtained by using the temperature dependent mass M^*. For instance the expectation value of the Polyakov loop is computed as

$$\langle\Omega\rangle = \frac{\int \mathrm{d}\Omega\,\rho_0(A_3, A_8)\,e^{-\Gamma_Q(M^*,\Omega,T)}\,\Omega}{\int \mathrm{d}\Omega\,\rho_0(A_3, A_8)\,e^{-\Gamma_Q(M^*,\Omega,T)}}. \quad (12)$$

The integral in $\mathrm{d}\Omega$ in the case $N_c = 3$ and in the Polyakov gauge can be computed numerically. At the mean-field level the Polyakov distribution becomes a delta function, but this picture is obviously only valid at high enough temperature, where quantum effects are negligible. At low temperatures the mean-field approximation can lead to wrong results. For example, for the expectation value of the Polyakov loop in the adjoint representation one obtains $-1/(N_c^2 - 1)$, instead of zero, as has been observed in lattice data [19]. The color group integration solves this problem [11]. Finally, we have taken into account here the local effects in the Polyakov loop by considering a confinement correlation volume $V_\sigma = 8\pi T^3/\sigma^3$, which follows from the simplest correlation of two Polyakov loops

$$\frac{1}{T}\int \mathrm{d}^3 x \int \mathrm{d}\Omega\,\operatorname{tr}_c \Omega(\boldsymbol{x})\,\operatorname{tr}_c \Omega^\dagger(\boldsymbol{y}) = \frac{1}{T}\int \mathrm{d}^3 x\, e^{-\sigma|\boldsymbol{x}-\boldsymbol{y}|/T}. \quad (13)$$

Higher correlation functions are expected to have a smaller contribution [11].

The result for the chiral condensate and Polyakov loop expectation value is presented in fig. 1. We consider $\sqrt{\sigma} = 425\,\text{MeV}$ and $m_u = m_d \equiv m_q = 5.5\,\text{MeV}$. As we can see the agreement is reasonable although the quark model predicts a shift in the critical temperature of about $50\,\text{MeV}$. This discrepancy is compatible with the model uncertainties discussed in our previous work [11] where

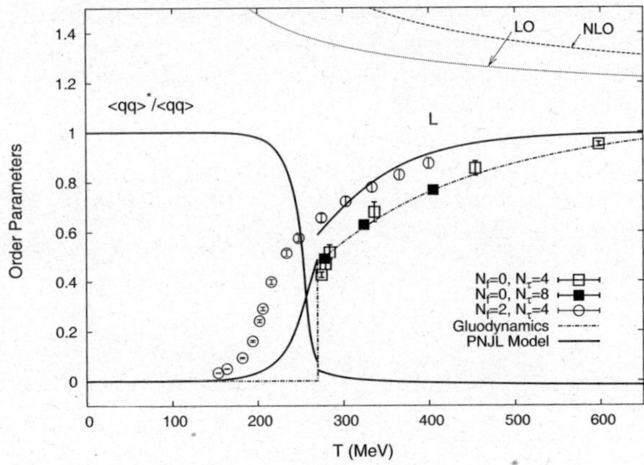

Fig. 1. Temperature dependence of chiral condensate $\langle\bar{q}q\rangle^*$ and Polyakov loop expectation value $L = \langle\operatorname{tr}_c\Omega\rangle/N_c$ in the 2-flavor PNJL model. We use the Gaussian distribution reproducing the Polyakov loop in the case of pure gluodynamics. Lattice data from [16] for gluodynamics ($N_f = 0$) and [17] for $N_f = 2$ full QCD. Perturbative LO and NLO results for the Polyakov loop with $N_f = 2$ are shown for comparison.

the explicit center symmetry breaking of the pure gluonic action was disregarded. It is also compatible with the large rescaling advocated in [12]. Basically, the shift in the critical temperature reflects our lack of knowledge on the gluon action at intermediate temperatures, in particular, the feedback of quarks on the gluons. It would be most useful to develop a theoretical framework where this feedback is taken into account, as it is done, *e.g.* in the mean-field approximation [12], while retaining the fundamentally quantum and local nature of the Polyakov loop [11].

This work is supported by the Spanish DGI and FEDER funds grant FIS2005-00810, Junta de Andalucía grant FQM-225, and EURIDICE Grant No. HPRN-CT-2002-00311.

References

1. J.B. Kogut *et al.*, Phys. Rev. Lett. **50**, 393 (1983).
2. M. Fukugita, A. Ukawa, Phys. Rev. Lett. **57**, 503 (1986).
3. F. Karsch, E. Laermann, Phys. Rev. D **50**, 6954 (1994) `hep-lat/9406008`.
4. F. Karsch, PoS (Corfu98) 008 (1999).
5. R.D. Pisarski, `hep-ph/0203271` (2002).
6. A. Mocsy, F. Sannino, K. Tuominen, Phys. Rev. Lett. **92**, 182302 (2004) `hep-ph/0308135`.
7. P.N. Meisinger, M.C. Ogilvie, Phys. Lett. B **379**, 163 (1996) `hep-lat/9512011`.
8. K. Fukushima, Phys. Lett. B **591**, 277 (2004) `hep-ph/0310121`.
9. E. Megías, E. Ruiz Arriola, L.L. Salcedo, *Mini-Workshop on Quark Dynamics, Bled 2004, Bled, Slovenia, 12-19 Jul. 2004*, in *Bled Workshops in Physics*, Vol. **5**, No. 1 (2004) pp. 1-6, `hep-ph/0410053`.
10. E. Megías, E. Ruiz Arriola, L.L. Salcedo, AIP Conf. Proc. **756**, 436 (2005) `hep-ph/0411293`.

11. E. Megías, E. Ruiz Arriola, L.L. Salcedo, Phys. Rev. D **74**, 065005 (2006) `hep-ph/0412308`.
12. C. Ratti, M.A. Thaler, W. Weise, Phys. Rev. D **73**, 014019 (2006) `hep-ph/0506234`.
13. S.K. Ghosh, T.K. Mukherjee, M.G. Mustafa, R. Ray, Phys. Rev. D **73**, 114007 (2006) `hep-ph/0603050`.
14. E. Gava, R. Jengo, Phys. Lett. B **105**, 285 (1981).
15. F. Zantow, `hep-lat/0301014` (2003).
16. O. Kaczmarek, F. Karsch, P. Petreczky, F. Zantow, Phys. Lett. B **543**, 41 (2002) `hep-lat/0207002`.
17. O. Kaczmarek, F. Zantow, Phys. Rev. D **71**, 114510 (2005) `hep-lat/0503017`.
18. E. Megías, E. Ruiz Arriola, L.L. Salcedo, JHEP **01**, 073 (2006) `hep-ph/0505215`.
19. A. Dumitru, Y. Hatta, J. Lenaghan, K. Orginos, R.D. Pisarski, Phys. Rev. D **70**, 034511 (2004) `hep-th/0311223`.

Eur. Phys. J. A **31**, 557–559 (2007)

DOI 10.1140/epja/i2006-10215-5

THE EUROPEAN
PHYSICAL JOURNAL A

Special Article – QNP 2006

Scalar isoscalar part of the hyperon-nucleon interaction

K. Sasaki[a], E. Oset, and M.J. Vicente Vacas

Departamento de Física Teórica and IFIC, Centro Mixto Universidad de Valencia-CSIC, Spain

Received: 8 November 2006
Published online: 5 March 2007 – © Società Italiana di Fisica / Springer-Verlag 2007

Abstract. We study the central part of the ΛN potential by considering the correlated and uncorrelated two-meson exchange besides the ω exchange contribution. The correlated two-meson is evaluated in a chiral unitary approach. We find that a short-range repulsion is generated by the correlated two-meson potential which also produces an attraction in the intermediate distance region. The uncorrelated two-meson exchange produces a sizeable attraction in all cases which is counterbalanced by ω exchange contribution.

PACS. 13.75.Ev Hyperon-nucleon interactions – 12.39.Fe Chiral Lagrangians

1 Introduction

The scalar-isoscalar potential plays an important role in the nucleon-nucleon interaction providing an intermediate-range attraction in all channels which is demanded by the data. In models of the NN interaction using one-boson exchange (OBE) this part of the interaction was traditionally accounted for by allowing the exchange of a "σ" particle whose nature has been a source of controversy.

The picture of the σ as a dynamically generated resonance called for a interpretation of the σ exchange in the NN interaction and this work was performed in [1]. In this work the traditional σ exchange was substituted by the exchange of two interacting mesons within the chiral unitary framework of [2], and an intermediate attraction was found together with a repulsion at short distances, which makes the picture qualitatively different from the ordinary, always attractive, σ exchange.

The work of [1] was complemented in [3], where in addition to the interacting two-pion exchange, the contribution of the uncorrelated two-pion exchange and the repulsive contribution of the ω exchange were considered, leading altogether to a good reproduction of the empirical scalar isoscalar interaction of [4,5].

The purpose of this work is to extend this to the strange sector evaluating the ΛN scalar isoscalar interaction, as shown in fig. 1.

Empirical evaluations of the YN scalar isoscalar interaction are done in several works allowing the exchange of a scalar meson and making fits to data [6–9]. Our approach evaluates directly the correlated two-pion exchange by explicitly using the chiral unitary approach to deal with the

pion-pion interaction and using appropriate triangle diagrams to account for the coupling of the two pions to the baryons. The success of this approach providing the scalar isoscalar NN interaction provides solid grounds to extend these ideas to the case of the ΛN interaction, which we present in this work.

2 Formalism

We define the correlated two-meson exchange potential in momentum space as

$$V_{\Lambda N}^{Cor}(q) = \sum_{ij}^{\pi\pi, K\overline{K}} N_{ij}\Delta_{\Lambda}^{i} t_{i\rightarrow j}^{(I=0)}\Delta_{N}^{j}, \tag{1}$$

where Δ indicates the triangle scalar loop contribution of two-meson for baryon, which is defined in [1], and N_{ij} is a factor from the isospin summation, concretely $N_{\pi\pi,\pi\pi} = 6$, $N_{\pi\pi,K\overline{K}} = N_{K\overline{K},\pi\pi} = 2\sqrt{3}$, $N_{K\overline{K},K\overline{K}} = 2$. Concretely, the Δ-function for the Λ with the correlated two-pion is given by

$$\Delta_{\Lambda}^{(\pi\pi)} = \left(\frac{D}{\sqrt{3}f_{\pi}}\right)^{2} V_{\Sigma\Lambda}^{(\pi\pi)} + \frac{2}{3}\left(\frac{f_{\pi N\Delta}^{*}}{\sqrt{2}m_{\pi}}\right)^{2} V_{\Sigma^{*}N}^{(\pi\pi)}, \tag{2}$$

where $V_{B'B}^{(m_1 m_2)}$ is the vertex function which is already evaluated in [1] and given in a generalized form as

$$V_{B'B}^{(m_1 m_2)} = \int \frac{\mathrm{d}^3 p}{(2\pi)^3} \frac{M_{B'}}{E_{B'}} \frac{(\boldsymbol{p}+\boldsymbol{q})\cdot\boldsymbol{p}}{2\omega_1\omega_2(\omega_1 + \omega_2)}$$
$$\times \frac{\omega_1 + \omega_2 + E_{B'} - M_B}{(\omega_1 + E_{B'} - M_B)(\omega_2 + E_{B'} - M_B)} \tag{3}$$

[a] e-mail: Kenji.Sasaki@ific.uv.es

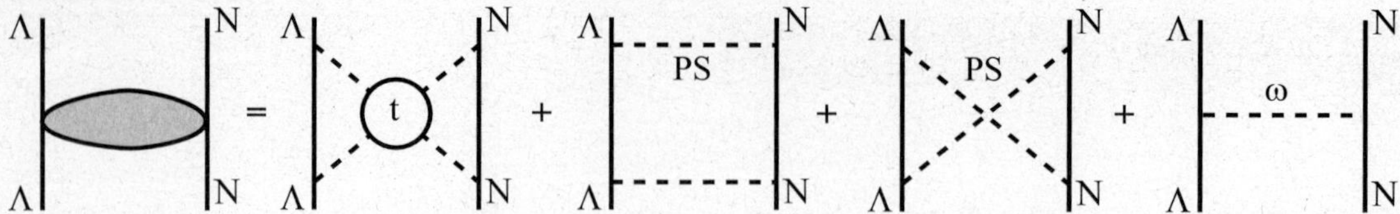

Fig. 1. Diagrammatic representation for the scalar part of ΛN potential. The first diagram stands for the correlated two-meson exchange potential and the last diagram is the contribution of the ω exchange. The others are the ladder and crossed two-meson exchange potentials.

with

$$E_B = \sqrt{p^2 + M_B^2}; \quad \omega_1 = \sqrt{\mu_1^2 + p^2}; \quad \omega_2 = \sqrt{\mu_2^2 + (p+q)^2}. \quad (4)$$

As in [1], we have put the initial momentum at rest. We introduce a static form factor in order to regularize the triangle loop function. The form factor employed in this calculation is

$$F(p)F(p+q) = \frac{\Lambda^2}{\Lambda^2 + p^2} \frac{\Lambda^2}{\Lambda^2 + (p+q)^2}, \quad (5)$$

where the cutoff is chosen as $\Lambda = 1.0\,\mathrm{GeV}$.

The t is the unitarized amplitude of meson-meson scattering which reproduces the π-π phase shift up to $1.2\,\mathrm{GeV}$ quite well.

The uncorrelated two-meson potential is given after performing analytically the p^0 integration in terms of the integrals

$$V_{\Lambda N}^{(i,\mu)}(q) = -\int \frac{\mathrm{d}^3 p}{(2\pi)^3} \frac{M_\Lambda}{E_\Lambda} \frac{M_N}{E_N} \frac{\left(4p^2 - q^2\right)^2}{32\omega_1\omega_2(\omega_1 + \omega_2)} R_i(\cdot), \quad (6)$$

where M_Λ and M_N are the Λ and nucleon mass, respectively and i stands for the direct (D) or crossed (C) terms and

$$R_D(\cdot) = \frac{\Omega(\epsilon^2 + 2\omega_1\omega_2) + (\Omega^2 + \omega_1\omega_2 + E_1'E_2')\epsilon}{\epsilon(\omega_1 + E_1')(\omega_1 + E_2')(\omega_2 + E_1')(\omega_2 + E_2')}, \quad (7)$$

$$R_C(\cdot) = \frac{\Omega\epsilon + \Omega^2 - \omega_1\omega_2 + E_1'E_2'}{(\omega_1 + E_1')(\omega_1 + E_2')(\omega_2 + E_1')(\omega_2 + E_2')}, \quad (8)$$

with $E_i' = E_i - M_i$, $\epsilon = E_1' + E_2'$ and $\Omega = \omega_1 + \omega_2$. It is worth mentioning that, if we compare without coupling constants, the crossed contribution is much smaller than the box-type contribution.

The ω exchange potential for ΛN in momentum space is given by

$$V_{\Lambda N}^{\omega}(q) = \frac{g_{\omega\Lambda\Lambda}g_{\omega NN}}{q^2 + m_\omega^2} \left(\frac{\Lambda_\omega^2 - m_\omega^2}{\Lambda_\omega^2 + q^2}\right)^2, \quad (9)$$

where we choose $g_{\omega NN} = 13$ and $\Lambda_\omega = 1.4\,\mathrm{GeV}$ [3]. The ideal mixing for ω and ϕ leads the relation, $g_{\omega\Lambda\Lambda} = \frac{2}{3}g_{\omega NN}$, which is deduced from the quark contents of the hadrons. For simplicity we assume the same form factor for both the ωNN and $\omega\Lambda\Lambda$ vertices.

The potential in configuration space is given by

$$V(r) = \frac{1}{2\pi^2 r} \int_0^\infty q\sin(qr)V(q)\mathrm{d}q \quad (10)$$

with the relative distance.

3 Results

The left panel of fig. 2 shows the ΛN potential in momentum space. The correlated two-meson contribution has a peak around $q = 400\,\mathrm{MeV}/c$. A similar peak in position and magnitude was found for the NN case in refs. [1,3]. It is worth discussing this shape because it is impossible to parameterize it by a single-meson exchange with usual form factors, like monopole or Gaussian. This contribution could be decomposed in, at least, two parts: a strong repulsive part and a weak attraction. In any case, this contribution is much smaller than the other ones, so that the main contribution comes from the uncorrelated two-meson exchange and from the ω exchange. These two potentials have opposite sign, and the uncorrelated two-meson potential is slightly stronger than the ω exchange potential in the whole range of q. Thus, the sum of these two potentials is always negative.

The total potential has positive strength around $q = 600\,\mathrm{MeV}/c$, which is pushed up by the correlated two-meson potential. The correlated two-meson potential plays a more important role in this region.

We notice the effect of the two-kaon contribution, suppressed due to its heavy mass, is very small and does not change the pionic potential much.

In the right panel of fig. 2, we can see a similar correlated two-meson potential as for the NN case, see fig. 10 in [1]. This should be expected since the definition of the potential is quite similar to the NN case except for the masses of the baryons and coupling constants. In fact both the NN and ΛN potential generated by the correlated two-pion exchange contribution pass through zero at $r \simeq 0.9\,\mathrm{fm}$ and have a minimum at $r \simeq 1.3\,\mathrm{fm}$. This potential is repulsive in the short-range region and, on the other hand, it is attractive beyond $1\,\mathrm{fm}$. The strength of the correlated two-meson potential is much smaller than the other contributions.

The right panel of fig. 2 shows that the uncorrelated two-meson generates a strong attraction and the ω produces a repulsion in the short-range region. The sum of these two potentials produces a relatively strong attraction around $1\,\mathrm{fm}$ and leads to large cancellations in the short-range region. We do not give the results below $0.5\,\mathrm{fm}$ since there the overlap of the baryons and quark exchange mechanisms can lead the sizeable corrections.

Although the attraction in the total potential is mainly generated by the uncorrelated two-meson potential, part of the repulsion is generated by the correlated two-meson

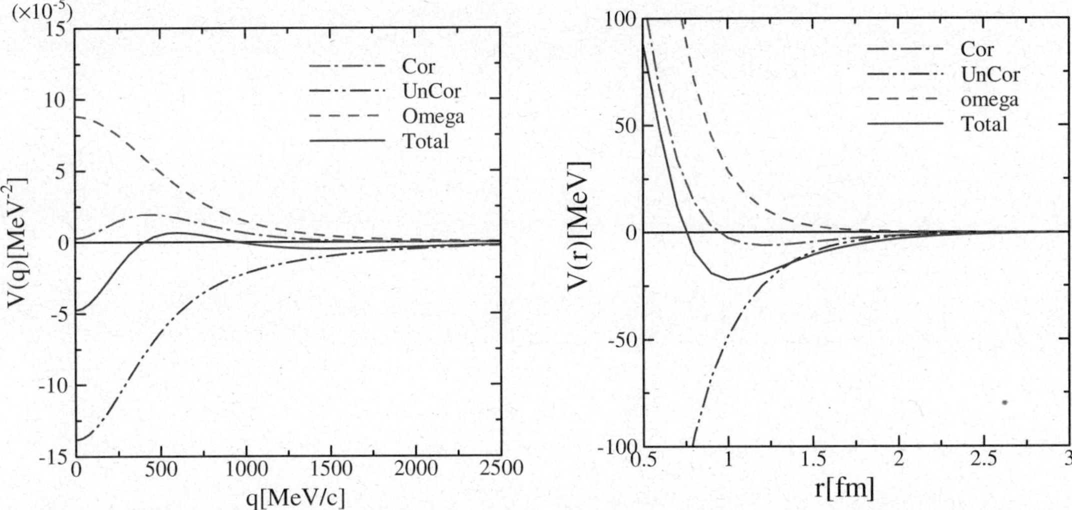

Fig. 2. The left (right) panel shows the ΛN potential in momentum (configuration) space, respectively. In both panels the dot-dashed, double-dot–dashed and dashed lines indicate the correlated two-meson, uncorrelated two-meson and omega exchange potential, respectively. The solid line stands for the total potential of the ΛN interaction in our approach.

potential. This is interesting because the correlated two-meson potential is considered as a σ-meson exchange in other papers and, there, the interaction would be always attractive (see eq. (3.19) of [1]).

The two-kaon contribution slightly enhances the magnitude of both the correlated and uncorrelated two-meson potential, and it makes the total potential a little deeper than the pionic potential.

4 Conclusions

We have evaluated the scalar channel potential between the Λ and nucleon. We have considered the correlated and uncorrelated two-meson exchange contributions in this channel besides ω exchange. The correlated two-meson exchange contribution was calculated by using a chiral unitary approach which reproduces very well the experimental meson-meson phase shift up to 1.2 GeV.

A strong attraction is produced by the uncorrelated two-meson exchange contribution. The ω exchange contribution makes a short-range repulsion, also similar to the NN case, but its strength is two-thirds of the NN case due to the simple counting of non-strange quarks in the baryons. These two contributions drive the attractive potential in the medium-range region and almost cancel each other at shorter distances.

The correlated two-meson exchange contribution is relatively smaller than the other two contributions. This potential produces some attraction at medium range distances and some strong repulsion in the short-range region. This behavior is quite similar to the NN case which is already calculated in [1]. The striking effect is the repulsion in the short-range region where the strong attraction generated by the uncorrelated two-meson potential is cancelled by the repulsion produced by the ω-meson. Thus the correlated two-meson potential plays an impor-

tant role for both the medium range attraction and the short-range repulsion in the ΛN.

We have also checked the contribution of two-kaon exchange diagrams. We have found that the two-kaon contribution is rather weak and it slightly enhances the magnitude of the potential without changing its main behavior for the ΛN potential. Therefore the two-pion exchange contribution plays a crucial role in the scalar isoscalar ΛN potential as well.

One of us, K.S., wishes to acknowledge support from the Ministerio de Educación y Ciencia and program of estancias de jóvenes doctores y tecnólogos extranjeros en España. This work is partly supported by DGICYT contract number BFM2003-00856, and the E.U. EURIDICE network contract no. HPRN-CT-2002-00311. This research is part of the EU Integrated Infrastructure Initiative Hadron Physics Project under contract number RII3-CT-2004-506078.

References

1. E. Oset, H. Toki, M. Mizobe, T.T. Takahashi, Prog. Theor. Phys. **103**, 351 (2000) [arXiv:nucl-th/0011008].
2. J.A. Oller, E. Oset, Nucl. Phys. A **620**, 438 (1997); **652**, 407 (1999)(E) [arXiv:hep-ph/9702314].
3. D. Jido, E. Oset, J.E. Palomar, Nucl. Phys. A **694**, 525 (2001) [arXiv:nucl-th/0101051].
4. R.B. Wiringa, R.A. Smith, T.L. Ainsworth, Phys. Rev. C **29**, 1207 (1984).
5. R.B. Wiringa, V.G.J. Stoks, R. Schiavilla, Phys. Rev. C **51**, 38 (1995) [arXiv:nucl-th/9408016].
6. T.A. Rijken, Y. Yamamoto, arXiv:nucl-th/0603042.
7. J. Haidenbauer, U.G. Meissner, Phys. Rev. C **72**, 044005 (2005) [arXiv:nucl-th/0506019].
8. Z.Y. Zhang, Y.W. Yu, P.N. Shen, L.R. Dai, A. Faessler, U. Straub, Nucl. Phys. A **625**, 59 (1997).
9. Y. Fujiwara, C. Nakamoto, Y. Suzuki, M. Kohno, K. Miyagawa, Prog. Theor. Phys. Suppl. **156**, 17 (2004).

Resonance Form-Factors: L_8 determination at Next-to-Leading Order in $1/N_C$

J.J. Sanz-Cillero[a]

Department of Physics, Peking University, Beijing 100871, PRC

Abstract. This talk presents a systematic procedure for the computation of the SS-PP correlator beyond the large–N_C limit. The present calculation is carried on within a perturbative $1/N_C$ framework . By constraining the meson form-factors at leading order in $1/N_C$, one obtains a one-loop spectral function well behaved at short distances. The Weinberg sum-rules get modified, gaining an extra contribution suppressed by $1/N_C$. This leads to a prediction for the low energy chiral perturbation theory coupling $L_8^r(\mu)$ at the one-loop level, i.e., up to next-to-leading order in $1/N_C$.

1 The large–N_C limit

Quantum Chromodynamics (QCD) has provided a wide understanding of hadronic processes. At long distances, the theory becomes non-perturbative. An approach that has been found to be useful is to consider QCD in the limit of an infinite number of colors $N_C \to \infty$, keeping $N_C \alpha_s$ finite [1]. Assuming confinement, large–N_C QCD results equivalent to a theory with an infinite number of narrow-width mesons, with the matrix elements given by the tree-level amplitudes. Hadronic loops are suppressed by $1/N_C$.

A crucial ingredient in light-meson interactions is chiral symmetry. The QCD lagrangian with n_f massless flavors is invariant under the chiral group $SU(n_f)_L \otimes SU(n_f)_R$, which gets spontaneously broken into the vector subgroup $SU(n_f)_{L+R}$ [2]. Actually, in the large–N_C limit the symmetry group gets enlarged into $U(n_f)_L \otimes U(n_f)_R$ and it is spontaneously broken into $U(n_f)_{L+R}$, producing n_f^2 Nambu-Goldstone bosons [3]. Below the $\rho(770)$ vector resonance multiplet, the Goldstones are the only hadrons in the spectrum. Their interaction can be described through an effective field theory based on chiral symmetry, namely chiral perturbation theory (χPT) [2, 4]. Symmetry imposes stringent constraints on the structure of the low energy interaction, so it is an essential ingredient to recover the proper low-energy QCD behavior. Hence, the interactions between mesons (Goldstones and resonance) must be given by a chiral theory of resonances (RχT) [5,6].

Following Refs. [7–10], this talk proposes a systematic program to aboard the analysis of hadronic amplitudes beyond the leading order in $1/N_C$ (LO). The example of

the isovector SS-PP correlator is studied:

$$\Pi(t) = i \int dx^4 e^{iqx} \langle T\{J(x)J(0)^\dagger - J_5(x)J_5(0)^\dagger\} \rangle, \quad (1)$$

with $J = \bar{d}u$ and $J_5 = i\bar{d}\gamma_5 u$, and $t = q^2$. The calculation is carried on within the chiral limit.

2 A program for calculations at NLO in $1/N_C$

2.1 RχT lagrangian

Although the large–N_C spectrum contains an infinite number of hadrons, the Green-functions that are chiral order parameters are mainly governed the lightest states. In general, one truncates the tower of states and works under a minimal hadronical approximation (MHA) [11], keeping the minimal number of resonance multiplets enough to fulfill the short-distance constraints. Though the truncation of the spectrum induces uncertainties [12,13], these can be estimated through the last absorptive channel included the calculation [7].

The particles included in our meson lagrangian are the chiral Goldstones and the lightest 1^{--}, 1^{++}, 0^{++}, 0^{--} resonance multiplets. Since we work within the large N_C framework, the hadrons are classified into $U(n_f)$ multiplets.

The hadronic lagrangian contains all the available operators consistent with chiral symmetry. Our building blocks are the resonance fields and the chiral tensors containing the Goldstones [5,6]. For the spin–1 fields we use the antisymmetric tensor formalism [5,14]. In addition, in order to avoid a wrong growing behavior of the Green-functions at high energies, the operators only contain tensors up to $\mathcal{O}(p^2)$. The operators of the lagrangian can be

---

[a] e-mail: `cillero@th.phy.pku.edu.cn`

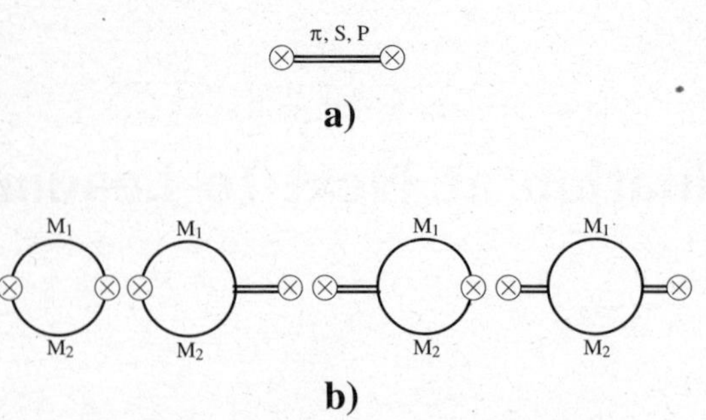

Fig. 1. Absorptive contributions for the SS-PP correlator: Tree-level meson exchange (a), and one-loop diagrams (b).

organized on the number of resonance fields:

$$\mathcal{L}_{R\chi T} = \mathcal{L}_{\chi PT}^{(2)} + \sum_{R_1} \mathcal{L}_{R_1} + \sum_{R_1,R_2} \mathcal{L}_{R_1,R_2} + \sum_{R_1,R_2,R_3} \mathcal{L}_{R_1,R_2,R_3} + ...$$

(2)

where the first term on the r.h.s. is the $\mathcal{O}(p^2)$ χPT lagrangian [2] and the second term, linear in the resonance fields, was long ago constructed in Ref. [5]. For the form-factors we are interested on (those with two mesons in the final state), only the operators with three or less resonance fields are relevant. They extra terms contain new couplings $\lambda_{R_1R_2}$, $\lambda_{R_1R_2R_3}$ [6,7].

2.2 Form-factors at large–N_C

In the large N_C limit the correlator is given by the tree-level exchange of Goldstones and resonances. At the one loop level one may find the two-meson absorptive diagrams shown in Fig. (1). The calculation is carried on within perturbation theory and no Dyson-Schwinger resummation is performed. Hence, all the lines in Fig. (1) stand for tree-level meson propagators.

At LO, the only absorptive cuts are one-particle cuts. At NLO, one may also have intermediate two-meson states. For a particular two-particle cut M_1M_2, its contribution to the spectral function is in general proportional to some form factor squared:

$$\mathrm{Im}\Pi(t)_{M_1M_2} \propto \left| \mathcal{F}_{M_1M_2}(t) \right|^2 .$$

(3)

By means of quark-counting rule arguments [15], it is usually accepted that the pion scalar form factor vanishes at infinite momentum. It accepts an unsubtracted dispersion relation which leads to the usual monopolar expression $\mathcal{F}_{\pi\pi}(t) = \frac{M_S^2}{M_S^2 - t}$ and to a contribution $\mathrm{Im}\Pi(t)_{\pi\pi}$ to the spectral function which vanishes at $t \to \infty$..

Demanding each separate absorptive contribution $\mathrm{Im}\Pi(t)_{M_1M_2}$ to vanish at least as fast as $\mathrm{Im}\Pi(t)_{\pi\pi}$ leads to a series of constraints for the form-factors at LO in $1/N_C$. Furthermore, the $\pi\pi$ and $R\pi$ form factors become determined in terms of the resonance masses [7].

2.3 Derivation of Π(t)

It is known from the operator product expansion (OPE) [16] that the SS-PP correlator accepts an unsubtracted dispersion relation:

$$\Pi(t) = \frac{1}{2\pi i} \oint dt' \frac{\Pi(t')}{t' - t} = \frac{1}{\pi} \int_0^\infty dt' \frac{\mathrm{Im}\Pi(t')}{t' - t} .$$

(4)

At LO, the spectral function is given by a sum of delta functions centered on the meson masses. Within the MHA, this gives the large–N_C correlator

$$\Pi(t) = 2B_0^2 \left[\frac{8c_m^2}{M_S^2 - t} - \frac{8d_m^2}{M_P^2 - t} + \frac{F^2}{t} \right] .$$

(5)

At NLO, the spectral function contains as well finite two-particle contributions $\mathrm{Im}\Pi(t)_{M_1M_2}$, related to the two-meson form-factors $\mathcal{F}_{M_1M_2}(t)$. By means of Eq. (4), one finds that the two-meson cuts contribute to the correlator with a finite part, $\Delta\Pi(t)_{M_1M_2}$, and a NLO renormalization of the scalar and pseudo-scalar masses and couplings. Hence, the whole correlator up to NLO shows the general structure:

$$\Pi(t) = 2B_0^2 \left[\frac{8c_m^{r\,2}}{M_S^{r\,2} - t} - \frac{8d_m^{r\,2}}{M_P^{r\,2} - t} + \frac{F^2}{t} \right] + \sum_{M_1M_2} \Delta\Pi(t)_{M_1M_2} ,$$

(6)

with the finite contribution from the M_1M_2 cut,

$$\Delta\Pi(t)_{M_1M_2} = \lim_{\epsilon \to 0^+} \left[\int_{\mathcal{R}_\epsilon} \frac{dt}{\pi} \frac{\mathrm{Im}\Pi(t')_{M_1M_2}}{t' - t} - \frac{2}{\pi\epsilon} \lim_{t' \to M_R^2} \left(\frac{(M_R^2 - t')^2 \, \mathrm{Im}\Pi(t')_{M_1M_2}}{t' - t} \right) \right] ,$$

(7)

with $R = S, P$ the corresponding s–channel resonance, and the interval $\mathcal{R}_\epsilon = [0, M_R^{r\,2} - \epsilon] \cup [M_R^{r\,2} + \epsilon, +\infty)$

It is interesting to recall that no new couplings are required after demanding an unsubtracted dispersion relation for $\Pi(t)$. To fix the correlator at NLO, one just needs to specify the value of the renormalized masses M_S^r, M_P^r and couplings c_m^r, d_m^r.

2.4 Matching OPE up to NLO in $1/N_C$

In the high energy limit, the two–meson contribution is found to behave as

$$\Delta\Pi(t) = \frac{F^2}{t} \delta_{\mathrm{NLO}}^{(1)} + \frac{F^2 M_S^2}{t^2} \left(\delta_{\mathrm{NLO}}^{(2)} + \widetilde{\delta}_{\mathrm{NLO}}^{(2)} \ln \frac{-t}{M_S^2} \right) ,$$

(8)

where the NLO constants $\delta_{\mathrm{NLO}}^{(1)}$, $\delta_{\mathrm{NLO}}^{(2)}$ and $\widetilde{\delta}_{\mathrm{NLO}}^{(2)}$ depend on the decay constant F and the resonance masses M_R [7].

The one-loop RχT correlator can be now matched to OPE in the deep euclidian region, finding similar expressions to the Weinberg sum-rules, but now containing extra terms, NLO in $1/N_C$:

$$-8c_m^{r\,2} + 8d_m^2 + F^2\left(1 + \delta_{\mathrm{NLO}}^{(1)}\right) = 0, \qquad (9)$$

$$-8c_m^{r\,2}M_S^{r\,2} + 8d_m^2 M_P^{r\,2} + F^2 M_S^2\, \delta_{\mathrm{NLO}}^{(2)} \simeq 0, \qquad (10)$$

where the dimension–4 OPE condensate is much smaller than each single term in Eq. (10) and it can be safely neglected [17]. The matching is fulfilled by demanding that the $\frac{1}{t^2}ln\frac{-t}{M_S^2}$ term also vanishes, this is, $\widetilde{\delta}_{\mathrm{NLO}}^{(2)} = 0$.

These relations allow fixing the resonance couplings up to NLO:

$$c_m^{r\,2} = \frac{F^2}{8}\,\frac{M_P^{r\,2}}{M_P^{r\,2} - M_S^{r\,2}}\left[1 + \delta_{(1)} - \frac{M_S^2}{M_P^2}\delta_{(2)}\right], \qquad (11)$$

$$d_m^{r\,2} = \frac{F^2}{8}\,\frac{M_S^{r\,2}}{M_P^{r\,2} - M_S^{r\,2}}\left[1 + \delta_{(1)} - \delta_{(2)}\right]. \qquad (12)$$

When considering just $\pi\pi$ and $R\pi$ cuts, one finds that after imposing the QCD short distance conditions everything becomes determined in terms of the renormalized masses M_R^r.

At low energies, the contribution from higher and higher thresholds becomes more and more suppressed. The two-resonance cuts are neglected in the present work. The uncertainty from the truncation is estimated from the $P\pi$ contribution, the higher threshold under consideration.

2.5 Recovery of χPT at low energies

One of the main advantages of working within a chiral invariant framework is the recovery of χPT at low energies even at the loop level. The one–loop RχT calculation exactly reproduces the one–loop χPT expression. This provides a prediction for the value of the renormalized low energy constant (LEC), $L_8^r(\mu)$, in terms of RχT parameters. The two expressions match at any μ and RχT generates the exact $L_8^r(\mu)$ running found in χPT [2]. There is not a specific saturation scale but a relation between renormalized LECs and renormalized RχT parameters.

In the low energy limit, the RχT expression can be expanded in powers of t:

$$\Pi(t) = B_0^2\left\{\frac{2F^2}{t} + 32\bar{L}_8^{U(3)} + \frac{3}{16\pi^2}\left(1 - \ln\frac{-t}{M_S^2}\right) + \mathcal{O}(t)\right\}, \qquad (13)$$

with the constant

$$\bar{L}_8^{U(3)} = \frac{F^2}{16}\left[\frac{1}{M_S^{r\,2}} + \frac{1}{M_P^{r\,2}}\right] \qquad (14)$$

$$\times \left\{1 + \delta_{\mathrm{NLO}}^{(1)} - \frac{M_S^{r\,2}}{M_S^{r\,2} + M_P^{r\,2}}\delta_{\mathrm{NLO}}^{(2)}\right\} - \frac{3\,\Delta}{256\pi^2}.$$

The $\mathcal{O}(1)$ constant Δ is given in Ref. [7]. It comes from the two-particle contribution $\Delta\Pi(t)_{\mathrm{M_1 M_2}}$ and is a function of F and M_R.

Comparing this result with $U(3) - \chi$PT [4], one gets a prediction for the renormalized LEC $L_8^r(\mu)$:

$$L_8^r(\mu)_{U(3)} = \bar{L}_8^{U(3)} - \frac{3}{512\pi^2}\ln\frac{\mu^2}{M_S^2}. \qquad (15)$$

The last step is to integrate out the chiral singlet η_0. In the large–N_C limit, the η_0 is the ninth Goldstone and it is massless [3]. However, it gains mass due to higher order corrections. Naively, one would expect that the effect of the η_0 mass in the one-loop diagrams would be next-to-next-to-leading order, this is, suppressed by $\frac{1}{N_C^2}$. Actually, since we study an energy limit below the η_0 threshold ($t \ll M_0^2$), the first effect from the η_0 mass appears at order $\frac{1}{N_C}\ln\frac{1}{N_C}$. Thus, the $SU(3) - \chi$PT constant is finally related to the $U(3)$ prediction through [4,7]

$$L_8^r(\mu)_{SU(3)} = L_8^r(\mu)_{U(3)} - \frac{1}{384\pi^2}\ln\frac{M_{\eta_0}^2}{\mu^2}. \qquad (16)$$

3 Conclusions

One of the main advantages of a chirally invariant theory of resonances is that the symmetry properties ensures the right recovery of the QCD low energy limit, χPT, even at the loop level.

The absorptive χPT logarithms are exactly reproduced by our result at long distances. This removes the large–N_C ambiguity about the renormalization scale of saturation of $L_8^r(\mu)$. The renormalized chiral coupling is given in terms of the renormalized resonance effective parameters c_m^r, d_m^r, M_S^r, M_P^r.

The systematic $1/N_C$ expansion within the RχT framework allows to derive Weinberg sum-rules beyond the leading order. This fixes the value of the renormalized scalar and pseudo-scalar couplings in terms of the renormalized resonance masses.

Talk given at QNP'06, Madrid, Spain. Further details can be found in Ref. [7]. Work done in collaboration with I. Rosell and A.Pich. It has been partially supported by EU RTN Contract CT2002-0311 and China National Natural Science Foundation under grant number 10575002 and 10421503. I want to acknowledge the organizers of the conference for their work and all the attentions received.

References

1. G. 't Hooft, *Nucl. Phys.* B **72** (1974) 461; **75** (1974) 461; E. Witten, *Nucl. Phys.* B **160** (1979) 57.
2. J. Gasser and H. Leutwyler, it Nucl. Phys. B **250** (1985) 465, 517, 539.
3. S. R. Coleman and E. Witten, *Phys. Rev. Lett.* **45** (1980) 100.
4. R. Kaiser and H. Leutwyler, Eur. Phys. J. C **17** (2000) 623; P. Herrera-Siklody *et al.*, Nucl. Phys. B **497** (1997) 345.

5. G. Ecker *et al.*, *Nucl. Phys.* B **321** (1989) 311.

6. V. Cirigliano *et al.*, *Nucl. Phys.* B **753** (2006) 139-177.

7. I. Rosell *et al.*, [ArXiv:hep-ph/0610290]; I. Rosell, J.J. Sanz-Cillero and A. Pich, to appear.

8. O. Catà and S. Peris, *Phys. Rev.* D **65** (2002) 056014.

9. I. Rosell *et al.*, *JHEP* **0408**, 042 (2004).

10. I. Rosell *et al.*, *JHEP* **0512**, 020 (2005).

11. M. Knecht and E. de Rafael, Phys. Lett. B **424**, 335 (1998).

12. J.J. Sanz-Cillero, *Nucl. Phys.* B **732** (2006) 136-168.

13. M. Golterman and S. Peris, [ArXiv:hep-ph/0607152].

14. G. Ecker *et al.*, *Phys. Lett.* B **223** (1989) 425.

15. G.P. Lepage and S.J. Brodsky, Phys. Lett. B **87** (1979) 359; Phys. Rev. D **22** (1980) 2157; Phys. Rev. D **24** (1981) 1808.

16. M. A. Shifman *et al.*, *Nucl. Phys.* B **147** (1979) 385.

17. A. Pich, [ArXiv:hep-ph/0205030].

Some comments on calculations of the scalar radius of the pion and the chiral constant $\bar{l}_4$

F. J. Ynduráin

Departamento de Física Teórica, C-XI, Universidad Autónoma de Madrid, Spain

Abstract. The pion scalar radius is given by $\langle r_S^2 \rangle = (6/\pi) \int_{4M_\pi^2}^{\infty} dt\, \delta_S(t)/t^2$, with δ_S the phase of the scalar form factor. Below $\bar{K}K$ threshold, $\delta_S = \delta_0$, δ_0 being the isoscalar, S-wave $\pi\pi$ phase shift. Between $\bar{K}K$ threshold and $t^{1/2} \sim 1.5\,\mathrm{GeV}$ I argued, in two previous letters [Ynduráin, Phys. Lett. B **578**, 99 and (E) **586**, 439 (2004); Phys. Lett. B **612**, 245 (2005)], that one can approximate $\delta_S \sim \delta_0$, because inelasticity is small, compared with the errors. This gives $\langle r_S^2 \rangle = 0.75 \pm 0.07\,\mathrm{fm}^2$ and the value $\bar{l}_4 = 5.4 \pm 0.5$ for the one-loop chiral perturbation theory constant, compared with the values given by Leutwyler and collaborators, $\langle r_S^2 \rangle = 0.61 \pm 0.04\,\mathrm{fm}^2$ and $\bar{l}_4 = 4.4 \pm 0.3$. At high energy, $t^{1/2} > 1.5\,\mathrm{GeV}$, I remarked that the value of δ_S that follows from perturbative QCD agrees with my interpolation and disagrees with that of Leutwyler and collaborators. In a recent article, Caprini, Colangelo and Leutwyler [Int. J. Mod. Phys. A **21**, 954 (2006)] claim that my estimate of the asymptotic phase δ_S is incorrect as it neglects higher twist contributions. Here I remark that, when correctly calculated, higher twist contributions are likely negligible. I also show that chiral perturbation theory gives $\bar{l}_4 = 6.60 \pm 0.43$, compatible with my estimate but widely off the value $\bar{l}_4 = 4.4 \pm 0.3$ of Leutwyler and collaborators. The results referred to have been published in F. J. Ynduráin, *The theory of quark and gluon interactions* (Springer-Verlag, August 2006) p. 239 and pp. 336, 337. The full text of the contribution may be found, with full references, in hep-ph/0510317.

PACS.

Eur. Phys. J. A **31**, 560–565 (2007)

DOI 10.1140/epja/i2006-10239-9

THE EUROPEAN
PHYSICAL JOURNAL A

Special Article – QNP 2006

Nucleon elastic form factors

Current status of the experimental effort

D. Day[a]

Department of Physics, University of Virginia, Charlottesville, VA 22904, USA

Received: 8 November 2006

Published online: 6 March 2007 – © Società Italiana di Fisica / Springer-Verlag 2007

Abstract. The nucleon form factors are still the subject of active investigation even after an experimental effort spanning 50 years. This is because they are of critical importance to our understanding of the electromagnetic properties of nuclei and provide a unique testing ground for QCD motivated models of nucleon structure. Progress in polarized beams, polarized targets and recoil polarimetry have allowed an important and precise set of data to be collected over the last decade. I will review the experimental status of elastic electron scattering from the nucleon along with an outlook for future progress.

PACS. 14.20.Dh Protons and neutrons – 13.40.Gp Electromagnetic form factors – 24.70.+s Polarization phenomena in reactions

1 Introduction

The experimental and theoretical study of the nucleon elastic form factors that began more than 50 years ago has returned to an examination of its roots —the Rosenbluth formula and the validity of one-photon approximation on which it depends. Deviations from this approximation are being examined to understand how they might alter our analysis of past, current and future electron scattering data. The Rosenbluth formula, which describes the elastic electro-nucleon cross-section in terms of the Pauli and Dirac form factors and is valid only in the one-photon approximation, had been considered unassailable until the appearance of high-Q^2 polarization transfer data on the proton from Jefferson Lab.

In the single-photon exchange, the Rosenbluth formula for the elastic cross-section is written (with F_1 and F_2 the Dirac and Pauli form factors, respectively, and which are functions of momentum transfer, $Q^2 = 4E_0 E' \sin^2(\theta/2)$ alone) as

$$\frac{\mathrm{d}\sigma}{\mathrm{d}\Omega} = \sigma_{\mathrm{Mott}} \frac{E'}{E_0} \left\{ (F_1)^2 + \tau \left[2 (F_1 + F_2)^2 \tan^2(\theta) + (F_2)^2 \right] \right\}, \quad (1)$$

where $\tau = \frac{Q^2}{4M^2}$, θ is the electron scattering angle, E_0, E' are the incident and final electron energies, respectively, and σ_{Mott} is the Mott cross-section. Because of their direct relation (in the Breit frame) to the Fourier transforms of the charge and magnetization distributions in the nucleon, the Sachs form factors are commonly used. They

are linear combinations of F_1 and F_2: $G_E = F_1 - \tau F_2$ and $G_M = F_1 + F_2$. Early measurements of the form factors established a scaling law relating three of the four nucleon elastic form factors and the dipole law describing their common Q^2-dependence, $G_E^p(Q^2) \approx \frac{G_M^p(Q^2)}{\mu_p} \approx \frac{G_M^n(Q^2)}{\mu_n} \approx G_D \equiv \left(1 + Q^2/0.71\right)^{-2}$. This behavior is known as form factor scaling. The neutron electric form factor has been usefully parametrized by $G_E^n = -\mu_N G_D \frac{\tau}{1+5.6\tau}$ [1].

2 Proton form factors

The proton form factors have been, until recently, only separated through the Rosenbluth technique, which can be understood by re-writing eq. (1) using the Sachs form factors,

$$\frac{\mathrm{d}\sigma}{\mathrm{d}\Omega} = \sigma_{\mathrm{NS}} \left[\frac{G_E^2 + \tau G_M^2}{1 + \tau} + 2\tau G_M^2 \tan^2(\theta/2) \right], \quad (2)$$

and rearranging, with $\epsilon^{-1} = 1 + 2(1 + \tau)\tan^2(\theta/2)$ and $\sigma_{\mathrm{NS}} = \sigma_{\mathrm{Mott}} E'/E_0$, to give

$$\sigma_{\mathrm{R}} \equiv \frac{\mathrm{d}\sigma}{\mathrm{d}\Omega} \frac{\epsilon(1 + \tau)}{\sigma_{\mathrm{NS}}} = \tau G_M^2(Q^2) + \epsilon G_E^2(Q^2). \quad (3)$$

By making measurements at a fixed Q^2 and variable $\epsilon(\theta, E_0)$, the reduced cross-section σ_{R} can be fit with a straight line with slope G_E^2 and intercept τG_M^2. Figure 1 gives an example of the reduced cross-section plotted in this way. The Rosenbluth formula holds only for single-photon exchange and it has been assumed (until recently) that any two-photon contribution is small.

[a] e-mail: dbd@virginia.edu

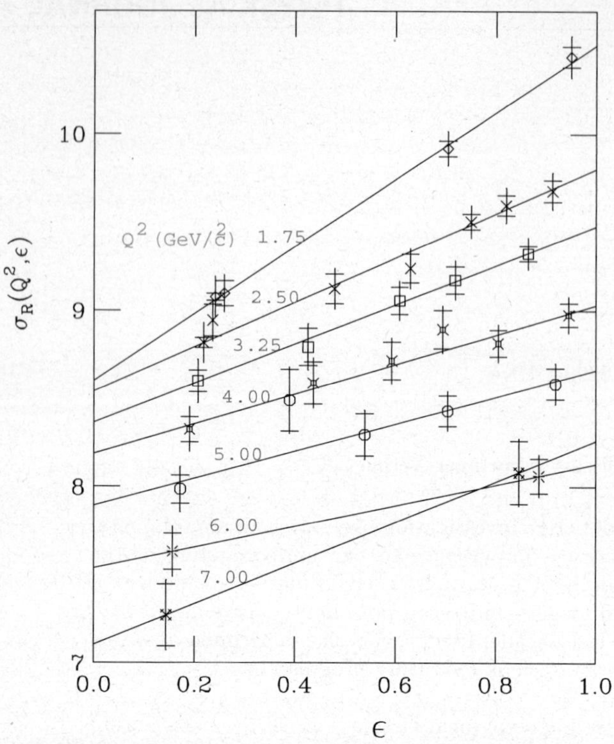

Fig. 1. Reduced cross-section plotted against ϵ for a range of fixed momentum transfers. The data is fit with a straight line with a slope G_E^2 and an intercept τG_M^2. The linear dependence of the reduced cross-section on ϵ assumes that any two-photon exchange effects are small. The data is from ref. [2].

The Rosenbluth method is problematic —it requires the measurement of absolute cross-sections and at large Q^2 the cross-section is insensitive to G_E and the error propagation is not favorable. Figure 2 presents the Rosenbluth data set. See ref. [3] for a compilation of the data, references, and useful fits. G_M^p has been successfully measured out to 30 $(\text{GeV}/c)^2$, while G_E^p begins to endure large errors at much smaller Q^2.

2.1 Neutron form factors

The neutron form factor data set (setting aside recent progress) has been inadequate in both quality and extent due to the lack of a free neutron target and the dominance of G_M^n over G_E^n. The traditional techniques (restricted to the use of unpolarized beams and targets) used to extract information about G_M^n and G_E^n have been: elastic scattering from the deuteron (D): $D(e, e')D$; inclusive quasielastic scattering: $D(e, e')X$; scattering from deuteron with the coincident detection of the scattered electron and recoiling neutron: $D(e, e'n)p$; a ratio method which minimizes uncertainties in the deuteron wave function and the role of FSI: $\frac{D(e, e'n)}{D(e, e'p)}$. The G_M^n data is shown in fig. 3. New data at large momentum transfers is available from an experiment [4–6] at Jefferson Lab, which used the ratio method to measure G_M^n with small errors out to nearly 5 $(\text{GeV}/c)^2$ in the CLAS.

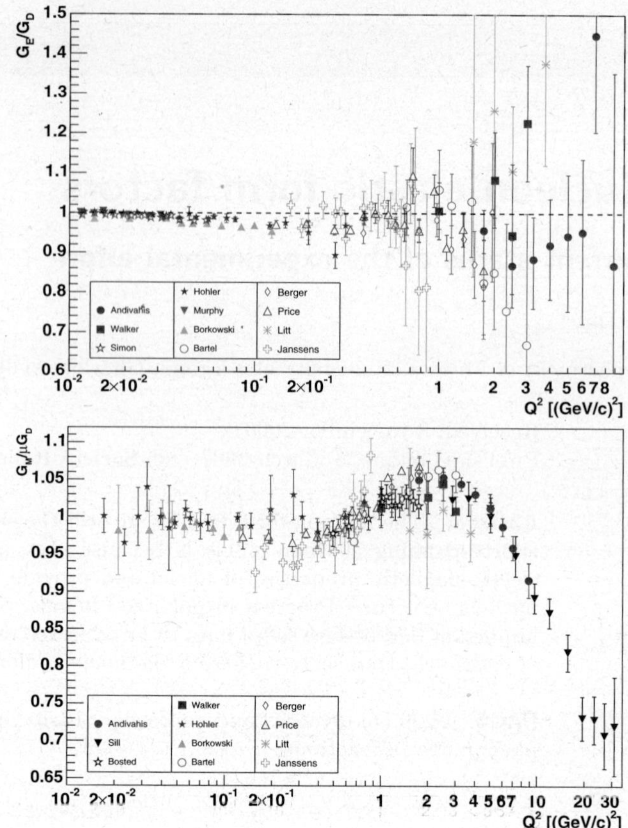

Fig. 2. G_E^p/G_D and $G_M^p/\mu_p/G_D$ *versus* Q^2 from the Rosenbluth method. The scaling law and the dipole law hold to a good approximation ($\approx 10\%$) for both form factors out to $Q^2 = 8$ $(\text{GeV}/c)^2$.

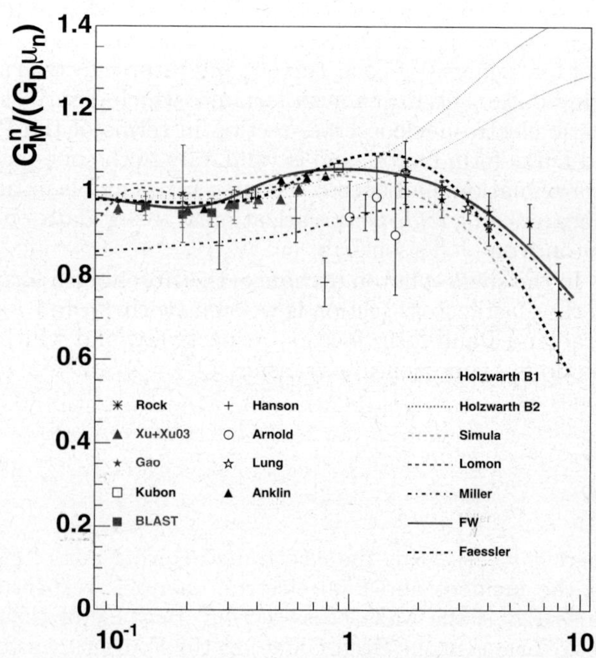

Fig. 3. G_M^n from unpolarized scattering [7–12] and polarized scattering [13–15] along with some theoretical predictions.

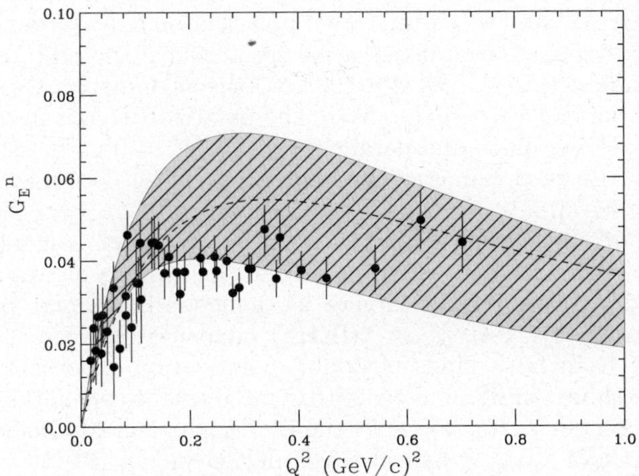

Fig. 4. G_E^n from elastic e-D [16]. The band represents the theoretical error associated with the extraction. The dashed line is the familiar Galster parametrization [1].

Until the early 1990s the extraction of G_E^n was done most successfully through either small-angle elastic e-D scattering [1,16] or by quasielastic e-D scattering [7]. In the Impulse Approximation (IA) the elastic electron-deuteron cross-section is the sum of proton and neutron responses with deuteron wave function weighting.

Experiments have been able to achieve small statistical errors but remain very sensitive to deuteron wave function model leaving a significant residual dependence on the NN potential. The most precise data on G_E^n from elastic e-D scattering are shown in fig. 4 from an experiment at Saclay, published in 1990 [16]. The band, a measure of the theoretical uncertainty, arises from the use of different NN potentials in the extraction.

3 Spin-dependent measurements

The nucleon electromagnetic form factors can be measured through spin-dependent elastic scattering from the nucleon, accomplished either through a measurement of the scattering asymmetry of polarized electrons from a polarized nucleon target [17,18] or equivalently by measuring the polarization transferred to the nucleon [19,20]. In the scattering of polarized electrons from a polarized target, an asymmetry appears in the elastic scattering cross-section when the beam helicity is reversed. In contrast, in scattering a polarized electron from an unpolarized target, the transferred polarization to the nucleon produces an azimuthal asymmetry in the secondary scattering of the nucleon (in a polarimeter) due to its dependence on the polarization. In both cases, the asymmetry is sensitive to the product $G_E G_M$. In the last decade experiments exploiting these spin degrees of freedom have become possible.

Extraction of the neutron form factors (necessarily from a nuclear target) using polarization observables is complicated by the need to account for Fermi motion, MEC, and FSI, complications that are absent when scattering from a proton target. Fortunately, it has been found for the deuteron that in kinematics that emphasize quasifree neutron knockout both the transfer polarization P_t [21] and the beam-target asymmetry A_V^{eD} [22] are especially sensitive to G_E^n and relatively insensitive to the NN potential describing the ground state of the deuteron and other reaction details. Calculations [23] of the beam-target asymmetry from a polarized ^{3}He target (which can be approximated as a polarized neutron) showed modest model dependence.

3.1 Recoil polarization

In elastic scattering of polarized electrons from a nucleon, the nucleon obtains (is transferred) a polarization whose components, P_l (along the direction of the nucleon momentum) and P_t (perpendicular to the nucleon momentum) are proportional to G_M^2 and $G_E G_M$, respectively. The recoil polarization technique has allowed precision measurements of G_E^p to nearly 6 $(\text{GeV}/c)^2$ [24–26] and of G_E^n out to $Q^2 = 1.5$ $(\text{GeV}/c)^2$ [27–30]. Polarimeters are sensitive only to the perpendicular polarization components so precession of the nucleon spin before the polarimeter in the magnetic field of the spectrometer (for the proton) or a dipole (inserted in the path of neutron) allows a measurement of the ratio P_t/P_l and the form factor ratio: $\frac{G_E}{G_M} = -\frac{P_t}{P_l}\frac{(E_0+E')}{2M_N}\tan(\theta/2)$.

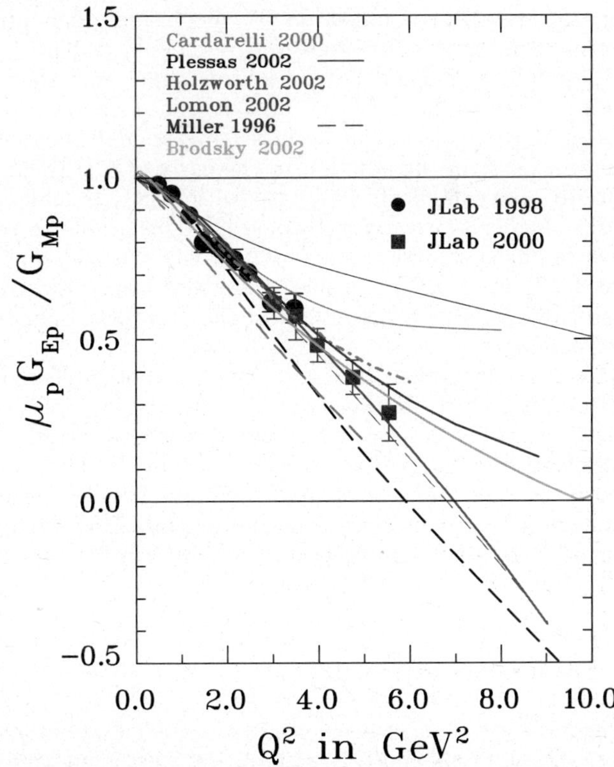

Fig. 5. (Color online) Comparison of theoretical model calculations with the data from ref. [24] (solid circles) and from [26] (solid squares). The curves are, black thin solid [31], green solid, dot-dashed and dashed [32], black dashed [33], red solid [34], yellow solid [35] and magenta dashed and solid [36].

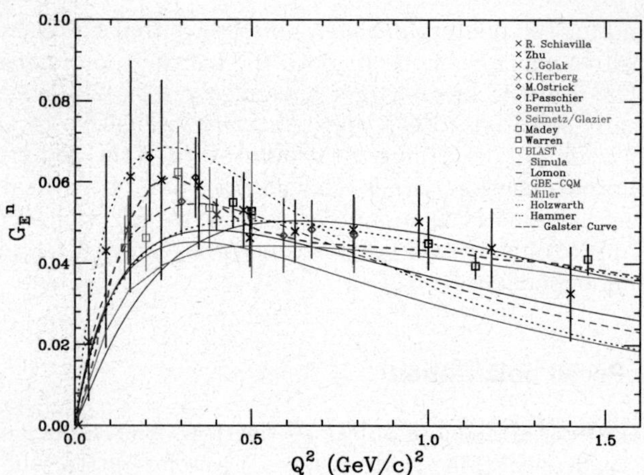

Fig. 6. (Color online) Comparison of selected theoretical model calculations with the data on G_E^n from polarized experiments. Starting at the top of the legend the data are from [37–39,28,29,40–42,30,43,44]. The neutron, at large momentum transfers, has the same Q^2-dependence as the proton. The red dashed line is the Friedrich and Walcher [45] fit to the data.

The results from Jefferson Lab for the proton are shown in fig. 5 where it is seen that the ratio of $\mu_p G_E^p / G_M^p$ does not follow the scaling law obtained from Rosenbluth separation, rather showing a steep decline with increasing Q^2. This suggests that the distribution of magnetization and charge densities in the proton are dissimilar. Shown with the data are a collection of calculations including several relativistic constituent-quark models (rQCM), a VMD-pQCD model and a chiral soliton model. Also shown is a pQCD calculation [35]. The same data, in terms of $Q^2 F_2 / F_1$, gives no indication of scaling at high momentum transfer, in contradiction to the early pQCD prediction [46]. Recent efforts [35], still within pQCD and including higher twist contributions, have been able to reproduce this behavior. Other pQCD calculations which consider quark angular orbital momentum are also successful [47,48]. For a discussion of the theoretical curves see ref. [49].

Recoil polarization has been used at both Jefferson Lab and Mainz to extract G_E^n / G_M^n when scattering polarized electrons from an unpolarized deuteron target in quasielastic kinematics. At both labs a dipole magnet was used to precess the neutron spin thereby allowing a measurement of both polarization components. The results from [30] are especially precise, extending our knowledge of G_E^n out to 1.5 GeV/c^2. See fig. 6.

3.2 Beam-target asymmetry

Polarized targets have been used to extract G_E^n [50,40, 51,41,42,39,38,43] and G_M^n [52,13–15]. The beam-target asymmetry can be written schematically (a, b, c, and d are known kinematic factors) as

$$A = \frac{a \cos \Theta^\star \left(G_M \right)^2 + b \sin \Theta^\star \cos \Phi^\star G_E G_M}{c \left(G_M \right)^2 + d \left(G_E \right)^2} \quad (4)$$

where $\Theta^\star$ and $\Phi^\star$ fix the target polarization axis. With the target polarization axis in the scattering plane and perpendicular to q, $(\Theta^\star, \Phi^\star = 90°, 0°)$ the asymmetry A_{TL} is proportional to $G_E G_M$. With the polarization axis in the scattering plane and parallel to q $(\Theta^\star, \Phi^\star = 0°, 0°)$, measuring the asymmetry A_T allows G_M to be determined. See fig. 3.

G_E^n has been extracted from beam-target asymmetry measurements using polarized ^{3}He targets at Mainz and polarized ND_3 targets at Jefferson Lab, and polarized gas targets at NIKHEF and Bates. Data for G_E^n from both kinds of double-polarization experiments are shown in fig. 6 along with some relevant calculations. The models, starting with the first in the legend, include a rCQM [53], a hybrid VMD-pQCD model [34], a relativistic CQM calculation [31], a light-front cloudy bag model [54], a soliton model [36], and a dispersion theory calculation [55]. While most of these calculations describe the Q^2-dependence, several badly fail to reproduce the slope at $Q^2 = 0$, $\frac{dG_E^n(0)}{dQ^2} = -\frac{1}{6} \langle r_E^2 \rangle$. The neutron charge radius, r_E, has been determined through neutron electron scattering [56].

4 Pion cloud

The neutron has long been thought to exist, part of the time, as a proton core surrounded by a negative pion cloud. This idea has appeared in models describing the nucleon form factors, ref. [57] and ref. [58], and has been recently re-emphasized by Friedrich and Walcher [45]. These authors fit all form factors consistently as a sum of a broad distribution and a "bump", where the "bump" is attributed to a π-cloud. The bump shows up in all 4 form factors at $Q^2 \simeq 0.25$, and is seen most clearly in G_E^n and in fig. 6 as the red online dashed line. This feature also shows up in the densities determined by Kelly in ref. [59].

Reinforcing this view this is a recent calculation within a chiral quark model by the Tuebingen group [60] that treats the baryons as bound states of constituent quarks, dressed by a cloud of mesons and which gives an excellent description of all four of the electromagnetic form factors.

5 The G_E^p / G_M^p discrepancy and two-photon effects

The recoil polarization measurements of the form factor ratio $\mu_p G_E^p / G_M^p$ contradict the Rosenbluth measurements and it has been suggested that the earlier experiments might have underestimated systematic errors or suffer from normalization problems. The Rosenbluth measurements have been re-examined [61]. This global reanalysis could find no systematic or normalization problems that could account for the discrepancy and concluded that a 5–6% linear ϵ-dependence correction (of origin yet unknown) to the cross-section measurements is required to explain the difference. Several investigators [62–65] have explored the possibility of two-photon exchange corrections (which would be less important in the direct ratio

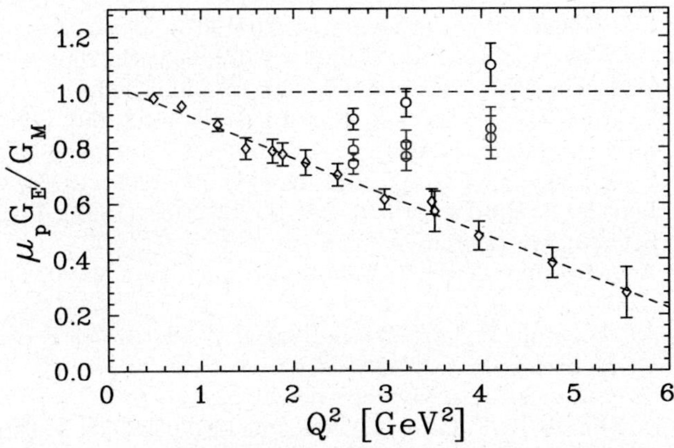

Fig. 7. Corrections to the three data points from the super Rosenbluth data set move those data towards (*i.e.* reduce the value of the form factor ratio) the recoil polarization data set shown with fit (dashed line). Starting with the largest value at each Q^2 is the measured ratio, the ratio with the two-photon corrections of ref. [65] applied and the ratio with the two-photon and with Coulomb corrections [66] applied [67].

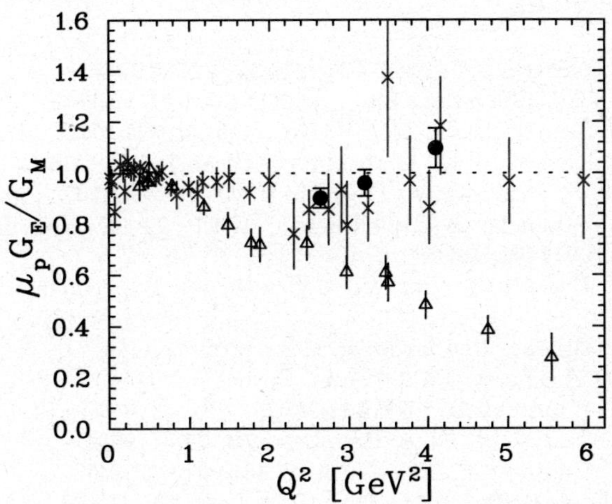

Fig. 8. (Color online) Proton form factor ratio where the (blue) triangles are from recoil polarization [24,26], (red) crosses from the reanalysis of the world's Rosenbluth data set [61] and the filled circles from the recent super Rosenbluth experiment [69].

measurement of recoil polarization) to explain the discrepancy. While only incomplete calculations exist, the results of refs. [62,65] account for part of the difference.

The most recent work by Chen *et al.* [65] employed a different approach than that of ref. [62] in that they describe the process in terms of hard scattering from a quark and use GPDs to describe the quark emission and absorption. They argue that when taking the recoil polarization form factors as input, the addition of the two-photon corrections reproduces the Rosenbluth data. However, Arrington [67] has shown that when the corrections of Chen *et al.* are applied to the new Jefferson Lab Rosenbluth data, which have small errors (see below and fig. 7) only one-half of the discrepancy is explained.

The Rosenbluth-polarization transfer discrepancy has been recently confirmed by a "super" Rosenbluth measurement [68] at Jefferson Lab that was designed to minimize the systematic errors that handicap Rosenbluth measurements. This was achieved by detecting the proton rather than the electron in elastic kinematics. In doing so many of the extreme rate variations and cross-section sensitivities that are normally encountered were avoided. The results [69] show that the discrepancy still exists. See fig. 8.

Direct tests for the existence of two-photon exchange include measurements of the ratio $\frac{\sigma(e^+p)}{\sigma(e^-p)}$, where the real part of the two-photon exchange amplitude leads to an enhancement, and in Rosenbluth data where it can lead to non-linearities in ϵ. There is no experimental evidence of non-linearities in the Rosenbluth data and the e^+/e^- ratio data [70] are of only modest precision, making it difficult to absolutely confirm the presence of two-photon effects in these processes.

It is the imaginary part of the two-photon amplitude that can lead to single-spin asymmetries but again the existing data [71,72] are of insufficient precision to allow one to make a statement. There is, however, one observable that has provided unambiguous evidence for a two-photon effect in ep elastic scattering. Groups in both the US [73] and Europe [74] have measured the transverse polarized beam asymmetry. These measurements are significant but have limited utility in solving the G_E^p discrepancy. The reader interested in more detail about the existence of two-photon effects and their role on the form factor measurements should refer to ref. [67].

Fortunately experiments are planned at Jefferson Lab to look for non-linearities in the Rosenbluth data, for the presence of induced recoil polarization and for an enhancement in the e^+/e^- ratio. We can expect that a concentrated effort in both experiment and theory will reveal the full extent of two-photon effects in the not too distant future.

6 Outlook

A recently completed experiment [75] at Jefferson Lab using a polarized ^{3}He target will provide data on G_E^n out to 3.5 $(\text{GeV}/c)^2$. Next year, the recoil technique will extend the measurement [76] of G_E^p/G_M^p out to 9 $(\text{GeV}/c)^2$. With the 12 GeV upgrade and improvements in targets and recoil polarimeters it anticipated that these quantities can be measured out to 8 and 12 $(\text{GeV}/c)^2$, respectively. Similarly data on G_M^n as high as 14 $(\text{GeV}/c)^2$ can be measured in the upgraded CLAS.

The capabilities of high duty factor accelerators, polarized beams and targets, and polarimeters have produced precision data out to large momentum transfer on the proton and neutron form factors. These new data, and that which will be earned in the next generation of experiments will continue to challenge our view of the structure of the proton and neutron and provide rigorous tests for any QCD-inspired model of nucleon structure.

References

1. S. Galster *et al.*, Nucl. Phys. B **32**, 221 (1971).
2. L. Andivahis *et al.*, Phys. Rev. D **50**, 5491 (1994).
3. P. Bosted, Phys. Rev. C **51**, 409 (1995); E.J. Brash, A. Kozlov, S. Li, G.M. Huber, Phys. Rev. C **65**, 05100 (2002); J.J. Kelly, Phys. Rev. C **70**, 068202 (2004); V.V. Ezhela, B.V. Polishchuk, IHEP 99-48, arXiv:hep-ph/9912401, 1999.
4. W. Brooks, M. Vineyard, Jefferson Lab Experiment E94-107.
5. W. Brooks, J.D. Lachniet, these proceeedings.
6. J.D. Lachniet, PhD Thesis, Carnegie Mellon University, 2005, unpublished UMI-31-86035.
7. A. Lung *et al.*, Phys. Rev. Lett. **70**, 718 (1993).
8. H. Anklin *et al.*, Phys. Lett. B **336**, 313 (1994).
9. E.E.W. Bruins *et al.*, Phys. Rev. Lett. **75**, 21 (1995).
10. H. Anklin *et al.*, Phys. Lett. B **428**, 248 (1998).
11. G. Kubon, H. Anklin *et al.*, Phys. Lett. B **524**, 26 (2002).
12. P. Markowitz *et al.*, Phys. Rev. C **48**, R5 (1993).
13. W. Xu *et al.*, Phys. Rev. Lett. **85**, 2900 (2000).
14. W. Xu *et al.*, Phys. Rev. C **67**, 012201 (2003).
15. N. Meitanis, PhD Thesis, MIT, March 2006, unpublished.
16. S. Platchkov *et al.*, Nucl. Phys. A **510**, 740 (1990).
17. N. Dombey, Rev. Mod. Phys. **41**, 236 (1969).
18. T.W. Donnelly, A.S. Raskin, Ann. Phys. (N.Y.) **169**, 247 (1986); **191**, 81 (1989).
19. A.I. Akhiezer, M.P. Rekalo, Sov. J. Part. Nucl. **3**, 277 (1974).
20. R.G. Arnold, C. Carlson, F. Gross, Phys. Rev. C **23**, 363 (1981).
21. H. Arenhövel, Phys. Lett. B **199**, 13 (1987).
22. H. Arenhövel, W. Leidemann, E.L. Tomusiak, Z. Phys A **331**, 123 (1988); **334**, 363 (1989)(E), Phys. Rev. C **46**, 455 (1992).
23. J. Golak *et al.*, Phys. Rev. C **63**, 034006 (2001); S. Ishikawa, J. Golak *et al.*, Phys. Rev. C **57**, 39 (1998).
24. M.K. Jones *et al.*, Phys. Rev. Lett. **84**, 1398 (2000).
25. O. Gayou *et al.*, Phys. Rev. C **64**, 038292 (2001).
26. O. Gayou *et al.*, Phys. Rev. Lett. **88**, 092301 (2002).
27. T. Eden *et al.*, Phys. Rev. C **50**, R1749 (1994).
28. C. Herberg *et al.*, Eur. Phys. J. A **5**, 131 (1999).
29. M. Ostrick *et al.*, Phys. Rev. Lett. **83**, 276 (1999).
30. R. Madey *et al.* Phys. Rev. Lett. **91**, 122002 (2003).
31. S. Boffi, L.Ya. Glozman, W. Klink, W. Plessas, M. Radici, R.F. Wagenbrunn, Eur. Phys. J. A **14**, 17 (2002).
32. E. Pace, G. Salme, F. Cardarelli, S. Simula, Nucl. Phys. A **666**, 33 (2000).
33. M.R. Frank, B.K. Jennings, G.A. Miller, Phys. Rev. C **54**, 920 (1996).
34. E.L. Lomon, Phys. Rev. C **66**, 045501 (2002).
35. S.J. Brodsky, arXiv:hep-ph/0208158, SLAC-PUB-9281.
36. G. Holzwarth, hep-ph/0201138; Z. Phys. A **356**, 339 (1996) (Fit B2 is shown for the neutron).
37. R. Schiavilla, I. Sick, Phys. Rev. C **64**, 041002 (2001).
38. H. Zhu *et al.*, Phys. Rev. Lett. **87**, 081801 (2001).
39. J. Golak *et al.*, Phys. Rev. C **63**, 034006 (2001).
40. I. Passchier *et al.*, Phys. Rev. Lett. **82**, 4988 (1999).
41. J. Bermuth *et al.*, Phys Lett. B. **564**, 199 (2003); D. Rohe *et al.*, Phys. Rev. Lett. **83**, 4257 (1999).
42. D.I. Glazier *et al.*, arXiv:nucl-ex/0410026.
43. G. Warren *et al.*, Phys. Rev. Lett. **92**, 042301 (2004).
44. V. Ziskin, PhD Thesis, MIT, 2005, unpublished.
45. J. Friedrich, T. Walcher, Eur. Phys. J. A **17**, 607 (2003) (arXiv:hep-ph/0303054).
46. S. Brodsky, G. Farrar, Phys. Rev. D **11**, 1309 (1975); G. Lepage, S. Brodsky, Phys. Rev. D **22**, 2157 (1980); Phys. Scr. **23**, 945 (1981).
47. A.V. Belitsky, X. Ji, F. Yuan, Phys. Rev. Lett. **91**, 092003 (2003).
48. J.P. Ralston, P. Jain, *Proceedings of Workshop on Testing QCD through Spin Observables in Nuclear Targets* (World Scientific, 2003) arXiv:hep-ph/0207129.
49. V. Punjabi, C.F. Perdrisat *et al.*, nucl-ex/0307001; V. Punjabi *et al.*, Phys. Rev. C **71**, 055202 (2005); **71**, 069902 (2005)(E) (arXiv:nucl-ex/0501018).
50. M. Meyerhoff *et al.*, Phys. Lett. B **327**, 201 (1994).
51. J. Becker *et al.*, Eur. Phys. J. A **6**, 329 (1999).
52. H. Gao *et al.*, Phys. Rev. C **50**, R546 (1994).
53. F. Cardarelli, S. Simula, Phys. Rev. C **62**, 065201 (2000); Phys. Lett. B **467**, 1 (1999).
54. G.A. Miller, Phys. Rev. C **66**, 032201(R) (2002).
55. H.W. Hammer, U.G. Meissner, Eur. Phys. J. A **20**, 469 (2004).
56. S. Kopecki *et al.*, Phys. Rev. Lett. **74**, 2427 (1995).
57. M.M. Kaskulov, P. Grabmayr, Eur. Phys. J. A **19**, 157 (2004) (arXiv:nucl-th/0308015).
58. G.A. Miller, Phys. Rev. C **66**, 032201 (2002) (arXiv:nucl-th/0207007).
59. J.J. Kelly, Phys. Rev. C **66**, 065203 (2002) (arXiv:hep-ph/0204239); AIP Conf. Proc. **698**, 393 (2004).
60. A. Faessler, T. Gutsche, V.E. Lyubovitskij, K. Pumsa-ard, Phys. Rev. D **73**, 114021 (2006) (arXiv:hep-ph/0511319).
61. J. Arrington, Phys. Rev. C **68**, 034325 (2003).
62. P.G. Blunden, W. Melnitchouk, J.A. Tjon, Phys. Rev. Lett. **91**, 142304 (2003).
63. M.P. Rekalo, E. Tomasi-Gustafsson, Eur. Phys. J. A **22**, 331 (2004).
64. P.A.M. Guichon, M. Vanderhaeghen, Phys. Rev. Lett. **91**, 142303 (2003).
65. Y.C. Chen, A. Afanasev, S.J. Brodsky, C.E. Carlson, M. Vanderhaeghen, Phys. Rev. Lett. **93**, 122301 (2004).
66. J. Arrington, I. Sick, Phys. Rev. C **70**, 028203 (2004).
67. J. Arrington, Phys. Rev. C **71**, 015202 (2005) (arXiv:hep-ph/0408261).
68. J. Arrington, R. Segel, Jefferson Lab Experiment E01-001.
69. I.A. Qattan *et al.*, Phys. Rev. Lett. **94**, 142301 (2005) (arXiv:nucl-ex/0410010).
70. J. Mar *et al.*, Phys. Rev. Lett. **21**, 482 (1968).
71. H.C. Kirkman *et al.*, Phys. Lett. B **32**, 519 (1970).
72. T. Powell *et al.*, Phys. Rev. Lett. **24**, 753 (1970).
73. SAMPLE Collaboration (S.P. Wells *et al.*), Phys. Rev. C **63**, 064001 (2001).
74. F.E. Maas *et al.*, Phys. Rev. Lett. **94**, 082001 (2005) (arXiv:nucl-ex/0410013).
75. B. Wojtsekhowski, G. Cates, spokespersons, Jefferson Lab Proposal PR02-013.
76. E. Brash, M. Jones, C. Perdrisat, V. Punjabi, spokespersons, Jefferson Lab Proposal PR04-108.

Eur. Phys. J. A **31**, 566–571 (2007)

DOI 10.1140/epja/i2006-10178-5

THE EUROPEAN
PHYSICAL JOURNAL A

Structure functions at HERA

D.H. Saxon[a]

Faculty of Physical Sciences, University of Glasgow, Glasgow, G12 8QQ, UK

Received: 24 September 2006
Published online: 16 February 2007 – © Società Italiana di Fisica / Springer-Verlag 2007

Abstract. HERA provides the key facility for the measurement of proton structure functions. Formalism and methods are outlined for the measurement and interpretation of inclusive structure functions, including the use of polarised $e^{\pm}$ beams. The measurement of charm, beauty and photon structure functions is discussed, together with special runs at low proton energy for measurement of the longitudinal structure function. Finally, the functions accessed using polarised beams on polarised targets are indicated.

PACS. 13.60.-r Photon and charged-lepton interactions with hadrons – 13.88.+e Polarisation in interactions and scattering – 14.65.-q Quarks – 14.70.Dj Gluons

1 Introduction

Deep inelastic scattering of $e^{\pm}$ on protons takes place via the exchange of a gauge boson, either neutral-current, (NC: γ or Z^0 exchange), or charged-current (CC: $W^{\pm}$ exchange, in which case the outgoing lepton is a neutrino or antineutrino). Defining k and k' as the incident and outgoing lepton four-momenta, then the exchanged boson carries four-momentum $q = k - k'$. We define the $Q^2 = -q^2 = -(k - k')^2$. If the incident proton four-momentum is p, then the centre-of-mass energy squared, $s = (p + k)^2 = 4E_e E_p$ (neglecting the rest masses). For $27.5 \,\mathrm{GeV}$ $e^{\pm}$ incident on $920 \,\mathrm{GeV}$ protons, $s = 101200 \,\mathrm{GeV}^2$ and the c.m. energy is $318 \,\mathrm{GeV}$. When $Q^2 \simeq 0$ the exchanged photon is quasi-real and the process is refered to as photoproduction. The allowed range is $0 < Q^2 < s$. For virtual photons $Q^2 \neq 0$ and the process is called deep inelastic scattering (DIS).

The reaction is seen as a hard collision between the exchanged boson and a parton within the proton carrying a fraction $x = Q^2/2p \cdot q$ of the proton's momentum. x lies between 0 and 1, and the distribution of partons as a function of x is unknown *a priori*, although its evolution as a function of Q^2 is predicted [1]. The final state in lowest order contains an outgoing lepton, a jet from the struck parton and a forward-going proton remnant.

It is useful to introduce the quantity $y = p \cdot q/p \cdot k$. Then $y = \sin^2 \theta/2$, where θ is the scattering angle in the lepton-parton centre of mass. Distributions in y are therefore influenced by the helicity structure of the process. Note that $0 < y < 1$ and that $Q^2 = sxy$.

For unpolarised $e^{\pm}p$ NC DIS the differential cross-section is given by

$$\frac{\mathrm{d}^2\sigma}{\mathrm{d}x\mathrm{d}Q^2} = \frac{2\pi\alpha^2}{xQ^4}\big[Y_+ F_2(x, Q^2) - y^2 F_L(x, Q^2) \mp Y_- xF_3(x, Q^2)\big],$$

where $Y_{\pm} = 1 \pm (1 - y)^2$. The structure functions, F_i, are interpreted in terms of the quark $(q, \bar{q})$, and gluon (g), parton structure at a given (x, Q^2) as follows:

- $F_2 \simeq \Sigma x(q + \bar{q})$: dominates;
- $xF_3 \simeq \Sigma x(q - \bar{q})$: contributes at high Q^2;
- $F_L \simeq \alpha_s xg$: contributes at high y, absent in zeroth order QCD.

Thus measurement of the DIS rate gives direct sensitivity to quark distributions but only indirect sensitivity to gluons —F_L contributes typically a few percent at the highest y-bins. Gluons in the proton are accessed via the Q^2 evolution ($\mathrm{d}F_2/\mathrm{d}\ln Q^2$) or through measurements of the dijet rate, caused by photon-gluon fusion, $\gamma^* g \to q\bar{q}$, where the gluon is part of the incident photon structure (as well as by hard gluon bremsstrahlung from outgoing quarks, $\gamma^* q \to qg$).

The precision measurement of structure functions is interesting in its own right, and also provides the input needed for the calculation of every hard process at the LHC. All such interpretations of data at LHC will depend on the accuracy achieved at HERA:

2 Measurement of F$_2$ and parton densities

F_2 has been measured over the range $(0.00005 < x < 1, \; 0.5 < Q^2 < 30000)$ using NC data. (See fig. 1.) As

---

[a] e-mail: `d.saxon@physics.gla.ac.uk`

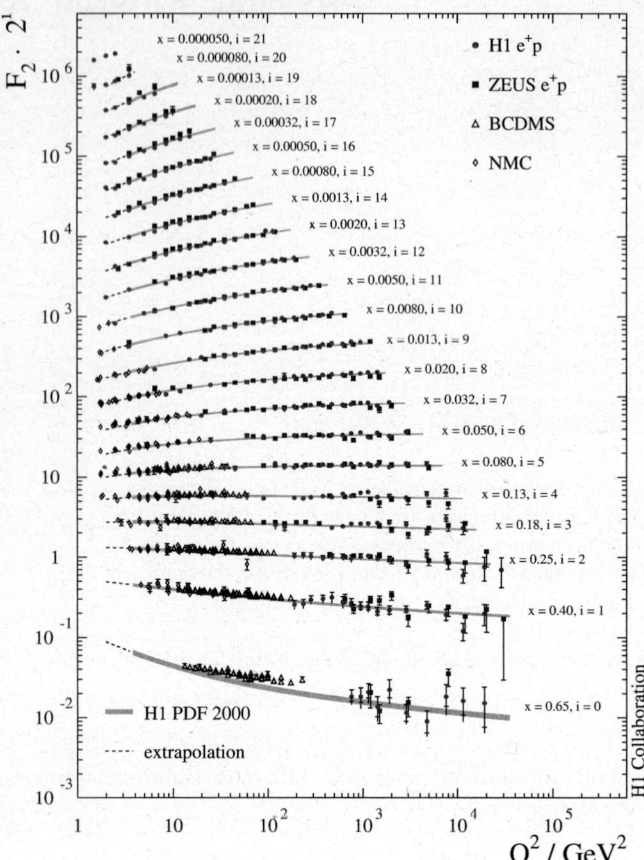

Fig. 1. Measurements of F_2.

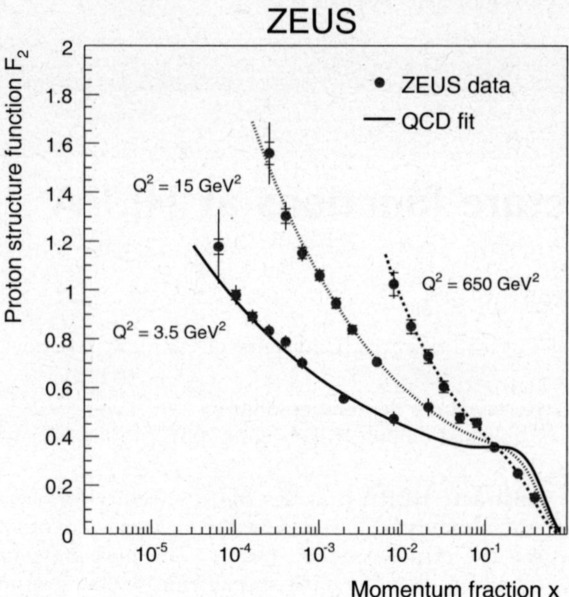

Fig. 2. Low-x behaviour of F_2 at different values of Q^2.

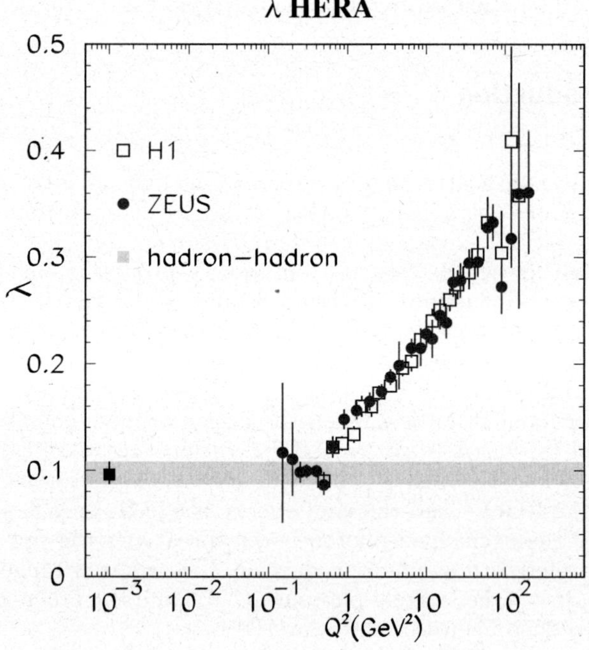

Fig. 3. Power λ for $F_2 \simeq x^{-\lambda}$ low-x fit.

a function of Q^2, F_2 falls for $x > 0.15$ and rises steeply for $x < 0.01$. (See fig. 2.) The high-x shape relates to the valence quark structure of the proton. At low x we are observing the universal structure of QCD radiation: at fixed Q^2 and low x, $F_2(x, Q^2) \simeq x^{-\lambda}$, with λ varying linearly with $\log Q^2$ from 0.1 for $Q^2 = 1$ to 0.35 for $Q^2 \simeq 100\,\mathrm{GeV}^2$. (See fig. 3.)

The full NC plus CC data set can be fitted to parametric forms to extract valence quark $(u_v(x), d_v(x))$ sea quark and gluon distributions at a reference value of $Q^2 = 10\,\mathrm{GeV}^2$. The 11-parameter analytic fit imposes factors such as the total number of valence quarks. The small F_L effects are input from NLO DGLAP calculations. For the reasons stated above the gluon distributions are relatively poorly determined. ZEUS have therefore made a combined fit to DIS rates and to dijet data [2]. This leads to reduced errors on fitted gluon distributions (see fig. 4) and to a major improvement in precision on α_s:

$$\alpha_s(M_Z) = 0.1183 \pm 0.0007(\text{stat}) \pm 0.0027(\text{syst})$$
$$\pm 0.0008(\text{model}) \pm 0.005(\text{theory})$$

to be compared to the 2004 world average value, 0.1182 ± 0.0027. This measurement was made using data from one experiment alone. A study by the same authors of combining results from H1 and ZEUS showed movements outside the errors if the two data sets are simply combined. Care is needed, therefore, over systematic error estimates [1].

HERA II has now provided high statistics for both e^+p and e^-p. This permits extraction of xF_3 from the differences between the two data sets. The xF_3 NC term arises from the interference of γ and Z^0 exchange and hence is significant at high Q^2. Figure 5 shows early HERA II results.

3 Charged current results

HERA has now delivered over $200\,\mathrm{pb}^{-1}$ of e^-p data, including beam polarisations of $+32\%$ and -41%. The rest

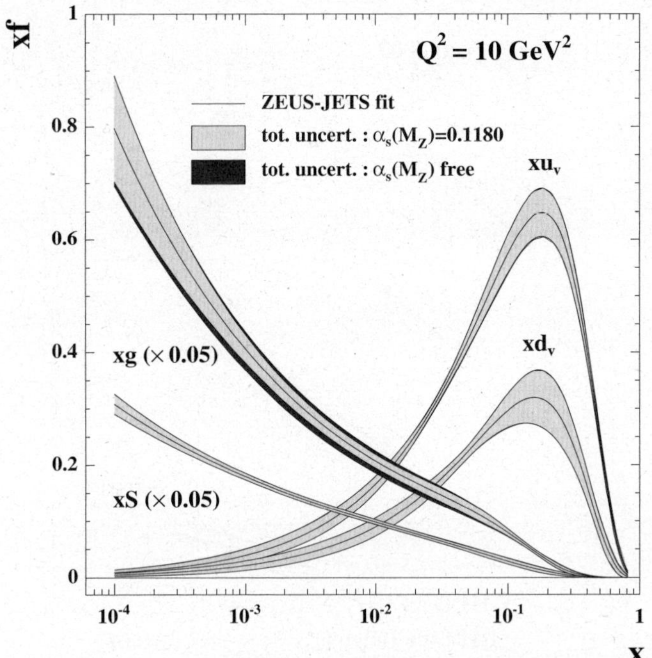

Fig. 4. Fitted parton distributions in the ZEUS jets fit.

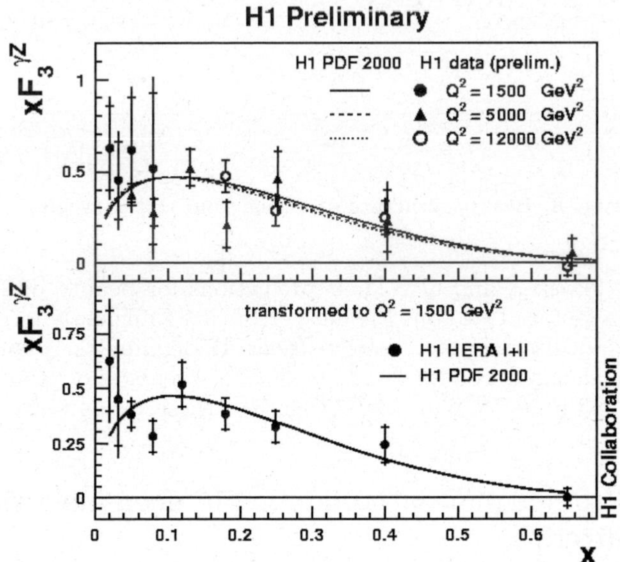

Fig. 5. xF_3 from differences of e^+p: $150\,\mathrm{pb}^{-1}$ and e^-p: $135\,\mathrm{pb}^{-1}$ data.

of the running will be for e^+p (where polarisations of $+33\%$ and -27% have been achieved). The high luminosity allows us to access high-statistics CC data. The CC cross-sections are sensitive to quark and antiquark flavours as follows:

$$\frac{\mathrm{d}^2\sigma}{\mathrm{d}x\mathrm{d}Q^2}(e^+p) = \frac{G_F^2}{2\pi x}\frac{M_W^4}{(Q^2+M_W^2)^2}x[\bar{u}+\bar{c}+(1-y)^2(d+s)],$$

$$\frac{\mathrm{d}^2\sigma}{\mathrm{d}x\mathrm{d}Q^2}(e^-p) = \frac{G_F^2}{2\pi x}\frac{M_W^4}{(Q^2+M_W^2)^2}x[u+c+(1-y)^2(\bar{d}+\bar{s})],$$

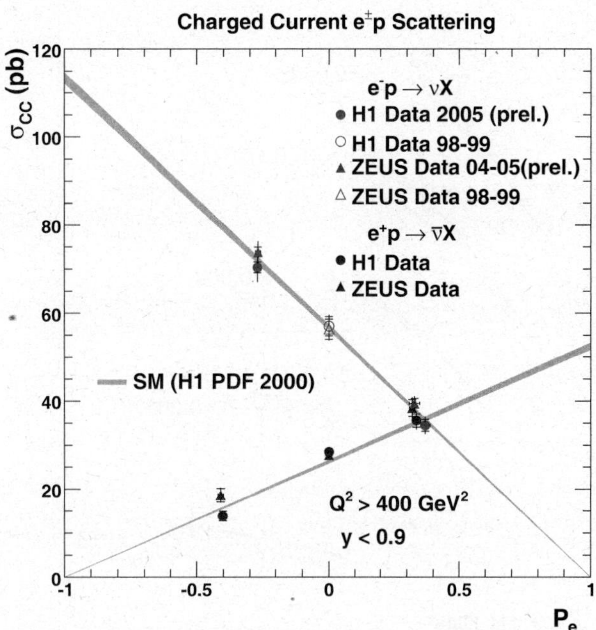

Fig. 6. The linear dependence of charged-current cross-sections on lepton beam polarisation.

where the quark flavour distributions are functions of x and Q^2. Both e^+p and e^-p are needed to separate the different quark flavours. Running at positive and negative beam helicities will assist this activity.

The first step in this process is to verify the helicity dependence of the cross-section. For pure $W^\pm$ exchange we expect

$$\sigma_{CC}^{e^\pm p}(P_e) = (1 \pm P_e)\sigma_{CC}^{e^\pm p}(P_e = 0).$$

Figure 6 shows a standard model fit to the data, with zero intercepts at $P_e = \pm 1$ [3]. This is a good fit to the data and so there is no need for W_R exchange and we can set lower limits on the W_R mass (for standard couplings) of order $200\,\mathrm{GeV}$. Bin-by-bin comparisons of $\sigma(RH)/\sigma(LH)$ show Q^2-dependence consistent with the standard model.

Polarised data also allow us to extract the vector and axial weak couplings of the u and d quarks. Precisions are comparable to measurements at LEP and resolve the LEP sign ambiguity (where only the sign of the product of vector and axial couplings is measured).

4 Heavy flavour production

Charm and beauty production NC cross-sections are used to define structure functions as follows (ignoring xF_3 for moderate Q^2):

$$\frac{\mathrm{d}^2\sigma}{\mathrm{d}x\mathrm{d}Q^2}(c\bar{c}) = \frac{2\pi\alpha^2}{xQ^4}[Y_+F_2^{c\bar{c}} - y^2F_L^{c\bar{c}}]$$

and similarly for b-production. Production is dominated by photon-gluon fusion. Both ZEUS and H1 have operational vertex detectors. Results are available using vertex

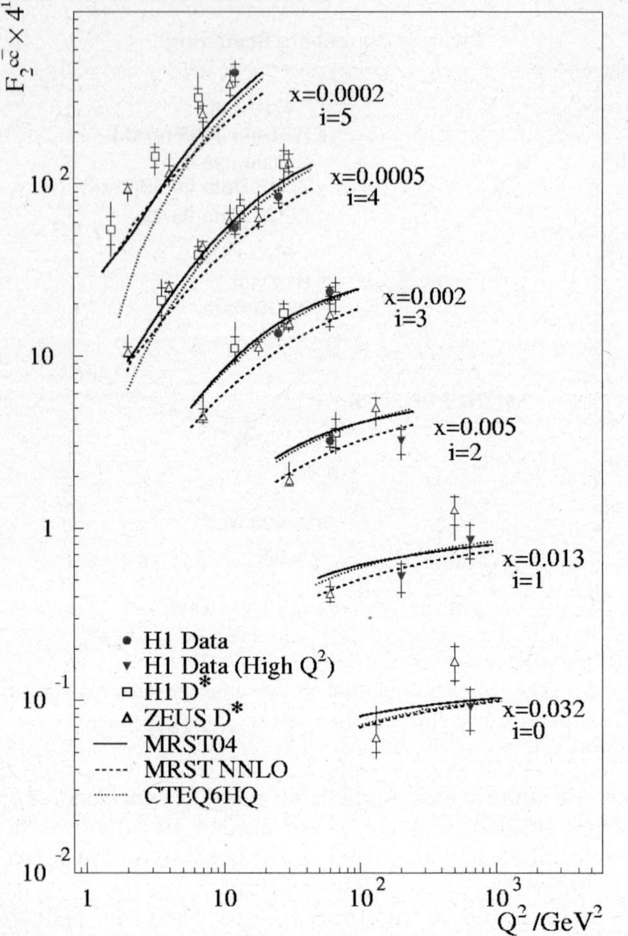

Fig. 7. Charm structure functions from vertex and $D^{*\pm}$ tagging.

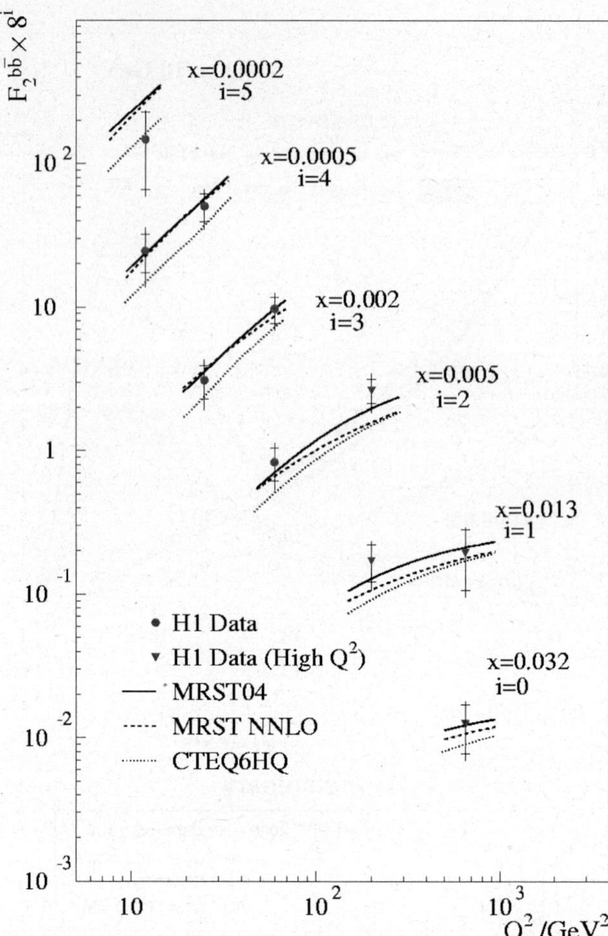

Fig. 8. Beauty structure functions from vertex tagging.

detector decay time tagging in H1 [4]. The detector acceptance is good at modest laboratory rapidity, providing measurements for $0.0002 < x < 0.032$ and Q^2 up to $650\,\text{GeV}^2$.

The H1 vertex detector provides trajectory measurements with accuracy $\sigma_{r\phi} = 12\,\mu\text{m}$, $\sigma_z = 25\,\mu\text{m}$ at $5.7\,\text{cm}$ radius. Tracks are extrapolated back to the primary production vertex. The miss distance (called the impact parameter, δ) arises from measuring error (positive or negative) or from parent flight before decay. $\delta \simeq 0.7c\tau$, so a b lifetime of $1.5\,\text{ps}$ gives a mean impact parameter of $300\,\mu\text{m}$. We define the significance $S = \delta/\sigma(\delta)$. Raw data distributions are dominated by measuring error. Algorithms are devised using the second highest significance for charm and the third highest for beauty. The negative-S tail is subtracted from the positive data since the light-quark background is symmetric in S and this reduces the sensitivity to normalisation and to the knowledge of the resolution. Results are shown in figs. 7 and 8, together with earlier charm results based on $D^{*\pm}$ tagging. F_L corrections are about 3% at the highest y. Scaling violations are large, as expected for a gluon-driven process. Theoretical predictions for charm are rather different for NLO and NNLO. MRST and CTEQ agree with the charm results except at the lowest x and Q^2. Their predictions for beauty differ by a factor of two but the data errors are sufficiently large not to discriminate between them. Depending on x and Q^2, charm accounts for 15 to 30% of events and beauty for 0.3 to 3.5%. NLO QCD describes these ratios well.

5 Prompt photon production in deep inelastic scattering

Photons emitted at high E_T and well isolated in rapidity-azimuth space are regarded as arising from hard emission from an electron or quark line in the primary Feynman diagram and not as the secondary products of jet fragmentation. ZEUS data on inclusive prompt photons and photon-plus-jet disagreed with the predictions of different (divergent) Monte Carlo models and agreed only modestly with NLO calculations available at the time [5]. The process has two hard scales, Q^2 and E_T and NLO for photon-plus-jet is of order $(\alpha^3 \alpha_s)$, so it is quite challenging to calculate.

This stimulated MRST to look at the process as the DIS of an electron off a photon which is itself a constituent of the photon, the electron emitting hard non-collinear bremsstrahlung [6]. Such constituent photons must be expected from the QED analogue of DGLAP evolution, and

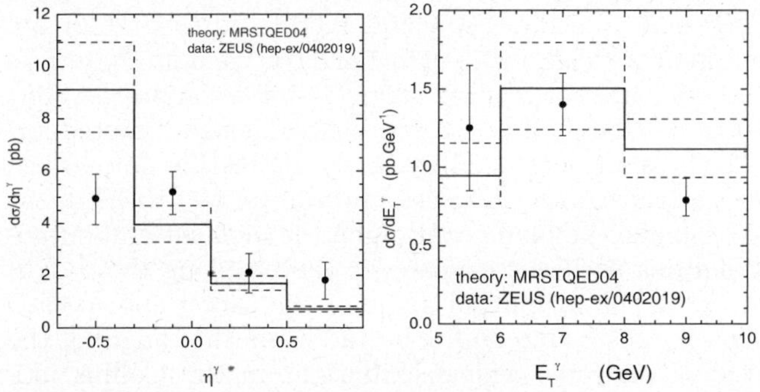

Fig. 9. Inclusive DIS prompt photon production compared to MRST predictions (absolute normalisation).

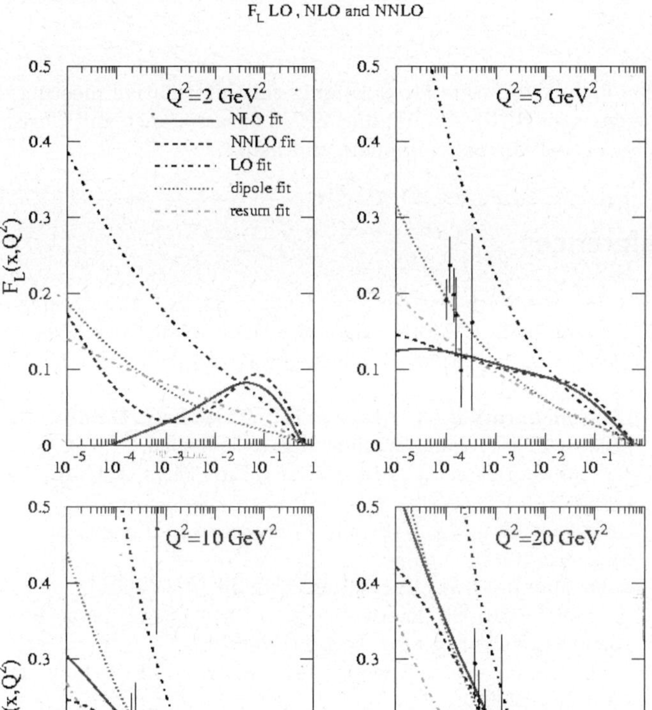

Fig. 10. Different theoretical predictions for F_L together with hypothetical measurements with errors of the expected size.

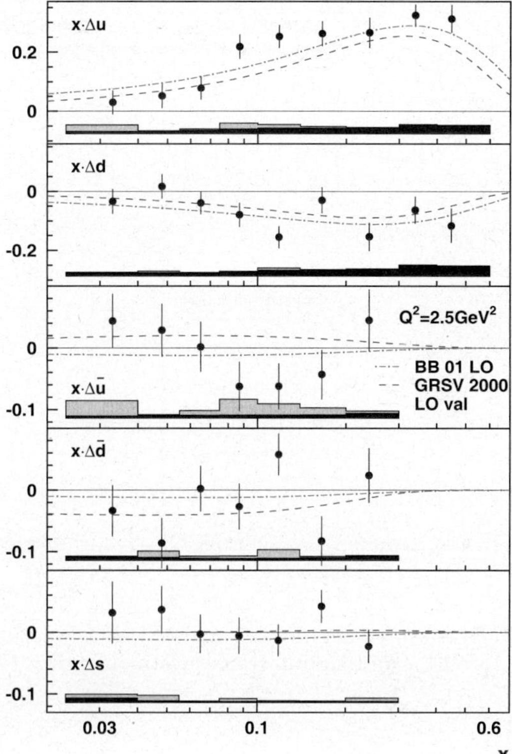

Fig. 11. HERMES measurements of nucleon longitudinal spin structure.

emission from quarks and jet production. Logically, this ought to exist.

6 Measurement of F_L

Extraction of F_L is difficult for several reasons. Recall we need high y to see its effect, which is normally only a few percent even then. Since $Q^2 = sxy$, at fixed x and Q^2 this means low s. Special runs have therefore been proposed for 2007 at reduced proton beam energy, for example three months at 50% of nominal beam energy. F_L is then extracted by comparing the same (x, Q^2) bins at different values of y. High y is experimentally quite challenging as it demands low outgoing electron energy, giving challenges to triggering, energy resolution and to backgrounds from photoproduction.

The present knowledge of F_L is scant and theories are divergent at low Q^2 and low x, due to large uncertainties in gluon distributions. There are major differences between LO, NLO and NNLO calculations [8] (see fig. 10). Measuring F_L is therefore both a test for perturbative QCD and a calibration for the LHC. $5\,\mathrm{pb}^{-1}$ of data at $460\,\mathrm{GeV}$ proton energy plus $3.5\,\mathrm{pb}^{-1}$ at $575\,\mathrm{GeV}$ give errors of order 0.05 to 0.1 (F_L is typically about 0.3) in a total of 18 (x, Q^2) bins.

7 Measurements with polarised targets

The HERMES experiment collides the electron beam with a proton target, either both polarised longitudinally or

one can calculate a generated $F_2^\gamma(x, Q^2)$. Predictions for absolute rate, Q^2 distribution, photon E_T and rapidity are all close to the data, though the rapidity distribution in the data is flatter than the theory (see fig. 9) suggesting that a photon radiation from quarks needs to be added. Subsequent to this work, new NLO calculations for conventional DIS have been produced close to the ZEUS data and also to recent H1 results [7]. There is therefore not yet an integrated approach combining $F_2^\gamma(x, Q^2)$ with photon

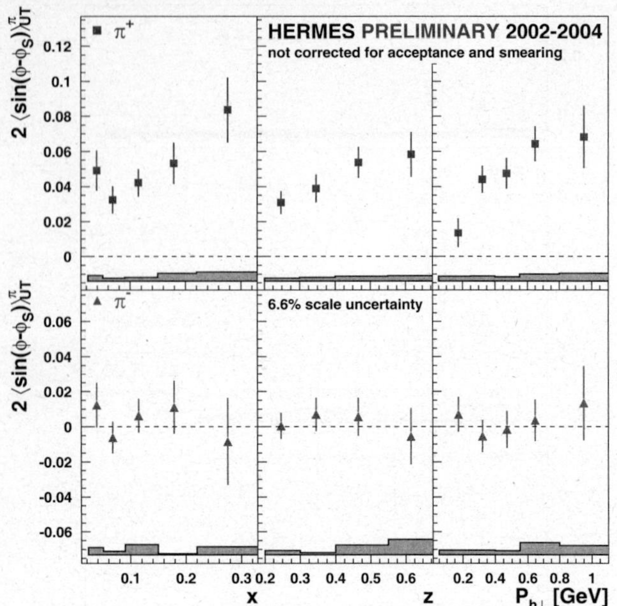

Fig. 12. Sivers moments for π^+ and π^- as a function of x, $z = E_{\mathrm{hadron}}/E_{\mathrm{jet}}$, and hadron transverse momentum $P_{h\perp}$.

unpolarised electron hitting transversely polarised proton. For a proton gas jet target $|P_z| \simeq 0.85$. The difference in rates for ep spins parallel or antiparallel allows us to access the spin structure of the proton. Choosing a proton with its spin pointing in the z-direction,

$$\langle S_z^N \rangle = \frac{1}{2} = \frac{1}{2}\Delta\Sigma + L_q + J_g \,,$$

$\Delta\Sigma$ here is the quark spin asymmetry. This should be 1 if all the proton spin is due to aligning the spin of the quarks. Other possible contributions arise from the orbital angular momentum of the quarks, L_q, or polarisation of the gluons, J_g. Events are flavour-tagged by identifying a leading $\pi^\pm$ or $K^\pm$. This allows decomposition of the spin asymmetry into $u, \overline{u}, d, \overline{d}$ and $(s + \overline{s})$ [9]. Figure 11 shows the results as a function of x. Integrating over the range $0.2 < x < 1$ HERMES find $\Delta\Sigma(u + \overline{u} + d + \overline{d}) = 0.286 \pm 0.026 \pm 0.011$,

$\Delta\Sigma(s + \overline{s}) = 0.006 \pm 0.029 \pm 0.007$. $\Delta\Sigma(\overline{u}$ and $\overline{d})$ are consistent with zero with errors of 0.04. We note that there is no evidence for a large strange-quark component, but that more than half of the proton spin is unaccounted for.

This question is addressed by HERMES data taken using a transversely polarised proton target [10]. Take as the xz-plane the plane containing the incident and outgoing leptons with the exchanged photon along the z-axis. Let ϕ_s be the azimuthal angle of the target spin polarisation in this frame and ϕ be the azimuthal angle of the outgoing hadrons. The distributions in the Collins moments, $\langle\sin(\phi + \phi_s)\rangle$, and Sivers moments $\langle\sin(\phi - \phi_s)\rangle$, together with $\langle\sin(2\phi - \phi_s)\rangle$, and $\langle\sin(\phi_s)\rangle$ are measured. The Sivers moments are consistent with zero for outgoing π^- but significant for π^+. This results requires L_q to be non-zero (see fig. 12).

I thank the organisers for arranging this informative meeting, colleagues on HERMES, H1 and ZEUS for assistance and Chris Collins-Tooth for his comments and help.

References

1. Robin Devenish and Amanda Cooper-Sarkar, *Deep Inelastic Scattering* (Oxford University Press, New York, 2004).
2. ZEUS Collaboration (S. Chekanov *et al.*), Eur. Phys. J. C **42**, 1 (2005).
3. H1 Collaboration (A. Aktar *et al.*), Phys. Lett. B **634**, 173 (2006); ZEUS Collaboration, hep-ex/0602026.
4. H1 Collaboration (A. Aktar *et al.*), Eur. Phys. J. C **45**, 23 (2006).
5. ZEUS Collaboration (S. Chekanov *et al.*), Phys. Lett. B **595**, 86 (2004).
6. A.D. Martin *et al.*, Eur. Phys. J. C **39**, 155 (2005).
7. A. Gehrmann-De Ridder *et al.*, Phys. Rev. Lett. **96**, 132002 (2006). See also, hep-ph/0605222.
8. A.D. Martin, W.J. Stirling, R.S. Thorne, Phys. Lett. B **635**, 305 (2006).
9. HERMES Collaboration (A. Airepetian *et al.*), Phys. Rev. Lett. **92**, 012005 (2004).
10. HERMES Collaboration (A. Airepetian *et al.*), Phys. Rev. Lett. **94**, 012002 (2005).

Eur. Phys. J. A **31**, 572–574 (2007)
DOI 10.1140/epja/i2006-10198-1

Special Article – QNP 2006

Analysis of the quadrupole deformation of $\Delta(1232)$ within an effective Lagrangian model for pion photoproduction from the nucleon

C. Fernández-Ramírez[1,2,a], E. Moya de Guerra[1,3], and J.M. Udías[3]

[1] Instituto de Estructura de la Materia, CSIC, Serrano 123, E-28006 Madrid, Spain
[2] Departamento de Física Atómica, Molecular y Nuclear, Universidad de Sevilla, Apdo. 1065, E-41080 Sevilla, Spain
[3] Departamento de Física Atómica, Molecular y Nuclear, Facultad de Ciencias Físicas, Universidad Complutense de Madrid, Avda. Complutense s/n, E-28040 Madrid, Spain

Received: 8 October 2006
Published online: 22 February 2007 – © Società Italiana di Fisica / Springer-Verlag 2007

Abstract. We present an extraction of the $E2/M1$ ratio of the $\Delta(1232)$ from experimental data applying an effective Lagrangian model. We compare the result obtained with different nucleonic models and we reconcile the experimental results with the lattice QCD calculations.

PACS. 14.20.Gk Baryon resonances with $S = 0$ – 25.20.Lj Photoproduction reactions – 13.60.Le Meson production

The deformation of the nucleon and its first excitation, the $\Delta(1232)$, is a topic that has focused the attention of many researchers in the last years from both the experimental and the theoretical sides [1]. The possibility of such deformation has been studied using the $E2/M1$ ratio (EMR) of the $\gamma N \to \Delta(1232)$ transition [2]. The emission (absorption) of a photon by a spin-3/2 particle involves a magnetic dipole ($M1$) multipolarity and an electric quadrupole ($E2$) multipolarity. From experiments it is found that $E2$ is small but not zero, which evokes a deformed nucleon picture. A deviation from zero of the EMR is a clear indication of the existence of such deformation and allows to quantify it. This ratio is mainly obtained in two different ways, from nucleonic models such as quark models or lattice QCD, and from experimental data. Reconciliation of both extractions is significant in order to understand the structure of the nucleon. This work is concerned with the extraction of the intrinsic $E2/M1$ ratio of the $\Delta(1232)$ from experimental data using a reaction model.

In [3,4] we have developed a pion photoproduction model up to 1 GeV of photon energy based upon effective Lagrangians. The model follows closely the work of Garcilazo and Moya de Guerra [5] and has similarities with the work of Sato and Lee [6] as well as with other works [7–9] based on the seminal work of Peccei [10]. The reaction model allows us to isolate the contribution of the $\Delta(1232)$ —taking into account the high-energy behaviour

of the tail of the resonance—, to calculate its EMR, and to compare it with the values provided by nucleonic models. In addition to Born terms (those which involve only photons, nucleons, and pions) and vector meson exchange terms (ρ and ω exchanges), the model includes all the four star resonances in the Particle Data Group (PDG) [11] up to 1.7 GeV mass and up to spin-3/2: $\Delta(1232)$, $N(1440)$, $N(1520)$, $\Delta(1620)$, $N(1650)$, and $\Delta(1700)$.

The model displays chiral symmetry, gauge invariance, and crossing symmetry, as well as a consistent treatment of the interaction with spin-3/2 particles that avoids well-known pathologies present in previous models [3,4]. The dressing of the resonances is considered by means of a phenomenological width which takes into account decays into one π, one η, and two π. The width fulfills crossing symmetry and contributes to both direct and crossed channels of the resonances.

We assume that the final-state interactions (FSI) factorize (πN rescattering) and can be included through the distortion of the πN final-state wave function. The calculation of the distortion requires one to calculate higher-order pion loops or to develop a phenomenological potential FSI model. Both approaches are far from the one we apply. The first one is overwhelmingly complex and the second would introduce additional model dependences, which are to be avoided in the present analysis, because we are mainly interested in the bare properties of the resonances. We rather include FSI in a phenomenological way by adding a phase $\delta_{\rm FSI}$ to the electromagnetic multipoles. We determine this phase so that the total phase of the.

[a] e-mail: cesar@nuc2.fis.ucm.es

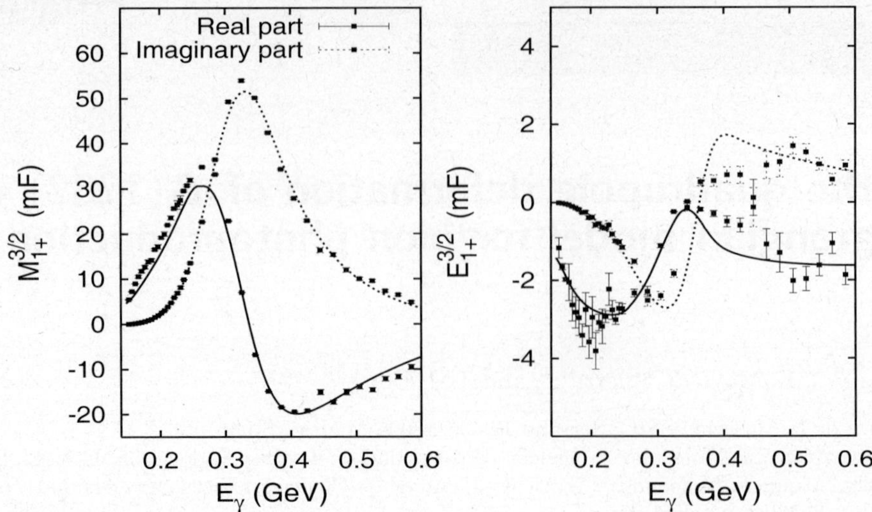

Fig. 1. $M_{1+}^{3/2}$ and $E_{1+}^{3/2}$ electromagnetic multipoles. Curve conventions: solid, real part of the multipole; dashed, imaginary part of the multipole. Data taken from [12].

electromagnetic multipole is identical to the one of the energy-dependent solution of SAID [12]. In this way we are able to isolate the electromagnetic vertex and remove the FSI effects.

In order to assess the parameters of the model we had to minimize the function χ^2 defined by

$$\chi^2 = \sum_{j=1}^{m} \left[\frac{\mathcal{M}_j^{exp} - \mathcal{M}_j^{th}(\lambda_1, \ldots, \lambda_n)}{\Delta \mathcal{M}_j^{exp}} \right]^2, \quad (1)$$

where $\mathcal{M}^{exp}$ stands for the current energy-independent extraction of the multipole analysis of SAID up to $1\,\mathrm{GeV}$ for E_{0+}, M_{1-}, E_{1+}, M_{1+}, E_{2-}, and M_{2-} multipoles in the three isospin channels $I = \frac{3}{2}, p, n$ for the $\gamma p \to \pi^0 p$ process [12]. $\Delta \mathcal{M}^{exp}$ is the error and $\mathcal{M}^{th}$ is the multipole given by the model which depends on the parameters $\lambda_1, \ldots, \lambda_n$, which stand for the electromagnetic coupling constants of the resonances and the cutoff Λ which regularizes the high-energy behaviour of the Born terms. The masses and the widths of the resonances have been taken from the multichannel analysis of Vrana, Dytman, and Lee [13] and from a *speed plot* calculation [4]. The EMR we present has been obtained as an average of the results obtained with both sets of masses and widths. Electromagnetic multipoles are complex quantities and we have taken into account 763 data for the real part of the multipoles and the same amount for the imaginary part. Thus, $m = 1526$ data points have been used in the fits.

In order to fit the data and determine the best parameters of the resonances we have written a genetic algorithm combined with the E04FCF routine from NAG libraries [14]. Although genetic algorithms are computationally more expensive than other algorithms, in a minimisation problem it is much less likely for them to get stuck at local minima than for other methods, namely gradient-based minimisation methods. Thus, in a multiparameter minimisation like the one we face here it is probably the best possibility to search for the minimum [3,15].

Table 1. Comparison of EMR^{physical} values from experiments compared to the values obtained with reaction models.

	EMR^{physical}	Ref.
Experiments		
LEGS Collaboration	$(-3.07 \pm 0.26 \pm 0.24)\%$	[17]
A1 Collaboration	$(-2.28 \pm 0.29 \pm 0.20)\%$	[18]
A2 Collaboration	$(-2.74 \pm 0.03 \pm 0.30)\%$	[19]
Particle Data Group	$(-2.5 \pm 0.5)\%$	[11]
Reaction models		
Fernández-Ramírez *et al.*	$(-3.9 \pm 1.1)\%$	[16]
Pascalutsa and Tjon	$(-2.4 \pm 0.1)\%$	[8]
Sato and Lee	-2.7%	[6]
Fuda and Alharbi	-2.09%	[20]

In fig. 1 we show the fits to $M_{1+}^{3/2}$ and $E_{1+}^{3/2}$ electromagnetic multipoles.

Different definitions of the EMR have been employed in the literature. We should distinguish between the *intrinsic* or *bare* EMR of the $\Delta(1232)$ and the directly measured value in experiments which is often called *physical* or *dressed* EMR value [6,8,16]. The physical EMR is obtained as the ratio between the imaginary parts of $E_{1+}^{3/2}$ and $M_{1+}^{3/2}$ electromagnetic multipoles at the invariant mass (photon energy) at which $\mathrm{Re}[M_{1+}^{3/2}] = 0 = \mathrm{Re}[E_{1+}^{3/2}]$. Since all the reaction models are fitted to the experimental electromagnetic multipoles, they generally reproduce the physical EMR value within the error bars (see table 1). We obtain

$$\mathrm{EMR}^{\mathrm{physical}} = \frac{\mathrm{Im}\left[E_{1+}^{3/2}\right]}{\mathrm{Im}\left[M_{1+}^{3/2}\right]} \times 100\% = (-3.9 \pm 1.1)\%. \quad (2)$$

However, this measured EMR value is not directly available from theoretical models of the nucleon and its

Table 2. Comparison of EMR^{bare} values extracted from experiments through reaction models compared to the values obtained with nucleonic models.

	EMR^{bare}	Ref.
Reaction models		
Fernández-Ramírez *et al.*	$(-1.30 \pm 0.52)\%$	[16]
Pascalutsa and Tjon	$(3.8 \pm 1.6)\%$	[8]
Sato and Lee	-1.3%	[6]
Davidson *et al.*	-1.45%	[7]
Garcilazo and Moya de Guerra	-1.42%	[5]
Vanderhaeghen *et al.*	-1.43%	[9]
Nucleonic models		
Non-relativistic quark model	0%	[21]
Constituent quark model	-3.5%	[22]
Skyrme model	$(-3.5 \pm 1.5)\%$	[23]
Lattice QCD (Leinweber *et al.*)	$(3 \pm 8)\%$	[24]
Lattice QCD (Alexandrou *et al.*)		[25]
$(Q^2 = 0.1\,\text{GeV}^2,\ m_\pi = 0)$	$(-1.93 \pm 0.94)\%$	
$(Q^2 = 0.1\,\text{GeV}^2,\ m_\pi = 370\,\text{MeV})$	$(-1.40 \pm 0.60)\%$	

resonances. Instead, if we want to compare to models of nucleonic structure, it is necessary to extract the bare EMR value of $\Delta(1232)$ which is defined as

$$\text{EMR}^{\text{bare}} = \frac{G_E^{\Delta(1232)}}{G_M^{\Delta(1232)}} \times 100\% = (-1.30 \pm 0.52)\%, \quad (3)$$

The EMR defined in this way depends only on the intrinsic characteristics of the $\Delta(1232)$ and can thus be compared directly to predictions from nucleonic models. It is not, however, directly measurable but must be inferred (in a model-dependent way) from reaction models.

The intrinsic quadrupole deformation of the $\Delta(1232)$ is found to be EMR $= (-1.30 \pm 0.52)\%$, indicative of a small oblate deformation. In tables 1 and 2 we compare our EMR values (bare and physical) to the ones extracted by other authors using other models for pion photoproduction, as well as to predictions of nucleonic models.

Reconciliation of the experimental value of the $E2/M1$ $\gamma N \to \Delta(1232)$ transition ratio (EMR^{physical}) with the one obtained using lattice QCD (EMR^{bare}) within a consistent and sound framework (besides our analysis only in [8] a consistent treatment of the spin-3/2 interaction is performed) is one of the goals of this work [16]. Our results also indicate that quark models need improvements in order to reproduce the value obtained from experiment.

C.F.-R.'s work has been developed under the Spanish Government grant UAC2002-0009. This work has been supported in part under contracts of Ministerio de Educación y Ciencia (Spain) FIS2005-00640 and BFM2003-04147-C02-01.

References

1. B. Krusche, S. Schadmand, Prog. Part. Nucl. Phys. **51**, 399 (2003).
2. A.M. Bernstein, Eur. Phys. J. A **17**, 349 (2003).
3. C. Fernández-Ramírez, *Electromagnetic production of light mesons*, PhD dissertation, Universidad Complutense de Madrid (2006).
4. C. Fernández-Ramírez, E. Moya de Guerra, J.M. Udías, Ann. Phys. (N.Y.) **321**, 1408 (2006).
5. H. Garcilazo, E. Moya de Guerra, Nucl. Phys. A **562**, 521 (1993).
6. T. Sato, T.-S.H. Lee, Phys. Rev. C **63**, 055201 (2001).
7. R.M. Davidson, N.C. Mukhopadhyay, R.S. Wittman, Phys. Rev. D **43**, 71 (1991).
8. V. Pascalutsa, J.A. Tjon, Phys. Rev. C **70**, 035209 (2004).
9. M. Vanderhaeghen, K. Heyde, J. Ryckebusch, M. Waroquier, Nucl. Phys. A **595**, 219 (1995).
10. R.D. Peccei, Phys. Rev. **181**, 1902 (1969).
11. W.-M. Yao *et al.*, J. Phys. G **33**, 1 (2006), http://pdg.lbl.gov.
12. R.A. Arndt, W.J. Briscoe, I.I. Strakovsky, R.L. Workman, Phys. Rev. C **66**, 055213 (2002); SAID database, http://gwdac.phys.gwu.edu.
13. T.P. Vrana, S.A. Dytman, T.-S.H. Lee, Phys. Rep. **328**, 181 (2000).
14. Numerical Algorithms Group Ltd., Wilkinson House, Jordan Hill Road, Oxford OX2-8DR, UK, http://www.nag.co.uk/.
15. D.G. Ireland, S. Janssen, J. Ryckebusch, Nucl. Phys. A **740**, 147 (2004).
16. C. Fernández-Ramírez, E. Moya de Guerra, J.M. Udías, Phys. Rev. C **73**, 042201(R) (2006).
17. G. Blanpied *et al.*, Phys. Rev. C **64**, 025203 (2001).
18. S. Stave *et al.*, Eur. Phys. J. A **30**, 471 (2006).
19. J. Ahrens *et al.*, Eur. Phys. J. A **21**, 323 (2004).
20. M.G. Fuda, H. Alharbi, Phys. Rev. C **68**, 064002 (2003).
21. C.M. Becchi, G. Morpurgo, Phys. Lett. **17**, 352 (1965).
22. A.J. Buchmann, E. Hernández, A. Faessler, Phys. Rev. C **55**, 448 (1997).
23. A. Wirzba, W. Weise, Phys. Lett. B **188**, 6 (1987).
24. D.B. Leinweber, T. Draper, R.M. Woloshyn, Phys. Rev. D **48**, 2230 (1993).
25. C. Alexandrou, Ph. de Forcrand, H. Neff, J.W. Negele, W. Schroers, A. Tsapalis, Phys. Rev. Lett. **94**, 021601 (2005).

Eur. Phys. J. A **31**, 575–577 (2007)

DOI 10.1140/epja/i2006-10231-5

THE EUROPEAN
PHYSICAL JOURNAL A

Special Article – QNP 2006

Longitudinal and transverse target-spin asymmetries associated with DVCS on the proton at HERMES

M. Kopytin[a]

On behalf of the HERMES Collaboration
DESY, 15738 Zeuthen, Germany

Received: 8 November 2006
Published online: 14 March 2007 – © Società Italiana di Fisica / Springer-Verlag 2007

Abstract. Measurements of azimuthal cross-section asymmetries from deeply virtual Compton scattering on transversely and longitudinally polarized hydrogen and longitudinally polarized deuterium targets at HERMES are reported. By comparing the HERMES results on the transverse target-spin asymmetry with theoretical calculations based on a phenomenological model of generalized parton distributions, a model-dependent constraint on the total angular momentum carried by quarks in the nucleon is obtained.

PACS. 13.60.-r Photon and charged-lepton interactions with hadrons – 24.85.+p Quarks, gluons, and QCD in nuclei and nuclear processes – 13.60.Fz Elastic and Compton scattering – 13.88.+e Polarization in interactions and scattering

1 Introduction

Generalized parton distributions (GPDs) were introduced a decade ago as a unified description of hard exclusive processes in the Bjorken regime [1,2]. As a generalization of the usual Parton distribution functions (PDFs) they give additional information about quark and gluon properties in the nucleon. Because of their off-forward nature GPDs contain information about both PDFs and nucleon form factors.

Strong interest in GPDs was triggered by the work of Ji [3] who demonstrated that in the forward limit GPDs can give information about the total angular momentum carried by quarks (and gluons) in the nucleon.

The presently cleanest way to access GPDs is to study deeply virtual Compton scattering (DVCS), the hard exclusive electroproduction of a real photon. This process occurs always together with the Bethe-Heitler (BH) process where the photon is radiated from one of the involved leptons. Both processes have identical final states, making them experimentally indistinguishable, and leading to interference between amplitudes.

The transverse target-spin asymmetry (TTSA) associated with DVCS on a *transversely* polarized target and an unpolarized lepton beam is defined as:

$$A_{UT}(\phi, \phi_S) = \frac{\mathrm{d}\sigma(\phi, \phi_S) - \mathrm{d}\sigma(\phi, \phi_S + \pi)}{\mathrm{d}\sigma(\phi, \phi_S) + \mathrm{d}\sigma(\phi, \phi_S + \pi)}, \quad (1)$$

where ϕ denotes the azimuthal angle between the plane containing the incoming and outgoing lepton momenta

and the plane defined by the virtual and the real photons, and ϕ_S the azimuthal angle of the target polarization vector with respect to the lepton plane. There are two azimuthal amplitudes $A_{UT}^{\sin(\phi-\phi_S)\cos\phi}$ and $A_{UT}^{\cos(\phi-\phi_S)\sin\phi}$ of the TTSA that appear to leading order in α_s and $1/Q$. With the assumption of the dominance of the BH cross-section w.r.t. that of DVCS, they can be approximated as

$$A_{UT}^{\sin(\phi-\phi_S)\cos\phi} \propto \mathrm{Im}\left[F_2\mathcal{H} - F_1\mathcal{E}\right],$$
$$A_{UT}^{\cos(\phi-\phi_S)\sin\phi} \propto \mathrm{Im}\left[F_2\widetilde{\mathcal{H}} - F_1\xi\widetilde{\mathcal{E}}\right]. \quad (2)$$

Here $\mathcal{H}$, $\mathcal{E}$, $\widetilde{\mathcal{H}}$ and $\widetilde{\mathcal{E}}$ denote convolutions of the respective GPDs H, E, $\widetilde{H}$ and $\widetilde{E}$ with hard scattering kernels [4], the skewness parameter is given in the Bjorken limit by $\xi \simeq \frac{x_B}{2-x_B}$, and F_1 and F_2 are the Dirac and Pauli form factors, respectively.

The longitudinal target-spin asymmetry (LTSA) associated with DVCS on a *longitudinally* polarized target and an unpolarized lepton beam, is defined as

$$A_{UL}(\phi) = \frac{\sigma^{\Rightarrow}(\phi) - \sigma^{\Leftarrow}(\phi)}{\sigma^{\Rightarrow}(\phi) + \sigma^{\Leftarrow}(\phi)}, \quad (3)$$

where $\Leftarrow (\Rightarrow)$ denotes target spin antiparallel (parallel) to the beam direction. The azimuthal amplitude $A_{UL}^{\sin\phi}$ of the LTSA appears to leading order in α_s and $1/Q$ and can be approximated as

$$A_{UL}^{\sin\phi} \propto \mathrm{Im}\left[F_1\widetilde{\mathcal{H}} + \xi(F_1 + F_2)\mathcal{H}\right], \quad (4)$$

with the same notations and assumptions as for eq. (2).

[a] e-mail: Mikhail.Kopytin@desy.de

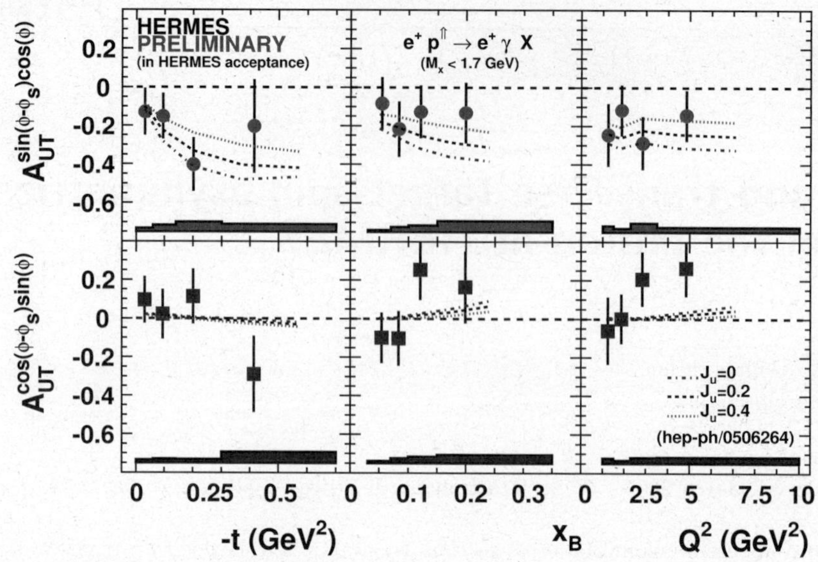

Fig. 1. The $A_{UT}^{\sin(\phi-\phi_S)\cos\phi}$ and $A_{UT}^{\cos(\phi-\phi_S)\sin\phi}$ amplitudes of the transverse target-spin asymmetry associated with DVCS on the proton, shown as a function of $-t$, x_B and Q^2 for the exclusive sample after background correction. The error bars (bands) represent the statistical (systematic) uncertainties. The curves are the predictions from a GPD model with different u-quark total angular momentum J_u and fixed d-quark total angular momentum $J_d = 0$ [5].

2 DVCS at HERMES

HERMES is a fixed-target experiment using the 27.6 GeV electron or positron beam of HERA [6]. Its internal gas target provides polarized H, D, ^{3}He as well as unpolarized H, D, Ne, Kr and Xe targets. In this paper results for H and D targets are discussed. HERMES took data with longitudinally polarized hydrogen (deuteron) target in the years 1996–1997 (1999–2000) and a transversely polarized hydrogen target in 2002–2005.

The selected DVCS events are required to have a detected photon in addition to one charged track identified as the scattered lepton. The kinematic requirements imposed on the scattered lepton are $Q^2 > 1\,\mathrm{GeV}^2$, $W^2 > 8\,\mathrm{GeV}^2$ and $\nu < 23\,\mathrm{GeV}$. The polar angle $\theta_{\gamma^*\gamma}$ between the virtual and the real photon is required to be between 5 mrad and 45 mrad. Since the recoiling proton was not detected the exclusive events were selected with the requirement that the missing mass M_x of the reaction $ep \to e\gamma X$ corresponds to the proton mass. Due to the limited detector resolution the missing-mass range $-1.5 < M_x < 1.7\,\mathrm{GeV}$ was selected based on a Monte Carlo simulation for the optimum separation of exclusive events from the semi-inclusive background. The resulting contribution from semi-inclusive π^0 production to the exclusive sample is about 5%. The contribution from the associated exclusive reaction, where the nucleon is excited to a resonant state, amounts to approximately 10%.

3 Results

The amplitudes $A_{UT}^{\sin(\phi-\phi_S)\cos\phi}$ and $A_{UT}^{\cos(\phi-\phi_S)\sin\phi}$ of the TTSA as a function of $-t$, x_B and Q^2, as extracted at HERMES, are shown in fig. 1 [7]. The analysis is based on data accumulated in 2002–2004 with the positron beam and a transversely polarized hydrogen target. Corrections

for background and smearing were applied. The main contributions to the systematic uncertainty are those from the determination of the target polarization and the background correction. Also shown in fig. 1 are theoretical predictions based on a phenomenological model of GPDs [8]. The curves represent the TTSA amplitudes evaluated with different u-quark total angular momentum J_u as a model parameter, while fixing the d-quark total angular momentum $J_d = 0$ [5].

In fig. 2 the dependences of the $A_{UL}^{\sin\phi}$ and $A_{UL}^{\sin 2\phi}$ amplitudes of the LTSA on proton and deuteron are shown as a function of $-t$ derived from separate fits for every $-t$ bin [9]. A sizable $A_{UL}^{\sin 2\phi}$ amplitude is found which is expected to be kinematically suppressed w.r.t. $A_{UL}^{\sin\phi}$. It can be sensitive to the next-to-leading contributions to the asymmetry (e.g., twist-3 GPDs H^3, $\widetilde{H}^3$) [4]. No difference is observed between the asymmetries on H and D targets in the first t bin, where effects from coherent scattering on the deuteron could be expected. The difference between the two targets in higher t bins might be due to incoherent scattering on the neutron. Calculations based on a GPD model developed in ref. [8] were carried out at the average kinematics of every t bin for the proton. Although the model describes the $A_{UL}^{\sin\phi}$ amplitude well, it does not agree with the data for $A_{UL}^{\sin 2\phi}$. This can be due to the fact that the twist-3 GPDs are modeled only by the Wandzura-Wilczek term, hence no information on quark-gluon correlations is included.

4 Constraint on the total angular momentum of quarks in nucleon

In order to constrain the total angular momentum of quarks a parameterization of GPDs developed in ref. [8]

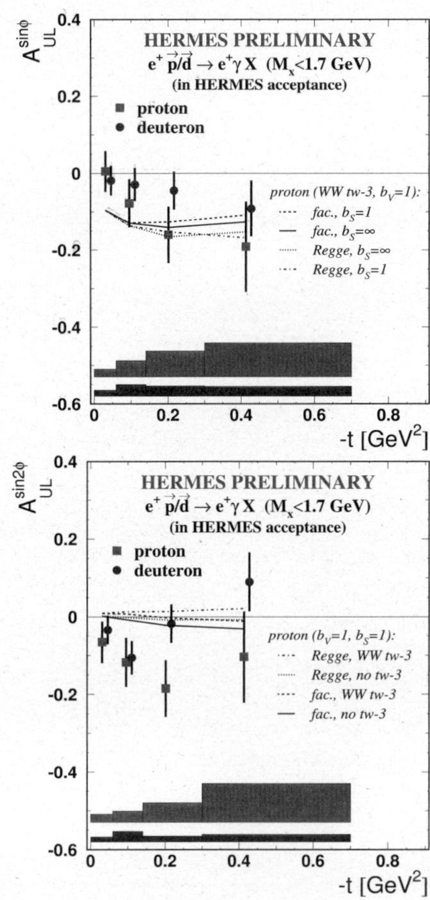

Fig. 2. The $A_{UL}^{\sin\phi}$ (top) and $A_{UL}^{\sin 2\phi}$ (bottom) amplitudes of the longitudinal target-spin asymmetry on the proton and the deuteron as a function of $-t$. The GPD model calculations for the proton use a factorized or a Regge-inspired t-dependence with or without a Wandzura-Wilczek (WW) term [8].

was used where the GPD E is modeled using J_u and J_d as free parameters. The reduced χ^2 value, defined as

$$\Delta\chi^2 = \chi^2 - \chi^2_{minimum} =$$
$$[A^{exp.} - A^{VGG}(J_u, J_d)]^2 / [\delta A^2_{stat.} + \delta A^2_{syst.}], \qquad (5)$$

is calculated for different values of J_u and J_d. Here $A^{exp.}$ denotes the measured (integrated) TTSA amplitude, $\delta A_{stat.}$ ($\delta A_{syst.}$) its statistical (systematic) uncertainty, and A^{VGG} are the values calculated at the average kinematics by a code [10] based on the GPD model [8]. As theoretical predictions on $A_{UT}^{\cos(\phi-\phi_S)\sin\phi}$ show minor changes (see fig. 1) with variations in J_u and J_d, only the contribution from $A_{UT}^{\sin(\phi-\phi_S)\cos\phi}$ to the reduced χ^2 value is included in eq. (5). The integrated result used in eq. (5) is $\langle A_{UT}^{\sin(\phi-\phi_S)\cos\phi}\rangle = -0.149 \pm 0.058(\text{stat.}) \pm 0.033(\text{syst.})$, extracted from the 2002–2004 data at the average kinematics $\langle -t \rangle = 0.12\,\text{GeV}^2$, $\langle x_B \rangle = 0.095$, $\langle Q^2 \rangle = 2.5\,\text{GeV}^2$.

The area in the (J_u, J_d)-plane, in which the reduced χ^2 value is not larger than unity, is defined as one-standard-deviation constraint on J_u vs. J_d. One obtains $J_u + J_d/2.9 = 0.42 \pm 0.21 \pm 0.06$ (see fig. 3). The first

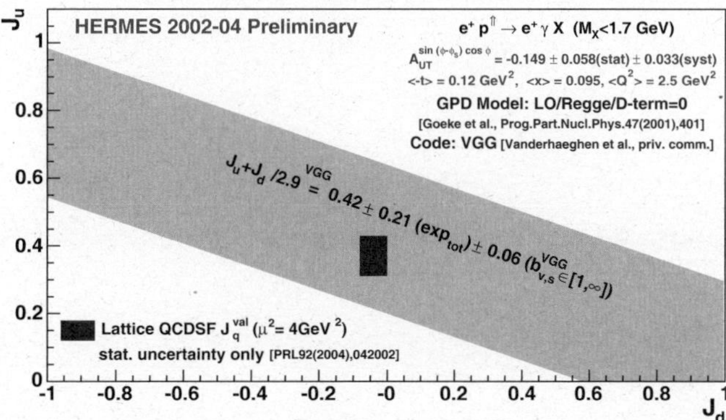

Fig. 3. Model-dependent constraint on u-quark total angular momentum J_u vs. d-quark total angular momentum J_d, obtained by comparing the experimental result and theoretical predictions on the TTSA amplitude $A_{UT}^{\sin(\phi-\phi_S)\cos\phi}$ [11]. Also shown is a lattice result from the QCDSF Collaboration, obtained at the scale $\mu^2 = 4\,\text{GeV}^2$ for valence quark contributions only.

uncertainty is due to the experimental uncertainty in the measured TTSA amplitude. The second one is a model uncertainty, obtained by varying from unity to infinity the unknown profile parameter b which controls the skewness of GPDs [8]. The t-dependence of GPDs is modeled using the Regge ansatz [8]. The impact of using it or its alternative —the factorized ansatz— on the theoretical prediction on the TTSA amplitudes was found to be negligible [5]. The D-term contribution to the GPDs H and E is set to zero, as suggested by HERMES results on the beam charge asymmetry [12]. If the D-term is modeled according to the chiral quark soliton model [8], the resulting constraint is $J_u + J_d/2.9 = 0.53 \pm 0.21 \pm 0.06$.

References

1. D. Mueller *et al.*, Fortschr. Phys. **42**, 101 (1994) `hep-ph/9812448`.
2. A. Radyushkin, Phys. Lett. B **385**, 333 (1996) `hep-ph/9605431`.
3. X. Ji, Phys. Rev. Lett. **78**, 610 (1997) `hep-ph/9603249`.
4. A. Belitsky, D. Müller, A. Kirchner, Nucl. Phys. B **629**, 323 (2002) `hep-ph/0112108`.
5. F. Ellinghaus, W.D. Nowak, A.V. Vinnikov, Z. Ye, Eur. Phys. J. C **46**, 729 (2006) `hep-ph/0506264`.
6. HERMES Collaboration (K. Ackerstaff *et al.*), Nucl. Instrum. Methods A **417**, 230 (1998) `hep-ex/9806008`.
7. HERMES Collaboration (Z. Ye), PoS (HEP2005) 120 (2006) `hep-ex/0512010`.
8. K. Goeke, M. Polyakov, M. Vanderhaegen, Prog. Part. Nucl. Phys. **47**, 401 (2001) `hep-ph/0106012`.
9. HERMES Collaboration (M. Kopytin), AIP Conf. Proc. **792**, 424 (2005).
10. M. Vanderhaeghen, P. Guichon, M. Guidal, *Computer code for the calculation of DVCS and BH processes in the reaction $ep \to ep\gamma$* (2001), private communication.
11. HERMES Collaboration (Z. Ye), `hep-ex/0606061` (2006).
12. HERMES Collaboration (A. Airapetian *et al.*), `hep-ex/0605108` (2006).

Eur. Phys. J. A **31**, 578–584 (2007)
DOI 10.1140/epja/i2006-10258-6

Special Article – QNP 2006

Implications of the nuclear EMC effect

G.A. Miller[a]

Physics Department, University of Washington, Seattle, WA 98195-1560, USA

Received: 23 November 2006
Published online: 16 March 2007 – © Società Italiana di Fisica / Springer-Verlag 2007

Abstract. The discovery more than twenty years ago, by the EMC Collaboration, that the deep-inelastic-scattering DIS structure functions are influenced by the nuclear environment stunned the nuclear physics community and brought quarks and gluons into the field with great impact. A great length of time has passed, but despite a semi-infinite number of papers on the subject, there is no explanation that is universally accepted. Many models (related in one way or another to QCD) have been successful in reproducing data for deep inelastic scattering on nuclear targets, but fewer have described both the DIS and nuclear Drell-Yan experiments. Although there are some positive indications, no model has been used to predict correctly and unambiguously new independent phenomena. We review the history and discuss the best experimental prospects for future discovery.

PACS. 12.39.-x Phenomenological quark models – 21.30.Fe Forces in hadronic systems and effective interactions – 24.85.+p Quarks, gluons, and QCD in nuclei and nuclear processes – 25.30.Mr Muon scattering (including the EMC effect)

1 Introduction

I begin by discussing the nuclear EMC effect and the related Drell-Yan (DY) reaction with the aim of defining these terms and showing that Conventional nuclear dynamics using only hadronic physics fails. This means that the structure of a nucleon is modified when it is placed inside nucleus. I will discuss three popular models of this modification may occur and the resulting implications for new experiments. Then I will close by discussing a favorite topic–the shape of the nucleon.

2 Nuclear EMC effect and related Drell-Yan process

The European Muon Collaboration (EMC) effect in which the structure function of a nucleus, measured in deep inelastic scattering at values of Bjorken $x \geq 0.4$ corresponding to the valence quark regime, was found to be reduced compared with that of a free nucleon was discovered almost twenty years ago [1]. Figure 1 shows the SLAC E139 data [1] for the ratio of the nuclear (per nucleon) to free nucleon structure function. This, contary to the expectations of many, is not unity. Despite much experimental and theoretical progress [2,3], no unique and universally accepted explanation of the depletion has emerged.

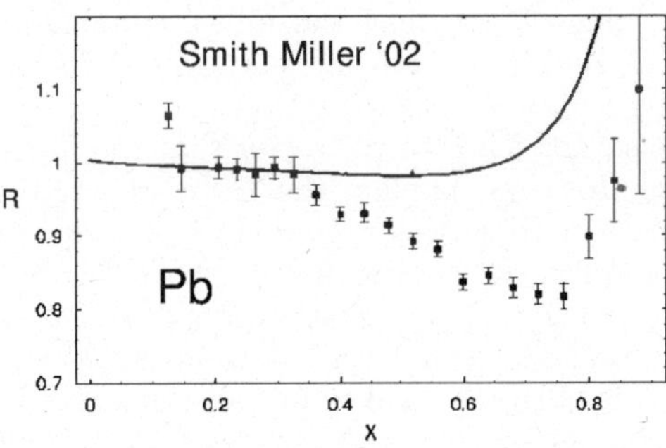

Fig. 1. EMC effect.

The immediate parton model interpretation that the nucleon bound in a nucleus carries less momentum than in free space seems uncontested, but determining the underlying origin remains an elusive goal. I intend to discuss some extant explanations, but will not discuss the issue of A-dependence. This is because Chen and Detmold [4] used the reasonable assumption that the EMC effect is dominated by two-nucleon effect to show in a model-independent manner that the A- and x-dependence factorizes. Thus, understanding the EMC effect for one heavy nucleus is a great start to understanding the entire effect.

The publication of the EMC data and the SLAC data led to the origination of a great number models that could account for the data. This caused me to write in 1985 that "EMC means Everyone's Model is Cool".

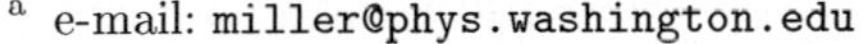
[a] e-mail: miller@phys.washington.edu

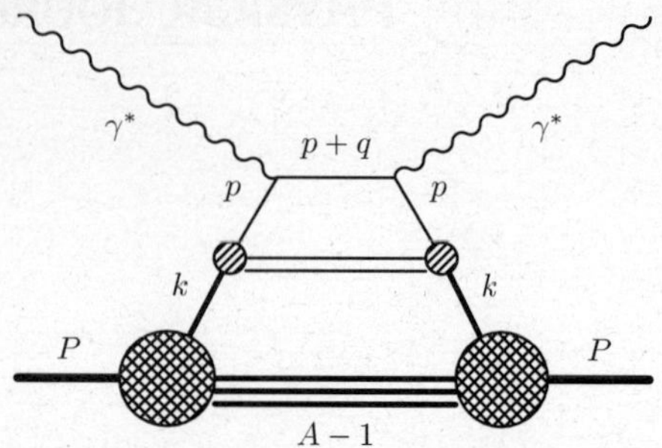

Fig. 2. Deep-inelastic-scattering diagram.

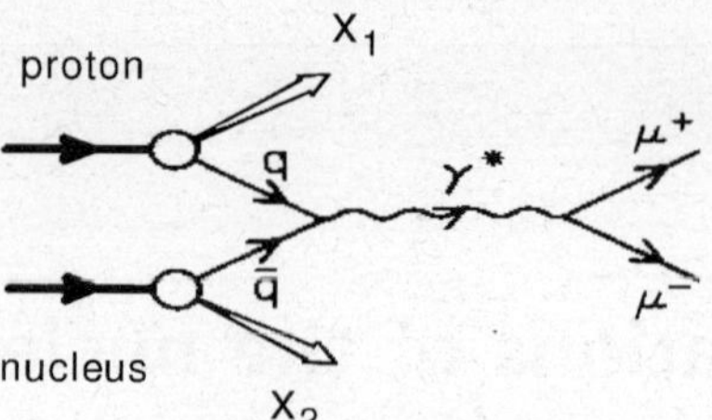

Fig. 3. DY effect.

π fails

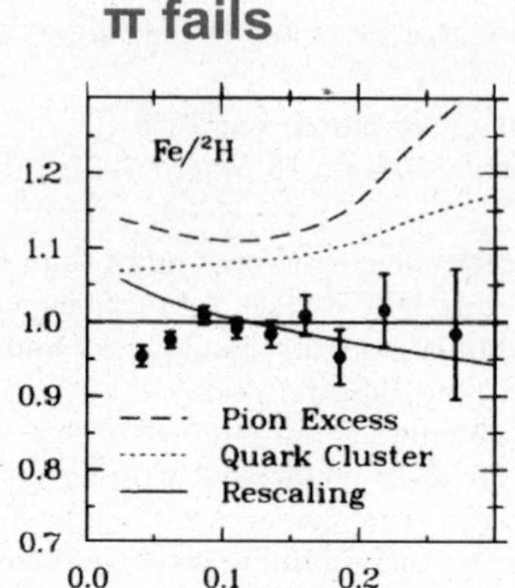

Fig. 4. DY data [8] and theory [7].

2.1 Nucleons only

It is important to rule out the possibility that the simple kinematic effects of binding energy and Fermi motion do not account for the EMC effect. If one starts with the assumption that the structure function is not modified and is the same on and off the energy shell (nucleon-only hypothesis) then evaluation of the diagram of fig. 2 leads to the simple formula

$$\frac{F_{2A}(x_A)}{A} = \int_{x_A}^{A} \mathrm{d}y\, f_N(y) F_{2N}(x_A/y), \qquad (1)$$

where P is the total four momentum of the nucleus, and

$$x_A \equiv Q^2 A/2P \cdot q = x A M/M_A, \qquad (2)$$

with M as the free nucleon mass. The variable $y = Ap^+/P^+$ is the fraction of the nuclear momentum (per nucleon) carried by a single nucleon, and $f_N(y)$ is the corresponding probability distribution. The formula (1) and simple reasoning leads to the result that the nucleon-only hypothesis does not explain the EMC effect. Under the Hugenholz van Hove theorem nuclear stability (pressure balance) implies (in the rest frame) the $P^P = P^- = M_A$. But to an excellent approximation $P^+ = A(M_N - 8\,\mathrm{MeV})$. Thus an average nucleon has $p^+ = M_N - 8\,\mathrm{MeV}$. The function $f_N(y)$ is narrowly peaked because the Fermi momentum is much smaller than the nucleon mass. This means that the value of y in the integral of (1) is contrained to be very near unity. Thus F_{2A}/A is well approximated by F_{2N} and one gets no EMC effect this way [5,6]. This is shown as the solid curve in fig. 1.

2.2 Nucleons and pions

The next thing to try is to assert that the nuclear pion cloud is enhanced, and that pions carry an excess of plus-momentum. In this case, $P^+ = P_N^+ + P_\pi^+ = M_A$. Many authors [2,3] found that using $P_\pi^+/M_A = 0.04$ is sufficient to account for the EMC effect. However, an excess of nuclear pions implies that the nuclear sea is enhanced and that enhancement should be observable in a nuclear Drell-Yan experiment [7]. The idea, see fig. 3, is that a quark from an incident proton (defined by large value of x_1) annihilates an anti-quark from the target nucleus defined by a smaller value of x_2. A significant enhancement of pions would enhance the anti-quarks and enhance the nuclear Drell-Yan reaction. But no such enhancement was observed [8] as shown in fig. 4. This caused Bertsch *et al.* [9] to announce "a crisis in nuclear theory", because conventional theory does not work.

3 Nucleon modification by nuclei

We now consider the modification of the structure of a single nucleon caused by its presence in the nuclear medium. The root cause of any such modification is the interaction between nucleons, so that one needs to consider whether or not the entire concept of single-nucleon modification makes any sense at all. Our belief is that if the kinematics of a given experiment selects single-nucleon properties, such as in quasi-elastic scattering, it does make sense to consider how a single nucleon is modified.

Actually, there is a well-known example of the nuclear modification of nucleon properties. It is an experimental fact that a single bound neutron is different than a free one. In particular, the lifetime is changed from about fifteen minutes to forever. This is because binding effects modify the propagator (energy denominator) that relates the $|pe^-\nu\rangle$ component of the neutron to its dominant $|n\rangle$ one. The increase in the energy denominator suppresses the component induced by the weak interaction. Changing the energy denominator changes the wave function.

Here we shall be mainly concerned with changes in a nucleon wave function induced by external strong fields supplied by other nucleons. These strong fields polarize nucleons and are an analog of the Stark effect in which an electric field induces a dipole moment of an atom. In nuclei

The European Physical Journal A

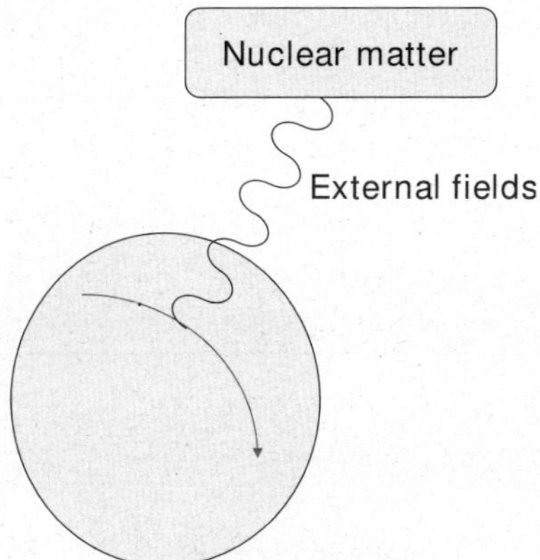

Fig. 5. Nuclear modification of nucleon properties through external fields.

there is no preferred spatial direction so we are examing phenomena related to monopole polarization.

3.1 Requirements and goals

Models aimed at explaining the EMC effect and Drell Yan experiment are subject to meeting a severe set of requirements. One needs to be able to model the free distributions for both the valence and sea quark components, satisfying the problem of maintaining good support. That is many models do not handle the requirements of momentum conservation exactly and using them can lead to obtaining structure functions that do not vanish for Bjorken x values greater than unity or less than zero. One must model the nuclear modifications in a manner that is consistent with known nuclear properties, and be able to describe the deep inelastic and di-muon production data. Finally, given all of this, one needs to predict new independent phenomena.

3.2 Three models

I discuss a set of three models, each built upon the common theme that in nuclei the quark wave functions of a single nucleon are modified by external fields provided by the surrounding nucleons, fig. 5.

In the quark meson coupling model [10] quarks in nucleons confined in an MIT bag exchange scalar and vector mesons with the nuclear medium. Later work [11] replaces the MIT bag with the NJL model. One may also immerse the chiral quark soliton model nucleon [12] in the medium [13]. In this case, the quarks exchange infinite pairs of pions as well as vector mesons with the surroundings. In both of these models the attraction needed to produce a bound state is generated by the exchange of scalar quantum numbers and the repulsion necessary to obtain nuclear saturation is caused by exchange of vector mesons. A third model involves the suppression of point-like configurations of the nucleon [14,15].

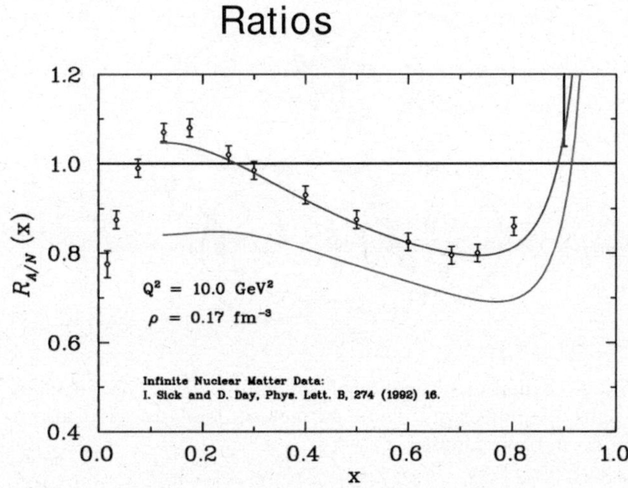

Fig. 6. (Color online) Ratio of spin-independent (blue and near data) and spin-dependent g_1 structure function (red). (I thank I. Cloet for this figure.)

3.2.1 QMC

The specific dynamical mechanism of interest is that the external scalar field enhances the lower component of the quark's Dirac wave function. This effect is exploited in a variety of interesting applications. The nuclear enhancement of the spin structure function [11] is of particular interest and is shown as fig. 6. The EMC effect of spin-dependent structure functions is significantly larger than that for the usual F_2 structure function.

3.2.2 CQSM

This model is based on the instanton-dominated nature of the vacuum [16]. The coupling of quarks couple to vacuum instantons generates spontaneously a constituent-quark mass of about 400 MeV. These quarks interact with pions through the effective Lagrangian The CQSM model Lagrangian with (anti)quark fields $\overline{\psi}, \psi$, and profile function $\Theta(r)$ representing the pion field is given by

$$\mathcal{L} = \overline{\psi}(i\partial\!\!\!/ - M e^{i\gamma_5 \boldsymbol{n}\cdot\boldsymbol{\tau}\Theta(r)})\psi, \tag{3}$$

where $\Theta(r \to \infty) = 0$ and $\Theta(0) = -\pi$ to produce a soliton with unit winding number. The nucleon is made of three such quarks. This model is known to reproduce good nucleon properties, as well as reasonable structure functions with excellent support properties [12]. One special feature worth noting is that this model nucleon contains an intrinsic sea (at a low momentum transfer scale) generated by the modification of the infinite Dirac sea by the quark interactions with the pionic field.

This model is applied to nuclear physics [13] by allowing collections of such nucleons to exchange scalar and vector mesons as cartooned in fig. 7. Obtaining the wave function for infinite nuclear matter involves solving a double self-consistency relation. The mass and internal wave

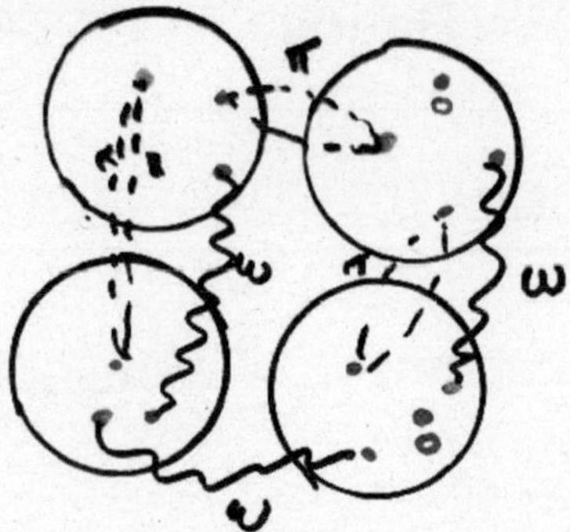

Fig. 7. Schematic display of the model [13]. The exchange of pairs of pions and of omega mesons leads to saturation of infinite nuclear matter.

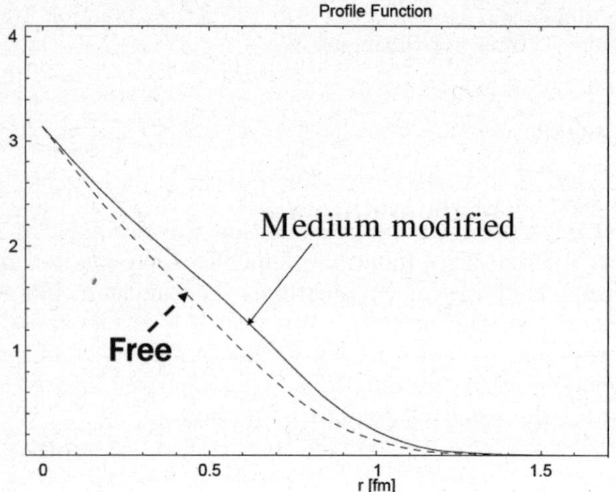

Fig. 8. Nuclear modification of the profile function.

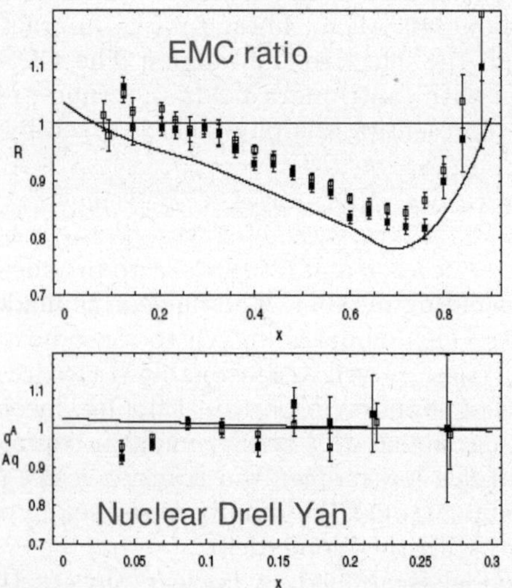

Fig. 9. Nuclear ratios of deep inelastic and Drell-Yan scattering. Nuclear matter data from Day and Sick. Drell-Yan data from [8].

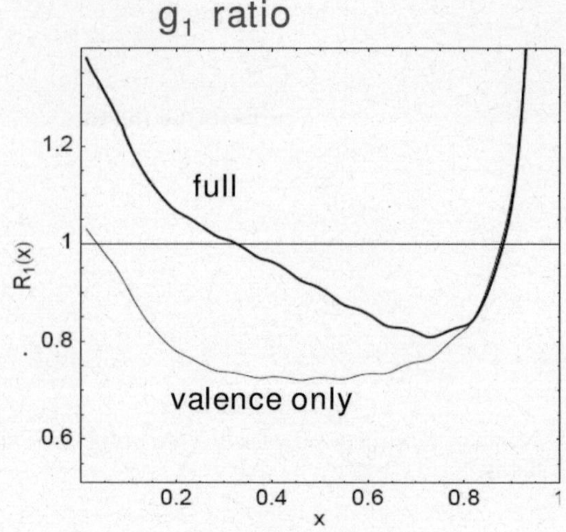

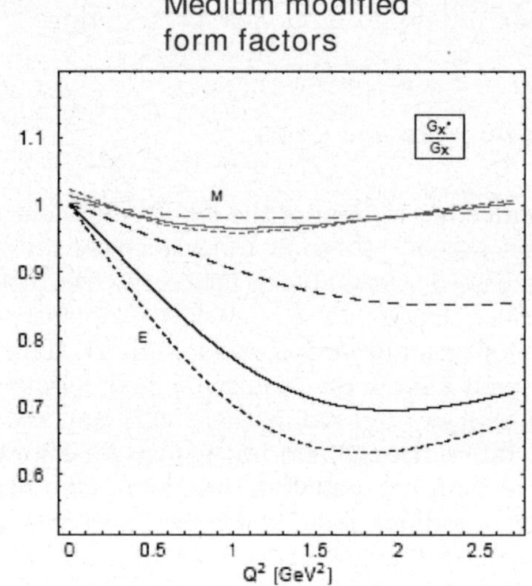

Fig. 10. Top panel: ratio of nuclear to nucleon spin-dependent g_1 structure function. The heavy curve is full calculation, the light curve is valence only. Bottom panel: medium modified electric (isoscalar) and magnetic (isovector) form factors. The three different sets of curves represent the effect of using three different nuclear densities.

function of a bound nucleon depends on the Fermi momentum. For a given set of parameters (the strength of the ωN coupling and the magnitude of the quark condensate) one minimizes the energy of the nucleus (per nucleon). We obtain excellent saturation properties. The resulting change in the profile function shown in fig. 8 is vey small, but corresponds to a slight broadening if the effective potential that binds the quarks in the nucleon. The use of the medium modified wave function to compute structure functions leads to the results shown in fig. 9. We are able to account for the EMC effect, and produce results that are in agreement with the Drell-Yan data. This indicates that the sea is not very much modified, and this can be seen by looking at the details of the results.

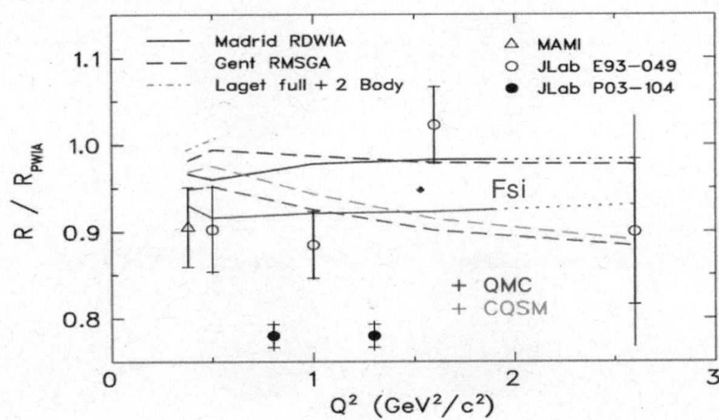

Fig. 11. Medium to free ratio of ratios G_E/G_M. Data for various measurements are compared with various theories. (I thank S. Strauch for preparing this figure.)

Fig. 12. Configurations of the nucleon.

The interesting extension is to the computation of other observables. The medium modified structure function is displayed in fig. 10. If one includes the effects of the valence quark only, the results are similar to those of Cloet *et al.*, but the full calculation including the effects of the sea quarks shows that the nuclear modification of g_1 is rather similar to the nuclear modification of F_2 of the usual EMC effect. This arises because the nucleon sea is not modified much. The modification of the nucleon electromagnetic form factors is shown as the bottom panel of fig. 10. The electric form factor is modified much more than the magnetic one. This effect can be studied using polarization transfer in the $^4\mathrm{He}(e, e'p)$ reactions [17]. These have the potential to measure the ratio of free to medium modified ratios of the ratio G_E/G_M. Some results are shown in fig. 11. Three calcluations that do include medium modifications line generally above the data. Both the QMC and CQSM calculations are closer to the data. The effects of a newer calculation [18] can reproduce the trend of the data without medium modifications. But the combination of medium modifications and this new final state interaction calculation would also agree with the data, within the large error bars. The crucial charge-exchange part of the final-state interaction used by [18] has not been constrained by data. This is a very exciting subject that needs to be pursued with further, more accurate experiments.

3.2.3 Supression of point-like configurations

The nucleon comes in a variety of configurations, which can be defined by the quark-gluon content of each Fock state component. See fig. 12. Suppose such a system is placed in the medium. Those of normal size feel the nuclear attraction and their energy goes down. The point-like-configurations PLC of very small size are prevented

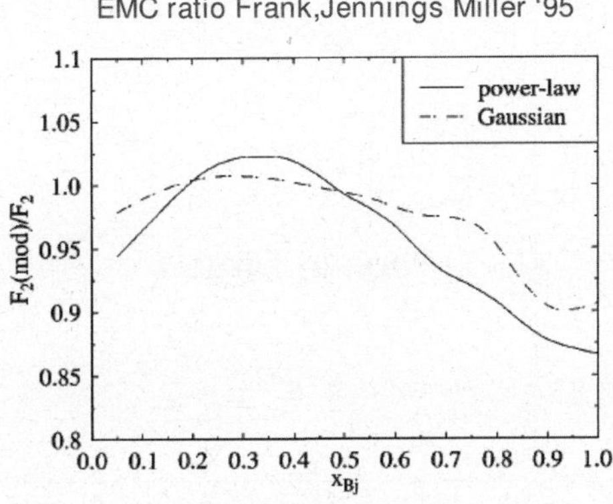

Fig. 13. EMC ratio due to PLC suppression. Only the curve labelled "power law" has meaning.

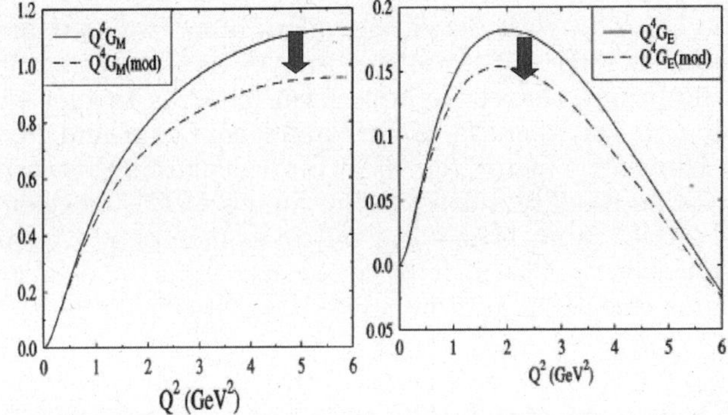

Fig. 14. Medium modifications of electric and magnetic form factors.

from feeling the attraction by the effects of color screening. The energy difference between the configurations of normal size and the PLCs increases. Then quantum mechanics tells us that the magnitude of the PLC is reduced and the PLC is suppressed. One consequence is that quarks in the medium have reduced momentum, in agreement with the EMC effect.

To make calculations, one needs a free nucleon wave function. Ours is of the relativistic constituent quark model. It is a three-quark antisymmetric wave function, that is frame independent and an eigenstate of the spin operator [19,15,20]. The wave function includes a product of three quark Dirac spinors, each evaluated at a relative momentum.

One places this wave function in the medium subject to a mean field that vanishes for configurations in which the three quarks are close together [15]. The resulting medium modifications of the structure function are shown in fig. 13. This calculation is made for a nucleon at rest, so there is no rise at large x. However, the suppression of the valence quark momentum distribution is seen.

The same model predicts medium modifications of the nucleon form factor as shown in fig. 14. The important modifications shown by the red arrow occur at larger values of momentum transfer than currently accessible experimentally. The result here show a more spectacular effect than medium modifications. The famous decrease in the

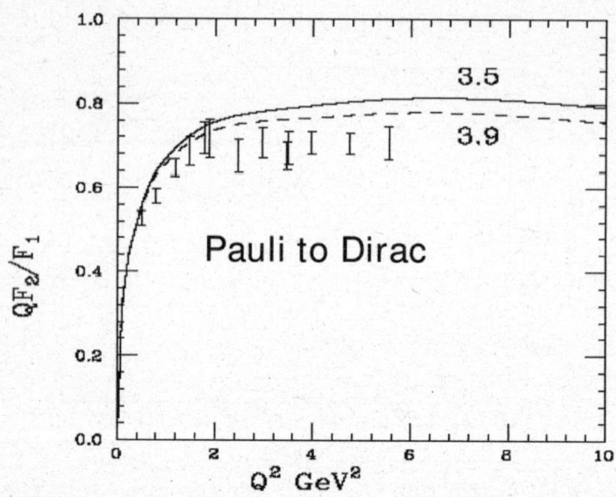

Fig. 15. QF_2/F_1, data from [21]. Theory from [20].

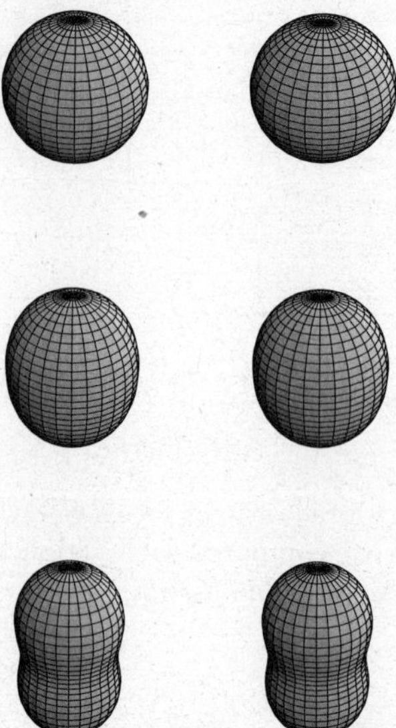

Fig. 16. Shapes of the matter distribution (eq. (4)) without the charge operator. First row: $K \doteq 250\,\mathrm{MeV}/c$, second row: $K = 1\,\mathrm{GeV}$, third row: $K \to \infty$. Left column: **n** parallel to the proton angular momentum, right column: anti-parallel.

ratio G_E/G_M with increasing values of Q^2 was our prediction in 1995 [15]. Another view the interesting nature of the proton electromagnetic form factor is to plot the ratio QF_2/F_1 where F_2 is the Pauli form factor and F_1 is the Dirac form factor, fig. 15. In our calculation this turns out to be roughly a constant for values of Q^2 between 2 and 20 GeV^2 [20]. There is a simple explanation in terms of the lower component of the quark Dirac spinor [20]. Quarks within the proton move relativistically.

4 Shapes of the nucleon

At one point I was asked to discuss the implications of the new Jlab data for the shape of the nucleon. Relativistically moving objects have a pancake shape when viewed from a frame at rest. This is not related to the intrinsic nature of the wave function. I defined the spin-dependent density [22] to probe the relativistic nature of the quark motion within the nucleon and to reveal the shape of the nucleon. One can compute the spin-dependent density which is the probability that a quark has a given momentum **K** and spin in a direction **n**. This operator is given by the expression

$$\widehat{\rho}(\mathbf{K},\mathbf{n}) = \int \frac{\mathrm{d}^3 r}{(2\pi)^3} e^{i\mathbf{K}\cdot\mathbf{r}} \bar{\psi}(\mathbf{r}) \frac{\widehat{Q}}{e} \left(\gamma^0 + \boldsymbol{\gamma}\cdot\mathbf{n}\gamma_5\right) \psi(0). \quad (4)$$

Taking matrix elements of this operator using model nucleon wave functions reveals that sometimes the proton is shaped like a bagel and sometimes like a pretzel [22]. See fig. 16. It now also seems possible to go beyond model calculations and use the lattice to determine whether or not the proton is always spherical. It is amusing to make a final speculation. We have found that lower components account for the flatness of the ratio QF_2/F_1 and also for the non-spherical shapes of the proton. We also known, through the EMC effect that effects of the medium modify the nucleon wave function. If the medium modifies the lower components, we might then expect that the medim will also modify the shape. A possible result is shown in fig. 17. At present, it remains a challenge to experiment to measure either shape.

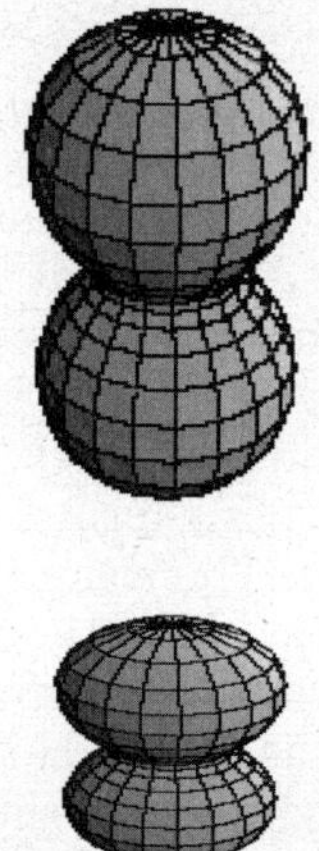

Fig. 17. Shapes of the charge distribution. Top: extreme shape of a rare configuration, bottom: possible medium modification of the same shape.

5 Summary

The predominance of the experimental evidence points to the conclusion that the structure of a nucleon is modified by immersing it in a nucleus. At the moment the use of models is the only way to address this physics. There are a set of minimum requirements for any model. The model should account for or be consistent with nuclear saturation as well as explaining the existing EMC and DY data. After this is achieved, the model should predict independent new phenomena. At the moment, the best hopes for measuring such phenomena lies with measurements of how the electromagnetic form factor is modified by the presence of the medium and in using spectator nucleon tagging in $eA \to e'XN$ experiments. Other possibilities involve trying to constrain the nuclear gluon distribution and measurements of how the nucleus might modify the Callan-Gross relation. There are a variety of possibilities for how new experiments, performed at Jefferson Laboratory, might reveal how quarks work inside the nucleus.

References

1. J. Aubert *et al.*, Phys. Lett. B **123**, 275 (1982); R.G. Arnold *et al.*, Phys. Rev. Lett. **52**, 727 (1984); A. Bodek *et al.*, Phys. Rev. Lett. **51**, 534 (1983).
2. For example, see the reviews: M. Arneodo, Phys. Rep. **240**, 301 (1994); D.F. Geesaman, K. Saito, A.W. Thomas, Annu. Rev. Nucl. Part. Sci. **45**, 337 (1995); G. Piller, W. Weise, Phys. Rep. **330**, 1 (2000).
3. L.L. Frankfurt, M.I. Strikman, Phys. Rep. **160**, 235 (1988).
4. J.W. Chen, W. Detmold, Phys. Lett. B **625**, 165 (2005).
5. G.A. Miller, J.R. Smith, Phys. Rev. C **65**, 015211 (2002); **66**, 049903 (2002)(E).
6. J.R. Smith, G.A. Miller, Phys. Rev. C **65**, 055206 (2002).
7. R.P. Bickerstaff, M.C. Birse, G.A. Miller, Phys. Rev. Lett. **53**, 2532 (1984); M. Ericson, A.W. Thomas, Phys. Lett. B **148**, 191 (1984).
8. D.M. Alde *et al.*, Phys. Rev. Lett. **64**, 2479 (1990).
9. G.F. Bertsch, L. Frankfurt, M. Strikman, Science **259**, 773 (1993).
10. P.A.M. Guichon, Phys. Lett. B **200**, 235 (1988); K. Saito, A.W. Thomas, Nucl. Phys. A **574**, 659 (1994).
11. I.C. Cloet, W. Bentz, A.W. Thomas, Phys. Rev. Lett. **95**, 052302 (2005).
12. D. Diakonov, V.Y. Petrov, arXiv:hep-ph/0009006.
13. J.R. Smith, G.A. Miller, Phys. Rev. Lett. **91**, 212301 (2003); Phys. Rev. C **70**, 065205 (2004); **72**, 022203 (2005); J.R. Smith, arXiv:nucl-th/0508036.
14. L.L. Frankfurt, M.I. Strikman, Nucl. Phys. B **250**, 143 (1985).
15. M.R. Frank, B.K. Jennings, G.A. Miller, Phys. Rev. C **54**, 920 (1996).
16. J.W. Negele, Nucl. Phys. Proc. Suppl. **73**, 92 (1999).
17. S. Strauch *et al.*, Phys. Rev. Lett. **91**, 052301 (2003).
18. R. Schiavilla, O. Benhar, A. Kievsky, L.E. Marcucci, M. Viviani, Phys. Rev. Lett. **94**, 072303 (2005).
19. F. Schlumpf, hep-ph/9211255.
20. G.A. Miller, M.R. Frank, Phys. Rev. C **65**, 065205 (2002); G.A. Miller, Phys. Rev. C **66**, 032201 (2002).
21. O. Gayou *et al.*, Phys. Rev. Lett. **88**, 092301 (2002).
22. G.A. Miller, Phys. Rev. C **68**, 022201 (2003); A. Kvinikhidze, G.A. Miller, Phys. Rev. C **73**, 065203 (2006).

Eur. Phys. J. A **31**, 585–587 (2007)

DOI 10.1140/epja/i2006-10227-1

Special Article – QNP 2006

The transparency of nuclei to nucleons and pions in a relativistic Glauber approximation

J. Ryckebusch[a], W. Cosyn, B. Van Overmeire, and C. Martínez[b]

Ghent University, Ghent, Belgium

Received: 8 November 2006
Published online: 8 March 2007 – © Società Italiana di Fisica / Springer-Verlag 2007

Abstract. We present a selection of results obtained within the context of a relativistic eikonal model. First, results of relativistic Glauber calculations for the nuclear transparency extracted from photon-induced pion production are presented. Second, computed differential cross-sections for the ^{12}C$(p, 2p)$ are compared to data.

PACS. 24.10.Jv Relativistic models – 11.80.-m Relativistic scattering theory – 25.40.-h Nucleon-induced reactions – 25.20.Lj Photoproduction reactions

1 Introduction

Electron scattering facilities like Jefferson lab probe nuclei at the femtometer and sub-femtometer scale and provide observables to look for deviations from predictions of models based on traditional nuclear physics. The interpretation of those experiments heavily depends on the availability of traditional nuclear-physics models which can compute the attenuation of fast nucleons and pions when they travel through nuclei. The ultimate goal of many experiments involving nuclei is mapping the transition from the non-perturbative (mesons and hadrons) to the perturbative (quarks and gluons) regime of QCD.

In this short report, a selection of results obtained with a relativistic extension to Glauber multiple-scattering theory is presented. The framework is dubbed the relativistic multiple-scattering Glauber approximation (RMSGA). The RMSGA model allows to compute the probability that a high-energy nucleon or pion will escape from an excited finite nucleus. The RMSGA model provides a unified framework to compute nuclear transparencies as they can for example be extracted from $A(e, e'p)$, $A(p, 2p)$, $A(p, pn)$ and $A(\gamma, \pi N)$ reactions. Here we concentrate on those channels with two hadrons in the final state.

2 Relativistic multiple-scattering Glauber approximation

Glauber theory is an exactly solvable multiple-scattering framework which allows one to compute the attenuation

effects on an energetic particle when it moves through a medium. The framework is valid under circumstances whereby, $\lambda < r_s < R$, with λ the wavelength of the particle, R the radius of the medium and r_s the typical interaction range between the energetic particle and the spectator particles in the target nucleus. The RMSGA framework introduced in ref. [1] is a relativistic extension of the Glauber model. It adopts the mean-field approximation with bound-state wave functions from the Serot-Walecka model. The framework involves an involving multiple integral which tracks the effect of all collisions of an energetic nucleon or pion with the remaining nucleons in the target nucleus. Thereby, each of the target nucleons acts as a scattering center and is represented by its own relativistic wave function.

3 Results and discussion

First, results of RMSGA calculations for the ^{4}He$(\gamma, \pi^- p)$ reaction are presented. Second, we turn to $A(p, pN)$ which pose a real challenge to models, as they involve three nucleons subject to attenuation effects.

At Jefferson Lab there are ongoing efforts to measure transparencies in $A(\gamma, \pi^- p)$ and $A(e, e'\pi^+ n)$. The combined presence of a pion and a nucleon in the final state will enhance non-perturbative QCD mechanisms, like color transparency (CT). There is one published result from experiment E91013 [2] for ^{4}He$(\gamma, \pi^- p)$ for $1.6 \leq E_\gamma \leq 4.5$ GeV and $\theta_\pi^{c.m.} = 70, 90°$. The choice for ^{4}He as target is considered to optimize medium effects. Not only is it a dense system, the ratio of the hadron formation length to the nuclear radius is large. In ref. [2] the data

---

[a] e-mail: `jan.ryckebusch@rug.ac.be`
[b] *Present address:* Universidad Complutense, Madrid, Spain.

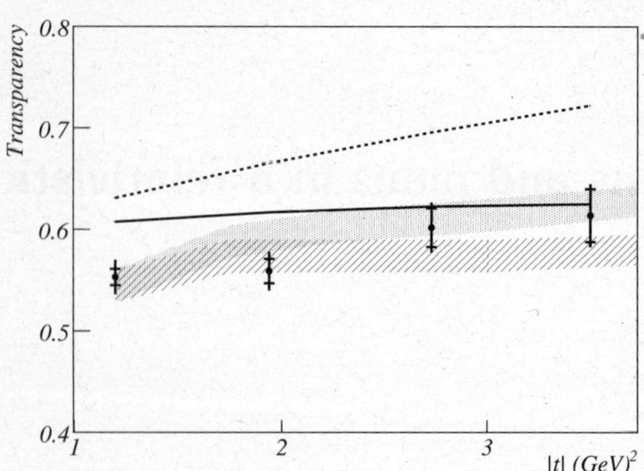

Fig. 1. The nuclear transparency extracted from ^{4}He$(\gamma, p\pi^-)$ *versus* the squared momentum transfer $|t|$ at $\theta^\pi_{c.m.} = 90°$. The solid (dashed) curve is the result of the RMSGA calculations without (with) color transparency. The semi-classical model [3] results are presented by the shaded areas: the hatched (dotted) area is a calculation without (with) CT. Data from [2].

are compared with calculations within the context of the semi-classical transport model by Gao, Holt and Pandharipande [3]. An improved description of the data was obtained after including CT effects. We have developed a quantum-mechanical and relativistic framework to extract nuclear transparencies from $\gamma + A \longrightarrow A - 1 + N + \pi$ reactions. In contrast to the semi-classical models, in the RMSGA model the effect of final-state interactions (FSI) is treated at the amplitude level. Every nucleon in the forward path of the outgoing N and π adds a phase to their respective wave function. Technically, this is achieved by means of the following Glauber phase operator ($\boldsymbol{r}(\boldsymbol{b}, z)$):

$$\mathcal{S}_{\mathrm{FSI}}(\boldsymbol{r}, \boldsymbol{r}_2, \ldots, \boldsymbol{r}_A) = \prod_{j=2}^{A} [1 - \Gamma_{N'N}(\boldsymbol{b} - \boldsymbol{b}_j)\theta(z - z_j)]$$

$$\times \left[1 - \Gamma_{\pi N}\left(\boldsymbol{b}'(\boldsymbol{b}, z) - \boldsymbol{b}'_j(\boldsymbol{b}_j, z_j)\right)\theta(z' - z'_j)\right].$$

The numerical computation of the effect of the above Glauber phase operator requires knowledge about $\pi N \to \pi N$ and $N'N \to N'N$ cross-sections, as well as a set of relativistic mean-field wave functions for the target nucleus. There are no free parameters to tune. The above-mentioned procedure results in involving multidimensional integrals which are performed numerically. The effect of CT is implemented through the model of ref. [4]. In fig. 1 we present results of our transparency calculations together with data and the predictions of the semi-classical model of ref. [3]. The parameters used to compute the CT are such that they maximize (or even, overestimate) the effect. The computed RMSGA nuclear transparencies are systematically about 10% larger than the semi-classical results. The RMSGA predictions overestimate the measured transparancies at small $|t|$ but

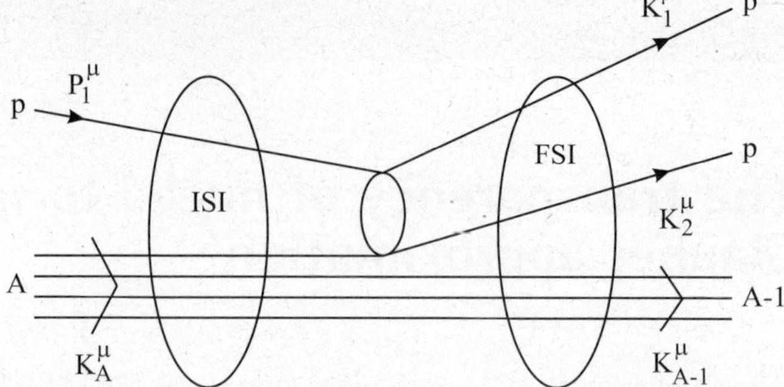

Fig. 2. Schematic representation of the $A(p, 2p)$ reaction. The incoming proton undergoes "soft" initial-state interactions with the target before knocking out a bound proton through the primary high-momentum-transfer pp scattering. Both the scattered and the ejected proton are subject to final-state interactions while leaving the nucleus. The scattered and the ejected proton are detected in coincidence.

do reasonably well for the higher values of $|t|$. It is clear that $|t|$ defines the hard scale and that CT mechanisms grow with this parameter.

As a second type of reaction which can be treated within the context of the relativistic eikonal approximation, we consider $A(p, pp)$ and $A(p, pn)$. A sketch is shown in fig. 2. We consider quasielastic processes: the impinging proton scatters from a single bound nucleon in the target and knocks it out of the target. The "hard" nucleon-nucleon collision is obscured by the "soft" initial- and final-state interactions (IFSI) of the incident and two outgoing nucleons with the nuclear medium. We adopt the cross-section factorized form for the $A(p, pN)$ cross-section

$$\frac{\mathrm{d}^5\sigma}{\mathrm{d}E_{k1}\,\mathrm{d}\Omega_1\,\mathrm{d}\Omega_2} \approx \frac{sM_{A-1}}{M_p M_A}\frac{k_1 k_2}{p_1} f^{-1}_{rec}\, \rho^D_{\alpha_1}\,(\boldsymbol{p}_m) \left(\frac{\mathrm{d}\sigma^{pp}}{\mathrm{d}\Omega}\right)_{\mathrm{c.m.}},$$

where, p_1 is the momentum of the incident proton, (k_1, k_2) those of the two ejectiles and α_1 the quantum number of the bound nucleon on which the hard scattering takes place. In computing $\rho^D_{\alpha_1}$ we consider phase factors from the two ejected nucleons and the impinging proton. Very often, in $A(p, pN)$ reactions one faces situations in which one of the ejectiles is relatively slow and a Glauber multiple-scattering approach is not applicable. To this end, our theoretical framework provides the flexibility to adopt an optical potential and a Glauber approach within the context of the relativistic eikonal approximation. Besides the RMSGA, this gives rise to the so-called relativistic optical model eikonal approximation (ROMEA).

The PNPI $A(p, pN)$ experiments [5] were carried out with an incident proton beam of energy 1 GeV. The scattered proton was detected at $\theta_1 = 13.4°$ with a kinetic energy between 800 and 950 MeV, while the knocked-out nucleon N was observed at $\theta_2 = 67°$ having a kinetic energy below 200 MeV.

Figure 3 displays differential cross-sections for ^{12}C$(p, 2p)$ as a function of the kinetic energy of the most

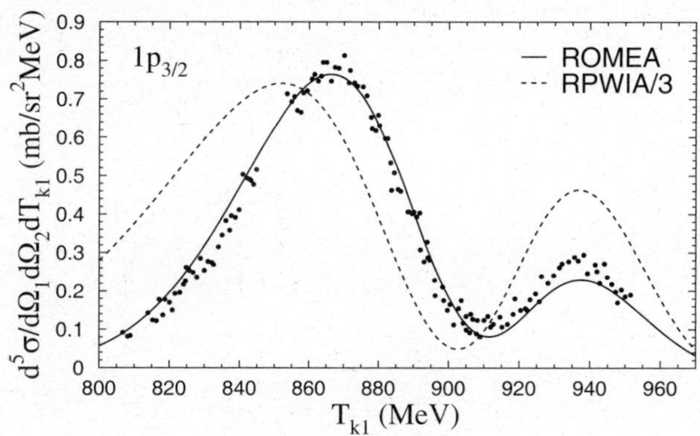

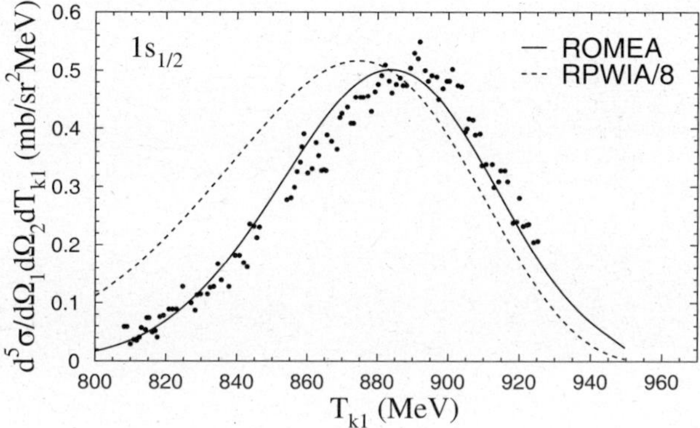

Fig. 3. Differential cross-sections for the ^{12}C$(p, 2p)$ reaction. The solid curve represents the ROMEA calculation, whereas the dashed curve is the plane-wave result reduced by the indicated factor. The ROMEA results are normalized to the data. Data points are from ref. [5]. The magnitude of the experimental error bars is estimated to be of the order of 5–10%.

energetic nucleon in the final state. The EDAI optical potential [6] was used for the ROMEA calculations. The RMSGA approach fails to give an adequate description of the data because of the low kinetic energy of the ejected nucleon. Since the experiment of ref. [5] only measured relative cross-sections, the ROMEA results were normalized to the experimental data.

The ROMEA calculations reproduce the shapes of the measured differential cross-sections. Furthermore, comparison of the relativistic plane-wave impulse approximation (RPWIA) and the ROMEA calculations shows that the effect of the IFSI is twofold. First, IFSI result in a reduction of the RPWIA cross-section that is both level and A-dependent. From fig. 3 it is clear that ejection of a nucleon from a deeper-lying level leads to stronger initial- and final-state distortions. This reflects the fact that the incoming and outgoing nucleons encounter more obstacles when a deeper-lying bound nucleon is probed. Besides the attenuation, the IFSI also make the measured missing momentum different from the initial momentum of the struck nucleon.

4 Conclusions

A relativistic eikonal framework to model the propagation of fast nucleons and pions through the nuclear medium has been developed. The framework adopts a mean-field approach to determine the wave functions of the bound nucleons and can be applied to even-even target nuclei with $A \geq 4$. Relativity is accommodated in both the dynamics and the kinematics of the reactions under study. The effect of FSI can be computed with the aid of optical potentials or in the Glauber approach. The model provides a common framework to describe a variety of nuclear reactions with electroweak and hadronic probes. A profound study [7] of nuclear transparencies extracted from $A(e, e'p)$ processes taught that there is a relatively smooth transition between the typical low-energy optical-potential description of FSI and the high-energy Glauber approach. This conclusion was drawn on the basis of comparable transparency predictions in an intermediate-energy regime where both models can be considered realistic.

In this contribution, results for the ^{4}He$(\gamma, \pi^- p)$ and the ^{12}C$(p, 2p)$ are presented. The computed transparencies for photon-induced pion production ^{4}He$(\gamma, \pi^- p)$ are larger than those from semi-classical models. The model has been extended to electroproduction processes for comparison with the forthcoming data from Jefferson lab. The relativistic eikonal method has also been applied to $A(p, pN)$ reactions. With three nucleons subject to attenuation effects, this reaction provides an excellent testing ground for the adopted assumptions. A fair description of the data for quasielastic proton scattering from ^{12}C, ^{16}O, and ^{40}Ca at 1 GeV and ^{4}He$(p, 2p)$ at 250 MeV is obtained [8]. The model has also been used to study $A(p, 2p)$ transparencies at high energies [9].

References

1. J. Ryckebusch, D. Debruyne, P. Lava, S. Janssen, B. Van Overmeire, T. Van Cauteren, Nucl. Phys. A **728**, 226 (2003).
2. D. Dutta *et al.*, Phys. Rev. C **68**, 052501 (2003).
3. H. Gao, R.J. Holt, V.R. Pandharipande, Phys. Rev. C **54**, 2779 (1996).
4. G.R. Farrar, H. Liu, L.L. Frankfurt, M.I. Strikman, Phys. Rev. Lett. **61**, 686 (1988).
5. S.L. Belostotsky, Yu.V. Dotsenko, N.P. Kuropatkin, O.V. Miklukho, V.N. Nikulin, O.E. Prokofiev, Yu.A. Scheglov, V.E. Starodubsky, A.Yu. Tsaregorodtsev, A.A. Vorobyov, M.B. Zhalov, in *Proceedings of the International Symposium on Modern Developments in Nuclear Physics, Novosibirsk, 1987*, p. 191.
6. E.D. Cooper, S. Hama, B.C. Clark, R.L. Mercer, Phys. Rev. C **47**, 297 (1993).
7. P. Lava, M.C. Martínez, J. Ryckebusch, J.A. Caballero, J.M. Udías, Phys. Lett. B **595**, 177 (2004).
8. B. Van Overmeire, W. Cosyn, P. Lava, J. Ryckebusch, Phys. Rev. C **73**, 0603013 (2006).
9. B. Van Overmeire, J. Ryckebusch, Phys. Lett. B **644**, 304 (2007).

Eur. Phys. J. A **31**, 588–592 (2007)

DOI 10.1140/epja/i2006-10268-4

Special Article – QNP 2006

Nucleon form factors and the BLAST experiment

R. Alarcon[a] and the BLAST Collaboration

Department of Physics, Arizona State University, Tempe, AZ 85287, USA

Received: 8 December 2006
Published online: 16 March 2007 – © Società Italiana di Fisica / Springer-Verlag 2007

Abstract. Measurements of the electric and magnetic form factors of the nucleon present a sensitive test of nucleon models and QCD-inspired theories. A precise knowledge of the neutron form factors at low Q^2 is also essential to reduce the systematic errors of parity violation experiments. At the MIT-Bates Linear Accelerator Center, the nucleon form factors have been measured by means of scattering of polarized electrons from vector-polarized hydrogen and deuterium. The experiment used the longitudinally polarized stored electron beam of the MIT-Bates South Hall Ring along with an isotopically pure, highly vector-polarized internal atomic hydrogen and deuterium target provided by an atomic beam source. The measurements have been carried out with the symmetric Bates Large Acceptance Spectrometer Toroid (BLAST) with enhanced neutron detection capability.

PACS. 25.30.-c Lepton-induced reactions – 25.30.Rw Electroproduction reactions – 29.25.-t Particle sources and targets – 29.30.-h Spectrometers and spectroscopic techniques

1 Introduction

The electromagnetic structure of the nucleon is traditionally described in terms of two form factors and it has been extensively studied by the scattering of electrons on nucleons. Experiments on electron-proton scattering have yielded abundant information on the magnetic G_M^p and charge G_E^p form factors of the proton in the range of momentum transfers squared, Q^2, up to about $30\,(\mathrm{GeV}/c)^2$. The information on the neutron form factors comes almost entirely from the scattering of electrons on deuterons or ^{3}He nuclei, and it is much less definite than that on the proton. The difficulty is due mainly to final-state interactions, mesonic currents, and the fact that the form factors of nucleons inside the deuteron or ^{3}He may differ from those of free particles.

Measurements of the nucleon form factors present a sensitive test of nucleon models and QCD-inspired theories. At low momentum transfer, $Q^2 \leq 1\,(\mathrm{GeV}/c)^2$, the pion cloud of the nucleon may play a significant role in the quantitative description of the form factors [1,2], in particular for the electric form factor of the neutron G_E^n in the absence of a net charge. Accurate measurements of the form factors at low Q^2 are required to reduce systematic uncertainty in the extraction of the strange-quark contribution to the nucleon electromagnetic structure as studied in parity violation electron scattering experiments [3].

In recent years the advent of polarized beams, targets, and polarimetry have made possible new classes of ex-

periments aimed at extracting the nucleon form factors utilizing spin degrees of freedom. The general form for the differential cross-section in the exclusive scattering of longitudinally polarized electrons from a polarized target is given by [4]

$$\frac{\mathrm{d}\sigma}{\mathrm{d}e'\mathrm{d}\Omega_e'\mathrm{d}\Omega_N} = \Sigma + h\Delta\,, \tag{1}$$

where h is the helicity of the incident electron, and Σ and Δ are the helicity-sum and helicity-difference cross-sections, respectively. The Σ and Δ cross-sections for the case of elastic scattering from a polarized nucleon can be written as

$$\Sigma = c\left(\rho_L G_E^2 + \rho_T \frac{q^2}{2M^2} G_M^2\right) \tag{2}$$

and

$$\Delta = -c\left(\rho_{LT}' \frac{q}{2^{3/2}M} G_E G_M P_x + \rho_T' \frac{q^2}{2M^2} G_M^2 P_z\right), \tag{3}$$

where c is a kinematical factor, the ρ's are the virtual photon densities, and P_j indicates the polarization of the nucleon along each of the three coordinate axes, of which the z-axis has been chosen parallel to the momentum transfer $\boldsymbol{q}$. The terms containing $G_E G_M$ and G_M^2 can be completely isolated by tuning the target spin polarization. Experimentally one measures spin asymmetries defined as

$$A_{exp} = p_e p_T \frac{\Delta}{\Sigma}\,, \tag{4}$$

[a] e-mail: ralarcon@asu.edu

where p_e and p_T are the electron beam and target polarization, respectively. To separate both terms of the Δ cross-section the target spin is oriented perpendicular (parallel) to the direction of $\boldsymbol{q}$ (*i.e.*, selecting P_x (P_z)).

This work reports preliminary results from new measurements of the nucleon form factors over a range of four-momentum transfer Q^2 between 0.12 and $0.70 \, (\mathrm{GeV}/c)^2$ with the BLAST experiment at the MIT-Bates Linear Accelerator Center. The technique makes use of elastic ep scattering of polarized electrons from polarized hydrogen to access the proton form factors, quasi-elastic $(e, e'n)$ scattering from vector-polarized deuterium to get at the charge form factor of the neutron, and inclusive (e, e') scattering from vector-polarized deuterium for the magnetic form factor of the neutron.

2 The BLAST experiment

The BLAST experiment has been designed to measure spin-dependent electron scattering at intermediate energies from polarized targets in the elastic, quasi-elastic and resonance region. Based on the internal-target technique BLAST optimizes the use of a longitudinally polarized electron beam stored in the South Hall Ring of the MIT-Bates Linear Accelerator Center, in combination with an isotopically pure, highly polarized internal target for both hydrogen or deuterium. In case of deuterium the target was both vector and tensor polarized. The polarized target is provided by an Atomic Beam Source (ABS) [5]. The ejected gas molecules are first dissociated into atoms before they pass sections of sextupole magnets and RF transition units to populate the desired single-spin states through Stern-Gerlach beam splitting and induced transitions between hyperfine states (see fig. 1).

This selection process is highly efficient and thus provides nuclear polarizations of more than 70%. The spin state selection was altered every five minutes in a random sequence to minimize systematics. The polarized atoms are injected into a 60 cm long cylindrical target cell with open ends through which the stored electron beam passes. As there are no target windows the experiment is very clean with negligibly small background rates of only a few percent in the prominent channels.

The direction of the target spin can be freely chosen within the horizontal plane using magnetic holding fields. During BLAST data taking, the spin direction pointed at 32° and 47° to the left side of the beam axis in the 2004 and 2005 runs, respectively. At Bates beam currents of up to 225 mA were stored in the ring at 65% polarization and beam lifetimes of 20–30 minutes. The electron beam energy was 850 MeV throughout the BLAST program. The relatively thin target in combination with the high beam intensity yields a luminosity of about $5 \times 10^{31}/(\mathrm{cm}^2 \, \mathrm{s})$ at an average current of 175 mA. The Bates storage ring contains a Compton polarimeter to monitor the longitudinal beam polarization in real time and without affecting the beam. The electron spin precession is compensated with a spin rotator (Siberian snake) in the ring section oppo-

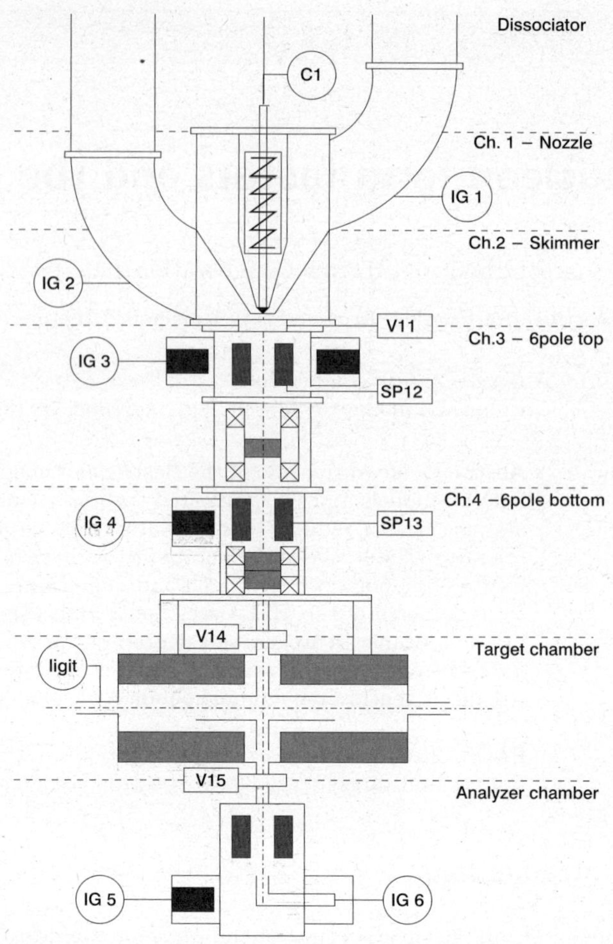

Fig. 1. Schematics of the Atomic Beam Source and the target region. The ABS was embedded in the strong, spatially varying magnetic field of the BLAST toroid [5].

site of BLAST. The helicity of the beam was flipped once before every ring fill.

The BLAST detector is schematically shown in fig. 2. It was built as a toroidal spectrometer consisting of eight normal-conducting copper coils producing a maximum field of 3800 G. The two in-plane sectors opposing each other are symmetrically equipped with drift chambers for the reconstruction of charged tracks, aerogel-Cerenkov detectors for e/π discrimination and $1''$ thick plastic scintillators for timing, triggering and particle identification. The angular acceptance covers scattering angles between 20° and 80° as well as $\pm 15°$ out of plane. The symmetric detector core is surrounded by thick large-area walls of plastic scintillators for the detection of neutrons using the time-of-flight method. The thin scintillators in combination with the voluminous wire chambers in front of the neutron detectors were used as a highly efficient veto for charged tracks, making the selection of $(e, e'n)$ events extremely clean. The neutron detectors are enhanced in the right sector with $\approx 30\%$ neutron detection efficiency ($\approx 10\%$ in the left sector). The reason for this is because

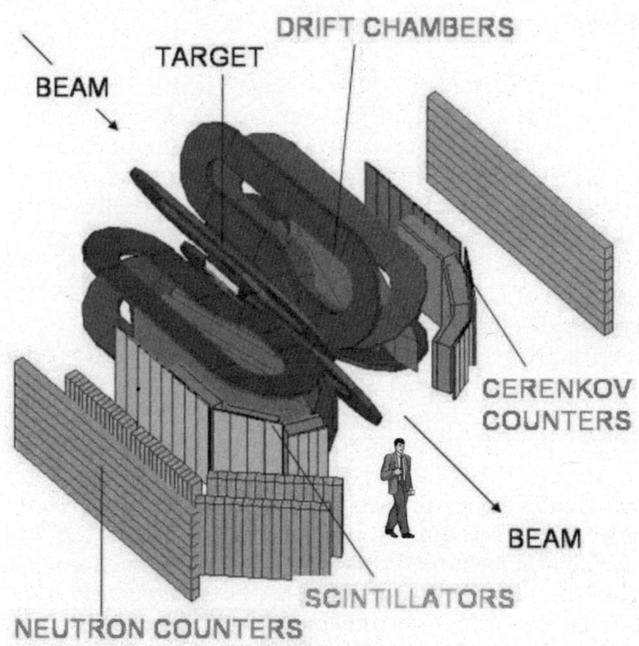

Fig. 2. Schematic, isometric view of the BLAST detector showing the main detector elements.

of the choice of the target spin orientation. The setup allows to simultaneously measure the inclusive and exclusive channels (e, e'), $(e, e'p)$, $(e, e'n)$, $(e, e'd)$ elastic or quasi-elastic, respectively, as well as $(e, e'\pi)$ in the excitation region of the Δ-resonance. By measuring many reaction channels at the same time over a broad range of momentum transfer, the systematic errors are minimal.

3 Results

3.1 Proton electric to magnetic form factor ratio

The polarized hydrogen data were divided into eight Q^2 bins and the yield distributions were in good agreement with results from a Monte Carlo simulation, including all detector efficiencies measured from the data. The BLAST detector configuration is symmetric about the incident electron beam and the target polarization angle is oriented $\sim 45°$ to the left of the beam. Two independent asymmetries of electrons scattered into the beam-left and beam-right sectors, respectively, can be measured simultaneously. The ratio G_E^p/G_M^p can then be determined, independent of p_e and p_T, from the ratio of these experimental asymmetries measured at the same Q^2 value but corresponding to different spin orientations [6].

Preliminary results for G_E^p/G_M^p are shown in fig. 3 with the error bars due to statistical and systematic contributions added in quadrature. Also shown in fig. 3 are published recoil polarization data [7–13], together with a few selected models: a soliton model [14], an extended vector meson dominance model [15], an updated dispersion model [16], a relativistic Constituent-Quark Model

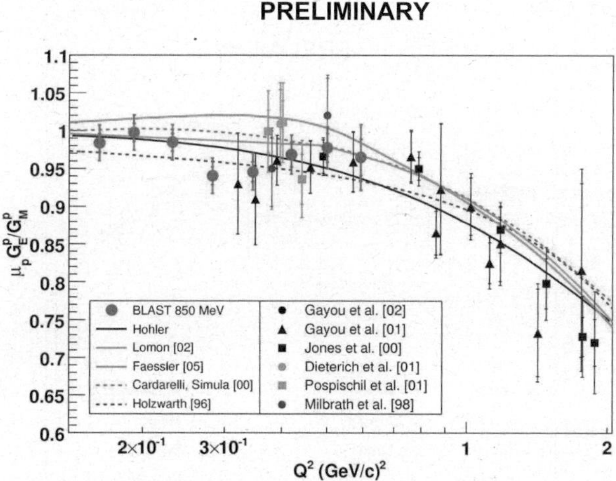

Fig. 3. Preliminary results of $\mu_p G_E^p/G_M^p$ shown with the world polarized data and several models described in the text.

(CQM) with $SU(6)$ symmetry breaking and a constituent-quark form factor [17], and a Lorentz covariant chiral quark model [2]. Using these results and the world differential cross-section data on e-p elastic scattering at the same Q^2 values, the proton electric and magnetic form factors can be extracted and work to this extent is in progress.

3.2 Neutron magnetic form factor

The first experiments to measure G_M^n used electron scattering off unpolarized deuterium in elastic or quasi-elastic kinematics [18–20]. These methods involved significant uncertainties due to the required proton contribution subtraction. Alternatively, the ratio of the cross-sections from the reactions ^{2}H$(e, e'n)p$ and ^{2}H$(e, e'p)n$ in quasi-elastic kinematics was used by [21–25]. By measuring the neutron in coincidence, the theoretical uncertainties related to the proton subtraction are eliminated. Furthermore, using the cross-section ratio reduces the dependence of the measurement on nuclear structure. However, these experiments require accurate knowledge of neutron detection efficiency, which is difficult to establish.

Exploiting the recent advances in polarization techniques, the inclusive quasi-elastic reaction ^{3}He(e, e') has been used for measurements of G_M^n [26–28]. Since in the ground state of ^{3}He the proton spins anti-align and cancel, the spin of the nucleus is carried by the unpaired neutron [29]. By the simultaneous use of beam and target polarization the experimental asymmetry is related to the ratio of the form factors of the neutron. The systematic uncertainties for these measurements stem mainly from Final-State Interaction (FSI) and reaction mechanism (Meson Exchange Currents (MEC) and Isobar Configurations (IC)) corrections.

With BLAST an alternative measurement of G_M^n has been carried out by using inclusive electron scattering with polarized beam and vector-polarized deuterium target, ^{2}H(e, e'), in quasi-elastic kinematics. The two-body

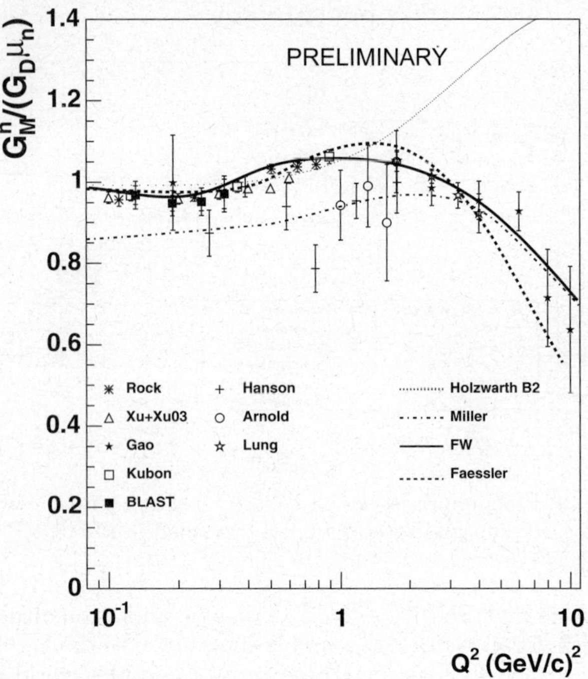

Fig. 4. Sample of the world's data on G_M^n along with the results of the BLAST experiment. The systematic and statistical errors for the BLAST data are added in quadrature. Holzwarth B2 is a soliton model [30], Miller [31] is a cloudy bag model, Faessler [2] is the chiral perturbation theory calculation and FW is the Friedrich and Walcher parametrization [1]. The data are taken from [18–20, 32, 24, 26–28] and the present work.

nature of the deuteron allows analytic solutions to its nuclear structure. Additionally, in the quasi-elastic regime, the FSI, MEC and IC are fairly small. Two data sets were collected, one with target polarization angle of $32°$ in 2004 ($320\,\mathrm{kC}$) and one with $47°$ in 2005 ($550\,\mathrm{kC}$). The asymmetries, the ratio and the sensitivity to G_M^n vary with the target polarization angle. Therefore, the form factor was extracted separately and the results were combined, weighted by their respective error bars. The G_M^n results from this work are shown in fig. 4 along with a selection of the world's data and theoretical calculations.

The systematic uncertainties are dominated by the uncertainty in the target polarization angle which was known to $0.5°$. This uncertainty amounted to $5\%/$degree for the $32°$ and $3\%/$degree for the $47°$ data sets. The second most important uncertainty was due to reconstruction and resolution which amounted to about 1.5%. Radiative effects on the asymmetries in both perpendicular and parallel kinematics were calculated using the MASCARAD software package by approximating the inclusive electrodisintegration channel as an incoherent sum of a proton and a neutron. The uncertainty due to radiative effects associated with this measurement was less than 1%.

The BLAST data agree fairly well with the recent parametrization of [1] and the calculations of [2]. Both of these show a dip in the form factor centered at $Q^2 \approx 0.2\,(\mathrm{GeV}/c)^2$ with a width of $\approx 0.2\,(\mathrm{GeV}/c)^2$.

3.3 Neutron electric form factor

The low-Q^2 region of G_E^n is an ideal testing ground for QCD– and pion-cloud–inspired and other effective nucleon models. Among the four nucleon electromagnetic form factors, G_E^n is experimentally the least known one with uncertainties of typically 15–20%. Significant improvement of the experimental uncertainty is highly desirable and is setting strong constraints for nucleon models. A precise knowledge of G_E^n at low Q^2 is also essential to reduce the systematic errors of parity violation experiments.

With BLAST measurements of G_E^n have been carried out by means of $(e, e'n)$ quasi-elastic scattering using polarized electrons and a vector-polarized deuterium target. The experimental double-spin asymmetry is formed from the measured $(e, e'n)$-yields in each beam-target spin state combination, properly normalized to the collected deadtime-corrected beam charge. For five bins in Q^2, the experimental asymmetry as a function of missing momentum is compared with the full BLAST Monte Carlo result based on deuteron electrodisintegration cross-section calculations by H. Arenhövel [33] with consistent inclusion of reaction mechanism and deuteron structure effects. The electric form factor of the neutron is varied as an input parameter to the Monte Carlo simulation and its measured value is extracted by a χ^2 minimization for each Q^2 bin.

Figure 5 shows the preliminary result for G_E^n from the 2004 run of BLAST along with the world data from polarization experiments [34]. Also shown is the parameterization by Galster *et al.* [35] $G_E^n = 1.91\tau/(1 + 5.6\tau)G_{Dipole}$, where $\tau = Q^2/(4M_n^2)$. The excess of the data over the

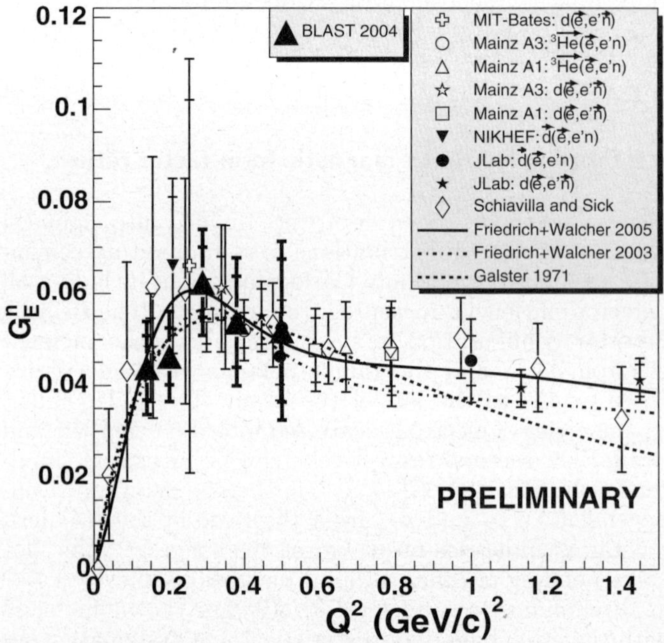

Fig. 5. Electric form factor of the neutron from polarization experiments [34] along with preliminary results from BLAST. The curves are the original parameterization by Galster *et al.* [35] and the recent parameterizations by Friedrich and Walcher [1].

Galster curve at high and at low Q^2 is better accounted for by the more recent parameterization by Friedrich and Walcher [34,1] (FW), who describe all four nucleon form factors as sums of a smooth and a bump part, where the latter is attributed to the role of the pion cloud around the nucleon. The new preliminary BLAST data are quite consistent with both the bulk of existing data as well as with the parameterizations shown in fig. 5, of which the FW parameterization appears slightly favored. Note that the BLAST data is preliminary and based on only about half of the statistics that were acquired in the total run in 2004 and 2005. The preliminary results shown here comprise parts of a PhD Thesis [36] based on the BLAST data taken in 2004. Analysis of the full 2004–2005 data set is in progress [37].

4 Summary

Preliminary results for the electromagnetic form factors of the nucleon have been obtained with the BLAST experiment. The technique utilizes a combination of polarized electron beam, polarized internal gas targets, and a large-acceptance detector. These new measurements cover a range of four-momentum transfer Q^2 between 0.12 and 0.70 $(\text{GeV}/c)^2$. The proton electric to magnetic form factor ratio was measured by elastic ep scattering of polarized electrons from polarized hydrogen. The neutron electric and magnetic form factors were extracted from simultaneous measurements of quasi-elastic $(e, e'n)$ scattering and inclusive (e, e') scattering from vector-polarized deuterium, respectively. The systematic uncertainties are small due to the use of spin degrees of freedom.

References

1. J. Friedrich, T. Walcher, Eur. Phys. J. A **17**, 607 (2003).
2. A. Faessler, Th. Gutsche, V.E. Lyubovitskij, K. Pumsaard, Phys. Rev. D **73**, 114021 (2006).
3. D.S. Armstrong *et al.*, Phys. Rev. Lett. **95**, 092001 (2005).
4. T.W. Donnelly, A.S. Raskin, Ann. Phys. **169**, 247 (1986).
5. D. Cheever *et al.*, Nucl. Instrum. Methods A **556**, 410 (2006); L.D. van Buuren *et al.*, Nucl. Instrum. Methods A **474**, 209 (2001).
6. C.B. Crawford, PhD Thesis, Massachusetts Institute of Technology (2005).
7. M. Jones *et al.*, Phys. Rev. Lett. **84**, 1398 (2000).
8. O. Gayou *et al.*, Phys. Rev. Lett. **88**, 092301 (2002).
9. V. Punjabi *et al.*, Phys. Rev. C **71**, 055202 (2005).
10. O. Gayou *et al.*, Phys. Rev. C **64**, 038202 (2001).
11. S. Dieterich *et al.*, Phys. Lett. B **500**, 47 (2001).
12. T. Pospischil *et al.*, Eur. Phys. J. A **12**, 125 (2001).
13. B. Milbrath *et al.*, Phys. Rev. Lett. **80**, 452 (1998); **82**, 2221 (1999)(E).
14. G. Holzwarth, Z. Phys. A **356**, 339 (1996).
15. E.L. Lomon, Phys. Rev. C **66**, 045501 (2002).
16. H.-W. Hammer, Ulf-G. Meissner, Eur. Phys. J. A **20**, 469 (2004).
17. F. Cardarelli, S. Simula, Phys. Rev. C **62**, 065201 (2000); S. Simula, e-print nucl-th/0105024.
18. S. Rock *et al.*, Phys. Rev. Lett. **49**, 1139 (1982).
19. R.G. Arnold *et al.*, Phys. Rev. Lett. **61**, 806 (1988).
20. A. Lung *et al.*, Phys. Rev. Lett. **70**, 718 (1993).
21. P. Markowitz *et al.*, Phys. Rev. C **48**, R5 (1993).
22. H. Anklin *et al.*, Phys. Lett. B **336**, 313 (1994).
23. E.E.W. Bruins *et al.*, Phys. Rev. Lett. **75**, 21 (1995).
24. G. Kubon *et al.*, Phys. Lett. B **524**, 26 (2002).
25. H. Anklin *et al.*, Phys. Lett. B **428**, 248 (1998).
26. H. Gao *et al.*, Phys. Rev. C **50**, R546 (1994).
27. W. Xu *et al.*, Phys. Rev. Lett. **85**, 2900 (2000).
28. W. Xu *et al.*, Phys. Rev. C **67**, R012201 (2003).
29. J.L. Friar *et al.*, Phys. Rev. C **42**, 2310 (1990).
30. G. Holzwarth, Z. Phys. A **356**, 339 (1996).
31. G.A. Miller, Phys. Rev. C **66**, 032201 (2002).
32. K.M. Hanson *et al.*, Phys. Rev. D **8**, 753 (1973).
33. H. Arenhövel, W. Leidemann, E.L. Tomusiak, Eur. Phys. J. A **23**, 147 (2005).
34. D.I. Glazier *et al.*, Eur. Phys. J. A **24**, 101 (2005) and references therein.
35. S. Galster *et al.*, Nucl. Phys. B **32**, 221 (1971).
36. V. Ziskin, PhD Thesis, Massachusetts Institute of Technology (2005).
37. E. Geis, PhD Thesis, Arizona State University, in preparation.

Eur. Phys. J. A **31**, 593–596 (2007)
DOI 10.1140/epja/i2006-10196-3

Special Article – QNP 2006

Opportunities with Drell-Yan scattering: Probing sea quarks in the nucleon and nuclei

P.E. Reimer[a]

Representing the Fermilab E866/NuSea and E906/Drell-Yan Collaborations
Physics Division, Argonne National Laboratory, Argonne, IL 60439, USA

Received: 8 October 2006
Published online: 22 February 2007 – © Società Italiana di Fisica / Springer-Verlag 2007

Abstract. The large $\bar{d}(x)/\bar{u}(x)$ ratio observed by Fermilab E866/NuSea convincingly demonstrated that the sea is not simply a result of pQCD. Moreover, meson cloud models also failed to explain fully the observed kinematic dependence. The Drell-Yan mechanism offers a unique, selective probe of antiquarks in the nucleon. Fermilab has approved a new Drell-Yan experiment, E906, that will exploit this feature to probe $\bar{d}(x)/\bar{u}(x)$ by measuring the ratio of cross-sections for the proton-induced Drell-Yan process on hydrogen to deuterium. When the nucleon is contained in a nucleus, the nucleon's parton distributions should to be modified; although this effect was not seen in the sea quark distributions obtained by Fermilab E772 with Drell-Yan scattering. The upcoming E906 Drell-Yan experiment will provide much more precise measurements over a wider kinematic range in order to guide and challenge the theoretical models.

PACS. 14.20.Dh Protons and neutrons – 14.65.Bt Light quarks

1 Introduction

The quark-level structure of the nucleon and the nucleus has been studied extensively with deep inelastic scattering (DIS). The electromagnetic DIS probe, while yielding a remarkable amount of information on this structure, lacks the basic ability to distinguish between quark and antiquark (or alternatively valence and sea) distributions of the target nucleon, thus leaving unanswered questions about sea quark distributions, their origins and their modifications in a nucleus. The Drell-Yan process provides a probe that is sensitive to the antiquark distributions of the interacting hadrons.

In leading order, the Drell-Yan process [1] is the annihilation of a quark in one hadron with an antiquark in a second hadron to form a virtual photon. The virtual photon decays into a lepton-antilepton pair which is detected. The cross-section is dependent on the charge-weighted sum of the distributions of quarks and antiquarks in the interacting hadrons:

$$\frac{\mathrm{d}^2\sigma}{\mathrm{d}x_1\mathrm{d}x_2} = \frac{4\pi\alpha^2}{9sx_1x_2} \sum_i e_i^2 \left[q_{1i}(x_1,Q^2)\bar{q}_{2i}(x_2,Q^2) \right.$$
$$\left. + \bar{q}_{1i}(x_1,Q^2)q_{2i}(x_2,Q^2) \right], \quad (1)$$

where q_{1i} (q_{2i}) are the beam (target) quark distributions, the sum is over all quark flavors (u, d, s, c, b, t) and e_i is the

quark charge. The fraction of the longitudinal momentum of the beam (target) carried by the participating quarks is $x_{1(2)}$. For a fixed target experiment, the squared total energy of the beam-target system is $s = 2m_2E_1 + m_1^2 + m_2^2$ with beam energy E_1 and $m_{1(2)}$ the rest mass of the beam (target) hadron.

In a fixed target environment, Drell-Yan scattering has a unique sensitivity to the antiquark distribution of the target hadron. The decay leptons are boosted far forward. This, combined with the acceptance of the typical dipole-based spectrometer, restricts the kinematic acceptance of the detector to $x_F \gtrsim 0$ and, consequently, to very high values of x_1 where the antiquark distributions are suppressed by several orders of magnitude relative to the quark distributions. These beam valence quarks must then annihilate with an antiquark in the target, thus preferentially selecting the first term in (1). This feature has been used by several recent experiments to study the sea quark distributions in the nucleon and in nuclei.

2 Isospin symmetry of the light-quark sea

For many years, it was believed that the proton's sea quark distributions were $\bar{d}$-$\bar{u}$ symmetric, because of approximately equal splitting of gluons into $d\bar{d}$ and $u\bar{u}$ pairs. The observation of a violation of the Gottfried Sum Rule [2] in muon DIS by the New Muon Collaboration [3,4] forced this belief to be reconsidered. The sensitivity of Drell-Yan

---
[a] e-mail: reimer@anl.gov

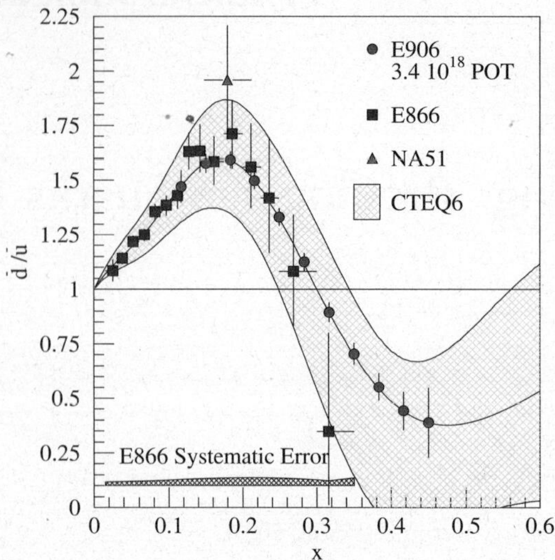

Fig. 1. (Colour on-line) Measurement of $\bar{d}(x)/\bar{u}(x)$ by E866/NuSea [10,11] (blue squares) and NA51 [6] (green triangle) are shown. The central curve in the cross filled band shows the $\bar{d}/\bar{u}$ ratio and uncertainty from the CTEQ5M fit, which included the E866/NuSea and NA51 data. The red circles represent the expected statistical uncertainties of the E906 experiment. The expected systematic uncertainty is approximately 1% [12].

scattering to antiquark distributions makes it an ideal probe of this asymmetry [5], and this was used by the CERN NA51 experiment [6] to verify, at $x = 0.18$, the inequality of $\bar{d}$ and $\bar{u}$ suggested by the Gottfried Sum Rule violation. In leading order, assuming $x_1 \gg x_2$ and the dominance of the $u\bar{u}$ annihilation term, the ratio (per nucleon) of the proton-proton to proton-deuterium Drell-Yan yields can be expressed as

$$\left.\frac{\sigma_{pd}}{2\sigma_{pp}}\right|_{x_1 \gg x_2} = \frac{1}{2}\left[1 + \frac{\bar{d}(x_2)}{\bar{u}(x_2)}\right]. \qquad (2)$$

The next–to–leading-order terms in the cross-section provide a small correction to this *ratio* and were considered in the analysis of the data, as well as the deviation from the $x_1 \gg x_2$ limit.

The Fermilab E866/NuSea experiment used this sensitivity to measure the x-dependence of the $\bar{d}/\bar{u}$ ratio. The E866/NuSea spectrometer consisted of two dipole magnets, the first primarily focused the muons into the spectrometer while the second performed a momentum measurement. This experiment used 800 GeV protons extracted from the Fermilab Tevatron incident on hydrogen and deuterium targets. The remainder of the beam that did not interact in the targets was intercepted by a copper beam dump contained within the first magnet. By using three different magnetic-field settings in the first two spectrometer magnet, the experiment was able to collect data over a broad range of kinematics covering $0.015 \leq x_2 \leq 0.35$. From the measured ratio of Drell-Yan yields, $\sigma^{pd}/(2\sigma^{pp})$, E866/NuSea was able to extract the ratio $\bar{d}(x)/\bar{u}(x)$ shown in fig. 1. The inclusion of the

measured cross-section ratios in global parton distribution fits [7–9] validated this extraction and completely changed the perception of the sea quark distributions in the nucleon.

The E866/NuSea data present an interesting picture of the sea quark distributions of the nucleon that may shed light on the origins of the sea quarks. At moderate values of x the data show more than 60% excess of $\bar{d}$ over $\bar{u}$, but as x grows larger, this excess disappears and the sea appears to be symmetric again. If the sea's origins are purely perturbative, then it is expected to exhibit only a very small asymmetry between $\bar{d}$ and $\bar{u}$. Non-perturbative explanations for the origin of the sea including meson cloud models, chiral perturbation theory or instantons can explain a large asymmetry, but not the return to a symmetric sea seen as $x \to 0.3$. (For a brief review of these models see [13] and referenced therein.) None of the models predicts an excess of $\bar{u}$ over $\bar{d}$ as shown by the CTEQ [7], MRST [8] or GRV [9] global parton distribution *fits*.

Unfortunately, as x increases beyond 0.25, the data become less precise and the exact trend of $\bar{d}/\bar{u}$ is not clear. To help understand this region better, the Fermilab E906 experiment has been approved to make collect Drell-Yan data in this region. The E906 experiment will use a 120 GeV proton beam rather than the 800 GeV beam used by E866. The Fermilab E906/Drell-Yan spectrometer [12] is modeled after its predecessors, Fermilab E772 and E866/NuSea. Experimentally, the lower beam energy has two significant advantages. First, the primary background in the experiment comes from J/ψ decays, the cross-section of which scales roughly with s, the square of the center-of-mass energy. The lower beam energy implies less background rate in the spectrometer and allows for a correspondingly higher instantaneous luminosity. Second, as seen in (1) the Drell-Yan cross-section is inversely proportional to s; thus, the lower beam energy provides a larger cross-section. The muons produced in a 120 GeV collision have a significantly smaller boost, which forces the apparatus to be shortened considerably in order to maintain the same transverse momentum acceptance. The expected statistical uncertainties of the E906/Drell-Yan experiment are shown in fig. 1. Systematic uncertainty in $\bar{d}/\bar{u}$ is expected to be approximately 1%.

3 Antiquark distributions of nuclei

The distributions of partons within a free nucleon differ from those of a nucleon bound within a heavy nucleus, an effect first discovered by the European Muon Collaboration (EMC) in 1983 [14]. (For a review of the Nuclear EMC effect, see [15].) Almost all of the data on nuclear dependencies is from charged lepton DIS experiments, which are sensitive only to the charge-weighted sum of all quark and antiquark distributions. This type of experiment is unable to distinguish between valence and sea effects. Sea quark nuclear effects may be entirely different from those in the valence sector [16], but an electron or muon DIS experiment would not be sensitive to this.

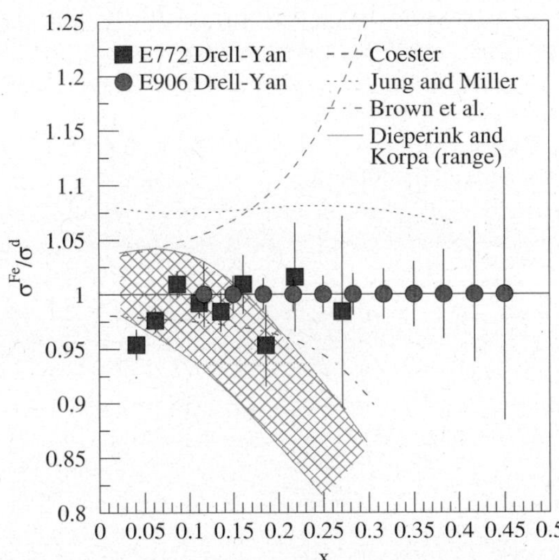

Fig. 2. (Colour on-line) Ratio of iron to deuterium Drell-Yan cross-sections measured by Fermilab E772 (blue squares) [17]. The expected sensitivity of the E906 experiment is shown by red circles. To illustrate the level of effects expected, curves based on several different representative models are also plotted.

In fact, no modification to the sea quark distributions, aside from shadowing at low x, was observed by Fermilab E772 [17] which studied the nuclear dependence of Drell-Yan scattering. (See fig. 2.) Because the widely accepted models of nuclear binding rely on the exchange of virtual mesons [18], a significant enhancement in the antiquark distribution in nuclei compared with deuterium was expected from the antiquarks in the virtual mesons. The non-observation of this enhancement calls these models into question. The expected enhancement of $\bar{u}$ quarks in iron relative to deuterium based on the nuclear convolution model calculations by Coester [19–21] is illustrated in fig. 2.

The lack of sea quark nuclear effects prompted a number of newer models. Jung and Miller [22] revisited the calculations of Berger and Coester [19,20] and examined the effect of the quantization of the pions on the light cone *versus* at "equal time", finding a roughly flat 8% increase in the Drell-Yan iron cross-section over deuterium. Brown *et al.* [23] argue that with the partial restoration of chiral symmetry, the mass of light-quark hadrons decreases with density. This rescaling leads to altered couplings and to an overall *decrease* in the Drell-Yan cross-section in nuclei. Based on a particle- and delta-hole model, which results in a strong distortion of the free-pion structure function, Dieperink and Korpa [24] also find cross-section decrease in nuclei. Finally, Smith and Miller [25] in a chiral quark-soliton model, find no nuclear dependence in Drell-Yan scattering. This model is significant in that it is able to simultaneously describe the EMC effect.

For $x > 0.2$, the E772 statistical uncertainties allow some freedom for these models and the data are not able to distinguish between them. To understand these models and nuclear binding better, higher precision data at larger

x are needed. E906/Drell-Yan will be able to provide these data with the statistical precision shown in fig. 2.

4 Conclusions

Previous Drell-Yan experiments have contributed greatly to the understanding of the antiquark distributions of the nucleon and their modifications by a nuclear environment. The Fermilab E906/Drell-Yan experiment will revisit many of these measurements with improved statistical precision and greater kinematic reach. The E906/Drell-Yan experiment has been approved by the Fermilab PAC and should begin collecting data in 2009. These data will allow E906/Drell-Yan to extend the measurement of $\bar{d}/\bar{u}$ to $x = 0.45$, thereby probing the region in which the sea appears to become flavor symmetric. They will also determine the nuclear dependence of the sea over the same range in x, with a statistical precision that will challenge the current models of nuclear binding.

The E866/NuSea Collaboration would like to thank W.K. Tung of CTEQ for providing the code to calculate the next-to-leading-order Drell-Yan cross-sections. This work was supported in part by the U.S. Department of Energy, Office of Nuclear Physics, under Contract No. W-31-109-ENG-38.

References

1. S.D. Drell, T.M. Yan, Phys. Rev. Lett. **25**, 316 (1970).
2. K. Gottfried, Phys. Rev. Lett. **18**, 1174 (1967).
3. P. Amaudruz, M. Arneodo, A. Arvidson, B. Badelek, G. Baum, J. Beaufays, I.G. Bird, M. Botje, C. Broggini, W. Brückner *et al.*, Phys. Rev. Lett. **66**, 2712 (1991).
4. M. Arneodo, A. Arvidson, B. Badelek, M. Ballintijn, G. Baum, J. Beaufays, I.G. Bird, P. Björkholm, M. Botje, C. Broggini *et al.*, Phys. Rev. D **50**, R1 (1994).
5. S.D. Ellis, W.J. Stirling, Phys. Lett. B **256**, 258 (1991).
6. NA51 Collaboration (A. Baldit *et al.*), Phys. Lett. B **332**, 244 (1994).
7. J. Pumplin *et al.*, JHEP **07**, 012 (2002) `hep-ph/0201195`.
8. A.D. Martin, R.G. Roberts, W.J. Stirling, R.S. Thorne, Eur. Phys. J. C **39**, 155 (2005) `hep-ph/0411040`.
9. M. Gluck, E. Reya, A. Vogt, Eur. Phys. J. C **5**, 461 (1998) `hep-ph/9806404`.
10. FNAL E866/NuSea Collaboration (E.A. Hawker *et al.*), Phys. Rev. Lett. **80**, 3715 (1998) `hep-ex/9803011`.
11. FNAL E866/NuSea Collaboration (R.S. Towell *et al.*), Phys. Rev. D **64**, 052002 (2001) `hep-ex/0103030`.
12. E906/Drell-Yan Collaboration (P. Reimer, D. Geesaman *et al.*), Experiment update to Fermilab PAC (2006).
13. E866/NuSea Collaboration (J.C. Peng *et al.*), Phys. Rev. D **58**, 092004 (1998) `hep-ph/9804288`.
14. European Muon Collaboration (J.J. Aubert *et al.*), Phys. Lett. B **123**, 275 (1983).
15. D.F. Geesaman, K. Saito, A.W. Thomas, Annu. Rev. Nucl. Part. Sci. **45**, 337 (1995).
16. S.A. Kulagin, R. Petti, Nucl. Phys. A **765**, 126 (2006) `hep-ph/0412425`.
17. D.M. Alde *et al.*, Phys. Rev. Lett. **64**, 2479 (1990).

18. J. Carlson, R. Schiavilla, Rev. Mod. Phys. **70**, 743 (1998).
19. E.L. Berger, F. Coester, R.B. Wiringa, Phys. Rev. D **29**, 398 (1984).
20. E.L. Berger, F. Coester, Phys. Rev. D **32**, 1071 (1985).
21. F. Coester, private communication (2001).
22. H. Jung, G.A. Miller, Phys. Rev. C **41**, 659 (1990).
23. G.E. Brown, M. Buballa, Z.B. Li, J. Wambach, Nucl. Phys. A **593**, 295 (1995) `nucl-th/9410049`.
24. A.E.L. Dieperink, C.L. Korpa, Phys. Rev. C **55**, 2665 (1997) `nucl-th/9703025`.
25. J.R. Smith, G.A. Miller, Phys. Rev. Lett. **91**, 212301 (2003) `nucl-th/0308048`.

Eur. Phys. J. A **31**, 597–599 (2007)
DOI 10.1140/epja/i2006-10207-5

Special Article – QNP 2006

Strange quarks in the nucleon sea

Results from HAPPEX II

K.A. Aniol[1,a] and HAPPEX Collaboration[2]

[1] Physics and Astronomy Department, California State University, Los Angeles, CA 90032, USA
[2] Jefferson Lab, 12000 Jefferson Avenue, Newport News, VA 23606, USA

Received: 25 October 2006
Published online: 23 February 2007 – © Società Italiana di Fisica / Springer-Verlag 2007

Abstract. The HAPPEX Collaboration measured parity-violating electron scattering from $^4\mathrm{He}(e,e)$ and $\mathrm{H}(e,e)$ in 2004 and 2005 for $Q^2 \leq 0.11\,\mathrm{GeV}^2$. Results for the strange-quark contributions to the electromagnetic form factors of the nucleon from the 2004 data will be reviewed. Preliminary results from the 2005 data, which have significantly greater statistical precision, are $G_E^s = 0.004 \pm 0.014_{stat} \pm 0.013_{syst}$ for $Q^2 = 0.0772\,\mathrm{GeV}^2$ from the helium data and $G_E^s + 0.088 G_M^s = 0.004 \pm 0.011_{stat} \pm 0.005_{syst} \pm 0.004_{FF}$ for $Q^2 = 0.1089\,\mathrm{GeV}^2$ from the hydrogen data.

PACS. 14.20.Dh Protons and neutrons – 13.40.Gp Electromagnetic form factors

1 Introduction

The structure of the nucleon is of fundamental interest. Almost all the non-dark-matter mass in the Universe is contained within the nucleon. The nucleon is unique among systems of ordinary matter in that most of its mass is not due to the masses of its constituents. For example, QCD calculations of nucleon mass [1] propose that most of the nucleon's mass is due to the energy in the gluon fields. Such strong gluon fields are expected to give rise to significant numbers of virtual quark and antiquark pairs. Indeed, the importance of this sea of $q\bar{q}$ pairs has been demonstrated [2] in the analysis of νN scattering. Since the nucleon contains no net strangeness, any effect of strange quarks on the structure of the nucleon should be attributable to the strange-quark sea. Hints of the importance of the strange quark sea to the mass of the nucleon [3] or to the spin structure of the nucleon [4] raise the question of whether static properties of the nucleon ground state, such as the electromagnetic form factors, also depend on the strength of the strange-quark sea.

2 Parity-violating electron scattering

Measurements of the electromagnetic form factors of the proton and neutron provide two pieces of data, as a function of Q^2, from which a two-component quark flavor separation is possible. If, however, a third quark flavor, that is strangeness, needs to be extracted, additional experimental data are needed. This third piece of

a e-mail: kaniol@calstatela.edu

data can be provided by parity-violating electron scattering [5]. Interference between photon and Z^0 exchange gives rise to a helicity-dependent asymmetry in the elastic scattering cross-section. For the Q^2 range, $0.01\,\mathrm{GeV}^2 \leq Q^2 \leq 1\,\mathrm{GeV}^2$ the experimentally measured asymmetry A_{PV} is expected to lie between 10^{-7} and 10^{-4}, where $A_{PV} = (\sigma_+ - \sigma_-)/(\sigma_+ + \sigma_-)$, and σ_+ and σ_- refer to positive and negative electron helicity states. In these experiments one measures G_E^s, and G_M^s, the strangeness analogs of the usual Sachs form factors. For the spin zero $^4\mathrm{He}$ target the asymmetry is sensitive to G_E^s, whereas for hydrogen one extracts a linear combination of G_E^s and G_M^s. The asymmetry has been exploited by several other groups (G0) [6],(A4) [7], (SAMPLE) [8] besides HAPPEX to obtain nucleon strange-quark electromagnetic form factors. Given the experimentally challengingly small asymmetries it has been a boon that different groups with different techniques have attacked the same question in overlapping Q^2 regions.

2.1 HAPPEX II, 2004 results

We have measured [9] the parity-violating electroweak asymmetry in the elastic scattering of polarized electrons from $^4\mathrm{He}$ at an average scattering angle $\theta_{lab} = 5.7°$ and a four-momentum transfer $Q^2 = 0.091\,\mathrm{GeV}^2$. From these data, for the first time, the strange electric form factor of the nucleon G_E^s has been isolated. The measured asymmetry of $A_{PV} = 6.72 \pm 0.84_{stat} \pm 0.21_{syst} \times 10^{-6}$ yields a value of $G_E^s = -0.038 \pm 0.042_{stat} \pm 0.010_{syst}$, consistent with zero. This data set consists of about 3 million helicity window pairs. A helicity window is 33.3 ms long. A pair consists of opposite helicity windows, with the helicity of the first window in each pair chosen at random. The details of the measurement are described in the paper [9].

We report [10] the most precise measurement to date of a parity-violating asymmetry in elastic electron-proton scattering. The measurement was carried out with a beam energy of $3.03\,\mathrm{GeV}$ and a scattering angle $\theta_{lab} = 6°$, with the result $A_{PV} = -1.14 \pm 0.24_{stat} \pm 0.06_{syst} \times 10^{-6}$. From this we extract, at $Q^2 = 0.099\,\mathrm{GeV}^2$, the strange form factor combination $G_E^s + 0.080 G_M^s = 0.030 \pm 0.025_{stat} \pm 0.006_{syst} \pm 0.012_{FF}$, where the first two errors are experimental and the last error is due to the uncertainty in the neutron electromagnetic form factor. This result significantly improves current knowledge of G_E^s and G_M^s at $Q^2 \approx 0.1\,\mathrm{GeV}^2$. A consistent picture emerges when several measurements at about the same Q^2 value are combined: G_E^s is consistent with zero while G_M^s prefers positive values though $G_E^s = G_M^s = 0$ is compatible with the data at 95% C.L. This data set consists of about 9 million helicity window pairs.

2.2 HAPPEX II, 2005 preliminary results

The 2005 experimental procedure was the same as that in 2004 except for some improvements to the shielding of the septa magnets that allow us to go to 6°. In 2005 we recorded 35 million helicity window pairs for the ^{4}He data, and 25 million window pairs for the hydrogen data. Our preliminary results from the 2005 data sets are: $G_E^s = 0.004 \pm 0.014_{stat} \pm 0.013_{syst}$ for $Q^2 = 0.0772\,\mathrm{GeV}^2$ from the helium data and $G_E^s + 0.088 G_M^s = 0.004 \pm 0.011_{stat} \pm 0.005_{syst} \pm 0.004_{FF}$ for $Q^2 = 0.1089\,\mathrm{GeV}^2$ from the hydrogen data.

3 Comparison of HAPPEX and world data

In fig. 1 we show a comparison of the HAPPEX 2005 preliminary results to the combined world's data for $Q^2 \approx 0.1\,\mathrm{GeV}^2$. The outer two ovals show the 68% and 95% probability contours. The HAPPEX 2004 data are included in the probability contours calculation.

We show in fig. 2 a comparison of the HAPPEX data with the world's data over a range of Q^2. For Q^2 around $0.1\,\mathrm{GeV}^2$ data from G0 and A4 indicate a nonzero value for the combined form factors of the nucleon. The HAPPEX 2004 results are consistent with these data but also with zero. The preliminary HAPPEX 2005 data favor a value much closer to zero. A value of zero would eliminate the need to explain the otherwise seeming cancellation of terms to produce a zero in the result for $Q^2 = 0.2\,\mathrm{GeV}^2$. A value of zero at $Q^2 = 0.1\,\mathrm{GeV}^2$ is also consistent with an empirical global fit [11] of the world's data for $Q^2 \leq 0.3\,\mathrm{GeV}^2$. These authors parameterized the form factors as: $G_E^s = \rho_s Q^2$ and $G_M^s = \mu_s$. With $\rho_s = -0.06 \pm 0.41\,\mathrm{GeV}^{-2}$ and $\mu_s = 0.12 \pm 0.55 \pm 0.07$. This fit includes the data from G0, A4, SAMPLE and HAPPEX II-2004. A recent lattice calculation [12] also

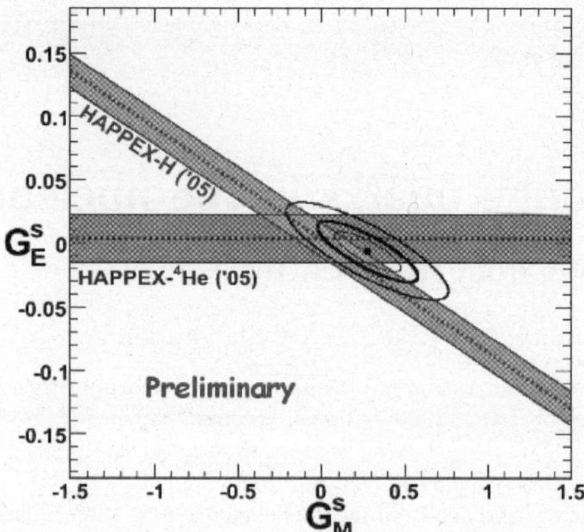

Fig. 1. A comparison of the HAPPEX II 2005 preliminary results with the world's data. The world's data includes HAPPEX II 2004 data. The second and third ovals are the 68% and 95% probability contours.

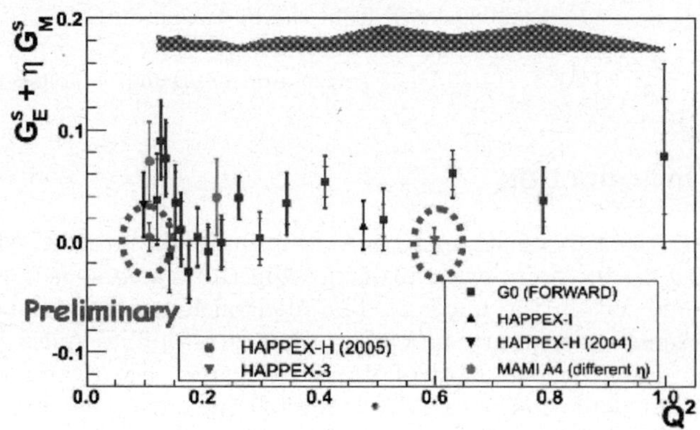

Fig. 2. A comparison of the HAPPEX data with the world's data. The future HAPPEX 3 data point is an estimate of the expected error. The band at the top of the figure represents correlated systematic uncertainty in the G0 (FORWARD) data.

predicts a very small value for the strange magnetic moment, $G_M^s = -0.046 \pm 0.19\,\mu_N$, consistent with the preliminary HAPPEX II 2005 result.

The G0 results in fig. 2 indicate that the combined strange electromagnetic form factors for the nucleon are positive for larger Q^2. The HAPPEX Collaboration has been approved to extend the measurements to $Q^2 = 0.6\,\mathrm{GeV}^2$. This is an important check on the G0 results. The future point is placed on zero in the figure and shows the expected error bar. We expect to be able to make the measurement in 2008.

References

1. X. Ji, Phys. Rev. Lett. **74**, 1071 (1995).
2. A.D. Martin, R.G. Roberts, W.J. Stirling, R.S. Thorne, Phys. Lett. B **604**, 61 (2004).

3. J. Gasser, H. Leutwyler, M.E. Sainio, Phys. Lett. B **253**, 252 (1991).
4. B. Adeva *et al.*, Phys. Rev. Lett. **71**, 959 (1993).
5. R.D. McKeown, Phys. Lett. B **219**, 140 (1989).
6. D. Armstrong *et al.*, Phys. Rev. Lett. **95**, 092001 (2005).
7. F.E. Maas *et al.*, Phys. Rev. Lett. **94**, 152001 (2005).
8. T.M. Ito *et al.*, Phys. Rev. Lett. **92**, 102003 (2004).
9. K.A. Aniol *et al.*, Phys. Rev. Lett. **96**, 022003 (2006).
10. K.A. Aniol *et al.*, Phys. Lett. B **265**, 275 (2006).
11. R.D. Young *et al.*, Phys. Rev. Lett. **97**, 102002 (2006).
12. D.B. Leinweber *et al.*, Phys. Rev. Lett. **94**, 212001 (2005).

Eur. Phys. J. A **31**, 600–602 (2007)

DOI 10.1140/epja/i2006-10250-2

THE EUROPEAN
PHYSICAL JOURNAL A

Special Article – QNP 2006

Meson cloud and nucleon strangeness: An update

F. Carvalho, F.S. Navarra[a], and M. Nielsen

Instituto de Física, Universidade de São Paulo, C.P. 66318, 05315-970 São Paulo, SP, Brazil

Received: 23 November 2006
Published online: 16 March 2007 – © Società Italiana di Fisica / Springer-Verlag 2007

Abstract. We use a version of the meson cloud model, including kaon, κ and K^* contributions, to estimate the electric and magnetic strange form factors of the nucleon. We compare our results with the recent measurements of the strange-quark contribution to parity-violating asymmetries in electron-proton scattering experiments.

PACS. 14.20.Dh Protons and neutrons – 12.40.-y Other models for strong interactions – 24.85.+p Quarks, gluons, and QCD in nuclei and nuclear processes

The strange component of the nucleon sea has been intensely investigated in the last years and there are new experimental and theoretical results. On the experimental side we have the low-energy parity-violating experiments carried out at TJNAF, where it was possible to measure the strange electric and magnetic form factors of the nucleon. The first measurements of these quantities (and combinations of them) were performed by the SAMPLE [1] and HAPPEX [2] Collaborations at TJNAF. The very recent results from SAMPLE [3], HAPPEX [4], A4 (at MAMI) [5] and G0 [6] Collaborations, provide information on the nucleon strange vector form factors over the range of momentum transfers $0.12 \leq Q^2 \leq 1.0\,\mathrm{GeV}^2$. The data indicate non-trivial, Q^2-dependent, strange-quark distributions inside the nucleon, and present a challenge to models of the nucleon structure.

On the theoretical side, there were new lattice QCD calculations of the strange magnetic [7] and electric [8] form factors. There are discrepancies between theory and experiment which deserve further investigation. For example, the measured strange magnetic moment is positive, whereas the lattice calculations point to a negative value. In this context, a further theoretical insight on the problem might come from chiral perturbation theory. Unfortunately, as it was shown in [9], in ChPT the matrix elements of the strangeness current depend on a new low-energy constant, which cannot be determined from fitting other data. Thus, for the observables discussed here, ChPT has no predictive power.

In the low-energy regime the strange component of the nucleon sea is expected to have a nonperturbative origin. One of the possible nonperturbative mechanisms of strangeness production is given by the meson cloud [10]. In this model the physical nucleon contains vir-

tual meson-baryon components, which "dress" the bare nucleon. The meson cloud mechanism provides a natural explanation for symmetry breaking in parton distributions [11]. Already in the first kaon-cloud models the nucleon strangeness distribution was generated by fluctuations of the "bare" nucleon into kaon-hyperon intermediate states which were described by the corresponding one-loop Feynman graphs [12]. Since then, some concerns have been raised in the literature regarding the implementation of the loop model of the nucleon. In particular, regarding the omission of contributions from higher-lying intermediate states in the meson-hyperon fluctuations [13,14]. While the effects of heavier hyperons, such as the Σ^*, have been shown to be negligible [14], the contribution of the K^*-Y pairs were found to be large [13]. Nevertheless, the results of [15] pointed to a "slow convergence" of the intermediate-state sum.

In view of the recent data mentioned above we think that it is interesting to update our previous calculations of the strange form factors with the meson cloud model. In a recent work [16] we have computed the momentum dependence of the strange vector form factors in the loop model at momentum transfers $0 \leq Q^2 \leq 3\,\mathrm{GeV}^2$ and compared our results with the measurements from the G0 Collaboration [6]. In the present work we extend our previous calculation, including, for the first time, the recently observed [17] strange scalar meson, κ, in the cloud and analyze new data on the strange form factors.

The nucleon matrix element of the strangeness current is parametrized by two invariant amplitudes, the Dirac and Pauli strangeness form factors $F_{1,2}^{(s)}$:

$$\langle N | \bar{s}\gamma_\mu s | N \rangle = \bar{U} \left[F_1^{(s)}(Q^2)\gamma_\mu + i\frac{\sigma_{\mu\nu}q^\nu}{2m_N} F_2^{(s)}(Q^2) \right] U ,$$

[a] e-mail: navarra@if.usp.br

$$(1)$$

where $U(p)$ denotes the nucleon spinor and $F_1^{(s)}(0) = 0$, due to the absence of an overall strangeness charge of the nucleon. The electric and magnetic form factors are defined through $G_E^{(s)}(Q^2) = F_1^{(s)}(Q^2) - \frac{Q^2}{4m_N^2} F_2^{(s)}(Q^2)$ and $G_M^{(s)}(Q^2) = F_1^{(s)}(Q^2) + F_2^{(s)}(Q^2)$.

We consider a hadronic one-loop model containing K, K^* and κ mesons as the dynamical framework for the calculation of these form factors. This model is based on the meson-baryon effective Lagrangians

$$\mathcal{L}_{KB} = -g_{ps}\bar{B}i\gamma_5 BK, \tag{2}$$

$$\mathcal{L}_{K^*B} = -g_v\left[\bar{B}\gamma_\alpha BK^{*\alpha} - \frac{\kappa_{tv}}{2m_N}\bar{B}\sigma_{\alpha\beta}B\partial^\alpha K^{*\beta}\right], \tag{3}$$

$$\mathcal{L}_{\kappa B} = -g_s\bar{B}B\kappa, \tag{4}$$

where B $(= N, \Lambda, \Sigma)$, K, $K^{*\alpha}$ and κ are the baryon, kaon, K^* (vector meson) and the strange scalar meson, respectively. In the above expression $m_N = 939\,$MeV is the nucleon mass and κ_{tv} is the ratio of tensor to vector coupling, $\kappa_{tv} = g_t/g_v$. In order to account for the finite extent of the above vertices, the model includes form factors from the Nijmegen NY potential [18] at the hadronic KNY and K^*NY $(Y = \Lambda, \Sigma)$ vertices, which have the monopole form

$$F(k^2) = \frac{m_M^2 - \Lambda_M^2}{k^2 - \Lambda_M^2} \tag{5}$$

with meson momenta k and the physical meson masses $m_K = 495\,$MeV, $m_{K^*} = 895\,$MeV and $m_\kappa = 800\,$MeV.

Since the nonlocality of the meson-baryon form factors (5) gives rise to vertex currents, gauge invariance was maintained in [13] by introducing the photon field via minimal substitution in the momentum variable k [19]. The resulting nonlocal seagull vertices are given explicitly in [13,15].

The diagonal couplings of $\bar{s}\gamma_\mu s$ to the strange mesons and baryons in the intermediate states are straightforwardly determined by current conservation, *i.e.* they are given by the net strangeness charge of the corresponding hadron. The nondiagonal (*i.e.* spin-flipping) couplings of the strange current in the vertices γ-K-K^* and γ-κ-K^* have been discussed in [13] and, more recently, in [20]. The other couplings in the effective Lagrangians are taken from the Nijmegen NY potential [18]: $g_{ps}/\sqrt{4\pi} = -4.005$, $g_v/\sqrt{4\pi} = -1.45$, $g_s = -10.0$, $\kappa_{tv} = 2.43$. In phenomenological potential models, as the Nijmegen [18] or Bonn-Jülich [21], the couplings and cut-off parameters are determined simultaneously. In fact, some coupling constants may be calculated with QCD sum rules, as, for example, in [22], where a smaller value of g_{ps} was found. In our numerical analysis, we shall fix the couplings to the numbers given above but we shall vary all the cut-off parameters in the interval $0.9\,\text{GeV} \le \Lambda \le 1.1\,\text{GeV}$, which encompasses the numbers obtained in refs. [18] and [21]. After some calculations we discovered that the contribution from κ to all the observables considered here tends to cancel the contribution of K and K^*, which add together. Since increasing the cut-off leads to an enhancement in the contribution of the corresponding meson, there are two extreme

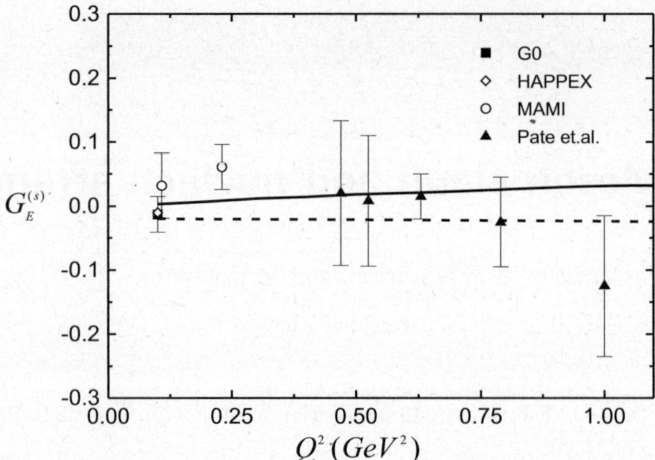

Fig. 1. The strange electric form factor G_E^s. The solid line shows our results with $\Lambda_K = \Lambda_{K^*} = 0.9\,$GeV and $\Lambda_\kappa = 1.1\,$GeV. The dashed line represents the choice $\Lambda_K = \Lambda_{K^*} = 1.1\,$GeV and $\Lambda_\kappa = 0.9\,$GeV.

choices: I) $\Lambda_K = \Lambda_{K^*} = 0.9\,$GeV and $\Lambda_\kappa = 1.1\,$GeV and II) $\Lambda_K = \Lambda_{K^*} = 1.1\,$GeV and $\Lambda_\kappa = 0.9\,$GeV. The first one maximizes the role played by κ, whereas the last one maximizes the joint contribution of K and K^*.

In fig. 1 we show the strange electric form factor G_E^s obtained with the loop model obtained by using choice I (solid line) and choice II (dashed line). Our final result for G_E^s is the area between these two "boundary" lines, since we cannot be more precise about the value of the Λ's. We see that the agreement with data points from refs. [4–6] and also from ref. [23] is reasonable.

One should keep in mind that the cut-off value $\Lambda_{K^*} = 0.9\,$GeV, is very close to the K^* mass. As a consequence, for this cut-off value the contributions from the K^* and the K/K^* and κ/K^* transitions are completely negligible relative to the kaon contribution. Using a bigger value for Λ_{K^*} makes the agreement with data worse. It is interesting to observe that the authors of [20], studying K^* photoproduction with a meson exchange model with the same Nijmegen form factors reached the same conclusion!

In fig. 2 we show the strange magnetic form factor G_M^s obtained with the loop model using choice I (solid line) and choice II (dashed line). Our final result for G_M^s is the area between these two lines. We see that the agreement with data points from refs. [4–6,23] is acceptable when the κ component is large but is very poor when it is small. In any case, however, our model cannot reproduce the strange magnetic moment $G_M^s(0)$. In order to have a better idea of the role played by the meson κ, in fig. 3 we replot the solid line of fig. 2 splitting it into its components given by the κ contribution (dashed line) and the $K + K^*$ contribution (dotted line). We clearly see that the inclusion of κ in the calculation compensates the $K + K^*$ contribution and improves the agreement with data.

In summary, we have calculated the electric and magnetic strange form factors of the nucleon with a version of the meson cloud model, which includes kaon, κ and K^* contributions. In contrast to other situations, in the

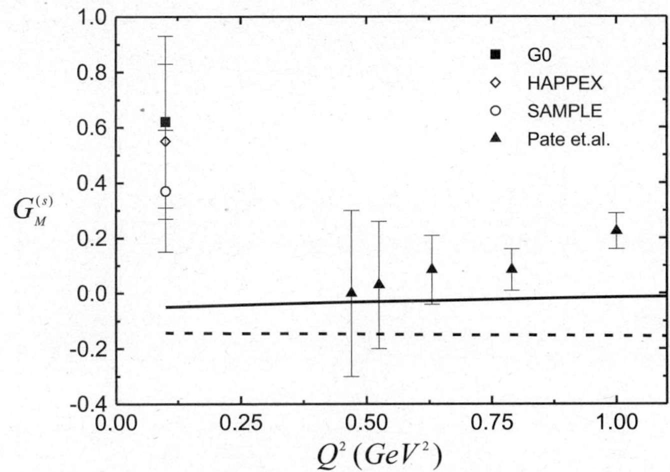

Fig. 2. The strange magnetic form factor G_M^s. The solid line shows our results with $\Lambda_K = \Lambda_{K^*} = 0.9\,\mathrm{GeV}$ and $\Lambda_\kappa = 1.1\,\mathrm{GeV}$. The dashed line represents the choice $\Lambda_K = \Lambda_{K^*} = 1.1\,\mathrm{GeV}$ and $\Lambda_\kappa = 0.9\,\mathrm{GeV}$.

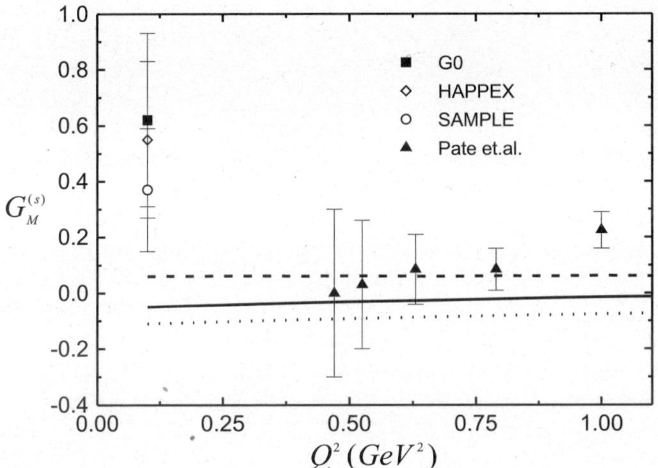

Fig. 3. The dashed (dotted) line shows the contribution of κ $(K + K^*)$ to the strange magnetic form factor G_M^s. The solid line shows the sum, which also corresponds to the solid line of fig. 2.

present case the K^* contribution did not cancel the kaon contribution. The κ contribution gives opposite results to the other strange mesons and improves the description of data.

References

1. SAMPLE Collaboration (D.T. Spayde *et al.*), Phys. Rev. Lett. **84**, 1106 (2000); B. Mueller *et al.*, Phys. Rev. Lett. **78**, 3824 (1997).
2. HAPPEX Collaboration (K.A. Aniol *et al.*), Phys. Rev. Lett. **82**, 1096 (1999).
3. SAMPLE Collaboration (D.T. Spayde *et al.*), Phys. Rev. Lett. **92**, 102003 (2004); Phys. Lett. B **583**, 79 (2004).
4. HAPPEX Collaboration (K.A. Aniol *et al.*), Phys. Rev. C **69**, 065501 (2004); nucl-ex/0506011.
5. A4 Collaboration (F.E. Mass *et al.*), Phys. Rev. Lett. **93**, 022002 (2004); **94**, 152001 (2005).
6. G0 Collaboration (D.S. Armstrong *et al.*), Phys. Rev. Lett. **95**, 092001 (2005); nucl-ex/0506021.
7. D.B. Leinweber *et al.*, Phys. Rev. Lett. **94**, 212001 (2005); hep-lat/0406002; Eur. Phys. J. A **24**, s2, 79 (2005); hep-lat/0502004.
8. D.B. Leinweber *et al.*, Phys. Rev. Lett. **97**, 022001 (2006); hep-lat/0601025.
9. B. Kubis, Eur. Phys. J. A **24**, s2, 97 (2005); M.J. Musolf, H. Ito, Phys. Rev. C **55**, 3066 (1997).
10. A.W. Thomas, Phys. Lett. B **126**, 97 (1983); C. Boros, A.W. Thomas, Phys. Rev. D **60**, 074017 (1999); A.I. Signal, A.W. Thomas, Phys. Lett. B **191**, 205 (1987); H. Holtmann, A. Szczurek, J. Speth, Nucl. Phys. A **596**, 631 (1996); F.S. Navarra, M. Nielsen, C.A.A. Nunes, M. Teixeira, Phys. Rev. D **54**, 842 (1996); S. Paiva, M. Nielsen, F.S. Navarra, F.O. Durães, L.L. Barz, Mod. Phys. Lett. A **13**, 2715 (1998).
11. E.M. Henley, G.A. Miller, Phys. Lett. B **251**, 453 (1990); F. Carvalho, F.O. Durães, F.S. Navarra, M. Nielsen, F.M. Steffens, Eur. Phys. J. C **18**, 127 (2000).
12. W. Koepf, E.M. Henley, Phys. Rev. C **49**, 2219 (1994); W. Koepf, S.J. Pollock, E.M. Henley, Phys. Lett. B **288**, 11 (1992); T.D. Cohen, H. Forkel, M. Nielsen, Phys. Lett. B **316**, 1 (1993); M.J. Musolf, M. Burkardt, Z. Phys. C **61**, 433 (1994); H. Forkel, M. Nielsen, X. Jin, T.D. Cohen, Phys. Rev. C **50**, 3108 (1994).
13. L.L. Barz, H. Forkel, H.-W. Hammer, F.S. Navarra, M. Nielsen, M.J. Ramsey-Musolf, Nucl. Phys. A **640**, 259 (1998).
14. W. Meltnichouk, M. Malheiro, Phys. Rev. C **55**, 431 (1997); **56**, 2373 (1997).
15. H. Forkel, F.S. Navarra, M. Nielsen, Phys. Rev. C **61**, 055206 (2000).
16. F. Carvalho, F.S. Navarra, M. Nielsen, Phys. Rev. C **72**, 068202 (2005).
17. E791 Collaboration (E.M. Aitala *et al.*), Phys. Rev. Lett. **89**, 121801 (2002); FOCUS Collaboration (J.M. Link *et al.*), Phys. Lett. B **535**, 43 (2002).
18. M.N. Nagels, T.A. Rijken, J.J. de Swart, Phys. Rev. D **15**, 2547 (1977); Th.A. Rijken, V.G.J. Stoks, Y. Yamamoto, Phys. Rev. C **59**, 21 (1999).
19. K. Ohta, Phys. Rev. D **35**, 785 (1987).
20. Yongseok Oh, Hungchong Kim, Phys. Rev. C **73**, 065202 (2006); hep-ph/0605105.
21. J. Haidenbauer, U.G. Meissner, Phys. Rev. C **72**, 044005 (2005); J. Haidenbauer, W. Melnitchouk, J. Speth, nucl-th/9805014.
22. M.E. Bracco, F.S. Navarra, M. Nielsen, Phys. Lett. B **454**, 346 (1999).
23. S.F. Pate, G.A. MacLachlan, D.W. McKee, V. Papavassiliou, hep-ex/0512032.

Eur. Phys. J. A **31**, 603–605 (2007)

DOI 10.1140/epja/i2006-10246-x

THE **EUROPEAN**
PHYSICAL JOURNAL A

Special Article – QNP 2006

Leading and higher twists in proton, neutron and deuteron unpolarized structure functions F_2

S. Simula[a]

Istituto Nazionale di Fisica Nucleare, Sezione di Roma 3, Via della Vasca Navale 84, I-00146 Roma, Italy

Received: 23 November 2006

Published online: 9 March 2007 – © Società Italiana di Fisica / Springer-Verlag 2007

Abstract. We summarize the results of a recent global analysis of proton and deuteron F_2 structure function world data performed over a large range of kinematics, including recent measurements done at JLab with the CLAS detector. From these data the lowest moments ($n \leq 10$) of the unpolarized structure functions are determined with good statistics and systematics. The Q^2 evolution of the extracted moments is analyzed in terms of an OPE-based twist expansion, taking into account soft-gluon effects at large x. A clean separation among the leading- and higher-twist terms is achieved. By combining proton and deuteron measurements the lowest moments of the neutron F_2 structure function are determined and its leading-twist term is extracted. Particular attention is paid to nuclear effects in the deuteron, which become increasingly important for the higher moments. Our results for the non-singlet, isovector (p-n) combination of the leading-twist moments are used to test recent lattice simulations. We also determine the lowest few moments of the higher-twist contributions, and find these to be approximately isospin independent, suggesting the possible dominance of ud correlations over uu and dd in the nucleon.

PACS. 13.60.Hb Total and inclusive cross sections (including deep-inelastic processes) – 12.38.Cy Summation of perturbation theory – 12.38.Qk Experimental tests – 12.38.Lg Other nonperturbative calculations

In the past few years, thanks to the performance of the CLAS detector of the Hall B at Jefferson lab, the unpolarized structure functions $F_2(x, Q^2)$ of both proton and deuteron have been measured precisely in a wide continuous interval of values of the Bjorken variable x and of the squared four-momentum transfer Q^2 [1,2]. In particular, the F_2 structure functions have been extracted over the whole resonance region ($W \leq 2.5\,\text{GeV}$) below $Q^2 \simeq 4.5\,(\text{GeV}/c)^2$. These measurements, together with existing world data, have allowed the determination of the moments $M_n(Q^2)$ of the structure function F_2 up to $n = 10$, drastically reducing the uncertainties related to data interpolation and extrapolation as well as providing the most accurate evaluation of the Q^2-dependence of the moments starting from $Q^2 \simeq 0.1\,(0.5)\,(\text{GeV}/c)^2$ for proton (deuteron) up to $Q^2 \simeq 100\,(\text{GeV}/c)^2$.

A very powerful tool for analyzing in QCD the Q^2-dependence of the moments $M_n(Q^2)$ is represented by the Operator Product Expansion (OPE). Through OPE the moments are given as a series expansion in terms of matrix elements of local operators, namely

$$M_n(Q^2) = \sum_{\tau=2}^{\infty} E_{n\tau}(\mu, Q^2)\, O_{n\tau}(\mu) \left(\frac{1}{Q^2}\right)^{\frac{1}{2}(\tau-2)}, \quad (1)$$

where μ is the renormalization scale, $O_{n\tau}(\mu)$ is the reduced matrix element of local operators with definite spin n and twist τ, related to the non-perturbative structure of the target, and $E_{n\tau}(\mu, Q^2)$ is a short-distance coefficient calculable in pQCD thanks to the asymptotic freedom.

The first term of the series (1) corresponds to the Leading Twist (LT) and is determined by single-parton distributions in the target. Subsequent terms are the so-called Higher Twists (HTs), which depend on the interactions among partons (*i.e.*, on multi-parton correlations). Their determination is considerably more challenging both experimentally and theoretically.

In refs. [1] and [2] both the LT and the HTs (up to twist-6) have been extracted from the proton and deuteron moments, respectively. Two important features of our twist analysis should be mentioned: 1) the analyzed moments are Nachtmann moments, and not Cornwall-Norton ones; 2) the perturbative evolution of the LT is calculated using soft-gluon resummation techniques.

We have adopted the Nachtmann definition of the moments in order to get rid of the target-mass corrections, which are power corrections of kinematical origin and affect the Cornwall-Norton (CN) definition of the moments. The use of Nachtmann moments allows to extract in a clean way dynamical HTs, which are the only ones related to multi-parton correlations. The target-mass corrections

[a] e-mail: simula@roma3.infn.it

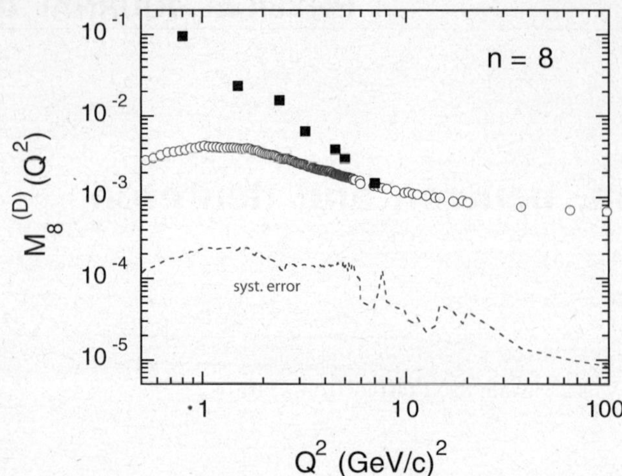

Fig. 1. Moment $M_8^{(D)}(Q^2)$ of the deuteron *versus* Q^2. Open dots are the results of ref. [2] using the Nachtmann definition of the moments, while full squares correspond to the CN moment as determined in ref. [3]. The reported systematic errors have been precisely estimated only in ref. [2].

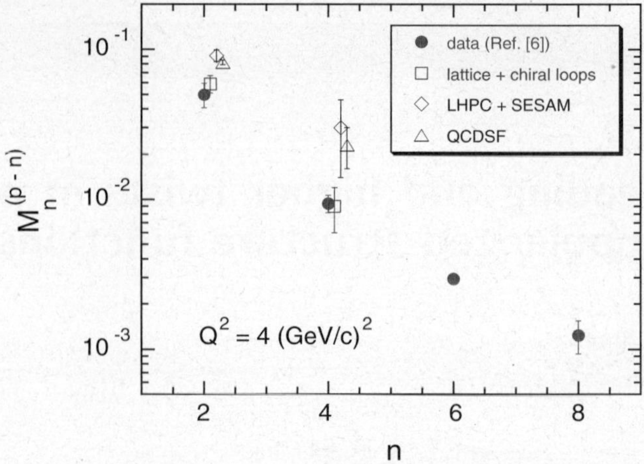

Fig. 2. Isovector $(p$-$n)$ moments of the nucleon F_2 structure function compared with lattice QCD simulations at $Q^2 = 4\,(\mathrm{GeV}/c)^2$. Full dots are the data from ref. [6], open diamonds and triangles correspond to the lattice results from refs. [7] and [8], obtained assuming a linear extrapolation in the quark masses, and open squares are the results of ref. [9], where chiral loop effects are considered.

can be subtracted in principle from the CN moments, but this procedure requires a precise knowledge of the LT moments at large n. Moreover, the contribution of target-mass corrections is sizable just in the region of $Q^2 \simeq$ few $(\mathrm{GeV}/c)^2$, where HTs are expected to show up. This point is clearly illustrated in fig. 1, where the Nachtmann moment $M_8^{(D)}(Q^2)$ is compared with the corresponding CN one as determined in ref. [3]. It is also worthwhile to note the much larger number of very accurate data points obtained thanks to the CLAS detector.

As for the perturbative evolution of the LT, the presence of large logarithms in the coefficient function at large n requires to go beyond any fixed-order approximation. As shown in ref. [4] for a reliable extraction of both LT and HTs it is crucial to apply the resummation of soft gluons. Moreover, for internal consistency the value of $\alpha_s(M_Z^2)$ has been extracted in ref. [5] by analyzing the large-Q^2 behavior of the proton moments of ref. [1] at the next-to-leading logarithmic accuracy used for the soft-gluon resummation.

The extracted LT components of proton and deuteron moments can be combined to form the LT moments of the neutron F_2 structure function. In the impulse approximation the deuteron structure function can be written as the sum of two terms: $F_2^{(D)} = F_2^{(D,conv.)} + F_2^{(D,off)}$, where the first term is given by a convolution of the nucleon structure function and the nucleon momentum distribution function in the deuteron, $f^{(D)}$, while the second term represents nuclear off-shell plus relativistic corrections[1]. Thus, the neutron moments can be obtained from

$$M_n^{(n)}(Q^2) = 2M_n^{(D)}(Q^2)\frac{1 - \Delta_n^{(off)}}{N_n^{(D)}} - M_n^{(p)}(Q^2)\,, \quad (2)$$

where $N_n^{(D)}$ is the moment of the function $f^{(D)}$ and $\Delta_n^{(off)}$ represents the nuclear off-shell correction. Thus, the quantity $N_n^{(D)}/[1 - \Delta_n^{(off)}]$ is the global nuclear correction factor, *i.e.* the EMC effect of the deuteron in moment space. The nuclear corrections and their uncertainties have been carefully estimated in ref. [6].

The extracted neutron LT moments can be combined with the proton LT ones to form the non-singlet, isovector $(p$-$n)$ LT moments of the nucleon F_2 structure function. Our findings are compared against lattice QCD simulations in fig. 2. While a linear extrapolation of the lattice results to the physical pion mass overestimates our data significantly, the results obtained using the extrapolation method of ref. [9] are in much better agreement. Our results for higher moments exhibit a remarkable precision, and it would be valuable to compare them with corresponding lattice moments, because the effects of chiral loops are expected to be suppressed at large n.

Another interesting issue is the behavior of the neutron to proton structure function ratio $R(x) \equiv F_2^n(x)/F_2^p(x)$ in the limit $x \to 1$. This ratio can be easily related to the corresponding d/u ratio of parton distributions. The "standard" value $R(x \doteq 1) = 1/4$ is used in most of the parton distribution fits and corresponds to a vanishing d/u ratio, while the value $R(x = 1) = 3/7$ is expected from pQCD arguments and corresponds to a limiting value of $1/5$ for the d/u ratio. As shown in ref. [6] the ratio of neutron to proton moments, M_n^n/M_n^p, is largely independent of Q^2 and is very sensitive at large n to the large-x behavior of $R(x)$. Our results are shown in fig. 3. The trend is clearly towards the "standard" value $1/4$ as n increases, although the precision of the data does not exclude higher limiting values. From fig. 3 one can also see the impact of the nuclear corrections on the deuteron moments at large n: for $n = 12$, which corresponds to an average value of

---

[1] For the moment with $n = 2$ shadowing and meson exchange current should be also considered (see ref. [6]).

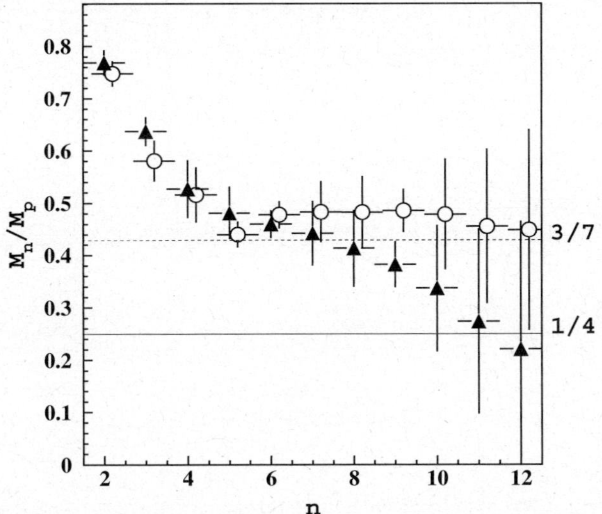

Fig. 3. Ratio of neutron to proton moments *versus* n: full triangles and open dots corresponds to the results of ref. [6] obtained with and without the nuclear corrections, respectively. The solid (dashed) line indicates the scenario where $d/u \to 0$ $(1/5)$ for $x \to 1$.

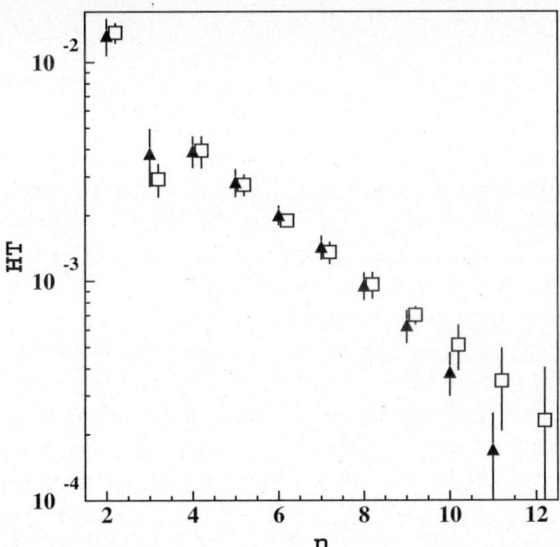

Fig. 4. Total higher twist contribution to the proton (filled triangles) and deuteron (open squares) moments evaluated at $Q^2 = 2\,(\mathrm{GeV}/c)^2$.

x around 0.75, the nuclear correction corresponds to a reduction factor of $\simeq 2$, though within large uncertainties.

Once the LT contribution to moments is determined, one can study the isospin dependence of the HT contribution. Assuming that final-state interactions and meson exchange currents have negligible impact on the HTs above $Q^2 \simeq 1\,(\mathrm{GeV}/c)^2$, the same nuclear corrections used for the LT can be applied to the HTs. In this way the total HT contributions to proton moments and to the corrected deuteron moments can be compared, as done in fig. 4. The comparison indicates clearly that the total HT contribution is almost independent of isospin, suggesting the possible dominance of ud correlations over uu and dd in the nucleon [6]. Moreover, the isovector combination $(p\text{-}n)$ of the F_2 structure functions should be almost free from HT contributions within the present uncertainties.

In conclusion, we have analyzed experimental data on proton and deuteron F_2 structure functions in order to determine their moments. Using the OPE we have extracted both leading- and higher-twist contributions. By combining proton and deuteron moments we have determined the LT moments of the neutron F_2 structure function, paying particular attention to the issue of nuclear effects in the deuteron, which are increasingly important at large n.

The main results of our analysis can be summarized as follows:

- the ratio of neutron to proton moments is consistent with $F_2^n/F_2^p \to 1/4$ as $x \to 1$, although the higher value of $3/7$, suggested by pQCD arguments, cannot be excluded;
- the non-singlet moments are in good agreement with the lattice data [7,8] only when the latter are extrapolated to physical quark masses after having taken into account chiral loops [9];
- the total contribution of HTs is found to be isospin independent, suggesting the possible dominance of ud correlations over uu and dd in the nucleon. This implies that in the isovector combination $(p\text{-}n)$ of F_2 structure functions the HTs are almost consistent with zero within the present uncertainties.

References

1. CLAS Collaboration (M. Osipenko *et al.*), Phys. Rev. D **67**, 092001 (2003) [arXiv:hep-ph/0301204]; M. Osipenko *et al.*, arXiv:hep-ex/0309052.
2. CLAS Collaboration (M. Osipenko *et al.*), Phys. Rev. C **73**, 045205 (2006) [arXiv:hep-ex/0506004]; CLAS Collaboration (M. Osipenko *et al.*), arXiv:hep-ex/0507098.
3. I. Niculescu, J. Arrington, R. Ent, C.E. Keppel, Phys. Rev. C **73**, 045206 (2006) [arXiv:hep-ph/0509241].
4. S. Simula, Phys. Lett. B **493**, 325 (2000) [arXiv:hep-ph/0005315].
5. S. Simula, M. Osipenko, Nucl. Phys. B **675**, 289 (2003) [arXiv:hep-ph/0306260].
6. M. Osipenko, W. Melnitchouk, S. Simula, S. Kulagin, G. Ricco, Nucl. Phys. A **766**, 142 (2006) [arXiv:hep-ph/0510189]; see also, arXiv:hep-ph/0610043.
7. LHPC Collaboration (D. Dolgov *et al.*), Phys. Rev. D **66**, 034506 (2002) [arXiv:hep-lat/0201021].
8. QCDSF Collaboration (M. Gockeler, R. Horsley, D. Pleiter, P.E.L. Rakow, G. Schierholz), Phys. Rev. D **71**, 114511 (2005) [arXiv:hep-ph/0410187].
9. W. Detmold, W. Melnitchouk, J.W. Negele, D.B. Renner, A.W. Thomas, Phys. Rev. Lett. **87**, 172001 (2001) [arXiv:hep-lat/0103006].

Eur. Phys. J. A **31**, 606–609 (2007)
DOI 10.1140/epja/i2006-10175-8

THE EUROPEAN
PHYSICAL JOURNAL A

Special Article – QNP 2006

Spin structure function of deuteron g_1^d from COMPASS

A. Korzenev[a]

For the COMPASS Collaboration
Mainz University, Institute of Nuclear Physics, D-55099, Mainz, Germany

Received: 24 September 2006
Published online: 16 February 2007 – © Società Italiana di Fisica / Springer-Verlag 2007

Abstract. New results on the longitudinal inclusive spin asymmetry A_1^d in the range $1 < Q^2 < 100\,(\mathrm{GeV}/c)^2$ and $0.004 < x < 0.7$ are presented. From these results we derive the spin-dependent structure function g_1^d which we use to evaluate the scale-invariant flavor-singlet axial charge $\hat{a}_0$. The contribution of the measured region is evaluated by a QCD fit of the world data. The data were obtained by the COMPASS experiment at CERN using a 160 GeV polarized muon beam scattered off a large double-cell polarized ^{6}LiD target. The results are in agreement with those from previous experiments and improve considerably the statistical accuracy in the region $x < 0.03$.

PACS. 13.60.Hb Total and inclusive cross-sections (including deep-inelastic processes) – 13.88.+e Polarization in interactions and scattering

1 Introduction

In view of the spin crisis in the quark parton model [1] the investigation of the spin structure of the nucleon is one of the most interesting issues in particle physics and it is the primary goal of the COMPASS experiment [2] at CERN SPS. The analysis of deep inelastic scattering (DIS) of 160 GeV longitudinally polarized muons off a longitudinally polarized ^{6}LiD target gives access to the spin-dependent structure function of the deuteron g_1^d.

The longitudinal virtual-photon deuteron asymmetry, A_1^d, is defined via the asymmetry of absorption cross-sections of transversely polarized photon as

$$A_1^d = (\sigma_0^T - \sigma_2^T)\,/\,(\tfrac{2}{3}(\sigma_0^T + \sigma_1^T + \sigma_2^T)),\tag{1}$$

where σ_J^T is the γ^*-deuteron absorption cross-section for a total spin projection J. The relation between the experimentally measured muon-deuteron asymmetry, A^d, and A_1^d is

$$A^d = \frac{\sigma^{\uparrow\downarrow} - \sigma^{\uparrow\uparrow}}{\sigma^{\uparrow\downarrow} + \sigma^{\uparrow\uparrow}} = D(A_1^d + \eta A_2^d),\tag{2}$$

where D and η depend on the event kinematics. Arrows indicate the direction of beam and target spins. The transverse asymmetry A_2^d has been measured at SLAC and found to be small [3]. In view of this, in our analysis, eq. (2) has been reduced to $A_1^d \simeq A^d/D$. The longitudinal

spin structure function g_1^d relates to A_1^d as

$$g_1^d = \frac{F_2^d}{2\,x\,(1+R)}A_1^d,\tag{3}$$

where F_2^d is the spin-independent deuteron structure function and R is the ratio of cross-sections for absorption of longitudinal and transverse photons $R = \sigma^L/\sigma^T$.

2 Data sample and results on A_1^d

In the analysis we require for all events to have a reconstructed primary interaction vertex defined by the incoming and the scattered muons. The energy of the beam muon is constrained in the interval $140 < E_\mu < 180\,\mathrm{GeV}$. To cancel out the muon flux normalization and thus minimize the systematic error, the trajectory of the incoming muon is required to cross the two cells of the target which are polarized in opposite direction. The kinematic region is defined by cuts on a photon virtuality Q^2 and a fractional energy y transfered from the beam muon to the virtual photon. The requirement $Q^2 > 1\,(\mathrm{GeV}/c)^2$ defines the region of DIS. The cut $0.1 < y < 0.9$ from one side removes events which are problematic from the reconstruction point of view due to a small energy transfer. From the other side it eliminates events at large y which are the most affected by radiative corrections. These cuts define the x interval accessible by COMPASS: $0.004 < x < 0.7$.

The counting rate asymmetry relates to A_1^d as

$$A^{exp} = P_t P_b f D\, A_1^d,\tag{4}$$

---

[a] On leave from JINR, Dubna, Russia;
e-mail: `Alexander.Korzenev@cern.ch`

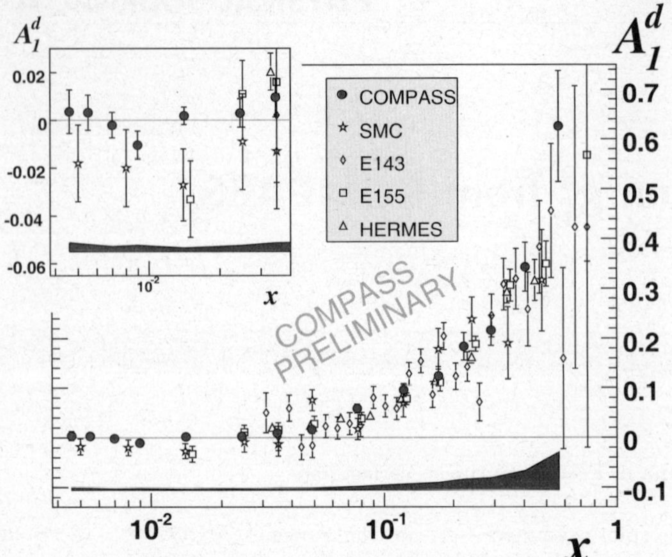

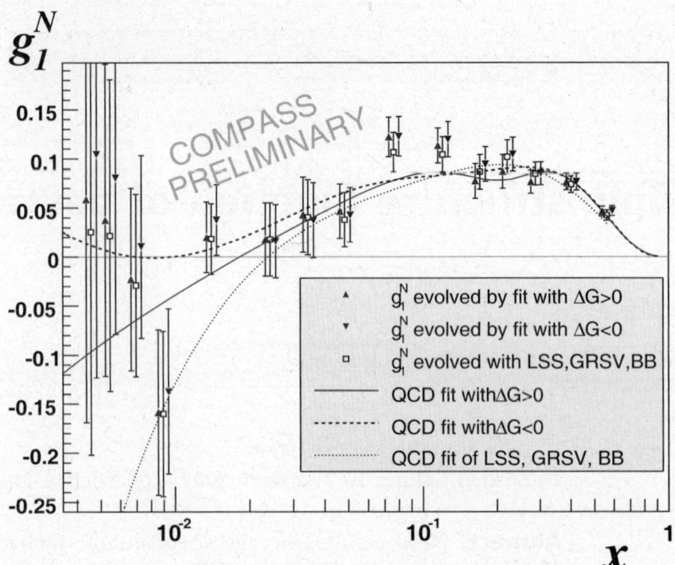

Fig. 1. The asymmetry $A_1^d(x)$ as measured in COMPASS and previous results from SMC [5], HERMES [6], E143 [7] and E155 [8] at $Q^2 > 1\,(\text{GeV}/c)^2$. Only statistical errors are shown with the data points. The shaded areas show the size of the COMPASS systematic errors.

Fig. 3. The COMPASS values of g_1^N evolved to $Q^2 = 3\,(\text{GeV}/c)^2$. In addition to our fits the curve obtained with three published parameterizations (BB, GRSV and LSS05) [10] is shown. These parameterizations lead almost to the same values of $g_1^N(x, Q^2 = 3\,(\text{GeV}/c)^2)$ and have been averaged. For clarity the data points evolved with different fits are shifted in x with respect to each other. Only statistical errors are shown.

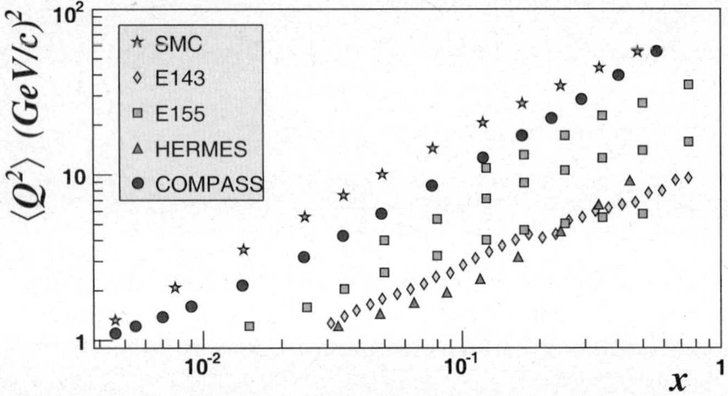

Fig. 2. The Q^2 values of COMPASS data compared to other experiments which performed measurement of g_1^d in the DIS region.

The average values of Q^2 *versus* x are shown in fig. 2. One can see that COMPASS points are slightly lower compared to SMC. However, they are a factor of 2–3 higher than E143 and HERMES data.

There are several sources of systematics uncertainties in the determination of A_1^d. The measurement error of the target polarization and parametrization of the beam polarization amount to 5% each. The uncertainty related to the dilution factor f is 6% over the full range of x. The ratio $R = \sigma^L/\sigma^T$ used to calculate the depolarization factor D gives an error of 2–3%. If added in quadrature all uncertainties of multiplicative factors (eq. (4)) amount to 10% systematic error. The main contributor to the systematic error additive with respect to A_1^d is the false asymmetry which could be generated by instabilities in some components of the spectrometer. It was evaluated as a fraction of the statistical error [4]: $\sigma_{syst} < 0.4\sigma_{stat}$. The total systematic uncertainty is shown as a band at the bottom of fig. 1.

where P_b is the incoming-muon polarization, P_t is the target polarization and f is the dilution factor which is the ratio of the absorption cross-section on the deuteron and on all nuclei in the target material. It also comprises corrections for radiative events. For the description of the method of asymmetry extraction and quality cuts for the sample, see [4]. The data were collected during the years 2002–2004. The resulting sample consists of 89×10^6 events.

The A_1^d as a function of x is shown in fig. 1, where it is superposed to data of previous experiments. One can observe the good agreement over the full range of x. COMPASS contributes significantly to the low-x region previously covered only by SMC. For the 6 points at $x < 0.03$ the statistical precision of COMPASS is a factor of 3–4 better. Moreover, the negative trend at low x which was observed by SMC is not confirmed by the present data.

3 Extraction of the singlet axial charge $\hat{a}_0$

With COMPASS data alone we evaluate the first moment $\Gamma_1^d(Q^2) = \int_0^1 g_1^d(x, Q^2)\mathrm{d}x$ from which we derive the scale-invariant matrix element of the flavor-singlet axial current $\hat{a}_0$ [9]. The evaluation of $\Gamma_1^d(Q^2)$ requires the evolution of all g_1 measurements to a common Q_0^2. This is most conveniently done by using a fitted parametrization $g_1^{fit}(x, Q^2)$:

$$g_1(x, Q_0^2) = g_1(x, Q^2) + \left[g_1^{fit}(x, Q_0^2) - g_1^{fit}(x, Q^2)\right]. \quad (5)$$

We have used several fits of g_1 from the Durham data base [10]: BB [11], GRSV [12] and LSS05 [13]. We have chosen $Q_0^2 = 3\,(\mathrm{GeV}/c)^2$ as reference Q^2 because it is close to the average Q^2 of the COMPASS DIS data. The three parameterizations are quite similar in the range of the COMPASS data and have been averaged. The resulting values of $g_1^N = (g_1^p + g_1^n)/2$ are shown as open squares in fig. 3. For clarity we now use g_1^N which is g_1^d corrected for the D-wave state of the deuteron:

$$g_1^N(x, Q^2) = g_1^d(x, Q^2)/(1 - 1.5\omega_D) \tag{6}$$

with $\omega_D = 0.05 \pm 0.01$ [14]. It can be seen in fig. 3 that the curve representing the average of the three fits does not reproduce the trend of our data for $x < 0.02$ and therefore cannot be used to estimate the unmeasured part of g_1^N at low x. In view of this, we have performed a new NLO QCD fit of all g_1 DIS data from the deuteron target [5–8] including the COMPASS data, proton [5–7,15,16] and ^{3}He [17] targets. In total 230 data points are used. The fit is performed in the $\overline{\mathrm{MS}}$ renormalization and factorization scheme. The fits have been performed with two different programs, the first one working in the (x, Q^2)-space [18], the other one in the space of moments [19]. Both programs give consistent values of the fitted PDFs and similar χ^2-probabilities. Each program yields two solutions, one solution with $\Delta G > 0$, the other with $\Delta G < 0$ (fig. 3).

The integral of g_1^N in the measured region is obtained from the experimental values evolved to a fixed Q^2. Taking into account the contributions from the fits in the unmeasured regions at low and high x we obtain

$$\left.\Gamma_1^N\right|_{Q^2=3\,(\mathrm{GeV}/c)^2} = 0.050 \pm 0.003(\text{stat.})$$
$$\pm 0.002(\text{evol.}) \pm 0.005(\text{syst.}). \tag{7}$$

The quoted value is the sum of a measured value plus two small corrections at high and low x. The measured part (0.049 ± 0.003) represents 98% of the total and the corrections only 2%. This is very different from the SMC analysis [5] where the measured value (over the same x range as COMPASS) was 0.037 ± 0.006 and the estimated correction -0.018 (*i.e.* about 50%). The measured parts are not expected to be the same in COMPASS and SMC because they are evaluated at different values of Q^2 (3 and $10\,(\mathrm{GeV}/c)^2$, respectively) but are compatible within $2\,\sigma$'s. The huge difference in the estimated low-x contribution reflects the fact that the COMPASS data do not support the fast decrease of $g_1^d(x, Q^2 = 3\,(\mathrm{GeV}/c)^2)$ at low x which was assumed in the SMC correction.

Γ_1^N is an important quantity because it gives access to the invariant matrix element of the singlet axial current $\hat{a}_0$ [9]. The relation between $\hat{a}_0$ and Γ_1^N has been calculated up to the third order in α_s:

$$\Gamma_1^N(Q^2) = \frac{1}{9}\, C_1^S(Q^2)\, \hat{a}_0 + \frac{1}{36}\, C_1^{NS}(Q^2)\, a_8 \tag{8}$$

with the coefficients [9]:

$$C_1^S(Q^2) = 1 - 0.33333\left(\frac{\alpha_s}{\pi}\right) - 0.54959\left(\frac{\alpha_s}{\pi}\right)^2$$
$$-4.44725\left(\frac{\alpha_s}{\pi}\right)^3,$$
$$C_1^{NS}(Q^2) = 1 - \left(\frac{\alpha_s}{\pi}\right) - 3.5833\left(\frac{\alpha_s}{\pi}\right)^2 - 20.2153\left(\frac{\alpha_s}{\pi}\right)^3.$$

The coupling constant at $Q^2 = 3\,(\mathrm{GeV}/c)^2$ is obtained by evolving the PDG value of $\alpha_s(M_z) = 0.1187 \pm 0.005$:

$$\left.\frac{\alpha_s}{\pi}\right|_{Q^2=3\,(\mathrm{GeV}/c)^2} = 0.084 \pm 0.003.$$

The value of a_8 is derived from hyperon β decay, assuming $SU(3)_f$ flavor symmetry: $a_8 = 0.585 \pm 0.025$ [20]. The resulting value of $\hat{a}_0$ is

$$\hat{a}_0 = 0.33 \pm 0.03(\text{stat.}) \pm 0.05(\text{syst.}). \tag{9}$$

This value is independent of Q^2 and directly comparable to a_8. The Ellis-Jaffe sum rule was based on the assumption of unpolarized strangeness $(\hat{a}_0 = a_8)$. Since the EMC experiment, it is known that this assumption is not valid. From eq. (9) we can derive the first moment of the strange quark spin distribution:

$$(\Delta s + \Delta \bar{s})_{Q^2 \to \infty} = \frac{1}{3}(\hat{a}_0 - a_8) =$$
$$-0.08 \pm 0.01(\text{stat.}) \pm 0.02(\text{syst.}). \tag{10}$$

4 Conclusion

The new measurement of the deuteron longitudinal spin asymmetry A_1^d and its spin-dependent structure function g_1^d at $Q^2 > 1\,(\mathrm{GeV}/c)^2$ over the range $0.004 < x < 0.7$ have been presented. Good agreement with the world data is observed over the full range of x. For $x < 0.03$ our points are consistent with zero and their statistical errors have been reduced by a factor 3–4 compared to SMC results. Thus, the negative trend of g_1^d at low x observed by SMC is not confirmed. The measured values have been evolved to a common Q^2 by a fit of the world g_1 data and the first moment Γ_1^N has been evaluated at $Q^2 = 3\,(\mathrm{GeV}/c)^2$. From Γ_1^N we have derived the invariant matrix element of the singlet axial current $\hat{a}_0$, which can be interpreted as a quark polarization in the nucleon at $Q^2 \to \infty$. Thus, at the order α_s^3, the COMPASS data alone give $\hat{a}_0 = 0.33 \pm 0.03(\text{stat.}) \pm 0.05(\text{syst.})$ and the first moment of the strange-quark distribution $(\Delta s + \Delta \bar{s})_{Q^2 \to \infty} = -0.08 \pm 0.01(\text{stat.}) \pm 0.02(\text{syst.})$. The quoted systematic errors account for the uncertainty from the evolution and for the experimental systematic error, combined in quadrature.

References

1. S.D. Bass, Rev. Mod. Phys. **77**, 1257 (2005).
2. COMPASS proposal CERN/SPSLC 96-14 (Geneva, 1996).
3. E155 Collaboration (P.L. Anthony *et al.*), Phys. Lett. B **553**, 18 (2003).
4. COMPASS Collaboration (E.S. Ageev *et al.*), Phys. Lett. B **612**, 154 (2005).
5. SMC Collaboration (B. Adeva *et al.*), Phys. Rev. D **58**, 112001 (1998).
6. HERMES Collaboration (A. Airapetian *et al.*), Phys. Rev. D **75**, 012003 (2005).
7. E143 Collaboration (K. Abe *et al.*), Phys. Rev. D **58**, 112003 (1998).
8. E155 Collaboration (P.L. Anthony *et al.*), Phys. Lett. B **463**, 339 (1999).
9. S.A. Larin *et al.*, Phys. Lett. B **404**, 153 (1997).
10. HEP Databases, `http://durpdg.dur.ac.uk/HEPDATA/pdf.html`.
11. J. Blümlein, H. Böttcher, Nucl. Phys. B **636**, 225 (2002).
12. M. Glück *et al.*, Phys. Rev. D **63**, 094005 (2001).
13. E. Leader, A.V. Sidorov, D.B. Stamenov, Phys. Rev. D **73**, 034023 (2006).
14. R. Machleidt *et al.*, Phys. Rep. **149**, 1 (1987).
15. EMC Collaboration (J. Ashman *et al.*), Nucl. Phys. B **328**, 1 (1989).
16. E155 Collaboration (P.L. Anthony *et al.*), Phys. Lett. B **493**, 19 (2000).
17. E142 Collaboration (P.L. Anthony *et al.*), Phys. Rev. D **54**, 6620 (1996); E154 Collaboration (K. Abe *et al.*), Phys. Rev. Lett. **79**, 26 (1997); JLAB/Hall A Collaboration (X. Zheng *et al.*), Phys. Rev. Lett. **92**, 012004 (2004); HERMES Collaboration (K. Ackerstaff *et al.*), Phys. Lett. B **404**, 383 (1997).
18. SMC Collaboration (B. Adeva *et al.*), Phys. Rev. D **58**, 112002 (1998).
19. A.N. Sissakian, O. Yu. Shevchenko, O.N. Ivanov, Phys. Rev. D **70**, 074032 (2004).
20. Y. Goto *et al.*, Phys. Rev. D **62**, 034017 (2000).

Eur. Phys. J. A **31**, 610–612 (2007)

DOI 10.1140/epja/i2006-10208-4

THE EUROPEAN
PHYSICAL JOURNAL A

Special Article – QNP 2006

Measurement of the generalized polarizabilities of the proton in virtual Compton scattering at MAMI

P. Janssens[a,b,c]

Department of Subatomic and Radiation Physics, Ghent University, 9000 Ghent, Belgium

Received: 25 October 2006
Published online: 26 February 2007 – © Società Italiana di Fisica / Springer-Verlag 2007

Abstract. Virtual Compton scattering off the proton has been studied at $Q^2 = 0.33$ $(\mathrm{GeV}/c)^2$ at the MAMI accelerator in Mainz (Germany). The goal of the experiment is to measure the generalized polarizabilities (GPs) of the proton using the double polarized $ep \to e'p'\gamma$ reaction. This paper reports the measurement of the unpolarized photon electroproduction cross-section for the new data and the extraction of two linear combinations of GPs, which are essential for the analysis of the double spin asymmetry.

PACS. 13.60.Fz Elastic and Compton scattering – 14.20.Dh Protons and neutrons

1 Introduction

Virtual Compton scattering (VCS) off the proton ($\gamma^* + p \to \gamma + p$, where γ^* is the incoming virtual photon and γ is the outgoing real photon) is an interesting reaction to study the internal structure of the proton. Below the pion production threshold, a double polarized VCS experiment allows to measure the six lowest-order generalized polarizabilities (GPs). Two of them are an extension of the static polarizabilities, α_E and β_M, measured in real Compton scattering (RCS). α_E and β_M quantify the deformation of the charge and current distributions inside the proton caused by an external electric or magnetic field, respectively. The GPs, which are functions of Q^2, measure the polarizability locally inside the nucleon on a distance scale given by Q^2 [1]. In the real-photon limit ($Q^2 \to 0$) two of the GPs are proportional to α_E and β_M.

VCS (see fig. 1) is accessed through photon electroproduction ($ep \to e'p'\gamma$). The five-fold differential cross-section[1], $\mathrm{d}^5\sigma/\mathrm{d}k'\mathrm{d}\Omega_{e'}\mathrm{d}\Omega_{p',\mathrm{cm}}$, depends on the virtual photon momentum q_{cm} and its polarization parameter ε, the real (outgoing) photon momentum q'_{cm} and the polar and azimuthal angle of the real photon with respect to the virtual photon direction, $\theta_{\gamma\gamma,\mathrm{cm}}$ and φ.

The photon electroproduction reaction contains three contributions (see fig. 2): the reaction is dominated by the Bethe-Heitler and Born (BH+B) contributions, where

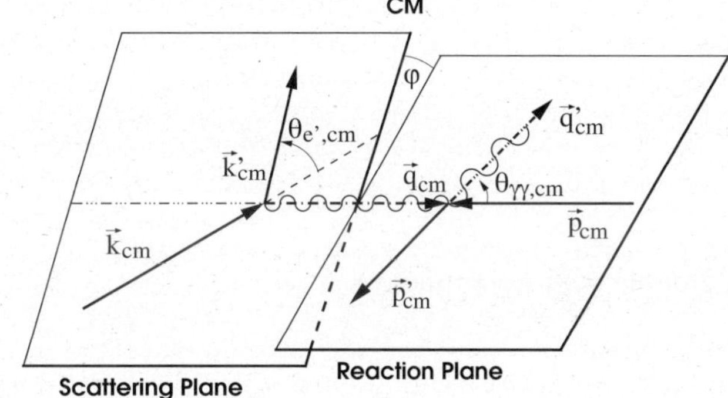

Fig. 1. Schematic drawing of the $ep \to e'p'\gamma$ reaction in the center of mass of the virtual photon and the target proton.

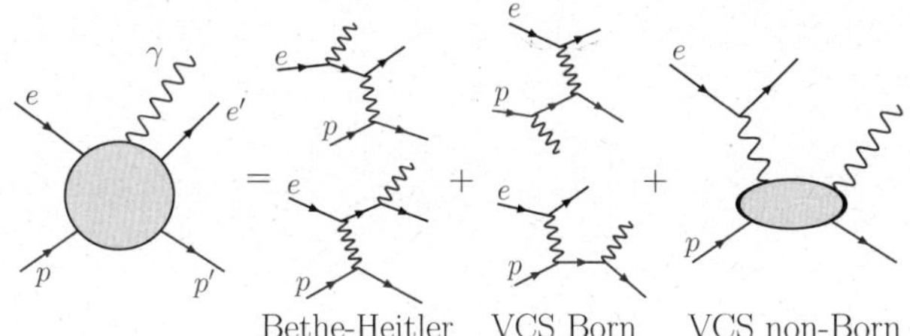

Fig. 2. Decomposition of the $ep \to e'p'\gamma$ reaction.

[a] e-mail: peter@inwfsun1.ugent.be
[b] On behalf of the VCS Collaboration at MAMI-A1.
[c] Aspirant of the FWO.

[1] All variables defined in the center of mass of the virtual photon and target proton have an index "cm". If no index is given the variable is defined in the laboratory system.

the outgoing photon is produced due to Bremsstrahlung of the electron or proton, respectively. The contribution of the BH+B process can be calculated exactly based on the proton form factors. The GPs contribute to the VCS non-Born part of the reaction.

2 Experimental determination of the GPs

The GPs cannot be measured directly. In the physical observables (cross-sections and asymmetries) they appear in specific linear combinations, the structure functions. Since there are six independent GPs, there are six independent structure functions [2].

2.1 Unpolarized cross-section

The low-energy theorem (LET) can be used to expand the unpolarized cross-section in powers of q'_{cm} ([1] and [2]):

$$d^5\sigma = d^5\sigma^{BH+B} + \phi q'_{cm}\Psi_0 + \mathcal{O}\big(q'^2_{cm}\big), \qquad (1)$$

where ϕ is a known phase space factor, $d^5\sigma^{BH+B}$ is the five-fold differential cross-section for the BH+B processes and Ψ_0, which contains the information about the GPs, is defined by

$$\frac{\Psi_0}{v_{LT}} = \frac{v_{LL}}{v_{LT}}(P_{LL} - P_{TT}/\varepsilon) + P_{LT}. \qquad (2)$$

In this equation v_{LT} and v_{LL} are kinematical coefficients (containing, e.g., $\theta_{\gamma\gamma,cm}$), and P_{LL}, P_{TT} and P_{LT} are the structure functions one wants to measure. The higher-order terms $\mathcal{O}(q'^2_{cm})$ can be neglected for low q'_{cm}, a condition which holds below the pion production threshold. In eqs. (1) and (2) two linear combinations of structure functions appear ($P_{LL} - P_{TT}/\varepsilon$ and P_{LT}), which can be separated by measuring the unpolarized cross-section in an appropriate kinematical region (see sect. 4).

2.2 Double spin asymmetry

More information about the GPs can be extracted from polarized observables. One wants to use the low-energy theorem to extract the GPs from the data, for which the experiment should be performed below the pion production threshold. Since in that regime all single spin asymmetries for VCS disappear [2], one has to perform a double polarized experiment (here $ep \rightarrow e'p'\gamma$). The double spin asymmetry (DSA) along the axis i in the center of mass is defined by

$$P_i = \frac{d^5\sigma_i^{\uparrow} - d^5\sigma_i^{\downarrow}}{d^5\sigma_i^{\uparrow} + d^5\sigma_i^{\downarrow}} = \frac{\Delta d^5\sigma_i}{2d^5\sigma} \qquad (i = x, y, z). \quad (3)$$

$d^5\sigma_i^{\uparrow}$ represents the differential cross-section, where the outgoing proton is polarized along the i-axis and the helicity, h, of the incoming electron equals $+\frac{1}{2}$, indicated by $\uparrow$ ($\downarrow$ corresponds to $h = -\frac{1}{2}$). The axes of the reference frame in the center of mass are defined as in [2].

The denominator is two times the unpolarized cross-section (see sect. 2.1). Similarly to the unpolarized cross-section, the low-energy theorem can be used to write the numerator as

$$\Delta d^5\sigma_i = \Delta d^5\sigma_i^{BH+B} + \phi q'_{cm}\Delta\Psi_{0,i} + \mathcal{O}\big(q'^2_{cm}\big), \quad (4)$$

where the terms $\Delta\Psi_{0,i}$ ($i = x, y, z$) are linear combinations of the structure functions:

$$\Delta\Psi_{0,z} = v_1^z P_{TT} + v_2^z P_{LT}^z + v_3^z P_{LT}'^z, \qquad (5)$$

and similar expressions for $\Delta\Psi_{0,x}$ and $\Delta\Psi_{0,y}$ (see [2]). In these three terms $\Delta\Psi_{0,i}$ the six independent structure functions appear, thus one can determine all six GPs by measuring the DSA for VCS.

3 Experimental setup

For the present experiment the standard setup of the A1 hall at MAMI was used [3]: the longitudinally polarized electron beam from the MAMI accelerator impinges with an energy of 854.6 MeV on a liquid-hydrogen target. The polarization of the beam is about 75%. The scattered electron and recoiling proton are detected in the high resolution magnetic spectrometers.

Real photon production events are identified by missing-mass reconstruction. The central momenta and angles of the spectrometers were optimized to detect real-photon production events with $q_{cm} = 600\,\text{MeV}/c$, $q'_{cm} = 90\,\text{MeV}/c$, $\varepsilon = 0.645$ and $\varphi = 0°$.

To measure the polarization of the recoiling proton a polarimeter was mounted behind the tracking detectors of spectrometer A. It consists of a 7 cm carbon scattering block and two double planes of horizontal drift chambers ([4] and [5]) and it allows to measure the transverse components of the proton polarization in the focal plane.

The magnetic field of the spectrometer rotates the spin of the protons on their way to the focal plane. This precession mixes the different components of the proton polarization, dependent on the magnetic field along the proton track in the spectrometer. This allows to measure the three components of the proton polarization in the center-of-mass system.

4 Preliminary results

The first step in the analysis of the experiment is the determination of the absolute unpolarized cross-section of the $ep \rightarrow e'p'\gamma$ reaction. Since the effect of the GPs is about 10% on the total cross-section, the solid angle of the detector acceptance has to be calculated accurately. This is performed using a Monte Carlo simulation [6], which incorporates all relevant resolution deteriorating effects. The events in the simulation are generated according to the BH+B cross-section behaviour, which is a very good approximation of the real cross-section for the kinematics of the experiment.

The measured cross-section is shown in fig. 3. The dotted line shows the BH+B contribution; the full line shows a fit to the data points and includes the effect of the GPs. This fit is shown in fig. 4, where Ψ_0/v_{LT} is plotted versus v_{LL}/v_{LT} (see eq. (2)). The results of the fit are shown in table 1. The first error indicates the statistical error.

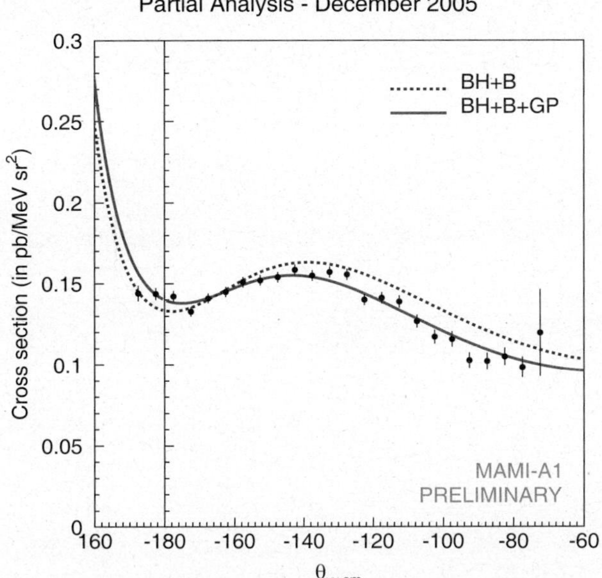

Fig. 3. The unpolarized $ep \rightarrow e'p'\gamma$ cross-section $\mathrm{d}^5\sigma/\mathrm{d}k'\,\mathrm{d}\Omega_{e'}\,\mathrm{d}\Omega_{p',\mathrm{cm}}$ measured at $q'_{\mathrm{cm}} = 90$ MeV/c, $q_{\mathrm{cm}} = 600$ MeV/c, $\varepsilon = 0.645$ and $\varphi = 0°$. The dotted line shows the BH+B cross-section, which deviates from the measured five-fold differential cross-section due to the contribution of the GPs. The full line shows the result of the fit in fig. 4. Negative values of $\theta_{\gamma\gamma,\mathrm{cm}}$ correspond to $\varphi = 180°$.

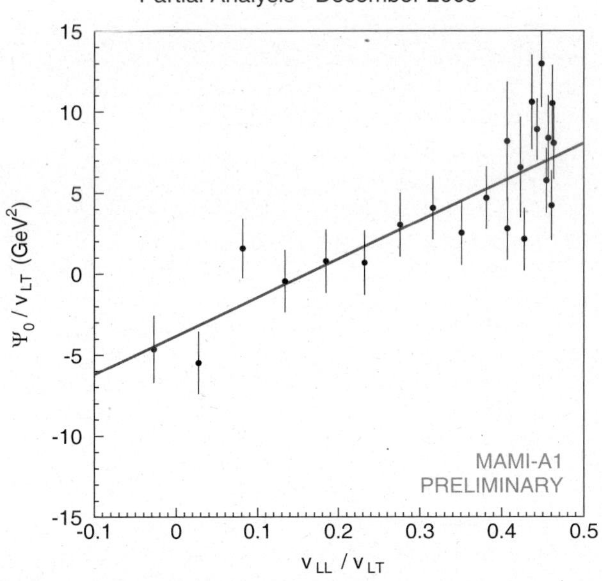

Fig. 4. A linear fit of Ψ_0/v_{LT} *versus* $v_{\mathrm{LL}}/v_{\mathrm{LT}}$ yields two linear combinations of GPs: $P_{\mathrm{LL}} - P_{\mathrm{TT}}/\varepsilon$ and P_{LT} (see eq. (2)). The reduced χ^2 of the fit is 1.35.

Table 1. Results for $P_{\mathrm{LL}} - P_{\mathrm{TT}}/\varepsilon$ and P_{LT} for $q_{\mathrm{cm}} = 600$ MeV/c. The meaning of the different errors is explained in the text. ε in reference [7] ($\varepsilon = 0.620$) was slightly different from the present experiment ($\varepsilon = 0.645$).

	$P_{\mathrm{LL}} - P_{\mathrm{TT}}/\varepsilon$ (GeV^{-2})	P_{LT} (GeV^{-2})
This work	$23.8 \pm 2.9 \pm 0.3 \pm 3.5$	$-3.8 \pm 1.0 \pm 0.9 \pm 1.1$
Ref. [7]	$23.7 \pm 2.2 \pm 0.6 \pm 4.3$	$-5.0 \pm 0.8 \pm 1.1 \pm 1.4$

The two other errors are the contributions of the absolute normalization of the cross-section and the absolute calibration of the spectrometers' central momentum to the systematic error, respectively. The result is in good agreement with the previous experiment [7].

In the next step the DSA will be determined. The DSA is defined in the center of mass (see sect. 2.2). Similar to the analysis of polarized pion electroproduction the Wigner rotation of the proton spin due to the Lorentz boost from the center of mass to the laboratory system has to be taken into account [8].

The GPs will be fitted to the data via the DSA along the x-, y- and z-axis using a maximum-likelihood method. The values for $P_{\mathrm{LL}} - P_{\mathrm{TT}}/\varepsilon$ and P_{LT} obtained in the unpolarized analysis are used as a constraint for this fit. This analysis is in progress.

This work was supported in part by the FWO-Flanders (Belgium), the French CEA and CNRS/IN2P3, the Deutsche Forschungsgemeinschaft (SFB 201 and SFB 443) and by the Federal State of Rhineland-Palatinate.

References

1. P.A.M. Guichon *et al.*, Nucl. Phys. A **591**, 606 (1995).
2. P.A.M. Guichon *et al.*, Prog. Part. Nucl. Phys. **41**, 125 (1998).
3. K.I. Blomqvist *et al.*, Nucl. Instrum. Methods A **403**, 263 (1998).
4. Th. Pospischil *et al.*, Nucl. Instrum. Methods A **483**, 713 (2002).
5. Th. Pospischil *et al.*, Nucl. Instrum. Methods A **483**, 726 (2002).
6. P. Janssens *et al.*, Nucl. Instrum. Methods A **566**, 675 (2006).
7. J. Roche *et al.*, Phys. Rev. Lett. **85**, 708 (2000).
8. H. Schmieden *et al.*, Eur. Phys. J. A **1**, 427 (1998).

Eur. Phys. J. A **31**, 613–615 (2007)

DOI 10.1140/epja/i2006-10265-7

THE EUROPEAN
PHYSICAL JOURNAL A

Special Article – QNP 2006

Electromagnetic properties of strange baryons in a relativistic quark model

T. Van Cauteren[1,a], J. Ryckebusch[1], B. Metsch[2], and H.-R. Petry[2]

[1] Department of Subatomic and Radiation Physics, Ghent University, Proeftuinstaat 86, B-9000 Gent, Belgium
[2] HISKP, Bonn University, Nußallee 14-16, D-53115 Bonn, Germany

Received: 23 November 2006
Published online: 16 March 2007 – © Società Italiana di Fisica / Springer-Verlag 2007

Abstract. We present some of our results for the electromagnetic properties of excited Σ hyperons, computed within the framework of the Bonn constituent-quark model, which is based on the Bethe-Salpeter approach. The seven parameters entering the model are fitted against the best-known baryon masses. Accordingly, the results for the form factors and helicity amplitudes are genuine predictions. We compare with the scarce experimental data available and discuss the processes in which Σ^*'s may play an important role.

PACS. 11.10.St Bound and unstable states; Bethe-Salpeter equations – 12.39.Ki Relativistic quark model – 13.40.Gp Electromagnetic form factors – 14.20.Jn Hyperons

1 Introduction

Strangeness physics is gaining more and more momentum in present-day hadronic physics. Due to the upgrading of existing and the commisioning of new ones, more experimental facilities than ever can perform measurements of hyperon structure and interactions.

Reactions such as the electromagnetic production of kaons or radiative kaon capture are sensitive to the electromagnetic (EM) properties of the particles involved. While for the low-lying nonstrange hadrons, a reasonable amount of data is available, the EM properties of hyperons and hyperon resonances are almost completely unknown. Yet modelling the above-mentioned reactions requires knowledge of the strength with which $Y^{(*)}$'s couple to photons. The work presented here and in refs. [1,2] tries to remedy this gap of knowledge by computing the EM form factors of $Y^{(*)}$'s in the framework of the covariant quark model developed by the Bonn group [3]. This constituent-quark (CQ) model is based on the Bethe-Salpeter approach and uses instantaneous interactions: confinement is ensured by a linearly rising Δ-shaped potential and hyperfine splittings are accounted for by an instanton-induced interaction. It should be mentioned that lattice-QCD calculations show a preference for a Y-shaped confinement potential between three static quarks at distances larger than ~ 0.8 fm [4]. However, the effects of assuming a different confinement ansatz (and subsequently rescaling the confinement parameters) seem to be rather small and at this

a e-mail: Tim.VanCauteren@UGent.be

moment beyond experimental verification, both for the spectrum as for EM form factors. The seven parameters in the model are fitted to the masses of the best-known baryon resonances, but no extra parameters are used in computing the EM form factors. The calculations are performed consistently at first order to obtain the vertexfunctions $\Gamma(p_\xi, p_\eta)$ as amputated Bethe-Salpeter amplitudes describing the baryon states in terms of the Jacobi momenta p_ξ and p_η. The EM current matrix elements are then computed in lowest order in the rest frame of the incoming state according to [1,2,5]

$$
\begin{aligned}
\left\langle \overline{P} | j^\mu(0) | M \right\rangle \simeq (-3) \int & \frac{\mathrm{d}^4 p_\xi}{(2\pi)^4} \frac{\mathrm{d}^4 p_\eta}{(2\pi)^4} \, \overline{\Gamma}_{\overline{P}}^{\Lambda}(p_\xi, p_\eta) \\
\times S_F^1 & \left(\frac{M}{3} + p_\xi + \frac{p_\eta}{2} \right) \otimes S_F^2 \left(\frac{M}{3} - p_\xi + \frac{p_\eta}{2} \right) \\
\otimes & \left[S_F^3 \left(\frac{M}{3} - p_\eta + q \right) \hat{q}\gamma^\mu S_F^3 \left(\frac{M}{3} - p_\eta \right) \right] \\
\times \Gamma_M^{\Lambda} & \left(p_\xi, p_\eta + \frac{2}{3} q \right),
\end{aligned}
\tag{1}
$$

where $\overline{P}$ is the total four-momentum of the outgoing on-shell state, M is the mass of the incoming state, $S_F^i(p)$ denotes the propagator for the i-th CQ, and $\hat{q}$ is the CQ-charge operator. The operator $\otimes$ denotes a direct product in the (qqq)-space. From this matrix element, the EM form factors or helicity amplitudes (HAs) can be derived. The limits of the current matrix elements for $Q^2 \to 0$ give the response of the baryon to real photons and are needed to compute EM decay widths.

2 Helicity amplitudes

The helicity amplitudes presented here are defined by [2]

$$A_{1/2}\left(\Sigma^* \to Y\right) =$$
$$\mathcal{D}\left\langle Y, \overline{P}, \frac{1}{2}\middle| j^1(0) + i\,j^2(0)\middle|\Sigma^*, \overline{P}^*, -\frac{1}{2}\right\rangle, \quad \text{(2a)}$$

$$A_{3/2}\left(\Sigma^* \to Y\right) =$$
$$\mathcal{D}\left\langle Y, \overline{P}, -\frac{1}{2}\middle| j^1(0) + i\,j^2(0)\middle|\Sigma^*, \overline{P}^*, -\frac{3}{2}\right\rangle, \quad \text{(2b)}$$

$$C_{1/2}\left(\Sigma^* \to Y\right) = \mathcal{D}\left\langle Y, \overline{P}, \frac{1}{2}\middle| j^0(0)\middle|\Sigma^*, \overline{P}^*, \frac{1}{2}\right\rangle, \quad \text{(2c)}$$

for the EM transitions between excited (Σ^*) and ground-state (Y) hyperons. Here, $\mathcal{D} = \sqrt{\frac{\pi\alpha}{2m(m^{*2}-m^2)}}$ depends on the specific normalization used for the vertex functions. The EM decay width for the process $\Sigma^* \to Y + \gamma$ is then given by

$$\Gamma_\gamma = \frac{|\mathbf{q}|^2}{4\pi^2\alpha} \frac{2m}{(2J^*+1)\,m^*} \left[|A_{1/2}|^2 + |A_{3/2}|^2\right], \quad \text{(3)}$$

where $|\mathbf{q}| = \frac{m^{*2}-m^2}{2m^*}$ is the photon three-momentum in the rest frame of the initial baryon resonance, and $\alpha = \frac{e^2}{4\pi} \simeq \frac{1}{137}$ is the EM fine-structure constant. Remark that this definition differs from the PDG [6] by a factor of $e^2 = 4\pi\alpha$. Wherever we can, we will compare our calculations with experiment. However, since we only compute EM current matrix elements, and not *e.g.* pion decay, only the relative signs between the three HAs of a certain $\Sigma^* \to Y$ decay are fixed. If the ground-state hyperon is a Σ, the relative HAs for the different isospin channels are also determined.

3 Results for the $J = 3/2$ Σ^*-resonances

Due to limited space, we only show the HAs of the $\Sigma^{*-} + \gamma^* \to \Sigma^-$ decays for lowest-lying spin $J = 3/2$ resonances in fig. 1. A few observations can be made here. First, in contrast to most elastic EM form factors [1], some HAs are larger at non-zero Q^2 than at the real-photon point. This property is more general than in the specific decays shown here and also turns up in other HAs both in the nonstrange [5] and in the hyperon sector [2]. It is therefore necessary to include electroproduction processes in the search for baryon resonances, and not only photo-production reactions.

Another observation one can infer from fig. 1 is that there are no HAs with a magnitude larger than $100 \times 10^{-3}\,\text{GeV}^{-1/2}$. Therefore, it is better to look for signals of spin $J = 3/2$ Σ-resonances in the positively charged isospin channel than in the negatively charged one. This

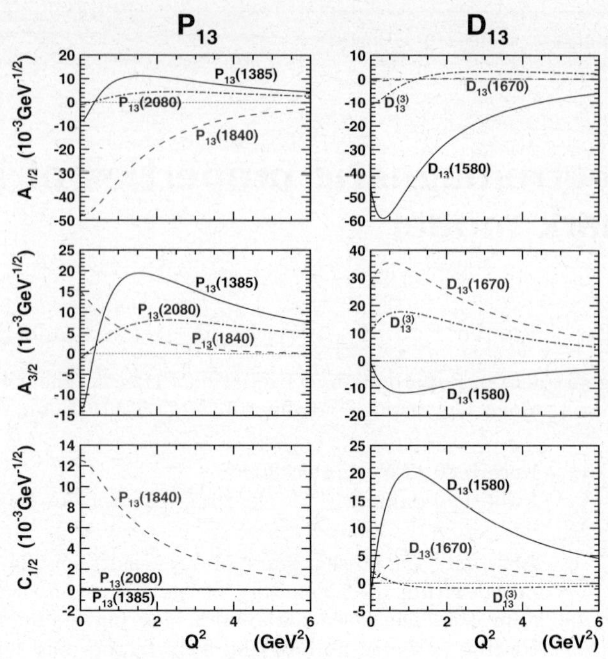

Fig. 1. The Q^2-dependence for the $\Sigma^{*-} + \gamma^* \to \Sigma^-$ decays for spin $J = 3/2$ resonances: left (right) panels show the results for the positive (negative) parity Σ^*-resonances.

Table 1. Calculated masses, photo-amplitudes and EM decay widths for the $\Sigma^* + \gamma \to \Sigma(1193)$ transitions for $J = 3/2$ Σ^*-resonances. The charge of the Σ^* isospin-triplet member is indicated by the superscripts $0, +, -$. Masses and decay widths are given in units of MeV, photo-amplitudes are given in units of $10^{-3}\,\text{GeV}^{-1/2}$. The value in parentheses is the experimental upper limit quoted by the PDG [6], with a 90% confidence level.

Resonance	M_{calc}	$A_{1/2}$	$A_{3/2}$	Γ_{calc}
$P^0_{13}(1385)$	1409	27.8	48.0	0.181
$P^0_{13}(1840)$	1902	15.4	-5.25	0.0960
$P^0_{13}(2080)$	1950	14.3	23.7	0.303
$D^0_{13}(1580)$	1675	-2.82	-32.9	0.230
$D^0_{13}(1670)$	1727	-6.77	-6.45	0.0214
$D^{(3)0}_{13}$	1780	24.4	25.5	0.349
$P^+_{13}(1385)$	1409	62.6	108.2	0.920
$P^+_{13}(1840)$	1902	80.0	-25.6	2.559
$P^+_{13}(2080)$	1950	29.7	48.5	1.280
$D^+_{13}(1580)$	1675	40.0	-65.2	1.235
$D^+_{13}(1670)$	1727	-13.1	-40.3	0.440
$D^{(3)+}_{13}$	1780	59.8	40.8	1.468
$P^-_{13}(1385)$	1409	-7.06	-12.2	0.0117
				$(< 0.0095 \pm 0.0006)$
$P^-_{13}(1840)$	1902	-47.1	15.1	0.887
$P^-_{13}(2080)$	1950	-1.20	-1.05	0.00101
$D^-_{13}(1580)$	1675	-45.7	-0.588	0.441
$D^-_{13}(1670)$	1727	-0.397	27.4	0.184
$D^{(3)-}_{13}$	1780	-10.9	10.3	0.0630

is better illustrated in table 1 where the computed decay widths are shown for all the isospin channels in the $\Sigma^*(J = 3/2) \to \Sigma$ EM decay. The predicted EM decays of the positively charged resonances are clearly larger than for the other channels.

Unfortunately, the experimental data on EM properties of Σ-resonances is very scarce. In fact, the only data mentioned in the PDG tables [6] are the EM decay width of the $\Sigma^{*0}(1385) \to \Lambda(1116)$ ($479 \pm 120^{+81}_{-100}$ keV) and an upper limit for the EM decay width of the $\Sigma^{*-}(1385) \to \Sigma^{-}(1193)$ process ($< 9.5 \pm 0.6$ keV at 90% confidence level). The former was extracted from looking at $p\gamma \to K^+\Lambda\gamma$ processes [7] at Jefferson Lab. Therefore, one may hope to extract information on the $\Sigma^{*+}(1385) \to \Sigma^+(1193)$ EM decay width from $p\gamma \to K^0\Sigma^+\gamma$ reactions. Our computed value of the $\Sigma^{*0} \to \Lambda$ EM decay width is 1.53 MeV, which is a factor of three larger than the measured value. In table 1, one can see that also our value for the $\Sigma^{*-} \to \Sigma^-$ EM decay width is slightly larger than the upper limit given in the PDG. This overestimation of the EM decay widths could be due to meson loop or exchange current contributions to the process not included in the present model.

4 Isospin asymmetries for the $J = 3/2$ Σ^{*0}-resonances

The *isospin asymmetries* are defined as

$$\mathcal{T}_{1/2} = \frac{|A^\Lambda_{1/2}|^2 - |A^\Sigma_{1/2}|^2}{|A^\Lambda_{1/2}|^2 + |A^\Sigma_{1/2}|^2}, \tag{4a}$$

$$\mathcal{T}_{3/2} = \frac{|A^\Lambda_{3/2}|^2 - |A^\Sigma_{3/2}|^2}{|A^\Lambda_{3/2}|^2 + |A^\Sigma_{3/2}|^2}, \tag{4b}$$

$$\mathcal{T}_{0} = \frac{|C^\Lambda_{1/2}|^2 - |C^\Sigma_{1/2}|^2}{|C^\Lambda_{1/2}|^2 + |C^\Sigma_{1/2}|^2}. \tag{4c}$$

Here, the superscript Λ (Σ) stands for the decay of the resonance to the Λ (Σ^0) ground state. It is clear that a positive (negative) value indicates that the resonance will preferentially decay to the Λ (Σ^0) ground state.

In fig. 2 the computed isospin asymmetries are displayed for the lowest-lying $J = 3/2$ Σ^{*0}-resonances. With the exception of the $P_{13}(1840)$ and the $D_{13}(1670)$, the shown asymmetries tend to drop with increasing Q^2. Therefore the Σ^{*0}-resonances will couple mostly to the $\gamma^*\Lambda$ channel at low Q^2, and to the $\gamma^*\Sigma^0$ channel at intermediate to high Q^2. The $\mathcal{T}_{1/2}$- and $\mathcal{T}_{3/2}$-asymmetries of the $P_{13}(1840)$ behave in the opposite way. The $D_{13}(1670)$ decays preferentially to the $\gamma^{(*)}\Lambda$ channel at vanishing and small Q^2. The mass of this resonance is in the kinematic region of the $K^-p \to \gamma Y^0$ experiments investigated by

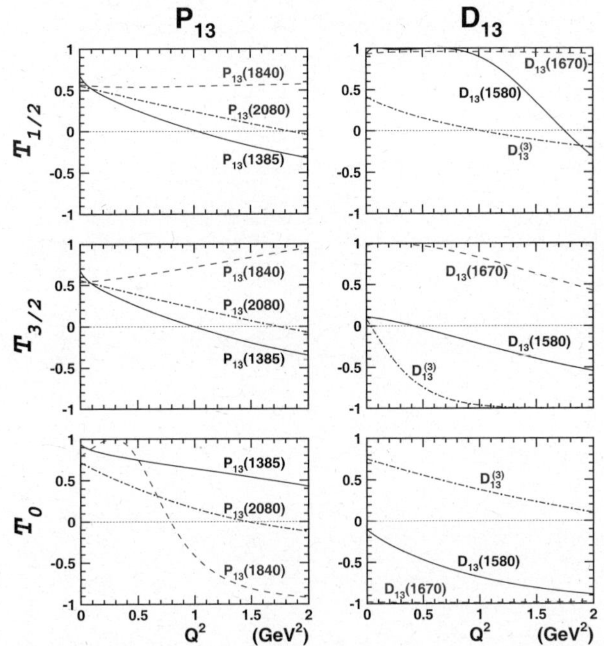

Fig. 2. The isospin asymmetries for $0 < Q^2 < 2.0$ GeV^2 as defined in eqs. (4) for the EM decays to different isospin channels for the three lowest-lying ($J = 3/2$, $S = -1$, $I = 1$) Σ^{*0}-resonances with positive-parity (left panels) and negative-parity (right panels).

the Crystal Ball Collaboration [8]. From our calculations one would expect that this resonance may only be seen in the $\gamma\Lambda$ channel and not in the $\gamma\Sigma^0$ channel.

References

1. T. Van Cauteren, D. Merten, J. Ryckebusch, T. Corthals, S. Janssen, B. Metsch, H.-R. Petry, Eur. Phys. J. A **20**, 283 (2004).
2. T. Van Cauteren, J. Ryckebusch, B. Metsch, H.-R. Petry, Eur. Phys. J. A **26**, 339 (2005).
3. U. Löring, K. Kretzschmar, B. Metsch, H.-R. Petry, Eur. Phys. J. A **10**, 309 (2001); U. Löring, B. Metsch, H.-R. Petry, Eur. Phys. J. A **10**, 395; 447 (2001).
4. Ph. de Forcrand, O. Jahn, Nucl. Phys. A **755**, 475c (2005).
5. D. Merten, U. Löring, K. Kretzschmar, B. Metsch, H.-R. Petry, Eur. Phys. J. A **14**, 477 (2002).
6. W.-M. Yao *et al.*, J. Phys. G **33**, 1 (2006).
7. S. Taylor *et al.*, Phys. Rev. C **71**, 054609 (2005); **72**, 039902 (2005)(E).
8. S. Prakhov, *Progress on Study of K^--Proton Reactions* (UCLA, USA, 2001) http://bmkn8.physics.ucla.edu/ Crystalball/Docs/documentation.html, unpublished; N. Phaisangittisakul, *First Measurement of the Radiative Process $K^-p \to \Lambda\gamma$ at Beam Momenta 520–750 MeV/c Using the Crystal Ball Detector*, PhD Thesis, University of California, Los Angeles, USA (2001).

Eur. Phys. J. A **31**, 616–619 (2007)

DOI 10.1140/epja/i2006-10182-9

THE EUROPEAN
PHYSICAL JOURNAL A

Special Article – QNP 2006

Retardation effects in the rotating string model

F. Buisseret[a], C. Semay[b], and V. Mathieu[c]

Groupe de Physique Nucléaire Théorique, Université de Mons-Hainaut, Académie universitaire Wallonie-Bruxelles, Place du Parc 20, BE-7000 Mons, Belgium

Received: 8 October 2006
Published online: 16 February 2007 – © Società Italiana di Fisica / Springer-Verlag 2007

Abstract. A new method to study the retardation effects in mesons is presented. It is based on a generalized rotating string model, in which a nonzero value of the relative time between the quark and the antiquark is allowed. This approach leads to a retardation term in the Hamiltonian which behaves as a perturbation of the nonretarded Hamiltonian and preserves the Regge trajectories for light mesons. The straight-line ansatz is used to describe the string, and the relevance of this approximation is tested. It is shown that the string is actually curved because of retardation, but this bending does not bring a relevant contribution to the energy spectrum of the model.

PACS. 12.39.Ki Relativistic quark model – 12.39.Pn Potential models – 14.40.-n Mesons

1 Introduction

The retardation effect between two interacting particles is a relativistic phenomenon, due to the finiteness of the interaction speed. Light mesons are typical systems in which these effects can significantly contribute to the dynamics, since the light quarks can move at a speed close to the speed of light. We present here a generalization of the rotating string model (RSM) [1,2] developed in refs. [3,4], which aims to take into account these effects in mesons.

2 Rotating string model with a nonzero relative time

The RSM is an effective model derived from the QCD Lagrangian, describing a meson by a quark and an antiquark linked by a straight string. Both particles are considered as spinless because spin interactions are sufficiently small to be added in perturbation. Our method to treat the retardation effects must be considered as a first trial to take into account such contributions in mesons. It relies on the hypothesis that the relative time between the quark and the antiquark must have a nonzero value. Consequently, in our approach, the evolution parameter of the system is not the common proper time of the quark, the antiquark and the string, but the time coordinate of the centre of mass which plays the role of an "average" time. The RSM with a nonzero relative time has been studied in detail in ref. [3]. So, we simply recall here the main points of this work.

Starting from the QCD Lagrangian and particularising it to the case of an interacting quark-antiquark pair, an effective Lagrangian can be derived. In units where $\hbar = c = 1$, it reads [1]

$$\mathcal{L} = -m_1\sqrt{\dot{\boldsymbol{x}}_1^2} - m_2\sqrt{\dot{\boldsymbol{x}}_2^2}$$
$$-a\int_0^1 \mathrm{d}\theta\sqrt{(\dot{\boldsymbol{w}}\boldsymbol{w}')^2 - \dot{\boldsymbol{w}}^2\boldsymbol{w}'^2}. \qquad (1)$$

The first two terms are the kinetic energy operators of the quark and the antiquark, whose current masses are m_i. These two particles are attached by a Nambu-Goto string with a tension a. $\boldsymbol{x}_i$ and $\boldsymbol{w}$ are the coordinates of the quark i and of the string, respectively. We defined $\dot{\boldsymbol{x}}_i = \partial_\tau \boldsymbol{x}_i$, $\dot{\boldsymbol{w}} = \partial_\tau \boldsymbol{w}$, and $\boldsymbol{w}' = \partial_\theta \boldsymbol{w}$. The string is generally assumed to be a straight line linking the quark to the antiquark, that is

$$\boldsymbol{w} = \theta\boldsymbol{x}_1 + (1-\theta)\boldsymbol{x}_2, \quad \theta \in [0,1]. \qquad (2)$$

Such an ansatz is suggested in particular by lattice QCD calculations, which show that the chromoelectric field between the quark and the antiquark appears to be roughly constant on a straight line joining the two particles [5].

Starting from Lagrangian (1) with a straight string given by eq. (2), one usually makes the equal-time ansatz, i.e.

$$x_1^0 = x_2^0 = w^0 = \tau = \bar{t}. \qquad (3)$$

____

[a] FNRS Research Fellow;
e-mail: `fabien.buisseret@umh.ac.be`
[b] FNRS Research Associate;
e-mail: `claude.semay@umh.ac.be`
[c] IISN Scientific Research Worker;
e-mail: `vincent.mathieu@umh.ac.be`

Then, we have $r = (0, r)$, and $R = (t, R)$. This procedure considerably simplifies the equations, but neglects the relativistic retardation effects due to a possible nonzero value of the relative time σ. That is why we made in ref. [3] a less restrictive hypothesis: We identified the temporal coordinate of the centre of mass with the evolution parameter, $\bar{t} = \tau$, and we allowed a nonvanishing relative time σ.

It is then possible to derive from the Lagrangian (1) a set of three equations for the RSM with a nonzero relative time,

$$0 = \mu_1 y_1 - \mu_2 y_2 - \frac{ar}{y_t}\left(\sqrt{1 - y_1^2} - \sqrt{1 - y_2^2}\right), \tag{4a}$$

$$\frac{L}{r} = \frac{1}{y_t}(\mu_1 y_1^2 + \mu_2 y_2^2) + \frac{ar}{y_t^2}(F(y_1) + F(y_2)), \tag{4b}$$

$$H = \frac{1}{2}\left[\frac{p_r^2 + m_1^2}{\mu_1} + \frac{p_r^2 + m_2^2}{\mu_2} + \mu_1(1 + y_1^2) + \mu_2(1 + y_2^2)\right]$$
$$+ \frac{ar}{y_t}(\arcsin y_1 + \arcsin y_2) + \Delta H, \tag{4c}$$

with

$$F(y_i) = \frac{1}{2}\left[\arcsin y_i - y_i\sqrt{1 - y_i^2}\right]. \tag{5}$$

p_r is the radial momentum and y_i is the transverse velocity of the quark i, and $y_t = y_1 + y_2$. The first relation gives the cancellation of the total momentum in the centre-of-mass frame, while the two last ones define, respectively, the angular momentum and the Hamiltonian of the system. Two auxiliary fields, denoted as μ_i, are introduced to simplify the computation. They can however be interpreted as dynamical masses of the quarks whose current masses are m_i [6]. They can be eliminated by minimising the energies with respect to them.

Equations (4) are identical to those of the usual RSM (see, for example, ref. [6]), but a perturbation of the Hamiltonian, denoted ΔH, is now present. It contains the contribution of the retardation effects and is given by [3]

$$\Delta H = -\frac{\Sigma^2}{2a_3} + \frac{c_2}{a_3}\Sigma\sigma - c_1\sigma - \frac{c_2^2}{2a_3}\sigma^2 + \frac{1}{2}a_4\sigma^2, \tag{6}$$

where Σ is the canonical momentum associated with the relative time σ. c_2, a_3, and a_4 are complicated functions of the spatial variables [6].

Let us now consider the quantized version of our model: $L \to \sqrt{\ell(\ell+1)}$, $[r, p_r] = i$, $[\sigma, \Sigma] = -i$. The Hamiltonian (4c) has then the following structure:

$$H(\sigma, r) = H_0(r) + \Delta H(\sigma, r). \tag{7}$$

The relative time only appears in the perturbation, and H_0 depends only on the radial variables. So, we can assume that the total wave function reads

$$|\psi(r)\rangle = |A(\sigma)\rangle \otimes |R(r)\rangle \otimes |Y_{\ell m}(\theta, \phi)\rangle, \tag{8}$$

where $|R(r)\rangle$ is a solution of the eigenequation

$$H_0(r)|R(r)\rangle = M_0|R(r)\rangle. \tag{9}$$

Such a problem can be numerically solved, for instance, by the Lagrange mesh technique [7]. The total mass is thus given by

$$M = M_0 + \langle A(\sigma)| \otimes \langle R(r)| \Delta H(r, \sigma) |R(r)\rangle \otimes |A(\sigma)\rangle$$
$$= M_0 + \Delta M. \tag{10}$$

As an excited state with respect to the relative time is irrelevant, the contribution ΔM is then given by the fundamental state of the eigenequation

$$\Delta\mathcal{H}(\sigma)|A(\sigma)\rangle = \Delta M |A(\sigma)\rangle, \tag{11}$$

where

$$\Delta\mathcal{H}(\sigma) = \langle R(r)| \Delta H(r, \sigma) |R(r)\rangle.$$
$$= -\frac{1}{2\langle a_3\rangle}\left[\Sigma^2 + \langle c_2^2 - a_4 a_3\rangle\sigma^2\right] \tag{12}$$

in the case $m_1 = m_2$. This constraint eliminates the unphysical degree of freedom due to the introduction of the relative time. Let us mention two consequences of eq. (11).

Firstly, the retardation contribution ΔM is negative,

$$\Delta M = -\frac{1}{2}\sqrt{\langle c_2^2 - a_4 a_3\rangle/\langle a_3\rangle^2}, \tag{13}$$

and it thus decreases the meson mass. It has been shown for a long time that quark models were able to roughly fit the hadron spectrum [8]. However, an arbitrary negative constant is still needed to shift the energies to the correct values. Our retardation term could thus be a good physical candidate to replace this arbitrary constant.

Secondly, the temporal part of the wave function reads

$$A(\sigma) = \left(\frac{\beta}{\pi}\right)^{1/4} \exp\left(-\frac{\beta}{2}\sigma^2\right), \tag{14}$$

with

$$\beta = \sqrt{\langle c_2^2 - a_4 a_3\rangle}. \tag{15}$$

It is a Gaussian function centered around $\sigma = 0$. This provides an interpretation of the equal-time ansatz (3) as the most probable configuration of the system.

A simple calculation shows that ΔM preserves the Regge trajectories of light mesons [3]. Moreover, our model, supplemented by a one-gluon-exchange potential and quark self-energy term can rather well reproduce the spin averaged experimental meson masses of light and heavy mesons.

3 String shape beyond the straight-line ansatz

It is worth noting that the use of a nonvanishing relative time is not really compatible with the straight-line ansatz. This can be seen by the following simple considerations: Let us assume that the world sheet of the system in the centre-of-mass frame is a helicoid area in the case of exactly circular quark orbits. The shape of the string is then

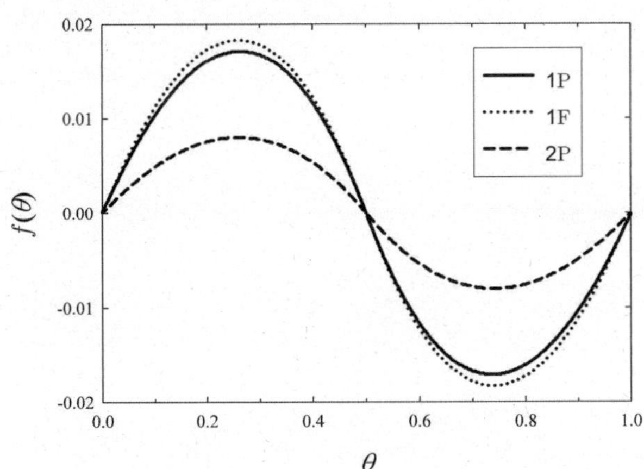

Fig. 1. The curved string linking the quark to the antiquark for different states.

a straight line for a slice at constant time and a curve for a slice not at constant time. This problem was investigated in ref. [4], whose main points are summed up in this section.

As the meson evolves in a plane, we can use for the string the complex coordinates (w^0, w, w^*) defined by

$$w = \frac{1}{\sqrt{2}}(w^1 + iw^2). \tag{16}$$

With these coordinates, a curved string rotating at a constant angular speed ω can be described by the ansatz

$$w^0(\theta, \tau) = \tau + \theta\,\sigma(\tau)/2, \tag{17a}$$

$$w(\theta, \tau) = \frac{r(\tau)}{2\sqrt{2}}\,[\theta + if(\theta)]\exp[i\omega\tau], \tag{17b}$$

where the spatial deformation f has been introduced to take into account the relative time σ. Equations (17) clearly describe a curved string, as can be seen by rewriting it when $\tau = 0$. Then we have simply $w^1 \propto \theta$ and $w^2 \propto f(\theta)$. As σ is assumed to be arbitrary, only f has to be determined thanks to the equations of motion of the Nambu-Goto Lagrangian, supplemented by the requirement that f vanishes at the centre of mass and at the ends of the string. The trivial solution $f = 0$ is only valid if $\sigma = 0$. This is the straight line without retardation. We already gave in ref. [3] arguments showing that the bending was small. Consequently, we can linearize the equations in f and solve them numerically. The conclusion of ref. [4] is that the string is curved, with a state-dependent amplitude. The bending is maximal for the $1F$ state and decreases with the quantum numbers, as illustrated in fig. 1. Moreover, the string shape between the centre of mass and one quark can be roughly approximated by the expression

$$f(\rho) \approx -\frac{\sigma\omega}{2}\,\rho\,(1 - \rho), \tag{18a}$$

with $\rho \in [0, 1]$, $\rho = 0$ being the centre of mass and $\rho = 1$ the quark. The other part of the string is readily obtained by a central symmetry with respect to the centre of mass. It is worth mentioning that in the case of a vanishing angular momentum, $\omega = 0$, the solution is trivially $f(\rho) = 0$. Even when the retardation is included, the string is straight when the angular momentum is zero.

Since the string brings an energetic contribution to the meson which is proportional to its length, it is interesting to evaluate the ratio between the lengths of both the curved and the straight strings. It is given by [4]

$$\Delta L/L \le 2 \times 10^{-3}. \tag{19}$$

The length of the string is only modified by some tenths of percent, because of the bending induced by retardation effects. As the typical mass scale for mesons is around 1–2 GeV, the correction due to the curved string is around 1–4 MeV. Such an order of magnitude was also obtained in a previous study of the string deformation due to nonuniform rotation [9]. The contribution of the bending of the string to the mass spectrum seems thus very small. This is also small compared with the retardation contribution, which can be around 100 MeV for massless quarks.

4 Conclusion

By allowing the relative time between the quark and the antiquark in a meson to be nonzero, we have been able to obtain a generalized rotating string model in which the retardation effects contribute as a perturbation of the nonretarded Hamiltonian. This contribution does not destroy the Regge trajectories and can lead to a good agreement with the experimental meson spectrum. We also showed that the string was curved due to retardation, but this bending does not influence significantly the mass spectrum of the model. In conclusion, our study reveals that the usually neglected retardation effects could be a relevant physical mechanism, in particular in the light meson sector where the retardation can give a contribution around 100 MeV. It should be interesting to study the dependence in the relative time of other observables, like the decay width of mesons, for example, in order to make further comparisons with experiment. We leave this for future investigations.

References

1. A.Yu. Dubin, A.B. Kaidalov, Yu.A. Simonov, Phys. At. Nucl. **56**, 1745 (1993); Yad. Fiz. **56**, 213 (1993) [hep-ph/9311344].
2. E.L. Gubankova, A.Yu. Dubin, Phys. Lett. B **334**, 180 (1994) [hep-ph/9408278].
3. F. Buisseret, C. Semay, Phys. Rev. D **72**, 114004 (2005) [hep-ph/0505168].
4. F. Buisseret, V. Mathieu, C. Semay, Eur. Phys. J. A **31**, 213 (2007) [hep-ph/0605328].
5. Y. Koma, E.M. Ilgenfritz, T. Suzuki, H. Toki, Phys. Rev. D **64**, 014015 (2001) [hep-ph/0011165].

6. F. Buisseret, C. Semay, Phys. Rev. D **70**, 077501 (2004) [hep-ph/0406216].

7. F. Buisseret, C. Semay, Phys. Rev. E **71**, 026705 (2005) [hep-ph/0409033].

8. A. De Rújula, H. Georgi, S.L. Glashow, Phys. Rev. D **12**, 147 (1975); W. Celmaster, Phys. Rev. D **15**, 1391 (1977).

9. T.J. Allen, M.G. Olsson, S. Veseli, Phys. Rev. D **59**, 094011 (1999) [hep-ph/9810363]; **60**, 074026 (1999) [hep-ph/9903222].

Eur. Phys. J. A **31**, 620–625 (2007)
DOI 10.1140/epja/i2006-10251-1

Special Article – QNP 2006

Spin physics at COMPASS

J.M. Friedrich[a]

For the COMPASS Collaboration
Physik Department, Technische Universität München, Germany

Received: 23 November 2006
Published online: 15 March 2007 – © Società Italiana di Fisica / Springer-Verlag 2007

Abstract. Results for the spin structure of the nucleon from the COMPASS data taking periods 2002 to 2004 are presented. The quark contribution to the nucleon spin, following from a QCD fit to the new data, turns out to be significantly larger than it was derived from the previous world data. The new data favour, on the other side, a comparatively small gluon polarisation in the range $x_g \approx 0.1$. In the data taken with the deuteron target polarised transversely the related asymmetries are found to be small on the level of accuracy reached so far, indicating a cancellation of the proton and neutron contributions. This is in agreement, for both the Collins and the Sivers asymmetry, with recent theoretical calculations. Also, a step towards the understanding of angular-momentum contributions with COMPASS is taken by the evaluation of asymmetries in exclusive vector meson production.

PACS. 25.30.Mr Muon scattering (including the EMC effect) – 21.10.Hw Spin, parity, and isobaric spin – 24.85.+p Quarks, gluons, and QCD in nuclei and nuclear processes – 13.60.-r Photon and charged-lepton interactions with hadrons

1 Introduction

Since many years the understanding of the structure and the spectrum of hadrons is a central goal of particle physics. As initialised by Hofstadter's results on the proton radius, dating now 50 years ago, a complete description of the underlying strong interaction is still a challenging task on both the experimental and theoretical sides. While in the sector of high-energetic collisions, perturbative quantum chromodynamics (QCD) is a powerful tool, for the low-energy processes and bound-state properties the effective degrees of freedom cannot be directly derived from the presumably underlying theory.

The COmmon Muon and Proton Apparatus for Structure and Spectroscopy (COMPASS) at the Super Proton Synchrotron (SPS) at CERN is devoted to this field with a broad variety of measurements, using different beams and spectrometer setups. While measurements with pion and proton beams will explore hadron spectroscopy, the data taking up to now has been focused on spin physics with a muon beam in deep inelastic scattering (DIS, for momentum transfer $Q^2 > 1\,\mathrm{GeV}^2/c^2$) as well as in quasi-real photo-production ($Q^2 < 1\,\mathrm{GeV}^2/c^2$) processes. During the data taking periods 2002–2004, the muon momentum was $160\,\mathrm{GeV}/c$ and the beam intensity $4 \cdot 10^7$ muons/s.

2 The COMPASS spectrometer

COMPASS [1] is a 2-stage magnetic spectrometer equipped with modern detector and data acquisition technology for large-acceptance measurements with high-intensity beams and high interaction rates. The tracking system is based on silicon microstrip detectors, together with scintillating fiber detectors, for high-precision tracking in the target region, the novel technologies of Micromegas and GEM detectors are employed for the small-area tracking, and drift chambers, straw tubes and wire chambers are used for the large-area tracking. A large-acceptance ring-imaging Cherenkov (RICH) detector in the first spectrometer stage provides charged-particle identification. Both stages are equipped with hadronic calorimeters, providing further particle identification. Their signals are also used in the trigger system, which is based on scintillator hodoscopes selecting the signals from scattered muons. The muons are identified throughout the apparatus in detectors behind hadron absorbers.

The target system [2] is equipped with two 60 cm long cells filled with solid ^{6}LiD. The cells are polarised oppositely by separate RF cavities. The target magnet provides a 2.5 T longitudinal solenoid field and a 0.4 T transverse dipole field, where the latter is used during the adiabatic field rotations and also as holding field during running with the target spin oriented transverse to the beam. The polarisable part of the ^{6}Li nucleus represents with good

---

[a] e-mail: Jan.Friedrich@ph.tum.de

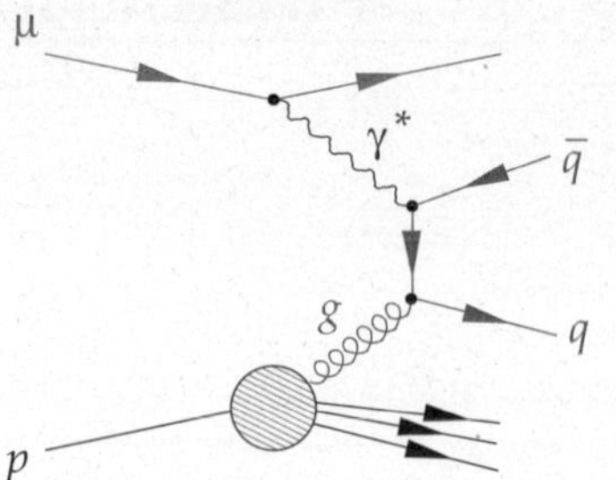

Fig. 1. The photon-gluon fusion process. The dependence of the production amplitude on the relative polarisation of the muon and the nucleon gives access to the gluon polarisation in the nucleon.

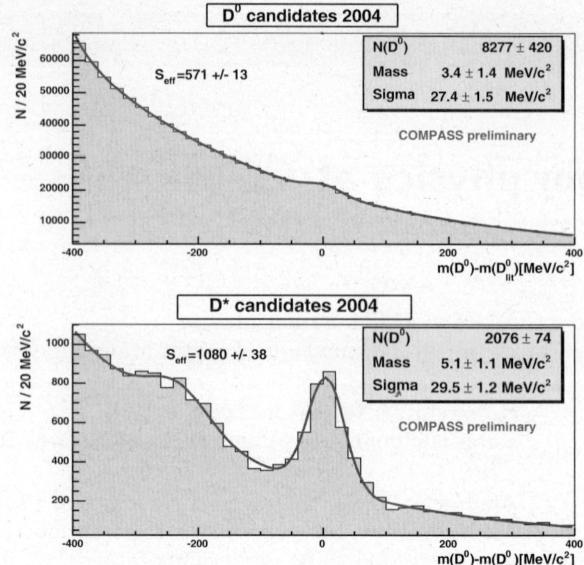

Fig. 2. D-meson signals from the COMPASS 2002–2004 data. The upper plot shows all reconstructed D^0-mesons, the lower plot the D^0 tagged by a $D^* \to D\pi$ decay.

accuracy an additional deuteron, so the presented asymmetries are effectively measurements on the deuteron, and the target material has a favorable dilution factor of ~ 0.4. Polarisations to about 50% are reached.

The resulting trigger rate of about 20 kHz is handled by custom-made frontend electronics for the 250000 readout channels and transfered through a farm of event builder computers to tape at a rate of about 5 TB per day.

3 Results on spin physics

3.1 Composition of the nucleon spin

In QCD, the spin of the nucleon can have four contributions,

$$\frac{1}{2} = \frac{1}{2}\Delta\Sigma + \Delta G + L_q + L_g, \tag{1}$$

where $\Delta\Sigma$ is the quark contribution, ΔG the gluon contribution, and L_q, L_g are the respective angular momenta.

Using the $\overline{MS}$ renormalisation scheme, the contribution of the quarks is given by the observable singlet axial current, $\Delta\Sigma = a_0$. This value has been extracted, using the new 2002–2004 COMPASS data only,

$$a_0(Q^2 = 3\,\mathrm{GeV}^2/c^2) = 0.35 \pm 0.03_{\mathrm{stat}} \pm 0.05_{\mathrm{syst}}, \tag{2}$$

the details of this analysis being given in a separate contribution to these proceedings [3] and also in [4]. The integral over g_1^N is performed by summing the obtained data points, with a small correction for the $x_{Bj} \to 0$ and 1 limits taken from a QCD fit to the world data , and the evolution to a fixed Q^2. The difference with respect to the former SMC analysis [5], leading to a significantly smaller value, mainly stems from the values at small Bjorken $x < 0.02$, where the new COMPASS data are well compatible with zero.

This method, even allowing the sign of the gluon contribution ΔG to be positive or negative within its systematic error [3], restricting only the absolute value to 0.2–0.3. With COMPASS this quantity can be accessed by the study of the photon-gluon fusion (PGF) process, depicted in fig. 1. The photon carries a well-determined fraction of the known incoming muon polarisation, while

the gluon polarisation, as related to the incoming proton polarisation, is the quantity of interest. There are two ways of extracting the process in COMPASS: The production of a $c\bar{c}$ pair not forming a bound system (open charm), which is intrinsically a small-scale process due to the high charm mass, and by this the theoretically favorable channel. The second possibility is the production of light quarks, $q = u, d, s$, where the PGF process is enriched by choosing a small spatial scale by either selecting $Q^2 > 1\,\mathrm{GeV}^2/c^2$, or by ensuring that the outgoing hadrons have high transverse momenta, $\Sigma p_T^2 > 2.5\,\mathrm{GeV}^2/c^2$.

3.2 Open-charm analysis

The production of charm quarks via photon-gluon fusion is observed through D-mesons, which are reconstructed from the tracks of their decay particles. The channels investigated so far are the decay $D^0 \to \pi^+ K^-$, as well as the same decay preceded by an original $D^* \to D\pi_{\mathrm{slow}}$ decay. The statistics and purity of the two channels as measured in the runs 2002 to 2004 is presented in fig. 2. While the available statistics is much higher for the D^0, the D^* sample is of higher purity, resulting in comparable uncertainties on the extracted asymmetry in the two cases.

The photon-nucleon asymmetry $A^{\mu N}$ is obtained by correcting the obtained experimental asymmetry A^{exp}:

$$A^{\mu N} = \frac{1}{P_B P_T f D} \frac{S+B}{S} A^{exp} \tag{3}$$

for the beam and target polarisations P_B, P_T, for the material dilution factor f, which accounts for the fact that not all nucleons in the target are polarised, the depolarisation factor D for the virtual photon, and the dilution factor due to the admixture of unpolarised combinatorial background B under the signal S. A possible background asymmetry due to other spin-dependent processes is assumed to be much smaller than the statistical uncertainty obtained so far. For the extraction of ΔG, the respective analysing power $a_{LL} = \Delta\sigma(\gamma^* g \to c\bar{c}X)/\sigma(\gamma^* g \to c\bar{c}X)$, only accessible via a model-dependent simulation, must be used, leading to

$$\frac{\Delta G}{G} = \frac{1}{P_B P_T f D\, a_{LL}} \frac{S+B}{S} A^{exp}. \qquad (4)$$

Since part of the factors have a different value for each event, the asymmetry is finally obtained by the method of event weighting. Combining the values for the two examined channels, the preliminary result from open charm is

$$\frac{\Delta G}{G} = -0.57 \pm 0.41_{\text{stat}} \qquad (5)$$

for the gluon momentum fraction centered at $x_g \approx 0.15$ with an RMS width of 0.08, the scale of the process being centered around $13\,\text{GeV}^2/c^2$.

3.3 High-p_T pairs

The leading deep inelastic process is given by the virtual photon colliding with a quark, which recoils and fragments into a hadron mainly in the direction of the initial photon. Choosing two hadrons with both having a large transverse momentum $p_{T,i} > 0.7\,\text{GeV}/c$ and $\Sigma_i p_{T,i}^2 > 2.5\,\text{GeV}^2/c^2$, enhances strongly the contribution where the photon hits a virtual $q\bar{q}$ pair carrying the information of a single gluon; in COMPASS kinematics this reaches a contribution up to 30%. The sample is purified by taking only hadrons identified in the hadron calorimeters, and further restricted to masses for the 2-hadron system $> 1.5\,\text{GeV}$, suppressing contributions from resonances.

The first option in order to obtain a result for $\Delta G/G$ is to choose events with $Q^2 > 1\,\text{GeV}^2/c^2$, since then the only competing processes to PFG are the leading-order deep inelastic process and QCD-Compton interaction, as shown in fig. 3 on the left side. These two competing processes are sensitive to the quark polarisation, whose contribution to the asymmetry can be neglected after further restricting the events to $x < 0.05$, given the smallness of A_1 in this region. Then, the asymmetry in 2-hadron production with high p_T can be connected with $\Delta G/G$,

$$A^{\gamma^* N \to hh} = R_{PGF}\, a_{LL}^{PGF} \frac{\Delta G}{G}. \qquad (6)$$

The ratio $R_{PGF} = 0.34\pm0.07$ is taken from the COMPASS Monte Carlo simulation using the LEPTO 6.5.1 generator, which is suited for physics at high Q^2, and from the 2002–2003 data the preliminary value

$$\frac{\Delta G}{G} = 0.06 \pm 0.31_{\text{stat}} \pm 0.06_{\text{syst}} \qquad (7)$$

is obtained at scale $\mu^2 = 3\,\text{GeV}^2/c^2$ and an average $x_g \approx 0.13$ with an RMS of 0.08.

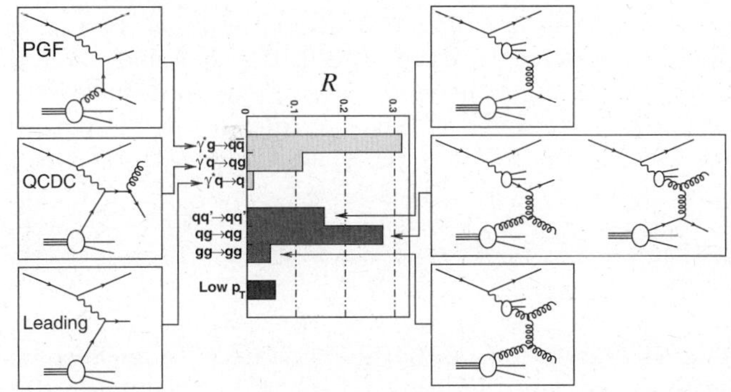

Fig. 3. Contributions of different processes as in the PYTHIA sample of events with $Q^2 < 1\,\text{GeV}^2/c^2$.

The selection of quasi-real photo-production with $Q^2 < 1\,\text{GeV}^2/c^2$, has about a factor of 10 more statistics, but now the asymmetry is more difficult to interpret, since also processes with resolved photons, as presented in fig. 3 on the right side, have to be taken into account. Those processes involving the hadronic part of the photon are estimated in [6] and used in the PYTHIA 6.2 [7] Monte Carlo generator for incorporation of these effects. The details of this analysis for the 2002–2003 data is given in [8], including the 2004 data the new preliminary value

$$\frac{\Delta G}{G} = 0.016 \pm 0.058_{\text{stat}} \pm 0.055_{\text{syst}} \qquad (8)$$

is extracted at $x_g = 0.095$ and the scale $\mu^2 = 3\,\text{GeV}^2/c^2$.

3.4 Results for $\Delta G/G$

The three preliminary COMPASS values for $\Delta G/G$ are summarized in fig. 4, together with the data previously obtained by SMC [9] and HERMES [10]. They are compared with the parametrisation given by GRSV [11] in

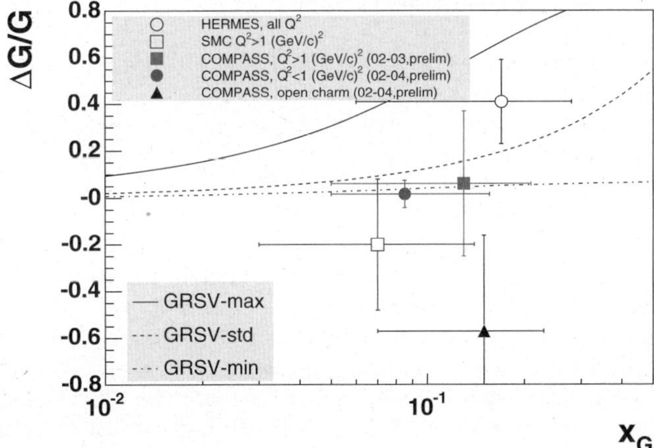

Fig. 4. Gluon polarisation of the nucleon as obtained in the different COMPASS analysis, compared to earlier measurements and theoretical distributions from [11] as discussed in the text.

the scenarios of maximal, best fit and minimal integrated value for ΔG being 2.5, 0.62 and 0.16, respectively. While the experimental values of the high-p_T analysis have been obtained at the same scale as the calculation was carried out, *i.e.* $Q^2 = 3\,\mathrm{GeV}^2/c^2$, the open-charm data refer to a much higher scale $\mu^2 = 13\,\mathrm{GeV}^2/c^2$.

3.5 Transversity

For the description of the quark state of a nucleon at twist-two level, in addition to the $q(x)$ and $\Delta q(x)$ distributions also the transverse spin distributions $\Delta_T q(x)$ contribute [12,13]. The sparseness of available data is explained by the fact that these *transverse distributions* are not observables in inclusive deep inelastic scattering due to their chiral-odd nature.

COMPASS runs about 20% of the beam time with the target transversely polarised in order to measure transversity asymmetries in semi-inclusive deep inelastic muon scattering. The produced hadrons exhibit azimuthal asymmetries with respect to the transverse target polarisation. Two possible angular correlations are discussed, depending on whether the hadron azimuthal angle is observed with respect to the azimuthal angle of the initial or fragmenting quark spin as depicted in fig. 5, defining the Sivers $(\phi_h - \phi_s)$ or Collins $(\phi_h - \phi_s')$ angle, respectively.

In case of the Collins asymmetry, the chiral-odd distribution $\Delta_T^0 D_q^h$, describing the spin dependence of the hadronization of the transversely polarised quark q into the hadron h, contributes together with the transverse-distribution function $\Delta_T q$ to the asymmetry

$$A_{Coll} = \frac{\sum_q e_q^2 \cdot \Delta_T q \cdot \Delta_T^0 D_q^h}{\sum_q e_q^2 \cdot q \cdot D_q^h}, \qquad (9)$$

where the Collins angle Φ_C is contained in the azimuthal dependence of $\Delta_T^0 D_q^h$ [14].

In case the initial nucleon spin direction is correlated to an intrinsic transverse momentum k_T of an unpolarised quark in the transversely polarised nucleon, the quark distribution could be written [15] as

$$q_T(x, \mathbf{k}_T) = q(x, k_T) + \Delta q_0^T(x, k_T) \cdot \sin \Phi_S \qquad (10)$$

and leads to a Sivers asymmetry

$$A_{Siv} = -\frac{\sum_q e_q^2 \cdot \Delta q_0^T \cdot D_q^h}{\sum_q e_q^2 \cdot q \cdot D_q^h}, \qquad (11)$$

a convolution of the Sivers distribution function Δq_0^T and the unpolarised fragmentation function D_q^h. So, the Sivers asymmetry is not connected to transversity, *i.e.* the transverse-distribution function $\Delta_T q$.

The analysis of these quantities from the COMPASS 2002–2004 transversity data is performed with DIS cuts $(Q^2 > 1\,\mathrm{GeV}^2/c^2,\ W > 5\,\mathrm{GeV},\ 0.1 < y < 0.9)$ and a cut on the transverse momentum of the hadrons, $p_T > 0.1\,\mathrm{GeV}/c$, ensuring the angles Φ_C and Φ_S to be well defined. The asymmetry is extracted either from all hadrons

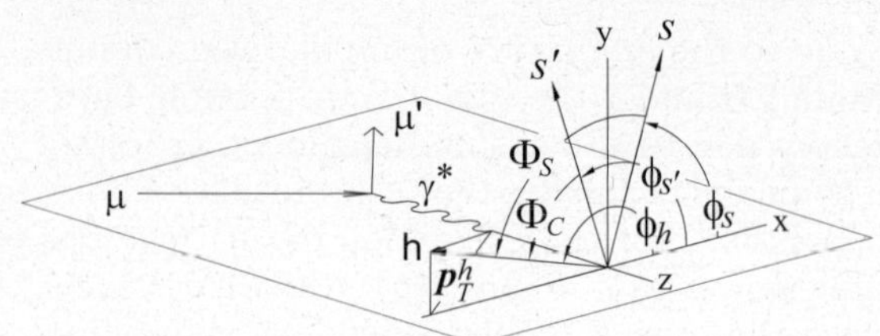

Fig. 5. Azimuthal angles used in the extraction of Collins and Sivers asymmetries.

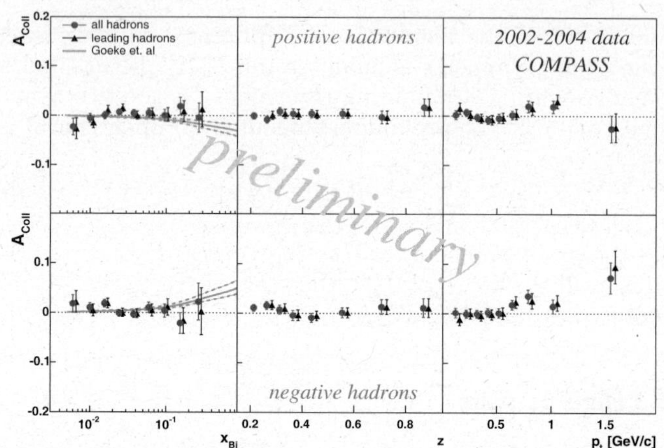

Fig. 6. Collins asymmetry from the COMPASS 2002–2004 data, compared in the x_{Bj}-dependence to a recent theoretical calculation [18].

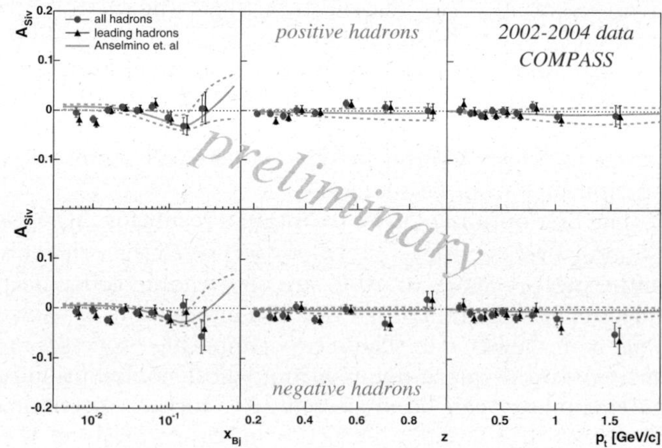

Fig. 7. Sivers asymmetry from COMPASS 2002–2004 data, shown together with a theoretical description [19] as discussed in the text.

with $0.2 < z < 1$, z being the fraction of the photon energy carried by the hadron, or from only the hadrons with highest momentum if they have $z > 0.25$. Further details of the analysis and systematic studies can be found in [16] and [17].

The results are presented for both options, in fig. 6 for the Collins asymmetry and in fig. 7 for the Sivers asymmetry. In case of the Collins asymmetry, the x_{Bj}-dependence is compared to a recent theoretical calcula-

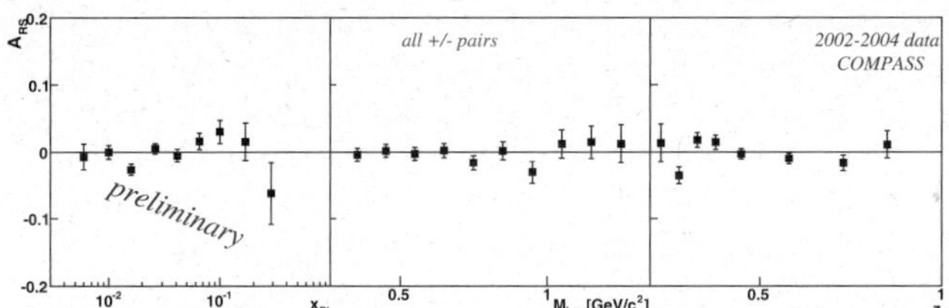

Fig. 8. Transverse asymmetry determined from hadron pairs.

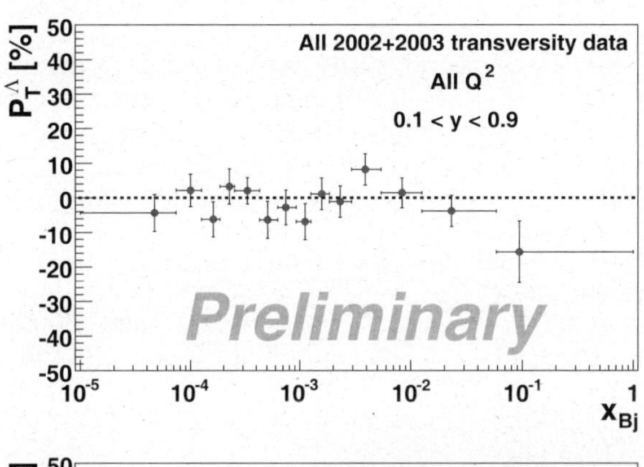

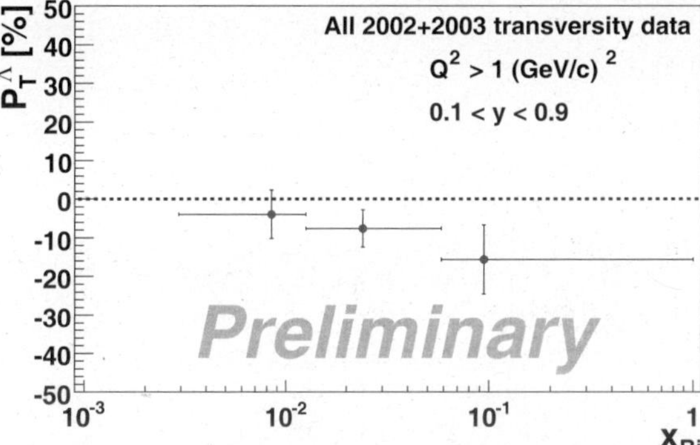

Fig. 9. Transversity from Λ polarisation.

tion [18] using transversity distributions from the chiral quark-soliton model, which has been found in agreement also with other experiments. In order to disentangle the transversity distributions from the transverse fragmentation function $\Delta_T^0 D_q^h$, this quantity must be determined by another method, as for example done with the analysis of spin asymmetries in $e^+ e^- \to 2h$ reactions studied by the BELLE Collaboration.

The measured Sivers asymmetry is compared to the theoretical analysis [19] in dependence on the kinematical variables x_{Bj}, z and p_T. This analysis takes into account these new COMPASS data together with the HERMES data on protons within the leading-order parton model including transverse quark motions.

Compared to the sizable asymmetries measured on the proton [20], the presented asymmetries on the deuteron are much smaller, which would follow if $\Delta_T u$ and $\Delta_T d$ had opposite sign like the helicity distributions, and so the contributions from the proton and neutron mainly cancel.

Also other ways of accessing the transverse spin distribution $\Delta_T q(x)$ are studied on the COMPASS data.

In the production of hadron pairs, the angle of the two-hadron plane with respect to the fragmenting quark spin may be defined analogously to the Collins case, giving rise to an asymmetry proportional to the convolution of the transverse spin distribution $\Delta_T q(x)$ and a two-hadron fragmentation function, referred to as interference fragmentation function $H_q^{\angle h}(z, M_h)$. The resulting transverse asymmetry is shown in fig. 8 in dependence of x_{Bj}, the invariant mass of the 2-hadron system, and z. The asymmetry is found to be small, so either both or one of the functions $\Delta_T q(x)$ and $H_q^{\angle h}(z, M_h)$ are small on the achieved level of accuracy, as in the case of single-hadron studies.

The last channel presented here in order to access transversity is the study of the transverse polarisation of Λ hyperons. Here, the measured Λ polarisation is expected to depend on $\Delta_T q(x)$. In fig. 9 the dependence of x_{Bj} for all measured Q^2 and with a cut on deep inelastic kinematics $Q^2 > 1 \, \text{GeV}^2/c^2$ is presented. The data from the 2002–2003 data taking for DIS kinematics may indicate a non-zero value, however at the statistical limit of data analysed so far. Including the 2004 data will increase the statistics by a factor of 2.

3.6 Exclusive ρ-production

Incoherent exclusive ρ^0 production on polarised deuterons, $\mu + N \to \mu + N + \rho$, is also studied at COMPASS at

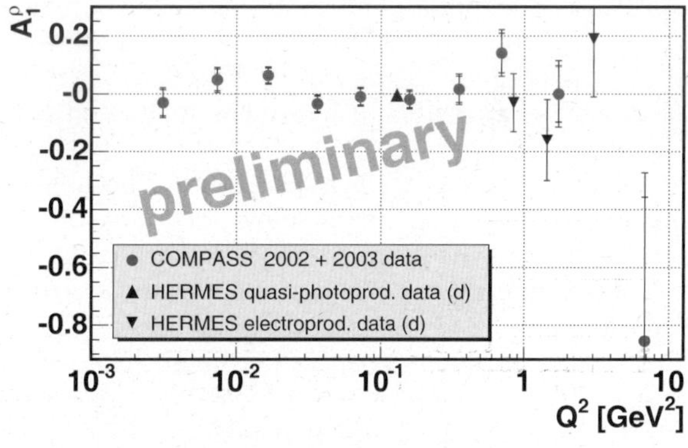

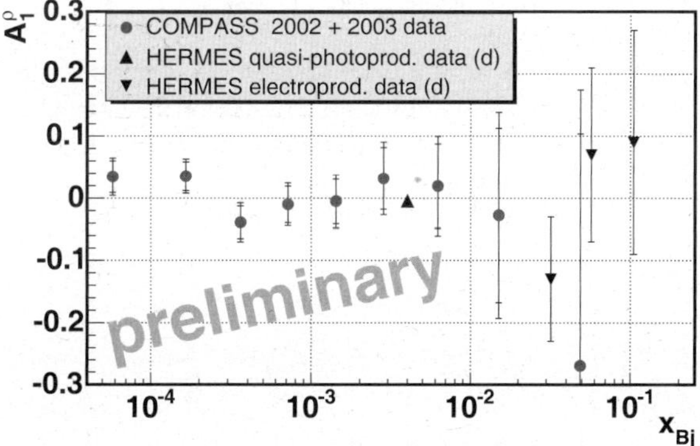

Fig. 10. Exclusive ρ production asymmetry, shown together with the results from HERMES [22].

$\langle W \rangle = 10\,\text{GeV}$ over a wide kinematical range: $3 \times 10^{-3} < Q^2/(\text{GeV}^2/c^2) < 7$ and $5 \times 10^{-5} < x_{Bj} < 0.05$.

For the analysis two hadron tracks are required, which correspond to charged pions from the ρ^0 decay. A cut on the invariant mass is applied: $0.5 < M_{\pi\pi}/\text{GeV} < 1$. As slow recoil target particles are not detected in the current setup, in order to select exclusive events we used cuts on the missing energy, $-2.5 < E_{\text{miss}}/\text{GeV} < 2.5$, and on the transverse momentum of ρ^0 with respect to the virtual-photon direction, $p_t^2 < 0.5\,\text{GeV}^2/c^2$. Coherent interactions on the target nuclei are suppressed by a cut $p_t^2 > 0.15\,\text{GeV}^2/c^2$.

The longitudinal double-spin asymmetry A_1^ρ from the COMPASS 2002 and 2003 data [21] is reported as a function of Q^2 (upper panel) and x_{Bj} (lower panel) in fig. 10. Comparison with the HERMES results [22] is also shown. For the whole kinematical range A_1^ρ is consistent with zero. This indicates that the role of unnatural-parity exchanges, like π- or A_1-Reggeon exchange, is small in that kinematical domain, which is to be expected if diffraction is the dominant process. At $Q^2 > 1\,\text{GeV}^2/c^2$ a non-zero asymmetry would indicate sensitivity to spin-dependent generalized parton distributions [23] but more data are needed to clarify this issue.

4 Summary and outlook

COMPASS contributes to a broad variety of questions around the spin riddle of the nucleon, where possibly the gluon contribution will turn out to be smaller than it was speculated at some point. The transversity phenomena are studied with dedicated data-taking periods, and have delivered first results for the deuteron. The ongoing analysis will increase the statistics in many channels, and lead to more precise information.

Apart from the aspects presented here, the huge data set collected with muon beam allows to also contribute to aspects of hadron spectroscopy, as the exclusion of a $\Xi^{--}(1860)$ pentaquark [24] on a high statistical level demonstrates.

The 2004 data taken with a pion beam have revealed the possibilities that the very flexible setup of the experiment offers, switching to soft hadronic and Primakoff reactions within a few days.

The preparation for the 2007 run is ongoing, where for the first time a polarised hydrogen target will allow at COMPASS transversity measurements on the proton, before the first measurements with hadron beam for central production are started.

In longer terms, an upgraded COMPASS spectrometer with even higher beam flux is discussed, aiming at the still unknown orbital angular momenta of the partons in the nucleon. This will be addressed, for example, by the determination of GPDs via deep virtual Compton scattering and hard exclusive vector meson production.

References

1. COMPASS proposal, CERN/SPSLC96-14 (1996).
2. J. Ball *et al.*, Nucl. Instrum. Methods A **498**, 101 (2003).
3. A. Korzenev (for the COMPASS Collaboration), these proceedings.
4. COMPASS Collaboration (V. Alexakhin *et al.*), Phys. Lett. B **647**, 8 (2007).
5. SMC Collaboration (B. Adeva *et al.*), Phys. Rev. D **58**, 112001 (1998).
6. M. Glück, E. Reya, C. Sieg, Eur. Phys. J. C **20**, 210 (2001).
7. T. Sjöstrand *et al.*, Comput. Phys. Commun. **138**, 238 (2001).
8. COMPASS Collaboration (E. Ageev *et al.*), Phys. Lett. B **633**, 25 (2006).
9. SMC Collaboration (B. Adeva *et al.*), Phys. Rev. D **70**, 012002 (2004).
10. HERMES Collaboration (A. Airapetian *et al.*), Phys. Rev. Lett. **84**, 2584 (2000).
11. M. Glück, E. Reya, M. Stratmann, W. Vogelsang, Phys. Rev. D **63**, 094005 (2001).
12. J.P. Ralston, D.E. Soper, Nucl. Phys. B **152**, 109 (1979).
13. V. Barone *et al.*, Phys. Rep. **359**, 1 (2002).
14. J. Collins, Nucl. Phys. **396**, 161 (1993).
15. D. Sivers, Phys. Rev. D **41**, 83 (1990).
16. COMPASS Collaboration (V. Alexakhin *et al.*), Phys. Rev. Lett. **94**, 202002 (2005).
17. COMPASS Collaboration (E. Ageev *et al.*), Nucl. Phys. B **765**, 31 (2007).
18. A.V. Efremov, K. Goeke, P. Schweitzer, Phys. Rev. D **73**, 094025 (2006).
19. M. Anselmino *et al.*, *Proceedings to Transversity 2005, Como, Italy* (World Scientific Publishing, 2006) ISBN 981-256-846-8, hep-ph/0511249.
20. HERMES Collaboration (A. Airapetian *et al.*), Phys. Rev. Lett. **94**, 012002 (2005).
21. A. Sandacz (for the COMPASS Collaboration), Nucl. Phys. B, Proc. Suppl. **146**, 58 (2005).
22. HERMES Collaboration (A. Airapetian *et al.*), Eur. Phys. J. C **29**, 171 (2003).
23. S.V. Goloskokov, P. Kroll, Eur. Phys. J. C **42**, 281 (2005).
24. COMPASS Collaboration (E. Ageev *et al.*), Eur. Phys. J. C **41**, 469 (2005).

Eur. Phys. J. A **31**, 626–629 (2007)

DOI 10.1140/epja/i2006-10179-4

THE EUROPEAN
PHYSICAL JOURNAL A

Special Article – QNP 2006

Nucleon sigma term and quark condensate in nuclear matter

K. Tsushima[1,a], K. Saito[2], A.W. Thomas[3], and A. Valcarce[1]

[1] Universidad de Salamanca, E-37008, Salamanca, Spain
[2] Tokyo University of Science, Noda 278-8510, Japan
[3] Jefferson Lab, 12000 Jefferson Avenue, Newport News, VA 23606, USA

Received: 24 September 2006
Published online: 19 February 2007 – © Società Italiana di Fisica / Springer-Verlag 2007

Abstract. We study the bound-nucleon sigma term and the quark condensate in nuclear matter. In the quark-meson coupling (QMC) model the nuclear correction to the sigma term is small and negative, *i.e.*, it decelerates the decrease of the quark condensate in nuclear matter. However, the quark condensate in nuclear matter is controlled primarily by the scalar-isoscalar σ field. Compared to the leading term, it moderates the decrease more than that of the nuclear sigma term alone at densities around and larger than the normal nuclear-matter density.

PACS. 24.85.+p Quarks, gluons, and QCD in nuclei and nuclear processes – 11.30.Rd Chiral symmetries – 21.65.+f Nuclear matter

1 Introduction

One of the most challenging questions in nuclear physics is whether or not the properties of nucleon and hadrons are modified in nuclei [1,2]. A number of pieces of evidence, such as the nuclear EMC effect [3], the quenching [4–6] (enhancing [7]) of the space (time) component of the axial coupling constant in nuclear β decays, the missing strength of the response functions and the suppression of the Coulomb sum rule [8], have stimulated investigations into this problem. In particular, to focus on the change of the internal structure of the nucleon (hadrons) is very appealing in the light of an approach based on QCD.

In addition to the recent experimental evidence for the vector meson mass shifts in nuclei [9,10], some important hints concerning the change of the internal structure of a bound nucleon have been reported in measurements of the electromagnetic form factors of a proton in ^{4}He [11]. The analyses suggest a reduction of the bound proton's electric-to-magnetic form factor ratio, where most of the conventional, sophisticated approaches employing the free-proton form factors [12] fail to explain the observed effect. (See ref. [13] for an exploration of a conventional mechanism which might explain at least some of the effect. We note, however, that for the moment that approach is hampered by a lack of experimental data to pin down the input parameters.) Indeed, excellent agreement with the data is achieved when the small correction associated with the change of the internal structure of the bound proton (which had been predicted [14] long before the experiments) is included.

Up to now, some models such as the quark-meson coupling (QMC) model [15–17] and Nambu–Jona-Lasinio model [18], have opened possibilities to understand the change of the nucleon internal structure in nuclei based on quark degrees of freedom. However, it is important to explore other ways to study the change in the internal structure of the bound nucleon through various nuclear phenomena.

In this paper, we study the bound-nucleon sigma term, which has a direct connection with the QCD Hamiltonian via Feynman-Hellmann theorem [19]. First, we study the density dependence of the nuclear correction to the sigma term in the QMC model [15,16]. Next, we study the influence of the nuclear correction that arises from the sigma term on the quark condensate.

2 Bound-nucleon sigma term

For the ground state of uniform, nucleon density ρ_B with A nucleons and volume V ($\rho_B = A/V$), the nuclear sigma term σ_A and the bound-nucleon sigma term σ_N^* may be defined by [20–22]

$$\sigma_A \equiv A\sigma_N^*(\rho_B),$$

$$= \frac{1}{3}\sum_{a=1}^{3}[\langle A(\rho_B)|[Q_5^a,[Q_5^a,H_{\mathrm{QCD}}]]|A(\rho_B)\rangle$$

$$- \langle 0|[Q_5^a,[Q_5^a,H_{\mathrm{QCD}}]]|0\rangle], \qquad (1)$$

$$= V\rho_B 2m_q[\langle A(\rho_B)|\bar{q}q|A(\rho_B)\rangle - \langle 0|\bar{q}q|0\rangle], \qquad (2)$$

where Q_5^a is the weak axial charge with isospin a, H_{QCD} the QCD Hamiltonian with current quark mass, $m_q \equiv$

---

[a] e-mail: `tsushima@usal.es`

$\frac{1}{2}(m_u+m_d)$, and $\bar{q}q \equiv \frac{1}{2}(\bar{u}u+\bar{d}d)$. Applying the Feynman-Hellmann theorem [19], we get

$$A\sigma_N^*(\rho_B) = m_q \frac{\partial}{\partial m_q}[\langle A(\rho_B)|H_{\rm QCD}|A(\rho_B)\rangle$$
$$-\langle 0|H_{QCD}|0\rangle], \tag{3}$$
$$= V\rho_B m_q \frac{\partial}{\partial m_q}[m_N(m_q) + \varepsilon_B(m_q,\rho_B)], \tag{4}$$
$$\equiv A[\sigma_{N\rm free} + \delta\sigma_N^*(\rho_B)], \tag{5}$$

where $\sigma_{N\rm free} \equiv m_q \frac{\partial m_N}{\partial m_q}$ is the free-nucleon sigma term (empirical value $\simeq$ 45 MeV [23]), $\delta\sigma_N^*(\rho_B) \equiv m_q \frac{\partial \varepsilon_B(m_q,\rho_B)}{\partial m_q}$ the nuclear correction to the sigma term, and $\varepsilon_B(m_q,\rho_B)$ the binding energy per nucleon. At present, $\varepsilon_B(m_q,\rho_B)$ which contains the contributions from the nucleon kinetic energy and N-N interaction etc., can only be calculated in a model-dependent way. Note that our focus concerns the dependence on the current quark mass m_q [19], which has a direct connection with the QCD Hamiltonian through eqs. (2)-(5). It may be contrasted with the calculation of the quark condensate in nuclear matter [24–27], which will also be studied later. For later convenience we recall the saturation conditions for nuclear matter:

$$\left.\frac{\partial\varepsilon_B(m_q,\rho_B)}{\partial\rho_B}\right|_{\rho_B=\rho_0} = 0, \tag{6}$$
$$\varepsilon_B(m_q,\rho_B=\rho_0) = -15.7\,\text{MeV}. \tag{7}$$

Now we study the density dependence of the nuclear correction to the nucleon sigma term $\delta\sigma_N^*(\rho_B)$ in eq. (5). Generally, models for nuclear matter based on quark degrees of freedom contain coupling constants, $g_j(m_q)(j = 1,2,\ldots)$, which depend on the current quark mass m_q. These coupling constants are determined so as to satisfy the saturation conditions eqs. (6) and (7) for a chosen value of m_q. This also fixes the function $\varepsilon(m_q,\rho_B)$ as a function of m_q. Of course, since in nature we have only one set of quark masses, which lead to nuclear saturation at the correct place, it is not appropriate to include the variation of g_j with m_q in calculating the nuclear sigma term.

In the following we use different versions of the QMC model (QMC-I [15], and QMC-II [16] with parameter set B), and study the density dependence of the nuclear correction $\delta\sigma_N^*(\rho_B)$. We use the standard parameters, $m_q = 5$ MeV, and the free-nucleon bag radius $R_N = 0.8$ fm (all the corresponding parameters, values for the coupling constants can be found in refs. [2,15,16]). Thus, we suppose that $m_q = 5$ MeV is the value in the QCD Hamiltonian.

First, we show in fig. 1 the nuclear correction to the sigma term, $\delta\sigma_N^*(\rho_B)$, together with the binding energy per nucleon, $\varepsilon_B = E/A - m_N$ (saturation curve) calculated in QMC-I and QMC-II. The negative contribution of the nuclear correction to the sigma term agrees with the result [22] based on the 2-body N-N interaction using the Skyrme model, although the correction is small

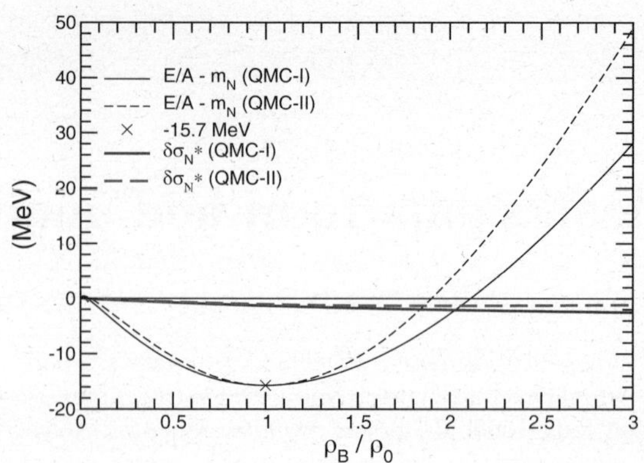

Fig. 1. Nuclear correction to the bound-nucleon sigma term ($\delta\sigma_N^*$), and the binding energy per nucleon ($\varepsilon_B = E/A - m_N$), where $\rho_0 = 0.15\,\text{fm}^{-3}$.

in the present case. In refs. [25,26], an effective density-dependent sigma term was introduced. Although they estimated the m_q-dependence of the meson masses appearing in the N-N interaction, in order to calculate the in-medium quark condensate, the calculation was not based on QCD but instead was rather qualitative. We do not attempt to make such an estimate for the variation of the masses of the meson fields.

3 Quark condensate in nuclear matter

Next, we discuss the quark condensate in nuclear matter. The quark condensate and the nuclear sigma term are related through eqs. (3)-(5). Using the Gell-Mann–Oakes–Renner (GOR) relation, the nuclear matter-to-vacuum quark condensate ratio is given by

$$\frac{Q^*}{Q_0} \equiv \frac{\langle \rho_B|\bar{q}q|\rho_B\rangle}{\langle 0|\bar{q}q|0\rangle} \simeq 1 - \frac{\sigma_{N\rm free} + \delta\sigma_N^*(\rho_B)}{m_\pi^2 f_\pi^2}\rho_B. \tag{8}$$

Without $\delta\sigma_N^*(\rho_B)$ above, it is the usual leading-term result.

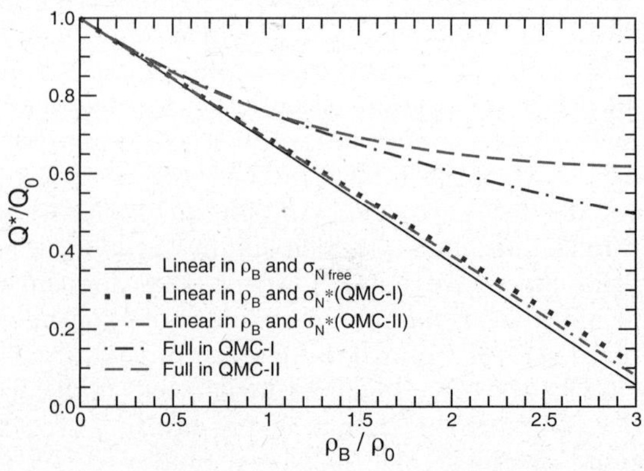

Fig. 2. The nuclear matter-to-vacuum quark condensate ratio. All results are obtained with $\sigma_{N\rm free} = 45$ MeV [23].

628 The European Physical Journal A

We show in fig. 2 the nuclear matter-to-vacuum quark condensate ratio calculated via eq. (8). The solid line is the leading term with the free-nucleon sigma term, $\sigma_{N\text{free}} = 45\,\text{MeV}$ [23], while the dotted and double-dash–dotted lines are those calculated with the bound nucleon sigma term σ_N^* in QMC-I and QMC-II. As expected from the results for the bound-nucleon sigma term, the influence of $\delta\sigma_N^*$ on the ratio is very small. (The full results will be discussed below.)

In the QMC model, nucleons interact through the self-consistent exchanges of the scalar-isoscalar (σ), vector-isoscalar (ω), and vector-isovector (ρ) meson mean fields, which couple directly to the quarks inside the nucleons. In principle, although the masses of the meson fields in QMC will also depend on the current quark mass m_q, this is not a feature of the model. Thus, in the QMC model the other nuclear correction to the quark condensate, aside from the bound-nucleon sigma term, is expected to arise from the scalar-isoscalar σ field. Indeed, the full expression for the nuclear matter-to-vacuum quark condensate ratio in QMC-I is given by [27]

$$\frac{Q^*}{Q_0} = 1 - \frac{\sigma_{N\text{free}} + \delta\sigma_N^*(\rho_B)}{m_\pi^2 f_\pi^2} \left(\frac{m_\sigma}{g_\sigma}\right)^2 (g_\sigma\sigma), \qquad (9)$$

where σ is the mean value of the scalar-isoscalar σ field in nuclear matter. For QMC-II, one may replace in eq. (9) $m_\sigma \to m_\sigma^*$ which depends on density or σ but not on m_q explicitly. This expression, which is obtained using the self-consistent equation for the σ field, may be a characteristic feature of the QMC model. The quark condensate in nuclear matter is directly connected to the scalar-isoscalar σ field, which plays an important role in describing the properties of nuclear matter and nuclei. This may not be a trivial result. Note that the differences between the present expression and that in ref. [27] arise because the present one does not contain the terms involving the derivative of the quark-meson coupling constants with respect to m_q.

Next, we show in fig. 2 the full results of eq. (9) in QMC-I (the dash-dotted line) and QMC-II (the dashed line). In both versions of QMC, the scalar-isoscalar σ field largely moderates the decrease of the quark condensate relative to the leading term. In refs. [25,26], although the m_q-dependence of the exchanged mesons in the N-N interaction was estimated qualitatively, a similar concave shape of the density dependence for the ratio was also obtained.

4 Summary

To summarise, it is shown that the nuclear correction to the bound-nucleon sigma term is negative, and decelerates the decrease of the quark condensate in nuclear matter. However, the magnitude is small and so is its effect on the quark condensate. The quark condensate in nuclear matter is predominantly controlled by the scalar-isoscalar σ field in the QMC model. It moderates appreciably the decrease of the condensate relative to the leading term

at densities around and larger than the normal nuclear-matter density.

K.T. would like to acknowledge the warm hospitality at Jefferson Lab, where part of the work was carried out. This work was supported in part by the U.S. DOE Contract No. DE-AC05-06OR23177, under which Jefferson Science Associates operate Jefferson Lab.

References

1. G.E. Brown, M. Rho, Phys. Rev. Lett. **66**, 2720 (1991).
2. K. Saito, K. Tsushima, A.W. Thomas, Prog. Part. Nucl. Phys. **58**, 1 (2007) arXiv:hep-ph/0506314.
3. M. Arneodo, Phys. Rep. **240**, 301 (1994); D.F. Geesaman, K. Saito, A.W. Thomas, Annu. Rev. Nucl. Part. Sci. **45**, 337 (1995); R.P. Bickerstaff, A.W. Thomas, J. Phys. G **15**, 1523 (1989).
4. A. Arima *et al.*, Adv. Nucl. Phys. **18**, 1 (1987); I.S. Towner, Phys. Rep. **155**, 263 (1987); F. Osterfeld, Rev. Mod. Phys. **64**, 491 (1992).
5. K. Tsushima, D.O. Riska, Nucl. Phys. A **549**, 313 (1992).
6. D.H. Lu, A.W. Thomas, K. Tsushima, nucl-th/0112001.
7. K. Kubodera, J. Delorme, M. Rho, Phys. Rev. Lett. **40**, 755 (1978); M. Kirchbach, D.O. Riska, K. Tsushima, Nucl. Phys. A **542**, 616 (1992); I.S. Towner, Nucl. Phys. A **542**, 631 (1992).
8. J. Morgenstern, Z.-E. Meziani, Phys. Lett. B **515**, 269 (2001); K. Saito, K. Tsushima, A.W. Thomas, Phys. Lett. B **465**, 27 (1999).
9. CBELSA/TAPS Collaboration (D. Trnka *et al.*), Phys. Rev. Lett. **94**, 192303 (2005).
10. M. Naruki *et al.*, Phys. Rev. Lett. **96**, 092301 (2006).
11. S. Malov *et al.*, Phys. Rev. C **62**, 057302 (2000); S. Dieterich *et al.*, Phys. Lett. B **500**, 47 (2001); S. Strauch *et al.*, Phys. Rev. Lett. **91**, 052301 (2003).
12. J.J. Kelly, Phys. Rev. C **60**, 044609 (1999); J.M. Udias, J.R. Vignote, Phys. Rev. C **62**, 034302 (2000); J.M. Udias *et al.*, Phys. Rev. Lett. **83**, 5451 (1999).
13. R. Schiavilla *et al.*, Phys. Rev. Lett. **94**, 072303 (2005).
14. D.H. Lu, A.W. Thomas, K. Tsushima, A.G. Williams, K. Saito, Phys. Lett. B **417**, 217 (1998); D.H. Lu, K. Tsushima, A.W. Thomas, A.G. Williams, K. Saito, Phys. Rev. C **60**, 068201 (1999).
15. P.A.M. Guichon, Phys. Lett. B **200**, 235 (1989); P.A.M. Guichon, K. Saito, E.N. Rodionov, A.W. Thomas, Nucl. Phys. A **601**, 349 (1996); K. Saito, K. Tsushima, A.W. Thomas, Nucl. Phys. A **609**, 339 (1996); P.G. Blunden, G.A. Miller, Phys. Rev. C **54**, 359 (1996).
16. K. Saito, K. Tsushima, A.W. Thomas, Phys. Rev. C **55**, 2637 (1997).
17. P.A.M. Guichon, A.W. Thomas, Phys. Rev. Lett. **93**, 132502 (2004).
18. W. Bentz, A.W. Thomas, Nucl. Phys. A **696**, 138 (2001).
19. R.P. Feynman, Phys. Rev. **56**, 340 (1939).
20. R.L. Jaffe, Phys. Rev. D **21**, 3125 (1980).
21. J. Delome, G. Chanfray, M. Ericson, Nucl. Phys. A **603**, 239 (1996).
22. A. Gammal, T. Frederico, Phys. Rev. C **57**, 2830 (1998).
23. J. Gasser, H. Leutwyler, M. Sainio, Phys. Lett. B **253**, 252 (1991).

24. T.D. Cohen, R.J. Furnstahl, D.K. Griegel, Phys. Rev. D **45**, 1881 (1992).
25. G.Q. Li, C.M. Ko, Phys. Lett. B **338**, 118 (1994).
26. R. Brockmann, W. Weise, Phys. Lett. B **367**, 40 (1996).
27. K. Saito, K. Tsushima, A.W. Thomas, Mod. Phys. Lett. A **10**, 769 (1998).

Eur. Phys. J. A **31**, 630–637 (2007)

DOI 10.1140/epja/i2006-10271-9

Special Article – QNP 2006

Mind the gap

M.S. Bhagwat[1], A. Krassnigg[2], P. Maris[3], and C.D. Roberts[1,a]

[1] Physics Division, Argonne National Laboratory, Argonne, IL, 60439, USA
[2] Fachbereich Theoretische Physik, Universität Graz, A-8010 Graz, Austria
[3] Department of Physics and Astronomy, University of Pittsburgh, PA 15260, USA

Received: 8 December 2006
Published online: 22 March 2007 – © Società Italiana di Fisica / Springer-Verlag 2007

Abstract. In this summary of the application of Dyson-Schwinger equations to the theory and phenomenology of hadrons, some deductions following from a nonperturbative, symmetry-preserving truncation are highlighted, notable amongst which are results for pseudoscalar mesons. We also describe inferences from the gap equation relating to the radius of convergence of a chiral expansion, applications to heavy-light and heavy-heavy mesons, and quantitative estimates of the contribution of quark orbital angular momentum in pseudoscalar mesons; and recapitulate upon studies of nucleon electromagnetic form factors.

PACS. 24.85.+p Quarks, gluons, and QCD in nuclei and nuclear processes – 12.38.Lg Other nonperturbative calculations

1 Introduction

The world's experimental hadron physics facilities are providing data of unprecedented accuracy with an enormous potential to impact on our understanding of the basic features of the strong interaction; *e.g.*, [1–4]. The explanation of that data in terms of QCD's key elements is an important task for contemporary theory. It is likewise vital for theory to link current observations to a future path of discovery. Success with that requires flexible tools, which can rapidly provide an intuitive understanding of information in hand and simultaneously anticipate its likely consequences. Models, parametrisations and truncations of QCD play this role. Such theory can also identify and explore novel possibilities within hadron physics, whose experimental verification could test the foundations of QCD.

Prominent among such tools are QCD's Dyson-Schwinger equations (DSEs), truncations thereof, and models based on this complex of integral equations. Glancing back at the last proceedings volume produced for this series of conferences [5], it is apparent that considerable progress has been made in the interim. We will provide a snapshot of that herein, which is augmented by other contributions to this volume.

2 Gap equation

In the presence of dynamical chiral symmetry breaking (DCSB); *i.e.*, the generation of mass *from nothing*, a study

[a] e-mail: cdroberts@anl.gov

of hadrons must begin with QCD's gap equation:

$$S(p)^{-1} = Z_2 \left(i\gamma \cdot p + m^{\mathrm{bm}}\right) + \Sigma(p), \tag{1}$$

$$\Sigma(p) = Z_1 \int_q^\Lambda g^2 D_{\mu\nu}(p-q)\frac{\lambda^a}{2}\gamma_\mu S(q)\Gamma_\nu^a(q,p), \tag{2}$$

where $\int_q^\Lambda$ represents a Poincaré invariant regularisation of the integral, with Λ the regularisation mass-scale [6], $D_{\mu\nu}(k)$ is the dressed-gluon propagator, $\Gamma_\nu(q,p)$ is the dressed-quark-gluon vertex, and m^{bm} is the quark's Λ-dependent bare current mass. The vertex and quark wave-function renormalisation constants, $Z_{1,2}(\zeta^2, \Lambda^2)$, depend on the gauge parameter.

The solution of the gap equation can be written

$$S(p) = Z(p^2, \zeta^2)/[i\gamma \cdot p + M(p^2)], \tag{3}$$

wherein the mass function, $M(p^2)$, is momentum-dependent but independent of the renormalisation point. The solution is obtained from (1) augmented by the renormalisation condition

$$S(p)^{-1}\big|_{p^2=\zeta^2} = i\gamma \cdot p + m(\zeta^2), \tag{4}$$

where $m(\zeta^2)$ is the renormalised (running) mass:

$$Z_2(\zeta^2, \Lambda^2)\, m^{\mathrm{bm}}(\Lambda) = Z_4(\zeta^2, \Lambda^2)\, m(\zeta^2), \tag{5}$$

with Z_4 the Lagrangian-mass renormalisation constant. In QCD the chiral limit is strictly defined by [6]

$$Z_2(\zeta^2, \Lambda^2)\, m^{\mathrm{bm}}(\Lambda) \equiv 0, \; \forall \Lambda^2 \gg \zeta^2, \tag{6}$$

which states that the renormalisation-point–invariant current-quark mass $\hat{m} = 0$.

In connection with (1) and (2) it has recently been established [7] that on a bounded, measurable domain of non-negative current-quark mass, realistic models for the kernel can simultaneously admit two inequivalent DCSB solutions and a solution that is unambiguously connected with the realisation of chiral symmetry in the Wigner mode. The Wigner solution and one of the DCSB solutions are destabilised by a current-quark mass and both disappear when that mass exceeds a critical value. This critical value also bounds the domain on which the surviving DCSB solution possesses a chiral expansion and can therefore be viewed as an upper limit on the domain within which a perturbative expansion in the current-quark mass around the chiral limit is uniformly valid for physical quantities. This critical mass is typically $\hat{m}_{\mathrm{cr}} \sim 60\text{--}70\,\mathrm{MeV}$, which for a flavour-nonsinglet 0^- meson constituted of equal-mass quarks corresponds to a mass $m_{0^-} \sim 0.45\,\mathrm{GeV}$ [8]. In arguing this case, properties of the two DCSB solutions of the gap equation that enable a valid definition of $\langle \bar{q}q \rangle$ in the presence of a nonzero current mass were employed. The behaviour of this condensate indicates that the essentially dynamical component of chiral symmetry breaking decreases with increasing current-quark mass, following the trend predicted by the constituent-quark σ-term ([9], sect. 5.2.2).

3 Symmetry-preserving truncation

One is forced to begin with (1) and (2) by the axial-vector Ward-Takahashi identity; viz.,

$$P_\mu \Gamma_{5\mu}(k; P) = S(k_+)^{-1} i\gamma_5 + i\gamma_5 S(k_-)^{-1} - 2im(\zeta^2)\Gamma_5(k; P), \tag{7}$$

written here for a quark and antiquark of equal current mass, wherein the axial-vector vertex is determined by the inhomogeneous Bethe-Salpeter equation

$$[\Gamma_{5\mu}(k; P)]_{tu} = Z_2\, [\gamma_5\gamma_\mu]_{tu} + \int_q^\Lambda [\chi(q; P)]_{sr} K_{tu}^{rs}(q, k; P), \tag{8}$$

with $\chi(k; P) = S(k_+)\Gamma_{5\mu}(k; P)S(k_-)$, $k_\pm = k \pm P/2$ and the colour- and Dirac-matrix structure of the elements in the equation is denoted by the indices r, s, t, u. In (8), $K(q, k; P)$ is the fully-amputated quark-antiquark scattering kernel. If one knows the form of K, then the nature of the interaction between quarks in QCD is completely understood. The pseudoscalar vertex in (7) is prescribed by an analogue of (8).

Every pseudoscalar meson appears as a pole contribution to the axial-vector and pseudoscalar vertices [6]; viz.,

$$\Gamma_{5\mu}(k; P) \overset{P^2 \simeq -m_{\pi_n}^2}{=} \frac{f_{\pi_n} P_\mu}{P^2 + m_{\pi_n}^2} \Gamma_{\pi_n}(k; P) + \Gamma_{5\mu}^{\mathrm{reg}}(k; P), \tag{9}$$

$$i\Gamma_5(k; P) \overset{P^2 \simeq -m_{\pi_n}^2}{=} \frac{\rho_{\pi_n}(\zeta^2)}{P^2 + m_{\pi_n}^2} \Gamma_{\pi_n}(k; P) + i\Gamma_5^{\mathrm{reg}}(k; P), \tag{10}$$

where the Γ^{reg} are regular in the neighbourhood of this pole, $\Gamma_{\pi_n}(k; P)$ represents the bound state's canonically normalised Bethe-Salpeter amplitude:

$$\Gamma_{\pi_n}(k; P) = \gamma_5\, [iE_{\pi_n}(k; P) + \gamma \cdot P F_{\pi_n}(k; P) + \gamma \cdot k\, k \cdot P\, G_{\pi_n}(k; P) + \sigma_{\mu\nu} k_\mu P_\nu\, H_{\pi_n}(k; P)], \tag{11}$$

and

$$f_{\pi_n} P_\mu = Z_2\, \mathrm{tr} \int_q^\Lambda \gamma_5 \gamma_\mu\, \chi_{\pi_n}(q; P), \tag{12}$$

$$i\rho_{\pi_n}(\zeta^2) = Z_4\, \mathrm{tr} \int_q^\Lambda \gamma_5\, \chi_{\pi_n}(q; P), \tag{13}$$

wherein the Bethe-Salpeter wave function is

$$\chi_{\pi_n}(k; P) = S(k_+)\Gamma_{\pi_n}(k; P)S(k_-) \tag{14}$$
$$= \gamma_5\, [i\mathcal{E}_{\pi_n}(k; P) + \gamma \cdot P \mathcal{F}_{\pi_n}(k; P) + \gamma \cdot k\, k \cdot P\, \mathcal{G}_{\pi_n}(k; P) + \sigma_{\mu\nu} k_\mu P_\nu\, \mathcal{H}_{\pi_n}(k; P)]. \tag{15}$$

In (9)–(15), π_0 denotes the lowest-mass pseudoscalar and increasing n labels bound states of increasing mass.

Equation (7) is a statement of chiral symmetry and the manner by which it is broken, explicitly and dynamically. It relates the solution of the two-body problem, (8), in quantum field theory to the one-body problem, (1). QCD is violated in any approach that does not preserve (7). A weak-coupling expansion guarantees (7). However, one cannot study bound states in perturbation theory, nor QCD's fundamental emergent phenomena; viz., confinement and DCSB. Fortunately, as related in [5], there is at least one symmetry-preserving truncation of the DSEs that is nonperturbative in the coupling [10–13]. This fact has enabled the proof of exact results in QCD and importantly in addition their illustration using practical truncations to which the corrections are quantifiable.

An exemplar: for a flavour-nonsinglet pseudoscalar meson[1] [6,14]

$$f_{\pi_n} m_{\pi_n}^2 = 2m(\zeta^2) \rho_{\pi_n}(\zeta^2). \tag{16}$$

The Gell-Mann–Oakes–Renner relation is a corollary of (16). Moreover, in deriving this expression, no assumptions are made about the current-quark mass of the constituents. Hence an analogue is equally valid for a meson containing one heavy-quark, a heavy-light system, and also heavy-heavy mesons.

---

[1] For notational simplicity we have written this identity for mesons constituted from a quark and antiquark with the same current mass. The unequal-mass case is little different.

Furthermore, the validity of (16) is not restricted to the ground state. This entails [14] that in the chiral limit

$$f_{\pi_n}^0 \equiv 0, \ \forall n \geq 1; \tag{17}$$

viz., Goldstone modes are the only pseudoscalar mesons to possess a nonzero leptonic decay constant in the chiral limit when chiral symmetry is dynamically broken. The decay constants of all other pseudoscalar mesons on this trajectory, *e.g.*, radial excitations, vanish. This result is consistent with model studies; *e.g.*, [15–19]. On the flip side, in the absence of DCSB, the leptonic decay constant of each such pseudoscalar meson vanishes in the chiral limit; namely, (17) is true $\forall n \geq 0$.

From the perspective of quantum mechanics, (17) is a surprising fact. The leptonic decay constant for S-wave states is typically proportional to the wave function at the origin. Compared with the ground state, this is smaller for an excited state because the wave function is broader in configuration space and wave functions are normalised. However, it is a modest effect; *e.g.*, consider the e^+e^- decay of vector mesons, for which a calculation in relativistic quantum mechanics based on light-front dynamics [20] yields $f_{\rho_1}/f_{\rho_0} \approx 0.6$, consistent with the value inferred from experiment. (The expression for f_V is analogous to (12), see [21].) As apparent in a recent lattice simulation [22], (17) can be realised in a framework if, and only if, (7) holds true and is veraciously realised.

The two-photon decay of 0^{-+} mesons provides another important example. In the presence of DCSB the ground state neutral pseudoscalar meson decays predominantly into two photons. So long as one employs a nonperturbative truncation that preserves chiral symmetry and the pattern of its dynamical breaking, the ground state's two photon decay is described in the chiral limit by a coupling [23–26]

$$g_{\pi_0^0 \gamma\gamma} := \mathcal{T}_{\pi_n^0}(-m_{\pi_n}^2 = 0, Q^2 = 0) = \frac{1}{2}\frac{1}{f_{\pi_0}}, \tag{18}$$

where $\mathcal{T}$ is the scalar function appearing in the matrix element. Equation (18) is the most widely known consequence of the Abelian anomaly.

Now, given that $f_{\pi_n \neq 0} \equiv 0$ in the chiral limit, it is natural to ask whether the $\pi_{n \neq 0} \to \gamma\gamma$ transition is affected. Since rainbow-ladder is the leading order in a symmetry-preserving truncation, it was used in [27] to provide a model-independent analysis of this process for arbitrary n. Therein, amongst other things, (18) is generalised and it is proved that

$$\mathcal{T}_{\pi_n^0}(-m_{\pi_n}^2, Q^2) \stackrel{Q^2 \gg \Lambda_{\mathrm{QCD}}^2}{=}$$
$$\frac{4\pi^2}{3}\left[\frac{f_{\pi_n}}{Q^2} + F_n^{(2)}(-m_{\pi_n}^2)\frac{\ln^\gamma Q^2/\omega_{\pi_n}^2}{Q^4}\right], \tag{19}$$

where: γ is an anomalous dimension; ω_{π_n} is a mass scale that gauges the momentum space width of the pseudoscalar meson; and $F_n^{(2)}(-m_{\pi_n}^2)$ is a structure-dependent constant, similar but unrelated to $f_{\pi_n}{}^2$. It is now plain

____

[2] With the interaction described in [27], $F_1^{(2)}(-m_{\pi_n}^2) \simeq -\langle\bar{q}q\rangle^0$, and it is generally nonzero in the chiral limit.

that $\forall n \geq 1$

$$\lim_{\hat{m}\to 0} \mathcal{T}_{\pi_n^0}(-m_{\pi_n}^2, Q^2) \stackrel{Q^2 \gg \Lambda_{\mathrm{QCD}}^2}{=}$$
$$\frac{4\pi^2}{3} F_n^{(2)}(-m_{\pi_n}^2)\frac{\ln^\gamma Q^2/\omega_{\pi_n}^2}{Q^4}\bigg|_{\hat{m}=0}; \tag{20}$$

namely, in the chiral limit the leading-order power law in the transition form factor for excited state pseudoscalar mesons is $O(1/Q^4)$.

4 Heavy quarks

The implications of (16) for heavy-light mesons are treated in [21,28,29]. Therefore herein we focus on heavy-heavy mesons, for which (16) is equally valid. For quark flavour Q, a constituent-quark spectrum-mass is defined via

$$M_Q^S = M_Q(p^2 = \zeta_{\rho_0}^2), \ \zeta_{\rho_0}^2 = -\frac{1}{4}M_{\rho_0}^2, \tag{21}$$

where $M_Q(p^2)$ is the renormalisation-point–invariant dressed-quark mass function in (3) obtained as the solution of (1) and (2) with the Q-quark current mass in (4)[3]. As $\hat{m}_Q$ is increased, M_Q^S becomes equivalent to the so-called pole-mass in the effective field theory for quarkonium systems; *i.e.*, non-relativistic-QCD (NRQCD). Now, with

$$m_{\pi_n}^{\bar{Q}Q} = 2M_Q^S[1 + \varepsilon_{\pi_n}^{\bar{Q}Q}/M_Q^S], \tag{22}$$

where $\varepsilon_{\pi_n}^{\bar{Q}Q}$ is a binding energy that does not grow with M_Q^S, and using the renormalisation-group-invariance of $m(\zeta^2)\rho_{\pi_n}^{\bar{Q}Q}(\zeta^2)$, (16) predicts

$$\rho_{\pi_n}^{\bar{Q}Q}(\zeta_{\rho_0}^2) \stackrel{\hat{m}_Q \to \infty}{=} f_{\pi_n}^{\bar{Q}Q} m_{\pi_n}^{\bar{Q}Q}. \tag{23}$$

This is an identity between the pseudoscalar and pseudovector projections of the meson's Bethe-Salpeter wave function at the origin in configuration space. Each element in (23) is gauge invariant and renormalisation point independent. The identity is exhibited in the heavy-quark limit of potential models for quarkonia; *e.g.*, [30,31].

In heavy-light systems it can be shown algebraically [21,28,29] that (23) is realised via

$$\rho_{\pi_n}^{\bar{Q}q} \propto (m_{\pi_n}^{\bar{Q}q})^{1/2} \quad \text{and} \quad f_{\pi_n}^{\bar{Q}q} \propto 1/(m_{\pi_n}^{\bar{Q}q})^{1/2}. \tag{24}$$

These results follow for heavy-light mesons because the integrands in (12) and (13) can in this instance be accurately approximated via an expansion in $\varepsilon_{\pi_n}^{\bar{Q}q}/M_Q^S$ and $w_{\pi_n}^{\bar{Q}q}/M_Q^S$, where $w_{\pi_n}^{\bar{Q}q}$ is the width of the meson's Bethe-Salpeter wave function, which we define as the value of

____

[3] In solving the Bethe-Salpeter equation, an amplitude peaked at zero relative four-momentum weights this value of $M_Q(p^2)$ most heavily in the dressed-quark propagator. In calculations based on [5], $M_c^S = 2.00\,\mathrm{GeV}$ and $M_b^S = 5.34\,\mathrm{GeV}$.

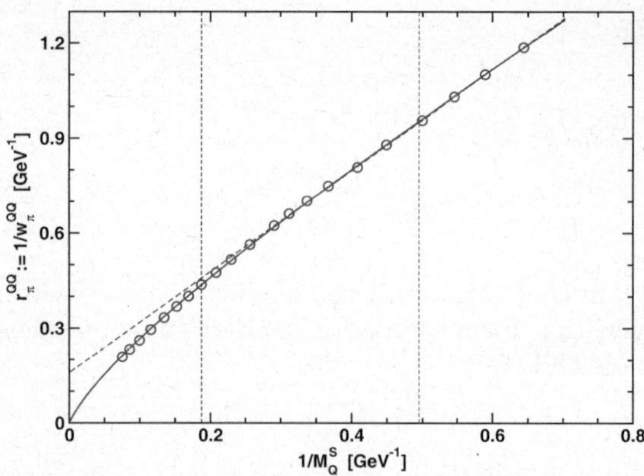

Fig. 1. $r_{\pi_n}^{\bar{Q}Q}$ *vs.* $1/M_Q^S$. Circles: $r_{\pi_n}^{\bar{Q}Q}$ calculated using the interaction model in [5]. Solid curve: described in (25). Dashed curve: linear fit to calculated result in neighbourhood of M_c^S. Dashed vertical lines mark, from left, $1/M_b^S$ and $1/M_c^S$. $r_{\pi_n}^{\bar{Q}Q}$, approaches zero as $M_S^Q \to \infty$.

the relative momentum whereat the first Chebyshev moment of the amplitude $\mathcal{E}_{\pi_n}^{\bar{Q}q}(k; P)$ falls to one-half of its maximum value. In the heavy-light meson $k \sim w_{\pi_n}^{\bar{Q}q}$ is the typical momentum of the light quark. Moreover, $w_{\pi_n}^{\bar{Q}q}$ obtains a finite nonzero value in the limit $M_Q^S \to \infty$. It follows that a heavy-light meson is always of nonzero spatial extent.

This is not true for heavy-heavy systems, as is apparent in fig. 1, which depicts the evolution of the spatial size of a heavy-heavy meson with constituent-quark spectrum mass[4]. The curve in the figure is

$$r_{\pi_n}^{\bar{Q}Q} = \frac{\gamma_M}{M_Q^S} \ln\left[\tau_M + \frac{M_Q^E}{\Lambda_{\mathrm{QCD}}}\right], \quad \gamma_M = 0.68, \quad \tau_M = 8.56,$$

$$(25)$$

with $\Lambda_{\mathrm{QCD}} = 0.234\,\mathrm{GeV}$ [5]. Plainly, with increasing constituent-mass a heavy-heavy system becomes "point-like" in configuration space and hence delocalised in momentum space. The evolution with M_Q^S of an observable such as $f_{\pi_n}^{\bar{Q}Q}$ may therefore be sensitive to the $\bar{Q}Q$ interaction over a wide range of momentum scales and hence a useful probe of that interaction.

In NRQCD the matrix elements for various spin states of a given quarkonium system are equal up to corrections of order $v_Q^2 \simeq (w_{\pi_n}^{\bar{Q}Q}/M_Q^S)^2$, where $k \sim w_{\pi_n}^{\bar{Q}Q}$ is the typical magnitude of the heavy-constituent's three-momentum in the meson's rest frame [32]. In this picture, 0^{-+} and 1^{--} mesons, which differ because spins are anti-aligned in the pseudoscalar and aligned in the vector, become degenerate in the limit $M_Q^S \to \infty$ and their leptonic decay constants become identical; *i.e.*, $f_{\pi_n}^{\bar{Q}Q} = f_{\rho_n}^{\bar{Q}Q}$. It is noteworthy, however, that (25) gives $v_c^2 \simeq 0.27$ and $v_b^2 \simeq 0.18$. Moreover, v_Q^2 falls only as $\alpha_s^2(M_Q^S)$. Hence, a quantitative discrep-

---

[4] The evolution was calculated using the renormalisation-group-improved rainbow-ladder DSE truncation with the interaction model in [5]. All corrections to this truncation vanish in the heavy-heavy limit; *e.g.*, see [13].

ancy between $f_{\pi_n}^{\bar{Q}Q}$ and $f_{\rho_n}^{\bar{Q}Q}$ can conceivably persist until rather large quark masses.

The evolution to mass degeneracy is exemplified in [13], which begins with a model for the dressed-quark-gluon vertex that appears in (2). Since the model's diagrammatic content is explicitly enumerable, a symmetry-preserving dressed-quark Bethe-Salpeter kernel could be explicitly constructed. The study showed that with rising current-quark mass the rainbow-ladder truncation provides an increasingly accurate estimate of the mass of a heavy-heavy system, and the mass splitting between vector and pseudoscalar meson masses vanishes. With the b-quark mass fitted to give $m_{\Upsilon(1S)} = 9.46\,\mathrm{GeV}$, the model predicts $m_{\eta_b} = 9.42\,\mathrm{GeV}$.

Little is known experimentally about heavy-heavy 0^{-+} mesons. However, 1^{--} mesons are readily produced experimentally and much studied. We therefore consider vector meson leptonic decays, for which it was noted [33,34] that

$$\frac{(f_{\rho_0}^{\bar{Q}Q})^2}{m_{\rho_0}^{\bar{Q}Q}} \propto \frac{\Gamma_{\rho_0}^{\bar{Q}Q} \to e^+e^-}{\langle e_Q\rangle^2} \approx 12.4\,\mathrm{keV}, \quad (26)$$

where $\langle e_Q\rangle$ is the mean electric charge of the valence-quark constituents in units of the electron's charge. (See fig. 2.) From this empirically based conjecture one might conclude that on the experimentally accessible domain of current-quark masses $f_{\rho_0}^{\bar{Q}Q} \propto (m_{\rho_0}^{\bar{Q}Q})^{1/2}$. This is a marked departure from the behaviour in heavy-light systems, (24), which is independent of the $\bar{Q}q$ interaction. Furthermore [34], Coulomb-potential models typically give $f_{\rho_0}^{\bar{Q}Q} \propto m_{\rho_0}^{\bar{Q}Q}$ whereas a linear potential produces $f_{\rho_0}^{\bar{Q}Q} \sim$ constant. Here is confirmation that the properties of quarkonia are a probe of the $\bar{Q}Q$ interaction.

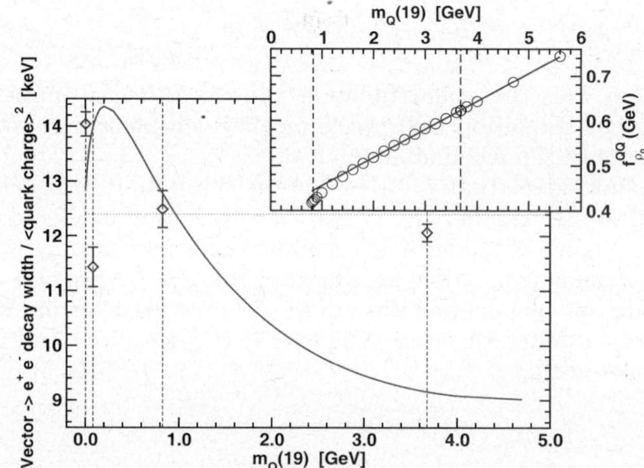

Fig. 2. Main panel: width/charge-squared ratio in (26). Solid line: result obtained in the approach of [5]; diamonds: data summarised in [34] and updated from [35]; the horizontal dotted line indicates the conjecture of (26); and the dashed vertical lines indicate, from left, the u, s, c, b current-quark masses fixed at a renormalisation scale $\zeta = 19\,\mathrm{GeV}$. Inserted panel: circles: calculated evolution of $f_{\rho_0}^{\bar{Q}Q}$. Solid line: straight-line fit to large-m_Q results. The c and b current-quark masses are indicated by vertical dashed lines.

In fig. 2 we depict the result obtained for the ratio in (26) using the renormalisation-group–improved rainbow-ladder DSE truncation with the interaction model of [5]. The interaction is precisely that of QCD in the ultraviolet; namely, colour-Coulomb. However, a single parameter is employed to express a model for the long-range part of the quark-quark scattering kernel. That parameter is a gluon mass scale $m_g = 720\,\mathrm{MeV}$. It was chosen in order to fit f_{π_0} and ρ_{π_0} and hence the results in fig. 2 are an untuned prediction. Quantitatively, that for the c-quark is good. Indeed, for ground-state vector mesons up to and including J/ψ the standard deviation between the calculated width and experiment is $\lesssim 15\%$. (NB. The calculated masses are used in determining the widths.) Adding the $\Upsilon(1S)$, that standard deviation rises to $\lesssim 25\%$. Qualitatively, however, the mass-dependence obtained with this interaction does not support (26). Our preliminary result, illustrated via the inset in fig. 2, is $f_{\rho_0}^{\bar{Q}Q} \propto m_{\rho_0}^{\bar{Q}Q}$ for masses in the neighbourhood of the b-quark and beyond. This is Coulomb-potential-like behaviour, which may be natural because the interaction employed is precisely that of QCD in the ultraviolet. We are continuing to examine the validity of the hypothesis in (26).

5 Quark orbital angular momentum

It is noteworthy that quark orbital angular momentum is not a Poincaré invariant. However, if absent in a particular frame, it will inevitably appear in another frame related via a Poincaré transformation. Nonzero quark orbital angular momentum is thus a necessary outcome of a Poincaré covariant description, which is why the Bethe-Salpeter wave function in (15) is a matrix-valued function with a rich structure.

A pseudoscalar meson naturally possesses total spin $J = 0$ and this is expressed in the fact that χ_{π_n} is an eigenstate of the Pauli-Lubanski operator with eigenvalue zero. Nonetheless, in the meson's rest frame one can straightforwardly decompose the Pauli-Lubanski operator into the sum of two terms: one measuring the angular momentum of the quarks and another measuring their spin. In this way one can show that the terms in (15) characterised by $\mathcal{E}$ and $\mathcal{F}$ are purely $L = 0$ in the rest frame, whereas the $\mathcal{G}$ and $\mathcal{H}$ terms are associated with $L = 1$. Thus a pseudoscalar meson Bethe-Salpeter wave function *always* contains both S- and P-wave components.

In this connection, it is worth recalling that E_{π_0} in (11) provides only a subleading contribution to the ultraviolet behaviour of the electromagnetic pion form factor [24]. The leading power law behaviour anticipated from perturbative QCD is produced by the amplitudes associated with F_{π_0} and G_{π_0} [25]. Moreover, it is an identity relating F_{π_n} and G_{π_n} in the ultraviolet that ensures f_{π_n} in (12) is gauge invariant, and cutoff- and renormalisation-point-independent [25,36,37].

We exhibit the rest-frame angular-momentum content of a 0^{-+} meson in the following way. A Bethe-Salpeter

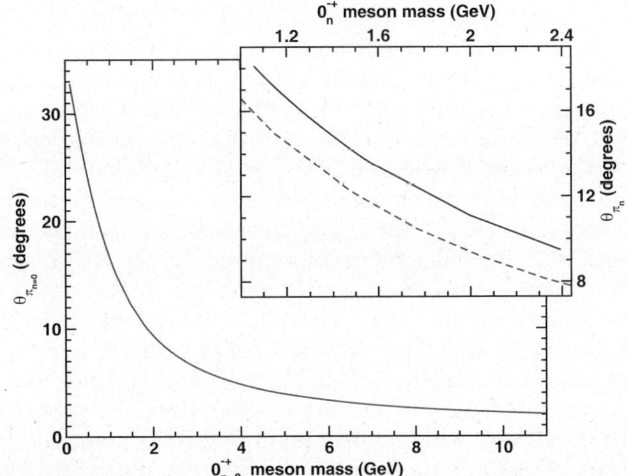

Fig. 3. Main frame: θ_{π_0}, which gauges the rest-frame admixture of $L = 1$ components in the ground-state 0^{-+} meson's Bethe-Salpeter wave function, plotted as a function of the meson's mass. Inset: solid curve: analogue, θ_{π_1}, for the first excited state; dashed curve: θ_{π_0} for comparison. In the chiral limit the ground state is naturally massless whereas the calculated mass of the first excited state is 1.04 GeV [27].

amplitude is canonically normalised. There are sixteen distinguishable terms in the associated sum; viz., an $\bar{\mathcal{E}}_{\pi_n}\mathcal{E}_{\pi_n}$ contribution plus an $\bar{\mathcal{E}}_{\pi_n}\mathcal{F}_{\pi_n}$ contribution, etc. In the sum of the squares of these terms we associate $\sin^2\theta_{\pi_n}$ with the nondiagonal contributions, in which case $\sin\theta_{\pi_n}$ gauges the role played by $L = 1$ components in the normalisation. In fig. 3, for both the ground and first radially excited state, we plot the bound-state mass-dependence of θ_{π_n} obtained in the renormalisation-group-improved rainbow-ladder truncation using the model interaction described in [5]. For both states, angular momentum is most significant in the neighbourhood of the chiral limit, and decreases with increasing current-quark mass. Notably, as measured by the angle θ_{π_n}, at a given bound-state mass the admixture of $L = 1$ components in the first radial excitation is roughly 15% greater than that in the ground state. Measured as a function of the current-quark mass, however, the situation is reversed: $\theta_1/\theta_0 \approx 0.5$ in the chiral limit and this ratio increases steadily to ≈ 0.9 at $m_Q(19) = 0.5\,\mathrm{GeV}$. Our analysis continues.

6 Baryons

Despite material progress with the study of mesons, the challenge of baryons remains. A nucleon appears as a pole in a six-point quark Green function. The pole's residue is proportional to the nucleon's Faddeev amplitude, which is obtained from a Poincaré covariant Faddeev equation that adds-up all possible quantum field-theoretical exchanges and interactions that can take place between three dressed quarks. In formulating and solving this problem, current expertise is approximately at the level it was for mesons ten years ago; *i.e.*, model building and phenomenology. However, we are a little ahead because a great deal has been learnt in applications to mesons. For example, we have acquired a veracious understanding of the structure of dressed quarks and gluons and therefore can straightforwardly incorporate effects owing to and arising from the strong infrared modification of the momentum dependence of these propagators. (See, *e.g.*, sect. 5.1 of [9].)

A tractable treatment of the Faddeev equation requires a truncation. One is based [38] on the observation that an interaction which describes colour-singlet mesons also generates quark-quark (diquark) correlations in the colour-$\bar{3}$ (antitriplet) channel [39]. The dominant correlations for ground state octet and decuplet baryons are 0^+ and 1^+ diquarks. This can be understood on the grounds that: the associated mass scales are smaller than the baryons' masses [40,41], with models giving (in GeV) $m_{[ud]_{0^+}} = 0.74\text{-}0.82$, $m_{(uu)_{1^+}} = m_{(ud)_{1^+}} = m_{(dd)_{1^+}} = 0.95\text{-}1.02$; the electromagnetic size of these correlations is less than that of the proton [42] —$r_{[ud]_{0^+}} \approx 0.7\,\mathrm{fm}$, from which we estimate $r_{(ud)_{1^+}} \sim 0.8\,\mathrm{fm}$ based on the ρ-meson/π-meson radius ratio [43,44]; and the positive parity of the correlations matches that of the baryons. Both 0^+ and 1^+ diquarks provide attraction in the Faddeev equation.

The truncation of the Faddeev equation's kernel is completed by specifying that the quarks are dressed, with two of the three dressed-quarks correlated always as a colour-$\bar{3}$ diquark. Binding is then effected by the iterated exchange of roles between the bystander and diquark-participant quarks. This ensures that the Faddeev amplitude exhibits the correct symmetry properties under fermion interchange. A Ward-Takahashi-identity-preserving electromagnetic current for the baryon thus constituted is subsequently derived [45]. It depends on the electromagnetic properties of the axial-vector diquark correlation: its magnetic and quadrupole moments; and the strength of electromagnetically induced axial-vector $\leftrightarrow$ scalar diquark transitions.

A Faddeev equation study of the nucleon's mass and the effect on this of a pseudoscalar meson cloud are detailed in [46]. The lessons learnt are employed in a series of studies of nucleon properties; *e.g.*, the nucleons' σ-term in [47,48], and nucleon form factors in [49–52]. Of particular contemporary interest is the ratio $\mu_p G_E^p(Q^2)/G_M^p(Q^2)$ [1,4]. This passes through zero at $Q^2 \approx 6.5\,\mathrm{GeV}^2$ [50]. The analogous ratio for the neutron is presented in [51]. In the neighbourhood of $Q^2 = 0$,

$$\mu_p \frac{G_E^n(Q^2)}{G_M^n(Q^2)} = -\frac{r_n^2}{6} Q^2, \tag{27}$$

where r_n is the neutron's electric radius. The calculation shows this to be a good approximation for $r_n^2 Q^2 \lesssim 1$, with which the data [53] are consistent. It is notable that, just as for the proton, the small-Q^2 behaviour of this ratio is materially affected by the neutron's pion cloud.

Pseudoscalar mesons are not pointlike and therefore their contributions to form factors diminish in magnitude with increasing Q^2. It follows therefore that the evolution of $\mu_p G_E^p(Q^2)/G_M^p(Q^2)$ and $\mu_n G_E^n(Q^2)/G_M^n(Q^2)$ on $Q^2 \gtrsim 2\,\mathrm{GeV}^2$ are both primarily determined by the quark-core

of the nucleon. While the proton ratio decreases uniformly on this domain [49,50], [51] predicts that the neutron ratio increases steadily until $Q^2 \simeq 8\,\mathrm{GeV}^2$.

As with mesons, sect. 5, in a Poincaré-covariant treatment the nucleon's quark core is necessarily described by a Faddeev amplitude with nonzero quark orbital angular momentum. The Faddeev amplitude is therefore a matrix-valued function that, in a baryons' rest frame, corresponds to a relativistic wave function with S-wave, P-wave and D-wave components [54]. In form factor studies [49–51] there is some quantitative sensitivity to the electromagnetic structure of the diquarks. However, the gross features of the form factors are primarily governed by correlations expressed in the nucleon's Faddeev amplitude and, in particular, by the amount of intrinsic quark orbital angular momentum [55]. The nature of the kernel in the Faddeev equation specifies just how much quark orbital angular momentum is present in a baryon's rest frame.

We see a baryon as composed primarily of a quark core, constituted of confined quark and confined diquark correlations, but augmented by 0^- meson cloud contributions that are sensed by long-wavelength probes. Short-wavelength probes pierce the cloud, and expose spin-isospin correlations and quark orbital angular momentum within the baryon. The veracity of this description makes plain that a picture of baryons as a bag of three constituent-quarks in relative S-waves is profoundly misleading.

7 Prospect

Two emergent phenomena are primarily responsible for the observed properties of hadrons: confinement and dynamical chiral symmetry breaking (DCSB). They can be viewed as an essential consequence of the presence and role of particle-antiparticle pairs in an asymptotically free theory and therefore can only be veraciously understood in relativistic quantum field theory. The Dyson-Schwinger equations (DSEs) provide a natural framework for the exploration of these phenomena. The DSEs are a generating tool for perturbation theory and thus give a clean connection with processes that are well understood. Moreover, they admit a systematic, symmetry preserving and nonperturbative truncation, and thereby give access to strong QCD in the continuum. On top of this, quantitative comparisons and feedback between DSE and lattice-QCD studies are today proving fruitful.

Dynamical chiral symmetry breaking (DCSB) is a singularly effective mass-generating mechanism. It is understood via QCD's gap equation, the solution of which delivers a quark mass function with a momentum-dependence that connects the perturbative and nonperturbative, constituent-quark domains. Despite the fact that light quarks are made heavy, the mass of the pseudoscalar mesons remains unnaturally small. That, too, owes to DCSB, expressed this time in a relationship between QCD's gap equation and those colour singlet Bethe-Salpeter equations which have a pseudoscalar projection. Goldstone's theorem is a natural consequence of this connection.

The existence of a sensible truncation scheme enables the proof of exact results using the DSEs. That the truncation scheme is also tractable provides a means by which the results may be illustrated, and furthermore a practical tool for the prediction of observables that are accessible at contemporary experimental facilities. The consequent opportunities for rapid feedback between experiment and theory brings within reach an intuitive understanding of nonperturbative strong-interaction phenomena.

There are indications that confinement may be expressed in the analyticity properties of the dressed propagators. (Section 4.2 in [9].) To build understanding, it is essential to work toward an accurate map of the confinement force between light quarks and elucidate how that evolves from the potential between two static quarks. Among the rewards are a clear connection between confinement and DCSB, an accounting of the distribution of mass within hadrons, and a realistic picture of hybrids and exotics.

CDR is grateful to the organisers for their hospitality and acknowledges fruitful conversations with O. Lakhina and P.C. Tandy. This work was supported by: US Department of Energy, Office of Nuclear Physics, contract nos. DE-AC02-06CH11357 and DE-FG02-00ER41135; Austrian Science Fund FWF, Schrödinger-Rückkehrstipendium R50-N08; and benefited from the facilities of the ANL Computing Resource Center and Pittsburgh's NSF Terascale Computing System.

References

1. H.-Y. Gao, Int. J. Mod. Phys. E **12**, 1 (2003) (**12**, 567 (2003)(E)).
2. V.D. Burkert, T.S.H. Lee, Int. J. Mod. Phys. E **13**, 1035 (2004).
3. E.J. Beise, M.L. Pitt, D.T. Spayde, Prog. Part. Nucl. Phys. **54**, 289 (2005).
4. J. Arrington, C.D. Roberts, J.M. Zanotti, *Nucleon electromagnetic form factors*, nucl-th/0611050.
5. P. Maris, A. Raya, C.D. Roberts, S.M. Schmidt, Eur. Phys. J. A **18**, 231 (2003).
6. P. Maris, C.D. Roberts, P.C. Tandy, Phys. Lett. B **420**, 267 (1998).
7. L. Chang, Y.-X. Liu, M.S. Bhagwat, C.D. Roberts, S.V. Wright, *Dynamical chiral symmetry breaking and a critical mass*, nucl-th/0605058.
8. A. Krassnigg, P. Maris, J. Phys. Conf. Ser. **9**, 153 (2005).
9. A. Höll, C.D. Roberts, S.V. Wright, *Hadron physics and Dyson-Schwinger equations*, nucl-th/0601071.
10. H.J. Munczek, Phys. Rev. D **52**, 4736 (1995).
11. A. Bender, C.D. Roberts, L. von Smekal, Phys. Lett. B **380**, 7 (1996).
12. A. Bender, W. Detmold, C.D. Roberts, A.W. Thomas, Phys. Rev. C **65**, 065203 (2002).
13. M.S. Bhagwat, A. Höll, A. Krassnigg, C.D. Roberts, P.C. Tandy, Phys. Rev. C **70**, 035205 (2004).
14. A. Höll, A. Krassnigg, C.D. Roberts, Phys. Rev. C **70**, 042203(R) (2004).
15. C.A. Dominguez, Phys. Rev. D **15**, 1350 (1977); **16**, 2313 (1977).
16. A. Le Yaouanc, L. Oliver, S. Ono, O. Pene, J.C. Raynal, Phys. Rev. D **31**, 137 (1985).
17. F.J. Llanes-Estrada, S.R. Cotanch, Phys. Rev. Lett. **84**, 1102 (2000).
18. M.K. Volkov, V.L. Yudichev, Phys. Part. Nucl. **31**, 282 (2000) (Fiz. Elem. Chast. At. Yad. **31**, 576 (2000)).
19. W. Lucha, D. Melikhov, S. Simula, Phys. Rev. D **74**, 054004 (2006).
20. J.P.B. de Melo, T. Frederico, E. Pace, G. Salme, Phys. Rev. D **73**, 074013 (2006).
21. M.A. Ivanov, Yu.L. Kalinovsky, C.D. Roberts, Phys. Rev. D **60**, 034018 (1999).
22. UKQCD Collaboration (C. McNeile, C. Michael), Phys. Lett. B **642**, 244 (2006).
23. M. Bando, M. Harada, T. Kugo, Prog. Theor. Phys. **91**, 927 (1994).
24. C.D. Roberts, Nucl. Phys. A **605**, 475 (1996).
25. P. Maris, C.D. Roberts, Phys. Rev. C **58**, 3659 (1998).
26. P. Maris, P.C. Tandy, Phys. Rev. C **65**, 045211 (2002).
27. A. Höll, A. Krassnigg, P. Maris, C.D. Roberts, S.V. Wright, Phys. Rev. C **71**, 065204 (2005).
28. M.A. Ivanov, Yu.L. Kalinovsky, P. Maris, C.D. Roberts, Phys. Lett. B **416**, 29 (1998).
29. M.A. Ivanov, Yu.L. Kalinovsky, P. Maris, C.D. Roberts, Phys. Rev. C **57**, 1991 (1998).
30. O. Lakhina, E.S. Swanson, Phys. Rev. D **74**, 014012 (2006).
31. O. Lakhina, private communication.
32. G.T. Bodwin, E. Braaten, G.P. Lepage, Phys. Rev. D **51**, 1125 (1995) (**55**, 5853 (1997)(E)).
33. D.R. Yennie, Phys. Rev. Lett. **34**, 239 (1975).
34. M.N. Achasov *et al.*, Phys. Rev. Lett. **86**, 1698 (2001).
35. Particle Data Group (W.M. Yao *et al.*), J. Phys. G **33**, 1 (2006).
36. P. Maris, C.D. Roberts, Phys. Rev. C **56**, 3369 (1997).
37. P. Maris, P.C. Tandy, Phys. Rev. C **60**, 055214 (1999).
38. R.T. Cahill, C.D. Roberts, J. Praschifka, Austral. J. Phys. **42**, 129 (1989).
39. R.T. Cahill, C.D. Roberts, J. Praschifka, Phys. Rev. D **36**, 2804 (1987).
40. C.J. Burden, L. Qian, C.D. Roberts, P.C. Tandy, M.J. Thomson, Phys. Rev. C **55**, 2649 (1997).
41. P. Maris, Few Body Syst. **32**, 41 (2002).
42. P. Maris, Few Body Syst. **35**, 117 (2004).
43. C.J. Burden, C.D. Roberts, M.J. Thomson, Phys. Lett. B **371**, 163 (1996).
44. F.T. Hawes, M.A. Pichowsky, Phys. Rev. C **59**, 1743 (1999).
45. M. Oettel, M. Pichowsky, L. von Smekal, Eur. Phys. J. A **8**, 251 (2000).
46. M.B. Hecht, M. Oettel, C.D. Roberts, S.M. Schmidt, P.C. Tandy, A.W. Thomas, Phys. Rev. C **65**, 055204 (2002).
47. V.V. Flambaum, A. Höll, P. Jaikumar, C.D. Roberts, S.V. Wright, Few Body Syst. **38**, 31 (2006).
48. A. Höll, P. Maris, C.D. Roberts, S.V. Wright, *Schwinger functions and light-quark bound states, and sigma terms*, nucl-th/0512048.
49. R. Alkofer, A. Höll, M. Kloker, A. Krassnigg, C.D. Roberts, Few Body Syst. **37**, 1 (2005).
50. A. Höll, R. Alkofer, M. Kloker, A. Krassnigg, C.D. Roberts, S.V. Wright, Nucl. Phys. A **755**, 298 (2005).
51. M.S. Bhagwat, A. Höll, A. Krassnigg, C.D. Roberts, *Theory and phenomenology of hadrons*, nucl-th/0610080.

52. A. Holl, C.D. Roberts, S.V. Wright, AIP Conf. Proc. **857**, 46 (2006).

53. E93-038 Collaboration (R. Madey *et al.*), Phys. Rev. Lett. **91**, 122002 (2003).

54. M. Oettel, G. Hellstern, R. Alkofer, H. Reinhardt, Phys. Rev. C **58**, 2459 (1998).

55. J.C.R. Bloch, A. Krassnigg, C.D. Roberts, Few-Body Syst. **33**, 219 (2003).

Eur. Phys. J. A **31**, 638–644 (2007)

DOI 10.1140/epja/i2006-10221-7

Special Article – QNP 2006

Hadron spectroscopy and structure from AdS/CFT

S.J. Brodsky[a]

Stanford Linear Accelerator Center, Stanford University, Stanford, CA 94309, USA

Received: 8 November 2006
Published online: 5 March 2007 – © Società Italiana di Fisica / Springer-Verlag 2007

Abstract. The AdS/CFT correspondence between conformal field theory and string states in an extended space-time has provided new insights into not only hadron spectra, but also their light-front wave functions. We show that there is an exact correspondence between the fifth-dimensional coordinate of anti-de Sitter space z and a specific impact variable ζ which measures the separation of the constituents within the hadron in ordinary space-time. This connection allows one to predict the form of the light-front wave functions of mesons and baryons, the fundamental entities which encode hadron properties and scattering amplitudes. A new relativistic Schrödinger light-front equation is found which reproduces the results obtained using the fifth-dimensional theory. Since they are complete and orthonormal, the AdS/CFT model wave functions can be used as an initial ansatz for a variational treatment or as a basis for the diagonalization of the light-front QCD Hamiltonian. A number of applications of light-front wave functions are also discussed.

PACS. 12.38.-t Quantum chromodynamics – 11.25.Tq Gauge/string duality – 12.38.Aw General properties of QCD (dynamics, confinement, etc.) – 12.40.Nn Regge theory, duality, absorptive/optical models

1 Hadron wave functions in QCD

One of the most important tools in atomic physics is the Schrödinger wave function; it provides a quantum-mechanical description of the position and spin coordinates of nonrelativistic bound states at a given time t. Clearly, it is an important goal in hadron and nuclear physics to determine the wave functions of hadrons in terms of their fundamental quark and gluon constituents.

Guy de Téramond and I have recently shown how one can use AdS/CFT to not only obtain an accurate description of the hadron spectrum for light quarks, but also how to obtain a remarkably simple but realistic model of the valence wave functions of mesons, baryons, and glueballs. As I review below, the amplitude $\Phi(z)$ describing the hadronic state in the fifth dimension of Anti-de Sitter space AdS$_5$ can be precisely mapped to the light-front wave functions $\psi_{n/h}$ of hadrons in physical space-time [1], thus providing a relativistic description of hadrons in QCD at the amplitude level. The light-front wave functions are relativistic and frame-independent generalizations of the familiar Schrödinger wave functions of atomic physics, but they are determined at fixed light-cone time $\tau = t + z/c$ —the "front form" advocated by Dirac— rather than at fixed ordinary time t.

Formally, the light-front expansion is constructed by quantizing QCD at fixed light-cone time [2] $\tau = t + z/c$ and forming the invariant light-front Hamiltonian $H_{LF}^{QCD} =$

$P^+ P^- - \vec{P}_\perp^2$, where $P^\pm = P^0 \pm P^z$ [3]. The momentum generators P^+ and $\vec{P}_\perp$ are kinematical; i.e., they are independent of the interactions. The generator $P^- = i\frac{\mathrm{d}}{\mathrm{d}\tau}$ generates light-front time translations, and the eigen-spectrum of the Lorentz scalar H_{LF}^{QCD} gives the mass spectrum of the color-singlet hadron states in QCD together with their respective light-front wave functions. For example, the proton state satisfies: $H_{LF}^{QCD} \, |\psi_p\rangle = M_p^2 \, |\psi_p\rangle$. Remarkably, the light-front wave functions are frame independent; thus knowing the LFWFs of a hadron in its rest frame determines the wave functions in all other frames.

Given the light-front wave functions $\psi_{n/H}(x_i, \vec{k}_{\perp i}, \lambda_i)$, one can compute a large range of hadron observables. For example, the valence and sea quark and gluon distributions which are measured in deep inelastic lepton scattering are defined from the squares of the LFWFS summed over all Fock states n. Form factors, exclusive weak transition amplitudes [4] such as $B \to \ell\nu\pi$ and the generalized parton distributions [5] measured in deeply virtual Compton scattering are (assuming the "handbag" approximation) overlaps of the initial and final LFWFS with $n = n'$ and $n = n' + 2$. The gauge-invariant distribution amplitude $\phi_H(x_i, Q)$ defined from the integral over the transverse momenta $\vec{k}_{\perp i}^2 \leq Q^2$ of the valence (smallest n) Fock state provides a fundamental measure of the hadron at the amplitude level [6,7]; they are the nonperturbative input to the factorized form of hard exclusive amplitudes and exclusive heavy-hadron decays in perturbative QCD. The resulting distributions obey the DGLAP and ERBL evo-

[a] e-mail: sjbth@slac.stanford.edu

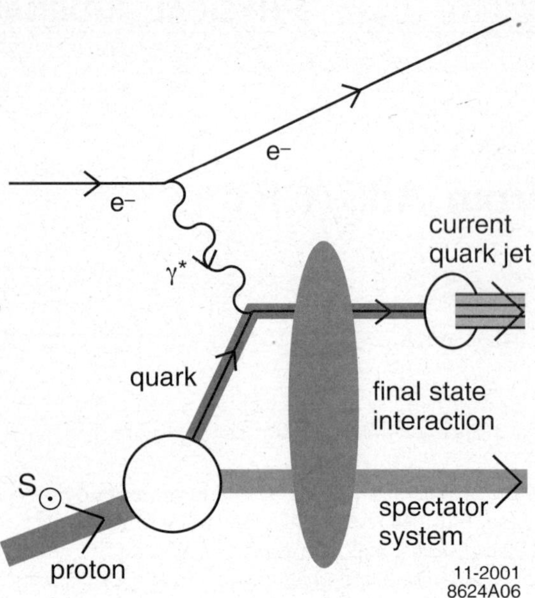

Fig. 1. A final-state interaction from gluon exchange in deep inelastic lepton scattering. The difference of the QCD Coulomb-like phases in different orbital states of the proton produces a single-proton spin asymmetry.

lution equations as a function of the maximal invariant mass, thus providing a physical factorization scheme [8]. In each case, the derived quantities satisfy the appropriate operator product expansions, sum rules, and evolution equations. However, at large x where the struck quark is far-off shell, DGLAP evolution is quenched [9], so that the fall-off of the DIS cross-sections in Q^2 satisfies inclusive-exclusive duality at fixed W^2.

The physics of higher Fock states such as the $|uudq\overline{Q}\rangle$ fluctuation of the proton is nontrivial, leading to asymmetric $s(x)$ and $\overline{s}(x)$ distributions, $\overline{u}(x) \neq \overline{d}(x)$, and intrinsic heavy quarks $c\overline{c}$ and $b\overline{b}$ which have their support at high momentum [10]. Color adds an extra element of complexity: for example there are five different color singlet combinations of six 3_C quark representations which appear in the deuteron's valence wave function, leading to the hidden color phenomena [11].

An important example of the utility of light-front wave functions in hadron physics is the computation of polarization effects such as the single-spin azimuthal asymmetries in semi-inclusive deep inelastic scattering, representing the correlation of the spin of the proton target and the virtual photon to hadron production plane: $\vec{S}_p \cdot \vec{q} \times \vec{p}_H$. Such asymmetries are time reversal odd, but they can arise in QCD through phase differences in different spin amplitudes. In fact, final-state interactions from gluon exchange between the outgoing quarks and the target spectator system lead to single-spin asymmetries in semi-inclusive deep inelastic lepton-proton scattering which are not power law suppressed at large photon virtuality Q^2 at fixed x_{bj} [12] (see fig. 1). In contrast to the SSAs arising from transversity and the Collins fragmentation function, the fragmentation of the quark into hadrons is not necessary; one predicts

a correlation with the production plane of the quark jet itself. Physically, the final-state interaction phase arises as the infrared-finite difference of QCD Coulomb phases for hadron wave functions with differing orbital angular momentum. The same proton matrix element which determines the spin-orbit correlation $\vec{S} \cdot \vec{L}$ also produces the anomalous magnetic moment of the proton, the Pauli form factor, and the generalized parton distribution E which is measured in deeply virtual Compton scattering. Thus, the contribution of each quark current to the SSA is proportional to the contribution $\kappa_{q/p}$ of that quark to the proton target's anomalous magnetic moment $\kappa_p = \sum_q e_q \kappa_{q/p}$ [12, 13]. The HERMES Collaboration has recently measured the SSA in pion electroproduction using transverse target polarization [14]. The Sivers and Collins effects can be separated using planar correlations; both processes are observed to contribute, with values not in disagreement with theory expectations [14,15]. The deeply virtual Compton amplitudes can be Fourier transformed to $b_\perp$ and $\sigma = x^- P^+/2$ space providing new insights into QCD distributions [16–19]. The distributions in the LF direction σ typically display diffraction patterns arising from the interference of the initial and final state LFWFs [18].

The final-state interaction mechanism provides an appealing physical explanation within QCD of single-spin asymmetries. Physically, the final-state interaction phase arises as the infrared-finite difference of QCD Coulomb phases for hadron wave functions with differing orbital angular momentum. An elegant discussion of the Sivers effect including its sign has been given by Burkardt [13]. As shown by Gardner and myself [20], one can also use the Sivers effect to study the orbital angular momentum of gluons by tagging a gluon jet in semi-inclusive DIS. In this case, the final-state interactions are enhanced by the large color charge of the gluons.

The final-state interaction effects can also be identified with the gauge link which is present in the gauge-invariant definition of parton distributions [21]. Even when the light-cone gauge is chosen, a transverse gauge link is required. Thus in any gauge the parton amplitudes need to be augmented by an additional eikonal factor incorporating the final-state interaction and its phase [22,23]. The net effect is that it is possible to define transverse-momentum–dependent parton distribution functions which contain the effect of the QCD final-state interactions.

A related analysis also predicts that the initial-state interactions from gluon exchange between the incoming quark and the target spectator system lead to leading-twist single-spin asymmetries in the Drell-Yan process $H_1 H_2^\uparrow \to \ell^+ \ell^- X$ [24,25]. Initial-state interactions also lead to a $\cos 2\phi$ planar correlation in unpolarized Drell-Yan reactions [26].

2 Diffractive deep inelastic scattering

A remarkable feature of deep inelastic lepton-proton scattering at HERA is that approximately 10% events are diffractive [27,28]: the target proton remains intact, and there is a large rapidity gap between the proton and the other hadrons in the final state. These diffractive deep inelastic scattering (DDIS) events can be understood most simply from the perspective of the color-dipole model: the $q\bar{q}$ Fock state of the high-energy virtual photon diffractively dissociates into a diffractive dijet system. The exchange of multiple gluons between the color dipole of the $q\bar{q}$ and the quarks of the target proton neutralizes the color separation and leads to the diffractive final state. The same multiple gluon exchange also controls diffractive vector meson electroproduction at large photon virtuality [29]. This observation presents a paradox: if one chooses the conventional parton model frame where the photon light-front momentum is negative $q+ = q^0 + q^z <$ 0, the virtual photon interacts with a quark constituent with light-cone momentum fraction $x = k^+/p^+ = x_{bj}$. Furthermore, the gauge link associated with the struck quark (the Wilson line) becomes unity in light-cone gauge $A^+ = 0$. Thus the struck "current" quark apparently experiences no final-state interactions. Since the light-front wave functions $\psi_n(x_i, k_{\perp i})$ of a stable hadron are real, it appears impossible to generate the required imaginary phase associated with pomeron exchange, let alone large rapidity gaps.

This paradox was resolved by Paul Hoyer, Nils Marchal, Stephane Peigne, Francesco Sannino and myself [30]. Consider the case where the virtual photon interacts with a strange quark —the $s\bar{s}$ pair is assumed to be produced in the target by gluon splitting. In the case of Feynman gauge, the struck s quark continues to interact in the final state via gluon exchange as described by the Wilson line. The final-state interactions occur at a light-cone time $\Delta\tau \simeq 1/\nu$ shortly after the virtual photon interacts with the struck quark. When one integrates over the nearly-on-shell intermediate state, the amplitude acquires an imaginary part. Thus the rescattering of the quark produces a separated color singlet $s\bar{s}$ and an imaginary phase. In the case of the light-cone gauge $A^+ = \eta \cdot A = 0$, one must also consider the final-state interactions of the (unstruck) $\bar{s}$ quark. The gluon propagator in the light-cone gauge $d_{LC}^{\mu\nu}(k) = (i/k^2 + i\epsilon)[-g^{\mu\nu} + (\eta^\mu k^\nu + k^\mu \eta^\nu/\eta \cdot k)]$ is singular at $k^+ = \eta \cdot k = 0$. The momentum of the exchanged gluon k^+ is of $\mathcal{O}(1/\nu)$; thus rescattering contributes at leading twist even in the light-cone gauge. The net result is gauge invariant and is identical to the color dipole model calculation. The calculation of the rescattering effects on DIS in Feynman and light-cone gauge through three loops is given in detail for an Abelian model in ref. [30]. The result shows that the rescattering corrections reduce the magnitude of the DIS cross-section in analogy to nuclear shadowing.

A new understanding of the role of final-state interactions in deep inelastic scattering has thus emerged. The multiple scattering of the struck parton via instantaneous interactions in the target generates dominantly imaginary diffractive amplitudes, giving rise to an effective "hard pomeron" exchange. The presence of a rapidity gap between the target and diffractive system requires that the target remnant emerges in a color singlet state; this is made possible in any gauge by the soft rescattering. The resulting diffractive contributions leave the target intact and do not resolve its quark structure; thus there are contributions to the DIS structure functions which cannot be interpreted as parton probabilities [30]; the leading-twist contribution to DIS from rescattering of a quark in the target is a coherent effect which is not included in the light-front wave functions computed in isolation. One can augment the light-front wave functions with a gauge link corresponding to an external field created by the virtual photon $q\bar{q}$ pair current [23,21]. Such a gauge link is process dependent [24], so the resulting augmented LFWFs are not universal [23,30,31]. We also note that the shadowing of nuclear structure functions is due to the destructive interference between multi-nucleon amplitudes involving diffractive DIS and on-shell intermediate states with a complex phase. In contrast, the wave function of a stable target is strictly real since it does not have on-energy-shell intermediate-state configurations. The physics of rescattering and shadowing is thus not included in the nuclear light-front wave functions, and a probabilistic interpretation of the nuclear DIS cross-section is precluded.

Rikard Enberg, Paul Hoyer, Gunnar Ingelman and I [32] have shown that the quark structure function of the effective hard pomeron has the same form as the quark contribution of the gluon structure function. The hard pomeron is not an intrinsic part of the proton; rather it must be considered as a dynamical effect of the lepton-proton interaction. Our QCD-based picture also applies to diffraction in hadron-initiated processes. The rescattering is different in virtual photon- and hadron-induced processes due to the different color environment, which accounts for the observed nonuniversality of diffractive parton distributions. This framework also provides a theoretical basis for the phenomenologically successful soft color interaction (SCI) model [33] which includes rescattering effects and thus generates a variety of final states with rapidity gaps.

The phase structure of hadron matrix elements is thus an essential feature of hadron dynamics. Although the LFWFs are real for a stable hadron, they acquire phases from initial-state and final-state interactions. In addition, the violation of CP invariance leads to a specific phase structure of the LFWFs. Dae Sung Hwang, Susan Gardner and I [34] have shown that this in turn leads to the electric dipole moment of the hadron and a general relation between the edm and anomalous magnetic moment Fock state by Fock state.

There are also leading-twist diffractive contributions $\gamma^* N_1 \to (q\bar{q})N_1$ arising from Reggeon exchanges in the t-channel [35]. For example, isospin-nonsinglet $C = +$ Reggeons contribute to the difference of proton and neutron structure functions, giving the characteristic Kuti-Weisskopf $F_{2p} - F_{2n} \sim x^{1-\alpha_R(0)} \sim x^{0.5}$ behavior at small x. The x-dependence of the structure functions reflects the Regge behavior $\nu^{\alpha_R(0)}$ of the virtual Compton amplitude at fixed Q^2 and $t = 0$. The phase of the diffractive amplitude is determined by analyticity and crossing to be proportional to $-1 + i$ for $\alpha_R = 0.5$, which together with the phase from the Glauber cut, leads to *constructive* interference of the diffractive and nondiffractive multi-step nuclear amplitudes. Furthermore, because of its x-dependence, the nuclear structure function is enhanced precisely in the domain $0.1 < x < 0.2$, where antishadowing is empirically observed. The strength of the Reggeon amplitudes is fixed by the fits to the nucleon structure functions, so there is little model dependence. Ivan Schmidt, Jian-Jun Yang, and I [36] have applied this analysis to the shadowing and antishadowing of all of the electroweak structure functions. Quarks of different flavors will couple to different Reggeons; this leads to the remarkable prediction that nuclear antishadowing is not universal; it depends on the quantum numbers of the struck quark. This picture leads to substantially different antishadowing for charged and neutral current reactions, thus affecting the extraction of the weak-mixing angle θ_W. We find that part of the anomalous NuTeV result [37] for θ_W could be due to the nonuniversality of nuclear antishadowing for charged and neutral currents. Detailed measurements of the nuclear dependence of individual quark structure functions are thus needed to establish the distinctive phenomenology of shadowing and antishadowing and to make the NuTeV results definitive. Antishadowing can also depend on the target and beam polarization.

3 The conformal approximation to QCD

One of the most interesting recent developments in hadron physics has been the use of Anti-de Sitter space holographic methods in order to obtain a first approximation to nonperturbative QCD. The essential principle underlying the AdS/CFT approach to conformal gauge theories is the isomorphism of the group of Poincaré and conformal transformations $SO(2,4)$ to the group of isometries of Anti-de Sitter space. The AdS metric is

$$\mathrm{d}s^2 = \frac{R^2}{z^2}(\eta^{\mu\nu}\mathrm{d}x_\mu \mathrm{d}x^\mu - \mathrm{d}z^2),$$

which is invariant under scale changes of the coordinate in the fifth dimension $z \to \lambda z$ and $\mathrm{d}x_\mu \to \lambda \mathrm{d}x_\mu$. Thus one can match scale transformations of the theory in $3+1$ physical space-time to scale transformations in the fifth dimension z. The amplitude $\phi(z)$ represents the extension of the hadron into the fifth dimension. The behavior of $\phi(z) \to z^\Delta$ at $z \to 0$ must match the twist dimension of the hadron at short distances $x^2 \to 0$. As shown by Polchinski and Strassler [38], one can simulate confinement by imposing the condition $\phi(z = z_0 = \frac{1}{\Lambda_{QCD}})$. This approach, has been successful in reproducing general properties of scattering processes of QCD bound states [38, 39], the low-lying hadron spectra [40,41], hadron couplings and chiral symmetry breaking [41,42], quark potentials in confining backgrounds [43] and pomeron physics [44].

It was originally believed that the AdS/CFT mathematical tool could only be applicable to strictly conformal theories such as $\mathcal{N} = 4$ supersymmetry. However, if one considers a semi-classical approximation to QCD with massless quarks and without particle creation or absorption, then the resulting β-function is zero, the coupling is constant, and the approximate theory is scale and conformal invariant. Conformal symmetry is of course broken in physical QCD; nevertheless, one can use conformal symmetry as a *template*, systematically correcting for its nonzero β-function as well as higher-twist effects. For example, "commensurate scale relations" [45] which relate QCD observables to each other, such as the generalized Crewther relation [46], have no renormalization scale or scheme ambiguity and retain a convergent perturbative structure which reflects the underlying conformal symmetry of the classical theory. In general, the scale is set such that one has the correct analytic behavior at the heavy-particle thresholds [47].

In a confining theory where gluons have an effective mass, all vacuum polarization corrections to the gluon self-energy decouple at long wavelength. Theoretical [48] and phenomenological [49] evidence is in fact accumulating that QCD couplings based on physical observables such as τ decay [50] become constant at small virtuality; *i.e.*, effective charges develop an infrared fixed point in contradiction to the usual assumption of singular growth in the infrared. The near-constant behavior of effective couplings also suggests that QCD can be approximated as a conformal theory even at relatively small momentum transfer. The importance of using an analytic effective charge [51] such as the pinch scheme [52,53] for unifying the electroweak and strong couplings and forces is also important [54]. Thus, conformal symmetry is a useful first approximant even for physical QCD.

4 Hadronic spectra in AdS/QCD

Guy de Téramond and I [1,40] have recently shown how a holographic model based on truncated AdS space can be used to obtain the hadronic spectrum of light quark $q\bar{q}, qqq$ and gg bound states. Specific hadrons are identified by the correspondence of the amplitude in the fifth dimension with the twist dimension of the interpolating operator for the hadron's valence Fock state, including its orbital angular-momentum excitations. An interesting aspect of our approach is to show that the mass parameter μR, which appears in the string theory in the fifth dimension, is quantized, and that it appears as a Casimir constant governing the orbital angular momentum of the hadronic constituents analogous to $L(L+1)$ in the radial Schrödinger equation.

As an example, the set of three-quark baryons with spin 1/2 and higher is described in AdS/CFT by the Dirac equation in the fifth dimension [1]:

$$\left[z^2 \, \partial_z^2 - 3z \, \partial_z + z^2 \mathcal{M}^2 - \mathcal{L}_\pm^2 + 4 \right] \psi_\pm(z) = 0. \quad (1)$$

The constants $\mathcal{L}_+ = L + 1$, $\mathcal{L}_- = L + 2$ in this equation are Casimir constants which are determined to match the twist dimension of the solutions with arbitrary relative

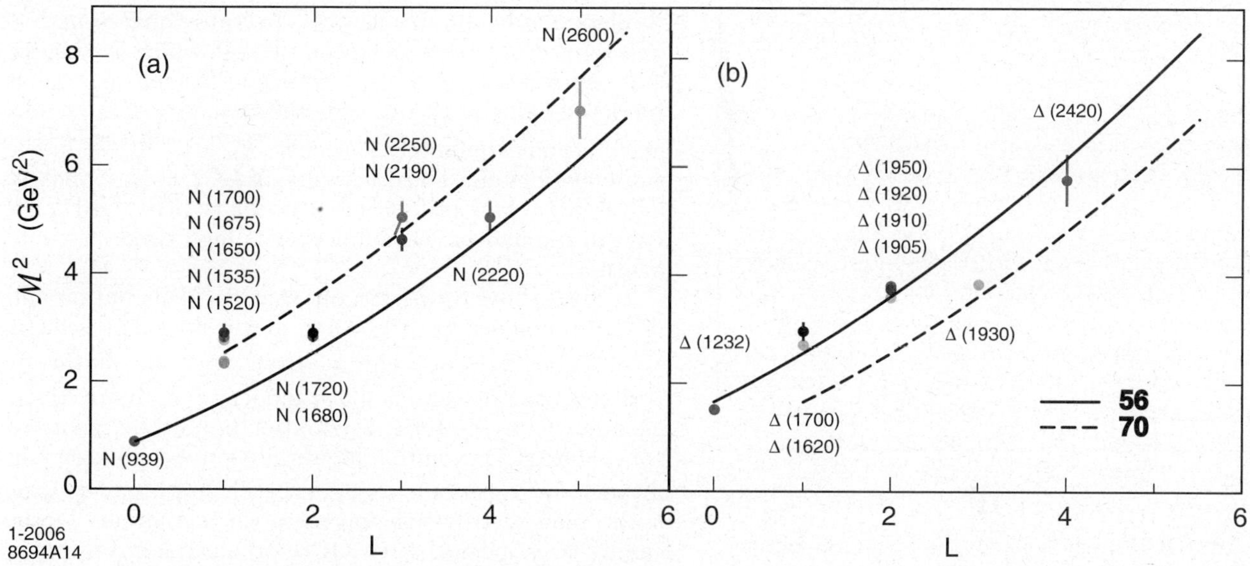

Fig. 2. Predictions for the light-baryon orbital spectrum for $\Lambda_{QCD} = 0.25$ GeV. The 56 trajectory corresponds to L even $P = +$ states, and the 70 to L odd $P = -$ states.

orbital angular momentum. The solution is

$$\Psi(x,z) = Ce^{-iP \cdot x} \left[\psi(z)_+ \, u_+(P) + \psi(z)_- \, u_-(P)\right], \quad (2)$$

with $\psi_+(z) = z^2 J_{1+L}(z\mathcal{M})$ and $\psi_-(z) = z^2 J_{2+L}(z\mathcal{M})$. The physical string solutions have plane waves and chiral spinors $u(P)_\pm$ along the Poincaré coordinates and hadronic invariant-mass states given by $P_\mu P^\mu = \mathcal{M}^2$. A discrete four-dimensional spectrum follows when we impose the boundary condition $\psi_\pm(z = 1/\Lambda_{QCD}) = 0$. One has $\mathcal{M}^+_{\alpha,k} = \beta_{\alpha,k}\Lambda_{QCD}$, $\mathcal{M}^-_{\alpha,k} = \beta_{\alpha+1,k}\Lambda_{QCD}$, with a scale-independent mass ratio [40]. The $\beta_{\alpha,k}$ are the first zeros of the Bessel eigenfunctions.

Figure 2(a) shows the predicted orbital spectrum of the nucleon states and fig. 2(b) the Δ orbital resonances. The spin-3/2 trajectories are determined from the corresponding Rarita-Schwinger equation. The data for the baryon spectra are from S. Eidelman *et al.* [55]. The internal parity of states is determined from the $SU(6)$ spin-flavor symmetry.

Since only one parameter, the QCD mass scale Λ_{QCD}, is introduced, the agreement with the pattern of physical states is remarkable. In particular, the ratio of Δ to nucleon trajectories is determined by the ratio of zeros of Bessel functions. The predicted mass spectrum in the truncated space model is linear, $M \propto L$, at high orbital angular momentum, in contrast to the quadratic dependence $M^2 \propto L$ in the usual Regge parametrization.

Our approach shows that there is an exact correspondence between the fifth-dimensional coordinate of anti-de Sitter space z and a specific impact variable ζ in the light-front formalism which measures the separation of the constituents within the hadron in ordinary space-time. The amplitude $\Phi(z)$ describing the hadronic state in AdS_5 can be precisely mapped to the light-front wave functions $\psi_{n/h}$ of hadrons in physical space-time [1], thus providing a relativistic description of hadrons in QCD at the amplitude

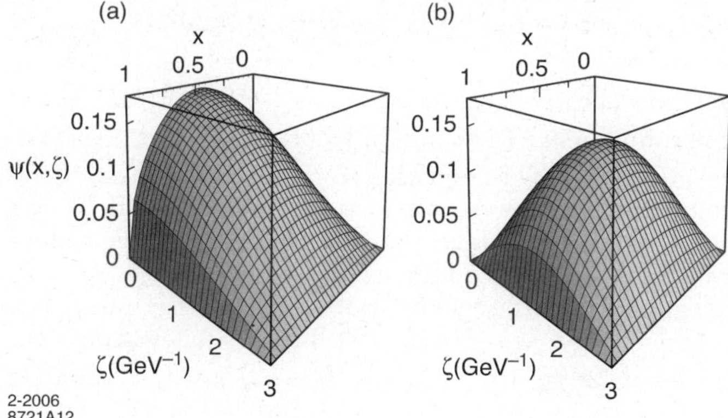

Fig. 3. AdS/QCD predictions for the $L = 0$ and $L = 1$ LFWFs of a meson.

level. We derived this correspondence by noticing that the mapping of $z \to \zeta$ analytically transforms the expression for the form factors in AdS/CFT to the exact Drell-Yan-West expression in terms of light-front wave functions. In the case of a two-parton constituent bound state the correspondence between the string amplitude $\Phi(z)$ and the light-front wave function $\widetilde{\psi}(x, \mathbf{b})$ is expressed in closed form [1]:

$$\left|\widetilde{\psi}(x,\zeta)\right|^2 = \frac{R^3}{2\pi} \, x(1-x) \, e^{3A(\zeta)} \, \frac{|\Phi(\zeta)|^2}{\zeta^4}, \qquad (3)$$

where $\zeta^2 = x(1-x)\mathbf{b}_\perp^2$. Here $b_\perp$ is the impact separation and Fourier conjugate to $k_\perp$. The variable ζ, $0 \leq \zeta \leq \Lambda_{QCD}^{-1}$, represents the invariant separation between point-like constituents, and it is also the holographic variable z in AdS; *i.e.*, we can identify $\zeta = z$. The prediction for the meson light-front wave function is shown in fig. 3. We can also transform the equation of motion in the fifth dimension using the z to ζ mapping to obtain an effective

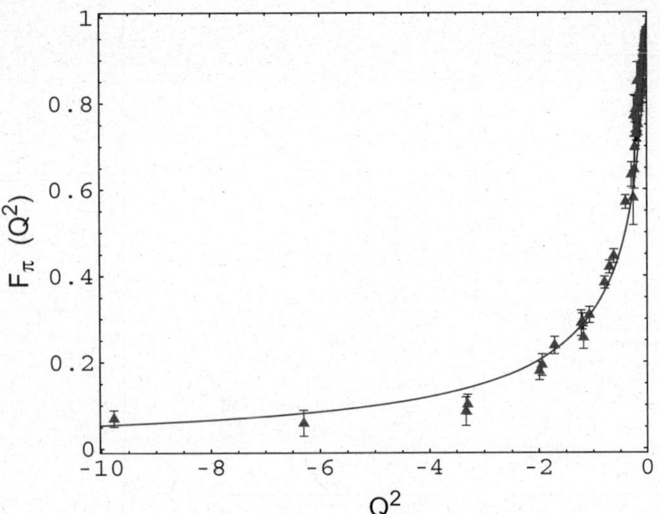

Fig. 4. AdS/QCD predictions for the pion form factor.

two-particle light-front radial equation

$$\left[-\frac{d^2}{d\zeta^2} + V(\zeta)\right]\phi(\zeta) = \mathcal{M}^2\phi(\zeta), \tag{4}$$

with the effective potential $V(\zeta) \to -(1-4L^2)/4\zeta^2$ in the conformal limit. The solution to (4) is $\phi(z) = z^{-\frac{3}{2}}\Phi(z) = Cz^{\frac{1}{2}}J_L(z\mathcal{M})$. This equation reproduces the AdS/CFT solutions. The lowest stable state is determined by the Breitenlohner-Freedman bound [56] and its eigenvalues by the boundary conditions at $\phi(z = 1/\Lambda_{QCD}) = 0$ and given in terms of the roots of the Bessel functions: $\mathcal{M}_{L,k} = \beta_{L,k}\Lambda_{QCD}$. Normalized LFWFs follow from (3):

$$\widetilde{\psi}_{L,k}(x,\zeta) = B_{L,k}\sqrt{x(1-x)}J_L\left(\zeta\beta_{L,k}\Lambda_{QCD}\right)$$
$$\times\theta(z \leq \Lambda_{QCD}^{-1}), \tag{5}$$

where $B_{L,k} = \pi^{-\frac{1}{2}}\Lambda_{QCD}J_{1+L}(\beta_{L,k})$. The resulting wave functions (see fig. 3) display confinement at large inter-quark separation and conformal symmetry at short distances, reproducing dimensional counting rules for hard exclusive processes in agreement with perturbative QCD.

The hadron form factors can be predicted from over-lap integrals in AdS space or equivalently by using the Drell-Yan-West formula in physical space-time. The prediction for the pion form factor is shown in fig. 4. The form factor at high Q^2 receives contributions from small ζ, corresponding to small $\vec{b}_\perp = \mathcal{O}(1/Q)$ (high relative $\vec{k}_\perp = \mathcal{O}(Q)$) as well as $x \to 1$. The AdS/CFT dynamics is thus distinct from endpoint models [57] in which the LFWF is evaluated solely at small transverse momentum or large impact separation.

The $x \to 1$ endpoint domain is often referred to as a "soft" Feynman contribution. In fact $x \to 1$ for the struck quark requires that all of the spectators have $x = k^+/P^+ = (k^0 + k^z)/P^+ \to 0$; this in turn requires high longitudinal momenta $k^z \to -\infty$ for all spectators

—unless one has both massless spectator quarks $m \equiv 0$ with zero transverse momentum $k_\perp \equiv 0$, which is a regime of measure zero. If one uses a covariant formalism, such as the Bethe-Salpeter theory, then the virtuality of the struck quark becomes infinitely spacelike: $k_F^2 \sim -\frac{k_\perp^2 + m^2}{1-x}$ in the endpoint domain. Thus, actually, $x \to 1$ corresponds to high relative longitudinal momentum; it is as hard a domain in the hadron wave function as high transverse momentum.

It is also interesting to note that the distribution amplitude predicted by AdS/CFT at the hadronic scale is $\phi_\pi(x, Q) = \frac{4}{\sqrt{3}\pi}f_\pi\sqrt{x(1-x)}$ from both the harmonic oscillator and truncated space models; it is quite different than the asymptotic distribution amplitude predicted from the PQCD evolution [6] of the pion distribution amplitude $\phi_\pi(x, Q \to \infty) = \sqrt{3}f_\pi x(1-x)$. The broader shape of the pion distribution increases the magnitude of the leading-twist perturbative QCD prediction for the pion form factor by a factor of $16/9$ compared to the prediction based on the asymptotic form, bringing the PQCD prediction close to the empirical pion form factor [58].

Since they are complete and orthonormal, the AdS/CFT model wave functions can be used as an initial ansatz for a variational treatment or as a basis for the diagonalization of the light-front QCD Hamiltonian. We are now in fact investigating this possibility with J. Vary and A. Harindranath. The wave functions predicted by AdS/QCD have many phenomenological applications ranging from exclusive B and D decays, deeply virtual Compton scattering and exclusive reactions such as form factors, two-photon processes, and two-body scattering. A connection between the theories and tools used in string theory and the fundamental constituents of matter, quarks and gluons, has thus been found.

The application of AdS/CFT to QCD phenomenology is now being developed in many new directions, incorporating finite quark masses, chiral symmetry breaking, asymptotic freedom, and finite-temperature effects. Some recent papers are given in refs. [59–68].

Work supported by the Department of Energy under contract number DE-AC02-76SF00515. The AdS/CFT results reported here were done in collaboration with Guy de Téramond.

References

1. S.J. Brodsky, G.F. de Teramond, Phys. Rev. Lett. **96**, 201601 (2006) [arXiv:hep-ph/0602252].
2. P.A.M. Dirac, Rev. Mod. Phys. **21**, 392 (1949).
3. S.J. Brodsky, H.C. Pauli, S.S. Pinsky, Phys. Rep. **301**, 299 (1998) [arXiv:hep-ph/9705477].
4. S.J. Brodsky, D.S. Hwang, Nucl. Phys. B **543**, 239 (1999) [arXiv:hep-ph/9806358].
5. S.J. Brodsky, M. Diehl, D.S. Hwang, Nucl. Phys. B **596**, 99 (2001) [arXiv:hep-ph/0009254].
6. G.P. Lepage, S.J. Brodsky, Phys. Lett. B **87**, 359 (1979).
7. A.V. Efremov, A.V. Radyushkin, Phys. Lett. B **94**, 245 (1980).
8. G.P. Lepage, S.J. Brodsky, Phys. Rev. D **22**, 2157 (1980).

9. S.J. Brodsky, G.P. Lepage, SLAC-PUB-2294, talk presented at the *Workshop on Current Topics in High Energy Physics, Cal Tech., Pasadena, CA, Feb. 13-17, 1979.*

10. S.J. Brodsky, arXiv:hep-ph/0004211.

11. S.J. Brodsky, C.R. Ji, G.P. Lepage, Phys. Rev. Lett. **51**, 83 (1983).

12. S.J. Brodsky, D.S. Hwang, I. Schmidt, Phys. Lett. B **530**, 99 (2002) [arXiv:hep-ph/0201296].

13. M. Burkardt, Nucl. Phys. Proc. Suppl. **141**, 86 (2005) [arXiv:hep-ph/0408009].

14. HERMES Collaboration (A. Airapetian *et al.*), Phys. Rev. Lett. **94**, 012002 (2005) [arXiv:hep-ex/0408013].

15. CLAS Collaboration (H. Avakian, L. Elouadrhiri), AIP Conf. Proc. **698**, 612 (2004).

16. M. Burkardt, Int. J. Mod. Phys. A **21**, 926 (2006) [arXiv:hep-ph/0509316].

17. X.d. Ji, Phys. Rev. Lett. **91**, 062001 (2003) [arXiv:hep-ph/0304037].

18. S.J. Brodsky, D. Chakrabarti, A. Harindranath, A. Mukherjee, J.P. Vary, arXiv:hep-ph/0604262.

19. P. Hoyer, arXiv:hep-ph/0608295.

20. S.J. Brodsky, S. Gardner, arXiv:hep-ph/0608219.

21. J.C. Collins, A. Metz, Phys. Rev. Lett. **93**, 252001 (2004) [arXiv:hep-ph/0408249].

22. X.d. Ji, F. Yuan, Phys. Lett. B **543**, 66 (2002) [arXiv:hep-ph/0206057].

23. A.V. Belitsky, X. Ji, F. Yuan, Nucl. Phys. B **656**, 165 (2003) [arXiv:hep-ph/0208038].

24. J.C. Collins, Phys. Lett. B **536**, 43 (2002) [arXiv:hep-ph/0204004].

25. S.J. Brodsky, D.S. Hwang, I. Schmidt, Nucl. Phys. B **642**, 344 (2002) [arXiv:hep-ph/0206259].

26. D. Boer, S.J. Brodsky, D.S. Hwang, Phys. Rev. D **67**, 054003 (2003) [arXiv:hep-ph/0211110].

27. H1 Collaboration (C. Adloff *et al.*), Z. Phys. C **76**, 613 (1997) [arXiv:hep-ex/9708016].

28. ZEUS Collaboration (J. Breitweg *et al.*), Eur. Phys. J. C **6**, 43 (1999) [arXiv:hep-ex/9807010].

29. S.J. Brodsky, L. Frankfurt, J.F. Gunion, A.H. Mueller, M. Strikman, Phys. Rev. D **50**, 3134 (1994) [arXiv:hep-ph/9402283].

30. S.J. Brodsky, P. Hoyer, N. Marchal, S. Peigne, F. Sannino, Phys. Rev. D **65**, 114025 (2002) [arXiv:hep-ph/0104291].

31. J.C. Collins, Acta Phys. Pol. B **34**, 3103 (2003) [arXiv:hep-ph/0304122].

32. S.J. Brodsky, R. Enberg, P. Hoyer, G. Ingelman, Phys. Rev. D **71**, 074020 (2005) [arXiv:hep-ph/0409119].

33. A. Edin, G. Ingelman, J. Rathsman, Phys. Lett. B **366**, 371 (1996) [arXiv:hep-ph/9508386].

34. S.J. Brodsky, S. Gardner, D.S. Hwang, Phys. Rev. D **73**, 036007 (2006) [arXiv:hep-ph/0601037].

35. S.J. Brodsky, H.J. Lu, Phys. Rev. Lett. **64**, 1342 (1990).

36. S.J. Brodsky, I. Schmidt, J.J. Yang, Phys. Rev. D **70**, 116003 (2004) [arXiv:hep-ph/0409279].

37. NuTeV Collaboration (G.P. Zeller *et al.*), Phys. Rev. Lett. **88**, 091802 (2002); **90**, 239902 (2003)(E) [arXiv:hep-ex/0110059].

38. J. Polchinski, M.J. Strassler, Phys. Rev. Lett. **88**, 031601 (2002) [arXiv:hep-th/0109174].

39. S.J. Brodsky, G.F. de Teramond, Phys. Lett. B **582**, 211 (2004) [arXiv:hep-th/0310227].

40. G.F. de Teramond, S.J. Brodsky, Phys. Rev. Lett. **94**, 201601 (2005) [arXiv:hep-th/0501022].

41. J. Erlich, E. Katz, D.T. Son, M.A. Stephanov, Phys. Rev. Lett. **95**, 261602 (2005) [arXiv:hep-ph/0501128].

42. S. Hong, S. Yoon, M.J. Strassler, arXiv:hep-ph/0501197.

43. H. Boschi-Filho, N.R.F. Braga, C.N. Ferreira, Phys. Rev. D **73**, 106006 (2006) [arXiv:hep-th/0512295].

44. R.C. Brower, J. Polchinski, M.J. Strassler, C.I. Tan, arXiv:hep-th/0603115.

45. S.J. Brodsky, H.J. Lu, Phys. Rev. D **51**, 3652 (1995) [arXiv:hep-ph/9405218].

46. S.J. Brodsky, G.T. Gabadadze, A.L. Kataev, H.J. Lu, Phys. Lett. B **372**, 133 (1996) [arXiv:hep-ph/9512367].

47. S.J. Brodsky, G.P. Lepage, P.B. Mackenzie, Phys. Rev. D **28**, 228 (1983).

48. R. Alkofer, C.S. Fischer, F.J. Llanes-Estrada, Phys. Lett. B **611**, 279 (2005) [arXiv:hep-th/0412330].

49. S.J. Brodsky, S. Menke, C. Merino, J. Rathsman, Phys. Rev. D **67**, 055008 (2003) [arXiv:hep-ph/0212078].

50. S.J. Brodsky, J.R. Pelaez, N. Toumbas, Phys. Rev. D **60**, 037501 (1999) [arXiv:hep-ph/9810424].

51. S.J. Brodsky, M.S. Gill, M. Melles, J. Rathsman, Phys. Rev. D **58**, 116006 (1998) [arXiv:hep-ph/9801330].

52. M. Binger, S.J. Brodsky, arXiv:hep-ph/0602199.

53. J.M. Cornwall, J. Papavassiliou, Phys. Rev. D **40**, 3474 (1989).

54. M. Binger, S.J. Brodsky, Phys. Rev. D **69**, 095007 (2004) [arXiv:hep-ph/0310322].

55. Particle Data Group (S. Eidelman *et al.*), Phys. Lett. B **592**, 1 (2004).

56. P. Breitenlohner, D.Z. Freedman, Ann. Phys. (N.Y.) **144**, 249 (1982).

57. A.V. Radyushkin, arXiv:hep-ph/0605116.

58. H.M. Choi, C.R. Ji, arXiv:hep-ph/0608148.

59. N. Evans, A. Tedder, arXiv:hep-ph/0609112.

60. Y. Kim, S.J. Sin, K.H. Jo, H.K. Lee, arXiv:hep-ph/0609008.

61. C. Csaki, M. Reece, arXiv:hep-ph/0608266.

62. J. Erdmenger, N. Evans, J. Grosse, arXiv:hep-th/0605241.

63. H. Boschi-Filho, N.R.F. Braga, arXiv:hep-th/0604091.

64. N. Evans, T. Waterson, arXiv:hep-ph/0603249.

65. M. Harada, S. Matsuzaki, K. Yamawaki, arXiv:hep-ph/0603248.

66. G.T. Horowitz, J. Polchinski, arXiv:gr-qc/0602037.

67. I.R. Klebanov, Int. J. Mod. Phys. A **21**, 1831 (2006) [arXiv:hep-ph/0509087].

68. E. Shuryak, arXiv:hep-th/0605219.

Measurement of ΔS in the proton

H. E. Jackson on behalf of the HERMES Collaboration

Argonne National Laboratory, Argonne, IL 60439, USA

Abstract. The helicity density of the strange sea in the proton has been measured in an "isoscalar" extraction from semi-inclusive deep-inelastic scattering on the deuteron. In the region of measurement, x > 0.02, the helicity density is zero within experimental error. The first moment of the strange quark helicity density over the measured region is determined to be 0.006±0.029(stat.)±0.007(sys.). The corresponding moment of the octet axial charge combination is sustantially less than the total axial charge extracted from hyperon semi-leptonic decays.

PACS. 13.88.+e Polarization in interactions and scattering – 14.65.Bt Light quarks

1 Introduction

The helicity distribution of the strange quark sea is of great interest as a probe of the spin properties of the quark sea in the nucleon. Its features reflect the reaction mechanism that is responsible for the formation of the quark sea. Because strange quarks carry no isospin, the total strange quark helicity density $\Delta S(x) \equiv \Delta s(x) + \Delta \bar{s}(x)$ can be extracted from polarization measurements in scattering on the deuteron alone which is isoscalar. Measurements of the inclusive double spin asymmetries provide an estimate of the helicity density $\Delta Q(x) \equiv \Delta u(x) + \Delta \bar{u}(x) + \Delta d(x) + \Delta \bar{d}(x)$ of the nonstrange sea. Using the spin asymmetries measured for the charged kaons as the second experimental data set, it is possible to extract $\Delta S(x)$. The fragmentation functions needed for the extraction process can be obtained without resorting to other experiments by measuring the charged kaon multiplicities with the same data set used for the helicity extraction. Aside from that of isospin symmetry between the proton and the neutron, the only assumption required in the analysis is charged-conjugation invariance in the fragmentation process. A precise "isoscalar" extraction of $\Delta S(x)$ using semi-inclusive DIS on the deuteron at HERMES [1] is presented here.

2 The experiment

In leading order (LO), the virtual photon double-spin asymmetry A_1^K in semi-inclusive production of a kaon is given by

$$A_1^K(x,z) = \frac{\sigma_{1/2}^K - \sigma_{3/2}^K}{\sigma_{1/2}^K + \sigma_{3/2}^K} = \frac{\sum_q e_q^2 \, \Delta q(x) \, D_q^K(z)}{\sum_q e_q^2 \, q(x) \, D_q^K(z)} . \quad (1)$$

The asymmetries in the analysis reported here were record ed by the HERMES experiment using a longitudinally polarized deuteron gas target internal to the E = 27.5 GeV HERA positron storage ring at DESY. The self-induced beam polarization is measured continuously with Compton backscattering of circularly polarized laser beams [2,3]. The open-ended target cell is fed by an atomic-beam source based on Stern-Gerlach separation [4] with hyperfine transitions. The nuclear polarization of the atoms is flipped at 90 s time intervals, while both this polarization and the atomic fraction inside the target cell are continuously measured [5,6]. Scattered beam leptons and coincident hadrons are detected by the HERMES spectrometer [7]. Leptons are identified with an efficiency exceeding 98% and a hadron contamination of less than 1% using an electromagnetic calorimeter, a transition-radiation detector, a preshower scintillation counter and a Čerenkov detector. Charged kaons are identified using a dual-radiator ring-imaging Čerenkov detector [8]. Events were selected subject to the kinematic requirements $Q^2 > 1\,\text{GeV}^2$, $W^2 > 10\,\text{GeV}^2$ and $y < 0.85$, where W is the invariant mass of the photon-nucleon system, and $y = \nu/E$. Coincident hadrons were accepted if $0.2 < z < 0.8$ and $x_F \approx 2p_L/W > 0.1$, where $z = E_K/\nu$ and p_L is the longitudinal momentum of the hadron with respect to the virtual photon direction in the photon-nucleon center of mass frame.

3 Charged-kaon multiplicities

For the deuteron, over a finite interval in z, Eq. 1 reduces to a simple form which reflects its isoscalar character

$$A_{1,d}^K(x) = \frac{\Delta Q(x) \int \mathcal{D}_Q^K(z)dz + \Delta S(x) \int \mathcal{D}_S^K(z)dz}{Q(x) \int \mathcal{D}_Q^K(z)dz + S(x) \int \mathcal{D}_S^K(z)dz} \quad (2)$$

where

$$\int \mathcal{D}_Q^K(z)dz = 4 \int D(z)_u^{K^\pm} dz + \int D(z)_d^{K^\pm} dz$$

$$\int \mathcal{D}_S^K(z)dz = 2 \int D(z)_s^{K^\pm} dz. \tag{3}$$

The corresponding inclusive asymmetry is given by

$$A_{1,d}(x) = \frac{5\Delta Q(x) + 2\Delta S(x)}{5Q(x) + 2S(x)}. \tag{4}$$

The fragmentation functions of Eq. 1 have been obtained directly from the same polarized DIS data set for the deuteron that has been used to extract the double spin asymmetries by averaging the two spin states. Assuming only charge conjugation invariance, i.e.

$$
\begin{aligned}
D_u^{K^+ + K^-} &= D_{\bar{u}}^{K^+ + K^-} \\
D_d^{K^+ + K^-} &= D_{\bar{d}}^{K^+ + K^-} \\
D_s^{K^+ + K^-} &= D_{\bar{s}}^{K^+ + K^-},
\end{aligned}
\tag{5}
$$

in leading order for charged kaon production in semi-inclusive DIS on the deuteron

$$\frac{dN^K(x)/dx}{dN^{dis}(x)/dx} = \frac{Q(x)\int \mathcal{D}_Q^K(z)dz + S(x)\int \mathcal{D}_S^K(z)dz}{5Q(x) + 2S(x)}. \tag{6}$$

The strange and non-strange fragmentation functions can be extracted by fitting the x dependence of this ratio using Eq. 6 together with parton distributions taken from the CTEQ6 compilation [9]. A fit to the $K^\pm$ multiplicity extracted from the HERMES data using Eq. 6 with the limits of integration $0.2<z<0.8$ is presented in Fig. 1. The fit gives $\int_{0.2}^{0.8} \mathcal{D}_Q^K(z)dz = 0.41 \pm 0.02(stat.) \pm 0.03(sys.)$ and $\int_{0.2}^{0.8} \mathcal{D}_S^K(z)dz = 1.41 \pm 0.29(stat.) \pm 0.15(sys.)$.

4 Helicity distributions

The helicity distributions $\Delta Q(x)$ and $\Delta S(x)$ have been extracted directly from the measured values of $A_{1,d}(x)$ and $A_{1,d}^{K^\pm}(x)$ using the fragmentation functions defined in Sect. 3 and the parton distributions $Q(x)$ and $S(x)$ taken from the latest CTEQ6L compilation [9]. The resulting strange and non-strange helicity distributions weighted by Bjorken x are presented in Fig. 2. The non-strange helicity distribution is in excellent agreement with that derived from the published five-component flavor decomposition [1] of the proton helicities. While of much improved precision and free of the systematic uncertainties in the fragmentation functions, the strange helicity distribution also agrees well with the results reported therein, and is consistent with zero over the measured range.

The first moments of the measured distributions in the measured range of Bjorken x are given in Table 1. The first moment over the measured region of $\Delta S(x)$ is consistent with zero. Because of the very small density of

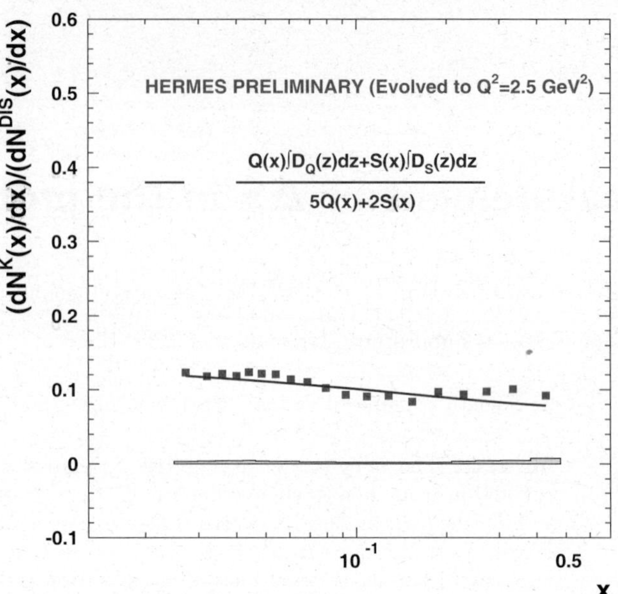

Fig. 1. Multiplicity in 4π of charged kaons in semi-inclusive DIS on a deuterium target as a function of Bjorken x. The statistical error bars are not visible, and the bands at the bottom represent the systematic uncertainties.

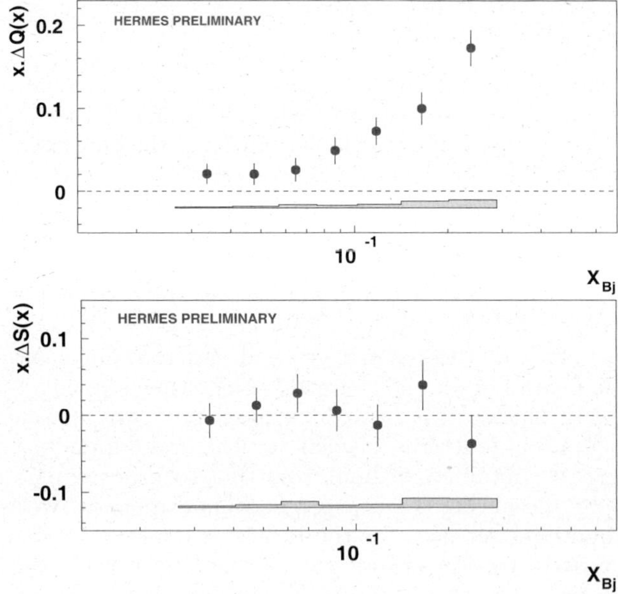

Fig. 2. Strange and non-strange quark helicity distributions at $\langle Q^2 \rangle = 2.5\,\mathrm{GeV}^2$ as a function of Bjorken x. The error bars are statistical, and the bands at the bottom represent the systematic uncertainties.

strange quarks above $x=0.3$ the contribution of any non-zero helicity density in this region is negligible compared to the systematic error in the measurement. Consequently, the value for ΔS can be safely taken as the moment over the Bjorken x range 0.02-1.0. The vanishing values reported recently [10] for $g_{1,d}(x)$ at lower values of x, suggest that any contribution to the first moment of $\Delta Q(x)$ below $x=0.02$ will be very small. This conclusion strongly contrasts those of the early inclusive measurements [11,12]

Table 1. First moments of various helicity distributions in the Bjorken range 0.02-0.6

	Moment in measured range
ΔQ	$0.286 \pm 0.026 \pm 0.011$
ΔS	$0.006 \pm 0.029 + 0.007$
Δq_8	$0.274 \pm 0.039 \pm 0.018$

which reported a substantial negative polarization for the strange quarks. It suggests that the violation of the Ellis-Jaffe sum rule observed in these inclusive measurements is not the result of a significant negative polarization of the strange sea.

It is also of interest to use the data on the first moments to test directly the assumption made in the analysis of inclusive DIS data, namely SU(3) symmetry in hyperon decays. The octet combination

$$\Delta q_8(x) = \Delta Q(x) - 2\Delta S(x) \qquad (7)$$

can be extracted from the first moments of $Q(x)$ and $S(x)$. The isoscalar result is included in Table 1. The result is to be compared with the value extracted under the assumption of SU(3) symmetry from the hyperon decay constants [13]

$$\Delta q_8 = (3F - D)C_{QCD} = 0.49. \qquad (8)$$

The difference from the isoscalar result of 0.274 must be found in the low x region x<0.02 if SU(3) symmetry is to be required.

In principle, an estimate of the integral of the strange quark helicity distribution at high Q^2 is also accessible from measurements of parity violation in elastic ep scattering. Such measurements provide estimates of the strange axial form factor at low momentum transfers. An estimate of the strange quark helicity distribution at high Q^2 follows from the relationship [14]

$$\Delta S = G_A^s(Q^2 = 0) = \int_0^1 \Delta S(x, Q^2 = \infty)dx. \qquad (9)$$

In a combined analysis of parity violation measurements in elastic ep scattering and of elastic νp scattering(see ref. [14]) $G_A^s(Q^2)$ has been extracted over the range 0.4 GeV^2<Q^2<1.0 GeV^2. While the data above 0.5 GeV^2 are consistent with zero the lowest measurement at Q^2=0.5 GeV^2 suggests that at very low momentum transfers the axial form factor may become increasingly negative as $Q^2 \to 0$. If an extrapolation of the axial form results do lead to a nonzero value, they will suggest that indeed, $\Delta S < 0$.

The semi-inclusive DIS result reported here appears to contradict that conclusion. The discrepency could be resolved by the onset of large values of $\Delta S(x,Q^2)$ at very low values of x_{bj}, $x \approx 10^{-2}$-10^{-3}. Alternatively, an "x=0" contribution to the singlet axial charge coming from a nonperturbative gluon topological contribution has been proposed [15]. Similarly, if SU(3) symmetry is preserved in hyperon decay, there must be missing strength in $\Delta q_8(x)$ in the low x region.

5 Conclusions

In conclusion, a direct measurement of the helicity density for strange quarks in the proton has been made using inclusive and semi-inclusive charged kaon spin asymmetries for a deuteron target. The helicity density for the strange quarks is consistent with zero in the measured range and the octet axial charge is observed to be substantially less than that extracted from hyperon decays under the assumption of SU(3) symmetry. Either earlier conclusions regarding $\Delta S(x,Q^2)$ are not correct, or some non-trivial features of $\Delta S(x,Q^2)$ at low x must emerge from experiment.

We gratefully acknowledge the DESY management for its support, the staff at DESY and the collaborating institutions for the significant effort, and our national funding agencies for their financial support.

References

1. A. Airapetian et al. (HERMES), Phys. Rev. D **71**, 012003 (2005)
2. D.P. Barber et al., Nucl. Inst. & Meth. **A 338**, 166 (1994)
3. M. Beckmann et al., Nucl. Inst. & Meth. **A 479**, 334 (2002)
4. F. Stock et al., Nucl. Inst. & Meth. **A 343**, 334 (1994)
5. C. Baumgarten et al., Nucl. Inst. & Meth. **A 482**, 606 (2002)
6. C. Baumgarten et al., Nucl. Inst. & Meth. **A 496**, 263 (2003)
7. K. Ackerstaff et al. (HERMES), Nucl. Inst. & Meth. **A 417**, 230 (1998)
8. N. Akopov et al., Nucl. Inst. & Meth. **A 479**, 511 (2002)
9. J. Pumplin et al., J. High Energy Phys. **7**, 12 (2002)
10. E.S. Ageev et al. (COMPASS), Phys. Lett. **B612**, 154 (2005)
11. J. Ashman et al. (EMC), Phys. Lett. **B206**, 364 (1988)
12. D. Adams et al. (SMC), Phys. Rev. D **56**, 5330 (1997)
13. P. Ratcliffe, Czech J. Phys. **54**, B11 (2004)
14. S. Pate et al., *in these proceedings* (2006)
15. S.D. Bass, Rev. Mod. Phys. **77**, 1257 (2005)

Eur. Phys. J. A **31**, 645–648 (2007)

DOI 10.1140/epja/i2006-10296-0

Special Article – QNP 2006

The E687 cryptoexotic vector meson candidate

R. Baldini[1,2], L. Benussi[2], M. Bertani[2], S. Bianco[2], F.L. Fabbri[2], S. Pacetti[2,a], and A. Zallo[2]

[1] Centro Studi e Ricerche Enrico Fermi, Roma, Italy
[2] Laboratori Nazionali di Frascati, INFN, Frascati, Italy

Received: 22 December 2006
Published online: 23 March 2007 – © Società Italiana di Fisica / Springer-Verlag 2007

Abstract. A new study of the dip observed at $1.9\,\mathrm{GeV}$ by the Fermilab experiment E687 in diffractive photoproduction of $3\pi^+3\pi^-$ is presented. The E687 and the BABAR data on the annihilation cross-section $\sigma(e^+e^- \to 3\pi^+3\pi^-)$, obtained with initial-state radiation, are fitted all together. The fit function is based on a simple mixing mechanism that explains the manifestation of a resonance as a dip. Possible interpretations in terms of hybrids and tetraquark states are considered.

PACS. 13.40.Hq Electromagnetic decays – 13.66.Bc Hadron production in e^-e^+ interactions – 25.20.Lj Photoproduction reactions

1 Introduction

In 2001 the E687 experiment at Fermilab confirmed [1,2], with a high statistical significance, the dip observed, in the $3\pi^+3\pi^-$ final state, with lower statistical significance, by the DM2 Collaboration [3] in 1998. A further confirmation, even if with larger width, has been provided by BABAR [4] (fig. 5), which observed such a dip, by means of initial-state radiation (ISR), not only in the $e^+e^- \to 3\pi^+3\pi^-$ channel but also in the $e^+e^- \to 2\pi^+2\pi^-2\pi^0$ one.

The DM2 and BABAR measurements give important hints on the quantum numbers of the dip, that, if interpreted as a resonance, has $J^{PC} = 1^{--}$ and, because of the six-pion decay channel, $G = +1$. Hence, as a resonance, the dip has the quantum numbers of a vector meson.

2 Photoproduction and e^+e^- annihilation

In principle it is quite difficult to obtain the spin and parity of a six-pion final state. However, supported also by other observations [5], if the transfer momentum between incident photon and nucleus target is small enough, the physical photon can be described as the superposition [6]

$$|\gamma\rangle \simeq \sqrt{1 - c^2\alpha_{em}}|\gamma_B\rangle + c\sqrt{\alpha_{em}}|h\rangle, \quad c = \mathcal{O}(1) \quad (1)$$

of a bare photon $|\gamma_B\rangle$, which does not interact with the hadrons, and a hadronic component $\sqrt{\alpha_{em}}|h\rangle$. The $|h\rangle$ state, which maintains the photon quantum numbers, in the low-energy regime $E_\gamma < 2\,\mathrm{GeV}$, can be described in

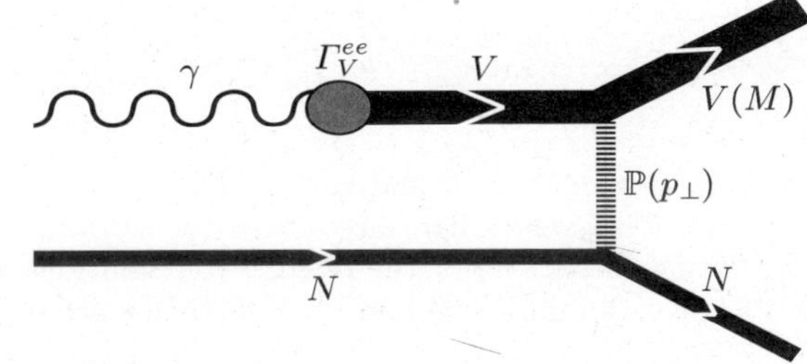

Fig. 1. Feynman diagram depicting the interaction, mediated by a pomeron $\mathbb{P}$, between the hadronic part of the incident photon and the nucleus target.

terms of vector meson dominance (VMD) [7] as [6]

$$c\sqrt{\alpha}|h\rangle = \sum_V^{\rho^0,\omega,\phi} \frac{e}{f_V}|V\rangle, \quad (2)$$

where f_V is the electromagnetic coupling to the meson V. By following the schematic diagram of fig. 1, the incident photon fluctuates (see eq. (1)) in a vector meson V (see eq. (2)) with a rate Γ_V^{ee} and then it is scattered, via a pomeron [8] exchange, by the nucleus target. The diffractive photoproduction cross-section (dPCS) for this process can be written as

$$\sigma^{\mathrm{diff}}_{\gamma N \to VN} \propto \Gamma_V^{ee} \cdot \sigma_{VN \to VN} \propto \int M^2 \sigma_{e^+e^- \to V}(M)\mathrm{d}M, \quad (3)$$

being $\Gamma_V^{ee} \propto \int M^2 \sigma_{e^+e^- \to V}(M)\mathrm{d}M$ and $\sigma_{VN \to VN}$ only weakly dependent on the produced hadronic mass M (fig. 1). Finally by differentiating eq. (3) with respect to

---

[a] e-mail: `simone.pacetti@lnf.infn.it`

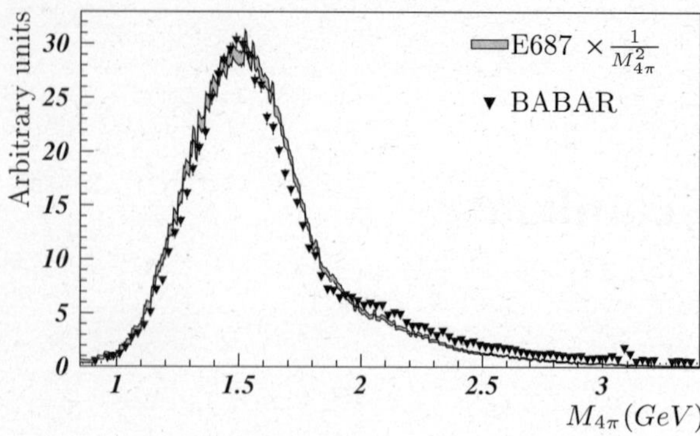

Fig. 2. BABAR (ISR) and E687 $2\pi^+2\pi^-$ invariant-mass distributions. The band represents the E687 data normalized to the BABAR cross-section via eq. (4).

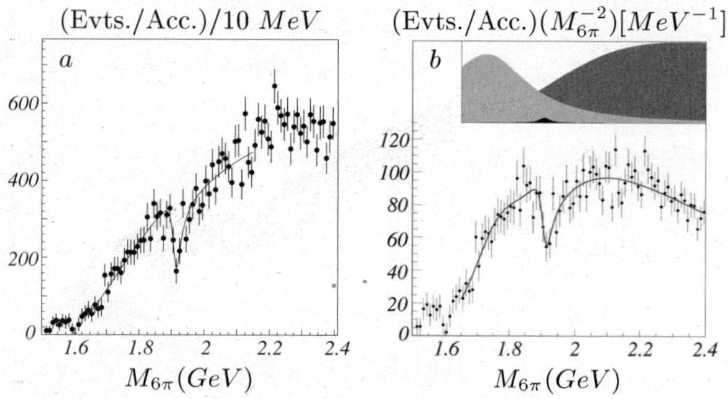

Fig. 3. Fits with one (a) [1] and two (b) [2] BWs interacting with a JS. In (b), where data are scaled by $M_{6\pi}^2$, the inset represents, in arbitrary units, the relative contributions of the two BWs (gray and dark gray) and the JS (black area).

M we get:

$$\frac{1}{M^2}\frac{\mathrm{d}\sigma^{\mathrm{diff}}_{\gamma N \to 6\pi N}}{\mathrm{d}M} \propto \sigma_{e^+e^- \to 6\pi}(M) \qquad (4)$$

and the proportionality between dPCS and annihilation cross-section is found. In fig. 2 the ISR BABAR data on the $e^+e^- \to 2\pi^+2\pi^-$ cross-section [9] are compared with the E687 dPCS for the same final state, scaled by $M_{4\pi}^2$, as prescribed in eq. (4). The good agreement in the low-energy region ($< 2\,\mathrm{GeV}$) verifies the claimed proportionality (eq. (4)). At high invariant masses the compatibility can be improved by considering a mild dependence on $M_{4\pi}$ of the elastic cross-section $\sigma_{VN \to VN}$.

3 Previous fit

There exist two previous analyses of the E687 data on $3\pi^+3\pi^-$ dPCS [1,2]. In both these cases, data have been described by means of Breit-Wigner formulae (BW) interacting with a real Jacob-Slansky (JS) continuum [10]. In fig. 3 these fits are shown with data superimposed, and in table 1 the parameters obtained for the BWs are reported.

Table 1. Fit parameters obtained in the two cases of refs. [1,2].

Ref.	Res.	Mass (MeV/c^2)	Width (MeV/c^2)	$\frac{B_{ee}B_{6\pi}}{M^2}$ ($\frac{\text{Yield}}{10\,\text{MeV}}$)	Phase (deg.)
[1]	V_0	1911 ± 4	29 ± 11	6 ± 1	62 ± 12
[2]	V_0	1910 ± 10	37 ± 13	5 ± 1	10 ± 30
	V_1	1710 ± 34	315 ± 100	17 ± 3	140 ± 10

Comparing these results we see that while the masses, found for the dip (V_0), are very similar, there is a discrepancy between the widths even if they are still compatible.

4 The mixing mechanism

If we assume that the structure V_0 is a resonance which can couple with both the e^+e^- initial state and the six-pion final state only through an intermediate broad vector meson V_1, then a mixing mechanism [11] to explain its appearing as a dip can be provided. We are considering mixing between the states V_0 and V_1 far from threshold, hence the mass-mixing formalism [12] may be used. In particular we adopt the 2×2 matrix propagator:

$$\mathcal{D}(s) = \frac{1}{\mathbb{M}^2 - s} + \sum_{j=0}^{1} a_j(s - m_0^2)^j e^{ib_j \sqrt{s-(6M_\pi)^2}}, \qquad (5)$$

where the mass matrix

$$\mathbb{M}^2 = \begin{bmatrix} m_0^2 - i\gamma_0 m_0 & \delta m^2 \\ \delta m^2 & m_1^2 - i\gamma_1 m_1 \end{bmatrix} \qquad (6)$$

is written in terms of "bare" masses and widths: m_i and γ_i ($i = 1, 2$) and the sum in eq. (5) is a Laurent series, around the pole V_0, up to the linear term, to account for the background. The non-diagonal term δm^2 induces the mixing and hence the coupling between the two states.

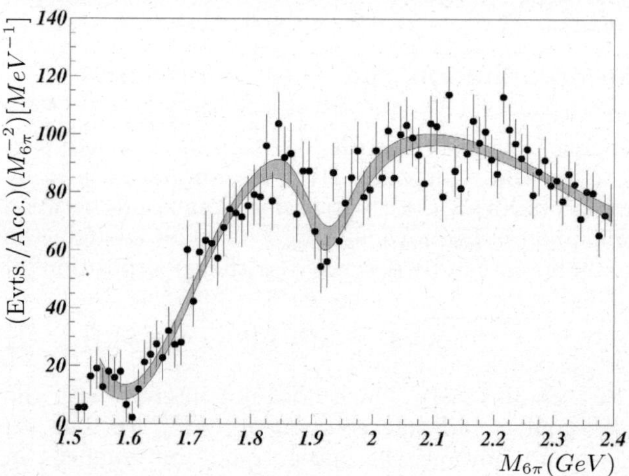

Fig. 4. Fit (gray band) of the E687 data scaled by $M_{6\pi}^2$.

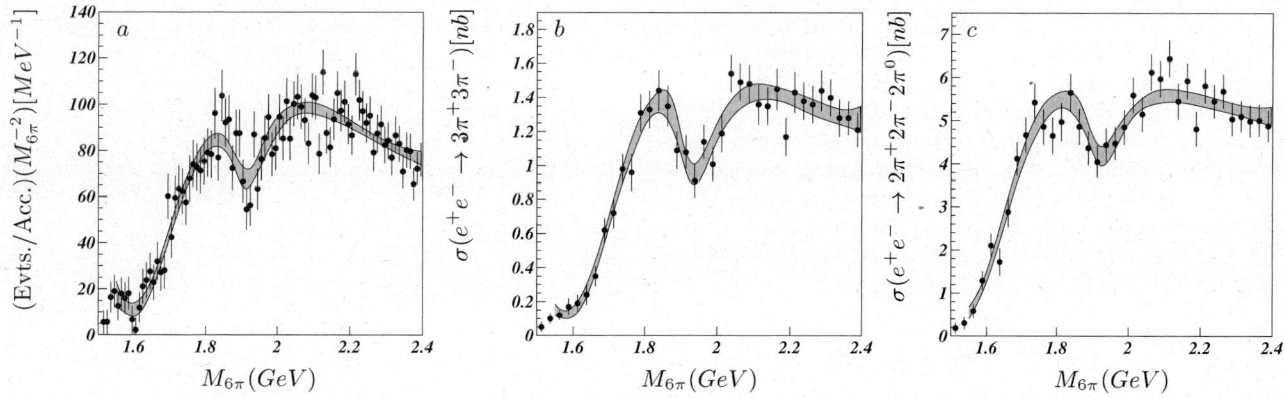

Fig. 5. Combined fit of E687 data scaled by $M_{6\pi}^2$ (a) and BABAR data on $3\pi^+3\pi^-$ (b) and $2\pi^+2\pi^-2\pi^0$ (c) final states.

Table 2. Best parameters for the fit shown in fig. 4. These values represent "real" masses and widths, obtained by diagonalizing the mass matrix of eq. (6).

Res.	Mass (MeV)	Width (MeV)	δm^2 (GeV^2)	χ^2
V_0	1905 ± 13	74 ± 44	0.17 ± 0.03	0.90
V_1	1540 ± 370	340 ± 70		

5 Fitting the E687 data alone

Figure 4 shows data and fit performed by using a function based on the propagator of eq. (5). In table 2 the obtained "real" masses and widths are reported. By comparing these parameters with those of the previous fits (table 1) we notice that the larger discrepancy concerns the V_0 width. It is about twice the previous values, even if, due to the large error, it is still compatible.

6 A combined fit

Since the larger width obtained for V_0 goes in the direction of reconciling the E687 and the ISR BABAR data in both $3\pi^+3\pi^-$ and $2\pi^+2\pi^-2\pi^0$, we perform a combined fit including these three sets of data. A matrix propagator (eq. (5)) with identical resonances, but with a free overall scale factor and independent background structures is used.

In table 3 the obtained parameters are reported and fig. 5 shows the three corresponding curves. Concerning the dip V_0, the mass is in agreement with previous estimates, while the width increases. In fact such a width is even higher than the previous fit, where only E687 data have been included. The broadening of the V_0 is an obvi-

ous consequence due to the BABAR data, which show a larger dip.

However, the good normalized $\chi^2 = 0.96$ allows to identify these dips as the same structure, that could be the manifestation of one or more interacting resonances.

7 Conclusions

A new fit procedure has been used to investigate the dip V_0, found at $\sim 1.9\,\text{GeV}$ by the E687 experiment. Such a procedure, based on a $V_0 - V_1$ mass-mixing mechanism, gives masses and widths, for the dip and the broad resonance V_1, in agreement with previous estimates. Nevertheless the central value of the V_0 width is increased by a factor of two.

We perform also a combined analysis, by considering the three sets of data: E687, BABAR ($3\pi^+3\pi^-$) and BABAR ($2\pi^+2\pi^-2\pi^0$). In this case, we obtain a larger width for V_0, which is incompatible with the previous result (only E687). However the good χ^2 seems to confirm that both E687 and BABAR are observing the same structure.

Concerning the possible interpretations, mass and decay modes are in agreement with many model-dependent predictions for vector meson cryptoexotic states [13].

In particular $J^{PC} = 1^{--}$ hybrids [$q\bar{q}g$] and tetraquark states [$qq\bar{q}\bar{q}$] seem favored because of their multi-particle decay channels.

We warmly thank the BABAR and E687 Collaborations for support and cooperation.

References

1. P.L. Frabetti *et al.*, Phys. Lett. B **514**, 240 (2001).
2. P.L. Frabetti *et al.*, Phys. Lett. B **578**, 290 (2004).
3. R. Baldini *et al.*, reported at the *FENICE Workshop, 1998, Frascati*.
4. B. Aubert *et al.*, Phys. Rev. D **73**, 052003 (2006).
5. E687 Collaboration (P. Lebrun), FERMILAB-CONF-97-387-E, talk presented at *Hadron 97, Upton, USA*.

Table 3. Best parameters for the combined fit shown in fig. 5.

Res.	Mass (MeV)	Width (MeV)	δm^2 (GeV^2)	χ^2
V_0	1925 ± 26	140 ± 50	0.16 ± 0.03	0.96
V_1	1620 ± 310	350 ± 40		

6. G.A. Schuler, T. Sjostrand, Nucl. Phys. B **407**, 539 (1993).
7. J.J. Sakurai, Ann. Phys. (N.Y.) **11**, 1 (1960); M. Gell-Mann, F. Zachariasen, Phys. Rev. **124**, 953 (1961).
8. A. Donnachie, H.G. Dosch, P.V. Landsho, O. Nachtmann, *Pomeron Physics and QCD* (Cambridge University Press, 2002); J.R. Forshaw, D.A. Ross, *Quantum Chromodynamics and the Pomeron* (Cambridge University Press, 1997).
9. B. Aubert *et al.*, Phys. Rev. D **71**, 052001 (2005).
10. M. Jacob, R. Slansky, Phys. Lett. B **37**, 408 (1971); Phys. Rev. D **5**, 1847 (1972).
11. P.J. Franzini, F.J. Gilman, Phys. Rev. D **32**, 237 (1985).
12. S.R. Coleman, H.J. Schnitzer, Phys. Rev. **134**, B863 (1964); F.M. Renard, Springer Tracts Mod. Phys. **63**, 98 (1972); Y. Dothan, D. Horn, Nucl. Phys. B **114**, 400 (1976).
13. F.E. Close, AIP Conf. Proc. **717**, 919 (2004); F. Iddir, L. Semlala, arXiv:hep-ph/0211289 and references therein; A. Donnachie, Yu.S. Kalashnikova, Z. Phys. C **59**, 621 (1993).

Eur. Phys. J. A **31**, 649–652 (2007)

DOI 10.1140/epja/i2006-10209-3

Special Article – QNP 2006

X(1859) Baryonium or something else?

D.R. Entem[a] and F. Fernández

Nuclear Physics Group and IUFFyM, University of Salamanca, E-37008 Salamanca, Spain

Received: 25 October 2006
Published online: 26 February 2007 – © Società Italiana di Fisica / Springer-Verlag 2007

Abstract. Using a constituent-quark model we study possible bound or resonance $N\bar{N}$ states. The model fits the $p\bar{p}$ and $p\bar{n}$ cross-sections and explains the large $^3P_0$ antiprotonium energy shift. Only a resonance is found in the $^3P_0$ $I = 0$ partial wave. The threshold enhancement in the $J/\Psi \to \gamma p\bar{p}$ decay can be explained with FSI effects in S-waves and no $N\bar{N}$ bound state is needed.

PACS. 12.39.Jh Non relativistic quark model – 13.75.Cs Nucleon-nucleon interactions (including antinucleons, deuterons, etc.) – 25.43.+t Antiproton-induced reactions – 21.10.Dr Binding energies and masses

1 Introduction

The recent observation of a near threshold narrow enhancement in the $p\bar{p}$ invariant-mass spectrum from radiative $J/\psi \to \gamma p\bar{p}$ decay by the BES Collaboration [1] has renewed the interest for the $N\bar{N}$ interaction and its possible baryonium bound states. If this enhancement is fitted with an S-wave Breit-Wigner resonance function the resulting peak mass is $M = 1859 \pm 6\,\mathrm{MeV}$, which is below the $p\bar{p}$ threshold, whereas a P-wave fit gives a peak mass very close to the threshold at $M = 1876 \pm 0.9\,\mathrm{MeV}$. The photon polar angle distribution is consistent with $1+\cos^2\theta$ which suggest that the total angular momentum is very likely to be $J = 0$. So this structure may have quantum numbers $J^{PC} = 0^{-+}$ or $J^{PC} = 0^{++}$ which, in principle, do not correspond to any known meson resonance. However, no similar signal was observed by BES in the $\pi^0 p\bar{p}$ channel which suggests that the enhancement is due to an isoscalar resonance.

The simplest interpretation of the experimental $J/\psi \to \gamma p\bar{p}$ is a baryonium bound state [2,3] although the result is currently being interpreted in several ways [4–6].

The study of the possible nucleon-antinucleon bound states has a very long history (see ref. [7] and references therein). Usually the real part of the $N\bar{N}$ interaction is derived by G-parity transformation of the NN potentials [7]. The derivation of the annihilation part is still a major challenge. Although some complicate microscopic models have been developed the best results are obtained by phenomenological treatments.

In the meson exchange picture the central force in the NN sector is provided basically by the σ and the ω which have opposite signs but adds in $N\bar{N}$. This gives a lot of attraction and usually generates many $N\bar{N}$ bound states. The ω-exchange contribution is replaced in quark based models by the antisymmetry which is not present in $N\bar{N}$. Therefore quark-based $N\bar{N}$ potentials may look different from the conventional ones. Furthermore, the NN one-pion exchange tensor interaction is attenuated also by antisymetrization and not by ρ-exchanges. This fact, which has observable consequences at the NN level [8], may also significantly change the whole $N\bar{N}$ interaction.

2 The N$\bar{\mathrm{N}}$ interaction

We use the constituent-quark model from refs. [9,10] as the microscopic model to get the $N\bar{N}$ interaction. The model takes into account spontaneous chiral symmetry-breaking through pseudo-goldstone boson modes and the constituent mass generation. Perturbative QCD corrections beyond the chiral symmetry-breaking scale are included through fluctuations of the one-gluon field which give gluon exchange diagrams.

In the $N\bar{N}$ system, it is not allowed any quark-antiquark exchange between N and $\bar{N}$. However quark-antiquark annihilation has to be considered. This makes important differences with the NN system in which exchange diagrams are important. For example the one-gluon exchange only contributes in the $N\bar{N}$ case through annihilation diagrams which are not present in the NN sector.

In order to get the $q\bar{q}$ interaction, we have to perform a G-parity transformation of the qq potentials. For the pseudo-scalar case only a -1 factor is needed while the scalar case remains unchanged. For the real part of the annihilation diagrams we use the derivation from [11].

---

[a] e-mail: `entem@usal.es`

Once the microscopic model is fixed we use the Resonating Group Method to derive the $N\bar{N}$ interaction in the same way as we did in the NN case. The N and $\bar{N}$ wave functions are the same provided that we use for the spin-isospin of the $\bar{N}$ the G-parity transformed.

In order to have a realistic model it is important to include the annihilation into mesons. Annihilation processes are very complicated, for a review see [12]. There are microscopic models at quark level that describes such processes. Usually these models are highly non-local and energy dependent, and instead of using one of these, we chose to describe such processes with an optical potential in order to simplify the calculation. This approximation was already used by Bryan and Phillips [13], Dover and Richard [14] and Kohno and Weise [15]. We chose the parametrization

$$V_{q\bar{q}}^{Anh}(\boldsymbol{q}) = i\, W_i\, e^{-\frac{q^2 b'^2}{3}}, \qquad (1)$$

where W_i gives the strength and b' the range. All but these two parameters of the model are fixed by the NN sector and we use those of [9]. b' and W_i are fixed fitting the total annihilation cross-section for the $p\bar{p}$ system, since the imaginary part of the potential dominates it. We use the values $W_i = -0.74\,\mathrm{GeV}^{-2}$ and $b' = 0.848\,\mathrm{fm}$. With these two parameters we are able to fit the annihilation data in the laboratory momentum range between 100 and 600 MeV as shown in fig. 1(a).

3 Results

Once the annihilation cross-section is fitted, we calculate the total cross-section, which comes in good agreement with the experimental data as can be seen in fig. 1(b). This indicates that the real part of the interaction should be fairly well.

We check the isospin dependence studying the $p\bar{n}$ reaction. Notice that we do not include any isospin dependence in the annihilation potential, so all the isospin dependence comes from the real part of the potential and is fixed from the NN sector. As can be seen in figs. 2 (a) and (b) the results, given by the solid line, are in agreement with the data for the total and annihilation cross-sections, although in both cases there are big uncertainties.

Now we are in the position to study possible $N\bar{N}$ bound states. If we do not include the imaginary potential, the real part of the potential predicts two bound states. One has quantum numbers $J^{PC} = 1^{--}$ which is a $^3S_1$-$^3D_1$ partial wave in $I = 0$ and the other is $J^{PC} = 0^{++}$ which is a $^3P_0$ state also with $I = 0$. Both states are very close to threshold being the binding energy 1.30 MeV for the first one and 1.32 MeV for the second one. We find the strongest interactions in S-waves with $I = 0$ which is much bigger than $I = 1$ and for the $^3P_0$ wave for $I = 0$ which is also much bigger than for $I = 1$.

When we include the annihilation potential the 1^{--} state is washed out and the 0^{++} state is transformed in a near-threshold resonance with a mass 18.5 MeV above threshold and a width of 33.6 MeV.

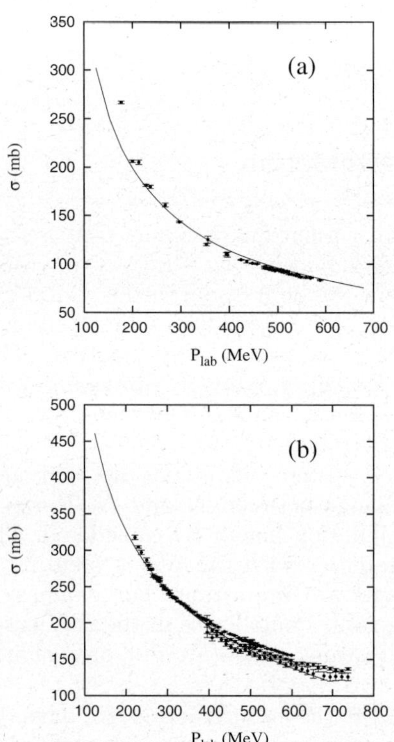

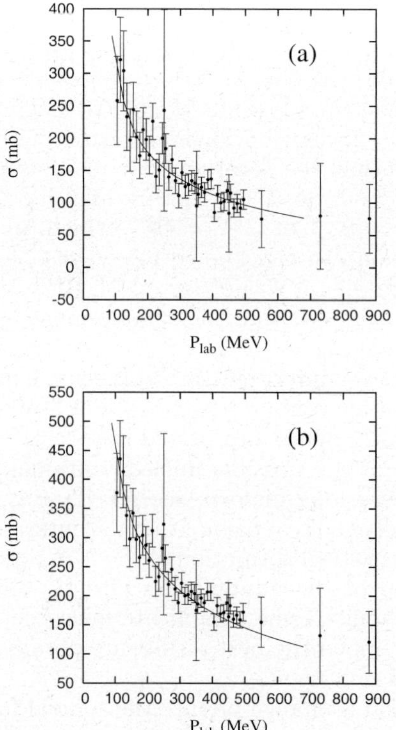

Fig. 1. $p\bar{p}$ annihilation (a) and total (b) cross-sections. Experimental data are from ref. [16].

Fig. 2. $p\bar{n}$ annihilation (a) and total (b) cross-sections. Experimental data are from ref. [17].

Table 1. $p\bar{p}$ energy shifts calculated from the improved Trueman formula (ΔE) and from perturbation theory including only $p\bar{p}$ states ($\Delta E_{p\bar{p}}$) and the coupling to $n\bar{n}$ ($\Delta E_f = \Delta E_{p\bar{p}} + \Delta E_{n\bar{n}}$). S-waves are in eV while P-waves are in meV. Experimental values are available on partial waves $\Delta E_{exp}(1\,^1S_0) = 440 \pm 75 - i(600 \pm 125)\,\mathrm{eV}$ [18] and $\Delta E_{exp}(2\,^3P_0) = -140 \pm 28 - i(60 \pm 12)\,\mathrm{meV}$ [19].

State	ΔE	$\Delta E_{p\bar{p}}$	ΔE_f
$1\,^1S_0$	$557.3 - i820.9$	$-587.5 - i2263.4$	
$2\,^1S_0$	$67.7 - i106.3$	$-73.4 - i28.3$	
$2\,^1P_1$	$-27.9 - i12.4$	$-27.0 - i10.4$	$-28.5 - i10.4$
$2\,^3P_0$	$-67.3 - i94.7$	$-71.8 - i10.4$	$-112.2 - i26.0$
$2\,^3P_1$	$36.5 - i10.9$	$40.1 - i10.4$	$29.6 - 12.1$

This resonance may explain the data from antiprotonic atoms which indeed are the only low energy data available. The protonium atomic levels are shifted and widened $\Delta E_{nl} = \delta E_{nl} - i\Gamma_{nl}/2$ by the strong interaction. One of the most interesting outcomings is that the $^3P_0$ protonium level shift is much bigger than any other. This result suggests the existence of a near threshold $N\bar{N}$ resonance which enhances the scattering amplitude. The energy shifts in hadronic atoms can be determined with the improved Trueman formula [20,7], using the scattering length (S-wave) or the scattering volume (P-waves):

$$\delta E_{nl} = -E_n \frac{4}{n} \frac{a_l}{a_B^{2l+1}} \alpha_{nl} \left(1 - \frac{a_l}{a_B^{2l+1}} \beta_{nl} + \dots\right). \quad (2)$$

An alternative approach is to use perturbation theory that for weak potentials reproduce the former result. However, the strong interaction in S-waves is big enough compared to the coulombic at short range that perturbation theory does not work. For P-waves the centrifugal barrier contributes to soften the strong interaction, and perturbation theory became useful.

We use the improved Trueman formula to calculate the energy shifts and half-width for the $p\bar{p}$ $L = 0$ and $L = 1$ partial waves. The results are shown in table 1. We also include the results of the perturbative calculation for the P-waves, which are compatibles with the one calculated with the Trueman formula. One can see from this table that the experimental results are well reproduced except for the $^3P_0$ energy shift which is underestimate by a factor two. These predictions coincide basically with those obtained with other potentials [21].

It is worth to note that $p\bar{p}$ and $n\bar{n}$ channels are coupled and it may have a significant influence in the theoretical predictions. So for P-waves we use coupled channel perturbation theory. Results are shown in table 1. As we can see the coupling with the $n\bar{n}$ channel does not affect very much to the energy shift except for the $^3P_0$ partial wave. The observed enhancement in this wave is due to the resonance located close to the $N\bar{N}$ threshold in this channel.

As we do not find any bound state we have studied FSI effects in the $J/\Psi \to \gamma p\bar{p}$ as a possible justification of the threshold enhancement. Using the Watson-Migdal

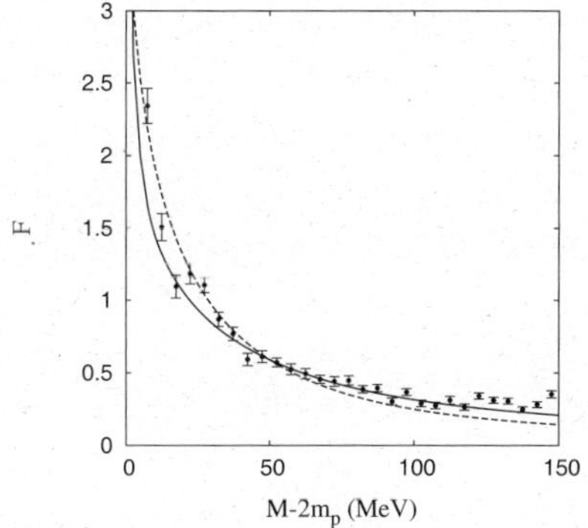

Fig. 3. $J/\Psi \to \gamma p\bar{p}$ invariant-mass distribution. Lines show the FSI factor for the $^1S_0$ $I = 0$ (solid) and $I = 1$ (dashed) channels. Experimental data are from [1,6].

prescription we find that it can be explained with FSI in S-waves. The result for the $^1S_0$ partial wave are shown in fig. 3. A more elaborate analysis is on preparation.

This work has been partially funded by MEC under Contract No. FPA2004-05616, and by Junta de Castilla y León under Contract No. SA-104/04.

References

1. BES Collaboration (J.Z. Bai *et al.*), Phys. Rev. Lett. **91**, 022001 (2003).
2. A. Datta, P.J. O'Donnell, Phys. Lett. B **567**, 273 (2003).
3. B. Loiseau, S. Wycech, Phys. Rev. C **72**, 011001(R) (2001).
4. Tao Huang, Shi-Lin Zhu, Phys. Rev. D **73**, 014023 (2006).
5. Bing An Li, Phys. Rev. D **74**, 034019 (2006).
6. A. Sibirtsev, J. Haidenbauer, S. Krewald, Ulf-G. Meissner, A.W. Thomas, Phys. Rev. D **71**, 054010 (2005).
7. E. Klempt, F. Bradamante, A. Martin, J.-M. Richard, Phys. Rep. **368**, 119 (2002).
8. F. Fernández, A. Valcarce, P. González, V. Vento, Phys. Lett. B **287**, 35 (1992).
9. D.R. Entem, F. Fernández, A. Valcarce, Phys. Rev. C **62**, 034002 (2000).
10. A. Valcarce, H. Garcilazo, F. Fernández, P. González, Rep. Prog. Phys. **68**, 965 (2005); J. Vijande, F. Fernández, A. Valcarce, J. Phys. G **19**, 2013 (2005).
11. A. Faessler, G. Lübeck, K. Shimizu, Phys. Rev. D **26**, 3280 (1982).
12. E. Klempt, C. Batty, J.-M. Richard, Phys. Rep. **413**, 197 (2005).
13. R.A. Bryan, R.J. Phillips, Nucl. Phys. B **5**, 201 (1986); **7**, 481 (1986) (E).
14. C.B. Dover, J.-M. Richard, Phys. Rev. C **21**, 1466 (1980); J.-M. Richard, M.E. Saino, Phys. Lett. B **110**, 349 (1982).
15. M. Kohno, W. Weise, Nucl. Phys. A **454**, 429 (1986).
16. W. Brückner *et al.*, Z. Phys. A **335**, 217 (1990); T. Kamae *et al.*, Phys. Rev. Lett. **44**, 1439 (1980); K. Nakamura *et al.*, Phys. Rev. D **29**, 349 (1984); A.S. Clough *et al.*, Phys. Lett. B **146**, 299 (1984); D.V. Bugg *et al.*, Phys. Lett. B **194**, 563 (1987).

17. B. Gunderson, J. Learned, J. Mapp, D.D. Reeder, Phys. Rev. D **23**, 587 (1981); T. Armstrong *et al.*, Phys. Rev. D **36**, 659 (1987).
18. M. Augsburger *et al.*, Nucl. Phys. A **658**, 149 (1999).
19. D. Gotta *et al.*, Nucl. Phys. A **660**, 283 (1999).
20. T.L. Trueman, Nucl. Phys. **26**, 57 (1961).
21. J. Carbonell, M. Mangin-Brinet, Nucl. Phys. A **692**, 11c (2001).

Eur. Phys. J. A **31**, 653–655 (2007)
DOI 10.1140/epja/i2006-10219-1

Special Article – QNP 2006

Search for exotic baryons at HERMES

W. Deconinck[a]

On behalf of the HERMES Collaboration
Randall Laboratory of Physics, University of Michigan, Ann Arbor, MI, USA

Received: 8 November 2006
Published online: 1 March 2007 – © Società Italiana di Fisica / Springer-Verlag 2007

Abstract. An experimental search for the $\Theta(1540)$- and $\Lambda(1520)$-resonance was performed in quasi-real photoproduction on deuterium at the HERMES experiment. While evidence for $\Theta(1540)$ was found in the decay channel $pK_S^0 \to p\pi^+\pi^-$, no evidence for the corresponding anti-particle was found. In some models it is expected that the $\Theta(1540)$ and the $\Lambda(1520)$ have similar production mechanisms. The photoproduction cross-sections for the $\Lambda(1520)$ in the decay channel $\Lambda(1520) \to pK^-$ and the corresponding anti-particle are determined. The partial photoproduction cross-sections for $\Lambda(1520)$ and $\bar{\Lambda}(1520)$ are obtained as $\sigma_{\Lambda(1520)} = 65.3\pm8.8(\text{stat})\pm6.9(\text{syst})$ nb and $\sigma_{\bar{\Lambda}(1520)} = 9.8\pm2.6(\text{stat})\pm0.9(\text{syst})$ nb, corresponding to a ratio $R_{\Lambda(1520)} = \sigma_{\bar{\Lambda}(1520)}/\sigma_{\Lambda(1520)} = 0.15 \pm 0.05(\text{stat}) \pm 0.02(\text{syst})$.

PACS. 12.39.Mk Glueball and nonstandard multi-quark/gluon states – 13.60.-r Photon and charged-lepton interactions with hadrons – 13.60.Rj Baryon production – 14.20.-c Baryons (including antiparticles)

1 Introduction

The theory of Quantum Chromodynamics describes the interactions of quarks and gluons as constituents of mesons and baryons. It does not explicitly prohibit exotic states consisting of more than three quarks. Recently, several experimental groups have reported evidence for a manifestly exotic baryon resonance Θ with a mass near 1540 MeV, whose quantum numbers can be explained only with five or more quarks. On the other hand, searches at several high-energy experiments show no evidence for an exotic resonance near 1540 MeV, casting serious doubt on the existence of the $\Theta(1540)$. For a review of the current experimental status of exotic baryons the reader is referred to ref. [1].

At HERMES, in a search for the $\Theta(1540)$ in quasi-real photoproduction on a deuterium target through the decay channel $\Theta(1540) \to pK_S^0 \to p\pi^+\pi^-$, a narrow peak at $1528 \pm 2.6(\text{stat}) \pm 2.1(\text{syst})$ MeV with 59 ± 16 events was observed [2]. The $M(p\pi^+\pi^-)$ invariant-mass spectrum is shown in fig. 1. Using the same data and event selection criteria, a search for the anti-particle $\overline{\Theta}(1540)$ was performed. The $M(\bar{p}\pi^+\pi^-)$ invariant-mass spectrum is shown in fig. 2. A similar fit as for the $\Theta(1540)$ but with a fixed Gaussian width results in 3±6 events, consistent with zero.

Recent results of the LEPS Collaboration [3] point to a similar production mechanism on deuterium for the exotic $\Theta(1540)$ and the hyperon $\Lambda(1520)$ with mass close to the observed $\Theta(1540)$ mass. Therefore the ratio of the

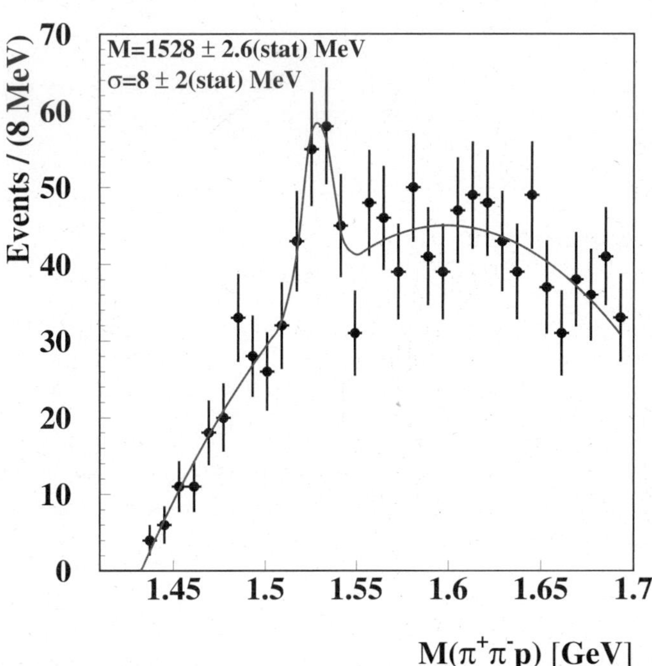

Fig. 1. Distribution of the invariant mass of the $p\pi^+\pi^-$ system. A fit to the data with a Gaussian plus a third-order polynomial is shown.

$\overline{\Theta}(1540)$ and $\Theta(1540)$ cross-sections is expected to be similar to that of the $\bar{\Lambda}(1520)$ and $\Lambda(1520)$. More generally, knowledge of the ratios of anti-hadrons to hadrons

[a] e-mail: wdconinc@umich.edu

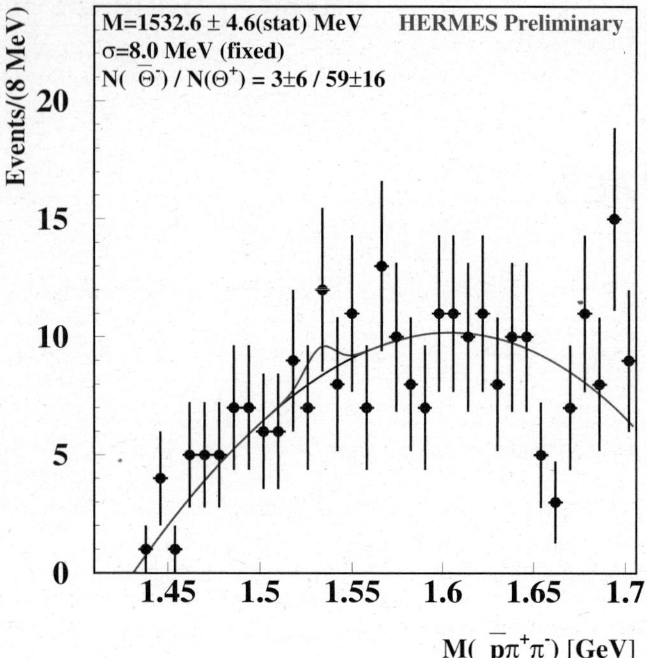

Fig. 2. Distribution of the invariant mass of the $\overline{p}\pi^+\pi^-$ system. A fit similar to fig. 1 but with fixed width for the $\overline{\Theta}(1540)$-resonance is shown.

could help understand the production mechanism of the $\Theta(1540)$ [4].

In this analysis the cross-section ratio of the $\bar{\Lambda}(1520)$ to $\Lambda(1520)$ is determined. For details about the $\Theta(1540)$ analysis, the reader is referred to ref. [2].

2 Experiment

For this analysis, data were obtained at the HERMES experiment with the 27.6 GeV positron beam of the HERA accelerator at DESY. An integrated luminosity of 209 pb^{-1} was collected on a longitudinally polarized deuterium gas target, and the yields were summed over the two spin orientations. The HERMES spectrometer [5] consists of two identical halves located above and below the positron beam pipe, and has an angular acceptance of ± 170 mrad horizontally and ± 40–140 mrad vertically.

The trigger was formed by a coincidence between scintillating hodoscopes, a preshower detector and a lead-glass calorimeter. Lepton and hadron tracks are distinguished by combining the signals from a transition radiation detector, a preshower detector, a ring-imaging Čerenkov detector and a lead-glass calorimeter. Identification of pions, kaons and protons is accomplished with the dual radiator ring-imaging Čerenkov (RICH) detector [6]. Data from simulations indicate that cross contaminations are negligible in a momentum range 1–15 GeV for pions, 2–15 GeV for kaons, and 4–9 GeV for protons, the kinematic restrictions subsequently used in this analysis.

3 Analysis

This analysis searched for inclusive photoproduction of the hyperon $\Lambda(1520)$ followed by the decay $\Lambda(1520) \to pK^-$, and of its anti-particle $\bar{\Lambda}(1520)$ followed by the decay $\bar{\Lambda}(1520) \to \bar{p}K^+$. Using the simulated effects of acceptance and efficiency the partial cross-sections for $\Lambda(1520)$ and $\bar{\Lambda}(1520)$ are determined.

To maximize the yield of the $\Lambda(1520)$ peak in the $M(pK^-)$ invariant-mass spectrum while minimizing its background, restrictions on the event topology of the pK system were applied. Based on the intrinsic tracking resolution, a distance of closest approach between the proton and kaon tracks less than 0.6 cm and a radial distance of the pK vertex to the HERA beam axis less than 0.4 cm are required. Because the $\Lambda(1520)$ has a mean path length much smaller than the longitudinal tracking resolution of 3 cm, a decay length of the $\Lambda(1520)$ less than 5 cm is required. Kaons mis-identified as protons were rejected when the $M(K^+K^-)$ invariant mass was within the reconstructed $\Phi(1020)$-resonance peak. The same event selection criteria were applied to the candidate $\bar{\Lambda}(1520)$ events.

The resulting $M(pK)$ invariant mass spectra for the pK^- and $\bar{p}K^+$ systems are shown in figs. 3 and 4. The invariant-mass spectra were fitted using an unbinned maximum-likelihood technique with the sum of a Breit-Wigner resonance shape convolved with the Gaussian de-

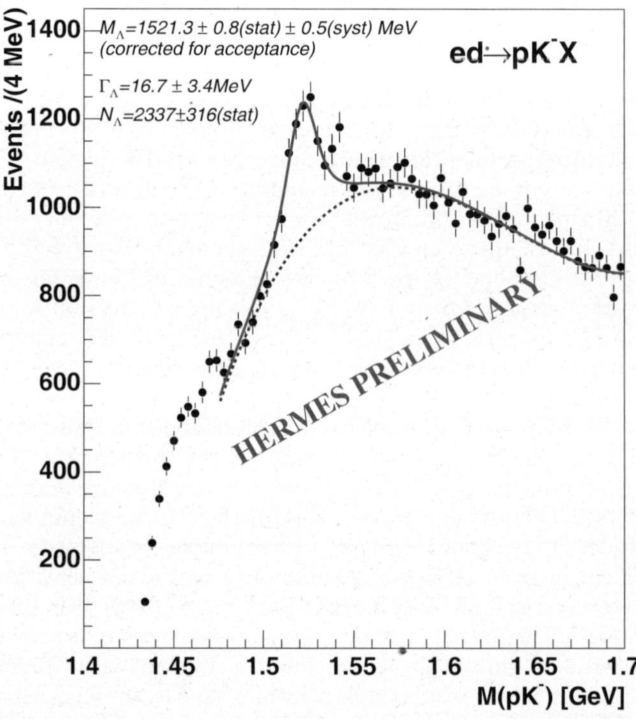

Fig. 3. Distribution of the invariant mass of the pK^- system. A fit to the data with a Breit-Wigner resonance shape convolved with the Gaussian resolution plus a third-order polynomial is shown. The position of the peak has been corrected for the effects of the HERMES acceptance.

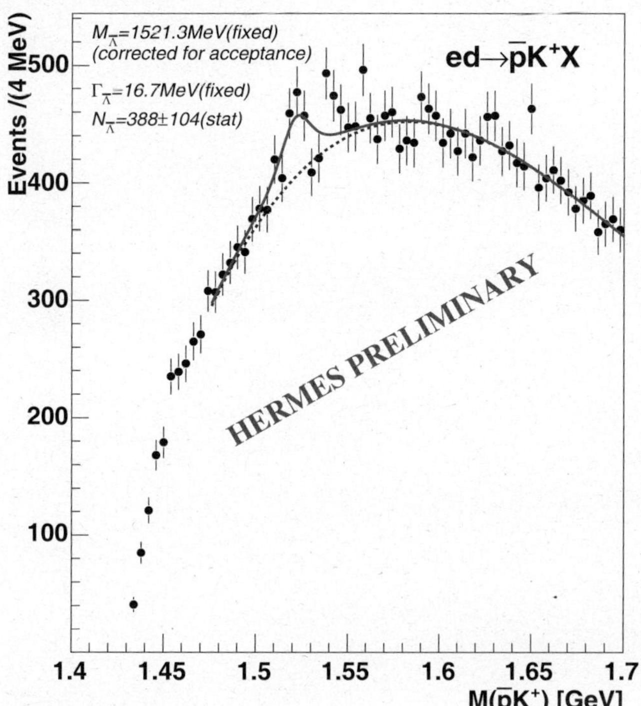

Fig. 4. Distribution of the invariant mass of the $\bar{p}K^+$ system. A fit similar to fig. 3 but with fixed width and position for the $\bar{\Lambda}(1520)$-resonance is shown.

tector resolution, in addition to a third-order polynomial for the description of the background. The resolution in $M(pK)$ was determined using a Monte Carlo simulation of the full spectrometer and is $\sigma = 4\,\text{MeV}$ for both the pK^- and $\bar{p}K^+$ system. For the $\Lambda(1520)$ a peak is observed at $1521.3 \pm 0.8(\text{stat}) \pm 0.5(\text{syst})\,\text{MeV}$ with an intrinsic width $\Gamma = 16.7 \pm 3.4\,\text{MeV}$, in agreement with the PDG estimates [7]. For the $\bar{\Lambda}(1520)$ the peak position and width were fixed to the results for the $\Lambda(1520)$. The extracted number of resonance events are $N_{\Lambda(1520)} = 2337 \pm 316$ and $N_{\bar{\Lambda}(1520)} = 388 \pm 104$.

The shoulder in the $M(pK^-)$ invariant-mass spectrum at a mass slightly below the $\Lambda(1520)$ mass is caused by kinematic effects and reproduced in the background shape obtained by event mixing. The excess of events near 1540 MeV can be removed by requiring more stringent cuts for correct particle identification, and is likely due to background of wrongly identified particles.

Monte Carlo simulations of $\Lambda(1520)$ events tracked through a full model of the spectrometer show that the variation of the detector acceptance in the region around the $\Lambda(1520)$ peak causes a shift in the resonance position of $1.2 \pm 0.5\,\text{MeV}$ towards higher masses. The reported masses and their systematic uncertainties have been corrected for this effect.

For the calculation of the cross-section, the combined effect of the geometrical and kinematic acceptance and of the event selection efficiency was determined using Monte Carlo simulations. The Pythia Monte Carlo generator does not include the $\Lambda(1520)$ hyperon, so the initial momen-

tum distribution for $\Lambda(1520)$ and $\bar{\Lambda}(1520)$ production is assumed to be similar to that of other hyperons available in Pythia (Λ, Σ, Ξ, $\Sigma(1385)$, $\Xi(1530)$). The acceptance for $\Lambda(1520)$ events exhibits a steady increase with the mass of the hyperon whose momentum distribution is used. Due to the kinematic acceptance of the HERMES spectrometer for $\Lambda(1520)$ decays, only candidates with a longitudinal momentum P_z greater than 6 GeV were considered. The determined acceptance for $\Lambda(1520)$ and $\bar{\Lambda}(1520)$ is $3.8 \pm 0.4\%$ and $4.2 \pm 0.4\%$, respectively. No $\Lambda(1520)$ polarization was taken into account for this calculation.

The resulting partial photoproduction cross-sections are $\sigma_{\Lambda(1520)} = 65.3 \pm 8.8(\text{stat}) \pm 6.9(\text{syst})\,\text{nb}$ and $\sigma_{\bar{\Lambda}(1520)} = 9.8 \pm 2.6(\text{stat}) \pm 0.9(\text{syst})\,\text{nb}$. The systematic uncertainties take into account the change of acceptance related to the uncertainty on the peak position, and the uncertainty in the choice of initial momentum distribution. The corresponding cross-section ratio $R_{\Lambda(1520)} = \sigma_{\bar{\Lambda}(1520)}/\sigma_{\Lambda(1520)}$ is $R_{\Lambda(1520)} = 0.15 \pm 0.05(\text{stat}) \pm 0.02(\text{syst})$.

4 Discussion

Assuming similar production processes for the $\Theta(1540)$ and the $\Lambda(1520)$, and equal ratios for the particle and antiparticle cross-sections of these resonances, the expected number of $\overline{\Theta}(1540)$ events at HERMES can be estimated as 10 ± 4. This is in agreement with the experimental result 3 ± 6 within statistical uncertainties.

At HERMES we are currently analyzing the data collected on a polarized hydrogen target between the years 2003 and 2005. Additional data on a high-density hydrogen and deuterium target are still being collected.

I would like to thank my colleagues in the HERMES Collaboration for the many suggestions and useful discussions, in particular Siguang Wang and Avetik Airapetian who contributed to this analysis. I highly acknowledge Wolfgang Lorenzon and Michael Düren for a critical reading of this manuscript. This research is supported by the U.S. National Science Foundation, Intermediate Energy Nuclear Science Division, under grants PHY-0244842 and PHY-055423.

References

1. For a review of the experimental situation, see M. Danilov, arXiv:hep-ex/0509012.
2. HERMES Collaboration (A. Airapetian *et al.*), Phys. Lett. B **585**, 213 (2004).
3. T. Nakano, presentation at *QCD05, Peking University, June 2005*.
4. M. Karliner, H.J. Lipkin, arXiv:hep-ph/0506084.
5. HERMES Collaboration (K. Ackerstaff *et al.*), Nucl. Instrum. Methods A **417**, 230 (1998).
6. N. Akopov *et al.*, Nucl. Instrum. Methods A **479**, 511 (2002).
7. W.-M. Yao *et al.*, J. Phys. G: Nucl. Part. Phys. **33**, 1 (2006).

Eur. Phys. J. A **31**, 656–661 (2007)

DOI 10.1140/epja/i2006-10234-2

Special Article – QNP 2006

QCD Coulomb gauge approach to exotic hadrons

S.R. Cotanch[a], I.J. General, and P. Wang

Department of Physics, North Carolina State University, Raleigh, NC 27695-8202, USA

Received: 8 November 2006
Published online: 8 March 2007 – © Società Italiana di Fisica / Springer-Verlag 2007

Abstract. The Coulomb gauge Hamiltonian model is used to calculate masses for selected J^{PC} states consisting of exotic combinations of quarks and gluons: ggg glueballs (oddballs), $q\bar{q}g$ hybrid mesons and $q\bar{q}q\bar{q}$ tetraquark systems. An odderon Regge trajectory is computed for the J^{--} glueballs with intercept much smaller than the pomeron, explaining its nonobservation. The lowest 1^{-+} hybrid-meson mass is found to be just above 2.2 GeV while the lightest tetraquark state mass with these exotic quantum numbers is predicted around 1.4 GeV consistent with the observed $\pi(1400)$.

PACS. 12.38.Lg Other nonperturbative calculations – 12.39.Ki Relativistic quark model – 12.39.Mk Glueball and nonstandard multi-quark/gluon states – 12.40.Yx Hadron mass models and calculations

1 Introduction

Establishing the existence of exotic hadrons (non $q\bar{q}$ or qqq structure) is of paramount importance and remains one of the key unsolved problems in hadronic physics. In particular, it is expected from general QCD principles that nonconventional color singlet states of gluons and quarks should exist such as glueballs (gg and ggg), hybrid mesons ($q\bar{q}g$) and multiquark states ($q\bar{q}q\bar{q}$ and $qqqq\bar{q}$). The present work addresses the structure of these hadrons and provides new information to assist experimental searches.

2 Coulomb gauge Hamiltonian model

In a series of publications [1–8] a realistic model for hadron structure has been developed and applied to the quark and gluon sectors. This field theoretical, relativistic many-body approach utilizes an effective QCD Hamiltonian, H_{eff}, formulated in the Coulomb gauge. It properly incorporates chiral symmetry using standard bare current quark masses but dynamically generates a constituent mass and spontaneous chiral symmetry breaking [2]. Through approximate many-body diagonalizations it successfully describes the meson spectrum [4,6] and is consistent [7] with lattice glueball predictions. It also yields a good description of the vacuum properties (quark and gluon condensates) within a minimal two-parameter theory.

The effective Hamiltonian, an approximation to the exact Coulomb gauge QCD Hamiltonian, is

$$H_{\mathrm{eff}} = H_q + H_g + H_{qg} + H_C, \tag{1}$$

[a] e-mail: cotanch@ncsu.edu

$$H_q = \int \mathrm{d}\mathbf{x}\Psi^\dagger(\mathbf{x})[-i\boldsymbol{\alpha}\cdot\boldsymbol{\nabla} + \beta m]\Psi(\mathbf{x}), \tag{2}$$

$$H_g = \frac{1}{2}\int \mathrm{d}\mathbf{x}\left[\boldsymbol{\Pi}^a(\mathbf{x})\cdot\boldsymbol{\Pi}^a(\mathbf{x}) + \mathbf{B}^a(\mathbf{x})\cdot\mathbf{B}^a(\mathbf{x})\right], \tag{3}$$

$$H_{qg} = g\int \mathrm{d}\mathbf{x}\,\mathbf{J}^a(\mathbf{x})\cdot\mathbf{A}^a(\mathbf{x}), \tag{4}$$

$$H_C = -\frac{1}{2}\int \mathrm{d}\mathbf{x}\mathrm{d}\mathbf{y}\rho^a(\mathbf{x})\hat{V}(|\mathbf{x}-\mathbf{y}|)\rho^a(\mathbf{y}), \tag{5}$$

with g the QCD coupling, Ψ the quark field, m the current quark mass, $\mathbf{A}^a$ the gluon fields satisfying the transverse gauge condition, $\boldsymbol{\nabla}\cdot\mathbf{A}^a = 0$, $a = 1, 2, \ldots, 8$, $\boldsymbol{\Pi}^a$ the conjugate fields and $\mathbf{B}^a$ the non-Abelian magnetic fields

$$\mathbf{B}^a = \nabla\times\mathbf{A}^a + \frac{1}{2}gf^{abc}\mathbf{A}^b\times\mathbf{A}^c. \tag{6}$$

The color densities, $\rho^a(\mathbf{x})$, and currents, $\mathbf{J}^a$, are

$$\rho^a(\mathbf{x}) = \Psi^\dagger(\mathbf{x})T^a\Psi(\mathbf{x}) + f^{abc}\mathbf{A}^b(\mathbf{x})\cdot\boldsymbol{\Pi}^c(\mathbf{x}), \tag{7}$$

$$\mathbf{J}^a = \Psi^\dagger(\mathbf{x})\boldsymbol{\alpha}T^a\Psi(\mathbf{x}), \tag{8}$$

with $T^a = \frac{\lambda^a}{2}$ and f^{abc} the SU_3 color matrices and structure constants, respectively. Confinement is described by a Cornell-type potential,

$$\hat{V}(r = |\mathbf{x}-\mathbf{y}|) = \hat{V}_C(r) + \hat{V}_L(r), \tag{9}$$

$$\hat{V}_C(r) = -\frac{\alpha_s}{r}, \tag{10}$$

$$\hat{V}_L(r) = \sigma r, \tag{11}$$

with previously determined string tension, $\sigma = 0.135\,\mathrm{GeV}^2$, and $\alpha_s = \frac{g^2}{4\pi} = 0.4$. Below we denote the Fourier transform of $\hat{V}$ by V.

The bare fields have the Fock operator expansions (quark spinors u, v, helicity, $\lambda = \pm 1$, and color vectors $\hat{\epsilon}_{C=1,2,3}$)

$$\Psi(\mathbf{x}) = \int \frac{d\mathbf{k}}{(2\pi)^3} \Psi_C(\mathbf{k}) e^{i\mathbf{k}\cdot\mathbf{x}} \hat{\epsilon}_C, \tag{12}$$

$$\Psi_C(\mathbf{k}) = u_\lambda(\mathbf{k}) b_{\lambda C}(\mathbf{k}) + v_\lambda(-\mathbf{k}) d^\dagger_{\lambda C}(-\mathbf{k}), \tag{13}$$

$$\mathbf{A}^a(\mathbf{x}) = \int \frac{d\mathbf{k}}{(2\pi)^3} \frac{1}{\sqrt{2k}} \left[\mathbf{a}^a(\mathbf{k}) + \mathbf{a}^{a\dagger}(-\mathbf{k}) \right] e^{i\mathbf{k}\cdot\mathbf{x}}, \tag{14}$$

$$\mathbf{\Pi}^a(\mathbf{x}) = -i \int \frac{d\mathbf{k}}{(2\pi)^3} \sqrt{\frac{k}{2}} \left[\mathbf{a}^a(\mathbf{k}) - \mathbf{a}^{a\dagger}(-\mathbf{k}) \right] e^{i\mathbf{k}\cdot\mathbf{x}}, \tag{15}$$

with quark, anti-quark and gluon Fock operators $b_{\lambda C}(\mathbf{k})$, $d_{\lambda C}(-\mathbf{k})$ and $a^a_\mu(\mathbf{k})$ ($\mu = 0, \pm 1$), respectively. The Coulomb gauge condition, $\mathbf{k}\cdot\mathbf{a}^a(\mathbf{k}) = (-1)^\mu k_\mu a^a_{-\mu}(\mathbf{k}) = 0$, produces transverse commutation relations,

$$\left[a^a_\mu(\mathbf{k}), a^{b\dagger}_{\mu'}(\mathbf{k}') \right] = (2\pi)^3 \delta_{ab} \delta^3(\mathbf{k} - \mathbf{k}') D_{\mu\mu'}(\mathbf{k}), \tag{16}$$

with

$$D_{\mu\mu'}(\mathbf{k}) = \delta_{\mu\mu'} - (-1)^\mu \frac{k_\mu k_{-\mu'}}{k^2}. \tag{17}$$

The ground state (model vacuum) is generated using the Bardeen-Cooper-Schrieffer (BCS) method, entailing rotated field operators (Bogoliubov-Valatin transformation),

$$B_{\lambda C}(\mathbf{k}) = \cos \frac{\theta_k}{2} b_{\lambda C}(\mathbf{k}) - \lambda \sin \frac{\theta_k}{2} d^\dagger_{\lambda C}(-\mathbf{k}),$$

$$D_{\lambda C}(-\mathbf{k}) = \cos \frac{\theta_k}{2} d_{\lambda C}(-\mathbf{k}) + \lambda \sin \frac{\theta_k}{2} b^\dagger_{\lambda C}(\mathbf{k}),$$

$$\boldsymbol{\alpha}^a(\mathbf{k}) = \cosh \Theta_k \mathbf{a}^a(\mathbf{k}) + \sinh \Theta_k \mathbf{a}^{a\dagger}(-\mathbf{k}), \tag{18}$$

producing the dressed, quasi-particle operators $\boldsymbol{\alpha}^a$, $B_{\lambda C}$ and $D_{\lambda C}$, respectively. The quasi-particle (BCS) vacuum, determined by $B_{\lambda C}|\Omega\rangle = D_{\lambda C}|\Omega\rangle = \alpha^a_\mu|\Omega\rangle = 0$, is built on the bare parton one, $b_{\lambda C}|0\rangle = d_{\lambda C}|0\rangle = a^a_\mu|0\rangle = 0$,

$$|\Omega_{\text{quark}}\rangle = e^{-\int \frac{d\mathbf{k}}{(2\pi)^3} \lambda \tan \frac{\theta_k}{2} b^\dagger_{\lambda C}(\mathbf{k}) d^\dagger_{\lambda C}(-\mathbf{k})} |0\rangle, \tag{19}$$

$$|\Omega_{\text{gluon}}\rangle = e^{-\int \frac{d\mathbf{k}}{(2\pi)^3} \frac{1}{2} \tanh \Theta_k D_{\mu\mu'}(\mathbf{k}) a^{a\dagger}_\mu(\mathbf{k}) a^{a\dagger}_{\mu'}(-\mathbf{k})} |0\rangle. \tag{20}$$

The composite BCS vacuum, $|\Omega\rangle = |\Omega_{\text{quark}}\rangle \otimes |\Omega_{\text{gluon}}\rangle$, contains quark and gluon condensates (correlated $q\bar{q}$ and gg Cooper pairs). A variational minimization of the vacuum expectation value of the Hamiltonian, $\delta\langle\Omega|H_{\text{eff}}|\Omega\rangle = 0$, yields the constituent quark and gluon gap equations

$$ks_k - mc_k = \frac{2}{3} \int \frac{d\mathbf{q}}{(2\pi)^3} (s_k c_q x - s_q c_k) V(|\mathbf{k} - \mathbf{q}|), \tag{21}$$

$$\omega_k^2 = k^2 - \frac{3}{4} \int \frac{d\mathbf{q}}{(2\pi)^3} V(|\mathbf{k} - \mathbf{q}|)[1 + x^2]\left(\frac{\omega_q^2 - \omega_k^2}{\omega_q} \right)$$

$$+ \frac{3}{4} g^2 \int \frac{d\mathbf{q}}{(2\pi)^3} \frac{1 - x^2}{\omega_q}, \tag{22}$$

with $s_k = \sin \phi_k$, $c_k = \cos \phi_k$ and $x = \mathbf{k}\cdot\mathbf{q}$. Here $\phi_k = \phi(k)$ is the quark gap angle related to the BCS angle θ_k by

$\tan(\phi_k - \theta_k) = m/k$, and $\omega_k = ke^{-2\Theta_k}$ is the effective gluon self-energy. The last term in eq. (22) is due to the non-Abelian component of the gluon kinetic energy. The quark gap equation is UV finite for the linear potential since $V_L(|\mathbf{p}|) = -8\pi\sigma/p^4$, but not for the Coulomb potential $V_C(|\mathbf{p}|) = -4\pi\alpha_s/p^2$. The gluon gap equation has both logarithmical and quadratical UV divergences and an integration cutoff, $\Lambda = 4\,\text{GeV}$, determined in previous studies is used in both equations.

3 Applications

Predictions for the low-lying spectra of glueballs, hybrid-mesons and tetraquark systems are now presented and discussed. Since these hadrons consist of 3 or more constituents, the masses for selected J^{PC} states are computed variationally,

$$M_{JPC} = \frac{\langle \Psi^{JPC} | H_{\text{eff}} | \Psi^{JPC} \rangle}{\langle \Psi^{JPC} | \Psi^{JPC} \rangle}. \tag{23}$$

The variational approximation has been comprehensively tested in two-body systems by comparison with exact diagonalization and found to be accurate to a few percent.

3.1 Glueballs

Previous work [5] has investigated gg glueballs which only have $C = 1$. For $C = -1$ glueballs (oddballs), Fock states with at least 3 gluons are necessary and the variational wave function is ($\mathbf{q}_{i=1,2,3}$ are the cm gluon momenta)

$$|\Psi^{JPC}_{ggg}\rangle = \int d\mathbf{q}_1 d\mathbf{q}_2 d\mathbf{q}_3 \delta(\mathbf{q}_1 + \mathbf{q}_2 + \mathbf{q}_3) \Phi^{JPC}_{\mu_1\mu_2\mu_3}(\mathbf{q}_1, \mathbf{q}_2, \mathbf{q}_3)$$

$$\times C^{abc} \alpha^{a\dagger}_{\mu_1}(\mathbf{q}_1) \alpha^{b\dagger}_{\mu_2}(\mathbf{q}_2) \alpha^{c\dagger}_{\mu_3}(\mathbf{q}_3) |\Omega_{gluon}\rangle, \tag{24}$$

with color tensor C^{abc} either totally anti-symmetric f^{abc} (for $C = 1$) or symmetric d^{abc} (for $C = -1$). Boson statistics thus requires the $C = -1$ oddballs to have a symmetric space-spin wave function. Using eq. (23) and a two-parameter variational radial wave function, the J^{--} oddball states have been calculated. Only the Abelian component of the magnetic fields is retained and the hyperfine interaction, H_{qg}, is suppressed. There are three terms contributing to the mass expectation value which are depicted in fig. 1 which correspond to the gluon self-energy (top), gluon-gluon scattering (middle) and annihilation (bottom).

The nine-dimensional variational calculation was performed using the Monte Carlo method with the adaptive sampling algorithm VEGAS [9] and numerical convergence required between 10^5 and 10^6 samples. The oddball mass predictions are compared in table 1 to available lattice gauge results [10,11] and a Wilson-loop–inspired model [12]. A study of the glueball mass sensitivity to both statistical and variational uncertainties yielded error bars at the few percent level.

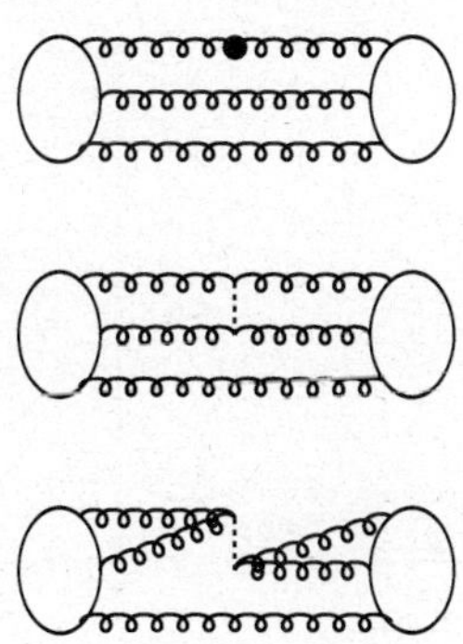

Fig. 1. Glueball diagrams for $\langle \Psi_{ggg}^{JPC} | H_{\mathrm{eff}} | \Psi_{ggg}^{JPC} \rangle$.

Table 1. Glueball quantum numbers and masses in MeV. Error in H_{eff} (from Monte Carlo only) is 100 MeV or less, the quoted lattice errors are typically 200–300 MeV.

Model	1^{--}	3^{--}	5^{--}	7^{--}
Coulomb gauge H_{eff}	3950	4150	5050	5900
Lattice [10]	3850	4130		
Lattice [11]	3100	4150		
Wilson loop [12]	3490	4030		

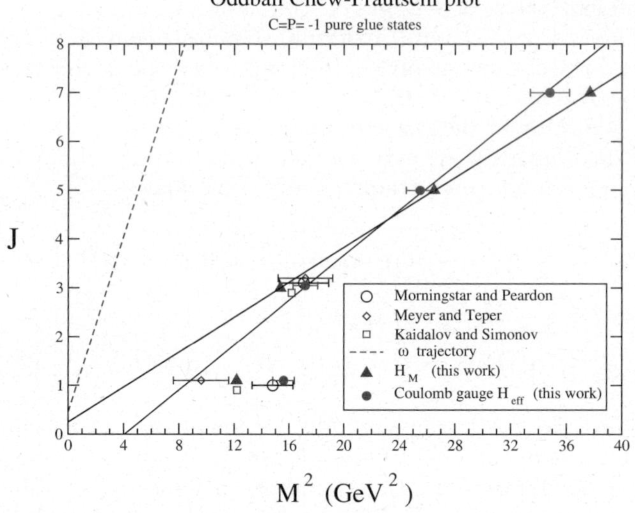

Fig. 2. Odderon trajectories from constituent gluon models and lattice compared to the ω-meson Regge trajectory.

Figure 2 displays predicted oddball Regge trajectories from the alternative approaches. Lattice results are depicted by open circles [10] and diamonds [11]. Constituent gluon predictions are represented by boxes, solid triangles and solid circles and correspond to a Wilson-loop–inspired potential model [12], a simpler harmonic-oscillator calculation [7] labeled H_{M} and the H_{eff} approach, respectively. The odderon trajectories for the latter two models are represented by the solid lines, $\alpha_O^{\mathrm{M}} = 0.18t + 0.25$ and $\alpha_O^{\mathrm{eff}} = 0.23t - 0.88$, which provide an overall theoretical

uncertainty. The much steeper dashed line is the ω trajectory.

Three key results follow. First, the predicted odderon has slope similar to the pomeron but intercept clearly lower than the ω value. Second, the odderon starts with the 3^{--} state and not the 1^{--} which is on a daughter trajectory. Note that there are no lattice 5^{--} glueball predictions which are necessary to confirm this point and we strongly recommend that future studies calculate higher J^{--} states. Third, all approaches agree that the 3^{--} mass is near 4 GeV.

3.2 Hybrid mesons

We denote the momenta of the dressed quark, anti-quark and gluon by $\mathbf{q}$, $\bar{\mathbf{q}}$ and $\mathbf{g}$, respectively, and work in the hybrid cm system. The color structure for a $q\bar{q}g$ hybrid is given by $SU_c(3)$ algebra, $(3 \otimes \bar{3}) \otimes 8 = (8 \otimes 8) \oplus (8 \otimes 1) = 27 \oplus 10 \oplus 10 \oplus 8 \oplus 8 \oplus 8 \oplus 1$. Hence for an overall color singlet the quarks must be in an octet state like the gluon which leads to a repulsive $q\bar{q}$ interaction, confirmed by lattice at short range, that raises the mass of the hybrid meson. The hybrid wave function has the general form

$$|\Psi_{q\bar{q}g}^{JPC}\rangle = \int \mathrm{d}\mathbf{q}\mathrm{d}\bar{\mathbf{q}}\mathrm{d}\mathbf{g}\, \delta(\mathbf{q}+\bar{\mathbf{q}}+\mathbf{g})\, \Phi_{\lambda\bar{\lambda}\mu}^{JPC}(\mathbf{q},\bar{\mathbf{q}},\mathbf{g})$$
$$\times T_{\mathcal{C}\bar{\mathcal{C}}}^a B_{\lambda\mathcal{C}}^\dagger(\mathbf{q}) D_{\bar{\lambda}\bar{\mathcal{C}}}^\dagger(\bar{\mathbf{q}}) \alpha_\mu^{a\dagger}(\mathbf{g}) |\Omega\rangle. \tag{25}$$

We have extended our previous hybrid study [3] by including the H_{qg} Hamiltonian term containing the $\mathbf{J}^a \cdot \mathbf{A}^a$ operators. Following [6], an effective quark hyperfine interaction with a $\mathbf{J}^a \cdot \mathbf{J}^a$ form is obtained using perturbation theory to second order in g and integrating over the gluonic degrees of freedom. This contribution to the hybrid mass is represented by the $q\bar{q}$ gluon exchange Feynman diagrams in fig. 3 (first two in the bottom row). The non-Abelian magnetic-field terms are also included and entail triple-gluon vertices (last two diagrams in fig. 3). The remaining diagrams represent the self-energy, scattering and quark annihilation mass contributions.

The hyperfine interaction from the H_{qg} term is

$$V_T = \frac{1}{2} \int \int \mathrm{d}\mathbf{x}\mathrm{d}\mathbf{y}\, J_i^a(\mathbf{x}) \hat{U}_{ij}(\mathbf{x},\mathbf{y}) J_j^a(\mathbf{y}), \tag{26}$$

Fig. 3. Hybrid-meson diagrams for $\langle \Psi_{q\bar{q}g}^{JPC} | H_{\mathrm{eff}} | \Psi_{q\bar{q}g}^{JPC} \rangle$.

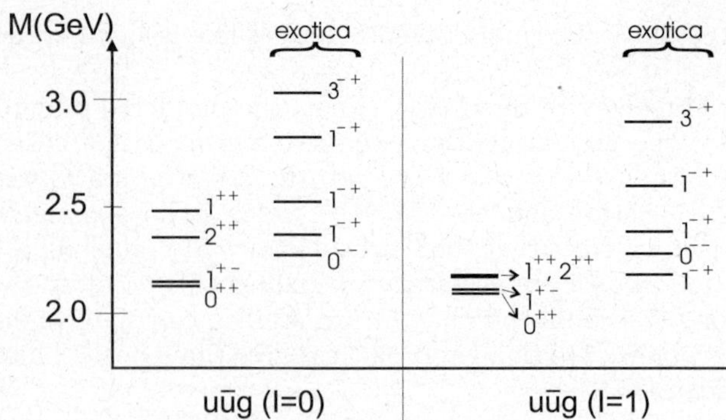

Fig. 4. Low-lying $u\bar{u}g$ spectra.

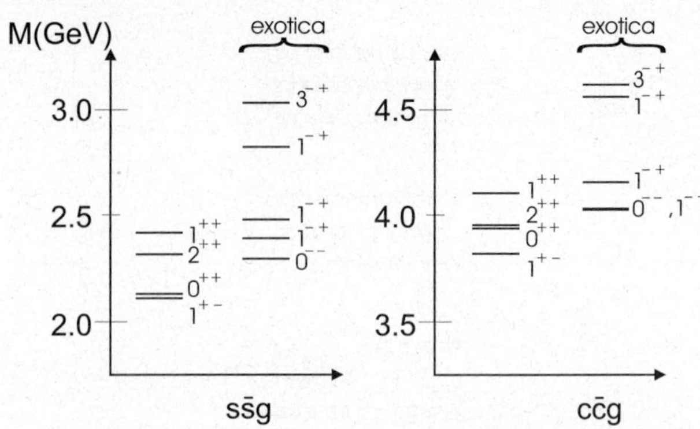

Fig. 5. Low-lying $s\bar{s}g$ and $c\bar{c}g$ spectra.

with kernel reflecting the transverse gauge

$$\hat{U}_{ij}(\mathbf{x},\mathbf{y}) = \left(\delta_{ij} - \frac{\nabla_i \nabla_j}{\nabla^2}\right)_{\mathbf{x}} \hat{U}(|\mathbf{x} - \mathbf{y}|). \qquad (27)$$

The potential $\hat{U}$ is a modified Yukawa with dynamical mass, $m_g = 600\,\text{MeV}$, for the exchanged gluon as explained in [6]. Its momentum space representation is

$$U(p) = \begin{cases} -\dfrac{8.04}{p^2} \dfrac{\ln^{-0.62}\left(\frac{p^2}{m_g^2}+0.82\right)}{\ln^{0.8}\left(\frac{p^2}{m_g^2}+1.41\right)}, & p > m_g, \\ -\dfrac{24.50}{p^2+m_g^2}, & p < m_g. \end{cases} \qquad (28)$$

The quark hyperfine interaction also generates additional terms in the quark gap equation [6].

The 9-dimensional integrals were calculated using the Monte Carlo method and repetitively evaluated with an increasing number of points until a weight-averaged result converged, typically involving about 50 million samples. The hybrid-mass error introduced by this procedure is about $\pm 50\,\text{MeV}$. For each J^{PC} hybrid state we optimized the two variational parameters.

Using standard current quark masses, $m_u = m_d = 5\,\text{MeV}$, $m_s = 80\,\text{MeV}$ and $m_c = 1000\,\text{MeV}$, the predicted low-lying mass spectra for light and heavy hybrid mesons are presented in figs. 4 and 5, respectively. Note that quark annihilation interactions increase the hybrid mass and this introduces isospin splitting since it only contributes in the $I_{q\bar{q}} = 0$ channel. More importantly, all hybrid masses, especially the lightest exotic 1^{-+} state, are clearly above $2\,\text{GeV}$. This is consistent with lattice [13–20] and Flux Tube model [21–23] results summarized in table 2. These composite predictions strongly suggest that the observed 1^{-+} $\pi(1600)$, and more clearly $\pi(1400)$, are not hybrid meson states.

The charmed $c\bar{c}g$ hybrid spectrum has a slightly different level order compared to the strange $s\bar{s}g$ and $u\bar{u}g$ spectra due to the hyperfine interaction. The predicted strange and charmed exotic 1^{-+} states are also in reasonable agreement with both lattice and Flux Tube results.

Table 2. Published predicted exotic 1^{-+} masses, in GeV, for light, strange and charmed hybrid mesons.

Model	u/d hybrid	s hybrid	c hybrid
Lattice QCD [13–20]	1.7–2.1	1.9	4.2–4.4
Flux Tube [21–23]	1.8–2.1	2.1–2.3	4.1–4.5

3.3 Tetraquark systems

This is the first four-body application using this approach. The $SU_c(3)$ color algebra for four quarks produces 81 color states, $3 \otimes \bar{3} \otimes 3 \otimes \bar{3} = 27 \oplus 10 \oplus \overline{10} \oplus 8 \oplus 8 \oplus 8 \oplus 8 \oplus 1 \oplus 1$, of which two are color singlets that can be obtained in four different ways, depending on the intermediate color coupling: singlet scheme (molecule), octet scheme, and two diquark schemes involving the triplet and the sextet representations (see fig. 6).

Working in the cm system and denoting the momenta of the quarks by $\mathbf{q}_1$ and $\mathbf{q}_3$, and those of the anti-quarks by $\mathbf{q}_2$ and $\mathbf{q}_4$, the tetraquark wave function is

$$|\Psi_{4q}^{JPC}\rangle = \int d\mathbf{q}_1 d\mathbf{q}_2 d\mathbf{q}_3 d\mathbf{q}_4 \delta(\mathbf{q}_1 + \mathbf{q}_2 + \mathbf{q}_3 + \mathbf{q}_4)$$
$$\times \Phi_{\lambda_1\lambda_2\lambda_3\lambda_4}^{JPC}(\mathbf{q}_1,\mathbf{q}_2,\mathbf{q}_3,\mathbf{q}_4)R_{\mathcal{C}_3\mathcal{C}_4}^{\mathcal{C}_1\mathcal{C}_2}$$
$$\times B_{\lambda_1\mathcal{C}_1}^\dagger(\mathbf{q}_1)D_{\lambda_2\mathcal{C}_2}^\dagger(\mathbf{q}_2)B_{\lambda_3\mathcal{C}_3}^\dagger(\mathbf{q}_3)D_{\lambda_4\mathcal{C}_4}^\dagger(\mathbf{q}_4)|\Omega_{\text{quark}}\rangle, \quad (29)$$

where the color elements $R_{\mathcal{C}_3\mathcal{C}_4}^{\mathcal{C}_1\mathcal{C}_2}$ depend on the specific color scheme chosen.

Contributions to the Hamiltonian expectation value are summarized in fig. 7 and correspond to 4 self-energy, 6 scattering, 4 annihilation and 70 exchange terms each of which can be reduced to 12-dimensional integrals that

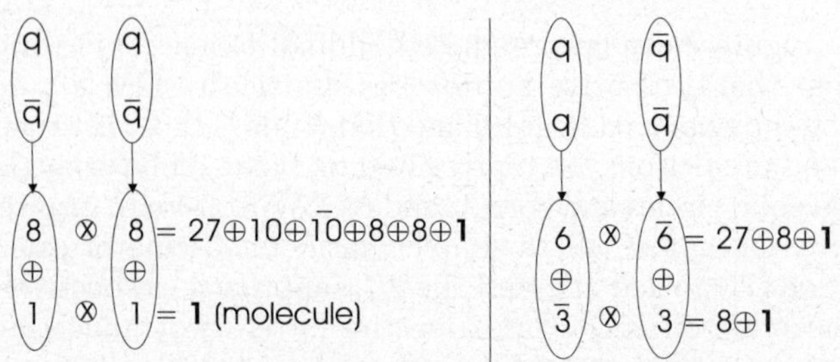

Fig. 6. Color singlets via four different representations.

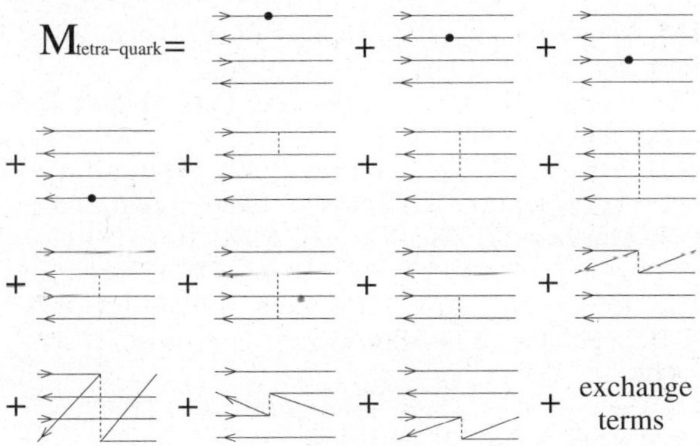

Fig. 7. Tetraquark diagrams for $\langle \Psi_{4q}^{JPC} | H_{\text{eff}} | \Psi_{4q}^{JPC} \rangle$.

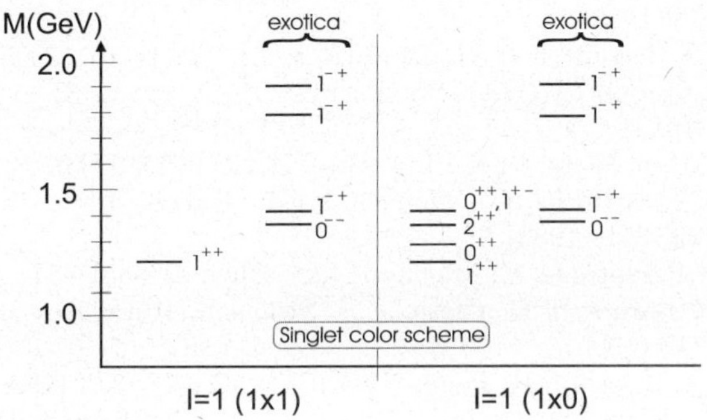

Fig. 8. Tetraquark $I = 1$ singlet scheme (molecule) spectra.

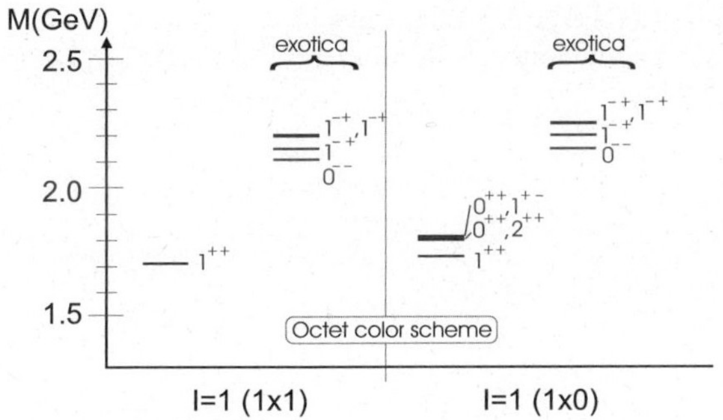

Fig. 9. Tetraquark $I = 1$ octet scheme spectra.

are evaluated in momentum space. Because of the computationally intensive nature of this analysis, the hyperfine interaction was not included. Performing large-scale Monte Carlo calculations (typically 50 million samples), has conclusively determined that the molecular representation (*i.e.* meson-meson) produces the lightest-mass state for a given J^{PC}. This is due to suppression of certain interactions (cancellation of color factors) in the singlet-singlet molecular representation and also the presence of repulsive forces in the other, more exotic, color schemes.

Using $m_u = m_d = 5\,\text{MeV}$, the predicted tetraquark ground state is the nonexotic vector 1^{++} state in the molecular representation with mass around $1.2\,\text{GeV}$. Figures 8 and 9 depict the predicted tetraquark spectra for states having conventional and exotic quantum numbers in both singlet and octet color representations. Due to quark annihilation interactions ($q\bar{q} \rightarrow g \rightarrow q\bar{q}$) in the $I_{q\bar{q}} = 0$ channel there are isospin splitting contributions,

up to several hundred MeV, in the octet but not singlet scheme as illustrated in the figures. The annihilation interaction terms are repulsion, yielding octet states with $I = 2$ lower than the $I = 1$ which are lower than the $I = 0$. The molecular-type states are all isospin degenerate and the lightest exotic molecule is the 0^{--} with mass $1.35\,\text{GeV}$. Because of the isospin degeneracy, there will be several molecular-type tetraquark states with the same J^{PC} in the 1 to $2\,\text{GeV}$ region. Further, these states can be observed in different electric charge channels (different I_z) at about the same energy, which is a useful experimental signature. There are 3 orbital angular momenta and the lightest 1^{-+} exotics have only 1 p-wave. The lightest 1^{-+} is predicted near $1.4\,\text{GeV}$ which is close to the observed $\pi(1400)$, suggesting this state has a meson-meson-type structure distinct from states having quarks in more exotic color representations (quark atoms). Related, the computed tetraquark mass for 1^{-+} states with these more exotic color configurations is above $2\,\text{GeV}$. This is consistent with model predictions [8] for exotic hybrid meson ($q\bar{q}g$) 1^{-+} states also lying above $2\,\text{GeV}$ due to repulsive color octet quark interactions. Finally, for the same J^{PC} states, including the 1^{-+}, the computed masses (not shown) in both the triplet and the sextet diquark color representations are all heavier than in the singlet representation and comparable to the octet scheme results.

4 Conclusions

Summarizing, using the Coulomb gauge Hamiltonian model we have performed large-scale Monte Carlo calculations for ggg glueballs, $q\bar{q}g$ hybrid mesons and $q\bar{q}q\bar{q}$ tetraquark systems. Our oddball spectrum, along with lattice and other glueball model predictions, clearly documents an odderon trajectory with slope similar to the pomeron but much lower intercept. This explains why the odderon has not been seen in reactions with pomeron exchange. If the odderon intercept is comparable to the ω value it may be possible to observe it in reactions where the pomeron is absent such as pseudoscalar [24] or tensor meson [25] electromagnetic production. However, if our prediction that the intercept is below 0.5 is correct, it is unlikely that the odderon will be seen. It is important therefore that lattice calculations for the 5^{--} be performed to confirm this.

For the hybrid-meson investigation, the Coulomb gauge Hamiltonian model predicts that all $u\bar{u}g$ and $u\bar{d}g$ states have mass above $2\,\text{GeV}$. In particular, lattice and constituent model predictions for light, strange and charmed exotic 1^{-+} hybrid mesons are collectively in agreement. This strongly suggests that $\pi_1(1400)$ and $\pi_1(1600)$ are not hybrid mesons but have an alternative structure as argued below.

Our tetraquark results clearly show that $[(q\bar{q})_1 \otimes (q\bar{q})_1]_1$ molecular states are lighter than the more exotic color octet $[(q\bar{q})_8 \otimes (q\bar{q})_8]_1$ atomic-like states. Most significantly, for the 1^{-+}, $I = 1$ channel our predicted lightest color octet state is above $2\,\text{GeV}$, while the lightest singlet scheme state is around 1400 MeV, close to $\pi(1400)$. Combining these results with several model 1^{-+} hybrid-meson predictions indicates that the observed π states are not hybrids or exotic color configurations of 4 quarks but rather more conventional meson-meson–type molecules.

Future work will address charmed ($c\bar{c}u\bar{u}$) and strange tetraquark ($c\bar{c}s\bar{s}$) systems to study the recently reported $X(3872)$ and $Y(4260)$ states. Dynamical mixing of glueball, hybrid and conventional meson and tetraquark states will also be investigated along with three-body forces [26].

The authors are very grateful to F.J. Llanes-Estrada and P. Bicudo for insightful discussions. Work supported by U.S. DOE grants DE-FG02-97ER41048 and DE-FG02-03ER41260.

References

1. A.P. Szczepaniak, E.S. Swanson, C.R. Ji, S.R. Cotanch, Phys. Rev. Lett. **76**, 2011 (1996).
2. F.J. Llanes-Estrada, S.R. Cotanch, Phys. Rev. Lett. **84**, 1102 (2000).
3. F.J. Llanes-Estrada, S.R. Cotanch, Phys. Lett. B **504**, 15 (2001).
4. F.J. Llanes-Estrada, S.R. Cotanch, Nucl. Phys. A **697**, 303 (2002).
5. F.J. Llanes-Estrada, S.R. Cotanch, P. Bicudo, J.E. Ribeiro, A.P. Szczepaniak, Nucl. Phys. A **710**, 45 (2002).
6. F.J. Llanes-Estrada, S.R. Cotanch, A.P. Szczepaniak, E.S. Swanson, Phys. Rev. C **70**, 035202 (2004).
7. F.J. Llanes-Estrada, P. Bicudo, S.R. Cotanch, Phys. Rev. Lett. **96**, 081601 (2006).
8. I.J. General, S.R. Cotanch, F.J. Llanes-Estrada, arXiv:hep-ph/0609115.
9. G.P. Lepage, J. Comput. Phys. **27**, 192 (1978); Cornell University Report CLNS 80-447, 1980, unpublished.
10. C.J. Morningstar, M. Peardon, Phys. Rev. D **60**, 034509 (1999).
11. H.B. Meyer, M. Teper, Phys. Lett. B **605**, 344 (2005).
12. A.B. Kaidalov, Y.A. Simonov, Phys. Lett. B **477**, 163 (2000).
13. C. Bernard *et al.*, Phys. Rev. D **56**, 7039 (1997).
14. C. Bernard *et al.*, Nucl. Phys. (Proc. Suppl.) B **73**, 264 (1999).
15. P. Lacock, K. Schilling, Nucl. Phys. (Proc. Suppl.) B **73**, 261 (1999).
16. J.N. Hedditch *et al.*, Phys. Rev. D **72**, 114507 (2005).
17. X.Q. Luo, Z.H. Mei, Nucl. Phys. (Proc. Suppl.) B **119**, 263 (2003).
18. Y. Liu, X. Q. Luo, Phys. Rev. D **73**, 054510 (2006).
19. L.A. Griffiths, C. Michael, P.E.L. Rakow, Phys. Lett. B **129**, 351 (1983).
20. S. Perantonis, C. Michael, Nucl. Phys. B **347**, 854 (1990).
21. T. Barnes, F.E. Close, E.S. Swanson, Phys. Rev. D **52**, 5242 (1995).
22. F.E. Close, P.R. Page, Nucl. Phys. B **443**, 233 (1995).
23. K. Waidelich, Diploma Thesis, North Carolina State University (2001).
24. E.R. Berger *et al.*, Eur. Phys. J. C **9**, 491 (1999).
25. E.R. Berger *et al.*, Eur. Phys. J. C **14**, 673 (2000).
26. A.P. Szczepaniak, P. Krupinski, Phys. Rev. D **73**, 116002 (2006).

Eur. Phys. J. A **31**, 662–664 (2007)
DOI 10.1140/epja/i2006-10192-7

Special Article – QNP 2006

Composite-meson model and meson exchange currents

R.Ya. Kezerashvili[a] and V.S. Boyko

Physics Department, New York City College of Technology, The City University of New York, Brooklyn, NY 11201, USA

Received: 8 October 2006
Published online: 22 February 2007 – © Società Italiana di Fisica / Springer-Verlag 2007

Abstract. We study the meson exchange currents (MEC) mechanism for the pion double-charge exchange (DCX) reaction in a composite-meson model. The model assumes that the mesons are two-quark systems and can interact with each other only through quark loops. The contributions of the ρ, σ, and f_0 mesons, the four-quark box diagram as well as a contact diagram has been taken into account. It is shown that the contribution of the ρ, σ, and f_0 mesons increases the forward scattering cross-section in an average by 25% and decreases with energy.

PACS. 25.80.Gn Pion charge-exchange reactions – 25.80.Hp Pion-induced reactions

The most important problem of the DCX of π-mesons on nuclei is the mechanism of the reaction. Normally, the reaction is dominated by the sequential mechanism (SQM), in which the incoming pion undergoes two sequential single-charge exchange scatterings on nucleons within a nucleus. In ref. [1] the MEC mechanism, was proposed and it consists of the assumption that the incident pion is scattered on the off-shell virtual pion exchanged by the nucleons within the nucleus and the DCX takes place at the $\pi\pi$ vertex as shown in fig. 1a. Later, ref. [2] introduced an additional diagram shown in fig. 1b and concluded that the MEC effects would be small for an analog DCX because this diagram partially cancels the contribution of the pole term. The MEC issue has been revived in refs. [3] and [4] using the Lovelace-Veneziano model and in refs. [5–9] based on an effective Lagrangian method. All calculations of the MEC in the effective Lagrangian formalism are performed in the lowest order which includes the contribution of the "trees" diagram and no pionic or baryonic closed loop diagrams are included. The tree diagrams correspond to the Born approximation for $\pi\pi$ scattering, and their contribution is defined by the first term of the expansion of the $\pi\pi$ amplitude in terms of $1/F_\pi^2$. In ref. [9], the authors showed that because of the small DCX cross-section produced by the SQM at high energies, the MEC mechanism could show up, even in the Born approximation. Below we study the MEC mechanism for the DCX reaction based on the composite-meson model. The model assumes that the mesons are two-quark systems and can interact with each other only through quark loops. We are considering the diagrams shown in fig. 2 which successfully describe $\pi\pi$ scattering, as well as the contact diagram in

a e-mail: rkezerashvili@citytech.cuny.edu

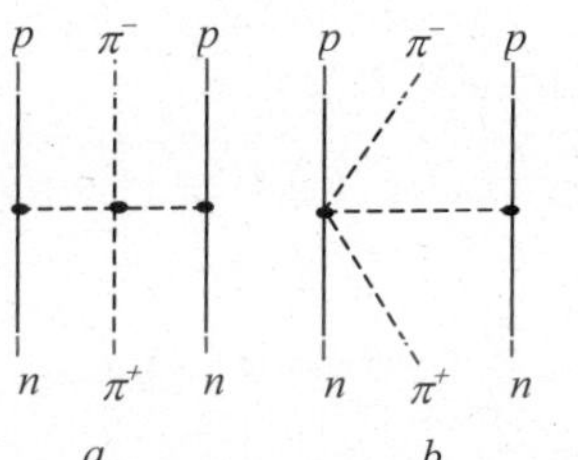

Fig. 1. The pole (a) and contact (b) diagrams.

fig. 1b. This approach allows to find the contribution of the σ, f_0 and ρ mesons to the MEC mechanism.

The chiral symmetry requires consideration of both diagrams in fig. 1, which describe the pion DCX on two nucleons through the MEC mechanism. We replaced the diagram 1a by the diagram on the left side in fig. 2, where the shaded vertex block corresponds to the $\pi\pi$ interaction and constructed the amplitude for the process $\pi^+ nn \to \pi^- pp$ using the amplitudes for $\pi\pi$ scattering and for the πN-$\pi\pi N$ process. The amplitude corresponding to the graph on the left side in fig. 2 in the nonrelativistic limit is

$$T_1 = (i\sqrt{2}g)^2 \frac{(\boldsymbol{\sigma}_1 \cdot \mathbf{q}_1)(\boldsymbol{\sigma}_2 \cdot \mathbf{q}_2)}{4m^2} \frac{M_{\pi^+\pi^-}(q_{1\mu}; q_{2\mu})}{(\mathbf{q}_1^2 + m_\pi^2)(\mathbf{q}_2^2 + m_\pi^2)}, \quad (1)$$

where g is the πN coupling constant, m is the mass of the nucleon, $\mathbf{q}_1 = \mathbf{p}_1' - \mathbf{p}_1$ and $\mathbf{q}_2 = \mathbf{p}_2' - \mathbf{p}_2$, m_π is the pion mass, and $q_{1\mu}$ and $q_{2\mu}$ are the 4-momenta of virtual mesons. In eq. (1) $M_{\pi^+\pi^-}$ is the transition matrix element for the $\pi^+\pi^- \to \pi^+\pi^-$ process. Similarly, we can find the amplitude for fig. 1b using the transition matrix for the πN-$\pi\pi N$ process.

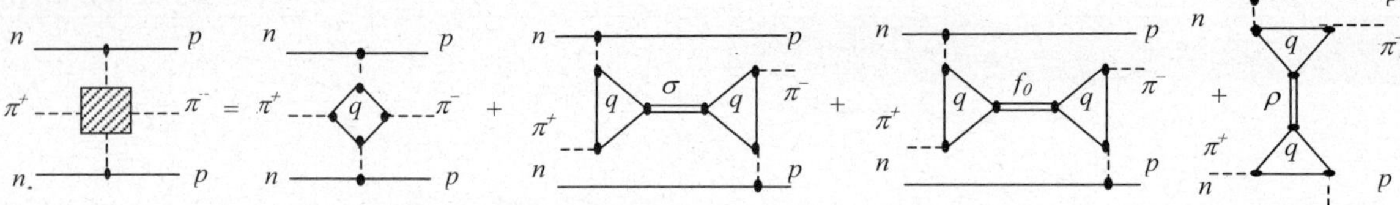

Fig. 2. The quark structure of the MEC diagram.

In the composite-meson model the shaded vertex box in fig. 2 can be considered at the quark level and includes the sum of the quark diagrams which successfully describe $\pi\pi$ scattering. Indeed, $\pi^+ = u\bar{d}$ and $\pi^- = \bar{d}u$ are two-quark systems, and can interact with each other only through quark loops. Therefore, in the single-loop approximation, the diagram on the left side in fig. 2 can be expanded as shown. The first quark box diagram actually corresponds to the pole diagram in fig. 1a for $\pi\pi$ scattering and a part of its contribution which is proportional to $1/F_\pi^2$ represents the Born approximation in the effective Lagrangian method. The choice of the other diagrams representing ρ, σ and f_0 mesons is based on the probability of the two-pion decay of mesons: $\rho(770) \to \pi\pi$, $\sigma(600) \to \pi\pi$ and $f_0(980) \to \pi\pi$. The inner quark blocks of the diagrams in fig. 2 can be easily obtained from the Lagrangian [10] and the set of the Lagrangians relevant to the processes of figs. 1b and 2 are

$$L = \overline{Q}\left[i\partial - m_q + \frac{m_q}{F_\pi}\left(\sigma\sin\alpha + f_0\cos\alpha + i\gamma_5\boldsymbol{\tau}\boldsymbol{\pi}\right) + \frac{g_\rho}{2}\boldsymbol{\tau}\boldsymbol{\rho}\right]Q,$$
$$(2)$$

$$L_{NN\pi} = \frac{g}{2m}\overline{\psi}\gamma^\mu\gamma_5\boldsymbol{\tau}\psi\cdot(\partial_\mu\boldsymbol{\pi}),$$
$$(3)$$

$$L_{NN\pi\pi} = -\frac{g}{2m}\frac{1}{4F_\pi^2}\overline{\psi}\gamma^\mu\gamma_5\boldsymbol{\tau}\psi\cdot(\partial_\mu\boldsymbol{\pi})\boldsymbol{\pi}^2,$$
$$(4)$$

where Q is the quark field, m_q is the mass of the quark, g_ρ is the decay constant of the ρ-meson, and α is the mixing angle. To find the contributions of the diagrams in fig. 2 we need to know all effective coupling constants and express them in terms of two decay constants g_ρ and F_π of the $\rho \to 2\pi$ and $\pi \to \mu\bar{\nu}$ decays. We shall impose a natural requirement that on the mass shell the form factors of the processes should coincide exactly with the corresponding physical coupling constants. The form factors for the $\rho \to 2\pi$, $\sigma \to 2\pi$ and $f_0 \to 2\pi$ decays can be obtained from the Lagrangian (2) and finally the results for the inner blocks of the diagrams in fig. 2 (for $\pi^+\pi^- \to \pi^+\pi^-$ amplitude) and the diagram in fig. 1b are

$$M_{\pi^+\pi^-} = -\frac{m_\pi^2}{F_\pi^2} + \frac{m_q^2}{(\pi F_\pi)^2}\left[\frac{s+t}{F_\pi^2} - \frac{m_q^2}{(4\pi F_\pi)^2}\frac{s+t}{F_\pi^2}\right]$$

$$+\frac{4m_q^2}{F_\pi^2}\left\{\frac{4m_q^2}{m_\sigma^2}\left(\frac{s}{(m_\sigma^2 - s)} + \frac{t}{(m_\sigma^2 - t)}\right)\sin^2\alpha\right.$$

$$+\frac{4m_q^2}{m_{f_0}^2}\left(\frac{s}{(m_{f_0}^2 - s)} + \frac{t}{(m_{f_0}^2 - t)}\right)\cos^2\alpha\right\}$$

$$+g_\rho\left\{\frac{t-u}{m_\rho^2 - s}\left(1 + \frac{s - m_\rho^2}{8\pi^2 F_\pi^2}\right)^2 + \frac{s-u}{m_\rho^2 - t}\left(1 + \frac{t - m_\rho^2}{8\pi^2 F_\pi^2}\right)^2\right\}, \quad (5)$$

$$T_2 = -\frac{g^2}{4m^2}\frac{2}{F_\pi^2}\frac{(\boldsymbol{\sigma}_1\cdot\mathbf{q}_1)[\boldsymbol{\sigma}_2\cdot\mathbf{q}_1 - \boldsymbol{\sigma}_2\cdot\mathbf{q}_2]}{(\mathbf{q}_1^2 + m_\pi^2)}. \quad (6)$$

In eq. (5) s, t, and u are the standard Mandelstam variables, the mixing angle α is determined from $\Gamma_{f_0\to 2\pi} = 26\,\mathrm{MeV}$ and it is $\alpha = 17°$ if $m_q = 280\,\mathrm{MeV}$ [11], and the decay constant of the ρ-meson g_ρ is $g_\rho^2/4\pi \approx 3$. The contribution of the quark box diagram depends on the undetermined parameter, whose value was fixed so that the experimental value of the pion-pion scattering length a_0^2 is obtained.

Let us compare the MEC amplitude in the composite-meson model given by (5) with one in the Born approximation for the Weinberg Lagrangian [12]: $M_{\pi^+\pi^-} = -m_\pi^2/F_\pi^2 + (s+t)/F_\pi^2$. It is easy to see that the constant term in eq. (5) is the same. The term in the square brackets is caused by the quark box diagram and inclusion of the so-called q^2 terms (the q^2 terms leads to convergent integrals [13]) in the quark box diagrams, and it is partially proportional to $(1/F_\pi^2)^2$. The factor in front of the square brackets $m_q^2/(\pi F_\pi)^2 = 1$ for $F_\pi = 89.2\,\mathrm{MeV}$ and $m_q = 280\,\mathrm{MeV}$ and a variation of this factor is nonsignificant when the pion decay constant varies from $87\,\mathrm{MeV}$ to $93\,\mathrm{MeV}$. Therefore, the constant term and the first term in the square brackets in eq. (5) give the Weinberg transition amplitude for the $\pi^+\pi^- \to \pi^+\pi^-$ process. The two terms in (5) related to the mixing angle α are representing the contribution of the σ and f_0 mesons. The term that follows the factor g_ρ is the contribution of the ϱ-meson and it has the parts which are also proportional to $(1/F_\pi^2)^2$. After the substitution of eq. (5) into (1) one can separate the part which corresponds to the amplitude of the pole diagram in fig. 1a for the Weinberg Lagrangian. The sum of this part and (6) gives the contribution of the quark box diagram, including only terms proportional to $1/F_\pi^2$ and the contact diagram and is the same as the sum of contributions of the pole (fig. 1a) and contact (fig. 1b) diagrams in the Born approximation. The rest of the contribution of the quark box diagram is proportional to $(1/F_\pi^2)^2$ and represents the deviation from the Born approximation.

We calculated the forward scattering cross-section for the $^{18}\mathrm{O}(\pi^+, \pi^-)^{18}\mathrm{Ne}$ reaction for the incident pion energy from $600\,\mathrm{MeV}$ to $1400\,\mathrm{MeV}$ and compare the contribution of the MEC evaluated in the composite-meson model with the calculations in the Born approximation. We use the shell model wave functions to describe $^{18}\mathrm{O}$ and $^{18}\mathrm{Ne}$ nuclei, and assuming that the DCX process takes place on the valence neutrons and leads to the double isobaric analog state. The wave function of the two odd neutrons in $^{18}\mathrm{O}$ has been used as in [3,6] with the harmonic-oscillator parameter $\alpha^2 = 0.32\,\mathrm{fm}^{-2}$. Since our main points are to

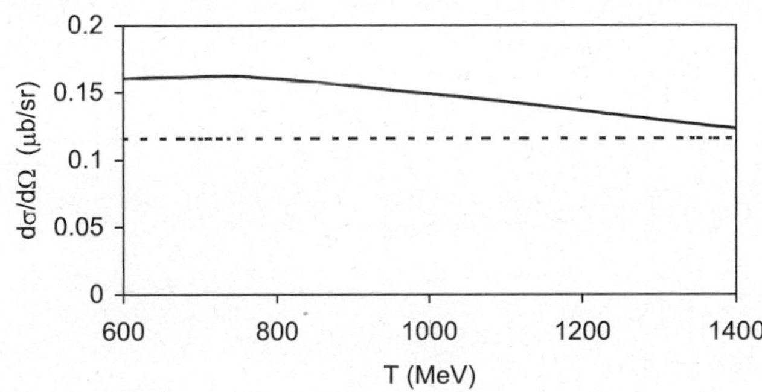

Fig. 3. Forward scattering cross-section for the reaction ^{18}O$(\pi^+,\pi^-)^{18}$Ne as a function of the incident pion energy. The contributions of the MEC in the composite-meson model with the contact diagram are shown by the solid curve, and the pole and contact diagrams in the Born approximation by the dashed line.

understand the contribution of three $\pi\pi$ resonances ρ, σ and f_0 to the MEC mechanism, establish the order of magnitude of this contribution and compare it to the conventional MEC mechanism, follow [7], for simplicity, we neglect the distortion of the pion waves, using the plane waves instead. In the plane-wave approximation integration of the sum of the contributions of the quark box diagram, including only terms proportional to $1/F_\pi^2$ and the contact diagram in fig. 1b over momentum q in the target, can be performed analytically and the result is

$$T^B = \frac{2}{F_\pi^2}\frac{1}{m_\pi}\frac{\partial}{\partial m_\pi}V(r), \qquad (7)$$

where $V(r)$ is a one-pion exchange potential.

The results of the calculations for the forward scattering cross-section as a function of the incoming pion kinetic energy are presented in fig. 3. The results of calculations for the composite-meson model with the contact diagram in fig. 3 are presented by a solid line. The cross-section for the MEC mechanism in the Born approximation, when both the pole term as well as the contact term are considered, is presented by the dashed line. The comparison of these results shows that the inclusion of the ρ, σ and f_0 mesons increases the cross-section and their contribution decreases with energy. The cross-section for the MEC in the composite-meson model with the contact diagram in fig. 1b is systematically larger in an average by 25% than that in the Born approximation. We checked the sensibility of the calculations to the changes of the masses of the σ and f_0 mesons. The cross-section is sensitive to the small

changes of the σ-meson mass and almost did not change with small changes of the f_0-meson mass. For example, the change of the σ-meson mass from 780 MeV to 700 MeV will increase the cross-section by about 14%. Let us also mention that the third term in eq. (5) related to the ϱ-meson contribution, appreciably changes the cross-section, and it shows the importance of the ϱ-meson at energies above 600 MeV. At energies above 1000 MeV the cross-section for the MEC mechanism with inclusion of the ρ, σ and f_0 mesons becomes larger than one for the SQM and that mechanism dominates in the reaction.

Thus, we can conclude that at the considered energy region the MEC mechanism in the composite-meson model with the contact diagram can reveal in the pion DCX, because it has a substantial contribution and the inclusion of the ρ, σ and f_0 mesons increases the contribution of the meson exchange currents in the MEC mechanism for the DCX reaction. It is important to mention that the distortion of the pion waves will generally reduce the cross-section in the composite-meson model as well as for the MEC in the Born approximation and for the SQM, but it will not change the conclusion of the importance of including the pion resonances into the MEC mechanism.

References

1. J.F. Germond, C. Wilkin, Lett. Nuovo Cimento **13**, 605 (1975).
2. M.R. Robilotta, C. Wilkin, J. Phys. G: Nucl. Phys. **4**, L115 (1978).
3. E. Oset, D. Strottman, M.J. Vicente Vacas, Ma Wei-hsing, Nucl. Phys. A **408**, 461 (1983).
4. N. Auerbach, W.R. Gibbs, J.N. Ginocchio, W.B. Kaufmann, Phys. Rev. C **38**, 1277 (1988).
5. R.I. Jibuti, R.Ya. Kezerashvili, Nucl. Phys. A **437**, 687 (1985).
6. Yu. Zi-quang, Cai Chong-hai, Ma Wei-hsing, Zhao Shuping, Phys. Rev. C **38**, 272 (1988).
7. M.F. Jiang, D.S. Koltun, Phys. Rev. C **42**, 2662 (1990); D.S. Koltun, M.F. Jiang, Phys. Lett. B **273**, 6 (1991).
8. M.B. Johnson, E. Oset, H. Sarafian, E.R. Siciliano, M.J. Vicente Vacas, Phys. Rev. C **44**, 2480 (1991).
9. L. Alvarez-Ruso, M.J. Vicente Vacas, J. Phys. G: Nucl. Part. Phys. **22**, L45 (1996).
10. M.K. Volkov, A.A. Osipov, Sov. J. Nucl. Phys. **39**, 440 (1984).
11. K. Kikkawa, Prog. Theor. Phys. **56**, 947 (1976)
12. S. Weinberg, Phys. Rev. Lett. **17**, 616 (1966); **18**, 188 (1967).
13. M.K. Volkov, D.V. Kreopalov, Theor. Math. Phys. **57**, 21 (1983).

Eur. Phys. J. A **31**, 665–671 (2007)
DOI 10.1140/epja/i2006-10283-5

Special Article – QNP 2006

Unraveling the f_0 nature by connecting KLOE and BABAR data through analyticity

S. Pacetti[a]

Laboratori Nazionali di Frascati, INFN, Frascati, Italy

Received: 18 December 2006
Published online: 23 March 2007 – © Società Italiana di Fisica / Springer-Verlag 2007

Abstract. We define a general procedure, based on analyticity and dispersion relations, to estimate low-energy amplitudes for processes like: $\phi \to e^+ e^- M$ and $\phi \to \gamma M$, starting from cross-section data on $e^+ e^- \to \phi M$, where M is a generic light scalar or pseudoscalar meson. In particular this procedure is constructed to obtain predictions on the radiative decay rate which are crucially linked on the assumed quark structure for the meson M under consideration. Three cases are analyzed: $M = \eta$, $M = f_0(q\bar{q})$ and $M = f_0(qq\bar{q}\bar{q})$. While in the η case the estimate of the branching fraction for the radiative decay $\phi \to \eta\gamma$ is in agreement with the data, in the case of f_0, such agreement is obtained only under the hypothesis of a tetraquark scalar meson.

PACS. 11.55.-n S-matrix theory; analytic structure of amplitudes – 13.25.-k Hadronic decays of mesons – 13.40.Gp Electromagnetic form factors

1 Introduction

1.1 The known properties of the $f_0(980)$ scalar meson

The $f_0(980)$ is the more extensively studied isospin zero scalar meson [1]. The properties of such a meson are very peculiar and, in some cases, in spite of the deep experimental investigation, they are still poorly known. In fact, while the mass is rather well measured $M_{f_0} = 980 \pm 10\,\text{MeV}$ [1], the width is known with a large uncertainty: $\Gamma_{f_0} = 40\text{–}100\,\text{MeV}$ [1]. This is a consequence of the scalar nature of this meson, which does not allow unambiguous interpretations when the resonance is studied by means of the usual Breit-Wigner analysis. The f_0 has the quantum number of the vacuum, the Higgs boson and also of the expected lowest-lying glueball [2].

Usually, this scalar meson is considered as a "bridge" between light (u and d) and strange quarks because it can be produced by both strange and non-strange particles.

1.2 Long outstanding questions

By following the standard quark model [1], where mesons are described as colorless $q\bar{q}$ bound states and classified in J^{PC} multiplets, we expect there should be the conventional flavor $SU(3)$ scalar nonet (0^{++}):

$$q^3 \otimes \bar{q}^3 = [\bar{q}q]^1 \oplus [\bar{q}q]^8. \tag{1}$$

In such a scenario the f_0 represents the singlet and, due to the large $s\bar{s}$ component, it should be the heaviest meson of this multiplet, in contradiction with the experimental data which provide: $M_{f_0} \simeq M_{a_0}$ and no degeneracy between a_0 and σ_0 ($M_{a_0} > M_{\sigma_0} \sim 450\,\text{MeV}$). Also the absolute values of the masses are not in agreement with the expectation, in fact, being $q\bar{q}$ bound in P-wave, these masses should lie in the $> 1\,\text{GeV}$ energy region of other P-wave states. This means that the $q\bar{q}$ bound state is only a small component of the f_0 Fock space.

Since some years, the interpretations in terms of tetraquark states (in different combinations) [3–9] is the object of deep theoretical and experimental investigation and also in this work we will focus on this possibility.

2 Probing the intrinsic nature of f_0

2.1 Common practice

At present, the main source of information on the f_0 structure is the radiative decay $\phi \to f_0\gamma$ [9], whose surprisingly high branching fraction is interpreted in terms of an abnormally strong affinity between the ϕ vector meson and the f_0 itself. Such an affinity has been described in different ways by different authors, even though, each of them put the cue on the key role played by the coupling $g_{KK}^{f_0}$. In particular, there are two schools of thought, the first one describes this coupling by introducing in the $\phi \to f_0\gamma$ amplitude a $K\overline{K}$-loop [3], while the second, in order to

[a] e-mail: simone.pacetti@lnf.infn.it

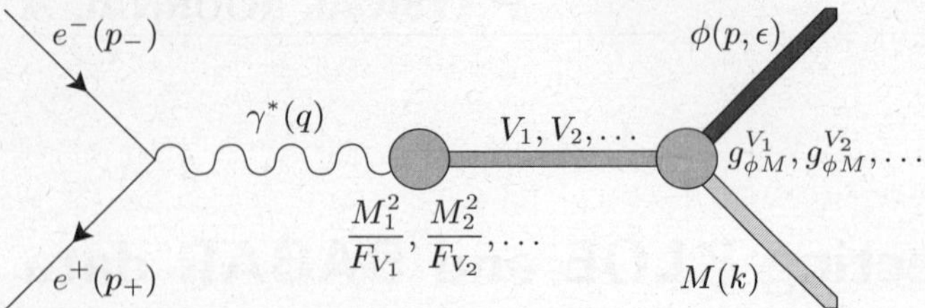

Fig. 1. Feynman diagram for the $e^+e^- \to \phi M$ in Born approximation, with intermediate vector mesons V_j.

avoid any model-dependent description, adopts a usual parameterization in terms of meson propagators with free couplings [6]. In both these descriptions, with different languages, the message is the same: the coupling $(g^{f_0}_{KK})^2$ is more than twice the $(g^{f_0}_{\pi\pi})^2$ [10].

2.2 The strategy

The idea of this work consists in probing the intrinsic nature of the f_0 by considering different processes to investigate the transition form factor (tff) which describes the coupling $\phi f_0 \gamma$. All the information on the tff are then used under the constraint of an assumed f_0 quark structure, which strongly affects every step of the procedure.

3 First step: parameterization in the resonance region $[3\mathrm{M}_\pi, \sim 3\,\mathrm{GeV}]$

The amplitude of $e^+e^- \to \phi M$ (fig. 1) is parameterized in terms of a tff, which describes the coupling between the virtual photon and the intermediate vector mesons V_j, i.e.:

$$\mathcal{M}_{\mathrm{Annihi}} = \sum_j^{N_V} \frac{g^{V_j}_{\phi M}}{M_j^2 - q^2 - i\Gamma_j M_j} \frac{M_j^2}{F_{V_j}} \epsilon_\nu \mathcal{T}_M^{\mu\nu} \frac{1}{q^2} J_\mu, \quad (2)$$

where $J_\mu = -ie\overline{v}(p_+)\gamma_\mu u(p_-)$, ϵ_ν is the ϕ polarization vector and, F_{V_j} and $g^{V_j}_{\phi M}$ are couplings with the initial and the final state. The second-rank tensors $\mathcal{T}_M^{\mu\nu}$ accounts for the dynamics of the meson M.

The formal expression of the ϕM tff, in the resonance region, as extracted from the amplitude of eq. (2), is

$$F^{\mathrm{Res}}_{\phi M}(q^2) = \sum_j^{N_V} \frac{M_j}{eF_{V_j}} \frac{g^{V_j}_{\phi M}}{\Gamma_j} \frac{\Gamma_j M_j}{M_j^2 - q^2 - i\Gamma_j M_j}. \quad (3)$$

This parameterization integrates all the information about the meson M structure[1]. In case of radiative decay, by using the previous arguments, the amplitude is:

$$\mathcal{M}_{\mathrm{Rad}} = F_{\phi M}(0)\,\mathcal{T}_M^{\mu\nu}\,\epsilon_\mu(p)\epsilon_\nu^*(q), \quad (4)$$

---

[1] E.g., for $M \equiv f_0$, a strong affinity of the f_0 with the $K\overline{K}$ intermediate-state (kaon loop) should manifest itself in an enhancement of the coupling $g^\phi_{\phi f_0}$, since in the s-channel the $K\overline{K}$ state is almost completely resonant in $\phi(1020)$.

it is constant and proportional to the value of the tff at $q^2 = 0$. To give the explicit expressions of the second-rank tensors $\mathcal{T}_M^{\mu\nu}$ we have to specify the nature of the mesons under consideration. We will study two cases

$$\mathcal{T}_\eta^{\mu\nu} = \varepsilon^{\mu\nu\rho\sigma} p_\rho q_\sigma, \qquad \mathcal{T}_{f_0}^{\mu\nu} = p^\mu q^\nu - g^{\mu\nu}(pq), \quad (5)$$

where p and q are 4-momenta as labeled in fig. 1.

The formal expressions for total annihilation cross-sections and decay rates are (eqs. (2), (5))

$$\sigma_{\phi\eta}(s) = \rho^{\mathrm{kin}}_{\phi\eta}(s)|F_{\phi\eta}(s)|^2, \quad \Gamma_{\phi\eta} = \tau^{\mathrm{kin}}_{\phi\eta} F_{\phi\eta}(0)^2,$$
$$\sigma_{\phi f_0}(s) = \rho^{\mathrm{kin}}_{\phi f_0}(s)|F_{\phi f_0}(s)|^2, \quad \Gamma_{\phi f_0} = \tau^{\mathrm{kin}}_{\phi f_0} F_{\phi f_0}(0)^2, \quad (6)$$

where $s = q^2$. These quantities are products between kinematic factors ($\rho^{\mathrm{kin}}(s)$ and τ^{kin}), which, depending on the nature of the meson M, are different in the two cases ($M \equiv \eta$ and $M \equiv f_0$), and the modulus squared of the corresponding tff's.

3.1 Step two: selection of the contributions

In this so-called resonance region, i.e., from the theoretical threshold, which is $s_0 = (3M_\pi)^2$ since the isoscalar final state, up to $\sim (3\,\mathrm{GeV})^2$, we adopt the parameterization of eq. (3). The first criterion used to select the vector mesons which contribute to the tff's is the quantum number conservation. For both $\phi\eta$ and ϕf_0 final state we have: $I^G(J^{PC}) = 0^-(1^{--})$, hence only contributions from the ω- and ϕ-family are expected. However, if we consider explicitly the structure of the mesons in terms of valence quarks, assuming the f_0 as $q\overline{q}$ bound state, we have

$$|\phi\rangle = s\overline{s} \qquad \begin{aligned} |\eta\rangle &= X_\eta|u,d;-\rangle + Y_\eta|s;-\rangle, \\ |f_0\rangle &= X_{f_0}|u,d;+\rangle + Y_{f_0}|s;+\rangle, \end{aligned} \quad (7)$$

with the normalization: $X^2_{\eta,f_0} + Y^2_{\eta,f_0} = 1$ and where the state $|q_1, q_2, \ldots; P\rangle$ indicates a normalized combination of $q_1\overline{q}_1$, $q_2\overline{q}_2$, etc., with parity P and $J^C = 0^+$.

While the ϕ-family contributions are allowed (fig. 2), the ω-family contributions are instead OZI-forbidden [11]. In fact, as it is shown in fig. 3, the corresponding Feynman diagrams have disconnected flavor lines in the vector sector.

It follows that, in light of the quantum number conservation and the OZI rule, only ϕ-family contributions are

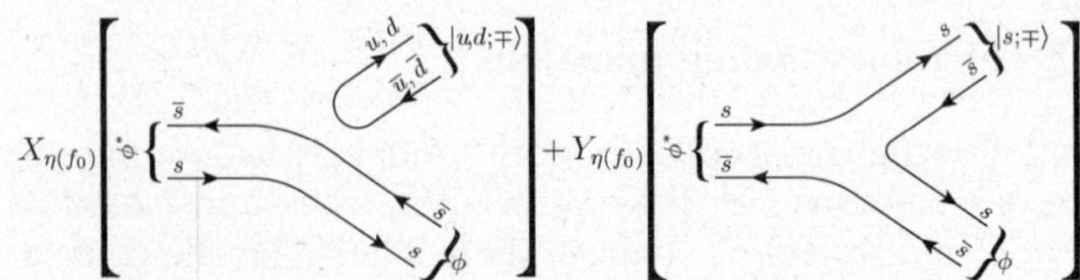

Fig. 2. Feynman diagram for $\phi^* \to \phi\eta(f_0)$ with: $|\phi\rangle = s\overline{s}$ and $|\eta(f_0)\rangle = X_{\eta,f_0}|u,d;\mp\rangle + Y_{\eta,f_0}|s;\mp\rangle$. The two components $|u,d;\mp\rangle$ and $|s;\mp\rangle$ are separately shown.

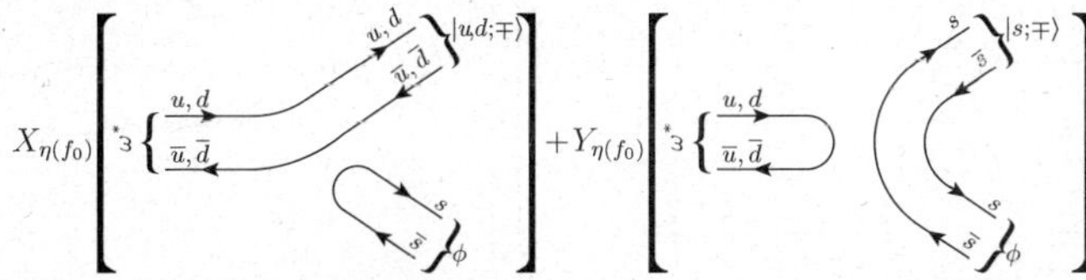

Fig. 3. Feynman diagram for $\omega^* \to \phi\eta(f_0)$ with: $|\phi\rangle = s\bar{s}$ and $|\eta(f_0)\rangle = X_{\eta,f_0}|u,d;\mp\rangle + Y_{\eta,f_0}|s;\mp\rangle$. The two components $|u,d;\mp\rangle$ and $|s;\mp\rangle$ are separately shown.

expected, not only for the $\phi\eta$, but also for ϕf_0 tff under the hypothesis of a $q\bar{q}$ f_0.

In addition, the uncertainty due to the unknown ω-ϕ mixing, allows only to reduce the $\phi\eta\gamma$ coupling with respect to $\phi f_0\gamma$. In fact, by putting $Y_\eta = 0$, it should remain the only $|u,d;-\rangle$ component, which, being OZI forbidden (see first diagram in fig. 2), should give $g_{\eta\gamma}^\phi = 0$. Contrary, by putting $Y_{f_0} = 0$, the remaining component $|u,d;+\rangle$ is a combination of u and d quarks with scalar quantum numbers ($J^{PC} = 0^{++}$) and then it is not OZI suppressed [12].

3.2 The asymptotic region

Another crucial point of this procedure is the knowledge of the time-like asymptotic behaviour of a hadronic form factor. To this purpose, the nominal power law, as predicted by the perturbative QCD (pQCD) [13], is used. The power to be considered, linearly depends on the number of elementary constituents, *i.e.* valence quarks, even though further sub-structures are present (*e.g.*: deuteron [14]).

In addition, in order to describe the asymptotic behaviour of the tff's under consideration, we have also to account for the flip of the hadronic helicity[2] [15], which passes from zero, in the leptonic initial state, to one in the $\phi\eta(f_0)$ final state ($\lambda_{\eta(f_0)} = 0$, $|\lambda_\phi| = 1$). This implies an additional suppression, and then the asymptotic behaviour to be considered, starting from a certain energy $s_{\text{asy}}^{\eta(f_0)}$, is [15,16]

$$F_{\phi\eta(f_0)}^{\text{asy}}(s) \propto \left(\frac{1}{s}\right)^{n_H + \frac{n_\lambda + l_q - 1}{2} = 2\left(\frac{5}{2}\right)} \qquad s > s_{\text{asy}}^{\eta(f_0)}, \qquad (8)$$

where $n_\lambda = |\lambda_{\eta(f_0)} + \lambda_\phi| = 1$, $n_H = 2$, $l_q = 0(1)$. The power is a function of the number of hadronic fields n_H, the modulus of the total helicity n_λ and the relative angular momentum of the quarks which constitute the meson η ($l_q = 0$) or f_0 ($l_q = 1$). The parameterization of eq. (8) provides different asymptotic powers for $\phi\eta$ and $\phi f_0(q\bar{q})$ tff.

____

[2] The virtual photon of the annihilation $e^+e^- \to \gamma^* \to \phi M$, at high energies, always has spin ± 1 along the beam axis. Angular-momentum conservation implies that the total angular momentum is 1. In the center-of-mass frame: $\boldsymbol{p}_\phi = -\boldsymbol{p}_M \equiv$

3.3 The region below the theoretical threshold s_0

Unitarity assures that a generic hadronic tff $F_H(s)$ is an analytic function in the s-complex plane with a cut, along the real axis, starting from the theoretical threshold s_0 up to infinity. This property, together with a power law vanishing asymptotic behaviour allows to use spectral representations for the tff's as, for instance, the dispersion relation (DR) for the imaginary part [17]. However, to use directly the experimental data and the asymptotic behaviour (eq. (8)) in the time-like region, we use the DR for the logarithm [18]:

$$\ln[F_H(t)] = \frac{\sqrt{s_0 - t}}{\pi} \int_{s_0}^\infty \frac{\ln|F_H(s)|}{(s-t)\sqrt{s-s_0}} ds, \qquad (10)$$

which relates the real values of the tff below the theoretical threshold ($t < s_0$) to its modulus over the cut. The DR for the logarithm can be used in the form of eq. (10) only if $F_H(s)$ has neither zeros nor poles on the physical sheet. The parameterization used in the resonance region (eq. (3)) guarantees the absence of singularities for the tff's in the physical sheet.

Since we know the modulus of the tff $F_{\phi\eta(f_0)}(s)$ over the cut (eqs. (3),(8)), the DR for the logarithm allows to perform an analytic continuation of the parameterization below the theoretical threshold and it gives:

$$F_{\phi\eta(f_0)}^{\text{an}}(t) = \exp\left[\frac{\sqrt{s_0 - t}}{\pi}\left(\int_{s_0}^{s_{\text{asy}}^{\eta(f_0)}} \frac{\ln|F_{\phi\eta(f_0)}^{\text{Res}}(s)|}{(s-t)\sqrt{s-s_0}} ds \right.\right.$$
$$\left.\left. + \int_{s_{\text{asy}}^{\eta(f_0)}}^\infty \frac{\ln|F_{\phi\eta(f_0)}^{\text{asy}}(s)|}{(s-t)\sqrt{s-s_0}} ds\right)\right], \quad t < s_0. \quad (11)$$

3.4 The overall parameterization

We are now ready to write down the parameterization which covers the whole s-real axis (in principle the whole s-complex plane), it is a three-fold definition:

$$F_{\phi\eta(f_0)}(s) = \begin{cases} F_{\phi\eta(f_0)}^{\text{an}}(s) & s \le s_0 & \text{(eq. (11))}, \\ F_{\phi\eta(f_0)}^{\text{Res}(\phi)}(s) & s_0 < s \le s_{\text{asy}}^{\eta(f_0)} & \text{(eq. (3))}, \\ F_{\phi\eta(f_0)}^{\text{asy}}(s) & s > s_{\text{asy}}^{\eta(f_0)} & \text{(eq. (8))}. \end{cases}$$
$$(12)$$

____

$\boldsymbol{p}$ it follows that

$$\left|\frac{\boldsymbol{p}_\phi \cdot \boldsymbol{s}_\phi}{|\boldsymbol{p}_\phi|} - \frac{\boldsymbol{p}_M \cdot \boldsymbol{s}_M}{|\boldsymbol{p}_M|}\right| = |\lambda_\phi - \lambda_M| = \left|\frac{\boldsymbol{p} \cdot \boldsymbol{s}_{\text{tot}}}{|\boldsymbol{p}|}\right| = 0, 1, \quad (9)$$

where λ_ϕ and λ_M are the helicities of the final mesons. But, since hadronic helicity conservation requires: $\lambda_\phi + \lambda_M = 0$, or $\lambda_\phi - \lambda_M = 2\lambda_\phi = -2\lambda_M$, eq. (9) holds, in case of mesons, only if $|\lambda_\phi| = |\lambda_M| = 0$. In the cases under consideration, the coupling $\gamma^*\phi M$ ($M = \eta, f_0$) is described in terms of only one tff $F_{\phi M}(s)$ and this requires $|\lambda_\phi| = 1$. Helicity zero would mean that the ϕ spin lies in the plane orthogonal to its 3-momentum, in this case we would need an additional degree of freedom (for instance: the azimuthal angle) and then an additional tff to describe this process. But since the tff is only one it follows that the ϕ spin lies in 3-momentum direction, *i.e.* $|\lambda_\phi| = 1$.

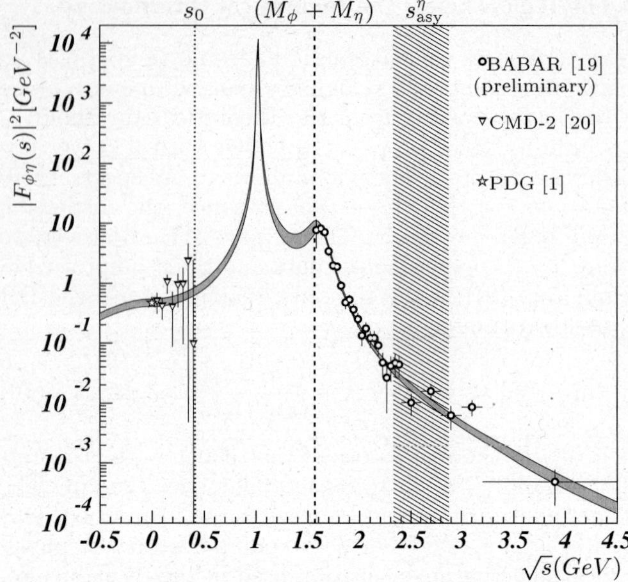

Fig. 4. Reconstructed $|F_{\phi\eta}(s)|^2$ tff. The grey band represents the error. The circles are the preliminary BABAR data used as input, while the triangles (CMD-2: $\phi \to \eta e^+ e^-$) and the star (PDG: $\phi \to \eta\gamma$) are not included in the computation. The dotted and the dashed vertical lines represent the theoretical and the physical thresholds, and the lined region in the obtained interval for s^η_{asy}.

Table 1. Parameters obtained for the $\phi\eta$ tff. The values without error are fixed.

Res	$g^V_{\phi\eta}(\mathrm{GeV}^{-1})$	$M_V(\mathrm{MeV})$	$\Gamma_V(\mathrm{MeV})$
ϕ	13 ± 1.4	1019.4	4.3
ϕ'	56 ± 6	1622 ± 15	203 ± 12

The real axis has been divided into three intervals to parameterize the tff. From higher energies: the asymptotic region, where the pQCD power law is used with power 2 for the η and $5/2$ for the $f_0(q\bar{q})$; the resonance region, where the tff is written in terms of vector propagators and coupling constants; the analytic region, where the tff is reconstructed by means of a DR.

4 The $\phi\eta$ transition form factor

In fig. 4 the reconstructed $\phi\eta$ tff is shown. This result has been obtained by considering only ϕ-family contributions in the resonance region, namely the $\phi(1020) \equiv \phi$ and $\phi(1680) \equiv \phi'$, and a power law asymptotic behaviour $\propto (1/s)^2$. Higher-mass resonances are not included because their effect in the low-energy region under consideration is negligible.

By using only the BABAR data (circles in fig. 4) and the analyticity to fix the free parameters of the procedure, the function defined in eq. (12) reproduces the low-energy behaviour of the tff in good agreement with the data. The obtained parameters are shown in table 1.

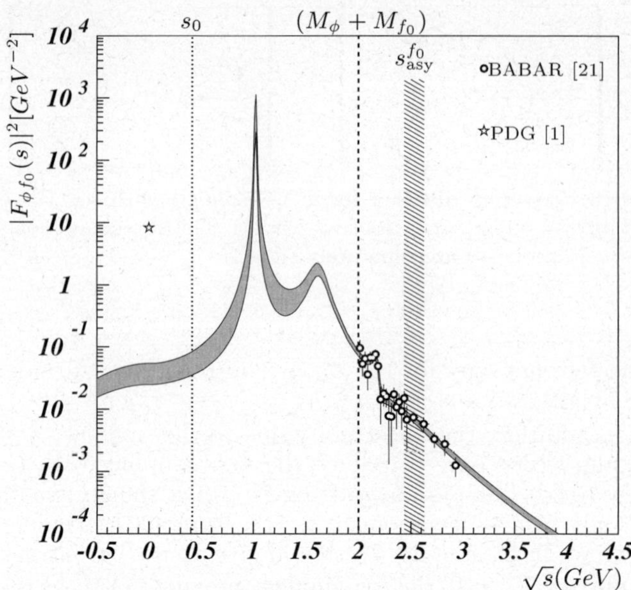

Fig. 5. Reconstructed $\phi f_0(q\bar{q})$ tff. The grey band represents the error. The circles are the BABAR data used as input, while the star (PDG: $\phi \to f_0\gamma$) is not included in the computation. The dotted and the dashed vertical lines represent the theoretical and the physical thresholds, and the lined region in the obtained interval for $s^{f_0}_{\mathrm{asy}}$.

Table 2. Parameters obtained in the case of the $\phi f_0(q\bar{q})$ tff. The additional resonance ϕ'' is included by hand [21]. The values without error are fixed.

Res	$g^V_{\phi f_0}(\mathrm{GeV}^{-1})$	$M_V(\mathrm{MeV})$	$\Gamma_V(\mathrm{MeV})$
ϕ	3.5 ± 0.9	1019.4	4.3
ϕ'	26 ± 3	1622	203
ϕ''	$F_{\phi''}(6.4 \pm 0.3) \cdot 10^{-3}/M_{\phi''}$	2176	51

By using eq. (6) we get: $BR(\phi \to \eta\gamma) = (1.3 \pm 0.2)\%$, to be compared with [1]: $BR_{\mathrm{PDG}}(\phi \to \eta\gamma) = (1.301 \pm 0.024)\%$ (see fig. 4). This good agreement provides an important positive check for two key points of the procedure i.e., the description in terms of resonances in the low-energy region and the power law asymptotic behaviour.

5 The $\phi f_0(q\bar{q})$ transition form factor

In fig. 5 is shown the reconstructed $\phi f_0(q\bar{q})$ tff, under the hypothesis of a $q\bar{q}$ f_0 and the corresponding parameters are reported in table 2. This result has been obtained with the same prescriptions used in the previous $\phi\eta$ case, i.e. by considering only ϕ-family contributions (here there is also the ϕ'' [21]) and a $(1/s)^{5/2}$ asymptotic power law (eq. (8)).

In this case we get: $BR(\phi \to f_0\gamma) = (2.1 \pm 0.8) \cdot 10^{-6}$, while: $BR_{\mathrm{PDG}}(\phi \to f_0\gamma) = (4.40 \pm 0.21) \cdot 10^{-4}$ [1]. We stress that this result has been obtained assuming for the f_0 a quark structure identical to that of the η (eq. (7)).

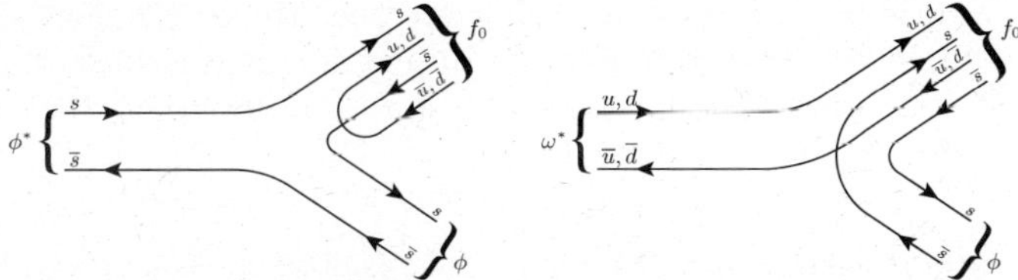

Fig. 6. Feynman diagrams for the ϕ^* and ω^* contribution to the ϕf_0 final state with a tetraquark f_0.

6 The $\phi f_0(qq\overline{qq})$ transition form factor

To account for an f_0 with a tetraquark structure like: $(ns\overline{ns})$, where $n = u, d$, two steps in our procedure need improvement.

The first upgrade concerns the resonance region, where the tetraquark structure allows both the ϕ- and ω-family contributions (see fig. 6), which were previously forbidden. The second is about the asymptotic behaviour. Since the pQCD asymptotic power depends on the number of hadronic fields, an additional valence quark implies a further factor $(1/s)$ and, having the two pairs qq and $\overline{qq}$ zero relative angular momentum and total spin [22], there is no more the $l_q = 1$ attenuation and the asymptotic power becomes: $p(n_H = 3, n_\lambda = 1, l_q = 0) = 3$.

The new overall parameterization is

$$F_{\phi f_0(qq\overline{qq})} = \begin{cases} F^{\mathrm{an}}_{\phi f_0}(s), & s \leq s_0, \\ F^{\mathrm{Res}(\phi,\omega)}_{\phi f_0}(s), & s_0 < s \leq s^{f_0}_{\mathrm{asy}}, \\ \propto s^{-3}, & s > s^{f_0}_{\mathrm{asy}}, \end{cases} \quad (13)$$

where the only $\omega(782)$ has been included in the region $]s_0, s^{f_0}_{\mathrm{asy}}]$ (higher mass ω-recurrences have no important effects) and a faster vanishing asymptotic behaviour is implemented for $s > s^{f_0}_{\mathrm{asy}}$.

6.1 Results for $f_0(qq\overline{qq})$

In fig. 7 is shown the ϕf_0 tff obtained under the hypothesis of a tetraquark f_0 and the corresponding parameters are reported in table 3. In this scenario our prediction for the tff, which gives: $BR(\phi \to f_0\gamma) = (3.9 \pm 0.9) \cdot 10^{-4}$ is in good agreement with the PDG value (mainly due to KLOE [23]), which is: $BR_{\mathrm{PDG}}(\phi \to f_0\gamma) = (4.40 \pm 0.21) \cdot 10^{-4}$.

This drastic change in the low-energy result, with respect to the previous case without the ω, is the consequence of a combined effect of the two modifications; the faster vanishing asymptotic behaviour enhances the resonance region and the additional ω contribution moves, down in energy, its mean value. The sum of these two effects is just a large enhancement of the tff at $s = 0$.

7 Duality

The pQCD nominal power law of eq. (8) behaves like a counter of valence quarks involved in a certain process.

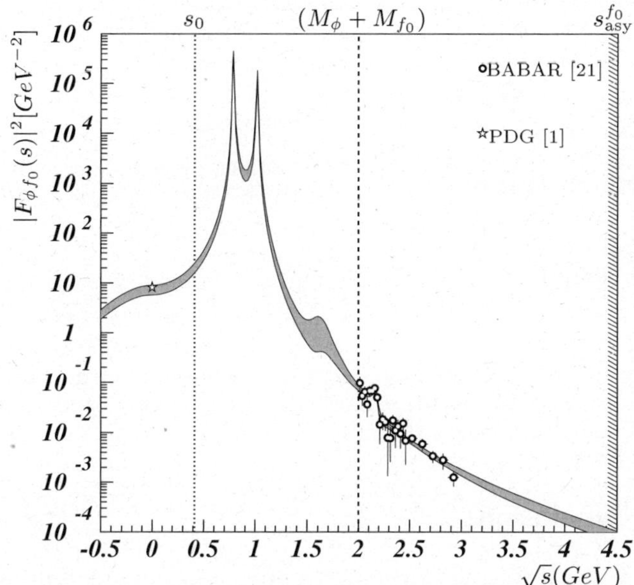

Fig. 7. Reconstructed $\phi f_0(qq\overline{qq})$ tff. The grey band represents the error. The circles, on the right, are the BABAR data used as input, while the star (PDG: $\phi \to f_0\gamma$) is not included in the computation. The dotted and the dashed vertical lines represent the theoretical and the physical thresholds, and the lined region in the obtained interval for $s^{f_0}_{\mathrm{asy}}$.

Table 3. Parameters obtained in the case of the $\phi f_0(qq\overline{qq})$ tff. The values without error are fixed.

Res	$g^V_{\phi f_0}(\mathrm{GeV}^{-1})$	$M_V(\mathrm{MeV})$	$\Gamma_V(\mathrm{MeV})$
ω	112 ± 13	783.7	8.5
ϕ	52 ± 7	1019.4	4.3
ϕ'	12 ± 7	1622	203
ϕ''	$F_{\phi''}(1.9 \pm 0.4) \cdot 10^{-3}/M_{\phi''}$	2176	51

Nevertheless, there is an objective difficulty in exploiting experimentally this power counting. In fact, the high-energy values of the tff's, due to their power law vanishing, are by definition very small quantities. In the low-energy region, where the tff's attain their higher values and then the experimental observation should be more feasible, the power law behaviour is covered by the resonances.

However, following the so-called quark-hadron duality [24], the asymptotic behaviour may be "restored" also at low energy by averaging the tff's and then by canceling the local effects of the resonances.

Figure 8 shows the asymptotic powers of the $\phi\eta$, ϕf_0 tff's and also of the pion form factor $F_{\pi\pi}(s)$, which is used as a check of the averaging procedure. The results are in agreement with the expectations and they confirm the further suppression due to the helicity flip in case of $F_{\phi\eta}(s)$ (first experimental observation) and the additional hadronic filed in case of $F_{\phi f_0}(s)$.

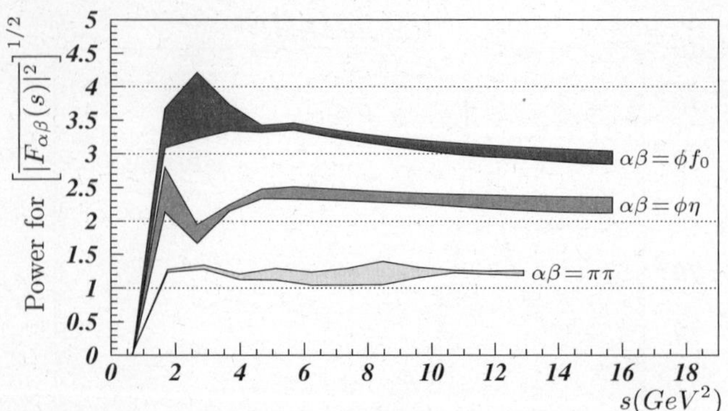

Fig. 8. Asymptotic powers for the mean values of $\pi\pi$ form factor (light-grey band) and, the $\phi\eta$ (medium-grey band) and ϕf_0 (dark-grey band) tff's.

8 Conclusions

We have defined a dispersive technique to construct, for a generic ϕM tff, an analytic parameterization which is defined in the whole s-complex plane (M is a pseudoscalar or scalar light meson). The main ingredients of this technique are: the pQCD counting rule [13] (and helicity rule [15]) to describe the asymptotic behaviour; data on the annihilation cross-section $\sigma(\phi M)$; Breit-Wigner parameterization in the resonance region and DR's for the logarithm.

In the case of the $\phi\eta$ tff, by considering only the ϕ-family contributions, assuming the asymptotic behaviour $(1/s^2)$, and using the BABAR data on $\sigma(\phi\eta)$, we achieved a parameterization $F_{\phi\eta}(s)$ (fig. 4), which is in good agreement with the low energy data, not used in the analysis.

In the case of the ϕf_0 tff, the same procedure, *i.e.*, an analysis under the hypothesis of a $q\bar{q}$ f_0 with the opportune $(1/s^{5/2})$ asymptotic behaviour, gives a prediction in huge disagreement with the low-energy data (fig. 5).

We modified the procedure in order to account for a tetraquark f_0. The modification concerns two steps: a new contribution coming from the ω-family is included; the asymptotic behaviour becomes $(1/s^3)$ because of the additional quarks in the final state and the disappearing of the angular momentum attenuation. These two upgrades, if individually considered, give opposite effects: the first reduces, while the second enhances the low-energy values of the tff. However, the combined effect gives an important enhancement of the tff, which reconciles the prediction with the data (fig. 7).

In addition, we have performed an analysis in terms of quark-hadron duality, which, by restoring the power law behaviour at low energy, allows to extract the asymptotic power from the data and the parameters of the resonances. The results are in good agreement with a tetraquark f_0 and with the helicity rule, which is for the first time experimentally confirmed.

Since, by means of this procedure we have tested two extreme hypotheses concerning the f_0 structure, *i.e.* the two- and the four-quark state, our conclusion is that: the available data, the analyticity requirement and the pQCD asymptotic behaviour prove that the tetraquark component in the f_0 is the dominant one.

A still in progress analysis, where both the components are included, gives for the $q\bar{q}$ content an upper limit of 12%. It follows that the four-quark state represents at least the 88% of the f_0 structure.

I warmly acknowledge R. Baldini, G. Isidori, L. Maiani, G. Pancheri and A. Polosa for precious discussions on the subject of this talk and, BABAR and KLOE Collaborations for support and cooperation.

References

1. W.-M. Yao *et al.*, J. Phys. G **33**, 1 (2006).
2. E. Klempt, arXiv:hep-ph/0404270.
3. N.N. Achasov, V.V. Gubin, Phys. Rev. D **63**, 094007 (2001); V.E. Markushin, Eur. Phys. J. A **8**, 389 (2000) (arXiv:hep-ph/0005164); J.A. Oller, Phys. Lett. B **426**, 7 (1998) (arXiv:hep-ph/9803214); E. Marco, S. Hirenzaki, E. Oset, H. Toki, Phys. Lett. B **470**, 20 (1999) (arXiv:hep-ph/9903217); J.A. Oller, Nucl. Phys. A **714**, 161 (2003) (arXiv:hep-ph/0205121).
4. M. Boglione, M.R. Pennington, Eur. Phys. J. C **30**, 503 (2003) (arXiv:hep-ph/0303200).
5. L. Maiani, F. Piccinini, A.D. Polosa, V. Riquer, Phys. Rev. Lett. **93**, 212002 (2004) (arXiv:hep-ph/0407017).
6. G. Isidori, L. Maiani, M. Nicolaci, S. Pacetti, JHEP **0605**, 049 (2006) (arXiv:hep-ph/0603241).
7. R.L. Jaffe, Phys. Rep. **409**, 1 (2005) (Nucl. Phys. Proc. Suppl. **142**, 343 (2005)) (arXiv:hep-ph/0409065).
8. F.E. Close, AIP Conf. Proc. **717**, 919 (2004) (arXiv:hep-ph/0311087).
9. N.N. Achasov, V.N. Ivanchenko, Nucl. Phys. B **315**, 465 (1989); S. Nussinov, T.N. Truong, Phys. Rev. Lett. **63**, 1349 (1989); **63**, 2002 (1989)(E); J.L. Lucio Martinez, J. Pestieau, Phys. Rev. D **42**, 3253 (1990); F.E. Close, N. Isgur, S. Kumano, Nucl. Phys. B **389**, 513 (1993) (arXiv:hep-ph/9301253); N. Brown, F.E. Close, *The DNE Physics Handbook*, edited by L. Maiani, G. Pancheri, N. Paver (INFN, Frascati, 1995) pp. 447-464.
10. KLOE Collaboration (F. Ambrosino *et al.*), Phys. Lett. B **634**, 148 (2006) (arXiv:hep-ex/0511031).
11. G. Zweig, CERN report S419/TH412 (1964) unpublished; S. Okubo, Phys. Lett. **5**, 165 (1963); I. Iizuka, K. Okuda, O. Shito, Prog. Theor. Phys. **35**, 1061 (1966).
12. P. Geiger, N. Isgur, Phys. Rev. D **47**, 5050 (1993); S. Descotes-Genon, L. Girlanda, J. Stern, JHEP **0001**, 041 (2000) (arXiv:hep-ph/9910537); N. Isgur, H.B. Thacker, Phys. Rev. D **64**, 094507 (2001) (arXiv:hep-lat/0005006).
13. S.J. Brodsky, G.R. Farrar, Phys. Rev. D **11**, 1309 (1975).
14. S.J. Brodsky, B.T. Chertok, Phys. Rev. D **14**, 3003 (1976).
15. S.J. Brodsky, G.P. Lepage, Phys. Rev. D **24**, 2848 (1981). S.J. Brodsky, G.F. de Teramond, Phys. Lett. B **582**, 211 (2004) (arXiv:hep-th/0310227).
16. V.L. Chernyak, A.R. Zhitnitsky, Phys. Rep. **112**, 173 (1984); V. Chernyak, arXiv:hep-ph/9906387; G.R. Farrar, D.R. Jackson, Phys. Rev. Lett. **35**, 1416 (1975).
17. See for instance: E.C. Tichmarsch, *The Theory of Functions* (Oxford University Press, London, 1939).
18. B.V. Geshkenbein, Yad. Fiz. **9**, 1232 (1969).
19. BABAR Collaboration (A. Zallo), private comunication.

20. M.N. Achasov, V.M. Aulchenko, K.I. Beloborodov, A.V. Berdyugin, Phys. Lett. B **504**, 275 (2001).

21. BABAR Collaboration (G. Solodov), *Initial state radiation study at BABAR and the application to the R measurement and hadron spectroscopy*, talk presented at *ICHEP 06, Moscow, Russia*.

22. R.L. Jaffe, Phys. Rev. D **15**, 267 (1977).

23. KLOE Collaboration (A. Aloisio *et al.*), Phys. Lett. B **537**, 21 (2002) (arXiv:hep-ex/0204013).

24. J.J. Sakurai, Phys. Lett. B **46**, 207 (1973).

Eur. Phys. J. A **31**, 672–675 (2007)

DOI 10.1140/epja/i2006-10278-2

Special Article – QNP 2006

New hadronic states observed at BES II

Ji Xiaobin[a]

For the BES Collaboration
Institute of High Energy Physics, CAS, Beijing 100049, China

Received: 18 December 2006
Published online: 20 March 2007 – © Società Italiana di Fisica / Springer-Verlag 2007

Abstract. The report includes the new observation of $X(1835)$ in $J/\psi \to \gamma\eta'\pi^+\pi^-$, the $\omega\phi$ threshold enhancement in $J/\psi \to \gamma\omega\phi$, the $\omega\omega$ structure in $J/\psi \to \gamma\omega\omega$ and the broad 1^{--} structure at the low K^+K^- invariant-mass spectrum in $J/\psi \to K^+K^-\pi^0$.

PACS. 12.39.Mk Glueball and nonstandard multi-quark/gluon states – 13.25.Gv Decays of J/ψ, Υ, and other quarkonia

1 Introduction

BES [1] is a large general purpose solenoidal detector at Beijing Electron Positron Collider (BEPC). The 5.8×10^7 J/ψ events were accumulated with BES II, which provides a good laboratory for the search of new hadronic states. All the results presented here are based on this data sample.

2 Observation of X(1835) in $J/\psi \to \gamma\eta'\pi^+\pi^-$

An anomalous enhancement near the mass threshold in the $p\bar{p}$ invariant-mass spectrum from $J/\psi \to \gamma p\bar{p}$ decays was reported by the BES II experiment [2]. This enhancement was fitted with a sub-threshold $\mathcal{S}$-wave Breit-Wigner (BW) resonance function with a mass $M = 1859^{+3+5}_{-10-25}\,\mathrm{MeV}/c^2$, a width $\Gamma < 30\,\mathrm{MeV}/c^2$ (at the 90% C.L.) and a product branching fraction $B(J/\psi \to \gamma X)\cdot B(X \to p\bar{p}) = (7.0\pm0.4^{+1.9}_{-0.8}) \times 10^{-5}$. Among various theoretical interpretations [3–8] of the $p\bar{p}$ mass threshold enhancement, the most intriguing one is that of a $p\bar{p}$ bound state, sometimes called baryonium [3,6,9], which has been the subject of many experimental searches [10]. The baryonium interpretation of the $p\bar{p}$ mass enhancement requires a new resonance with a mass around $1.85\,\mathrm{GeV}/c^2$ and it would be supported by the observation of the resonance in other decay channels. $\pi^+\pi^-\eta'$ [5,6] is a possible decay mode for a $p\bar{p}$ bound state.

The analysis of $J/\psi \to \gamma\eta'\pi^+\pi^-$ is described in detail in ref. [11]. Two decay modes of η' ($\eta' \to \pi^+\pi^-\eta$, $\eta \to \gamma\gamma$ and $\eta' \to \gamma\rho$) are used in the analysis. The combined

a e-mail: jixb@ihep.ac.cn

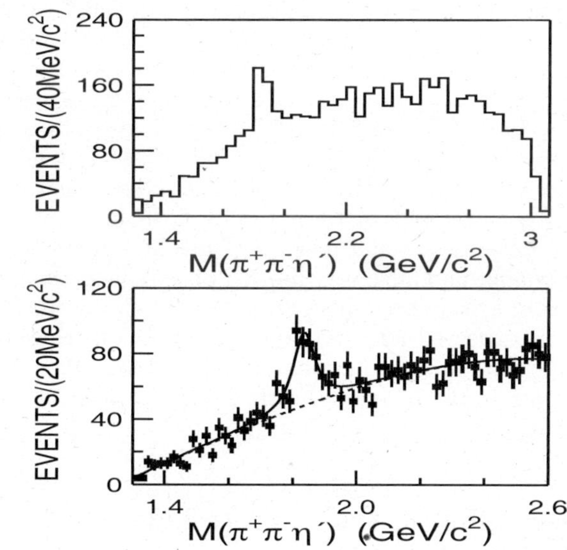

Fig. 1. The $\pi^+\pi^-\eta'$ invariant-mass distribution for selected events from both the $J/\psi \to \gamma\eta'\pi^+\pi^-$ ($\eta' \to \pi^+\pi^-\eta$, $\eta \to \gamma\gamma$) and $J/\psi \to \gamma\eta'\pi^+\pi^-$ ($\eta' \to \gamma\rho$) analyses. The bottom panel shows the fit (solid curve) to the data (points with error bars); the dashed curve indicates the background function.

$\pi^+\pi^-\eta'$ spectrum for the two decay modes of η' is fitted with a BW function convolved with a Gaussian mass resolution function ($\sigma = 13\,\mathrm{MeV}/c^2$) to represent the $X(1835)$ signal plus a smooth polynomial background function. The mass and width obtained from the fit (shown in the bottom panel of fig. 1) are $M = 1833.7\pm6.1\,\mathrm{MeV}/c^2$ and $\Gamma = 67.7\pm20.3\,\mathrm{MeV}/c^2$. The signal yield from the fit is 264 ± 54 events with a confidence level 45.5% ($\chi^2/\mathrm{d.o.f.} = 57.6/57$) and $-2\ln\mathcal{L} = 58.4$. A fit to the mass spectrum without BW signal function returns $-2\ln\mathcal{L} = 126.5$. The change in $-2\ln\mathcal{L}$ with $\Delta(\text{d.o.f}) = 3$ corresponds to a statistical significance of 7.7 σ for the signal.

Using MC-determined selection efficiencies of 3.72% and 4.85% for the $\eta' \to \pi^+\pi^-\eta$ and $\eta' \to \gamma\rho$ modes, respectively, we determine a product BF of

$$B(J/\psi \to \gamma X) \cdot B(X \to \pi^+\pi^-\eta') = (2.2 \pm 0.4) \times 10^{-4}.$$

The mass and width of the $X(1835)$ are not compatible with any known meson resonance [12]. We examined the possibility that the $X(1835)$ is responsible for the $p\bar{p}$ mass threshold enhancement observed in radiative $J/\psi \to \gamma p\bar{p}$ decay [2]. It has been pointed out that the S-wave BW function used for the fit in ref. [2] should be modified to include the effect of final-state interactions (FSI) on the shape of the $p\bar{p}$ mass spectrum [7,8]. Redoing the S-wave BW fit to the $p\bar{p}$ invariant-mass spectrum in ref. [2] including the zero isospin, S-wave FSI factor of ref. [8], yields a mass $M = 1831 \pm 7\,\text{MeV}/c^2$ and a width $\Gamma < 153\,\text{MeV}/c^2$ (at the 90% C.L.); these values are in good agreement with the mass and width of $X(1835)$ reported here. Moreover, according to ref. [6], the $\pi\pi\eta'$ mode is expected to be strong for a $p\bar{p}$ bound state. Thus, the $X(1835)$-resonance is a prime candidate for the source of the $p\bar{p}$ mass threshold enhancement in $J/\psi \to \gamma p\bar{p}$ process. In this case, the J^{PC} and I^G of the $X(1835)$ could only be 0^{-+} and 0^+, which can be tested in future experiments. Also in this context, the relative $p\bar{p}$ decay strength is quite strong: $B(X \to p\bar{p})/B(X \to \pi^+\pi^-\eta') \sim 1/3$ (The product BF determined from the fit that includes FSI effects on the $p\bar{p}$ mass spectrum is within the systematic errors of the result report in ref. [2]). Since decays to $p\bar{p}$ are kinematically allowed only for a small portion of the high-mass tail of the resonance and have very limited phase space, the large $p\bar{p}$ branching fraction implies an unusually strong coupling to $p\bar{p}$, as expected for a $p\bar{p}$ bound state [9,13]. However, other possible interpretations of the $X(1835)$ that have no relation to the $p\bar{p}$ mass threshold enhancement are not excluded.

3 Observation of the $\omega\phi$ threshold enhancement in $J/\psi \to \gamma\omega\phi$

Systems of two vector particles have been intensively studied for signatures of gluonic bound states. The radiative J/ψ decay $J/\psi \to \gamma\omega\phi$ is a double Okubo-Zweig-Iizuka (OZI) suppressed process, the measurement of this decay and the search for possible resonance states will provide useful information on two vector-meson systems. MARK III Collaboration [14] studied $J/\psi \to \gamma\omega\phi$ decays, but did not find clear structures in the $\omega\phi$ invariant-mass spectrum. The final states of $\omega\phi$ are also observed in photon-photon collisions by ARGUS [15,16] experiment and the cross-sections were measured [16].

The detailed analysis of $J/\psi \to \gamma\omega\phi$ is described in ref. [17]. The decay modes of $\omega \to \pi^+\pi^-\pi^0$ and $\phi \to K^+K^-$ are used in the analysis. Since the decays of $J/\psi \to \omega\phi$ and $J/\psi \to \omega\phi\pi^0$ are forbidden by C invariance, the 294 observed $\omega\phi$ events present direct evidence

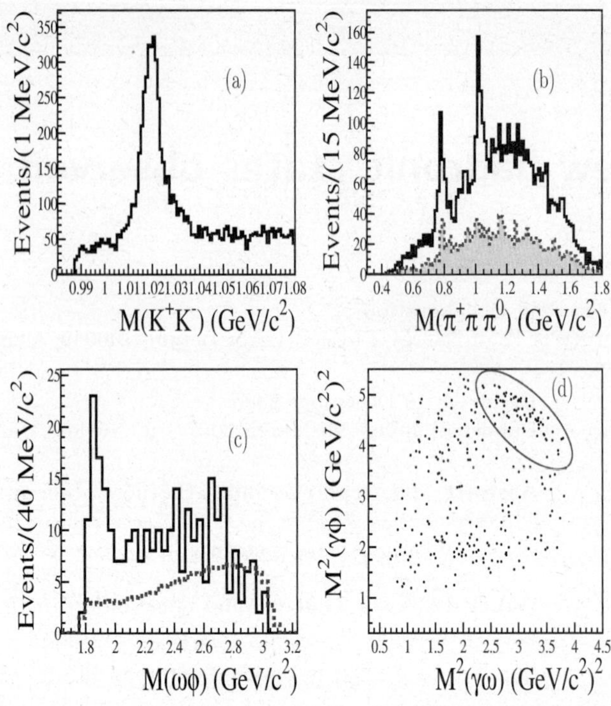

Fig. 2. (a) The K^+K^- invariant-mass distribution. (b) The $\pi^+\pi^-\pi^0$ invariant-mass distribution; the open histogram is for candidate events with $m_{K^+K^-}$ being in the ϕ range, and the shaded histogram is for events with $m_{K^+K^-}$ being in the ϕ sideband region. (c) The $K^+K^-\pi^+\pi^-\pi^0$ invariant-mass distribution for the $J/\psi \to \gamma\omega\phi$ candidate events. The dashed curve indicates the acceptance varying with the $\omega\phi$ invariant mass. (d) Dalitz plot.

for the radiative $J/\psi \to \gamma\omega\phi$ decay. The histogram in fig. 2(c) shows the $K^+K^-\pi^+\pi^-\pi^0$ invariant-mass distribution for events with $|m_{K^+K^-} - m_\phi| < 15\,\text{MeV}/c^2$ and $|m_{\pi^+\pi^-\pi^0} - m_\omega| < 30\,\text{MeV}/c^2$, and a structure peaked near the $\omega\phi$ threshold is observed. The dashed curve in the figure indicates how the acceptance varies with invariant mass. The peak is also evident as a diagonal band along the upper right-hand edge of the Dalitz plot in fig. 2(d). No evidence of an enhancement near the $\omega\phi$ threshold is observed from the sideband events of ω and ϕ. And the study from inclusive and exclusive MC shows that the $\omega\phi$ threshold enhancement is not from backgrounds.

The significance of the $\omega\phi$ threshold enhancement is more than 10σ. From a partial-wave analysis with covariant helicity coupling amplitudes, a spin-parity of $X = 0^{++}$ with an S-wave $\omega\phi$ system is favored. The mass and width of the enhancement are determined to be $M = 1812^{+19}_{-26}\,(\text{stat}) \pm 18\,(\text{syst})\,\text{MeV}/c^2$ and $\Gamma = 105 \pm 20\,(\text{stat}) \pm 28\,(\text{syst})\,\text{MeV}/c^2$, and the product branching fraction is $\mathcal{B}(J/\psi \to \gamma X) \cdot \mathcal{B}(X \to \omega\phi) = (2.61 \pm 0.27\,(\text{stat}) \pm 0.65\,(\text{syst})) \times 10^{-4}$. The mass and width of this state are not compatible with any known scalars listed in the Particle Data Group (PDG) [12]. It could be an unconventional state [18–22]. However, more statistics and further studies are needed to clarify this.

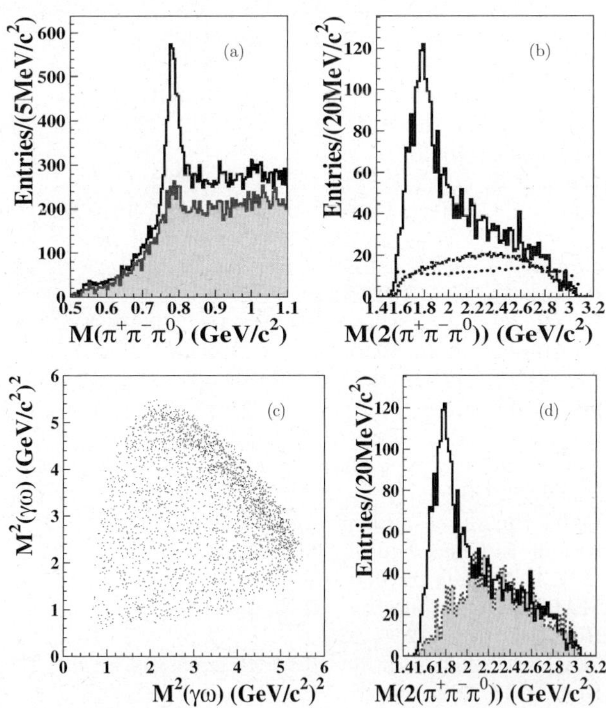

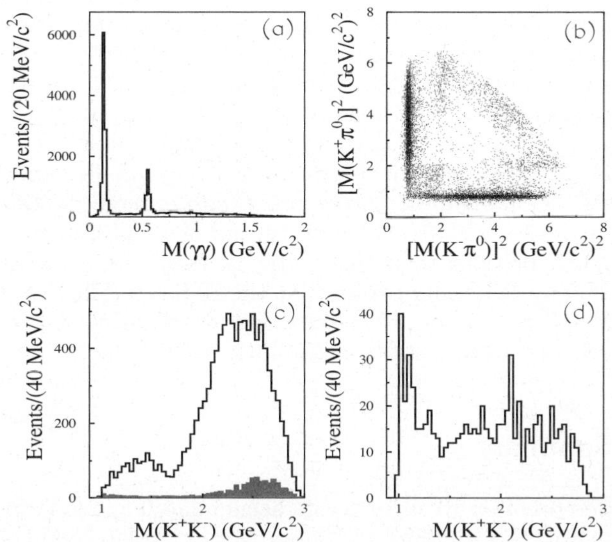

Fig. 4. (a) The $\gamma\gamma$ invariant-mass distribution. (b) The Dalitz plot for $K^+K^-\pi^0$ candidate events. (c) The K^+K^- invariant-mass distribution for $K^+K^-\pi^0$ candidate events; the solid histogram is data and the shaded histogram is the background (normalized to data). (d) The K^+K^- invariant-mass distribution for the π^0 mass sideband events (not normalized).

Fig. 3. (a) The $\pi^+\pi^-\pi^0$ mass distribution for the best $\omega\omega$ combination after the requirement that the invariant mass of the other $\pi^+\pi^-\pi^0$ is inside the ω region (open histogram), defined by $|m_{\pi^+\pi^-\pi^0} - m_\omega| < 40\,\mathrm{MeV}/c^2$, or in the sideband range (shaded histogram), defined by $40\,\mathrm{MeV}/c^2 < |m_{\pi^+\pi^-\pi^0} - m_\omega| < 80\,\mathrm{MeV}/c^2$. (b) The $2(\pi^+\pi^-\pi^0)$ invariant-mass distribution for candidate events. The dashed curve is the phase space invariant-mass distribution, and the dotted curve shows the acceptance *versus* the $\omega\omega$ invariant mass. (c) The Dalitz plot. (d) The $2(\pi^+\pi^-\pi^0)$ invariant mass of the inclusive Monte Carlo sample (shaded histogram).

4 Pseudoscalar production at the $\omega\omega$ threshold in $J/\psi \to \gamma\omega\omega$

The $\eta(1760)$ was reported by the MARK III Collaboration in J/ψ radiative decays and was found to decay to $\omega\omega$ [23] and $\rho\rho$ [24]. It was also observed by the DM2 [25, 26] and BES I [27]. The $\eta(1760)$ was suggested to be a 3^1S_0 pseudoscalar $q\bar{q}$ meson, but some authors suggest a mixture of glueball and $q\bar{q}$ or a hybrid [28,29]. In our analysis of $J/\psi \to \gamma\omega\omega$ ($\omega \to \pi^+\pi^-\pi^0$) [30], the presence of a signal around $1.76\,\mathrm{GeV}/c^2$ and its pseudoscalar character are confirmed, and the mass, width, and branching fraction are measured by partial-wave analysis.

The histogram of fig. 3(b) shows the $2(\pi^+\pi^-\pi^0)$ invariant-mass distribution of events with both $\pi^+\pi^-\pi^0$ masses within the ω range ($|m_{\pi^+\pi^-\pi^0} - m_\omega| < 40\,\mathrm{MeV}/c^2$). There are 3046 events with a clear peak at $1.76\,\mathrm{GeV}/c^2$. The phase space invariant-mass distribution and the acceptance *versus* $\omega\omega$ invariant mass are also shown in the figure. The corresponding Dalitz plot is shown in fig. 3(c). The shaded histogram of fig. 3(d) shows the $2(\pi^+\pi^-\pi^0)$ invariant-mass distribution of the inclusive sample. There is no peak at the $\omega\omega$ mass threshold

in the invariant-mass distribution at around $1.76\,\mathrm{GeV}/c^2$. The fitting of fig. 3(d) yields 1441 ± 50 background events within the $\omega\omega$ invariant-mass range from $1.6\,\mathrm{GeV}/c^2$ to $2.8\,\mathrm{GeV}/c^2$.

The partial-wave analysis shows that the structure around $1.76\,\mathrm{GeV}/c^2$ in the $\omega\omega$ invariant-mass spectrum is predominantly pseudoscalar, with small contributions from $f_0(1710)$, $f_2(1640)$, and $f_2(1910)$. The mass of the pseudoscalar is $M = 1744 \pm 10$ (stat) ± 15 (syst) MeV/c^2, the width $\Gamma = 244^{+24}_{-21}$ (stat) ± 25 (syst) MeV/c^2, and the product branching fraction is $\mathrm{Br}(J/\psi \to \gamma\eta(1760)) \cdot \mathrm{Br}(\eta(1760) \to \omega\omega) = (1.98 \pm 0.08$ (stat) ± 0.32 (syst)$) \times 10^{-3}$.

5 Observation of a broad 1^{--} resonant structure in the K^+K^- mass spectrum in $J/\psi \to K^+K^-\pi^0$

A broad peak is observed at low K^+K^- invariant mass in $J/\psi \to K^+K^-\pi^0$ decays, detailed analysis is described in ref. [31]. The Dalitz plot for the selected 10631 events is shown in fig. 4(b), where a broad K^+K^- band is evident in addition to the $K^*(892)$ and $K^*(1410)$ signals. This band corresponds to the broad peak observed around $1.5\,\mathrm{GeV}/c^2$ in the K^+K^- invariant mass projection shown in fig. 4(c).

A partial-wave analysis shows that the J^{PC} of this structure is 1^{--}. Its pole position is determined to be $(1576^{+49+98}_{-55-91})\,\mathrm{MeV}/c^2 - i(409^{+11+32}_{-12-67})\,\mathrm{MeV}/c^2$, and the branching ratio is $B(J/\psi \to X\pi^0) \cdot B(X \to K^+K^-) = (8.5 \pm 0.6^{+2.7}_{-3.6}) \times 10^{-4}$, where the first errors are statistical and the second are systematic. These parameters are not compatible with any known meson resonances [12].

To understand the nature of the broad 1^{--} peak, it is important to search for a similar structure in $J/\psi \to K_S K^{\pm} \pi^{\mp}$ decays to determine its isospin. It is also intriguing to search for $K^* K, K K \pi$ decay modes. In the mass region of the X, there are several other 1^{--} states, such as the $\rho(1450)$ and $\rho(1700)$, but the width of the X is much broader than the widths of these other mesons. This may be an indication that the X has a different nature than these other mesons. For example, very broad widths are expected for multiquark states [32].

6 Summary

Based on 5.8×10^7 J/ψ events accumulated at the BES II detector, the X(1835) in $J/\psi \to \gamma \eta' \pi^+ \pi^-$, the $\omega\phi$ threshold enhancement in $J/\psi \to \gamma \omega \phi$ and a broad 1^{--} resonant structure around $1.5\,\mathrm{GeV}/c^2$ in the $K^+ K^-$ mass spectrum in $J/\psi \to K^+ K^- \pi^0$ are observed. And partial-wave analysis shows that $\eta(1760)$ is dominant in the $\omega\omega$ in the invariant-mass spectrum in $J/\psi \to \gamma \omega \omega$.

References

1. BES Collaboration (J.Z. Bai *et al.*), Nucl. Instrum. Methods A **344**, 319 (1994); **458**, 627 (2001).
2. BES Collaboration (J.Z. Bai *et al.*), Phys. Rev. Lett. **91**, 022001 (2003).
3. A. Datta, P.J. O'Donnell, Phys. Lett. B **567**, 273 (2003); M.L. Yan *et al.*, hep-ph/0405087; B. Loiseau, S. Wycech, Phys. Rev. C **72**, 011001 (2005); G.J. Ding, R.G. Ping, M.L. Yan, Phys. Rev. D **74**, 014029 (2006).
4. J. Ellis, Y. Frishman, M. Karliner, Phys. Lett. B **566**, 201 (2003); J.L. Rosner, Phys. Rev. D **68**, 014004 (2003).
5. C.S. Gao, S.L. Zhu, Commun. Theor. Phys. **42**, 844 (2004).
6. G.J. Ding, M.L. Yan, Phys. Rev. C **72**, 015208 (2005).
7. B.S. Zou, H.C. Chiang, Phys. Rev. D **69**, 034004 (2003).
8. A. Sibirtsev *et al.*, Phys. Rev. D **71**, 054010 (2005).
9. I.S. Shapiro, Phys. Rep. **35**, 129 (1978); C.B. Dover, M. Goldhaber, Phys. Rev. D **15**, 1997 (1977).
10. E. Klempt *et al.*, Phys. Rep. **368**, 119 (2002); J.-M. Richard, Nucl. Phys. Proc. Suppl. **86**, 361 (2000).
11. BES Collaboration (M. Ablikim *et al.*), Phys. Rev. Lett. **95**, 262001 (2005).
12. W.-M. Yao *et al.*, J. Phys. G **33**, 1 (2006).
13. S.L. Zhu, C.S. Gao, Commun. Theor. Phys. **46**, 291 (2006).
14. J. Becker *et al.*, Contribution to the *23rd International Conference on High Energy Physics, Berkeley, CA, 1986*, Report No. SLAC-PUB-4242 (1987) p. 10.
15. H. Albrecht *et al.*, Phys. Lett. B **210**, 273 (1988).
16. H. Albrecht *et al.*, Phys. Lett. B **332**, 451 (1994).
17. BES Collaboration (M. Ablikim *et al.*), Phys. Rev. Lett. **96**, 162002 (2006).
18. Bing An Li, Phys. Rev. D **74**, 054017 (2006).
19. Xiao-Gang He, Xue-Qian Li *et al.*, Phys. Rev. D **73**, 051502 (2006).
20. Pedro Bicudo *et al.*, hep-ph/0602172.
21. Kuang-Ta Chao, hep-ph/0602190.
22. D.V. Bugg, hep-ph/0603018.
23. MARKIII Collaboration (R.M. Baltrusaitis *et al.*), Phys. Rev. Lett. **55**, 1723 (1985).
24. MARKIII Collaboration (R.M. Baltrusaitis *et al.*), Phys. Rev. D **33**, 1222 (1986).
25. DM2 Collaboration (D. Bisello *et al.*), Phys. Rev. D **39**, 701 (1989).
26. DM2 Collaboration (D. Bisello *et al.*), Phys. Lett. B **192**, 239 (1987).
27. BES Collaboration (J.Z. Bai *et al.*), Phys. Lett. B **446**, 356 (1999).
28. P.R. Page, X.Q. Li, Eur. Phys. J. C **1**, 579 (1998).
29. N. Wu *et al.*, Chin. Phys. **10**, 611 (2001).
30. BES Collaboration (M. Ablikim *et al.*), Phys. Rev. D **73**, 112007 (2006).
31. BES Collaboration (M. Ablikim *et al.*), Phys. Rev. Lett. **97**, 142002 (2006).
32. L. Kopke, N. Wermes, Phys. Rep. **174**, 67 (1989); R. Jaffe, K. Kohnson, Phys. Lett. B **60**, 201 (1976); R. Jaffe, Phys. Rev. D **15**, 267; 281 (1977); R. Jaffe, F. Low, Phys. Rev. D **19**, 2105 (1979), G.J. Ding, M.L. Yan, Phys. Lett. B **643**, 33 (2006).

Eur. Phys. J. A **31**, 676–678 (2007)

DOI 10.1140/epja/i2006-10176-7

Special Article – QNP 2006

$D_{s0}^+(2317)$ as an iso-triplet four-quark meson

K. Terasaki[a]

Yukawa Institute for Theoretical Physics, Kyoto University, Kyoto 606-8502, Japan and
Institute for Theoretical Physics, Kanazawa University, Kanazawa 920-1192, Japan

Received: 24 September 2006
Published online: 16 February 2007 – © Società Italiana di Fisica / Springer-Verlag 2007

Abstract. Although assigning $D_{s0}^+(2317)$ to the $I_3 = 0$ component $\hat{F}_I^+$ of iso-triplet four-quark mesons is favored by experiments, its neutral and doubly charged partners have not yet been observed. It is discussed why they were not observed in inclusive $e^+e^- \to c\bar{c}$ experiment and that they can be observed in B decays.

PACS. 14.40.Lb Charmed mesons

The charm-strange scalar meson $D_{s0}^+(2317)$ has been observed in inclusive e^+e^- annihilation [1,2]. It is very narrow ($\Gamma < 3.8\,\mathrm{MeV}$ [3]) and it decays dominantly into $D_s^+\pi^0$ while no signal of $D_s^{*+}\gamma$ decay has been observed. Therefore, the CLEO provided a severe constraint [2],

$$R(D_{s0}^+(2317)) < 0.059, \tag{1}$$

where $R(S) = \Gamma(S \to D_s^{*+}\gamma)/\Gamma(S \to D_s^+\pi^0)$ with $S = D_{s0}^+(2317)$. Similar resonances have been observed in B decays: $B \to \bar{D}\tilde{D}_{s0}^+(2317)[D_s\pi^0, D_s^{*+}\gamma]$ [4], $B \to \bar{D}$(or $\bar{D}^*$) $\tilde{D}_{s0}^+(2317)[D_s\pi^0]$ [5]. Here the new resonances have been denoted by $\tilde{D}_{s0}^+(2317)$ (observed channel(s)) to distinguish them from the above $D_{s0}^+(2317)$. It is because the resonance signals have been observed in the $D_s^{*+}\gamma$ channel in addition to the $D_s^+\pi^0$ [4]. It is quite different from the previous $D_{s0}^+(2317)$.

As will be seen later, assigning $D_{s0}^+(2317)$ to the $I_3 = 0$ component $\hat{F}_I^+$ [6] of scalar $\hat{F}_I \sim [cn][\bar{s}\bar{n}]_{I=1}$, $(n = u, d)$, is favored by eq. (1). In this case, its narrow width can be realized by a small overlap between wave functions of the initial and final states of its dominant decay $\hat{F}_I^+ \to D_s^+\pi^0$. Such a small overlap can be seen by decomposing a scalar $|[qq][\bar{q}\bar{q}]\rangle$ with $\mathbf{1}_s \times \mathbf{1}_s$ of spin $SU(2)$ and $\bar{\mathbf{3}}_c \times \mathbf{3}_c$ of color $SU_c(3)$ into a sum of $|\{q\bar{q}\}\{q\bar{q}\}\rangle$ states:

$$|[qq]_{\bar{\mathbf{3}}_c}^{\mathbf{1}_s}[\bar{q}\bar{q}]_{\mathbf{3}_c}^{\mathbf{1}_s}\rangle_{\mathbf{1}_c}^{\mathbf{1}_s} = -\sqrt{1/4}\sqrt{1/3}|\{q\bar{q}\}_{\mathbf{1}_c}^{\mathbf{1}_s}\{q\bar{q}\}_{\mathbf{1}_c}^{\mathbf{1}_s}\rangle_{\mathbf{1}_c}^{\mathbf{1}_s}$$
$$+\sqrt{3/4}\sqrt{1/3}|\{q\bar{q}\}_{\mathbf{1}_c}^{\mathbf{3}_s}\{q\bar{q}\}_{\mathbf{1}_c}^{\mathbf{3}_s}\rangle_{\mathbf{1}_c}^{\mathbf{1}_s} + \cdots, \tag{2}$$

where the above state is the lowest-lying four-quark state. However, another possible $\mathbf{6}_c \times \bar{\mathbf{6}}_c$ state has been ignored because it is much heavier [7]. The color and spin wave function overlap between $|\hat{F}_I^+\rangle$ and $\langle D_s^+\pi^0|$ is given by the first term of the right-hand side of eq. (2). To see its

narrow width more explicitly, it might be better to calculate directly its rate. However, it is difficult because no one knows spatial wave functions of particles included. Therefore, we estimate the rate for $\hat{F}_I^+ \to D_s^+\pi^0$ by comparing it with $\hat{\delta}^{s+} \to \eta\pi^+$. To this aim, we assign the observed $a_0(980)$, $f_0(980)$, $\kappa(800)$ and $f_0(600)$ [8] to scalar $[qq][\bar{q}\bar{q}]$ mesons, $\hat{\delta}^s$, $\hat{\sigma}^s$, $\hat{\kappa}$ and $\hat{\sigma}$ [7], although their assignment is still in controversy. However, the above overlap can be quite different from that of $\langle \eta\pi^+|$ and $|\hat{\delta}^{s+}\rangle$ at the scale of $m_{\hat{\delta}^s} \sim 1\,\mathrm{GeV}$, because a gluon exchange between $\{q\bar{q}\}$ pairs will reshuffle the above decomposition (while such a reshuffling will be rare at the 2 GeV or a higher energy scale, because it is known that the s-quark at the 2 GeV scale is much more slim ($m_s \simeq 90\,\mathrm{MeV}$) [9] than the s-quark in the constituent quark model, $i.e.$, the quark-gluon coupling at 2 GeV scale is much weaker than that at 1 GeV scale). With this in mind, we introduce a parameter β_0 describing the difference between overlaps of color and spin wave functions at the scale of $m_{\hat{\delta}^s}$ and at the scale of $m_{\hat{F}_I^+}$. In the limiting case of full reshuffling around 1 GeV but no reshuffling at the scale of $m_{\hat{F}_I}$, we have $|\beta_0|^2 = 1/12$ as seen in eq. (2). By using a hard-pion technique and the asymptotic $SU_f(4)$ symmetry, which have been reviewed comprehensively in ref. [10], we have

$$\Gamma(\hat{F}_I^+ \to D_s^+\pi^0)_{SU_f(4)} \simeq 5\text{--}10\,\mathrm{MeV}, \tag{3}$$

where the spatial wave function overlap is in the $SU_f(4)$ symmetry limit at this stage. Here, we have used $\Gamma(a_0(980) \to \eta\pi)_{\mathrm{exp}} = 50\text{--}100\,\mathrm{MeV}$ and the η-η' mixing angle $\theta_P \simeq -20°$ [8] as the input data. Noting that the above $SU_f(4)$ symmetry overestimates by 20–30% the amplitude, we have $\Gamma(\hat{F}_I^+) \simeq \Gamma(\hat{F}_I^+ \to D_s^+\pi^0) \sim 3.5$–$7\,\mathrm{MeV}$ [11], which is sufficiently narrow.

Next, we study the radiative decay of $D_{s0}^+(2317)$ to see that its assignment to $\hat{F}_I^+$ is consistent with eq. (1).

[a] e-mail: `terasaki@hep.s.kanazo`

Table 1. Rates for radiative decays of charm-strange mesons under the VMD, where the spatial wave function overlap is in the $SU_f(4)$ symmetry. The input data are taken from ref. [8].

Decay	Pole(s)	Input data	Rate (keV)
$D_s^{*+} \to D_s^+ \gamma$	ϕ, ψ	$\Gamma(\omega \to \pi^0 \gamma)_{\mathrm{exp}}$	0.8
$\hat{F}_I^+ \to D_s^{*+} \gamma$	ρ^0	$\Gamma(\phi \to a_0 \gamma)_{\mathrm{exp}}$	45
$\hat{F}_0^+ \to D_s^{*+} \gamma$	ω	$\Gamma(\phi \to a_0 \gamma)_{\mathrm{exp}}$	4.7
$D_{s0}^{*+} \to D_s^{*+} \gamma$	ϕ, ψ	$\Gamma(\chi_{c0} \to \psi \gamma)_{\mathrm{exp}}$	35

Table 2. Rates for isospin non-conserving decays, where the spatial wavefunction overlap is in the $SU_f(4)$ symmetry. Input data are taken from ref. [8].

Decay	Input data	Rate (keV)
$D_s^{*+} \to D_s^+ \pi^0$	$\Gamma(\rho \to \pi\pi)_{\mathrm{exp}}$	0.05
$\hat{F}_0^+ \to D_s^+ \pi^0$	$\Gamma(a_0 \to \eta\pi) \simeq 70\,\mathrm{MeV}$	0.7
$D_{s0}^{*+} \to D_s^+ \pi^0$	$\Gamma(K_0^{*0} \to K^+ \pi^-)_{\mathrm{exp}}$	0.6

For later convenience, we study the typical three cases, $D_{s0}^+(2317)$ as i) the iso-triplet $\hat{F}_I^+$, ii) the iso-singlet $\hat{F}_0^+ \sim [cn][\bar{s}\bar{n}]_{I=0}$ and iii) the conventional scalar $D_{s0}^{*+} \sim \{c\bar{s}\}$, under the vector meson dominance (VMD) hypothesis. To test our approach, we study $D_s^{*+} \to D_s^+ \gamma$ in the same way. Here, we take VVP and SVV couplings with spatial wave function overlap in the $SU_f(4)$ symmetry limit, where V, P and S denote a vector, a pseudoscalar and a scalar meson, respectively, and take the overlapping factor $|\beta_1|^2 = 1/4$ between wave functions of a scalar four-quark and two vector-meson states, as seen in eq. (2). The results are listed in table 1, where the input data are taken from ref. [8]. Comparing the rate for $\hat{F}_I^+ \to D_s^{*+} \gamma$ in table 1 with eq. (3), we obtain $R(\hat{F}_I^+) \sim (4.5–9) \times 10^{-3}$ [11], which is consistent with the constraint eq. (1). It implies that assigning $D_{s0}^+(2317)$ to $\hat{F}_I^+$ is favored by experiments.

In the cases of the above assignments ii) and iii), $D_{s0}^+(2317) \to D_s^+ \pi^0$ is isospin non-conserving. The isospin non-conservation is assumed to be caused by the η-π^0 mixing as usual. The mixing parameter ϵ has been estimated [12] as $\epsilon = 0.0105 \pm 0.0013$, i.e., $\epsilon \sim O(\alpha)$ with the fine-structure constant α. It implies that isospin non-conserving decays are much weaker than the radiative ones. By using the hard-pion approximation, the asymptotic $SU_f(4)$ symmetry and the above value of ϵ, the rates for the isospin non-conserving decays can be obtained as listed in table 2. The results on the decays of D_s^{*+} in tables 1 and 2 lead to the ratio of decay rates $R(D_s^{*+})^{-1} \simeq 0.06$. This reproduces well the experimental value [13] $R(D_s^{*+})_{\mathrm{BABAR}}^{-1} = 0.062 \pm 0.005 \pm 0.006$. This means that the present approach is sufficiently reliable. The corresponding ratios of the decay rates in the cases ii) and iii) are also obtained as ii) $R(\hat{F}_0^+) \simeq 7$ and iii) $R(D_{s0}^{*+}) \simeq 60$. They are much larger than the experi-

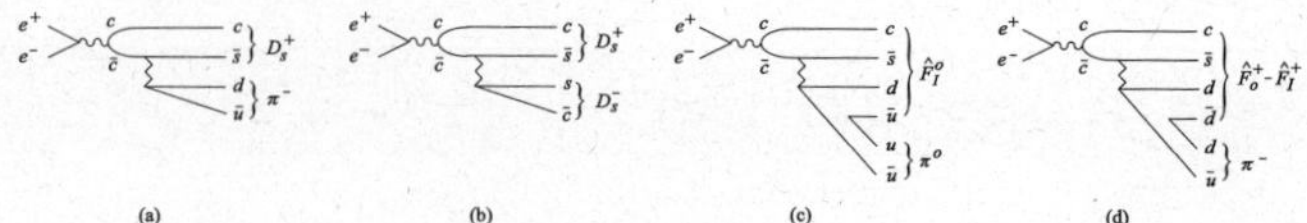

Fig. 1. Production of charm-strange scalar four-quark mesons through $e^+ e^- \to c\bar{c}$. (a) and (b) describe the productions $D_s^+ \pi^-$, $D_s^{*+} \pi^-$, etc., and $D_s^+ D_s^-$, $D_s^{*+} D_s^-$, etc., respectively. The production of $\hat{F}_I^0$, $\hat{F}_I^+$ and $\hat{F}_0^+$ is given by (c) and (d).

mental upper bound. It should be noted that the isospin non-conserving decays are much weaker than the radiative decays, as expected intuitively above. The assignment of $D_{s0}^+(2317)$ to the iso-singlet DK molecule [14] leads to $R(\{DK\}) \simeq 3$ which is much larger than the experimental upper bound in eq. (1). Hence, such an assignment should be rejected [15]. Thus, assigning $D_{s0}^+(2317)$ to an iso-singlet state ($\hat{F}_0^+$, D_{s0}^{*+} or DK molecule) is disfavored by experiments.

From the above considerations, it is natural to assign $D_{s0}^+(2317)$ to the iso-triplet $\hat{F}_I^+$. However, its neutral and doubly charged partners, $\hat{F}_I^{++}$ and $\hat{F}_I^0$, have not yet been observed [3]. With this in mind, we study the production of charm-strange scalar mesons ($\hat{F}_I^{++,+,0}$ and $\hat{F}_0^+$) by assigning $D_{s0}^+(2317)$ to $\hat{F}_I^+$, and discuss why experiments have observed $D_{s0}^+(2317)$ but not its neutral and doubly charged partners. To this aim, we consider their production through weak interactions as a possible mechanism, because the OZI-rule violating productions of multi-$q\bar{q}$-pairs and their recombinations into four-quark meson states are believed to be strongly suppressed at high energies. First, we recall that color-mismatched decays which include rearrangements of colors in weak-decay processes would be suppressed compared with color favored ones as long as non-factorizable contributions, which are actually small in B decays and are expected to be much smaller at higher energies, are ignored. Next, we draw quark-line diagrams within the minimal $q\bar{q}$-pair creation, noting the OZI rule. Because there is no diagram yielding $\hat{F}_I^{++}$ production in this approximation, as seen in fig. 1, it is easy to understand why no evidence of $\hat{F}_I^{++}$ was found in $e^+ e^- \to c\bar{c}$ experiments. As will be seen in productions of $\tilde{D}_{s0}^+(2317)[D_s^+ \pi^0]$ in the B decays, their observed rates are comparable with color-mismatched decays, so that rates for $\hat{F}_I^0$ and $\hat{F}_0^+$ productions in $e^+ e^- \to c\bar{c}$ annihilation will be expected to be comparable with those through the color-suppressed ones. Therefore, they would be much weaker (possibly by about two orders of magnitude) than productions of $D_s^+ \pi^-$, $D_s^{*+} \pi^-$, $D_s^{*+} \rho^-$, etc., and $D_s^+ D_s^-$, $D_s^{*+} D_s^-$, $D_s^{*+} D_s^{*-}$, etc., created through the reaction depicted by figs. 1(a) and (b). The $D_s^+ \pi^-$ produced through fig. 1(a) obscures the signal $\hat{F}_I^0 \to D_s^+ \pi^-$ events. In addition, the D_s^{*+} and γ from $D_s^{*-} \to D_s^- \gamma$ produced through the ordinary $e^+ e^- \to c\bar{c} \to D_s^{*+} D_s^{*-}$ (and through fig. 1 (b)) obscure the signal of $\hat{F}_0^+ \to D_s^{*+} \gamma$ events. Therefore, it is understood why the inclusive $e^+ e^-$ annihilation experiment found no signal of scalar resonance in the $D_s^+ \pi^-$ and $D_s^{*+} \gamma$ channels. In the case of

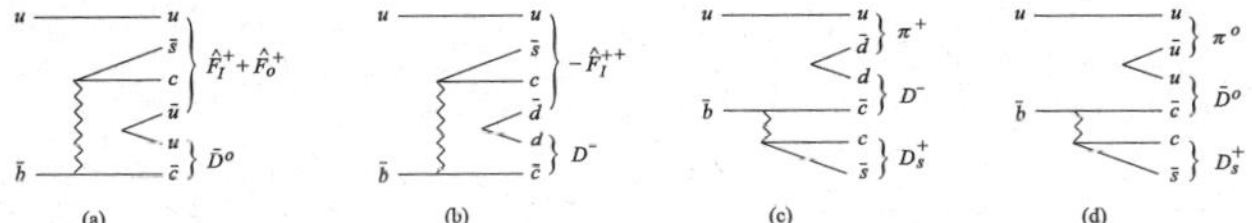

Fig. 2. Production of charm-strange scalar mesons in the B_u^+ decays. (c) and (d) describe the production of backgrounds of $\hat{F}_I^{++}$ and $\hat{F}_I^+$ signals, respectively.

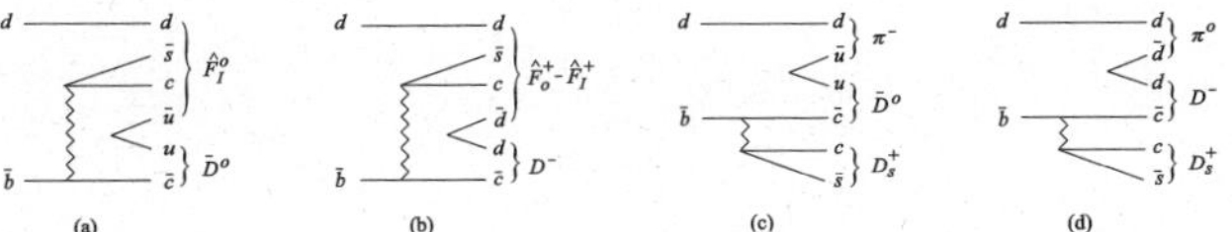

Fig. 3. Production of $\hat{F}_I^+$, $\hat{F}_0^+$ and $\hat{F}_I^0$ in the B_d^0 decays. (c) and (d) depict the production of backgrounds of $\hat{F}_I^0$ and $\hat{F}_I^+$.

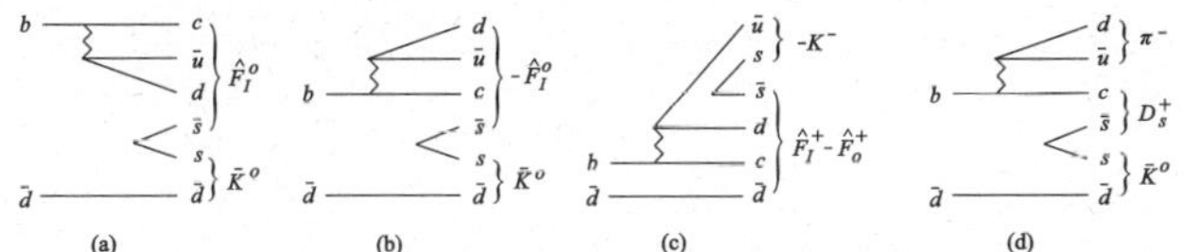

Fig. 4. Production of $\hat{F}_I^+$, $\hat{F}_I^0$ and $\hat{F}_0^+$ in the decays of $\bar{B}_d^0$. (d) describes the production of backgrounds of the $\hat{F}_I^0$ signal.

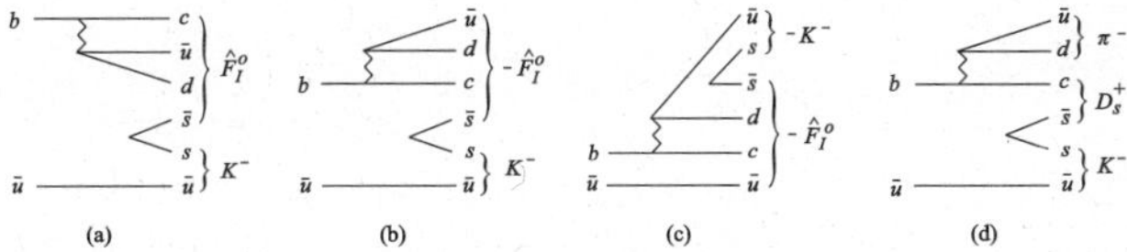

Fig. 5. Production of $\hat{F}_I^0$ in the $\bar{B}_u^-$ decays. (d) describes the production of backgrounds of its signals.

$\hat{F}_I^+$, however, there do not exist large numbers of background events described by figs. 1(a) and (b), because its main decay is $\hat{F}_I^+ \to D_s^+ \pi^0$. In fact, $D_{s0}^+(2317)$ has been observed in the $D_s^+ \pi^0$ channel. This seems to imply that the production of four-quark mesons in hadronic weak decays plays an essential role [16].

Because it is difficult to observe $\hat{F}_I^{++}$ and $\hat{F}_I^0$ in inclusive $e^+ e^- \to c\bar{c}$ experiments, we study their productions in B decays. First, we draw quark-line diagrams in the same way as the above. As expected in figs. 2(a) and 3(b), $\tilde{D}_{s0}^+(2317)[D_s^+ \pi^0]$, which can be identified to $\hat{F}_I^+$ because it decays dominantly into $D_s^+ \pi^0$ as seen above, has been observed. The production of $\hat{F}_I^{++}$ is given by fig. 2(b) which is of the same type as fig. 2(a). In addition, the production of $\hat{F}_I^0$ is given by fig. 3(a) which is again of the same type as figs. 2(a) and (b), so that their production rates are not very different from each other; $B(B_u^+ \to \hat{F}_I^{++} D^-) \sim B(B_d^+ \to \hat{F}_I^0 \bar{D}^0) \sim B(B_u^+ \to \hat{F}_I^+ \bar{D}^0)$, where $B(B_u^+ \to \hat{F}_I^+ \bar{D}^0)_{\exp} \sim 10^{-3}$ [4, 5]. Besides, BELLE Collaboration observed indications of $\tilde{D}_{s0}^+(2317)[D_s^{*+}\gamma]$ which are conjectured to be signals of $\hat{F}_0^+ \to D_s^{*+}\gamma$ because the production of $\hat{F}_0^+$ and $\hat{F}_I^+$ are depicted by the same diagrams and the $\hat{F}_I^+ \to D_s^{*+}\gamma$ is much weaker than the $\hat{F}_I^+ \to D_s^+ \pi^0$ while the $\hat{F}_0^+ \to D_s^{*+}\gamma$ is much stronger than the $\hat{F}_0^+ \to D_s^+ \pi^0$, as seen before. As expected in fig. 4(c), BELLE Collaboration [17] observed $\bar{B}_d \to \tilde{D}_{s0}^+(2317)[D_s^+ \pi^0]K^-$, and provided $B(\bar{B}_d \to \tilde{D}_{s0}^+(2317)K^-) \cdot B(\tilde{D}_{s0}^+(2317) \to D_s^+ \pi^0) = (5.3^{+1.5}_{-1.3} \pm 0.7 \pm 1.4) \times 10^{-5}$. If $\tilde{D}_{s0}^+(2317)[D_s^+ \pi^0]$ is identified to $\hat{F}_I^+$ and $B(\tilde{D}_{s0}^+(2317) \to D_s^+ \pi^0) \simeq 100\%$ is taken, $B(\bar{B}_d^0 \to \hat{F}_I^+ K^-) \sim 10^{-5} - 10^{-4}$ would be obtained. Using it as the input data and noting that figs. 4(c) and 5(c) are of the same type, we could estimate $B(B_u^- \to K^- \hat{F}_I^0) \sim 10^{-5} - 10^{-4}$, if contributions from the diagrams of figs. 5(a) and (b) cancel each other (because the phases of $\hat{F}_I^0$ in these diagrams have opposite signs arising from the antisymmetry property of its wave function).

In summary, we have seen that assigning $D_{s0}^+(2317)$ to an iso-triplet $\hat{F}_I^+$ is favored by experiments. In addition, we have discussed why inclusive $e^+ e^- \to c\bar{c}$ experiments observed no evidence for its neutral and doubly charged partners $\hat{F}_I^0$ and $\hat{F}_I^{++}$. $\tilde{D}_{s0}^+(2317)[D_s^+ \pi^0]$ which was observed in B decays has been identified to $D_{s0}^+(2317)$. Indications of $\hat{F}_0^+$ have also been observed as $\tilde{D}_{s0}^+(2317)[D_s^{*+}\gamma]$ in B decays. $\hat{F}_I^0$ and $\hat{F}_I^{++}$ will be observed in B decays.

The author would like to thank the organizers of QNP06 for supporting local expense during the conference. This work is supported in part by the Grant-in-Aid for Science Research, Ministry of Education, Culture, Sports, Science and Technology of Japan (No. 16540243).

References

1. BABAR Collaboration (B. Aubert *et al.*), Phys. Rev. Lett. **90**, 242001 (2003).
2. CLEO Collaboration (D. Besson *et al.*), Phys. Rev. D **68**, 032002 (2003).
3. BABAR Collaboration (B. Aubert *et al.*), hep-ex/0604030.
4. BELLE Collaboration (P. Krokovny *et al.*), Phys. Rev. Lett. **91**, 262002 (2003).
5. BABAR Collaboration (E. Robutti), Acta Phys. Pol. B **36**, 2315 (2005).
6. K. Terasaki, Phys. Rev. D **68**, 011501(R) (2003).
7. R.L. Jaffe, Phys. Rev. D **15**, 267 and 281 (1977).
8. Particle Data Group (S. Eidelman *et al.*), Phys. Lett. B **592**, 1 (2004).
9. R. Gupta, hep-lat/0502005.
10. S. Oneda, K. Terasaki, Prog. Theor. Phys. Suppl. **82**, 1 (1985).
11. A. Hayashigaki, K. Terasaki, Prog. Theor. Phys. **114**, 1191 (2005); K. Terasaki, hep-ph/0512285.
12. R.H. Dalitz, F. von Hippel, Phys. Lett. **10**, 153 (1964).
13. BABAR Collaboration (B. Aubert *et al.*), hep-ex/0508039.
14. T. Barns, F.E. Close, H.J. Lipkin, Phys. Rev. D **68**, 054006 (2003).
15. T. Mehen, R.P. Springer, Phys. Rev. D **70**, 0704014 (2004).
16. K. Terasaki, Prog. Theor. Phys. **116**, 435 (2006).
17. BELLE Collaboration (A. Drutskoy *et al.*), Phys. Rev. Lett. **94**, 061802 (2005).

Eur. Phys. J. A **31**, 679–683 (2007)

DOI 10.1140/epja/i2006-10211-9

Special Article – QNP 2006

New physics and technical challenges of $\overline{\text{P}}$ANDA

P. Gianotti[a]

LNF, Via E. Fermi 40, 00044 Frascati, Italy

Received: 25 October 2006
Published online: 1 March 2007 – © Società Italiana di Fisica / Springer-Verlag 2007

Abstract. An antiproton beam of unprecedented intensity and quality will soon be available at the HESR machine foreseen for the new FAIR accelerator complex of Darmstadt. This new facility, together with a properly designed new detector ($\overline{\text{P}}$ANDA), will be the ideal environment to study fundamental questions of hadron and nuclear physics and to carry out precise tests of the strong interaction.

PACS. 13.60.Le Meson production – 13.60.Rj Baryon production – 13.30.Eg Hadronic decays – 21.80.+a Hypernuclei

1 Introduction

The Gesellschaft für Schwerionenforschung (GSI) [1] of Darmstadt, Germany, is undergoing a major upgrade of the existing accelerator complex [2]. This upgrade foresees ion beams of higher intensity and better quality, and, first for GSI, an antiproton beam. At FAIR the antiprotons will be produced at the rate of 2×10^7/s and then, after accumulation and cooling, will be transferred inside the High Energy Storage Ring (HESR) for the experimental activity. The HESR machine will be equipped with an internal target surrounded by a general purpose detector: $\overline{\text{P}}$ANDA (Antiproton Annihilation at Darmstadt). The momentum of the antiprotons will vary between 1.5 and 15 GeV/c so that the maximum center-of-mass energy will reach 5.5 GeV, enough for the associate production of Ω_c which is the upper limit for the mass range of charmonium hybrid predictions.

The HESR foresees two different operation modes: the high intensity mode, where with a beam momentum spread $\delta p/p$ of 10^{-4} a luminosity of $10^{32}\,\text{cm}^{-2}\text{s}^{-1}$ will be available, and the high resolution mode, where the luminosity requirement will be released to $10^{31}\,\text{cm}^{-2}\text{s}^{-1}$ to have a maximum momentum precision of 10^{-5}.

$\overline{\text{P}}$ANDA will perform a complete program of hadron spectroscopy to test many aspects of Quantum Chromo Dynamics (QCD), the generally accepted theory of strong interactions. The aim is to investigate both the dynamics of the interaction of fundamental particles, and the existence of new forms of matter such as extra charmonium states, nuclei with an explicit strange-quark content. Furthermore, particle properties, when produced inside the nuclear medium, will be studied. The importance of these measurements is related to our capability of predicting,

confirming, and explaining the physical states of the theory, in other words, an exhaustive understanding of the strong-interaction mechanism.

In the past, experiments with antiprotons have proven to be a rich source of information in this field, and with the new GSI antiproton machine, all the above-mentioned topics will be addressed with a more powerful detector and dedicated experimental campaigns. $\overline{\text{P}}$ANDA will be the new general purpose detector optimized to accomplish this complete hadron physics program. In the following, the main topics of this investigation will be illustrated.

2 Hadron spectroscopy

Quantum Chromo Dynamics is extremely successful in describing phenomena at high energies where the interaction among quarks by gluon exchange can be treated by perturbation theory. Nevertheless, hadron properties cannot be directly derived from this approach, and phenomenological models, tuned on the experimental data, are used to describe lower-energy systems.

The experimental results obtained by past experiments at LEAR and Fermilab by studying $\bar{p}p$ annihilations have been excellent tools to perform hadron spectroscopy. The HESR antiprotons will allow continuation and expansion of this research field with the new possibilities offered by the higher beam intensity and the better energy resolution. Figure 1 gives an overlook on QCD-systems which could be studied in the HESR energy range.

The eight charmonium states lie exactly in the center of this mass region, and even if today all of them have been identified, open questions remain for their masses and widths. Concerning the ground state of charmonium, despite the abundance of measurements performed recently mainly at the B-factories, the knowledge of the η_c

---

[a] e-mail: `paola.gianotti@lnf.infn.it`

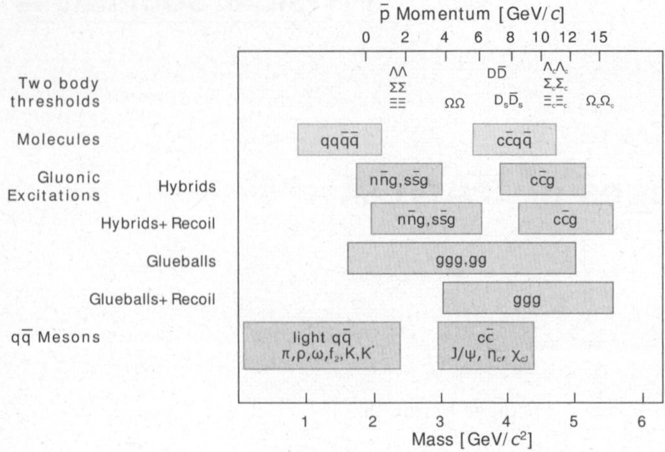

Fig. 1. QCD-systems accessible in the HESR energy range.

mass is not yet satisfactory [3]. Moreover, the different measurements of the width range between 7 and 34 MeV. New, high-precision determinations of these parameters are desirable.

The study of $\eta_c(2S)$ has just started. Discovered by the Belle Collaboration [4] it was then confirmed by CLEO [5] and BaBar [6], but the values of its mass and width are not compatible with an earlier observation of the Crystal Ball Collaboration [7]. Even the consistency with the model predictions is marginal.

The singlet-P (h_c) resonance of charmonium is of extreme importance to determine the spin dependent component of the $q\bar{q}$ confinement potential. The $^1P_1$ charmonium state has been discovered by the E760 Collaboration [8] that, nevertheless, could only set an upper limit for the width ($< 1\,\mathrm{MeV}$). Recently E835 [9] and CLEO [10] experiments confirmed the mass value quoted by E760, but the narrow width of such a state could only be measured by experiments able, like $\overline{\mathrm{P}}\mathrm{ANDA}$, to carry out a systematic study of its decay modes.

Concerning χ_{cJ} states, the angular distribution of radiative decays needs to be studied with high statistics, in order to clarify the discrepancies observed between the available measurements [11,12] and the theoretical predictions.

Above the $D\bar{D}$ threshold (3.73 GeV) very little is known positively. In this energy region narrow $^1D_2$, $^3D_2$ states are expected (narrow because they cannot decay to $D\bar{D}$) together with the first radial excitations of the singlet and triplet charmonium states. Actually, something unexpected in this region came out in 2003 when the Belle Collaboration reported the observation of a new narrow state with a mass of 3872.0 MeV/c^2 [13]. This new state, called X(3872), has been subsequently confirmed by other experiments, but its nature is still controversial. Speculations range from a $D^0\overline{D^{0*}}$ molecule to a $^3D_2$ state. All these interpretations are not satisfactory and only new data on different decay modes can shed light on the nature of this particle. The investigation of the energy region above $D\bar{D}$ threshold is a central part of the $\overline{\mathrm{P}}\mathrm{ANDA}$ charmonium spectroscopy program. At $\overline{\mathrm{P}}\mathrm{ANDA}$,

precise measurements would be possible thanks to the high yield of charmonium production in $\bar{p}p$ annihilation (*i.e.* BR($\bar{p}p \to \eta_c = (1.2 \pm 0.4) \cdot 10^{-3}$)). By detecting hadronic final states ($KK\pi\pi$, $4K$, 4π, $K\bar{K}\pi$, $\eta\pi\pi$,...) having branching fractions two orders of magnitude higher than the $\gamma\gamma$ decay mode studied so far, high-statistic samples could be easily collected.

The spectrum of fig. 1 also indicates where "gluonic hadrons", *i.e.* hadronic systems with explicit content of glue, have been predicted by theoretical models. These states fall in two general categories; hybrids and glueballs. Hybrids consist of a quark and an anti-quark pair plus a gluon; glueballs are pure gluonic states. The additional degrees of freedom carried by the gluons allow hybrids and glueballs to have J^{PC} quantun mumbers that are forbidden for ordinary $q\bar{q}$ systems. The study of exotics has been carried out systematically and intesively in the past years at LEAR, and the most promising candidates for gluonic hadrons have all been seen by antiproton-proton annihilation experiments. Two particles with exotic quantum mumbers ($J^{PC} = 1^{-+}$) $\pi_1(1400)$ [14] and $\pi_1(1600)$ [15], actually discovered in pion-induced reactions at BNL, have been deeply studied in $p\bar{p}$ annihilation. Up to now, the hunt for exotic hadrons has mainly been accomplished in the mass region below $2.2\,\mathrm{GeV}/c^2$; a region where a lot of broad ordinary states sit. Therefore, a precise and unambiguous identification of exotics is hardly possible. The idea of the $\overline{\mathrm{P}}\mathrm{ANDA}$ Collaboration is to extend this search into the charmonium energy region, where only eight narrow systems are present. Up to now, LQCD calculations have been centered around the lowest-lying charmonium hybrids. Among these, three are expected to have exotic quantum numbers so that mixing effects with nearby $c\bar{c}$ states are excluded, and their identification should be easier.

Regarding glueballs, the best candidate for the glueball ground state ($J^{PC} = 0^{++}$) is the f$_0$(1500) [3] discovered in $p\bar{p}$ annihilation. However, this state mixes up with other nearby ordinary $q\bar{q}$ systems, making its exotic nature not accepted worldwide. LQCD calculations [16] make rather detailed predictions for the glueball's mass spectrum. Figure 2 shows such spectrum. In the mass range that $\overline{\mathrm{P}}\mathrm{ANDA}$ will explore, about 15 states are predicted, some of them even with exotic quantum numbers (oddballs). The lightest oddball, with $J^{PC} = 2^{+-}$, is expected to have a mass of $4.3\,\mathrm{GeV}/c^2$.

In $\bar{p}p$ annihilation, particles can be studied in two different environments: formation or production reactions. In formation reactions the $\bar{p}p$ annihilation leads to the appearance in the final state of the desired particle, while in productions reactions an extra meson ($\pi, \eta, \ldots$) is generated together with that under investigation. Formation experiments would generate exotic states with ordinary quantum numbers, while production experiments which yield an exotic together with another particle allow access to non-$q\bar{q}$ quantum numbers. From previous experiments at LEAR, we learned that production rates of exotic states are similar to those of ordinary particles. Therefore, for exotics we estimate cross-sections of formation

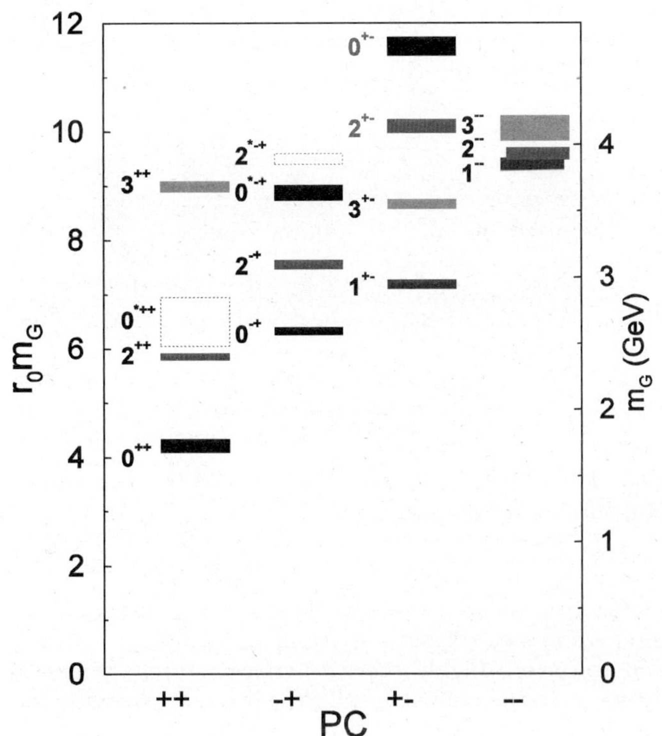

Fig. 2. Glueball prediction from LQCD calculation from ref. [16].

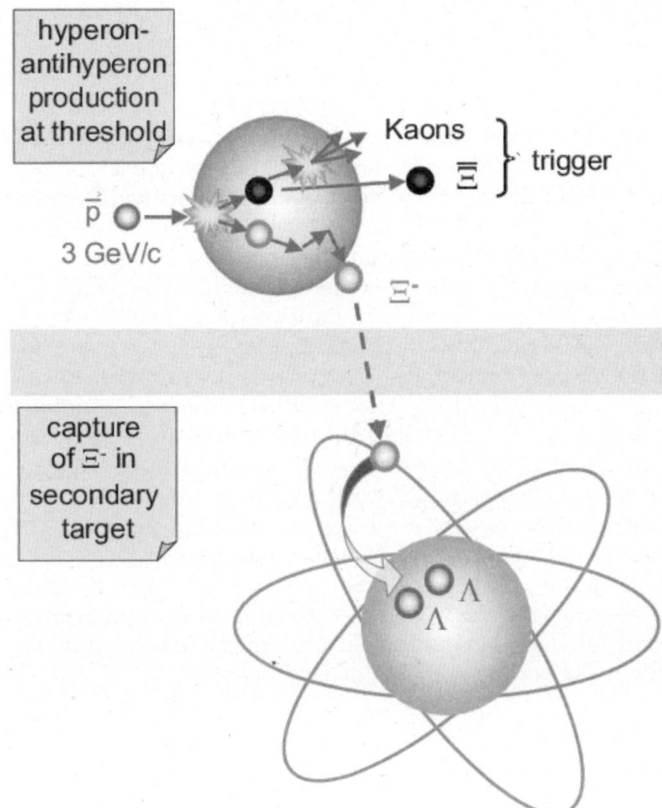

Fig. 3. The two-step reaction that will be used by $\overline{P}$ANDA to produce double Λ-hypernuclei.

and of production something similar to conventional charmonium states, *i.e.* of the order of 120 pb. The $\overline{P}$ANDA program for exotic search foresees to start with production measurements at the highest antiproton momentum ($15\,\mathrm{GeV}/c$) with the aim of studying all possible production channels of exotic and ordinary channels. The second step would consist of formation measurements by scanning the antiproton energy in small steps in the regions where promising candidates have been observed in production measurements. This will allow us to check J^{PC} assignment as well as masses and widths of the measured states.

3 Hadrons in nuclear matter

The study of the modification of hadronic properties in nuclear matter is one of the present research activities at GSI. This work is aimed at understanding the origin of hadron masses and the effect of chiral symmetry breaking on the process of mass generation. Actually, these studies, carried out on the light-quark sector [17], have showed sizeable mass splitting for both pions and kaons. A high-intensity $\overline{p}$ beam up to $15\,\mathrm{GeV}/c$ will allow to extend this research to the charm sector. .

For the low-lying charmonium states (η_c, J/ψ) theoretical calculations indicate [18] small in-medium mass reductions, of the order of 5–$10\,\mathrm{MeV}/c^2$, but since the effect is expected to scale with the volume occupied by the $c\bar{c}$ pair, the situation may change for excited charmonium states.

For the D-meson family the situation is different: made of a c-quark and of a light anti-quark, they represent QCD analogues of hydrogen atoms. Hence, they provide the unique opportunity of studying the in-medium dynamics of a system with a single light quark.

Experimentally, the study of in-medium mass modification of charmonium states will be performed measuring di-lepton decay branching ratios on different nuclear targets, whereas D-mesons will be identified via their hadronic decays to channels with kaons.

$\overline{P}$ANDA will be the first experiment carrying out a systematic study of the properties of D charmed mesons, and of charmonium $c\bar{c}$ states in nuclear matter.

4 Multi-strangeness systems

Hypernuclear physics is not a new field of nuclear research but, in spite of its age, it is experiencing a renewed interest thanks to the availability of better experimental conditions that could be used to clarify some long-standing problems. One of these is the precise evaluation of $\Lambda\Lambda$-hypernuclear binding energy. Up to now, only three $\Lambda\Lambda$-hypernuclei have been detected via their double pion decay, but if they could be produced at a reasonable rate, they could be a unique source of information on Λ-Λ interactions. $\overline{P}$ANDA is planning to produce copiously such systems making a $3\,\mathrm{GeV}/c\,\bar{p}$ beam interacting with a nucleus and giving rise to a $\Xi\bar{\Xi}$ hyperon pair. The $\bar{\Xi}$ will annihilate inside the residual nucleus, and since strangeness is conserved in strong interactions, this process will produce 2 anti-kaons that could be used as a tag for the reaction. The slow Ξ^- will be captured in a secondary target, and after an atomic cascade, could interact with a proton producing two Λs that eventually could be stuck to the nucleus. For the identification and the spectroscopy of the hypernucleus, Ge-detectors will be installed. Figure 3 is illustrating schematically the two-steps process to produce $\Lambda\Lambda$-hypernuclei in $\overline{\mathrm{PANDA}}$. Moreover, the existence of an $S = -2$ six-quark state, the H-particle [19], is another challenging topic that could be addressed studying $\Lambda\Lambda$-hypernuclei. Furthermore, one could also study the properties of the hyperatoms created during the capture process of hyperons. This will supply new information on fundamental properties of hyperons. As an example, the Ω-hyperon (sss) is particularly interesting because of its long lifetime (82 ps) and of its spin $= 3/2$. This is the only elementary baryon with a non-vanishing static quadrupole moment that therefore could be directly measured determining hyperfine splitting of Ω^- atoms.

5 Further topics

Other interesting topics could be accessed exploiting the unique features offered by the HESR machine running at full luminosity. Among those, $\overline{\mathrm{PANDA}}$ Collaboration will address new and old open questions regarding the strong interaction. A large number of D-meson pairs will be produced with antiprotons of momentum larger than $6.4\,\mathrm{GeV}/c$. B-factory experiments have discovered several particles in the D and D_s sector. It is important to verify these findings and to settle the open questions regarding the static parameters of these resonances. Threshold pair production can be employed to precisely measure the mass and the width of narrow excited D and D_s states, and to study rare decays of D-mesons, and direct CP violation.

Deeply Virtual Compton Scattering (DVCS) allows to access more details about generalized parton distribution functions. The measurements of the inverted process could provide complementary information. Furthermore, the study of Drell-Yan processes ($\bar{p}p \to l^+l^- X$) will allow access to nucleon transverse polarization functions.

Finally, in $\overline{\mathrm{PANDA}}$ the electromagnetic form factor of the proton in the time like region will be determined over the widest Q^2 range from threshold to $20\,\mathrm{GeV}^2/c^4$ and above. The high statistics collected will allow us to measure $|G_E|$ and $|G_M|$ separately.

6 The $\overline{\mathrm{PANDA}}$ detector

To investigate this wide physics program, a general purpose detector will be necessary. It should be able to deal with different targets, to cover an almost full solid angle, to identify charged and neutral particles over a wide

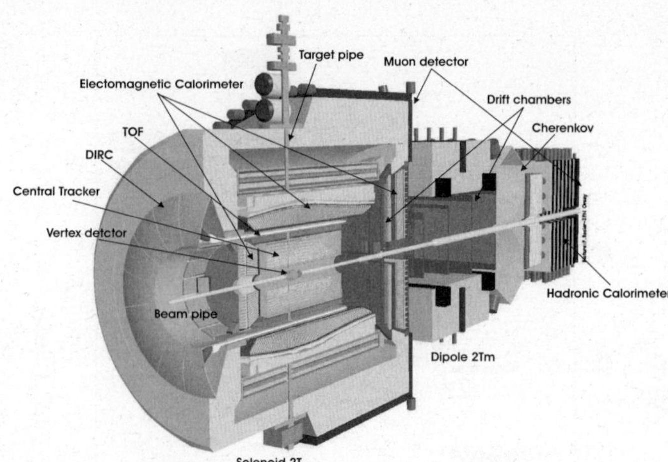

Fig. 4. Cross-sectional view of the $\overline{\mathrm{PANDA}}$ spectrometer. More details are given in the test.

momentum range, to be able to stand high rates ($\sim 10^7$ interaction per second), to resolve secondary vertexes, and to have a highly selective trigger system. Figure 4 shows a cross-sectional view of the detector presently under design [20]. The whole apparatus consists of a target spectrometer, surrounding the interaction region, and a forward spectrometer to detect particles emitted in the forward direction. The basic concept of the target spectrometer is a shell-like arrangement of various detectors surrounding the interaction point inside the field of a solenoidal magnet. The forward spectrometer consists of a large-gap dipole in combination with tracking detectors and calorimeters. The antiproton beam will interact with the target at the cross point with the target pipe located inside the solenoid. Both the target and the forward spectrometer will allow the detection, identification, and energy/momentum evaluation of charged and neutral particles. The combination of the two spectrometers takes into account the wide range of energies of the particles following antiproton annihilation. Furthermore, it is sufficiently flexible to make individual components exchangeable or insertable for specific measurements, *i.e.* hypernuclear measurements. In the target spectrometer, we find first a micro-vertex silicon detector, and then a charged-particle tracking system. Particle identification will be performed with a particular Cherenkov system (DIRC) similar to that constructed for the BaBar experiment. The forward region will be covered by a set of mini drift chambers and another Cherenkov detector. An electromagnetic calorimeter of about 19000 crystals will complete the setup inside the solenoid. Outside the return yoke of this magnet, muon detectors will be installed. Particles emitted with polar angles below 10° and 5° in the horizontal and vertical directions, respectively, are detected by the forward spectrometer. The current design foresees a 1 m gap dipole, tracking detectors for momentum evaluations of charged particles, a shashlyk-type calorimeter consisting of lead and scintillator sandwiches for photons detection, a hadron calorimeter and a muon detector. More details about the $\overline{\mathrm{PANDA}}$ detector can be found in ref. [20].

7 Conclusion

After the LEAR shutdown in 1996 and the end of the Fermilab fixed target program, a new challenging project involving antiprotons is started in Europe. The characteristics of the new beam, together with the high performance of the detector involved, will determine a step forward in the hadron physics sector, allowing to continue the investigations of charmed hadrons and their interaction with matter. This will deliver unique information on many aspect of non-perturbative QCD and nuclear physics.

References

1. http://www.gsi.de/.
2. http://www.gsi.de/fair/.
3. W-M Yao et al., J. Phys. G: Nucl. Part. Phys **33**, 1 (2006).
4. S.K. Choi et al., Phys. Rev. Lett. **89**, 102001 (2002).
5. D.M. Asner et al., Phys. Rev. Lett. **92**, 142001 (2004).
6. B. Aubert et al., Phys. Rev. Lett. **92**, 142002 (2004).
7. C. Edwards et al., Phys. Rev. Lett. **48**, 70 (1982).
8. T. Armstrong et al., Phys. Rev. Lett. **69**, 2337 (1992).
9. M. Andreotti et al., Phys. Rev. D **72**, 032001 (2005).
10. J.L. Rosner et al., Phys. Rev. Lett. **95**, 102003 (2005).
11. T. Armstrong et al., Phys. Rev. D **48**, 3037 (1993).
12. M. Ambrogiani et al., Phys. Rev. D **62**, 052002 (2000).
13. S.K. Choi et al., Phys. Rev. Lett. **91**, 262002 (2003).
14. A. Abele et al., Phys. Lett. B **446**, 349 (1999).
15. J. Reinnarth, Nucl. Phys. A **692**, 268c (2001).
16. C.J. Morningstar, M. Peardon, Phys. Rev. D **60**, 034509 (1999).
17. See A. Gillitzer, these proceedings.
18. F. Klingl et al., Phys. Rev. Lett. **82**, 3396 (1999).
19. R.L. Jaffe, Phys. Rev. Lett. **38**, 195 (1997).
20. The $\overline{\text{P}}$ANDA Collaboration, *Technical Progress Report*, http://www-panda.gsi.de/archive/public/panda_tpr.pdf/.

Eur. Phys. J. A **31**, 684–686 (2007)

DOI 10.1140/epja/i2006-10264-8

THE EUROPEAN
PHYSICAL JOURNAL A

Special Article – QNP 2006

Conventional view versus FINUDA claims of a deeply bound K^-pp state

A. Ramos[1,a], V.K. Magas[1], E. Oset[2], and H. Toki[3]

[1] Departament d'Estructura i Constituents de la Matèria, Universitat de Barcelona, Diagonal 647, 08028 Barcelona, Spain
[2] Departamento de Física Teórica and IFIC Centro Mixto Universidad de Valencia-CSIC, Institutos de Investigación de Paterna
     Apdo. correos 22085, 46071, Valencia, Spain
[3] Research Center for Nuclear Physics, Osaka University, Ibaraki, Osaka 567-0047, Japan

Received: 23 November 2006
Published online: 16 March 2007 – © Società Italiana di Fisica / Springer-Verlag 2007

Abstract. We critically revise the recent claims of a deeply bound K^-pp state associated to a peak seen in the Λp invariant-mass spectrum following nuclear K^- absorption reactions measured by the FINUDA Collaboration. An explicit theoretical simulation shows that the peak is simply generated from a two-nucleon absorption process, like $K^-pp \to \Lambda p$, followed by final-state interactions of the produced particles with the residual nucleus.

PACS. 25.80.Nv Kaon-induced reactions – 13.75.Jz Kaon-baryon interactions – 21.45.+v Few-body systems – 21.80.+a Hypernuclei

1 Introduction

Over the last years, the study of the interactions of antikaons with nuclei has revealed many interesting aspects that have triggered various speculations on the existence of new exotic phenomena.

The possible appearance of a condensate of antikaons in neutron stars was postulated after examining the sizable attractive properties of the chiral $\bar{K}N$ interaction at tree level [1]. Phenomenological fits to kaonic atoms fed the idea because a solution where antikaons would feel strongly attractive potentials of the order of $-200\,\mathrm{MeV}$ at the center of the nucleus [2] was preferred. However, a deeper understanding of the antikaon optical potential demands it to be linked to the elementary $\bar{K}N$ scattering amplitude which is dominated by the presence of a resonance, the $\Lambda(1405)$, located only 27 MeV below threshold. This makes the problem to be a highly non-perturbative one. In recent years, the scattering of $\bar{K}$ mesons with nucleons has been treated within the context of chiral unitary methods [3]. The explicit incorporation of medium effects, such as Pauli blocking, was shown to be important [4] and it was soon realized that the influence of the resonance demanded the in-medium amplitudes to be evaluated self-consistently [5]. The resulting antikaon optical potentials were quite shallower than the phenomenological one, with depths between -70 and $-40\,\mathrm{MeV}$ [6], but gave an acceptable description of the kaonic atom data [7].

More recently, variational calculations of few-body systems using a simplified phenomenological $\bar{K}N$ interaction predicted extremely deep kaonic states in ^{3}He and ^{4}He, reaching densities of up to ten times that of normal nuclear matter [8,9]. Motivated by this finding, experiment KEK-E471 used the ^{4}He(stopped K^-,p) reaction and reported [10] a structure in the proton momentum spectrum, which was denoted as the tribaryon $S^0(3115)$ with strangeness $S = -1$. If interpreted as a (K^-pnn) bound state, it would have a binding energy of 194 MeV. However, in a recent work [11] strong criticisms to the theoretical claims of refs. [8,9] have been put forward and, in addition, a reinterpretation of the KEK proton peak has been given in terms of two-nucleon absorption processes, $K^-pn \to \Sigma^-p, K^-pp \to \Sigma^0(\Lambda)p$, where the rest of the nucleus acts as a spectator. We note that deficiencies in the efficiency corrections were revealed in the last *HYP2006 Conference*, and the new reanalyzed KEK proton momentum spectrum does not show evidence of this peak [12], although the conditions of the analysis (cuts and coincidences with other charged particles) were not the same as in the previous one. In contrast, a FINUDA proton momentum spectrum from K^- absorption on ^{6}Li [13] does show a peak, which is interpreted with the two-nucleon absorption mechanism advocated in ref. [11].

Another experiment of the FINUDA Collaboration has measured the invariant-mass distribution of Λp pairs [14]. The spectrum shows a narrow peak at 2340 MeV, which corresponds to the same signal seen in the proton momentum spectrum, namely K^- absorption by a two-nucleon

a e-mail: ramos@ecm.ub.es

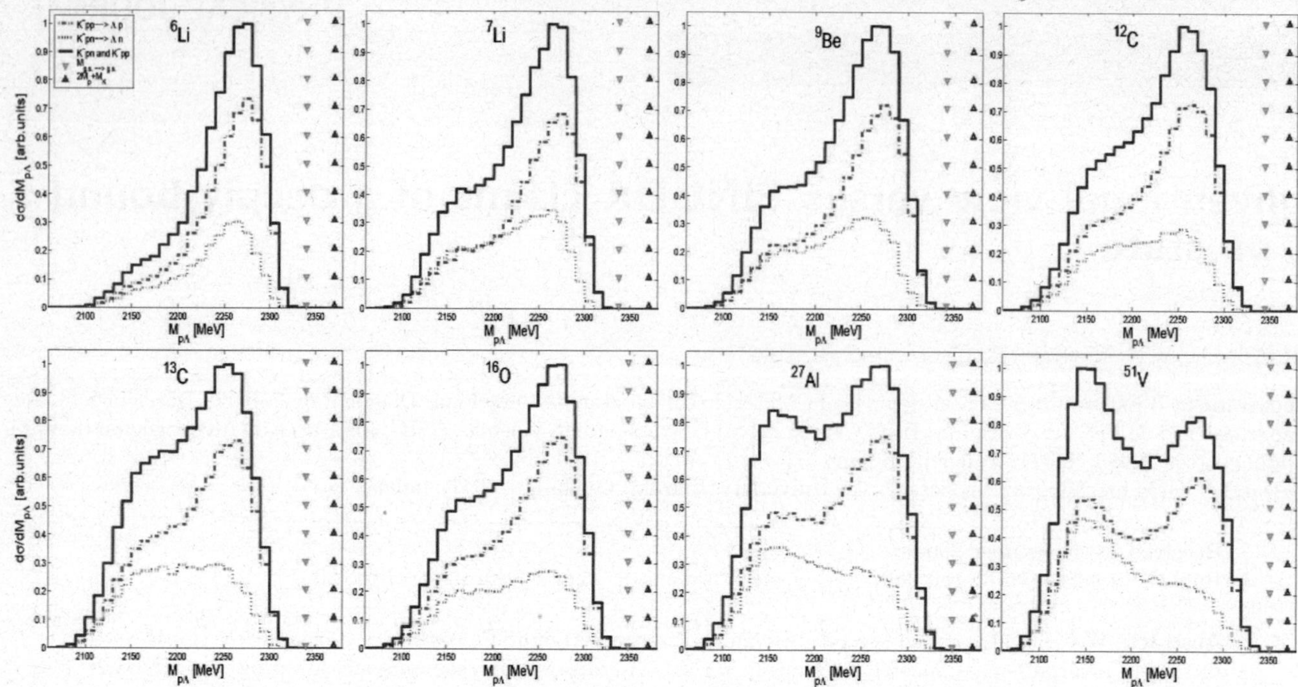

Fig. 1. Λp invariant-mass distribution after K^- absorption in several nuclei imposing $P_\Lambda > 300\,\mathrm{MeV}/c$ and $\cos\Theta_{\boldsymbol{p}_\Lambda\boldsymbol{p}_p} < -0.8$.

pair leaving the daughter nucleus in its ground state. Another wider peak is also seen at around 2255 MeV, which is interpreted in ref. [14] as being a K^-pp bound state with $B_{K^-pp} = 115^{+6}_{-5}(\text{stat})^{+2}_{-3}(\text{syst})\,\mathrm{MeV}$ and having a width of $\Gamma = 67^{+14}_{-11}(\text{stat})^{+2}_{-3}(\text{syst})\,\mathrm{MeV}$. In a recent work [15] we showed that this peak is generated from the interactions of the Λ and the nucleon, produced after K^- absorption, with the residual nucleus. We here present some additional results, improved by the use of more realistic ΛN scattering probabilities, and summarize the present status of the field.

2 Theoretical model

The reaction $(K^-)_{\text{stopped}}\, A \to \Lambda p A'$ proceeds by capturing a slow K^- in a high atomic orbit of the nucleus, which later cascades down till the K^- reaches a low-lying orbit, from where it is finally absorbed. We assume that the absorption of the K^- takes place from the lowest level where the energy shift for atoms has been measured, or, if it is not measured, from the level where the calculated shift [7] falls within the measurable range.

The width for K^- absorption from pN pairs in a nucleus with mass number A is given, in Local Density Approximation (LDA) by

$$\Gamma_A \propto \int \mathrm{d}^3r\,|\Psi_{K^-}(\boldsymbol{r})|^2 \int \frac{\mathrm{d}^3p_1}{(2\pi)^3} \int \frac{\mathrm{d}^3p_2}{(2\pi)^3}\,\Gamma_m(\boldsymbol{p}_1,\boldsymbol{p}_2,\boldsymbol{p}_K,\boldsymbol{r}),\tag{1}$$

where $|\Psi_{K^-}(\boldsymbol{r})|^2$ is the probability of finding the K^- in the nucleus, $|\boldsymbol{p}_1|, |\boldsymbol{p}_2| < k_F(r)$ with $k_F(r) = \left(3\pi^2\rho(r)/2\right)^{1/3}$ being the local Fermi momentum and $\Gamma_m(\boldsymbol{p}_1,\boldsymbol{p}_2,\boldsymbol{p}_K,\boldsymbol{r})$ is the in-medium decay width for the $K^-pN \to \Lambda N$ process.

The structure of the integrals determining Γ_m,

$$\Gamma_m \propto \int \mathrm{d}^3\boldsymbol{p}_\Lambda \mathrm{d}^3\boldsymbol{p}_N \ldots K(\boldsymbol{p}_\Lambda,\boldsymbol{r}) K(\boldsymbol{p}_N,\boldsymbol{r}),\tag{2}$$

allows us to follow the propagation of the produced nucleon and Λ through the nucleus after K^- absorption via the kernel $K(\boldsymbol{p},\boldsymbol{r})$. The former two equations describe the process in which a kaon at rest is absorbed by two nucleons (pp or pn) within the local Fermi sea emitting a nucleon and a Λ. The primary nucleon (Λ) is allowed to re-scatter with nucleons in the nucleus according to a probability per unit length given by $\sigma_{N(\Lambda)}\rho(r)$, where $\sigma_{N(\Lambda)}$ is the experimental $NN(\Lambda N)$ cross-section at the corresponding energy, while in [15] a simpler parameterization for the Λ, of the type $\sigma_\Lambda = 2\sigma_N/3$ was employed. The angular distribution of the re-scattered particles are also generated according to experimental differential cross-sections. We note that particles move under the influence of a mean-field potential, of Thomas-Fermi type. After several possible collisions, one or more nucleons and a Λ emerge from the nucleus and the invariant mass of all possible Λp pairs, as well as their relative angle, are evaluated for each Monte Carlo event. See ref. [15] for more details.

3 The Λp invariant-mass spectrum

Absorption of a K^- from a nucleus leaving the final daughter nucleus in its ground state gives rise to a narrow peak in the Λp invariant-mass distribution, as it is observed in the spectrum of [14]. We note that our local density formalism, in which the hole levels in the Fermi sea form a continuum of states, cannot handle properly transitions to discrete states of the daughter nucleus, in particular to the ground state. For this reason, we will remove in our calculations those events in which the p and Λ produced after K^- absorption leave the nucleus without having suffered a secondary collision. However, due to the small overlap between the two-hole initial state after K^- absorption and the residual $(A-2)$ ground state of the daughter nucleus, as well as to the limited survival probability for both the p and the Λ crossing the nucleus without any collision, this strength represents only a moderate fraction, estimated to be smaller than 15% in ^{7}Li [15]. This means, in practice, that the excitation of the nucleus will require the secondary collision of the p or Λ after the K^-pp absorption process, similarly as to what happens in (p,p') collisions, where the strength of the cross-section to elastic or bound excited states is very small compared to that of nuclear breakup producing the quasi-elastic peak.

Our invariant-mass spectra requiring at least a secondary collision of the $p(n)$ or Λ after the $K^-pp(np)$ absorption process are shown in fig. 1 for several nuclei, where we have applied the same cuts as in the experiment, namely $P_\Lambda > 300\,\mathrm{MeV}/c$ (to eliminate events from $K^-p \to \Lambda\pi$) and $\cos\Theta_{p_\Lambda p_p} < -0.8$ (to filter Λp pairs going back-to-back). Actually, the calculated angular distribution shown in ref. [15] demonstrates that, even after collisions, a sizable fraction of the events appear at the back-to-back kinematics. These events generate the main bump at 2260–2270 MeV in all the Λp invariant-mass spectra shown in fig. 1, about the same position as the main peak shown in the inset of fig. 3 of [14]. We note that, since one measures the Λp invariant mass, the main contribution comes from $K^-pp \to \Lambda p$ absorption (dot-dashed lines), although the contribution from the $K^-pn \to \Lambda n$ reaction followed by $np \to pn$ (dotted line) is non-negligible. It is interesting to observe that the width of the distribution gets broader with the size of the nucleus, while the peak remains in the same location, consistently to what one expects for the behavior of a quasi-elastic peak. Let us point out in this context that the work of [16] shows that the possible interpretation of the FINUDA peak as a bound state of the K^- with the nucleus, not as a K^-pp bound state, would unavoidably lead to peaks at different energies for different nuclei. We finally observe that the spectra of heavy nuclei develop a secondary peak at lower invariant masses due to the larger amount of re-scattering processes. It is slightly more pronounced than that shown in our earlier work [15], due to the use here of a realistic ΛN scattering cross-section.

In summary, we have seen how the experimental Λp invariant-mass spectrum of the FINUDA Collaboration [14] is naturally explained in our Monte Carlo simulation as a consequence of final-state interactions of the particles produced in nuclear K^- absorption as they leave the nucleus, without the need of resorting to exotic mechanisms like the formation of a K^-pp bound state. Together with the interpretation of the proton momentum spectrum given in refs. [11,13], it seems then clear that there is at present no experimental evidence for the existence of deeply bound kaonic states. Theoretically, a major step forward has been given by recent few-body calculations,

either solving Faddeev equations [17] or applying variational techniques [18], using realistic $\bar{K}N$ interactions and short-range nuclear correlations. These works predict few-nucleon kaonic states bound by 50–70 MeV but having large widths of the order of 100 MeV, thereby disclaiming the findings of refs. [8,9]. It remains to be explored whether the larger attraction expected in much heavier kaonic nuclear systems is enough to block the decaying channels, thus opening the possibility of narrow, hence measurable, deeply bound kaonic states.

This work is partly supported by contracts BFM2003-00856 and FIS2005-03142 from MEC (Spain) and FEDER, the Generalitat de Catalunya contract 2005SGR-00343, and the E.U. EURIDICE network contract HPRN-CT-2002-00311. This research is part of the EU Integrated Infrastructure Initiative Hadron Physics Project under contract number RII3-CT-2004-506078.

References

1. D.B. Kaplan, A.E. Nelson, Phys. Lett. B **175**, 57 (1986); **179**, 409 (1986)(E).
2. E. Friedman, A. Gal, C.J. Batty, Nucl. Phys. A **579**, 518 (1994).
3. N. Kaiser, P.B. Siegel, W. Weise, Nucl. Phys. A **594**, 325 (1995); E. Oset, A. Ramos, Nucl. Phys. A **635**, 99 (1998); J.A. Oller, U.G. Meissner, Phys. Lett. B **500**, 263 (2001); M.F.M. Lutz, E.M. Kolomeitsev, Nucl. Phys. A **700**, 193 (2002); C. Garcia-Recio, J. Nieves, E. Ruiz Arriola, M.J. Vicente Vacas, Phys. Rev. D **67**, 076009 (2003); D. Jido, J.A. Oller, E. Oset, A. Ramos, U.G. Meissner, Nucl. Phys. A **725**, 181 (2003); B. Borasoy, R. Nissler, W. Weise, Eur. Phys. J. A **25**, 79 (2005); J.A. Oller, J. Prades, M. Verbeni, Phys. Rev. Lett. **95**, 172502 (2005).
4. V. Koch, Phys. Lett. B **337**, 7 (1994).
5. M. Lutz, Phys. Lett. B **426**, 12 (1998).
6. A. Ramos, E. Oset, Nucl. Phys. A **671**, 481 (2000).
7. S. Hirenzaki, Y. Okumura, H. Toki, E. Oset, A. Ramos, Phys. Rev. C **61**, 055205 (2000); A. Baca, C. Garcia-Recio, J. Nieves, Nucl. Phys. A **673**, 335 (2000).
8. Y. Akaishi, T. Yamazaki, Phys. Rev. C **65**, 044005 (2002).
9. Y. Akaishi, A. Dote, T. Yamazaki, Phys. Lett. B **613**, 140 (2005).
10. T. Suzuki *et al.*, Phys. Lett. B **597**, 263 (2004).
11. E. Oset, H. Toki, Phys. Rev. C **74**, 015207 (2006).
12. M. Iwasaki, invited talk at the *IX International Conference on Hypernuclear and Strange Particle Physics, Mainz, October 10-14, 2006*.
13. FINUDA Collaboration (M. Agnello *et al.*), Nucl. Phys. A **775**, 35 (2006).
14. FINUDA Collaboration (M. Agnello *et al.*), Phys. Rev. Lett. **94**, 212303 (2005).
15. V.K. Magas, E. Oset, A. Ramos, H. Toki, Phys. Rev. C **74**, 025206 (2006).
16. J. Mares, E. Friedman, A. Gal, Nucl. Phys. A **770**, 84 (2006).
17. N.V. Shevchenko, A. Gal, J. Mares, Phys. Rev. Lett. **98**, 082301 (2007).
18. A. Dote, contributed talk at the *IX International Conference on Hypernuclear and Strange Particle Physics, Mainz, October 10-14, 2006*.

Eur. Phys. J. A **31**, 687–690 (2007)

DOI 10.1140/epja/i2006-10229-y

Special Article – QNP 2006

Excited baryons and heavy pentaquarks in large-N_c QCD

D. Pirjol[1] and C. Schat[2,3,a]

[1] Center for Theoretical Physics, Massachusetts Institute of Technology, Cambridge, MA 02139, USA
[2] CONICET and Departamento de Física, FCEyN, Universidad de Buenos Aires, Ciudad Universitaria, Pab. 1, (1428) Buenos Aires, Argentina
[3] Departamento de Física, Universidad de Murcia, E30071 Murcia, Spain

Received: 8 November 2006
Published online: 8 March 2007 – © Società Italiana di Fisica / Springer-Verlag 2007

Abstract. We briefly discuss the large-N_c picture for excited baryons, present a new method for the calculation of matrix elements and illustrate it by computing the strong decays of heavy exotic states.

PACS. 11.15.Pg Expansions for large numbers of components (*e.g.*, $1/N_c$ expansions) – 14.20.-c Baryons (including antiparticles)

1 Introduction

The $1/N_c$ expansion of QCD has turned out to be a fruitful approach to its non-perturbative regime, as is shown by many examples [1]. The successful applications to the study of ground-state baryons make the excited baryons and exotic states especially interesting because they provide a wider testing ground for the $1/N_c$ expansion.

It is useful to recall a few general facts that make the large number of colors limit interesting and useful:

- The $1/N_c$ expansion is the only candidate for a perturbative expansion of QCD at all energies.
- In the $N_c \to \infty$ limit baryons fall into irreducible representations of the *contracted* spin-flavor algebra $SU(2n_f)_c$, also known as $\mathcal{K}$-symmetry, that relates properties of states in different multiplets of flavor symmetry.
- The breaking of spin-flavor symmetry can be studied order by order in $1/N_c$ as an operator expansion.

It is important to stress that already at leading order in the large-N_c limit it is possible to obtain significant insights into the structure of excited baryons, among which we would like to highlight the following:

- The three towers [2–4] predicted by $\mathcal{K}$-symmetry for the $L = 1$ negative-parity N^* baryons, labeled by $\mathcal{K} = 0, 1, 2$ with $\mathcal{K}$ related to the isospin I and spin J of the N^*'s by $\mathcal{K} \geq |I - J|$.
- The vanishing of the strong-decay width $\Gamma(N^*_{\frac{1}{2}} \to [N\pi]_S)$ for $N^*_{\frac{1}{2}}$ in the $\mathcal{K} = 0$ tower, which provides a natural explanation for the relative supresion of pion decays for the $N^*(1535)$ [2,4,5].

- The order $\mathcal{O}(N_c^0)$ mass splitting of the $SU(3)$ singlets $\Lambda(1405)$-$\Lambda(1520)$ in the $[\mathbf{70}, 1^-]$ multiplet [6].

The general framework is based on the observation that at the fundamental level of QCD diagrams can be classified according to their scaling with N_c. Planar diagrams are the leading order, non-planar diagrams and quark loops are subleading in $1/N_c$. In order to obtain finite amplitudes the quark-gluon coupling constant must scale as $g \propto N_c^{-1/2}$. An m-body operator requires at least the exchange of $m - 1$ gluons which gives a suppression factor of N_c^{1-m}. However, the matrix elements of an operator can eventually be enhanced by coherence effects, as is the case of G^{ia} defined below[1]. In an explicit quark operator representation different hadronic operators like the masses, magnetic moments, axial currents, etc., can be expanded [7] in $1/N_c$. For example, for the mass operator we have schematically

$$\hat{M} = \sum_{k=0}^{N_c} \frac{1}{N_c^{k-1}} C_k \mathcal{O}_k \tag{1}$$

with $\mathcal{O}_k$ a k-body operator. Both the coefficients C_k (which correspond to reduced matrix elements of QCD operators) and the matrix elements of the quark operators on baryon states $\langle \mathcal{O}_k \rangle$ have power expansions in $1/N_c$ with coefficients determined by nonperturbative dynamics. The basic building blocks to construct the $\mathcal{O}_k$ are the

---

[a] e-mail: schat@df.uba.ar

[1] $\langle G^{ia} \rangle \propto N_c$ when restricted to the subspace of states with spin and isospin of order N_c^0, which are the ones that will correspond to the $N_c = 3$ physical states.

generators of $SU(2n_f)$, where n_f is the number of flavors

$$S^i = \sum_{\alpha=1}^{N_c} s^i_{(\alpha)}, \quad T^a = \sum_{\alpha=1}^{N_c} t^a_{(\alpha)}, \quad G^{ia} = \sum_{\alpha=1}^{N_c} s^i_{(\alpha)} t^a_{(\alpha)}. \quad (2)$$

In the large-N_c limit we can define $X^0_{ia} \equiv \lim_{N_c \to \infty} \frac{G_{ia}}{N_c}$, because the matrix elements of G_{ia} scale like N_c for the states of interest, which is the coherence effect mentioned before. In this way we obtain for $n_f = 2$ the contracted algebra $SU(4)_c$

$$\begin{aligned}
[S_i, S_j] &= i\epsilon_{ijk} S_k, & [S_i, X^0_{ja}] &= i\epsilon_{ijk} X^0_{ka}, \\
[T_a, T_b] &= i\epsilon_{abc} T_c, & [T_a, X^0_{ib}] &= i\epsilon_{abc} X^0_{ic}, \\
[X^0_{ia}, X^0_{jb}] &= 0.
\end{aligned} \quad (3)$$

The last commutation relations can also be obtained in a purely hadronic language. They are known as consistency relations [7] and are necessary to obtain finite amplitudes for pion-nucleon scattering. Consider the direct and crossed diagrams that contribute at tree level. The pion-nucleon coupling scales like $\sqrt{N_c}$, which makes each diagram separately to scale like N_c. To obtain a finite amplitude for the physical process we need a cancellation to happen. This requires $[X^0_{ia}, X^0_{jb}] = \mathcal{O}(1/N_c)$, which in the large-$N_c$ limit gives eq. (3). This symmetry structure gives rise to model-independent predictions like the three towers for excited baryons that was mentioned above. In an explicit quark operator representation this is manifest by the presence of two $\mathcal{O}(N_c^0)$ operators (that also involve the generator of $O(3)$ [8]) and has been checked by an explicit calculation [3,4].

2 Occupation number formalism

In this section we give an outline of the occupation number formalism [9] that we use to compute matrix elements for arbitrary N_c. In broken $SU(3)$, the $SU(6)$ spin-flavor generators can be decomposed into generators of the subgroup

$$SU(6)_{SF} \supset SU(4)_{SI} \otimes SU(2)_{J_s} \otimes U(1)_{n_s}$$

$$J^i, \quad I^a = T^a, \quad G^{ia} = G^{ia} \quad (i, a = 1, \ldots, 3),$$

$$J^i_s = s^\dagger \frac{\sigma^i}{2} s, \quad N_s = s^\dagger s$$

plus operators mediating transitions between sectors of different n_s

$$\tilde{t}^\alpha = q^{\dagger\alpha} s, \quad t_\alpha = s^\dagger q_\alpha \quad (\alpha = \pm 1/2),$$

$$\tilde{Y}^{i\alpha} = q^{\dagger\alpha} \frac{\sigma^i}{2} s, \quad Y^i_\alpha = s^\dagger \frac{\sigma^i}{2} q_\alpha.$$

We introduce the "$6n$-symbol" defined as $(N = \sum_{i=1}^6 n_i)$

$$\{n_1, n_2, n_3, n_4, n_5, n_6\} = \sqrt{\frac{n_1! n_2! n_3! n_4! n_5! n_6!}{N!}}$$
$$\times (u_\uparrow^{n_1} u_\downarrow^{n_2} d_\uparrow^{n_3} d_\downarrow^{n_4} s_\uparrow^{n_5} s_\downarrow^{n_6} + \text{perms}).$$

The nonstrange states in a $\mathcal{K} = 0$ tower have spin and isopin satisfying $I = J$. Their spin-flavor symmetric wave functions can be given in closed form as

$$|I I_3 J_3; N_{ud}\rangle = \sum_i \left(\begin{matrix} \frac{N_u}{2} & \frac{N_d}{2} \\ i & J_3 - i \end{matrix} \middle| \begin{matrix} I \\ J_3 \end{matrix} \right)$$
$$\times \left\{ \frac{N_u}{2} + i, \frac{N_u}{2} - i, \frac{N_d}{2} + J_3 - i, \frac{N_d}{2} - J_3 + i \right\},$$

where $N_{u,d}$ are the number of up and down quarks: $N_u = \frac{N_{ud}}{2} + I_3$, $N_d = \frac{N_{ud}}{2} - I_3$ with $N_{ud} = N_c - n_s$.

A few representative nonstrange $J_3 = +\frac{1}{2}$ states are

$$p_\uparrow = \sqrt{\frac{2}{3}} \{2, 0, 0, 1\} - \frac{1}{\sqrt{3}} \{1, 1, 1, 0\}, \quad \Delta_\uparrow^{++} = \{2, 1, 0, 0\}.$$

Acting with

$$\begin{aligned}
q_i\{\cdots, n_i, \cdots\} &= \sqrt{n_i}\{\cdots, n_i - 1, \cdots\}, \\
q_i^\dagger\{\cdots, n_i, \cdots\} &= \sqrt{n_i + 1}\{\cdots, n_i + 1, \cdots\}
\end{aligned} \quad (4)$$

we obtain the matrix elements of any operator for arbitrary N_c.

3 Pentaquark towers

For the exotic $q^{N_c+1}\bar{q}$ states with $N_c + 1$ quarks in a "$\bar{\mathbf{3}}$" of color, Fermi statistics implies the $SU(6) \otimes O(3)$ decomposition

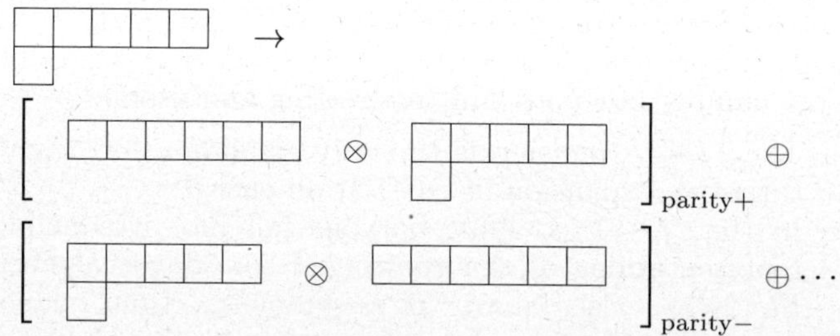

The negative-parity states were studied in [10]. Here we reconsider the positive-parity states [11], which are all members of the two towers

$$\begin{aligned}
\mathcal{K} = 1/2: \quad & \overline{\mathbf{10}}_{\frac{1}{2}}, \quad \mathbf{27}_{\frac{1}{2}, \frac{3}{2}}, \quad \mathbf{35}_{\frac{3}{2}, \frac{5}{2}}, \cdots \\
\mathcal{K} = 3/2: \quad & \overline{\mathbf{10}}_{\frac{3}{2}}, \quad \mathbf{27}_{\frac{1}{2}, \frac{3}{2}, \frac{5}{2}}, \quad \mathbf{35}_{\frac{1}{2}, \frac{3}{2}, \frac{5}{2}, \frac{7}{2}}, \cdots.
\end{aligned}$$

In [11] only states in the first tower were considered. In the heavy-quark limit $m_Q \to \infty$ these two towers become degenerate and the tower label for the light degrees of freedom becomes a good quantum number:

$$\mathcal{K}_{light} = 1: \quad \overline{\mathbf{6}}_1, \quad \mathbf{15}_{0,1,2}, \quad \mathbf{15}'_{1,2,3}, \quad \cdots. \quad (5)$$

On the other hand, the heavy pentaquarks considered in [11] belong to the tower

$$\mathcal{K}_{light} = 0: \quad \overline{\mathbf{6}}_0, \quad \mathbf{15}_1, \quad \mathbf{15}'_2, \quad \cdots, \quad (6)$$

Table 1. Reduced matrix elements Y and large-N_c width for the $\mathcal{K} = 1/2$ pentaquark $\Theta_{\bar{Q}J_\ell} \to NK, \Delta K$ decays.

Decay	$(I'J', IJ)$	$Y(I'J'\mathcal{K}', IJ\mathcal{K})$	$\frac{1}{p^3}\Gamma^{p\text{-wave}}_{N_c \to \infty}$
$\Theta_0(\frac{1}{2}) \to NK$	$(\frac{1}{2}\frac{1}{2}, 0\frac{1}{2})$	$-\frac{\sqrt{3}}{2}\sqrt{N_c + 1}$	1
$\Theta_1(\frac{1}{2}) \to NK$	$(\frac{1}{2}\frac{1}{2}, 1\frac{1}{2})$	$\frac{1}{2}\sqrt{N_c + 5}$	$\frac{1}{9}$
$\quad \to \Delta K$	$(\frac{3}{2}\frac{3}{2}, 1\frac{1}{2})$	$\frac{1}{\sqrt{2}}\sqrt{N_c - 1}$	$\frac{8}{9}$
$\Theta_1(\frac{3}{2}) \to NK$	$(\frac{1}{2}\frac{1}{2}, 1\frac{3}{2})$	$-\sqrt{2}\sqrt{N_c + 5}$	$\frac{4}{9}$
$\quad \to \Delta K$	$(\frac{3}{2}\frac{3}{2}, 1\frac{3}{2})$	$-\frac{1}{2}\sqrt{\frac{5}{2}}\sqrt{N_c - 1}$	$\frac{5}{9}$

which arises naturally in the Skyrme model.

As an example we compute the strong decays of the $\mathcal{K} = 1/2$ states in [11]. The reduced matrix elements of the transition operator are defined by

$$\langle I'I'_3, J'J'_3; n_s - 1|Y^{i\alpha}|II_3, JJ_3; n_s\rangle =$$
$$\begin{pmatrix} I & \frac{1}{2} & I' \\ I_3 & \alpha & I'_3 \end{pmatrix} \begin{pmatrix} J & 1 & J' \\ J_3 & i & J'_3 \end{pmatrix} Y(I'J'\mathcal{K}', IJ\mathcal{K}). \quad (7)$$

In the large-N_c limit we find [12]

$$Y_0(I'J'\mathcal{K}', IJ\mathcal{K}) \propto \sqrt{[I][J]} \begin{Bmatrix} \frac{1}{2} & 1 & \frac{1}{2} \\ I & J & \mathcal{K} \\ I' & J' & \mathcal{K}' \end{Bmatrix}. \quad (8)$$

The expressions for arbitrary N_c are found in table 1. Averaging over initial states and summing over final states the p-wave widths are obtained as

$$\Gamma(I'J'\mathcal{K}', IJ\mathcal{K}) \propto \frac{[I'][J']}{[I][J]}|Y(I'J'\mathcal{K}', IJ\mathcal{K})|^2. $$

In the large-N_c limit all pentaquark states in the same tower have the same total width. This leads to sum rules like

$$\Gamma\left(\Theta_0\left(\frac{1}{2}\right) \to NK\right) =$$
$$\Gamma\left(\Theta_1\left(\frac{1}{2}\right) \to NK\right) + \Gamma\left(\Theta_1\left(\frac{1}{2}\right) \to \Delta K\right) =$$
$$\Gamma\left(\Theta_1\left(\frac{3}{2}\right) \to NK\right) + \Gamma\left(\Theta_1\left(\frac{3}{2}\right) \to \Delta K\right)$$

as can be verified from table 1. The results for $N_{c.} = 3$ in [11] can also be verified from table 1.

4 Large-N_c and heavy-quark limit predictions

Heavy-quark symmetry predicts the amplitudes in terms of a few reduced matrix elements f_i (see table 2). The decay amplitude for $\Theta_{\bar{Q}}(IJJ_\ell) \to [NH_{\bar{Q}}^{(*)}(J'J'_\ell)]_{J_N}$, where $\mathbf{J}_N = \mathbf{S}_N + \mathbf{L}$ is the angular momentum carried by the final baryon, is given by [13]

$$A_i = \sqrt{(2J_\ell + 1)(2J' + 1)} \begin{Bmatrix} J_\ell & J'_\ell & J_N \\ J' & J & \frac{1}{2} \end{Bmatrix} f_i. \quad (9)$$

Table 2. Heavy-quark symmetry predictions for the decay amplitudes $\Theta_{\bar{Q}J_\ell} \to [NH_{\bar{Q}}^{(*)}]_{p\text{-wave}}$.

Decay	$J_N = 1/2$	$J_N = 3/2$
$\Theta_{\bar{Q}0}(\frac{1}{2}) \to NH_{\bar{Q}}$	$-\frac{1}{2}f_0$	$-$
$\Theta_{\bar{Q}1}(\frac{1}{2}) \to NH_{\bar{Q}}$	$\frac{\sqrt{3}}{2}f_1$	$-$
$\Theta_{\bar{Q}1}(\frac{3}{2}) \to NH_{\bar{Q}}$	$-$	$-\frac{1}{2}\sqrt{\frac{3}{2}}f_2$
$\Theta_{\bar{Q}0}(\frac{1}{2}) \to NH_{\bar{Q}}^*$	$\frac{\sqrt{3}}{2}f_0$	$-$
$\Theta_{\bar{Q}1}(\frac{1}{2}) \to NH_{\bar{Q}}^*$	$\frac{1}{2}f_1$	$-f_2$
$\Theta_{\bar{Q}1}(\frac{3}{2}) \to NH_{\bar{Q}}^*$	$-f_1$	$\frac{1}{2}\sqrt{\frac{5}{2}}f_2$

Table 3. Ratios of strong-decay widths for heavy pentaquarks $R^I(J) = \Theta_{\bar{Q}}^I(J) \to (NH_{\bar{Q}}) : (NH_{\bar{Q}}^*)$.

$I = 1$	$R^I(J = \frac{1}{2})$		$R^I(J = \frac{3}{2})$	
$\mathcal{K}_{light} = 1$	$1:3$	$(J_\ell = 0)$	$\frac{1}{2}:\frac{7}{2}$	$(J_\ell = 1)$
	$2:2$	$(J_\ell = 1)$	$\frac{5}{2}:\frac{3}{2}$	$(J_\ell = 2)$
$\mathcal{K}_{light} = 0$	$1:11$	$(J_\ell = 1)$	$4:8$	$(J_\ell = 1)$

Combining the heavy-quark symmetry predictions with the large-N_c amplitudes we can fix the reduced amplitudes f_i and obtain predictions for the ratios of decays widths, as summarized for the $I = 1$ states in table 3. More details will be given elsewhere [12].

5 Conclusions

The large-N_c limit reveals a structure of mass degeneracies and sum rules for decay widths that is not apparent at $N_c = 3$. This picture can be corrected systematically in $1/N_c$. We presented a new method for computing matrix elements for arbitrary N_c which is useful for this purpose. As an illustration, we showed how the combined large-N_c and heavy-quark limit allows to compute decay width ratios that discriminate between different heavy-pentaquark states. In the heavy-quark limit the spin of the light degrees of freedom is a conserved quantum number. When this is combined with the large-N_c limit we can label the states by the new quantum number $\mathcal{K}_{light}$. The states considered in [11] have $\mathcal{K}_{light} = 0$, while the states considered in this work have $\mathcal{K}_{light} = 1$. The predictions for their strong decays differ, as can be seen in table 3.

The work of C.S. was supported in part by Fundación Antorchas, Argentina and Fundación Séneca, Murcia, Spain. C.S. thanks for the hospitality of the Departamento de Física, Universidad de Murcia during the completion of this work.

References

1. J.L. Goity, R.F. Lebed, A. Pich, C.L. Schat, N.N. Scoccola (Editors), *Proceedings of the Workshop Large N(C) QCD 2004, Trento, Italy* (World Scientific, Singapore, 2005).

2. D. Pirjol, T.M. Yan, Phys. Rev. D **57**, 5434; 1449 (1998).
3. D. Pirjol, C. Schat, Phys. Rev. D **67**, 096009 (2003).
4. T.D. Cohen, R.F. Lebed, Phys. Rev. Lett. **91**, 012001 (2003).
5. D. Pirjol, C. Schat, AIP Conf. Proc. **698**, 548 (2004).
6. C.L. Schat, J.L. Goity, N.N. Scoccola, Phys. Rev. Lett. **88**, 102002 (2002).
7. R.F. Dashen, E. Jenkins, A.V. Manohar, Phys. Rev. D **49**, 4713 (1994) [**51**, 2489 (1995)(E)]; **51**, 3697 (1995).
8. J.L. Goity, Phys. Lett. B **414**, 140 (1997); C.E. Carlson, C.D. Carone, J.L. Goity, R.F. Lebed, Phys. Rev. D **59**, 114008 (1999).
9. D. Pirjol, C. Schat, Phys. Lett. B **638**, 340 (2006).
10. D. Pirjol, C. Schat, Phys. Rev. D **71**, 036004 (2005); M.E. Wessling, Phys. Lett. B **603**, 152 (2004).
11. E. Jenkins, A.V. Manohar, JHEP **0406**, 039 (2004).
12. D. Pirjol, C. Schat, hep-ph/0612314, to be published in Phys. Rev. D.
13. N. Isgur, M.B. Wise, Phys. Rev. Lett. **66**, 1130 (1991).

Inclusive pentaquark and strange baryons production in hadron beam experiments at high energy

I.M.Narodetskii, M.A.Trusov, and A.I.Veselov

ITEP, Moscow, 117218 Russia

Abstract. We calculate the cross sections for the inclusive production in the fragmentation region of $\Theta^+(1540)$ and $\Lambda(1520)$ in pp collisions and $\Lambda(1520)$ in Σp collisions at high energy using the K- and π-meson exchange diagrams, respectively. The contributions of these diagrams survive at asymptotically large energies and are energy independent in this region up to logarithmic and power corrections. We find that inclusive $\Theta^+(1540)$ production should be at the level of $1~\mu$b $\times~\Gamma_{\Theta KN}/1$ MeV. The ratio of the $\Theta^+(1540)$ over the $\Lambda(1520)$ yields is found to be $\sim 1\%$. The fraction of $\Lambda(1520)$ yields in Σp and pp collisions is ~ 2.7 that quantitatively agrees with the preliminary result of the Fermilab fixed target experiment E781.

PACS. 13.85.Ni

1 Introduction

The possible existence of the Θ^+ pentaquark remains one of the puzzling mysteries of recent years. To date there are more than 20 experiments with evidence for this state, but criticism for the Θ^+ claim arises because similar number of high energy experiments did not find any evidence for the Θ^+, even though the other "conventional" three-quark hyperons such as $\Lambda(1520)$ hyperon resonance are seen clearly [1].

Most of negative high energy experiments are high statistic hadron beam experiments. *E.g.* HERA-B, a fixed target experiment at the 920 GeV proton storage ring of DESY [2] finds no evidence for narrow signals in the $\bar{K}^0_S p$ channel and only sets modest upper limits for Θ^+ production of less than 16 μb/N and less than about 12% relative to $\Lambda(1520)$ in mid-rapidity region. This negative result would present serious rebuttal evidence to worry about. However, without obvious production mechanism of the Θ^+ (if it exists) or even $\Lambda(1520)$ the rebuttal is not very convincing.

In this paper we estimate the high-energy behavior of the Θ^+ and $\Lambda(1520)$ production cross sections in inclusive pp collisions using the K exchange diagram, which is known to survive at high energies in the beam/target fragmentation region. We show that the cross section of the Θ^+-production is suppressed compared to the production of $\Lambda(1520)$. This suppression is mainly due to the smallness of the coupling constant $G^2_{\Theta KN}$ compared to $G^2_{\Lambda KN}$ that in turn is related to the small width of the Θ^+. As a byproduct we also estimate the contribution of the π exchange diagram for the inclusive $\Lambda(1520)$ production in Σp collisions.

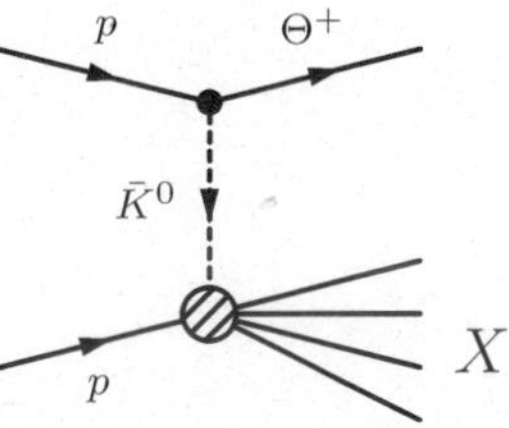

Fig. 1. The $\bar{K}^0$ exchange diagram for the $\Theta^+(1540)$ production in inclusive pp scattering.

2 Inclusive cross sections

We assume that the Θ^+ exists and $J^P(\Theta^+) = \frac{1}{2}^+$. Consider the Θ^+ production in the reaction $p + p \to \Theta^+ + X$, where X is unspecified inclusive final state carrying the strangeness -1. The $\bar{K}^0$ exchange diagram for $pp \to \Theta^+ X$ is shown in Fig. 1.

The standard expression for the contribution of this diagram written in terms of the 4-momentum transfer squared $t = q^2$ and the invariant mass W of the $\bar{K}^0 p$ system is well known [3]. At high energy, it is more convenient to convert this expression to an integral over the Feynman variable x_F, the fraction of the incident proton momentum carried by the Θ^+ in the initial direction of the proton (in the center-of-mass system), and $k_\perp$, the transverse momentum of Θ^+ relative to the initial proton direction. Then the contribution of the K meson exchange to the double differential cross section for the Θ^+ inclusive

"

production reads

$$\frac{d\sigma}{dx_F dk_\perp^2} = \frac{1}{4\pi} \frac{G_{\Theta KN}^2}{4\pi} J \frac{p}{E_\Theta} \Phi_{p\Theta}(t) F^4(t) \sigma_{\text{tot}}^{\bar{K}^0 p}(s_1), \quad (1)$$

where $G_{\Theta KN}$ is the coupling constant for the decay $\Theta^+ \to \bar{K}^0 p$, $E_\Theta = \sqrt{x_F^2 p^2 + k_\perp^2 + m_\Theta^2}$ is the Θ^+ energy in the center-of-mass system, $s_1 = W^2 = s + m_\Theta^2 - 2E_\Theta\sqrt{s}$, $s = 4(p^2 + m_p^2)$, p is the center-of-mass momentum, and $t = m_p^2 + m_\Theta^2 - 2E_\Theta\sqrt{p^2 + m_p^2} + 2x_F p^2$. The factor J is the ratio of flux factors in the pp and $\bar{K}^0 p$ reactions. To evaluate the cross sections away from the pole position $t = M_K^2$ we include the phenomenological form factor $F_K(t)$. The function $\Phi_{p\Theta}(t)$ is the squared product of the vertex function for $p \to \Theta^+ \bar{K}^0$ and the kaon propagator:

$$\Phi_{p\Theta}(t) = \frac{(m_p - m_\Theta)^2 - t}{(t - m_K^2)^2}. \quad (2)$$

In the high energy limit with accuracy $\mathcal{O}(1/p^2)$

$$J \cdot \frac{p}{E_\Theta} \approx \frac{1 - x_F}{x_F}, \quad (3)$$

$$s_1 \approx (1 - x_F)s, \quad t \approx m_\Theta^2 + m_p^2(1 - x_F) - \frac{m_\Theta^2 + k_\perp^2}{x_F}, \quad (4)$$

and the double differential cross section written in terms of x_F and $k_\perp^2$ reads

$$\frac{d\sigma}{dx_F dk_\perp^2} = \frac{1}{4\pi} \frac{G_{\Theta KN}^2}{4\pi} \cdot \frac{1 - x_F}{x_F} \Phi_{p\Theta}(t) F^4(t) \sigma_{\text{tot}}^{\bar{K}^0 p}(s_1). \quad (5)$$

3 The $\Theta^+ KN$ vertex

The $\Theta^+ KN$ vertex is

$$L_{\Theta KN} = iG_{\Theta KN}(K^\dagger \bar{\Theta} \gamma_5 N + \bar{N} \gamma_5 \Theta K), \quad (6)$$

with the operator γ_5 corresponding to positive Θ^+ parity. The Lagrangian (6) corresponds with the Θ^+ being a p-wave resonance in the $K^0 p$ system. The partial decay width $\Gamma_{\Theta \to K^0 p}$ is

$$\Gamma_{\Theta \to K^0 p} = \frac{G_{\Theta KN}^2}{4\pi} \cdot \frac{2p_K^3}{(m_\Theta + m_p)^2 - m_K^2}, \quad (7)$$

where $p_K = 260$ MeV/c is the kaon momentum in the rest frame of Θ^+. To extract the value for $G_{\Theta KN}$, we need the experimental information of the width $\Gamma_{\Theta KN}$, which is not known precisely but whose measurement is the subject of several planned dedicated experiments. To provide numerical estimates, we will use the value $\Gamma_{\Theta \to K^0 p} = 1$ MeV. This corresponds to the full width $\Gamma_{\Theta KN} = \Gamma_{\Theta \to K^0 p} + \Gamma_{\Theta \to K^+ n} = 2$ MeV, which is consistent with the upper limit for the width derived from elastic KN scattering [4] [1].

---

[1] An additional reason for the smallness of the pentaquark width arises in the string model, in which the pentaquark decay is accompanied by the annihilation of the two string junctions. Indeed, in this model, the pentaquark containing three string junctions dissociates "fall apart" into two minimal color singlets containing only one string junction.

Evaluating Eq. (7) with $\Gamma_{\Theta \to K^0 p} = 1$ MeV, we extract the value $G_{\Theta KN}$

$$\frac{G_{\Theta KN}^2}{4\pi} = 0.167 \cdot \frac{\Gamma_{\Theta \to K^0 p}}{1 \text{ MeV}}, \quad (8)$$

which will be used in the subsequent estimates for the inclusive cross section.

4 $\Lambda(1520) KN$ vertex

The $\Lambda(1520) KN$ vertex is

$$L_{\Lambda KN} = \frac{G_{\Lambda KN}}{m_K} \left(\bar{\Lambda}^\mu \gamma_5 N \partial_\mu K + \bar{N} \gamma_5 \Lambda^\mu \partial_\mu K^\dagger \right), \quad (9)$$

where Λ^μ is the vector spinor for the spin $3/2$ particle. The Lagrangian (9) corresponds with the $\Lambda(1520)$ being a d-wave resonance in the $K^- p$ system. The $\Lambda(1520) \to pK^-$ width is

$$\Gamma_{\Lambda \to K^- p} = \frac{G_{\Lambda KN}^2}{4\pi} \cdot \frac{2p_K^5}{3m_K^2} \cdot \frac{1}{(m_\Lambda + m_p)^2 - m_K^2}, \quad (10)$$

where $p_K = 246$ MeV/c is the kaon momentum in the rest frame of $\Lambda(1520)$. Using the PDG values of $\Gamma_{\text{tot}}(\Lambda(1520)) = 15.6$ MeV and $\text{Br}(\Lambda(1520) \to N\bar{K}) = 45\%$ we obtain

$$\frac{G_{\Lambda KN}^2}{4\pi} \approx 8.14. \quad (11)$$

The function $\Phi_{p\Lambda}(t)$ is

$$\Phi_{p\Lambda}(t) = \frac{(m_p + m_\Lambda)^2 - t}{6m_\Lambda^2 m_K^2} \cdot \frac{((m_p - m_\Lambda)^2 - t)^2}{(t - m_K^2)^2}. \quad (12)$$

The expression (12) includes the factor $1/m_K$ in (9).

5 Results

The total cross section for the general case of the fragmentation of a baryon a into a baryon b due to the exchange by the meson m is

$$\sigma_{ab} = \frac{G_{bma}^2}{4\pi} \int dx_F \int dk_\perp^2 \, K_{ab}(x_F, k_\perp^2) \, \sigma_{\text{tot}}^{mp}(s_1), \quad (13)$$

where $K_{ab}(x_F, k_\perp^2) = (1 - x_F) \Phi_{ab}(t) F^4(t)/x_F$. We employ two representative examples for the form factor $F(t)$:

$$A: \ F(t) = \frac{\Lambda^2 - m_K^2}{\Lambda^2 - t}, \qquad B: \ F(t) = \frac{\Lambda^4}{\Lambda^4 + (t - m_K^2)^2},$$

the cut-off parameter Λ being a typical hadronic scale $\Lambda = 1$ GeV.

Because of (3), (4) all the energy dependence of the right hand side of Eq. (5) is due to the factor $\sigma^{mp}(s_1)$. Since $\sigma^{mp}(s_1)$ is slow varying function of $s_1 = (1 - x_F)s$ everywhere, except the low energy region, we can take it

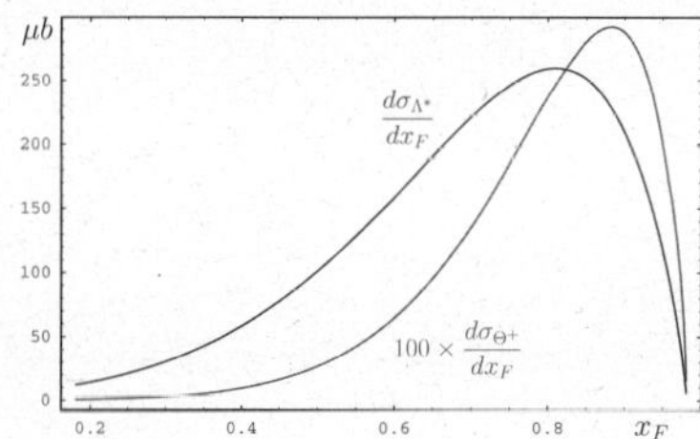

Fig. 2. x_F dependence of the inclusive $pp \to \Theta^+(1540)$ and $\Lambda(1520)$ cross sections.

out of the integral at the point $\hat{s}_1 = (1 - \hat{x}_F)s$, where $\hat{x}_F$ is the point at which $d\sigma^{mp}/dx_F$ reaches the maximum. The typical values of $\hat{x}_F$ are $\sim 0.8 - 0.9$, depending on the reaction considered. Therefore, at high energies in the fragmentation region the effective energy $\sqrt{\hat{s}_1}$ *is always much smaller* than $\sqrt{s}$, *is still large enough* to use the asymptotic of σ^{mp}. Then we obtain

$$\sigma_{ab} \approx \frac{G_{bma}^2}{4\pi} \sigma_{\text{tot}}^{mp}(\hat{s}_1) \hat{K}_{ab}, \qquad (14)$$

where the quantities

$$\hat{K}_{ab} = \int dx_F \int dk_\perp^2 \, K_{ab}(x_F, k_\perp^2) \qquad (15)$$

do not depend on energy, and $\sigma_{\text{tot}}^{mp}(\hat{s}_1)$ is a constant up to logarithmic and power corrections.

For estimation we take the total cross sections $\sigma_{\text{tot}}^{\bar{K}^0 p}$ and $\sigma_{\text{tot}}^{K^+ p}$ to be a constant ($\sigma_{\text{tot}}^{\bar{K}^0 p} \sim \sigma_{\text{tot}}^{K^+ p} \sim 20$ mb at $\sqrt{\hat{s}_1} > 10$ GeV). Then we obtain for the production cross sections[2]

$$\sigma(pp \to \Theta^+(1540)X) = 0.8\,(1.6) \times \frac{\Gamma_{\Theta \to K^0 p}}{1 \text{ MeV}} \, \mu\text{b}, \qquad (16)$$

$$\sigma(pp \to \Lambda^+(1520)X) = 106\,(126) \, \mu\text{b}, \qquad (17)$$

where the first values refer to the form factor (A) and the second ones to the form factor (B). The result for $\sigma(pp \to \Theta^+ X)$ matches well that of Ref. [6] for the inclusive $pp \to \Theta^+ X$ production at $\sqrt{s} < 10$ GeV. If $\Gamma_{\Theta KN} = 0.36 \pm 0.11$ MeV [7], our result for the Θ^+ production cross section should be correspondingly smaller.

In scattering hadronic probes at high energy from nuclear target the only positive signal for the Θ^+ decaying to $K_S^0 p$ was reported by the SVD Collaboration, using 70 GeV proton in a fixed target arrangement $pA \to \Theta^+ X$ at a center-of-mass energy of about 11.5 GeV [5]. Our prediction for $\sigma(pp \to \Lambda(1520)^+ X)$ agrees with the preliminary result of the SVD-2 collaboration, but $\sigma(pp \to \Theta^+ X)$ is lower than the cross section estimation (for $x_F > 0$) : $\sigma \cdot \text{Br}(\Theta^+ \to pK^0) \sim 6 \, \mu\text{b}$. The illustrative examples of the x_F distributions for the Θ^+ and $\Lambda(1520)$ are shown in Fig. 2 for the form factor A.

---

[2] The values in (16), (17) correspond to the region $x_F > 0$. For pp scattering the total cross sections are two times larger.

The ratio of Θ^+ to $\Lambda(1520)$ production cross-sections is $\sim 1\%$. Our estimation is a bit larger than that obtained in the fragmentation-recombination model [8] but still is rather small and probably can be useful to explain why the Θ^+ production is suppressed in some high energy experiments.

In the same way it is possible to make quantitative predictions for other type of colliding particles. As an example, we estimate the cross section for the inclusive $\Lambda(1520)$ production in $\Sigma^- p \to \Lambda(1520)$ collisions at 600 GeV/c studied in the fixed target Fermilab experiment E781 (SE-LEX). In the fragmentation region of the Σ-hyperon this reaction can proceed via the π-meson exchange. Using

$$\Gamma_{\Lambda \to \pi^- \Sigma^+} = \frac{1}{3} \cdot \text{Br}(\Lambda \to \pi\Sigma) \cdot \Gamma_{\text{tot}} = 2.18 \text{ MeV}, \qquad (18)$$

$G_{\Lambda\pi\Sigma}^2/4\pi \approx 0.353$, and $\sigma(\pi N) = 25 \, \mu\text{b}$ we get

$$\sigma(\Sigma p \to \Lambda(1520)X) = 314\,(340) \, \mu\text{b}, \qquad (19)$$

where, as before, the first value refers to the form factor (A) and the second ones to the form factor (B). For the ratio of inclusive Σp and pp cross sections we get

$$\frac{\sigma(\Sigma p \to \Lambda(1520)X)}{\sigma(pp \to \Lambda(1520)X)} \approx 2.9\,(2.7), \qquad (20)$$

that agrees with the preliminary experimental result ≈ 2.6 of the SELEX collaboration [9].

6 Conclusions

Let us recall that our estimations may somehow depend on specific assumptions regarding for instance the K-meson exchange dominance at forward direction, and on the choice of the form factor. As an outlook, it would be interesting to go beyond the present calculation and to perform a systematic study of K, K^* and π Regge exchanges into inclusive production of (anti)strange baryons in pp collisions. We plan to come back to these issues in a next publication.

This work was supported by RFBR grants 04-02-17263, 05-02-17869, 06-02-17120, and by the State Contract 02.445.11.7424, part 2006-RI-112.0/001/398.

References

1. For a recent review see T. Nakano, Pentaquark experimental searches, These Proceedings
2. I. Abt et al. Phys. Rev. Lett. **93** (2004) 212003
3. T. Yao, Phys. Rev. **125** (1962) 1048
4. R.N. Cahn and G.H. Trilling, Phys. Rev. D **69** (2004) 011501
5. A. Aleev et al., hep-ex/0401024, hep-ex/0509033
6. V.Yu. Grishina et al., Eur. Phys. J. A **25** (2005) 141
7. V. V. Barmin et al., hep-ex/0603017.
8. A. I. Titov et al., Phys. Rev. C **70** (2004) 042202
9. A. G. Dolgolenko, V. A. Matveev, private communication

$\Lambda(1405)$ resonance as a superposition of two states

V.K. Magas[1], E. Oset[2], and A. Ramos[1]

[1] Departament d'Estructura i Constituents de la Matèria, Universitat de Barcelona, Diagonal 647, 08028 Barcelona, Spain
[2] Departamento de Física Teórica and IFIC, Centro Mixto, Institutos de Investigación de Paterna - Universidad de Valencia-CSIC, Apdo. correos 22085, 46071, Valencia, Spain

Abstract. Chiral unitarity models have shown the existence of two states with the same quantum numbers in the vicinity of the $\Lambda(1405)$, both contributing to the final experimental invariant mass distribution. The $K^- p \to \pi^0\pi^0\Sigma^0$ reaction, discussed in detail in this work, gives maximal possible weight to the second $\Lambda(1405)$ state, which is narrower and of higher energy than the nominal $\Lambda(1405)$. The calculated distribution of $\pi^0\Sigma^0$ states forming the $\Lambda(1405)$ is in agreement with a recent experimental data, and shows a peak at 1420 MeV and a relatively narrow width of $\Gamma = 38$ MeV. In contrast, the $\pi^- p \to K^0\pi\Sigma$ reaction gives more weight to the pole at lower energy and with a larger width. The data of these two experiments, together with the present theoretical analysis, provides a clear experimental evidence of the two pole structure of the $\Lambda(1405)$. This excludes the possibility to describe the $\Lambda(1405)$ as a bound state of $\bar{K}N$.

PACS. 13.75.-n – 12.39.Fe – 14.20.Jn – 11.30.Hv

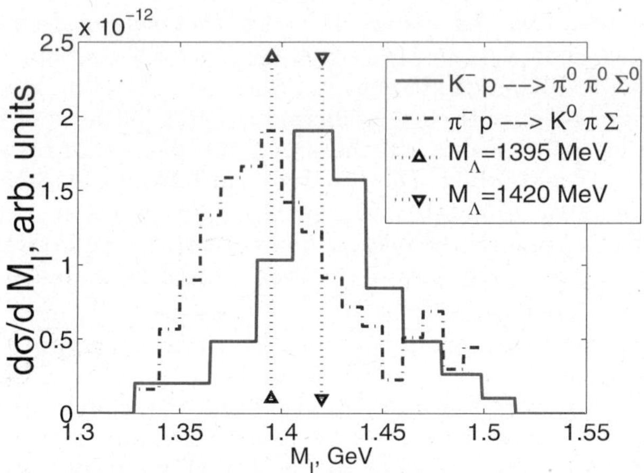

Fig. 1. Two experimental shapes of the $\Lambda(1405)$ resonance. See text for more details.

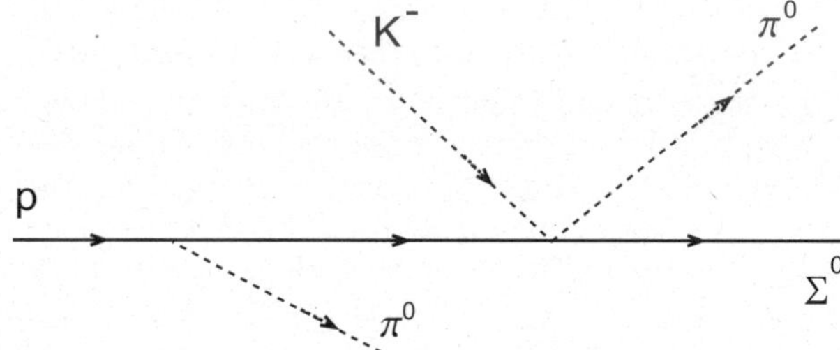

Fig. 2. Nucleon pole term for the $K^- p \to \pi^0\pi^0\Sigma$ reaction.

The study of the $\Lambda(1405)$ as a dynamically generated resonance has a long history [1–7]. It was shown, in the framework of unitary extensions of chiral perturbation theory ($U\chi PT$), that the $\Lambda(1405)$ is a superposition of two states, both contributing to the final experimental invariant mass distribution [5–11]. The properties of these two states are quite different, one has a mass around 1390 MeV, a large width of about 130 MeV and couples mostly to $\pi\Sigma$, while the second one has a mass around 1425 MeV, a narrow width of about 30 MeV and couples mostly to $\bar{K}N$ [8]. The two states are populated with different weights in different reactions and, hence, their superposition can lead to different distribution shapes. Since the $\Lambda(1405)$ resonance is always seen from the invariant mass of its only strong decay channel, the $\pi\Sigma$, hopes to see the second pole are tied to having a reaction where the $\Lambda(1405)$ is formed from the $\bar{K}N$ channel.

In this sense a calculation of the $K^- p \to \gamma\Lambda(1405)$ reaction [12], prior to the knowledge of the existence of the two $\Lambda(1405)$ poles, showed a narrow structure at about 1420 MeV. With the present perspective this is clearly interpreted as the reaction proceeding through the emission of the photon followed by the generation of the resonance from $K^- p$, thus receiving a large contribution from the second narrower state at higher energy. The same idea is used in the reaction $\gamma p \to K^*\Lambda(1405)$ [13] which proceeds via K-meson exchange and is now under investigation at

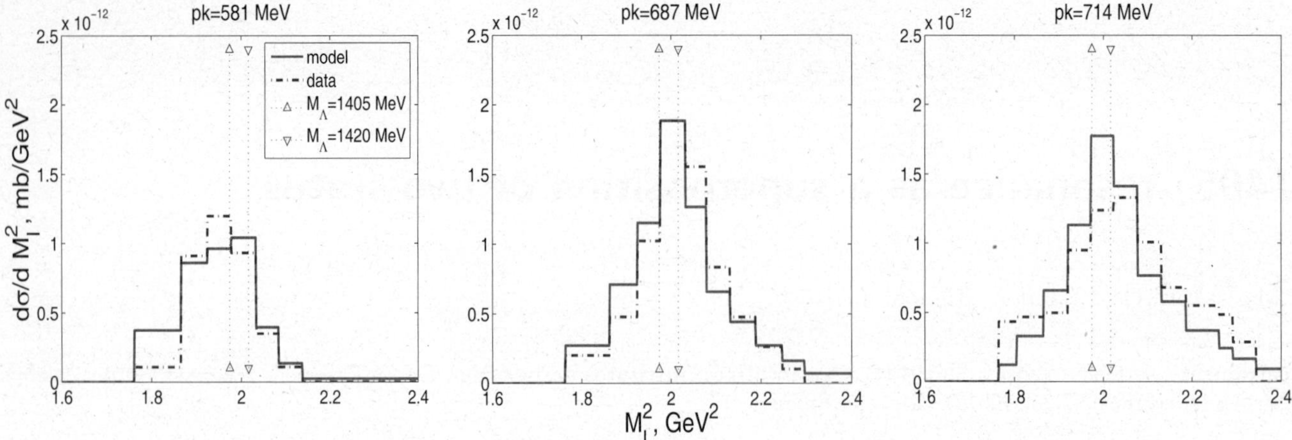

Fig. 3. The $(\pi^0\Sigma^0)$ invariant mass distribution for three different initial kaon momenta.

Spring8/Osaka [14]. Luckily, the recently measured reaction $K^-p \to \pi^0\pi^0\Sigma^0$ [15] allows us to test already the two-pole nature of the $\Lambda(1405)$. This process shows a strong similarity with the reaction $K^-p \to \gamma\Lambda(1405)$, where the photon is replaced by a π^0.

Thus, to test experimentally the two-pole nature of the $\Lambda(1405)$ we have to compare two experimental shapes of this resonance, seen in two reactions tied mostly to the first and the second $\Lambda(1405)$ poles correspondingly. In Fig. 1 we compare the experimental shapes of the two $\pi^0\Sigma^0$ invariant mass experimental distributions coming from the recently measured $K^-p \to \pi^0\pi^0\Sigma^0$ reaction [15] and the $\pi^-p \to K^0\pi\Sigma$ experiment [16]. The figure shows a clear signal of the two-pole structure of the $\Lambda(1405)$.

The experimental shapes of the $\Lambda(1405)$ shown in Fig. 1 are not only different from each other, but also in good agreement with chiral unitary theory calculations, which at the same time explain many other reactions involving these channels. The recent $K^-p \to \pi^0\pi^0\Sigma^0$ reaction [15], which leads to a narrower peak at higher energies, is studied in detail in [18]. The shape of the $\Lambda(1405)$ in the $\pi^-p \to K^0\pi\Sigma$ reaction is largely built from the $\pi\Sigma \to \pi\Sigma$ amplitude, which is dominated by the wider lower energy state [17]. Therefore this second experiment has found a distribution peaked at 1395 MeV with the apparent width of about 60 MeV. One of the important findings in the detailed study of [18] is that the reaction $K^-p \to \pi^0\pi^0\Sigma^0$ in the energy region of $p_{K^-} = 514$ to 750 MeV/c, as in the experiment [15], is largely dominated by the nucleon pole term shown in Fig. 2. As a consequence, the $\Lambda(1405)$ thus obtained comes mainly from the $K^-p \to \pi^0\Sigma^0$ amplitude which, as mentioned above, gives the largest possible weight to the second (narrower) state. In this paper we are going to briefly review our results of Ref. [18] and their comparison with the experimental data.

The main source of the background in this reaction is the indistinguishability of the two emitted pions. This requires the implementation of symmetrization, which is achieved in our calculations by summing two amplitudes evaluated with the two pion momenta exchanged, and in addition we have to include a factor of $1/2$ for indistinguishable particles in the total cross section.

In Fig. 3 our results for the invariant mass distribution for three different energies of the incoming K^- are compared to the experimental data. Symmetrization of the amplitudes produces a sizable amount of background. At a kaon laboratory momentum of $p_K = 581$ MeV/c this background distorts the $\Lambda(1405)$ shape producing cross section in the lower part of M_I, while at $p_K = 714$ MeV/c the strength of this background is shifted toward the higher M_I region. An ideal situation is found for momenta around 687 MeV/c, where the background sits below the $\Lambda(1405)$ peak distorting its shape minimally. The peak of the resonance shows up at $M_I^2 = 2.02$ GeV^2 which corresponds to $M_I = 1420$ MeV, larger than the nominal $\Lambda(1405)$, and in agreement with the predictions of Ref. [8] for the location of the peak when the process is dominated by the $t_{\bar{K}N \to \pi\Sigma}$ amplitude. The apparent width from experiment is about $40-45$ MeV, but a precise determination would require to remove the background mostly coming from the "wrong" $\pi^0\Sigma^0$ couples due to the indistinguishability of the two pions.

A theoretical analysis permits extracting the pure resonant part by not symmetrizing the amplitude. This is plotted in Fig. 4 as a function of M_I. The width of the resonant part is $\Gamma = 38$ MeV, which is smaller than the nominal $\Lambda(1405)$ width of 50 ± 2 MeV [19], obtained from the average of several experiments.

The invariant mass distributions shown here are not normalized, as in experiment. But we can also compare our absolute values of the total cross sections with those in Ref. [15]. As shown in Fig. 5, our results are in excellent agreement with the data, in particular for the three kaon momentum values whose corresponding invariant mass distributions have been displayed in Fig. 3.

Thus, this combined study of the $K^-p \to \pi^0\pi^0\Sigma^0$ [18] reaction together with $\pi^-p \to K^0\pi\Sigma$ [17], besides demonstrating once more the great predictive power of the chiral unitary theories, gives the first clear evidence of the two-pole nature of the $\Lambda(1405)$. So, we know the way the $\Lambda(1405)$ should be included into any simulation. And it is

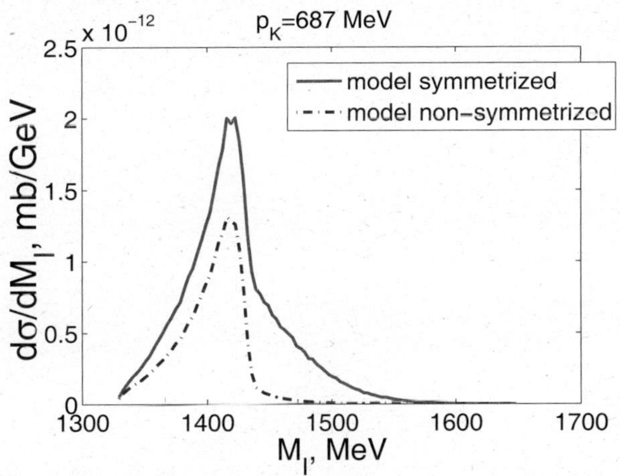

Fig. 4. Theoretical $(\pi^0 \Sigma^0)$ invariant mass distribution for an initial kaon lab momenta of 687 MeV. The non-symmetrized distribution also contains the factor 1/2 in the cross section.

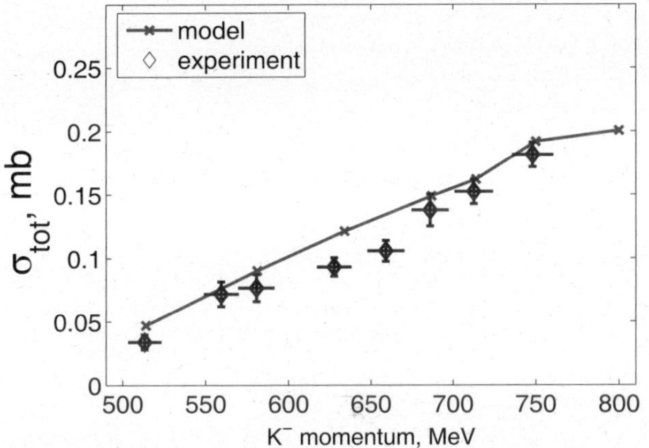

Fig. 5. Total cross section for the reaction $K^- p \to \pi^0 \pi^0 \Sigma^0$. Experimental data are taken from Ref. [15].

clear that the $\Lambda(1405)$ can not be taken as a bound state of $\bar{K}N$, what is done, for example, in [20]. In this paper, [20], due to several reasons, see refs. [21] for more details, in particular due to incorrect treatment of the $\Lambda(1405)$, the authors obtain an unreasonably deep $K^- N$ potential, much deeper than any realistic chiral model calculation.

In summary, the study of the $K^- p \to \pi^0 \pi^0 \Sigma^0$ reaction [18,15] together with the $\pi^- p \to K^0 \pi \Sigma$ reaction [17] shows that the quite different shapes of the $\Lambda(1405)$ resonance seen in these experiments can be interpreted in favour of the existence of two poles with the corresponding states having the characteristics predicted by the chiral theoretical calculations. Our model has proved to be accurate in reproducing both the invariant mass distributions and integrated cross sections seen in a recent experiment [15].

This work is partly supported by DGICYT contracts BFM2002-01868, BFM2003-00856, the Generalitat de Catalunya contract SGR2001-64, and the E.U. EURIDICE network contract HPRN-CT-2002-00311. This research is part of the EU Integrated Infrastructure Initiative Hadron Physics Project under contract number RII3-CT-2004-506078.

References

1. M. Jones, R. H. Dalitz and R. R. Horgan, Nucl. Phys. B **129** (1977) 45.
2. N. Kaiser, P. B. Siegel and W. Weise, Phys. Lett. B **362** (1995) 23.
3. N. Kaiser, T. Waas and W. Weise, Nucl. Phys. A **612** (1997) 297.
4. E. Oset and A. Ramos, Nucl. Phys. A **635** (1998) 99.
5. J. A. Oller and U. G. Meissner, Phys. Lett. B **500** (2001) 263.
6. D. Jido, A. Hosaka, J. C. Nacher, E. Oset and A. Ramos, Phys. Rev. C **66** (2002) 025203.
7. C. Garcia-Recio, J. Nieves, E. Ruiz Arriola and M. J. Vicente Vacas, Phys. Rev. D **67** (2003) 076009.
8. D. Jido, J. A. Oller, E. Oset, A. Ramos and U. G. Meissner, Nucl. Phys. A **725** (2003) 181.
9. C. Garcia-Recio, M. F. M. Lutz and J. Nieves, Phys. Lett. B **582** (2004) 49.
10. T. Hyodo, S. I. Nam, D. Jido and A. Hosaka, Phys. Rev. C **68** (2003) 018201.
11. S. I. Nam, H. C. Kim, T. Hyodo, D. Jido and A. Hosaka, arXiv:hep-ph/0309017.
12. J. C. Nacher, E. Oset, H. Toki and A. Ramos, Phys. Lett. B **461** (1999) 299.
13. T. Hyodo, A. Hosaka, M. J. Vicente Vacas and E. Oset, Phys. Lett. B **593** (2004) 75.
14. T. Nakano, private communication.
15. S. Prakhov et al. [Crystall Ball Collaboration], Phys. Rev. C **70** (2004) 034605.
16. D. W. Thomas, A. Engler, H. E. Fisk, and R. W. Kraemer, Nucl. Phys. B **56**, 15 (1973).
17. T. Hyodo, A. Hosaka, E. Oset, A. Ramos and M. J. Vicente Vacas, Phys. Rev. C **68** (2003) 065203.
18. V.K. Magas, E. Oset, A. Ramos, Phys. Rev. Lett. **95** (2005) 052301; hep-ph/0510122, also in Proceedings of the "New Trends in High-Energy Physics" (Crimea 2005), Yalta, Crimea, Ukraine, September 10-17, 2005, pp. 177-182; E. Oset, V.K. Magas, A. Ramos, nucl-th/0512090, also in Proceedings of the XI International Conference on Hadron Spectroscopy (Hadron 05), Rio de Janeiro, Brazil, Aug 21-26, 2005, American Institute of Physics (AIP) Conf. Proc. 814 (2006) 273-277; hep-ph/0512361.
19. K. Hagiwara et al. [Particle Data Group], Phys. Rev. D **66** (2002) 010001.
20. Y. Akaishi and T. Yamazaki, Phys. Rev. C **65** (2002) 044005.
21. E. Oset and H. Toki, Phys.Rev. C **74** (2006) 015207.

Casimir scaling, glueballs, and hybrid gluelumps[*]

Vincent Mathieu[1,a], Claude Semay[1,b], and Fabian Brau[2,c]

[1] Groupe de Physique Nucléaire Théorique, Université de Mons-Hainaut, Académie universitaire Wallonie-Bruxelles, Place du Parc 20, BE-7000 Mons, Belgium
[2] CWI, P.O. Box 94079, 1090 GB Amsterdam, The Netherlands

Abstract. Assuming that the Casimir scaling hypothesis is well verified in QCD, masses of glueballs and hybrid gluelumps are computed within the framework of the rotating string formalism. In our model, two gluons are attached by an adjoint string in a glueball, while the gluon and the colour octet $c\bar{c}$ pair are attached by two fundamental strings in a hybrid gluelump. Masses for such exotic hadrons are computed with very few free parameters. These predictions can serve as a guide for experimental searches. In particular, the ground state glueballs lie on a Regge trajectory and the lightest 2^{++} state has a mass compatible with some experimental candidates.

PACS. 12.39.Mk Glueball and nonstandard multiquark gluon states – 12.39.Ki Relativistic quark model – 12.39.Pn Potential model

1 Introduction

Lattice calculations [1] predict that the Casimir scaling hypothesis is well verified in QCD. This mechanism can be tested for instance in two-gluon glueballs. The Casimir scaling hypothesis can also be tested in another system whose colour-spin structure is similar to the one of a glueball. Let us consider an hybrid meson containing a colour octet spin one $c\bar{c}$ pair and a gluon. Due their very heavy masses, the charm quarks can be assumed nearly fixed at the centre of mass with the gluon orbiting around. A gluon attached to such a static colour octet is called a hybrid gluelump.

In our model, a quark and an antiquark are attached by a fundamental string while two gluons in a glueball are attached by an adjoint string. We assume here that the three-body nature of the confinement implies that the gluon and the colour octet pair are attached by two fundamental strings in a hybrid gluelump. Within these hypothesis, we compute the masses of glueballs and hybrid gluelumps with a simple effective model of QCD. This semirelativistic potential model is described in Sec. 2. It depends on few parameters which are fixed on well known mesons in Sec. 3. Masses of glueballs and hybrid gluelumps

are computed respectively in Sec. 4 and 5. Some concluding remarks are given in Sec. 6.

2 Hamiltonian

Starting from the QCD theory, a Lagrangian for a system of two confined spinless colour sources can be derived taking into account the dynamical degrees of freedom of the string, with tension σ, joining the two particles, the rotating string model (RSM) [2]. We will use an approximation of these effective QCD theories [3], which is simply given by the following spinless Salpeter Hamiltonian

$$H_0 = \sqrt{\mathbf{p}^2 + m_1^2} + \sqrt{\mathbf{p}^2 + m_2^2} + \sigma r, \tag{1}$$

completed by a perturbative part due to the motion of the string

$$\Delta H_{\mathrm{str}} = -\frac{\sigma L(L+1)}{r}$$
$$\times \frac{\left[4(\mu_1^2 + \mu_2^2 - \mu_1\mu_2) + (\mu_1 + \mu_2)\sigma r\right]}{2\mu_1\mu_2\left[12\mu_1\mu_2 + 4(\mu_1+\mu_2)\sigma r + (\sigma r)^2\right]}. \tag{2}$$

The quantity μ_i appearing in the above equation is a kind of constituent particle mass given by

$$\mu_i = \left\langle \sqrt{\mathbf{p}^2 + m_i^2} \right\rangle, \tag{3}$$

in which the average value is evaluated with an eigenstate of the Hamiltonian (1). The contribution ΔM_{str} to the

[*] The extended version of this paper has been published in Eur. Phys. J. A **27**, 225 (2006).

[a] IISN Scientific Research Worker;
e-mail: `vincent.mathieu@umh.ac.be`
[b] FNRS Research Associate;
e-mail: `claude.semay@umh.ac.be`
[c] e-mail: `f.brau@cwi.nl`

mass due to the string can be computed with a good precision by the following approximation [3]

$$\Delta M_{\mathrm{str}} = -\sigma L(L+1)\left\langle \frac{1}{r} \right\rangle$$

$$\times \frac{[4(\mu_1^2 + \mu_2^2 - \mu_1\mu_2) + (\mu_1 + \mu_2)\sigma\langle r\rangle]}{2\mu_1\mu_2\,[12\mu_1\mu_2 + 4(\mu_1 + \mu_2)\sigma\langle r\rangle + (\sigma\langle r\rangle)^2]}. \quad (4)$$

A significant contribution to hadron masses is also given by the one-gluon exchange (OGE) mechanism between colour sources. At the zero order, the interaction has the following form

$$\Delta H_{\mathrm{Coul}} = \kappa\frac{\alpha_S}{r}. \quad (5)$$

κ is a colour factor depending on the color Casimir operator given by $\kappa = (C_{12} - C_1 - C_2)/2$.

If M_0 is the solution of the eigenequation $H_0|\phi\rangle = M_0|\phi\rangle$, the total mass M for the system of the two colour sources is given by

$$M = M_0 + \Delta M_{\mathrm{str}} + \langle \Delta H_{\mathrm{Coul}}\rangle. \quad (6)$$

Recently, it was shown that the quark self-energy (QSE) contribution adds a negative constant to the hadron masses [4]. The QSE contribution ΔS_i for a quark of current mass m_i is given by

$$\Delta S_i = -f\frac{\sigma}{2\pi}\frac{\eta(m_i/\delta)}{\mu_i}. \quad (7)$$

δ is the inverse of the gluonic correlation length. We fix the value of δ at 1 GeV. The factor f has been computed by lattice calculations and is equal to $f = 3$ [5]. The η–function is given in [3].

3 Parameters

As we are interested in describing the main features of spectra and high orbital momemtum states, we can use a Hamiltonian of spinless particles since spin effects are generally an order of magnitude smaller than orbital or radial excitations in states with non zero angular momentum. Moreover, we will consider hadrons for which the spin-dependent part of the Hamiltonian is the weakest possible, that is to say, in the systems where the spin is maximum. For this reason, we will only consider systems with maximal value of the total spin S: $J = L + 1$ for mesons and $J = L + 2$ for glueballs and hybrid gluelumps.

For theoretical reasons, we have chosen to take $m_n = 0$, $m_g = 0$, $f = 3$, and $\delta = 1$ GeV. Now, we will fix the values of the remaining free parameters with well established meson states. As, the light $n\bar{n}$ spin 1 mesons can be isoscalar ($I = 0$) or isovector ($I = 1$), we will use an average mass defined, as usual, by

$$M_{n\bar{n}}(\mathrm{av.}) = \frac{M_{n\bar{n}}(I = 0) + 3\,M_{n\bar{n}}(I = 1)}{4}. \quad (8)$$

Table 1. Experimental masses in GeV for some light mesons ($n\bar{n}$), D^* mesons ($c\bar{n}$), and charmoniums ($c\bar{c}$).

$N = 0,\ J^{P(C)}$	$1^{-(-)}$	$2^{+(+)}$	$3^{-(-)}$	$4^{+(+)}$
$n\bar{n}\ I = 0$	0.783	1.275	1.667	2.034
$n\bar{n}\ I = 1$	0.775	1.318	1.689	2.001
$n\bar{n}$ av.	0.777	1.307	1.684	2.009
$c\bar{n}$	2.010	2.460		
$c\bar{c}$	3.097	3.556		

Table 2. Parameters of the model.

$m_n = 0$	$a = 0.175$ GeV^2
$m_g = 0$	$\alpha_s = 0.10$
$m_c = 1.300$ GeV	$f = 3$
$m_{c\bar{c}} = 2\,m_c$	$\delta = 1$ GeV

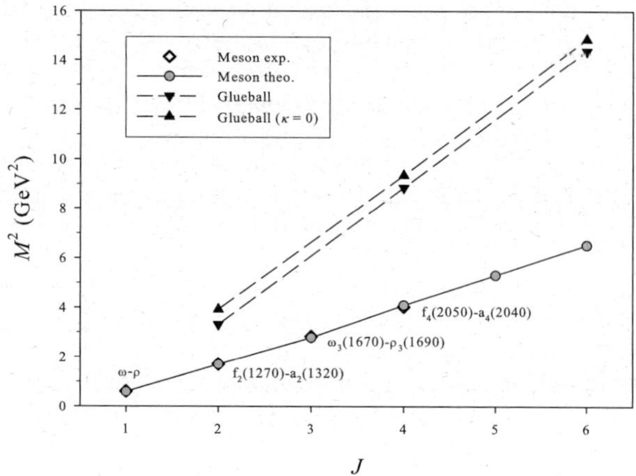

Fig. 1. Regge trajectories for light mesons and glueballs. These states are characterized by $N = 0$, S maximal, and $J = L + S$.

We fix the parameters by fitting the results, obtained with formula (6) and the exact solutions of Hamiltonian (1), on experimental data [6] (see Table 1). Using the Regge trajectory for light meson, we find $a = 0.175$ GeV^2 and $\alpha_s = 0.10$. The mass of the quark c is determined by computing the masses of D^* mesons instead of $c\bar{c}$ mesons. We consider that a hybrid gluelump is dynamically close to a D^* because, in both cases, a light particle orbits around a heavy one. All parameters are gathered in Table 2.

4 Glueballs

In this paper, we assume that the string tension between two colour sources is given by the Casimir scaling. In a two-gluon system, we have then $\sigma = 9\,a/4$. Moreover there is no self-energy contribution coming from gluons.

At the lowest order, the colour Coulomb factor in such a system is $\kappa = 3$. Nevertheless, in some model, it is found that the Coulomb force in glueballs could be damped [7]. So, we have considered both cases $\kappa = 0$ and $\kappa = 3$.

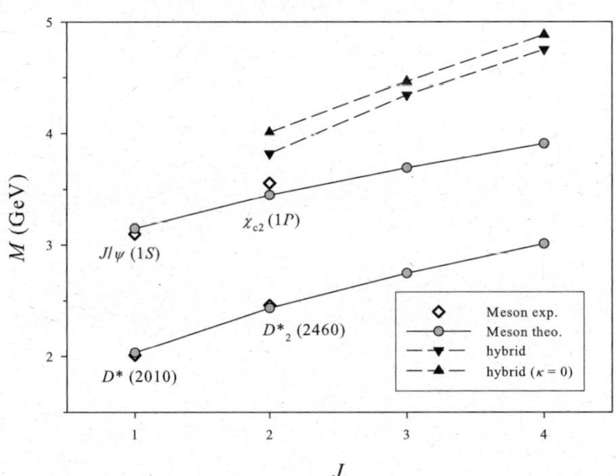

Fig. 2. Masses for $c\bar{n}$, $c\bar{c}$, and hybrid gluelumps (with or without Coulomb interaction). These states are characterized by $N = 0$, S maximal, and $J = L + S$.

The colour wave function of two gluons is symmetric and the spin wave function considered here is symmetric since $S = 2$. As gluons are bosons, the spatial wave function is also symmetric and L is even. The masses of the first three states are computed with those of the six lowest light mesons of the first Regge trajectory. Results, obtained with formula (6) and the exact solutions of Hamiltonian (1), are given in Table 3 and in Fig. 1. When $\kappa = 3$, the trajectory is

$$J = 0.36 M^2 + 0.80. \tag{9}$$

Our glueball masses are compatible with those of the model III.A of Ref. [8].

5 Hybrids

We find $\kappa = 3/2$ for a gluon-quark pair inside the hybrid. As the quark and the antiquark are assumed to be very close, the total colour factor for the Coulomb-like force between the gluon and the point-like $c\bar{c}$ pair is $2 \times 3/2 = 3$, which is the same as into a two gluon glueball. As in the previous case, we have considered both cases $\kappa = 0$ and $\kappa = 3$.

The situation is different for the confinement which is actually a three-body force. We can show, using the triangle inequalities, that the confining energy is equal to two times the confining energy in a meson [9], and we take $\sigma = 2a$.

A hydrid gluelump is then considered here as a gluon and a point-like $c\bar{c}$ pair attached by two fundamental strings. To compute its mass, we must determine the mass of the $c\bar{c}$ pair. For heavy quark, the constituent mass μ is around the current mass m. For a $c\bar{c}$, the contribution of the self-energy can be estimated around -40 MeV. So, if we neglect the kinetic energy of the quarks inside the point-like $c\bar{c}$ pair, we can take the point-like $c\bar{c}$ with a mass equal to $2\,m_c$. We estimate that this approximation

Table 3. Predicted masses in GeV for $n\bar{n}$, $c\bar{n}$, $c\bar{c}$, glueballs, and hybrid gluelumps. These states are characterized by $N = 0$, S maximal, and $J = L + S$. The asterisk indicates that the states is calculated with $\kappa = 0$.

J	1	2	3	4	5	6
$n\bar{n}$	0.767	1.306	1.670	2.024	2.304	2.555
Glueball	-	1.820	-	2.976	-	3.793
Glueball*	-	1.981	-	3.059	-	3.855
$c\bar{n}$	2.033	2.436	2.747	3.010		
$c\bar{c}$	3.146	3.449	3.693	3.908		
Hybrid	-	3.820	4.343	4.749		
Hybrid*	-	4.010	4.464	4.884		

can generate an error on hybrid glueball masses of about 100 MeV.

The masses of hybrid gluelumps, obtained with formula (6) and the exact solutions of Hamiltonian (1), are given up to $J = 4$ with the masses of $c\bar{n}$ and $c\bar{c}$ mesons (see Table 3 and Fig. 2).

6 Concluding remarks

We compute the masses of the lightest glueballs and hybrid gluelumps within the framework of a simple effective model of QCD derived from the rotating string model [2]. The corresponding semirelativistic Hamiltonian is dominated by a linear confinement, supplemented by a one gluon exchange interaction and a contribution from self-energy for the quarks only. It is assumed that the Casimir scaling hypothesis is well verified, and that two gluons are attached by an adjoint string in a glueball, while the gluon and the colour octet $c\bar{c}$ pair are attached by two fundamental strings in a hybrid gluelump.

In order to minimize the contributions of the spin dependent interactions, only the masses of states with maximal spin have been calculated. This strongly constraints the values of our three free parameters: the mass of the charm quark, the strong coupling constant and the string tension in a meson. With these parameters, we find glueballs and hybrid gluelumps with masses relatively small. The contribution of the Coulomb interaction is about 100-200 MeV, depending on the particular hadron. The ground state glueballs lie on a Regge trajectory and the lightest 2^{++} state has a mass compatible with some experimental candidates [10].

References

1. S. Deldar, Phys. Rev. D **62**, 034509 (2000).
2. A. Yu. Dubin, A. B. Kaidalov, Yu. A. Simonov, Phys. At. Nucl. **56**, 1745 (1993); [Yad. Fiz. **56**, 213 (1993)].
3. F. Buisseret, C. Semay, Phys. Rev. D **71**, 034019 (2005).
4. Yu. A. Simonov, Phys. Lett. B **515**, 137 (2001).
5. A. Di Giacomo, Yu. A. Simonov, Phys. Lett. B **595**, 368 (2004).

6. Particle Data Group, S. Eidelman *et al.*, Phys. Lett. B **592**, 1 (2004).
7. A. B. Kaidalov, Yu. A. Simonov, Phys. Atom. Nucl. **63**, 1428 (2000) [Yad. Fiz. **63**, 1428 (2000)].
8. F. Brau, C. Semay, Phys. Rev. D **70**, 014017 (2004).
9. V. Mathieu, C. Semay, F. Brau, Eur. Phys. J. A **27** (2006) 225.
10. D. V. Bugg, M. Peardon, B. S. Zou, Phys. Lett. B **486**, 49 (2000).

Three-gluon glueballs in a semirelativistic potential model

B. Silvestre-Brac[1], V. Mathieu[2], and C. Semay[2]

[1] LPSC, 53 Av. des martyrs, F-38026 Grenoble Cedex, France
[2] Université de Mons-Hainaut, place du parc 20, B-7000 Mons, Belgium

Abstract. The 3-gluon glueballs are studied in a semirelativistic model including a potential with long range and short range contributions. The long range is the confinement potential of traditional Y form coming from the flux tube model and the short range stems from the one gluon exchange contributions. The bare potentials are dressed by convoluting them with the phenomenological matter density for the gluons that cannot be considered as pointlike particles. Our model contains 3 parameters: the string tension, the strong coupling constant and the gluon size. They are determined from the 2-gluon glueball states and kept unchanged for the 3-gluon glueball calculations. Our results are in complete agreement with LQCD conclusions except for the 2^{--} state that shows a very puzzling behaviour.

PACS. 12.39.Pn Potential models – 12.39.Mk Glueball and non-standard multi quark/gluon states

1 Introduction

It is well known that QCD allows the coupling of gluons with themselves through 3-leg and 4-leg vertices. As a consequence, physical states made of valence gluons only and interacting exchanging gluons may exist; the required condition is that these states are colour singlets: the corresponding objects are named glueballs.

The simplest systems are made of 2 gluons, but in this case only states with positive C parity are possible; the lowest states have the quantum numbers J^{PC} equal to 0^{++}, 2^{++} and 0^{-+}. The systems studied here are made of 3 gluons; in this case both possibilities of C parity are allowed. They have been studied in several works [1,2]. Here we present results for glueballs with $L = 0$ orbital angular momentum.

Lattice QCD [3,4] predicts around 12 glueballs of both types in the energy range between 1.5 GeV and 4.8 GeV, and it is very likely that one 0^{++} resonance that was observed around 1.5-1.8 GeV has an important 2-gluon component.

2 The wave function

Let us now discuss the type of wave function that describes a 3-gluon system. The physical wave function is expanded on some basis states that are not orthogonal: $|\Psi_\alpha\rangle = \sum_i C_i^\alpha |\Psi_i\rangle$. The basis states are a tensor product of colour, spin and space degrees of freedom: $|\Psi_i\rangle = \mathcal{S}[|C_i\rangle |S_i\rangle |\phi_i\rangle]$. The symmetrizer $\mathcal{S}$ insures good symmetry properties for this 3-boson system.

Concerning colour, the gluon is an $SU(3)$ octet object. It exist 2 possible colour functions: $|C_1\rangle = [(8\,8)_{8_S}\,8]_1$ in which the intermediate coupling of the first two gluons is the symmetric octet and $|C_2\rangle = [(88)_{8_A}\,8]_1$ in which it is in the antisymmetric octet. $|C_1\rangle$ is totally symmetric under particle exchange, while $|C_2\rangle$ is totally antisymmetric.

Since a gluon has a spin 1, the only possibilities for the spin function are the following:

- only one state with total spin $S = 3$, $|S\rangle = [(1\,1)_2\,1]_3$ which is totally symmetric under particle exchange;
- 2 states for $S = 2$, $|S_1\rangle = [(1\,1)_1\,1]_2$, $|S_2\rangle = [(1\,1)_2\,1]_2$ which are of mixed symmetry;
- 3 states with $S = 1$, $|S_0'\rangle = [(11)_0\,1]_1$, $|S_1'\rangle = [(11)_1\,1]_1$, $|S_2'\rangle = [(1\,1)_2\,1]_1$, a combination of them being totally symmetric and the other two of mixed symmetry;
- and lastly only one state $S = 0$, $|S_0\rangle = [(1\,1)_1\,1]_0$, which is totally antisymmetric.

The space function is expanded on gaussian type functions [5] built with the natural Jacobi coordinates for the system: as usual $x = r_1 - r_2$ is the relative distance between gluons 1 and 2, and $y = (r_1 + r_2)/2 - r_3$ is the distance between the pair and the third gluon.

For $L = 0$, the spatial basis states are just gaussian functions built with the most general scalar function in terms of the Jacobi coordinates, namely

$$\langle x, y|\phi_i\rangle = \exp(-a_i\,x^2 - b_i\,y^2 - 2c_i\,x \cdot y). \qquad (1)$$

The mixed term $x \cdot y$ complicates the formalism, but, on the other hand, it allows a much faster convergence, because each pair of particles can form an arbitrary orbital angular momentum, and not only an S-wave.

Thus for each spatial basis state, there are 3 free non linear parameters a_i, b_i, c_i.

3 The dynamical equation

For solving the dynamical equation, we need to calculate 2 types of kernels. The norm kernel $\mathcal{N}_{ij} = \langle \Psi_i | \Psi_j \rangle$ which is simply the overlap of the non orthogonal basis states, and the energy kernel $\mathcal{H}_{ij} = \langle \Psi_i | H | \Psi_j \rangle$ which is the matrix element of the total hamiltonian on those states. The Schrödinger equation leads to a well known generalized eigenvalue problem

$$\sum_j C_j^\alpha [\mathcal{H}_{ij} - E \mathcal{N}_{ij}] = 0 \tag{2}$$

which provides the energy E and the components of the wave function C_j^α. The energy $E(a_i, b_i, c_i)$ depends on the free parameters appearing in the gaussian basis states, and these parameters are determined by a usual variational procedure minimizing the energy $\partial E / \partial a_i = 0 = \partial E / \partial b_i = \partial E / \partial c_i$. Once these parameters $\bar{a}_i, \bar{b}_i, \bar{c}_i$ are obtained, they are inserted in the kernels and the dynamical equation gives the resulting total wave function.

4 The hamiltonian

Let us now discuss the form of the Hamiltonian in a potential model. In this framework, the Hamiltonian $H = T + V$ is the sum of a kinetic energy T term and a potential interaction V.

Since the bare gluons have a null mass, one is forced to take the kinetic energy operator under a relativistic form like this

$$T = \sqrt{p_1^2} + \sqrt{p_2^2} + \sqrt{p_3^2}. \tag{3}$$

The potential interaction $V = V_l + V_s$ has long range and short range contributions.

The long range part V_l is the confinement potential. The model that sticks the best to the results obtained from Lattice QCD is the flux tube model. In this model, the energy is concentrated in a tube that connects a colour source to a corresponding anti-colour source, and the density energy is constant all along the tube, so that the total confining energy is proportional to the length of the tube. For a meson, made of 1 quark and 1 antiquark, the source is the quark itself and the anti-source is the antiquark, so that the form of the potential is

$$V_l = a'(\lambda^2/4) |r_1 - r_2|. \tag{4}$$

a' is a universal constant (more or less the energy density); the colour contribution is proportional to the Casimir of the colour representation $\lambda^2/4$, which equals 4/3 for a triplet, and the length of the tube is simply the relative distance. In this case, the potential has a well known linear form ar, the constant $a = (4/3)a'$ being called the string tension.

For a glueball made of 3 gluons, the situation is more complicated because the tube connects each gluon to the Steiner point Y which minimizes the total length of the tube. Moreover, the colour contribution is different, the gluon being an octet instead of a triplet. The corresponding contribution is thus $(9/4)a \sum_{i=1}^3 |r_i - r_Y|$, with a string tension multiplied by a 9/4 factor. The flux tube potential is a very complicated three-body operator. Instead, we used, in our calculations, an approximation of this potential that is much simpler and nevertheless still very accurate [6]. In this approximation, the Steiner point Y is replaced by the centre of mass G, so that the confining operator is now a one-body operator, and this difference is compensated by introducing a correction factor f, whose value is around 0.95 for 3 identical particles [6]. The long range potential that is finally used for practical calculations has the following form

$$V_l = \frac{9}{4} f a \sum_{i=1}^3 |r_i - r_G| . \tag{5}$$

The short range potential V_s is due to one gluon exchange. In contrast to the interaction between 2 quarks which is characterized by only one diagram (a gluon being exchanged between 2 quark lines), we have in addition the 4-leg vertex diagram (the annihilation diagram cancels in this peculiar case).

The corresponding two-body potential is due to Hou et al. [7], and takes this form

$$V_s = \sum_{i<j=1}^3 (\lambda_i \cdot \lambda_j/4)\alpha_s \left[(1/4 + S_{ij}^2/3)U(r_{ij}) \right.$$
$$\left. - (\pi^2/m^2)\delta(r_{ij})(\beta + 5S_{ij}^2/6)\right] . \tag{6}$$

It contains the usual $\lambda_i \cdot \lambda_j/4$ colour factor, a Yukawa term $U(r) = \exp(-m r)/r$ and a short range Dirac term. The spin dependence appears through the total spin of the pair $S_{ij} = S_i + S_i$. α_s is the traditional strong coupling constant, and β is a parameter which takes the value -1 if the total colour of the pair is a symmetric octet 8_s, and $+1$ if it is an antisymmetric octet 8_a. In principle, spin-orbit and tensor terms do exist, but since we are concerned only by a $L = 0$ states, they do not play any role.

As such, this potential presents two big problems.

The first one is related to the presence of the Dirac function which leads to a collapse. One can cure this problem by considering that the interacting gluons are already effective particles dressed by a lot of things. To simulate those complicated effects, one considers that the interacting gluons are no longer pointlike particles but rather they possess some matter density; this density $\rho(u)$ must be a function sharply peaked around the mean position of the gluon. We choose a function of Yukawa type $\rho(u) = (1/(4\pi\gamma^2)) \exp(-u/\gamma)/u$ that depends on some range γ. Now we must modify all the bare potentials by using instead their convolution with the density [8]. This means we must replace everywhere the bare potential V, by a dressed potential $\widetilde{V}$ which depends on γ. After this treatment, all the singularities disappear.

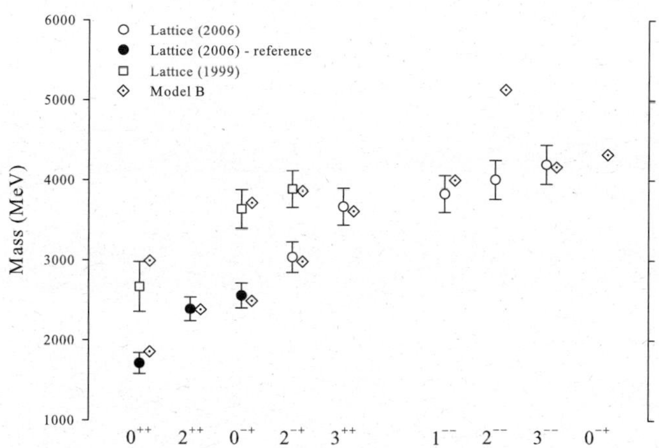

Fig. 1. Absolute masses for glueball states. On the left hand side we report the 2-gluon glueballs [10], and on the right hand side the 3-gluon glueballs. The dotted diamonds stand for our calculations based on a relativistic potential model, while circles and squares stand for LQCD results [3,4]. Black circles indicate the reference states taken as input for the determination of the parameters. The error bars for LQCD are computed by summing the two kinds of uncertainties.

The second problem comes from the fact that the bare gluons have a zero mass $m = 0$, and, again, this leads to singularities. To solve this problem, we follow a procedure suggested by the authors of Ref. [9], which works very well. They consider that confinement is the driving potential; in the confinement potential the gluon mass does not appear. So, one splits the Hamiltonian in a part H_0 which contains the kinetic energy and the confinement and which does not depend on the mass, and isolate for a second treatment the short range mass dependent potential. The first step consists in solving the dynamical equation with H_0 only and obtaining the corresponding wave function $H_0 |\Phi\rangle = E_0 |\Phi\rangle$. The second step is the calculation of the dynamical gluon mass $\mu_i = \langle \Phi | \sqrt{p_i^2} |\Phi\rangle$ as the mean value of the square root of the momentum. Then, in a last step, one inserts this dynamical value μ_i in the short range potential that is now perfectly defined, and solve the dynamical equation with the total dressed potential.

5 Results

The 2-gluon glueball spectrum was calculated in a previous paper [10]. Since experimentally the situation is not definitively clear, the parameters of the model of Ref. [10] were determined from LQCD calculations. They are the string tension a, the strong coupling constant α_s and the gluon size γ. The obtained parameter values are $a = 0.21$ GeV^2, $\alpha_s = 0.5$, $\gamma = 0.495$ GeV^{-1}.

We present in the figure 1 the absolute masses for glueballs obtained by our calculations with dotted diamonds and by LQCD with circles and squares for comparison [3, 4]. We keep unchanged the parameters of Ref. [10] to perform the 3-body calculations corresponding to 3-gluon glueballs. Thus, most of our calculations essentially do not need any free parameter.

On the left hand side, we compare the 2-gluon glueballs (with positive C parity), and on the right hand side the 3-gluon glueballs. One sees that the agreement between LQCD and our potential model is quite good in both C parity sectors. The notable exception concerns the 2^{--} state that is found too high in our model. The fact that this state is higher than the corresponding 3^{--} and 1^{--} can be understood by considering that the space function is of mixed symmetry while the wave functions of the two last states are completely symmetric. Why it is so high as compared to LQCD is nevertheless puzzling, and we do not have clear explanations of this feature. Since similar potential models give the same conclusion than ours, it seems that there is a discrepancy between LQCD and potential models for this peculiar state. But the rest of the spectrum looks quite nice.

6 Conclusions

We presented a phenomenological approach of 3-gluon glueballs, based on a potential model with a semi-relativistic kinetic energy. The long range potential stems from the flux tube model and the short range potential is due to one gluon exchange. To solve the 3-body problem, we used space functions of gaussian type which include a mixed term. The convergence is fast enough to have reliable results with 10 gaussians for each colour, spin degrees of freedom. The dynamical equation has a form of a generalized eigenvalue problem relying on a norm kernel and an energy kernel. A variational procedure based on minimization of the eigenenergy is used to determine the mass of the states and the corresponding wave function. Our calculations were performed without any free parameter for 3-gluon glueballs and comparison with LQCD gives a very good agreement, except for the peculiar 2^{--} state for which the status remains unclear.

References

1. A.B. Kaidalov and Y.A. Simonov, Phys. Lett B **636**, (2006) 101.
2. F.J Llanes-Estrada, P. Bicudo and S.R. Cotanch, Phys. Rev. Lett **96**, (2006) 081601.
3. C.J. Morningstar and M.J Peardon, Phys. Rev. D **60**, (1999) 034509.
4. Y. Chen *et al.*, Phys. Rev. D **73**, (2006) 014516.
5. Y. Suzuki and K. Varga, *Stochastic variational approach to quantum mechanical few-body problems* (Springer, Berlin 1998) 310.
6. B.Silvestre-Brac *et al.*, Eur. Phys. J C **32**, (2004) 385.
7. W.S. Hou, C.S. Luo and G.G Wong, Phys. Rev. D **64**, (2001) 014028.
8. F. Brau, C. Semay and B. Silvestre-Brac, Phys. Rev. C **66**, (2002) 055202.
9. B.O. Kerbikov and Yu.A. Simonov, Phys. Rev. D **62**, (2000) 093016.
10. F. Brau and C. Semay, Phys. Rev. D **70**, (2004) 014017.

Hadron spectrum and hadrons in the nuclear medium

Some recent developments

M. J. Vicente Vacas

Departamento de Física Teórica and IFIC, Centro Mixto Universidad de Valencia CSIC; Institutos de Investigación de Paterna, Aptdo. 22085, 46071 Valencia, Spain

Abstract. Some recent developments in chiral dynamics of hadrons and hadrons in a medium are presented. Unitary schemes based on chiral Lagrangians describe some hadronic states as being dynamically generated resonances. We discuss how standard quantum many body techniques can be used to calculate the properties of these dynamically generated and other hadrons in the nuclear medium. We present some results for vector mesons (ρ and ϕ), scalar mesons (σ, κ, $a_0(980)$, $f_0(980)$), the $\Lambda(1520)$ and for the in-medium baryon-baryon interaction.

PACS. 12.39.Fe – 12.40.Yx – 21.65.+f

1 Introduction

Coupled channels unitary models based on the chiral Lagrangians have provided a framework that allows to extend the calculations to higher energies than standard χPT and have been very successful in the description of hadronic phenomenology. These models reproduce well meson-meson phase shifts up to energies above 1 GeV [1] and are able to generate a number of well established baryonic resonances like $N^*(1535)$, $\Lambda(1405)$, $\Lambda(1670)$, $\Sigma(1620)$, $\Xi(1690)$ and others [2–6]. These resonances appear dynamically as poles of the meson baryon scattering amplitudes. The analysis of the residues of the poles in the different scattering channels gives the couplings and branching ratios of the resonances.

An important problem in nuclear physics is the understanding of the properties of hadrons at finite baryonic densities. Many different theoretical approaches to this problem can be found in the literature. A quite interesting question is if the very special nature of the dynamically generated resonances obtained in chiral unitary models will have consequences for their in medium properties. It looks clear that this description of these particles has some implications. For instance, the $\Lambda(1405)$ appears in these models as a resonant state in $\bar{K}N$ and/or $\pi\Sigma$ scattering. The changes in the properties of the constituents $\bar{K}$, N, π, Σ will also change the results for their scattering amplitude and thus for its resonant states like the $\Lambda(1405)$.

The basic idea of the approach we present here is to start from these models which describe adequately the properties of several hadronic resonances in vacuum and their couplings to pseudoscalar mesons. Then, we incorporate nuclear medium effects by including vertex corrections and a suitable modification of the pseudoscalar meson and the baryon propagators, which can be done in terms of density dependent selfenergies. The selfenergies are studied by means of a many body calculation accounting for their interactions with the nuclear medium. With this formulation we automatically consider the presence of additional decay channels which are not present in vacuum, since they require an interaction with one or more nucleons in the medium.

We will start discussing the π and K selfenergies. They experience a sizable renormalization at finite densities as a consequence of their interactions with the nucleons as it has been found in many theoretical calculations and experimental results. Then, using these selfenergies as ingredients, we study some medium effects on vector mesons (ρ and ϕ), dynamically generated scalar mesons (σ, κ, $a_0(980)$ and $f_0(980)$) and for the $\Lambda(1520)$ resonance. Finally, we will discuss briefly some recent results on the NN interaction in nuclear matter.

2 Pion and Kaons selfenergies in nuclear matter

The pion selfenergy in the nuclear medium at intermediate energies is relatively well known. It is mainly driven by its p-wave coupling to $p-h$ and $\Delta-h$ excitations as depicted in Fig. 1. Higher order terms obtained by the iteration of these excitations, considering also short range correlations, and the small contribution of s-wave scattering are also included in the calculations. Analytical expressions, as well as results can be found in Refs. [7]. In Fig. 2 we show the results for the pion propagator. Three wide peaks

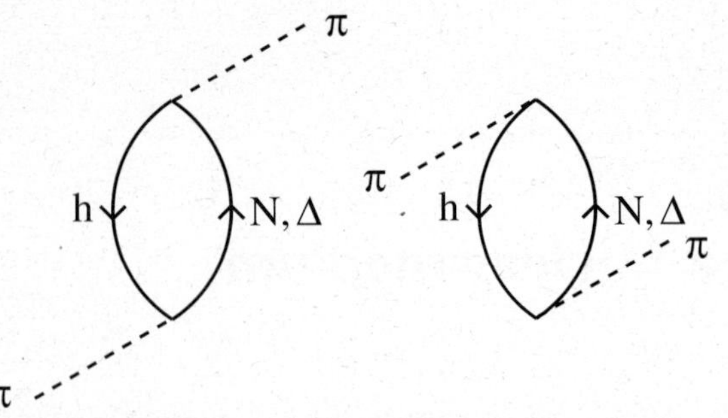

Fig. 1. Diagrams contributing to π selfenergy in the nuclear medium.

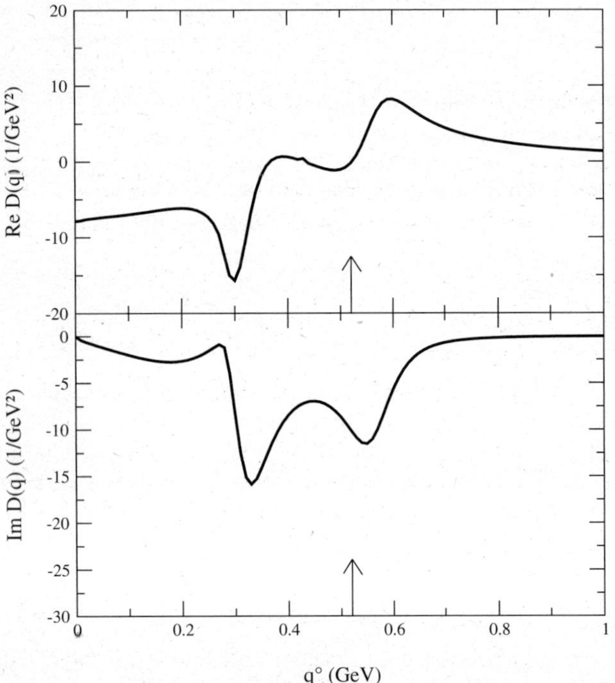

Fig. 2. Real and imaginary part of the π propagator at $\rho = \rho_0$ as a function of the energy for a pion with momentum $q = 500$ MeV. The arrow indicates the position of the pion pole in vacuum.

appear in its imaginary part corresponding to $p - h$ excitation, the pion peak that occurs at lower energies than in vacuum and $\Delta - h$ excitation respectively. This model has been widely tested and agrees well with pion-nucleus phenomenology from low to intermediate energies.

The kaon selfenergy is smooth at low energies and a simple $t\rho$ approximation can be used [8]. The situation is more controversial for the antikaons with recent experimental data that might be interpreted as very deep kaonic atoms [9,10] suggesting very strong potentials, incompatible with the calculations based on chiral models, although alternative interpretations of these results have been presented [11]. See also Ref. [12] and references therein.

The model used in what follows is based on the chiral unitary model in coupled channels for the s-wave $\bar{K}N$ scattering of Ref. [13]. This model is quite successful in the description of many observables, like threshold ratios

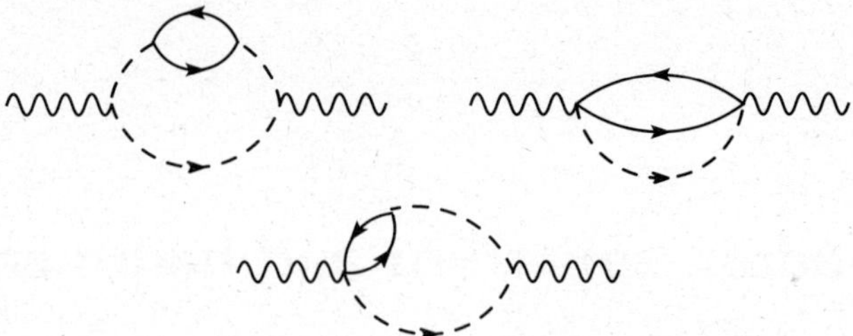

Fig. 3. Diagrams contributing to ρ selfenergy in the nuclear medium.

of K^-p to inelastic channels and K^-p cross sections in the elastic and inelastic channels. A selfconsistent calculation of the selfenergy in the nuclear medium was later carried out in Refs. [14,15] and includes the p-wave excitation of Λh, Σh and $\Sigma(1385)h$. The results of these works show an antikaon spectral function quite wide at normal nuclear densities and with the peak displaced toward low energies. The optical potential from Ref. [14] also compares well with kaonic atoms data [16] although other phenomenological potentials with a much deeper attraction also show a good agreement [17].

3 Vector mesons (ρ and ϕ)

In medium properties of the ρ meson have been extensively studied [18–26] after some predictions of scaling laws implying strong droppings of the ρ meson mass even at normal nuclear density. On the other hand, this mass reduction could be observed through the vector meson leptonic decays, free from strong final state distortions. Here, we will briefly present our approach to this topic.

The ρ meson decays in vacuum into two pions. The large modification of the pion properties in nuclei discussed in the previous section implies changes of the ρ properties, i.e., in the medium new decay channels are open because of the coupling of pions to baryon-hole excitations, see Fig. 3. Other new decay channels appear because of the direct coupling of the ρ to the nucleon, sometimes producing some baryon-hole excitation, as shown in Fig. 4. In particular, the excitation of the $N^*(1520)$ is very relevant. The contribution of these and other pieces is done using standard many body techniques. Details can be found in Refs. [27,28]. The obtained spectral function is depicted in Fig. 5. Although some strength appears at low masses, due to the coupling to $N^*(1520)$, and there is some widening, the ρ peak moves very little toward a mass higher than in vacuum. This is in contrast with popular dropping mass scenarios. Several microscopic calculations have obtained very similar results [29,30] which seem to have been confirmed by recent experimental data [31] when compared with various theoretical approaches [32].

The ϕ meson in medium properties have been studied along similar lines in Refs. [8,33]. The ϕ decays primarily in $K\bar{K}$ and in the medium the kaons have an spectral function quite different from vacuum, as discussed previously. The consideration of these effects, depicted in the

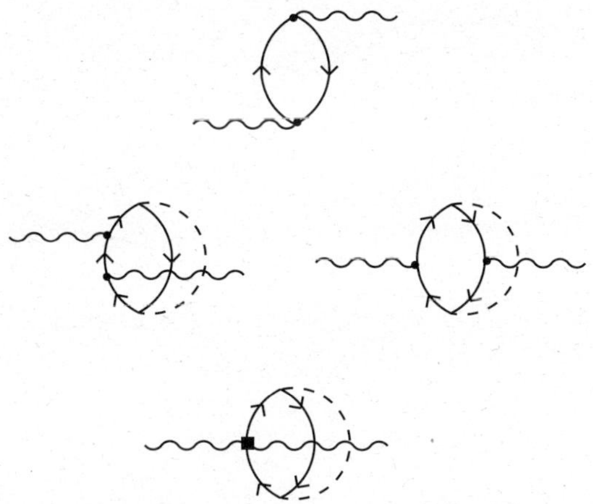

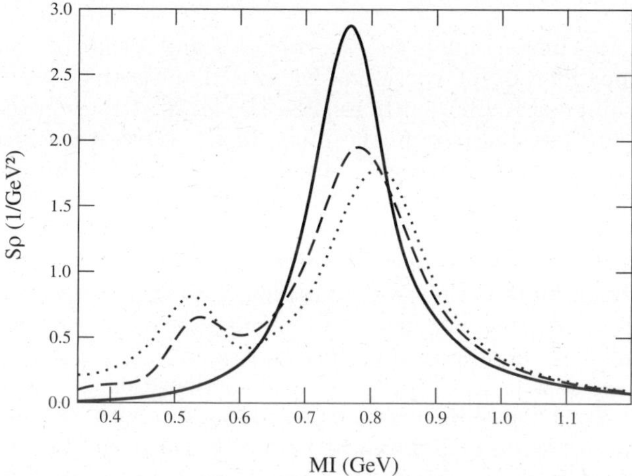

Fig. 4. Other diagrams contributing to ρ selfenergy in the nuclear medium.

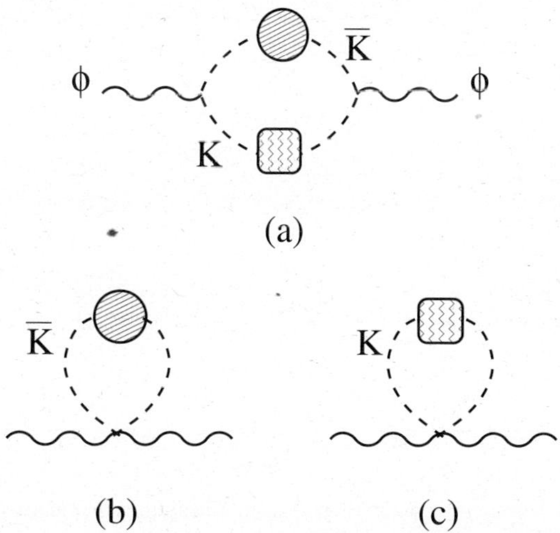

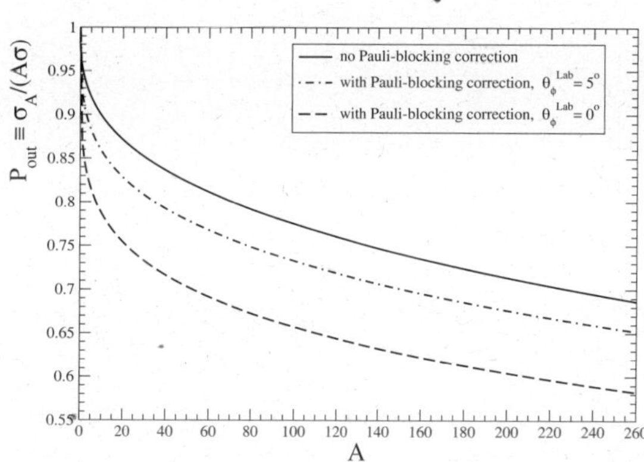

Fig. 6. Diagrams contributing to ϕ selfenergy.

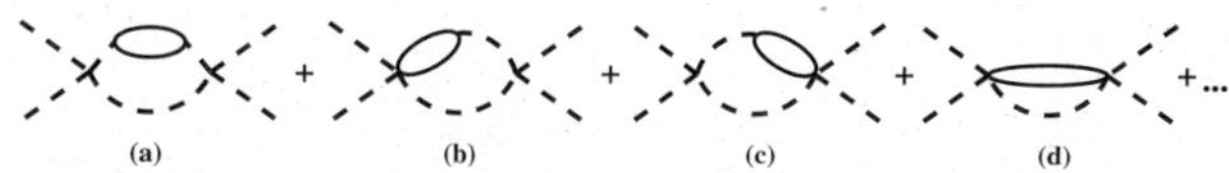

Fig. 5. ρ spectral function in vacuum: solid line, for $\rho = \rho_0/2$: dashed line and for $\rho = \rho_0$: dotted line.

Fig. 7. Ratio of Nucleus to A times nucleon cross sections, $\sigma_A/(A\sigma)$, for ϕ production as a function of A ($p_\phi = 2\,GeV$)

Fig. 8. Diagrams contributing to meson meson scattering in the nuclear medium.

diagrams shown in Fig. 6, leads to a width one order of magnitude larger than in vacuum and also to some small mass reduction. In [34], the loss of flux in photoproduction reactions was proposed as a possible experimental test of the model, see Fig. 7. Later experiments [35] have confirmed the importance of medium effects suggesting a ϕ width even larger than expected that might indicate a very large ϕN cross section [36]. The small reduction of the mass has also been confirmed recently [38] although a detailed comparison of the theoretical models with data is still to be done.

The theoretical study of the ω meson in the medium is technically more complex due to its strong coupling to three-body channels although its narrow width makes it very well suited for experimental observation. Recently, the CBELSA/TAPS Collaboration has published some interesting results showing medium effects on the ω meson [37]. A possible explanation of the data is the reduction of the ω mass. Other possibilities related to final state interactions of the detected π^0 are currently been studied.

4 Scalar mesons (σ, a_0, f_0 and κ)

One of the major successes of chiral unitary models is the description of meson-meson scattering amplitudes up to energies above 1 GeV. Some resonances appear that can be identified with the a_0 and f_0 mesons. Much further from the real axis appear the poles corresponding to the σ and κ mesons. The medium modifications of these resonant states can be obtained calculating meson meson scattering but now including the proper in medium selfenergies of the mesons as depicted in Fig. 8, apart from some vertex corrections that produce only minor effects.

The properties of the σ meson in medium are of particular interest because of theoretical predictions claiming that at normal nuclear densities it could become much lighter and narrower, due to chiral symmetry restoration, and also because of some experimental results [39,

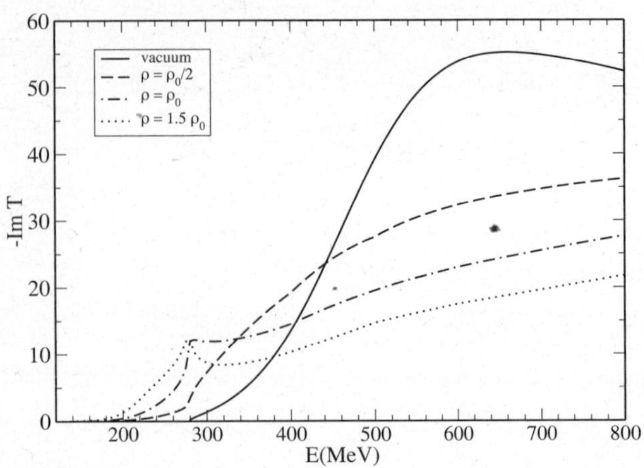

Fig. 9. Imaginary part of the pion pion scattering amplitude in the σ channel at several densities.

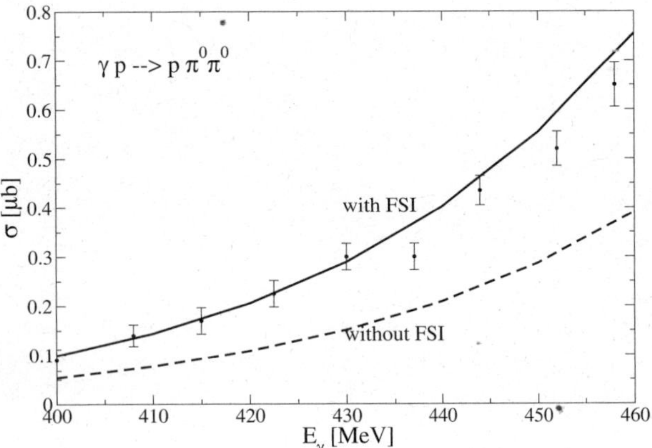

Fig. 10. Cross section for $2\,\pi^0$ photoproduction in nuclei as a function of the energy with and without final state interaction between the two pions in the I=0 channel. Experimental points from Ref. [44].

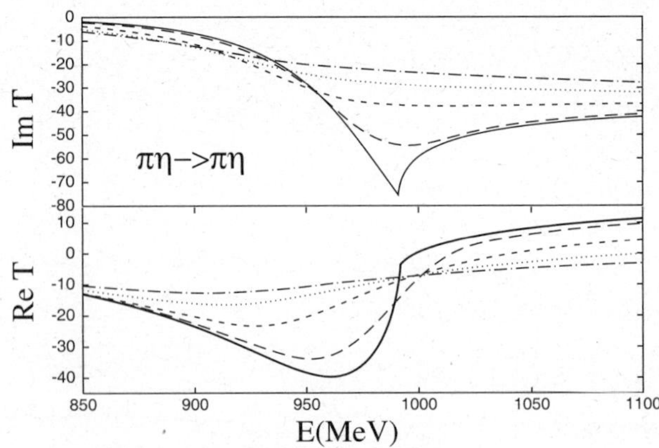

Fig. 11. Real Imaginary part of the $\pi\eta$ scattering amplitude in the I=1 channel at several densities. Solid line, free amplitude; long dashed line, $\rho = \rho_0/8$; short dashed line, $\rho = \rho_0/2$; dotted line, $\rho = \rho_0$; dashed dotted line, $\rho = 1.5\rho_0$.

40] showing an enhancement of two pion production in the isoscalar channel at very low invariant masses that might be due to the in medium σ meson. Indeed, in chiral unitary models the σ spectral function has a large enhancement at low masses [41], see Fig. 9 where the imaginary part of the $\pi\pi$ scattering amplitude is shown. This enhancement is related in these models to the position of the σ pole that at $\rho = \rho_0$ appears at low mass and closer to the real axis [42]. The calculation of two pion photoproduction making use of these techniques for the nuclear effects and starting from a good microscopical model for the reaction in vacuum, has found good agreement with experiment [43] for both total and differential cross sections.

However, the presence of additional medium effects makes difficult the isolation of possible σ effects [45,46], which are small due to the low effective density of these processes. Nonetheless, these results together with those studying pion induced two pion production [47,48] support the possibility of testing the calculations when experiments with better statistics and for heavier nuclei are carried out.

Also interesting is the case of the κ and $\bar{\kappa}$ that because of the different interaction of kaons and antikaons with the medium get a different mass[49]. The large attraction that the nucleus exerts over the pions makes these resonances lighter and also narrower. However, they do not become as narrow as to be clearly visible experimentally. Nonetheless, these results are very sensitive to the strength of the antikaon potential so controversial nowadays. If very deep antikaon potentials were confirmed that would automatically imply the existence of a low mass narrow $\bar{\kappa}$ in nuclei. In addition to the κ pole positions, also the $K\pi$ and $\bar{K}\pi$ s-wave scattering amplitudes at finite densities were studied. Strong modifications were found for both channels that also became different although they are equal in vacuum. The most noticeable effect is a peak found at low invariant masses of the $K\pi$ system that could be observed experimentally in reactions where a s-wave low energy $K\pi$ pair is produced.

Other dynamically generated mesons like the a_0 and f_0 have also been studied theoretically using the same method in Ref. [50]. These two resonances are relatively narrow and couple strongly to the $K\bar{K}$ channel. A large growth of their widths was obtained but without relevant changes in their masses. As an example, we show in Fig. 11 the $\pi\eta$ scattering amplitude. Whereas in vacuum there is a peak, related to the $a_0(980)$ pole, in medium, even at low densities the peak practically disappears.

5 Baryons

We will discuss first the case of the $\Lambda(1520)$ which is generated from the s-wave interaction of the decuplet of $3/2^+$ baryons with the octet of pseudoscalar mesons[51,52]. For a better description, the inclusion of the KN and $\pi\Sigma$ d-wave channels was also required to get a good agreement on the position of the peak and the width of the resonance[53,54], see Fig. 12.

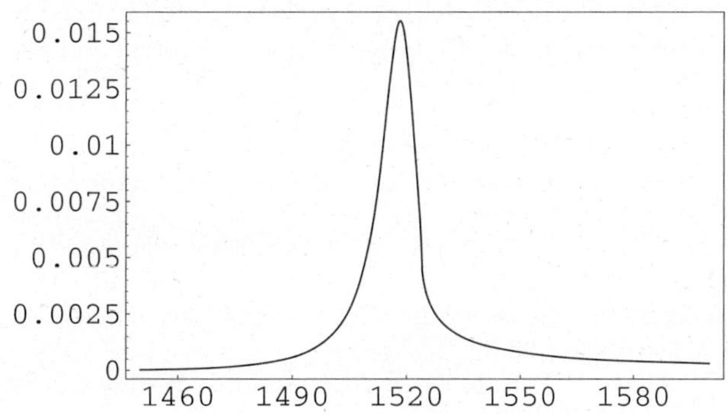

Fig. 12. $|T_{\bar{K}N\to\pi\Sigma^*}|^2$ in units of MeV^{-2} showing the $\Lambda(1520)$ peak.

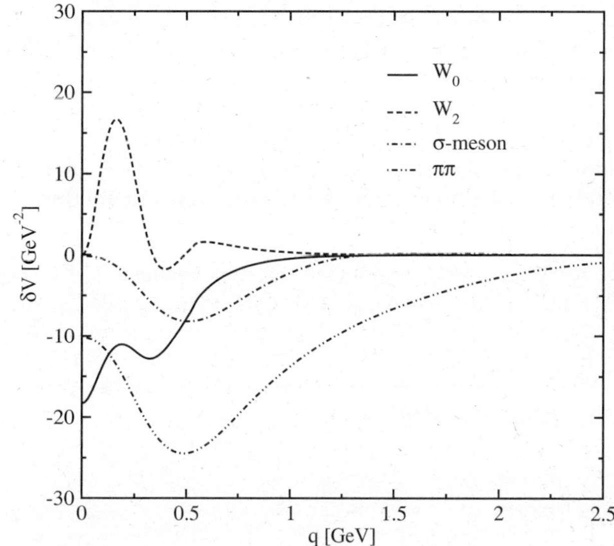

Fig. 13. Modification of the NN interaction at normal nuclear matter density in momentum space. Solid line: central part of OPEP, dashed line: tensor part of OPEP, dot-dashed: correlated σ-meson exchange and dot-dot-dashed curve: uncorrelated $\pi\pi$ exchange.

Its relatively small width makes this resonance suitable for the experimental verification of the predicted medium effects.

In vacuum, $\Lambda(1520)$ couples strongly to the $\pi\Sigma^*(1385)$ channel, even when the decay is largely suppressed because of the small phase space. However, in the nuclear medium, this decay channel is much enhanced due to both the pion and $\Sigma^*(1385)$ selfenergies, leading to a predicted width ~ 5 times larger than in vacuum at normal nuclear matter densities [55]. This large effect could be observable experimentally, studying the A-dependence of $\Lambda(1520)$ production reactions [56].

As mentioned above, the antikaon in medium selfenergy was calculated in a self-consistent way in Ref. [15]. This also implies the study of medium effects of several hyperons relevant for the calculation. Their results show a downwards shift for the Λ and Σ masses of -30 MeV at normal nuclear density and a large enhancement of the Σ^* width. The dynamically generated $\Lambda(1405)$ is strongly diluted at $\rho = \rho_0$ and remains only as a small bump close to its vacuum position.

6 In medium baryon baryon interaction

The scalar isoscalar channel is a very relevant part of the nucleon nucleon interaction, basically mediated by the exchange of two correlated (”σ”) or uncorrelated pions. The σ part of the channel is strongly modified in the nuclear medium, as shown in a previous section, and this could lead to some effects on the NN interaction in the medium. This topic was analyzed in Ref. [57]. The corrections for the NN interaction in the medium were found to be sizable, see Fig. 13. We also found that short range correlations are fundamental to produce only moderate changes. Further studies are still required to analyze the stability and properties of nuclear matter with potentials incorporating these results.

The central part of the ΛN and $\Lambda\Lambda$ potential has also been studied including the correlated, which in this case involves both σ and κ, and uncorrelated two-meson exchange[58]. We find a short range repulsion generated by the correlated two-meson potential which also produces an

attraction in the intermediate distances region. This interesting feature cannot be reproduced by the exchange of a σ meson. The uncorrelated two-meson exchange produces a sizable attraction in all cases.

7 Conclusions

In this work we have presented some recent results on the study of hadron properties. The use of χPT to incorporate low energy constraints on the scattering amplitudes and of coupled channels unitary models which generate dynamically meson and baryon resonances provides a suitable framework to study nuclear medium effects on those generated resonances. These medium effects are considered using standard techniques of many body Quantum Field Theory. In this approach, we have shown results for several scalar mesons: σ, κ, $a_0(980)$ and $f_0(980)$ and for some baryonic resonances like $\Lambda(1520)$ and $\Lambda(1405)$ and also for the controversial antikaon optical potential. Some of the results presented here already have good experimental support and others will soon be tested.

We have also discussed the scalar isoscalar part of the nucleon nucleon interaction and how this piece, partly mediated by the σ, is strongly modified in a nuclear medium. It must be considered that in our approach the σ is only a resonant two pions state and the interaction resulting from the exchange of this resonance cannot be fitted by a single meson exchange.

Finally, the pion and kaon in medium selfenergies have important consequences for the decay of vector mesons that have also been analysed with our methods providing an alternative picture to other calculations. We find quite small mass changes, consistent with many experimental data, but large width enhancements and even complicated spectral functions with several bumps produced by the strong coupling of the vector mesons to some particular excitation mode. This shows the need of detailed microscopic calculations that properly take into account the relevant degrees of freedom in the nuclear medium.

This work is partly supported by the Spanish CSIC and JSPS collaboration, the DGICYT contract number BFM 2003-00856, and the E.U. EURIDICE network contract no. HPRN - CT -2002-00311. This research is part of the EU Integrated Infrastructure Initiative Hadron Physics Project under contract number RII3 - CT -2004-506078.

References

1. J. A. Oller and E. Oset, Nucl. Phys. A **620** (1997) 438 [Erratum-ibid. A **652** (1999) 407]; J. A. Oller, E. Oset and J. R. Pelaez, Phys. Rev. D **59** (1999) 074001 [Erratum-ibid. D **60** (1999) 099906].
2. M. F. M. Lutz and E. E. Kolomeitsev, Nucl. Phys. A **700** (2002) 193.
3. E. Oset, A. Ramos and C. Bennhold, Phys. Lett. B **527** (2002) 99 [Erratum-ibid. B **530** (2002) 260].
4. C. Garcia-Recio, J. Nieves, E. Ruiz Arriola and M. J. Vicente Vacas, Phys. Rev. D **67**, 076009 (2003).
5. B. Borasoy, R. Nissler and W. Weise, Phys. Rev. Lett. **94** (2005) 213401.
6. T. Inoue, E. Oset and M. J. Vicente Vacas, Phys. Rev. C **65**, 035204 (2002).
7. Torleif E.O. Ericson and W. Weise, *Pions And Nuclei* Clarendon, Oxford, UK (1988). (The International Series of Monographs on Physics, 74); E. Oset, P. Fernandez de Cordoba, L. L. Salcedo and R. Brockmann, Phys. Rept. **188** (1990) 79.
8. E. Oset and A. Ramos, Nucl. Phys. A **679** (2001) 616.
9. T. Suzuki et al., Phys. Lett. B **597** (2004) 263.
10. M. Agnello et al., Phys. Rev. Lett. **94** (2005) 212303.
11. E. Oset and H. Toki, Phys. Rev. C **74**, 015207 (2006); V. K. Magas, E. Oset, A. Ramos and H. Toki, Phys. Rev. C **74**, 025206 (2006).
12. J. Mares, E. Friedman and A. Gal, Nucl. Phys. A **770** (2006) 84.
13. E. Oset and A. Ramos, Nucl. Phys. A **635** (1998) 99.
14. A. Ramos and E. Oset, Nucl. Phys. A **671** (2000) 481.
15. L. Tolos, A. Ramos and E. Oset, arXiv:nucl-th/0609046.
16. A. Baca, C. Garcia-Recio and J. Nieves, Nucl. Phys. A **673** (2000) 335.
17. A. Cieply, E. Friedman, A. Gal and J. Mares, Nucl. Phys. A **696** (2001) 173.
18. T. Hatsuda and S. H. Lee, Phys. Rev. **C46** (1992) 34.
19. S. Leupold, W. Peters and U. Mosel, Nucl. Phys. **A628** (1998) 311.
20. F. Klingl, N. Kaiser and W. Weise, Nucl. Phys. **A624** (1997) 527.
21. S. Mallik and A. Nyffeler, Phys. Rev. C **63** (2001) 065204.
22. M. Asakawa, C. M. Ko, P. Levai and X. J. Qiu, Phys. Rev. **C46** (1992) 1159; M. Asakawa and C. M. Ko, Phys. Rev. **C48** (1993) 526.
23. G. Chanfray and P. Schuck, Nucl. Phys. **A555** (1993) 329.
24. M. Herrmann, B. L. Friman and W. Norenberg, Nucl. Phys. **A560** (1993) 411.
25. W. Peters, M. Post, H. Lenske, S. Leupold and U. Mosel, Nucl. Phys. **A632** (1998) 109.
26. M. Lutz, B. Friman and G. Wolf, Nucl. Phys. **A661** (1999) 526.
27. D. Cabrera, E. Oset and M. J. Vicente-Vacas, Acta Phys. Polon. B **31** (2000) 2167.
28. D. Cabrera, E. Oset and M. J. Vicente Vacas, Nucl. Phys. A **705** (2002) 90.
29. M. Urban, M. Buballa, R. Rapp and J. Wambach, Nucl. Phys. A **641** (1998) 433.
30. M. Post, S. Leupold and U. Mosel, Nucl. Phys. A **689** (2001) 753.
31. S. Damjanovic et al., Nucl. Phys. A **774** (2006) 715.
32. H. van Hees and R. Rapp, arXiv:hep-ph/0604269.
33. D. Cabrera and M. J. Vicente Vacas, Phys. Rev. C **67** (2003) 045203.
34. D. Cabrera, L. Roca, E. Oset, H. Toki and M. J. Vicente Vacas, Nucl. Phys. A **733** (2004) 130.
35. J. K. Ahn et al., Phys. Lett. B **608** (2005) 215.
36. P. Muhlich and U. Mosel, Nucl. Phys. A **765** (2006) 188.
37. D. Trnka et al., Phys. Rev. Lett. **94** (2005) 192303.
38. R. Muto et al., arXiv:nucl-ex/0511019.
39. P. Camerini, N. Grion, R. Rui and D. Vetterli, Nucl. Phys. A **552** (1993) 451 [Erratum-ibid. A **572** (1994) 791].
40. J. G. Messchendorp et al., Phys. Rev. Lett. **89** (2002) 222302.
41. H. C. Chiang, E. Oset and M. J. Vicente-Vacas, Nucl. Phys. A **644** (1998) 77.
42. D. Cabrera, E. Oset and M. J. Vicente Vacas, Phys. Rev. C **72** (2005) 025207.
43. L. Roca, E. Oset and M. J. Vicente Vacas, Phys. Lett. B **541** (2002) 77.
44. M. Wolf et al., Eur. Phys. J. A **9** (2000) 5.
45. P. Muhlich, L. Alvarez-Ruso, O. Buss and U. Mosel, Phys. Lett. B **595** (2004) 216.
46. L. Alvarez-Ruso, P. Muhlich, O. Buss and U. Mosel, Int. J. Mod. Phys. A **20** (2005) 578.
47. R. Rapp et al., Phys. Rev. C **59** (1999) 1237.
48. M. J. Vicente Vacas and E. Oset, Phys. Rev. C **60** (1999) 064621.
49. D. Cabrera and M. J. Vicente Vacas, Phys. Rev. C **69** (2004) 065204.
50. E. Oset and M. J. Vicente Vacas, Nucl. Phys. A **678** (2000) 424.
51. E. E. Kolomeitsev and M. F. M. Lutz, Phys. Lett. B **585** (2004) 243.
52. S. Sarkar, E. Oset and M. J. Vicente Vacas, Nucl. Phys. A **750**, 294 (2005).
53. S. Sarkar, E. Oset and M. J. Vicente Vacas, Phys. Rev. C **72**, 015206 (2005).
54. L. Roca, S. Sarkar, V. K. Magas and E. Oset, Phys. Rev. C **73** (2006) 045208.
55. M. Kaskulov and E. Oset, Phys. Rev. C **73** (2006) 045213.
56. M. Kaskulov, L. Roca and E. Oset, Eur. Phys. J. A **28** (2006) 139.
57. M. M. Kaskulov, E. Oset and M. J. Vicente Vacas, Phys. Rev. C **73**, 014004 (2006).
58. K. Sasaki, E. Oset and M. J. Vicente Vacas, arXiv:nucl-th/0607068.

Eur. Phys. J. A **31**, 691–694 (2007)
DOI 10.1140/epja/i2006-10242-2

THE EUROPEAN
PHYSICAL JOURNAL A

Special Article – QNP 2006

Doubly heavy-quark baryon spectroscopy and semileptonic decay

C. Albertus[1], E. Hernández[2], J. Nieves[1], and J.M. Verde-Velasco[2,a]

[1] Departamento de Física Atómica, Molecular y Nuclear, Universidad de Granada, E-18071 Granada, Spain
[2] Grupo de Física Nuclear, Departamento de Física Fundamental e IUFFyM, Facultad de Ciencias, E-37008 Salamanca, Spain

Received: 8 November 2006
Published online: 12 March 2007 – © Società Italiana di Fisica / Springer-Verlag 2007

Abstract. Working in the framework of a nonrelativistic quark model we evaluate the spectra and semileptonic decay widths for the ground state of doubly heavy Ξ and Ω baryons. We solve the three-body problem using a variational ansatz made possible by the constraints imposed by heavy-quark spin symmetry. In order to check the dependence of our results on the inter-quark interaction, we have used five different quark-quark potentials which include Coulomb and hyperfine terms coming from one-gluon exchange, plus a confining term. Our results for the spectra are in good agreement with a previous calculation done using a Faddeev approach. For the semileptonic decay our results for the total decay widths are in good agreement with the ones obtained within a relativistic quark model in the quark-diquark approximation.

PACS. 12.39.Jh Nonrelativistic quark model – 12.40.Yx Hadron mass models and calculations – 13.30.Ce Leptonic, semileptonic, and radiative decays

1 Introduction

Even though only recently the mass of a baryon with two heavy quarks has been measured experimentally [1], these systems have been being studied for more than a decade. Working with a system with two heavy quarks one can take advantage of the constraints imposed by heavy-quark spin symmetry (HQSS). This symmetry amounts to the decoupling of the heavy-quark spins in the infinity heavy-quark mass limit. In that limit one can consider the total spin of the two heavy-quarks subsystem to be well defined. This result, that we shall assume to be valid for the actual heavy-quark masses, will simplify the solution of the three-body problem.

In this contribution we shall present results for masses and total semileptonic decay widths. We have also analyzed other static observables as well as form factors, differential decay widths and angular asymmetries of the weak decays. For a detailed account of the full calculation see ref. [2]

In table 1 we summarize the quantum numbers of the baryons considered in this study.

2 The model

Once the centre of mass (CM) motion has been removed, the intrinsic Hamiltonian that describes the inner dynam-

Table 1. Quantum numbers of doubly heavy baryons analyzed in this study. S, J^P are the strangeness and the spin parity of the baryon, I is the isospin, and S_h^π is the spin parity of the heavy degrees of freedom. l denotes a light u or d quark.

Baryon	S	J^P	I	S_h^π	Quark content
Ξ_{cc}	0	$\frac{1}{2}^+$	$\frac{1}{2}$	1^+	ccl
Ξ_{cc}^*	0	$\frac{3}{2}^+$	$\frac{1}{2}$	1^+	ccl
Ω_{cc}	-1	$\frac{1}{2}^+$	0	1^+	ccs
Ω_{cc}^*	-1	$\frac{3}{2}^+$	0	1^+	ccs
Ξ_{bb}	0	$\frac{1}{2}^+$	$\frac{1}{2}$	1^+	bbl
Ξ_{bb}^*	0	$\frac{3}{2}^+$	$\frac{1}{2}$	1^+	bbl
Ω_{bb}	-1	$\frac{1}{2}^+$	0	1^+	bbs
Ω_{bb}^*	-1	$\frac{3}{2}^+$	0	1^+	bbs
Ξ_{bc}'	0	$\frac{1}{2}^+$	$\frac{1}{2}$	0^+	bcl
Ξ_{bc}	0	$\frac{1}{2}^+$	$\frac{1}{2}$	1^+	bcl
Ξ_{bc}^*	0	$\frac{3}{2}^+$	$\frac{1}{2}$	1^+	bcl
Ω_{bc}'	-1	$\frac{1}{2}^+$	0	0^+	bcs
Ω_{bc}	-1	$\frac{1}{2}^+$	0	1^+	bcs
Ω_{bc}^*	-1	$\frac{3}{2}^+$	0	1^+	bcs

[a] e-mail: jmverde@usal.es

ics of the baryon is given by

$$H^{\text{int}} = \sum_{j=1,2} H_j^{sp} + V_{h_1 h_2}(\boldsymbol{r}_1 - \boldsymbol{r}_2, spin) - \frac{\boldsymbol{\nabla}_1 \cdot \boldsymbol{\nabla}_2}{m_q} + \overline{M},$$

$$H_j^{sp} = -\frac{\boldsymbol{\nabla}_j^2}{2\mu_j} + V_{h_j q}(\boldsymbol{r}_j, spin), \quad j = 1,2, \tag{1}$$

where $\boldsymbol{r}_1$, $\boldsymbol{r}_2$ are the relative positions of the h_1, h_2 heavy quarks with respect to the light quark q, $\overline{M} = m_{h_1} + m_{h_2} + m_q$, $\mu_j = (1/m_{h_j} + 1/m_q)^{-1}$ and $\boldsymbol{\nabla}_j = \partial/\partial_{\boldsymbol{r}_j}$, $j = 1,2$. $V_{h_j q}$ and $V_{h_1 h_2}$ are the heavy-light and heavy-heavy interaction potentials. Note the presence of the Hughes-Eckart term that results from the separation of the CM motion.

For the quark-quark interaction we have considered five different phenomenological potentials, one suggested by Bhaduri and collaborators [3] (BD) and four suggested by B. Silvestre-Brac and C. Semay [4,5] (AL1, AL2, AP1 and AP2). All of them include Coulomb and hyperfine terms coming from one-gluon exchange and a confining term, and differ in the form factor used for the hyperfine term, the use of a form factor in the one-gluon exchange Coulomb term or in the power of the confinement term. All free parameters had been adjusted to reproduce the light and heavy-light meson spectra. Details on the potentials can be found in refs. [3–5].

For the interactions considered, the total spin and internal orbital angular momentum commute with the intrinsic Hamiltonian, and thus are well defined. In this work we will study the ground state of baryons with total angular momentum $J = 1/2$, $3/2$ so we can assume the orbital angular momentum to be 0. This implies that the spatial wave function only depend on r_1, r_2 and $r_{12} = |\boldsymbol{r}_1 - \boldsymbol{r}_2|$. We will also assume that taking the total spin of the heavy degrees of freedom to be well defined, as obtained in the infinite heavy-quark mass limit, is a good approximation. That will allow us to write the wave function in a simple way (see ref. [2] for details).

The spatial part of wave function will be determined using a variational method in which we will assume the following functional form:

$$\Psi_{h_1 h_2}^B(r_1, r_2, r_{12}) = N\, F^B(r_{12})\, \phi_{h_1 q}(r_1)\, \phi_{h_2 q}(r_2), \tag{2}$$

where N is a normalization constant, $\phi_{h_j q}$ is the S-wave ground-state wave function $\varphi_j(r_j)$ of the single-particle Hamiltonian H_j^{sp} corrected at large distances:

$$\phi_{h_j q}(r_j) = (1 + \alpha_j r_j)\, \varphi_j(r_j), \quad j = 1,2. \tag{3}$$

The heavy-heavy Jastrow correlation function F^B will be given as a linear combination of Gaussians:

$$F^B(r_{12}) = \sum_{j=1}^{4} a_j e^{-b_j^2(r_{12}+d_j)^2}, \quad a_1 = 1, \tag{4}$$

where α_i, a_i $i \neq 1$, b_i and d_i are free variational parameters. The values that we get for the variational parameters are compiled in ref. [2].

We have also used the wave function obtained in this model to study different doubly $B(1/2^+) \rightarrow B'(1/2^+)$ baryon semileptonic decays involving a $b \rightarrow c$ transition at the quark level. We have worked in the spectator approximation with only one-body currents.

The differential decay width reads

$$d\Gamma = 8|V_{cb}|^2 m_{B'} G_F^2 \frac{d^3 p'}{(2\pi)^3 2E'_{B'}} \frac{d^3 k}{(2\pi)^3 2E_{\bar{\nu}_l}} \frac{d^3 k'}{(2\pi)^3 2E'_l} (2\pi)^4$$

$$\times \delta^4(p - p' - k - k')\, \mathcal{L}^{\alpha\beta}(k, k')\mathcal{H}_{\alpha\beta}(p, p'), \tag{5}$$

where $|V_{cb}|$ is the modulus of the corresponding Cabibbo-Kobayashi-Maskawa matrix element, $m_{B'}$ is the mass of the final baryon, G_F is the Fermi decay constant, p, p', k and k' are the four-momenta of the initial baryon, final baryon, final anti-neutrino and final lepton, respectively, and $\mathcal{L}$ and $\mathcal{H}$ are the lepton and hadron tensors.

The lepton tensor is given as

$$\mathcal{L}^{\mu\sigma}(k, k') = k'^\mu k^\sigma + k'^\sigma k^\mu - g^{\mu\sigma} k \cdot k' + i\epsilon^{\mu\sigma\alpha\beta} k'_\alpha k_\beta \tag{6}$$

where we use the convention $\epsilon^{0123} = -1$, $g^{\mu\mu} = (+,-,-,-)$.

The hadron tensor is given as

$$\mathcal{H}_{\mu\sigma}(p, p') =$$
$$\frac{1}{2} \sum_{r,r'} \left\langle B', r'\, \boldsymbol{p}' \,\middle|\, \overline{\Psi}^c(0)\gamma_\mu(I - \gamma_5)\Psi^b(0) \,\middle|\, B, r\, \boldsymbol{p} \right\rangle$$
$$\times \left\langle B', r'\, \boldsymbol{p}' \,\middle|\, \overline{\Psi}^c(0)\gamma_\sigma(I - \gamma_5)\Psi^b(0) \,\middle|\, B, r\, \boldsymbol{p} \right\rangle^* \tag{7}$$

with $|B, r\, \boldsymbol{p}\rangle$ ($|B', r'\, \boldsymbol{p}'\rangle$) representing the initial (final) baryon with three-momentum $\boldsymbol{p}$ ($\boldsymbol{p}'$) and spin index r (r'). The baryon states are normalized such that

$$\langle r\, \boldsymbol{p} | r'\, \boldsymbol{p}' \rangle = (2\pi)^3 (E(\boldsymbol{p})/m)\, \delta_{rr'}\, \delta^3(\boldsymbol{p} - \boldsymbol{p}'). \tag{8}$$

We compute the widths similarly as we did in ref. [6] for baryons with a heavy quark.

3 Results and discussion

The mass of the baryon is simply given by the expectation value of the intrinsic Hamiltonian. In table 2 we give our results for doubly heavy Ξ baryons, while in table 3 are the results for the doubly heavy Ω ones. Our central values correspond to the results obtained using the AL1 potential, while the errors quoted take into account the variations found when using the other potentials. That also applies to the quoted results for ref. [4], obtained with the same interaction potentials but within a Faddeev approach. When comparison with this work is possible we find an excellent agreement between the two calculations. Besides we give predictions for states not considered in the study of ref. [4]. We also compare with other theoretical models. All calculations give similar results that vary within a few percent. From the experimental side only the mass of the Ξ_{cc} has been measured. The experimental

Table 2. Doubly heavy Ξ masses in MeV.

	This work	[4]	Exp. [1]	Lattice [7]
Ξ_{cc}	3612^{+17}	3609^{+22}	3519 ± 1	3549 ± 95
Ξ_{cc}^*	3706^{+23}			3641 ± 97
Ξ_{bb}	10197^{+10}_{-17}	10194^{+10}_{-19}		
Ξ_{bb}^*	10236^{+9}_{-17}			
Ξ_{bc}	6919^{+17}_{-7}	6916^{+18}_{-9}		
Ξ_{bc}'	6948^{+17}_{-6}			
Ξ_{bc}^*	6986^{+14}_{-5}			

	This work	[8]	[9]	[10]	[11]	[12]	[13]
Ξ_{cc}	3612^{+17}	3620	3480	3740	3478	3660	3524
Ξ_{cc}^*	3706^{+23}	3727	3610	3860	3610	3740	3548
Ξ_{bb}	10197^{+10}_{-17}	10202	10090	10300	10093	10340	
Ξ_{bb}^*	10236^{+9}_{-17}	10237	10130	10340	10133	10370	
Ξ_{bc}	6919^{+17}_{-7}	6933	6820	7010	6820	7040	
Ξ_{bc}'	6948^{+17}_{-6}	6963	6850	7070	6850	6990	
Ξ_{bc}^*	6986^{+14}_{-5}	6980	6900	7100	6900	7060	

Table 3. Doubly heavy Ω masses in MeV.

	This work	[4]	Lattice [7]
Ω_{cc}	3702^{+41}	3711^{+30}_{-2}	3663 ± 97
Ω_{cc}^*	3783^{+22}		3734 ± 98
Ω_{bb}	10260^{+14}_{-34}		
Ω_{bb}^*	10297^{+5}_{-28}		
Ω_{bc}	6986^{+27}_{-17}	7003^{+20}_{-32}	
Ω_{bc}'	7009^{+24}_{-15}		
Ω_{bc}^*	7046^{+11}_{-9}		

	This work	[8]	[9]	[10]	[11]	[12]
Ω_{cc}	3702^{+41}	3778	3590	3760	3590	3740
Ω_{cc}^*	3783^{+22}	3872	3690	3900	3690	3820
Ω_{bb}	10260^{+14}_{-34}	10359	10180	10340	10180	10370
Ω_{bb}^*	10297^{+5}_{-28}	10389	10200	10380	10200	10400
Ω_{bc}	6986^{+27}_{-17}	7088	6910	7050	6910	7090
Ω_{bc}'	7009^{+24}_{-15}	7116	6930	7110	6930	7060
Ω_{bc}^*	7046^{+11}_{-9}	7130	6990	7130	6990	7120

Table 4. Semileptonic decay widths in units of 10^{-14} GeV. We have used $|V_{cb}| = 0.0413$. l stands for $l = e, \mu$.

	This work	[14]	[15]	[16]	[17]
$\Gamma(\Xi_{bb} \to \Xi_{bc}\, l\bar{\nu}_l)$	$3.84^{+0.49}_{-0.10}$	3.26		28.5	
$\Gamma(\Xi_{bc} \to \Xi_{cc}\, l\bar{\nu}_l)$	$5.13^{+0.51}_{-0.05}$	4.59	0.79	8.93	4.0
$\Gamma(\Xi_{bb} \to \Xi_{bc}'\, l\bar{\nu}_l)$	$2.12^{+0.26}_{-0.05}$	1.64		4.28	
$\Gamma(\Xi_{bc}' \to \Xi_{cc}\, l\bar{\nu}_l)$	$2.71^{+0.19}_{-0.05}$	1.76		7.76	

	This work	[14]	[16]
$\Gamma(\Omega_{bb} \to \Omega_{bc}\, l\bar{\nu}_l)$	$4.28^{+0.39}_{-0.03}$	3.40	28.8
$\Gamma(\Omega_{bc} \to \Omega_{cc}\, l\bar{\nu}_l)$	$5.17^{+0.39}$	4.95	
$\Gamma(\Omega_{bb} \to \Omega_{bc}'\, l\bar{\nu}_l)$	$2.32^{+0.26}$	1.66	
$\Gamma(\Omega_{bc}' \to \Omega_{cc}\, l\bar{\nu}_l)$	$2.71^{+0.17}$	1.90	

tions appear for the BD potential, with differences of the order of 7–12%. We compare our results with the predictions of different models. For that purpose we need to fix a value for $|V_{cb}|$ for which we take $|V_{cb}| = 0.0413$. Our results are in reasonable agreement with the ones in ref. [14] where they use a relativistic quark model evaluated in the quark-diquark approximation. For $\Gamma(\Xi_{bc} \to \Xi_{cc})$ we also agree with the value of ref. [17] obtained using heavy-quark effective theory. A much smaller value for the same width is obtained in the relativistic three-quark model calculation of ref. [15]. In ref. [16], where they use the Bethe-Salpeter equation applied to a quark-diquark system, they obtain much larger results for all transitions.

This research was supported by DGI and FEDER funds, under contracts FIS2005-00810, BFM2003-00856 and FPA2004-05616, by Junta de Andalucía and Junta de Castilla y León under contracts FQM0225 and SA104/04, and it is part of the EU integrated infrastructure initiative Hadron Physics Project under contract number RII3-CT-2004-506078. C.A. acknowledges a research contract with Universidad de Granada. J.M.V.-V. acknowledges an E.P.I.F. contract with Universidad de Salamanca.

value for $M_{\Xi_{cc}}$ obtained by the SELEX Collaboration [1] is 100 MeV smaller than our result. Note nevertheless that no account is given of the systematic error. There are also lattice calculations, by the UKQCD Collaboration [7], of the masses of the doubly charmed Ξ_{cc}, Ξ_{cc}^*, Ω_{cc} and Ω_{cc}^* baryons. Our results are within errors of the lattice determinations.

In table 4 we present our results for the semileptonic decay widths for the different processes under study. Our central values again correspond to the results obtained using the AL1 potential, while the errors show the variations when using the other four potentials. The biggest varia-

References

1. SELEX Collaboration (M. Mattson *et al.*), Phys. Rev. Lett. **89**, 112001 (2002).
2. C. Albertus, E. Hernández, J. Nieves, J.M. Verde-Velasco, hep-ph/0610030, submitted to Phys. Rev. D.
3. R.K. Bhaduri, L.E. Cohler, Y. Nogami, Nuovo Cimento A **65**, 376 (1981).
4. B. Silvestre-Brac, Few-Body Syst. **20**, 1 (1996).
5. C. Semay, B. Silvestre-Brac, Z. Phys. C **61**, 271 (1994).

6. C. Albertus, E. Hernandez, J. Nieves, Phys. Rev. D **71**, 014012 (2005).

7. UKQCD Collaboration (J.M. Flynn, F. Mescia, A.S.B. Tariq), JHEP **0307**, 066 (2003).

8. D. Ebert, R.N. Faustov, V.O. Galkin, A.P. Martynenko, Phys. Rev. D **66**, 014008 (2002).

9. V.V. Kiselev, A.K. Likhoded, Phys. Usp. **45**, 455 (2002) (Usp. Fiz. Nauk **172**, 497 (2002)) arXiv:hep-ph/0103169.

10. S.-P. Tong, Y.-B. Ding, X.-H. Guo, H.-Y. Jin, X.-Q. Li, P.-N. Shen, R. Zhang, Phys. Rev. D **62**, 054024 (2000).

11. S.S. Gershtein, V.V. Kiselev, A.K. Likhoded, A.I. Onishchenko, Phys. Rev. D **62**, 054021 (2000).

12. R. Roncaglia, D.B. Lichtenberg, E. Predazzi, Phys. Rev. D **52**, 1722 (1995); R. Roncaglia, A. Dzierba, D.B. Lichtenberg, E. Predazzi, Phys. Rev. D **51**, 1248 (1995).

13. J. Vijande, H. Garcilazo, A. Valcarce, F. Fernandez, Phys. Rev. D **70**, 054022 (2004).

14. D. Ebert, R.N. Faustov, V.O. Galkin, A.P. Martynenko, Phys. Rev. D **70**, 014018 (2004).

15. A. Faessler, Th. Gutsche, M.A. Ivanov, J.G. Körner, V.E. Lyubovitdkij, Phys. Lett. B **518**, 55 (2001).

16. X.-H. Guo, H.-Y. Jin, X.-Q. Li, Phys. Rev. D **58**, 114007 (1998).

17. M.A. Sanchis-Lozano, Nucl. Phys. B **440**, 251 (1995).

Eur. Phys. J. A **31**, 695–697 (2007)

DOI 10.1140/epja/i2006-10174-9

THE EUROPEAN
PHYSICAL JOURNAL A

Special Article – QNP 2006

Hadron spectroscopy at BABAR

S. Tosi[a]

On behalf of the BABAR Collaboration
Università di Genova, Dipartimento di Fisica and INFN, I-16146 Genova, Italy

Received: 24 September 2006
Published online: 16 February 2007 – © Società Italiana di Fisica / Springer-Verlag 2007

Abstract. The high integrated luminosity collected by the BABAR detector at the SLAC PEP-II e^+e^- B-Factory offers an excellent opportunity for the study of heavy-quark spectroscopy. A selection of the most recent results reported by BABAR will be presented, focussing on recently observed states with both open- and hidden-charm content.

PACS. 13.25.Gv Decays of J/ψ, Υ, and other quarkonia – 13.25.Ft Decays of charmed mesons – 13.66.Bc Hadron production in e^-e^+ interactions

1 Introduction

The B-Factory experiments offer an excellent opportunity for hadronic-spectroscopy studies. In fact, they accumulated huge data samples, more than $390\,\mathrm{fb}^{-1}$ in the case of BABAR, and allow to exploit various hadron-production mechanisms: not only B-meson decays, but also $e^+e^- \to q\bar{q}$ events ($q = u,d,s,c$), events with initial state radiation (ISR) and $\gamma\gamma$ collisions, for which all a high efficiency is maintained.

2 States with $c\bar{s}$ content

Recently BABAR [1] and CLEO [2] discovered two narrow mesons with $c\bar{s}$ content in $e^+e^- \to c\bar{c}$ events, around a mass of $2320\,\mathrm{MeV}/c^2$ ($D_{sJ}^*(2320)^+$ [3]) and $2460\,\mathrm{MeV}/c^2$ ($D_{sJ}(2460)^+$), respectively, later confirmed by Belle [4]. They could be the missing 0^+ and 1^+ levels of the D_s-meson spectrum, but their masses are quite below the expectations, namely they were expected to lie above the DK and D^*K thresholds, thus being very wide. In addition, they have been discovered in the $D_s^{(*)+}\pi^0$ decay modes, that violate isospin (I) conservation for D_s states. Less conventional interpretations have been proposed for these two particles, including tetraquarks [5] and molecular states [6]. BABAR carried on an extensive program aimed at improving the knowledge of their properties. Precision measurements of the masses of the two states have been reported, $(2319.6 \pm 0.2 \pm 1.4)$ and $(2460.2 \pm 0.2 \pm 0.8)\,\mathrm{MeV}/c^2$, as well as upper limits on the total widths of 3.8 and $3.5\,\mathrm{MeV}/c^2$, at the 95% C.L., respectively [7].

The decay channels $D_s^+\pi^0$, $D_s^+\gamma$, $D_s^+\pi^0\gamma$, $D_s^+\pi^+\pi^-$ and $D_s^+\pi^0\pi^0$ have been studied. The $D_{sJ}^*(2320)^+$ has only been observed to decay to $D_s^+\pi^0$, whereas the $D_{sJ}(2460)^+$ has been observed in the $D_s^+\gamma$, $D_s^+\pi^0\gamma$ and $D_s^+\pi^+\pi^-$ decay modes. No sign of $D_{sJ}(2460)^+$ has been observed in the $D_s^+\pi^0\pi^0$ final state: this is suggestive that the $\pi^+\pi^-$ pair in the $D_s^+\pi^+\pi^-$ final state has isospin 1, hence that isospin is not conserved in the $D_s^+\pi\pi$ decay. Also, isospin partners of $D_{sJ}^*(2317)^+$ with 0 or 2 charge, foreseen in molecular models, have been sought for in the $D_s^+\pi^\pm$ decay modes, but no hint of any signal has been reported, implying null isospin.

Also, B-meson decays to $D_{sJ}(2460)^+$ have been studied [8]. By fully reconstructing one of the two B-mesons produced in the process $e^+e^- \to \Upsilon(4S) \to B\overline{B}$, and a $D^{(*)}$-meson in its recoil, a clear signal of $D_{sJ}(2460)^+$ was reported in the missing-mass spectrum (fig. 1). Using previously measured products of branching fractions $\mathcal{B}(B \to D_{sJ}(2460)^+D^{(*)}) \times \mathcal{B}(D_{sJ}(2460)^+ \to f)$ [9], this technique allowed to derive absolute measurements of the branching fractions for the $D_{sJ}(2460)^+$ decay channels f: $\mathcal{B}(D_{sJ}(2460)^+ \to D_s^+\pi^+\pi^-) = (0.04 \pm 0.01)$, $\mathcal{B}(D_{sJ}(2460)^+ \to D_s^+\gamma) = (0.16 \pm 0.04 \pm 0.03)$, $\mathcal{B}(D_{sJ}(2460)^+ \to D_s^{*+}\pi^0) = (0.56\pm0.13\pm0.09)$: these add up to $76\pm17\%$ of the total $D_{sJ}(2460)^+$ decay width. In addition, the helicity distribution of $D_{sJ}(2460)^+ \to D_s^+\gamma$ in fully reconstructed $B \to D_{sJ}D$ decays favours the hypothesis that the $D_{sJ}(2460)^+$ angular momentum J be 1 [9].

Summing up all this information, one can conclude that $D_{sJ}^*(2320)^+$ has been established to have $I = 0$, $J = 0$, positive parity, and to mostly decay to $D_s^+\pi^0$, while $D_{sJ}(2460)^+$ is favoured to have $I = 0$, $J = 1$ and positive parity. In addition, its absolute branching fractions have been measured. These states indeed appear to

^{a} e-mail: tosi@ge.infn.it

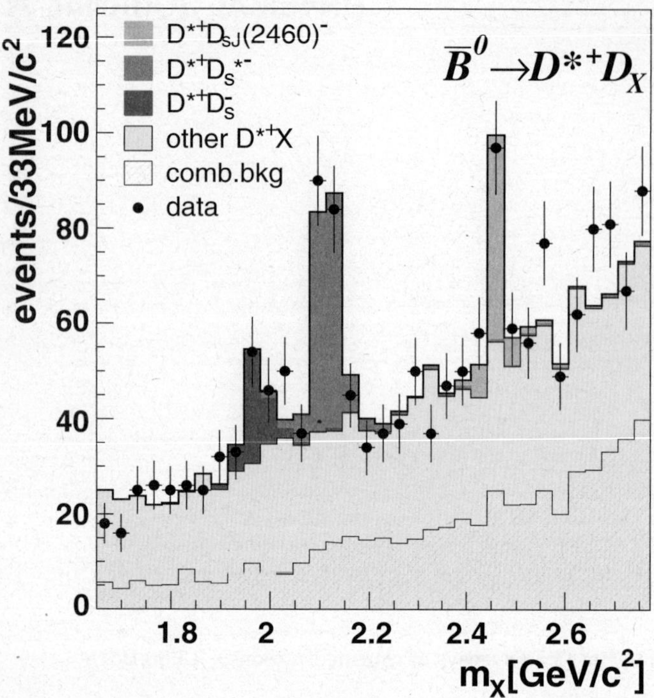

Fig. 1. Distribution of the missing mass (points) in events with a fully reconstructed $\bar{B}^0$- and D^{*+}-meson, overlaid to the result of a fit to a signal plus background hypothesis (histogram).

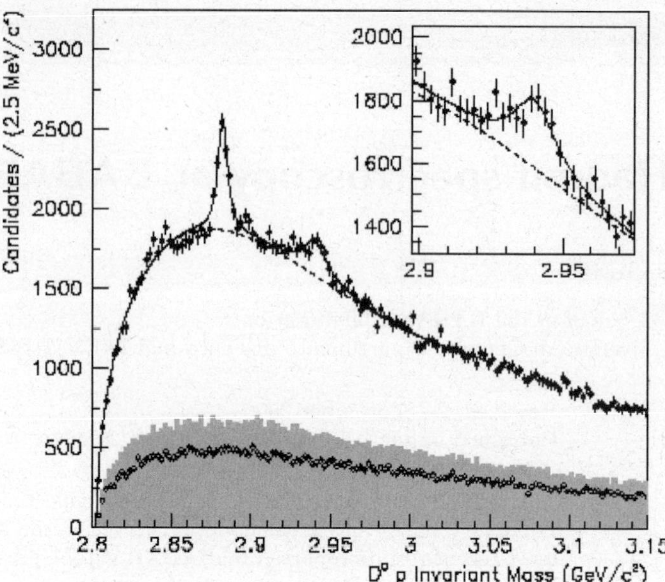

Fig. 2. $D^0 p$ invariant-mass distribution (solid points) overlaid to the result of a fit to a background component plus a $\Lambda_c(2880)^+$ and $\Lambda_c(2940)^+$ signal hypothesis (solid curve). The shaded histogram represents the background from D^0 mass sidebands, while the open points represent the invariant mass of wrong-sign $\bar{D}^0 p$ pairs.

be consistent with the missing 0^+ and 1^+ levels of the D_s spectrum, although the low masses remain to be better understood.

3 A new charm baryon

The study of the invariant-mass spectrum of D^0-proton pairs in $e^+e^- \to c\bar{c}$ events (fig. 2) led to the observation of a new baryon with mass $(2939.8 \pm 1.3 \pm 1.0)\,\mathrm{MeV}/c^2$ and width $(17.5 \pm 5.2 \pm 5.9)\,\mathrm{MeV}/c^2$ [10]. Also, the observation of the $\Lambda_c(2880)^+ \to D^0 p$ decay was reported, allowing the first measurement of its width $(5.8 \pm 1.5 \pm 1.1\,\mathrm{MeV}/c^2)$: previously, this state was only observed to decay to $\Lambda_c \pi^+ \pi^-$. These represent the first observations of a c-baryon–to–c-meson decay. The study of the $D^+ p$ invariant-mass spectrum did not reveal any charged partners of the two states, confirming the assignment of both to the Λ_c family.

4 States with $c\bar{c}$ content

The spectrum of $c\bar{c}$ bound states is well known below the open-charm threshold, and there good agreement is found with the potential model expectations. More charmonium states with $J^{PC} = 1^{--}$ are known above the open-charm threshold from the R scans [11]. On the other hand, many levels are still to be discovered. In addition, QCD predicts a richer spectroscopy: hybrids $(c\bar{c}g)$, $D^{(*)}\bar{D}^{(*)}$ molecules, tetraquarks; the recent discovery at the B-Factories of

many states with properties not well fitting within the potential model for conventional $c\bar{c}$ states, like $X(3870)$ [12] and $Y(4260)$ [13], led to a new interest in this spectroscopy. BABAR undertook a rich program of studies of both B decays and ISR events to try and improve the knowledge of these states.

The production of $X(3870)$ in both charged and neutral B decays to $J/\psi \pi^+ \pi^- K$ was studied [14]. If $X(3870)$ is a conventional charmonium state, a similar production ratio r_{0+} from B^0 and B^+ is expected; if, on the contrary, it is a molecule, several models predict that the B^0 production be suppressed [15]. Also, if it were a tetraquark, it is foreseen that the states produced by B^0 and B^+ be not the same, and have a mass difference of about $7\,\mathrm{MeV}/c^2$ [5]. A clear signal was observed in B^+ decays, and a 2.5 standard-deviations excess in B^0 decays; r_{0+} was measured to be $0.50 \pm 0.30 \pm 0.05$. The mass difference was measured to be $(2.7 \pm 1.3 \pm 0.2)\,\mathrm{MeV}/c^2$. The method appears therefore interesting, however more data are needed to distinguish between the various models.

Charged partners of $X(3870)$, foreseen in molecular models, were sought for in B^0 and B^- decays to $J/\psi \pi^- \pi^0 K$: no signal was reported, ruling out the isovector hypothesis for $X(3870)$ at the 10^{-4} C.L. [16].

The observation of the $J/\psi \gamma$ decay of $X(3870)$ was reported in the $B^+ \to J/\psi \gamma K^+$ process [17]. This decay mode implies positive charge parity (C) for $X(3870)$, and consequently that the $\pi^+ \pi^-$ pair in the $J/\psi \pi^+ \pi^-$ mode has isospin 1, hence that the decay to $J/\psi \pi^+ \pi^-$ violates isospin conservation, confirming BELLE [18] and CDF [19] results. The ratio of the branching fraction for $J/\psi \gamma$ to $J/\psi \pi^+ \pi^-$ was measured to be 0.34 ± 0.14.

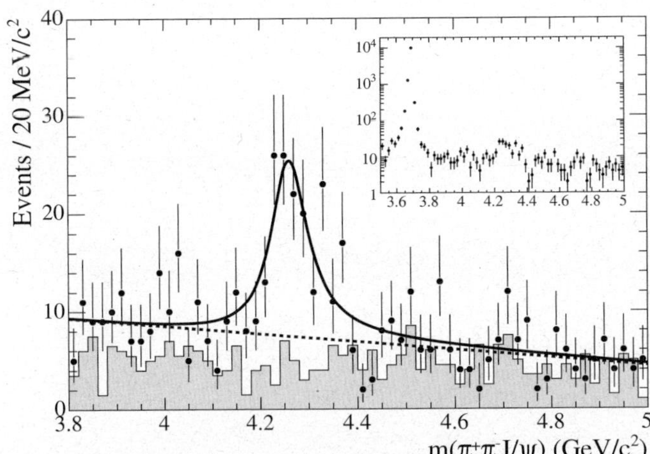

Fig. 3. The $J/\psi\pi^+\pi^-$ invariant-mass spectrum in ISR events in the 3.8–5.0 GeV/c^2 range and in a wider range (inset). The result of a fit to a single-resonance hypothesis plus background (solid curve) is superimposed to the data points; the shaded histogram represents the scaled data from J/ψ mass sidebands.

A recoil technique similar to the one described above was also used for the $X(3870)$, studying the missing mass recoiling against a K^+. An upper limit was set for $B^+ \to X(3870)K^+$ of 3.2×10^{-4}, at the 90% C.L., allowing to derive $\mathcal{B}(X(3870) \to J/\psi\pi^+\pi^-) > 4.2\%$ [20].

The study of $J/\psi\pi^+\pi^-$ states in ISR events did not reveal any signal for $X(3870)$, consistent with the $C = +$ assignement, however a structure ($Y(4260)$) was observed at higher masses (fig. 3): if interpreted as a single resonance, this has a mass of $(4259 \pm 8^{+2}_{-6})$ MeV/c^2 and a width of $(88 \pm 23^{+6}_{-4})$ MeV/c^2 [13]. Being well above the open-charm threshold and wide, its partial width to $J/\psi\pi^+\pi^-$ is expected to be very small in the conventional charmonium scenario, so alternative interpretations were proposed [21]. The search for other decay modes of $Y(4260)$ in ISR events led to negative results: no signals were reported for the $\phi\pi^+\pi^-$ [22], $p\bar{p}$ [23], $D\bar{D}$ [24] modes. A hint of a possible $Y(4260)$ signal was reported in $B^+ \to J/\psi\pi^+\pi^-K^+$ decays, leading to a branching fraction product $\mathcal{B}(B^+ \to Y(4260)K^+) \times \mathcal{B}(Y(4260) \to J/\psi\pi^+\pi^-) = (2.0 \pm 0.7 \pm 0.2) \times 10^{-5}$ [14].

Summing up all this information and the information reported by other experiments, mainly by BELLE, but also by CLEO, CDF and D0, it is apparent that $X(3870)$ and $Y(4260)$ are not easy to accommodate within the conventional charmonium picture. The $X(3870)$ is near the DD^* threshold, but it is very narrow, it decays to $J/\psi\pi^+\pi^-$ with isospin violation, and it was also observed to decay to $J/\psi\pi^+\pi^-\pi^0$, $D^0\bar{D}^{*0}$ and $J/\psi\gamma$. The $J^{PC} = 1^{++}$ and $I = 0$ quantum number assignments are favoured: the χ'_{c1} charmonium state would be consistent with this, but the mass would be too small. The $Y(4260)$ was only observed so far in the $J/\psi\pi\pi$ mode; since it was seen in ISR, it has $J^{PC} = 1^{--}$, however it was not observed in the R scans in the past, in contrast with the other 1^{--} charmonium states.

5 Conclusion

BABAR is greatly contributing to hadron spectroscopy. Significant progresses have been attained towards a better understanding of states with $c\bar{s}$ and $c\bar{c}$ content, including precision measurements of masses and widths, as well as quantum numbers of recently discovered states, and more decay modes and production mechanisms have been investigated. More data are being accumulated and new results are on the way in this field of renewed interest.

References

1. BABAR Collaboration (B. Aubert et al.), Phys. Rev. Lett. **90**, 242001 (2003).
2. CLEO Collaboration (D. Besson et al.), Phys. Rev. D **68**, 032002 (2003).
3. Charge conjugation is implied throughout.
4. BELLE Collaboration (K. Abe et al.), Phys. Rev. Lett. **92**, 012002 (2004).
5. See, for example, L. Maiani et al., Phys. Rev. D **71**, 014028 (2005).
6. See, for example, T. Barnes et al., Phys. Rev. D **68**, 054006 (2003).
7. BABAR Collaboration (B. Aubert et al.), Phys. Rev. D **74**, 032007 (2006).
8. BABAR Collaboration (B. Aubert et al.), Phys. Rev. D **74**, 031103 (2006).
9. BABAR Collaboration (B. Aubert et al.), Phys. Rev. Lett. **93**, 181801 (2004).
10. BABAR Collaboration (B. Aubert et al.), Phys. Rev. Lett. **98**, 012001 (2007), hep-ex/0603052.
11. W.-M. Yao et al., J. Phys. G **33**, 1 (2006).
12. BELLE Collaboration (S.K. Choi et al.), Phys. Rev. Lett. **91**, 262001 (2003).
13. BABAR Collaboration (B. Aubert et al.), Phys. Rev. Lett. **95**, 142001 (2005).
14. BABAR Collaboration (B. Aubert et al.), Phys. Rev. D **73**, 011101 (2006).
15. E. Braaten, M. Kusunoki, Phys. Rev. D **71**, 074005 (2005).
16. BABAR Collaboration (B. Aubert et al.), Phys. Rev. D **71**, 031501 (2005).
17. BABAR Collaboration (B. Aubert et al.), Phys. Rev. D **74**, 071101 (2006), hep-ex/0607050,
18. BELLE Collaboration (K. Abe et al.), hep-ex/0505037; hep-ex/0505038.
19. CDF Collaboration (A. Abulencia et al.), Phys. Rev. Lett. **96**, 102002 (2005).
20. BABAR Collaboration (B. Aubert et al.), Phys. Rev. Lett. **96**, 052002 (2006).
21. See, for example, L. Maiani et al., Phys. Rev. D **72**, 031502 (2005); X. Liu et al., Phys. Rev. D **72**, 054023 (2005) or F.E. Close, P.R. Page, Phys. Lett. B **628**, 215 (2005).
22. BABAR Collaboration (B. Aubert et al.), Phys. Rev. D **71**, 052001 (2005).
23. BABAR Collaboration (B. Aubert et al.), Phys. Rev. D **73**, 012005 (2006).
24. BABAR Collaboration (B. Aubert et al.), hep-ex/0607083.

Eur. Phys. J. A **31**, 698–700 (2007)

DOI 10.1140/epja/i2006-10269-3

Special Article – QNP 2006

Multichannel calculation of the very narrow $D_{s0}^*(2317)$ and the very broad $D_0^*(2300\text{--}2400)$

G. Rupp[1],[a] and E. van Beveren[2]

[1] Centro de Física das Interacções Fundamentais, Instituto Superior Técnico, Edifício Ciência, P-1049-001 Lisboa, Portugal
[2] Centro de Física Teórica, Departamento de Física, Universidade de Coimbra, P-3004-516 Coimbra, Portugal

Received: 8 December 2006
Published online: 15 March 2007 – © Società Italiana di Fisica / Springer-Verlag 2007

Abstract. The narrow $D_{s0}^*(2317)$ and broad $D_0^*(2300\text{--}2400)$ charmed scalar mesons and their radial excitations are described in a coupled-channel quark model that also reproduces the properties of the light scalar nonet. All two-meson channels containing ground-state pseudoscalars and vectors are included. The parameters are chosen fixed at published values, except for the overall coupling constant λ, which is fine-tuned to reproduce the $D_{s0}^*(2317)$ mass, and a damping constant α for subthreshold contributions. Variations of λ and $D_0^*(2300\text{--}2400)$ pole postions are studied for different α values. Calculated cross-sections for S-wave DK and $D\pi$ scattering, as well as resonance pole positions, are given for the value of α that fits the light scalars. The thus predicted radially excited state $D_{s0}^{*\prime}(2850)$, with a width of about $50\,\mathrm{MeV}$, seems to have been observed already.

PACS. 14.40.Lb Charmed mesons – 14.40.Ev Other strange mesons – 13.25.-k Hadronic decays of mesons – 12.39.Pn Potential models

The very narrow $D_{s0}^*(2317)$ charm-strange scalar meson first observed [1] three years ago has turned out to be the precursor of a series of new discoveries in hadron spectroscopy that have breathed new life into this field. The surprisingly low mass of the $D_{s0}^*(2317)$ itself has given rise to a flurry of theoretical work and speculations, mostly embracing nonstandard quark configurations (see, e.g., ref. [2] for a list of references). Moreover, the very broad charm-nonstrange partner meson $D_0^*(2300\text{--}2400)^1$ discovered [4] shortly afterwards further added to the confusion, as its Breit-Wigner mass seems of the same order as the mass of the $D_{s0}^*(2317)$, and perhaps even larger [5]. However, the large width ($\simeq 260\,\mathrm{MeV}$) of the $D_0^*(2300\text{--}2400)$ and the conflicting experimental mass determinations leave enough room for a possible reconciliation with the $D_{s0}^*(2317)$ mass.

In ref. [6], we described the quasi-bound $D_{s0}^*(2317)$- and the $D_0^*(2300\text{--}2400)$-resonance as P-wave $c\bar{s}$ and $c\bar{n}$ ($n = u, d$) states, respectively, strongly coupled to the lowest S-wave two-meson channel, i.e., DK, respectively $D\pi$. The framework of our calculation was a simple coupled-channel model previously used to fit the S-wave $K\pi$ phase shifts and predict the $K_0^*(800)$ (alias κ) meson [7]. As

a result, both the quasi-bound $D_{s0}^*(2317)$ *below* the DK threshold and the very broad $D_0^*(2300\text{--}2400)$-resonance *above* the $D\pi$ threshold were roughly reproduced, though with a too low-lying $D_0^*(2300\text{--}2400)$ pole. Scaling arguments from flavour invariance later allowed to somewhat improve [2, 8, 9] our predictions, with no new parameters. In the present study (also see ref. [10]), we ameliorate our coupled-channel description, by including all pseudoscalar-pseudoscalar (PP) and vector-vector (VV) channels, via a generalisation of our model that has recently been applied with success [11] to the whole light scalar nonet. The extension to PP and VV channels should allow us to make reliable predictions at least up to $\sim 3\,\mathrm{GeV}$.

Inclusion of all (ground-state) PP and VV channels implies that we couple the scalar $c\bar{s}$ states to DK, $D_s\eta$, $D_s\eta'$ in S-waves, and to D^*K^*, $D_s^*\phi$ in S- as well as D-waves, leading to a total number of 7 meson-meson channels. For the corresponding $c\bar{n}$ states, we need the coupling to $D\pi$, $D\eta$, $D\eta'$, D_sK in S-waves, and to $D\rho$, $D\omega$, $D_s^*K^*$ in S- and D-waves, thus totalling 10 channels. For the (bare) confined $c\bar{q}$ states, an infinite harmonic-oscillator spectrum is taken, as in previous work. These bare states are then coupled, via the $^3P_0$ mechanism, to the two-meson channels, assuming that transitions only occur at a certain distance r_0. The resulting T-matrix can be solved in closed form (see refs. [10, 11] for the formula).

---

[a] e-mail: `george@ist.utl.pt`

[1] We adopt here the designation $D_0^*(2300\text{--}2400)$, instead of the official PDG nomenclature $D_0^*(2400)$ [3], to roughly indicate the two [4,5] observed mass values.

Table 1. Pole positions of $D^*_0(2300\text{--}2400)$ in MeV. Parameter λ fitted to $D^*_{s0}(2317)$ mass.

λ (GeV$^{-3/2}$)	r_0 (GeV^{-1})	θ_{PS} (°)	α (GeV^{-2})	$M_1 M_2$	$D^*_{s0}(2317)$	$D^*_0(2300\text{--}2400)$
2.491	3.2	−13.5	0.0	PP	2.317	$2186 - i109$
2.497	3.2	−17.3	0.0	PP	2.317	$2185 - i108$
1.468	3.2	−13.5	0.0	PP + VV	2.317	$2233 - i35.1$
1.469	3.2	−17.3	0.0	PP + VV	2.317	$2233 - i35.0$
2.700	3.2	−13.5	2.0	PP	2.317	$2165 - i111$
2.714	3.2	−17.3	2.0	PP	2.317	$2163 - i110$
2.137	3.2	−13.5	2.0	PP + VV	2.317	$2205 - i66.7$
2.144	3.2	−17.3	2.0	PP + VV	2.317	$2203 - i66.3$
2.854	**3.2**	**−13.5**	**4.0**	**PP**	2.317	$\mathbf{2149 - i111}$
2.868	3.2	−17.3	4.0	PP	2.317	$2147 - i110$
2.617	**3.2**	**−13.5**	**4.0**	**PP + VV**	2.317	$\mathbf{2174 - i96.4}$
2.629	3.2	−17.3	4.0	PP + VV	2.317	$2172 - i95.3$
2.988	3.2	−13.5	6.0	PP	2.317	$2135 - i108$
3.001	3.2	−17.3	6.0	PP	2.317	$2133 - i107$
2.901	3.2	−13.5	6.0	PP + VV	2.317	$2145 - i105$
2.913	3.2	−17.3	6.0	PP + VV	2.317	$2143 - i104$

An important and very difficult issue when dealing with coupled channels is how to treat subthreshold contributions, *i.e.*, the effects of channels that are kinematically closed. Obviously, one cannot simply neglect channels as soon as the energy drops below threshold, which would be in gross violation of analyticity and even of common sense. However, it is also clear that, far below threshold, only a non-perturbative field-theoretic treatment of the Dyson-Schwinger type might provide a rigorous description, since constituent masses are inexorably subject to major self-energy corrections in deeply bound systems. Evidently, a non-covariant approach like our coupled-channel Schrödinger equation cannot account for such effects, despite the use of relativistic kinematics, as particles are manifestly on-mass-shell. Suppression of closed channels due to wave functions, which is naturally included in our Schrödinger fomalism, empirically turns out to be insufficient in relativistic systems, as recently observed in the mentioned application of our model to the light scalars [11]. Therefore, we adopt here the same remedy as employed in the latter paper, and also in many multichannel data analyses, namely the use of subthreshold form factors. Thus, for closed channels we multiply the squares of the individual channel couplings that show up in our closed-form T-matrix expression by an exponential $\exp(\alpha k_i^2)$, where k_i is the relativistic channel momentum (with $\Re e\, k_i^2 < 0$) and α is a positive parameter, assumed to be universal. Clearly, this ansatz is not fully analytic either, namely on top of a threshold, but at least it is continuous there.

Now we proceed by fine-tuning the overall coupling constant λ [10] so as to reproduce the mass of the nowadays firmly established $D^*_{s0}(2317)$. We do this for a variety of situations in which not only the parameter α is chosen at different values, but also a comparison is made between calculations with only PP channels included, and with VV channels accounted for as well. Moreover, we also choose two different but both frequently quoted values for the pseudoscalar mixing angle θ_{PS}, which intro-

duces slight variations in the predictions owing to the channels involving an η- or η'-meson. In each case, we determine the pole position of the ground-state scalar $c\bar{n}$ state. In table 1, these pole positions are given together with the values of the parameters λ, α, θ_{PS}, and r_0 (fixed). From the table, we first of all observe that the dependence of the pole positions on the pseudoscalar mixing angle is indeed very feeble. Then, we note that the inclusion of the VV channels has a very significant effect on the $D^*_0(2300\text{--}2400)$ pole positions, especially on the imaginary parts, which nevertheless becomes smaller for increasing α. Also this can be easily understood, as all VV channels are highly virtual at these pole energies, so that large values of α lead to a strong suppression of these channels. Finally, all pole positions come out too low when compared to both experimental [4,5] $D^*_0(2300\text{--}2400)$ masses, even when noticing that the cross-sections corresponding to these poles peak at somewhat higher masses. For instance, in the case $\alpha = 4.0\,\mathrm{GeV}^{-2}$ (boldface in table 1), which was the value used in ref. [11] for all light scalars, our $D^*_0(2300\text{--}2400)$ cross-section peaks lie at $2.18\,\mathrm{GeV}$ (PP) and $2.19\,\mathrm{GeV}$ (PP + VV). However, some words of caution are due here. Besides the mentioned 100 MeV discrepancy between the two central experimental masses, it should be realised that one cannot just compare our predicted cross-sections for elastic scattering with the Breit-Wigner fits of a very broad resonance observed in production processes, where other and more pronounced resonances like, *e.g.*, the $D^*_2(2460)$ show up as well. The resulting distributions for the $D^*_0(2300\text{--}2400)$ may be quite different (see, *e.g.*, fig. 2 of ref. [10]). Of course, pole positions must be the same in elastic scattering and production, but experiment does not extract any poles from the data. So better data on the $D^*_0(2300\text{--}2400)$ are definitely needed.

Focusing now our attention on the case $\alpha = 4.0\,\mathrm{GeV}^{-2}$, which fits the light scalars, we compute the elastic S-wave $D\pi$ and DK cross-sections up to $3\,\mathrm{GeV}$, which are then plotted in fig. 1, both for the PP only

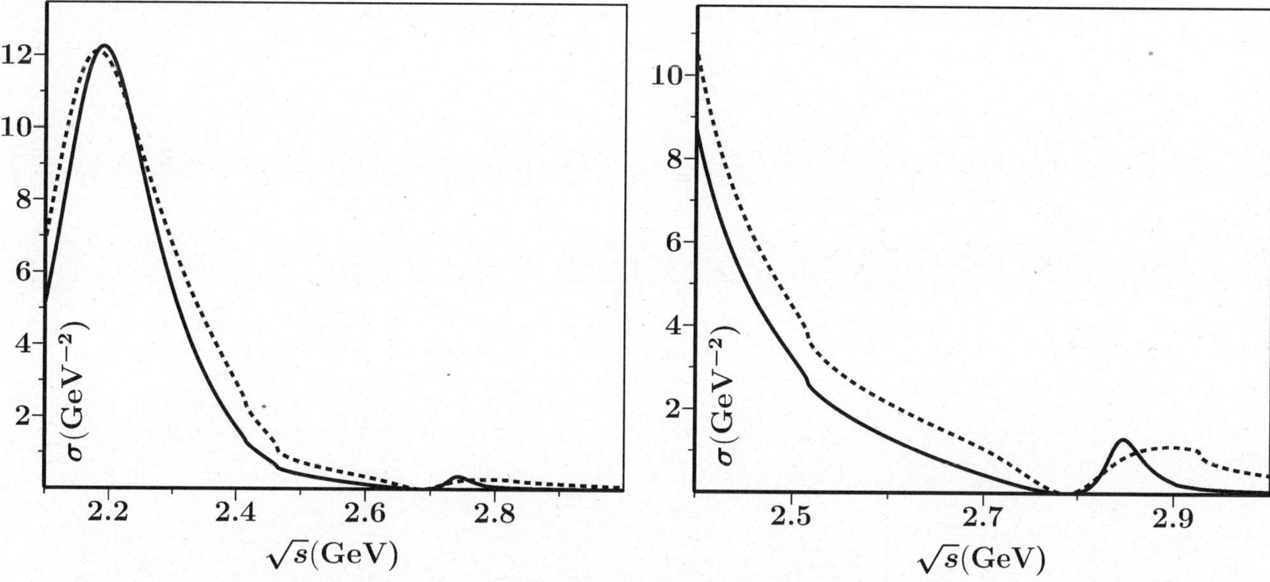

Fig. 1. S-wave $D\pi$ (left) and DK (right) cross-sections. Dashed curves: PP channels only; full curves: PP + VV.

and the full PP + VV cases. Besides the large bump in $D\pi$ due to the $D_0^*(2300$–$2400)$, and the steeply falling cross-section in DK owing to the $D_{s0}^*(2317)$ quasi-bound state, we observe additional structures at higher energies, which are more pronounced and narrower when all channels are included. Concretely, there is a tiny bump in $D\pi$ at about 2.74 GeV, with a peak width of roughly 50 MeV. We find the corresponding pole at $2737 - i24.0$ MeV, which can be traced back, for vanishing λ, to the first radial excitation of the bare confinement spectrum at $E = 2823$ MeV. Moreover, there is a very broad pole as well, *viz* at $2703 - i228$ MeV, which is connected to the confinement ground state, the $D_0^*(2300$–$2400)$ being a continuum pole [2,6]. In the DK case, there is a clear bump at about 2.85 GeV, with a peak width of again some 50 MeV. Besides a very broad continuum pole at $2779 - i233$ MeV, there is indeed a narrow pole at $2842 - i23.6$ MeV, originating from the bare confinement state at $E = 2925$ MeV. This resonance should thus correspond to the first radial excitation of the $D_{s0}^*(2317)$. Quite significantly, a new $c\bar{s}$ resonance denoted $D_{sJ}(2860)$, with a mass of $2856.6 \pm 1.5 \pm 5.0$ MeV and a width of $48 \pm 7 \pm 10$ MeV, was reported [12] by the BABAR Collaboration [13] very shortly after the presentation of the present results. The observation of the DK decay mode and the nonobservation of the D^*K mode, as reported by BABAR, are compatible with the radially excited scalar $c\bar{s}$ state predicted by us.

In the meantime, three other theoretical papers [14–16] on the $D_{sJ}(2860)$ have appeared. The first one argues in favour of a 3^- assignment, the second one supports a radially excited scalar as we do, and the third admits either option. So experimental confirmation of the $D_{sJ}(2860)$ is needed, as well as observation of another decay mode.

We are indebted to S. Tosi for drawing our attention to the $D_{sJ}(2860)$, right after its first public announcement [12]. This work was supported in part by the *Fundação para a Ciência e a Tecnologia* of the *Ministério da Ciência, Tecnologia e Ensino Superior* of Portugal, under contract POCI/FP/63437/2005.

References

1. BABAR Collaboration (B. Aubert *et al.*), Phys. Rev. Lett. **90**, 242001 (2003) [arXiv:hep-ex/0304021].
2. E. van Beveren, J.E.G.N. Costa, F. Kleefeld, G. Rupp, Phys. Rev. D **74**, 037501 (2006) [arXiv:hep-ph/0509351].
3. Particle Data Group (W.M. Yao *et al.*), J. Phys. G **33**, 1 (2006).
4. Belle Collaboration (K. Abe *et al.*), Phys. Rev. D **69**, 112002 (2004) [arXiv:hep-ex/0307021].
5. FOCUS Collaboration (J.M. Link *et al.*), Phys. Lett. B **586**, 11 (2004) [arXiv:hep-ex/0312060].
6. E. van Beveren, G. Rupp, Phys. Rev. Lett. **91**, 012003 (2003) [arXiv:hep-ph/0305035].
7. E. van Beveren, G. Rupp, Eur. Phys. J. C **22**, 493 (2001) [arXiv:hep-ex/0106077].
8. E. van Beveren, G. Rupp, Mod. Phys. Lett. A **19**, 1949 (2004) [arXiv:hep-ph/0406242].
9. G. Rupp, F. Kleefeld, E. van Beveren, AIP Conf. Proc. **756**, 360 (2005) [arXiv:hep-ph/0412078].
10. E. van Beveren, G. Rupp, Phys. Rev. Lett. **97**, 202001 (2006) [arXiv:hep-ph/0606110].
11. E. van Beveren, D.V. Bugg, F. Kleefeld, G. Rupp, Phys. Lett. B **641**, 265 (2006) [arXiv:hep-ph/0606022].
12. A. Palano, plenary talk given on 7 June 2006 at the *Charm2006 International Workshop, 5–7 June 2006, Beijing, China*, Int. J. Mod. Phys. A **21**, 5601 (2006).
13. BABAR Collaboration (B. Aubert *et al.*), Phys. Rev. Lett. **97**, 222001 (2006) [arXiv:hep-ex/0607082].
14. P. Colangelo, F. De Fazio, S. Nicotri, arXiv:hep-ph/0607245.
15. F.E. Close, C.E. Thomas, O. Lakhina, E.S. Swanson, arXiv:hep-ph/0608139.
16. B. Zhang, X. Liu, W.Z. Deng, S.L. Zhu, arXiv:hep-ph/0609013.

Eur. Phys. J. A **31**, 701–704 (2007)

DOI 10.1140/epja/i2006-10287-1

THE EUROPEAN
PHYSICAL JOURNAL A

Special Article – QNP 2006

Radial excitations of heavy-light mesons

T. Matsuki[1,a], T. Morii[2], and K. Sudoh[3]

[1] Tokyo Kasei University, 1-18-1 Kaga, Itabashi, Tokyo 173, Japan
[2] Graduate School of Science and Technology, Kobe University, Nada, Kobe 657-8501, Japan
[3] Institute of Particle and Nuclear Studies, High Energy Accelerator Research Organization, 1-1 Ooho, Tsukuba, Ibaraki 305-0801, Japan

Received: 18 December 2006
Published online: 20 March 2007 – © Società Italiana di Fisica / Springer-Verlag 2007

Abstract. The recent discovery of D_s states suggests the existence of radial excitations. Our semirelativistic quark potential model succeeds in reproducing these states within one to two percent of accuracy compared with the experiments, $D_{s0}(2860)$ and $D_s^*(2715)$, which are identified as 0^+ and 1^- radial excitations ($n = 2$). We also present calculations of radial excitations for B/B_s heavy mesons. The relation between our formulation and the modified Goldberger-Treiman relation is also described.

PACS. 12.39.Hg Heavy quark effective theory – 12.39.Pn Potential models – 12.40.Yx Hadron mass models and calculations – 14.40.Lb Charmed mesons

1 Introduction

BaBar has recently announced the discovery of a new D_s state, which seems to be the $c\bar{s}$ state [1], $D_{s0}(2860)$. Subsequent to this experiments Belle has observed a new state of $D_s^*(2715)$ whose spin and parity are determined to be 1^- [2].

We were the first in predicting 0^+ and 1^+ states of D_s and D particles, $D_{s0}(2317)$, $D_{s1}(2460)$, $D_0^*(2308)$, and $D_1'(2427)$, [3], and we have also succeeded in reproducing higher resonances of B/B_s particles, $B_1(5720)$, $B_2^*(5745)$, and $B_{s2}^*(5839)$ [4], using our semirelativistic quark potential model. Our model succeeds in lowering 0^+ and 1^+ states of D_s so that these mass values are below DK/D^*K threshold while other models do not. This model respects both heavy-quark symmetry and chiral symmetry in a certain limit of parameters [5], which relates our formulation with the idea of the modified Goldberger-Treiman relation proposed in refs. [6,7]. Hence it is natural to try to explain the newly discovered D_s states by using our semirelativistic model and we are again successful in reproducing these states discovered by BaBar and Belle.

To interpret the state $D_{s0}(2860)$, there are arguments that this $c\bar{s}$ state is explained to be a scalar by a coupled channel model [8], or that it is a $J^P = 3^-$ state [9], or that it can be explained by using a phenomenological interaction term like our quark potential model [10], or that it can be analyzed by using the $^3P_0$ model [11].

Starting from the astonishing discovery of D_{sJ} particles with narrow decay width by BaBar and CLEO, and

a e-mail: matsuki@tokyo-kasei.ac.jp

Table 1. Most optimal values of the parameters.

Parameters	$\alpha_s^{n=2}$	a (GeV^{-1})	b (GeV)
	0.344	1.939	0.0749

$m_{u,d}$ (GeV)	m_s (GeV)	m_c (GeV)	m_b (GeV)
0.0112	0.0929	1.032	4.639

confirmed by Belle, a series of successive experiments on the spectrum of a heavy-light system, *i.e.*, heavy mesons, heavy quarkonium, and heavy baryons, stimulates theorists to explain all these spectra as well as their decay modes. See the recent reviews of refs. [12,13]. It seems that a new era of spectroscopy is opening, which is challenging to theorists to solve these spectra at the same time.

2 Numerical calculation

Our model starts from a Hamiltonian with a scalar confining potential together with a Coulombic vector potential, we expand the whole system, *i.e.*, the Hamiltonian, the wave function, and the eigenvalue is expanded in $1/m_Q$, and we solve the equations order by order consistently. In the actual numerical calculation, we expand the wave function in a power series of relative coordinate with some weighting exponential times power functions. The wave function has positive components of a heavy quark due to the lowest-order constraint when expanding in $1/m_Q$

Table 2. $D_s(n=2)$ meson mass spectra (first order). Units are in MeV.

$^{2s+1}L_J(J^P)$	M_0	c_1/M_0	p_1/M_0	n_1/M_0	M_{calc}	M_{obs}
$^1S_0(0^-)$	2328	1.006×10^{-1}	0.919×10^{-1}	8.695×10^{-3}	2563	−
$^3S_1(1^-)$		1.830×10^{-1}	1.824×10^{-1}	5.744×10^{-4}	2755	2715
$^3P_0(0^+)$	2456	1.553×10^{-1}	1.470×10^{-1}	8.245×10^{-3}	2837	2856
$^{``3}P_1"(1^+)$		2.551×10^{-1}	2.543×10^{-1}	7.667×10^{-4}	3082	−
$^{``1}P_1"(1^+)$	2585	1.969×10^{-1}	1.966×10^{-1}	2.531×10^{-4}	3094	−
$^3P_2(2^+)$		2.209×10^{-1}	2.209×10^{-1}	5.070×10^{-7}	3157	−
$^3D_1(1^-)$	2391	8.605×10^{-1}	8.600×10^{-1}	5.016×10^{-4}	4449	−
$^{``3}D_2"(2^-)$		-4.287×10^{-1}	-4.287×10^{-1}	5.482×10^{-7}	1366	−

Table 3. $D(n=2)$ meson mass spectra (first order). Units are in MeV.

$^{2s+1}L_J(J^P)$	M_0	c_1/M_0	p_1/M_0	n_1/M_0	M_{calc}	M_{obs}
$^1S_0(0^-)$	2241	1.078×10^{-1}	0.975×10^{-1}	1.038×10^{-2}	2483	−
$^3S_1(1^-)$		1.917×10^{-1}	1.910×10^{-1}	6.882×10^{-4}	2671	−
$^3P_0(0^+)$	2418	1.540×10^{-1}	1.444×10^{-1}	9.621×10^{-3}	2791	−
$^{``3}P_1"(1^+)$		2.493×10^{-1}	2.488×10^{-1}	5.352×10^{-4}	3021	−
$^{``1}P_1"(1^+)$	2491	2.076×10^{-1}	2.070×10^{-1}	5.956×10^{-4}	3008	−
$^3P_2(2^+)$		2.319×10^{-1}	2.318×10^{-1}	1.101×10^{-4}	3069	−
$^3D_1(1^-)$	2280	1.869×10^{-1}	1.863×10^{-1}	5.418×10^{-4}	2706	−
$^{``3}D_2"(2^-)$		2.100×10^{-1}	2.099×10^{-1}	1.203×10^{-4}	2759	−

Table 4. $B(n=2)$ meson mass spectra (first order). Units are in MeV.

$^{2s+1}L_J(J^P)$	M_0	c_1/M_0	p_1/M_0	n_1/M_0	M_{calc}	M_{obs}
$^1S_0(0^-)$	5849	0.919×10^{-2}	0.831×10^{-2}	8.849×10^{-4}	5902	−
$^3S_1(1^-)$		1.634×10^{-2}	1.629×10^{-2}	5.867×10^{-5}	5944	−
$^3P_0(0^+)$	6025	1.375×10^{-2}	1.289×10^{-2}	8.590×10^{-4}	6108	−
$^{``3}P_1"(1^+)$		2.226×10^{-2}	2.221×10^{-2}	4.779×10^{-5}	6160	−
$^{``1}P_1"(1^+)$	6098	1.886×10^{-2}	1.881×10^{-2}	5.412×10^{-5}	6213	−
$^3P_2(2^+)$		2.108×10^{-2}	2.107×10^{-2}	1.001×10^{-5}	6227	−
$^3D_1(1^-)$	5888	1.610×10^{-2}	1.605×10^{-2}	4.668×10^{-5}	5982	−
$^{``3}D_2"(2^-)$		1.809×10^{-2}	1.808×10^{-2}	1.037×10^{-5}	5994	−

Table 5. $B_s(n=2)$ meson mass spectra (first order). Units are in MeV.

$^{2s+1}L_J(J^P)$	M_0	c_1/M_0	p_1/M_0	n_1/M_0	M_{calc}	M_{obs}
$^1S_0(0^-)$	5936	0.878×10^{-2}	0.802×10^{-2}	7.588×10^{-4}	5988	−
$^3S_1(1^-)$		1.597×10^{-2}	1.592×10^{-2}	5.013×10^{-5}	6031	−
$^3P_0(0^+)$	6063	1.399×10^{-2}	1.325×10^{-2}	7.429×10^{-4}	6148	−
$^{``3}P_1"(1^+)$		2.299×10^{-2}	2.292×10^{-2}	6.908×10^{-5}	6202	−
$^{``1}P_1"(1^+)$	6193	1.828×10^{-2}	1.826×10^{-2}	2.351×10^{-5}	6306	−
$^3P_2(2^+)$		2.052×10^{-2}	2.052×10^{-2}	4.709×10^{-8}	6320	−
$^3D_1(1^-)$	5999	7.631×10^{-2}	7.627×10^{-2}	4.449×10^{-5}	6456	−
$^{``3}D_2"(2^-)$		-3.802×10^{-2}	-3.802×10^{-2}	4.861×10^{-8}	5770	−

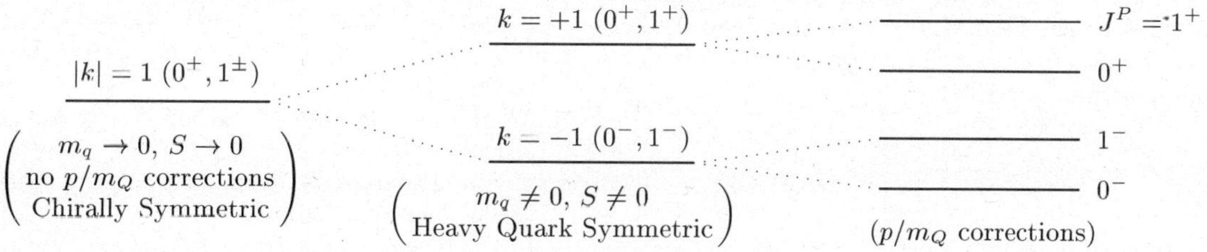

$$|k| = 1\ (0^+, 1^\pm)$$

$$k = +1\ (0^+, 1^+) \quad\cdots\cdots\quad J^P = 1^+$$
$$0^+$$

$$\left(\begin{array}{c} m_q \to 0,\ S \to 0 \\ \text{no } p/m_Q \text{ corrections} \\ \text{Chirally Symmetric} \end{array}\right)$$

$$k = -1\ (0^-, 1^-) \quad\cdots\cdots\quad 1^-$$
$$0^-$$

$$\left(\begin{array}{c} m_q \neq 0,\ S \neq 0 \\ \text{Heavy Quark Symmetric} \end{array}\right) \qquad (p/m_Q \text{ corrections})$$

Fig. 1. Procedure how the degeneracy is resolved in our model. A quantum number k is defined in the main text.

and four components of a light antiquark. Thus, the lowest eigenfunction is a two-by-four matrix. Other than the angular part, its radial part can be written as:

$$u_k(r),\ v_k(r) \sim w_k(r) \left(\frac{r}{a}\right)^\lambda \exp\left[-(m_q + b)r - \frac{1}{2}\left(\frac{r}{a}\right)^2\right],$$

where k is the quantum number of the operator, $-\beta_q(\boldsymbol{\Sigma}_q \cdot \boldsymbol{L} + 1)$ with $\boldsymbol{\Sigma}_q$ light-quark spin and $\boldsymbol{L}$ light-quark angular momentum, which distinguishes uniquely each state [3], for instance, $k = -1$ for a spin multiplet ($J^P = 0^-, 1^-$), $k = +1$ for $(0^+, 1^+)$, etc., $u_k(r)$ and $v_k(r)$ are the upper and lower components of the radial wave functions, and $\lambda = \sqrt{k^2 - (4\alpha_s/3)^2}$, where a and b are included in a scalar potential as $S(r) = r/a^2 + b$, α_s is the strong coupling, and $w_k(r)$ is the finite series of a polynomial in r, $w_k(r) = \sum_{i=0}^{N-1} a_i^k (r/a)^i$, which takes different coefficients for $u_k(r)$ and $v_k(r)$. In the actual calculation, we have used $N = 7$. Hence we can in principle obtain seven different radial excitations.

In this paper, we calculate the most optimal values of the parameters so that the recently discovered and known $n = 2$ (the first radial excitation) D_s particles are all fitted well around one percent of accuracy compared with the experiments. Here only the strong coupling α_s is modified and other parameters are kept the same as in [4], in which we have obtained $\alpha_s = 0.261$ both for D and D_s. These are presented in table 1 at the first order of calculation in p/m_Q with p being the internal quark momentum and m_Q the heavy-quark mass.

With these values of the parameters, we obtain $n = 2$ masses of D_s and D at the same time. The results are shown in table 2 for D_s. We also predict $n = 2$ states for D, B, and B_s states, which are shown in tables 3, 4, and 5 assuming the same strong coupling α_s for D_s, which may actually be different for B/B_s particles. In these tables, p_i and n_i are the i-th order positive and negative component contributions of a heavy quark, respectively, and $c_i = p_i + n_i$. When one carefully looks at these tables, one notices that the values of the higher states $^3D_1$ and "$^3D_2$" are not reliable even though we have listed them in the tables for consistency with the former calculations.

3 Mass gap

Noticing that the mass gaps between spin multiplets, $(0^-, 1^-)$ and $(0^+, 1^+)$ for D_{sJ} mesons, are almost equal

Table 6. Theoretical mass gap. Values in brackets are experiments. Units are in MeV.

Mass gap ($n = 1$)	D	D_s	B	B_s
0^+–0^-	414 (441)	358 (348)	322	239
1^+–1^-	410 (419)	357 (348)	320	242

($n = 2$)	D	D_s	B	B_s
0^+–0^-	308	274	206	160
1^+–1^-	350	327	216	171

to each other, the modified Goldberger-Treiman relation is proposed by Bardeen *et al.* [6,7] to understand the facts, *i.e.*, the underlying physics might be chiral physics. In other words the mass gap is due to the chiral symmetry breakdown which is expressed by this relation. Further they have assumed that the hyperfine splittings due to breakdown of heavy-quark symmetry (inclusion of $1/m_Q$ corrections) are the same within two spin multiplets, $(0^-, 1^-)$ and $(0^+, 1^+)$, so that the mass gap is not affected by this hyperfine splitting.

In our formulation this is explained in fig. 1 [5]. Heavy-quark symmetry reduces the original Hamiltonian into the one without spin structure after projecting wave functions into the positive and negative components of a heavy quark [3,5]. The chiral symmetry of a light quark is realized by taking a limit of $m_q \to 0$ and $S(r) \to 0$, in which case all the members of two spin multiplets, $(0^-, 1^-)$ and $(0^+, 1^+)$, are degenerate. When the light-quark mass and a scalar potential are turned on, then the degeneracy due to chiral symmetry is resolved and the mass gap corresponding to the modified Goldberger-Treiman relation is given by

$$\Delta M = M_0(k = +1) - M_0(k = -1),$$

where $M_0(k)$ is a degenerate mass for a quantum number k, which appears in tables 2–5 to distinguish states. This is described as $\Delta M = \tilde{g}_\pi f_\pi / G_A$ in [6]. They have assumed dominant interaction terms for hyperfine splitting due to $1/m_Q$ corrections so that mass gaps between 0^+–0^- and 1^+–1^- are almost equal to each other, which seems to hold in our formulation. Our hyperfine splittings due to $1/m_Q$ to this mass gap ΔM are calculated and we have added those to degenerate mass gaps between 0^+–0^- and 1^+–1^- for D, D_s, B, and B_s with $n = 1$, which are given in table 6. Our dynamical calculation supports the assumption that the mass gaps between $0^+\text{–}0^-$ and $1^+\text{–}1^-$ are almost equal to each other in the case of $n = 1$. In the second row of the same table values for $n = 2$ are given, which is not as good as the case for $n = 1$.

References

1. A. Palano, *New Spectroscopy with Charm Quarks at B Factories*, talk presented at *Charm 2006, Beijing, June 5-7, 2006*; BaBar Collaboration (B. Aubert *et al.*), Phys. Rev. Lett. **97**, 222001 (2006) hep-ex/0607082.
2. Belle Collaboration (K. Abe *et al.*), Belle report BELLE-CONF-0643, hep-ex/0608031.
3. T. Matsuki, T. Morii, Phys. Rev. D **56**, 5646 (1997); see also T. Matsuki, Mod. Phys. Lett. A **11**, 257 (1996); T. Matsuki, T. Morii, Aust. J. Phys. **50**, 163 (1997); T. Matsuki, T. Morii, K. Seo, Trends Appl. Spectrosc. **4**, 127 (2002).
4. T. Matsuki, T. Morii, K. Sudoh, hep-ph/0605019.
5. T. Matsuki, K. Mawatari, T. Morii, K. Sudoh, Phys. Lett. B **606**, 329 (2005).
6. W.A. Bardeen, E.J. Eichten, C.T. Hill, Phys. Rev. D **68**, 054024 (2003).
7. M. Harada, M. Rho, C. Sasaki, Phys. Rev. D **70**, 074002 (2004).
8. E. van Beveren, G. Rupp, Phys. Rev. Lett. **97**, 202001 (2006) hep-ph/0606110.
9. P. Colangelo, F. De Fazio, S. Nicotri, Phys. Lett. B **642**, 48 (2006) hep-ph/0607245.
10. F.E. Close, C.E. Thomas, O. Lakhina, E.S. Swanson, hep-ph/0608139.
11. B. Zhang, Z. Liu, W. Deng, S. Zhu, to be published in Eur. Phys. J. C, hep-ph/0609013.
12. J.L. Rosner, talk presented at the *2006 Conference on the Intersections of Particle and Nuclear Physics (CIPAN 2006), Rio Grande, Puerto Rico, May 30-June 3, 2006*, hep-ph/0606166.
13. E.S. Swanson, Phys. Rep. **429**, 243 (2006) hep-ph/0601110.

Eur. Phys. J. A **31**, 705–710 (2007)

DOI 10.1140/epja/i2006-10255-9

Special Article – QNP 2006

Overview of Non-Relativistic QCD

J. Soto[a]

Departament d'Estructura i Constituents de la Matèria, Universitat de Barcelona, Spain

Received: 23 November 2006
Published online: 12 March 2007 – © Società Italiana di Fisica / Springer-Verlag 2007

Abstract. An overview of recent theoretical progress on Non-Relativistic QCD (NRQCD) and related effective theories is provided.

PACS. 13.20.Gd Decays of J/ψ, Υ, and other quarkonia – 12.39.St Factorization – 14.40.Gx Mesons with $S = C = B = 0$, mass > 2.5 GeV (including quarkonia)

1 Introduction

Heavy-quarkonium systems have been a good laboratory to test our theoretical ideas since the early days of QCD (see [1] for an extensive review). Based on the pioneering work of Caswell and Lepage [2], a systematic approach to study such systems from QCD has been developed over the years which makes use of effective-field theory techniques and is generically known as Non-Relativistic QCD (NRQCD) [3]. NRQCD has been applied to spectroscopy, decay and production of heavy quarkonium. The NRQCD formalism for spectroscopy and decay is very well understood (this is also so for electromagnetic threshold production) so that NRQCD results may be considered QCD results up to a given order in the expansion parameters (usually $\alpha_s(m_Q)$ and $1/m_Q$, m_Q being the heavy-quark mass). I will restrict myself to review recent progress on these issues and refer the reader to the review [4] for earlier developments. The NRQCD formalism for production is more controversial. Important work has been carried out recently to which I will devote some time.

2 Heavy quarkonium

Heavy quarkonia are mesons made out of a heavy quark and a heavy antiquark (not necessarily of the same flavor), whose masses are larger than Λ_{QCD}, the typical hadronic scale. These include bottomonia ($b\bar{b}$), charmonia ($c\bar{c}$), B_c systems ($b\bar{c}$ and $c\bar{b}$) and would-be toponia ($t\bar{t}$). Baryons made out of two or three heavy quarks share some similarities with these systems (see [5,6]).

In the quarkonium rest frame the heavy quarks move slowly ($v \ll 1$, v being the typical heavy-quark velocity in the center-of-mass frame), with a typical momentum $m_Q v \ll m_Q$ and binding energy $\sim m_Q v^2$. Hence any study

of heavy quarkonium faces a multiscale problem with the hierarchies $m_Q \gg m_Q v \gg m_Q v^2$ and $m_Q \gg \Lambda_{QCD}$. The use of effective-field theories is extremely convenient in order to exploit these hierarchies.

3 Effective-Field Theories

Direct QCD calculations in multiscale problems are extremely difficult, no matter if one uses analytic approaches (*i.e.*, perturbation theory) or numerical ones (*i.e.*, lattice). We may try to construct a simpler theory (the Effective-Field Theory (EFT)), in which less scales are involved in the dynamics and which is equivalent to the fundamental theory (QCD) in the particular energy region where heavy-quarkonia states lie. The clues for the construction are: i) identify the relevant degrees of freedom, ii) enforce the QCD symmetries and iii) exploit the hierarchy of scales. Typically, one integrates out the higher-energy scales so that the Lagrangian of the EFT can be organized as a series of operators over powers of these scales. Each operator has a so-called matching coefficient in front, which encodes the remaining information on the higher-energy scales. The matching coefficients may be calculated by imposing that a selected set of observables coincide when are calculated in the fundamental and in the effective theory.

NRQCD is the (first) effective theory relevant for heavy quarkonium. Out of the four components of the relativistic Dirac fields describing the heavy quarks (antiquarks) only the upper (lower) are relevant for energies lower than m_Q (no pair production is allowed anymore). Hence, a two-component Pauli spinor field is used to describe the quark (antiquark). The hierarchy of scales exploited in NRQCD is $m_Q \gg m_Q v, m_Q v^2, \Lambda_{QCD}$. The remaining hierarchy $m_Q v \gg m_Q v^2$, may be exploited using a further effective theory called Potential NRQCD (pNRQCD) [7,8].

  [a] e-mail: `joan.soto@ub.edu`, `soto@ecm.ub.es`

3.1 Non-Relativistic QCD

The part of the NRQCD Lagrangian bilinear on the heavy-quark fields coincides with the one of Heavy-Quark Effective Theory (HQET) (see [9] for a review), and reads

$$\mathcal{L}_\psi = \psi^\dagger \left\{ iD_0 + \frac{1}{2m_Q}\mathbf{D}^2 + \frac{1}{8m_Q^3}\mathbf{D}^4 + \frac{c_F}{2m_Q}\boldsymbol{\sigma}\cdot g\mathbf{B} \right.$$

$$+ \frac{c_D}{8m_Q^2}\left(\mathbf{D}\cdot g\mathbf{E} - g\mathbf{E}\cdot\mathbf{D}\right) + i\frac{c_S}{8m_Q^2}\boldsymbol{\sigma}$$

$$\left. \cdot\left(\mathbf{D}\times g\mathbf{E} - g\mathbf{E}\times\mathbf{D}\right) \right\}\psi;$$

ψ is a Pauli spinor which annihilates heavy quarks. c_F, c_D and c_S are short-distance matching coefficients which depend on m_Q and μ (factorization scale). Analogous terms exist for χ, a Pauli spinor field which creates antiquarks. Unlike HQET, it also contains four fermion operators,

$$\mathcal{L}_{\psi\chi} = \frac{f_1(^1S_0)}{m_Q^2}O_1(^1S_0) + \frac{f_1(^3S_1)}{m_Q^2}O_1(^3S_1)$$

$$+ \frac{f_8(^1S_0)}{m_Q^2}O_8(^1S_0) + \frac{f_8(^3S_1)}{m_Q^2}O_8(^3S_1),$$

$$O_1(^1S_0) = \psi^\dagger\chi\,\chi^\dagger\psi, \qquad O_1(^3S_1) = \psi^\dagger\boldsymbol{\sigma}\chi\,\chi^\dagger\boldsymbol{\sigma}\psi, \quad (1)$$
$$O_8(^1S_0) = \psi^\dagger\mathbf{T}^a\chi\,\chi^\dagger\mathbf{T}^a\psi, \qquad O_8(^3S_1) = \psi^\dagger\mathbf{T}^a\boldsymbol{\sigma}\chi\,\chi^\dagger\mathbf{T}^a\boldsymbol{\sigma}\psi.$$

The f's are again short-distance matching coefficients which depend on m_Q and μ, which have imaginary parts when the quark and the antiquark are of the same flavor. This is due to the fact that hard gluons (of energy $\sim m_Q$), which remain accessible through annihilation of the quark and antiquark, have been integrated out. The fact that NRQCD is equivalent to QCD at any desired order in $1/m_Q$ and $\alpha_s(m_Q)$ makes the lack of unitarity innocuous. In fact, it is turned into an advantage: it facilitates the calculation of inclusive decay rates to light particles.

The Lagrangian above can be used as such for spectroscopy, inclusive decays and electromagnetic threshold production of heavy quarkonia. Spectroscopy studies have been carried out on the lattice and are discussed in [10] (see also [11] and references therein).

3.1.1 Inclusive decays

Let us just show, as an example, the NRQCD formula for inclusive decays of P-wave states to light hadrons at leading order

$$\Gamma(\chi_Q(nJS) \to LH) =$$
$$\frac{2}{m_Q^2}\left(\mathrm{Im}\, f_1(^{2S+1}P_J)\frac{\langle\chi_Q(nJS)|O_1(^{2S+1}P_J)|\chi_Q(nJS)\rangle}{m_Q^2} \right.$$
$$\left. + \mathrm{Im}\, f_8(^{2S+1}S_S)\langle\chi_Q(nJS)|O_8(^1S_0)|\chi_Q(nJS)\rangle \right),$$

f_1 and f_8 are short-distance matching coefficients which can be calculated in perturbation theory in $\alpha_s(m_Q)$. $O_1(^{2S+1}P_J)$ is a color singlet dimension-8 operator and $O_8(^1S_0)$ is the color octet operator given in (1). Earlier QCD factorization formulas were missing the color octet contribution. They are inconsistent because the color octet matrix element is necessary to cancel the factorization scale dependence which arises in one-loop calculations of $\mathrm{Im}\, f_1(^{2S+1}P_J)$ [12]. The matrix elements cannot be calculated in perturbation theory of $\alpha_s(m_Q)$. They can, however, be calculated on the lattice (see, for instance, [13]) or extracted from data. The imaginary parts of the matching coefficients of the dimension-6 and -8 operators are known at NLO [14], and those of the electromagnetic decays of dimension 10 at LO [15–17].

3.1.2 Problems

Unlike HQET, in which each term in the Lagrangian (1) has a definite size (by assigning Λ_{QCD}^n to any operator of dimension n), NRQCD does not enjoy a homogeneous counting. This is due to the fact that the scales m_Qv, m_Qv^2 and Λ_{QCD} are still entangled. Even though a counting was put forward in [3], the so-called NRQCD velocity counting, which has become the standard book-keeping for NRQCD calculations, there is no guarantee that it holds for all the states, and hence there is a problem on how to make a sensible organization of the calculation for a given state. There are two approaches which aim at disentangling these scales, and hence to facilitate the counting. The first one consist of constructing a further effective theory by integrating out energy scales larger m_Qv^2. This is the pNRQCD approach first proposed in [7], which will be described in the next section. The second one consist in decomposing the NRQCD fields in several modes so that each one has a homogeneous counting. This is the vNRQCD approach first proposed in [18] (see also [19]). The scale Λ_{QCD} has not been discussed so far in this approach, and, in fact, a consistent formulation has only become available recently [20] (see also [21,22]).

3.2 Potential NRQCD

As mentioned above, pNRQCD is the effective theory which arises from NRQCD after integrating out energy scales largen than m_Qv^2, namely than the typical binding energy. If $\Lambda_{QCD} \lesssim m_Qv^2$, then $m_Qv \gg \Lambda_{QCD}$ and the matching between NRQCD and pNRQCD can be carried out in perturbation theory in $\alpha_s(m_Qv)$. This is the so-called weak-coupling regime. If $\Lambda_{QCD} \gg m_Qv^2$, the matching cannot be carried out in perturbation theory in $\alpha_s(m_Qv)$ anymore, but one can still exploit the hierarchy $m_Q \gg m_Qv, \Lambda_{QCD} \gg m_Qv^2$. This is the so-called strong-coupling regime.

3.2.1 Weak-coupling regime

The Lagrangian in this regime reads

$$\mathcal{L}_{pNRQCD} =$$
$$\int d^3\mathbf{r}\, \mathrm{Tr}\left\{ S^\dagger \left(i\partial_0 - h_s(\mathbf{r},\mathbf{p},\mathbf{P_R},\mathbf{S}_1,\mathbf{S}_2,\mu) \right) S \right.$$
$$+ O^\dagger \left(iD_0 - h_o(\mathbf{r},\mathbf{p},\mathbf{P_R},\mathbf{S}_1,\mathbf{S}_2,\mu) \right) O \Big\}$$
$$+ V_A(r,\mu)\, \mathrm{Tr}\left\{ O^\dagger \mathbf{r}\cdot g\mathbf{E}\, S + S^\dagger \mathbf{r}\cdot g\mathbf{E}\, O \right\}$$
$$+ \frac{V_B(r,\mu)}{2}\, \mathrm{Tr}\left\{ O^\dagger \mathbf{r}\cdot g\mathbf{E}\, O + O^\dagger O\mathbf{r}\cdot g\mathbf{E} \right\},$$

where S and O are color singlet and color octet wave function fields respectively. h_s and h_o are color singlet and color octet Hamiltonians, respectively, which may be obtained by matching to NRQCD in perturbation theory in $\alpha_s(m_Q v)$ and $1/m_Q$ at any order of the multipole expansion ($\mathbf{r} \sim 1/m_Q v$). The static potential terms in h_s and h_o are known at two loops [23–25] (the logarithmic contributions at three loops and (for h_s only) at four loops are also known [26,8,27]). The renormalon singularities in the static potentials are also understood in some detail [28, 29]. The $1/m_Q$ and $1/m_Q^2$ terms in h_s are known at two and one loop, respectively [30]. $V_A, V_B = 1 + \mathcal{O}(\alpha_s^2)$ [27]. This Lagrangian has been used to carry our calculations at fixed order in α_s: an almost complete NNNLO (assuming $\Lambda_{QCD} \ll m_Q \alpha_s^2$) expression for the spectrum is available [31]. Most remarkably resummations of Logarithms can also be carried out using renormalization group techniques [32–34][1]. Thus, the η_b mass and its electromagnetic decay width have been predicted at NNLL and NLL, respectively [35,36].

3.2.2 Strong-coupling regime

The Lagrangian in this regime reads[2]

$$L_{pNRQCD} = \int d^3\mathbf{R} \int d^3\mathbf{r}\, S^\dagger (i\partial_0 - h_s(\mathbf{r},\mathbf{p},\mathbf{P_R},\mathbf{S}_1,\mathbf{S}_2))S,$$

$$h_s(\mathbf{r},\mathbf{p},\mathbf{P_R},\mathbf{S}_1,\mathbf{S}_2) = \frac{\mathbf{p}^2}{m_Q} + \frac{\mathbf{P_R}^2}{4m_Q} + V_s(\mathbf{r},\mathbf{p},\mathbf{P_R},\mathbf{S}_1,\mathbf{S}_2),$$

$$V_s = V_s^{(0)} + \frac{V_s^{(1)}}{m_Q} + \frac{V_s^{(2)}}{m_Q^2} + \cdots,$$

V_s cannot be calculated in perturbation theory of $\alpha_s(m_Q v)$ anymore, but can indeed be, and most of the terms have been, calculated on (quenched) lattice simulations [37–41] (see [42] for a large-N_c calculation). The fact that

$\Lambda_{QCD} \gg m_Q v^2$ can now be exploited to further factorize NRQCD matrix elements [43–45], for instance

$$\langle \Upsilon(n)|O_8(^1S_0)|\Upsilon(n)\rangle = C_A \frac{|R_n(0)|^2}{2\pi}$$
$$\times \left(-\frac{(C_A/2 - C_f)c_F^2 \mathcal{B}_1}{3m_Q^2} \right),$$

where $R_n(0)$ is the wave function at the origin, $\mathcal{B}_1 \sim \Lambda_{QCD}^2$ is a universal (independent of n) non-perturbative parameter, and c_F a (computable) short-distance matching coefficient. This further factorization allows to put forward new model-independent predictions, for instance, the ratios of hadronic decay widths of P-wave states in bottomonium were predicted from charmonium data [43], or the ratio of photon spectra in radiative decays of vector resonances [46], as we will see below.

3.2.3 Weak or strong (or else)?

Since m_Q, v and Λ_{QCD} are not directly observable, given a heavy-quarkonium state, it is not clear to which of the above regimes, if to any[3], it must be assigned to. The leading non-perturbative ($\sim \Lambda_{QCD}$) corrections to the spectrum in the weak-coupling regime scale as a large power of the principal quantum number [47,48], which suggests that only the $n = 1$ states of bottomonium and charmonium may belong to this regime. However, if one ignores this and proceeds with weak-coupling calculations one finds, for instance, that renormalon-based approaches at NNLO [49,50] and the NNNLO calculation [36] give a reasonable description of the bottomonium spectrum up to $n = 3$. It has recently been proposed that precise measurements of the photon spectra in radiative decays, as the ones carried out by CLEO [51], will clarify the assignments [46]. It turns out that in the strong-coupling regime one can work out the following formula, which holds at NLO:

$$\frac{\dfrac{d\Gamma_n}{dz}}{\dfrac{d\Gamma_r}{dz}} = \frac{\langle O_1(^3S_1)\rangle_n}{\langle O_1(^3S_1)\rangle_r} \left(1 + \frac{C_1'\,[^3S_1]\,(z)}{C_1\,[^3S_1]\,(z)} \frac{1}{m_Q}\,(E_n - E_r) \right),$$

$$\frac{\langle O_1(^3S_1)\rangle_n}{\langle O_1(^3S_1)\rangle_r} = \frac{\Gamma\left(\Upsilon(n) \to e^+e^- \right)}{\Gamma\left(\Upsilon(r) \to e^+e^- \right)}$$
$$\times \left[1 - \frac{\mathrm{Im}\,g_{ee}\,(^3S_1)}{\mathrm{Im}\,f_{ee}\,(^3S_1)} \frac{E_n - E_r}{m_Q} \right],$$

$C_1'[^3S_1](z)$, $C_1[^3S_1](z)$, $\mathrm{Im}\,g_{ee}(^3S_1)$, $\mathrm{Im}\,f_{ee}(^3S_1)$ are matching coefficients computable in perturbation theory, and $E_n - E_r$ the mass difference between the two states. If data follow this formula it will indicate that both n and r are in the strong-coupling regime. For the $n = 1,2,3$ of bottomonia, current data disfavor $n = 1$ in the strong-coupling regime and is compatible with $n = 2,3$ in it.

---

[1] The importance of log resummations was first emphasized in the vNRQCD approach [18]. However, the correct results for the spectrum and electromagnetic decay widths were first obtained in pNRQCD [32,33] and later on reproduced in vNRQCD [21].

[2] We ignore for simplicity pseudo-Goldstone bosons.

[3] States close or above the open flavor threshold are expected to belong neither the weak- nor the strong-coupling regimes.

3.3 Soft-Collinear Effective Theory

For exclusive decays and for certain kinematical endpoints of semi-inclusive decays, NRQCD must be supplemented with collinear degrees of freedom. This can be done in the effective theory framework of Soft-Collinear Effective Theory (SCET) [52,53]. Exclusive radiative decays of heavy quarkonium in SCET have been addressed in [54], where results analogous to those of traditional light-cone factorization formulas have been obtained [55]. Concerning semi-inclusive decays, let me focus on the photon spectrum of $\Upsilon(1S) \to \gamma X$ when the photon is very energetic. Describing experimental data in this region has been a challenge since the early days of QCD, and the best fits were obtained using a model with a finite gluon mass of the order of the GeV [56]. The naive inclusion of color octet contributions in the NRQCD framework seemed to further deepen the discrepancy [57]. In recent years a remarkable improvement in the understanding of this endpoint region has been achieved by combining SCET and NRQCD. The following factorization formula has been proved [58,59]:

$$\frac{\mathrm{d}\Gamma^e}{\mathrm{d}z} = \sum_\omega H(M,\omega,\mu) \int \mathrm{d}k S(k,\mu) \operatorname{Im} J_\omega(k+M(1-z),\mu),$$

where H is a hard matching coefficient computable in perturbation theory, J_w is a jet function (also computable in perturbation theory if z is not too close to 1) and S is a (soft) shape function which depends on the bound-state dynamics. M is the heavy-quarkonium mass and μ a factorization scale. Large (Sudakov) logs have been resummed using renormalization group equations both in the color octet [60] and color singlet [61] contributions. Furthermore, using pNRQCD the shape function S has been calculated [62]. When all this is put together, an excellent description of data is achieved [63], see fig. 1.

4 Production

Production of heavy quarkonium is a far more complicated issue than spectroscopy and decay. This is due to the fact that, in addition to the several scales which characterize the heavy quarkonium system, further scales due to the kinematics of the production process may also appear.

4.1 Electromagnetic threshold production

This is the simplest and best understood production process in the weak-coupling regime, which is relevant for a precise measurement of top quark mass in the future International Linear Collider (ILC). The cross-section at NNLO is known for some time [64] and the log resummation at NLL is also available [33,21]. The effort is now in calculations at NNNLO, where partial results already exist [65–70], and in the log resummation at NNLL, where partial results also exist [67,71]. At this level of precision the electroweak effects must also be taken into account [72–74] (see [75,76] for recent reviews).

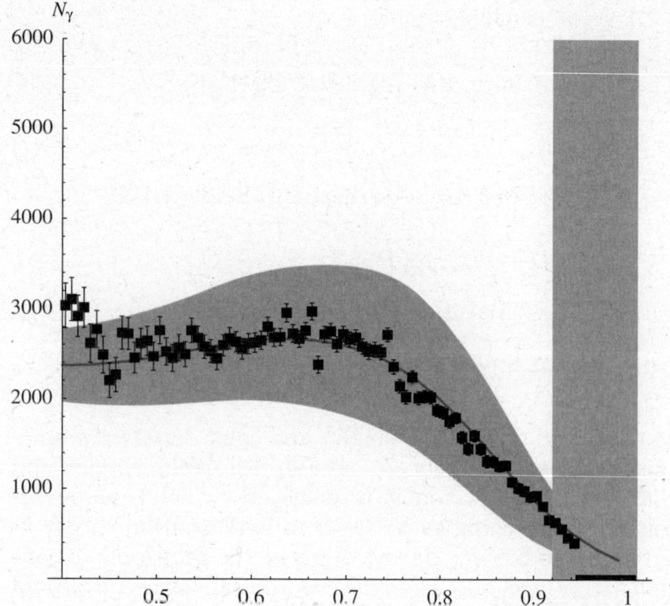

Fig. 1. N_γ is the number of photons and z the energy of the photons normalized to half the $\Upsilon(1S)$ mass. The black squares are CLEO data [51] and the (red on-line) solid curve the theoretical result of [63] with theoretical errors (grey band) displayed. The (green on-line) rectangle on the right displays the region where the theoretical results are not reliable. The plot is taken from [77].

4.2 Inclusive production

A factorization formula for the inclusive production of heavy quarkonium was put forward in [3], which was assumed to hold provided that the tranverse momentum $\mathbf{p}_\perp$ was larger or of the order of the heavy-quark mass,

$$\sigma(H) = \sum_n \frac{F_n(\mu)}{m_Q^{d_n-4}} \langle 0|\mathcal{O}_n^H(\mu)|0\rangle.$$

The formula contains a partonic level cross-section $(F_n(\mu)/m_Q^{d_n-4})$ in which the heavy-quark pair may be produced in a color singlet or a color octet state, and long distance production matrix elements $(\langle 0|\mathcal{O}_n^H(\mu)|0\rangle)$ which encode the evolution of the pair to the actual physical state. The production matrix elements have the generic form

$$\mathcal{O}_n^H = \chi^\dagger \mathcal{K}_n \psi \left(\sum_X \sum_{m_J} |H+X\rangle\langle H+X| \right) \psi^\dagger \mathcal{K}_n' \chi,$$

where the sums are over the $2J+1$ spin states of the quarkonium H and over all other final-state particles X. $\mathcal{K}_n$, $\mathcal{K}_n'$ are gluonic operators (including no gluon content and covariant derivatives). The production matrix elements were assigned sizes according to the velocity scaling rules of [3], which, as discussed above, correspond to the weak-coupling regime. A great success of this formula was the explanation of charmonium production at the Tevatron [78], and it has been applied to a large number of production processes (see [79–81] for reviews), including some NLO calculations in photoproduction [82,83].

This factorization formalism has received a closer look recently in the framework of fragmentation functions, and has been proved to be correct at NNLO in $\alpha_s(m_Q)$, provided a slight redefinition of the matrix elements is carried out [84–86]. The interplay of the scales $m_Q v$, $m_Q v^2$ and Λ_{QCD} has not been discussed in detail for production. This may be important in resolving the polarization puzzle at the Tevatron [87], since the NRQCD prediction that heavy quarkonium must be produced transversely polarized in fragmentation processes [88], sometimes phrased as a smoking gun for NRQCD, depends crucially on the velocity counting used. For instance, in the strong-coupling regime one would not expect a sizable polarization [89].

Near certain kinematical endpoints NRQCD production processes must be supplemented with collinear degrees of freedom, in analogy to semi-inclusive decays discussed above. This issue has recently been addressed using SCET [90–93].

4.3 Exclusive production

Although no detailed formalism has been worked out for exclusive production, the basic ideas of NRQCD factorization have indeed been applied to it [94,95], mostly after the surprisingly large double charmonium cross-section first measured at Belle [96]. The traditional light-cone factorization does not seem to have major problems to accommodate this cross-section [97–99], although some modeling of the light-cone distribution amplitudes has so far been required. Recently, a further factorization of these objects in NRQCD has been presented [100], and a detailed comparison of the NRQCD and light-cone approaches has been carried out [101] (see also [102]). The latter suggests that the actual cross-section may well be accommodated into the NRQCD factorization results once higher-order contributions in the velocity expansion are taken into account.

5 Conclusions

The NRQCD formalism provides a solid QCD-based framework where heavy-quarkonium spectroscopy and inclusive decays can be systematically described starting from QCD. An NRQCD factorization formalism has also been put forward for semi-inclusive decays, inclusive production and, more recently, exclusive production, which, in spite of its successes, is not so well understood theoretically. Nevertheless, remarkable progress has been done recently concerning the structure of the factorization formulas for inclusive processes and the relation to the light-cone formalism for exclusive ones. Certain kinematical endpoints both in semi-inclusive decays and production require the inclusion of collinear degrees of freedom in NRQCD. Progress has also occurred here by combining NRQCD and SCET.

I acknowledge financial support from MEC (Spain) grant CYT FPA 2004-04582-C02-01, the CIRIT (Catalonia) grant 2005SGR00564, and the RTNs Euridice HPRN-CT2002-00311 and Flavianet MRTN-CT-2006-035482 (EU).

References

1. N. Brambilla *et al.*, arXiv:hep-ph/0412158.
2. W.E. Caswell, G.P. Lepage, Phys. Lett. B **167**, 437 (1986).
3. G.T. Bodwin, E. Braaten, G.P. Lepage, Phys. Rev. D **51**, 1125 (1995); **55**, 5853 (1997)(E) [arXiv:hep-ph/9407339].
4. N. Brambilla, A. Pineda, J. Soto, A. Vairo, Rev. Mod. Phys. **77**, 1423 (2005) [arXiv:hep-ph/0410047].
5. N. Brambilla, A. Vairo, T. Rosch, Phys. Rev. D **72**, 034021 (2005) [arXiv:hep-ph/0506065].
6. S. Fleming, T. Mehen, Phys. Rev. D **73**, 034502 (2006) [arXiv:hep-ph/0509313].
7. A. Pineda, J. Soto, Nucl. Phys. Proc. Suppl. **64**, 428 (1998) [arXiv:hep-ph/9707481].
8. N. Brambilla, A. Pineda, J. Soto, A. Vairo, Nucl. Phys. B **566**, 275 (2000) [arXiv:hep-ph/9907240].
9. M. Neubert, Phys. Rep. **245**, 259 (1994) [arXiv:hep-ph/9306320].
10. A. Vairo, these proceedings, arXiv:hep-ph/0610251.
11. A. Gray, I. Allison, C.T.H. Davies, E. Dalgic, G.P. Lepage, J. Shigemitsu, M. Wingate, Phys. Rev. D **72**, 094507 (2005) [arXiv:hep-lat/0507013].
12. G.T. Bodwin, E. Braaten, G.P. Lepage, Phys. Rev. D **46**, 1914 (1992) [arXiv:hep-lat/9205006].
13. G.T. Bodwin, D.K. Sinclair, S. Kim, Phys. Rev. D **65**, 054504 (2002) [arXiv:hep-lat/0107011].
14. A. Petrelli, M. Cacciari, M. Greco, F. Maltoni, M.L. Mangano, Nucl. Phys. B **514**, 245 (1998) [arXiv:hep-ph/9707223].
15. N. Brambilla, E. Mereghetti, A. Vairo, JHEP **0608**, 039 (2006) [arXiv:hep-ph/0604190].
16. G.T. Bodwin, A. Petrelli, Phys. Rev. D **66**, 094011 (2002) [arXiv:hep-ph/0205210].
17. J.P. Ma, Q. Wang, Phys. Lett. B **537**, 233 (2002) [arXiv:hep-ph/0203082].
18. M.E. Luke, A.V. Manohar, I.Z. Rothstein, Phys. Rev. D **61**, 074025 (2000) [arXiv:hep-ph/9910209].
19. H.W. Griesshammer, Phys. Rev. D **58**, 094027 (1998) [arXiv:hep-ph/9712467].
20. A.V. Manohar, I.W. Stewart, arXiv:hep-ph/0605001.
21. A.H. Hoang, I.W. Stewart, Phys. Rev. D **67**, 114020 (2003) [arXiv:hep-ph/0209340].
22. A. Manohar, these proceedings.
23. M. Peter, Nucl. Phys. B **501**, 471 (1997) [arXiv:hep-ph/9702245].
24. Y. Schroder, Phys. Lett. B **447**, 321 (1999) [arXiv:hep-ph/9812205].
25. B.A. Kniehl, A.A. Penin, Y. Schroder, V.A. Smirnov, M. Steinhauser, Phys. Lett. B **607**, 96 (2005) [arXiv:hep-ph/0412083].
26. N. Brambilla, A. Pineda, J. Soto, A. Vairo, Phys. Rev. D **60**, 091502 (1999) [arXiv:hep-ph/9903355].
27. N. Brambilla, X. Garcia i Tormo, J. Soto, A. Vairo, arXiv:hep-ph/0610143.
28. A. Pineda, J. Phys. G **29**, 371 (2003) [arXiv:hep-ph/0208031].
29. G.S. Bali, A. Pineda, Phys. Rev. D **69**, 094001 (2004) [arXiv:hep-ph/0310130].
30. B.A. Kniehl, A.A. Penin, V.A. Smirnov, M. Steinhauser, Nucl. Phys. B **635**, 357 (2002) [arXiv:hep-ph/0203166].
31. A.A. Penin, M. Steinhauser, Phys. Lett. B **538**, 335 (2002) [arXiv:hep-ph/0204290].

32. A. Pineda, Phys. Rev. D **65**, 074007 (2002) [arXiv:hep-ph/0109117].
33. A. Pineda, Phys. Rev. D **66**, 054022 (2002) [arXiv:hep-ph/0110216].
34. A. Pineda, Phys. Rev. A **66**, 062108 (2002) [arXiv:hep-ph/0204213].
35. B.A. Kniehl, A.A. Penin, A. Pineda, V.A. Smirnov, M. Steinhauser, Phys. Rev. Lett. **92**, 242001 (2004) [arXiv:hep-ph/0312086].
36. A.A. Penin, A. Pineda, V.A. Smirnov, M. Steinhauser, Nucl. Phys. B **699**, 183 (2004) [arXiv:hep-ph/0406175].
37. Y. Koma, these proceedings.
38. G.S. Bali, K. Schilling, A. Wachter, Phys. Rev. D **56**, 2566 (1997) [arXiv:hep-lat/9703019].
39. S. Necco, R. Sommer, Nucl. Phys. B **622**, 328 (2002) [arXiv:hep-lat/0108008].
40. Y. Koma, M. Koma, H. Wittig, Phys. Rev. Lett. **97**, 122003 (2006) [arXiv:hep-lat/0607009].
41. Y. Koma, M. Koma, arXiv:hep-lat/0609078.
42. F. Jugeau, H. Sazdjian, Nucl. Phys. B **670**, 221 (2003).
43. N. Brambilla, D. Eiras, A. Pineda, J. Soto, A. Vairo, Phys. Rev. Lett. **88**, 012003 (2002) [arXiv:hep-ph/0109130].
44. N. Brambilla, D. Eiras, A. Pineda, J. Soto, A. Vairo, Phys. Rev. D **67**, 034018 (2003) [arXiv:hep-ph/0208019].
45. N. Brambilla, A. Pineda, J. Soto, A. Vairo, Phys. Lett. B **580**, 60 (2004) [arXiv:hep-ph/0307159].
46. X. Garcia i Tormo, J. Soto, Phys. Rev. Lett. **96**, 111801 (2006) [arXiv:hep-ph/0511167].
47. M.B. Voloshin, Nucl. Phys. B **154**, 365 (1979).
48. H. Leutwyler, Phys. Lett. B **98**, 447 (1981).
49. N. Brambilla, Y. Sumino, A. Vairo, Phys. Lett. B **513**, 381 (2001) [arXiv:hep-ph/0101305].
50. N. Brambilla, Y. Sumino, A. Vairo, Phys. Rev. D **65**, 034001 (2002) [arXiv:hep-ph/0108084].
51. CLEO Collaboration (D. Besson *et al.*), Phys. Rev. D **74**, 012003 (2006) [arXiv:hep-ex/0512061].
52. C.W. Bauer, S. Fleming, M.E. Luke, Phys. Rev. D **63**, 014006 (2001) [arXiv:hep-ph/0005275].
53. C.W. Bauer, S. Fleming, D. Pirjol, I.W. Stewart, Phys. Rev. D **63**, 114020 (2001) [arXiv:hep-ph/0011336].
54. S. Fleming, C. Lee, A.K. Leibovich, Phys. Rev. D **71**, 074002 (2005) [arXiv:hep-ph/0411180].
55. J.P. Ma, Nucl. Phys. B **605**, 625 (2001); **611**, 523 (2001)(E) [arXiv:hep-ph/0103237].
56. J.H. Field, Phys. Rev. D **66**, 013013 (2002).
57. S. Wolf, Phys. Rev. D **63**, 074020 (2001).
58. S. Fleming, A.K. Leibovich, Phys. Rev. Lett. **90**, 032001 (2003) [arXiv:hep-ph/0211303].
59. S. Fleming, A.K. Leibovich, Phys. Rev. D **67**, 074035 (2003) [arXiv:hep-ph/0212094].
60. C.W. Bauer, C.W. Chiang, S. Fleming, A.K. Leibovich, I. Low, Phys. Rev. D **64**, 114014 (2001).
61. S. Fleming, A.K. Leibovich, Phys. Rev. D **70**, 094016 (2004) [arXiv:hep-ph/0407259].
62. X. Garcia i Tormo, J. Soto, Phys. Rev. D **69**, 114006 (2004) [arXiv:hep-ph/0401233].
63. X. Garcia i Tormo, J. Soto, Phys. Rev. D **72**, 054014 (2005) [arXiv:hep-ph/0507107].
64. A.H. Hoang *et al.*, EPJdirect C **2**, C3, 1 (2000) doi:10.1007/s1010500c003.
65. B.A. Kniehl, A.A. Penin, Nucl. Phys. B **577**, 197 (2000) [arXiv:hep-ph/9911414].
66. B.A. Kniehl, A.A. Penin, M. Steinhauser, V.A. Smirnov, Phys. Rev. Lett. **90**, 212001 (2003).
67. A.H. Hoang, Phys. Rev. D **69**, 034009 (2004).
68. M. Beneke, Y. Kiyo, K. Schuller, Nucl. Phys. B **714**, 67 (2005) [arXiv:hep-ph/0501289].
69. A.A. Penin, V.A. Smirnov, M. Steinhauser, Nucl. Phys. B **716**, 303 (2005) [arXiv:hep-ph/0501042].
70. P. Marquard, J.H. Piclum, D. Seidel, M. Steinhauser, arXiv:hep-ph/0607168.
71. A. Pineda, A. Signer, arXiv:hep-ph/0607239.
72. A.H. Hoang, C.J. Reisser, Phys. Rev. D **71**, 074022 (2005) [arXiv:hep-ph/0412258].
73. A.H. Hoang, C.J. Reisser, Phys. Rev. D **74**, 034002 (2006) [arXiv:hep-ph/0604104].
74. D. Eiras, M. Steinhauser, Nucl. Phys. B **757**, 197 (2006) [arXiv:hep-ph/0605227].
75. A.H. Hoang, PoS (TOP2006) 032 (2006) [arXiv:hep-ph/0604185].
76. A. Signer, PoS (TOP2006) 033 (2006) [arXiv:hep-ph/0604032].
77. X. Garcia i Tormo, arXiv:hep-ph/0610145.
78. E. Braaten, S. Fleming, Phys. Rev. Lett. **74**, 3327 (1995) [arXiv:hep-ph/9411365].
79. M. Kramer, Prog. Part. Nucl. Phys. **47**, 141 (2001).
80. G.T. Bodwin, J. Lee, R. Vogt, arXiv:hep-ph/0305034.
81. G.T. Bodwin, Int. J. Mod. Phys. A **21**, 785 (2006).
82. M. Klasen, B.A. Kniehl, L.N. Mihaila, M. Steinhauser, Nucl. Phys. B **713**, 487 (2005) [arXiv:hep-ph/0407014].
83. M. Klasen, B.A. Kniehl, L.N. Mihaila, M. Steinhauser, Phys. Rev. D **71**, 014016 (2005) [arXiv:hep-ph/0408280].
84. G.C. Nayak, J.W. Qiu, G. Sterman, Phys. Lett. B **613**, 45 (2005) [arXiv:hep-ph/0501235].
85. G.C. Nayak, J.W. Qiu, G. Sterman, Phys. Rev. D **72**, 114012 (2005) [arXiv:hep-ph/0509021].
86. G.C. Nayak, J.W. Qiu, G. Sterman, Phys. Rev. D **74**, 074007 (2006) [arXiv:hep-ph/0608066].
87. CDF Collaboration (A.A. Affolder *et al.*), Phys. Rev. Lett. **85**, 2886 (2000) [arXiv:hep-ex/0004027].
88. P.L. Cho, M.B. Wise, Phys. Lett. B **346**, 129 (1995).
89. S. Fleming, I.Z. Rothstein, A.K. Leibovich, Phys. Rev. D **64**, 036002 (2001) [arXiv:hep-ph/0012062].
90. S. Fleming, A.K. Leibovich, T. Mehen, Phys. Rev. D **68**, 094011 (2003) [arXiv:hep-ph/0306139].
91. K. Hagiwara, E. Kou, Z.H. Lin, C.F. Qiao, G.H. Zhu, Phys. Rev. D **70**, 034013 (2004) [arXiv:hep-ph/0401246].
92. Z.H. Lin, G.h. Zhu, Phys. Lett. B **597**, 382 (2004).
93. S. Fleming, A.K. Leibovich, T. Mehen, arXiv:hep-ph/0607121.
94. E. Braaten, J. Lee, Phys. Rev. D **67**, 054007 (2003) **72**, 099901 (2005)(E) [arXiv:hep-ph/0211085].
95. Y.J. Zhang, Y.J. Gao, K.T. Chao, Phys. Rev. Lett. **96**, 092001 (2006) [arXiv:hep-ph/0506076].
96. Belle Collaboration (K. Abe *et al.*), Phys. Rev. Lett. **89**, 142001 (2002) [arXiv:hep-ex/0205104].
97. J.P. Ma, Z.G. Si, Phys. Rev. D **70**, 074007 (2004).
98. A.E. Bondar, V.L. Chernyak, Phys. Lett. B **612**, 215 (2005) [arXiv:hep-ph/0412335].
99. V.V. Braguta, A.K. Likhoded, A.V. Luchinsky, Phys. Rev. D **74**, 094004 (2006) [arXiv:hep-ph/0602232].
100. J.P. Ma, Z.G. Si, arXiv:hep-ph/0608221.
101. G.T. Bodwin, D. Kang, J. Lee, arXiv:hep-ph/0603185.
102. V.V. Braguta, A.K. Likhoded, A.V. Luchinsky, arXiv:hep-ph/0611021.

Eur. Phys. J. A **31**, 711–713 (2007)

DOI 10.1140/epja/i2006-10243-1

THE EUROPEAN
PHYSICAL JOURNAL A

Special Article – QNP 2006

Relativistic NJL model with light and heavy quarks

A.L. Mota[1,2,a] and E. Ruiz Arriola[2]

[1] Departamento de Ciências Naturais, Universidade Federal de São João del Rei, 36301-160, São João del Rei, Brazil
[2] Departamento de Física Atómica, Molecular y Nuclear, Universidad de Granada, E-18071 Granada, Spain

Received: 23 November 2006
Published online: 12 March 2007 – © Società Italiana di Fisica / Springer-Verlag 2007

Abstract. We study the Nambu-Jona-Lasinio model with light and heavy quarks in a relativistic approach. We emphasize relevant regularization issues as well as the transition from light to heavy quarks. The approach of the electromagnetic meson form factor to the Isgur-Wise function in the heavy-quark limit is also discussed.

PACS. 11.30.-j Symmetry and conservation laws – 11.30.Rd Chiral symmetries – 12.39.-x Phenomenological quark models – 12.39.Fe Chiral Lagrangians

The physics of light and heavy quarks and their corresponding effective field theories cannot be more disparate even though a smooth transition between both limits is expected. In the case of light up, down and strange quarks the spontaneous breaking of chiral symmetry is the dominant feature which explains the mass gap between pions and kaons and the rest of the hadronic spectrum enabling the use of Chiral Perturbation Theory (ChPT) for energies much smaller than the mass gap [1,2]. In the opposite limit of heavy charm, bottom and top quarks, spin symmetry largely explains the degeneracy between hadronic states which differ only in the spin of the heavy quark like *e.g.* $B(5280)$ *vs.* $B^*(5325)$, or $D(1870)$ *vs.* $D^*(2010)$ and a systematic Heavy Quark Effective Theory (HQET) [3–5] can be designed for masses much larger than the mass gap. Besides these two fairly known extreme limits, the understanding of the transition from light to heavy quarks is not only of theoretical interest, but may also provide some insight into lattice simulations where the putative light quarks are most frequently artificially heavy. Unfortunately, there is no general framework describing the heavy-light transition in a model independent way, even though ChPT and HQET describe the extreme cases.

In HQET the heavy quark limit is taken *before* implementing dimensional regularization because as is well known heavy particles do not decouple in this regularization scheme. For a heavy quark the relevant degrees of freedom are given by

$$\Psi_v(x) = \frac{1 + \not{v}}{2} e^{i m_0 v \cdot x} \Psi(x), \qquad (1)$$

where $\Psi(x)$ is the heavy-quark spinor, m_0 is the heavy-quark mass, and v^μ is a quadrivector where the spacial components v corresponds to the velocity of the heavy quark and the time component is chosen in order to have $v^2 = 1$. After integrating out the irrelevant degrees of freedom, the resulting effective Lagrangian is expanded in $1/m_0$, and the propagator for the heavy-quark effective field, in leading order, is given by

$$S(k) = \frac{1}{v \cdot k + i\epsilon}, \qquad (2)$$

k^μ being the residual momentum of a heavy quark with total momentum $k^\mu + m_0 v^\mu$.

In this work we discuss the heavy-light transition with the guidance of the Nambu–Jona-Lasinio (NJL) model for quarks (for reviews see, *e.g.*, [6–10]). The corresponding Lagrangian reads

$$\mathcal{L} = \bar{\psi}(i\not{\partial} - \hat{m}_0)\psi - \frac{G}{2}((\bar{\psi}\lambda\psi)^2 + (\bar{\psi}i\gamma_5\lambda\psi)^2), \qquad (3)$$

where λ are the $N_f^2 - 1$ flavour $SU(N_f)$ Gell-Mann matrices and $\hat{m}_0 = \mathrm{diag}(m_{u0}, m_{d0}, m_{s0}, \ldots, m_{n0})$ is a diagonal current mass matrix which explicitly breaks chiral invariance. With the exception of the mass term all flavours are treated on the *same footing* and Lagrangian (3) is invariant under the $SU_R(N_f) \otimes SU_L(N_f)$ chiral group and also under $SU(N_c)$ global transformations. Summation over color and flavour indexes is implicit. As is well known the NJL model is not renormalizable and a finite cut-off Λ is required to make sense of it. A technical, but crucial, issue is that of the finite cut-off regularization method and its consistency with the gauge and chiral symmetries.

Quark models have been studied in the last decade in connection to the heavy quark effective Lagrangian

a Spokesperson; e-mail: motaal@ufsj.edu.br

obtained from HQET and its interplay with chiral quark models [11–14]. The basic underlying assumption is that all quark species are assumed to have the same kind of contact interactions and the mass terms are indeed treated in a rather asymmetric fashion. However, mimicking HQET itself, in HQET models, the infinite heavy-quark limit is taken *before* applying a definite regularization scheme in the heavy sector. Whether or not the $m_0 \to \infty$ limit commutes with the regularization procedure is not obvious. In addition, due to the non-renormalizability of the models, as in the light quarks NJL model, the finite regularization procedure employed is a part of the model and different regularizations can lead to different results. The choice of a regularization procedure that does not violate symmetries of the model becomes relevant, and in particular naively shifting the internal momenta of the loops may be dangerous. There is no reason *a priori* why neglecting finite cut-off corrections is more justified in the heavy sector as it is in the light sector.

Following previous experience we use the Pauli-Villars method with two subtractions in the coincidence limit. The proper way to do this is by bosonization and separation of the effective action into normal and abnormal parity contributions [15]. After integrating out the fermions one gets the normal parity contribution to the effective action

$$S_{\text{even}} = -\frac{iN_c}{2}\sum_i c_i \operatorname{Tr} \log(\mathbf{D}_5 \mathbf{D} + \Lambda_i^2)$$
$$-\frac{1}{4G_S}\int \mathrm{d}^4 x \operatorname{tr}_f (S^2 + P^2), \tag{4}$$

where $\mathbf{D}$ and $\mathbf{D}_5$ are Dirac operators given by

$$\mathbf{D} = +i\slashed{\partial} - S - \hat{m}_0 - i\gamma_5 P + \slashed{v} + \slashed{a}\gamma_5,$$
$$\mathbf{D}_5 = -i\slashed{\partial} - S - \hat{m}_0 + i\gamma_5 P - \slashed{v} + \slashed{a}\gamma_5, \tag{5}$$

where $S = \lambda^a S^a$, $P = \lambda^a P^a$ are dynamical fields and $v^\mu = \lambda^a v_a^\mu$ and $a^\mu = \lambda^a a_a^\mu$ stand for external sources. The Pauli-Villars regulators fulfill $c_0 = 1$, $\Lambda_0 = 0$ and the conditions $\sum_i c_i = 0$, $\sum_i c_i \Lambda_i^2 = 0$ which render finite the logarithmic and quadratic divergences respectively. In practice, we take two cut-offs in the coincidence limit $\Lambda_1 \to \Lambda_2 = \Lambda$ and hence $\sum_i c_i f(\Lambda_i^2) = f(0) - f(\Lambda^2) + \Lambda^2 f'(\Lambda^2)$. For light quarks with small current masses, m_0, the dynamical breaking of chiral symmetry generates a constituent mass, M, for the quarks, and pion physics phenomenology yields values $\Lambda \sim 1\,\text{GeV}$ for $M \sim 300\,\text{MeV}$ and both the constituent as well as the current masses are much smaller than the cut-off $m_0 \ll M \ll \Lambda$.

Naively, one would expect that, as a matter of principle, processes involving scales above the cut-off cannot be reliably addressed by the model. However, this is not necessarily so. A compelling example is provided by the study of high energy processes which involve asymptotically large momenta $Q^2 \gg \Lambda^2$ which enable the determination of mesonic parton leading twist distributions and amplitudes [16]. The surprisingly good agreement found in such an analysis, at least for the pseudoscalar

bosons, when gluonic radiative corrections via QCD evolution equations are implemented, suggests not only that there is nothing fundamentally wrong in looking at high scales as compared to the model cut-off but also that a rather acceptable description of existing data may be achieved. An important lesson learned from these studies was that a sloppy treatment of the finite cut-off regularization violates significantly relevant constraints regarding gauge and relativistic invariances which control the normalization and momentum fraction shared by the constituents, respectively. With this insights in mind we dare to explore with the necessary provisos the NJL model for *any current quark masses* including as a particular case the heavy-quark limit, *i.e.* for current quark masses much larger than the cut-off $m_0 \gg \Lambda$.

The relevant observable quantities can be read off from the effective action, by collecting the coefficients of the corresponding terms. The electroweak decay constant appears as the coefficient of the term involving one axial vector current and one pseudoscalar meson field. From now on, we will use m to denote the total mass of a light quark and m_0 for the total mass of a heavy quark. For a given channel involving one light quark and one heavy quark, one has (PV regularization over-understood)

$$p^\mu f_M(p^2) = i\int \frac{\mathrm{d}^4 k}{(2\pi)^4}\operatorname{Tr}\left\{i\gamma_5 \frac{i}{(\slashed{k} + m_0\slashed{v}) - m_0}\right.$$
$$\left.\times \gamma^\mu \gamma_5 \frac{i}{(\slashed{k} - \slashed{p}) - m}\right\}, \tag{6}$$

and $f_M(M_\Phi^2)$ is the heavy-meson electroweak-decay constant, where M_Φ stands for the heavy-light meson mass.

In the heavy-quark limit, the Isgur-Wise function is a universal form factor, defined as the matrix elements of the electroweak heavy-to-heavy currents between two heavy mesons of different non-relativistic velocities [17]. Within the Nambu–Jona-Lasinio model with heavy quarks, it can be computed as the heavy-quark limit of the electromagnetic form factor for an arbitrary current quark mass, yielding

$$\Gamma^\mu(p, p') = -i(\Gamma_0^\mu(p, p') + \Gamma_0^\mu(-p', -p)), \tag{7}$$

where, formally

$$\Gamma_0^\mu(p, p') = \int \frac{\mathrm{d}^4 k}{(2\pi)^4} Tr\left\{\frac{\slashed{k} - m}{k^2 - m^2}\right.$$
$$\left.\times \frac{\slashed{k} + m_0\slashed{v} + \slashed{p} + m_0}{(k + m_0 v + p)^2 - m_0^2}\gamma^\mu \frac{\slashed{k} + m_0\slashed{v} + \slashed{p}' + m_0}{(k + m_0 v + p')^2 - m_0^2}\right\}. \tag{8}$$

The form factor is computed with on-shell mesons ($p^2 = p'^2 = (\Delta M)^2$), $\Delta M = M_\Phi - m_0$ and we choose the arbitrary parameter v such as $v = \frac{p}{\sqrt{p^2}}$. We also define $\omega = \frac{p \cdot p'}{(\Delta M)^2}$. As before, we are assuming a Pauli-Villars gauge invariant regularization scheme. The explicit result will be given elsewhere [18].

In fig. 1 we compare our result to that of Ebert *et al.* [11]. To see clearly the effect when comparing to, we

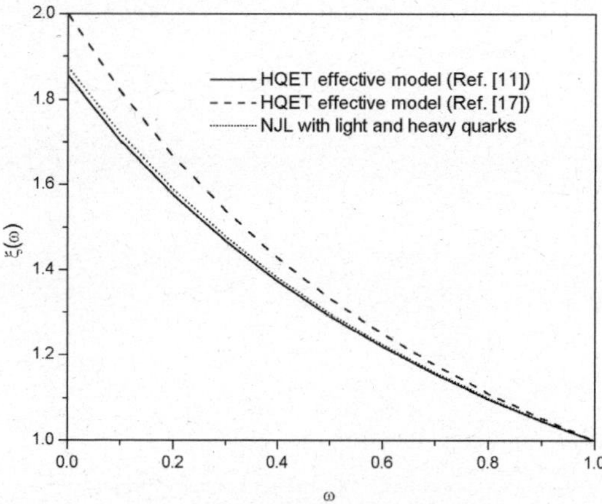

Fig. 1. The Isgur-Wise function for the NJL model and other HQET effective models. (Solid line: ref. [11], dashed line: ref. [19] (with $m = 0$), dotted line: present approach.)

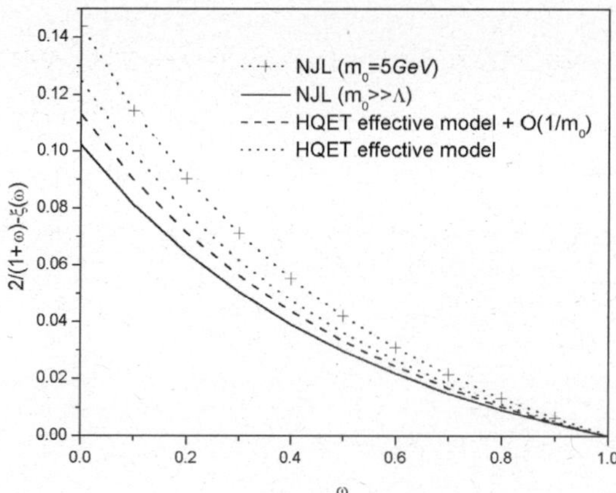

Fig. 2. Effects of $\frac{1}{m_0}$ corrections to the Isgur-Wise function. The present results are represented by the solid ($m_0 \to \infty$) and dotted with crosses ($m_0 = 5\,\text{GeV}$) lines. The HQET effective model [11] results are represented by the dotted ($m_0 \to \infty$) and dashed ($m_0 = 5\,\text{GeV}$) lines.

choose the set of parameters to reproduce $f_\pi = 93\,\text{MeV}$, $m_\pi = 140\,\text{MeV}$, $m_\rho = 770\,\text{MeV}$ and $M_B = 5.3\,\text{GeV}$, obtaining $m_u = m_d = 300\,\text{MeV}$, $\Lambda = 875\,\text{MeV}$, $m_s = 510\,\text{MeV}$, $m_B = 5.1\,\text{GeV}$ and $G = 2.9\,\text{GeV}^{-2}$. In contrast with [11], no independent coupling for the heavy-meson sector needs to be inserted on the model in order to reproduce the light-light and heavy-light mesons masses, although a very low B-meson weak decay constant is obtained ($f_B = 59\,\text{MeV}$). Nevertheless, other processes that could be important to the correct description of this decay constant, as mesons loops, are absent in the present treatment. In the same figure, we also compare to the HQET effective model presented in [19] where

$$\xi(\omega) = \frac{2}{1+\omega}. \qquad (9)$$

in both $m_0 \to \infty$ and $m \to 0$ limits. As we see, differences become more significant as ω goes to zero. The results obtained from the present model are very close to the results

obtained in [11]. The differences of both models to the results presented in [19] are due to finite light-*constituent*-quark mass effects: even in the $m_0 \to \infty$ limit, result (9) is only achieved when $m = 0$.

In fig. 2 we present the comparison between the behaviour of the Isgur-Wise function as m_0 goes from 5 GeV to 50 GeV ($m_0 \gg \Lambda$), computed on both present and [11] models (corrections of order $1/m_0$ on the HQET effective model of ref. [11] were included). As can be viewed from fig. 2, as m_0 increases the slope of the Isgur-Wise function on the present model increases, while the same slope decreases on the model presented in [11]. The magnitude of the changes on the slope of the Isgur-Wise function is also different in the two models.

Finally, let us mention that a derivative expansion of the bosonized version of the model can be employed to construct an effective mesonic Lagrangian, as was done for the light quark sector [20]. One problem is that such an expansion assumes small momenta for the corresponding meson bosonized fields, while for heavy mesons they are large. It is possible, however, to overcome this problem in a way that the treatment of light and heavy mesons is as symmetric as possible, so that at any stage the heavy pseudoscalars would become Goldstone bosons if the quarks were light. Further details will be further elaborated elsewhere [18].

This work is supported by CAPES-Brazil and Spanish DGI and FEDER funds with grant no. BFM2002-03218, Junta de Andalucía grant No. FQM-225, and EU RTN Contract CT2002-0311 (EURIDICE)

References

1. J. Gasser, H. Leutwyler, Ann. Phys. (N.Y.) **158**, 142 (1984).
2. J.F. Donoghue, E. Golowich, B.R. Holstein, Cambridge Monogr. Part. Phys. Nucl. Phys. Cosmol. **2**, 1 (1992).
3. H. Georgi, Phys. Lett. B **240**, 447 (1990).
4. M. Neubert, Phys. Rep. **245**, 259 (1994).
5. A.V. Manohar, M.B. Wise, Cambridge Monogr. Part. Phys. Nucl. Phys. Cosmol. **10**, 1 (2000).
6. U. Vogl, W. Weise, Prog. Part. Nucl. Phys. **27**, 195 (1991).
7. S.P. Klevansky, Rev. Mod. Phys. **64**, 649 (1992).
8. T. Hatsuda, T. Kunihiro, Phys. Rep. **247**, 221 (1994).
9. R. Alkofer, H. Reinhardt, H. Weigel, Phys. Rep. **265**, 139 (1996).
10. C.V. Christov *et al.*, Prog. Part. Nucl. Phys. **37**, 91 (1996).
11. D. Ebert, T. Feldmann, R. Friedrich, H. Reinhardt, Nucl. Phys. B **434**, 619 (1995).
12. A. Deandrea, N. Di Bartolomeo, R. Gatto, G. Nardulli, A.D. Polosa, Phys. Rev. D **58**, 034004 (1998).
13. A. Hiorth, J.O. Eeg, Phys. Rev. D **66**, 074001 (2002).
14. W.A. Bardeen, E.J. Eichten, C.T. Hill, Phys. Rev. D **68**, 054024 (2003).
15. C. Schuren, E. Ruiz Arriola, K. Goeke, Nucl. Phys. A **547**, 612 (1992).
16. E. Ruiz Arriola, Acta Phys. Pol. B **33**, 4443 (2002).
17. N. Isgur, M.B. Wise, Phys. Lett. B **232**, 113 (1989).
18. A.L. Mota, E. Ruiz Arriola, in preparation.
19. W.A. Bardeen, C.T. Hill, Phys. Rev. D **49**, 409 (1994).
20. D. Ebert, H. Reinhardt, Nucl. Phys. B **271**, 188 (1986).

Eur. Phys. J. A **31**, 714–717 (2007)

DOI 10.1140/epja/i2006-10270-x

Special Article – QNP 2006

Study of semileptonic and nonleptonic decays of the B_c^- meson

E. Hernández[1,a], J. Nieves[2], and J.M. Verde-Velasco[1]

[1] Grupo de Física Nuclear, Departamento de Física Fundamental e IUFFyM, Universidad de Salamanca, Spain
[2] Departamento de Física Atómica, Molecular y Nuclear, Universidad de Granada, Spain

Received: 8 December 2006
Published online: 21 March 2007 – © Società Italiana di Fisica / Springer-Verlag 2007

Abstract. We evaluate semileptonic and two-meson nonleptonic decays of the B_c^- meson in the framework of a nonrelativistic quark model. The former are done in spectator approximation using one-body current operators at the quark level. Our model reproduces the constraints of heavy-quark spin symmetry obtained in the limit of infinite heavy-quark mass. For the two-meson nonleptonic decays we work in factorization approximation. We compare our results to the ones obtained in different relativistic approaches.

PACS. 12.39.Hg Heavy quark effective theory – 12.39.Jh Nonrelativistic quark model – 13.20.Fc Decays of charmed mesons – 13.20.He Decays of bottom mesons

1 Introduction

In this work we have studied, in the framework of a nonrelativistic quark model, exclusive semileptonic and two-meson nonleptonic decays of the B_c^- meson driven by a $b \to c$ or $\bar{c} \to \bar{d}, \bar{s}$ transitions at the quark level. We have not considered semileptonic processes driven by the quark $b \to u$ transition to avoid known problems both at too high-q^2 transfers, where one might need to include the exchange of B_c^*-resonances, and at too low q^2 where recoil effects could be important [1].

In order to check the sensitivity of our results to the interquark interaction we have used five different quark-quark potentials taken from refs. [2,3]. All the potentials used have a confinement term plus Coulomb and hyperfine terms coming from one-gluon exchange, and differ from one another in the power of the confining term or in the use of different form factors in the Coulomb and hyperfine terms. Their free parameters had been adjusted to reproduce the light- and heavy-light meson spectra. Our central results have been obtained with the AL1 potential of ref. [3], while our errors show the spread of the results when using the other four potentials.

A detailed account of the full contents of this work is now available in ref. [4].

2 Semileptonic decays

We have made our calculations in the spectator approximation using one-body current operators. In table 1 we

Table 1. Branching ratios in % for semileptonic $B_c^- \to c\bar{c}$ and $B_c^- \to \overline{B}$ decays. In the first part of the table l stands for $l = e, \mu$.

	This work	[5]	[7]
$B_c^- \to \eta_c\, l^-\, \bar{\nu}_l$	$0.48^{+0.02}$	0.81	0.42
$B_c^- \to \eta_c\, \tau^-\, \bar{\nu}_\tau$	$0.17^{+0.01}$	0.22	
$B_c^- \to \chi_{c0}\, l^-\, \bar{\nu}_l$	$0.11^{+0.01}$	0.17	
$B_c^- \to \chi_{c0}\, \tau^-\, \bar{\nu}_\tau$	$0.013^{+0.001}$	0.013	
$B_c^- \to J/\Psi\, l^-\, \bar{\nu}_l$	$1.54^{+0.06}$	2.07	1.23
$B_c^- \to J/\Psi\, \tau^-\, \bar{\nu}_\tau$	$0.41^{+0.02}$	0.49	
$B_c^- \to \chi_{c1}\, l^-\, \bar{\nu}_l$	$0.066^{+0.003}_{-0.002}$	0.092	
$B_c^- \to \chi_{c1}\, \tau^-\, \bar{\nu}_\tau$	$0.0072^{+0.0002}_{-0.0003}$	0.0089	
$B_c^- \to h_c\, l^-\, \bar{\nu}_l$	$0.17^{+0.02}$	0.27	
$B_c^- \to h_c\, \tau^-\, \bar{\nu}_\tau$	$0.015^{+0.001}$	0.017	
$B_c^- \to \chi_{c2}\, l^-\, \bar{\nu}_l$	$0.13^{+0.01}$	0.17	
$B_c^- \to \chi_{c2}\, \tau^-\, \bar{\nu}_\tau$	$0.0093^{+0.0002}_{-0.0005}$	0.0082	
$B_c^- \to \Psi(3836)\, l^-\, \bar{\nu}_l$	$0.0043_{-0.0005}$	0.0066	
$B_c^- \to \Psi(3836)\, \tau^-\, \bar{\nu}_\tau$	$0.000083_{-0.000010}$	0.000099	

	This work	[6]	[8]
$B_c^- \to \overline{B}^0\, e\, \bar{\nu}_e$	$0.046^{+0.004}_{-0.007}$	0.071	0.042
$B_c^- \to \overline{B}^0\, \mu\, \bar{\nu}_\mu$	$0.044^{+0.005}_{-0.006}$		
$B_c^- \to \overline{B}_s^0\, e\, \bar{\nu}_e$	$1.06^{+0.05}_{-0.02}$	1.10	0.84
$B_c^- \to \overline{B}_s^0\, \mu\, \bar{\nu}_\mu$	$1.02^{+0.04}_{-0.02}$		
$B_c^- \to \overline{B}^{*0}\, e\, \bar{\nu}_e$	$0.11^{+0.01}_{-0.01}$	0.063	0.12
$B_c^- \to \overline{B}^{*0}\, \mu\, \bar{\nu}_\mu$	$0.11^{+0.01}_{-0.02}$		
$B_c^- \to \overline{B}_s^{*0}\, e\, \bar{\nu}_e$	$2.35^{+0.14}_{-0.10}$	2.37	1.75
$B_c^- \to \overline{B}_s^{*0}\, \mu\, \bar{\nu}_\mu$	$2.22^{+0.12}_{-0.10}$		

---

[a] e-mail: gajatee@usal.es

Table 2. Forward-backward asymmetry A_{FB} of the final charged lepton (e, μ or τ) measured in the leptons center-of-mass frame.

	$A_{FB}(e)$	$A_{FB}(\mu)$	$A_{FB}(\tau)$
$B_c \to \eta_c$			
This work	$0.60^{+0.01}\ 10^{-6}$	$0.13^{+0.01}\ 10^{-1}$	0.35
[5]	$0.953\ 10^{-6}$		0.36
$B_c \to \chi_{c0}$			
This work	$0.72^{+0.02}\ 10^{-6}$	$0.15\ 10^{-1}$	0.40
[5]	$1.31\ 10^{-6}$		0.39
$B_c \to J/\Psi$			
This work	-0.19	$-0.18_{-0.01}$	$-0.35^{+0.02}\ 10^{-1}$
[5]	-0.21		$-0.48\ 10^{-1}$
$B_c \to \chi_{c1}$			
This work	$-0.60_{-0.01}$	$-0.60_{-0.01}$	-0.46
[5]	0.19		0.34
$B_c \to h_c$			
This work	$-0.83_{-0.05}\ 10^{-2}$	$0.97^{+0.01}_{-0.05}\ 10^{-2}$	0.35
[5]	$-3.6\ 10^{-2}$		0.31
$B_c \to \chi_{c2}$			
This work	-0.14	-0.13	$0.55^{+0.02}\ 10^{-1}$
[5]	-0.16		$0.44\ 10^{-1}$
$B_c \to \Psi(3836)$			
This work	-0.59	-0.59	-0.42
[5]	0.21		0.41

	$A_{FB}(e)$	$A_{FB}(\mu)$
$B_c^- \to \overline{B}^0$		
This work	$0.67^{+0.02}\ 10^{-5}$	$0.82^{+0.01}\ 10^{-1}$
$B_c^- \to \overline{B}_s^0$		
This work	$0.89^{+0.01}\ 10^{-5}$	$0.96^{+0.01}\ 10^{-1}$
$B_c^- \to \overline{B}^{*0}$		
This work	$0.17_{-0.01}$	$0.19_{-0.01}$
$B_c^- \to \overline{B}_s^{*0}$		
This work	$0.14_{-0.01}$	$0.16^{+0.01}$

show our branching ratios for semileptonic B_c^- decays and compare them to the results obtained by Ivanov *et al.* [5,6], within a relativistic quark model calculation, and Ebert *et al.* [7,8], within the quasipotential approach to the relativistic quark model. The three calculations give similar results.

In table 2 we show the forward-backward asymmetry, with the forward direction being defined by the three-momentum of the final meson, of the charged lepton measured in the leptons center-of-mass frame. Our results are in reasonable agreement with the ones of ref. [5] with two exceptions corresponding to final χ_{c1} and $\Psi(3836)$, where not even the sign of the asymmetry is the same.

For the decay $B_c^- \to J/\Psi\, l\bar{\nu}_l$ with the J/Ψ decaying into a $\mu^+\mu^-$ pair one can evaluate another angular asymmetry. Calling x_μ the cosine of the polar angle of the $\mu^+\mu^-$ pair, measured in the $\mu^+\mu^-$ rest frame relative to the mo-

Table 3. α^* angular asymmetry for the $\mu^+\mu^-$ pair in the decay $B_c^- \to \mu^+\mu^- (J/\Psi)\, l\bar{\nu}_l$.

	$\alpha^*(e)$	$\alpha^*(\mu)$	$\alpha^*(\tau)$
$B_c \to J/\Psi$			
This work	$-0.29_{-0.01}$	-0.29	-0.19
[5]	-0.34		-0.24

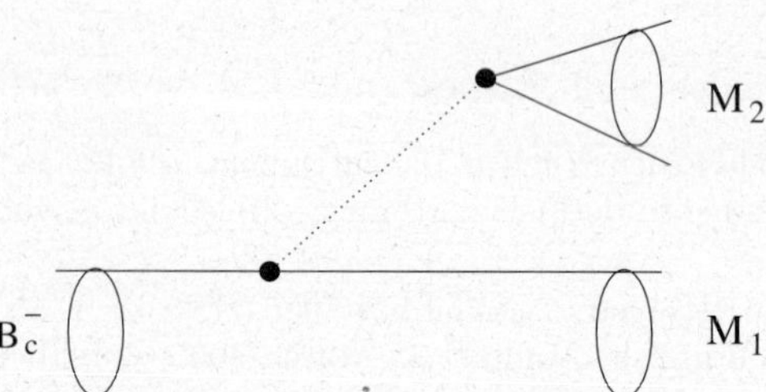

Fig. 1. Diagrammatic representation of the B_c^- two-meson nonleptonic decay in the factorization approximation.

mentum of the decaying J/Ψ, we have that the differential decay width $d\Gamma_{B_c^- \to \mu^+\mu^-(J/\Psi)l\bar{\nu}_l}/dx_\mu \equiv 1 + \alpha^* x_\mu$. Our results for the asymmetry parameter α^* are given in table 3. They are in reasonable agreement with the ones obtained by Ivanov *et al.* [5].

2.1 Heavy quark spin symmetry

While one cannot apply ordinary heavy-quark symmetry to hadrons with two heavy quarks, there is a symmetry that survives for those systems, which is the heavy-quark spin symmetry (HQSS). This symmetry reflects the fact that for infinite heavy-quark masses the spins of the two heavy quarks decouple.

We have checked that our model reproduces the constraints imposed by HQSS in the infinite heavy-quark mass limit. Those constraints relate the form factors for B_c semileptonic decays into final 0^- and 1^- mesons near the zero recoil point [9].

For the actual heavy-quark masses we find small deviations from the infinite heavy-quark mass limit relations for the $B_c \to \eta_c$, J/Ψ case, while corrections are large for some of the form factors in the $B_c \to \overline{B}^0, \overline{B}^{*0}$ and $B_c \to \overline{B}_s^0, \overline{B}_s^{*0}$ cases. See ref. [4] for details.

3 Two-meson nonleptonic decays

Neglecting penguin contributions, the effective Hamiltonians used for these calculations are given by local four-quark operators of the current-current type [7,10]. We work, as it is usually done, in factorization approximation in which the amplitude only contains contributions as the ones depicted in fig. 1. To evaluate the amplitude we need different meson decay constants that we take from experiment or lattice data. For the η_c we use our own theoretical result obtained with the model described in ref. [11].

Table 4. Branching ratios in % for exclusive two-meson non-leptonic decays of the B_c^- meson that include a $c\bar{c}$ meson in the final state.

	This work	[10]	[7]
$B_c^- \to \eta_c\, \pi^-$	$0.094^{+0.006}$	0.19	0.083
$B_c^- \to \eta_c\, \rho^-$	$0.24^{+0.01}$	0.45	0.20
$B_c^- \to \eta_c\, K^-$	$0.0075^{+0.0005}$	0.015	0.006
$B_c^- \to \eta_c\, K^{*-}$	$0.013^{+0.001}$	0.025	0.011
$B_c^- \to J/\Psi\, \pi^-$	$0.076^{+0.008}$	0.17	0.060
$B_c^- \to J/\Psi\, \rho^-$	$0.24^{+0.02}$	0.49	0.16
$B_c^- \to J/\Psi\, K^-$	$0.0060^{+0.0006}$	0.013	0.005
$B_c^- \to J/\Psi\, K^{*-}$	$0.014^{+0.002}$	0.028	0.010
$B_c^- \to \chi_{c0}\, \pi^-$	$0.026^{+0.003}$	0.055	
$B_c^- \to \chi_{c0}\, \rho^-$	$0.067^{+0.006}_{-0.001}$	0.13	
$B_c^- \to \chi_{c0}\, K^-$	$0.0020^{+0.0002}$	0.0042	
$B_c^- \to \chi_{c0}\, K^{*-}$	$0.0037^{+0.0005}$	0.0070	
$B_c^- \to \chi_{c1}\, \pi^-$	$0.00014^{+0.00001}$	0.0068	
$B_c^- \to \chi_{c1}\, \rho^-$	$0.010^{+0.001}_{-0.001}$	0.029	
$B_c^- \to \chi_{c1}\, K^-$	$1.1^{+0.1}\,10^{-5}$	$5.1\,10^{-4}$	
$B_c^- \to \chi_{c1}\, K^{*-}$	$0.00073^{+0.00007}_{-0.00002}$	0.0018	
$B_c^- \to h_c\, \pi^-$	$0.053^{+0.007}$	0.11	
$B_c^- \to h_c\, \rho^-$	$0.13^{+0.01}$	0.25	
$B_c^- \to h_c\, K^-$	$0.0041^{+0.0006}$	0.0083	
$B_c^- \to h_c\, K^{*-}$	$0.0071^{+0.0008}$	0.013	
$B_c^- \to \chi_{c2}\, \pi^-$	$0.022^{+0.002}$	0.046	
$B_c^- \to \chi_{c2}\, \rho^-$	$0.065^{+0.006}_{-0.002}$	0.12	
$B_c^- \to \chi_{c2}\, K^-$	$0.0017^{+0.0001}$	0.0034	
$B_c^- \to \chi_{c2}\, K^{*-}$	$0.0038^{+0.0003}_{-0.0002}$	0.0065	
$B_c^- \to \Psi(3836)\, \pi^-$	$4.1^{+0.03}_{-0.02}\,10^{-5}$	0.0017	
$B_c^- \to \Psi(3836)\, \rho^-$	$0.0020_{-0.0003}$	0.0055	
$B_c^- \to \Psi(3836)\, K^-$	$3.1^{+0.2}_{-0.2}\,10^{-6}$	0.00012	
$B_c^- \to \Psi(3836)\, K^{*-}$	$0.00014_{-0.00002}$	0.00032	

	This work	[10]	[12]	[13]
$B_c^- \to \eta_c\, D^-$	$0.014^{+0.001}$	0.019	0.014	0.015
$B_c^- \to \eta_c\, D^{*-}$	$0.012^{+0.001}$	0.019	0.013	0.010
$B_c^- \to J/\Psi\, D^-$	$0.0083^{+0.0005}$	0.015	0.009	0.009
$B_c^- \to J/\Psi\, D^{*-}$	$0.031^{+0.001}$	0.045	0.028	0.028
$B_c^- \to \eta_c\, D_s^-$	$0.44^{+0.02}$	0.44	0.26	0.28
$B_c^- \to \eta_c\, D_s^{*-}$	$0.24^{+0.02}$	0.37	0.24	0.27
$B_c^- \to J/\Psi\, D_s^-$	$0.24^{+0.02}$	0.34	0.15	0.17
$B_c^- \to J/\Psi\, D_s^{*-}$	$0.68^{+0.03}$	0.97	0.55	0.67

Table 5. Branching ratios in % for exclusive two-meson non-leptonic decays of the B_c^- meson that include a $\overline{B}$, B meson in the final state.

	This work	[10]	[8]
$B_c^- \to \overline{B}^0\, \pi^-$	$0.11^{+0.01}_{-0.01}$	0.20	0.10
$B_c^- \to \overline{B}^0\, \rho^-$	$0.14^{+0.02}_{-0.02}$	0.20	0.13
$B_c^- \to \overline{B}^0\, K^-$	$0.010^{+0.001}_{-0.001}$	0.015	0.009
$B_c^- \to \overline{B}^0\, K^{*-}$	$0.0039^{+0.0003}_{-0.0005}$	0.0048	0.004
$B_c^- \to \overline{B}^{*0}\, \pi^-$	$0.072^{+0.012}_{-0.012}$	0.057	0.026
$B_c^- \to \overline{B}^{*0}\, \rho^-$	$0.58^{+0.05}_{-0.08}$	0.30	0.67
$B_c^- \to \overline{B}^{*0}\, K^-$	$0.0048^{+0.0007}_{-0.0008}$	0.0036	0.004
$B_c^- \to \overline{B}^{*0}\, K^{*-}$	$0.030^{+0.002}_{-0.004}$	0.013	0.032
$B_c^- \to \overline{B}_s^0\, \pi^-$	$3.51^{+0.19}_{-0.06}$	3.9	2.46
$B_c^- \to \overline{B}_s^0\, \rho^-$	$2.34^{+0.05}_{-0.06}$	2.3	1.38
$B_c^- \to \overline{B}_s^0\, K^-$	$0.29^{+0.01}_{-0.01}$	0.29	0.21
$B_c^- \to \overline{B}_s^0\, K^{*-}$	$0.013_{-0.001}$	0.011	0.0030
$B_c^- \to \overline{B}_s^{*0}\, \pi^-$	$2.34^{+0.19}_{-0.14}$	2.1	1.58
$B_c^- \to \overline{B}_s^{*0}\, \rho^-$	$13.4^{+0.5}_{-0.6}$	11	10.8
$B_c^- \to \overline{B}_s^{*0}\, K^-$	$0.13^{+0.01}_{-0.01}$	0.13	0.11

	This work	[10]	[8]
$B_c^- \to B^-\, \pi^0$	$0.0038^{+0.0005}_{-0.0006}$	0.007	0.004
$B_c^- \to B^-\, \rho^0$	$0.0050^{+0.0004}_{-0.0007}$	0.0071	0.005
$B_c^- \to B^-\, K^0$	$0.25^{+0.03}_{-0.04}$	0.38	0.24
$B_c^- \to B^-\, K^{*0}$	$0.093^{+0.006}_{-0.013}$	0.11	0.09
$B_c \to B^{*-}\, \pi^0$	$0.0025^{+0.0004}_{-0.0005}$	0.0020	0.001
$B_c \to B^{*-}\, \rho^0$	$0.020^{+0.002}_{-0.003}$	0.011	0.024
$B_c \to B^{*-}\, K^0$	$0.12^{+0.02}_{-0.02}$	0.088	0.11
$B_c \to B^{*-}\, K^{*0}$	$0.73^{+0.06}_{-0.10}$	0.32	0.84

In table 4 we show results for decays that include a $c\bar{c}$ meson in the final state. For decays with a π^-, ρ^-, K^-, K^{*-} meson in the final state we find good agreement with the data by Ebert *et al.* [7] while our results are roughly a factor of two smaller than the ones obtained by Ivanov *et al.* [10]. For decays with a D^-, D^{*-}, D_s^-, D_s^{*-} meson in the final state we are in good agreement with the results by El-Hady *et al.* [12], obtained using the Bethe-Salpeter equation, and the ones by Kiselev [13], obtained within the three point sum rules of QCD and nonrelativistic QCD.

In table 5 we show results for decays that include a meson with a b quark. Our branching ratios for decays with a $\overline{B}^0$, $\overline{B}^{*0}$ meson in the final state are in very good agreement, being $B_c^- \to \overline{B}^{*0}\pi^-$ the exception, with the results by Ebert *et al.* [8], while for the case with a $\overline{B}_s^0$, $\overline{B}_s^{*0}$ meson in the final state we are in very good agreement with the results by Ivanov *et al.* [10]. Finally, for the case with a B^-, B^{*-} meson in the final state we find, with just one exception ($B_c^- \to B^{*-}\pi^0$), very good agreement with the results by Ebert *et al.* [8].

This research was supported by DGI and FEDER funds, under contracts FIS2005-00810, BFM2003-00856 and FPA2004-05616, by Junta de Andalucía and Junta de Castilla y León under contracts FQM0225 and SA104/04, and it is part of the EU integrated infrastructure initiative Hadron Physics Project under contract number RII3-CT-2004-506078. J.M.V.-V. acknowledges a contract E.P.I.F. with the University of Salamanca.

References

1. C. Albertus, J.M. Flynn, E. Hernández, J. Nieves, J.M. Verde-Velasco, Phys. Rev. D **72**, 033002 (2005).
2. R.K. Bhaduri, L.E. Cohler, Y. Nogami, Nuovo Cimento A **65**, 376 (1981).
3. B. Silvestre-Brac, Few-Body Syst. **20**, 1 (1996).
4. E. Hernández, J. Nieves, J.M. Verde-Velasco, Phys. Rev. D **74**, 074008 (2006).
5. M.A. Ivanov, J.G. Körner, P. Santorelli, Phys. Rev. D **71**, 094006 (2005); **75**, 019901 (2007)(E).
6. M.A. Ivanov, J.G. Körner, P. Santorelli, Phys. Rev. D **63**, 074010 (2001).
7. D. Ebert, R.N. Faustov, V.O. Galkin, Phys. Rev. D **68**, 094020 (2003).
8. D. Ebert, R.N. Faustov, V.O. Galkin, Eur. J. Phys. C **32**, 29 (2003).
9. E. Jenkins, M. Luke, A.V. Manohar, M.J. Savage, Nucl. Phys. B **390**, 463 (1993).
10. M.A. Ivanov, J.G. Körner, P. Santorelli, Phys. Rev. D **73**, 054024 (2006).
11. C. albertus, E. Hernández, J. Nieves, J.M. Verde-Velasco, Phys. Rev. D **71**, 113006 (2005).
12. A. Abd El-Hady, J.M. Munoz, J.P. Vary, Phys. Rev. D **62**, 014019 (2000).
13. V.V. Kiselev, hep-ph/0211021.

Eur. Phys. J. A **31**, 718–721 (2007)

DOI 10.1140/epja/i2006-10256-8

Special Article – QNP 2006

Heavy quarkonia from classical SU(3) Yang-Mills configurations

R.A. Coimbra[a] and O. Oliveira

Centro de Física Computacional, Universidade de Coimbra, 3004 516 Coimbra, Portugal

Received: 23 November 2006

Published online: 14 March 2007 – © Società Italiana di Fisica / Springer-Verlag 2007

Abstract. A generalized Cho-Faddeev-Niemi ansatz for $SU(3)$ Yang-Mills is investigated. The corresponding classical field equations are solved for its simplest parametrization. From these solutions it is possible to define a confining non-relativistic central potential used to study heavy quarkonia. The associated spectra reproduces the experimental spectra with an error of less than 3% for charmonium and 1% for bottomonium. Moreover, the recently discovered new charmonium states can be accomodate in the spectra, keeping the same level of precision. The leptonic widths show good agreement with the recent measurements. The charmonium and bottomonium $E1$ electromagnetic transitions widths are computed and compared with the experimental values.

PACS. 12.39.Pn Potential models – 12.38.Lg Other nonperturbative calculations – 12.38.-t Quantum chromodynamics

1 Introduction

Charmonium and bottomonium are essentially non relativistic systems. Their dynamics can be understood in terms of a potential. Ideally, we would be able to derive such a potential directly from QCD. However, the confining potentials currently used to describe such systems, although related to the fundamental theory, are not directly derived from QCD.

In this paper, we report on a class of classical solutions of the Yang-Mills theory, computed after the introduction of a generalized Cho-Faddeev-Niemi [1,2] ansatz for the gluon field. The solutions allow a definition of a confining potential which is then used to study charmonium and bottomonium spectra, their leptonic decays and charmonium electromagnetic transitions. For interquark distance between 0.2 fm to 1 fm, the new potential is essentially the lattice singlet potential. For larger distances it grows exponentially with the interquark distance and for smaller distances it is Coulomb like. Concerning the spectra, the new potential is able to reproduce charmonium and bottomonium spectra with an error of less than 3% and 1%, respectively, including the recently discovered new charmonium states $X(3872)$, $X(3940)$, $Y(3940)$, $Z(3940)$ and $Y(4260)$. If the spectra is reasonably well described, the hyperfine splittings $^3S_1$-$^1S_0$ turn out to be too small. This can be due to a missing scalar confining potential or the need for including contributions from other configurations, namely those related with the short-distance behaviour of the the-

ory. As for the leptonic decays and the charmonium electromagnetic transitions, the new potential describes well the experimental data. Part of this work is described in [3].

2 Classical configuration and a heavy quark potential

Following a procedure suggested in [4], we introduce a real field n^a and we require the field n^a to be a covariant constant

$$D_\mu n^a = \partial_\mu n^a + g f_{abc} n^b A^c_\mu = 0 \,. \tag{1}$$

Multiplying this equation by n^a, we realize that n^a can be unitary. Defining the color projected field C_μ such that $A^a_\mu = n^a C_\mu + X^a_\mu$, with $n^a X^a_\mu = 0$ and replacing this definition in (1), X^a_μ can be related to n^a:

$$A^a_\mu = n^a C_\mu + \frac{3}{2g} f_{abc} n^b \partial_\mu n^c + Y^a_\mu \,, \quad n^a Y^a_\mu = 0 \,. \tag{2}$$

For the simplest parametrization of $n^a = \delta^{a1} \sin\theta + \delta^{a2} \cos\theta$ and for the simplest gluon configuration (with $Y_\mu = C_\mu = 0$), the classical field equations in Landau gauge become

$$\partial^\mu \partial_\mu \theta = 0 \,, \tag{3}$$

with $g A^a_\mu = -\delta^{a3} \partial_\mu \theta$. Note that there are no boundary conditions for θ. Equation (3) can be solved by the usual

---

[a] e-mail: `rita@teor.fis.uc.pt`

method of separation of variables. The solution considered here is

$$A_0^3 = \Lambda\left(e^{\Lambda t} - b_T e^{-\Lambda t}\right)V_0(r), \tag{4}$$

$$\mathbf{A}^3 = -\left(e^{\Lambda t} + b_T e^{-\Lambda t}\right)\nabla V_0(r), \tag{5}$$

$$V_0(r) = A\frac{\sinh(\Lambda r)}{r} + B\frac{e^{-\Lambda R}}{r}. \tag{6}$$

Assuming that the spatial part of A_0^3 can be identified with a non relativistic potential for heavy quarkonia (see [3] for details), the squared difference between V_0 and the lattice singlet potential integrated between 0.2–1 fm was minimized to define the various parameters; for r in MeV^{-1}, $A = 11.254$, $B = -0.701$, $\Lambda = 228.026$ MeV. The difference between V_0 and the singlet potential is less than 50 MeV for $r \in [0.2{-}1]$ fm. From now on, we assume that V_0 describes the interaction between the heavy quarks and we include perturbatively a spin-dependent contribution supposing that the potential is of pure vectorial type.

2.1 Charmonium and bottomonium spectra

The charm quark mass, $m_c = 1870$ MeV, was adjusted to reproduce the experimental value of the $\chi_{c2}(1P) - J/\psi(1S)$ mass difference, whereas the bottom quark mass, $m_b = 4185.25$ MeV, was defined to reproduce the experimental value for the $\Upsilon(2S) - \Upsilon(1S)$ mass difference.

In table 1 we report the low-lying charmonium states. The masses were shifted ($\Delta_{cc} = -3223$ MeV) to reproduce the $J/\psi(1S)$ experimental value ($M = E_{nr} + 2m_Q + \Delta_{QQ}$). In what concerns the spectrum, there is good agreement between the theoretical and the measured values. The only particles which do not fit well in the spectrum are $\psi(4040)$ and $\psi(4415)$. About these two states, the experimental information is scarse and the particle data book comments that the "interpretation of these states as a single resonance is unclear because of the expectation of substantial threshold effects in this energy region". Curiously, both particle masses are essentially the sum of J/ψ with $J^{PC} = 0^{++}$ known mesons.

As for the new charmonium states, the theoretical predictions which are close to the experimental values are: $Z(3930)$, mass of $3929 \pm 5 \pm 2$ MeV, $J = 2$, is a 1^3F_2 state with 3932 MeV or a 2^3P_2 state with 4048 MeV; $X(3940)$, mass of $3942 \pm 11 \pm 13$ MeV, and $Y(3940)$, mass of $3943 \pm 11 \pm 13$ MeV, can be any of the following states 2^3P_0, 3831 MeV, 2^3P_1, 3938 MeV, or 2^1P_1, 3959 MeV; $Y(4260)$, mass of 4260 MeV, and with quantum numbers 1^{--} as established by the experience, can be identified as a 3^3S_1 state with 4164 MeV. For charmonium $X(3872)$, mass of 3871.2 ± 0.5 MeV, the experimental data favours the following quantum numbers $J^{PC} = 1^{++}, 2^{-+}$. For the potential considered here, the closest states with such quantum numbers are 2^3P_1, 3938 MeV, and 2^3P_0, 3832 MeV.

In table 2 we report the low-lying bottomonium states. The masses were shifted ($\Delta_{bb} = -1136$ MeV) to reproduce the $\Upsilon(1S)$ experimental value. The discrepancy between

Table 1. Charmonium spectra. The right columns show the difference, in MeV, between the theoretical prediction and the experimental values taken from [5].

Theor. (MeV)	Expt. (MeV)		Theor. (MeV)	Expt. (MeV)	
$J^{PC} = 0^{-+}$			[1P]		
3075	$\eta_c(1S)$	95	3556	$\chi_{c2}(1P)$	0
3632	$\eta_c'(2S)$	-5	3462	$\chi_{c1}(1P)$	-49
4135	—	—	3372	$\chi_{c0}(1P)$	-43
			3478	$h_c(1P)$	-48
$J^{PC} = 1^{--}$ $[^3S_1]$			$J^{PC} = 1^{--}$ $[^3D_1]$		
3097	$J/\psi(1S)$	0			
3659	$\psi(2S)$	-27	3688	$\psi(3770)$	-83
4164	$Y(4260)$	-95	4155	—	—

$$M[J/\psi(1S) - \eta_c(1S)] = 22|_{\text{Theo}} = 117|_{\text{Exp}} \text{ MeV}$$
$$M[\psi(2S) - \eta_c'(2S)] = 26|_{\text{Theo}} = 48|_{\text{Exp}} \text{ MeV}$$

Table 2. Bottomonium spectra. The right columns report the difference, in MeV, between theory and experimental values.

Theor. (MeV)	Expt. (MeV)		Theor. (MeV)	Expt. (MeV)	
$J^{PC} = 0^{-+}$			$J^{PC} = 1^{--}$ $[^3S_1]$		
9457	$\eta_b(1S)$	157	9460	$\Upsilon(1S)$	0
10018	$\eta_b'(2S)$	—	10023	$\Upsilon(2S)$	0
10380	—	—	10385	$\Upsilon(3S)$	30
10721	—	—	10727	$\Upsilon(4S)$	148
11059	—	—	11065	$\Upsilon(11020)$	46
[1P]			[2P]		
9983	$\chi_{b2}(1P)$	71	10326	$\chi_{b2}(2P)$	57
9941	$\chi_{b1}(1P)$	48	10283	$\chi_{b1}(2P)$	28
9894	$\chi_{b0}(1P)$	35	10234	$\chi_{b0}(2P)$	2
9955	$h_b(1P)$	—	10296	$h_b(2P)$	—
$[^3D_1]$			$[^3D_2]$		
10159	—	—	10179	$\Upsilon(1D)$	18
10476	$\Upsilon(4S)$ (?)	103			
10796	$\Upsilon(10860)$	69			

the predicted and experimental value for $\Upsilon(4S)$ is too large when compared with other results in the table. It is more natural to assign the $\Upsilon(4S)$ to a 2^3D_1 state, just taking into account the predicted masses.

In conclusion, the non-relativistic potential V_0 is able to explain the charmonium spectrum with an error of less than 3% (~ 100 MeV) and the bottomonium spectrum with an error of less than 1.5% (~ 100 MeV), i.e., the level of accuracy achieved is within other potential model calculations (see, for example, [6] and references therein).

Table 3. Charmonium and bottomonium leptonic widths in keV. The experimental figures are from [5]. The limit for $Y(4260)$ is from [7].

		Theor.	Expt.		Theor.	Expt.
2^3S_1	$\psi(2S)$	2.84	2.48(6)	$\Upsilon(2S)$ 0.426	0.612(11)
3^3S_1	$Y(4260)$	2.16	< 0.40	$\Upsilon(3S)$ 0.356	0.443(8)
4^3S_1	—	—	—	$\Upsilon(4S)$ 0.335	0.272(29)
5^3S_1	—	—	—	$\Upsilon(5S)$ 0.311	0.130(30)

2.2 Leptonic decays of charmonium and bottomonium

For the leptonic decays we follow the van Royen-Weisskopf [8] approach and assume that QCD corrections factorize in the calculation of the widths, *i.e.*

$$\Gamma_{e^+e^-}(^3S_1) = \frac{4\,e_q^2\,\alpha^2}{M^2}\,|R_{nS}(0)|^2\,, \tag{7}$$

$$\Gamma_{e^+e^-}(^3D_1) = \frac{25\,e_q^2\,\alpha^2}{2\,m_q^4\,M^2}\,\left|R_{nD}^{(2)}(0)\right|^2\,, \tag{8}$$

times QCD corrections; e_q is the quark electric charge, α the fine structure constant, m_q the quark mass, M the meson mass, $R_{nS}(0)$ the S-wave meson radial wave function at the origin and $R_{nD}^{(2)}(0)$ the D-wave meson second derivative of the radial wave function at the origin. In order to avoid estimating the QCD corrections, in table 3 we report the theoretical widths computed relatively to $J/\psi(1S)$ for charmonium and $\Upsilon(1S)$ for bottomonium.

For charmonium, $\Gamma_{e^+e^-}$ for $\psi(2S)$ is slightly larger than the experimental value. For bottomonium, the witdh for the $\Upsilon(2S)$ is below the experimental value although the witdhs for $\Upsilon(3S)$ and $\Upsilon(4S)$ agree with the experimental figure within two standard deviations. For $\Upsilon(5S)$, the theoretical prediction is a factor of 3 higher than the experimental figure. The theoretical prediction for $\Gamma_{e^+e^-}$ for $Y(4260)$ is much larger than the experimental limit. However, the 3^3S_1 and 2^3D_1 states are almost degenerate in mass and we expect a large mixing for the leptonic width to become compatible with the experimental value. A similar situation is seen in $\Psi(3770)$ and $\Psi(2S)$; see [3] for details.

2.3 E1 electromagnetic transitions

In relation to the computation of the $E1$ electromagnetic charmonium transitions we follow [9]. Table 4 reports the theoretical estimates of the decay widths known experimentaly. Overall, the agreement between theory and experiment is good. The exception being the transitions involving the scalar meson $\chi_{c0}(1P)$. Remember that for the meson spectra, the larger deviations from the experimental numbers occured for the scalar mesons.

Given the good agreement with the known experimental radiative transitions of the charmonium, we have computed the $E1$ electromagnetic widths for the theoretical

Table 4. $E1$ electromagnetic charmonium widths in keV. In the calculation for the meson mass it was used the experimental one. Experimental values are from [5].

		Theor.	Expt.
$\psi(2S) \longrightarrow$	$\chi_{c2}(1P) + \gamma$	30	27 ± 2
	$\chi_{c1}(1P) + \gamma$	43	29 ± 2
	$\chi_{c0}(1P) + \gamma$	51	31 ± 2
$\chi_{c2}(1P) \longrightarrow$	$J/\psi(1S) + \gamma$	414	416 ± 32
$\chi_{c1}(1P) \longrightarrow$	$J/\psi(1S) + \gamma$	308	317 ± 25
$\chi_{c0}(1P) \longrightarrow$	$J/\psi(1S) + \gamma$	146	135 ± 15
$\psi(3770)(1^3D_1) \longrightarrow$	$\chi_{c2}(1P) + \gamma$	4	< 21
	$\chi_{c1}(1P) + \gamma$	110	75 ± 18
	$\chi_{c0}(1P) + \gamma$	359	172 ± 30

Table 5. $E1$ electromagnetic charmonium widths, in keV, for the new charmonium states.

		E_γ	Γ
$X(3872)[2^3P_1] \longrightarrow$	$J/\psi(1S) + \gamma$	697	48
	$\psi(2S) + \gamma$	181	63
$X(3940),\ Y(3940)$			
2^3P_1 or $2^3P_0 \longrightarrow$	$J/\psi(1S) + \gamma$	755	61
	$\psi(2S) + \gamma$	249	165
$2^3P_0 \longrightarrow$	$\psi(3770) + \gamma$	168	70
$2^1P_1 \longrightarrow$	$\eta_c(1S) + \gamma$	845	86
	$\eta_c'(2S) + \gamma$	293	271
$Z(3930)[2^3P_2] \longrightarrow$	$J/\psi(1S) + \gamma$	744	59
	$\psi(2S) + \gamma$	235	140

Table 6. $E1$ electromagnetic bottomonium widths in keV. The experimental values are from [10].

		Theor.	Expt.
$\Upsilon(2S) \longrightarrow$	$\chi_{b2}(1P) + \gamma$	3.30 ± 0.04	2.21 ± 0.16
	$\chi_{b1}(1P) + \gamma$	3.21 ± 0.04	2.11 ± 0.16
	$\chi_{b0}(1P) + \gamma$	2.10 ± 0.02	1.14 ± 0.16
$\Upsilon(3S) \longrightarrow$	$\chi_{b2}(2P) + \gamma$	3.22	2.95 ± 0.21
	$\chi_{b1}(2P) + \gamma$	2.95	2.71 ± 0.20
	$\chi_{b0}(2P) + \gamma$	1.82	1.26 ± 0.14

states which can describe the new charmonium states. In table 5 we report those widths which are larger than $\sim 40\,\mathrm{keV}$. In principle, the electromagnetic transitions allows us to distinguish not only the various theoretical models but also the different meson states.

To conclude, in table 6 we report the bottomonium $E1$ electromagnetic transitions.

RAM acknowledges FCT for financial support, grant SFRH/BD/8736/2002. This work was partly supported by FCT under contract POCI/FP/63436/2005.

References

1. L. Faddeev, A.J. Niemi, Phys. Rev. Lett. **82**, 1624 (1999); Phys. Lett. B **464**, 90 (1999).
2. Y.M. Cho, Phys. Rev. D **21**, 1080 (1980); **23**, 2415 (1981).
3. O. Oliveira, R.A. Coimbra, hep-ph/0603046.
4. Y.M. Cho, Phys. Rev. D **62**, 074009 (2000).
5. W.-M. Yao *et al.*, Particle Data Book, J. Phys. G **33**, 1 (2006).
6. S.S. Gershtein, A.K. Likhoded, A.V. Luchinsky, Phys. Rev. D **74**, 016002 (2006).
7. X.H. Mo *et al.*, Phys. Lett. B **640**, 182 (2006).
8. R. van Royen, V.F. Weisskopf, Nuovo Cimento **50**, 617 (1967); **51**, 583 (1967).
9. W. Kwong, J.L. Rosner, Phys. Rev. D **38**, 279 (1988).
10. CLEO Collaboration (M. Artuso *et al.*), Phys. Rev. Lett. **94**, 032001 (2005).

Eur. Phys. J. A **31**, 722–724 (2007)

DOI 10.1140/epja/i2006-10199-0

Special Article – QNP 2006

The puzzle of the D and D$_s$ mesons

J. Vijande[1,2,a], F. Fernández[1], and A. Valcarce[1]

[1] Grupo de Física Nuclear and IUFFyM, Universidad de Salamanca, Salamanca, Spain
[2] Departamento de Física Teórica and IFIC, Universidad de Valencia - CSIC, Valencia, Spain

Received: 25 October 2006
Published online: 22 February 2007 – © Società Italiana di Fisica / Springer-Verlag 2007

Abstract. We present a theoretical framework that accounts for the new D_J and D_{sJ} mesons measured in the open-charm sector. These resonances are properly described if considered as a mixture of conventional P-wave quark-antiquark states and four-quark components. The narrowest states are basically P-wave quark-antiquark mesons, while the dominantly four-quark states are shifted above the corresponding two-meson threshold. We study the electromagnetic decay widths as basic tools to scrutiny their nature.

PACS. 14.40.Lb Charmed mesons – 14.40.Ev Other strange mesons

During the last few years, there has been a renewed interest in heavy-meson spectroscopy due to the discovery of several new charmed mesons. Three years ago the BABAR Collaboration reported the observation of a charm-strange state, the $D^*_{sJ}(2317)$ [1], that was later on confirmed by CLEO [2] and Belle Collaborations [3]. Besides, BABAR had also pointed out to the existence of another charm-strange meson, the $D_{sJ}(2460)$ [1]. This resonance was measured by CLEO [2] and confirmed by Belle [3]. Belle results are consistent with the assignments of $J^P = 0^+$ for the $D^*_{sJ}(2317)$ and $J^P = 1^+$ for the $D_{sJ}(2460)$. However, although these states are well established, they present unexpected properties quite different from those predicted by quark potential models. If they would correspond to standard P-wave mesons made of a charm quark, c, and a strange antiquark, $\bar{s}$, their masses would be larger, around 2.48 GeV for the $D^*_{sJ}(2317)$ and 2.55 GeV for the $D_{sJ}(2460)$. They would be therefore above the DK and D^*K thresholds, respectively, becoming broad resonances. However, the states observed by BABAR and CLEO are very narrow, $\Gamma < 4.6$ MeV for the $D^*_{sJ}(2317)$ and $\Gamma < 5.5$ MeV for the $D_{sJ}(2460)$.

The intriguing situation of the charm-strange mesons has been extended to the nonstrange sector with the Belle observation [4] of a nonstrange broad scalar resonance, D^*_0, with a mass of $2308 \pm 17 \pm 15 \pm 28$ MeV/c^2 and a width $\Gamma = 276 \pm 21 \pm 18 \pm 60$ MeV. A state with similar properties has been suggested by the FOCUS Collaboration at Fermilab [5] during the measurement of masses and widths of excited charm mesons D^*_2. This state generates for the open-charm nonstrange mesons a very similar problem to the one arising in the strange sector with the $D^*_{sJ}(2317)$.

If the $D^*_0(2308)$ would correspond to a standard P-wave meson made of a charm quark, c, and a light antiquark, $\bar{n}$, its mass would have to be larger, around 2.46 GeV. In this case, the quark potential models prediction and the measured resonance are both above the $D\pi$ threshold, the large width observed being expected although not its low mass.

The difficulties in identifying the D_J and D_{sJ} states with conventional $c\bar{q}$ mesons are rather similar to those appearing in the light-scalar meson sector [6] and may be indicating that other configurations are playing a role. $q\bar{q}$ states are more easily identified with physical hadrons when virtual quark loops are not important. This is the case of the pseudoscalar and vector mesons, mainly due to the P-wave nature of this hadronic dressing. In contrast, in the scalar sector the $q\bar{q}$ pair is the one in a P-wave state, whereas quark loops may be in an S-wave. In this case the intermediate hadronic states that are created may play a crucial role in the composition of the resonance, in other words unquenching is important. This has been shown to be relevant for the proper description of the low-lying scalar mesons [7].

In this work we have explored the same ideas for the understanding of the properties of the D_J and D_{sJ} meson states. In nonrelativistic quark models the wave function of a hadron with baryon number equal to zero may be written as $|B = 0\rangle = \Omega_1|q\bar{q}\rangle + \Omega_2|qq\bar{q}\bar{q}\rangle + \ldots$, where q stands for quark degrees of freedom and the coefficients Ω_i take into account the mixing of four- and two-quark states. The Hamiltonian considering the mixing between both configurations could be described using the $^3P_0$ model, however, since this model depends on the vertex parameter, we prefer in a first approximation to parametrize this coefficient by looking to the quark pair that is annihilated and not

a e-mail: `javier.vijande@uv.es`

Table 1. $c\bar{s}$ and $c\bar{n}$ masses (QM), in MeV. Experimental data (Exp.) are taken from ref. [9], except for the state denoted by a dagger that has been taken from ref. [4].

$nL\ J^P$	State	QM ($c\bar{s}$)	Exp.
$1S\ 0^-$	D_s	1981	1968.5 ± 0.6
$1S\ 1^-$	D_s^*	2112	2112.4 ± 0.7
$1P\ 0^+$	$D_{sJ}^*(2317)$	2489	2317.4 ± 0.9
$1P\ 1^+$	$D_{sJ}(2460)$	2578	2459.3 ± 1.3
$1P\ 1^+$	$D_{s1}(2536)$	2543	2535.3 ± 0.6
$1P\ 2^+$	$D_{s2}(2573)$	2582	2572.4 ± 1.5
$1S\ 0^-$	D	1883	1867.7 ± 0.5
$1S\ 1^-$	D^*	2010	2008.9 ± 0.5
$1P\ 0^+$	$D_0^*(2308)$	2465	$2308 \pm 17 \pm 15 \pm 28^\dagger$
$1P\ 1^+$	$D_1(2420)$	2450	2422.2 ± 1.8
$1P\ 1^+$	$D_1^0(2430)$	2546	$2427 \pm 26 \pm 25$
$1P\ 2^+$	$D_2^*(2460)$	2496	2459 ± 4

Table 2. Probabilities (P), in %, of the wave function components and masses (QM), in MeV, of the open-charm mesons once the mixing between $q\bar{q}$ and $qq\bar{q}\bar{q}$ configurations is considered. Experimental data are taken from ref. [9] except for the state denoted by a dagger that has been taken from ref. [4].

$I = 0\ J^P = 0^+$		
QM	2339	2847
Exp.	2317.4 ± 0.9	–
$P(cn\bar{s}\bar{n})$	28	55
$P(c\bar{s}_{1^3P})$	71	25
$P(c\bar{s}_{2^3P})$	~ 1	20
$I = 0\ J^P = 1^+$		
QM	2421	2555
Exp.	2459.3 ± 1.3	2535.3 ± 0.6
$P(cn\bar{s}\bar{n})$	25	~ 1
$P(c\bar{s}_{1^1P})$	74	~ 1
$P(c\bar{s}_{1^3P})$	~ 1	98
$I = 1/2\ J^P = 0^+$		
QM	2241	2713
Exp.	$2308 \pm 17 \pm 15 \pm 28^\dagger$	–
$P(cn\bar{n}\bar{n})$	46	49
$P(c\bar{n}_{1P})$	53	46
$P(c\bar{n}_{2P})$	~ 1	5

to the spectator quarks that will form the final $q\bar{q}$ state. Therefore, we have taken $V_{q\bar{q}\leftrightarrow qq\bar{q}\bar{q}} = \gamma$. Further details about the formalism and the constituent quark model used are given in refs. [7,8].

A thorough study of the full meson spectrum in this model has been presented in ref. [8]. The results for the open-charm mesons are summarized in table 1. It can be seen how the open-charm states are easily identified with standard $c\bar{q}$ mesons except for the cases of the $D_{sJ}^*(2317)$, the $D_{sJ}(2460)$, and the $D_0^*(2308)$. This is a common feature of almost all quark potential model calculations [10]. In a similar manner, quenched lattice NRQCD predicts for the $D_{sJ}^*(2317)$ a mass of 2.44 GeV [11], while using relativistic charm quarks the mass obtained is 2.47 GeV [12]. Unquenched lattice QCD calculations of $c\bar{s}$ states do not find a window for the $D_{sJ}^*(2317)$ [6], supporting the difficulty of a P-wave $c\bar{s}$ interpretation.

Using for the qq interaction the parametrization of ref. [7], the results obtained for the $cn\bar{s}\bar{n}$ configuration are 2731 and 2699 MeV for the $J^P = 0^+$ with $I = 0$ and $I = 1$, and 2841 and 2793 MeV for the $J^P = 1^+$ with $I = 0$ and $I = 1$. For the $cn\bar{n}\bar{n}$ configuration with $I = 1/2$ the energy is 2505 MeV. The $I = 1$ and $I = 0$ states are far above the corresponding strong decay thresholds and therefore should be broad, what rules out a pure four-quark interpretation of the new open-charm mesons.

As outlined above, for P-wave mesons the hadronic dressing is in an S-wave, thus physical states may correspond to a mixing of two- and four-body configurations. In the isoscalar sector, the $cn\bar{s}\bar{n}$ and $c\bar{s}$ states mix, as it happens with $cn\bar{n}\bar{n}$ and $c\bar{n}$ for the $I = 1/2$ case. The parameter γ has been fixed to reproduce the mass of the $D_{sJ}^*(2317)$ meson, $\gamma = 240$ MeV. The results obtained are shown in table 2. Let us first analyze the nonstrange sector. The $^3P_0$ $c\bar{n}$ pair and the $cn\bar{n}\bar{n}$ have a mass of 2465 MeV and 2505 MeV, respectively. Once the mixing is considered one obtains a state at 2241 MeV with 46% of four-quark component and 53% of $c\bar{n}$ pair. The lowest state, representing the $D_0^*(2308)$, is above the isospin-

preserving threshold $D\pi$, becoming broad as it is observed experimentally. The mixed configuration compares much better with the experimental data than the pure $c\bar{n}$ state. The orthogonal state appears higher in energy, at 2713 MeV, with an important four-quark component.

Concerning the strange sector, the $D_{sJ}^*(2317)$ and the $D_{sJ}(2460)$ are dominantly $c\bar{s}$ $J = 0^+$ and $J = 1^+$ states, respectively, with almost 30% of four-quark component. Such component is responsible for the shift of the mass of the unmixed states to the experimental values below the DK and D^*K thresholds. Being both states below their isospin-preserving two-meson threshold, the only allowed strong decays to $D_s^*\pi$ would violate isospin and are expected to have small widths $O(10)$ keV [13,14]. As a consequence, they should be narrower than the $D_{s2}(2573)$ and $D_{s1}(2536)$, opposite to what it is expected from heavy-quark symmetry. The second isoscalar $J^P = 1^+$ state, with an energy of 2555 MeV and 98% of $c\bar{s}$ component, corresponds to the $D_{s1}(2536)$. Regarding the $D_{sJ}^*(2317)$, it has been argued that a possible DK molecule would be preferred with respect to an $I = 0$ $cn\bar{s}\bar{n}$ tetraquark, what would anticipate an $I = 1$ $cn\bar{s}\bar{n}$ partner nearby in mass [15]. Our results confirm the last argument, the vicinity of the isoscalar and isovector tetraquarks, however, the restricted coupling to the $c\bar{s}$ system allowed only for the $I = 0$ four-quark states opens the possibility of a mixed nature for the $D_{sJ}^*(2317)$ while the $I = 1$ $J = 0^+$ and $J = 1^+$ four-quark states appear above 2700 MeV and cannot be shifted to lower energies.

Apart from the masses, the structure of the $D_{sJ}^*(2317)$ and the $D_{sJ}(2460)$ mesons could be investigated also

Table 3. Electromagnetic decay widths, in keV, for the $D^*_{sJ}(2317)$ and $D_{sJ}(2460)$ (QM), compared to the results of two different quark models based only on $q\bar{q}$ states. To compare with the experimental data by CLEO and Belle we have assumed $\Gamma(D^{*+}_s\pi^0) \approx \Gamma(D^+_s\pi^0) \approx 10\,\mathrm{keV}$ as estimated in ref. [14].

	Quark models		
Transition	QM	Ref. [13]	Ref. [14]
$D^*_{sJ}(2317) \to D^{*+}_s\gamma$	1.6	0.8	1.9
$D_{sJ}(2460) \to D^{*+}_s\gamma$	0.06	2.2	5.5
$D_{sJ}(2460) \to D^+_s\gamma$	6.7	2.4	6.2

	Experiments	
Transition	CLEO [2]	Belle [3]
$D^*_{sJ}(2317) \to D^{*+}_s\gamma$	< 0.59	< 1.8
$D_{sJ}(2460) \to D^{*+}_s\gamma$	< 1.6	< 3.1
$D_{sJ}(2460) \to D^+_s\gamma$	< 4.9	$5.5 \pm 1.3 \pm 0.8$

Table 4. Masses (QM), in MeV, of the recently measured charmonium and B_c states obtained within the model of ref. [8] used in this work.

Name	Mass	Ref.	$n^{2S+1}L_J$	QM
$X(3940)$	$3943 \pm 6 \pm 6$	[18]	2^1P_1	3923
$-$	$-$		2^3P_0	3878
$Y(3940)$	$3943 \pm 11 \pm 13$	[19]	2^3P_1	3915
$X'_{c2}(3940)$	$3931 \pm 4 \pm 2$	[20]	2^3P_2	3936
$Y(4260)$	$4260 \pm 8 \pm 2$	[21]	4^3S_1	4307
$B_c(6287)$	$6287 \pm 4.8 \pm 1.1$	[22]	1^1S_0	6277

through the study of their electromagnetic decay widths. In table 3 we compare our results with different theoretical approaches and the experimental limits reported by Belle and CLEO. The main difference is the suppression predicted for the $D_{sJ}(2460) \to D^{*+}_s\gamma$ decay relative to the $D_{sJ}(2460) \to D^+_s\gamma$. A ratio

$$D_{sJ}(2460) \to D^+_s\gamma/D_{sJ}(2460) \to D^{*+}_s\gamma \approx 1\text{–}2$$

has been obtained assuming a $q\bar{q}$ structure for both states [13,14] (which seems incompatible with their properties). We find a much larger value,

$$D_{sJ}(2460) \to D^+_s\gamma/D_{sJ}(2460) \to D^{*+}_s\gamma \approx 100,$$

due to the small 1^3P_1 $c\bar{s}$ probability of the $D_{sJ}(2460)$. A similar enhancement has been obtained in ref. [16] in the framework of light-cone QCD sum rules in contrast to a previous calculation of the same authors using vector meson dominance [17].

Let us finally mention that the difficulties encountered for the interpretation of the new open-charm states as two-quark systems do not appear for the case of the recent charmed and B_c states measured at different facilities. They do nicely agree with the predictions of the $q\bar{q}$ model, see table 4, giving confidence to the results obtained in the present work.

In summary, we have obtained a rather satisfactory description of the positive-parity open-charm mesons in terms of two- and four-quark configurations. The mixing between these two components is responsible for the unexpected low mass and widths of the $D^*_{sJ}(2317)$, $D_{sJ}(2460)$, and $D^*_0(2308)$. The same mechanism has been used to account for the spectroscopic properties of the light-scalar mesons. The obtained electromagnetic decay widths give hints that would help in distinguishing the nature of these states. We predict a ratio

$$D_{sJ}(2460) \to D^+_s\gamma/D_{sJ}(2460) \to D^{*+}_s\gamma$$

much larger than the one obtained in a pure $q\bar{q}$ scheme, as a consequence of the small $^3P_1$ $c\bar{s}$ component of the $D_{sJ}(2460)$. We encourage experimentalists to measure the electromagnetic decay widths of the $D^*_{sJ}(2317)$ and the $D_{sJ}(2460)$, which would help to clarify the exciting situation of the open-charm mesons.

References

1. BABAR Collaboration (B. Aubert *et al.*), Phys. Rev. Lett. **90**, 242001 (2003).
2. CLEO Collaboration (D. Besson *et al.*), Phys. Rev. D **68**, 032002 (2003).
3. Belle Collaboration (Y. Mikani *et al.*), Phys. Rev. Lett. **92**, 012002 (2004).
4. Belle Collaboration (K. Abe *et al.*), Phys. Rev. D **69**, 112002 (2004).
5. FOCUS Collaboration (J.M. Link *et al.*), Phys. Lett. B **586**, 11 (2004).
6. G.S. Bali, Phys. Rev. D **68**, 071501(R) (2003).
7. J. Vijande, A. Valcarce, F. Fernández, B. Silvestre-Brac, Phys. Rev. D **72**, 034025 (2005).
8. J. Vijande, F. Fernández, A. Valcarce, J. Phys. G **19**, 2013 (2005).
9. S. Eidelman *et al.*, Phys. Lett. B **592**, 1 (2004).
10. S. Godfrey, R. Kokoski, Phys. Rev. D **43**, 1679 (1991).
11. J. Hein *et al.*, Phys. Rev. D **62**, 074503 (2000).
12. UKQCD Collaboration (P. Boyle), Nucl. Phys. B (Proc. Suppl.) **63**, 314 (1998); **53**, 398 (1997).
13. W.A. Bardeen, E.J. Eichten, C.T. Hill, Phys. Rev. D **68**, 054024 (2003).
14. S. Godfrey, Phys. Lett. B **568**, 254 (2003).
15. T. Barnes, F.E. Close, H.J. Lipkin, Phys. Rev. D **68**, 054006 (2003).
16. P. Colangelo, F. de Fazio, A. Ozpineci, Phys. Rev. D **72**, 074004 (2005).
17. P. Colangelo, F. de Fazio, Phys. Lett. B **570**, 180 (2003).
18. Belle Collaboration (K. Abe *et al.*), hep-ex/0507019.
19. Belle Collaboration (S.-K. Choi *et al.*), Phys. Rev. Lett. **94**, 182002 (2005).
20. Belle Collaboration (K. Abe *et al.*), Phys. Rev. Lett. **96**, 082003 (2006).
21. BABAR Collaboration (B. Aubert *et al.*), Phys. Rev. Lett. **95**, 142001 (2005).
22. CDF Collaboration (M.D. Corcoran), hep-ex/0506061.

Eur. Phys. J. A **31**, 725 727 (2007)

DOI 10.1140/epja/i2006-10247-9

Special Article – QNP 2006

Study of the semileptonic decays B → π, D → π and D → K

C. Albertus[1,a], J.M. Flynn[1], E. Hernández[2], J. Nieves[3], and J.M. Verde-Velasco[2]

[1] School of Physics and Astronomy, University of Southampton, Southampton SO17 1BJ, UK
[2] Grupo de Física Nuclear, Departamento de Física Fundamental e IUFFyM, Facultad de Ciencias, E-37008 Salamanca, Spain
[3] Departamento de Física Atómica, Molecular y Nuclear, Universidad de Granada, E-18071 Granada, Spain

Received: 23 November 2006
Published online: 12 March 2007 – © Società Italiana di Fisica / Springer-Verlag 2007

Abstract. The semileptonic decay $B \to \pi$ is studied starting from a simple quark model that takes into into account the effect of the B^*-resonance. A novel, multiply subtracted, Omnès dispersion relation has been implemented to extend the predictions of the quark model to all q^2 values accessible in the physical decay. By comparison to the experimental data, we extract $|V_{ub}| = (3.4 \pm 0.2(\text{exp.}) \pm 0.7(\text{theory}))\, 10^{-3}$. As a further test of the model, we have also studied $D \to \pi$ and $D \to K$ decays for which we get good agreement with experiment.

PACS. 12.15.Hh Determination of Kobayashi-Maskawa matrix elements – 11.55.Fv Dispersion relations – 12.39.Jh Nonrelativistic quark model – 13.20.He Decays of bottom mesons

1 Introduction

The exclusive semileptonic decay $B \to \pi l^+ \nu_l$ provides an important alternative to inclusive reactions $B \to X_u l^+ \nu_l$ in the determination of the Cabibbo-Kobayashi-Maskawa (CKM) matrix element $|V_{ub}|$.

This reaction has been studied in different approaches like lattice QCD (both in the quenched and unquenched approximations), light-cone sum rules (LCSR) and constituent-quark models (CQM), each of them having a limited range of applicability: LCSR are suitable for describing the low momentum transfer square (q^2) region, while lattice-QCD provides results only in the high-q^2 region. CQM can in principle provide form factors in the whole q^2 range but they are not directly connected to QCD. A combination of different methods seems to be the best strategy.

The use of Watson's theorem for the $B \to \pi l^+ \nu_l$ process allows one to write a dispersion relation for each of the form factors entering in the hadronic matrix element. This procedure leads to the so-called Omnès representation, which can be used to constrain the q^2-dependence of the form factors using the elastic $\pi B \to \pi B$ scattering amplitudes. The problem posed by the unknown $\pi B \to \pi B$ scattering amplitudes at high energies can be dealt with by using a multiply subtracted dispersion relation. The latter will allow for the combination of predictions from various methods in different q^2 regions.

[a] e-mail: c.albertus@mail.phys.soton.ac.uk,
albertus@ugr.es

In this work we study the semileptonic $B \to \pi l^+ \nu_l$ decay. The use of a multiply subtracted Omnès representation of the form factors will allow us to use the predictions of LCSR calculations at $q^2 = 0$ in order to extend the results of a simple nonrelativistic constituent-quark model (NRCQM) from its region of applicability, near the zero recoil point, to the whole physically accessible q^2 range. To test our model we shall also study the $D \to \pi$ and $D \to K$ semileptonic decays for which the relevant CKM matrix elements are well known and there is precise experimental data.

2 B → πl⁺ν̄

The matrix element for the semileptonic $B^0 \to \pi^- l^+ \nu_l$ decay can be parametrized in terms of two dimensionless form factors,

$$\langle \pi(p_\pi) | V^\mu | B(p_B) \rangle = \left(p_B + p_\pi - q \frac{m_B^2 - m_\pi^2}{q^2} \right)^\mu f^+(q^2)$$

$$+ q^\mu \frac{m_B^2 - m_\pi^2}{q^2} f^0(q^2), \qquad (1)$$

where $q^\mu = p_B - p_\pi$ is the four-momentum transfer and $m_B = 5279.4\,\text{MeV}$ and $m_\pi = 139.57\,\text{MeV}$ are the B^0 and π^- masses. For massless leptons, the total decay width is given by

$$\Gamma(B^0 \to \pi^- l^+ \nu_l) = \frac{G_F^2 |V_{ub}|^2}{192\pi^3 m_B^3} \int_0^\infty dq^2 [\lambda(q^2)]^{\frac{3}{2}} |f^+(q^2)|^2 \quad (2)$$

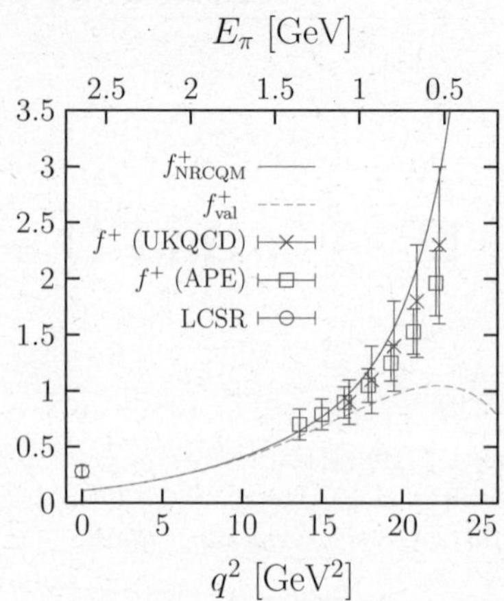

Fig. 1. f^+ form factor obtained with the valence quark (val) contribution alone and with the valence quark plus B^* contribution (NRCQM). We also plot lattice QCD results by the UKQCD [1] and APE [2] Collaborations, and LCSR [3] f^+ results.

with $q^2_{\max} = (m_B - m_\pi)^2$, $G_F = 1.16637 \times 10^{-5}\,\mathrm{GeV}^{-2}$ and $\lambda(q^2) = (m_B^2 + m_\pi^2 - q^2)^2 - 4m_B^2 m_\pi^2 = 4m_B^2 |\boldsymbol{p}_\pi|^2$, with $\boldsymbol{p}_\pi$ the pion three-momentum in the B rest frame.

2.1 Nonrelativistic constituent-quark model: valence quark and B*-resonance contributions

Figure 1 shows how the naive NRCQM valence quark description of the f^+ form factor fails in the whole q^2 range (see ref. [4] for details on the calculation). In the region close to $q^2_{\max}$, where a nonrelativistic model should work best, the influence of the B^*-resonance pole is evident. Close to $q^2 = 0$ the pion is ultra relativistic, and thus predictions from a nonrelativistic model are unreliable.

As first pointed out in ref. [5], the effects of the B^*-resonance pole dominate the $B \to \pi l^+ \nu_l$ decay near the zero recoil point $(q^2_{\max})$. Those effects must be added coherently as a distinct contribution to the valence result. The hadronic amplitude from the B^*-pole contribution is given by

$$-iT^\mu =$$
$$-i\hat{g}_{B^*B\pi}(q^2)p_\pi^\nu \left(i\frac{-g_\nu^\mu + q^\mu q_\nu/m_{B^*}^2}{q^2 - m_{B^*}^2} \right) i\sqrt{q^2}\hat{f}_{B^*}(q^2) \quad (3)$$

with $m_{B^*} = 5325\,\mathrm{MeV}$. $\hat{f}_{B^*}$ and $\hat{g}_{B^*B\pi}$ are, respectively, the off-shell B^* decay constant and off-shell strong $B^*B\pi$ coupling constant. Details on their calculation are given in ref. [4] and references therein. From the above equation one can easily obtain the B^*-pole contribution to f^+ which is given by

$$f^+_{\mathrm{pole}}(q^2) = \frac{1}{2}\hat{g}_{B^*B\pi}(q^2)\frac{\sqrt{q^2}\hat{f}_{B^*}(q^2)}{m_{B^*}^2 - q^2} . \quad (4)$$

The inclusion of the B^*-resonance contribution to the form factor improves the simple valence quark prediction down to q^2 values around $15\,\mathrm{GeV}^2$. Below that the description is still poor.

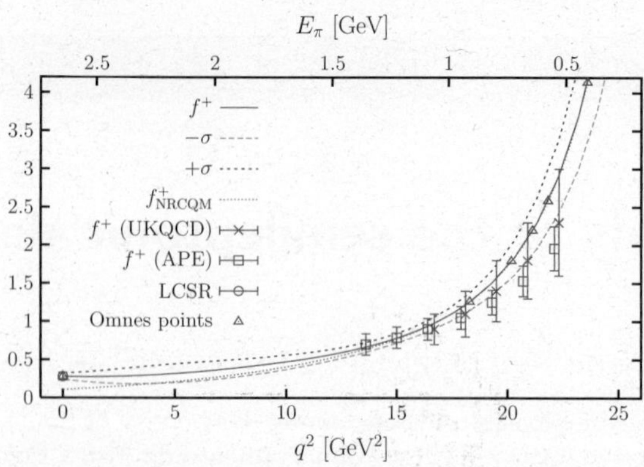

Fig. 2. Omnès improved form factor (solid line). The subtraction points are denoted by triangles. The $\pm\sigma$ lines show the theoretical-uncertainty band.

2.2 Omnès representation

Now one can use the Omnès representation to combine the NRCQM predictions at high q^2 with the LCSR at $q^2 = 0$. This representation requires as an input the elastic $B\pi \to B\pi$ phase shift $\delta(s)$ in the $J^P = 1^-$ and isospin $I = 1/2$ channel, plus the form factor at different q^2 values below the πB threshold where the subtractions will be performed. For a large enough number of subtractions, only the phase shift at or near threshold is needed. In that case one can approximate $\delta(s) \approx \pi$, arriving at the result that

$$f^+(q^2) \approx \frac{1}{s_{\mathrm{th}} - q^2} \prod_{j=0}^n [f^+(q_j^2)(s_{\mathrm{th}} - q_j^2)]^{\alpha_j(q^2)}, \; n \gg 1 \quad (5)$$

with $s_{\mathrm{th}} = m_B + m_\pi$ and $\alpha_j(q^2) = \prod_{j\neq k=0} \frac{q^2 - q_k^2}{q_j^2 - q_k^2}$.

Figure 2 shows with a solid line the form factor obtained using the Omnès representation with six subtraction points: we take five q^2 values between $18\,\mathrm{GeV}^2$ and $q^2_{\max}$ for which we use the f^+ NRCQM predictions (valence $+ B^*$ pole), plus the LCSR prediction at $q^2 = 0$. The $\pm\sigma$ lines enclose a 68% confidence level region that we have obtained from an estimation of the theoretical uncertainties. The latter have two origins: i) uncertainties in the quark-antiquark nonrelativistic interaction and ii) uncertainties on the product $g_{B^*B\pi}f_{B^*}$, and on the input to the multiply subtracted Omnès representation. See ref. [4] for details.

By comparison with the world average (w.a.) value for the decay width by the Heavy Flavor Averaging Group (HFAG) [6], we obtain

$$|V_{ub}| = (3.4 \pm 0.2(\mathrm{exp.}) \pm 0.7(\mathrm{theo.}))\,10^{-3} \quad (6)$$

to be compared to the exclusive and inclusive w.a. [6]

$$|V_{ub}| = (3.80 \pm 0.27 \pm 0.47)\,10^{-3} \; \text{exclusive w.a.},$$
$$|V_{ub}| = (4.39 \pm 0.19 \pm 0.27)\,10^{-3} \; \text{inclusive w.a.} \quad (7)$$

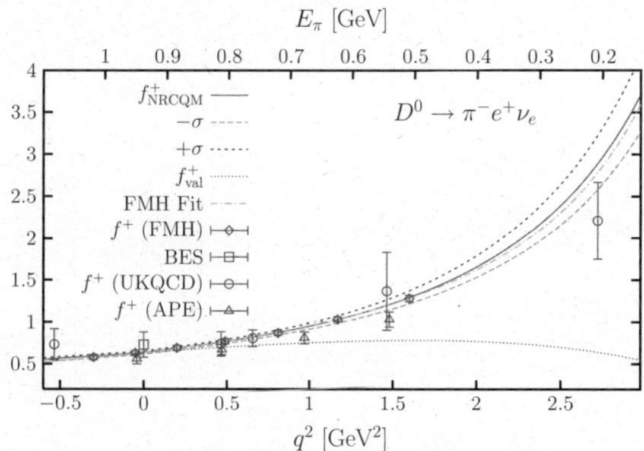

Fig. 3. The solid line denotes our determination of the f^+ form factor (f^+_{NRCQM}) for the $D^0 \to \pi^- e^+ \nu_e$ decay. The $\pm\sigma$ lines denote the theoretical-uncertainty band on the form factor. We compare with experimental data by the BES Collaboration [7] and with lattice results by the Fermilab-MILC-HPQCD [8], UKQCD [9] and APE [2] Collaborations.

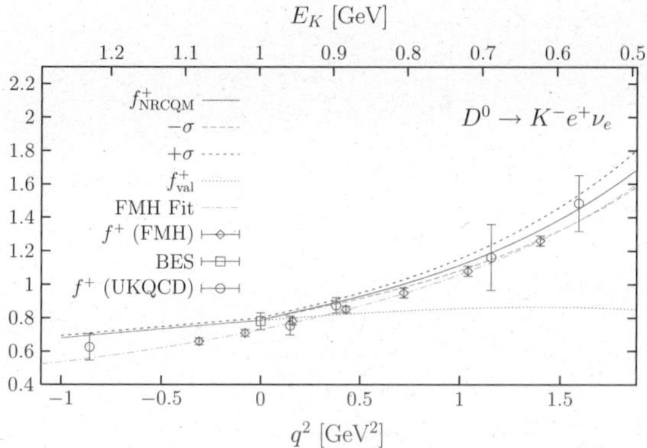

Fig. 4. Same as fig. 3 for the decay $D^0 \to K^- e^+ \nu_e$

3 $D \to \pi l \bar{\nu}_l$ and $D \to K l \bar{\nu}_l$

Our results for the f^+ form factor are depicted in figs. 3 and 4. As before we have considered valence quark plus resonant pole contributions (D^* and D^*_s, respectively). In both cases, we obtain a good description in the physical region of the experimental data [7] and previous lattice results [8,9,2], without using the Omnès dispersion relation. In the case of the $D \to K$ decay, our predictions for negative q^2 values could have been improved by the Omnès representation.

In fig. 5 we compare our results for the $f^+(q^2)/f^+(0)$ with experimental results by the FOCUS Collaboration [10]. We find very good agreement with the data.

Besides we have found for the decay widths

$$\Gamma(D^0 \to \pi^- e^+ \nu_e) = (5.2 \pm 0.1(\mathrm{exp.}) \pm 0.5(\mathrm{theo.}))$$
$$\times 10^{-12}\,\mathrm{MeV},$$
$$\Gamma(D^0 \to K^- e^+ \nu_e) = (66 \pm 3(\mathrm{theo.})) \times 10^{-12}\,\mathrm{MeV}. \quad (8)$$

For $D \to \pi$ we are in good agreement with experimental data while for $D \to K$ our result is two standard deviations higher.

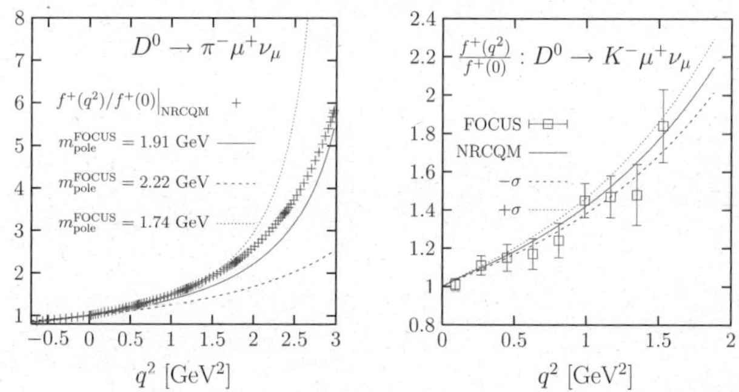

Fig. 5. NRCQM predictions for the ratio $f^+(q^2)/f^+(0)$ for $D \to \pi$ and $D \to K$ decays. We compare with experimental results by the FOCUS Collaboration [10] (a pole fit ($m_{\mathrm{pole}} = 1.91^{+0.31}_{-0.17}$ GeV) to data in the $D \to \pi$ case). For the $D \to K$ case we show the theoretical-uncertainty band.

4 Concluding remarks

We have shown the limitations of a pure valence quark model to describe the $B \to \pi$, $D \to \pi$ and $D \to K$ semileptonic decays. As a first correction, we have included vector resonance pole contributions which dominate the relevant f^+ form factor at high-q^2 transfers. Subsequently, for the $B \to \pi$ decay, we have applied a multiply subtracted Omnès dispersion relation. This has allowed us to extend the results of the NRCQM model to the whole q^2 range. Our result for $|V_{ub}|$ agrees within errors with the exclusive and inclusive w.a. by HFAG [6]. For $f^+(q^2)$ of the $D \to \pi$ and $D \to K$ decays and q^2 in the physical region we have found good agreement with experimental and lattice data.

This work was supported by DGI and FEDER funds, under contracts Nos. FIS2005-00810, BFM2003-00856 and FPA2004-05616, by the Junta de Andalucía and Junta de Castilla y León under contracts No. FQM0225 and No. SA104/04, and it is a part of the EU integrated infrastructure initiative Hadron Physics Project under contract No. RII3-CT-2004-506078. J.M.V.-V. acknowledges an E.P.I.F contract with the University of Salamanca. C.A. acknowledges a research contract with the University of Granada.

References

1. UKQCD Collaboration (K.C. Bowler *et al.*), Phys. Lett. B **486**, 111 (2000).
2. A. Abada *et al.*, Nuc. Phys. B **619**, 565 (2001).
3. A. Khodjamiriam *et al.*, Phys. Rev. D **62**, 114002 (2000).
4. C. Albertus, J.M. Flynn, E. Hernández, J. Nieves, J.M. Verde-Velasco, Phys. Rev. D **72**, 033002-1 (2005).
5. N. Isgur, M.B. Wise, Phys. Rev. D **41**, 151 (1990).
6. Heavy Flavor Averaging Group (E. Barbeiro *et al.*), hep-ex/0603003.
7. BES Collaboration (M. Abilikim *et al.*), Phys. Lett. B **597**, 39 (2004).
8. Fermilab MILC and HPQCD Collaborations (C. Aubin *et al.*), Phys. Rev. Lett. **94**, 011601 (2005).
9. UKQCD Collaboration (K.C. Bowler *et al.*), Phys. Rev. D **51**, 4905 (1995).
10. FOCUS Collaboration (J.M. Link *et al.*), Phys. Lett. B **607**, 233 (2005).

Eur. Phys. J. A **31**, 728–733 (2007)

DOI 10.1140/epja/i2006-10200-0

Special Article – QNP 2006

Heavy-quarkonium physics from effective-field theories

A. Vairo[a]

Dipartimento di Fisica dell'Università di Milano and INFN, via Celoria 16, 20133 Milano, Italy

Received: 25 October 2006
Published online: 22 February 2007 – © Società Italiana di Fisica / Springer-Verlag 2007

Abstract. I review recent progress in heavy-quarkonium physics from an effective-field theory perspective. In this unifying framework, I discuss advances in perturbative calculations for low-lying quarkonium observables and in lattice calculations for high-lying ones, and progress and lasting puzzles in quarkonium production.

PACS. 12.38.-t Quantum chromodynamics – 12.39.Hg Heavy quark effective theory – 13.20.Gd Decays of J/ψ, Υ, and other quarkonia

1 Introduction

Heavy quarkonium is most frequently and successfully studied as a non-relativistic bound state of QCD [1]. Non-relativistic bound states made of heavy quarks are characterized by a hierarchy of energy scales: m, mv, mv^2, ..., m being the heavy-quark mass and v the heavy-quark velocity in the centre-of-mass frame. A way to disentangle rigorously these scales is by substituting QCD, scale by scale, with simpler but equivalent Effective-Field Theories (EFTs). Modes of energy and momentum of order m may be integrated out from QCD in a perturbative manner ($m \gg \Lambda_{\rm QCD}$, by definition of heavy quark) leading to an EFT known as non-relativistic QCD (NRQCD) [2,3]; integrating out gluons of energy or momentum of order mv leads to an EFT known as potential NRQCD (pNRQCD) [4], see fig. 1. pNRQCD is close to a Schrödinger-like description of the bound system and, therefore, as simple. The bulk of the interaction is carried by potential-like terms, but non-potential interactions, associated with the propagation of low-energy dynamical degrees of freedom, are generally present as well. For a review on the subject we refer to [5]. EFTs for quarkonium were discussed at the conference by J. Soto.

QCD is characterized by an intrinsic scale $\Lambda_{\rm QCD}$. Let us consider states below threshold. In the perspective of non-relativistic EFTs, it may be important to distinguish between low-lying quarkonium resonances, for which we assume $mv^2 \gtrsim \Lambda_{\rm QCD}$, and high-lying states, for which we assume $mv \sim \Lambda_{\rm QCD}$. In the first case, the system is weakly coupled and the potential is perturbative, in the second case, the system is strongly coupled and the potential must be determined non-perturbatively, for instance,

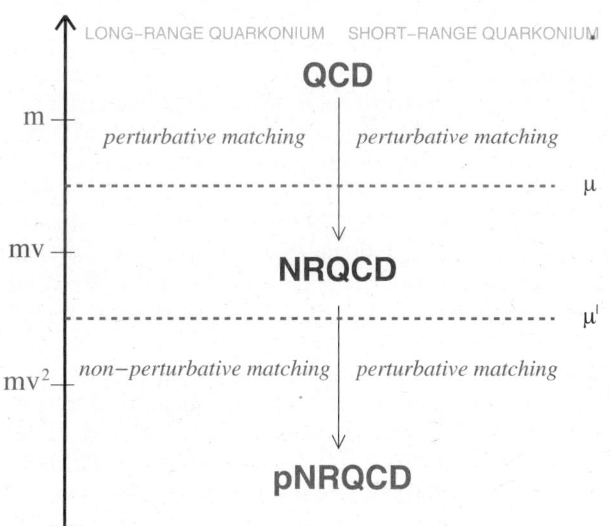

Fig. 1. EFTs for quarkonium.

on the lattice. This difference is not appreciated in potential models that typically describe the whole spectrum with the same interaction. Besides potential terms, which encode the dynamics at the soft scale mv, EFTs contain also terms that describe the dynamics of lower-energy degrees of freedom. These terms are typically absent in potential models. For what concerns systems close or above the open flavor threshold, a complete and satisfactory understanding of their dynamics has not yet been achieved. Therefore, the study of these systems, despite their phenomenological interest, is on a less secure ground than the study of states below threshold. Indeed, it largely relies on phenomenological models.

In the following, I will outline some recent (and very recent) progress in our theoretical understanding of heavy-

---

[a] e-mail: antonio.vairo@mi.infn.it

quarkonium physics. The perspective is quite personal, not only because the framework, which I will privilege, is that one of QCD EFTs, but also because I can only present a selection of results. It is important, when discussing about progress, to keep in mind the distinction made above about low-lying, high-lying and threshold states. In the first case, we may rely on perturbation theory: the challenge is in performing higher-order calculations and the goal is precision physics. In the second case, we have to rely on non-perturbative methods: the challenge is in providing a consistent framework where to perform lattice calculations and progress is measured by the progress in lattice computations. In the third case, one of the major challenges is to interpret the new charmonium states discovered at the B-factories in the last few years (for an updated list, see http://www.qwg.to.infn.it/). The distinction may help to better appreciate the quality of the progress made.

2 Low-lying $Q\bar{Q}$

Low-lying $Q\bar{Q}$ states are assumed to realize the hierarchy: $m \gg mv \gg mv^2 \gtrsim \Lambda_{\mathrm{QCD}}$, where mv is the typical scale of the inverse distance between the heavy quark and antiquark and mv^2 the typical scale of the binding energy. At a scale μ such that $mv \gg \mu \gg mv^2$ the effective degrees of freedom are $Q\bar{Q}$ states (in color singlet and octet configurations), low-energy gluons and light quarks.

The lowest-lying quarkonium states are η_b (not yet detected), $\Upsilon(1S)$, η_c and J/ψ. The $\Upsilon(1S)$ and J/ψ masses may be used to determine the bottom and charm quark masses. These determinations are competitive with other ones (for the b mass see, e.g., [6]). We report some recent determinations in table 1.

Once the heavy-quark masses are known, one may use them to extract other quarkonium ground-state observables. At NNLO the B_c mass was calculated in [7] ($M_{B_c} = 6326 \pm 29\,\mathrm{MeV}$), [8] ($M_{B_c} = 6324 \pm 22\,\mathrm{MeV}$) and [9] ($M_{B_c} = 6307 \pm 17\,\mathrm{MeV}$). These values agree well with the unquenched lattice determination of [10] ($M_{B_c} = 6304 \pm 12^{+18}_{-0}\,\mathrm{MeV}$), which shows that the B_c mass is not very sensitive to non-perturbative effects. This is confirmed by a recent measurement of the B_c in the channel $B_c \to J/\psi\,\pi$ by the CDF Collaboration at the Tevatron; they obtain with $360\,\mathrm{pb}^{-1}$ of data $M_{B_c} = 6285.7 \pm 5.3 \pm 1.2\,\mathrm{MeV}$ [11], while the latest available figure based on $1.1\,\mathrm{fb}^{-1}$ of data is $M_{B_c} = 6276.5 \pm 4.0 \pm 2.7\,\mathrm{MeV}$ (see http://www-cdf.fnal.gov/physics/new/bottom/060525.blessed-bc-mass/).

The bottomonium (and charmonium) ground-state hyperfine splitting has been calculated at NLL in [12]. Combining it with the measured $\Upsilon(1S)$ mass, this determination provides a quite precise prediction for the η_b mass: $M_{\eta_b} = 9421 \pm 10^{+9}_{-8}\,\mathrm{MeV}$, where the first error is an estimate of the theoretical uncertainty and the second one reflects the uncertainty in α_s. Note that the discovery of the η_b may provide a very competitive source of α_s at the bottom mass scale with a projected error at the

Table 1. Different recent determinations of $\overline{m}_b(\overline{m}_b)$ and $\overline{m}_c(\overline{m}_c)$ in the $\overline{\mathrm{MS}}$ scheme from the bottomonium and the charmonium systems. The displayed results either use direct determinations or non-relativistic sum rules. Here and in the text, the * indicates that the theoretical input is only partially complete at that order.

Reference	Order	$\overline{m}_b(\overline{m}_b)$ (GeV)
[13]	NNNLO*	$4.210 \pm 0.090 \pm 0.025$
[9]	NNLO + charm	$4.190 \pm 0.020 \pm 0.025$
[14]	NNLO	4.24 ± 0.10
[15]	NNNLO*	4.346 ± 0.070
[16]	NNNLO*	4.20 ± 0.04
[17]	NNNLO*	4.241 ± 0.070
[18]	NNLL*	4.19 ± 0.06

Reference	Order	$\overline{m}_c(\overline{m}_c)$ (GeV)
[8]	NNLO	1.24 ± 0.020
[14]	NNLO	1.19 ± 0.11

M_Z scale of about 0.003. Similarly, in [19], the hyperfine splitting of the B_c was calculated at NLL accuracy: $M_{B_c^*} - M_{B_c} = 65 \pm 24^{+19}_{-16}\,\mathrm{MeV}$.

The ratio of electromagnetic-decay widths was calculated for the ground state of charmonium and bottomonium at NNLL order in [20]. In particular, they report: $\Gamma(\eta_b \to \gamma\gamma)/\Gamma(\Upsilon(1S) \to e^+e^-) = 0.502 \pm 0.068 \pm 0.014$, which is a very stable result with respect to scale variation. A partial NNLL* order analysis of the absolute width of $\Upsilon(1S) \to e^+e^-$ can be found in [21].

Allowed magnetic dipole transitions between charmonium and bottomonium ground states have been considered at NNLO in [22,23]. The results are: $\Gamma(J/\psi \to \gamma\,\eta_c) = (1.5 \pm 1.0)\,\mathrm{keV}$ and $\Gamma(\Upsilon(1S) \to \gamma\,\eta_b) = (k_\gamma/39\,\mathrm{MeV})^3\,(2.50 \pm 0.25)\,\mathrm{eV}$, where the errors account for uncertainties (which are large in the charmonium case) coming from higher-order corrections. The width $\Gamma(J/\psi \to \gamma\,\eta_c)$ is consistent with [24]. Concerning $\Gamma(\Upsilon(1S) \to \gamma\,\eta_b)$, a photon energy $k_\gamma = 39\,\mathrm{MeV}$ corresponds to a η_b mass of $9421\,\mathrm{MeV}$.

The radiative transition $\Upsilon(1S) \to \gamma\,X$ has been considered in [25,26]. The agreement with the CLEO data of [27] is very satisfactory. J. Soto has reported about this at the conference.

3 Low-lying QQq

The SELEX Collaboration at Fermilab reported evidence of five resonances that may possibly be identified with doubly charmed baryons [28]. Although these findings have not been confirmed by other experiments (notably by FOCUS, BELLE and BABAR) they have triggered a renewed theoretical interest in doubly heavy-baryon systems.

Low-lying QQq states are assumed to realize the hierarchy: $m \gg mv \gg \Lambda_{\mathrm{QCD}}$, where mv is the typical inverse distance between the two heavy quarks and Λ_{QCD}

is the typical inverse distance between the centre of mass of the two heavy quarks and the light quark. At a scale μ such that $mv \gg \mu \gg \Lambda_{\mathrm{QCD}}$ the effective degrees of freedom are QQ states (in color antitriplet and sextet configurations), low-energy gluons and light quarks. The most suitable EFT at that scale is a combination of pN-RQCD and HQET [29,30]. The hyperfine splittings of the doubly heavy-baryon lowest states have been calculated at NLO in α_s and at LO in Λ_{QCD}/m by relating them to the hyperfine splittings of the D and B mesons (this method was first proposed in [31]). In [29], the obtained values are: $M_{\Xi_{cc}^*} - M_{\Xi_{cc}} = 120 \pm 40\,\mathrm{MeV}$ and $M_{\Xi_{bb}^*} - M_{\Xi_{bb}} = 34 \pm 4\,\mathrm{MeV}$, which are consistent with the quenched lattice determinations of [32–35]. Chiral corrections to the doubly heavy-baryon masses, strong-decay widths and electromagnetic-decay widths have been considered in [36].

Also low-lying QQQ baryons can be studied in a weak-coupling framework. Three quark states can combine in four color configurations: a singlet, two octets and a decuplet, which lead to a rather rich dynamics [29]. Masses of various QQQ ground states have been calculated with a variational method in [37]: since baryons made of three heavy quarks have not been discovered so far, it may be important for future searches to remark that the baryon masses turn our to be lower than those generally obtained in strong-coupling analyses.

4 High-lying Q$\bar{\text{Q}}$

High-lying $Q\bar{Q}$ states are assumed to realize the hierarchy: $m \gg mv \sim \Lambda_{\mathrm{QCD}} \gg mv^2$. A first question is where the transition from low-lying to high-lying takes place. This is not obvious, because we cannot measure directly mv. Therefore, the answer can only be indirect and, so far, there is no clear agreement in the literature. A weak-coupling treatment for the lowest-lying bottomonium states ($n = 1$, $n = 2$ and also for the $\Upsilon(3S)$) appears to give positive results for the masses at NNLO in [8] and at N^{3}LO* in [38]. The result is more ambiguous for the fine splittings of the bottomonium $1P$ levels in the NLO analysis of [39] and positive only for the $\Upsilon(1S)$ state in the N^{3}LO* analysis of [40]. In the weak-coupling regime, the magnetic-dipole hindered transition $\Upsilon(2S) \to \gamma\eta_b$ at leading order [22] does not agree with the experimental upper bound [41], while the ratios for different n of the radiative decay widths $\Gamma(\Upsilon(nS) \to \gamma X)$ are better consistent with the data if $\Upsilon(1S)$ is assumed to be a weakly coupled bound state and $\Upsilon(2S)$ and $\Upsilon(3S)$ strongly coupled ones [42].

Masses of high-lying quarkonia may be accessed by lattice calculations. A recent unquenched QCD determination of the charmonium spectrum below the open flavor threshold with staggered sea quarks may be found in [43]. At present, bottomonium is too heavy to be implemented directly on the lattice. A solution is provided by NRQCD [44]. Since the heavy-quark mass scale has been integrated out, for NRQCD on the lattice, it is sufficient to have a lattice spacing a as coarse as $m \gg 1/a \gg mv$.

A price to pay is that, by construction, the continuum limit cannot be reached. Another price to pay is that the NRQCD Lagrangian has to be supplemented by matching coefficients calculated in lattice perturbation theory, which encode the contributions from the heavy-mass energy modes that have been integrated out. A recent unquenched determination of the bottomonium spectrum with staggered sea quarks can be found in [45]. Note that all matching coefficients of NRQCD on the lattice are taken at their tree level value. This induces a systematic effect of order $\alpha_s v^2$ for the radial splittings and of order α_s for the fine and hyperfine splittings. In [45], also the ratio $\Gamma(\Upsilon(2S) \to e^+e^-)/\Gamma(\Upsilon(1S) \to e^+e^-) \times M_{\Upsilon(2S)}^2/M_{\Upsilon(1S)}^2$ has been calculated. The result on the finest lattice compares well with the experimental one.

In order to describe electromagnetic and hadronic inclusive decay widths of heavy quarkonia, many NRQCD matrix elements are needed. The specific number depends on the order in v of the non-relativistic expansion to which the calculation is performed and on the power counting. At order mv^5 and within a conservative power counting, S- and P-wave electromagnetic and hadronic decay widths for bottomonia and charmonia below threshold depend on 46 matrix elements [46]. More are needed at order mv^7 [47–49]. Order mv^7 corrections are particularly relevant for P-wave quarkonium decays, since they are numerically as large as NLO corrections in α_s, which are known since long time [50] and to which the most recent data are sensitive [51,1]. NRQCD matrix elements may be fitted to the experimental decay data [52,53] or calculated on the lattice [54–56]. The matrix elements of color-singlet operators can be related at leading order to the Schrödinger wave functions at the origin [3] and, hence, may be evaluated by means of potential models [57] or potentials calculated on the lattice [58]. However, most of the matrix elements remain poorly known or unknown. We refer to [1] for a summary of results.

At a scale μ such that $mv \sim \Lambda_{\mathrm{QCD}} \gg \mu \gg mv^2$, confinement sets in. Far from threshold, the effective degrees of freedom are $Q\bar{Q}$ states (in color singlet configuration) and light quarks. Neglecting light quarks, the $Q\bar{Q}$ propagation is simply described by a non-relativistic potential [59,60]. This will be in general a complex valued function admixture of perturbative terms, inherited from NRQCD, which encode high-energy contributions, and non-perturbative ones. The latter may be expressed in terms of Wilson loops and, therefore, are well suited for lattice calculations.

The real part of the potential has been one of the first quantities to be calculated on the lattice (for a review see [58]). In the last year, there has been some remarkable progress. In [61], the $1/m$ potential has been calculated for the first time. The existence of this potential was first pointed out in the pNRQCD framework [59]. A $1/m$ potential is typically missing in potential model calculations. The lattice result shows that the potential has a $1/r$ behaviour, which, in the charmonium case, is of the same size as the $1/r$ Coulomb tail of the static potential and, in the bottomonium one, is about 25%. Therefore, if the

$1/m$ potential has to be considered part of the leading-order quarkonium potential together with the static one, as the pNRQCD power counting suggests and the lattice seems to show, then the leading-order quarkonium potential would be, somewhat surprisingly, a flavor-dependent function. In [62], spin-dependent potentials have been calculated with unprecedented precision. In the long range, they show, for the first time, deviations from the flux-tube picture of chromoelectric confinement [63] (for a review see, for instance, [64]). The knowledge of the potentials in pNRQCD could provide an alternative to the direct determination of the spectrum in NRQCD lattice simulations: the quarkonium masses would be determined by solving the Schrödinger equation with the lattice potentials. The approach may present some advantages: the leading-order pNRQCD Lagrangian, differently from the NRQCD one, is renormalizable, the potentials are determined once for ever for all quarkonia, and the solution of the Schrödinger equation provides also the quarkonium wave functions, which enter in many quarkonium observables: decay widths, transitions, production cross-sections, ….

The imaginary part of the potential provides the NRQCD decay matrix elements in pNRQCD. They typically factorize in a part, which is the wave function in the origin square (or its derivatives), and in a part which contains gluon tensor-field correlators [65,46,66,67]. This drastically reduces the number of non-perturbative parameters needed; in pNRQCD, these are wave functions at the origin and universal gluon tensor-field correlators, which can be calculated on the lattice. Another approach may consist in determining the correlators on one set of data (*e.g.*, in the charmonium sector) and use them to make predictions for another (*e.g.*, in the bottomonium sector). Following this line in [65,68], at NLO in α_s, but at leading order in the velocity expansion, it was predicted $\Gamma_{\text{had}}(\chi_{b0}(2P))/\Gamma_{\text{had}}(\chi_{b2}(2P)) \approx 4.0$ and $\Gamma_{\text{had}}(\chi_{b1}(2P))/\Gamma_{\text{had}}(\chi_{b2}(2P)) \approx 0.50$. Both determinations turned out to be consistent, within large errors, with the CLEO III data [1].

5 Threshold states

For states near or above threshold a general systematic treatment does not exist so far. Also lattice calculations are inadequate. Most of the existing analyses rely on models (*e.g.*, the Cornell coupled-channel model [69] or the $^3P_0$ model [70]).

However, in some cases, one may develop an EFT owing to special dynamical conditions. An example is the $X(3872)$ discovered by BELLE [71] and seen also by CDF [72], D0 [73] and BABAR [74]. If interpreted as a loosely bound $D^0\bar{D}^{*0}$ and $\bar{D}^0 D^{*0}$ molecule, one may take advantage of the hierarchy of scales $\Lambda_{\text{QCD}} \gg m_\pi \gg m_\pi^2/(2m_{\text{red}}) \approx 10\,\text{MeV} \gg E_{\text{binding}}$. Indeed, the binding energy, E_{binding}, which may be estimated from $M_{X(3872)} - (M_{D^{0*}} + M_{D^0})$, is very close to zero, *i.e.* much smaller than the natural scale $m_\pi^2/(2m_{\text{red}})$. Systems with a short-range interaction and a large scattering length have

universal properties that may be exploited; in particular, production and decay amplitudes factorize in a short-range and a long-range part, where the latter only depends on one single parameter, the scattering length [75–77].

Another interesting case is provided by the $Y(4260)$, discovered by BABAR [78] and seen also by CLEO [79] and BELLE [80]. If interpreted as a heavy charmonium hybrid (see, *e.g.*, [81]), one may rely on the heavy-quark expansion and on lattice calculations to study its properties. In particular, analogously to the energy of a static quark-antiquark pair, the energies of static hybrids have been calculated in quenched approximation on the lattice in [82]. One may expect that the static energy is related to the static potential of the system and that this may be a relevant quantity for the dynamics of the system, but to substantiate this a suitable EFT for heavy hybrids, like pNRQCD for heavy quarkonium, needs to be formulated. Such a formulation does not exist yet[1].

6 Production

Although a formal proof of the NRQCD factorization formula for heavy-quarkonium production has not yet been developed, NRQCD factorization has proved to be very successful to explain a large variety of quarkonium production processes (for a review, see the production chapter in [1]). In the last year, there has been a noteworthy progress toward an all order proof. In [83,84], it has been shown that a necessary condition for factorization to hold at NNLO is that the conventional octet NRQCD production matrix elements must be redefined by incorporating Wilson lines that make them manifestly gauge invariant.

Differently from decay processes, a pNRQCD treatment does not exist so far for quarkonium production. The difficulty of such a formulation may be linked to that of providing a full proof of factorization at the level of NRQCD and a consistent definition of the NRQCD production matrix elements at the level of pNRQCD. A pNRQCD formulation of quarkonium production may present potentially the same advantages as that of quarkonium decay: a sensible reduction in the number of parameters and hence more predictive power.

In the last years, two main problems have plagued our understanding of heavy-quarkonium production: 1) double charmonium production in e^+e^- collisions and 2) charmonium polarization at the Tevatron.

In [85], BELLE measures $\sigma(e^+e^- \rightarrow J/\psi + \eta_c)\,\text{Br}(c\bar{c} \rightarrow> 2\ \text{charged}) = 25.6 \pm 2.8 \pm 3.4\,\text{fb}$ and in [86], BABAR finds $\sigma(e^+e^- \rightarrow J/\psi + \eta_c)\,\text{Br}(c\bar{c} \rightarrow> 2\ \text{charged}) = 17.6 \pm 2.8^{+1.5}_{-2.1}\,\text{fb}$. When these cross-sections first appeared, they were about one order of magnitude

---

[1] It is suggestive, however, that the lowest hybrid state is expected to contain a pseudoscalar color-octet quark-antiquark pair and gluons, whose quantum numbers are those of the electric cloud in a diatomic Π_u molecule, so that the system has J^{PC} numbers 1^{--}, like the Y. Moreover, its mass, obtained by solving the Schrödinger equation, is consistent, within 100 MeV, with the experimental range 4.25–4.30 GeV.

above theoretical expectations. In the meantime, some errors have been corrected in some of the theoretical determinations, and, more important, NLO corrections in α_s have been calculated in [87] and higher-order v^2 corrections in [88]. All these improvements have shifted the theoretical value much closer to the experimental one. In [89], a preliminary estimate of $\sigma(e^+e^- \to J/\psi + \eta_c)$ that includes the above corrections has been presented, it reads: $16.7 \pm 4.2\,\mathrm{fb}$. It appears that, within the uncertainties, the discrepancy has been resolved. Still open is the issue of the inclusive double charm production in the presence of a J/ψ. BELLE measures a ratio $\sigma(e^+e^- \to J/\psi + c\bar{c})/\sigma(e^+e^- \to J/\psi + X)$ that is about 80%, to be compared with theoretical estimates, which are about 10% (see [1] for a detailed discussion). New theoretical analyses are timely.

Charmonium polarization has been measured at the Tevatron by the CDF Collaboration at run I on $110\,\mathrm{pb}^{-1}$ of data [90] and recently at run II on $800\,\mathrm{pb}^{-1}$ of data [91]. The data of the two runs do not seem consistent with each other in the 7–12 GeV region of transverse momentum, p_T, and both are not with NRQCD expectations. For large p_T, NRQCD predicts that the main mechanism of charmonium production is via color-octet gluon fragmentation, the gluon is transversely polarized and most of the gluon polarization is expected to be transferred to the charmonium. For an analysis of the NRQCD prediction and its dependence on the adopted power counting, we refer to [92]. The CDF data do not show any sign of transverse polarization at large p_T. Before drawing definite conclusions, a polarization study from the D0 experiment would be most welcome, at least to settle the possible discrepancy between the run-I and run-II data.

7 Conclusions

Many new data on heavy-quark bound states are provided in these years by the B-factories, CLEO, BES, HERA and the Tevatron experiments. Many more will come in the future from the LHC and GSI. They will show new (perhaps exotic) states, new production and decay mechanisms. What makes all this interesting is that we may investigate a wide range of heavy-quarkonium observables in a controlled and systematic fashion and, therefore, learn about one of the most elusive sectors of the Standard Model: low-energy QCD. The tools for this systematic investigation are provided by EFTs and lattice gauge theories. Still challenging remains for both the description of threshold states.

I acknowledge the financial support obtained inside the Italian MIUR program "incentivazione alla mobilità di studiosi stranieri e italiani residenti all'estero".

References

1. N. Brambilla *et al.*, *Heavy quarkonium physics*, CERN-2005-005 (CERN, Geneva, 2005) [arXiv:hep-ph/0412158].
2. W.E. Caswell, G.P. Lepage, Phys. Lett. B **167**, 437 (1986).
3. G.T. Bodwin, E. Braaten, G.P. Lepage, Phys. Rev. D **51**, 1125 (1995) [**55**, 5853 (1997)(E)] [hep-ph/9407339].
4. A. Pineda, J. Soto, Nucl. Phys. Proc. Suppl. **64**, 428 (1998) [arXiv:hep-ph/9707481]; N. Brambilla, A. Pineda, J. Soto, A. Vairo, Nucl. Phys. B **566**, 275 (2000) [arXiv:hep-ph/9907240].
5. N. Brambilla, A. Pineda, J. Soto, A. Vairo, Rev. Mod. Phys. **77**, 1423 (2005) [arXiv:hep-ph/0410047].
6. A.X. El-Khadra, M. Luke, Annu. Rev. Nucl. Part. Sci. **52**, 201 (2002) [arXiv:hep-ph/0208114].
7. N. Brambilla, A. Vairo, Phys. Rev. D **62**, 094019 (2000) [arXiv:hep-ph/0002075].
8. N. Brambilla, Y. Sumino, A. Vairo, Phys. Lett. B **513**, 381 (2001) [arXiv:hep-ph/0101305].
9. N. Brambilla, Y. Sumino, A. Vairo, Phys. Rev. D **65**, 034001 (2002) [arXiv:hep-ph/0108084].
10. HPQCD Collaboration (I.F. Allison, C.T.H. Davies, A. Gray, A.S. Kronfeld, P.B. Mackenzie, J.N. Simone), Phys. Rev. Lett. **94**, 172001 (2005) [arXiv:hep-lat/0411027].
11. CDF Collaboration (D. Acosta *et al.*), Phys. Rev. Lett. **96**, 082002 (2006) [arXiv:hep-ex/0505076].
12. B.A. Kniehl, A.A. Penin, A. Pineda, V.A. Smirnov, M. Steinhauser, Phys. Rev. Lett. **92**, 242001 (2004) [arXiv:hep-ph/0312086].
13. A. Pineda, JHEP **0106**, 022 (2001) [arXiv:hep-ph/0105-008].
14. M. Eidemüller, Phys. Rev. D **67**, 113002 (2003) [arXiv:hep-ph/0207237].
15. A.A. Penin, M. Steinhauser, Phys. Lett. B **538**, 335 (2002) [arXiv:hep-ph/0204290].
16. T. Lee, JHEP **0310**, 044 (2003) [arXiv:hep-ph/0304185].
17. C. Contreras, G. Cvetic, P. Gaete, Phys. Rev. D **70**, 034008 (2004) [arXiv:hep-ph/0311202].
18. A. Pineda, A. Signer, Phys. Rev. D **73**, 111501 (2006) [arXiv:hep-ph/0601185].
19. A.A. Penin, A. Pineda, V.A. Smirnov, M. Steinhauser, Phys. Lett. B **593**, 124 (2004) [arXiv:hep-ph/0403080].
20. A.A. Penin, A. Pineda, V.A. Smirnov, M. Steinhauser, Nucl. Phys. B **699**, 183 (2004) [arXiv:hep-ph/0406175].
21. A. Pineda, A. Signer, arXiv:hep-ph/0607239.
22. N. Brambilla, Y. Jia, A. Vairo, Phys. Rev. D **73**, 054005 (2006) [arXiv:hep-ph/0512369].
23. A. Vairo, arXiv:hep-ph/0608327.
24. Particle Data Group (W.M. Yao *et al.*), J. Phys. G **33**, 1 (2006).
25. S. Fleming, A.K. Leibovich, Phys. Rev. D **67**, 074035 (2003) [arXiv:hep-ph/0212094].
26. X. Garcia i Tormo, J. Soto, Phys. Rev. D **72**, 054014 (2005) [arXiv:hep-ph/0507107].
27. CLEO Collaboration (B. Nemati *et al.*), Phys. Rev. D **55**, 5273 (1997) [arXiv:hep-ex/9611020].
28. SELEX Collaboration (A. Ocherashvili *et al.*), arXiv:hep-ex/0406033; SELEX Collaboration (M. Mattson *et al.*), Phys. Rev. Lett. **89**, 112001 (2002) [arXiv:hep-ex/0208014]; SELEX Collaboration (M.A. Moinester *et al.*), Czech. J. Phys. **53**, B201 (2003) [arXiv:hep-ex/0212029]; see also the SELEX web page http://www-selex.fnal.gov/Welcome.html.
29. N. Brambilla, A. Vairo, T. Rösch, Phys. Rev. D **72**, 034021 (2005) [arXiv:hep-ph/0506065].
30. S. Fleming, T. Mehen, Phys. Rev. D **73**, 034502 (2006) [arXiv:hep-ph/0509313].
31. M.J. Savage, M.B. Wise, Phys. Lett. B **248**, 177 (1990).

32. UKQCD Collaboration (J.M. Flynn, F. Mescia, A.S.B. Tariq), JHEP **0307**, 066 (2003) [arXiv:hep-lat/0307025].

33. R. Lewis, N. Mathur, R.M. Woloshyn, Phys. Rev. D **64**, 094509 (2001) [arXiv:hep-ph/0107037].

34. A. Ali Khan *et al.*, Phys. Rev. D **62**, 054505 (2000) [arXiv:hep-lat/9912034].

35. N. Mathur, R. Lewis, R.M. Woloshyn, Phys. Rev. D **66**, 014502 (2002) [arXiv:hep-ph/0203253].

36. J. Hu, T. Mehen, Phys. Rev. D **73**, 054003 (2006) [arXiv:hep-ph/0511321].

37. Y. Jia, arXiv:hep-ph/0607290.

38. A.A. Penin, V.A. Smirnov, M. Steinhauser, Nucl. Phys. B **716**, 303 (2005) [arXiv:hep-ph/0501042].

39. N. Brambilla, A. Vairo, Phys. Rev. D **71**, 034020 (2005) [arXiv:hep-ph/0411156].

40. M. Beneke, Y. Kiyo, K. Schuller, Nucl. Phys. B **714**, 67 (2005) [arXiv:hep-ph/0501289].

41. CLEO Collaboration (M. Artuso *et al.*), Phys. Rev. Lett. **94**, 032001 (2005) [hep-ex/0411068].

42. X. Garcia i Tormo, J. Soto, Phys. Rev. Lett. **96**, 111801 (2006) [arXiv:hep-ph/0511167].

43. S. Gottlieb *et al.*, PoS **LAT2005**, 203 (2006) [arXiv:hep-lat/0510072].

44. G.P. Lepage, L. Magnea, C. Nakhleh, U. Magnea, K. Hornbostel, Phys. Rev. D **46**, 4052 (1992) [arXiv:hep-lat/9205007].

45. A. Gray, I. Allison, C.T.H. Davies, E. Gulez, G.P. Lepage, J. Shigemitsu, M. Wingate, Phys. Rev. D **72**, 094507 (2005) [arXiv:hep-lat/0507013].

46. N. Brambilla, D. Eiras, A. Pineda, J. Soto, A. Vairo, Phys. Rev. D **67**, 034018 (2003) [arXiv:hep-ph/0208019].

47. G.T. Bodwin, A. Petrelli, Phys. Rev. D **66**, 094011 (2002) [arXiv:hep-ph/0205210].

48. J.P. Ma, Q. Wang, Phys. Lett. B **537**, 233 (2002) [arXiv:hep-ph/0203082].

49. N. Brambilla, E. Mereghetti, A. Vairo, JHEP **0608**, 039 (2006) [arXiv:hep-ph/0604190].

50. R. Barbieri, M. Caffo, R. Gatto, E. Remiddi, Phys. Lett. B **95**, 93 (1980); A. Petrelli, M. Cacciari, M. Greco, F. Maltoni, M.L. Mangano, Nucl. Phys. B **514**, 245 (1998) [arXiv:hep-ph/9707223].

51. A. Vairo, AIP Conf. Proc. **756**, 101 (2005) [arXiv:hep-ph/0412331].

52. M.L. Mangano, A. Petrelli, Phys. Lett. B **352**, 445 (1995) [arXiv:hep-ph/9503465].

53. F. Maltoni, arXiv:hep-ph/0007003.

54. G.T. Bodwin, D.K. Sinclair, S. Kim, Phys. Rev. Lett. **77**, 2376 (1996) [arXiv:hep-lat/9605023].

55. G.T. Bodwin, D.K. Sinclair, S. Kim, Phys. Rev. D **65**, 054504 (2002) [arXiv:hep-lat/0107011].

56. G.T. Bodwin, J. Lee, D.K. Sinclair, Phys. Rev. D **72**, 014009 (2005) [arXiv:hep-lat/0503032].

57. E.J. Eichten, C. Quigg, Phys. Rev. D **52**, 1726 (1995) [arXiv:hep-ph/9503356].

58. G.S. Bali, Phys. Rep. **343**, 1 (2001) [arXiv:hep-ph/0001312].

59. N. Brambilla, A. Pineda, J. Soto, A. Vairo, Phys. Rev. D **63**, 014023 (2001) [arXiv:hep-ph/0002250].

60. A. Pineda, A. Vairo, Phys. Rev. D **63**, 054007 (2001) [**64**, 039902 (2001)(E)] [arXiv:hep-ph/0009145].

61. Y. Koma, M. Koma, H. Wittig, Phys. Rev. Lett. **97**, 122003 (2006) [arXiv:hep-lat/0607009].

62. Y. Koma, M. Koma, arXiv:hep-lat/0609078.

63. W. Buchmüller, Phys. Lett. B **112**, 479 (1982).

64. N. Brambilla, A. Vairo, arXiv:hep-ph/9904330.

65. N. Brambilla, D. Eiras, A. Pineda, J. Soto, A. Vairo, Phys. Rev. Lett. **88**, 012003 (2002) [arXiv:hep-ph/0109130].

66. N. Brambilla, A. Pineda, J. Soto, A. Vairo, Phys. Lett. B **580**, 60 (2004) [arXiv:hep-ph/0307159].

67. A. Vairo, Mod. Phys. Lett. A **19**, 253 (2004) [arXiv:hep-ph/0311303].

68. A. Vairo, Nucl. Phys. Proc. Suppl. **115**, 166 (2003) [arXiv:hep-ph/0205128].

69. E. Eichten, K. Gottfried, T. Kinoshita, K.D. Lane, T.M. Yan, Phys. Rev. D **17**, 3090 (1978) [**21**, 313 (1980)(E)].

70. A. Le Yaouanc, L. Oliver, O. Pene, J.C. Raynal, Phys. Rev. D **8**, 2223 (1973).

71. Belle Collaboration (S.K. Choi *et al.*), Phys. Rev. Lett. **91**, 262001 (2003) [arXiv:hep-ex/0309032].

72. CDF II Collaboration (D. Acosta *et al.*), Phys. Rev. Lett. **93**, 072001 (2004) [arXiv:hep-ex/0312021].

73. D0 Collaboration (V.M. Abazov *et al.*), Phys. Rev. Lett. **93**, 162002 (2004) [arXiv:hep-ex/0405004].

74. BABAR Collaboration (B. Aubert *et al.*), Phys. Rev. D **71**, 071103 (2005) [arXiv:hep-ex/0406022].

75. E. Braaten, M. Kusunoki, Phys. Rev. D **69**, 074005 (2004) [arXiv:hep-ph/0311147].

76. E. Braaten, M. Kusunoki, Phys. Rev. D **72**, 014012 (2005) [arXiv:hep-ph/0506087].

77. M.T. AlFiky, F. Gabbiani, A.A. Petrov, Phys. Lett. B **640**, 238 (2006) [arXiv:hep-ph/0506141].

78. BABAR Collaboration (B. Aubert *et al.*), Phys. Rev. Lett. **95**, 142001 (2005) [arXiv:hep-ex/0506081].

79. CLEO Collaboration (T.E. Coan *et al.*), Phys. Rev. Lett. **96**, 162003 (2006) [arXiv:hep-ex/0602034].

80. Belle Collaboration, arXiv:hep-ex/0608018.

81. E. Kou, O. Pene, Phys. Lett. B **631**, 164 (2005) [arXiv:hep-ph/0507119].

82. K.J. Juge, J. Kuti, C. Morningstar, Phys. Rev. Lett. **90**, 161601 (2003) [arXiv:hep-lat/0207004].

83. G.C. Nayak, J.W. Qiu, G. Sterman, Phys. Rev. D **72**, 114012 (2005) [arXiv:hep-ph/0509021].

84. G.C. Nayak, J.W. Qiu, G. Sterman, arXiv:hep-ph/0608066.

85. Belle Collaboration (K. Abe *et al.*), Phys. Rev. D **70**, 071102 (2004) [arXiv:hep-ex/0407009].

86. BABAR Collaboration (B. Aubert *et al.*), Phys. Rev. D **72**, 031101 (2005) [arXiv:hep-ex/0506062].

87. Y.J. Zhang, Y.J. Gao, K.T. Chao, Phys. Rev. Lett. **96**, 092001 (2006) [arXiv:hep-ph/0506076].

88. G.T. Bodwin, D. Kang, J. Lee, Phys. Rev. D **74**, 014014 (2006) [arXiv:hep-ph/0603186].

89. J. Lee, talk at the *Quarkonium Working Group meeting (BNL, 2006)*, http://www.qwg.to.infn.it/WS-jun06 /WS4talks/Thursday_PM/Lee.pdf.

90. CDF Collaboration (A.A. Affolder *et al.*), Phys. Rev. Lett. **85**, 2886 (2000) [arXiv:hep-ex/0004027].

91. M.J. Kim, talk at the *Quarkonium Working Group meeting (BNL, 2006)*, http://www.qwg.to.infn.it/WS-jun06 /WS4talks/Thursday_AM/Kim.pdf.

92. S. Fleming, I.Z. Rothstein, A.K. Leibovich, Phys. Rev. D **64**, 036002 (2001) [arXiv:hep-ph/0012062].

Eur. Phys. J. A **31**, 734 738 (2007)
DOI 10.1140/epja/i2006-10286-2

THE EUROPEAN
PHYSICAL JOURNAL A

Special Article – QNP 2006

Superfluidity in many fermion systems: Exact renormalisation group treatment

B. Krippa[a]

School of Physics and Astronomy, The University of Manchester, M13 9PL, Manchester, UK

Received: 18 December 2006
Published online: 20 March 2007 – © Società Italiana di Fisica / Springer-Verlag 2007

Abstract. The application of the exact renormalisation group to symmetric as well as asymmetric many-fermion systems with a short-range attractive force is studied. Assuming an ansatz for the effective action with effective bosons, describing pairing effects, a set of approximate flow equations for the effective coupling including boson and fermionic fluctuations has been derived. The phase transition to a phase with broken symmetry is found at a critical value of the running scale. The mean-field results are recovered if boson-loop effects are omitted. The calculations with two different forms of the regulator are shown to lead to similar results. We find that, being quite small in the case of the symmetric many-fermion system the corrections to mean-field approximation become more important with increasing mass asymmetry.

PACS. 21.65.+f Nuclear matter – 21.60.-n Nuclear structure models and methods – 68.18.Jk Phase transitions – 73.22.Gk Broken symmetry phases

1 Introduction

There is a growing interest in applying the exact renormalisation group (ERG) formalism to few- and many-body systems [1–3] when the underlying interaction is essentially nonperturbative. Regardless of the details all ERG-based approaches share the same distinctive feature, a successive elimination/suppression of some modes, resulting in effective interaction between the remaining degrees of freedom. One specific way of implementing such a procedure is to eliminate modes by applying a momentum-space blocking transformation with some physically motivated cutoff. The effect of varying a cutoff is described by nonlinear ERG evolution equations, which include the effect of the eliminated modes. In the following we will use the variant of ERG based on the concept of average effective action (AEA) [4]. The corresponding evolution equation can be written in the following general form:

$$\partial_k \Gamma = -\frac{i}{2} \mathrm{Tr}\left[(\partial_k R)\left(\Gamma^{(2)} - R\right)^{-1}\right].\qquad(1)$$

Here $\Gamma^{(2)}$ is the second functional derivative of the AEA taken with respect to all types of field included in the action, and R is a regulator which should suppress the contributions of states with momenta less than or of the order of the running scale k. To recover the full effective action we require $R(k)$ to vanish as $k \to 0$, in other respects its form is rather arbitrary. The concrete functional form of the regulator has no effect on physical results provided no approximations/truncations were made. By solving the ERG equations one can find a scale dependence of the coupling constants and thus determine the ERG flow. The whole approach is nonperturbative so that some physically motivated assumption about the functional form of the effective action should be made It is well known that the many-fermion system with attractive interaction favours the formation of the correlated fermion pairs leading to the symmetry breaking and related phenomena. Since we expect the appearance of the correlated fermion pairs in a physical ground state, we need to parametrise our effective action in a way that can describe the qualitative change in the physics when this occurs. A natural way to do this is to introduce a boson field whose vacuum expectation value (VEV) describes this correlated pair [5] and study the evolution of this effective degrees of freedom. At the start of the RG evolution, the boson field is not dynamical and is introduced through a Hubbard-Stratonovich transformation of the four-point interaction. As we integrate out more and more of the fermion degrees of freedom by running the cutoff scale k to lower values, we generate dynamical terms in the bosonic effective action. In this paper we treat both symmetric and asymmetric many-fermion systems. The corresponding ansatz for the boson-fermion effective action consists of the kinetic terms for boson and fermions and the interaction term and for two types of fermions it can be written as

$$\Gamma[\mu, k] = \int \mathrm{d}^4 x \left(\Gamma_B[\mu, k] + \Gamma_F[\mu, k] + \Gamma_I[k]\right).\qquad(2)$$

[a] e-mail: krippa@theoserv.phy.umist.ac.uk

Here $\Gamma_{B(F)}$ is the boson (fermion) part of AEA,

$$\Gamma_B = \phi^\dagger \left(Z_\phi(i\partial_t + \mu_a + \mu_b) + \frac{Z_m}{2m}\nabla^2 \right) \phi - U(\phi, \phi^\dagger), \quad (3)$$

$$\Gamma_F = \sum_{i=a}^b \psi_i^\dagger \left(Z_{\psi,i}(i\partial_t + \mu_i) + \frac{Z_{M,i}}{2M_i}\nabla^2 \right) \psi_i, \quad (4)$$

and Γ_I is the interaction term,

$$\Gamma_I = -Z_g \left(\frac{i}{2}\psi_b^T \sigma_2 \psi_a \phi^\dagger - \frac{i}{2}\psi_a^\dagger \sigma_2 \psi_b^{\dagger T}\phi \right). \quad (5)$$

M is the reduced mass of the fermion in vacuum and the factor $1/2m$ with $m = M_a + M_b$ in the boson kinetic term is chosen simply to make Z_m dimensionless. The coupling Z_g, the wave function renormalisations factors $Z_{\phi,\psi}$ and the kinetic-mass renormalisations factors $Z_{m,M}$ all run with k, the scale of the regulator. Having in mind the future applications to the crossover from BCS to BEC (where the chemical potential becomes negative) we also let the chemical potentials μ_a and μ_b run, thus keeping the corresponding densities (and Fermi momenta $p_{F,i}$) constant. The bosons are, in principle, coupled to the chemical potentials via a quadratic term in ϕ, but this can be absorbed into the potential by defining $\overline{U} = U - (\mu_1 + \mu_2)Z_\phi\phi^\dagger\phi$. The evolution equations include running of chemical potentials, effective potential and all couplings ($Z_\phi, Z_m, Z_{M,i}, Z_{\psi,i}, Z_g$). However, in this paper we allow to run only Z_ϕ, parameters in the effective potential ($u's$ and ρ_0) and chemical potentials since this is the minimal set needed to include the effective boson dynamics. The system with one type of fermion corresponds to the limit $M_a = M_b$.

We expand the effective potential about its minimum, $\phi^\dagger\phi = \rho_0$, so that the coefficients u_i are defined at $\rho = \rho_0$,

$$\overline{U}(\rho) = u_0 + u_1(\rho-\rho_0) + \frac{1}{2}u_2(\rho-\rho_0)^2 + \frac{1}{6}u_3(\rho-\rho_0)^3 + \cdots, \quad (6)$$

where we have introduced $\rho = \phi^\dagger\phi$. A similar expansion can be written for the renormalisation factors. The coefficients of the expansion run with the scale. The phase of the system is determined by the coefficient u_1. We start evolution at high scale where the system is in the symmetric phase so that $u_1 > 0$. When the running scale becomes comparable with the pairing scale (close to average Fermi momentum) the system undergoes the phase transition to the phase with broken symmetry, energy gap etc. The point of the transition corresponds to the scale where $u_1 = 0$. The bosonic excitations in the gapped phase are gapless Goldstone bosons. Note, that in this phase the minimum of the potential will also run with the scale k so that the value $\rho_0(k \to 0)$ determines the physical gap.

The important part of any ERG treatment is the choice of the regulator. Ideally, the physical results should not depend on this choice. However, some sort of truncations and approximations should always be made in real calculations to render the system of the resulting evolution equations solvable so that the convenient choice of the regulator is the question of significant practical importance.

In our approach the boson regulator has the structure

$$\mathbf{R}_B = R_B \operatorname{diag}(1,1), \quad (7)$$

and the fermion regulator for both types of fermions has the structure

$$\mathbf{R}_{F,i} = \operatorname{sgn}(\epsilon_i(q) - \mu_i)R_{F,i}(q,\mu_i,k)\operatorname{diag}(1,-1). \quad (8)$$

Note that this regulator is positive for particle states above the Fermi surface and negative for the hole states below the Fermi surface. The function R_F should suppress the contributions of states with momenta near the Fermi surface, $|q - p_F| \sim k$. Once a large gap has appeared in the fermion spectrum, there are no low-energy fermion excitations and so the fermionic regulator plays little further role. However, while the gap is zero or small, it is crucial that the sign of the regulator matches that of the energy, $q^2/2M - \mu$, and hence it is μ which appears in the sign functions.

The other important part of the ERG approach is fixing the boundary conditions which define the form of the AEA at some initial scale so that at some large starting scale $k = K$ we demand that the Lagrangian be equivalent to a purely fermionic theory with the contact interaction,

$$\mathcal{L}_i = -\frac{1}{4}C_0 \left(\psi^\dagger \sigma_2 \psi^{\dagger T} \right) \left(\psi^T \sigma_2 \psi \right). \quad (9)$$

Here $C_0(K)$ is the strength of the energy-independent term in the effective NN interaction in vacuum. This is evaluated at the scale K, using the same regularisation procedure as we apply in matter. This equation implies that u_1 and g at the this scale are related by

$$C_0(K) = -\frac{g(K)^2}{u_1(K)}. \quad (10)$$

Using this boundary condition we can relate the pairing phenomena in a many-body environment with the fermion-fermion interaction in vacuum.

The boson potential $\overline{U}$ is obtained by evaluating the effective action for uniform boson fields. It evolves according to

$$\partial_k \overline{U} = -\frac{1}{\mathcal{V}_4}\partial_k \Gamma, \quad (11)$$

where $\mathcal{V}_4$ is the volume of spacetime. Substituting our expansion of $\overline{U}$, eq. (6), on the left-hand side leads to a set of ordinary differential equations for the u_n.

For the boson wave function renormalisation factor, Z_ϕ, we need to consider a time-dependent background field. Taking

$$\phi(x) = \phi_0 + \eta e^{-ip_0 t}, \quad (12)$$

where η is a constant, we can get the evolution of Z_ϕ from

$$\partial_k Z_\phi = \frac{1}{\mathcal{V}_4}\frac{\partial}{\partial p_0}\left(\frac{\partial^2}{\partial\eta\partial\eta^\dagger}\partial_k\Gamma \right)\bigg|_{\eta=0}\bigg|_{p_0=0}. \quad (13)$$

Let us first consider the results of the calculations in the case of symmetric fermion matter. We solve the evolution equations numerically with two types of cutoff. First, we use the smoothed step-function type of regulator (called hereafter as R_{1F}):

$$R_{1F} = \frac{k^2}{2M}\theta_1(q-p_F,k,\sigma); \quad R_{1B} = \frac{k^2}{2m}\theta_1(q,k,\sigma), \quad (14)$$

where

$$\theta_1(q,k,\sigma) = \frac{1}{2\,\mathrm{erf}(1/\sigma)}\left[\mathrm{erf}\left(\frac{q+k}{k\sigma}\right) + \mathrm{erf}\left(\frac{q-k}{k\sigma}\right)\right] \quad (15)$$

with σ being a parameter determining the sharpness of the step. Second, we use the sharp-cutoff function R_{2F} which is somewhat similar to the one suggested in ref. [6] for the pure boson case and chosen in a rather peculiar way to make the calculations as simple as possible

$$R_{2F} = \frac{k^2}{2M}\big[((k+p_\mu)^2 - q^2)\theta(p_\mu + k - q) \\ +((k+p_\mu)^2 + q^2 - 2p_\mu^2)\theta(q - p_\mu + k)\big], \quad (16)$$

$$R_{2B} = \frac{k^2}{2m}(k^2 - q^2)\theta(k - q), \quad (17)$$

where $p_\mu = (2M\mu)^{1/2}$. The fermion sharp cutoff consists of two terms which result in the modification of the particle and hole propagators, respectively. The hole term is further modified to suppress the contribution from the surface terms, which may bring in the dangerous dependence of the regulator on the cutoff scale even at the vanishingly small k. As an example, we focus on the parameters relevant to neutron matter: $M = 4.76\,\mathrm{fm}^{-1}$, $p_F = 1.37\,\mathrm{fm}^{-1}$. Let us first discuss the results obtained with the smooth cutoff R_1. We found that the value of the physical gap is practically independent of either the values of the width parameter σ (varied within some range) or the starting scale K provided $K > 5\,\mathrm{fm}^{-1}$. The results of the calculations are shown in fig. 1. At the starting scale the system is in the symmetric phase and remains in this phase until u_1 hits zero at $k_{crit} \simeq 1.2\,\mathrm{fm}^{-1}$ where the artificial second-order phase transition to a broken phase occurs and the energy gap is formed. Already at $k \simeq 0.5$ the running scale has essentially no effect on the gap. It is worth mentioning that we found very small (on the level of 1%) contribution to the gap from the boson loops, due to cancellations between the direct contributions to the running of the gap and indirect ones via u_2. The boson loops play a much more important role in the evolution of u_2 and Z_ϕ. In fact, they drive both couplings to zero at $k \to 0$. We note however, that the effect of the boson loops for the gap may still be more visible if the evolution of the other couplings is included.

The results obtained with the sharp-cutoff regulator are shown in fig. 2. One immediate observation is that the results become starting scale independent as long as $K > 5\,\mathrm{fm}^{-1}$ similarly to the case with the smooth cutoff. However, the artificial phase transition occurs at lower values of the running scale $k \simeq 0.7\,\mathrm{fm}^{-1}$. At approximately $k \simeq 0.2\,\mathrm{fm}^{-1}$ the value of the gap becomes scale independent. One notes that the curves obtained with different

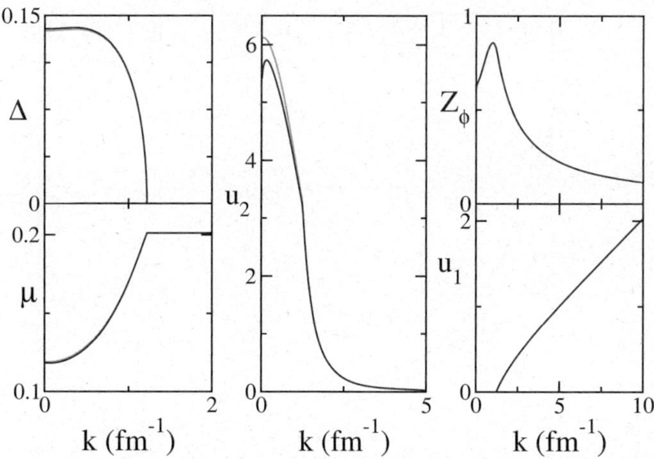

Fig. 1. (Colour on-line) Numerical solutions to the evolution equations for infinite a_0 and $p_F = 1.37\,\mathrm{fm}$, starting from $K = 16\,\mathrm{fm}^{-1}$. We show the evolution of all relevant parameters for the cases of fermion loops only (orange/grey lines), and of bosonic loops with a running Z_ϕ (blue/black lines). All quantities are expressed in appropriate powers of fm^{-1}.

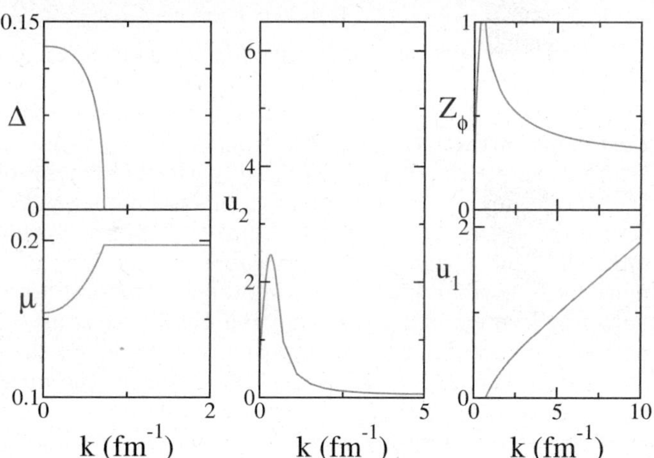

Fig. 2. (Colour on-line) Evolution of the parameters when the sharp cutoff is used.

regulators and describing the evolution of the gap, being rather different at intermediate scales, approach each other with decreasing scale resulting in very close values for the physical gap. This is an encouraging result taking into account that, although the hypothetical exact results must be independent of the choice of the regulator in practice it is not guaranteed given the assumed ansatz for the effective action and truncations made. The same conclusion also holds for the other quantities. The couplings Z_ϕ and u_2 first grow with scale and then start decreasing eventually coming to zero. The chemical potential begins to decrease at the point of phase transition and becomes scale independent at $k \simeq 0.2\,\mathrm{fm}^{-1}$. However, in this case the numerical values of the chemical potentials obtained with different regulators differ by approximately 20% so that this quantity is more sensitive to the details of effective action and to the trancations made.

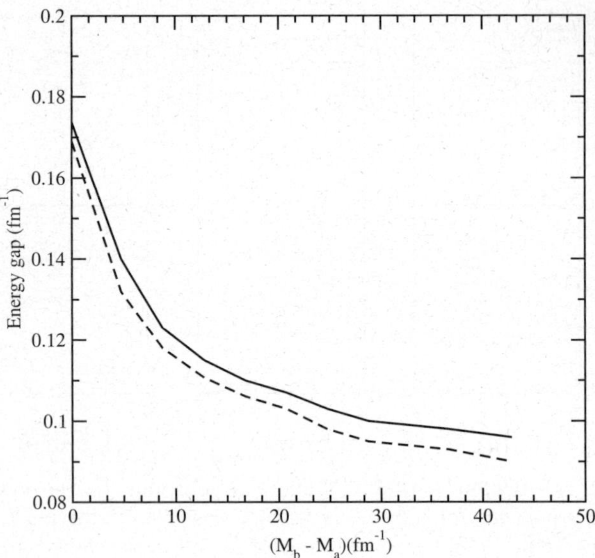

Fig. 3. Evolution of the gap in the MF approach (dashed curve) and with boson loops (solid curve) in the unitary regime $a = -\infty$ as a function of the mass asymmetry.

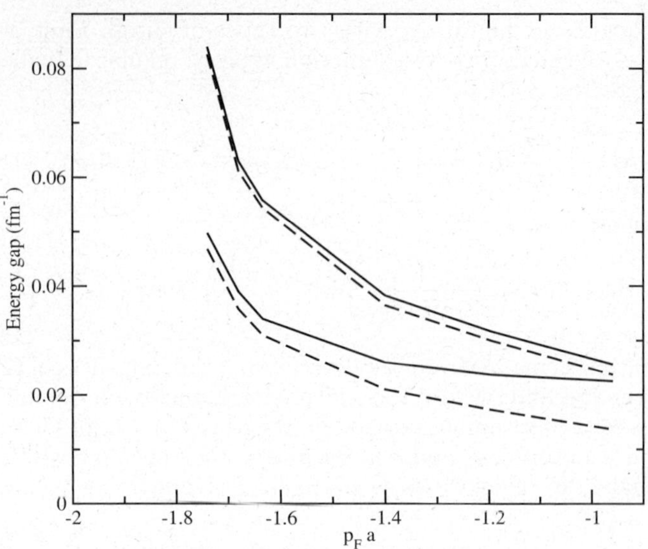

Fig. 4. Evolution of the gap as a function of the parameter $p_F a$. The upper pair of curves corresponds to the calculations with no asymmetry in the MF approach (dashed curve) and with boson loops (solid curve) and the lower pair of curves describes the results of calculations with the maximal asymmetry when $M_b = 10M_a$.

Let us now discuss the results for the many-fermion system with two fermion species. For simplicity we consider the case of the hypothetical "nuclear" matter with short-range attractive interaction between two types of fermions, light and heavy, and study the behaviour of the energy gap as a function of the mass asymmetry. We choose the Fermi momentum to be $p_F = 1.37\,\mathrm{fm}^{-1}$. One notes that the formalism is applicable to any type of a many-body system with two fermion species from quark matter to fermionic atoms so that the hypothetical asymmetrical "nuclear" matter is simply chosen as a study case. We assume that $M_a < M_b$, where M_a is always the mass of the physical nucleon.

First we consider the case of the unitary limit with the infinite scattering length. The results of our calculations for the gap are shown in fig. 3.

We see from this figure that increasing mass asymmetry leads to a decreasing gap that seems to be a natural result. However, the effect of the boson loops is found to be small. We found essentially no effect in the symmetric phase, 2–4% corrections for the value of the gap in the broken phase and even smaller corrections for the chemical potential so that one can conclude that the mean-field approach (MFA) indeed provides the reliable description in the unitary limit for both small and large mass asymmetries. It is worth mentioning that, similar to the above-considered case of the fermion matter with one type of fermion, the boson contributions are more important for the evolution of u_2 where they drive u_2 to zero as $k \to 0$ making the effective potential convex in agreement with the general expectations. This tendency retains in the unitary regime regardless of the mass asymmetry.

We have also considered the behaviour of the gap as a function of the parameter $p_F a$ for the cases of the zero asymmetry $M_a = M_b$ and the maximal asymmetry $M_b = 10M_a$. The results are shown in fig. 4.

One can see from fig. 4 that in the case of zero (or small) asymmetry the corrections stemming from boson loops are small at all values of the parameter $p_F a$ considered here (down to $p_F a = 0.94$). On the contrary, when $M_b = 10M_a$ these corrections, being rather small at $p_F a \geq 2$ become significant ($\sim 40\%$) when the value of $p_F a$ decreases down to $p_F a \sim 1$. We found that at $p_F a \sim 1$ the effect of boson fluctuations becomes $\sim 10\%$ already for $M_b = 5M_a$. One can therefore conclude that the regime of large mass asymmetries, which starts approximately at $M_b > 5M_a$, moderate scattering length and/or the Fermi momenta is the one where the MF description becomes less accurate so that the calculations going beyond the MFA are needed. One might expect that the deviation from the mean-field results could even be stronger in a general case of a large mass asymmetry and the mismatched Fermi surfaces but the detailed conclusion can only be drawn after the actual calculations are performed.

We were not able to follow the evolution of the system at small gap (or small $p_F a$) because of the nonanalyticity of the effective action in this case. This nonanalyticity of the effective action can explicitly be demonstrated in the mean-field approximation. The flow equations can be solved analytically in this case and one can see from the solution, which has a closed-form expression in terms of an associated Legendre function, $P_l^m(y)$ at $k = 0$, that the fermion loops contain a term $\phi^\dagger \phi \log(\phi^\dagger \phi)$. It remains to be seen whether, within the given ansatz, the full solution of the system of the partial differential equations for the effective potential and running couplings is required to trace the evolution of the system in the case of small gaps.

In summary, we have studied the pairing effect for the asymmetric fermion matter with two fermion species as a function of fermion mass asymmetry. We found that regardless of the size of the fermion mass asymmetry the boson loop corrections are small at large enough values of $p_F a$ so that the MFA provides a consistent description of the pairing effect in this case. However, when $p_F a \sim 1$ these corrections become significant at large asymmetries ($M_b > 5M_a$) making the MFA inadequate. In this case it seems to be necessary to go beyond the mean-field description.

There are several ways where this approach can further be developed. In the case of asymmetric systems the next natural step would be to consider the case of the mismatched Fermi surfaces taking into account the possibility of formation of Sarma [7], mixed [8,9] and/or LOFF [10] phases, exploring the importance of the boson loop for the stability of those phases and applying the approach to the real physical systems, for example fermionic atoms. Work in this direction is in progress. The other important extension of this approach would be to take into account running of all couplings of the average effective action, use different types of cutoff function, preferably the smooth one and include both particle-hole channel and long-range forces. The three-body force effects [11], when the correlated pair interact with the unpaired fermion may also be important, especially for nondilute systems.

The author would like to thank Mike Birse, Niels Walet and Judith McGovern for very useful discussions.

References

1. M.C. Birse, J.A. McGovern, K.G. Richardson, Phys. Lett. B **464**, 169 (1999); Meyer *et al.*, Phys. Rev. C **61**, 035202 (2000); B.J. Schaefer, O. Bohr, J. Wambach, Phys. Rev. D **65**, 105008 (2002).
2. M.C. Birse, B. Krippa, N.R. Walet, J.A. McGovern, Nucl. Phys. A **749**, 134 (2005); Int. J. Mod. Phys. A **20**, 596 (2005); Phys. Lett. B **605**, 287 (2005); B. Krippa, J. Phys. A **39**, 8075 (2006).
3. B. Krippa, nucl-th/0605071.
4. J. Berges, N. Tetradis, C. Wetterich, Phys. Rep. **363**, 223 (2002).
5. S. Weinberg, Nucl. Phys. B **413**, 567 (1994).
6. D. Litim, JHEP **0111**, 059 (2001) hep-th/0111159.
7. W.V. Liu, F. Wilczek, Phys. Rev. Lett. **90**, 047002 (2003); E. Gubankova, W.V. Liu, F. Wilczek, Phys. Rev. Lett., **91**, 032001 (2003); G. Sarma, J. Phys. Chem. Solids **24**, 1029 (1963).
8. P.F. Bedaque *et al.*, Phys. Rev. Lett. **91**, 247002 (2003).
9. H. Caldas, Phys. Rev. A **69**, 063602 (2004).
10. A.I. Larkin, Yu.N. Ovchinnikov, JETP **20**, 762 (1965); P. Fulde, R.A. Ferrell, Phys. Rev. A **135**, 550 (1964).
11. M.C. Birse, B. Krippa, N.R. Walet, J.A. McGovern, Phys. Rev. C **67**, 031301 (2003).

Eur. Phys. J. A **31**, 739–741 (2007)

DOI 10.1140/epja/i2006-10184-7

THE EUROPEAN
PHYSICAL JOURNAL A

Special Article – QNP 2006

Dimension-2 condensates, ζ-regularization and large-N_c Regge models

E. Ruiz Arriola[1,a] and W. Broniowski[2,3]

1 Departamento de Física Atómica, Molecular y Nuclear, Universidad de Granada, E-18071 Granada, Spain
2 The H. Niewodniczański Institute of Nuclear Physics, Polish Academy of Sciences, PL-31342 Kraków, Poland
3 Institute of Physics, Świętokrzyska Academy, PL-25406 Kielce, Poland

Received: 25 October 2006
Published online: 19 February 2007 – © Società Italiana di Fisica / Springer-Verlag 2007

Abstract. Dimension-2 and -4 gluon condensates are re-analyzed in large-N_c Regge models with the ζ-function regularization which preserves the spectrum in any $\bar{q}q$ channel separately. We demonstrate that the signs and magnitudes of both condensates can be properly described within the framework.

PACS. 12.38.Lg Other nonperturbative calculations – 12.38.-t Quantum chromodynamics

The dimension-2 gluon condensate, corresponding to the vacuum expectation value of a gauge-invariant non-perturbative and non-local operator and generating the lowest $1/Q^2$ power corrections, was proposed long ago [1] and has been determined in instanton model studies [2], phenomenological QCD sum rule re-analyses [3,4], theoretical considerations [5–9], non-local quark models [10], and lattice simulations at zero [11,12] and finite [13] temperatures. In the present paper we confirm our recent findings [14], namely, that within the large-N_c expansion these $1/Q^2$ corrections appear naturally within the Regge framework, in the light of ζ-regularization. We show that the signs and magnitudes of *both* the dimension-2 and -4 condensates can be accommodated comfortably with reasonable values of the parameters of the hadronic spectra. Many works compare Regge models to the Operator Product Expansion (OPE) [15–19] but besides [19] the dimension-2 condensate has been ignored.

We begin with a simple quantum-mechanical derivation of the (radial) Regge spectrum. For two relativistic scalar quarks of mass m interacting via a linear confining potential the mass operator in the CM frame is given by

$$M = 2\sqrt{\boldsymbol{p}^2 + m^2} + \sigma_S r, \qquad (1)$$

where σ_S is the (scalar) string tension and $\boldsymbol{p}$ and r are the relative momentum and distance, respectively. Squaring $\sqrt{\boldsymbol{p}^2 + m^2}$ yields an equivalent Schrödinger operator, such that $\boldsymbol{p}^2 = p_r^2 + L^2/r^2$. For excited radial states the Bohr-Sommerfeld semiclassical quantization condition holds,

$$2\pi(n + \alpha) = 2 \int_0^a p_r(r)\mathrm{d}r, \qquad (2)$$

where $p_r(r)$ is the local classical radial momentum defined by $p_r(r)^2 + L^2/r^2 + m^2 = (M - \sigma_S r)^2/4$, a is the turning point and α is of order of unity. Taking $L = 0$ and $m = 0$ for simplicity, one has $p_r(r) = |M - \sigma_S r|/2$ and $a = M/\sigma_S$. The integral is trivial leading to

$$M_n^2 = 4\pi\sigma_S(n + \alpha) = 2\pi\sigma(n + \alpha), \qquad (3)$$

a (radial) Regge mass spectrum which in terms of the spinor string tension $\sigma = 2\sigma_S$ (the factor depends on the type of interaction [20]) seems fulfilled experimentally for mesons [21] and signals confinement for the quarks. For large meson masses, the level density becomes

$$\rho(M^2) = \sum_n \delta(M^2 - M_n^2) \to \int \mathrm{d}n\delta(M^2 - M_n^2)$$
$$= \frac{\mathrm{d}n}{\mathrm{d}M^2} = \frac{1}{\pi} \int_0^a \frac{\mathrm{d}r}{p_r(r)} = \frac{1}{2\pi\sigma} \qquad (4)$$

which is a constant. Inclusion of finite quark mass corrections is straightforward, yielding

$$\frac{\mathrm{d}n}{\mathrm{d}M^2} = \frac{1}{2\pi\sigma}\sqrt{1 - \frac{4m^2}{M^2}} \quad \text{for} \quad M^2 \geq 4m^2 . \qquad (5)$$

Note that this corresponds to the two-body phase space factor appearing in the absorptive part of two-point correlators. Thus, at large energies the WKB approximation holds and $\rho(M^2)$ looks like the phase space of two free particles, featuring the *quark-hadron duality*.

The best way to look at the $q\bar{q}$ level density is to consider two-point correlation functions in different channels,

$$i\Pi^{\mu a, \nu b}(q) = \int \mathrm{d}^4 x e^{-iq \cdot x} \langle 0|T\left\{J^{\mu a}(x)J^{\nu b}(0)\right\}|0\rangle$$
$$= \Pi(q^2)\left(q^\mu q^\nu - g^{\mu\nu}q^2\right)\delta^{ab}, \qquad (6)$$

a Speaker; e-mail: earriola@ugr.es

where current conservation for both the vector and axial currents has explicitly been used. At high Euclidean momentum OPE can be performed. Equivalently, in the Euclidean coordinate space one has

$$\langle J_\mu(x) J_\nu(0)\rangle = \left(g^{\mu\nu}\partial^2 - \partial^\mu\partial^\nu\right)\Pi(x). \qquad (7)$$

The function $\Pi(x)$ has dimension x^{-4}. Thus, at short distances one expects to have (up to possible logarithms)

$$\Pi(x) = \frac{O_0}{x^4} + \frac{O_2}{x^2} + \frac{O_4}{x^0} + O_6 x^2 + \dots. \qquad (8)$$

We digress that the term containing O_2, an operator of dimension-2, is very special, since it yields a contribution of the form $(g^{\mu\nu}\partial^2 - \partial^\mu\partial^\nu)x^{-2} = (g^{\mu\nu}x^2 - 4x^\mu x^\nu)x^{-4}$ which, in addition to being conserved, is also traceless. Thus, contracting and taking the derivative ∂^α do no commute. There is no traceless and transverse term in momentum space, since conservation implies the form $A(q^\mu q^\nu - q^2 g^{\mu\nu})$ but tracelessness requires $A = 0$. This problem appears in chiral quark models; the dimension-2 object is the constituent-quark mass squared, $O_2 \sim \langle\sigma^2 + \pi^2\rangle \sim M^2$.

Recent discussions incorporate O_2 in OPE. From [9] we get up to dimension 6 in the chiral limit,

$$\Pi_{V+A}(Q^2) = \frac{1}{4\pi^2}\left\{ -\left(1 + \frac{\alpha_S}{\pi}\right)\log\frac{Q^2}{\mu^2}\right.$$
$$\left. -\frac{\alpha_S}{\pi}\frac{\lambda^2}{Q^2} + \frac{\pi}{3}\frac{\langle\alpha_S G^2\rangle}{Q^4} + \frac{256\pi^3}{81}\frac{\alpha_S\langle\bar q q\rangle^2}{Q^6}\right\},$$
$$\Pi_{V-A}(Q^2) = -\frac{32\pi}{9}\frac{\alpha_S\langle\bar q q\rangle^2}{Q^6}, \qquad (9)$$

where λ^2 corresponds to the dimension-2 condensate.

In the large-N_c limit one has (up to subtractions) [22]

$$\Pi_V(Q^2) = \sum_V \frac{F_V^2}{M_V^2 + Q^2} + \text{c.t.}, \qquad (10)$$

where the sum involves infinitely many resonances. This function satisfies a dispersion relation of the form

$$\Pi_V(Q^2) = \int_0^\infty ds \frac{Q^2}{s}\frac{\rho_V(s)}{s + Q^2}, \qquad (11)$$

with one subtraction, $\Pi_V(0) = 0$, due to Coulomb's law and the spectral function in the vector channel given by

$$\rho_V(s) = \frac{1}{\pi}\operatorname{Im}\Pi_V(s) = \sum_V F_V^2\delta(s - M_V^2). \qquad (12)$$

At large values of the squared CM energy s, it becomes

$$\rho_V(s) \to \int_0^\infty F_V^2\delta(s - M_V^2)dn = F_V(n)^2\frac{dn}{dM_V^2}\bigg|_{M_V^2 = s}. \qquad (13)$$

Matching to the free massless quark result $\rho_V(s) = N_c/(12\pi^2)\theta(s)$ gives at large values of n

$$F_V(n)^2\frac{dn}{dM_V^2} \to \frac{N_c}{3}\frac{1}{4\pi^2}. \qquad (14)$$

For constant F_V this implies the asymptotic spectrum $M_V^2 \to 2\pi\sigma n$ with the string tension given by $2\pi\sigma = 24\pi^2 F_V^2/N_c$. If we identify $F_V = 154\,\text{MeV}$ from the $\rho \to 2\pi$ decay [23] which corresponds to only one resonance, we get $\sqrt\sigma = 546\,\text{MeV}$. Lattice calculations [24] provide $\sqrt\sigma = 420\,\text{MeV}$.

In dimensional regularization the coupling of the resonance to the current acquires an additional dimension $F_V \to F_V \mu^\epsilon$ with $\epsilon = d - 4$. By choosing $\mu = M_V$ one gets $F_V^2 \to F_V^2 M_V^{2\epsilon}$. The regularized correlator is an analytic function for $Q^2 < m_\rho^2$ (the lowest mass), so we can Taylor-expand at small Q^2. We can then regularize the finite coefficients of the expansion and proceed by analytic continuation both in Q^2 and ϵ. The regularization only acts for truly *infinitely* many resonances. At large Euclidean momenta one gets

$$\Pi_V(Q^2) = \sum_V F_V^2 M_V^{2\epsilon} - \sum_V F_V^2\frac{M_V^{2+2\epsilon}}{Q^2} + \dots. \qquad (15)$$

The coefficients of powers in $1/Q^2$ of the expansion are convergent provided one computes the sum first and then takes the limit $\epsilon \to 0$ corresponding to the use of the ζ-function regularization (see, *e.g.*, [25]),

$$\sum_V F_V^2 M_V^{2n} \equiv \lim_{s\to n}\sum_V F_V^2 M_V^{2s}. \qquad (16)$$

In other words, one may expand formally at large Q^2 and re-interpret the result by means of the ζ-function regularization. Using the axial-axial correlator at large N_c,

$$\Pi_A(Q^2) = \frac{f^2}{Q^2} + \sum_A \frac{F_A^2}{M_A^2 + Q^2} + \text{c.t.}, \qquad (17)$$

and matching to (9) yields the two Weinberg sum rules:

$$f^2 = \sum_A F_A^2 - \sum_V F_V^2 \qquad (\text{WSR I}),$$

$$0 = \sum_A F_A^2 M_A^2 - \sum_V F_V^2 M_V^2 \qquad (\text{WSR II}).$$

These sums are assumed to be ζ-regularized, see eq. (16).

The simplest Regge model is given by

$$M_{V,n}^2 = M_V^2 + 2\pi\sigma n, \qquad M_{A,n}^2 = M_A^2 + 2\pi\sigma n, \qquad (18)$$

$n = 0, 1, 2 \dots$, which seems fulfilled [21] experimentally. The corresponding couplings are constant, $F_V = F_A = F$ and the ζ-function regularized sums follow from

$$\sum_{n=0}^\infty (na + M^2)^s = a^s\zeta\left(-s, \frac{M^2}{a}\right). \qquad (19)$$

This function is analytic in the complex plane $\operatorname{Re}(s) \le -1$ with the exception of $s = -1$, admitting analytic continuation to any s. Actually, for positive powers one gets the Bernoulli polynomials $\zeta(-k, z) = -B_{k+1}(z)/(k+1)$,

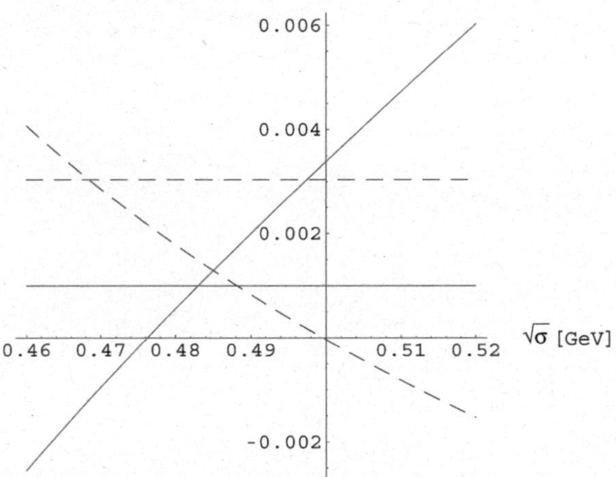

Fig. 1. Dimension-2 (solid line, in GeV^2) and -4 (dashed line, in GeV^4) condensates plotted as functions of $\sqrt{\sigma}$. The horizontal lines indicate estimates from the literature.

where $B_0 = 1$, $B_1 = x - 1/2$, $B_2 = x^2 - x + 1/6$, etc. An important feature of the ζ-function is that it regulates each spectrum separately, *i.e.* under regularization one cannot apply the distributive property. For instance,

$$\sum_{n=0}^{\infty} (an + M_V^2)^0 - \sum_{n=0}^{\infty} (an + M_A^2)^0 \neq 0. \tag{20}$$

In other words, the difference of the regularized sums does not coincide with the regularized difference. The finite terms in the difference have to do with preserving the spectra in the vector and axial channels separately, and hence a chiral asymmetry is generated. All these ζ-function results reproduce the direct asymptotic expansion of exact sums in terms of the digamma functions, but allow to discuss cases where the sums cannot be carried out *before* expanding in large Q^2.

The strict linear Regge model does not generate condensates with the proper signs. In [14] (see also [26]) we consider the following simple modification:

$$M_{V,0} = \mathring{m}_\rho, \quad M_{V,n}^2 = M_V^2 + 2\pi\sigma n, \quad n \geq 1,$$
$$M_{A,n}^2 = M_A^2 + 2\pi\sigma n, \quad n \geq 0. \tag{21}$$

Here the lowest ρ mass is shifted, otherwise all is kept "universal", including constant residues for all states. With (21) the Weinberg sum rules are (we set $N_c = 3$)

$$M_A^2 = M_V^2 + 8\pi^2 f^2,$$
$$2\pi\sigma = 8\pi^2 F^2 = \frac{8\pi^2 f^2 \left(4\pi^2 f^2 + M_V^2\right)}{4\pi^2 f^2 - m_\rho^2 + M_V^2}. \tag{22}$$

When $m_\rho = 0.77\,\mathrm{GeV}$ is fixed, the model has only one free parameter left. We may take it to be M_V; however, it is more convenient to express it through the string tension σ, which is then treated as a free parameter. Thus

$$M_V^2 = \frac{-16\pi^3 f^4 + 4\pi^2 \sigma f^2 - m_\rho^2 \sigma}{4f^2\pi - \sigma}, \tag{23}$$

and the condensates, obtained by matching to (9), are

$$-\frac{\alpha_S \lambda^2}{4\pi^3} = \frac{16\pi^3 f^4 - \pi\sigma^2 + m_\rho^2 \sigma}{16 f^2 \pi^3 - 4\pi^2 \sigma}$$
$$\frac{\alpha_S \langle G^2 \rangle}{12\pi} = 2\pi^2 f^4 - \pi\sigma f^2$$
$$+ \frac{3\sigma m_\rho^2}{8\pi^2} \left(\frac{m_\rho^2 \sigma}{(\sigma - 4f^2\pi)^2} - 2\pi \right) + \frac{\sigma^2}{12}. \tag{24}$$

Despite the strong dependence on $\sqrt{\sigma}$ (see fig. 1) the correct signs, $-\lambda^2 > 0$, $\langle G^2 \rangle > 0$, and reasonable numbers are reproduced in the range $0.48 \leq \sqrt{\sigma} \leq 0.5\,\mathrm{GeV}$ [14].

Work supported by the Polish Ministry of Education and Science grants 2P03B 02828 by the Spanish Ministerio de Asuntos Exteriores and the Polish Ministry of Education and Science project 4990/R04/05, by the Spanish DGI and FEDER funds grant BFM2002-03218, Junta de Andalucía grant FQM-225, and EU RTN Contract CT2002-0311 (EURIDICE).

References

1. L.S. Celenza, C.M. Shakin, Phys. Rev. D **34**, 1591 (1986).
2. M. Hutter, arXiv:hep-ph/9501335.
3. C.A. Dominguez, Phys. Lett. B **345**, 291 (1995).
4. K.G. Chetyrkin, S. Narison, V.I. Zakharov, Nucl. Phys. B **550**, 353 (1999).
5. F.V. Gubarev, L. Stodolsky, V.I. Zakharov, Phys. Rev. Lett. **86**, 2220 (2001).
6. F.V. Gubarev, V.I. Zakharov, Phys. Lett. B **501**, 28 (2001).
7. K.I. Kondo, Phys. Lett. B **514**, 335 (2001).
8. H. Verschelde, K. Knecht, K. Van Acoleyen, M. Vanderkelen, Phys. Lett. B **516**, 307 (2001).
9. S. Narison, V.I. Zakharov, Phys. Lett. B **522**, 266 (2001).
10. A.E. Dorokhov, W. Broniowski, Eur. Phys. J. C **32**, 79 (2003).
11. P. Boucaud, A.Le Yaouanc, J.P. Leroy, J. Micheli, O. Pene, J. Rodriguez-Quintero, Phys. Rev. D **63**, 114003 (2001).
12. E. Ruiz Arriola, P.O. Bowman, W. Broniowski, Phys. Rev. D **70**, 097505 (2004).
13. E. Megias, E. Ruiz Arriola, L.L. Salcedo, JHEP **0601**, 073 (2006).
14. E. Ruiz Arriola, W. Broniowski, Phys. Rev. D **73**, 097502 (2006).
15. M. Golterman, S. Peris, JHEP **0101**, 028 (2001); Phys. Rev. D **67**, 096001 (2003).
16. S.R. Beane, Phys. Rev. D **64**, 116010 (2001).
17. Y.A. Simonov, Phys. At. Nucl. **65**, 135 (2002) (Yad. Fiz. **65**, 140 (2002)).
18. S.S. Afonin, Phys. Lett. B **576**, 122 (2003).
19. S.S. Afonin, A.A. Andrianov, V.A. Andrianov, D. Espriu, JHEP **0404**, 039 (2004).
20. C. Goebel et al., Phys. Rev. D **41**, 2917 (1990).
21. A.V. Anisovich, V.V. Anisovich, A.V. Sarantsev, Phys. Rev. D **62**, 051502 (2000).
22. A. Pich, arXiv:hep-ph/0205030.
23. G. Ecker, J. Gasser, A. Pich, E. de Rafael, Nucl. Phys. B **321**, 311 (1989).
24. O. Kaczmarek, F. Zantow, Phys. Rev. D **71**, 114510 (2005).
25. L.L. Salcedo, E. Ruiz Arriola, Ann. Phys. (N.Y.) **250**, 1 (1996)
26. S.S. Afonin, D. Espriu, JHEP **0609**, 047 (2006).

Eur. Phys. J. A **31**, 742–745 (2007)
DOI 10.1140/epja/i2006-10281-7

Special Article – QNP 2006

On dynamical gluon mass generation

A.C. Aguilar[1] and J. Papavassiliou[2,a]

[1] Instituto de Física Teórica, Universidade Estadual Paulista, Rua Pamplona 145, 01405-900, São Paulo, SP, Brazil
[2] Departamento de Física Teórica and IFIC, Universidade de Valencia-CSIC, E-46100, Burjassot, Valencia, Spain

Received: 18 December 2006
Published online: 20 March 2007 – © Società Italiana di Fisica / Springer-Verlag 2007

Abstract. The effective gluon propagator constructed with the pinch technique is governed by a Schwinger-Dyson equation with special structure and gauge properties, that can be deduced from the correspondence with the background field method. Most importantly the non-perturbative gluon self-energy is transverse order-by-order in the dressed loop expansion, and separately for gluonic and ghost contributions, a property which allows for a meanigfull truncation. A linearized version of the truncated Schwinger-Dyson equation is derived, using a vertex that satisfies the required Ward identity and contains massless poles. The resulting integral equation, subject to a properly regularized constraint, is solved numerically, and the main features of the solutions are briefly discussed.

PACS. 12.38.Lg Other nonperturbative calculations – 12.38.Aw General properties of QCD (dynamics, confinement, etc.)

It is well known that one of the main theoretical problems when dealing with Schwinger-Dyson (SD) equations is that they are built out of unphysical off-shell Green's functions; thus, the extraction of reliable physical information depends crucially on delicate all-order cancellations, which may be inadvertently distorted in the process of the truncation. The truncation scheme based on the pinch technique (PT) [1,2] implements a drastic modification at the level of the building blocks of the SD series. The PT enables the construction of new, effective Green's functions endowed with very special properties; most importantly, they are independent of the gauge-fixing parameter, and satisfy QED-like Ward identities (WI) instead of the usual Slavnov-Taylor identities. The upshot of this approach would then be to trade the conventional SD series for another, written in terms of the new Green's functions, and then truncate this new series, by keeping only a few terms in a "dressed-loop" expansion, maintaining exact gauge-invariance. Of central importance in this context is the connection between the PT and the Background Field Method (BFM), a special gauge-fixing procedure that preserves the symmetry of the action under ordinary gauge transformations with respect to the background (classical) gauge field $\widehat{A}_\mu^a$. As a result, the background n-point functions satisfy QED-like all-order WIs. The connection between PT and BFM, which is known to persist to all orders (last two articles in [2]), affirms that the (gauge-independent) PT effective n-point functions coincide with

the (gauge-dependent) BFM n-point functions provided that the latter are computed in the Feynman gauge. In this paper we report recent progress on the issue of gluon mass generation in the PT-BFM scheme [3].

We first define some basic quantities. There are two gluon propagators appearing in this problem, $\widehat{\Delta}_{\mu\nu}(q)$ and $\Delta_{\mu\nu}(q)$, denoting the background and quantum gluon propagator, respectively. Defining $P_{\mu\nu}(q) = g_{\mu\nu} - \frac{q_\mu q_\nu}{q^2}$, we have that $\widehat{\Delta}_{\mu\nu}(q)$, in the Feynman gauge is given by

$$\widehat{\Delta}_{\mu\nu}(q) = -i\left[P_{\mu\nu}(q)\widehat{\Delta}(q^2) + \frac{q_\mu q_\nu}{q^4}\right], \tag{1}$$

The gluon self-energy, $\widehat{\Pi}_{\mu\nu}(q)$, has the form $\widehat{\Pi}_{\mu\nu}(q) = P_{\mu\nu}(q)\,\widehat{\Pi}(q^2)$, and $\widehat{\Delta}^{-1}(q^2) = q^2 + i\widehat{\Pi}(q^2)$. Exactly analogous definitions relate $\Delta_{\mu\nu}(q)$ with $\Pi_{\mu\nu}(q)$.

As is widely known, in the conventional formalism the inclusion of ghosts is instrumental for the transversality of $\Pi_{\mu\nu}^{ab}(q)$, already at the level of the one-loop calculation. On the other hand, in the PT-BFM formalism, due to new Feynman rules for the vertices, the one-loop gluon and ghost contribution are individually transverse [4].

As has been shown in [3], this crucial feature persists at the non-perturbative level, as a consequence of the simple WIs satisfied by the full vertices appearing in the diagrams of fig. 1, defining the BFM SD equation for $\widehat{\Delta}_{\mu\nu}(q)$ [5]. Specifically, the gluonic and ghost sector are separately transverse, within each individual order in the dressed-loop expansion.

a e-mail: Joannis.Papavassiliou@uv.es

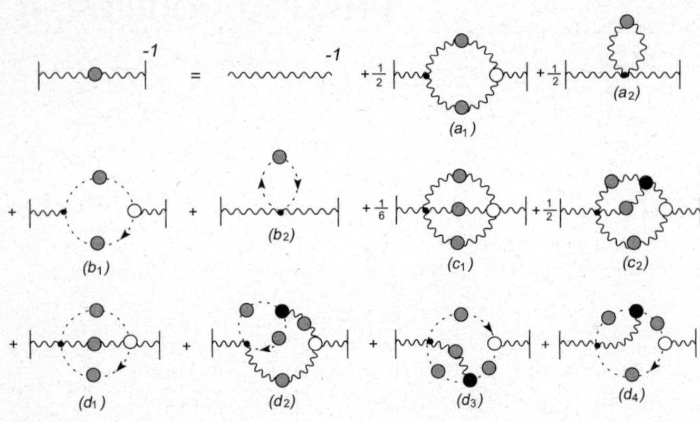

Fig. 1. The SD equation for the gluon propagator in the BFM. All external legs (ending with a vertical line) are background gluons, wavy lines with grey blobs denote full-quantum gluon propagators, dashed lines with grey blobs are full-ghost propagators, black dots are the BFM tree-level vertices, black blobs are the full conventional vertices, and white blobs denote full three- or four-gluon vertices with one external background leg.

Let us demonstrate this property for graphs (a_1) and (a_2), given by

$$\widehat{\Pi}^{ab}_{\mu\nu}(q)\Big|_{a_1} = \frac{1}{2}\int [\mathrm{d}k]\, \widetilde{\Gamma}^{aex}_{\mu\alpha\beta}\Delta^{\alpha\rho}_{ee'}(k)\widetilde{\Pi}^{be'x'}_{\nu\rho\sigma}\Delta^{\beta\sigma}_{xx'}(k+q),$$

$$\widehat{\Pi}^{ab}_{\mu\nu}(q)\Big|_{a_2} = \frac{1}{2}\int [\mathrm{d}k]\, \widetilde{\Gamma}^{abex}_{\mu\nu\alpha\beta}\Delta^{\alpha\beta}_{ex}(k), \qquad (2)$$

where $[\mathrm{d}k] = \mathrm{d}^d k/(2\pi)^d$ with $d = 4 - \epsilon$ the dimension of space-time. By virtue of the BFM all-order WI

$$q_1^\mu\,\widetilde{\Pi}^{abc}_{\mu\alpha\beta}(q_1, q_2, q_3) = g f^{abc}\left[\Delta^{-1}_{\alpha\beta}(q_2) - \Delta^{-1}_{\alpha\beta}(q_3)\right], \quad (3)$$

and using the tree-level $\widetilde{\Gamma}_{\mu\nu\alpha\beta}$ given in [4], we have

$$q^\nu\,\widehat{\Pi}^{ab}_{\mu\nu}(q)\Big|_{a_1} = C_A\, g^2 \delta^{ab}\, q_\mu \int [\mathrm{d}k]\, \Delta^\rho_\rho(k),$$

$$q^\nu\,\widehat{\Pi}^{ab}_{\mu\nu}(q)\Big|_{a_2} = -C_A\, g^2 \delta^{ab}\, q_\mu \int [\mathrm{d}k]\, \Delta^\rho_\rho(k), \qquad (4)$$

and thus, $q^\nu\big(\widehat{\Pi}^{ab}_{\mu\nu}(q)\big|_{a_1} + \widehat{\Pi}^{ab}_{\mu\nu}(q)\big|_{a_2}\big) = 0$. The importance of this transversality property in the context of the SD equation is that it allows for a meaningful first approximation: instead of the system of coupled equations involving gluon and ghost propagators, one may consider only the subset containing gluons, without compromising the crucial property of transversality. We will therefore study as the first non-trivial approximation for $\widehat{\Pi}_{\mu\nu}(q)$ the diagrams (a_1) and (a_2). Of course, we have no *a priori* guarantee that this particular subset is numerically dominant. Actually, as has been argued in a series of SD studies, in the context of the conventional Landau gauge it is the ghost sector that furnishes the leading contribution [6] Clearly, it is plausible that this characteristic feature may persist within the PT-BFM scheme as well, and we will explore this crucial issue in the near future.

The equation given in (2) is not a genuine SD equation, in the sense that it does not involve the unknown quantity $\widehat{\Delta}$ on both sides. Substituting $\Delta \to \widehat{\Delta}$ on the RHS of (2) (see discussion in [3]), we obtain

$$\widehat{\Pi}_{\mu\nu}(q) = \frac{1}{2} C_A\, g^2 \int [\mathrm{d}k]\, \widetilde{\Gamma}^{\alpha\beta}_\mu \widehat{\Delta}(k)\widetilde{\Pi}_{\nu\alpha\beta}\widehat{\Delta}(k+q)$$

$$- C_A\, g^2\, d\, g_{\mu\nu} \int [\mathrm{d}k]\, \widehat{\Delta}(k), \qquad (5)$$

with $\widetilde{\Gamma}_{\mu\alpha\beta} = (2k+q)_\mu g_{\alpha\beta} - 2q_\alpha g_{\mu\beta} + 2q_\beta g_{\mu\alpha}$, and

$$q^\nu\widetilde{\Pi}_{\nu\alpha\beta} = \left[\widehat{\Delta}^{-1}(k+q) - \widehat{\Delta}^{-1}(k)\right] g_{\alpha\beta}. \qquad (6)$$

We can then linearize the resulting SD equation, by resorting to the Lehmann representation for the scalar part of the gluon propagator [1],

$$\widehat{\Delta}(q^2) = \int \mathrm{d}\lambda^2\, \frac{\rho(\lambda^2)}{q^2 - \lambda^2 + i\epsilon}, \qquad (7)$$

and setting on the first integral of the RHS of eq. (5):

$$\widehat{\Delta}(k)\widetilde{\Pi}_{\nu\alpha\beta}\widehat{\Delta}(k+q) = \int \frac{\mathrm{d}\lambda^2\, \rho(\lambda^2)\, \widetilde{\Gamma}^{\mathrm{L}}_{\nu\alpha\beta}}{[k^2 - \lambda^2][(k+q)^2 - \lambda^2]}, \qquad (8)$$

where $\widetilde{\Gamma}^{\mathrm{L}}_{\nu\alpha\beta}$ must be such as to satisfy the tree-level WI

$$q^\nu\widetilde{\Gamma}^{\mathrm{L}}_{\nu\alpha\beta} = \left[(k+q)^2 - \lambda^2\right] g_{\alpha\beta} - (k^2 - \lambda^2)g_{\alpha\beta}. \qquad (9)$$

We propose the following form for the vertex:

$$\widetilde{\Gamma}^{\mathrm{L}}_{\nu\alpha\beta} = \widetilde{\Gamma}_{\nu\alpha\beta} + c_1\left((2k+q)_\nu + \frac{q_\nu}{q^2}\left[k^2 - (k+q)^2\right]\right) g_{\alpha\beta}$$

$$+ \left(c_3 + \frac{c_2}{2\,q^2}\left[(k+q)^2 + k^2\right]\right)(q_\beta g_{\nu\alpha} - q_\alpha g_{\nu\beta}), \quad (10)$$

which, due to the presence of the massless poles, allows the possibility of infrared finite solution.

Due to the QED-like WIs satisfied by the PT Green's functions, $\widehat{\Delta}^{-1}(q^2)$ absorbs all the RG-logs. Consequently, the product $\widehat{d}(q^2) = g^2 \widehat{\Delta}(q^2)$ forms a RG-invariant (μ-independent) quantity. Notice however that eq. (5) does not encode the correct RG behavior: when written in terms of $\widehat{d}(q^2)$ it is not manifestly g^2-independent, as it should. In order to restore the correct RG behavior, we use the simple prescription proposed in [1], whereby we substitute every $\widehat{\Delta}(z)$ appearing on the RHS of the SD equation by

$$\widehat{\Delta}(z) \to \frac{g^2\,\widehat{\Delta}(z)}{\bar{g}^2(z)} \equiv [1 + \tilde{b}g^2 \ln(z/\mu^2)]\widehat{\Delta}(z). \qquad (11)$$

Then, setting $\tilde{b} \equiv \frac{10\,C_A}{48\pi^2}$, $\sigma \equiv \frac{6(c_1+c_2)}{5}$, $\gamma \equiv \frac{4+4\,c_1+3\,c_2}{5}$, we finally obtain

$$\widehat{d}^{-1}(q^2) = q^2\left\{K' + \tilde{b}\int_0^{q^2/4} \mathrm{d}z\left(1 - \frac{4z}{q^2}\right)^{1/2} \frac{\widehat{d}(z)}{\bar{g}^2(z)}\right\}$$

$$+ \gamma\tilde{b}\int_0^{q^2/4} \mathrm{d}z\, z\left(1 - \frac{4z}{q^2}\right)^{1/2} \frac{\widehat{d}(z)}{\bar{g}^2(z)} + \widehat{d}^{-1}(0), \quad (12)$$

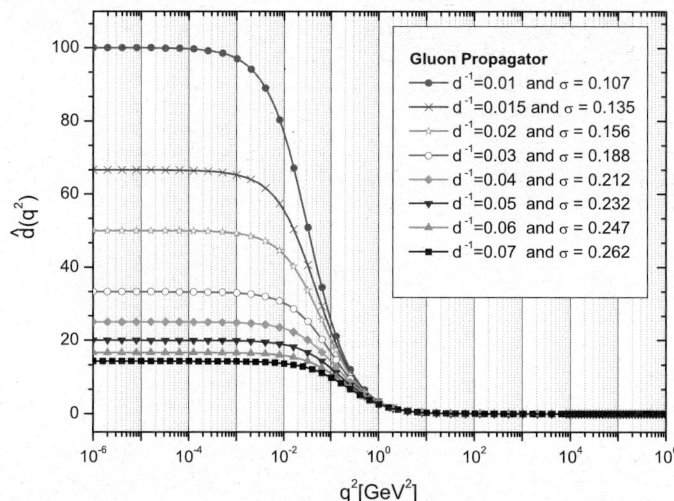

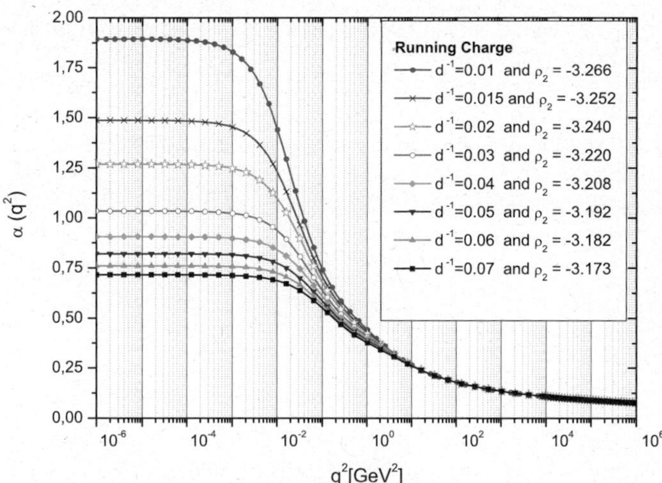

Fig. 2. Results for $\widehat{d}(q^2)$, for different values for $\widehat{d}^{-1}(0)$ (all in GeV^2), and the corresponding values for σ.

Fig. 3. $\alpha(q^2) = \bar{g}^2_{\mathrm{NP}}(q^2)/4\pi$ from the propagators of fig. 2.

$$K' = \frac{1}{g^2} - \tilde{b} \int_0^{\mu^2/4} \mathrm{d}z \left(1 + \gamma \frac{z}{\mu^2}\right) \left(1 - \frac{4z}{\mu^2}\right)^{1/2} \frac{\widehat{d}(z)}{\bar{g}^2(z)}, \tag{13}$$

and

$$\widehat{d}^{-1}(0) = -\frac{\tilde{b}\sigma}{\pi^2} \int \mathrm{d}^4 k \, \frac{\widehat{d}(k^2)}{\bar{g}^2(k^2)}. \tag{14}$$

It is easy to see now that eq. (12) yields the correct UV behavior, *i.e.* $\widehat{d}^{-1}(q^2) = \tilde{b}\, q^2 \ln(q^2/\Lambda^2)$.

When solving (12) we will be interested in solutions that are qualitatively of the general form [1]

$$\widehat{d}(q^2) = \frac{\bar{g}^2_{\mathrm{NP}}(q^2)}{q^2 + m^2(q^2)}, \tag{15}$$

where

$$\bar{g}^2_{\mathrm{NP}}(q^2) = \left[\tilde{b} \ln \left(\frac{q^2 + f(q^2, m^2(q^2))}{\Lambda^2}\right)\right]^{-1}, \tag{16}$$

$\bar{g}^2_{\mathrm{NP}}(q^2)$ represents a non-perturbative version of the RG-invariant effective charge of QCD: in the deep UV it goes over to $\bar{g}^2(q^2)$, while in the deep IR it "freezes" [1,7], due to the presence of the function $f(q^2, m^2(q^2))$, whose form will be determined by fitting the numerical solution. The function $m^2(q^2)$ may be interpreted as a momentum-dependent "mass". In order to determine the asymptotic behavior that eq. (12) predicts for $m^2(q^2)$ at large q^2, we replace eq. (15) on both sides, set $(1 - 4z/q^2)^{1/2} \to 1$, obtaining self-consistency provided that

$$m^2(q^2) \sim m_0^2 \ln^{-a}\left(q^2/\Lambda^2\right), \quad \text{with} \quad a = 1 + \gamma > 0. \tag{17}$$

The seagull-like contributions, defining $\widehat{d}^{-1}(0)$ in (14), are essential for obtaining IR finite solutions. However, the integral in (14) should be properly regularized, in order to ensure the finiteness of such a mass term. Recalling that in dimensional regularization $\int [\mathrm{d}k]/k^2 = 0$, we rewrite

eq. (14) (using (15)) as

$$\begin{aligned}
\widehat{d}^{-1}(0) &\equiv -\frac{\tilde{b}\sigma}{\pi^2} \int [\mathrm{d}k] \left(\frac{\bar{g}^2_{\mathrm{NP}}(k^2)}{[k^2 + m^2(k^2)]\bar{g}^2(k^2)} - \frac{1}{k^2}\right) \\
&= \frac{\tilde{b}\sigma}{\pi^2} \int [\mathrm{d}k] \frac{m^2(k^2)}{k^2[k^2 + m^2(k^2)]} \\
&\quad + \frac{\tilde{b}^2 \sigma}{\pi^2} \int [\mathrm{d}k] \, \widehat{d}(k^2) \ln\left(1 + \frac{f(k^2, m^2(k^2))}{k^2}\right).
\end{aligned} \tag{18}$$

The first integral converges provided that $m^2(k^2)$ falls asymptotically as $\ln^{-a}(k^2)$, with $a > 1$, while the second requires that $f(k^2, m^2(k^2))$ should drop asymptotically at least as fast as $\ln^{-c}(k^2)$, with $c > 0$. Notice that perturbatively $\widehat{d}^{-1}(0)$ vanishes, because $m^2(k^2) = 0$ to all orders, and, in that case, $f = 0$ also.

Solving numerically eq. (12), subject to the constraint of eq. (14), we obtain solutions shown in fig. 2; they can be fitted perfectly by means of a running coupling that freezes in the IR, shown in fig. 3, and a running mass that vanishes in the UV [3]. σ is treated as a free parameter, whose values are fixed in such a way as to achieve compliance between eqs. (12)–(14).

This research was supported by Spanish MEC under the grant FPA 2005-01678 and by Fundação de Amparo à Pesquisa do Estado de São Paulo (FAPESP-Brazil) through the grant 05/04066-0. J.P. thanks the organizers of *QNP 2006* for their hospitality.

References

1. J.M. Cornwall, Phys. Rev. D **26**, 1453 (1982).
2. J.M. Cornwall, J. Papavassiliou, Phys. Rev. D **40**, 3474 (1989); J. Papavassiliou, Phys. Rev. D **41**, 3179 (1990); D. Binosi, J. Papavassiliou, Phys. Rev. D **66**, 111901 (2002); J. Phys. G **30**, 203 (2004).
3. A.C. Aguilar, J. Papavassiliou, JHEP **0612**, 012 (2006).
4. L.F. Abbott, Nucl. Phys. B **185**, 189 (1981).

5. R.B. Sohn, Nucl. Phys. B **273**, 468 (1986); A. Hadicke, JENA-N-88-19.
6. L. von Smekal, R. Alkofer, A. Hauck, Phys. Rev. Lett. **79**, 3591 (1997); R. Alkofer, L. von Smekal, Phys. Rep. **353**, 281 (2001); C.S. Fischer, J. Phys. G **32**, R253 (2006).
7. A.C. Aguilar, A.A. Natale, P.S. Rodrigues da Silva, Phys. Rev. Lett. **90**, 152001 (2003); A.C. Aguilar, A. Mihara, A.A. Natale, Phys. Rev. D **65**, 054011 (2002); Int. J. Mod. Phys. A **19**, 249 (2004).

Eur. Phys. J. A **31**, 746–749 (2007)

DOI 10.1140/epja/i2006-10244-0

THE EUROPEAN
PHYSICAL JOURNAL A

Special Article – QNP 2006

Aspects of quark mass generation on a torus

C.S. Fischer[1,a] and M.R. Pennington[2]

[1] Institut für Kernphysik, Technical University of Darmstadt, Schlossgartenstraße 9, 64289 Darmstadt, Germany
[2] IPPP, Durham University, Durham DH1 3LE, UK

Received: 23 November 2006
Published online: 9 March 2007 – © Società Italiana di Fisica / Springer-Verlag 2007

Abstract. In this paper we report on recent results for the quark propagator on a compact manifold. The corresponding Dyson-Schwinger equations on a torus are solved on volumes similar to the ones used in lattice calculations. The quark-gluon interaction is fixed such that the lattice results are reproduced. We discuss both the effects in the infinite volume/continuum limit as well as effects when the volume is small.

PACS. 12.38.Aw General properties of QCD (dynamics, confinement, etc.) – 12.38.Gc Lattice QCD calculations – 12.38.Lg Other nonperturbative calculations – 14.65.Bt Light quarks

1 Introduction

In the low-energy sector of QCD dynamical chiral symmetry breaking is the dominant nonperturbative effect with respect to hadron phenomenology. Thus any framework attempting to explain the spectra and decay properties of light hadrons has to satisfy all constraints related to the correct pattern of chiral symmetry breaking in QCD. In this respect lattice simulations have to deal with two types of problems: on the one hand, commutation relations between γ-matrices are affected by the finite lattice spacing. On the other hand, there is the finite volume. Since continuous symmetries cannot be spontaneously broken at a finite volume V, chiral symmetry is restored in the limit of zero current quark mass, $m \to 0$ (see, *e.g.*, [1]). Thus one first has to perform both the continuum and infinite volume limit before one can investigate the chiral limit. In turn, studies with physical up- and down-quark masses necessarily require large volumes to avoid chiral restoration effects.

It is certainly desirable to study volume effects in chiral symmetry breaking in different frameworks. Chiral perturbation theory has turned out to be a reliable tool for both volume and chiral extrapolations. On the other hand, chiral perturbation theory has nothing to say about volume effects in the underlying quark and gluon substructure. For this the Green's function approach employing Dyson-Schwinger equations (DSEs) [2,3] provides a suitable alternative. In this paper we present results for the quark propagator in Landau gauge QCD. Based on ideas developed in [4,5], we use lattice results to fix the quark-gluon

[a] Spokesperson;
e-mail: `christian.fischer@physik.tu-darmstadt.de`

interaction at quark masses accessible on the lattice. To this end we solve the DSEs on tori with similar volumes. We then change current quark masses and volumes to explore the corresponding effects on the quark propagator.

2 Quark-gluon interaction

The Dyson-Schwinger equations (DSEs) for the ghost, gluon and quark propagators $D_G, D_{\mu,\nu}, S(p)$

$$D_G(p) = -\frac{G(p^2)}{p^2}, \tag{1}$$

$$D_{\mu\nu}(p) = \left(\delta_{\mu\nu} - \frac{p_\mu p_\nu}{p^2}\right)\frac{Z(p^2)}{p^2}, \tag{2}$$

$$S(p) = \frac{Z_f(p^2)}{i\not{p} + M(p^2)}, \tag{3}$$

in quenched approximation are given diagrammatically in fig. 1. This system of equations is closed but for the dressed vertices, which have to be approximated by suitable ansätze. The truncation scheme for the Yang-Mills sector is discussed, *e.g.*, in [3,6]. The corresponding numerical solutions for the ghost and gluon propagator can be represented accurately by

$$R(p^2) = \frac{c\,(p^2/\Lambda_{YM}^2)^\kappa + d\,(p^2/\Lambda_{YM}^2)^{2\kappa}}{1 + c\,(p^2/\Lambda_{YM}^2)^\kappa + d\,(p^2/\Lambda_{YM}^2)^{2\kappa}},$$

$$Z(p^2) = \left(\frac{\alpha(p^2)}{\alpha(\mu^2)}\right)^{1+2\delta} R^2(p^2), \tag{4}$$

$$G(p^2) = \left(\frac{\alpha(p^2)}{\alpha(\mu^2)}\right)^{-\delta} R^{-1}(p^2), \tag{5}$$

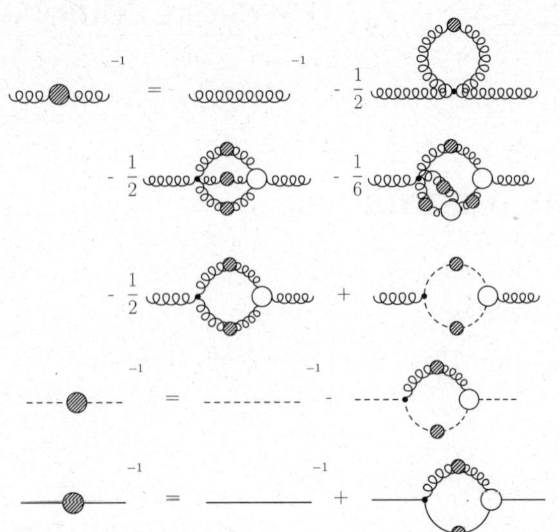

Fig. 1. Dyson-Schwinger equations for the gluon (curly lines), ghost (dashed lines) and quark (straight lines) propagators. Filled circles denote dressed propagators and empty circles denote dressed vertex functions.

Table 1. Parameters used in the vertex model, eqs. (7)-(10).

	h	Λ_g (GeV)	Λ_{QCD} (GeV)	a_1	a_2	a_3
Staggered	1.48	1.50	0.35	18.39	5.25	-0.21

cutoff for the logarithms. The quark mass dependence of the vertex is parametrised by

$$a(M) = \frac{a_1}{1 + a_2 M(\zeta^2)/\Lambda_{QCD} + a_3 M^2(\zeta^2)/\Lambda_{QCD}^2}, \quad (11)$$

where $M(\zeta^2)$ is determined during the iteration process at $\zeta = 2.9\,\text{GeV}$.

In [5] the parameters have been fitted such that lattice results for the quark propagator [8,9] are reproduced on similar compact manifolds. To this end, solutions for the ghost and gluon propagators on similar manifolds have been used [10]. These solutions have been improved in [11] with the effect that the correct infinite-volume limits can be seen on tori with volumes larger than $V \approx (10\,\text{fm})^4$. Here we use the improved solutions to update the results for the quark propagator reported in [5]. The resulting modified values for the parameters (staggered quarks [8] only) are given in fig. 1. The numerical results, shown below, are discussed in the next section.

with the scale $\Lambda_{YM} = 0.658\,\text{GeV}$, the coupling $\alpha(\mu^2) = 0.97$ and the parameters $c = 1.269$ and $d = 2.105$ in the auxiliary function $R(p^2)$. The quenched anomalous dimension of the ghost is given by $\delta = -9/44$ and related via $\gamma = -1 - 2\delta$ to the one of the gluon. The infrared exponent κ is determined in an analytical infrared analysis [7]: $\kappa = (93 - \sqrt{1201})/98 \approx 0.595$. The running coupling $\alpha(p^2)$, defined via the nonperturbative ghost-gluon vertex $\alpha(p^2) = \alpha(\mu^2)\, G^2(p^2)\, Z(p^2)$, can be represented by

$$\alpha(p^2) = \frac{1}{1 + p^2/\Lambda_{YM}^2}\left[\alpha(0) + p^2/\Lambda_{YM}^2\right.$$
$$\left. \times \frac{4\pi}{\beta_0}\left(\frac{1}{\ln(p^2/\Lambda_{YM}^2)} - \frac{1}{p^2/\Lambda_{YM}^2 - 1}\right)\right], \quad (6)$$

where $\alpha(0) \approx 8.915/N_c$ is also known analytically [7].

Apart from the gluon propagator the other missing piece in the quark-DSE is the quark-gluon vertex. We use an ansatz of the form [5]

$$\Gamma_\nu(k, \mu^2) = \gamma_\nu\, \Gamma_1(k^2)\, \Gamma_2(k^2, \mu^2)\, \Gamma_3(k^2, \mu^2) \quad (7)$$

with the components

$$\Gamma_1(k^2) = \frac{\pi\gamma_m}{\ln(k^2/\Lambda_{QCD}^2 + \tau)}, \quad (8)$$

$$\Gamma_2(k^2, \mu^2) = G(k^2, \mu^2)\, G(\zeta^2, \mu^2)\, \widetilde{Z}_3(\mu^2)$$
$$\times\, h\left[\ln(k^2/\Lambda_g^2 + \tau)\right]^{1+\delta} \quad (9)$$

$$\Gamma_3(k^2, \mu^2) = Z_2(\mu^2)\, \frac{a(M) + k^2/\Lambda_{QCD}^2}{1 + k^2/\Lambda_{QCD}^2}, \quad (10)$$

where $\gamma_m = 12/33$ is the quenched anomalous dimension of the quark and $\tau = e - 1$ acts as a convenient infrared

3 Numerical results

Our results for the quark mass function $M(p^2)$ and the wave function $Z_f(p^2)$ for renormalised current quark masses of $m(2.9\,\text{GeV}) = 44, 80, 151\,\text{MeV}$ are displayed in fig. 2. The data for the quark mass function from the staggered lattice action can be very well reproduced with the simple vertex ansatz specified in the last section. The results for the wave functions agree less with each other: the lattice data for the three different masses are smaller spread than the DSE-results. This may be an indication for the importance of tensor structures in the quark-gluon vertex different from the simple γ_μ-term used in eqs. (7)-(10). In fig. 2 we also show results in the infinite volume/continuum limit. For the quark mass function these are very close to the results on the compact manifold, indicating that the used lattice volume, $V = (2.04\,\text{fm})^4$ is indeed large enough such that the correct pattern of chiral symmetry breaking can be observed for the quark masses shown. Small effects are only seen in the renormalisation point dependent and therefore unphysical quark wave function $Z_f(p^2)$. This result agrees with conclusions from volume studies on the lattice [12][1].

The resulting quark mass and wave functions in the infinite volume, continuum and chiral limit are also shown in

---

[1] The larger volume effects found in [5] also for the mass function are due to an overestimation of the effects in the Yang-Mills sector. An improved analysis of volume effects in the ghost and gluon propagators can be found in [11].

     The European Physical Journal A

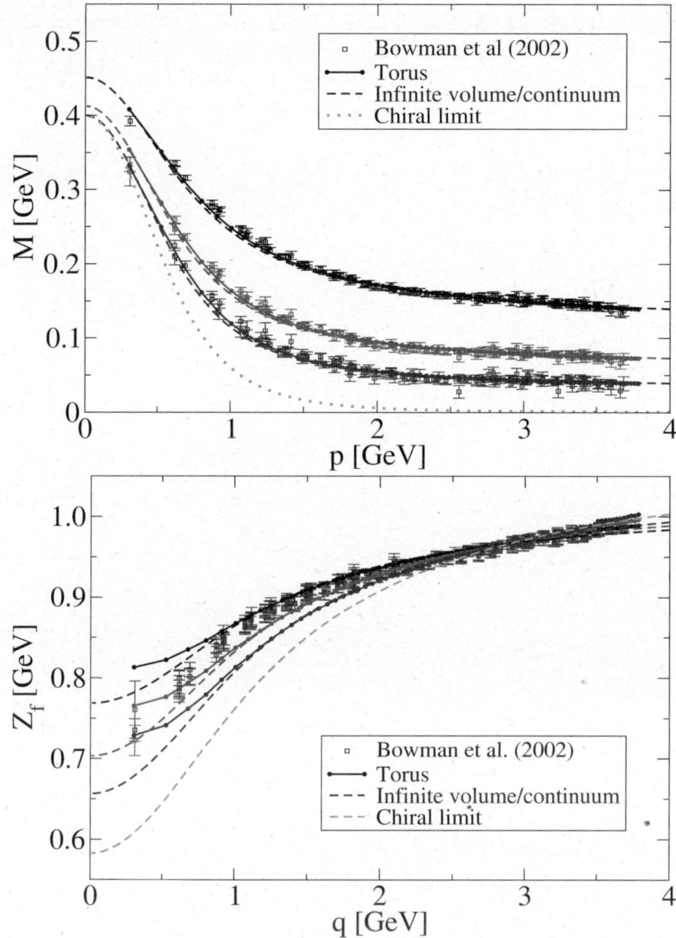

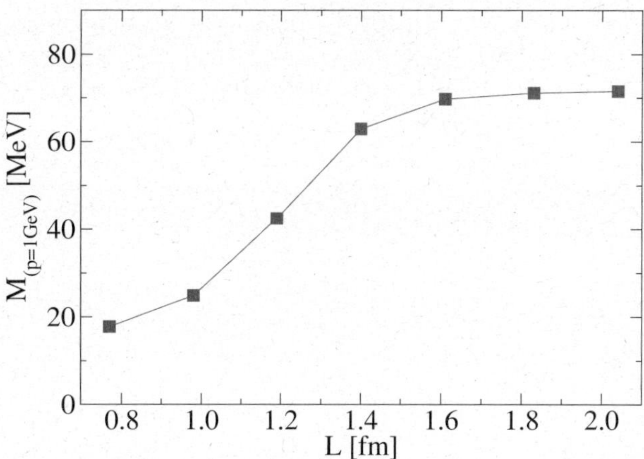

Fig. 3. The quark mass function for an up/down quark at a given momentum $p = 1\,\mathrm{GeV}$ plotted as a function of the box length $L = V^{1/4}$.

Fig. 2. The quark mass function $M(p^2)$ and the wave function $Z_f(p^2)$ for renormalised current quark masses of $m(2.9\,\mathrm{GeV}) = 44, 80, 151\,\mathrm{MeV}$. Compared are the lattice results of [8] with DSE-solutions on a torus (straight lines with filled circles) and in the infinite volume/continuum limit (dashed lines).

fig. 2. The corresponding quark mass at zero momentum is $M(0) = 400\,\mathrm{MeV}$. In the complex momentum plane the propagator has a leading analytic structure given by a pair of conjugate poles at $m = 416(20) \pm i\,304(20)\,\mathrm{MeV}$. The corresponding spectral function violates positivity and is therefore characteristical for a confined quark. The chiral condensate is $|\langle \bar{q}q \rangle|_{\overline{MS}}^{2\,\mathrm{GeV}} = (235 \pm 5\,\mathrm{MeV})^3$, using the quenched scale $\Lambda_{\overline{MS}} = 0.225(21)\,\mathrm{GeV}$ in the $\overline{MS}$ scheme. The pion decay constant, as determined from solutions of the pion Bethe-Salpeter equation, is $f_\pi = 72\,\mathrm{MeV}$. Both values are considerably smaller than the experimental result. Taken at face value this means that the quenched theory underestimates the amount of chiral symmetry breaking by about 10–20%. A similar conclusion has been drawn in [4].

Finally we study the behaviour of the quark mass function at small volumes. In fig. 3 we plot $M(p = 1\,\mathrm{GeV})$ as a function of the box length L. The corresponding current quark mass is of the order of a typical up/down-quark mass, $M(p = 2.9\,\mathrm{GeV}) = 10\,\mathrm{MeV}$. We clearly see that the quark mass function grows rapidly in the range

$1.0 < L < 1.6\,\mathrm{fm}$ signalling the onset of dynamical chiral symmetry breaking. Above $L = 1.6\,\mathrm{fm}$, a plateau is reached. This picture does not change when we extract the mass function $M(p^2)$ at smaller momenta p^2 or when we employ even smaller quark masses. The value is also not dependent on the type of quark-gluon interaction used. The result indicates that a safe volume to reproduce the correct pattern of chiral symmetry breaking on a compact manifold is at least

$$L_{\chi SB} \simeq 1.6\,\mathrm{fm}. \tag{12}$$

In particular this gives a (surprisingly small) miminal box length below which chiral perturbation theory cannot safely be applied.

We thank the organisers of *QNP 2006* for all their efforts. CF has been supported by the Deutsche Forschungsgemeinschaft (DFG) under contract Fi 970/7-1.

References

1. H. Leutwyler, A. Smilga, Phys. Rev. D **46**, 5607 (1992).
2. R. Alkofer, L. von Smekal, Phys. Rept. **353**, 281 (2001) [arXiv:hep-ph/0007355].
3. C.S. Fischer, J. Phys. G **32**, R253 (2006) [arXiv:hep-ph/0605173].
4. M.S. Bhagwat *et al.*, Phys. Rev. C **68**, (2003) 015203 [arXiv:nucl-th/0304003].
5. C.S. Fischer, M.R. Pennington, Phys. Rev. D **73**, 034029 (2006) [arXiv:hep-ph/0512233].
6. C.S. Fischer, R. Alkofer, Phys. Lett. B **536**, 177 (2002) [arXiv:hep-ph/0202202]; Phys. Rev. D **67**, 094020 (2003) [arXiv:hep-ph/0301094].
7. C. Lerche, L. von Smekal, Phys. Rev. D **65**, 125006 (2002) [arXiv:hep-ph/0202194].
8. P.O. Bowman, U.M. Heller, A.G. Williams, Phys. Rev. D **66**, 014505 (2002) [arXiv:hep-lat/0203001].
9. J.B. Zhang *et al.*, Phys. Rev. D **71**, 014501 (2005) [arXiv:hep-lat/0410045].

10. C.S. Fischer, R. Alkofer, H. Reinhardt, Phys. Rev. D **65**, 094008 (2002) [arXiv:hep-ph/0202195]; C.S. Fischer, B. Gruter, R. Alkofer, Ann. Phys. (N.Y.) **321**, 1918 (2006) [arXiv:hep-ph/0506053].

11. C.S. Fischer, A. Maas, J.M. Pawlowski, L. von Smekal, to be published in Ann. Phys., arXiv:hep-ph/0701050.

12. J.B. Zhang *et al.*, Nucl. Phys. Proc. Suppl. **141**, 15 (2005).

Eur. Phys. J. A **31**, 750–753 (2007)

DOI 10.1140/epja/i2006-10295-1

THE EUROPEAN
PHYSICAL JOURNAL A

Special Article – QNP 2006

Constraints on the IR behaviour of gluon and ghost propagator from Ward-Slavnov-Taylor identities

Ph. Boucaud[1], J.P. Leroy[1], A. Le Yaouanc[1], A.Y. Lokhov[2], J. Micheli[1], O. Pène[1], J. Rodríguez-Quintero[3,a], and C. Roiesnel[2]

[1] Laboratoire de Physique Théorique et Hautes Energies, Université de Paris XI, Bâtiment 211, F-91405 Orsay Cedex, France
[2] Centre de Physique Théorique de l'Ecole Polytechnique, F-91128 Palaiseau cedex, France
[3] Departamento Física Aplicada, Facultad de Ciencias Experimentales, Universidad de Huelva, E-21071 Huelva, Spain

Received: 22 December 2006
Published online: 20 March 2007 – © Società Italiana di Fisica / Springer-Verlag 2007

Abstract. We consider the constraints of the Slavnov-Taylor identity of the IR behaviour of gluon and ghost propagators and their compatibility with solutions of the ghost Dyson-Schwinger equation and with the lattice picture.

PACS. 12.38.Aw General properties of QCD (dynamics, confinement, etc.) – 12.38.Gc Lattice QCD calculations – 11.15.-q Gauge field theories – 11.15.Ha Lattice gauge theory

1 Introduction

In ref. [1] we have considered the constraints on the propagator dressing functions which can be derived from the Ward-Slavnov-Taylor identity (WSTI) —supplemented with some minimal assumptions on the analytic behaviour of the former and of the vertex form factors— and we were confronted with a contradiction between them and the ones that stem from the Dyson-Schwinger equation. The analysis of the ghost propagator Dyson-Schwinger equation seems to indicate that only a non-divergent gluon can match the lattice picture for the infrared behaviour of Landau gauge Green functions. On the other hand, WSTI seems to require that the gluon propagator diverges while the ghost dressing function should be finite and non-vanishing. In that ref. [1] we proposed, as a possible way out, that the ghost-gluon vertex function was singular (which does not contradict Taylor's theorem contrary to frequent claims). That hypothesis did not look very natural and the further work of [2] made it even less plausible.

In view of the very general validity of the WSTI, this situation is rather embarrassing and we wish to reconsider the problem. In the following, we will re-analyse the problem and clarify the working hypotheses to conclude either that the gluon propagator diverges[1] or that some of these hypoteses should fail.

<hr>

[a] e-mail: `jose.rodriguez@dfaie.uhu.es`

[1] Although softly enough as not to contradict the apparent finiteness previously stated from lattice data.

2 Notations and main hypotheses

We use the following notations [1]:

$$\left(F^{(2)}\right)^{ab}(k^2) = -\delta^{ab}\,\frac{F(k^2)}{k^2},$$

$$\left(G^{(2)}_{\mu\nu}\right)^{ab}(k^2) = \delta^{ab}\,\frac{G(k^2)}{k^2}\left(\delta_{\mu\nu} - \frac{k_\mu k_\nu}{k^2}\right), \tag{1}$$

where $G^{(2)}$ and $F^{(2)}$ are, respectively, the gluon and ghost propagators, G and F are, respectively, the gluon and ghost dressing functions. The ghost-gluon vertex $\widetilde{\Gamma}_\mu(p,k;q)$ (k and $-p$ are the momenta of the incoming and outgoing ghosts and q the gluon momentum) is defined as follows:

$$\Gamma^{abc}_\mu(p,k;q) = g_0(-ip_\nu)f^{abc}\widetilde{\Gamma}_{\nu\mu}(p,k;q)$$

$$= ig_0 f^{abc}\,\widetilde{\Gamma}_\mu(p,k;q). \tag{2}$$

It will be also useful to define the following scalars H_1 and H_2:

$$\widetilde{\Gamma}_\mu(-q,k;q-k) = q_\mu H_1(q,k) + (q-k)_\mu H_2(q,k) \tag{3}$$

that, after applying the standard tensor decomposition [3],

$$\widetilde{\Gamma}_{\nu\mu}(p,k;q) = \delta_{\nu\mu}a(p,k;q) - q_\nu k_\mu b(p,k;q)$$

$$+ p_\nu q_\mu c(p,k;q) + q_\nu p_\mu d(p,k;q) + p_\nu p_\mu e(p,k;q), \tag{4}$$

could be written as follows:

$$H_1(q,k) = a(-q,k;q-k) - q^2\,(b(-q,k;q-k)$$

$$+ d(-q,k;q-k) + e(-q,k;q-k))$$

$$H_2(q,k) = q^2\,(b(-q,k;q-k) - c(-q,k;q-k)). \tag{5}$$

We can at this point make our first hypothesis: the scalar factors present in that decomposition are regular when one of their arguments goes to zero while the others are kept finite. Thus, we suppose that

$$a(-r, r-p; p) = a_1(p^2) + \mathcal{O}(p \cdot r), \qquad (6)$$

and the same for the other scalars in the particular kinematic configurations we shall encounter. We adopt the notations $a_i(p^2)$, $b_i(p^2)$, $c_i(p^2)$ and so on, where the subindex i means that their i-th argument is a zero momentum.

The most general tensorial decomposition of the three-gluon vertex, $\Gamma_{\lambda\mu\nu}$ (of course, the antisymmetric color tensor f^{abc} is factorised) is given in ref. [4]. We will be interested in the limit of one vanishing gluon momentum while the two others remain finite. Such a limit deserves a careful analysis in the framework of WST identities because of the interplay of gluon and ghost propagator singularities and those of scalar functions in the decomposition [5]. When one of the momenta is zero the three-gluon vertex reduces to (cf. ref. [3]):

$$\Gamma_{\lambda\mu\nu}(q, -q, 0) =$$
$$(2\delta_{\lambda\mu}q_\nu - \delta_{\lambda\nu}q_\mu - \delta_{\nu\mu}q_\lambda)\, T_1(q^2)$$
$$- \left(\delta_{\lambda\mu} - \frac{q_\lambda q_\mu}{q^2}\right) q_\nu T_2(q^2) + q_\lambda q_\mu q_\nu T_3(q^2). \qquad (7)$$

For our purposes here, we will only assume that the limit of one vanishing gluon momentum can be safely taken, *i.e.*[2]:

$$\Gamma_{\lambda\mu\nu}(q - r, -q, r) = \Gamma_{\lambda\mu\nu}(q, -q, 0) + o(1). \qquad (8)$$

3 WSTI and IR propagators

The Ward-Slavnov-Taylor [6] identity for the three-gluon function reads

$$p^\lambda \Gamma_{\lambda\mu\nu}(p, q, r) = \frac{F(p^2)}{G(r^2)}(\delta_{\lambda\nu}r^2 - r_\lambda r_\nu)\widetilde{\Gamma}_{\lambda\mu}(r, p; q)$$
$$- \frac{F(p^2)}{G(q^2)}(\delta_{\lambda\mu}q^2 - q_\lambda q_\mu)\widetilde{\Gamma}_{\lambda\nu}(q, p; r). \qquad (9)$$

We shall now study the behaviour when $r \to 0$ while keeping q and p finite and apply decompositions (2), (7) and the hypotheses (6), (8) to replace the vertices in eq. (9). Then, if one only retains the leading terms, STI reads

$$T_1(q^2)\,(q_\mu q_\nu - q^2 \delta_{\mu\nu}) + q^2 q_\mu q_\nu\, T_3(q^2) + o(1) =$$
$$\frac{F(q^2)}{G(r^2)}\big[a_1(q^2)\,(r^2\delta_{\mu\nu} - r_\mu r_\nu)$$
$$+ b_1(q^2)q_\mu\,(r^2 q_\nu - (q \cdot r)r_\nu) + o(r^2)\big]$$
$$+ \frac{F(q^2)}{G(q^2)}\big[a_3(q^2)\,(q_\mu q_\nu - q^2 \delta_{\mu\nu}) + o(1)\big]. \qquad (10)$$

Thus, if one multiplies both l.h.s. and r.h.s. of this eq. (10) by r_ν, we obtain:

$$T_1(q^2)\,(q_\mu\,(q \cdot r) - q^2 r_\mu)$$
$$+ q^2 q_\mu(q \cdot r)\, T_3(q^2) + o(r \cdot q) =$$
$$\frac{F(q^2)}{G(q^2)}\, a_3(q^2)\,(q_\mu(q \cdot r) - q^2 r_\nu) + o(r \cdot q), \qquad (11)$$

where the first term of the r.h.s. of eq. (10) vanishes because it is transverse to r_ν. Thus, by identifying both r.h.s. and l.h.s. of eq. (11), one is led to the familiar relations [3]:

$$T_1(q^2) \doteq \frac{F(q^2)}{G(q^2)}a_3(q^2),$$
$$T_3(q^2) = 0. \qquad (12)$$

Now, let us multiply both r.h.s. and l.h.s. of eq. (10) by q_μ and apply that T_3 has been seen to be exactly 0 in eq. (12) and we obtain then

$$\frac{F(q^2)}{G(r^2)}\, r^2 \left[\big(a_1(q^2) + q^2 b_1(q^2)\big)\right.$$
$$\left. \times \left(q_\nu - \frac{(q \cdot r)}{r^2}\, r_\nu\right) + o(1)\right] = o(1). \qquad (13)$$

Thus, if $a_1(q^2) \neq 0$ or $b_1(q^2) \neq 0$ (and, indeed, one knows from perturbation theory that at large momenta $a_1 = 1$, cf. [3,6]) eq. (9) implies

$$\lim_{r \to 0} \frac{G(r^2)}{r^2} \to \infty, \qquad (14)$$

or, in other words, that *the gluon propagator diverges in the infrared limit*. If we stick to the commonly accepted idea that G behaves as a power in the infrared ($G(p^2) \sim (p^2)^{\alpha_G}$), then $\alpha_G < 1$ is to be concluded. Another attractive possibility would be to suppose an infrared behaviour less divergent than any power as, for instance, that of the form $G(p^2) \sim p^2 \log^\nu(p^2)$ for some positive ν. This will be considered in more detail in a forthcoming paper [5].

We can also, instead of letting $r \to 0$, study now the behaviour when $p \to 0$ of eq. (9) as is done in [3]. The dominant part of the l.h.s. of (9) reads

$$(2\delta_{\mu\nu}p \cdot q - p_\mu q_\nu - p_\nu q_\mu)\, a_3(q^2)\frac{F(q^2)}{G(q^2)}$$
$$- \left(\delta_{\mu\nu} - \frac{q_\mu q_\nu}{q^2}\right)(p \cdot q)T_2(q^2), \qquad (15)$$

where the results in eq. (12) have been implemented. Let us now multiply both sides with q^μ and keep only the leading tems in p and one obtains

$$(q_\nu(p \cdot q) - q^2 p_\nu)a_3(q^2)F(q^2) =$$
$$(q_\nu(p \cdot q) - q^2 p_\nu)F(q^2)(a_2(q^2) - q^2 d_2(q^2))$$
$$+ \mathcal{O}(p^2) \qquad (16)$$

---

[2] It is shown in [4], on a perturbative basis, that the vertex remains finite when one takes the limit $r \to 0$ while keeping the two other momenta fixed. Our hypothesis amounts to assuming that this result survives beyond perturbation theory.

that of course can be true only if $F(p^2)$ goes to some finite limit when $p^2 \to 0$ and whence, in terms of scalars,

$$F(q^2) \underset{q^2 \to 0}{\sim} F(0)\,\frac{a_2(q^2) - q^2 d_2(q^2)}{a_3(q^2)}, \qquad (17)$$

where $a_2/a_3 \to 1$ as $q^2 \to 0$ [3].

Let us repeat here that all these considerations are valid only when our regularity hypotheses about the ghost-gluon scalar factors and about the three-gluon vertex (see (6), (8)) are satisfied. *Under those hypotheses* one obtains important constraints on the gluon and ghost propagators —namely that they are divergent in the zero-momentum limit. Let us now briefly analyze the ghost propagator Dyson-Schwinger equation (GPDSE).

4 Ghost DSE: the case $\alpha_F = 0$

In a previous paper (ref. [1]) we studied all the classes of solutions for the GPDSE, that can be pictured, in diagrammatic form, as:

$$\left(F^{ab}(q^2)\right)^{-1} = \left(F^{ab}_{\text{tree}}(q^2)\right)^{-1} -$$

Let us first recall that the *unsubtracted* GPDSE is actually meaningless since the integral in its right-hand side is UV-divergent, behaving as $\int \mathrm{d}q^2 \frac{1}{q^2}(1 + 11\alpha_s/(2\pi)\log(q/\mu))^{-35/44}$. A way out of this difficulty would be to renormalise the equation to deal properly with its UV divergencies. Instead of that, we preferred to study the following subtracted version of the bare GPDSE equation for two scales, λk and $\kappa\lambda k$ (see eq. (14) of ref. [1]) with k the external ghost momentum and κ some fixed number (< 1). λ is an extra parameter that we shall ultimately let go to 0 in order to study the infared behaviour of the GPDSE. This subtracted version of the GPDSE reads (see eq. (14) of ref. [1]):

$$\frac{1}{F(\lambda k)} - \frac{1}{F(\kappa\lambda k)} =$$

$$g_B^2 N_c \int \frac{\mathrm{d}^4 q}{(2\pi)^4} \left(\frac{F(q^2)}{q^2}\left(\frac{(k \cdot q)^2}{k^2} - q^2\right) \right.$$

$$\left. \times \left[\frac{G((q - \lambda k)^2) H1(q, \lambda k)}{(q - \lambda k)^2} - (\lambda \to \kappa\lambda) \right] \right), \qquad (18)$$

where H_1 is the particular combination of the scalars defined in eq. (2) playing the GPSDE game. Furthermore, a proper dimensional analysis of eq. (18) requires to cut the integration domain in its r.h.s. into two pieces by introducing some additional scale q_0^2 (of the order of Λ^2_{QCD}). Clearly, the external momentum is not the only relevant

Table 1. Constraints imposed by the GPDSE to the critical behaviour of ghost and gluon propagators for the case $\alpha_F \neq 0$. The second column shows the behaviour on λ ($\lambda \to 0$) of eq. (18)'s r.h.s., while the l.h.s. behaves as $(\lambda^2)^{-\alpha_F}$.

$\alpha_F \neq 0$		
$\alpha_F + \alpha_G$	r.h.s.	Constraint
> 1	λ^2	$\alpha_F = -1$
$= 1$	$\lambda^2 \log \lambda$	excluded
< 1	$(\lambda^2)^{\alpha_F + \alpha_G}$	$2\alpha_F + \alpha_G = 0$

Table 2. The same constraints analysed in table 1 but here for $\alpha_F = 0$. The l.h.s. of eq. (18) behaves now as the next-to-leading term of the deep infrared expansion of $F(q^2)$ (third column).

$\alpha_F = 0$		
$\alpha_F + \alpha_G$	r.h.s.	Constraint
> 1	λ^2	$F(q^2) = A + Bq^2$
$= 1$	$\lambda^2 \log \lambda$	$F(q^2) = A + Bq^2 \log q^2$
< 1	$(\lambda^2)^{\alpha_G}$	$F(q^2) = A + Bq^{2\alpha_G}$

scale in the problem and Λ_{QCD}, without which it would not be understandable that the UV behaviour differs drastically from the IR one, must be taken into account. A careful dimensional analysis of the integrals extended over both domains, $q^2 > q_0^2$ and $q^2 < q_0^2$, is mandatory [1]. In the second one —and only there— we will initially use the common, convenient, but not really justified assumption of a power law behaviour of the propagators in the deep infrared:

$$F(k^2) \sim \left(\frac{k^2}{q_0^2}\right)^{\alpha_F}, \qquad G(k^2) \sim \left(\frac{k^2}{q_0^2}\right)^{\alpha_G}. \qquad (19)$$

We shall not repeat here the details of our scaling analysis of eq. (18)[3] and simply summarize our conclusions in the following 2 tables. It is often claimed, after the study of the GPDSE, that $2\alpha_F + \alpha_G = 0$. In fact, as can be seen in the next table 1, this results emerges only[4] after assuming $\alpha_F \neq 0$ and discarding (reasonably) $\alpha_F = -1$.

However, if $\alpha_F = 0$ other solutions are also compatible with GPDSE (see table 2).

Some recent lattice results seem to exclude the standard ($2\alpha_G + \alpha_F = 0$)-solution [1,2]. If one admits these results (lattice also discards $\alpha_F = -1$), then one is led to conclude that GPDSE implies $\alpha_F = 0$.

Furthermore, it was shown in ref. [7] that the r.h.s. of eq. (18) is the sum of two terms behaving, respectively, as $\lambda^{2\text{Min}(\alpha_F + \alpha_G + \alpha_\Gamma, 1)}$ and λ^2 when $\lambda \to 0$. So, it behaves as λ^2 when $\alpha_F = 0$. Then, one can prove that for any κ there

---

[3] The analysis done in [1] missed some possible solutions (for instance, the case $\alpha_F = 0$, $\alpha_G < 1$) mainly because of the fact that we had rejected the possibility of non-analytic sub-dominant terms in the dressing functions.

[4] The regularity of the ghost-gluon vertex is also needed as was discussed in [1].

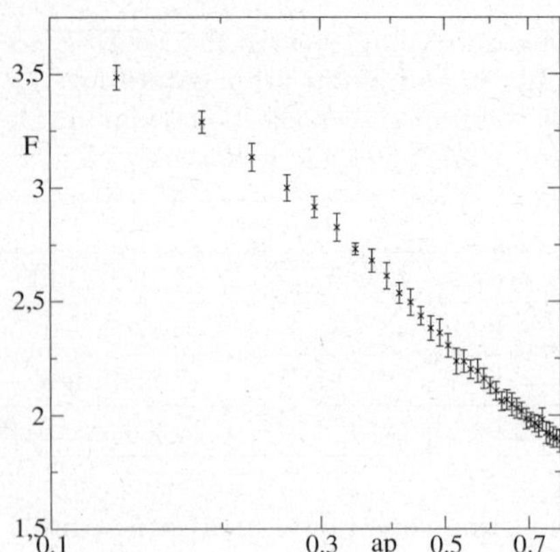

Fig. 1. $F(p)$ from a $SU(2)$ simulations on a 48^4 lattice at $\beta = 2.3$. (From ref. [7].)

is a value of λ and c such that

$$\left| \frac{1}{F(\lambda k)} - \frac{1}{F(\kappa^n \lambda k)} \right| \leq c \frac{1 - \kappa^{2n}}{1 - \kappa^2} \lambda^2. \qquad (20)$$

So $F \to \infty$ when $\lambda \to 0$ is excluded because taking the limit of the above expression when $n \to \infty$ we should have $\left| \frac{1}{F(\lambda k)} \right| \leq c \frac{1}{1-\kappa^2} \lambda^2$ and F would diverge as or more rapidly than $\frac{1}{\lambda^2}$ implying $\alpha_F \leq -1$ in contradiction with the hypothesis $\alpha_F = 0$. Let us remark that $F \to 0$ is also excluded: equation (20) implies $\left| \frac{1}{F(\kappa^n \lambda k)} \right| \leq \left| \frac{1}{F(\lambda k)} \right| + c \frac{1-\kappa^{2n}}{1-\kappa^2} \lambda^2$ and $\frac{1}{F(\kappa^n \lambda k)}$ cannot tend to infinity when $n \to \infty$. It should be emphasized that the dimensional analysis driving to eq. (20) is also valid if $F(q^2)$ is admitted to behave in a way other than a power. Thus, *if a leading power behaviour is discarded for the ghost dressing function, it has to be finite and $\neq 0$ in the IR limit.*

5 Conclusion

We derive from the Ward-Slavnov-Taylor identity for the three-gluon and ghost-gluon vertices, after assuming their regularity that gluon propagator diverges and ghost dressing function remains finite as the momentum goes to zero. A dimensional analysis of the GPDSE, provided that we trust the lattice results excluding $2\alpha_F + \alpha_G = 0$ [1,2] and $\alpha_F = -1$, leads to conclude independently that the ghost dressing function remains finite at zero momentum [7] (see tables 1, 2). Both GPDSE and WSTI constraints will offer compatible solutions provided that one admits non-analytic sub-leading terms for the low momentum expansion of dressing functions.

On the other hand, such a solution respecting WSTI and GPDSE constraints still match in the present picture of lattice knowledge about the IR behavior of propagators and vertices. The current simulations of ghost-gluon vertex seem to discard $2\alpha_F + \alpha_G = 0$ but those of ghost and gluon propagators cannot yet exclude or confirm the smooth divergences we propose as a way out [2,8,9] (as

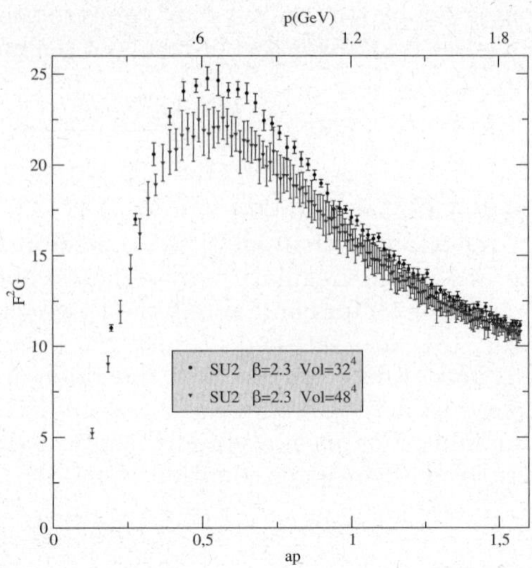

Fig. 2. $F^2 G$ from lattice simulation for $SU(2)$ (32^4 and 48^4, $\beta_{SU(2)} = 2.3$) gauge groups. $2\alpha_F + \alpha_G = 0$ implies a constant in the infrared domain. (From ref. [1].)

an example, see figs. 1 and 2, from refs. [7] and [1], respectively). A non-power behaviour (logarithmic, for instance) could be specially elusive for lattice extrapolations at infinite volume. Of course, new simulation results on bigger lattice volumes (or with twisted boundary conditions [10]) and careful extrapolations will be very welcome to dig into this matter.

This is a very interesting task to be acomplished, because either such a logarithmic (or similar) behaviour is found or one is led to conclude that the tensorial decomposition of ghost-gluon or three-gluon vertex admits non-regularities.

References

1. Ph. Boucaud *et al.*, arXiv:hep-ph/0507104.
2. A. Sternbeck, E.M. Ilgenfritz, M. Muller-Preussker, A. Schiller, Nucl. Phys. Proc. Suppl. **153**, 185 (2006).
3. K.G. Chetyrkin, A. Retey, arXiv:hep-ph/0007088; A.I. Davydychev, P. Osland, O.V. Tarasov, Phys. Rev. D **54**, 4087 (1996); **59**, 109901 (1999)(E).
4. J.S. Ball, T.W. Chiu, Phys. Rev. D **22**, 2550 (1980).
5. Ph. Boucaud *et al.*, in preparation.
6. J.C. Taylor, Nucl. Phys. B **33**, 436 (1971); A.A. Slavnov, Theor. Math. Phys. **10**, 99 (1972).
7. Ph. Boucaud *et al.*, JHEP **0606**, 001 (2006).
8. F.D.R. Bonnet, P.O. Bowman, D.B. Leinweber, A.G. Williams, J.M. Zanotti, Phys. Rev. D **64**, 034501 (2001).
9. D. Becirevic, P. Boucaud, J.P. Leroy, J. Micheli, O. Pene, J. Rodriguez-Quintero, C. Roiesnel, Phys. Rev. D **61**, 114508 (2000).
10. T. Tok, K. Langfeld, H. Reinhardt, L. von Smekal, *The gluon propagator in lattice Landau gauge with twisted boundary*, PoS (LAT2005) 334 (2006) [arXiv:hep-lat/0509134].

Eur. Phys. J. A **31**, 754–757 (2007)

DOI 10.1140/epja/i2007-10008-4

Special Article – QNP 2006

Finite-temperature ϕ^4 theory from the 2PI effective action: Two-loop truncation

A. Arrizabalaga[1,a] and U. Reinosa[2]

[1] NIKHEF, Kruislaan 409, 1098 SJ, Amsterdam, The Netherlands
[2] Institut für Theoretische Physik, Universität Heidelberg, Philosophenweg 16, 69120 Heidelberg, Germany

Received: 8 January 2007
Published online: 28 March 2007 – © Società Italiana di Fisica / Springer-Verlag 2007

Abstract. Using resummation techniques based on the 2PI effective action we study the scalar ϕ^4 theory at finite temperature. We present an analytical as well as numerical study for a renormalized two-loop truncation of the action. Both the spectral properties and critical behaviour of the theory are investigated. Within the truncation, we explicitly check that the physical observables are UV-finite.

PACS. 11.10.Wx Finite-temperature field theory

1 Introduction

The thermodynamical properties of a physical system can be extracted from the effective potential, which is given as a function $\gamma[\phi]$ of a condensate field ϕ. In the thermodynamic limit, its stationary point defines the free energy $\mathcal{F}$ as

$$\mathcal{F} = \lim_{V \to \infty} \gamma[\bar{\phi}], \quad \text{with} \quad \left. \frac{\delta \gamma[\phi]}{\delta \phi} \right|_{\phi = \bar{\phi}} = 0. \quad (1)$$

Once the free energy is known, all other thermodynamic quantities, such as the entropy or energy density, can be easily derived. At finite temperature collective phenomena are known to modify substantially the properties of the elementary excitations, so the use of a perturbative expansion around the free theory to calculate the effective potential becomes questionable and often leads to inconsistencies. One needs then to consider nonperturbative schemes. For situations where the effects of the collective phenomena can be conveniently captured by a suitable modification of the two-point functions, resummation schemes based on the 2PI effective action can be very useful [1,2]. The 2PI effective action consists of a reorganization of the perturbative expansion around dressed two-point functions, which are determined self-consistently for a given approximation/truncation. The technique is systematic and allows to go beyond mean-field and Hartree-type approximations, which are also included at lowest orders in the truncation. In this work we study the thermodynamics of the scalar ϕ^4 theory using a two-loop truncation of the 2PI effective action.

[a] e-mail: arrizaba@nikhef.nl

2 Two-loop truncation of the 2PI effective action

For the scalar ϕ^4 theory the 2PI effective action is usually parametrized as [3]

$$\Gamma_{2\mathrm{PI}}[\phi, G] = \frac{1}{2} \phi \cdot G_0^{-1} \cdot \phi + \frac{1}{2} \mathrm{Tr} \left[\ln G^{-1} + \left(G_0^{-1} - G^{-1} \right) \cdot G \right] + \Gamma_{\mathrm{int}}[\phi, G], \quad (2)$$

with ϕ and G generic one- and two-point functions and $A \cdot B$ a shorthand notation for the convolution of A and B. The term Γ_{int} contains the interactions and can be written as an expansion in terms of two-particle-irreducible (2PI) diagrams. Up to two-loops it is given by

$$-\Gamma_{\mathrm{int}}[\phi, G] = \frac{1}{4!} \;\; + \frac{1}{4} \;\; + \frac{1}{8} \;\; + \frac{1}{12} \;\;, \quad (3)$$

where the Feynman rules are:

$$\mathsf{X} = -\lambda, \quad - = G, \quad -\!\otimes = \phi.$$

Within the approximation, "physical" one- and two-point functions $\bar{\phi}$ and $\bar{G}$ are determined self-consistenly as the stationary points of the 2PI effective action, *i.e.*

$$\left. \frac{\delta \Gamma_{2\mathrm{PI}}}{\delta G} \right|_{\bar{G}[\phi], \phi} = 0 \quad \text{and} \quad \left. \frac{\delta \Gamma_{2\mathrm{PI}}}{\delta \phi} \right|_{\bar{G}[\phi], \bar{\phi}} = 0. \quad (4)$$

The stationary conditions (4) turn into a set of coupled implicit equations for $\bar{G}$ and $\bar{\phi}$ which have to be solved in order to calculate any physical quantity. In particular, the knowledge of the dressed two-point function $\bar{G}$ allows one to calculate the effective potential as $\gamma[\phi] = TV^{-1}\Gamma_{2\mathrm{PI}}[\phi, \bar{G}[\phi]]$.

One of the main complications that arises when dealing with truncations of the 2PI effective action is the fact that two- and higher n-point functions are *not uniquely defined* [4]. In particular, a given truncation defines two possible two-point functions and three four-point functions. One of the two-point functions is given by the stationary value $\bar{G}$, while the other is related to the inverse curvature of the effective potential as $\hat{G}^{-1} = TV^{-1}[\delta^2\gamma[\phi]/\delta\phi^2]$. The three four-point functions and their corresponding vertices are given in terms of coupled Bethe-Salpeter–like equations [4]. Concerning *renormalization*, the ambiguity in the definition of the vertex functions implies that there are more than one counterterm of a given type [5]. A given counterterm is determined by applying a renormalization condition to the corresponding vertex. For consistency, identical renormalization conditions are applied to all counterterms of the same type. It is simpler to consider renormalization conditions applied at a reference temperature $T_\star$ for which the physical field configuration is $\bar{\phi}_\star = 0$. This requires that $T_\star > T_c$ if a critical temperature T_c exists. Defining the self-energy from Dyson's equation $\bar{\Sigma} = \bar{G}^{-1} - G_0^{-1}$, the renormalized equations for $\bar{\Sigma}$ and $\bar{\phi}$ are

$$\bar{\Sigma}(P) = \delta m_0^2 + \frac{\lambda + \delta\lambda_2}{2}\phi^2 + \frac{\lambda + \delta\lambda_0}{2}\int_K^T \bar{G}(K)$$
$$- \frac{\lambda^2}{2}\phi^2\Theta(P) \tag{5}$$

and

$$\left(m^2 + \delta m_2^2\right)\bar{\phi} = \frac{\lambda + \delta\lambda_4}{6}\bar{\phi}^3 + \frac{\lambda + \delta\lambda_2}{2}\bar{\phi}\int_K^T \bar{G}(K)$$
$$- \frac{\lambda^2}{6}\bar{\phi}\int_K^T \bar{G}(K)\Theta(K), \tag{6}$$

with $\int_K^T$ a shorthand notation for the standard sum-integral over momentum K at temperature T, and

$$\Theta(P) = \int_K^T \bar{G}(K)\bar{G}(K+P). \tag{7}$$

The bar in ϕ has been omitted in the first equation to stress the fact that it can be solved for any value. This is needed, for instance, to calculate the effective potential $\gamma[\phi]$. The explicit expressions for the counterterms δm_0^2, δm_2^2, $\delta\lambda_0$, $\delta\lambda_2$ and $\delta\lambda_4$ can be found in ref. [4]. With those counterterms it can be shown that, for any value of T and ϕ, the results for $\bar{G}$, $\bar{\phi}$ and $\gamma[\phi]$ are UV-finite.

3 Numerical analysis

The gap and field equations (5) and (6) are solved numerically in Minkowski space. We split the self-energy into a local and a non-local part as $\bar{\Sigma}(P) = \bar{\Sigma}^l + \bar{\Sigma}^{nl}(P)$. The

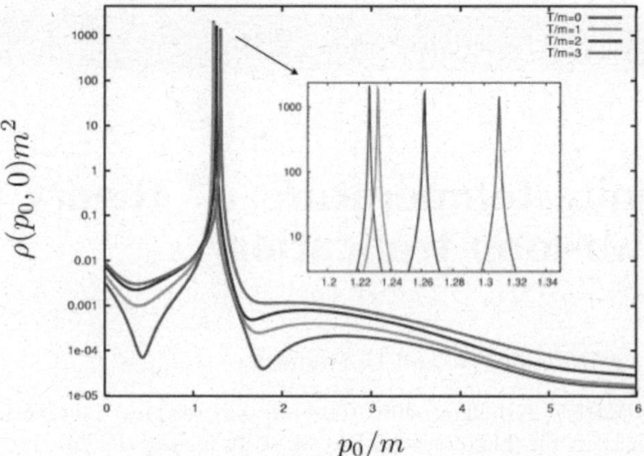

Fig. 1. Spectral function $\rho(p_0, 0)$ for $\lambda = 1$, $\phi/m = 1$ and $T_\star/m = 1$ and several temperatures.

momentum-dependent part is expanded in both p_0 and $|\mathbf{p}|$ using N Chebyshev polynomials as

$$\bar{\Sigma}^{nl}(p_0, |\mathbf{p}|) = \sum_i^N \sum_j^N c_{ij} T_i(p_0) T_j(|\mathbf{p}|). \tag{8}$$

The solution to the gap and field equations is obtained by solving the matrix equation for the N^2 coefficients c_{ij} plus the local terms Σ^l and/or $\bar{\phi}^2$ by means of a multi-dimensional Newton-Raphson method. This numerical algorithm is fairly stable and converges after few iterations. For a given N, the numerical solution oscillates around the "full" solution, which is convenient for the calculation of integrated quantities, such as the effective potential. Unlike lattice-based methods, this technique also allows the use of large (small) UV (IR) cutoffs, which is useful for the study of *renormalization* and/or *critical phenomena*. Although numerically more expensive, the advantage of solving the equations in Minkowski space is that one can obtain directly spectral properties of the system without the need of analytic continuation. In particular, the spectral function $\rho(p_0, |\mathbf{p}|)$ can be constructed from the knowledge of the real and imaginary parts of the retarded self-energy, which come directly from solving eq. (5) (see fig. 1 for an example result at several temperatures). The width, the position of the quasiparticle pole and the effect of multiparticle contributions can be adequately studied with the presented algorithm.

We can also look at the critical behaviour in a system with broken symmetry. A simple quantity to compute is the critical temperature which can be read off the two-point functions at vanishing effective mass M. As we discussed previously there are two possible two-point functions ($\bar{G}$ and $\hat{G}$), for which the corresponding effective masses are (in the symmetric phase)

$$M(T, \lambda)^2 = m^2 + \Sigma^l(T, \lambda), \tag{9}$$

$$\hat{M}(T, \lambda)^2 = \hat{G}^{-1}(p=0, T, \lambda) = \frac{\delta^2\Gamma[\phi, \bar{G}[\phi]]}{\delta\phi^2}\bigg|_{\bar{\phi}}(p=0, T, \lambda). \tag{10}$$

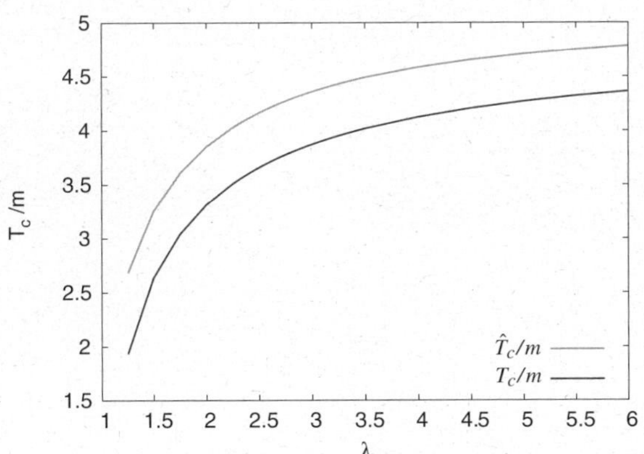

Fig. 2. Critical temperatures for several couplings ($T_\star/m = 5$).

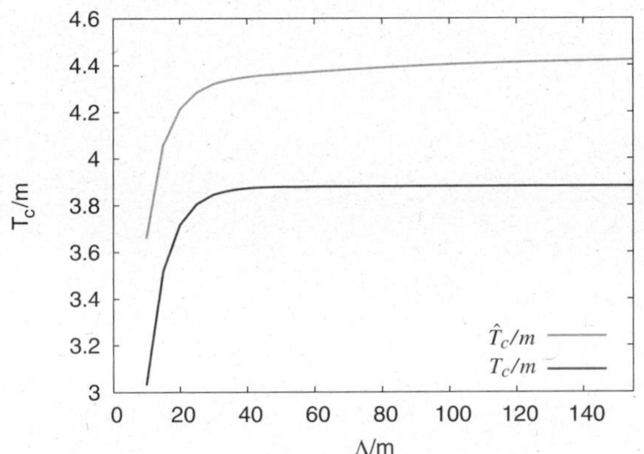

Fig. 3. UV cutoff dependence for $\lambda = 3$ ($T_\star/m = 5$).

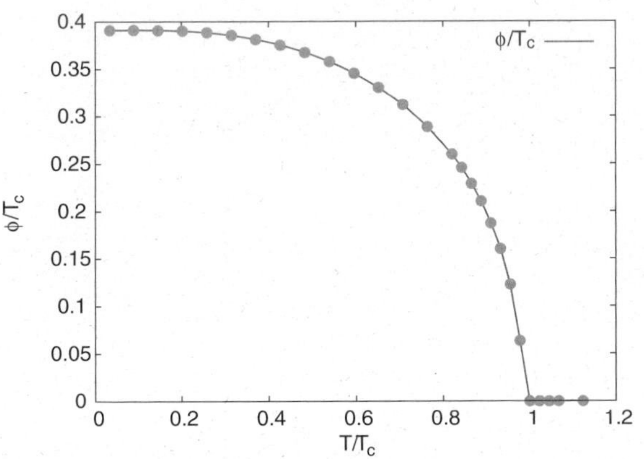

Fig. 4. Field expectation value $\bar\phi^2$ as a function of T for $\lambda = 3$ and $T_\star/m = 5$.

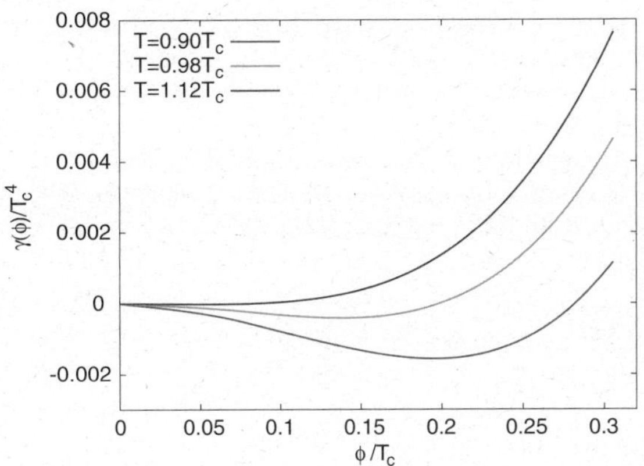

Fig. 5. Effective potential for $\lambda = 3$ and $T_\star/m = 5$.

of the order parameter $\bar\phi$ as a function of the temperature. We see that the transition is clearly of second order. This agrees with lattice studies as well as results from other non-perturbative approaches [6]. So, unlike mean-field or Hartree approximations, which predict the wrong order for the transition, methods based on the 2PI effective action give the correct result. A similar conclusion is reached by computing directly the effective potential $\gamma[\phi]$ (see fig. 5). Finally, we checked that $\gamma[\bar\phi]$, from which a renormalized free energy and pressure can be extracted, is indeed cutoff independent.

4 Conclusions and prospects

We have shown for a simple scalar ϕ^4 theory that the resummation methods based on the 2PI effective actions can be used to obtain physically meaningful (*i.e.*, renormalized) information about the thermodynamics, spectral properties and phase structure of a given system beyond mean-field or Hartree approximations. The natural next step is to apply the technique to theories physically more relevant, specially to those for which most non-perturbative finite-temperature studies are limited to mean-field and/or Hartree-type approximations. Of particular importance to heavy-ion phenomenology are chiral effective theories such as sigma or Nambu-Jona-Lasinio–type models (some work in that direction is underway [2]) and gauge theories[1].

References

1. B. Vanderheyden, G. Baym, J. Stat. Phys. **93**, 843 (1998); J.-P. Blaizot, E. Iancu, A. Rebhan, Phys. Rev. Lett. **83**, 2906 (1999); J.T. Lenaghan, D.H. Rischke, J. Phys. G **26**, 431 (2000); Y. Nemoto, K. Naito, M. Oka, Eur. Phys. J. A **9**, 245 (2000); Phys. Rev. D **63**, 065003 (2001); J. Berges, Sz. Borsányi, U. Reinosa, J. Serreau, Phys. Rev. D **71**, 105004 (2005); J.-P. Blaizot, A. Ipp, A. Rebhan, U. Reinosa, Phys. Rev. D **72**, 125005 (2005).

---

[1] In gauge theories, however, the application of 2PI effective action techniques suffers from a residual gauge dependence [7].

Starting from $T_\star$ and reducing the temperature, the critical values T_c and $\hat{T}_c$ are found when M^2 and $\hat{M}^2$ vanish. The results for several couplings are shown in fig. 2. We can also vary the UV cutoff in the calculation of the critical temperatures and therefore check the validity of the renormalization procedure (see fig. 3). We find that both quadratic and logarithmic divergences (present, respectively, in T_c and $\hat{T}_c$) are properly renormalized and hence the renormalization is satisfactory.

Solving in addition the field equation (6) allows us to study the phase transition. In fig. 4 we show the behavior

2. D. Roder, J. Ruppert, D.H. Rischke, Nucl. Phys. A **775**, 127 (2006); D. Roder, arXiv:hep-ph/0509232.
3. J.M. Cornwall, R. Jackiw, E. Tomboulis, Phys. Rev. D **10**, 2428 (1974).
4. A. Arrizabalaga, U. Reinosa, in preparation.
5. J. Berges, Sz. Borsányi, U. Reinosa, J. Serreau, Ann. Phys. (N.Y.) **320**, 344 (2005).
6. G. Smet, T. Vanzielighem, K. Van Acoleyen, H. Verschelde, Phys. Rev. D **65**, 045015 (2002).
7. A. Arrizabalaga, J. Smit, Phys. Rev. D **66**, 065014 (2002); M.E. Carrington, G. Kunstatter, H. Zaraket, Eur. Phys. J. C **42**, 253 (2005).

Eur. Phys. J. A **31**, 758–760 (2007)

DOI 10.1140/epja/i2007-10007-5

THE EUROPEAN
PHYSICAL JOURNAL A

Special Article – QNP 2006

Goldstone boson counting in relativistic systems at finite density

T. Brauner[a]

Department of Theoretical Physics, Nuclear Physics Institute, Academy of Sciences of the Czech Republic, 250 68 Řež, Czech Republic

Received: 8 January 2007

Published online: 22 March 2007 – © Società Italiana di Fisica / Springer-Verlag 2007

Abstract. We study the effects of finite chemical potential on the pattern of symmetry breaking within the relativistic linear sigma model. In accordance with previous works we show that type-II Goldstone bosons may appear whose dispersion relation is quadratic in momentum in the long-wavelength limit. We show that their presence is tightly connected with nonzero densities of non-Abelian Noether charges, and formulate a general counting rule for the number of the Goldstone bosons. Working at tree level, we conclude with the discussion of the loop effects. Our results find an application in particular to cold dense quark matter, where a type-II Goldstone boson has been found, *e.g.*, in the phases with kaon condensation.

PACS. 11.30.Qc Spontaneous and radiative symmetry breaking

1 Introduction

Spontaneous symmetry breaking plays a key role in understanding vastly different physical phenomena in several branches of physics, ranging from current high-energy and particle physics to condensed matter. One of its general consequences is the existence of the so-called Goldstone bosons —soft fluctuations of the order parameter(s)— guaranteed by the Goldstone theorem [1,2]. The number and properties of the Goldstone bosons (in particular, their dispersion relations) are essential for the low-energy dynamics of a system with spontaneously broken symmetry. Moreover, they significantly affect the thermodynamical properties of the system such as the heat capacity or the transport coefficients.

As written in any textbook on relativistic quantum field theory, in Lorentz-invariant theories the number of Goldstone bosons associated with a spontaneously broken *internal* symmetry (the case of a broken spacetime symmetry is treated, for instance, in ref. [3]) is always equal to the number of broken symmetry generators. On the other hand, in Lorentz-noninvariant systems[1], the situation is more complicated. The basic result in this respect was achieved by Nielsen and Chadha [4]. They showed that under certain technical assumptions, the energy of the Goldstone boson is proportional to some power of momen-

tum in the long-wavelength limit. The Goldstone boson is then classified as type I, if this power is odd, or type II, if it is even, respectively. The Nielsen-Chadha counting rule states that *the number of type-I Goldstone bosons plus twice the number of type-II Goldstone bosons is greater or equal to the number of broken generators.*

In the past decade, the number of Goldstone bosons was proven to be connected with the possibility that some of the conserved (Noether) charges develop nonzero density in the ground state. In particular, Leutwyler analyzed spontaneous symmetry breaking in nonrelativistic systems within the framework of low-energy effective field theory [5]. He showed that nonzero density of a non-Abelian charge induces a term in the effective Lagrangian with a single time derivative, which, in turn, gives rise to a type-II Goldstone boson with a quadratic dispersion relation.

To summarize, the Nielsen-Chadha theorem clarifies the connection of the number of the Goldstone bosons and their dispersion relations. These are related, by Leutwyler's work, to the Noether charge densities. However, to the best of the author's knowledge, a direct connection of the Goldstone boson counting and the charge densities is still missing. A partial result in this respect was achieved by Schaefer *et al.* [6]: *The Goldstone boson counting is as usual, provided the densities of commutators of all pairs of broken generators vanish.*

In this contribution, we investigate the spontaneous symmetry breaking within the relativistic linear sigma model at finite chemical potential. This model was used to describe kaon condensation in the so-called Color-Flavor-Locked phase of dense quark matter [6,7]. Within this restricted framework, we are able to convert the theorem

---

[a] e-mail: `brauner@ujf.cas.cz`

[1] In the following, these will be collectively called *nonrelativistic* in order to simplify the nomenclature. One should, however, keep in mind that this term may denote both intrinsically nonrelativistic systems and relativistic systems with Lorentz invariance explicitly broken, *e.g.*, by nonzero density.

of Schaefer *et al.* and show that the existence of type-II Goldstone bosons is unavoidable once some of the Noether charges develop nonzero density. Furthermore, we show that the Nielsen-Chadha inequality for the number of the Goldstone bosons is saturated. In the concluding section, we discuss the extension of our results to other relativistic systems at finite density.

2 Linear sigma model at finite chemical potential

2.1 Model with SU(2) × U(1) symmetry

We start with the model of Schaefer *et al.* [6], and Miransky and Shovkovy [7]. It is defined by the Lagrangian,

$$\mathcal{L} = D_\mu \phi^\dagger D^\mu \phi - M^2 \phi^\dagger \phi - \lambda (\phi^\dagger \phi)^2, \qquad (1)$$

where $D_\nu \phi = (\partial_\nu - i\delta_{\nu 0}\mu)\phi$. The scalar field ϕ transforms into the doublet representation of the global $SU(2)$ symmetry group, and μ is the chemical potential associated with the global $U(1)$ symmetry (particle number).

This model describes relativistic Bose-Einstein condensation: When $\mu > M$, the scalar field develops nonzero vacuum expectation value. Consequently, the $SU(2) \times U(1)$ symmetry of the Lagrangian (1) is spontaneously broken to its $U(1)$ subgroup (different from the original $U(1)$). Thus, *three* of the symmetry generators are spontaneously broken. However, only *two* Goldstone bosons appear, one type I and one type II. Their low-energy dispersion relations read

$$E = \sqrt{\frac{\mu^2 - M^2}{3\mu^2 - M^2}}|\mathbf{p}|, \qquad E = \frac{\mathbf{p}^2}{2\mu}, \qquad (2)$$

respectively.

2.2 Properties of the type-II Goldstone boson

To get more insight into the nature of the type-II Goldstone boson, we now investigate the corresponding plane-wave solutions of the classical equations of motion. With the standard choice of the vacuum expectation value v, the scalar ϕ is reparametrized as

$$\phi = \frac{1}{\sqrt{2}} e^{i\pi_k \tau_k / v} \begin{pmatrix} 0 \\ v + H \end{pmatrix}.$$

The type-II Goldstone boson is then annihilated by the complex field $\psi = \frac{1}{\sqrt{2}}(\pi_2 + i\pi_1)$. The relevant bilinear part of the Lagrangian reads

$$\mathcal{L}_\psi = 2i\mu\psi^\dagger \partial_0 \psi + \partial_\mu \psi^\dagger \partial^\mu \psi. \qquad (3)$$

At the leading order of the power expansion in energy and momentum, it is evidently of the Schroedinger type, which is due to the fact that the field ψ carries nonzero charge

of the unbroken $U(1)$ symmetry, generated by the matrix $\frac{1}{2}(1 + \tau_3)$.

The Lagrangian (3) describes a free particle with exact dispersion relation $E = \sqrt{\mathbf{p}^2 + \mu^2} - \mu$, whose low-momentum limit is given by eq. (2). The corresponding classical plane wave is simply $\psi = \psi_0 e^{-ip \cdot x}$. The $SU(2) \times U(1)$ symmetry of the Lagrangian (1) gives rise to four conserved currents, the isospin current and the particle number current, $j^\nu = -2\,\mathrm{Im}\,\phi^\dagger T\partial^\nu \phi + 2\mu\delta^{\nu 0}\phi^\dagger T\phi$, where $T = \{\tau, 1\}$, respectively. For the two broken generators, τ_1 and τ_2, that create the type-II Goldstone boson, we find

$$j_1^\nu = +(p^\nu + 2\delta^{\nu 0}\mu)v\sqrt{2}\,\mathrm{Re}\,\psi,$$
$$j_2^\nu = -(p^\nu + 2\delta^{\nu 0}\mu)v\sqrt{2}\,\mathrm{Im}\,\psi.$$

It is apparent that the type-II Goldstone boson corresponds to an isospin wave, circularly polarized in the plane perpendicular to the vacuum density of the isospin. Note also that the other circular polarization corresponds to an excitation with a gap 2μ so that there is indeed a single Goldstone boson which couples to the two broken generators.

The unbroken $U(1)$ symmetry generates the current $j^\nu = 2(p^\nu + \delta^{\nu 0}\mu)|\psi|^2$. This uniform current proves that the isospin wave transfers the unbroken charge. In other words, the type-II Goldstone boson carries the unbroken charge, which seems to be a generic feature of type-II Goldstone bosons.

2.3 General bilinear Lagrangians

At tree level, the spectrum of the linear sigma model follows from the bilinear part of the Lagrangian upon a proper reparametrization of the scalar field. Once the chemical potential is introduced, this bilinear Lagrangian attains new terms with a single time derivative, which communicate the effects of Lorentz violation by the dense medium to the excitation spectrum. The generic form of the bilinear Lagrangian one encounters is (see ref. [8] for details)

$$\mathcal{L}_{\mathrm{bilin}} = \frac{1}{2}(\partial_\mu \pi)^2 + \frac{1}{2}(\partial_\mu h)^2 - \frac{1}{2}f^2(\mu)h^2 - g(\mu)h\partial_0\pi. \qquad (4)$$

As long as at least one of the functions $f(\mu), g(\mu)$ is nonzero, the Lagrangian (4) describes a massive mode, with a mass gap $\sqrt{f^2(\mu) + g^2(\mu)}$, and a Goldstone boson with a dispersion relation

$$E^2 = \frac{f^2(\mu)}{f^2(\mu) + g^2(\mu)}\mathbf{p}^2 + \frac{g^4(\mu)}{[f^2(\mu) + g^2(\mu)]^3}\mathbf{p}^4 + \mathcal{O}(\mathbf{p}^6). \qquad (5)$$

Equation (5) shows that when $f(\mu) \neq 0$, *i.e.*, the chemical potential mixes a Goldstone field with a Higgs field (as is the case of π_3 and H in the simple model (1)), one finds the expected result: One massive and one massless excitation. On the other hand, mixing of two Goldstone fields (such as π_1 and π_2 above) gives rise to just *one* Goldstone boson, which is type-II —its dispersion relation is quadratic at low momentum.

2.4 Model with arbitrary symmetry

In ref. [8] we analyzed Bose-Einstein condensation in the linear sigma model with an arbitrary symmetry breaking pattern. The generic Lagrangian reads $\mathcal{L} = D_\mu \phi^\dagger D^\mu \phi - V(\phi)$. The covariant derivative, $D_\mu \phi = (\partial_\mu - iA_\mu)\phi$, involves a constant external field A_μ that accounts for the chemical potential.

To find the spectrum, one minimizes the potential $V(\phi)$ and determines the vacuum expectation value ϕ_0. Next the scalar field is reparametrized as $\phi = e^{i\Pi}[\phi_0 + H]$, where the matrix field Π factorizes out the Goldstone degrees of freedom, while H represents the radial (Higgs) modes. The resulting bilinear Lagrangian reads

$$
\begin{aligned}
\mathcal{L}_{\text{bilin}} = {} & \partial_\mu H^\dagger \partial^\mu H - V_{\text{bilin}}(H) \\
& -2\,\text{Im}\,H^\dagger A^\mu \partial_\mu H - 4\,\text{Re}\,H^\dagger A^\mu \partial_\mu \Pi \phi_0 \\
& +\phi_0^\dagger \partial_\mu \Pi \partial^\mu \Pi \phi_0 - \text{Im}\,\phi_0^\dagger A^\mu [\Pi, \partial_\mu \Pi]\phi_0,
\end{aligned} \tag{6}
$$

where V_{bilin} is the bilinear part of the potential which depends explicitly only on H.

The bilinear Lagrangian in eq. (6) contains three terms with a single derivative, proportional to the chemical potential. It can be shown that, with a proper choice of the basis of the symmetry generators, every excitation mode appears in exactly one of these terms so that the analysis of the simple two-field Lagrangian (4) applies. The most notable result is that the Goldstone-Goldstone mixing term in the last line of eq. (6), which according to eq. (5) gives rise to type-II Goldstone bosons, is proportional to the ground-state expectation value of the commutator of two broken generators. This proves the assertion made in the Introduction that *a nonzero density of a commutator of two broken generators gives rise to one type-II Goldstone boson with a quadratic dispersion relation*. Moreover, it is also obvious that the Nielsen-Chadha inequality for the number of the Goldstone bosons is saturated. (The only exception to this saturation known to the author, is the case of phase transitions where the phase velocity of a type-I Goldstone boson may vanish, thus making it an "accidental" type-II Goldstone boson.)

3 Summary and outlook

In this contribution, we investigated spontaneous symmetry breaking within the relativistic linear sigma model at finite chemical potential. We clarified the connection of Goldstone boson counting and their dispersion relations with nonzero densities of the Noether charges. In particular, we proved that nonzero density of a commutator of two broken charges produces one type-II Goldstone boson with a quadratic dispersion relation. It should be stressed, however, that all the results were achieved at the classical, tree level. Nevertheless, in ref. [9] it was shown that they are not altered by the one-loop radiative corrections.

Besides the radiative corrections to the linear sigma model, it would also be desirable to extend the results to other relativistic systems at finite density. In such systems, the Lorentz invariance is broken in a very particular way by the presence of the dense medium. It is, however, manifest on the microscopic level and hence could serve to constrain the patterns of symmetry breaking and the properties of the Goldstone bosons. One could thus hopefully strengthen the Nielsen-Chadha counting rule for this restricted class of systems. Based on the results achieved so far, we conjecture that generally an equality holds instead of the inequality, and that the Goldstone boson dispersion relation is either linear or quadratic, depending on the Lagrangian. (Recall that Nielsen and Chadha just distinguish odd and even powers of momentum.)

A preliminary argument in this direction was already given [8]. It was shown that when the symmetry group is non-Abelian, the charge density itself may serve as an order parameter for symmetry breaking. As a consequence, a single Goldstone boson couples to the two broken charges whose commutator yields the order parameter. In fact, by a proper analysis one may even show that such a Goldstone boson then necessarily has a quadratic dispersion relation. Using Leutwyler's effective Lagrangian approach, the coefficients in the dispersion relation can be related to the amplitude for the annihilation of the Goldstone boson by the broken current. This leads to a convenient model-independent parametrization of the Goldstone boson dispersion relations, which can be used as a check on models of spontaneous symmetry breaking such as that of Nambu and Jona-Lasinio. This issue will be investigated in detail in our future work.

The author wishes to thank the organizers of the *International Conference on Quarks and Nuclear Physics 2006* for giving him an opportunity to present this material, and to J. Hošek for fruitful discussions. This work was supported in part by the Institutional Research Plan AV0Z10480505, and by the GACR grants Nos. 202/06/0734 and 202/05/H003.

References

1. J. Goldstone, Nuovo Cimento **19**, 154 (1961).
2. J. Goldstone, A. Salam, S. Weinberg, Phys. Rev. **127**, 965 (1962).
3. I. Low, A.V. Manohar, Phys. Rev. Lett. **88**, 101602 (2002).
4. H.B. Nielsen, S. Chadha, Nucl. Phys. B **105**, 445 (1976).
5. H. Leutwyler, Phys. Rev. D **49**, 3033 (1994).
6. T. Schaefer, D.T. Son, M.A. Stephanov, D. Toublan, J.J.M. Verbaarschot, Phys. Lett. B **522**, 67 (2001).
7. V.A. Miransky, I.A. Shovkovy, Phys. Rev. Lett. **88**, 111601 (2002).
8. T. Brauner, Phys. Rev. D **72**, 076002 (2005).
9. T. Brauner, Phys. Rev. D **74**, 085010 (2006).

Eur. Phys. J. A **31**, 761–765 (2007)

DOI 10.1140/epja/i2006-10289-y

THE EUROPEAN
PHYSICAL JOURNAL A

Special Article – QNP 2006

Nonequilibrium quasi-classical effective meson gas: Thermalization

R.F. Alvarez-Estrada[a]

Departamento de Física Teórica I, Facultad de Ciencias Físicas, Universidad Complutense, 28040 Madrid, Spain

Received: 18 December 2006

Published online: 20 March 2007 – © Società Italiana di Fisica / Springer-Verlag 2007

Abstract. We consider a gas of interacting relativistic effective mesons (qualitatively, like those produced in a heavy-ion collision), regarded as an out-of-equilibrium statistical system. We suppose large occupation numbers, temperature somewhat below typical critical temperatures and the quasi-classical regime. At some initial time t_0, let the gas be in a nonequilibrium state, with spatial inhomogeneities. The time evolution of the gas for $t > t_0$ is studied by a moment method, and appropriate long-time approximations, which could yield the approach to global thermal equilibrium, are discussed.

PACS. 11.10.Wx Finite-temperature field theory – 11.25.Db Properties of perturbation theory – 11.10.Gh Renormalization – 11.90.+t Other topics in general theory of fields and particles

1 Introduction

In heavy-ion collisions, an important issue is how thermal equilibrium is created [1,2]. After a time less than some fm/c, quarks and gluons have thermalized locally, temperatures have fallen a bit below the critical temperatures, and the confinement and chiral phase transitions have occurred. Then, large numbers of pions are produced and, for a time interval between about some fm/c and a few tens of fm/c, before "freeze-out", the pions can be regarded, as least approximately, as a nonequilibrium interacting relativistic quantum gas. We leave aside many important physical features [2] and focus on the time evolution of the pion gas and the eventual formation of a state of more global thermal equilibrium, say, its thermalization. Even leaving aside gauge degrees of freedom and half-integral spin, the analysis is still difficult due to many degrees of freedom out of thermal equilibrium, relativity and quantum aspects. For accounts of nonequilibrium relativistic quantum field theory, see [3,4]. We shall analyze a meson gas, qualitatively similar to, but far simpler than, the pion gas, in the quasi-classical approximation: in so doing, we disregard quantum features, leaving aside their possible or potential relevance, for instance, for Hanbury-Brown and Twiss interferometry [2]. The latter simplification could perhaps be not entirely unreasonable, at least qualitatively, for describing some gross features, in suitably large time and spatial scales, of a nonequilibrium many-meson system, with mesons distributed with large occupation numbers over their quantum states and for a restricted temperature range, analogue to that between the pion mass and the critical temperatures. We accept the possibility that interactions in the meson gas, with an infinite number of degrees of freedom, could give rise to thermalization, and will focus on approximations which could lead to it. We shall deal with the time evolution of the meson gas after some initial time $t = t_0$, analogue of some fm/c, and its eventual approach towards approximate global thermalization for long time, analogue of a few tens of fm/c.

This work is organized as follows. Section 2 deals, for illustrative purposes, with the nonequilibrium statistical mechanics for one degree of freedom in an external "heat bath", and presents the moment technique and the long-time approximation. Section 3 treats one generalization of sect. 2 to an infinite number of degrees of freedom: an effective neutral scalar field. Section 4 outlines some generalization for effective nonlinear chiral fields. Section 5 contains some conclusions and discussions.

2 Oscillator in a "heat bath"

We shall outline the nonequilibrium statistical mechanics of one particle of mass m and momentum p, in one spatial dimension x, with Hamiltonian

$$H = p^2/(2m) + V, \quad V = 2^{-1}m\omega^2 x^2 + (4!)^{-1}gx^4 \quad (2.1)$$

in the presence of a "heat bath" at thermal equilibrium at absolute temperature β_{eq}^{-1}. ω and g are positive constants. The classical probability distribution $W = W(x,p;t)$ for

---

[a] e-mail: `rfa@fis.ucm.es`

the particle fulfills the reversible Liouville equation:

$$\frac{\partial W}{\partial t} = \{H, W\} = -\frac{p}{m}\frac{\partial W}{\partial x} + \frac{\partial V}{\partial x}\frac{\partial W}{\partial p} \quad (2.2)$$

at time t. $\{H, W\}$ is the classical Poisson bracket. The initial condition at $t = t_0$ is W_{in}. W seems to qualify not just as a classical probability distribution, but also as a quasi-classical one, say, it also accounts for the first correction in Planck's constant, $\hbar$. In fact, let H_Q be the quantum Hamiltonian associated to (2.1), let ρ and $[,]$ be the density operator representing the quantum particle and the commutator. The Schrödinger equation yields:

$$\frac{\partial \rho}{\partial t} = \frac{1}{i\hbar}[H_Q, \rho]. \quad (2.3)$$

Also, let W_Q be the quantum Wigner function determined by ρ [5]. The quantum evolution equation for $\partial W_Q/\partial t$, implied by (2.3), includes in its right-hand-side additive terms of order $\hbar^2$, but no corrections of order $\hbar$ [5]. As $\hbar \to 0$, W_Q and the quantum evolution equation for $\partial W_Q/\partial t$ become W and (2.2), respectively [5]. We shall concentrate on W and (2.2).

Any integration will be performed in $(-\infty, +\infty)$. We shall introduce the following moments W_n $(n = 0, 1, 2, \ldots)$ of W regarding the p-dependence:

$$W_n = W_n(x; t) = \int \mathrm{d}p \frac{H_n((\beta_{eq}/2m)^{1/2}p)}{(\pi^{1/2}2^n n!)^{1/2}} W \quad (2.4)$$

which incorporate the temperature of the "heat bath". H_n is the Hermite polynomial of order n. Equations (2.4) and (2.2) imply the following infinite three-term linear recurrence relation for all W_n's $(n = 0, 1, 2, \ldots, W_{-1} = 0)$:

$$\frac{\partial W_n}{\partial t} = -M_{n,n+1}W_{n+1} - M_{n,n-1}W_{n-1}, \quad (2.5)$$

$$M_{n,n+1}W_{n+1} \equiv \left[\frac{(n+1)}{m\beta_{eq}}\right]^{1/2}\frac{\partial W_{n+1}}{\partial x}, \quad (2.6)$$

$$M_{n,n-1}W_{n-1} \equiv \left[\frac{n}{m\beta_{eq}}\right]^{1/2}$$
$$\times \left(\frac{\partial W_{n-1}}{\partial x} + \beta_{eq}\frac{\partial V}{\partial x}W_{n-1}\right). \quad (2.7)$$

The initial condition $W_{n,in}$ is obtained by replacing W by W_{in} in (2.4). A t-independent solution of eq. (2.2) is

$$W_{eq} = \exp[-\beta_{eq}(p^2/(2m) + V)]$$

and, through (2.4), it yields $W_{0,eq}$ proportional to $\exp[-\beta_{eq}V]$ and $W_{n,eq} = 0$, $n = 1, 2, \ldots$. Equation (2.5) implies exactly for any $n \geq 0$

$$\sum_{n'=0}^{n} \int \mathrm{d}x (W_{0,eq})^{-1}[2^{-1}(\partial W_{n'}^2/\partial t)$$
$$+ W_n M_{n,n+1}W_{n+1}] = 0. \quad (2.8)$$

We introduce the Laplace transform:

$$\tilde{W}_n(s) \equiv \int_0^{+\infty} \mathrm{d}t W_n \exp(-st). \quad (2.9)$$

The Laplace transform of (2.5) can be solved formally. That yields all $\tilde{W}_n(s)$, for any $n = 1, \ldots$, in terms of sums of products of s-dependent linear operators $D[n'; s]$, $n' \geq n$, acting upon $\tilde{W}_{n-1}(s)$ and upon all $W_{n',in}$'s, with $n' \geq n$. The $D[n; s]$'s are infinite continued fractions of products of linear operators, generated by iterating

$$D[n; s] = [s - M_{n,n+1}D[n+1; s]M_{n+1,n}]^{-1}. \quad (2.10)$$

For $g = 0$, the harmonic oscillator, the operator $D[n; s]$ (2.10) can be evaluated in closed form, and we shall outline the result. Let

$$y = [m\beta_{eq}/2]^{1/2}\omega x, \qquad A = -\frac{1}{2}\frac{\mathrm{d}}{\mathrm{d}y}\left(\frac{\mathrm{d}}{\mathrm{d}y} + 2y\right), \quad (2.11)$$

$D[n; s] \equiv D[n; s; A]$ is given by the following fraction:

$$D[n; s; A] = [s + (n+1)\omega^2 A D[n+1; s; A-1]]^{-1}. \quad (2.12)$$

Let $f_{n'} = H_{n'}(y)\exp[-2^{-1}y^2]$, $n' = 0, 1, 2 \ldots$. Then, the eigenfunctions of A and of $D[n; s; A]$ are $\exp[-2^{-1}y^2]f_{n'}$. The eigenvalues of A are n' which, through iteration of (2.12), yield directly those of $D[n; s; A]$ as finite fractions, due to the structure $A - 1$. The $D[n; s]$'s cannot be evaluated in closed form for $g \neq 0$.

Thus far, no long-time approximation has been made. We shall analyze the irreversible evolution of the oscillator towards thermal equilibrium, say, its thermalization, induced by the "heat bath". We choose some $n_0(\geq 1)$ and, for $n \geq n_0$, fix $s = \epsilon > 0$ in any $D[n; s]$, ϵ being suitably small. A crucial property, for any $g \geq 0$, is the following: the s-independent $W_{0,eq}^{-1/2}D[n; \epsilon]W_{0,eq}^{1/2}$'s, $n \geq n_0$, are Hermitian operators, have denumerably infinite discrete spectra, without singularities for suitable $\epsilon > 0$, and their eigenvalues have non-negative real parts. Then, the long-time approximation for $n \geq n_0$ is as follows: we replace any $D[n'; s]$ yielding $\tilde{W}_n(s)$, $n \geq n_0$, in terms of $\tilde{W}_{n-1}(s)$ and of $W_{n'',in}$'s, $n'' \geq n$, by $D[n'; \epsilon]$. That approximation is not done for $n < n_0$, and it is the better fulfilled the larger n_0 is. It constitutes a necessary ingredient for the approach towards equilibrium. For a simpler analysis, we also neglect all $W_{n',in}$'s for any $n' \geq n_0$ and set, for small s,

$$\tilde{W}_{n_0}(s) \simeq -D[n_0; \epsilon]M_{n_0,n_0-1}\tilde{W}_{n_0-1}(s). \quad (2.13)$$

The hierarchy becomes closed, by using (2.5), as they stand, for $n = 0, 1, \ldots, n_0 - 1$, and the inverse Laplace transform of (2.13). Its t-independent solution is $W_{0,eq}$ and $W_{n,eq} = 0$, $n = 1, 2, \ldots, n_0$. The solutions of the closed hierarchy relax irreversibly, for $t \gg t_0$ and any reasonable W_{in}, towards the t-independent solution: thermalization of the oscillator due to the "heat bath". For $g \neq 0$ and $\epsilon > 0$, a rough estimate of the matrix elements of $D[n_0; \epsilon]$

over some finite part of its discrete spectrum has been performed: it indicates that, in some average sense, the relaxation times for W_{n_0} are adequately small. Equation (2.8) becomes

$$\sum_{n'=0}^{n_0-1} \int dx[(2W_{0,eq})^{-1}(\partial W_{n'}^2/\partial t)] \leq 0 \qquad (2.14)$$

which also expresses irreversibility due to the "heat bath". In the simplest case, $n_0 = 1$, approximate the linear operator $D[1; \epsilon]$ by a real constant (> 0), and come to (2.5) for $n = 0$. The resulting (irreversible) Fokker-Planck equation for the quasi-classical probability distribution function W_0 is

$$\frac{\partial W_0}{\partial t} = \frac{D[1; \epsilon]}{\beta_{eq}} \frac{\partial}{\partial x}\left[\frac{\partial}{\partial x} + \beta_{eq}\frac{\partial V}{\partial x}\right] W_0 \qquad (2.15)$$

with the quasi-classical initial condition $W_{0,in}$ at $t = t_0$. For $g = 0$, possible values for the constant $D[1; \epsilon]$ may be estimated qualitatively from (2.12), but a choice for its most reasonable value is open to discussion. The same happens for $g \neq 0$. Equation (2.15) also follows from other different methods [6], and a comparison with them could perhaps help to assess $D[1; \epsilon]$ adequately. This section extends [7], for a discretized spectrum.

3 Quasi-classical effective scalar fields

We shall consider a large statistical system, the dynamics of which is described by a relativistic real scalar classical field χ, having mass parameter m and quartic coupling, in three-dimensional space. m^2 is real. The classical Hamiltonian is, with $\mathbf{x} = (x_1, x_2, x_3)$,

$$H = \int d^3\mathbf{x}\frac{\pi^2}{2} + V_1 , \qquad (3.1)$$

$$V_1 = \int d^3\mathbf{x}\left\{\frac{1}{2}\sum_{i=1}^{3}\left(\frac{\partial\chi}{\partial x_i}\right)^2 + \frac{m^2\chi^2}{2} + V(\chi)\right\} , \qquad (3.2)$$

$$\pi = \frac{\partial\chi}{\partial t} , \qquad V(\chi) = \frac{g\chi^4}{4!} . \qquad (3.3)$$

χ and π are $\mathbf{x}$-dependent, but t-independent, fields. g is the dimensionless coupling constant. An ultraviolet cut-off, Λ, is included in H. χ, m and g are unrenormalized classical quantities. Now, there is no external "heat bath", but the infinite number of degrees of freedom of the classical field will give rise to statistical effects. Let $W = W[\chi, \pi; t]$ be the quasi-classical probability distribution for the system to be described, at time t, by the field configuration χ with momentum π. W fulfills the quasi-classical reversible Liouville equation. The latter reads

$$\frac{\partial W}{\partial t} = \int d^3\mathbf{x}\left[\left(\frac{\delta V_1}{\delta\chi}\right)\frac{\delta}{\delta\pi} - \pi\frac{\delta}{\delta\chi}\right] W ; \qquad (3.4)$$

δ/δ denotes the functional derivative. Equation (3.4) generalizes eq. (2.2) to the actual classical-field system. The initial condition is $W_{in} = W_{in}[\chi, \pi]$ at t_0. Let W_{in} be a

nonequilibrium state, with spatial inhomogeneities characterized by some function $\beta(\mathbf{x})$: $\beta^{-1}(\mathbf{x})$ could be interpreted, at least qualitatively, as the absolute temperature of the infinitesimal volume $d^3\mathbf{x}$ at $\mathbf{x}$. With

$$p_0 = p_0(\mathbf{x}) = (2/\beta(\mathbf{x}))^{1/2} ,$$

we shall introduce the functional Hermite polynomials H_n through the Rodrigues-like (functional differentiation) formula:

$$H_n \equiv (-)^n \exp\left[\int d^3\mathbf{x}\frac{\pi^2}{p_0(\mathbf{x})^2}\right]\frac{\delta}{\delta(\pi(\mathbf{x}_1)/p_0(\mathbf{x}_1))}$$

$$\times \cdots \frac{\delta}{\delta(\pi(\mathbf{x}_n)/p_0(\mathbf{x}_n))}\exp\left[-\int d^3\mathbf{x}\frac{\pi^2}{p_0(\mathbf{x})^2}\right] \qquad (3.5)$$

thereby generalizing the ordinary Hermite polynomials. H_n depends on $\pi(\mathbf{x}_1)/p_0(\mathbf{x}_1), \ldots, \pi(\mathbf{x}_n)/p_0(\mathbf{x}_n)$. Let $\int[d\pi]$ denote the functional integration over the classical momentum $\pi(\mathbf{x})$. We shall introduce the moments W_n of W:

$$W_n \equiv \frac{1}{(n!2^n)^{1/2}}$$

$$\times \int[d\pi]H_n(\pi(\mathbf{x}_1)/p_0(\mathbf{x}_1), \ldots, \pi(\mathbf{x}_n)/p_0(\mathbf{x}_n))W . \qquad (3.6)$$

The W_n's depend on $\mathbf{x}_1, \ldots, \mathbf{x}_n$ and, although not written explicitly, also on χ and t. Like (2.2) via (2.4), (3.4) gives, through (3.6), the following (reversible) infinite linear hierarchy for all W_n's $(n = 0, 1, \ldots, W_{-1} \equiv 0)$:

$$\frac{\partial W_n}{\partial t} = -M_{n,n+1}W_{n+1} - M_{n,n-1}W_{n-1} , \qquad (3.7)$$

$$M_{n,n+1}W_{n+1} \equiv \left[\frac{n+1}{2}\right]^{1/2}\int d^3\mathbf{x}p_0(\mathbf{x})\frac{\delta}{\delta\chi(\mathbf{x})}$$

$$\times W_{n+1}(\mathbf{x}, \mathbf{x}_1, \ldots, \mathbf{x}_n), \qquad (3.8)$$

$$M_{n,n-1}W_{n-1} \equiv \int d^3\mathbf{x}\frac{p_0(\mathbf{x})}{(2n)^{1/2}}FP(\mathbf{x})\left[\sum_{i=1}^{n}\delta^{(3)}(\mathbf{x} - \mathbf{x}_i)\right.$$

$$\left.\times W_{n-1}(\mathbf{x}_1, \ldots, \mathbf{x}_{i-1}, \mathbf{x}_{i+1}, \ldots, \mathbf{x}_n)\right], \qquad (3.9)$$

$$FP(\mathbf{x}) = \frac{\delta}{\delta\chi(\mathbf{x})} + \frac{2}{p_0(\mathbf{x})^2}\frac{\delta V_1}{\delta\chi(\mathbf{x})} . \qquad (3.10)$$

The initial condition $W_{n,in}$ for (3.7) is obtained by replacing W by W_{in} in (3.6). In particular, the equations in (3.7) for $n = 0, 1, 2$ can be shown to be exactly consistent with the balance equations for momentum, energy and angular momentum.

Like in sect. 2, we introduce the Laplace transform $\tilde{W}_n(s)$ of W_n, solve formally the Laplace transform of (3.7) and get all $\tilde{W}_n(s)$, for any $n = 1, \ldots$, in terms of similar structures and new s-dependent linear operators $D[n'; s]$. The $D[n; s]$'s are infinite continued fractions formally similar to (2.10).

3.1 Initial state not far from equilibrium

Let W_{in}, representing also nonequilibrium, be not far from global thermal equilibrium at constant absolute temperature β_{eq}^{-1}. For instance: i) $\beta(\mathbf{x}) = \beta_{eq} + \delta\beta(\mathbf{x})$, where the function $\delta\beta(\mathbf{x}) \to 0$ for $|\mathbf{x}| \to +\infty$ (and $|\delta\beta(\mathbf{x})| \le \beta_{eq}$), or ii) the larger parts of the gas are at global thermal equilibrium at β_{eq}^{-1}, while the remaining, smaller, parts are, still, out of equilibrium, with spatial inhomogeneities. β_{eq}^{-1} is not imposed by any external "heat bath", but by the infinite number of degrees of freedom, not far from thermal equilibrium, of the whole field system. Then, we replace $p_0(\mathbf{x})$ by the constant $p_0 = (2/\beta_{eq})^{1/2}$ in (3.5)–(3.10), and until otherwise stated. Then, $W_{0,eq} = \exp[-\beta_{eq} \int \mathrm{d}^3\mathbf{x} V_1(\chi)]$ and $W_{n,eq} = 0$, $n \ge 1$ yield a t-independent solution of (3.7). Equation (3.7) implies exactly the analogue of (2.8). With the actual W_{in}, the long-time approximation, with $D[n_0; s] \simeq D[n_0; \epsilon]$ for $n \ge n_0$, proceeds formally like in sect. 2, and it is the better fulfilled the larger n_0 is. The approximation seems justified due to the integrations and functional dependences involved in (3.8) and (3.9), say, to the infinite number of degrees of freedom of the large statistical system. Then, $W_{0,eq}^{-1/2} D[n; \epsilon] W_{0,eq}^{1/2}$'s, $n \ge n_0$, now have continuous spectra. In principle, the operator $D[n'; \epsilon]$ depends on g and on the dimensionful quantities ϵ, m, Λ and β_{eq}^{-1}. The analogue of (2.14) also holds. For $n_0 = 1$, by proceeding like in sect. 2, the resulting (irreversible) Fokker-Planck equation for the quasi-classical probability distribution functional W_0 is

$$\frac{\partial W_0}{\partial t} = \frac{D[1; \epsilon]}{\beta_{eq}} \int \mathrm{d}^3\mathbf{x} \frac{\delta}{\delta\chi(\mathbf{x})} FP(\mathbf{x}) W_0 \qquad (3.11)$$

with the quasi-classical initial condition $W_{0,in}$ at $t = t_0$. $\chi = \chi(\mathbf{x})$ plays the role of an order parameter. The t-independent solution of (3.11) is $W_{0,eq}$. The ansatz has been made of interpreting $D[1; \epsilon]$ as a positive constant, instead of as an operator. Then, the solutions of (3.11) relax irreversibly, for $t \gg t_0$ and any reasonable $W_{0,in}$, towards $W_{0,eq}$: thermalization. Physically, the larger parts of the gas, already at, or close to, global thermal equilibrium at β_{eq}^{-1} at t_0, iron out all spatial inhomogeneities and drive the remaining, smaller, parts also to the same global equilibrium distribution at β_{eq}^{-1}, $W_{0,eq}$, for long times.

3.2 Removing the cut-off

With the same W_{in} as in subsect. 3.1, let us now remove the ultraviolet cut-off in (3.11): $\Lambda \to +\infty$. We apply results in [8,9] to (3.11). With the quartic self-coupling in eq. (3.3), $W_{0,eq}$ characterizes a superenormalizable theory in three spatial dimensions. In $W_{0,eq}$ and in the dynamical theory described by (3.11) for $t > t_0$, mass renormalization is necessary but neither χ nor g require ultraviolet renormalization. The question then arises whether the constant $D[1; \epsilon]$ would require it: if that were the case, that would signal some physical failure in the long-time approximation. It turns out that $D[1; \epsilon]$ does not require any ultraviolet renormalization either. Then, the dynamical theory

described by (3.11) for $t > t_0$ is also superenormalizable. Physically, $D[1; \epsilon]$, being related to long-time and large-distance behaviours, should remain finite as $\Lambda \to +\infty$. This indicates, *a posteriori*, that the long-time approximation does not run into conflict with the ultraviolet behaviour, at least in the actual three-dimensional classical model. The above analysis for $n_0 = 1$ suggests the following, for $n_0 > 1$. The approximate dynamical theory given by the analogue of (2.13) with $D[n_0; \epsilon]$ interpreted as a positive constant, and by (3.7) for $n < n_0$ would be superenormalizable and would relax for a long time towards the same $W_{0,eq}$ as above. $D[n_0; \epsilon]$ would require no ultraviolet renormalization, only m being in need of it.

3.3 Initial state far from equilibrium

Now, let the gas be at t_0 in a state W_{in} quite appreciably out of global thermal equilibrium, say, with spatial inhomogeneities so that $\beta(\mathbf{x})$ is neatly different from any β_{eq}. Now, we deal with (3.7) with $p_0 = p_0(\mathbf{x})$. The long-time approximation also proceeds (formally, at least) by setting $s = \epsilon$ in any $D[n'; s]$ for $n' \ge n \ge n_0(\ge 1)$. The hermiticity and positivity properties of $D[n'; \epsilon]$ in subsect. 3.1 no longer hold necessarily, due to the $\mathbf{x}$-dependence of p_0. Equations (2.8) and (2.14) do not hold necessarily. Anyway, simplifying assumptions similar to those yielding (3.11) now give (at least, formally and for fixed Λ) the generalized Fokker-Planck equation:

$$\frac{\partial W_0}{\partial t} = D[1; \epsilon] \int \mathrm{d}^3\mathbf{x} \frac{p_0(\mathbf{x})}{2} \frac{\delta}{\delta\chi(\mathbf{x})}$$
$$\times p_0(\mathbf{x}) FP(\mathbf{x}) W_0 \qquad (3.12)$$

with the quasi-classical initial condition $W_{0,in}$ and a real and positive constant $D[1; \epsilon]$, so that irreversibility holds. As $p_0 = p_0(\mathbf{x})$, the actual equilibrium distribution $W_{0,eq}$ is not given by $FP(\mathbf{x}) W_{0,eq} = 0$, but as the limit of the solution W_0 of (3.12) for $t \to +\infty$: such $W_{0,eq}$ would require longer times to be reached and depend on some global equilibrium temperature ($\ne p_0(\mathbf{x})^2/2$). See [8], for instance, for functional integral representations of $W_{0,eq}$.

4 Quasi-classical effective chiral fields

Let the meson gas be a nonequilibrium (quasi-classical) large statistical system, described by the simplest ($O(N)$-invariant) nonlinear σ model [8]. Let $\chi_i = \chi_i(\mathbf{x})$, $i = 1, \ldots, N - 1$, be classical effective nonlinear chiral fields for mesons and $\pi_i = \pi_i(\mathbf{x})$ be their associated momenta. The classical Hamiltonian is

$$H = \int \mathrm{d}^3\mathbf{x} h_\sigma + V_1 , \qquad (4.1)$$

$$h_\sigma = \frac{g_\sigma^2}{2} \sum_{i,j=1}^{N-1} \pi_i (G^{-1})_{i,j} \pi_j , \qquad (4.2)$$

$$V_1 = \frac{g_\sigma^2}{2} \int \mathrm{d}^3\mathbf{x} \sum_{i,j=1}^{N-1} \left[G_{ij} \sum_{l=1}^{3} \left(\frac{\partial\chi_i}{\partial x_l} \frac{\partial\chi_j}{\partial x_l} \right) \right] , \qquad (4.3)$$

$$G_{ij} = \delta_{i,j} + \frac{\chi_i \chi_j}{1 - \bar{\chi}^2} . \qquad (4.4)$$

The coupling constant g_σ is now dimensionful. G^{-1} is the inverse of the $(N-1) \times (N-1)$ matrix formed by all G_{ij}. An ultraviolet cut-off about or somewhat larger than g_σ is supposed. The role played in sect. 3 by the functional Hermite polynomials H_n will now be played by new functional polynomials $H_{\sigma,n}$ which, by definition, are orthogonalized with respect to the functional measure

$$\int \prod_{i=1}^{N-1} [\mathrm{d}\pi_i] \exp\left[-\int \mathrm{d}^3\mathbf{x}\, 2p_0(\mathbf{x})^{-2} h_\sigma \right].$$

$H_{\sigma,n}$ would allow to introduce moments and to generalize formally the developments and results in sect. 3. Thus, if the initial nonequilibrium distribution at t_0 is not far from global thermal equilibrium at constant absolute temperature β_{eq}^{-1}, the counterpart of (3.11) reads for long times

$$\frac{\partial W_0}{\partial t} = \frac{D[1;\epsilon]}{\beta_{eq}} \sum_{l=1}^{N-1} \int \mathrm{d}^3\mathbf{x}\, \frac{\delta}{\delta\chi_i(\mathbf{x})} \sum_{j=1}^{N-1} (G^{-1})_{i,j}$$
$$\times FP_j(\mathbf{x}) W_0 , \tag{4.5}$$

$$FP_j(\mathbf{x}) = \frac{\delta}{\delta\chi(\mathbf{x})_j} + \beta_{eq}\frac{\delta V_1}{\delta\chi(\mathbf{x})_j} + \frac{\delta \ln[\det G^{-1}]^{1/2}}{\delta\chi(\mathbf{x})_j} . \tag{4.6}$$

$[\det G^{-1}]$ denotes the determinant of G^{-1} (arising from $\int \prod_{i=1}^{N-1}[\mathrm{d}\pi_i]$), both as a $(N-1) \times (N-1)$ matrix determinant and as a functional determinant. $D[1;\epsilon]$ is a positive constant. Equation (4.6) has the t-independent solution

$$W_{0,eq} = [\det G^{-1}]^{-1/2} \exp\left[-\beta_{eq} \int \mathrm{d}^3\mathbf{x}\, V_1 \right] ,$$

to which W_0 relaxes (at least, formally) for long times. For fixed ultraviolet cut-off, the analysis in subsect. 3.1 could apply in principle, but $[\det G^{-1}]$, $W_{0,eq}$ and (4.6) would require a detailed regularization and study, through techniques related to those employed for the quantum nonlinear σ model [8]. See [10] for the dynamics of a relaxational nonlinear σ model, for cooperative phenomena.

5 Conclusion and discussion

We have treated a nonequilibrium meson gas, through quasi-classical effective fields, as a caricature of the pion gas produced in a heavy-ion collision, say, between the phase transitions and "freeze-out". For simplicity, we started out with one degree of freedom and, later, turned to an infinite number of degrees of freedom: a scalar field and nonlinear chiral fields. At some initial time t_0, the meson gas is out of global thermal equilibrium. Neither transport theory nor Kubo formulae have been invoked. The temporal evolution of the gas has been analyzed through three-term linear hierarchies for moments of probability distributions. They provide an adequate framework to infer that, due to the infinite number of degrees of freedom involved and under suitable long-time approximations: a) higher-order moments follow adiabatically the

dynamics of the lower ones, and b) lower-order moments drive the relaxation towards approximate global thermal equilibrium. Not far from global thermal equilibrium, certain Hermiticity and positivity properties of the operators $D[n;\epsilon]$ play a crucial role. Those techniques seem, so far, less efficient for direct quantitative estimates of relaxation times: for this reason, one would require explicit approximations for the $D[n;\epsilon]$'s, outside our scope here.

In [3], the nonequilibrium quantum generating functional Z associated to (3.1)–(3.3) has been analyzed for long times, with several quasi-classical assumptions and approximations. Then, Z becomes approximately the nonequilibrium generating functional for a purely dissipative quasi-classical functional Fokker-Planck process, for a suitable order parameter. The nonperturbative quasi-classical and long-time approximations in sect. 3 are quite different from those in [3], but both [3] and the present work lead, essentially and consistently, to equivalent Fokker-Planck dynamics. The time evolution and long-time thermalization in classical-field theories has been investigated by other different methods: see [4,11–15].

The author acknowledges the financial support of Ministerio de Educacion y Ciencia (Project FPA2004-02602), Spain. He is grateful to Mr D. Fernandez-Fraile and Drs A. Gomez Nicola and F.J. Llanes-Estrada, for discussions on Thermal RQFT and heavy-ion collisions, and to Dr G.F. Calvo about the orthogonalized polynomials for (4.2). He thanks Dr F.J. Llanes-Estrada for a critical reading of the manuscript and kind facilities.

References

1. D. Boyanovsky, *Phase transitions in the early and the present universe: from the Big Bang to heavy ion collisions*, arXiv:hep-ph/0102120 v2 21 Feb. (2001).
2. U. Heinz, *Concepts in heavy-ion physics*, arXiv:hep-ph/0407360 v1 30 July (2004).
3. K.-C. Chou, Z.-B. Su, B.-L. Hao, L. Yu, Phys. Rep. **118**, 1 (1985).
4. J. Berges, *Introduction to nonequilibrium quantum field theory*, arXiv:hep-ph/0409233 v1 20 Sep. (2004).
5. E. Wigner, Phys. Rev. **40**, 749 (1932).
6. R. Kubo, M. Toda, N. Hashitsume, *Statistical Mechanics II*, second edition (Springer, Berlin, 1998).
7. R.F. Alvarez-Estrada, Ann. Phys. (Leipzig) **11**, 357 (2002).
8. J. Zinn-Justin, *Quantum Field Theory and Critical Phenomena*, fourth edition (Clarendon Press, Oxford, 2002).
9. A. Munoz Sudupe, R.F. Alvarez-Estrada, J. Phys. A: Math. Gen. **16**, 3049 (1983).
10. R. Bausch, H.K. Janssen, Y. Yamazaki, Z. Phys. B **37**, 163 (1980).
11. G. Aarts, G.F. Bonini, C. Wetterich, Nucl. Phys. B **587**, 403 (2000).
12. G.F. Bonini, C. Wetterich, Phys. Rev. D **60**, 105026 (1999).
13. F. Cooper, A. Khare, H. Rose, Phys. Lett. B **515**, 463 (2001).
14. G. Aarts, J. Smit, Nucl. Phys. B **511**, 451 (1998).
15. W. Buchmuller, A. Jakovac, Phys. Lett. B **407**, 39 (1997).

Eur. Phys. J. A **31**, 766–768 (2007)
DOI 10.1140/epja/i2006-10238-x

Special Article – QNP 2006

Behavior of the topological susceptibility at finite T and μ and signs of restoration of chiral symmetries

M.C. Ruivo[a], P. Costa, and C.A. de Sousa

Centro de Física Teórica, Departamento de Física, Universidade de Coimbra, P-3004-516 Coimbra, Portugal

Received: 8 November 2006
Published online: 12 March 2007 – © Società Italiana di Fisica / Springer-Verlag 2007

Abstract. We investigate the possible restoration of chiral and axial symmetries across the phase transition at finite temperature and chemical potential, by analyzing the behavior of several physics quantities, such as the quark condensates and the topological susceptibility, the respective derivatives with respect to the chemical potential, and the masses of meson chiral partners. We discuss whether only chiral symmetry or both chiral and axial symmetries are restored and what the role of the strange quark is. The results are compared with recent lattice results.

PACS. 11.30.Rd Chiral symmetries – 11.10.Wx Finite-temperature field theory – 14.40.Aq π, K, and η mesons

Understanding the rich content of the QCD phase diagram is a major challenge nowadays. Phase transitions, associated to deconfinement, restoration of chiral and axial $U_A(1)$ symmetries are expected to occur at high density and/or temperature. A question that has attracted a lot of attention is whether these phase transitions take place simultaneously and which observables could signal their occurrence.

The $U_A(1)$ symmetry is explicitly broken at the quantum level by the axial anomaly, that may be described at the semiclassical level by instantons, giving a mass to η' in the chiral limit, which implies that, in the real world, this meson is not a remnant of a Goldstone boson. The $U_A(1)$ anomaly causes flavor mixing, which has the effect of lifting the degeneracy between several mesons. So, the effective restoration of this symmetry should have relevant consequences on the meson masses as well as on the phenomenology of meson mixing angles. In particular, the η' mass should decrease and this meson should degenerate with other Goldstone bosons. There are several reasons to expect that the singlet axial symmetry might be restored. In fact, large instantons are supposed to be suppressed at high densities or temperatures, and interactions between instantons contribute to eliminate fluctuations of the topological charge, which implies that the effects of the anomaly could disappear [1]. The topological susceptibility, χ, is related to the η' mass through the Witten-Veneziano formula, and the behavior of χ and its slope are relevant to understand the possible restoration of the $U_A(1)$ symmetry [2–5].

The topological susceptibility is defined as

$$\chi = \int \mathrm{d}^4 x \, \langle T\{Q(x)Q(0)\}\rangle, \tag{1}$$

where $Q(x)$ is the topological charge density. Several lattice calculations (see [6] and references therein) indicate a sharp decrease of this quantity with temperature at zero density. Model calculations also give indications in this direction. Chiral models include an anomaly term that breaks the $U_A(1)$ symmetry. In order to simulate the fate of the anomaly it is usually assumed that the anomaly coefficient is a dropping function of temperature, whether the approach is phenomenological [7] or lattice inspired [3–5]. Recent lattice calculations with two colors and eight flavors [8] show that, at a fixed T, and varying the chemical potential μ, a critical μ is found, where the quark condensate and the topological susceptibility drop and the Polyakov loop rises; its derivatives vary sharply. The topological susceptibility and the quark condensate go to zero at much higher μ.

Another subject that deserves attention is the role played by the strange quark regarding the restoration of chiral and axial symmetries. In fact, chiral symmetry could still be broken for strange quarks, while it is already restored for light quarks. Lattice calculations show that the critical temperature for light $\langle \bar{q}q \rangle$ is lower than for $\langle \bar{s}s \rangle$ [9], a result also found in chiral perturbation theory and model calculations [10,11]. However the critical temperatures could become closer at high chemical potential [11].

Here we report a study on the behavior of the topological susceptibility at finite temperature, T, and baryonic chemical potential, $\mu_B = \mu_q = \mu_s$, and the possible

---

[a] e-mail: `maria@teor.fis.uc.pt`

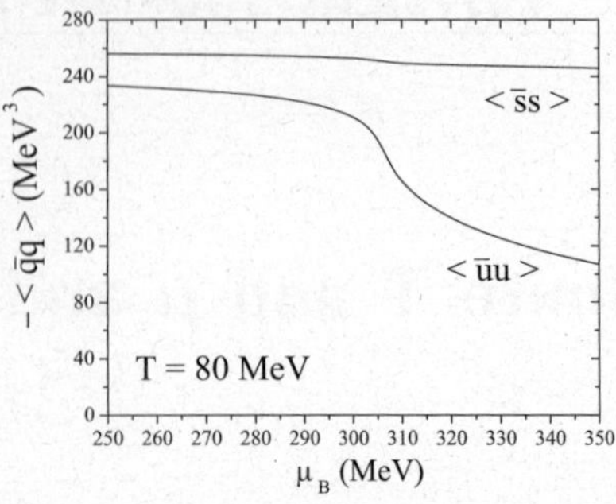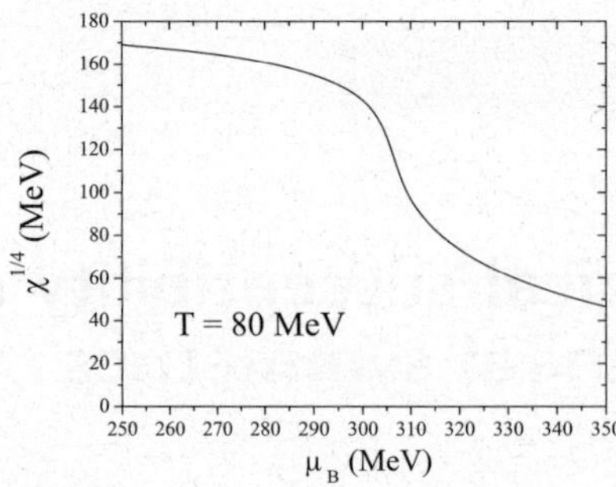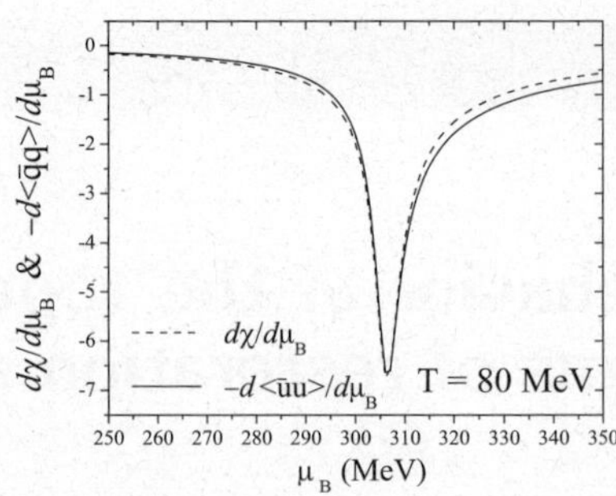

Fig. 1. Quark condensates (left panel), the topological susceptibility (middle panel) and the respective derivatives (right panel) (Case I).

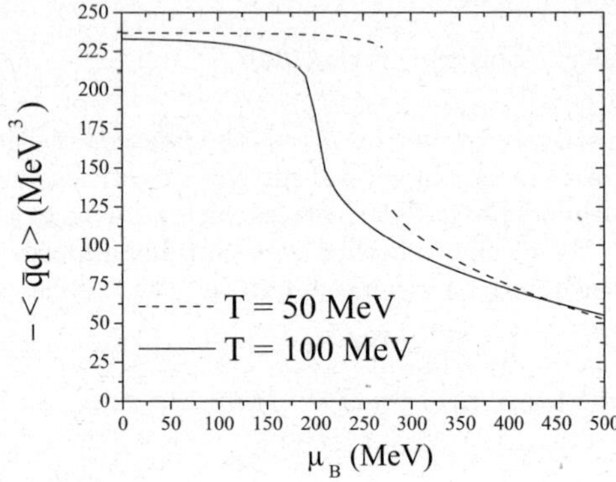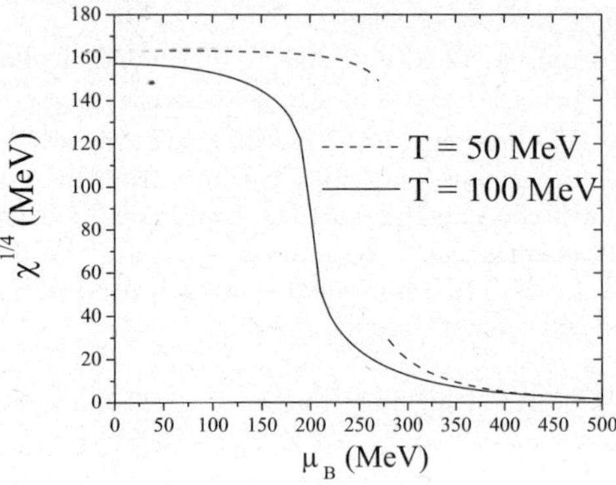

Fig. 2. Quark condensates (left panel) and topological susceptibility (right panel), above the CEP (Case II).

restoration of chiral and axial symmetries, discussing, in particular, the role played by the strange quark in this concern.

We perform our calculations in the framework of the three-flavor NJL model, including the determinantal 't Hooft interaction that breaks the $U_A(1)$ symmetry, which has the following Lagrangian:

$$\mathcal{L} = \bar{q}\left(i\partial \cdot \gamma - \hat{m}\right)q + \frac{g_S}{2}\sum_{a=0}^{8}\left[\left(\bar{q}\lambda^a q\right)^2 + \left(\bar{q}(i\gamma_5)\lambda^a q\right)^2\right]$$

$$+ g_D\left[\det\left[\bar{q}(1+\gamma_5)q\right] + \det\left[\bar{q}(1-\gamma_5)q\right]\right]. \quad (2)$$

Here $q = (u, d, s)$ is the quark field with $N_f = 3$ and $N_c = 3$. $\hat{m} = \mathrm{diag}(m_u, m_d, m_s)$ is the current quark mass matrix and λ^a are the Gell-Mann matrices, $a = 0, 1, \ldots, 8$, $\lambda^0 = \sqrt{\frac{2}{3}}\,\mathbf{I}$. The model is fixed by the coupling constants g_S, g_D, the cutoff parameter Λ, which regularizes the divergent integrals, and the current quark masses m_i. We use the parameter set: $m_u = m_d = 5.5\,\mathrm{MeV}$, $m_s = 140.7\,\mathrm{MeV}$, $g_S\Lambda^2 = 3.67$, $g_D\Lambda^5 = -12.36$ and $\Lambda = 602.3\,\mathrm{MeV}$ (for details see [12]).

In a previous work [5], we studied the possible effective restoration of axial symmetry, at finite temperature and zero chemical potential, by modeling the anomaly coefficient as a Fermi function from lattice results for χ; a similar approach was extrapolated for quark matter at zero temperature and finite density. For the sake of comparison with our present study, we summarize here the main conclusions for $\mu_B = 0$ and $T \neq 0$. It was found that at $T \approx 250\,\mathrm{MeV}$ the chiral partners (π^0, σ) and (a_0, η) become degenerate, which is a manifestation of the effective restoration of $SU(2)$ chiral symmetry; at $T \approx 350\,\mathrm{MeV}$, $\chi \to 0$, the pair (a_0, σ) becomes degenerate with (π^0, η) and the mixing angles go to the ideal values, indicating an effective restoration of axial symmetry. The strange-quark condensate decreases slightly but, in the range of temperatures considered, chiral symmetry in the strange sector remains broken. The mesons η' and f_0 become purely strange and do not show a tendency to converge. Therefore there is no restoration of the full $U(3) \otimes U(3)$ symmetry.

We would like to discuss questions such as whether the restoration of chiral symmetry contributes to the restoration of the singlet axial symmetry $U_A(1)$, what is the role of the strange quark and the combined effect of finite T, μ on the restoration of symmetries. For this purpose we will study the relevant observables as functions of the chemical potential, $\mu_B = \mu_q = \mu_s$, at a fixed temperature (below/above the critical end point (CEP)) in two cases: Case I: $m_q \neq m_s$; Case II: $m_q = m_s = 0.5\,\mathrm{MeV}$. g_D is kept constant in both cases. The main difference between results at a fixed T, below and above the CEP is that, in the first case, the observables present a discontinuity at the critical μ_B.

Concerning Case I, we plot in fig. 1 only the results above the CEP ($\mu_B = 318.5\,\mathrm{MeV}$, $T = 67.7\,\mathrm{MeV}$). It can be seen that there is a pronounced decrease of $\langle \bar{u}u \rangle$, but not of $\langle \bar{s}s \rangle$; $\frac{\partial \langle \bar{u}u \rangle}{\partial \mu_B}$, $\frac{\partial \chi}{\partial \mu_B}$ vary sharply at the same critical

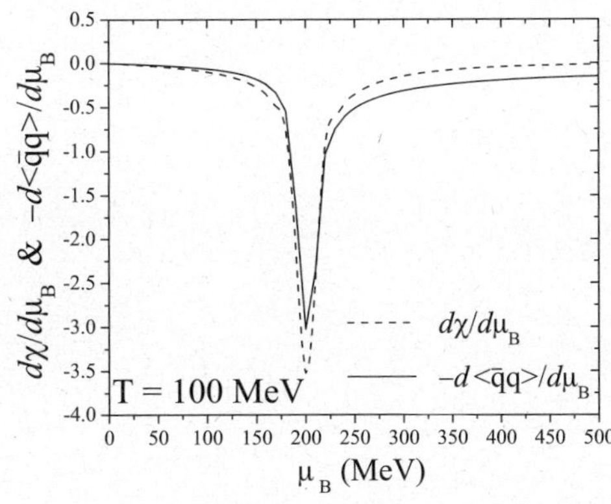
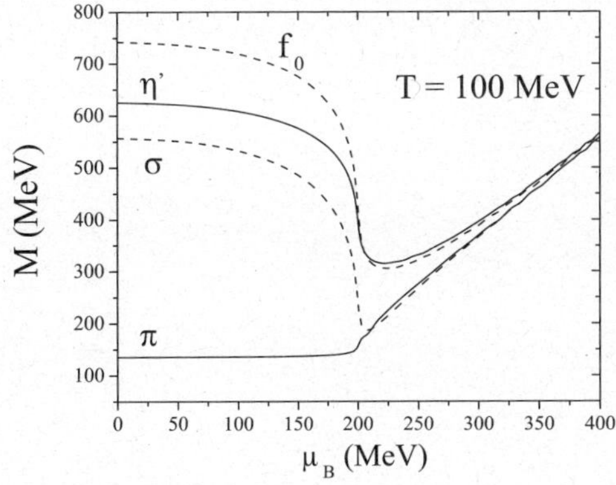

Fig. 3. Derivatives of $\langle \bar{q}q \rangle$ and χ (left panel) and meson masses (right panel), above the CEP (Case II).

μ_B. We interpret this result as indicating a phase transition associated to partial restoration of chiral $SU(2)$. Using as a criterion for the effective restoration of symmetries the convergence of the respective chiral partners, we verified that the effective restoration of chiral $SU(2)$ symmetry occurs at a higher μ_B, but there is no effective restoration of $U_A(1)$ symmetry: the respective chiral partners do not converge and χ does not vanish. So, when g_D is kept constant, the combined effect of finite T and μ_B is not sufficient to lead to effective restoration of axial symmetry.

In order to compare our results with those from lattice calculations [8] and to discuss the role of the strange quark, we will consider Case II (CEP: $\mu_B = 219$ MeV, $T = 87$ MeV). Since all the quark masses are equal, a_0 is degenerate with σ and η with π^0, even in the vacuum.

In fig. 2 we see that the quark condensate and the topological susceptibility have a behavior that compares well, at a qualitative level, with the results of [8], as well as the derivatives $\frac{\partial \langle \bar{q}q \rangle}{\partial \mu_B}$, $\frac{\partial \chi}{\partial \mu_B}$, that vary sharply at the same critical μ_B ($\mu_B \simeq 250(200)$ MeV for $T \simeq 50(100)$ MeV) (fig. 3, left panel). This critical μ_B is certainly the onset of a new phase with partial restoration of chiral symmetry. The information from the mesonic behavior (fig. 3, right panel) shows that, differently from Case I, the effective and partial restorations of chiral symmetry occur at about the the same critical chemical potential (convergence of the chiral partners (π^0, σ)). However, the effective restoration of $U_A(1)$ symmetry takes place at a higher chemical potential ($\mu_B \simeq 350$ MeV for $T \simeq 100$ MeV): the pair (f_0, η') degenerates with (π^0, σ) and χ vanishes. For higher temperatures, the phase transition occurs earlier.

In conclusion, we have studied the behavior of the topological susceptibility across the phase transition and discussed its relation with the restoration of chiral symmetries. We studied the observables as functions of the chemical potential, $\mu_B = \mu_q = \mu_s$, at a fixed temperature

(below/above the critical end point (CEP)) for two cases. In Case I, the only way of obtaining effective restoration of axial symmetry is by assuming that g_D is a dropping function of $T(\mu)$. In Case II, we have effective restoration of axial symmetry with $g_D = $ const: χ vanishes and all the mesons converge; the difference between the results for the two cases is due to the behavior of the strange-quark condensate.

Work supported by grant SFRH/BPD/23252/2005 from FCT (P.C.).

References

1. T. Schaefer, hep-ph/0412215.
2. A. Di Giacomo, E. Meggiolaro, H. Panagopoulos, Phys. Lett. B **277**, 491 (1992).
3. J. Schafner-Bielich, Phys. Rev. Lett. **84**, 3261 (2000).
4. K. Fukushina, K. Ohnishi, K. Ohta, Phys. Rev. C **514**, 045203 (2001).
5. P. Costa, M.C. Ruivo, Y.L. Kalinovsky, C.A. de Sousa, Phys. Rev. D **70**, 116013 (2004) hep-ph/0408177; **71**, 116002 (2005) hep-ph/0503258.
6. B. Allés, M.D. Elia, A. Di Giacomo, Nucl. Phys. B **494**, 281 (1997).
7. R. Alkofer, P.A. Amundsen, H. Reinhardt, Phys. Lett. B **218**, 75 (1989); T. Kunihiro, Phys. Lett. B **219**, 363 (1989).
8. B. Allés, M.D. Elia, M.P. Lombardo, Nucl. Phys. B **752**, 124 (2006).
9. C. Bernard et al., Phys. Rev. D **71**, 034504 (2005).
10. J.R. Pelaez, Phys. Rev. D **66**, 096007 (2002).
11. A. Tawfik, D. Toublan, Phys. Lett. B **623**, 48 (2005).
12. P. Costa, M.C. Ruivo, Y.L. Kalinovsky, C.A. de Sousa, Phys. Rev. C **70**, 025204 (2004) hep-ph/0304025.

Eur. Phys. J. A **31**, 769–772 (2007)
DOI 10.1140/epja/i2006-10222-6

Special Article – QNP 2006

2 + 1 flavor simulations of QCD with improved staggered quarks

U.M. Heller[a]

On behalf of the MILC Collaboration
American Physical Society, One Research Road, Ridge, NY 11961-9000, USA

Received: 8 November 2006
Published online: 6 March 2007 – © Società Italiana di Fisica / Springer-Verlag 2007

Abstract. The MILC Collaboration has been performing realistic simulations of full QCD with 2+1 flavors of improved staggered quarks. Our simulations allow for controlled continuum and chiral extrapolations. I present results for the light pseudoscalar sector: masses and decay constants, quark masses and Gasser-Leutwyler low-energy constants. In addition I will present some results for heavy-light mesons, decay constants and semileptonic form factors, obtained in collaboration with the HPQCD and Fermilab lattice Collaborations. Such calculations will help in the extraction of CKM matrix elements from experimental measurements.

PACS. 12.38.Gc Lattice QCD calculations – 12.39.Fe Chiral Lagrangians – 12.15.Hh Determination of Kobayashi-Maskawa matrix elements – 13.20.Fc Decays of charmed mesons

1 Introduction

QCD simulations with the effects of three light quark flavors fully included have been a long-standing goal of lattice gauge theorists. The MILC Collaboration has made significant advances towards this goal, employing an improved staggered formalism, "asqtad" fermions [1]. This formalism, and the ever-increasing computational resources available today, have enabled simulations with the strange-quark mass near its physical value, and the up and down quarks, taken to be of equal mass, as light as 1/10 the strange-quark mass, corresponding to pion masses as low as 240 MeV. To control the extrapolation to the continuum limit, simulations have been done, and are still ongoing, at multiple lattice spacings, $a \approx 0.15$, 0.12, 0.09 and 0.06 fm. These are referred to as coarser, coarse, fine and super-fine lattices, respectively.

All the lattices generated by the MILC Collaboration are made available to other researchers at the "NERSC Gauge Connection". They have been widely used, for example by the LHPC and NPLQCD Collaborations for the study of nucleon properties, form factors and scattering lengths, and by the Fermilab lattice, HPQCD and UKQCD Collaborations for the study of heavy quarkonia and heavy-light mesons. The last three joined efforts with the MILC Collaboration to validate these QCD simulations. They compared selected, "gold-plated", *i.e.* well controlled, quantities computed on the lattice with their experimentally well-known values. Agreement within errors of 1–3% was found [2].

Such a validation is, of course, valuable for any lattice simulation. It is, in particular, needed for simulations with staggered fermions, which rely on the so-called "fourth-root trick" to eliminate the extra species of fermions, called "tastes", present with staggered fermions, and thus to generate the correct number of sea quark flavors. A nice review of the validity or possible problems with this fourth-root trick can be found in [3].

While the MILC Collaborations has made simulations with lighter up and down quark masses, m_l, than have been reached by other groups, simulations at the physical light quark mass are still too costly, even with the improved "asqtad" quarks. Therefore, extrapolations in the light-quark mass —chiral extrapolations— are still needed to reach the physical value. Such extrapolations are well understood, theoretically, based on chiral perturbation theory, χPT. For results from lattice QCD, it is important to include a treatment of lattice effects into the chiral extrapolations. This holds, in particular, for simulation results with staggered fermions, because of the taste symmetry-breaking effects. For this purpose, χPT was adapted to staggered quarks in staggered chiral perturbation theory, SχPT [4].

MILC's first set of accurate results for light pseudoscalars, based on two lattice spacings (coarse and fine, *i.e.* 0.12 and 0.09 fm), one simulation strange quark mass m_s' and several simulation light-quark masses $m_l'^{1}$, with the SχPT formalism for joint chiral and continuum extrapolations, was published in 2004 [5]. To obtain quark masses, mass renormalization constants are needed. The

---

[a] e-mail: `heller@aps.org`

[1] We denote simulation masses by a prime, m'.

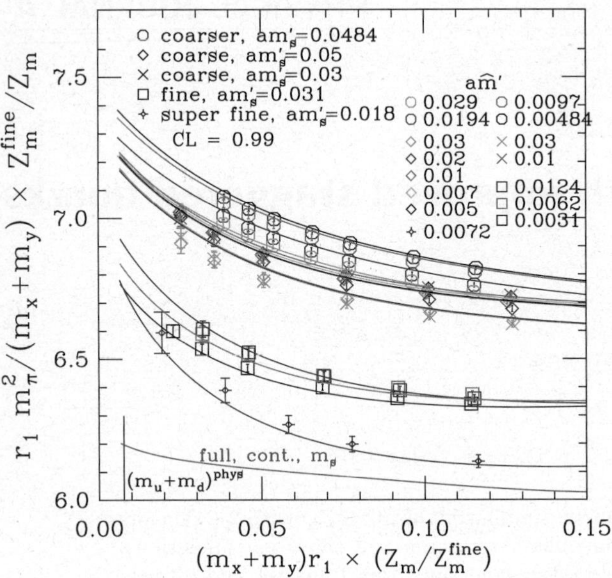

Fig. 1. (Color on-line) The square of the pion mass divided by the sum of the valence quark masses as a function of the sum of the valence quark masses in units of r_1.

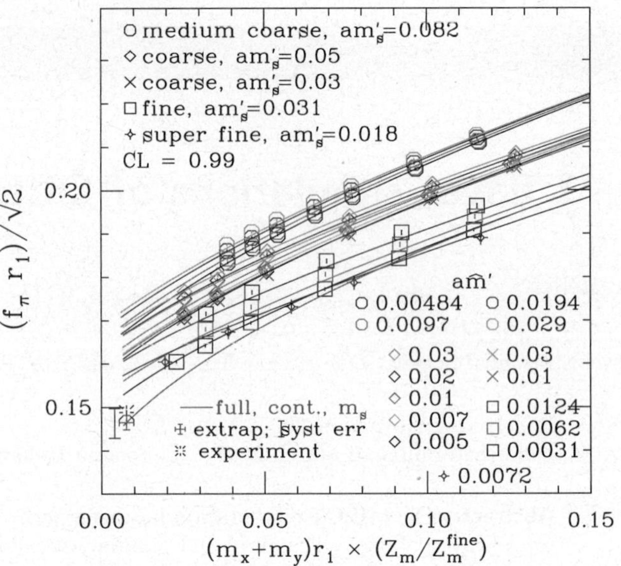

Fig. 2. (Color on-line) The pion decay constant as a function of the sum of the valence quark masses in units of r_1.

values with one-loop Z-factors, computed by members of the HPQCD and UKQCD Collaborations, appeared in [6].

Since 2004, our simulations expanded in several ways. On the coarse, $a = 0.12\,\mathrm{fm}$, lattice a second simulation strange-quark mass m_s' has been used, allowing interpolation to the physical strange-quark mass. On the fine, $a = 0.09\,\mathrm{fm}$, lattice a simulation with a lighter light-quark mass, $m_l' \simeq 0.1 m_s$, has been done. A coarser lattice ensemble, with $a = 0.15\,\mathrm{fm}$, has been added, increasing the lever arm for continuum extrapolations. And, finally, a superfine-lattice ensemble, with $a = 0.06\,\mathrm{fm}$, has been started. A run with $m_l' = 0.4 m_s'$, corresponding to $m_\pi \approx 430\,\mathrm{MeV}$ is half-finished, and lighter-mass simulations have begun.

2 The light pseudoscalar sector

The SχPT fitting is illustrated in figs. 1 and 2. Dimensionful quantities are given in units of $r_1 = 0.318(7)\,\mathrm{fm}$, a scale related to the heavy-quark potential [7]. The quark masses are renormalized (at one loop) relative to those on the fine lattice. The data is accurate enough and at small enough quark masses so that the effects of chiral logs are evident. The red on-line lines are the fit functions in "full continuum QCD" (valence and sea quark masses set equal) after extrapolation of parameters to the continuum limit.

Preliminary numerical results for f_π and f_K obtained from our most recent data are[2]

$$f_\pi = 128.6 \pm 0.4 \pm 3.0\,\mathrm{MeV} \quad [129.5 \pm 0.9 \pm 3.5\,\mathrm{MeV}], \quad (1)$$

$$f_K = 155.3 \pm 0.4 \pm 3.1\,\mathrm{MeV} \quad [156.6 \pm 1.0 \pm 3.6\,\mathrm{MeV}], \quad (2)$$

$$f_K/f_\pi = 1.208(2)\left(^{+7}_{-14}\right) \quad [1.210(4)(13)]. \quad (3)$$

---

[2] The numbers here are updated as of Lattice 2006 [8].

Here the numbers on the left are the new values, and those on the right in square brackets are from ref. [5]. In each case the first error is statistical, and the second systematic.

The lattice QCD value of f_K/f_π can be combined with experimental data to extract the CKM matrix element V_{us} [9]. We obtain $|V_{us}| = 0.2223(^{+26}_{-14})$, with an accuracy comparable to, and a value compatible with, the latest (2006) PDG value, $V_{us} = 0.2257(21)$ [10].

The up-, down- and strange-quark masses can be determined from the masses of the π and K mesons using the SχPT fits. To relate these masses to the experimental ones requires some continuum input for isospin breaking and electromagnetic effects [11]. Details can be found in ref. [5]. Preliminary results from our current data set, evaluated at scale $\mu = 2\,\mathrm{GeV}$, are

$$m_s^{\overline{\mathrm{MS}}} = 90(0)(5)(4)(0)\,\mathrm{MeV} \quad [76(0)(3)(7)(0)\,\mathrm{MeV}], \quad (4)$$

$$m_u^{\overline{\mathrm{MS}}} = 2.0(0)(1)(1)(1)\,\mathrm{MeV} \quad [1.7(0)(1)(2)(2)\,\mathrm{MeV}], \quad (5)$$

$$m_d^{\overline{\mathrm{MS}}} = 4.6(0)(2)(2)(1)\,\mathrm{MeV} \quad [3.9(0)(1)(4)(2)\,\mathrm{MeV}], \quad (6)$$

$$m_s/m_l = 27.2(0)(4)(0)(0) \quad [27.4(1)(4)(0)(1)], \quad (7)$$

$$m_u/m_d = 0.42(0)(1)(0)(4) \quad [0.43(0)(1)(0)(8)]. \quad (8)$$

Again, new results are on the left, and earlier ones from refs. [5,6] are in square brackets on the right. Errors are from statistics, simulation systematics, perturbation theory and electromagnetic effects. The main difference between the new and old results comes from the use of a two-loop mass renormalization constant [12], compared to the one-loop one used previously. A non-perturbative mass renormalization calculation is in progress. For further results, including Gasser-Leutwyler low-energy constants, see ref. [8].

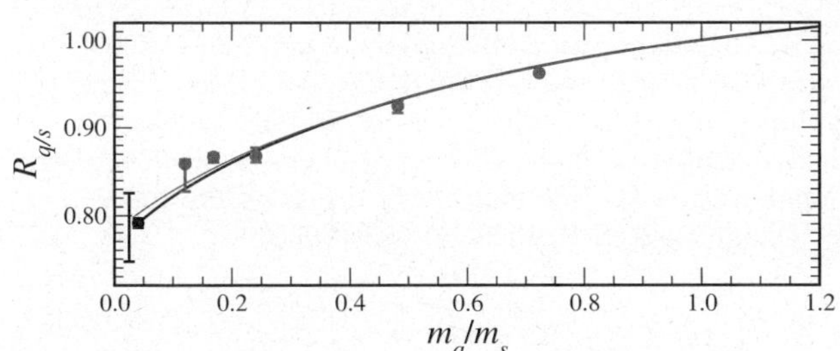

Fig. 3. (Color on-line) The ratio $R_{q/s} = f_D \sqrt{m_D}/f_{D_s}\sqrt{m_{D_s}}$ as function of m_q/m_s together with the SχPT fit. The black line and extrapolated point are obtained after removing $\mathcal{O}(a^2)$ effects from the fit, representing a continuum estimate.

3 Heavy-light meson physics

In a joint effort with the Fermilab lattice and HPQCD Collaborations, we used our full QCD ensembles to study properties of heavy-light mesons. For the heavy c and b quarks we used clover fermions with the Fermilab interpretation [13], while "asqtad" fermions where used for the light valence quarks. This allows to go much closer to the physical light-quark mass than was achieved previously. Use of SχPT, adopted to heavy-light mesons [14], makes the necessary remaining chiral extrapolation well controlled, as illustrated in fig. 3.

We found the lattice predictions [15,16]

$$f_{D_s} = 249 \pm 3 \pm 16\,\text{MeV}, \tag{9}$$

$$f_D = 201 \pm 3 \pm 17\,\text{MeV}, \tag{10}$$

$$f_{D_s}/f_D = 1.21 \pm 0.01 \pm 0.04, \tag{11}$$

compared to the later experimentally measured values from leptonic decays [17,18]

$$f_{D_s} = 283 \pm 17 \pm 7 \pm 14\,\text{MeV}, \tag{12}$$

$$f_{D^+} = 222.6 \pm 16.7^{+2.8}_{-3.4}\,\text{MeV}. \tag{13}$$

With the Fermilab lattice and HPQCD Collaborations, we are also computing form factors for semileptonic $D \to \pi/K$ and $B \to \pi/D$ decays. The heavy-to-light decay amplitudes are parametrized as

$$\langle P|V^\mu|H\rangle = f_+(q^2)(p_H + p_P - \Delta)^\mu + f_0(q^2)\Delta^\mu, \tag{14}$$

where $\Delta^\mu = (m_H^2 - m_P^2)q^\mu/q^2$. We can compare our calculation of $f_+^{(K)}(q^2)$ for $D^0 \to K^- l^+ \nu$ [19] with the recent measurement by Belle [20] (see fig. 4).

4 Conclusions

The MILC Collaboration's QCD simulations with $2+1$ flavors have been designed to have all errors, in particular from chiral and continuum extrapolations, well controlled. For many quantities they have reached hitherto unprecedented accuracy, so that the lattice QCD calculations are starting to have an impact on current experimental physics programs. For example, using only our lattice

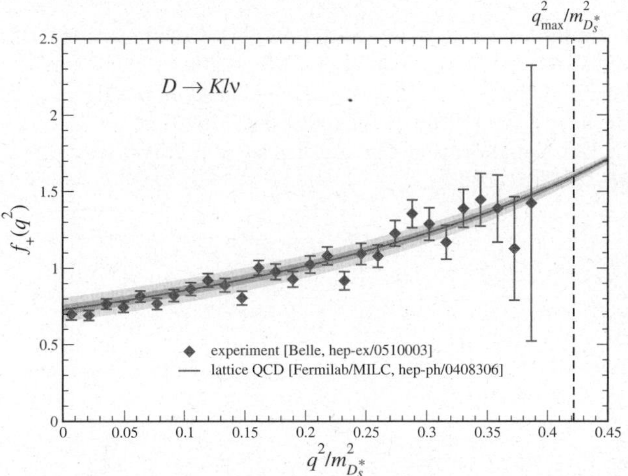

Fig. 4. (Color on-line) The form factor $f_+^{(K)}(q^2)$ for $D^0 \to K^- l^+ \nu$ compared to recent measurements by Belle [20].

calculation for f_K/f_π as a theoretical input, the CKM matrix V_{us} can be determined with an accuracy comparable to the latest PDG result. A determination of the entire CKM matrix with lattice QCD results based on the MILC ensembles of gauge configurations as the only theoretical input has been given in ref. [21].

References

1. MILC Collaboration (T. Blum *et al.*), Phys. Rev. D **55**, 1133 (1997); K. Orginos, D. Toussaint, Phys. Rev. D **59**, 014501 (1999); K. Orginos, D. Toussaint, R.L. Sugar, Phys. Rev. D **60**, 054503 (1999); G.P. Lepage, Phys. Rev. D **59**, 074502 (1999); J.F. Lagäe, D.K. Sinclair, Nucl. Phys. (Proc. Suppl.) **63**, 892 (1998); MILC Collaboration (C. Bernard *et al.*), Phys. Rev. D **58**, 014503 (1998).
2. Fermilab Lattice, HPQCD and MILC Collaborations (C.T.H Davies *et al.*), Phys. Rev. Lett. **92**, 022001 (2004).
3. S. Sharpe, PoS (LAT2006) 022 (2006).
4. C. Bernard, Phys. Rev. D **65**, 054031 (2002); C. Aubin, C. Bernard, Phys. Rev. D **68**, 034014; 074011 (2003).
5. MILC Collaboration (C. Aubin *et al.*), Phys. Rev. D **70**, 114501 (2004).
6. HPQCD, MILC and UKQCD Collaborations (C. Aubin *et al.*), Phys. Rev. D **70**, 031504(R) (2004).
7. MILC Collaboration (C. Bernard *et al.*), Phys. Rev. D **62**, 034503 (2000).
8. MILC Collaboration (C. Bernard *et al.*), PoS (LAT2006) 163 (2006).
9. W.J. Marciano, Phys. Rev. Lett. **93**, 231803 (2004).
10. W.M. Yao *et al.*, J. Phys. G **33**, 1 (2006).
11. J. Bijnens, J. Prades, Nucl. Phys. B **490**, 239 (1997); J.F. Donoghue, A.F. Perez, Phys. Rev. D **55**, 7075 (1997); B. Moussallam, Nucl. Phys. B **504**, 381 (1997).
12. Q. Mason, H. Trottier, R. Horgan, PoS (LAT2005) 011 (2005); Q. Mason *et al.*, Phys. Rev. D **73**, 114501 (2006).
13. A.X. El-Khadra, A.S. Kronfeld, P.B. Mackenzie, Phys. Rev. D **55**, 3933 (1997).
14. C. Aubin, C. Bernard, Nucl. Phys. B (Proc. Suppl.) **140**, 491 (2005); Phys. Rev. D **73**, 014515 (2006).

15. Fermilab lattice, MILC and HPQCD Collaborations (C. Aubin *et al.*), Phys. Rev. Lett. **95**, 122002 (2005).
16. Fermilab lattice, MILC and HPQCD Collaborations (C. Bernard *et al.*), PoS (LAT2006) 094 (2006).
17. CLEO Collaboration (M. Artuso *et al.*), Phys. Rev. Lett. **95**, 251801 (2005).
18. BABAR Collaboration (B. Aubert *et al.*), hep-ex/0607094.
19. Fermilab lattice, MILC and HPQCD Collaborations (C. Aubin *et al.*), Phys. Rev. Lett. **94**, 011601 (2005).
20. Belle Collaboration (K. Abe *et al.*), hep-ex/0510003; Belle Collaboration (L. Widhalm *et al.*), hep-ex/0604049.
21. M. Okamoto, PoS (LAT2005) 013 (2005).

Eur. Phys. J. A **31**, 773–776 (2007)

DOI 10.1140/epja/i2006-10245-y

Special Article – QNP 2006

Lattice QCD with a twisted mass term and a strange quark

A.M. Abdel-Rehim[1,a], R. Lewis[1,b], R.M. Woloshyn[2], and J.M.S. Wu[2]

[1] Department of Physics, University of Regina, Regina, SK, S4S 0A2, Canada
[2] TRIUMF, 4004 Wesbrook Mall, Vancouver, BC, V6T 2A3, Canada

Received: 23 November 2006
Published online: 12 March 2007 – © Società Italiana di Fisica / Springer-Verlag 2007

Abstract. There are three quarks with masses at or below the characteristic scale of QCD dynamics: up, down and strange. However, twisted mass lattice QCD relies on quark doublets. Various options for including three quark flavors within the twisted mass approach are explored by studying the kaon masses, both analytically (through chiral Lagrangians) and numerically (through lattice simulations). Advantages and disadvantages are revealed for each "strange and twisted" option.

PACS. 11.15.Ha Lattice gauge theory – 12.39.Fe Chiral Lagrangians – 14.40.Aq π, K, and η mesons

1 Two-flavor twisted mass lattice QCD

Consider a hypercubic lattice in four-dimensional space-time, and a two-flavor system:

$$\psi(x) = \begin{pmatrix} u(x) \\ d(x) \end{pmatrix}. \tag{1}$$

In the $m_u = m_d \equiv m_q$ limit, the fermion action for twisted mass lattice QCD is

$$S_{\text{fermion}} = a^4 \sum_x \bar{\psi}(x) \left[D(\omega) + m_q\right] \psi(x), \tag{2}$$

with Dirac and Wilson terms,

$$D(\omega) = \gamma \cdot \nabla + \exp(-i\omega\gamma_5\tau_3)W, \tag{3}$$

$$W\psi(x) \equiv M_{cr}\psi(x) - \frac{a}{2}\Box\,\psi(x). \tag{4}$$

As usual, the covariant derivative ∇_μ and d'Alembertian $\Box$ contain the gauge field $U_\mu(x) \equiv \exp(iagT^b A_\mu^b(x))$.

Two key features of twisted mass lattice QCD are a) the existence of a firm lower bound for the eigenvalues of $D(\omega)+m_q$, as long as quark mass m_q and twist angle ω are nonzero [1] and b) the absence of $O(a)$ errors in observable quantities at maximal twist [2], $\omega(\text{renormalized}) = \pm\pi/2$. Another significant consequence of $\omega \neq 0$ is that parity and flavor symmetries are broken; they only get restored in the continuum limit.

[a] *Present address:* Department of Physics, Baylor University, Waco, TX, 76798-7316, USA.
[b] Spokesperson; e-mail: `randy.lewis@uregina.ca`

2 Options for including the strange quark

Given the phenomenological importance of u, d, s physics, how should the strange quark be added to the two-flavor twisted mass theory of the previous section?

One option is to leave the strange quark untwisted. It will not be protected from near-zero modes, but this is not a practical problem for simulations at the physical strange-quark mass. There will be $O(a)$ errors in the theory, but one could remove their effects on observables through addition of a Sheikholeslami-Wohlert term [3]. Kaon operators, containing one twisted and one untwisted fermion, would appear as parity mixtures, requiring extra effort for the extraction of continuum physics.

Another option is to twist the strange quark by using the charm quark as its twisting partner. To accommodate $m_c \neq m_s$, the action for this "heavy" doublet will need one additional term relative to the "light" doublet action used in the previous section for up and down. The quark mass terms in the twisted basis are

$$\begin{aligned}
\mathcal{L} &= \bar{\psi}_l \left(m_l + i\gamma_5\mu_l\tau_3\right)\psi_l \\
&\quad + \bar{\psi}_h \left(m_h + i\gamma_5\mu_h\tau_3 + \epsilon\tau_a\right)\psi_h,
\end{aligned} \tag{5}$$

where τ_a is a Pauli matrix in flavor space. Should we choose $\tau_a = \tau_3$ as in [4] or $\tau_a \neq \tau_3$ as in [5]? An advantage of the parallel choice, *i.e.* $\tau_a = \tau_3$, is that flavors do not mix, but an important disadvantage is that the fermion determinant is not real. Therefore this parallel choice corresponds to fermions that can be used as valence quarks but not as sea quarks. The perpendicular choice, $\tau_a \neq \tau_3$, leads to a real fermion determinant so those quarks can be both sea and valence, but the flavors in the action mix, as is evident from (5). To elucidate the flavor and isospin structure of this choice, we study kaon masses in this work.

The notation $(c,s)_\parallel$ and $(c,s)_\perp$ will respectively denote twisted doublets with the parallel and perpendicular choices for τ_a relative to τ_3. A degenerate doublet is denoted by $(u,d)_0$. The three scenarios to be explored are

scenario 1: $(u,d)_0 + s$ [$+c$ if desired],
scenario 2: $(u,d)_0 + (c,s)_\parallel$,
scenario 3: $(u,d)_0 + (c,s)_\perp$.

3 Kaon mass splittings and lattice artifacts

3.1 Chiral perturbation theory

The chiral Lagrangian for two twisted doublets is [6]

$$\mathcal{L}_{\chi PT} = \mathcal{L}^{(2)} + \mathcal{L}^{(4)} + \dots , \tag{6}$$

$$\mathcal{L}^{(2)} = \frac{f^2}{4}\,\mathrm{Tr}(D_\mu \Sigma D_\mu \Sigma^\dagger) - \frac{f^2}{4}\,\mathrm{Tr}(\chi^\dagger \Sigma + \Sigma^\dagger \chi)$$

$$-\frac{f^2}{4}\,\mathrm{Tr}(\hat{A}^\dagger \Sigma + \Sigma^\dagger \hat{A}), \tag{7}$$

$$\mathcal{L}^{(4)} = -W_8'\,\mathrm{Tr}\left[(\hat{A}^\dagger \Sigma + \Sigma^\dagger \hat{A})^2\right] + \dots , \tag{8}$$

where

$$\Sigma = \begin{pmatrix} e^{i\omega_l \tau_3/2} & 0 \\ 0 & e^{i\omega_h \tau_3/2} \end{pmatrix} e^{i\Phi/f} \begin{pmatrix} e^{i\omega_l \tau_3/2} & 0 \\ 0 & e^{i\omega_h \tau_3/2} \end{pmatrix} \tag{9}$$

for meson matrix Φ. The quark mass matrix appears as

$$\chi = 2B \begin{pmatrix} m_l + i\mu_l \tau_3 & 0 \\ 0 & m_h + i\mu_h \tau_3 + \epsilon_h \tau_a \end{pmatrix}, \tag{10}$$

and the lattice spacing as $\hat{A} = 2W_0 a$, using notation familiar from [7–9].

To interpret $(u,d)_0 + (c,s)_\perp$, we diagonalize the quark mass matrices,

$$m_l + i\mu_l \tau_3 = \sqrt{m_l^2 + \mu_l^2}\, e^{i\tau_3 \omega_l}, \tag{11}$$

$$m_h + i\mu_h \tau_3 + \epsilon_h \tau_1 = e^{i\tau_3 \omega_h/2} Y^\dagger M Y e^{i\tau_3 \omega_h/2}, \tag{12}$$

where $Y = \frac{1}{\sqrt{2}} \begin{pmatrix} 1 & 1 \\ -1 & 1 \end{pmatrix}$ is a standard $\pi/4$ rotation, and the diagonal matrix M is

$$M = \begin{pmatrix} \sqrt{m_h^2 + \mu_h^2} + \epsilon_h & 0 \\ 0 & \sqrt{m_h^2 + \mu_h^2} - \epsilon_h \end{pmatrix}. \tag{13}$$

The resulting mass terms for kaons and D-mesons have the form

$$(\mathbb{K}^+\ \bar{\mathbb{D}}^0\ \mathbb{K}^0\ \mathbb{D}^-) \begin{pmatrix} m-\delta & \alpha & 0 & 0 \\ \alpha & m+\delta & 0 & 0 \\ 0 & 0 & m-\delta & -\alpha \\ 0 & 0 & -\alpha & m+\delta \end{pmatrix} \begin{pmatrix} \mathbb{K}^- \\ \mathbb{D}^0 \\ \bar{\mathbb{K}}^0 \\ \mathbb{D}^+ \end{pmatrix}, \tag{14}$$

where m, δ and α contain various parameters from the chiral Lagrangian. Here, mesons are labelled by their identities in the untwisted limit. They are clearly not mass

Table 1. Eigenvalues and eigenstates of the matrix in (14). Note that, for general θ, none of these mass eigenstates are isospin partners.

Mass eigenstates	Mass eigenvalues
$\psi_{u-} = \mathbb{K}^+ \cos\theta - \bar{\mathbb{D}}^0 \sin\theta$	
$\psi_{d-} = \mathbb{K}^0 \cos\theta + \mathbb{D}^- \sin\theta$	$m_{u-} = m_{d-} = m - \sqrt{\delta^2 + \alpha^2}$
$\psi_{u+} = \mathbb{K}^+ \sin\theta + \bar{\mathbb{D}}^0 \cos\theta$	
$\psi_{d+} = \mathbb{K}^0 \sin\theta - \mathbb{D}^- \cos\theta$	$m_{u+} = m_{d+} = m + \sqrt{\delta^2 + \alpha^2}$

eigenstates for nonzero twist, $\alpha \neq 0$. To determine meson eigenstates for a general twist, we diagonalize the matrix in (14) through a simple change of basis, and obtain the eigenstates and eigenvalues of table 1. The rotation (not twist!) angle, $0 \leq \theta < \pi/2$, is given by

$$\tan(2\theta) = \left|\frac{\alpha}{\delta}\right|, \tag{15}$$

where $\alpha \sim \sin\omega_l \sin\omega_h$ vanishes if either (u,d) or (c,s) is not twisted, and $\delta \sim \epsilon_h$ vanishes when (c,s) has no explicit mass splitting term.

Is there a correspondence between these eigenstates and the physical mesons? As a special case, consider $\alpha = 0$. This implies $\theta = 0$ and leads to scenario 1: $(u,d)_0 + s + c$. Quark flavors do not mix; the physical kaons are a (lighter) degenerate isospin doublet, and the physical D-mesons are a (heavier) degenerate isospin doublet. This scenario is also discussed in [10].

As a different special case, consider $\delta = 0$. This implies $\theta = \pi/4$ and leads to $(u,d)_0 + (c,s)_0$. Quark flavors can now be diagonalized, and mixings are thus avoided. In fact, this special case then becomes identical to $(u,d)_0 + (c,s)_\parallel$ with $m_c = m_s$. The mass-degenerate doublets are (K^+, D^-) and $(K^0, \bar{D}^0)$. The physical kaons only become degenerate in the continuum limit.

In general, when $\alpha \neq 0$ and $\delta \neq 0$, table 1 corresponds to $(u,d)_0 + (c,s)_\perp$. The light quark doublet has no flavor mixing, but the heavy doublet does. Interestingly, none of the mass eigenstates form isospin doublets unless isospin is somehow defined to involve (c,s) as well as (u,d). Our findings are sketched in fig. 1. Scenario 1 has no isospin splittings among kaons, and none among D-mesons either, but there are $O(a)$ splittings between the kaons and D-mesons. In scenario 2 all splittings begin at $O(a^2)$, but twist artifacts do cause unphysical $O(a^2)$ isospin splittings between charged and neutral kaons, and also between charged and neutral D-mesons. In scenario 3, flavors mix. There are two degenerate pairs of eigenstates, but the symmetry relating the degenerate states involves both quark doublets. Identification of these states with physical particles only becomes clear upon extrapolation to the continuum limit.

3.2 Lattice QCD simulations

According to the chiral Lagrangian discussion above, the splitting $m^2(K^0) - m^2(K^\pm)$ that arises from twist artifacts in the $(u,d)_0 + (c,s)_\parallel$ scenario should be, at leading order,

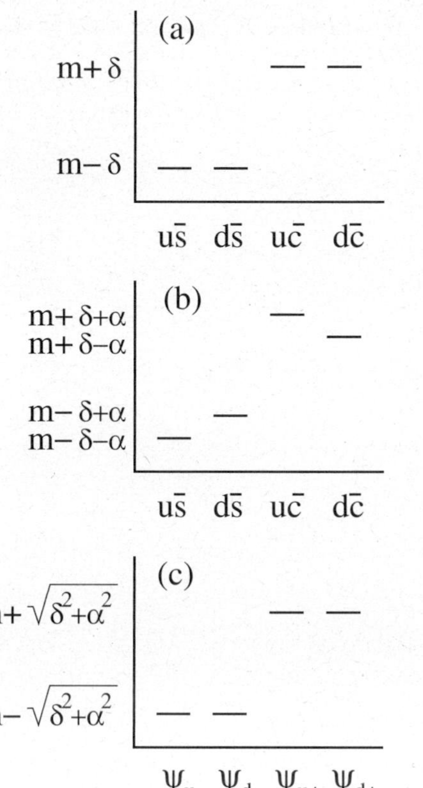

Fig. 1. Meson mass eigenstates obtained from three scenarios, (a) scenario 1: $(u,d)_0 + s + c$, (b) scenario 2: $(u,d)_0 + (c,s)_\parallel$, (c) scenario 3: $(u,d)_0 + (c,s)_\perp$.

i) independent of quark mass, ii) linear in a^2, and iii) vanishing as $a^2 \to 0$. Quenched numerical simulations for this scenario [6,11] are consistent with these expectations, though the vanishing of the mass splitting as $a^2 \to 0$ is not as clear in the data as one might have hoped. Even at the smallest lattice spacing, $a \approx 0.068\,\mathrm{fm}$, the kaon mass splitting is found to be sizeable: $m(K^0) - m(K^\pm) \sim 50\,\mathrm{MeV}$.

Reference [6] also contains quenched results for the $(u,d)_0 + s$ scenario, and these confirm the chiral Lagrangian claim of $m(K^0) = m(K^\pm)$, *i.e.* the kaon masses are not split by twist artifacts.

For scenario 3, $(u,d)_0 + (c,s)_\perp$, conventional isospin is only restored in the continuum limit. Therefore meson mass eigenstates must be identified by computing the full meson correlation matrix and then diagonalizing it numerically. With quark mass terms

$$
\mathcal{L} = (\bar{u}\;\bar{d}) \begin{pmatrix} m_l + i\gamma_5\mu_l & 0 \\ 0 & m_l - i\gamma_5\mu_l \end{pmatrix} \begin{pmatrix} u \\ d \end{pmatrix}
$$
$$
+ (\bar{q}_1\;\bar{q}_2) \begin{pmatrix} m_h + i\gamma_5\mu_h & \epsilon_h \\ \epsilon_h & m_h - i\gamma_5\mu_h \end{pmatrix} \begin{pmatrix} q_1 \\ q_2 \end{pmatrix} \tag{16}
$$

and operators $O_{ui} = \bar{u}\gamma_5 q_i$ and $O_{di} = \bar{d}\gamma_5 q_i$, numerical results give

$$
C_{ij}^{(u)} \equiv \left\langle O_{ui} O_{uj}^\dagger \right\rangle \Rightarrow C^{(u)} = \begin{pmatrix} a & c \\ c & b \end{pmatrix}, \tag{17}
$$

$$
C_{ij}^{(d)} \equiv \left\langle O_{di} O_{dj}^\dagger \right\rangle \Rightarrow C^{(d)} = \begin{pmatrix} b & c \\ c & a \end{pmatrix}, \tag{18}
$$

where a, b and c are real-valued. Diagonalization of $C^{(u)}$ and $C^{(d)}$ provides the meson mass eigenstates. Simulation results [12,13] are consistent with our chiral Lagrangian discussion.

4 Summary

Twisted mass lattice QCD removes $O(a)$ errors and unphysical near-zero modes. The up and down quarks fit naturally into a twisted doublet but the strange quark has no natural partner. Three scenarios for including the strange quark were considered within chiral perturbation theory, and found to have the following features:

1) $(u,d)_0 + s + c$
   - no unphysical mass splittings within isospin multiplets,
   - the strange quark does not benefit from twisting,
   - kaon operators (*i.e.*, twisted+untwisted) require care;
2) $(u,d)_0 + (c,s)_\parallel$
   - all quarks benefit from twisting,
   - suitable for valence quarks but not for sea quarks,
   - flavors do not mix and isospin is easily managed,
   - discretization effects appear as kaon mass splittings;
3) $(u,d)_0 + (c,s)_\perp$
   - all quarks benefit from twisting,
   - dynamical simulations can use a single action,
   - flavors mix; isospin is not straightforward for $a \neq 0$.

These features have also been observed in explicit numerical simulations, and may be helpful when choosing which scenario is optimal for a particular phenomenological application of lattice QCD.

The authors thank István Montvay for helpful communications. RL is grateful to the organizers of *QNP 2006* for the opportunity to participate in such an excellent conference. The work was supported in part by the Natural Sciences and Engineering Research Council of Canada, the Canada Foundation for Innovation, the Canada Research Chairs Program, and the Government of Saskatchewan.

References

1. Alpha Collaboration (R. Frezzotti, P.A. Grassi, S. Sint, P. Weisz), JHEP **0108**, 058 (2001).
2. R. Frezzotti, G.C. Rossi, JHEP **0408**, 007 (2004).
3. B. Sheikholeslami, R. Wohlert, Nucl. Phys. B **259**, 572 (1985).
4. C. Pena, S. Sint, A. Vladikas, JHEP **0409**, 069 (2004).
5. R. Frezzotti, G.C. Rossi, Nucl. Phys. B (Proc. Suppl.) **128**, 193 (2004).
6. A.M. Abdel-Rehim, R. Lewis, R.M. Woloshyn, J.M.S. Wu, Phys. Rev. D **74**, 014507 (2006).
7. G. Rupak, N. Shoresh, Phys. Rev. D **66**, 054503 (2002).
8. O. Bär, G. Rupak, N. Shoresh, Phys. Rev. D **67**, 114505 (2003).

9. O. Bär, G. Rupak, N. Shoresh, Phys. Rev. D **70**, 034508 (2004).
10. G. Münster, T. Sudmann, JHEP **0608**, 085 (2006).
11. A.M. Abdel-Rehim, R. Lewis, R.M. Woloshyn, Phys. Rev. D **71**, 094505 (2005).
12. F. Farchioni *et al.*, PoS (LAT2005) 072 (2006).
13. T. Chiarappa *et al.*, hep-lat/0606011.

Eur. Phys. J. A **31**, 777–780 (2007)

DOI 10.1140/epja/i2007-10005-7

Special Article – QNP 2006

Yukawa model on a lattice: Two-body states

F. De Soto[1,2,a], J. Carbonell[2], C. Roiesnel[3], Ph. Boucaud[4], J.P. Leroy[4], and O. Pène[4]

[1] Departamento de Sistemas Físicos Químicos y Naturales, Universidad Pablo de Olavide, 41013 Sevilla, Spain
[2] Laboratoire de Physique Subatomique et Cosmologie, 53 av. des Martyrs, 38026 Grenoble, France
[3] Centre de Physique Théorique Ecole Polytechnique, UMR7644, 91128 Palaiseau, France
[4] Laboratoire de Physique Théorique, UMR8627, Université Paris-XI, 91405 Orsay, France

Received: 8 January 2007
Published online: 22 March 2007 – © Società Italiana di Fisica / Springer-Verlag 2007

Abstract. We present some Quantum Field Theory (QFT) results concerning the Yukawa model, solved non-perturbatively with the help of lattice techniques. In particular we focus on the possibility of generating a two-nucleon bound state, as compared to the non-relativistic limit of the same model. Preliminary results show the appearance of zero modes of the Dirac operator. They limit the numerical solution of the model to values of the coupling constant which are too small to allow binding of the two-nucleon system.

PACS. 13.75.Cs Nucleon-nucleon interactions – 11.10.-z Field theory

1 Introduction

The nucleon-nucleon (NN) interaction is one of the most widely studied problems in theoretical physics. From meson exchange models [1, 2] to effective chiral Lagrangians [3], a huge effort has been devoted to developing suitable NN potentials that could reproduce the nuclear binding energies and scattering properties. Using Green's function Monte Carlo methods one can nowadays compute the nuclear spectrum up to ~ 12 nucleons [2].

Most potential models are inspired by an underlying QFT, from which only a very particular kind of diagrams is taken into account when solving the dynamical equations: in practice the solution is currently possible only in the ladder approximation.

For the Wick-Cutkosky (WC) model, all crossed-ladder graphs were summed up in [4], and the resulting binding energies are much bigger than the ones obtained within the ladder approximation. This strong disagreement is one of the most important motivations for the present work. As the WC model is not consistent as a field theory [5], we will study the simplest renormalizable QFT involving fermions, where one species of fermions interacts with a scalar meson via a Yukawa coupling.

The interest of this approach is manifold. On the one hand, it allows a comparison with the results of the ladder approximation in various relativistic and non-relativistic (NR) equations. On the other hand, and including other couplings, it could provide a relativistic description of nuclear ground states in terms of the traditional degrees of

freedom —mesons and nucleons— with no other restriction than those arising from the assumed structureless character of the constituents.

2 The model

We consider a system of two identical fermions (ψ) interacting through the exchange of a scalar meson (ϕ) described by the Lagrangian density,

$$\mathcal{L} = \overline{\psi} D \psi + \mathcal{L}_{\mathrm{KG}}(\phi) + g_0 \overline{\psi} \phi \psi \,, \tag{1}$$

where $D = \gamma_\mu \partial_\mu - M_0$ is the Dirac operator with a bare fermion mass M_0 and $\mathcal{L}_{\mathrm{KG}}$ is the Klein-Gordon Lagrangian for the scalar field. This model —with an additional $\lambda \phi^4$ term— has been widely studied in the framework of the Higgs phenomenology [6]. In the NR limit (1), it gives rise to the potential

$$V(r) = -\frac{g_0^2}{4\pi} \frac{e^{-m_s r}}{r} \,, \tag{2}$$

where m_s is the meson mass. The NR model depends on a unique parameter, $G = \frac{g_0^2}{4\pi} \frac{M}{m_s}$, and the first bound state appears for $G \approx 1.68$. The existence of this unique scaling parameter holds in the Schrödinger equation but it is no longer true in a relativistic or a QFT description.

In order to study the bound states, one needs to take into account contributions to all orders in the coupling. A perturbative approach is therefore not suitable. Instead, a genuinely non-perturbative tool will be used: the lattice

---

[a] e-mail: `fcsotbor@upo.es`

field theory. The Yukawa model is solved on a Euclidean space-time lattice, where vacuum expectation values are computed in the Feynman path integral approach. For the dressed nucleon propagator one has, for instance,

$$G^{\alpha\beta}(x,y) = \langle 0|\psi^\alpha(x)\overline{\psi}^\beta(y)|0\rangle =$$
$$\frac{1}{Z}\int [\mathrm{d}\overline{\psi}][\mathrm{d}\psi][\mathrm{d}\phi]\psi^\alpha(x)\overline{\psi}^\beta(y)e^{-S_E(\overline{\psi},\psi,\phi)}, \qquad (3)$$

where the Euclidean action acts as a probability distribution, allowing for a Monte Carlo integration.

We have chosen the following discretization of scalar fields:

$$S_{\mathrm{KG}} = \frac{1}{2}\sum_x\left[\left(8 + a^2 m_s^2\right)\phi_x^2 - 2\sum_\mu \phi_{x+\mu}\phi_x\right] \qquad (4)$$

and for fermion ones:

$$S = \sum_{xy}\overline{\psi}_x D_{xy}\psi_y\,,$$

where D_{xy} is the Wilson-Dirac operator

$$D_{xy} = (1 + g_L\phi_x)\,\delta_{x,y}$$
$$-\kappa\sum_\mu [(1 - \gamma_\mu)\delta_{x+\hat\mu,y} + (1 + \gamma_\mu)\delta_{x-\hat\mu,y}]\,, \qquad (5)$$

in which the hopping parameter, $\kappa = 1/(8 + 2aM_0)$, and $g_L = 2\kappa g_0$ have been introduced.

Fermion fields, being Grassmann variables, have to be integrated out in an algebraic way, resulting in

$$G^{\alpha\beta}(x,y) = \frac{1}{Z}\int [\mathrm{d}\phi]D^{-1}{}_{xy}^{\alpha\beta}\det(D)e^{-S_{\mathrm{KG}}}\,. \qquad (6)$$

This calculation is rather demanding in computing time due to the determinant. The task is considerably simplified in the "quenched" approximation, which consists in neglecting all virtual nucleon-antinucleon pairs originating from the meson field $\phi \to \overline{\psi}\psi$. Thanks to the heaviness of the nucleon, this appears to be a good approximation for the problem at hand and has been adopted all along this work. Note that this is not *a priori* justified for QCD, where quarks are very light. Nevertheless, the quenched approximation gives there qualitatively good results. Mathematically it amounts to setting $\det(D) = 1$. The main numerical task in calculating (6) is the inversion of the Dirac operator D_{xy}.

In the quenched approximation, and in the absence of meson self-interaction terms, the meson field is free, and ϕ field configurations can be independently generated by a Gaussian probability distribution in momentum space.

3 Spectrum of the Dirac operator

The spectrum of the free Dirac operator (5) lies inside a circle centered on $\lambda_0 = (1,0)$ with radius 8κ. In QCD when the interaction is turned on, the eigenvalues are modified,

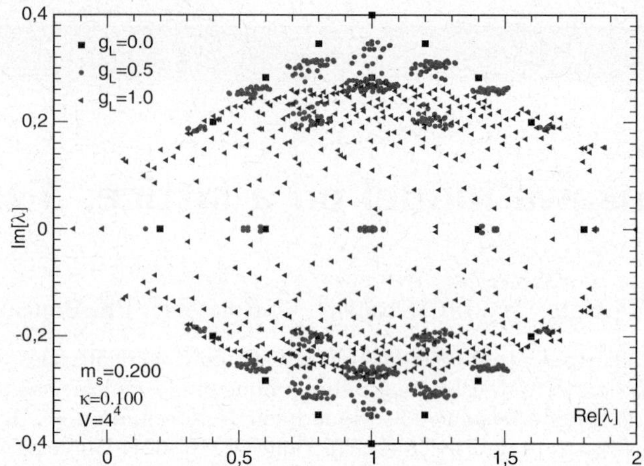

Fig. 1. Spectrum of the Wilson-Dirac operator for a small lattice and several values of the coupling.

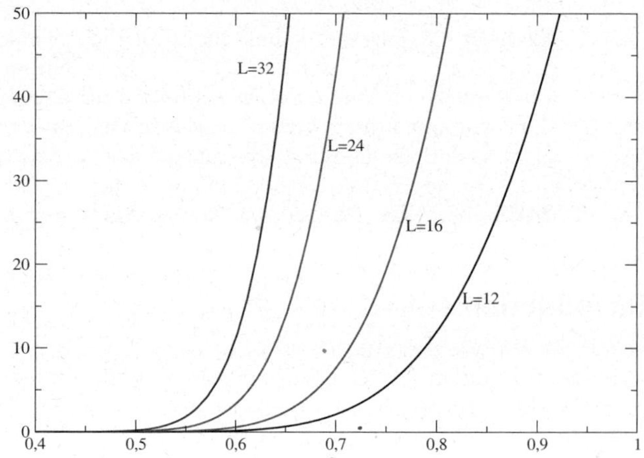

Fig. 2. Average number of negative eigenvalues for several lattices with $Lam_s = 5$ as a function of g_L.

but their real part is always positive. In the Yukawa model, on the contrary, the coupling term plays the role of a mass: as the coupling constant grows, the spectrum spreads out and some eigenvalues move to the $\mathrm{Re}(\lambda) < 0$ half-plane (fig. 1).

Eigenvalues with a negative real part spoil the convergence of most iterative algorithms, but this is not a fundamental problem. Nevertheless, for large values of the coupling and large lattices the probability of having very small eigenvalues grows dramatically. A simplified but significant picture can help to estimate the appearance of those small eigenvalues. The diagonal terms in (5) have the form $1 + g_L\phi_x$, which vanish for $\phi_x = -1/g_L$. According to (4), the values of ϕ_x are distributed around zero with a width:

$$\sigma^2 = \sum_{k\in V}\frac{1}{\hat{k}^2 + m_s^2}\,, \qquad \hat{k}^2 = 2\sum_\mu(1 - \cos(k_\mu))\,. \qquad (7)$$

The average number of negative eigenvalues quickly increases with the coupling and the lattice size, as is shown in fig. 2. The non-diagonal terms in (5) sligthly modify this picture. In practice, for $L \sim 16$ the small eigenvalues

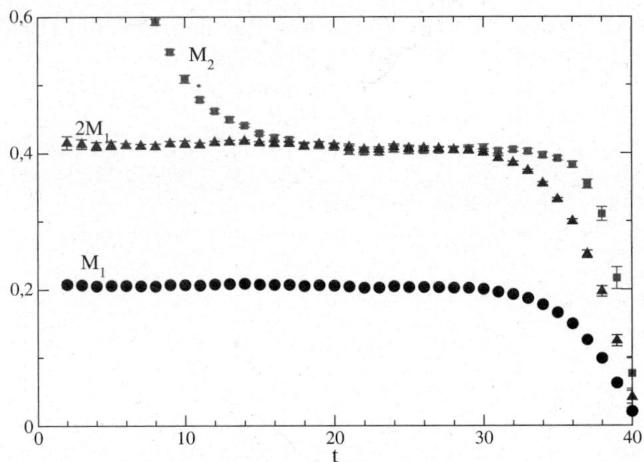

Fig. 3. One- (M_1) and two-body (M_2) effective masses *versus* the Euclidean time, for a $20^3 \times 80$ lattice with $g_0 \approx 1.6$ and $am_s = 0.200$. Two times fermion mass is plotted ($2M_1$) to compare with mass of two-fermion 0^+ state.

hinder the numerical solution of the linear system for $g_L \gtrsim 0.8$. This implies the existence of a maximum value of the coupling constant that can be used. The problem could perhaps be solved in the unquenched case, as the fermionic determinant would eliminate the configurations with very small $\det(D)$.

4 One- and two-body masses

The one-body mass is computed from the time-dependence of the Euclidean correlators:

$$C_1(t) = \sum_{\boldsymbol{x}} \text{Tr}\left[G(\boldsymbol{x}, t)\right] \sim e^{-M_1 t}, \qquad (8)$$

for large values of t, determining the fermion renormalized mass, M_1. Preliminary results on one-body masses for both scalar and pseudoscalar coupling were already presented in [7].

The two-body mass, M_2, is obtained in a similar way, from the time evolution of the $J(x) = \Gamma_{\alpha\beta}\psi_\alpha(x)\psi_\beta(x)$ operator, creating a nucleon pair

$$C_2(t) = \sum_{\boldsymbol{x}} \text{Tr}\langle J(\boldsymbol{x}, t) J^\dagger(\boldsymbol{0}, 0)\rangle \sim e^{-M_2 t}. \qquad (9)$$

For large values of t, it projects on the lowest energy state with the quantum numbers of $J(x)$. The matrix Γ determines the spin and parity of the state ($\Gamma = i\gamma_2\gamma_0\gamma_5$ for $J^\pi = 0^+$ —ground state— and $\Gamma = \gamma_2\gamma_0$ for $J^\pi = 0^-$). The exponential behavior is reached only at large values of t. It is useful to define an effective mass as

$$M_{\text{eff}}(t) = \ln\left(\frac{C(t)}{C(t+1)}\right)$$

which, for large t, goes to the mass of the state and helps to find the adequate fitting window. Some results for one- and two-body masses can be found in fig. 3.

With this set of parameters, the two-fermion mass value is not distinguishable from twice the fermion mass. This is a common picture for the whole set of parameters tested and no signal of the existence of a bound state has been found below the critical value of the coupling constant[1].

5 Discussion

Renormalization effects have been analyzed for one-body masses, and the renormalization issues concerning the coupling constant were discussed in [7].

The existence of a maximum value of the coupling in a QFT treatment of the Yukawa model has been established. This critical value is smaller than the one needed to form a bound state in the NR limit, and no signal of such a bound state for lower couplings has been observed. The limitation on the coupling constant value is a consequence of the QFT approach. This is in contrast with the potential models where the coupling can usually take arbitrary large values.

The physical meaning of this result needs to be clarified. It may be related to the quenched approximation, or to the fact that we have neglected meson self-interactions. Note however that both approximations are performed in the non-relativistic treatment as well. For a given value of the lattice spacing, there are other ways to discretize the Yukawa model which do not have these zero modes. This has to be further studied and it is not clear that it would allow to reach larger renormalized coupling constants and in particular the bound-state regime. We are not yet in a position to decide whether the bound on the coupling constant we encounter is a lattice artefact or whether it really casts a doubt on the Yukawa theory itself.

It is known that the Yukawa theory is infrared free and, as such, encounters the "triviality problem" *i.e.* the fact that the ultraviolet cut-off cannot be driven to infinity without the theory becoming free. This means that it can only be considered as an effective theory with a physical ultraviolet cut-off. In addition, the problem encountered in inverting the Dirac operator appears also to put a restriction on the continuum limit. It is not clear whether both problems are related or just happen both to hinder the continuum limit.

Whether we manage or not to overcome the difficulty of reaching the domain where bound states appear, the connection between the QFT treatment and the Schrödinger approach can still be performed by calculating the low-energy scattering parameters thanks to the method proposed by Luscher [8] and recently considered in [9].

---

[1] There might exist bound states for very light mesons, near the Coulomb limit, but this regime is difficult to reach on the lattice due to the hierarchy of scales that appear.

References

1. V.G. Stoks *et al.*, Phys. Rev. C **49**, 2950 (1994); R.B. Wiringa *et al.*, Phys. Rev. C **51**, 38 (1995); R. Machleidt, Phys. Rev. C **63**, 0240041 (2001).
2. S.C. Pieper *et al.*, Phys. Rev. C **66**, 044310 (2002).
3. S. Weinberg, Nucl. Phys. B **363**, 3 (1991); C. Ordonez *et al.*, Phys. Rev. C **53**, 2086 (1996); E. Epelbaum, W. Gockle, U.G. Meissner, Nucl. Phys. A **671**, 295 (2000).
4. T. Nieuwenhuis, J.A. Tjon, Phys. Rev. Lett. **77**, 814 (1996).
5. G. Baym, Phys. Rev. **117**, 886 (1960).
6. I. Montvay, Nucl. Phys. Proc. Suppl. **26**, 57 (1992).
7. F. De Soto *et al.*, hep-lat/0511009.
8. M. Luscher, Commun. Math. Phys. **104**, 177 (1986); **105**, 153 (1986); Nucl. Phys. B **354**, 531 (1991); M. Luscher, Nucl. Phys. B **364**, 237 (1991).
9. F. De Soto *et al.*, hep-lat/0610040, hep-lat/0610086.

Eur. Phys. J. A **31**, 781–783 (2007)

DOI 10.1140/epja/i2006-10202-x

Special Article – QNP 2006

WChPT analysis of twisted mass lattice data

S. Aoki[1,2] and O. Bär[3,a]

[1] Graduate School of Pure and Applied Sciences, University of Tsukuba, Tsukuba 305-8571, Japan
[2] Riken BNL Research Center, Brookhaven National Laboratory, Upton, NY 11973, USA
[3] Institute of Physics, Humboldt University Berlin, Newtonstrasse 15, 12489 Berlin, Germany

Received: 25 October 2006
Published online: 28 February 2007 – © Società Italiana di Fisica / Springer-Verlag 2007

Abstract. We perform a Wilson Chiral Perturbation Theory (WChPT) analysis of quenched twisted mass lattice data. The data were generated by two independent groups with three different choices for the critical mass. For one choice, the so-called pion mass definition, one observes a strong curvature for small quark masses in various mesonic observables ("bending phenomenon"). Performing a combined fit to the next-to-leading (NLO) expressions, we find that WChPT describes the data very well and the fits provide very reasonable values for the low-energy parameters.

PACS. 12.38.Gc Lattice QCD calculations – 11.15.Ha Lattice gauge theory

1 Introduction

Twisted mass lattice QCD [1,2] has many advantages for numerical lattice simulations, with automatic $O(a)$ improvement at maximal twist [3–5] being probably the most striking one (for a recent review see ref. [6]). Maximal twist is achieved by tuning the bare untwisted mass m_0 to a particular (critical) value such that some matrix element vanishes. The condition for maximal twist is not unique and various choices have been employed in quenched simulations [7–11].

A puzzle observed in early quenched simulations is the so-called "bending phenomenon" [7]. Employing the pion mass definition for maximal twist (*i.e.* tuning m_0 to the value where the pion mass would vanish without a twisted mass term), one observed a strong curvature in the quark mass dependence of many observables (m_π, f_π, m_ρ) for small twisted quark masses μ. This unexpected observation spurred further numerical simulations with other definitions for maximal twist [8–10] as well as theoretical studies. Nowadays the bending phenomenon seems well understood both in terms of the Symanzik effective theory [12] as well as in Wilson Chiral Perturbation theory (WChPT) [13–15] (for introductions to lattice ChPT see also refs. [16,17]).

It is a pleasant side effect of this effort to understand the bending phenomenon that there are lots of data available for five different lattice spacings and three definitions of maximal twist. Moreover, light quark masses could be reached with values for m_π/m_ρ down to ~ 0.3, where one

might expect WChPT to provide an effective description of the data. In particular, the characteristic curvature of the bending phenomenon provides a distinctive test for WChPT to pass if it is the correct low-energy effective theory for twisted mass lattice QCD. Provided WChPT describes the data very well we may also obtain estimates for some combinations of low-energy constants of the effective theory. This was sufficient motivation for us to carry out a WChPT analysis of the existing lattice data. Preliminary results involving data for two definitions of maximal twist only can be found in ref. [18].

2 Fitting the data

We analyzed quenched lattice data generated by two different groups [7–10] with the Wilson plaquette action at $\beta = 5.85$ ($a \approx 0.123\,\mathrm{fm}$) and $\beta = 6.0$ ($a \approx 0.093\,\mathrm{fm}$). The standard Wilson fermion action with a twisted mass term was employed to calculate a variety of mesonic observables. The twisted quark mass covers the range $m_\pi \approx$ 270–1200 MeV, or, alternatively, $m_\pi/m_\rho \approx 0.3$–0.8. Besides the pion mass there is data available for the pseudoscalar decay constant, the angle $\cot\omega_{\mathrm{WT}}$, as well as some more observables which we did not analyze. The untwisted bare quark mass was tuned according to three different definitions of maximal twist: the pion mass definition, the PCAC mass definition and the parity conservation definition. In total there are at most 52 data points available for each lattice spacing.

There exists a vast literature on WChPT for twisted mass lattice QCD [19,20,13,21,14], which contains expres-

---

[a] Spokesperson; e-mail: `obaer@physik.hu-berlin.de`

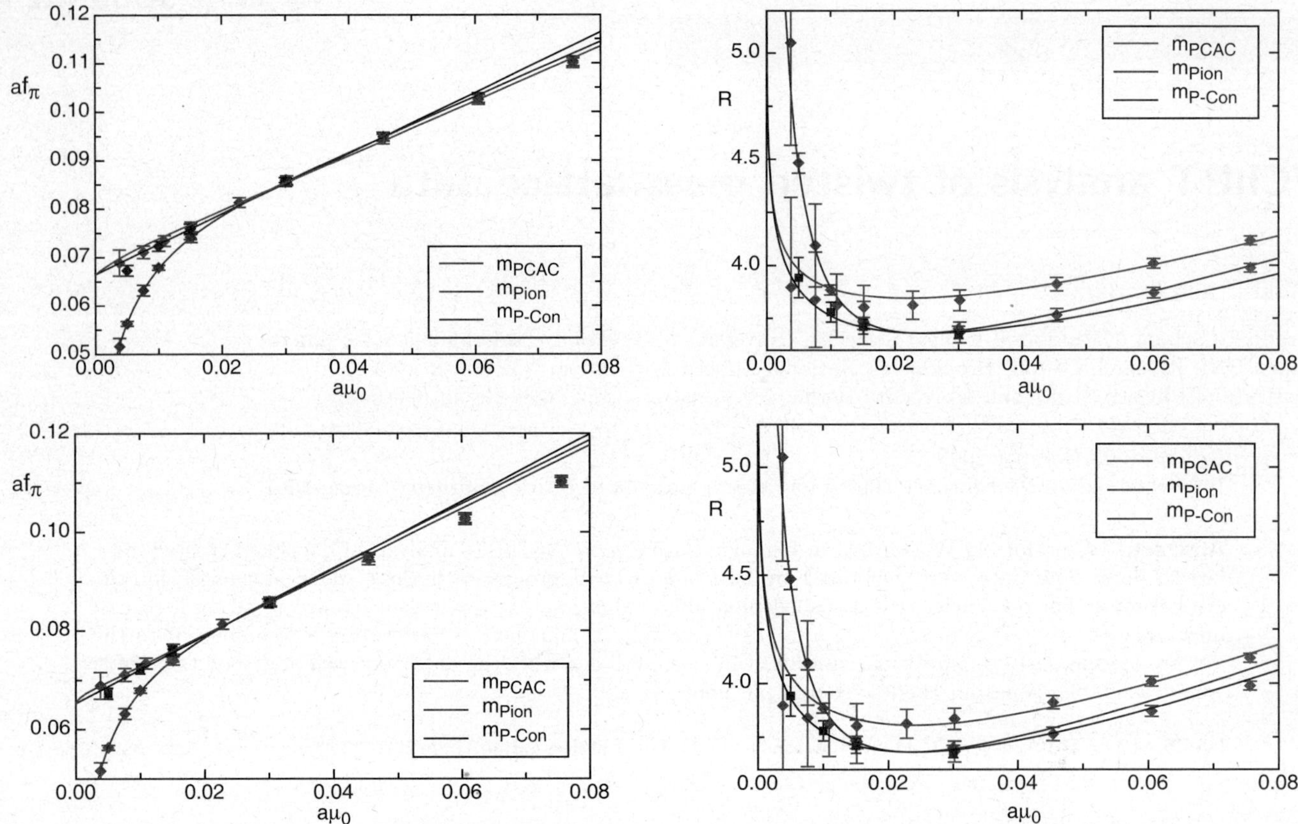

Fig. 1. (Colour on-line) Results of a combined fit for f_π and R at $\beta = 6.0$ ($\chi^2/$d.o.f. $= 0.23$ with d.o.f. $= 26$). The upper plots show the results where all data points are included in the fit, while data points with $a\mu_0 \leq 0.0302$ only are included in the fits shown in the lower plots.

sions for various mesonic observables up to next-to-leading order (NLO). The lattice artifacts are included through order $O(a^2)$ for different power countings and definitions of maximal twist. Here we use the NLO formulae given in ref. [15], which include all NLO terms consistently for the two regimes where either $\mu \sim a$ or $\mu \sim a^2$. The quenched chiral logarithm [22,23] is also included in the formulae.

We performed combined fits of the WChPT formulae at NLO to the data for three observables: f_π,

$$R = \frac{(a\,m_\pi)^2}{a\mu} \, , \tag{1}$$

$$\cot \omega_{\mathrm{WT}} = \frac{\langle \partial_\mu A_\mu^1 P^1 \rangle}{\langle \partial_\mu V_\mu^2 P^1 \rangle} \, , \tag{2}$$

where V_μ^a and A_μ^a denote the (nonsinglet) vector and axial vector current, respectively. At NLO we have in total thirteen free fit parameters. Even though this is a fairly large number it is still small compared to the number of data points.

We performed various fits, starting with all data points included and then successively remove the data points at high quark masses. In all cases we obtain good fit results with $\chi^2/$d.o.f. ≈ 0.2–0.5, even if all data points up to $m_\pi/m_\rho \approx 0.8$ are included[1]. Since we do not trust ChPT

---

[1] Note that the χ^2 value is underestimated since the data is highly correlated.

to work at such high masses we prefer to drop the data for the highest three masses. The fit results for f_π and R at $\beta = 6.0$ are shown in fig. 1. Even in this fit the heaviest point corresponds to $m_\pi/m_\rho \approx 0.63$, which is still heavy. Dropping more data points, however, makes the fit more and more unstable, so we cannot reduce the number of data points much further.

Apparently, WChPT describes the data very well. In particular, the bending for small masses in case of the pion mass definition is very well reproduced. This feature is independent of the number of data points included in the fit, even though the values for the fit parameters are different (see below). Note that the curvature in the data for R with $a\mu_0 \geq 0.3$ is also well described even though the heavier data points are excluded from the fit.

The fits give reasonable values for the fit parameters. For the quenched chiral log parameter δ_0, for example, we find

$$\delta_0 = \begin{cases} 0.10 \pm 0.03, & \beta = 6.0, \\ 0.054 \pm 0.011, & \beta = 5.85, \end{cases} \tag{3}$$

which is in very good agreement with the results obtained by other groups (for a summary, see ref. [24]).

We also obtain an estimate for the low-energy constant c_2. This parameter was first introduced in ref. [25] and enters the chiral Lagrangian according to[2]

$$L_\chi = \ldots + \frac{c_2}{16} \left\{ \mathrm{tr}(\Sigma + \Sigma^\dagger) \right\}^2 \ldots . \qquad (4)$$

The sign of c_2 determines the phase diagram of the lattice theory and the pion mass splitting $\Delta m_\pi^2 = m_{\pi^0}^2 - m_{\pi^\pm}^2$ in the chiral limit is given by c_2/f_π^2 [25].

From our fits we obtain for c_2 the value

$$c_2 = \left[291\,\mathrm{MeV}^{+4\%}_{-5\%} \right]^4, \quad \beta = 5.85. \qquad (5)$$

This is the first determination of this low-energy constant, so we cannot compare with other results. However, the value seems reasonable based on dimensional analysis arguments. The fit for the smaller lattice spacing with $\beta = 6.0$ does not determine c_2 very well; for the mean value we obtain approximately $[170\,\mathrm{MeV}]^4$, but the error is of the same size.

The physical parameters, on the other hand, are very well determined by the fit. For the pseudoscalar decay constant in the chiral limit we find

$$f_0 = \begin{cases} 141.2\,\mathrm{MeV} \pm 1\%, & \beta = 6.0, \\ 141.4\,\mathrm{MeV} \pm 1\%, & \beta = 5.85, \end{cases} \qquad (6)$$

which is in very good agreement with earlier determinations. For the low-energy constant α_5^q [26], entering the NLO expression for the decay constant, we obtain

$$\alpha_5^q = \begin{cases} 1.03(5), & \beta = 6.0, \\ 0.97(4), & \beta = 5.85. \end{cases} \qquad (7)$$

Also these values agree very well with previous results in ref. [26]. Note that the results for f_0 and α_5^q do not show any significant dependence on the lattice spacing. This is expected if WChPT works, since the main dependence on a is captured by other terms in the chiral expansion, being directly proportional to powers of a.

We emphasize that the errors we quoted so far are only the statistical errors given by MINUIT which we used to perform the fits. These errors are underestimated due to the highly correlated data and the true statistical error can be substantially larger. A second error source are systematic uncertainties, induced, for example, by the number of data points included in the fit. It is not simple to give a precise estimate for this error but we observed that the central value for f_0 changes by roughly 3 percent for the different fits we performed, while α_5^q varies by about 7 percent.

3 Conclusions

We performed fits of WChPT to quenched twisted mass data for m_π^2, f_π and the Ward-Takahashi angle $\cot \omega_{\mathrm{WT}}$. We find that the NLO expressions describe the data very

well with small χ^2 values and reasonable values for the low-energy parameters. In particular, the bending phenomenon in case of the pion mass definition is very well reproduced.

The bending phenomenon is a very characteristic feature of twisted mass lattice QCD. It is encouraging that WChPT describes this distinct curvature very well. This indicates that WChPT, *i.e.* ChPT for lattice QCD, seems to work. Previous studies [27–29], using untwisted Wilson fermions, came to contradicting results and were not conclusive at all.

So far we performed separate fits for each lattice spacing. In a next step it would be very interesting to perform a combined fit to the entire data set and take the continuum limit. These results should be compared to the results one obtains after a standard continuum extrapolation where one assumes a polynomial lattice spacing dependence. This would partly answer the question whether WChPT is not only able to describe the lattice data but also necessary to extract the correct continuum physics from the data.

References

1. R. Frezzotti, S. Sint, P. Weisz, JHEP **07**, 048 (2001).
2. R. Frezzotti, P.A. Grassi, S. Sint, P. Weisz, JHEP **08**, 058 (2001).
3. R. Frezzotti, G.C. Rossi, JHEP **08**, 007 (2004).
4. R. Frezzotti, G.C. Rossi, JHEP **10**, 070 (2004).
5. R. Frezzotti, G. Rossi, hep-lat/0507030.
6. A. Shindler, PoS (LAT2005) 014 (2005).
7. W. Bietenholz *et al.*, JHEP **12**, 044 (2004).
8. A.M. Abdel-Rehim, R. Lewis, R.M. Woloshyn, Phys. Rev. D **71**, 094505 (2005).
9. K. Jansen *et al.*, Phys. Lett. B **619**, 184 (2005).
10. K. Jansen *et al.*, JHEP **09**, 071 (2005).
11. A.M. Abdel-Rehim, R. Lewis, R.M. Woloshyn, J.M.S. Wu, Phys. Rev. D **74**, 014507 (2006).
12. R. Frezzotti, G. Martinelli, M. Papinutto, G.C. Rossi, JHEP **04**, 038 (2006).
13. S. Aoki, O. Bär, Phys. Rev. D **70**, 116011 (2004).
14. S.R. Sharpe, Phys. Rev. D **72**, 074510 (2005).
15. S. Aoki, O. Bär, hep-lat/0604018.
16. O. Bär, Nucl. Phys. Proc. Suppl. **140**, 106 (2005).
17. S.R. Sharpe, hep-lat/0607016.
18. S. Aoki, O. Bär, PoS (LAT2005) 046 (2005).
19. G. Münster, C. Schmidt, Europhys. Lett. **66**, 652 (2004).
20. L. Scorzato, Eur. Phys. J. C **37**, 445 (2004).
21. S.R. Sharpe, J.M.S. Wu, Phys. Rev. D **71**, 074501 (2005).
22. C.W. Bernard, M.F.L. Golterman, Phys. Rev. D **46**, 853 (1992).
23. S.R. Sharpe, Phys. Rev. D **46**, 3146 (1992).
24. H. Wittig, Nucl. Phys. Proc. Suppl. **119**, 59 (2003).
25. S.R. Sharpe, J. Singleton, Robert, Phys. Rev. D **58**, 074501 (1998).
26. J. Heitger, R. Sommer, H. Wittig, Nucl. Phys. B **588**, 377 (2000).
27. F. Farchioni, I. Montvay, E. Scholz, Eur. Phys. J. C **37**, 197 (2004).
28. S. Aoki, Phys. Rev. D **68**, 054508 (2003).
29. Y. Namekawa *et al.*, Phys. Rev. D **70**, 074503 (2004).

---

[2] The definition of c_2 is not unique and it is sometimes defined differently, for example in ref. [13].

Eur. Phys. J. A **31**, 784–786 (2007)
DOI 10.1140/epja/i2006-10285-3

Special Article – QNP 2006

Understanding nucleon structure using lattice simulations

Recent progress on three different structural observables

W. Schroers[a]

John von Neumann-Institut für Computing NIC/DESY, 15738 Zeuthen, Germany

Received: 18 December 2006
Published online: 20 March 2007 – © Società Italiana di Fisica / Springer-Verlag 2007

Abstract. This review focuses on the discussion of three key results of nucleon structure calculations on the lattice. These three results are the quark contribution to the nucleon spin, J_q, the nucleon-Δ transition form factors, and the nucleon axial coupling, g_A. The importance for phenomenology and experiment is discussed and the requirements for future simulations are pointed out.

PACS. 12.38.Gc Lattice QCD calculations

1 Introduction

In recent years lattice gauge theory has become a mature and reliable way to investigate the structure of strong interactions. It provides a model-independent way to do calculations in QCD. However, contemporary lattice computations become extremely costly at quark masses corresponding to pion masses below 500 MeV. Nature, however, has chosen the pion mass to be only 140 MeV. The lightness of the pseudoscalar mesons is due to the mechanism of spontaneous chiral symmetry breaking. If, however, we can investigate only the regime of heavy quarks, where chiral symmetry is broken explicitly by the quark mass, we might not describe physics accurately at light-quark masses.

To address and overcome this challenge, three different procedures have been proposed and are actively pursued: i) pushing existing simulations with Wilson-type quarks down to smaller quark masses by relying on improved algorithms and faster computers [1], ii) using a hybrid action approach by using different formulations for sea and valence quarks [2], and iii) doing simulations using dynamical Ginsparg-Wilson formulations, such as domain-wall fermions [3] or overlap fermions [4]. The last approach is certainly the most challenging and demanding one since the entire parameter space has to be explored again. This applies also to heavy quarks, a regime in which Ginsparg-Wilson fermions are about 30 to 100 times more expensive than standard Wilson-type fermions.

The hybrid action ansatz is an excellent compromise between quark mass and performance, but suffers from conceptual problems. First of all, the hybrid theory breaks unitarity at finite lattice spacing. Thus, it cannot act as

an effective theory at finite lattice spacing, and the existence of the continuum limit is crucial. Furthermore, usually staggered-type quarks are being used for the sea with the square-root being taken of the determinant. It is not clear if the procedure of taking the square root commutes with taking the continuum limit, see, *e.g.*, [5] for a recent review: Finally, the matching of sea and valence quark masses is prescription dependent, and particular choices may give rise to additional possibly large $\mathcal{O}(a^2)$ artifacts [6].

In this review we focus on three observables with relevance to phenomenological and experimental applications. The first one is the quark contribution to the nucleon spin, J_q. The second one is the transition form factors of the nucleon-Δ transition. The third one is the nucleon axial coupling, g_A. The former two of these quantities have so far been understood qualitatively, but a precise matching between the light-quark regime and the lattice —possibly by chiral perturbation theory or an effective model of the strong interaction like [7]— still remains to be done. For the latter observable it has been shown that lattice data can in fact be consistent with experiment when fitting it using the leading logarithmic chiral perturbation theory expression. This achievement marks a milestone in the field of nucleon structure.

2 Quark contribution to nucleon spin

The quark contribution to the spin of the nucleon has been under intense scrutiny after the observation that only about $(20 \pm 15)\%$ of the nucleon spin arises from the quark spin [8]. Recently, it has been realized how the use of GPDs [9–11] provides the means to directly compute the quark contribution to the nucleon spin via the

---

[a] e-mail: `Wolfram.Schroers@Field-theory.org`

energy momentum tensor [10],

$$J_q = \lim_{t \to 0} \left(A_{20}^{u+d}(t) + B_{20}^{u+d}(t) \right) . \qquad (1)$$

The virtuality t is given by $t \equiv (p' - p)^2$, where p' and p are the nucleon's incoming and outgoing momenta. The generalized form factors, $A_{20}^{u+d}(t)$ and $B_{20}^{u+d}(t)$, show up in the parameterization of the nucleon's energy-momentum tensor. For further details and the exact definition, consult [10]. The challenge is to understand which fraction of the nucleon spin, $J_N = 1/2$, arises from the quark spin, $1/2\Sigma_q$, the quark orbital angular momentum, L_q, and which fraction comes from gluon contributions, J_g:

$$J_N = 1/2 = J_q + J_g = 1/2\Sigma_q + L_q + J_g . \qquad (2)$$

The value of Σ_q has been known before [12]. The new ingredient is the ability to directly calculate J_q, and thus also L_q. To this end, there is no experimental determination of that quantity. The first computation of J_q on the lattice has been done in [12]. This calculation only utilizes quenched Wilson fermions, but features a calculation of the disconnected contribution using noisy estimators. A later calculation [13] calculates all generalized form factors in the energy-momentum tensor separately and at the same time a publication [14] features full QCD and introduces an improved technology to extract form factors from matrix elements. Higher moments of GPDs have also been computed [15].

As of today, the understanding gained from the world of pions weighing 500 MeV and beyond is that the quark contribution to the nucleon spin is about 70%, all of which comes from the quark spin alone. The remaining 30% comes from the gluons. The quark orbital angular momentum is negligibly small due to a cancellation between the contributions of u- and d-quarks [2].

This result differs from the finding outlined above which indicates that this quantity can be expected to substantially depend on the pion mass. The cancellation of the orbital angular momentum for u- and d-quarks is an interesting qualitative feature. The insight that the nucleon in the heavy-pion world receives a larger fraction of its spin from quarks rather than gluons is compatible with expectations from the non-relativistic quark model, but the exact interpolation between the heavy-quark and the light-quark regime can give further insight into how the strong interaction operates.

The extrapolation to the chiral regime, however, has not yet been possible and hence a precise quantitative matching with nature has not yet been established. Although ref. [16] suggests a rather flat expression it is yet unclear whether the same straight line is to be used for the light-quark regime as the one fitting the simulations. In this situation it is inevitable to perform similar calculations at smaller pion masses before a matching between lattice and small-scale expansion schemes can be established and a definitive prediction from the lattice can be provided.

Further investigations from several groups are underway and all three different paths outlined in sect. 1 are taken to resolve this important question. We can conclude, however, that the technology and understanding of how to compute these matrix elements are available and can be deployed easily once sufficiently light pion masses are available.

3 N-Δ transition form factors

A key question is whether the baryon states of QCD are spherical or deformed. Although the nucleon is easily accessible in exclusive and inclusive scattering experiments, it cannot have a spectroscopic quadrupole moment since it has spin $J_N = 1/2$. The excited states with spin $3/2$ and above can have a quadrupole moment, but these are not easily accessible in experiments. The only way to learn about deformations of the low-lying baryon spectrum is to consider transitions between the nucleon and the first-excited state, the $\Delta(1232)$-resonance. Experimentally, a flurry of activity has recently led to several important and exciting results [17].

The nucleon-Δ transition can be parameterized using three form factors —the dominant magnetic dipole form factor, $\mathcal{G}_{M1}$, the electric quadrupole, $\mathcal{G}_{E2}$, and the Coulomb quadrupole, $\mathcal{G}_{C2}$. Should the nucleon-Δ system be deformed, the latter two form factors will not vanish. Should the system be spherical, only the magnetic dipole form factor will be non-zero.

On the lattice, publications reporting the successful computation of these transition form factors are given in [18]. This set of calculations used unquenched Wilson fermions with pion masses beyond 600 MeV and quenched Wilson fermions with pion masses larger than 370 MeV. Later it has been attempted to apply these techniques also for hybrid actions [19], but to this end the statistical error bars on the quadrupole form factors turn out to be too large.

From these studies it has been clearly established, that the nucleon-Δ system is indeed deformed. The sign and the order of magnitude of the quadrupole form factors $\mathcal{G}_{E2}$ and $\mathcal{G}_{C2}$ were extracted successfully. The heavy-pion world in fact is similar to nature for these observables.

However, the extrapolation to the physical pion masses yields an inconsistency for $\mathcal{G}_{C2}$ at values below $Q^2 < 0.2\,\text{GeV}^2$. This discrepancy has been addressed recently in the framework of chiral perturbation theory [20]. It appears plausible that the discrepancy in fact arises from the inadequacy of a linear chiral extrapolation —it still remains to be seen if lattice data at smaller pion masses can indeed verify the pion mass dependence suggested in [20].

4 Nucleon axial coupling g_A

The investigation of the nucleon axial coupling has a long history on the lattice, see [21] for recent reviews. Several groups have performed investigations using a wide array of different lattice actions, spacings, volumes, and pion masses.

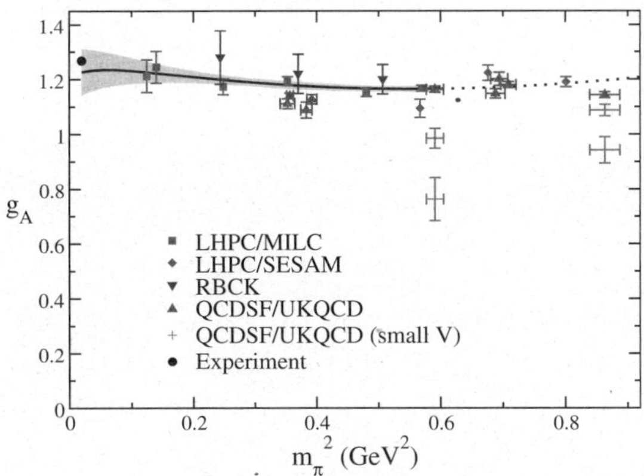

Fig. 1. Full QCD computations of the nucleon axial coupling, g_A. The line shows the fit of the leading logarithmic χPT expression to the hybrid lattice data from the LHPC Collaboration. Results from other groups are plotted, but not included in the fit. Figure taken from ref. [22], QCDSF data updated from ref. [23].

Recently, two independent papers, [22] and [23], have appeared showing how current lattice data can in fact be combined with chiral perturbation theory to arrive at the experimental value. Figure 1 shows the application of the leading logarithmic expression from χPT to the hybrid data computed by the LHPC Collaboration in [22]. The gray-shaded error band shows the error arising from statistical uncertainties only. The fit yields quantitative agreement with the experimental value.

However, the applicability of the leading-order chiral perturbation theory expression to the pion masses available has been questioned in [24]. The flat behavior at pion masses beyond 300 MeV is attributed to fine-tuning between different terms in the expansion. On the other hand, the expansion can still be consistent with experiment when applied to lattice calculations employing pion masses as large as 600 MeV [23]. It is perhaps fair to say that the exact range of applicability of χPT is under debate. Nonetheless, the striking agreement between the fit of lattice data and the experimental value mark an important milestone for the lattice treatment of nucleon structure.

5 Summary

We have given three examples of recent lattice calculations which are of great interest to both phenomenologists and experimentalists alike. The limiting factor of all these lattice results, however, is their limitation to rather large quark masses. Currently, the question of chiral extrapolations is under debate and the applicability depends strongly on the observable. While some groups successfully apply fits to pion masses as large as 600 MeV, other groups believe that pion masses lower than 300 MeV are essential. While the latter mass regime has not been reachable so far, we believe that the upcoming generation of lattice calculations will be able to settle the debate.

This work was supported in part by the DFG, contract FOR 465 (FG Gitter-Hadronen-Phänomenologie), and in part by the EU Integrated Infrastructure Initiative Hadron Physics (I3HP), contract No. RII3-CT-2004-506078.

References

1. K. Jansen *et al.*, these proceedings; M. Göckeler *et al.*, arXiv:hep-lat/0610066.
2. J.W. Negele *et al.*, Nucl. Phys. Proc. Suppl. **128**, 170 (2004); LHP Collaboration (D.B. Renner *et al.*), Nucl. Phys. Proc. Suppl. **140**, 255 (2005); LHPC Collaboration (Ph. Hägler, J.W. Negele, D.B. Renner, W. Schroers, T. Lippert, K. Schilling), Eur. Phys. J. A **24**, s01, 29 (2005); LHPC Collaboration (R.G. Edwards *et al.*), PoS LAT2005, 056 (2006); R.G. Edwards *et al.*, arXiv:hep-lat/0610007.
3. UKQCD and RBC Collaborations (R. Tweedie *et al.*), PoS LAT2005, 096 (2006).
4. N. Cundy, S. Krieg, A. Frommer, T. Lippert, K. Schilling, Nucl. Phys. Proc. Suppl. **140**, 841 (2005); N. Cundy, S. Krieg, T. Lippert, PoS LAT2005, 107 (2006); S. Schäfer, arXiv:hep-lat/0609063.
5. S. Dürr, PoS LAT2005, 021 (2006).
6. O. Bär, C. Bernard, G. Rupak, N. Shoresh, Phys. Rev. D **72**, 054502 (2005).
7. K. Goeke, J. Ossmann, P. Schweitzer, A. Silva, Eur. Phys. J. A **27**, 77 (2006).
8. R.D. Ball, S. Forte, G. Ridolfi, Phys. Lett. B **378**, 255 (1996).
9. D. Müller, D. Robaschik, B. Geyer, F.M. Dittes, J. Horejsi, Fortsch. Phys. **42**, 101 (1994).
10. X.D. Ji, Phys. Rev. Lett. **78**, 610 (1997).
11. A.V. Radyushkin, Phys. Rev. D **56**, 5524 (1997).
12. N. Mathur, S.J. Dong, K.F. Liu, L. Mankiewicz, N.C. Mukhopadhyay, Phys. Rev. D **62**, 114504 (2000).
13. QCDSF Collaboration (M. Göckeler, R. Horsley, D. Pleiter, P.E.L. Rakow, A. Schäfer, G. Schierholz, W. Schroers), Phys. Rev. Lett. **92**, 042002 (2004).
14. LHPC Collaboration (P. Hägler, J. Negele, D.B. Renner, W. Schroers, T. Lippert, K. Schilling), Phys. Rev. D **68**, 034505 (2003).
15. LHPC Collaboration (P. Hägler, J.W. Negele, D.B. Renner, W. Schroers, T. Lippert, K. Schilling), Phys. Rev. Lett. **93**, 112001 (2004).
16. J.W. Chen, X.d. Ji, Phys. Rev. Lett. **88**, 052003 (2002).
17. LEGS Collaboration (G. Blanpied *et al.*), Phys. Rev. Lett. **76**, 1023 (1996); R. Beck *et al.*, Phys. Rev. C **61**, 035204 (2000); C. Mertz *et al.*, Phys. Rev. Lett. **86**, 2963 (2001); OOPS Collaboration (N.F. Sparveris *et al.*), Phys. Rev. Lett. **94**, 022003 (2005); CLAS Collaboration (K. Joo *et al.*), Phys. Rev. Lett. **88**, 122001 (2002); A.M. Bernstein, Eur. Phys. J. A **17**, 349 (2003).
18. C. Alexandrou *et al.*, Phys. Rev. D **69**, 114506 (2004); C. Alexandrou, Ph. de Forcrand, H. Neff, J.W. Negele, W. Schroers, A. Tsapalis, Phys. Rev. Lett. **94**, 021601 (2005).
19. C. Alexandrou *et al.*, PoS LAT2005, 091 (2006).
20. V. Pascalutsa, M. Vanderhaeghen, Phys. Rev. Lett. **95**, 232001 (2005).
21. W. Schroers, Nucl. Phys. A **755**, 333 (2005); Nucl. Phys. Proc. Suppl. **153**, 277 (2006).
22. LHPC Collaboration (R.G. Edwards *et al.*), Phys. Rev. Lett. **96**, 052001 (2006).
23. A. Ali Khan *et al.*, arXiv:hep-lat/0603028.
24. V. Bernard, U.G. Meissner, Phys. Lett. B **639**, 278 (2006).

Eur. Phys. J. A **31**, 787–789 (2007)

DOI 10.1140/epja/i2006-10173-x

THE EUROPEAN
PHYSICAL JOURNAL A

Special Article – QNP 2006

Quark matter in QC$_2$D

S. Hands[1,a], S. Kim[2], and J.-I. Skullerud[3]

[1] Department of Physics, Swansea University, Singleton Park, Swansea SA2 8PP, UK
[2] Department of Physics, Sejong University, Gunja-Dong, Gwangjin-Gu, Seoul 143-747, South Korea
[3] School of Mathematics, Trinity College, Dublin 2, Ireland

Received: 24 September 2006
Published online: 16 February 2007 – © Società Italiana di Fisica / Springer-Verlag 2007

Abstract. Results are presented from a numerical study of lattice QCD with gauge group $SU(2)$ and two flavors of Wilson fermion at non-zero quark chemical potential $\mu \gg T$. Studies of the equation of state, the superfluid condensate, and the Polyakov line all suggest that in addition to the low-density phase of Bose-condensed diquark baryons, there is a deconfined phase at higher quark density in which quarks form a degenerate system, whose Fermi surface is only mildly disrupted by Cooper pair condensation.

PACS. 11.15.Ha Lattice gauge theory – 21.65.+f Nuclear matter

1 Introduction

The phase structure of QCD at large baryon density is one of the most fascinating areas of strong-interaction physics, and yet a systematic calculational approach to this problem remains elusive. Lattice QCD simulation, the usual non-perturbative approach of choice, fails dismally because in Euclidean metric the quark action $\bar{q}M(\mu)q$, where $M = \not{D}[A] + \mu\gamma_0 + m$ with μ the quark chemical potential, results in a complex-valued path integral measure det M when $\mu \neq 0$. Since $\mu > 0$ promotes baryon current flow in the positive t-direction, the fundamental reason for this *Sign Problem* can be traced to the explicit breaking of time reversal symmetry. Because the measure no longer has an interpretation as a probability distribution, Monte Carlo importance sampling, the mainstay of lattice simulations, is completely ineffective in the thermodynamic limit.

It is instructive to ask what goes wrong when simulations are performed with a measure det $M^\dagger M$ which is positive definite by construction, as is the case for all practical fermion algorithms. It turns out that while M describes a color-triplet quark $q \in \mathbf{3}$, $M^\dagger$ describes a *conjugate quark* $q^c \in \bar{\mathbf{3}}$. The model's spectrum thus contains gauge-singlet qq^c states, indistinguishable from mesons at $\mu = 0$, but carrying non-zero baryon number. As μ rises, baryonic matter first appears in the ground state (*i.e.* $n_q > 0$) at an onset $\mu_o \sim \frac{1}{2}m_\pi$, *i.e.* with an energy per quark comparable with the lightest baryon in the spectrum, which is degenerate with the pion, rather than the physically expected $\mu_o \sim \frac{1}{3}m_{nucleon}$. Only calculations performed with the correct measure det^{N_f}M have cancellations among configurations, due to the fluctuating

^{a} e-mail: s.hands@swan.ac.uk

phase of the determinant, which ensure that n_q vanishes for $\frac{1}{2}m_\pi < \mu < \frac{1}{3}m_{nucleon}$.

For Two-Color QCD (QC$_2$D), *i.e.* for the gauge group $SU(2)$, this bug is actually a feature. Since q and $\bar{q}$ live in equivalent representations of the color group, hadron multiplets contain both $q\bar{q}$ mesons and qq baryons. It is correspondingly straightforward to show that the quark determinant is positive definite for even N_f [1]. QC$_2$D is thus the simplest model of dense strongly interacting matter amenable to study with orthodox lattice techniques. Additionally, if there is a separation of scales $m_\pi \ll m_\rho$ in the spectrum, then at low densities attention may be focussed on the Goldstone bosons of the system (both mesons and baryons) using chiral perturbation theory (χPT) [2]. The key result is that for $\mu \geq \mu_o = \frac{1}{2}m_\pi$, a non-vanishing quark density $n_q > 0$ develops, along with a superfluid diquark condensate $\langle qq \rangle \neq 0$. Just above onset, the system is thus a textbook Bose Einstein Condensate (BEC) formed from tightly bound scalar diquarks.

Using the χPT prediction for $n_q(\mu)$ [2], it is simple to develop the full equation of state, *i.e.* pressure p and energy density ε_q, at $T = 0$ [3]:

$$n_q = 8N_f f_\pi^2 \mu \left(1 - \frac{\mu_o^4}{\mu^4}\right);$$

$$p = \int_{\mu_o}^{\mu} n_q d\mu = 4N_f f_\pi^2 \left(\mu^2 + \frac{\mu_o^4}{\mu^2} - 2\mu_o^2\right); \quad (1)$$

$$\varepsilon_q = -p + \mu n_q = 4N_f f_\pi^2 \left(\mu^2 - 3\frac{\mu_o^4}{\mu^2} + 2\mu_o^2\right);$$

$$\langle qq \rangle \propto \sqrt{1 - \frac{\mu_o^4}{\mu^4}}.$$

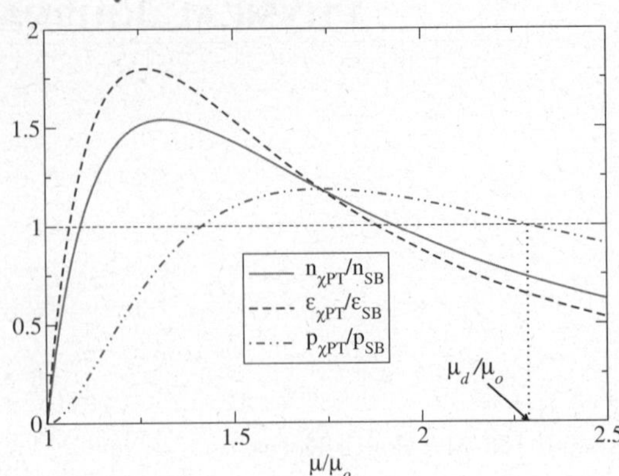

Fig. 1. Model equation of state for $f_\pi^2 = N_c/6\pi^2$.

Here, f_π is a parameter of the model. Contrast this with another paradigm for cold dense matter, namely a degenerate system of weakly-interacting massless quarks populating a Fermi sphere up to some maximum momentum $k_F \approx \mu$:

$$n_q = \frac{N_f N_c}{3\pi^2}\mu^3; \qquad \varepsilon_q = 3p = \frac{N_f N_c}{4\pi^2}\mu^4. \qquad (2)$$

Superfluidity in this scenario arises from the condensation of quark Cooper pairs within a layer of thickness Δ centred on the Fermi surface, so that $\langle qq \rangle \propto \Delta\mu^2$.

Figure 1 plots n_q, p and ε_q from (1), each divided by the free field results (2), as functions of μ. On equating pressures, this naive model, which ignores all non-Goldstone and gluonic degrees of freedom, predicts a first-order deconfining transition from BEC to "quark matter" at $\mu_d \approx 2.3\mu_o$ with the choice $f_\pi^2 = N_c/6\pi^2$.

2 Simulation

To test whether this prediction holds in a more systematic calculation we have performed simulations of $SU(2)$ lattice gauge theory with $N_f = 2$ Wilson fermions with $\mu \neq 0$ [3]. The Wilson formulation is not obviously a stupid choice: Wilson fermions retain a conserved baryon charge; any problems with chiral symmetry should dominate in the low-k region of the quark dispersion curve, which lies at the bottom of the Fermi sea and is hence inert; moreover, studies with free fermions show that saturation artifacts due to the complete filling of the first Brillouin zone actually set in at higher values of μ than is the case for staggered fermions [4]. Most importantly, the eigenvalue spectrum of the Wilson-Dirac operator has the same symmetries as that of continuum QC$_2$D. As shown in [3], this fact permits an exact ergodic hybrid Monte Carlo algorithm for $N_f = 2$, with no requirement to take a fourth root, which may be problematic for $\mu \neq 0$ [5]. The only novelty of our simulation is the inclusion of a diquark source term

$$jqq \equiv j\kappa(-\bar\psi_1(x)C\gamma_5\tau_2\bar\psi_2^{tr}(x) + \psi_2^{tr}(x)C\gamma_5\tau_2\psi_1(x)) \quad (3)$$

in the dynamics, where subscripts label flavor and the Pauli matrix acts on color. As well as making the algorithm ergodic, setting $j \neq 0$ mitigates the effect of IR fluctuations due to Goldstone modes in any superfluid phase, and of course enables direct estimation of the $\langle qq \rangle$ condensate.

Our initial study has been performed on an $8^3 \times 16$ lattice using a standard Wilson gauge action, with parameters $\beta = 1.7$, $\kappa = 0.178$, and $j = 0.04$ (with a few points taken at $j = 0.02, 0.06$). Studies of the static quark potential and the hadron spectrum at $\mu = 0$ yield $a = 0.220\,\mathrm{fm}$, $m_\pi a = 0.79(1)$, and $m_\pi/m_\rho = 0.80(1)^1$. We thus expect the onset of baryonic matter at $\mu_o a \approx 0.4$. Thermodynamic observables are calculated as follows: quark density is given by a local operator via

$$n_q = -\frac{\partial \ln \mathcal{Z}}{\partial \mu}. \qquad (4)$$

As a component of a conserved current, it is immune from quantum corrections, but may be affected by artifacts due to $a > 0$, $V < \infty$. We therefore prefer to quote our results in terms of $n_q/n_{SB}^{\mathrm{latt}}$, where $n_{SB}^{\mathrm{latt}}(\mu)$ is evaluated for free massless quarks on the same lattice. The pressure follows from an integral formula

$$\frac{p}{p_{SB}} = \int_{\mu_o}^{\mu} \frac{n_{SB}^{\mathrm{cont}}}{n_{SB}^{\mathrm{latt}}} n_q \mathrm{d}\mu \bigg/ \int_{\mu_o}^{\mu} n_{SB}^{\mathrm{cont}} \mathrm{d}\mu. \qquad (5)$$

Note that although p is calculated purely in terms of quark observables, it is in principle the pressure of the system as a whole, although both continuum and thermodynamic limits must eventually be taken. Finally, quark energy density is also estimated by a local operator

$$\varepsilon_q = \kappa\left\langle \bar\psi_x(\gamma_0 - 1)e^\mu U_{0x}\psi_{x+\hat0} - \bar\psi_x(\gamma_0 + 1)e^{-\mu}U_{0x-\hat0}^\dagger\psi_{x-\hat0}\right\rangle; \qquad (6)$$

this requires both subtraction of the $\mu = 0$ vacuum contribution, and a μ-independent but as yet unknown multiplicative renormalisation. In what follows, therefore, the shape of the curve is in principle correct, but the overall scale still undetermined.

Figure 2 summarises our results. Both n_q and p start to rise from zero at $\mu a \approx 0.3$, although a careful $j \to 0$ extrapolation will be needed to pinpoint the onset with any precision. By $\mu a \approx 0.5$ both quantities scale with μ in general accordance with free-field predictions, but with approximately twice the expected value. One explanation of this mismatch is that the system has formed a Fermi sphere with $\mu = E_F < k_F \propto n_q^{\frac{1}{3}}$, which could be attributed to a negative binding contribution to E from interactions. The quark energy density, by contrast, increases more slowly than free-field expectations up to $\mu a \approx 0.65$, whereupon free-field scaling sets in rather abruptly. Another intriguing result [3] is that for $0.4 \leq \mu a < 1.0$ the gluon energy density ε_g (identically zero in free-field theory) scales to quite high precision as μ^4, the only physically sensible possibility once $\mu/T \gg 1$. Note that $\varepsilon_g > 0$

---

^{1} This corrects the value erroneously given in [3].

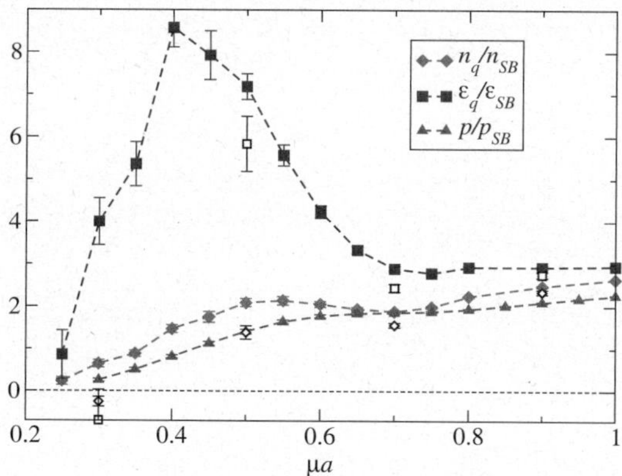

Fig. 2. Lattice equation of state for $j = 0.04$ (open symbols give $j \to 0$ extrapolation).

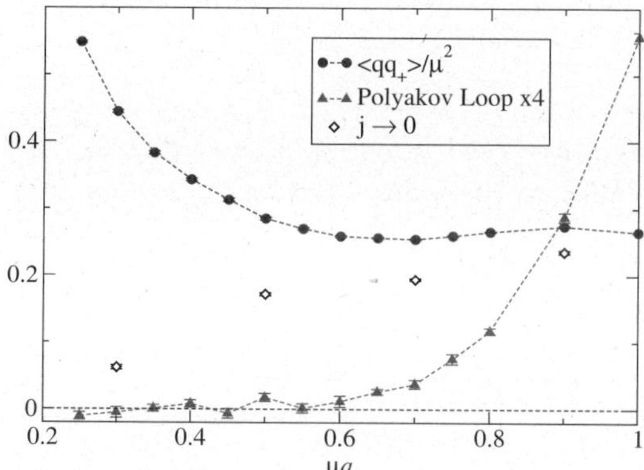

Fig. 3. Order parameters $\langle qq \rangle$ and L vs. μ.

entirely as a result of interactions with the background quark density, since this is the only means by which μ-dependence can arise.

To elucidate what is happening, fig. 3 plots both the superfluid order parameter $\langle qq \rangle$ divided by μ^2, and the Polyakov line L. For $\mu a \geq 0.5$ it is clear that the system is in a superfluid phase, but what is remarkable is that at $\mu a \approx 0.6$ there is a sudden transition to a regime where $\langle qq \rangle \propto \mu^2$, as expected for BCS pairing at a Fermi surface. At roughly the same point L rises from zero; although for theories with fundamental matter L is not strictly an order parameter, this is suggestive that at $\mu a \approx 0.65$ there is a *deconfining* transition, beyond which the effective degrees of freedom are best thought of as quarks (or even *quasiquarks*) and not the scalar diquarks of χPT.

3 Discussion

Our initial study of thermodynamic quantities, and of the properties of the ground state, strongly suggests that QC$_2$D at low temperature has at least two transitions as the chemical potential μ is raised. The first is between the vacuum and a phase of Bose-condensed tightly bound diquarks; the second, a relativistic analogue of the BEC/BCS crossover currently discussed in both strongly

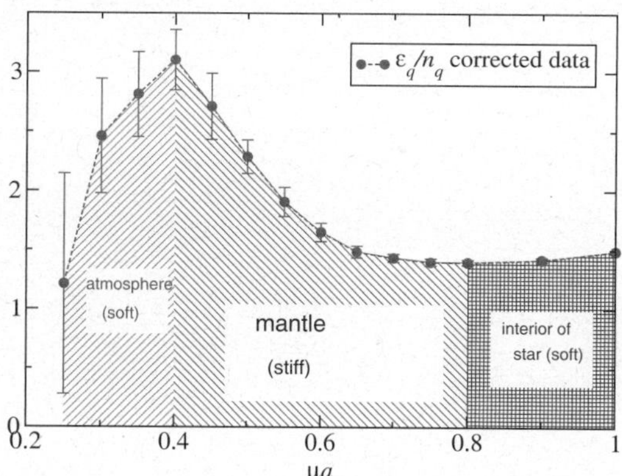

Fig. 4. Energy per quark ε_q/n_q vs. μ.

correlated electron and cold atom systems, is a deconfining transition to a system of degenerate quarks; the Fermi surface being mildly disrupted by a Cooper pair condensate. Although QC$_2$D clearly models nuclear matter unrealistically, its description of quark matter may well prove to have much in common with that of QCD. We are currently extending our study to the hadron spectrum, and to finer lattice spacings to check that this conclusion is not due to lattice artifacts. Interesting results obtained from a study of the gluon propagator on the current system will be discussed elsewhere [6,3].

Meanwhile it is hard to resist the temptation to speculate on what a two-color star might look like. Figure 4 plots the energy per quark ε_q/n_q vs. μ using the data of fig. 2. The most striking feature of this plot is the pronounced minimum at $\mu a \approx 0.8$, which is both robust (since it occurs even if corrections for $a > 0$, $V < \infty$ are left out), and unexpected (since it does not occur for the model EoS of fig. 1). We infer that any large object assembled from a fixed number of QC$_2$D quarks, such as a star, will have the bulk of its interior in the neighbourhood of this minimum, which as fig. 3 shows, means that the object would in effect be a quark star formed from deconfined matter. Somewhat speculatively, we have labelled the different regions of the μ-axis with the corresponding layers of the star, although a quantitative solution for the radial profile must await correctly normalised calculations of the energy densities ε_q and ε_g.

References

1. S. Hands, I. Montvay, S. Morrison, M. Oevers, L. Scorzato, J.I. Skullerud, Eur. Phys. J. C **17**, 285 (2000).
2. J.B. Kogut, M.A. Stephanov, D. Toublan, J.J.M. Verbaarschot, A. Zhitnitsky, Nucl. Phys. B **582**, 477 (2000).
3. S. Hands, S. Kim, J.I. Skullerud, Eur. Phys. J. C **48**, 193 (2006).
4. W. Bietenholz, U.J. Wiese, Phys. Lett. B **426**, 114 (1998).
5. M. Golterman, Y. Shamir, B. Svetitsky, Phys. Rev. D **74**, 071501 (2006).
6. J.I. Skullerud, these proceedings.

Eur. Phys. J. A **31**, 790–792 (2007)

DOI 10.1140/epja/i2006-10210-x

Special Article – QNP 2006

Infrared gluon and ghost propagators from lattice QCD

Results from large asymmetric lattices

O. Oliveira[a] and P.J. Silva

Centro de Física Computacional, Universidade de Coimbra, 3004 516 Coimbra, Portugal

Received: 25 October 2006
Published online: 1 March 2007 – © Società Italiana di Fisica / Springer-Verlag 2007

Abstract. We report on the infrared limit of the quenched lattice Landau gauge gluon and ghost propagators as well as the strong-coupling constant computed from large asymmetric lattices. The infrared lattice propagators are compared with the pure power law solutions from Dyson-Schwinger equations (DSE). For the gluon propagator, the lattice data is compatible with the DSE solution. The preferred measured gluon exponent being ~ 0.52, favouring a vanishing propagator at zero momentum. The lattice ghost propagator shows finite-volume effects and, for the volumes considered, the propagator does not follow a pure power law. Furthermore, the strong-coupling constant is computed and its infrared behaviour investigated.

PACS. 12.38.-t Quantum chromodynamics – 11.15.Ha Lattice gauge theory – 12.38.Gc Lattice QCD calculations – 14.70.Dj Gluons

1 Introduction

In the pure gauge $SU(3)$ Yang-Mills theory, a number of authors has been using first principles approaches, *i.e.* Dyson-Schwinger equations (DSE) and lattice QCD methods, to investigate the infrared gluon and ghost propagators in Landau gauge, respectively,

$$D_{\mu\nu}^{ab}(k) = \delta^{ab} \left(\delta_{\mu\nu} - \frac{k_\mu k_\nu}{k^2} \right) D(k^2) , \tag{1}$$

$$G_{\mu\nu}^{ab}(k) = -\delta^{ab} G(k^2) , \tag{2}$$

and the strong-coupling constant [1] defined as

$$\alpha_S(k^2) = \alpha_S(\mu^2) \, Z_{ghost}^2(k^2) \, Z_{gluon}(k^2) ; \tag{3}$$

$Z_{ghost}(k^2) = k^2 G(k^2)$ and $Z_{gluon}(k^2) = k^2 D(k^2)$ are the ghost and gluon dressing functions. In part, these studies have been trigerred by the solution of the DSE [2] which assuming infrared ghost dominance and infrared finiteness of the loop-integrals predicts pure power laws for the dressing functions

$$Z_{gluon}(k^2) = A\left(k^2\right)^{\kappa'} , \quad Z_{ghost}(k^2) = B\left(k^2\right)^{-\kappa} . \tag{4}$$

Moreover, the DSE solution relates the exponents of the two propagators, $\kappa' = 2\kappa$, and predicts a finite strong-coupling constant at zero momentum. The infrared solution predicts a vanishing gluon propagator and an infinite ghost propagator at zero momentum ($\kappa > 0.5$). The

[a] e-mail: `orland@teor.fis.uc.pt`

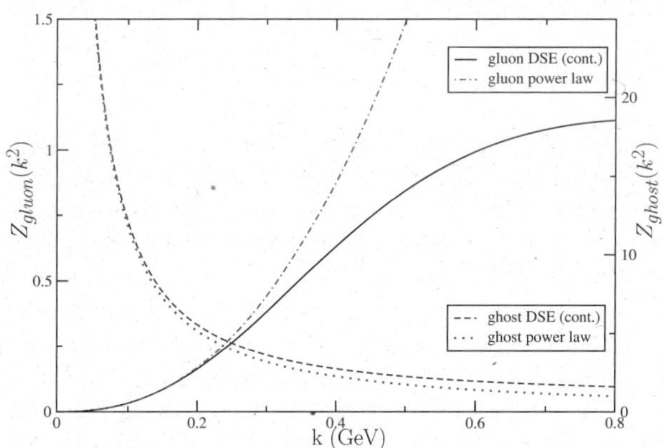

Fig. 1. DSE gluon and ghost dressing functions *versus* the pure power law solutions. The data is taken from [3].

infrared behaviour of the propagators can be related to gluon confinement criteria [4]. Looking at this particular DSE solution for the gluon propagator, the comparison with the pure power law, see fig. 1, shows that the pure power law is valid only for momenta $k < 200$ MeV. Notice that for the ghost, the deviations from a pure power law start earlier.

One should have in mind that the solution discussed above is a particular solution of the DSE. Indeed, there are, in the literature, different types of solutions for the DSE [5]. Moreover, the authors of [6] have investigated a generalization of the conditions assumed in [2] and found alternative behaviours for the infrared gluon and ghost propagators. Given the different scenarios, it would be important if lattice QCD can provide additional input, helping to check if any of the proposed solutions reproduces the lattice data. For a recent discussion on the infrared behaviour of the gluon and ghost propagators see [7].

For the lattice it is a challenge to investigate such low momenta. A possible way out is to consider large asymmetric lattices, which allows to access the momenta required in such an investigation [8–11]. Of course, there are lattice effects that have to be carefully estimated. Here we report on the gluon, ghost and strong-coupling constant computed from large asymmetric lattices.

2 Gluon propagator

In [11] we have investigated the infrared lattice Landau gauge gluon propagator. For notation and definitions see the above cited article. The simulation uses the Wilson action with $\beta = 6.0$ and combines a number of asymmetric lattices $L^3 \times 256$, $L = 8, 10, 12, 14, 16, 18$ to allow for $L \to +\infty$ extrapolation. In what concerns the time direction, the $16^3 \times 256$ and $16^3 \times 128$ gluon propagators were compared and no deviations were observed. This suggests that we have a sufficient number of points in the time direction.

For the lattices with the largest time direction, the minimum momentum being accessed is 47 MeV, while for the $16^3 \times 128$ the minimum momentum is 98 MeV. For the largest lattices with $T = 256$ a pure power law in the infrared is observed. In contrast, for the lattice of $16^3 \times 128$, power law fits are very poor with $\chi^2/\text{d.o.f} > 10$. This result is not surprising, given the smallest range of momenta available for this lattice (97–294 MeV) and the validity of the pure power law. Nevertheless, the $16^3 \times 128$ will allow us to estimate Gribov copies effects, with reasonable computational effort.

Let us summarize the results reported in [11] using the simulations with the largest time extension. Asymmetric lattices display finite size effects (see figs. 4 and 5 of [11]). However, the extrapolation towards $L \to +\infty$ is smooth. The extrapolated gluon propagator was well reproduced by a pure power law, with measured $\kappa = 0.49$–0.52. Clearly, the lattice data favors $\kappa \sim 0.52$, in agreement with other theoretical estimates. Unfortunately, one cannot yet provide a definitive answer concerning the value of the gluon propagator at zero momentum. We are currently engaged in trying to give such an answer.

For the $16^3 \times 128$ lattice, Gribov copies effects were estimated by comparing the gluon propagator computed from 164 configurations gauge fixed as in [11] (SD in the figures) and the same gauge configurations gauge fixed using the method described in [12] (CEASD in the figures), which aims to estimate the absolute maximum of the gauge fixing function. The gluon propagator given by the two methods is, within the statistical precision of the simulation, the same (see fig. 2), with the various fits to the lattice data (infrared, ultraviolet, all momenta) reproducing similar results. Therefore, no visible effects of Gribov copies are observed on the gluon propagator.

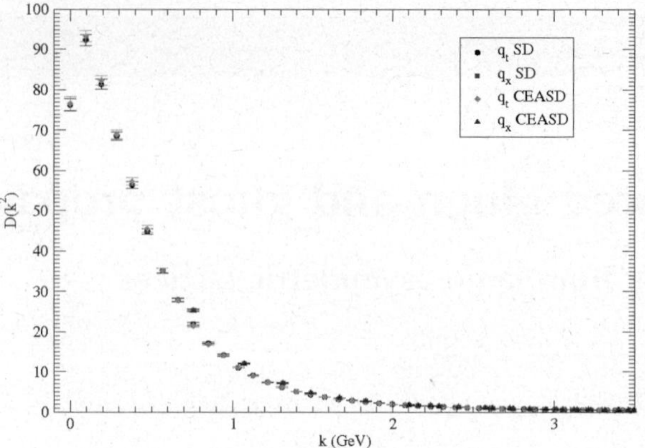

Fig. 2. Bare gluon propagator for $16^3 \times 128$.

3 Ghost propagator

The ghost propagator [13] was computed using two different methods [14,15]. The first allows, by solving a linear system, to access all momenta. Its drawback being that, by computing $G(x,0)$, the statistical errors in the propagator are much larger. The second method requires an inversion of the same linear system per momentum, which is more computationally demanding. However, for the momenta considered, it determines the propagator with better statistical accuracy. Computing the propagator using two methods provides a valuable cross-check.

In the following, for the ghost propagator the colour average was always performed. For the propagator computed with the [14] method, in order to reduce the statistical error, seven different sources for the linear system were considered and their result averaged.

For the ghost propagator, one observes sizeable finite-size effects (see fig. 3), in agreement with the $SU(2)$ study [16], and a very sign of clear Gribov copies effects for a large range of momenta.

In what concerns finite volume effects, fig. 3 shows that the ghost dressing function is enhanced when the lattice volume is increased. Moreover, the larger volumes have a larger dressing function and, for the smallest momenta,

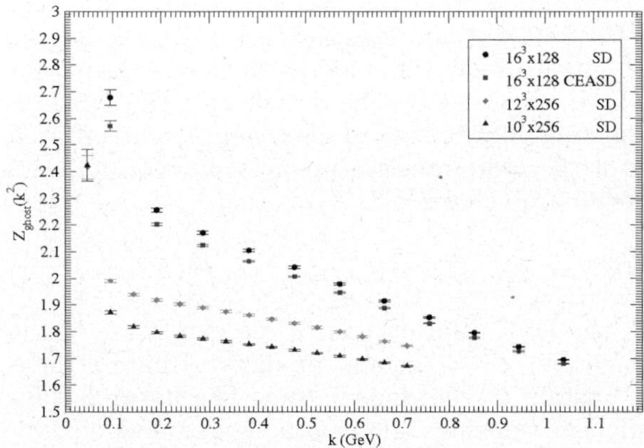

Fig. 3. Bare ghost dressing functions for a $16^3 \times 128$ and $L^3 \times 256$, $L = 10, 12$ lattices.

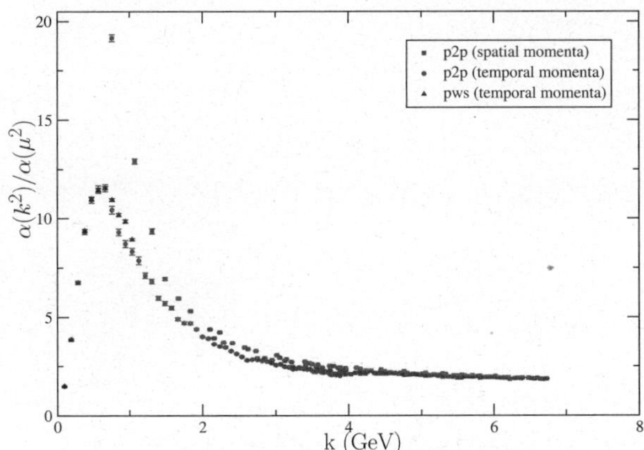

Fig. 4. Strong-coupling constant for $16^3 \times 128$ configurations for the CEASD gauge fixing method; "p2p" ("pws") stands for the ghost computed using [14] ([15]) method.

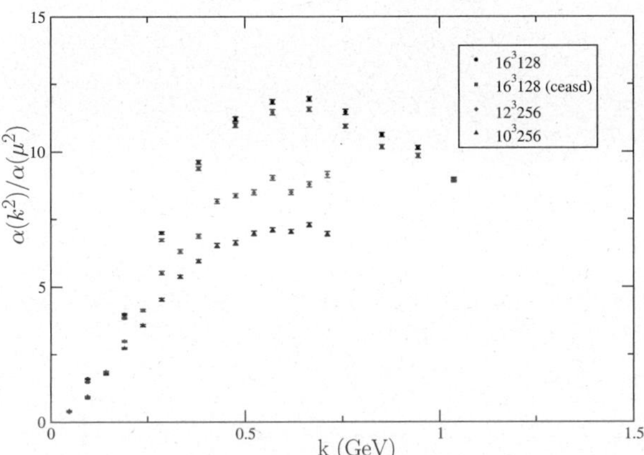

Fig. 5. Small-momenta strong coupling for all the simulations.

the derivatives of the dressing function seem to become smaller when the volume increases. The effect of Gribov copies, seen in fig. 3, cannot be eliminated by the renormalization of the propagator.

To study the compatibility of the ghost propagator with the DSE solutions, the lattice data was fitted to pure power laws and naïve corrections to the pure power law. It turns out that the data are not described by any of the functions considered. Given the results of a similar study for the $16^3 \times 128$ gluon propagator, this result is not a complete surprise.

4 Strong-coupling constant

The strong-coupling constant as defined by eq. (3) is displayed in fig. 4 for the $16^3 \times 128$ lattice and CEASD gauge fixing method. For the other simulations and the SD method, the measured coupling constants are, qualitatively, similar. Figure 5 shows the strong-coupling constant for small momenta, for all the simulations considered previously.

The simulations show a decreasing lattice strong-coupling constant for small momenta, in agreement with previous lattice calculations [17,18]. From our simulation, it is not clear if $\alpha_S(0)$ is vanishing. We have tried a number of fits to the strong-coupling constant as a function of the momentum, some requiring $\alpha_S(0) = 0$ and some leaving $\alpha_S(0)$ as a free parameter, and the data can be equally described by the two situations. For small momenta, only for the $16^3 \times 128$ data and the CEASD gauge fixing algorithm, is α_S described by a pure power law $\alpha_S(q^2) = (q^2)^{\kappa_\alpha}$, with $\kappa_\alpha = 0.69$ suggesting a vanishing strong-coupling constant at zero momentum. However, notice that the data show that α_S increases with lattice volume as seen in fig. 5 and is affected by Gribov copies. Further studies will hopefully lead to definitive conclusions.

References

1. For a review on strong-coupling constant see, for example, G.M. Prosperi, M. Raciti, C. Simolo, hep-ph/0607209.
2. L. von Smekal, A. Hauck, R. Alkofer, Phys. Rev. Lett. **79**, 3591 (1997); Ann. Phys. (N.Y.) **267**, 1 (1998).
3. C.S. Fischer, M.R. Pennington, Phys. Rev. D **73**, 034029 (2006).
4. For a review on DSE gluon and ghost propagators see, for example, C.S. Fischer, J. Phys. G **32**, R253 (2006) [hep-ph/0605173].
5. A.C. Aguillar, A.A. Natale, P.S. Rodrigues da Silva, Phys. Rev. Lett. **90**, 152001 (2003); A.C. Aguillar, A.A. Natale, JHEP **08**, 057 (2004).
6. Ph. Boucaud, J.P. Leroy, A. Le Yaouanc, A.Y. Lokhov, J. Micheli, O. Pène, J. Rodríguez-Quintero, C. Roiesnel, hep-ph/0507104.
7. C.S. Fischer, J.M. Pawlowski, hep-th/0609009.
8. O. Oliveira, P.J. Silva, AIP Conf. Proc. **756**, 290 (2005).
9. P.J. Silva, O. Oliveira, PoS (LAT2005) 286 (2006).
10. O. Oliveira, P.J. Silva, PoS (LAT2005) 287 (2006).
11. P.J. Silva, O. Oliveira, Phys. Rev. D **74**, 034513 (2006) [hep-lat/0511043].
12. O. Oliveira, P.J. Silva, Comput. Phys. Commun. **158**, 73 (2004) [hep-lat/0309184].
13. The definitions concerning the ghost propagator can be found in the articles where the lattice ghost propagator has been computed so far.
14. H. Suman, K. Schilling, Phys. Lett. B **373**, 314 (1996) [hep-lat/9512003].
15. A. Cucchieri, Nucl. Phys. B **508**, 353 (1997) [hep-lat/9705005].
16. A. Cucchieri, T. Mendes, Phys. Rev. D **73**, 071502 (2006) [hep-lat/0602012].
17. S. Furui, H. Nakajima, PoS (LAT2005) 291 [hep-lat/0509035]; hep-lat/0609024.
18. A. Sternbeck, E.-M. Ilgenfritz, M. Muller-Preussker, Phys. Rev. D **73**, 014502 (2006) [hep-lat/0510109]; **72**, 014507 (2005) [hep-lat/0506007].

Eur. Phys. J. A **31**, 793–798 (2007)
DOI 10.1140/epja/i2006-10177-6

Special Article – QNP 2006

Hadronic decays from the lattice

C. Michael[a]

Theoretical Physics Division, Department of Mathematical Science, University of Liverpool, Liverpool L69 3BX, UK

Received: 24 September 2006
Published online: 16 February 2007 – © Società Italiana di Fisica / Springer-Verlag 2007

Abstract. I review the lattice QCD approach to determining hadronic-decay transitions. Examples considered include $\rho \to \pi\pi$; $b_1 \to \pi\omega$; hybrid meson decays and scalar meson decays. I discuss what lattices can provide to help understand the composition of hadrons.

PACS. 13.25.-k Hadronic decays of mesons – 12.38.Gc Lattice QCD calculations – 12.39.Mk Glueball and nonstandard multi-quark/gluon states

1 Introduction

Relatively few hadronic states are stable to strong decays (*i.e.*, via QCD with degenerate u and d quarks). Among the mesons, we have [1]

Stable:	π, K, η, D, D_s, B, B_s, B_c, D_s^*, B^*, B_s^*, $D_s(0^+)$, $B_s(0^+)$;
$\Gamma < 1\,\mathrm{MeV}$:	η', D^*, $\psi(1S)$, $\psi(2S)$, χ_1, χ_2, $\Upsilon(1S)$, $\Upsilon(2S)$, $\Upsilon(3S)$;
$\Gamma < 10\,\mathrm{MeV}$:	ω, ϕ, χ_0, $X(3872)$;
$\Gamma > 10\,\mathrm{MeV}$:	ρ, f_0, a_0, h_1, b_1, a_1, f_2, f_1, a_2, etc., including η_c.

The mass of an unstable state is usually defined as the energy corresponding to a 90° phase shift. This definition[1] seems to accord with simple mass formulae: For example,

– $\rho(776)$ and $\omega(783)$ are close in mass despite having widths of 150 and 8 MeV, respectively.
– The baryon decuplet ($\Delta(1232)$, $\Sigma(1385)$, $\Xi(1530)$, $\Omega(1672)$) is roughly equally spaced in mass despite having widths of $(120, 37, 9, 0)\,\mathrm{MeV}$, respectively.

So, on the one hand, unstable particles seem to fit in well with stable ones; on the other hand, the presence of open decay channels will have an influence in lattice studies.

Some of the motivations to study hadronic decays on the lattice are:

– What are hadrons made of?
Is a meson made predominantly of $\bar{q}q$, $\bar{q}\bar{q}qq$ or meson-meson?

– A state that can decay strongly (resonance) necessarily has a meson-meson component —is this important?
Are unstable particles different from stable ones?
– What is the nature of light-light scalar mesons?
Where is the glueball?
– Are there hybrid mesons?

Lattice QCD is a first-principles method of attack. But we use unphysical (too heavy) quark masses, we have Euclidean time. So what can we learn?

2 Decays in Euclidean time

NO GO. At large spatial volume, the two-body continuum *masks* any resonance state. The extraction of the spectral function from the correlator $C(t)$ is ill-posed unless a model is made [3,4], since the low-energy continuum dominates at large t.

GO. For finite spatial volume (L^3), the two-body continuum is *discrete* and Lüscher showed [5–7] how to use the small energy shifts with L of these two-body levels to extract the elastic-scattering phase shifts. The phase shifts then determine the resonance mass and width, see ref. [8] for a review. Thus, a relatively broad resonance such as the ρ appears as a distortion of the $\pi_n\pi_{-n}$ energy levels where the pion momentum $q = 2\pi n/L$.

As a check of this approach to unstable particles on a lattice, the coupling of ρ to $\pi\pi$ has been determined from first principles [9]. The method is to arrange the ρ and $\pi\pi$ state (with definite relative momentum) to be approximately degenerate in energy on a lattice. Then, several independent methods allow to determine the transition amplitude x and, hence, the effective coupling constant $\bar{g}$ from the lattice (where decay does not proceed) and compare with experiment:

---

[a] e-mail: c.michael@liv.ac.uk

[1] Note that defining the mass as the real part of the pole will cause a downward shift of masses for wider states, *e.g.*, 22 MeV less [2] for the $\Delta(1232)$ pole, and this prescription will fit the equal mass rule less well.

Method	m_{val}	m_{sea}	$\bar{g}$
Lattice xt	s	s	1.40^{+47}_{-23}
Lattice ρ shift	s	s	1.56^{+21}_{-13}
$\phi \to K\bar{K}$	s	u, d	1.5
$K^* \to K\pi$	$u, d/s$	u, d	1.44
$\rho \to \pi\pi$	u, d	u, d	1.39

Note that the lattice has heavier sea quarks than experiment. Nevertheless, the level of agreement between first-principles lattice evaluation and experiment is very encouraging.

2.1 Hadronic transitions from the lattice

Here I summarise the steps that allow a fairly direct determination of hadronic transition amplitudes from the lattice.

i) Consider a lattice study of the off-diagonal correlator: from a ρ-meson to $\pi\pi$. Diagrammatically,

$$
\begin{array}{l}
\rho \to \pi\pi \\
0 \text{———} X \text{———} 0 \\
0 \quad\quad t \quad\quad\quad T
\end{array}
$$

ii) Now to evaluate this contribution, since the intermediate point marked X at time t is not observed on a lattice, it must be summed over:

$$
\sum_{t=0}^{T} e^{-m(\rho)t} x e^{-m(\pi\pi)(T-t)} \to xT e^{-mT}
$$

if $m(\rho) \approx m(\pi\pi)$.

iii) So a plot of the (normalised) transition from the lattice *versus* T has slope of x, which can be related [9] to the continuum coupling g^2.

Note this approach is like measuring the wave function overlap —but with no model for the wave functions. In order to control excited-state contributions, however, it is subject to the restriction that initial and final states have similar energies on a lattice. A more rigorous approach is possible by using the method of determining the two-body energy *versus* lattice volume, described above, but in practice sufficient precision is not available in general to study resonance decays, although results for scattering lengths have been obtained.

3 Hybrid meson decay

One of the characteristic predictions of QCD is that there can be mesons in which the gluonic degrees of freedom are non-trivially excited. The simplest example is a hybrid meson with spin-exotic $J^{PC} = 1^{-+}$ which is a J^{PC} combination not available to a $\bar{q}q$ state. The spin-exotic quantum numbers then require a non-trivial gluonic contribution to the state. These hybrid mesons have been studied extensively on the lattice, here I discuss their decay.

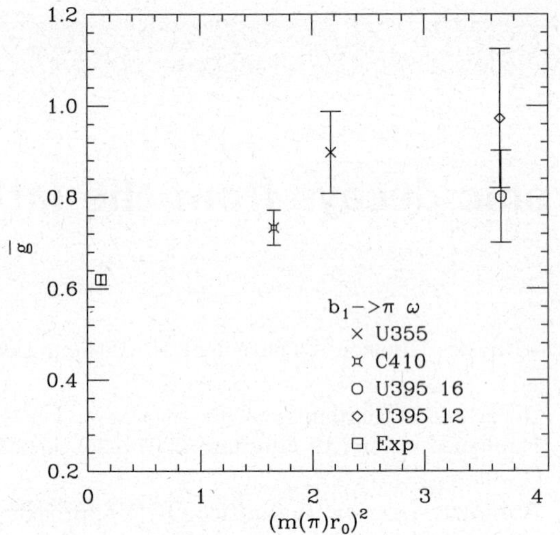

Fig. 1. First test of S-wave decays from the lattice for $b_1 \to \pi\omega$ from ref. [10]. The effective coupling $\bar{g}$ is plotted *versus* the quark mass, as determined by the pseudoscalar mass squared where the scale is set by $r_0 \approx 0.5$ fm.

3.1 Light quarks

The S-wave decay of the $J^{PC} = 1^{-+}$ spin-exotic hybrid meson($\hat{\rho}$) to πb_1 has been studied recently on the lattice [10]. Since S-wave decays have not previously been studied in this way, a check was made by extracting the decay strength for the S-wave component of the transition $b_1 \to \pi\omega$. As shown in fig. 1, the lattice determination, though using heavier quarks than experiment, fits in well. This gives confidence that the hybrid decay prediction will be reliable. Since the experimental effective coupling constant lies lower than the lattice results, this also shows that lattice results, with heavier than experimental quark masses, may overestimate the coupling somewhat. This may be interpreted, phenomenologically, as arising from form factor effects [11].

From lattices with $N_f = 2$ flavours of sea quark, a recent study [10] obtains a spin-exotic hybrid meson state at $2.2(2)$ GeV. The S-wave decay transitions are then evaluated, as shown in fig. 2, obtaining a partial width to πb_1 of $400(120)$ MeV and to πf_1 of $90(60)$ MeV. These results indicate that the decay width of this hybrid meson may be large and, hence, more difficult to extract experimentally.

For a recent preliminary lattice study of decay of a $J^{PC} = 1^{-+}$ hybrid meson to πa_1 using the Lüscher method which obtains a width of around 60 MeV, see ref. [12]. This decay channel implies that the hybrid meson considered has $I = 0$, rather than $I = 1$ as above.

3.2 Heavy quarks

The cleanest environment in which to study such hybrid states on a lattice is in the limit of very heavy quarks which is relevant to $\bar{b}b$. This can be approximated by using static quarks and the gluonic excitation arises as an

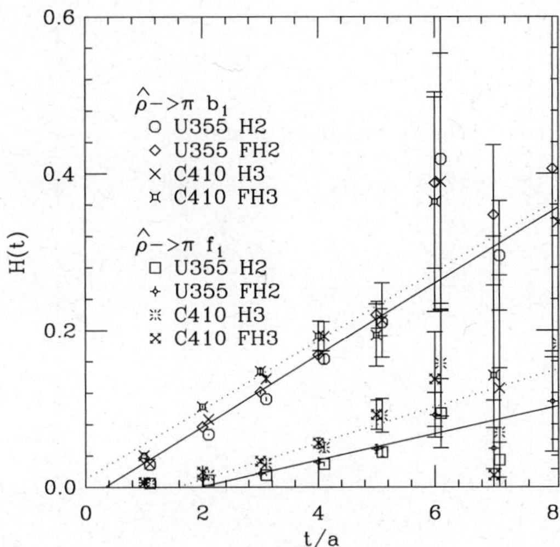

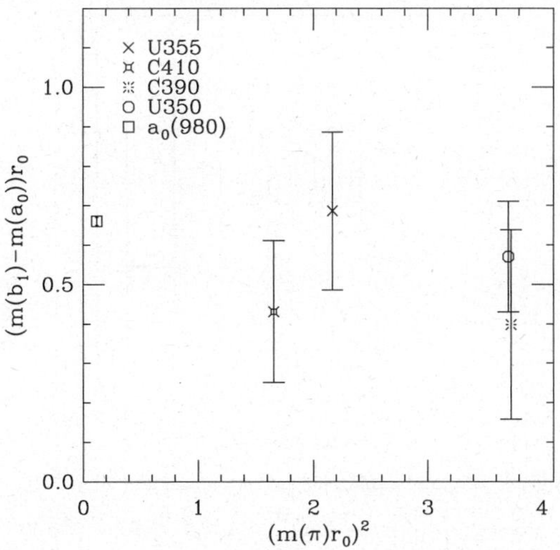

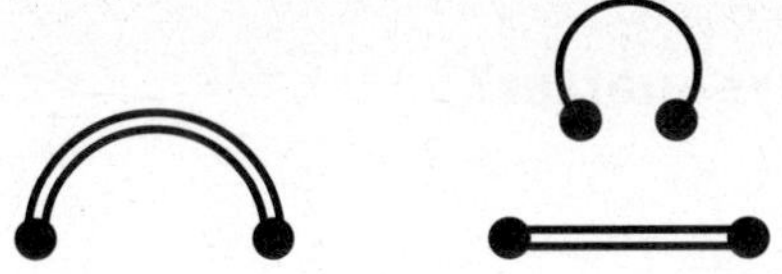

Fig. 2. Spin-exotic hybrid decay transition from ref. [10]. Here the decay transition strength is given by the slope.

Fig. 3. Initial and final states relevant for the decay of the heavy-quark spin-exotic hybrid meson to $\chi_b f_0$.

Fig. 4. Mass difference of b_1 and a_0 mesons from ref. [19] plotted *versus* the quark mass, as determined by the pseudoscalar mass squared with the scale is set by $r_0 \approx 0.5\,\mathrm{fm}$. The open box is the experimental point if the a_0-meson is at 980 MeV.

- 0^{++} glueball decay $\to \pi\pi$: quenched study [14,15].
- Glueball mixing with $q\bar{q}$-meson. This hadronic transition has been studied using quenched [16] and dynamical lattices [17].

A full lattice study is needed which includes glueball, $\bar{q}q$ and $\pi\pi$ channels but the disconnected diagram for $f_0 \to \pi\pi$ is very noisy in practice —as shown in ref. [18].

To reduce the contribution from disconnected diagrams, one can study flavour non-singlet scalar mesons. The simplest case is a_0 which has a decay $\eta\pi$ and this has been explored in quenched studies which have an anomalous behaviour: since the η itself is unphysical (appearing as a double pole degenerate in mass with the pion). Rather than try to correct for this anomaly which gives a wrong sign to the a_0 correlator at larger t, it is preferable to use a ghost-free theory. With two flavours of sea quark ($N_f = 2$), this problem is avoided. A recent study [19] of the a_0-meson concentrates on the mass difference between it and the b_1-meson, as shown in fig. 4. This study concludes that the $\bar{q}q$ non-singlet scalar meson lies around 1 GeV —which is considerably lighter than some previous lattice studies (see ref. [19] for a summary).

As well as determining the mass values, this study evaluates the $a_0 \to \eta\pi$ and $a_0 \to KK$ transitions. Results from the connected contribution to these transitions are shown in fig. 5. The resulting coupling constant is determined to be of similar value to that obtained by some phenomenological studies of decays of the $a_0(980)$-meson. This again points to the possibility that the $a_0(980)$ may be substantially a $\bar{q}q$ state.

A full study of flavour singlet scalar mesons will need to take into account the $\bar{q}q$, gluonic and meson-meson channel and their mixing. This has not yet been achieved, for a summary of the current state of lattice studies see ref. [18].

excited string state between these static quarks with nontrivial gluonic angular momentum. Lattice studies have long predicted the spectrum of such states.

To guide experiment, however, it is important to know the expected decay mechanism and associated width. In the static quark limit, several symmetries can be used which imply [13] that the dominant decay will be string de-excitation (rather than string breaking) as illustrated in fig. 3. Lattice study [13] shows that the dominant decay of the hybrid meson H_b is string de-excitation to $\chi_b f_0$. The width is predicted to be around 80 MeV.

This estimate from first principles of the decay width is of significance in guiding experimental searches for such hybrid states.

4 Scalar mesons

4.1 Light quarks

Since $u\bar{u} + d\bar{d}$, $s\bar{s}$, glueball, and meson-meson components are all possible for flavour-singlet scalar mesons, this is a difficult area to study both on a lattice, and in interpreting experimental data. For scalar mesons the lowest mass decay channels are $\pi\pi$ (flavour singlet: f_0) or $\eta\pi$ (flavour non-singlet: a_0) and these decay channels are open in many dynamical lattice studies. The history of lattice attempts to study the complex mixing between these different contributions is

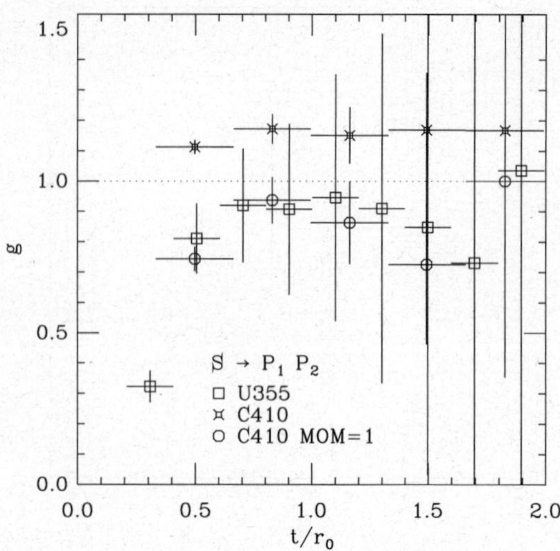

Fig. 5. Effective coupling from connected contribution to decay transition of scalar meson $\to$ 2 pseudoscalars from ref. [19]. The t region around 0.5 to 1.5 is expected to be relevant (in units with $r_0 \approx 0.5\,\text{fm}$).

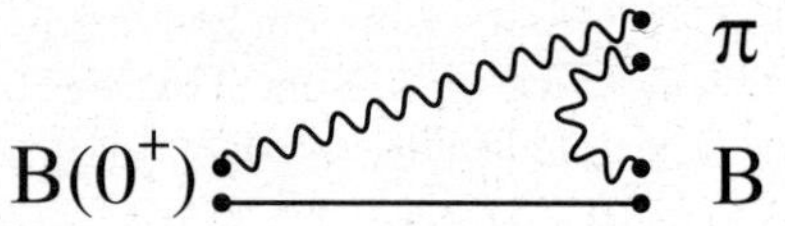

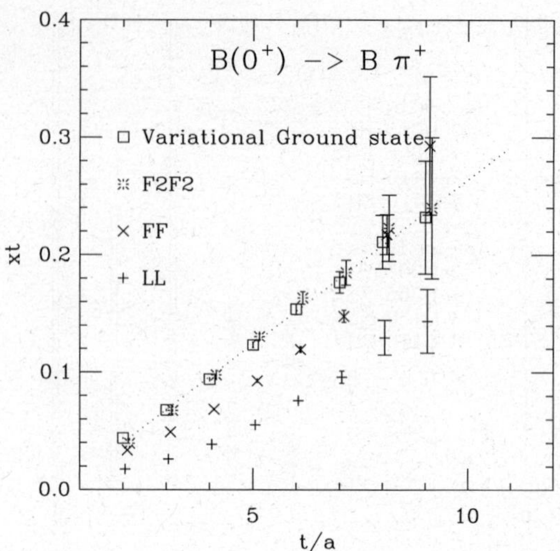

Fig. 7. $B(0^+) \to B\pi$ transition strength (given as slope) on a lattice from ref. [21].

$B(0^+)$ $\qquad$ π $\qquad$ B

Fig. 6. Diagram for the $B(0^+) \to B\pi$ transition.

4.2 Heavy-light quarks

One of the most promising ways to study scalar mesons on lattice is through heavy-light mesons. The scalar meson with $\bar{c}s$ quantum number is known experimentally [1] to be very narrow (it decays only via the isospin-violating channel $D_s\pi$ or electromagnetically), while the scalar meson with $\bar{b}s$ quantum number is predicted to be similarly narrow from a lattice study of its energy [20].

Here we consider the heavy-light scalar meson in the limit of a static heavy quark —which will be relevant to mesons with a b-quark. Lattice studies indicate [20] that the $\bar{b}s$ scalar meson is expected to be stable to strong decay while the $\bar{b}n$ scalar meson (where $n = u, d$) will decay to $B\pi$. Evaluating the diagram shown in fig. 6, a lattice estimate, shown in fig. 7, of the decay rate of $B(0^+) \to B(0^-)\pi$ gives a width predicted [21] as $162(30)\,\text{MeV}$. This state has not been observed experimentally yet, but the experimental results for the corresponding $\bar{c}n$ state, $D(0^+)$, are that the width is $270 \pm 50\,\text{MeV}$. Although significant $1/m_Q$ effects are expected in the HQET in extrapolating to charm quarks, this is indeed a similar magnitude to that predicted for B mesons. It will be interesting so see how the lattice prediction of the mass and width of the $B(0^+)$ fares when experimental results are available.

5 Do decays matter?

The previous discussion of decays on a lattice emphasises that $q\bar{q}$ states do mix with two-body states with the same quantum numbers. In the real world, the two-body states are a continuum and nearby states have a predominant influence. Since there is a suppression in the amplitude near threshold from the factor q^L for an L-wave transition, for S-wave transitions ($L = 0$) the threshold will turn on most abruptly and hence will have stronger mixing.

For bound states there is an influence of nearby many-body states (*e.g.*, $N\pi$ on N or $\pi\pi\pi$ on π) which mix to reduce the mass. The nearest such thresholds will be those with pionic channels since pions are the lightest mesons. This is the province of low-energy effective theories, especially chiral perturbation theory which is discussed in other talks. This then provides a reliable guide in extrapolating lattice results to the physical light quark masses.

For unstable states (resonances) the influence of the two-body continuum is less clear since the two-body states are both lighter and heavier. In the continuum at large volume, effective field theories can again be used to explore this. On a lattice, however, the signal for a particle becomes obscured as the quark mass is reduced so that it becomes unstable. Techniques, such as those discussed above, are needed to extract the elastic-scattering phase shift and hence the mass and width.

For quenched QCD, however, where these two-body states are not coupled (or have the wrong sign as in $a_0 \to \eta\pi$), then the unstable states will be distorted compared to full QCD. For instance, in existing quenched QCD studies, the ρ will be *too heavy* since it is not repelled by the heavier $\pi\pi$ states. Indeed an example of this effect was seen above in the study of ρ decay including dynamical sea quarks, where the ρ mass decreased [9] when it could couple to $\pi\pi$ compared to when it could not.

6 Molecular states?

Can lattice QCD provide evidence about possible molecular states: hadrons made predominantly of two hadrons?

The prototype is the deuteron: $n\,p$ bound in a relative S-wave (with some D-wave admixture) by π exchange.

There are many states close to two-body thresholds. Since S-wave thresholds are the most abrupt, it is usually in this case that the influence of the threshold on the state has been discussed. Some of these cases are

$$
\begin{aligned}
f_0(980)\ a_0(980) &\leftrightarrow K\overline{K}, \\
D_s(0^+) &\leftrightarrow D(0^-)K, \\
B_s(0^+) &\leftrightarrow B(0^-)K, \\
X(3872) &\leftrightarrow D^*\overline{D}, \\
\Lambda(1405) &\leftrightarrow \overline{K}N, \\
N(1535) &\leftrightarrow \eta N.
\end{aligned}
$$

Some of these states ($D_s(0^+)$, $B_s(0^+)$) are stable (in QCD in the isospin-conserving limit) whereas the rest have other channels open. There is a large literature, stretching over 40 years, discussing the consequences of the nearby threshold on these states. One definite implication is that isospin breaking is enhanced by mass splittings in thresholds (e.g., $\overline{K^0}K^0$ compared to K^+K^- is 8 MeV higher and this induces isospin mixing between the states at 980 MeV). This level of detail is not accessible in lattice studies at present, but lattice QCD should be able to address the issue of the influence of thresholds on these states.

The observation of a state near a 2-body threshold implies that there is an attractive interaction between the two bodies. But this is a topic like that of whether the chicken or egg was created first: an attractive interaction implies and is implied by a nearby state. What can lattice QCD offer here? We are in the position of being able to vary the quark masses and this is a very useful tool. A two-body threshold will move in general in a different way with changing quark mass than a $\overline{q}q$ state. We can also move the strange and non-strange masses separately and this can be helpful too.

Another line of investigation is that lattice studies can explore the wave function of a state —either the Bethe-Salpeter wave function or the charge or matter spatial distribution. One can also explore the coupling of a state to a 2-body channel, as was discussed above.

The prototype of a molecular state is the deuteron: it has a tiny binding energy (2.2 MeV) and a very extended spatial wave function. Pion exchange between neutron and proton gives a mechanism for this long-range attraction. In general, it is difficult to reproduce such small binding energies in lattice studies.

Another case where a long-range pion exchange can give binding is in the BB system. Here lattice results indicate [22] the possibility of molecular bound states in some quantum number channels which have an attractive interaction from pion exchange, but also the possibility of bound multi-quark states which are not described as hadron-hadron but where the two heavy quarks form a colour triplet and the light quarks are arranged as in a

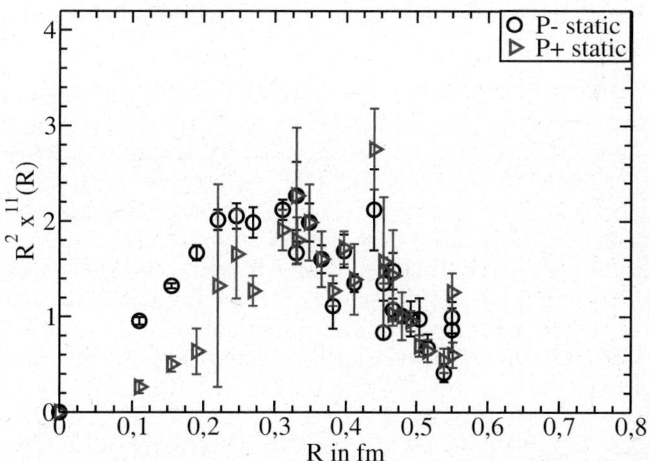

Fig. 8. Charge distributions of $B_s(0^+)$ and $B_s(2^+)$ mesons from the lattice. The heavy quark is static and the light quark is at a distance R from it. The $J^P = 0^+$ meson is P_-, while the $J^P = 2^+$ meson is P_+.

heavy-light-light baryon. This BB example illustrates the rich structure available to multi-quark systems.

One case where lattice studies have been able to shed considerable light is for the $\overline{b}s$ and $\overline{c}s$ scalar mesons. Since, in the isospin-conserving limit, these states cannot couple to $B_s\pi$, $D_s\pi$, they have a lightest open decay channel BK, DK. Lattice studies [20,23] indicate that, in both cases, the scalar meson lies below the open threshold, so the states should be stable.

Because simple quark model expectations were that these scalar mesons were unstable, theorists have suggested that a BK, DK molecular composition was responsible. This can be explored on a lattice by measuring the spatial distribution of the heavy-light meson. A study [24] of the charge distribution of the light quark in a $B_s(0^+)$-meson is illustrated in fig. 8. This shows that the light quark spatial distribution is similar to that of other $\overline{Q}q$ states (e.g., $J^P = 2^+$) for which no molecular interpretation is proposed. This reinforces the conclusion that the $B_s(0^+)$ is predominantly a $\overline{b}s$ state. The decay transition from $B_s(0^+)$ to BK has also been determined (see above) and it has an effective coupling constant which is consistent with that found for other (non-molecular) decays. Overall, lattice evidence does not support the hypothesis that $B_s(0^+)$ is a molecular state.

7 Conclusions

Lattice can address hadronic structure:

- Form factors (e.g., charge wave functions) can be evaluated;
- Decay transitions (and mixing transitions) can be evaluated;
- Structure function moments can be evaluated (steady but slow progress here);
- Hadronic matrix elements are needed to interpret experiment (e.g., f_B relates the B-meson to the b-quark,

etc.) in searches for signs of physics beyond the standard model.

Hadronic physics involves *unstable* states and lattice techniques are being developed to study these as we have summarised. These techniques have been tested against experiment for $\rho \to \pi\pi$ and $b_1 \to \pi\omega$. In particular, we presented evidence that the spin exotic hybrid meson (made of light quarks) is at a mass around $2\,\mathrm{GeV}$ and has a wide width. We discussed scalar mesons, and presented evidence that the $a_0(980)$-meson is basically a $q\bar{q}$ state and that the $B_s(0^+)$-meson is also predominantly a $b\bar{q}$ state.

There is a lot to be learnt from the lattice beyond mass spectra.

References

1. Particle Data Group (S. Eidelman *et al.*), Phys. Lett. B **592**, 1 (2004).
2. C. Michael, Phys. Rev. **156**, 1677 (1967).
3. C. Michael, Nucl. Phys. B **327**, 515 (1989).
4. L. Maiani, M. Testa, Phys. Lett. B **245**, 585 (1990).
5. M. Luscher, Commun. Math. Phys. **104**, 177 (1986).
6. M. Luscher, Commun. Math. Phys. **105**, 153 (1986).
7. M. Luscher, Nucl. Phys. B **354**, 531 (1991).
8. M. Luscher, Nucl. Phys. B **364**, 237 (1991).
9. UKQCD Collaboration (C. McNeile, C. Michael), Phys. Lett. B **556**, 177 (2003) hep-lat/0212020.
10. UKQCD Collaboration (C. McNeile, C. Michael), Phys. Rev. D **73**, 074506 (2006) hep-lat/0603007.
11. T. Burns, F.E. Close, Phys. Rev. D **74**, 034003 (2006) hep-ph/0604161.
12. M.S. Cook, H.R. Fiebig, Phys. Rev. D **74**, 034509 (2006) hep-lat/0606005.
13. UKQCD Collaboration (C. McNeile, C. Michael, P. Pennanen), Phys. Rev. D **65**, 094505 (2002) hep-lat/0201006.
14. J. Sexton, A. Vaccarino, D. Weingarten, Phys. Rev. Lett. **75**, 4563 (1995) hep-lat/9510022.
15. J. Sexton, A. Vaccarino, D. Weingarten, Nucl. Phys. Proc. Suppl. **47**, 128 (1996) hep-lat/9602022.
16. W.J. Lee, D. Weingarten, Phys. Rev. D **61**, 014015 (2000) hep-lat/9910008.
17. UKQCD Collaboration (C. McNeile, C. Michael), Phys. Rev. D **63**, 114503 (2001) hep-lat/0010019.
18. A. Hart, C. McNeile, C. Michael, J. Pickavance, hep-lat/0608026 (2006).
19. UKQCD Collaboration (C. McNeile, C. Michael), Phys. Rev. D **74**, 014508 (2006) hep-lat/0604009.
20. UKQCD Collaboration (A.M. Green, J. Koponen, C. McNeile, C. Michael, G. Thompson), Phys. Rev. D **69**, 094505 (2004) hep-lat/0312007.
21. UKQCD Collaboration (C. McNeile, C. Michael, G. Thompson), Phys. Rev. D **70**, 054501 (2004) hep-lat/0404010.
22. UKQCD Collaboration (C. Michael, P. Pennanen), Phys. Rev. D **60**, 054012 (1999) hep-lat/9901007.
23. UKQCD Collaboration (A. Dougall, R.D. Kenway, C.M. Maynard, C. McNeile), Phys. Lett. B **569**, 41 (2003) hep-lat/0307001.
24. UKQCD Collaboration (A.M. Green *et al.*), PoS (LAT2005) 205 (2006) hep-lat/0509161.

Eur. Phys. J. A **31**, 799–803 (2007)

DOI 10.1140/epja/i2006-10267-5

THE EUROPEAN
PHYSICAL JOURNAL A

Special Article – QNP 2006

Lattice QCD and nuclear physics

Computations of hadron-hadron scattering

K. Orginos[a]

Department of Physics, College of William and Mary, Williamsburg, VA 23187-8795, USA and
Jefferson Laboratory, 12000 Jefferson Avenue, Newport News, VA 23606, USA

Received: 8 December 2006
Published online: 23 March 2007 – © Società Italiana di Fisica / Springer-Verlag 2007

Abstract. A steady stream of developments in lattice QCD have made it possible today to begin to address the question of how nuclear physics emerges from the underlying theory of strong interactions. A central role in this understanding play both the effective-field theory description of nuclear forces and the ability to perform accurate non-perturbative calculations in low-energy QCD. Here I present some recent results that attempt to extract important low-energy constants of the effective-field theory of nuclear forces from lattice QCD.

PACS. 21.30.-x Nuclear forces – 13.75.Cs Nucleon-nucleon interactions (including antinucleons, deuterons, etc.) – 12.38.Gc Lattice QCD calculations – 12.38.-t Quantum chromodynamics

1 Introduction

Computing from first principles the properties of hadron interactions that lead to the formation of atomic nuclei is a major challenge for lattice QCD. The fact that the QCD dynamics is at scales of 1 GeV while the nuclear forces are resulting effects of the order a few MeV creates a classic two-scale problem. Effective-field theory techniques are a powerful tool to tackle such problems. The matching though of the effective-field theory to the high-energy scale theory, namely QCD, requires non-perturbative calculations within QCD. Lattice QCD is the only way to perform such calculations.

For realistic lattice calculations dynamical fermions with pion masses below 400 MeV are needed. This allows chiral effective-field theories to be used with some reliability. In addition, a dynamical strange quark is required in order to guarantee that the low-energy constants of the chiral Lagrangian match those of the physical theory. Large physical volumes are also needed so that finite volume systematic errors are under control. Although this task seems formidable, in the last several years there were developments in lattice QCD calculations that permit the performance of phenomenologically interesting calculations that address these questions.

The emergence of fermions that respect chiral symmetry [1–4] on the lattice was one of the major recent developments in lattice QCD. These formulations of lattice fermions allow us to reduce the lattice spacing errors and approach the continuum limit in a smoother manner.

In addition, the development of improved Kogut-Susskind fermion actions [5,6] that significantly reduce the $O(a^2)$ errors, allowed for cheap inclusion of quark loop effects in the QCD correlation functions computed on the lattice. With this formulation we can work at volumes as large as 3.5 fm and quark masses as low as 1/10th of the strange-quark mass.

In the work I present here, we used domain wall fermions for the valence sector and Kogut-Susskind fermions to represent the quark loops. This mixed action calculation allows us to both take advantage of the chiral symmetry properties of domain wall fermions and have quark loops with masses close to the physical regime. Although one might think that this mixed action scheme is a complication difficult to control, in practice it has been shown that the effects of the mismatch between the sea and the valence sectors are small in the case of flavor non-singlet quantities. In theory all these complications can be taken care of in the context of mixed action chiral perturbation theory [7,8]. Another problem with Kogut-Susskind fermions is that there is still a theoretical issue of the validity of computations with number of flavors not an integer multiple of four. However, recent theoretical work indicates that the troublesome non-localities of the lattice action are going away in the continuum limit [9–12]. Hence these effects are most likely taken care of in the continuum limit.

Given the available technology for lattice QCD calculations, there is a variety of physical observables with direct impact to nuclear physics that can be computed. The nucleon mass spectrum, decay constants and axial couplings

---

[a] e-mail: `kostas@wm.edu`

are now standard lattice calculations that can be done with very good precision with reliable control of the systematic errors involved. In addition to these types of calculations, more recently, lattice QCD calculations of scattering lengths and phase shifts have began to emerge [13–15] In the following, I am presenting a few of such calculations that I have worked on with my collaborators in LHPC and NPLQCD.

2 Two particles in a box

Although the energy levels of single-particle states can be easily computed from Euclidean two-point correlation functions, scattering amplitudes cannot be extracted from Euclidean four-point functions in infinite volume except on kinematic thresholds, as stated by the Maiani-Testa no-go theorem. However, Lüscher showed [16,17] that the s-wave scattering amplitude for two particles below inelastic thresholds can be determined using the measurements of one or more energy levels of the two-particle system in a finite volume. In particular, he showed that in the center-of-mass frame, the s-wave energy levels of two particles of identical mass m in a finite volume are shifted from those of two non-interacting particles by an amount that is related to the scattering amplitude. This energy level shift is $\Delta E_n \equiv E_n - 2m = 2\sqrt{p_n^2 + m^2} - 2m$, where p_n is defined by this equation and satisfies

$$p_n \cot \delta(p_n) = \frac{1}{\pi L} \mathbf{S}\left(\frac{p_n^2 L^2}{4\pi^2}\right) = \frac{1}{a} + \frac{1}{2} r p_n^2 + \cdots, \quad (1)$$

where $\delta(p_n)$ is the elastic-scattering phase shift, a and r are the scattering length and effective range, respectively, L is the length of the spatial dimension in a cubically symmetric lattice, and

$$\mathbf{S}(\eta) \equiv \lim_{\Lambda \to \infty}\left[\sum_{\mathbf{j}}^{|\mathbf{j}|<\Lambda} \frac{1}{|\mathbf{j}|^2 - \eta} - 4\pi\Lambda\right]. \quad (2)$$

This definition is equivalent to the analytic continuation of zeta-functions presented by Lüscher [18].

3 The lattice calculation

In our computation we use the mixed-action scheme developed by LHPC [19,20] using domain-wall valence quarks on $N_f = 2 + 1$ asqtad-improved [5,6] MILC configurations generated with rooted[1] staggered sea quarks [21]. For more details on the numerical calculations see [19, 20,15,14]. The essential feature of our calculation is that the pion masses ranged between 300 MeV and 770 MeV and that the strange-quark mass was fixed near its physical mass. The domain wall fermion mass was tuned so

---

[1] The "legality" of the rooting procedure has been questioned and investigated in recent lattice literature. Here, due to space limitations, I cannot address the issues raised fully.

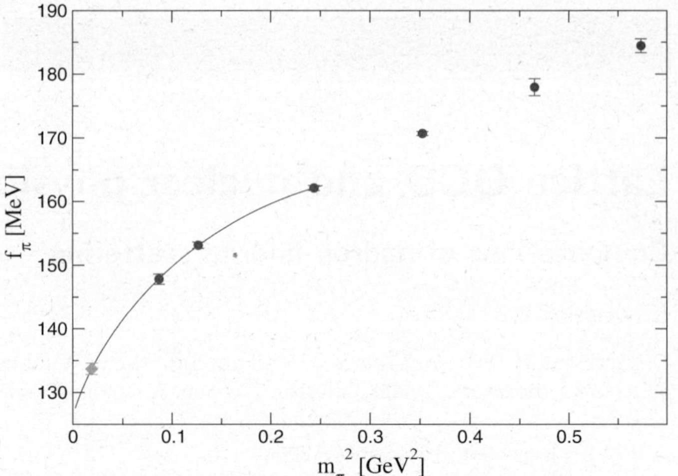

Fig. 1. The pion decay constant. Only the lighter three points are used in the fit. The diamond is the extrapolated value with its statistical error.

that the domain wall pions matched the Kogut-Susskind Goldston pion mass. The scattering length is extracted by computing on the lattice the energy level shift ΔE_0. In order to do this, the ratio of correlators of the two-particle state to the product of the single-particle correlators is fitted to a single exponential [15,14]. The decay rate of this exponential is the energy level shift we need.

3.1 Decay constants

The pseudoscalar decay constants are important parameters of the chiral Lagrangian which we would like to compute from lattice QCD. These constants have been computed lately to great precision by the MILC Collaboration [22,23] using staggered fermions. The dominant error (3%) in these calculations is systematic (due to scale setting, chiral extrapolations, continuum limit) rather than statistical. Within all the errors the results are in remarkable agreement with experiment. We have repeated these calculations using the mixed action scheme described above. Our results for the ratio of the kaon to pion decay constant ratio are presented in detail in [24]. We find that $f_K/f_\pi = 1.218(2)^{+0.011}_{-0.024}$, where the first error is statistical and the second is systematic due to the chiral extrapolation. In addition, by fitting to leading-order chiral perturbation theory formulas [25] we obtain $f_\pi = 133.7(9)(3.0)$ MeV, where the second error is systematic due to the chiral fit. The fit shown in fig. 1 results in a low-energy constant $\bar{l}_4^{phys} = 4.39(3)$ compatible with the experimental expectations and the lower statistics result obtained in [15]. The lower four solid points plotted in fig. 1 are high-statistics NPLQCD data while the rest are LHPC data. One remarkable feature of this plot is the fact that for pion masses above 400 MeV the data are almost linear while below there is a clear indication of the chiral log curvature.

Although in principle one needs to use the effective theory that takes into account the taste breaking effects in

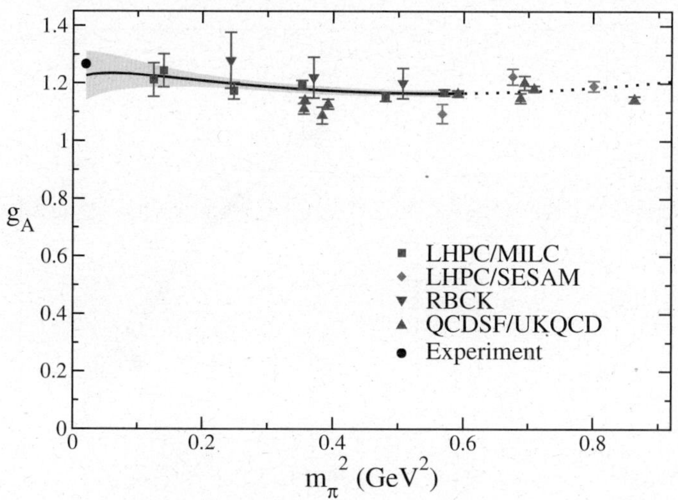

Fig. 2. The nucleon axial coupling.

the sea sector [7], in practice it turns out that the regular continuum chiral perturbation theory formulas fit our data quite well, leading us to believe that the taste breaking effects on the decay constant are rather small.

The compatibility of our results with those obtained by the MILC Collaboration is very encouraging for the mixed-action scheme we have adopted.

3.2 Nucleon axial coupling

Another low-energy constant useful for the effective-field theories needed to describe the nuclear forces is the nucleon axial coupling. The ability to reproduce the well-known experimental result has been a challenge for lattice QCD. Recently though, the use of light pion masses together with the understanding of the finite-volume corrections [26–28] has led to remarkable improvement in the lattice results. In fig. 2 the results by LHPC presented in [27] are shown. Our chiral extrapolation using the finite-volume formulas of [29] obtain a value of $g_A = 1.22(8)$ which is in agreement with the experimental result. This result still has unknown systematic errors due to continuum and chiral extrapolation, but we expect both these errors to be smaller, or at least comparable to our statistical error.

Recent work by Bernard and Meissner [30], going to two-loop chiral perturbation theory, indicates that one may need pion masses below 300 MeV before reliable chiral extrapolations can be obtained. Certainly, the lattice results shown here indicate that improving the control of the chiral extrapolation would definitely benefit from lower masses. In fact it seems that this is the only way one can improve the precision on the g_A calculation both from the point of view of statistical and systematic errors. LHPC is currently working in pushing closer to the chiral limit. Using the Kogut-Susskind lattices produced by MILC it is expected to be able to perform calculations in the range of 250 MeV pions. In addition, calculations on a smaller

lattice spacing is in the pipe line in order to control the continuum extrapolation error.

3.3 Charge symmetry breaking

Recently [31], using partially quenched chiral perturbation theory, we have calculated the proton-neutron mass splitting due to charge symmetry breaking. Although this splitting is small one can obtain an accurate estimate in lattice calculations taking advantage of the statitsical correlations between quantities computed on the same ensemble. The neutron-proton mass difference due to the up-down quark mass splitting ($m_u - m_d$) is related to this splitting by

$$M_n - M_p\big|^{d-u} = \frac{2}{3}\left(2\overline{\alpha} - \overline{\beta}\right)\left(\frac{1-\eta}{1+\eta}\right) m_\pi^2, \qquad (3)$$

where $\eta = m_u/m_d$ and $2\overline{\alpha} - \overline{\beta}$ is the strong iso-spin breaking term in the chiral expansion of the proton (neutron) mass. Partially quenched calculations with non-degenerate up and down quarks, *i.e.* keeping the sea quarks degenerate, allow us to compute from lattice calculations the term $2\overline{\alpha} - \overline{\beta}$ to great precision. Using $\eta = 0.43(1)(8)$ determined by MILC [23] we can obtain

$$M_n - M_p\big|^{d-u} m_\pi^{phys} = 2.26(57)(42)\,\mathrm{MeV} \qquad (4)$$

using the $O(m_q^{3/2})$ partially quenched formulas. This is to be compared to the experimental value of

$$M_n - M_p\big|_{exp}^{d-u} = 2.05(30)\,\mathrm{MeV} \qquad (5)$$

after correcting for electromagnetic effects [32].

3.4 Pion scattering

The scattering length is the observable that is directly related to the hadron-hadron interactions. Using Luscher's finite-volume technique described above we computed the pion-pion scattering length in the $I = 2$ channel. Our results are plotted in fig. 3 together with the lightest CP-PACS point [13]. This calculation is the first dynamical fermion calculation at these light pion masses. The lattice results are fitted using the one-loop chiral perturbation formula [25]

$$m_\pi a_2 = -\frac{m_\pi^2}{8\pi f_\pi^2}\left[1 + \frac{3m_\pi^2}{16\pi^2 f_\pi^2}\left(\log\frac{m_\pi^2}{f_\pi^2} + l_{\pi\pi}(f_\pi)\right)\right], \quad (6)$$

where $l_{\pi\pi}(f_\pi)$ is the Gasser-Leutwyler low-energy constant which scales as

$$l_{\pi\pi}(\mu) = l_{\pi\pi}(f_\pi) + 2\log(\mu/f_\pi). \qquad (7)$$

The fitted result at the physical point is

$$m_\pi a_2 = -0.0422(3)(18), \qquad (8)$$

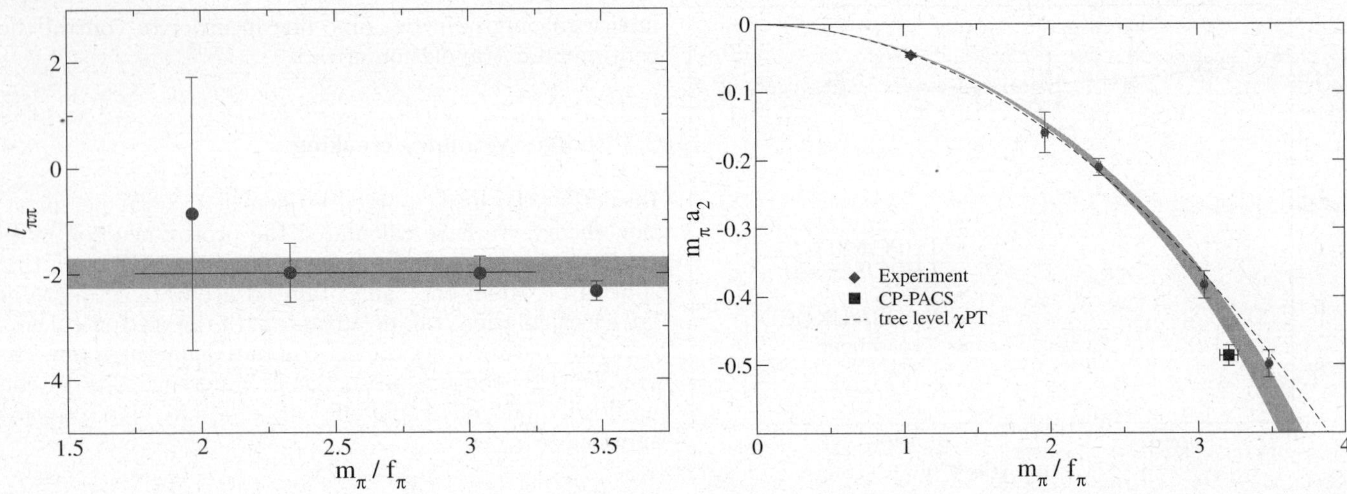

Fig. 3. The $I = 2$ π-π scattering length. The fit to the lattice results to extract the only fitted parameter $l_{\pi\pi}$ (left). The resulting fit together with the tree level result and the lightest CP-PACS point (right).

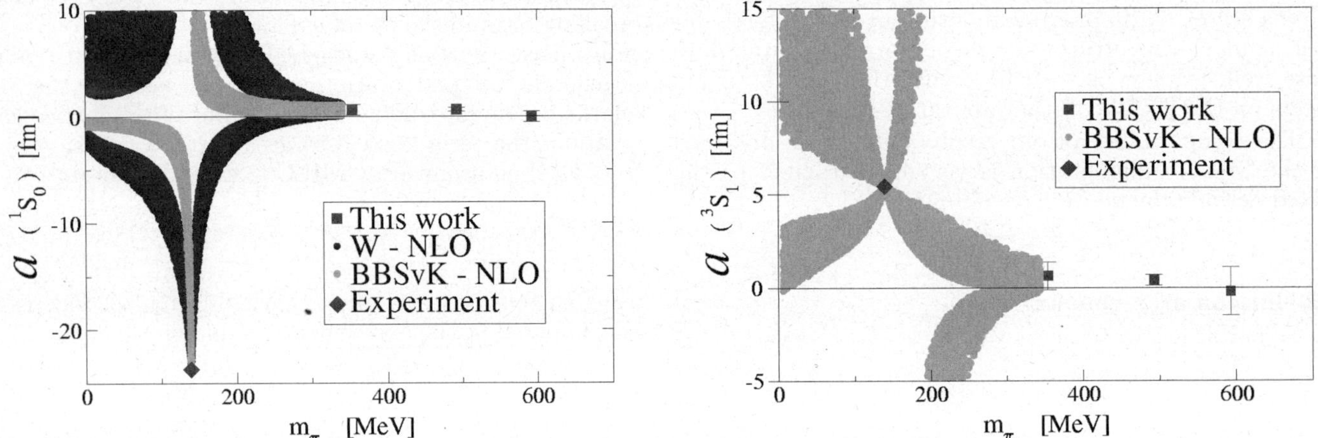

Fig. 4. The nucleon-nucleon scattering lengths using the opposite-sign convention from the π-π case.

where the first error is statistical and the second is an estimated systematic error due to chiral extrapolation. This result is slightly different from the one in [15] due to increased statistics. After scaling to $1\,\mathrm{GeV}$ using eq. (7), the low-energy constant $l_{\pi\pi}$ is $l_{\pi\pi}(1\,\mathrm{GeV}) = 2.1(3)$. In the above fits only the lightest three points were used. This is in good agreement with the best experimental result

$$m_\pi a_2 = -0.0454(31). \tag{9}$$

Mixed-action χ-PT [8] has insignificant effect on the final result. The authors of [8] estimated it to be -0.003. We are currently extending our studies of meson-meson scattering to the kaon-pion [33] and kaon-kaon systems.

3.5 Nucleon-nucleon scattering

We also computed the scattering length in the nucleon-nucleon system where very little lattice work has been done [34,35]. Our results are plotted in fig. 4. The shaded areas are the allowed regions that various effective-field

theory approaches predict for these scattering lengths, if we use the experimental points together with the lightest lattice point to constrain their parameters [36–41,14]. Clearly, the lightest lattice point is on the boundary of applicability of these effective-field theories, indicating that we need significantly lighter pion masses in order to be able to predict from QCD the physical scattering lengths in these processes. In addition to lighter pion masses, we need to study the volume dependence of the energy levels in order establish the scattering nature of the states we observe. Certainly, this calculation is far from definitive. It is an attempt to understand what it takes to perform such calculations. In the next few years we hope to improve substantially the quality of our results. Nonetheless, even at such an early stage, our calculation undoubtedly shows this very valuable piece of information: in the range of pion masses used here, the scattering lengths in the nucleon-nucleon channel are very different from their values at the physical point.

We are currently working on significantly enhancing our statistics in order to be able to reduce our error bars and obtain unambiguous signals for the scattering lengths.

Also we are investigating techniques to enhance our signal and alternative lattice formulations that will allow us to push towards the chiral limit given the computational resource available to us.

4 Conclusions

In this talk I have presented a number of lattice calculations that are relevant to nuclear physics. The ultimate goal is to obtain an understanding of nuclear forces from QCD. Together with effective-field theory techniques, lattice QCD is valuable tool in this endeavor. It is the only way to compute from QCD the low-energy constants needed by the low-energy effective-field theories. Recent theoretical, algorithmic and hardware developments have made it possible today to perform phenomenologically interesting calculations with direct impact in our understanding of the nature and phenomenology of strong interactions. Calculations directly related to the nucleon-nucleon interactions are just starting indicating that within the next few years one can expect significant results in this area.

I would like to thank my collaborators in NPLQCD, and LHPC. In particular I would like to acknowledge Silas Beane, Paulo Bedaque, and Martin Savage for their contributions in every aspect of the calculations done within NPLQCD. All calculations were done with the QDP++/ Chroma programming environment [42] at the JLab LQCD cluster. This work was supported in part by DOE contract DE-AC05-84ER40150, under which SURA operated the Thomas Jefferson National Accelerator Facility (JLab) and in part by DOE contract DE-AC05-06OR23177 under which Jefferson Science Associates, LLC, currently operates JLab.

References

1. D.B. Kaplan, Phys. Lett. B **288**, 342 (1992) `hep-lat/9206013`.
2. R. Narayanan, H. Neuberger, Nucl. Phys. B **412**, 574 (1994) `hep-lat/9307006`.
3. V. Furman, Y. Shamir, Nucl. Phys. B **439**, 54 (1995) `hep-lat/9405004`.
4. H. Neuberger, Phys. Lett. B **417**, 141 (1998) `hep-lat/9707022`.
5. MILC Collaboration (K. Orginos, D. Toussaint, R.L. Sugar), Phys. Rev. D **60**, 054503 (1999) `hep-lat/9903032`.
6. MILC Collaboration (K. Orginos, D. Toussaint), Phys. Rev. D **59**, 014501 (1999) `hep-lat/9805009`.
7. O. Bar, C. Bernard, G. Rupak, N. Shoresh, Phys. Rev. D **72**, 054502 (2005) `hep-lat/0503009`.
8. J.W. Chen, D. O'Connell, R.S. Van de Water, A. Walker-Loud, Phys. Rev. D **73**, 074510 (2006) `hep-lat/0510024`.
9. S.R. Sharpe, PoS (LAT2006) (2006) `hep-lat/0610094`.
10. Y. Shamir, `hep-lat/0607007` (2006).
11. C. Bernard, M. Golterman, Y. Shamir, Phys. Rev. D **73**, 114511 (2006) `hep-lat/0604017`.
12. C. Bernard, Phys. Rev. D **73**, 114503 (2006) `hep-lat/0603011`.
13. CP-PACS Collaboration (T. Yamazaki et al.), Phys. Rev. D **70**, 074513 (2004) `hep-lat/0402025`.
14. S.R. Beane, P.F. Bedaque, K. Orginos, M.J. Savage, `hep-lat/0602010` (2006).
15. NPLQCD Collaboration (S.R. Beane, P.F. Bedaque, K. Orginos, M.J. Savage), Phys. Rev. D **73**, 054503 (2006) `hep-lat/0506013`.
16. M. Luscher, Commun. Math. Phys. **104**, 177 (1986).
17. M. Luscher, Commun. Math. Phys. **105**, 153 (1986).
18. S.R. Beane, P.F. Bedaque, A. Parreno, M.J. Savage, Phys. Lett. B **585**, 106 (2004) `hep-lat/0312004`.
19. LHP Collaboration (D.B. Renner et al.), Nucl. Phys. Proc. Suppl. **140**, 255 (2005) `hep-lat/0409130`.
20. LHPC Collaboration (R.G. Edwards et al.), PoS (LAT2005) 056 (2005) `hep-lat/0509185`.
21. C.W. Bernard et al., Phys. Rev. D **64**, 054506 (2001) `hep-lat/0104002`.
22. C. Aubin, C. Bernard, Phys. Rev. D **68**, 074011 (2003) `hep-lat/0306026`.
23. MILC Collaboration (C. Aubin et al.), Phys. Rev. D **70**, 114501 (2004) `hep-lat/0407028`.
24. S.R. Beane, P.F. Bedaque, K. Orginos, M.J. Savage, `hep-lat/0606023` (2006).
25. J. Gasser, H. Leutwyler, Ann. Phys. **158**, 142 (1984).
26. The RIKEN-BNL-Columbia-KEK Collaboration (S. Sasaki, K. Orginos, S. Ohta, T. Blum), Phys. Rev. D **68**, 054509 (2003) `hep-lat/0306007`.
27. LHPC Collaboration (R.G. Edwards et al.), Phys. Rev. Lett. **96**, 052001 (2006) `hep-lat/0510062`.
28. A.A. Khan et al., `hep-lat/0603028` (2006).
29. S.R. Beane, M.J. Savage, Phys. Rev. D **70**, 074029 (2004) `hep-ph/0404131`.
30. V. Bernard, U.G. Meissner, `hep-lat/0605010` (2006).
31. S.R. Beane, K. Orginos, M.J. Savage, `hep-lat/0605014` (2006).
32. J. Gasser, H. Leutwyler, Phys. Rep. **87**, 77 (1982).
33. S.R. Beane et al., `hep-lat/0607036` (2006).
34. M. Fukugita, Y. Kuramashi, H. Mino, M. Okawa, A. Ukawa, Phys. Rev. Lett. **73**, 2176 (1994) `hep-lat/9407012`.
35. M. Fukugita, Y. Kuramashi, M. Okawa, H. Mino, A. Ukawa, Phys. Rev. D **52**, 3003 (1995) `hep-lat/9501024`.
36. S. Weinberg, Phys. Lett. B **251**, 288 (1990).
37. S. Weinberg, Nucl. Phys. B **363**, 3 (1991).
38. C. Ordonez, L. Ray, U. van Kolck, Phys. Rev. C **53**, 2086 (1996) `hep-ph/9511380`.
39. D.B. Kaplan, M.J. Savage, M.B. Wise, Nucl. Phys. B **534**, 329 (1998) `nucl-th/9802075`.
40. D.B. Kaplan, M.J. Savage, M.B. Wise, Phys. Lett. B **424**, 390 (1998) `nucl-th/9801034`.
41. S.R. Beane, P.F. Bedaque, M.J. Savage, U. van Kolck, Nucl. Phys. A **700**, 377 (2002) `nucl-th/0104030`.
42. SciDAC Collaboration (R.G. Edwards, B. Joo), Nucl. Phys. Proc. Suppl. **140**, 832 (2005) `hep-lat/0409003`.

Eur. Phys. J. A **31**, 804–809 (2007)

DOI 10.1140/epja/i2006-10290-6

THE EUROPEAN
PHYSICAL JOURNAL A

Special Article – QNP 2006

Exploring the details of the QCD phase diagram

Ph. de Forcrand[1,2] and O. Philipsen[3,a]

[1] Institut für Theoretische Physik, ETH Zürich, CH-8093 Zürich, Switzerland
[2] CERN, Physics Department, TH Unit, CH-1211 Geneva 23, Switzerland
[3] Institut für Theoretische Physik, Westfälische Wilhelms-Universität Münster, Germany

Received: 22 December 2006
Published online: 22 March 2007 – © Società Italiana di Fisica / Springer-Verlag 2007

Abstract. We summarize our recent results on the phase diagram of QCD with $N_f = 2 + 1$ quark flavors, as a function of temperature T and quark chemical potential μ. Using staggered fermions, lattices with temporal extent $N_t = 4$, and the exact RHMC algorithm, we first determine the critical line in the quark mass plane $(m_{u,d}, m_s)$ where the finite-temperature transition at $\mu = 0$ is second order. We confirm that the physical point lies on the crossover side of this line. Our data are consistent with a tricritical point at $(m_{u,d}, m_s) = (0, \sim 500)$ MeV. Then, using an imaginary chemical potential, we determine in which direction this second-order line moves as the chemical potential is turned on. Contrary to standard expectations, we find that the region of first-order transitions shrinks in the presence of a chemical potential, which is inconsistent with the presence of a QCD critical point at small chemical potential. The emphasis is put on clarifying the translation of our results from lattice to physical units, and on discussing the apparent contradiction of our findings with earlier lattice studies. Finally, we review related results obtained via simulations at fixed baryon number via the canonical ensemble.

PACS. 12.38.Gc Lattice QCD calculations – 12.38.Mh Quark-gluon plasma

1 Introduction

In recent years, considerable efforts have been devoted to the determination of the phase diagram of QCD at finite temperature and density [1]. At zero chemical potential, the nature of the quark hadron phase transition depends on the quark masses $m_{u,d}$ and m_s, and the qualitative expectations are summarized in fig. 1. In the limits of zero and infinite quark masses (lower left and upper right corners), order parameters corresponding to the breaking of a symmetry can be defined, and one finds numerically that a first-order transition takes place at a finite temperature T_c. On the other hand, one observes an analytic crossover at intermediate quark masses. Hence, each corner must be surrounded by a region of first-order transition, bounded by a second-order line as in fig. 1. The line in the heavy-quark corner has been studied in [2]. Here, we want to determine the chiral critical line.

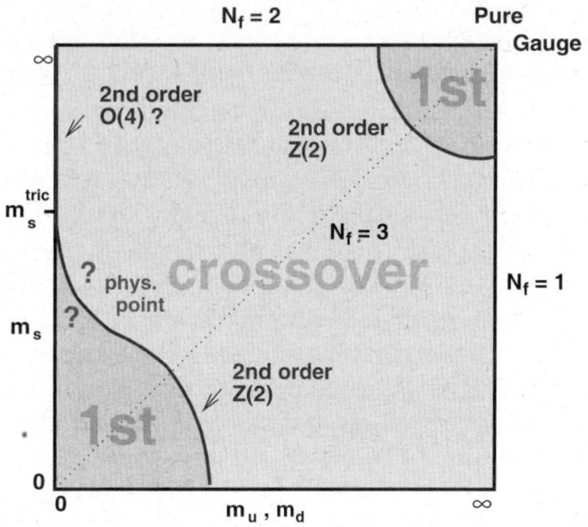

Along both lines, the universality class has been numerically determined to be that of the $3d$ Ising model. With this knowledge at hand, a powerful observable to determine the critical couplings is the Binder cumulant $B_4 \equiv \langle \delta X^4 \rangle / \langle \delta X^2 \rangle^2$, where $\delta X = X - \langle X \rangle$, and we take for X the u, d quark condensate $\bar{\psi}\psi$. In the infinite volume

Fig. 1. Schematic phase transition behaviour of $N_f = 2 + 1$ flavor QCD for different choices of quark masses $(m_{u,d}, m_s)$, at $\mu = 0$ (from [1]).

limit and when evaluated at the (pseudo-)critical temperature, this observable takes on the values 1 or 3 when the phase change corresponds to a first-order transition or a crossover, respectively, while it assumes the value $B_4 = 1.604$ characteristic for the $3d$ Ising universality class on a second-order critical point.

[a] e-mail: ophil@uni-muenster.de

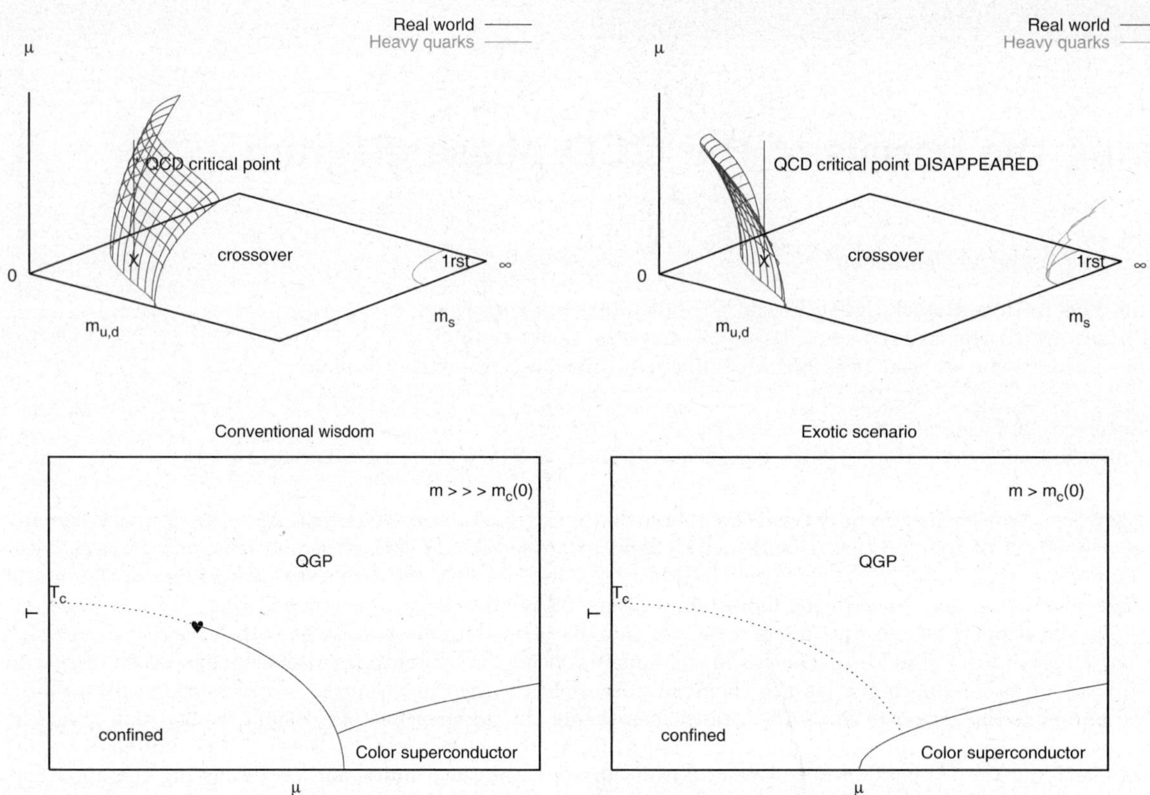

Fig. 2. Upper panel: the chiral critical surface in the case of positive (left) and negative (right) curvature. If the physical point is in the crossover region for $\mu = 0$, a finite μ phase transition will only arise in the scenario (left) with positive curvature, where the first-order region expands with $|\mu|$. Note that for heavy quarks, the first-order region shrinks with $|\mu|$ (right) [5]. Lower panel: phase diagrams for fixed quark mass (here $N_f = 3$) corresponding to the two scenarios depicted above.

On lattices $8^3, 12^3$ and $16^3 \times 4$, we thus estimate the critical couplings as those for which $B_4 = 1.604$. For each mass point $(m_{u,d}, m_s)$, we accumulate at least 200k RHMC trajectories, and interpolate among 4 or more $m_{u,d}$ values to find the critical $m_{u,d}$ mass m^c at previously fixed m_s. We obtain the set of points in fig. 4, left.

We then consider the effect of a baryonic chemical potential, $\mu_B = 3\mu$. As a function of quark chemical potential μ, represented vertically in fig. 2, the critical line determined at $\mu = 0$ now spans a surface. The standard expectation for the QCD phase diagram is depicted in fig. 2 left. The first-order region expands as μ is turned on, so that the physical point, initially in the crossover region, eventually belongs to the critical surface. At that chemical potential μ_E, the transition is second order: that is the QCD critical point. Increasing μ further makes the transition first order. Drawn in the (T, μ)-plane, this corresponds to the standard expected diagram fig. 2, left. A completely different scenario arises if instead the first-order region shrinks as μ is turned on. In that case (fig. 2, right), the physical point remains in the crossover region for any μ.

Since the phenomenologically interesting question is whether a QCD critical point (μ_E, T_E) exists at small $\mu_E, \mu_E/T_E \lesssim 1$, *i.e.* for $\mu_B \lesssim 500\,\mathrm{MeV}$, this question can be addressed by simulations with an imaginary chemical potential, followed by an analytic continuation based on a Taylor expansion [3,4]. The benefit of using an imaginary chemical potential is that the fermion determinant is positive in this case, hence there is no sign problem and simulations are technically equally feasible as those for $\mu = 0$. Using this approach, we determine the curvature $\frac{dm^c}{d\mu^2}|_{\mu=0}$ of the critical surface at $\mu = 0$. We find that it is negative, so that the first-order region shrinks as in fig. 2, right. Note that in the opposite corner, the first-order region also shrinks [5].

Section 2 tests our methodology in the $N_f = 3$ case. Section 3 describes the $N_f = 2 + 1$ study. Section 4 compares our results with earlier lattice studies and discusses the various limitations of our approach.

2 $N_f = 3$

We first check our methodology in the case of 3 degenerate flavors. This is basically a repeat of ref. [6], this time using the RHMC algorithm [7] instead of the R algorithm.

The RHMC algorithm eliminates the stepsize error of the R algorithm, which differs in magnitude in the chirally symmetric and broken phases [8]. As a result, the value of $m^c(\mu = 0)$ is considerably different: $(am^c(\mu = 0))$ moves from 0.033(1) (R algorithm) [6] to 0.0260(5) (see fig. 3, left). We have checked, by performing zero-temperature

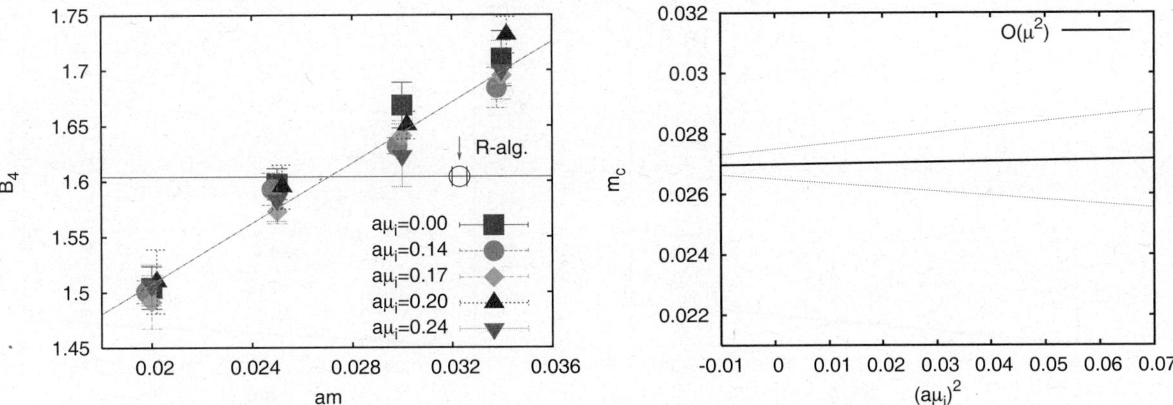

Fig. 3. Left: $B_4(am, a\mu_I)$ for different imaginary chemical potentials. Right: one-sigma error band for the critical mass $am^c(a\mu_I)$ resulting from a linear fit.

simulations at this quark mass, that this is not a simple renormalization effect, but that the physical ratio m_π/T_c is lowered by about 10%. Therefore, an exact algorithm appears mandatory for the study of the $N_f = 2 + 1$ critical line. Moreover, RHMC turns out to be vastly more efficient, by up to a factor 20 in our case for the smallest quark masses [9].

We now turn on an imaginary chemical potential $\mu = i\mu_I$, and for each μ_I monitor the Binder cumulant B_4 as a function of the quark mass. Our results are summarized in fig. 3, left. The chemical potential has almost no influence on B_4. A lowest-order fit, linear in am and $(a\mu)^2$, gives the error band in fig. 3, right, corresponding to

$$am^c(a\mu) = 0.0270(5) - 0.0024(160)(a\mu)^2. \quad (1)$$

Care must be taken for the conversion to physical units. The crucial point is that, as we increase the chemical potential μ_I, we tune the gauge coupling β upwards to maintain criticality, so that $a(\beta)$ decreases: our observation that $am^c(\mu_I) \approx$ const does *not* mean that $m^c(\mu_I) \approx$ const, but that $m^c(\mu_I)$ increases with μ_I, or *decreases* with a real chemical potential μ. If we express

$$\frac{am^c(\mu)}{am^c(0)} = 1 + \frac{c_1'}{am^c(0)}(a\mu)^2 + \dots, \quad (2)$$

$$\frac{m^c(\mu)}{m^c(0)} = 1 + c_1 \left(\frac{\mu}{\pi T}\right)^2 + \dots \quad (3)$$

then c_1 and c_1' are related by

$$c_1 = \frac{\pi^2}{N_t^2} \frac{c_1'}{am^c(0)} + \left(\frac{1}{T_c(m,\mu)} \frac{dT_c(m,\mu)}{d(\mu/\pi T)^2}\right)_{\mu=0}, \quad (4)$$

where $m = m^c(\mu)$ in the second term. Writing the transition temperature as

$$\frac{T_c(m,\mu)}{T_c(m_0^c,0)} = 1 + A\frac{m - m_0^c}{\pi T} + B\left(\frac{\mu}{\pi T}\right)^2 + \dots \quad (5)$$

one obtains

$$c_1 = \left(B + \frac{\pi^2}{N_t^2} \frac{c_1'}{am^c(0)}\right)\left(1 - A\frac{m_0^c}{\pi T}\right)^{-1}. \quad (6)$$

c_1' and $\frac{m_0^c}{\pi T}$ are both small, so that c_1 is nearly equal to B. Estimates of B and A can be obtained by converting our result for the pseudo-critical gauge coupling

$$\beta_0(am, a\mu) = 5.1369(3) + 1.94(3)(am - am_0^c) + 0.781(7)(a\mu)^2 \quad (7)$$

to physical units. Using the 2-loop β-function gives $A = 2.111(17)$, $B = -0.667(6)$ so that finally

$$\frac{m^c(\mu)}{m^c(0)} = 1 - 0.7(4)\left(\frac{\mu}{\pi T}\right)^2 + \dots \quad (8)$$

The error above is conservative and includes the uncertainty from using different fitting forms (see [10], table 2). The main source of systematic error comes from using the 2-loop β-function to obtain B. The non-perturbative β-function varies more steeply and may increase A and B, in magnitude, by up to a factor 2. This will make c_1 more negative.

We thus have clear evidence that, in the $N_f = 3$ theory on an $N_t = 4$ lattice, the region of first-order transitions *shrinks* as a baryon chemical potential is turned on, and the "exotic scenario" of fig. 2, right, is the correct one. This result is further supported by recent simulations of the same theory, under an isospin chemical potential [11].

3 $N_f = 2 + 1$

We now proceed to the non-degenerate case. First, at $\mu = 0$, we map out the line of second-order transitions in the $(am_{u,d}, am_s)$-plane. Our results, shown fig. 4, left, are in qualitative agreement with expectations fig. 1. In particular, they are consistent with the possible existence of a tricritical point $(m_{u,d} = 0, m_s = m_s^{tric})$. Using its known, Gaussian exponents, our data favor (blue on-line line in fig. 4, left) a heavy $m_s^{tric} \sim 2.8T_c$.

A more immediate issue is whether the QCD physical point lies on the crossover side of the critical line as expected. For that purpose, we have performed spectrum calculations at $T \sim 0$, at the parameters corresponding to

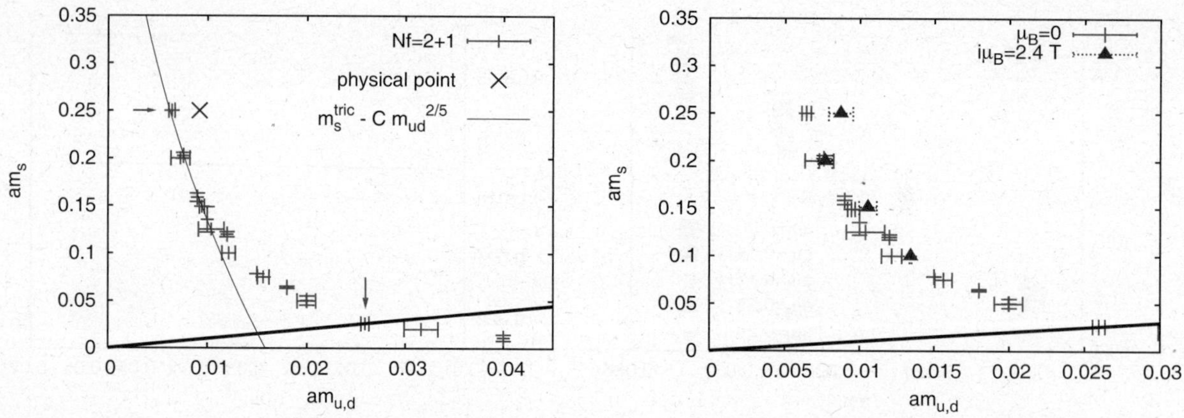

Fig. 4. Left: the chiral critical line in the bare quark mass plane at $\mu = 0$. $N_f = 3$ is shown by the thick solid line. Also shown are the physical point according to [12], and a fit corresponding to a tricritical point $m_s^{tric} \sim 2.8 T_c$. Right: comparison of the critical line at $\mu = 0$ and $a\mu_I = 0.2$.

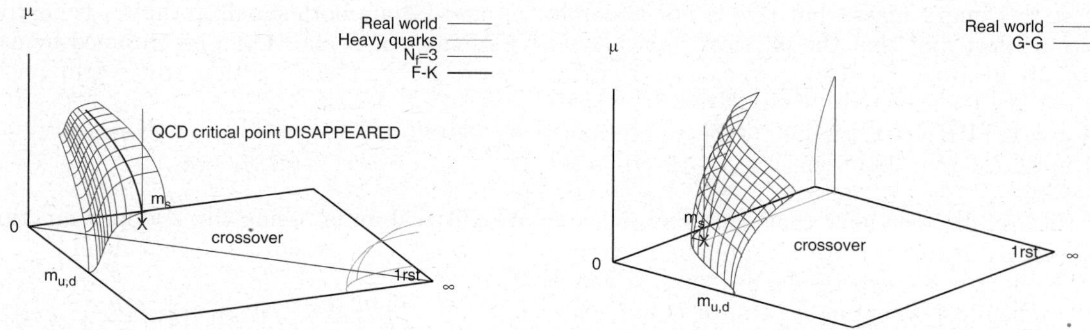

Fig. 5. Left: effect of keeping the quark mass fixed in lattice units in [12]. Right: comparison at finite μ between the $N_f = 2 + 1$ and the $N_f = 2$ theory considered in [16].

the horizontal arrow in fig. 4, left ($am_{u,d} = 0.005, am_s = 0.25, \beta = 5.1857$). They show that m_s is approximately tuned to its physical value ($\frac{m_K}{m_\rho} \sim \frac{m_K}{m_\rho}|_{phys}$), while the pion is lighter than in QCD ($\frac{m_\pi}{m_\rho} = 0.148(2) < 0.18$). This confirms that the physical point lies on the right of the critical line, $i.e.$ in the crossover region[1]. This conclusion has been confirmed by very recent calculations on finer lattices [13]. Also, we find T_c to vary little along the critical line, in accordance with model calculations [14].

We now couple an imaginary chemical potential $a\mu_I = 0.2$ to the two light flavors, and measure the change in the critical mass $am_{u,d}$ as in the $N_f = 3$ case. Figure 4, right, shows the same trend as for $N_f = 3$: the critical mass is constant or slightly increasing, $in\ lattice\ units$. The conversion to physical units proceeds as in eqs. (2)–(8). Since the critical gauge coupling $\beta_0(a\mu_I)$ increases with μ_I, the coefficient B, which is the dominant contribution to c_1, is negative. Together with a very small or slightly negative value for c_1', it implies again that the first-order region $shrinks$ as the baryon chemical potential is turned on, and the "exotic scenario" of fig. 2, right, is the correct one.

---

[1] In fact, our estimate of the lattice parameters corresponding to the physical point is consistent with that of Fodor and Katz using the same action, but the R algorithm [12].

This statement comes with several caveats: i) our lattice is very coarse ($a \sim 0.3$ fm); ii) as we consider lighter $m_{u,d}$, our box becomes small ($m_\pi L \sim 1.7$ for the worst case); iii) we use "rooting" of the staggered determinant to simulate 1 and 2 flavors, albeit our measure is positive with an imaginary μ, so that we avoid the pitfalls of [15].

4 Discussion

Our results appear in qualitative contradiction with those of Fodor and Katz [12] and of Gavai and Gupta [16], which both conclude for the existence of a critical point (μ_E, T_E) at small chemical potential $\mu_E/T_E \lesssim 1$. Let us consider the reasons for such disagreement.

– Fodor and Katz obtain Monte Carlo results at $\beta = \beta_c, \mu = 0$, and perform a double reweighting in (β, μ) along the pseudo-critical line $\beta_c(a\mu)$. By construction, this reweighting is performed at a quark mass fixed $in\ lattice\ units$: $am_{u,d} = \frac{m_{u,d}}{T_c} = $ const. Since the critical temperature T_c decreases as they turn on μ, so does their quark mass. This decrease of the quark mass pushes the transition towards first order, which might be the reason why they find a critical point at small μ. This effect is illustrated in the sketch fig. 5, left, where the bent trajectory

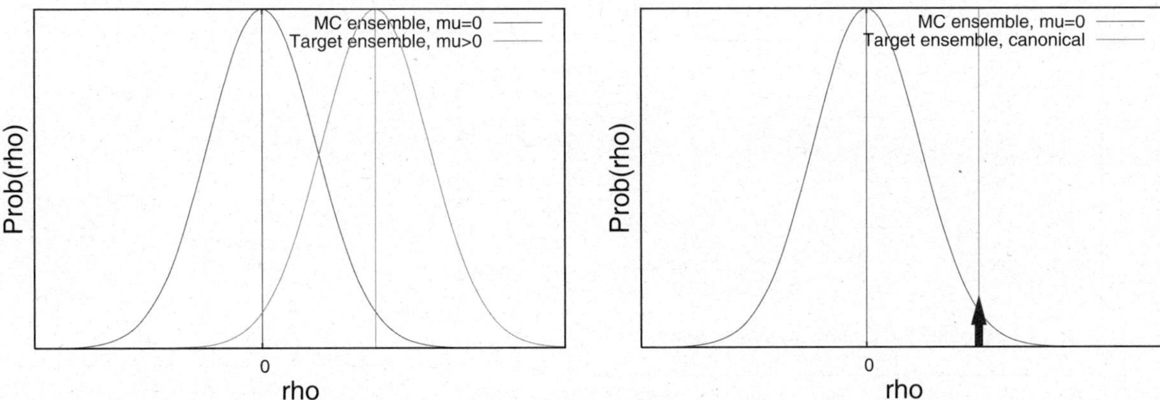

Fig. 6. (Colour on-line) Left: the Monte Carlo ensemble produces a distribution of baryon densities ρ (red) centered about $\rho = 0$, to be reweighted according to the desired, target ensemble (green) centered about $\rho > 0$. Very little large-ρ information is available. Right: by chosing the target ensemble to be canonical, one alleviates the need for large-ρ information, thus increasing the reliability of the results.

à la Fodor and Katz intersects the critical surface, while the vertical line of constant physics does not.

Put another way, Fodor and Katz measure the analogue of eq. (2) instead of (3). From their fig. 1 (ref. [12]), the coefficient c_1' which one would extract would be essentially zero like ours. As in our case, the variation of T_c with μ makes a dominant contribution, which might change the results qualitatively.

– Gavai and Gupta try to infer the location of the critical point by estimating the radius of convergence of the Taylor expansion of the free energy in $(\mu/\pi T)^2$. Regardless of the systematic error attached to such an estimate when only 4 Taylor coefficients are available, we want to point out that they consider a theory without strange quark, *i.e.* $N_f = 2$ only. The (μ, T) phase diagram of such a theory is qualitatively different from that of $N_f = 2 + 1$ QCD. At $\mu = 0$, the order of the finite-temperature transition as $m_{u,d} \to 0$ is not settled [17]. Assuming a second-order $O(4)$ transition, one expects then a tricritical point at $(m_{u,d} = 0, \mu = \mu^{tric})$, beyond which a non-zero critical mass $m_{u,d}^c(\mu)$ can be defined, as sketched in fig. 5, right. The quantitative relevance of results, even accurate, for this $N_f = 2$ theory to QCD is unclear to us.

Therefore, we find no inconsistency between our results and those above. We conclude that the existence of a critical point (μ_E, T_E) in QCD at small chemical potential $\mu_E/T_E \lesssim 1$ is an open question. Our numerical evidence, with the caveats mentioned in sect. 3, is that the curvature of the critical surface is as illustrated fig. 2, right. Our main systematic error comes from our coarse lattice spacing $a \sim 0.3\,\mathrm{fm}$ [18]. If confirmed on a finer lattice, the implications of our finding would be as follows. In the region where a leading Taylor expansion of the critical surface is a good approximation, *i.e.* $\mu/T \lesssim 1$, corresponding to the experimentally accessible regime, no critical point exists which is analytically connected to $\mu = 0$. Of course, we cannot exclude that the QCD phase diagram is more complex, and partly inaccessible to our imaginary μ + Taylor expansion strategy. In particular

this leaves open the possibility of a critical point *not* analytically connected to the one at $\mu = 0$.

5 Canonical approach

The canonical ensemble, where the baryon density is fixed instead of the conjugate chemical potential, provides a potentially fruitful approach to the study of QCD at finite density. Its theoretical advantage is illustrated in fig. 6. The ensemble of configurations sampled from the Monte Carlo ensemble, whatever this ensemble is chosen to be, has a distribution of baryon densities ρ centered around $\rho = 0$, illustrated by the Gaussian-like red on-line curve in fig. 6. By reweighting with a factor $\propto \exp(\mu/TV\rho)$, one attempts to obtain information on the target, $\mu \neq 0$ ensemble, which has a distribution of baryon densities illustrated by the Gaussian-like green on-line curve, centered around $\rho \neq 0$. When the volume V is increased, the red and green on-line curves both become narrower Gaussians, and an *overlap* problem appears: the Monte Carlo ensemble contains very little information relevant to the target ensemble, and correct results can only be maintained by increasing statistics exponentially with V. The problem is particularly acute for the large-density tail of the target ensemble. By changing the target ensemble to canonical, *i.e.* with a fixed baryon density as illustrated by the blue on-line delta-function in fig. 6, one eliminates the need to include high-density information. Thus, while the canonical ensemble does not provide a cure to the sign and overlap problems, there is some reason to expect that it delivers more reliable results than the usual reweighting to the grand-canonical ensemble, up to larger volumes or average baryon densities.

Reweighting of the $\mu = 0$ ensemble to the canonical one has been performed in ref. [19], in a pilot project with $N_f = 4$ degenerate flavors of staggered quarks, on a small $6^3 \times 4$ lattice. A selection of results is presented in fig. 7.

Figure 7 (left) shows, as a function of the baryon density, the derivative of the free-energy density with

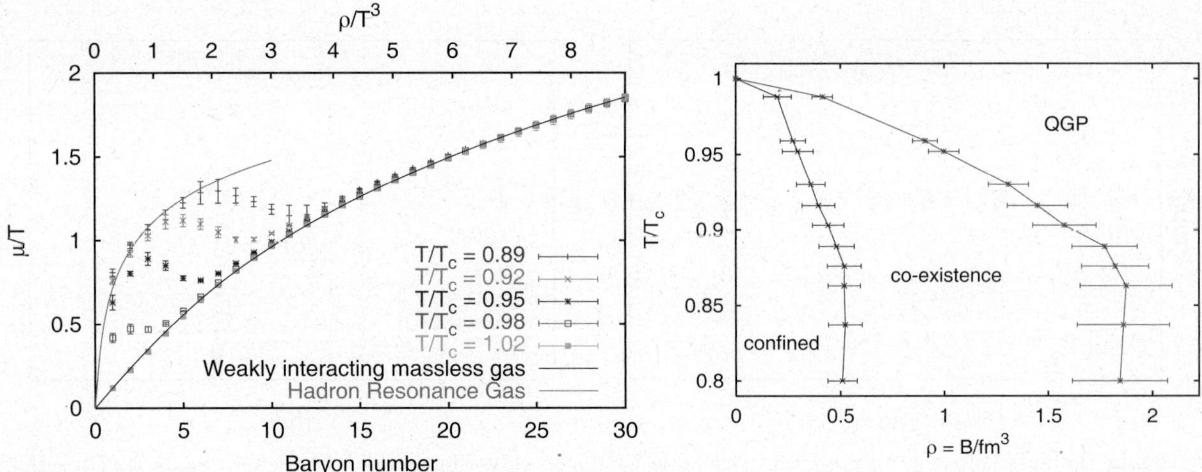

Fig. 7. Results from the canonical-ensemble approach [19]. Left: equivalent chemical potential *versus* baryon density. Right: boundaries of the coexistence region, obtained from the left panel by a Maxwell construction.

respect to the baryon density. This quantity is equal to the baryon chemical potential, in the saddle point approximation which is exact in the thermodynamic limit. Note that reliable results are obtained for systems containing up to 30 baryons, and chemical potentials up to $\mu/T \sim 2$. For the chosen quark mass $m_q/T = 0.2$, a first-order transition takes place between the confining and the plasma phases. The transition is visible in the S-shape of the data, indicating multiple solutions for the baryon density at a given chemical potential. A Maxwell construction reveals the critical chemical potential $\mu_c(T)$, and the limits $\rho_1(T), \rho_2(T)$ of the coexistence region. The latter are shown in fig. 7 (right) as a function of temperature. One sees that the pure hadronic phase has a maximum baryon density of about 0.5 baryon/fm³, which is not too far from the real-world nuclear density of 0.17 baryon/fm³. The minimum density in the plasma phase is about 6 quarks/fm³.

By choosing a heavier quark mass, one can ensure that the $\mu = 0$ transition becomes a crossover. If a critical point exists at some finite $\mu = \mu_E$, the transition will become first-order for $\mu > \mu_E$, which will be evidenced by the appearance of an S-shaped curve in fig. 7 (left) for $\mu > \mu_E$ only. While these simulations are still performed on a small volume, which causes important systematic errors in the estimate of the critical couplings and of the order of the phase transition, comparison with the imaginary-μ approach of the previous sections, and with the reweighting approach of Fodor and Katz [12] will provide a valuable crosscheck.

References

1. E. Laermann, O. Philipsen, Annu. Rev. Nucl. Part. Sci. **53**, 163 (2003) [arXiv:hep-ph/0303042]; O. Philipsen, PoS (LAT2005) 016 (2006); PoS (JHW2005) 012 (2006) [arXiv:hep-lat/05100677].

2. P. Hasenfratz, F. Karsch, I.O. Stamatescu, Phys. Lett. B **133**, 221 (1983); C. Alexandrou *et al.*, Phys. Rev. D **60**, 034504 (1999) [arXiv:hep-lat/9811028]; A. Dumitru, D. Roder, J. Ruppert, Phys. Rev. D **70**, 074001 (2004) [arXiv:hep-ph/0311119].

3. P. de Forcrand, O. Philipsen, Nucl. Phys. B **642**, 290 (2002) [arXiv:hep-lat/0205016].

4. M. D'Elia, M.P. Lombardo, Phys. Rev. D **67**, 014505 (2003) [arXiv:hep-lat/0209146].

5. S. Kim *et al.*, PoS (LAT2005) 166 (2006) [arXiv:hep-lat/0510069].

6. P. de Forcrand, O. Philipsen, Nucl. Phys. B **673**, 170 (2003) [arXiv:hep-lat/0307020].

7. M.A. Clark *et al.*, Nucl. Phys. Proc. Suppl. **140**, 835 (2005) [arXiv:hep-lat/0409133].

8. J.B. Kogut, D.K. Sinclair, arXiv:hep-lat/0504003.

9. M.A. Clark *et al.*, PoS (LAT2005) 115 (2006) [arXiv:hep-lat/0510004].

10. P. de Forcrand, O. Philipsen, arXiv:hep-lat/0607017.

11. D.K. Sinclair, J.B. Kogut, arXiv:hep-lat/0609041.

12. Z. Fodor, S.D. Katz, JHEP **0404**, 050 (2004) [arXiv:hep-lat/0402006].

13. Y. Aoki, G. Endrodi, Z. Fodor, S.D. Katz, K.K. Szabo, Nature **443**, 675 (2006).

14. Z. Szep, PoS (JHW2005) 017 (2006) [arXiv:hep-ph/0512241].

15. M. Golterman, Y. Shamir, B. Svetitsky, Phys. Rev. D **74**, 071501 (2006) [arXiv:hep-lat/0602026].

16. R.V. Gavai, S. Gupta, Phys. Rev. D **71**, 114014 (2005) [arXiv:hep-lat/0412035].

17. J.B. Kogut, D.K. Sinclair, Phys. Rev. D **73**, 074512 (2006) [arXiv:hep-lat/0603021]; M. D'Elia, A. Di Giacomo, C. Pica, Phys. Rev. D **72**, 114510 (2005) [arXiv:hep-lat/0503030].

18. F. Karsch *et al.*, Nucl. Phys. Proc. Suppl. **129**, 614 (2004) [arXiv:hep-lat/0309116].

19. P. de Forcrand, S. Kratochvila, PoS (LAT2005) 167 (2006) [arXiv:hep-lat/0509143]; Nucl. Phys. Proc. Suppl. **153**, 62 (2006) [arXiv:hep-lat/0602024].

Static-light mesons on a dynamical anisotropic lattice

Justin Foley[1], Alan ó Cais[2], Mike Peardon[2], Sinéad M. Ryan[2]

[1] Department of Physics, Swansea University, Singleton Park, Swansea SA2 8PP, UK
[2] School of Mathematics, Trinity College, Dublin 2, Ireland

Abstract. We present results for the spectrum of static-light mesons from $N_f = 2$ lattice QCD. These results were obtained using all-to-all light quark propagators on an anisotropic lattice, yielding an improved signal resolution when compared to more conventional lattice techniques. In particular, we consider the inversion of orbitally-excited multiplets with respect to the 'standard ordering', which has been predicted by some quark models.

PACS. 11.15.Ha – 12.38.Gc

1 Introduction

It has long been known that hadrons containing a single charm or bottom quark exhibit approximate heavy quark spin and flavour symmetries and can be described to a reasonable approximation by an idealised system in which the heavy quark is taken to be infinitely massive [1]. In this limit the heavy quark symmetries become exact and the system consists of light degrees of freedom bound to a static colour point source. The static approximation can be systematically improved using a power-counting rule in Λ_{QCD}/m_Q, which forms the basis for heavy quark effective theory (HQET). For hadrons containing a bottom quark this approach is particularly suitable. In this case one expects leading-order corrections to be already quite small, at the level of ten percent.

On the lattice, mass-dependent errors which arise from the discretisation of the Dirac operator can limit the accuracy of simulations of heavy-light systems. In principle, one can avoid such errors by using the HQET lagrangian to describe the heavy quark dynamics. An alternative approach is to perform a number of simulations both at unphysically light heavy quark masses using the lattice Dirac operator and in the static limit and to interpolate the results to the physical heavy quark mass guided by the predictions of HQET. Therefore, precise calculations in the static limit are of considerable interest to the lattice community. In addition, little is known experimentally of the excited state spectrum of heavy-light hadrons and accurate simulations of their simpler static-light counterparts will have considerable phenomenological impact. However, traditionally, simulations of static-light systems have been extremely noisy. Ultimately, this stems from the fact that the use of conventional point-to-all propagators for the light quarks restricts interpolating operators for the static-light hadron to a single spatial lattice site.

This problem could be overcome if it were possible to compute all elements of the light quark propagator. This would allow source and sink operators for the static-light correlators to be placed at every spatial lattice site, yielding a dramatic increase in statistics. Indeed, simulations of static-light mesons are the natural testing ground for all-to-all techniques since they require a single light quark propagator per configuration while the static quark propagator is easy to evaluate. It is therefore not surprising that a number of groups have attempted to tackle this particular problem [2,3].

In this contribution, we describe a preliminary study of the excited state spectrum of static-light mesons using the all-to-all propagator technique introduced in Ref. [4]. In this approach, the important contribution of the low-lying eigenmodes of the Dirac operator to the fermion propagator is computed exactly. A noisy estimate is then used to correct for the contribution of the remaining eigenmodes. This estimate is obtained by inverting the Dirac operator on a set of Z_4 noise sources which have been partitioned in some subset of quarkfield indices. This partitioning or 'dilution' yields a substantial reduction in the variance of the stochastic estimate. This study has been performed on a 3+1 anisotropic lattice, where the temporal lattice spacing a_t is much finer than the spatial lattice spacing a_s. The fine temporal lattice spacing proves particularly useful when determining a plateau in the effective masses of the higher excited states. The correlators have been computed on $N_f = 2$ background configurations.

2 Interpolating operators

Due to heavy quark spin symmetry, states which lie in the same hyperfine multiplet are degenerate and in the continuum it is conventional to label these multiplets by the

Lattice irrep	Dimension	J
G_1	2	$\frac{1}{2}, \frac{7}{2}\ldots$
G_2	2	$\frac{5}{2}, \frac{7}{2}\ldots$
H	4	$\frac{3}{2}, \frac{5}{2}, \frac{7}{2}\ldots$

Table 1. Irreducible representations of the group of 24 proper spatial lattice rotations with low-lying angular momentum content.

angular momentum and parity of the light degrees of freedom [5]. For example, in the static limit the pseudoscalar and vector correspond to $J_\ell^{P_\ell} = \frac{1}{2}^-$.

At finite lattice spacing, it makes sense to classify states according to the irreducible representations (irreps) of the octahedral point group O_h [6]. This 48 element group is the direct product of the 24 proper spatial rotations allowed by the lattice with the group consisting of the identity and the space inversion operator. The relationship between these representations and continuum quantum numbers is not difficult to determine. Restricting the continuum irreps to the elements of the lattice symmetry group generates representations for O_h which are in general reducible and identification of the constituent irreps is a straight-forward exercise in group theory. Here, we take advantage of the degeneracies which arise from the heavy quark spin symmetry and construct interpolating operators specifically for the light degrees of freedom. Therefore, our operators transform according to the double-valued or spinorial representations of O_h. There are 6 such representations

$$G_{1u}, G_{1g}, G_{2u}, G_{2g}, H_u, H_g. \tag{1}$$

where the subscripts u and g denote representations which are even and odd under spatial inversion respectively.

Table 1 lists the irreps of the subgroup of proper rotations and the angular momenta of their lower-lying constituent states. From this table we see that states with the same angular momentum but with different J_z may be scattered across the lattice irreps. For example, states which transform according to the rows of the 6-dimensional spin $\frac{5}{2}$ representation are divided between the G_2 and H irreps. In a numerical study we should therefore expect to observe energy levels in these irreps which are degenerate, up to lattice artifacts, corresponding to the $J_\ell = \frac{5}{2}$ states. In this study, we were particularly interested in orbital excitations of the static-light meson. There is a single S-wave state with $J_\ell^{P_\ell}$ quantum numbers $\frac{1}{2}^-$, the P-wave doublet is $\frac{1}{2}^+, \frac{3}{2}^+$ and the D-wave states are labelled $\frac{3}{2}^-, \frac{5}{2}^-$. Of these, only the S-wave and the $\frac{1}{2}^+$ P-wave which lie in the G_{1u} and G_{1g} irreps can be accessed using local operators, while all other irreps require the use of extended interpolating operators. It is worth noting that when one uses all-to-all light quark propagators the construction of extended interpolating fields requires no additional fermion matrix inversions and the extra computational cost incurred is minimal. In order to determine

accurately the excited state spectrum it is important to choose operators which couple strongly to the states of interest and we tested a number of operators across the lattice representations. Further details of our choice of interpolating operators will be presented in Ref. [7].

3 Simulation details

The lattice Dirac and gauge actions used in this study have previously been described in detail in Refs. [8,9]. Both are specifically formulated for 3+1 anisotropic lattices. The Wilson-like fermion action is accurate to $\mathcal{O}(a_t, a_s^3)$ and employs stout-smeared spatial links to reduce radiative corrections. The gauge action incorporates tree-level Symanzik and tadpole improvement and is accurate to $\mathcal{O}(a_t^2, a_s^4, \alpha_s a_s^2)$.

The anisotropic lattice breaks hypercubic symmetry and introduces two new parameters: the bare quark and gauge anisotropies, denoted ξ_q^0 and ξ_g^0, which must be tuned such that the measured anisotropy, a_s/a_t, assumes a consistent value. In dynamical QCD the tuning procedure is quite complicated and is detailed in Ref. [10]. In this study we use bare anisotropy values taken from that paper of $\xi_q^0 = 7.52$ and $\xi_g^0 = 8.06$ which yields a measured anisotropy of 6 with a three percent error.

We employ the standard Eichten-Hill action [11] for the static quark and the corresponding static quark propagator is a product of unsmeared temporal links.

Measurements have been performed on 249 background configurations on an $8^3 \times 80$ lattice with a spatial lattice spacing of about 0.2fm. The mass of the sea quarks and the light valence quark was found to be approximately the strange quark mass. For the light quark propagator, we computed the contribution of 100 low-lying eigenvectors of the Dirac operator. Two independent Z_4 noise sources which were diluted in time and colour were used in the stochastic estimate.

To extract excited state energies we applied the standard variational analysis to correlation matrices which were obtained by applying 5 or 6 levels of Jacobi smearing to the light quark fields. This was facilitated by the use of all-to-all propagators which meant that no additional inversions were required to smear the quark fields.

4 Results

Fig. 1 shows effective masses for the lowest-lying states across a number of the lattice irreps. In this case the lowest effective mass corresponds to the S-wave $\frac{1}{2}^-$ state and the higher-lying states are orbital excitations. The highest-lying states lie in D-wave channels. Clear plateaux are evident in each channel.

In addition, we obtain strong signals for a number states in each of the lattice irreps. Fig. 2 shows the ground state and the first two excited states in the G_{1u} (S-wave) irrep. These excited states shown here are almost certainly radial excitations in the S-wave channel.

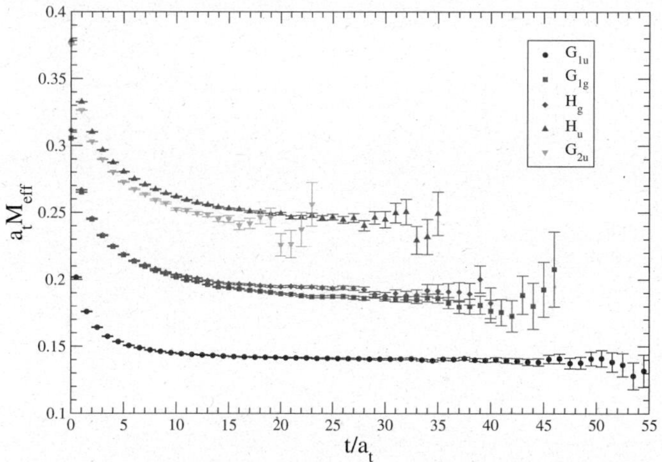

Fig. 1. Effective masses for the lowest-lying states in 5 lattice irreps. These include P-wave and D-wave excitations.

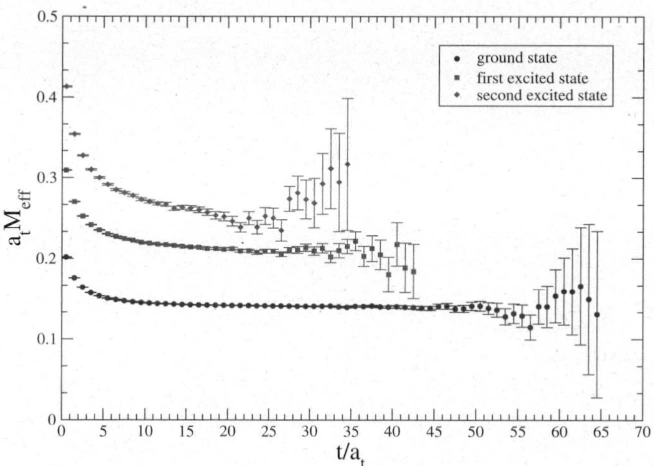

Fig. 2. Effective masses for the lowest-lying states in the G_{1u} (S-wave) irrep.

Since we are working in the static limit the energies determined by fitting to the effective masses contain an unphysical shift. This shift cancels in energy differences, which are therefore physically meaningful. Fig. 3 plots the energy differences between the lowest-lying state in the G_{1u} representation (i.e. the S-wave multiplet) and a number of excited states. We have not presented these preliminary results in physical units because we have performed our simulation on quite a small spatial volume which might distort the energies of the orbitally and radially excited states. The main point here is the precision to which we can compute the spectrum. Nevertheless, it is interesting to note the position of the lowest-lying states in the H_u and G_{2u} irreps. Naively, one expects the lightest state in the H_u representation to correspond to the $\frac{3}{2}^-$ D-wave while the $\frac{5}{2}^-$ ought to be the lightest state in the G_{2u} irrep. However, there appears to be no significant splitting between these states which may indicate that the $\frac{5}{2}^-$ multiplet is lighter than the $\frac{3}{2}^-$. Such a scenario has been predicted by quark models [12,13].

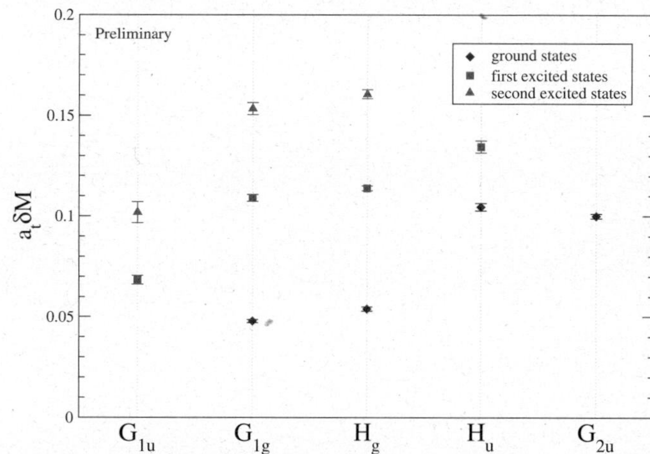

Fig. 3. Mass differences between the ground state S-wave multiplet and higher lying states.

5 Conclusions and outlook

We have presented preliminary results for the spectrum of static-light mesons in $N_f = 2$ QCD. We have been able to compute energy differences between the S-wave multiplet and a number of orbital and radial excitations. The use of all-to-all propagators allows us to minimise the statistical uncertainty in our measurements and facilitates the use of spatially-extended interpolating operators and the variational approach. The results presented here have allowed us to make some tentative remarks about the nature of the spectrum; however, we have not yet made a serious attempt to identify the continuum quantum numbers of the observed states. More recently we have performed similar simulations using larger operator bases on two different spatial volumes. These runs should allow us to assess finite volume effects and identify multi-particle states. We are currently analysing the results of these simulations and we intend to present our findings in the near future [7].

References

1. N. Isgur and M.B. Wise, Phys. Lett. **B232**, (1989) 113.
2. A. M. Green et al., Phys. Rev. **D69**, (2004) 094505.
3. T. Burch and C. Hagen, hep-lat/0607029.
4. J. Foley et al., Comput. Phys. Commun. **172**, (2005) 145-162.
5. N. Isgur and M. B. Wise, Phys. Rev. Lett. **66**, (1991) 1130-1133.
6. S. Basak et al, Phys. Rev. **D72**, (2005) 094506; Phys. Rev. **D72**, (2005) 074501.
7. J. Foley et al., in preparation.
8. J. Foley, A. Ó Cais, M. Peardon and S. M. Ryan, Phys. Rev. **D72**, (2005) 074503.
9. C. Morningstar and M. J. Peardon, Nucl. Phys. Proc. Suppl. **83**, (2000) 887.
10. R. Morrin, A. Ó Cais, M. Peardon, S. M. Ryan and J.I. Skullerud, Phys. Rev. **D74**, (2006) 014505.
11. E. Eichten and B. Hill, Phys. Lett. **B240**, (1990) 193.
12. H. J. Schnitzer, Phys. Lett. **B226** (1989) 171.
13. N. Isgur, Phys. Rev. **D57**, (1998) 4041-4053.

Eur. Phys. J. A **31**, 810–815 (2007)
DOI 10.1140/epja/i2006-10266-6

THE EUROPEAN
PHYSICAL JOURNAL A

Special Article – QNP 2006

Neutron stars as cosmic laboratories to explore hadronic matter at ultra-high densities

I. Bombaci[a]

Dipartimento di Fisica "E. Fermi", Università di Pisa, and INFN, Sezione di Pisa, largo B. Pontecorvo 3, 56127 Pisa, Italy

Received: 23 November 2006
Published online: 19 March 2007 – © Società Italiana di Fisica / Springer-Verlag 2007

Abstract. We examine the present status of the theoretical calculations for the internal structure of neutron stars, and the connection with the microscopic properties of ultradense hadronic matter. We discuss the possibility to have quark deconfinement phase transition in the core of neutron stars, and we explore some of its astrophysical implications as the quark-deconfinement nova model for gamma-ray bursts.

PACS. 97.60.Jd Neutron stars – 26.60.+c Nuclear matter aspects of neutron stars – 24.10.Cn Many-body theory – 25.75.Nq Quark deconfinement, quark-gluon plasma production, and phase transitions

1 Introduction

Neutron stars (NS) are the final product of the evolution of massive stars ($M > 8M_\odot$, being $M_\odot \simeq 2 \times 10^{33}$ g the mass of the Sun). These stars at the end of their *lives* experience a catastrophic gravitational collapse of their core, which triggers a type-II supernova (SN) explosion and leaves behind a hot and lepton-rich compact remnant (protoneutron star), which within a few minutes evolves to a cold catalyzed configuration (neutron star)[1].

The bulk properties and the internal constitition of NS primarily depend upon the equation of state (EOS) of dense hadronic matter, which is the main ingredient in solving the stellar-structure equations in general relativity [4–6]. Different models for the EOS of dense matter predict a neutron star maximum mass (M_{max}) in the range of 1.4–2.4 $M_\odot$, and a corresponding central density n_c in range of 4–10 times the saturation density ($n_0 = 0.16\,\mathrm{fm}^{-3}$) of nuclear matter. In the case of a star with $M \sim 1.4\,M_\odot$, different EOS models predict a radius in the range of 7–16 km. Neutron stars are thus the densest macroscopic objects in the Universe. They represent the limit beyond which gravity overwhelm all the other forces of nature and lead to the formation of a black hole.

Due to the large value of the stellar central densities, various regimes (particle species and phases) of dense hadronic matter are expected in the interiors of NS. Consequently, different types of "neutron stars" are hypothesized, as schematically summarized in fig. 1.

---

[a] e-mail: bombaci@df.unipi.it

[1] Under some circumstances the protoneutron star could collapse to a black hole [1–3].

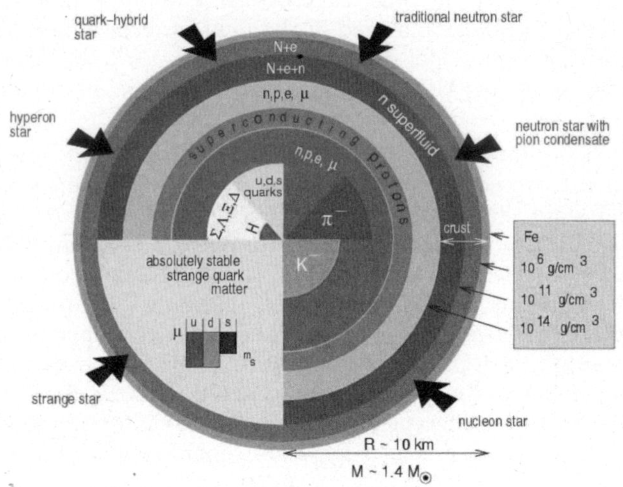

Fig. 1. Schematic cross-section of a neutron star [6].

In the simplest and conservative picture the core of a neutron star is modeled as a uniform fluid of neutron-rich nuclear matter in equilibrium with respect to the weak interaction: these are the so-called "traditional" neutron stars. However, due to the large value of the stellar central density and to the rapid increase of the nucleon chemical potentials with density, hyperons (Λ, Σ^-, Σ^0, Σ^+, Ξ^- and Ξ^0 particles) are expected to appear in the inner core of the star. Other *exotic* phases of hadronic matter such as a Bose-Einstein condensate of negative pion (π^-) or negative kaon (K^-) could be present in the inner part of the star. The core of the more massive NS is also one of the best candidates in the Universe where a phase transition from hadronic matter to a deconfined quark phase could occur. Compact stars which possess a "quark matter core" either as a mixed phase of deconfined quarks and hadrons or as a pure quark matter (QM) phase are called *Hybrid Stars* (HyS) [5]. In the following, the more *conventional* neutron stars in which no fraction of QM is present, will be referred to as *pure Hadronic Stars* (HS).

Even more challenging than the existence of a quark core in a neutron star, is the possible existence of a new type of compact stars consisting completely of a deconfined charge neutral mixture of *up*, *down*, *strange* quarks and electrons, satisfying the hypothesis on the absolute stability [7] of strange-quark matter (SQM). Such compact stars have been called *strange stars* (SS). The analysis of different type of observational data has given indirect evidence for the possible existence of SS (see, *e.g.*, ref. [8]). In the following, we will refer to hybrid stars and strange stars collectively as *Quark Stars* (QS).

Recently, there has been a considerable advance in our understanding of the properties of quark matter. In QCD any attractive quark-quark interaction will lead to pairing and color superconductivity, a subject already addressed in the late 1970s and early 1980s which came back a few years ago since the realization that the typical superconducting gaps in QM may be larger ($\Delta \sim 100\,\mathrm{MeV}$) than those predicted in these early works. The phase diagram of QCD has been analyzed in the light of color superconductivity and model calculations suggest that the phase structure is very rich at high densities (see, *e.g.*, [9,10] and references therein quoted).

Since 1967, the year in which the first radio pulsar was discovered and it was interpreted as a rotating neutron star, the masses of many NS have been measured. The calculated maximum mass, for a given EOS, must be larger than the values of all the measured neutron star masses. Thus, an accurate determination of the mass of NS give a stringent constraint to the global properties of dense matter EOS, and if coupled to additional observable (*i.e.* stellar-radius determination) could allow to discriminate between different possibilities for the internal stellar constitution (presence of hyperons, quark matter, etc.).

Binary stellar systems in which at least one component is a neutron star represent the most reliable way to measure the mass of the compact star. One of the most accurate mass determination is that of the neutron star associated to the pulsar PSR 1913 +16, which is a member of a tight (orbital period equal to 7 h 45 min) neutron star - neutron star system. The mass of PSR 1913 +16 is $1.4408 \pm 0.0003 M_\odot$. Such impressive accuracy is made possible by measuring general relativistic effects, such as the orbital decay due to gravitational radiation, the advance of periastron, the Shapiro delay, etc. Neutron star masses in NS-NS binary systems lie in the range 1.18 to 1.44 $M_\odot$ [11]. However, in at least two accreting X-ray binaries it has been found evidence for compact stars with higher masses. The first of these star is Vela X-1, with a reported mass $1.88 \pm 0.13 M_\odot$ [12], the second is Cygnus X-2, with a reported mass [13] of $1.78 \pm 0.23 M_\odot$. Unfortunately, mass determinations in X-ray binaries are affected

Table 1. Properties of the maximum-mass configuration obtained for different EOS models.

EOS	$M/M_\odot$	R (km)	n_c/n_0
BBB1	1.80	9.70	8.37
BBB2	1.94	9.54	8.31
WFF	2.13	9.40	7.81
APR	2.20	10.10	7.13
BPAL32	1.95	10.54	7.58
KS	2.24	10.79	6.30

by large uncertainties [14], therefore the previous quoted "high mass values" should always be handled with care.

Recently Nice *et al.* [15] have determined the mass of the neutron star associated to the millisecond pulsar PSR J0751 +1808, which is a member of a binary system with a helium white dwarf secondary. Measuring general relativistic effects, the authors of ref. [15] have obtained for the mass of PSR J0751 +1808 the value $2.1 \pm 0.2 M_\odot$ at 68% confidence level ($2.1^{+0.4}_{-0.5} M_\odot$ at 95% confidence level). This is the largest measured value for the mass of any neutron star.

2 Traditonal neutron stars

As we said before, in "traditional" neutron stars one assumes the stellar core to be made of an uncharged mixture of neutrons, protons, electrons and muons in equilibrium with respect to the weak interactions (β-stable nuclear matter). Even in this simplified picture, the determination of the EOS remains a formidable theoretical problem. In fact, one has to extrapolate the EOS to extreme conditions of high density and high neutron-proton asymmetry, *i.e.* in a regime where the EOS is poorly constrained by nuclear data and experiments. In the last decade there has been a substantial progress in the microscopic numerical methods for solving the nuclear many-body problem both within the non-relativistic and relativistic approaches. For example, the convergence of the Brueckner-Bethe-Goldstone (BBG) hole line expansion has beed inspected up to the three-hole line contribution [16] in the continuous choice for the auxiliary single-particle potential. The introduction of nuclear three-body forces (TBF) constrained by nuclear data [17] has permitted to solve the *saturation problem* of nuclear matter in the non-relativistic approach [18–20].

In table 1, we report the maximum-mass configuration of NS using different EOS. The first two stellar models (BBB) have been obtained [19] within the Brueckner-Hartree-Fock (BHF) approximation of the BBG theory. Model BBB1 makes use of the Argonne v14 (Av14) nucleon-nucleon (NN) interaction implemented by the Urbana TBF. Model BBB2, considers the Paris NN interaction and the Urbana TBF. We next consider two stellar models (WFF, APR) which are based on microscopic EOSs derived by variational chain summation methods. The WFF model [18] makes use of the Av14 interaction plus the Urbana (model VII) TBF. The APR model [20] is

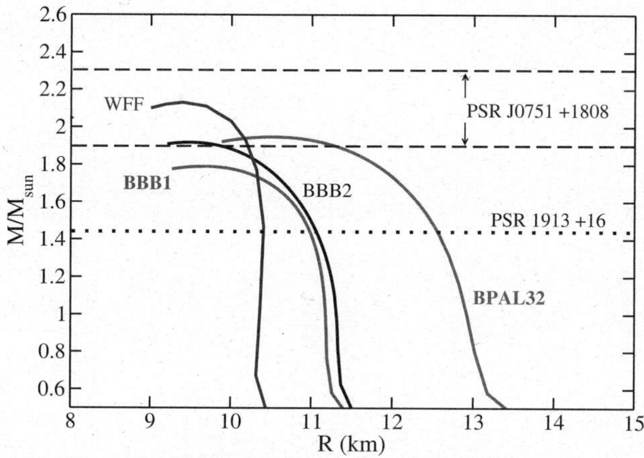

Fig. 2. The mass-radius relation for neutron stars for different EOS for β-stable nuclear matter. The dotted line gives the value of the mass of PSR 1913 +16, the band between the long-dashed lines the mass of PSR J0751 +1808 (at 68% c.l.).

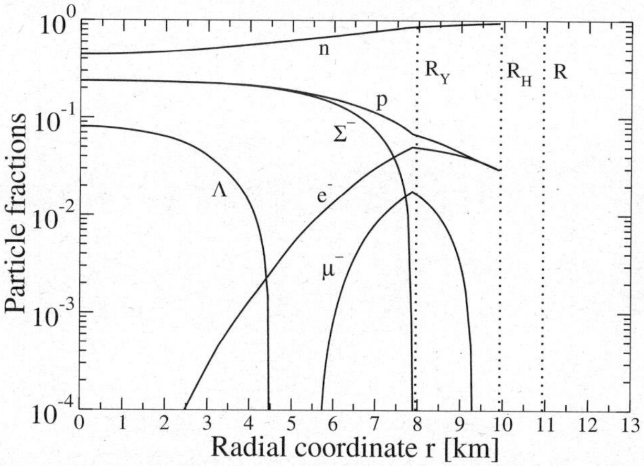

Fig. 3. The internal composition of a neutron star with hyperonic-matter core. R_Y is the radius of the hyperonic core. The nuclear-matter layer extend between R_Y and R_H. The stellar crust extend between R_H and R.

based on the charge-dependent Argonne v18 (Av18) two-body potential plus the Urbana (model IX) TBF. In addition, the APR model includes boost corrections to the NN interaction, which give the leading relativistic corrections to the EOS. The BPAL 32 neutron star configurations have been calculated using a phenomenological EOS for asymmetric nuclear matter [21] (see also [3,22]) derived from a momentum- and density-dependent effective NN interaction. The BPAL model is an extension of the EOS of refs. [23,24]. An important feature of the BPAL EOS is the possibility to have different forms for the density dependence of the potential part of the nuclear symmetry energy, reproducing different results predicted by various microscopic calculations. The nuclear symmetry energy E_{sym} (particularly its density dependence) plays an important role in the physics of NS. In fact, $E_{sym}(n)$ affects the values of the stellar radius [22,25], the values of the threshold densities for the onset of various new particle species (*e.g.*, K^- condensation [26], hyperons, QM phase [27]) and the value of the proton fraction in the stellar core [24,19]. The latter quantity has a strong influence on the thermal evolution of neutron stars. Finally, we consider a stellar model (KS) based on a miscroscopic EOS derived [28] from the Dirac-BHF approach using the Bonn-B nuclear interaction. The mass radius relations for some of these stellar models are reported in fig. 2, together with the measured values of the masses of PRS 1913 +16 and PSR J0751 +1808.

3 Hyperon stars

The reason why hyperons are expected in the high dense core of a neutron star is very simple, and it is mainly a consequence of the fermionic nature of nucleons, which makes the nucleon chemical potentials a very rapidly increasing function of density. As soon as the chemical potential of neutrons becomes sufficiently large, the most energetic neutrons (*i.e.*, those on the Fermi surface) can

decay via the weak interactions into Λ hyperons and form a Fermi sea of this new hadronic species with $\mu_\Lambda = \mu_n$. The Σ^- can be produced via the process $e^- + n \rightarrow \Sigma^- + \nu_e$ when the Σ^- chemical potential fulfill the condition[2] $\mu_{\Sigma^-} = \mu_n + \mu_e$. Hyperons appear at a relatively moderate density of about 2 times n_0. Notice that the Σ^- hyperon appears at a lower density than the Λ, even though the Σ^- is more massive than the Λ. This is due to the contribution of the electron chemical potential μ_e to the threshold condition for the Σ^- (*i.e.*, $M_{\Sigma^-} = \mu_n + \mu_e$, for free hyperons) and to the fact that μ_e in dense matter is large and can compensate for the mass difference $M_{\Sigma^-} - M_\Lambda = 81.76\,\text{MeV}$.

In fig. 3, we show the radial profile of a typical hyperon star [29]. As we see the hyperonic-matter inner core of the star extend for about 8 km. This radius has to be compared with the total stellar radius $R \sim 11\,\text{km}$, and with the thickness of the nuclear-matter layer (outer core) which is about 2 km.

The influence of hyperons on neutron stars properties has been investigated using different approaches to determine the EOS of hyperonic matter. One of the most popular approaches, to solve this problem, is the relativistic mean-field model [30,5]. Some of the parametrizations of the Lagrangian of the theory have tried to reconcile measured values of neutron star masses with the binding energy of the Λ particle in hypernuclei [31]. Considerable progress has been done in the last few years in microscopic calculations of hyperonic matter. This method is based on an extension of the BBG theory to include hyperonic degrees of freedom [32,33,29]. In particular, the study of ref. [29] focus on the properties of a newborn neutron star, and explore the consequences of neutrino trapping in dense matter on the structural properties and on the early evolution of neutron stars [3].

---

[2] Except from the very initial stage soon after neutron star birth, neutrinos freely escape the star and thus the neutrino chemical potentials can be put equal to zero.

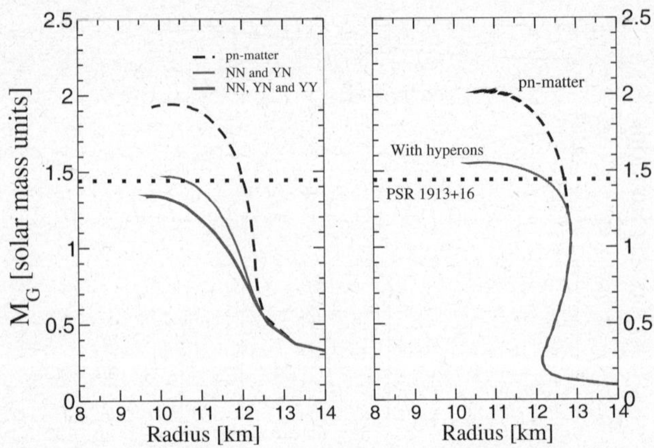

Fig. 4. Mass-radius relation for "traditional" neutron stars and hyperon stars calculated [33] within the BHF approach with the NSC9e interaction (left panel) and with the relativistic mean-field EOS GM3 of ref. [31] (rigth panel). The dotted horizontal line indicates the mass of PSR1913+16.

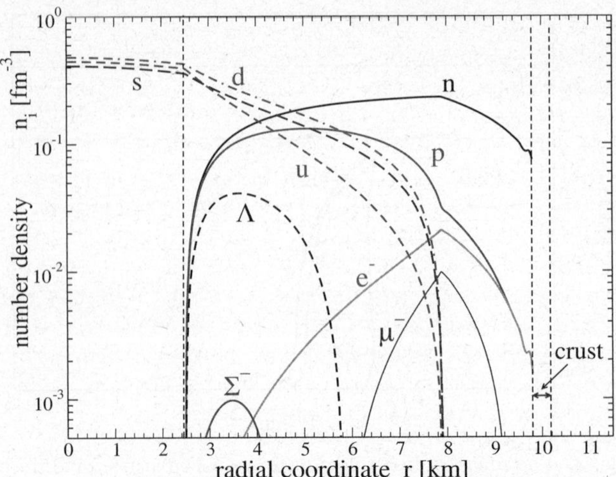

Fig. 5. Internal composition of a hybrid star with $M = M_{max} = 1.448\ M_\odot$. The GM3 EOS [31] has been used for the hadronic phase, and the bag model EOS, with $B = 136.6\ \mathrm{MeV/fm}^3$ and $m_s = 150\ \mathrm{MeV}$, for the quark phase.

As expected, the presence of hyperons reduces in a sizeable manner the value of the pressure of β-stable hyperonic matter with respect to the case of β-stable nuclear matter at the same density. This softening of the EOS has important consequences on many macroscopic properties of the star: the maximum mass is reduced by $\Delta M_{max} \sim$ 0.5–0.8 $M_\odot$, and the corresponding central density is increased. Also, hyperon stars are more compact (*i.e.* they have a smaller radius) with respect to traditional neutron stars. This is illustrated in fig. 4, where we show the mass-radius relation for traditional neutron stars and for hyperon stars obtained with the microscopic EOS of ref. [33] (left panel) and with the relativistic mean-field EOS (GM3 model) given in ref. [31]. The results depicted in fig. 4 clearly demonstrate that to neglect hyperons leads to an overstimate of M_{max}.

It is important to notice the "low" value of the stellar maximum mass, predicted within the approach of ref. [33], which is in contrast with the measured mass of PSR 1913 +16. The prediction of a value for M_{max} below some of the measured neutron star masses is a common feature of all the present microscopic EOS of hyperonic matter based on G-matrix BHF calculations [32,33,29]. For example, the authors of ref. [32], in case of the Argonne v_{18} NN interaction, found $M_{max} = 2.00\ M_\odot$, a corresponding radius of $R = 10.54\ \mathrm{km}$ and a central density $\rho_c = 1.11\ \mathrm{fm}^{-3}$ for neutron stars with a pure nucleonic core. When hyperons are considered as possible stellar constituents, they found [32] $M_{max} = 1.22\ M_\odot$, a corresponding radius of $R = 10.46\ \mathrm{km}$ and a central density $\rho_c = 1.25\ \mathrm{fm}^{-3}$. Therefore the current EOS for hyperonic matter, deduced from microscopic G-matrix BHF calculations, are "too soft" to explain observed neutron star masses.

Clearly, one should try to trace the origin of this problem back to the underlying YN and YY two-body interactions or to the possible repulsive three-body baryonic forces involving one or more hyperons, not included in the work of refs. [32,33,29]. Presently, this is a subject of very active research by people working in this field. Therefore, the use of microscopic EOS of hyperonic matter in the contest of neutron star physics is of fundamental importance for our understanding of the strong interactions involving hyperons, and to learn how these interactions behave in dense many-body systems.

4 Hybrid stars

As we have seen before, different sophisticated approaches, based on advanced many-body thecniques, have been utilized to derive the EOS for the hadronic phase. This is possible, to a large extent, thanks to the rich body of experimantal data at density $n \sim n_0$. The situation is drastically different in the case of quark matter. Presently, only indirect experimental evidence has been found for the existence of this new phase of matter. Moreover, lattice QCD calculations at finite density, to derive the EOS of quark matter, are still in an early stage. Thus, simple phenomenological (*e.g.*, MIT bag, Nambu–Jona-Lasinio) models have been used to describe QM for hybrid-star calculations.

In fig. 5, we show the typical internal composition of a hybrid star. The cross-section of the star is relative to the maximum-mass configuration ($M = M_{max} = 1.448 M_\odot$, with radius $R = 10.2\ \mathrm{km}$, and central density $\rho_c = 28.1 \times 10^{14}\ \mathrm{g/cm}^3$) for the EOS described in the figure caption. This star has a pure QM core which extend for about 2.5 km, next it has a hadron-quark mixed phase layer with a thickness of about 5.5 km, followed by a nuclear matter layer about 2 km thick. On the top we have the usual neutron star crust. The presence of quarks makes the EOS softer with respect to the corresponding pure hadronic-matter EOS. The stellar sequence associated to the latter EOS has the maximum-mass configuration: $M_{max} = 1.552\ M_\odot$, $R = 10.7\ \mathrm{km}$, $\rho_c = 25.4 \times 10^{14}\ \mathrm{g/cm}^3$.

Many possible astrophysical signals for the presence of a quark core in neutron stars have been proposed (see [5,6] and references quoted therein). Particularly, pulse timing properties of pulsars have attracted much attention since they are a manifestation of the rotational properties of the associated neutron star. The onset of quark-deconfinement in the core of the star, will cause a change in the stellar moment of inertia [34]. This change will produce a peculiar evolution of the stellar rotational period ($P = 2\pi/\Omega$) which will cause large deviations of the so-called pulsar braking index $n(\Omega) = (\Omega\ddot{\Omega}/\dot{\Omega}^2)$ from the *canonical* value $n = 3$, derived within the magnetic-dipole model for pulsars and assuming a constant moment of inertia for the star. The possible measurement of a value of the braking index very different from the canonical value (*i.e.* $|n| \gg 3$) has been proposed [34] as a signature for the occurrence of the quark-deconfinement phase transition in a neutron star. However, it must be stressed that a large value of the braking index could also results from the pulsar magnetic-field decay or alignment of the magnetic axis with the rotation axis [35].

5 Metastability of hadronic stars and GRBs

In bulk matter the quark-hadron mixed phase begins at the *static transition point* defined according to the Gibbs' criterion for phase equilibrium

$$\mu_H = \mu_Q \equiv \mu_0, \qquad P_H(\mu_0) = P_Q(\mu_0) \equiv P_0, \qquad (1)$$

where $\mu_H = (\varepsilon_H + P_H)/n_{b,H}$ and $\mu_Q = (\varepsilon_Q + P_Q)/n_{b,Q}$ are the chemical potentials for the hadron and quark phase respectively, ε_H (ε_Q), P_H (P_Q) and $n_{b,H}$ ($n_{b,Q}$) denote respectively the total (*i.e.*, including leptonic contributions) energy density, the total pressure and baryon number density for the hadron (quark) phase.

Consider now the more realistic situation in which one takes into account the energy cost due to finite-size effects in creating a drop of QM in the hadronic environment. As a consequence of these effects, the formation of a critical-size drop of QM is not immediate and it is necessary to have an overpressure $\Delta P = P - P_0$ with respect to the static transition point. Thus, above P_0, hadronic matter is in a metastable state, and the formation of a real drop of QM occurs via a quantum nucleation mechanism. Quark flavor must be conserved during the deconfinement transition [36]. We will call this form of deconfined matter, in which the flavor content is equal to that of the β-stable hadronic system at the same pressure, as the Q*-phase. Soon afterwards a critical-size drop of QM is formed the weak interactions will have enough time to act, changing the quark flavor fraction of the deconfined droplet to lower its energy, and a droplet of β-stable SQM is formed (hereafter the Q-phase).

In the scenario proposed in ref. [37], one considers a pure HS whose central pressure is increasing due to spin-down or due to mass accretion, *e.g.*, from the material left by the SN explosion, or from a companion star. As the central pressure exceeds the threshold value P_0^* at the static transition point, a virtual drop of quark matter in the Q*-phase can be formed in the center of the star. As

soon as a real drop of Q*-matter is formed, it will grow very rapidly and the original HS will be converted to and hybrid star or to a strange star, depending on the detail of the EOS for quark matter employed to model the phase transition. The nucleation time τ (*i.e.*, the time needed to form the first critical droplet of QM), can be calculated for different values of the stellar central pressure P_c, and thus for different values of the mass $M(P_c)$ of the corresponding hadronic star. The nucleation time τ dramatically depends on [37] the value of the stellar mass (central pressure). A metastable hadronic star can have a mean lifetime many orders of magnitude larger than the age of the Universe. As the star accretes a small amount of mass (of the order of a few per cent of the mass of the Sun), the consequential increase of the central pressure leads to a huge reduction of the nucleation time and, as a result, to a dramatic reduction of the HS *mean lifetime*.

To summarize, pure hadronic stars having a central pressure larger than the static transition pressure P_0^* for the formation of the Q*-phase are metastable to the *decay* (conversion) to a more compact stellar configuration in which deconfined QM is present (HyS or SS). These metastable HS have a *mean lifetime*[3] which is related to the nucleation time to form the first critical-size drop of deconfined matter in their interior. We define as *critical mass* M_{cr} of the metastable HS, the value of the gravitational mass for which the nucleation time is equal to one year: $M_{cr} \equiv M_{HS}(\tau = 1\,\text{y})$. Pure hadronic stars with $M_H > M_{cr}$ are very unlikely to be observed. M_{cr} plays the role of an *effective maximum mass* [36] for the hadronic branch of compact stars. While the Oppenheimer-Volkoff maximum mass [4] $M_{HS,max}$ is determined by the overall stiffness of the EOS for hadronic matter, the value of M_{cr} will depend in addition on the bulk properties of the EOS for quark matter and on the properties at the interface between the confined and deconfined phases of matter (*e.g.*, the droplet surface tension σ).

In fig. 6, we show the mass-radius (MR) curve for pure HSs within the GM1 [31] EOS for the hadronic phase, and that for hybrid or strange stars for different values of the bag constant B. The configuration marked with an asterisk on the hadronic MR curves represents the HS for which the central pressure is equal to P_0^* and $\tau = \infty$. The full circle on the HS sequence represents the critical-mass configuration, in the case of $\sigma = 30\,\text{MeV/fm}^2$. The full circle on the HyS (SS) mass-radius curve represents the hybrid (strange) star which is formed from the conversion of the hadronic star with $M_{HS} = M_{cr}$. We assume that during the stellar-conversion process the total number of baryons in the star (*i.e.*, the stellar baryonic mass) is conserved. Thus, the total energy liberated in the stellar conversion is given [38] by the difference between the gravitational mass of the initial hadronic star ($M_{in} \equiv M_{cr}$) and that of the final hybrid or strange stellar configuration $M_{fin} \equiv M_{QS}(M_{cr}^b)$ with the same baryonic mass: $E_{conv} = (M_{in} - M_{fin})c^2$.

---

[3] The actual *mean lifetime* of the HS depends on the mass accretion or on the spin-down rate which modifies τ via an explicit time dependence of the stellar central pressure.

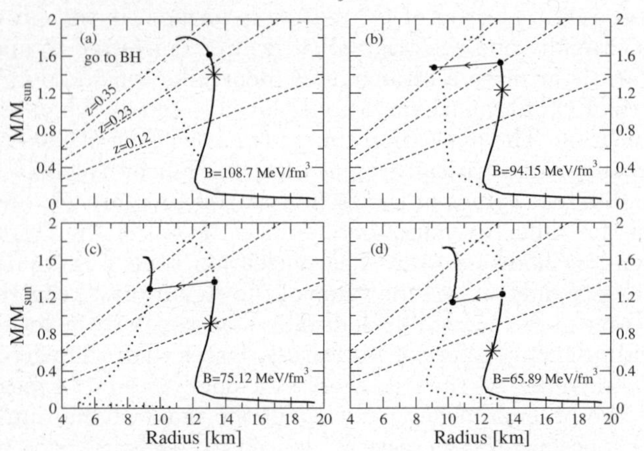

Fig. 6. Mass-radius relation for pure HSs [36] and for HyS or SS configurations for several values of the bag constant. The configuration marked with an asterisk represents the HS for which the central pressure is equal to P_0^* and $\tau = \infty$. The conversion process of the HS, with a mass equal to M_{cr}, into a final QS is denoted by the full circles connected by an arrow. In all cases $\sigma = 30\,\mathrm{MeV/fm^2}$. The dashed lines show the gravitational red shift deduced for the X-ray compact sources EXO 0748-676 ($z = 0.35$) and 1E 1207.4-5209 ($z = 0.12$–0.23).

The stellar-conversion process, described so far, will start to populate the new branch of quark stars (the part of the QS sequence plotted as a continuous curve in fig. 6). Long-term accretion on the QS can next produce stars with masses up to the limiting mass $M_{QS,max}$ for the quark star configurations.

The stellar conversion energy E_{conv} is in the range [36] 0.5–1.7×10^{53} erg. This mechanism has been proposed as a possible energy source for Gamma Ray Bursts (GRBs). The model of ref. [37] accounts for the association between SN explosions and GRBs, in fact, the stellar conversion represents a second "explosion" (the *Quark Deconfinement Nova* [36]) which occurs after the first explosion (the SN) which form the hadronic star. This model is also able to explain, in a natural way, the possibility to have a *long*-time delay ΔT between the SN explosion and the associated GRB, as inferred, for example, in the case of GRB990705 ($\Delta T \sim$ a few years) [39], GRB011211 ($\Delta T \sim$ a few days) [40].

References

1. G.E. Brown, H.A. Bethe, Astrophys. J. **423**, 659 (1994).
2. I. Bombaci, Astron. Astrophys. **305**, 871 (1996).
3. M. Prakash, I. Bombaci, M. Prakash, P.J. Ellis, R. Knorren, J.M. Lattimer, Phys. Rep. **280**, 1 (1997).
4. J.R. Oppenheimer, G.M. Volkoff, Phys. Rev. **62**, 035801 (2000).
5. N.K. Glendenning, *Compact Stars: Nuclear Physics, Particle Physics, and General Relativity* (Springer Verlag, 1996).
6. F. Weber, *Pulsars as Astrophysical Laboratories for Nuclear and Particle Physics* (IoP Publishing, 1999).
7. A.R. Bodmer, Phys. Rev. D **4**, 1601 (1971); H. Terazawa, INS Report **336** (Tokio University, 1979); E. Witten, Phys. Rev. D **30**, 272 (1984).
8. X.-D Li, I. Bombaci, M. Dey, J. Dey, E.P.J. van den Heuvel, Phys. Rev. Lett. **83**, 3776 (1999).
9. M.G. Alford, Annu. Rev. Nucl. Part. Sci. **51**, 131 (2001).
10. G. Nardulli, Riv. Nuovo Cimento **25**, 1 (2001).
11. S.E. Thorsett, D. Chakrabarty, Astrophys. J. **512**, 288 (1999).
12. H. Quaintrell *et al.*, Astron. Astrophys. **401**, 313 (2003).
13. J.A. Orosz, E. Kuulkers, Mon. Not. R. Astron. Soc. **305**, 1 (1999).
14. M.H. van Kerkwijk *et al.*, Astron. Astrophys. **303**, 483 (1995).
15. D.J. Nice *et al.*, Astrophys. J. **634**, 1242 (2005).
16. H.Q. Song *et al.*, Phys. Rev. Lett. **81**, 1584 (1998).
17. B.S. Pudliner *et al.*, Phys. Rev. Lett. **74**, 4396 (1995).
18. R.B. Wiringa, V. Ficks, A. Fabrocini, Phys. Rev. C **38**, 1010 (1988).
19. M. Baldo, I. Bombaci, G.F. Bugio, Astron. Astrophys. **328**, 274 (1997).
20. A. Akmal, V.R. Pandharipande, D.G. Ravenhall, Phys. Rev. C **58**, 1804 (1998).
21. I. Bombaci, in *Perspectives on Theoretical Nuclear Physics, Proceedings of the Conference Problems in Theoretical Nuclear Physics, October 1995, Cortona (Italy)*, edited by I. Bombaci *et al.* (ETS, Pisa, 1996) p. 223.
22. I. Bombaci, in *Isospin Physics in Heavy-Ion Collisions at Intermediate Energies*, edited by B.-A. Li, W.U. Schröder (Nova Science Publisher, New York, 2001) p. 35.
23. C. Gale, G, Bertsch, S. Das Gupta, Phys. Rev. C **35**, 1666 (1997).
24. M. Prakash, T.L. Ainsworth, J.M. Lattimer, Phys. Rev. Lett. **61**, 2518 (1998).
25. B.-A. Li, A.W. Steiner, Phys. Lett. B **642**, 436 (2006).
26. W. Zuo *et al.*, Phys. Rev. C **70**, 055802 (2004).
27. M. Di Toro *et al.*, Nucl. Phys. A **775**, 102 (2006).
28. P.G. Krastev, F. Sammarruca, Phys. Rev. C **74**, 025808 (2006).
29. I. Vidaña, I. Bombaci, A. Polls, A. Ramos, Astron. Astrophys. **399**, 687 (2003).
30. N.K. Glendenning, Astrophys. J. **293**, 470 (1985).
31. N.K. Glendenning, S.A. Moszkowski, Phys. Rev. Lett. **67**, 2414 (1991).
32. M. Baldo, G.F. Burgio, H.J. Schulze, Phys. Rev. C **61**, 055801 (2000).
33. I. Vidaña, A. Polls, A. Ramos, L. Engvik, M. Hjorth-Jensen, Phys. Rev. C **62**, 035801 (2000).
34. N.K. Glendenning, S. Pei, F. Weber, Phys. Rev. Lett. **79**, 1603 (1997).
35. T.M. Tauris, S. Konar, Astron. Astrophys. **376**, 543 (2001).
36. I. Bombaci, I. Parenti, I. Vidaña, Astrophys. J. **614**, 314 (2004).
37. Z. Berezhiani, I. Bombaci, A. Drago, F. Frontera, A. Lavagno, Astrophys. J. **586**, 1250 (2003).
38. I. Bombaci, B. Datta, Astrophys. J. **530**, L72 (2000).
39. L. Amati *et al.*, Science **290**, 953 (2000).
40. J.N. Reeves *et al.*, Nature **414**, 512 (2002).

Eur. Phys. J. A **31**, 816–823 (2007)

DOI 10.1140/epja/i2006-10206-6

THE EUROPEAN
PHYSICAL JOURNAL A

Special Article – QNP 2006

Low-x QCD physics from RHIC and HERA to the LHC

D. d'Enterria[a]

CERN, PH-EP, CH-1211 Geneva 23, Switzerland

Received: 25 October 2006

Published online: 27 February 2007 – © Società Italiana di Fisica / Springer-Verlag 2007

Abstract. We present a summary of the physics of gluon saturation and non-linear QCD evolution at small values of the parton momentum fraction x in the proton and nucleus in the context of recent experimental results at HERA and RHIC. The rich physics potential of low-x studies at the LHC, especially in the forward region, is discussed and some benchmark measurements in pp, pA and AA collisions are introduced.

PACS. 12.38.-t Quantum chromodynamics – 24.85.+p Quarks, gluons, and QCD in nuclei and nuclear processes – 25.75.-q Relativistic heavy-ion collisions

1 Introduction

1.1 Parton structure and evolution

The partonic structure of the proton (nucleus) can be probed with high precision in deep inelastic scattering (DIS) electron-proton ep (electron-nucleus, eA) collisions. The inclusive DIS hadron cross-section, $\mathrm{d}^2\sigma/\mathrm{d}x\,\mathrm{d}Q^2$, is a function of the virtuality Q^2 of the exchanged gauge boson (*i.e.* its "resolving power"), and the Bjorken-x fraction of the total nucleon momentum carried by the struck parton. The differential cross-section for the neutral-current (γ, Z exchange) process can be written in terms of the target structure functions as

$$\frac{\mathrm{d}^2\sigma}{\mathrm{d}x\,\mathrm{d}Q^2} = \frac{2\pi\alpha^2}{x\,Q^4}\left[Y_+ \cdot F_2 \mp Y_- \cdot xF_3 - y^2 \cdot F_L\right], \quad (1)$$

where $Y_\pm = 1 \pm (1-y)^2$ is related to the collision inelasticity y, and the structure functions $F_{2,3,L}(x,Q^2)$ describe the density of quarks and gluons in the hadron: $F_2 \propto e_q^2\, x\, \Sigma_i(q_i + \bar{q}_i)$, $xF_3 \propto x\, \Sigma_i(q_i - \bar{q}_i)$, $F_L \propto \alpha_s\, xg$ ($xq_i, x\bar{q}_i$ and xg are the corresponding parton distribution functions, PDF). F_2, the dominant contribution to the cross-section over most of phase space, is seen to rise strongly for decreasing Bjorken-x at HERA (fig. 1). The growth in F_2 is well described by $F_2(x,Q^2) \propto x^{-\lambda(Q^2)}$, with $\lambda \approx 0.1$–0.3 logarithmically rising with Q^2 [1]. The F_2 scaling violations evident at small x in fig. 1 are indicative of the increasing gluon radiation from sea quarks. The $xg(x,Q^2)$ distribution itself can be indirectly determined (fig. 2) from the F_2 slope:

$$\frac{\partial F_2(x,Q^2)}{\partial \ln(Q^2)} \approx \frac{10\,\alpha_s(Q^2)}{27\pi}\, xg(x,Q^2). \quad (2)$$

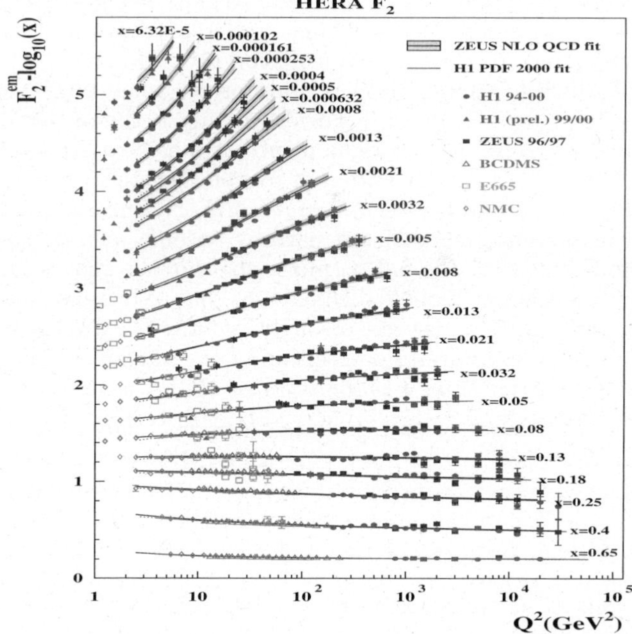

Fig. 1. $F_2(x,Q^2)$ measured in proton DIS at HERA ($\sqrt{s} = 320\,\mathrm{GeV}$) and fixed-target ($\sqrt{s} \approx 10$–$30\,\mathrm{GeV}$) experiments.

Although the PDFs are non-perturbative objects obtained from fits to the DIS data, once measured at an input scale $Q_0^2 \gtrsim 2\,\mathrm{GeV}^2$ their value at any other Q^2 can be determined with the Dokshitzer-Gribov-Lipatov-Altarelli-Parisi (DGLAP) evolution equations which govern the probability of parton branchings (gluon splitting, q, g-strahlung) in QCD [2].

The DGLAP parton evolution, however, only takes into account the Q^2-dependence of the PDFs, effectively

[a] e-mail: dde@cern.ch

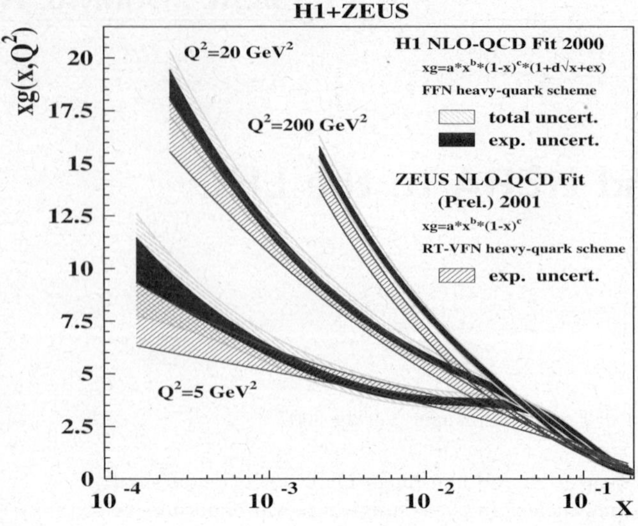

Fig. 2. Gluon distributions extracted at HERA (H1 and ZEUS) as a function of x in three bins of Q^2 [3].

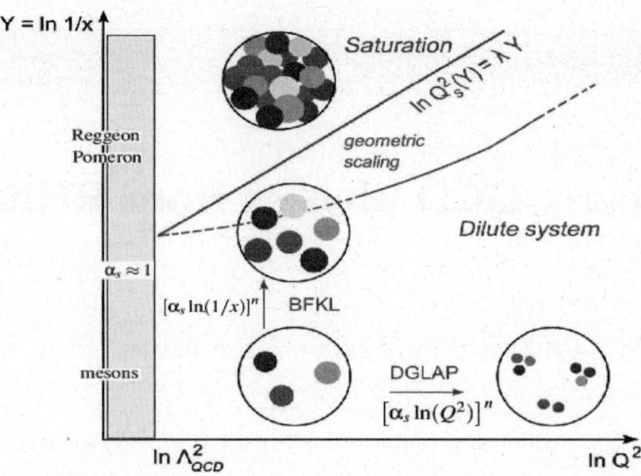

Fig. 3. QCD "phase diagram" in the $1/x, Q^2$ plane (each dot represents a parton with transverse area $\sim 1/Q^2$ and fraction x of the hadron momentum). The different evolution regimes (DGLAP, BFKL, saturation) as well as the "saturation scale" and "geometric scaling" curves between the dense and dilute domains are indicated. Adapted from [8].

summing leading powers of $[\alpha_s \ln(Q^2)]^n$ ("leading twist") generated by parton cascades in a region of phase space where the gluons have strongly ordered transverse momenta towards the hard subcollision $Q^2 \gg k_{nT}^2 \gg \cdots \gg k_{1T}^2$. Such a resummation is appropriate when $\ln(Q^2)$ is much larger than $\ln(1/x)$. For decreasing x, the probability of emitting an extra gluon increases as $\propto \alpha_s \ln(1/x)$. In this regime, the evolution of parton densities proceeds over a large rapidity region, $\Delta y \sim \ln(1/x)$, and the finite transverse momenta of the partons become increasingly important. Here the full k_T phase space of the gluons (including scattering of off-shell partons) has to be taken into account and not just the strongly ordered DGLAP part. Thus, the appropriate description of the parton distributions is in terms of k_T-*unintegrated* PDFs, $xg(x, Q^2) = \int^{Q^2} dk_T^2 G(x, k_T^2)$ with $G(x, k_T^2) \sim h(k_T^2) x^{-\lambda}$, described by the Balitski-Fadin-Kuraev-Lipatov (BFKL) equation [4] which governs parton evolution in x at fixed Q^2. Hints of extra BFKL radiation have been recently found at HERA in the enhanced production of forward jets compared to DGLAP expectations [5,6]. At large Q^2, a description resumming over both $\alpha_s \ln(Q^2)$ and $\alpha_s \ln(1/x)$ is given by the Ciafaloni-Catani-Fiorani-Marchesini (CCFM) evolution equation [7].

1.2 Parton saturation and non-linear evolution at low x

As shown in fig. 2, the gluon density rises very fast for decreasing x. Eventually, at some small enough value of x ($\alpha_s \ln(1/x) \gg 1$) the number of gluons is so large that non-linear (gg fusion) effects become important, taming the growth of the parton densities. In such a high-gluon density regime three things are expected to occur: i) the standard DGLAP and BFKL *linear* equations should no longer be applicable since they only account for single parton branchings ($1 \rightarrow 2$ processes) but not for non-linear

($2 \rightarrow 1$) gluon recombinations; ii) pQCD (collinear and k_T) factorization should break due to its (now invalid) assumption of *incoherent* parton scattering; and, as a result, iii) standard pQCD calculations lead to a *violation of unitarity* even for $Q^2 \gg \Lambda_{QCD}^2$. Figure 3 schematically depicts the different parton evolution regimes as a function of $y = \ln(1/x)$ and Q^2. For small enough x values and for virtualities below an energy-dependent "saturation momentum", Q_s, intrinsic to the *size* of the hadron, one expects to enter the regime of saturated PDFs. Since $xg(x, Q^2)$ can be interpreted as the number of gluons with transverse area $r^2 \sim 1/Q^2$ in the hadron wave function, an increase of Q^2 effectively diminishes the "size" of each parton, partially compensating for the growth in their number (*i.e.* the higher Q^2 is, the smaller the x at which saturation sets in). Saturation effects are, thus, expected to occur when the size occupied by the partons becomes similar to the size of the hadron, πR^2. In the case of nuclear targets with A nucleons (*i.e.* with gluon density $xG = A \cdot xg$), this condition provides a definition for the saturation scale [9, 10]:

$$Q_s^2(x) \simeq \alpha_s \frac{1}{\pi R^2} xG(x, Q^2) \sim A^{1/3} x^{-\lambda}$$
$$\sim A^{1/3} (\sqrt{s})^\lambda \sim A^{1/3} e^{\lambda y}, \tag{3}$$

with $\lambda \approx 0.25$ [11]. Equation (3) tell us that Q_s grows with the number of nucleons in the target and with the energy of the collision, $\sqrt{s}$, or equivalently, the rapidity of the gluon $y = \ln(1/x)$. The nucleon number dependence implies that, at equivalent energies, saturation effects will be enhanced by factors as large as $A^{1/3} \approx 6$ in heavy nuclear targets ($A = 208$ for Pb) compared to protons. In the last fifteen years, an effective field theory of QCD in the high-energy (high density, small x) limit has been developed —the Colour Glass Condensate (CGC) [12]— which describes the hadrons in terms of classical fields (saturated gluon wave functions) below the saturation scale Q_s. The saturation momentum Q_s introduces a (semi-)hard scale, $Q_s \gg \Lambda_{QCD}$, which not only acts as an infrared cut-off to unitarize the cross-sections but allows weak-coupling perturbative calculations ($\alpha_s(Q_s) \ll 1$) in a strong $F_{\mu\nu}$ colour field background. In the CGC framework, hadronic and nuclear collisions are seen as collisions of classical wave functions which "resum" all gluon recombinations and multiple scatterings. The quantum evolution in the CGC approach is given by the JIMWLK [13] non-linear equations (or by their mean-field limit for $N_c \to \infty$, the Balitsky-Kovchegov equation [14]) which reduce to the standard BFKL kernel at higher x values.

2 Parton saturation: experimental studies

The main source of information on the PDFs is obtained from hard processes as they involve outgoing particles directly coupled to the partonic scattering vertices. Figure 4 summarizes the variety of measurements at different experimental facilities which are sensitive to the gluon density and their approximate x coverage. xG enters directly at LO in hadron-hadron collisions with i) prompt photons, ii) jets, and iii) heavy quarks in the final state, as well as in the (difficult) DIS measurement[1] of iv) the longitudinal structure function F_L. In addition, v) heavy vector mesons ($J/\psi, \Upsilon$) from diffractive photoproduction processes[2] are a valuable probe of the gluon density since their cross-sections are proportional to the *square* of xG [16,17]:

$$\left. \frac{d\sigma_{\gamma p, A \to V p, A}}{dt} \right|_{t=0} = \frac{\alpha_s^2 \Gamma_{ee}}{3\alpha M_V^5} 16\pi^3 \left[xG(x, Q^2) \right]^2 , \quad (4)$$

$$\text{with} \quad Q^2 = M_V^2/4 \quad \text{and} \quad x = M_V^2/W_{\gamma p, A}^2. \quad (5)$$

The main source of information on the *quark* densities is obtained from measurements of i) the structure functions $F_{2,3}$ in lepton-hadron scattering, and ii) lepton pair (Drell-Yan, DY) production in hadron-hadron collisions. In hadronic collisions, one commonly measures the perturbative probes at central rapidities ($y = 0$), where $x = x_T = Q/\sqrt{s}$, and $Q \sim p_T, M$ is the characteristic scale of the hard scattering. However, one can probe smaller x_2 values in the target by measuring the corresponding cross-sections in the *forward* direction. Indeed, for a $2 \to 2$ parton scattering the *minimum* momentum fraction probed in a process with a particle of momentum p_T produced at pseudo-rapidity η is [18]

$$x_2^{min} = \frac{x_T\, e^{-\eta}}{2 - x_T\, e^{\eta}}, \quad \text{where} \quad x_T = 2p_T/\sqrt{s}, \quad (6)$$

---

[1] xG can also be (in)directly extracted from F_2 through the derivative in eq. (2) as well as from the F_2^{charm} data [15].

[2] Diffractive γp (γA) processes are characterized by a quasi-elastic interaction —mediated by a Pomeron or two gluons in a colour singlet state— in which the p (A) remains intact (or in a low excited state) and separated by a rapidity gap from the rest of final-state particles.

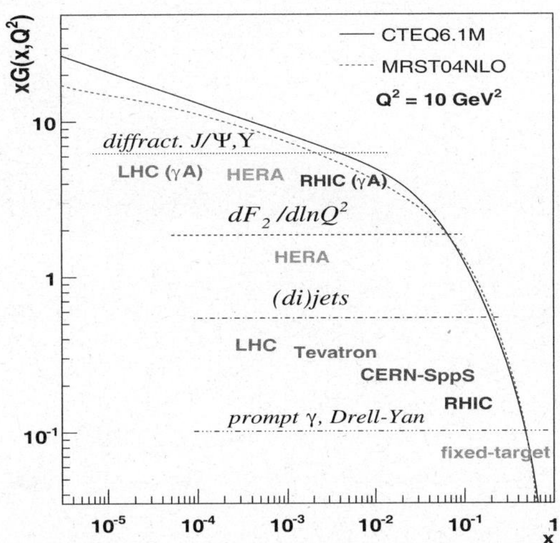

Fig. 4. Examples of experimental measurements at various facilities providing information on the gluon PDF in the range $x \sim 10^{-5}$–0.8.

i.e. x_2^{min} decreases by a factor of ~ 10 every 2 units of rapidity. Though eq. (6) is a lower limit at the end of phase-space (in practice the $\langle x_2 \rangle$ values in parton-parton scatterings are at least 10 larger than x_2^{min} [18]), it provides the right estimate of the typical $x_2 = (p_T/\sqrt{s})\, e^{-\eta}$ values reached in non-linear $2 \to 1$ processes (in which the momentum is balanced by the gluon "medium") as described in parton saturation models [19,20].

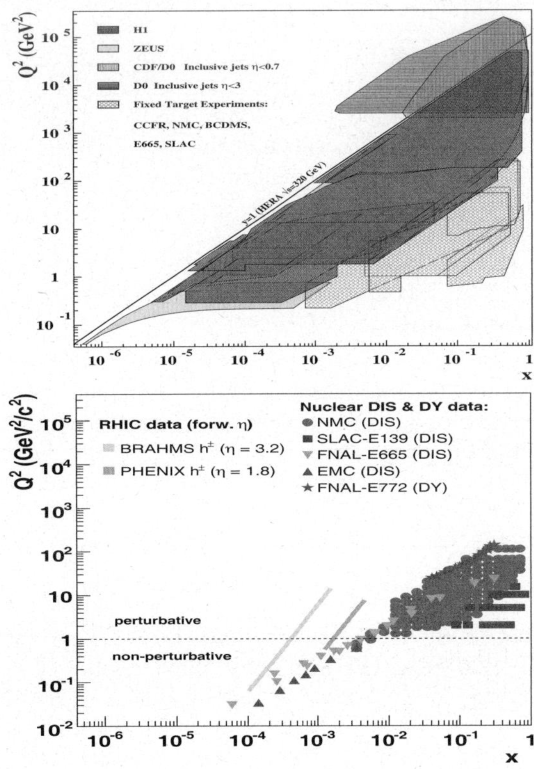

Fig. 5. Available measurements in the (x, Q^2) plane used for the determination of the proton [21] (top) and nuclear [22] (bottom) PDFs.

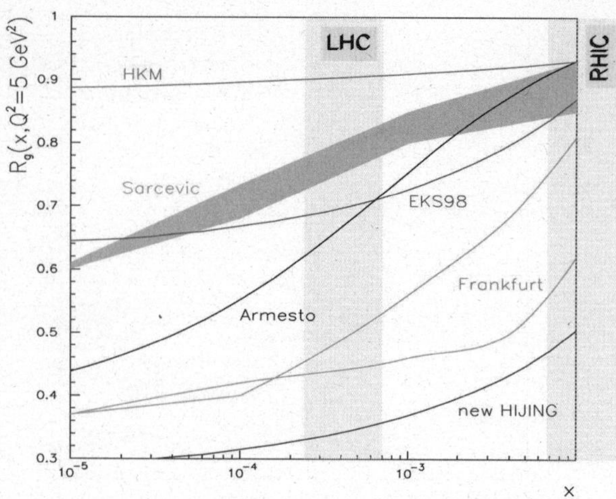

Fig. 6. Ratios of the Pb over proton gluon PDFs *versus* x from different models at $Q^2 = 5\,\mathrm{GeV}^2$. Figure taken from [23].

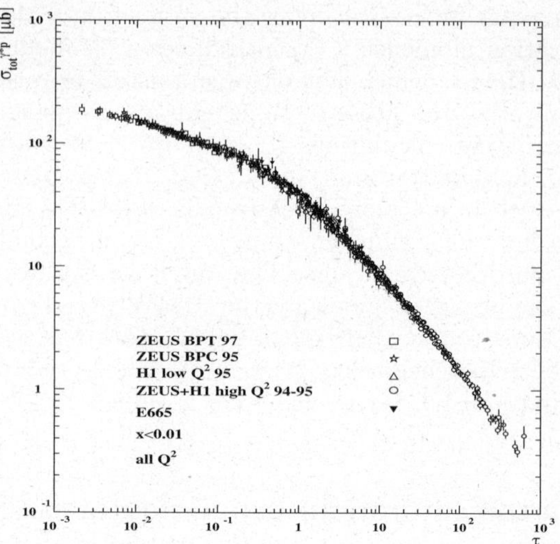

Fig. 7. Geometric scaling in the DIS $\gamma^\star p$ cross-sections plotted *versus* $\tau = Q^2/Q_s^2$ in the range $x < 0.01$, $0.045 < Q^2 < 450\,\mathrm{GeV}^2$ [25].

Figure 5 shows the kinematical map in (x, Q^2) of the existing DIS, DY, direct γ and jet data used in the PDF fits. Results from HERA and the Tevatron cover a substantial range of the proton structure ($10^{-4} \lesssim x \lesssim 0.8$, $1 \lesssim Q^2 \lesssim 10^5\,\mathrm{GeV}^2$) but the available measurements are much rarer in the case of nuclear targets (basically limited to fixed-target studies, $10^{-2} \lesssim x \lesssim 0.8$ and $1 \lesssim Q^2 \lesssim 10^2\,\mathrm{GeV}^2$). As a matter of fact, the nuclear parton distributions are basically unknown at low x ($x < 0.01$) where the only available measurements are fixed-target data in the *non-perturbative* range ($Q^2 < 1\,\mathrm{GeV}^2$) dominated by Regge dynamics rather than quark/gluon degrees of freedom. An example of the current lack of knowledge of the nuclear densities at low x is presented in fig. 6 where different available parametrizations of the ratio of Pb to proton gluon distributions, consistent with the available n DIS data at higher x, show differences as large as a factor of three at $x \sim 10^{-4}$ [23,24].

2.1 HERA results

Though the large majority of ep DIS data collected during the HERA-I phase are successfully reproduced by standard DGLAP predictions, more detailed and advanced experimental and theoretical results in the recent years have pointed to interesting hints of non-linear QCD effects in the data. Arguably, the strongest indication of such effects is given by the so-called "geometric scaling" property observed in inclusive σ_{DIS} for $x < 0.01$ [25] as well as in various diffractive cross sections [26,27]. For inclusive DIS events, this feature manifests itself in a total cross-section at small x ($x < 0.01$) which is only a function of $\tau = Q^2/Q_s^2(x)$, instead of being a function of x and Q^2/Q_s^2 separately (fig. 7). The saturation momentum follows $Q_s(x) = Q_0(x/x_0)^\lambda$ with parameters $\lambda \sim 0.3$, $Q_0 = 1\,\mathrm{GeV}$, and $x_0 \sim 3 \cdot 10^{-4}$. Interestingly, the scaling is valid up to very large values of τ, well above the saturation scale, in an "extended scaling" region (see fig. 3), where

$Q_s^2 < Q^2 < Q_s^4/\Lambda_{QCD}^2$ [28,8]. The saturation formulation is suitable to describe not only inclusive DIS, but also inclusive diffraction $\gamma^\star p \to X\,p$. The very similar energy dependence of the inclusive diffractive and total cross-sections in $\gamma^\star p$ collisions at a given Q^2 is easily explained in the Golec-Biernat Wüsthoff model [25] but not in standard collinear factorization. Furthermore, geometric scaling has been also found in different diffractive DIS cross-sections (inclusive, vector mesons, deeply-virtual Compton scattering DVCS) [26,27]. All the observed scalings are suggestive manifestations of the QCD saturation regime. Unfortunately, the values of $Q_s^2 \sim 0.5$–$1.0\,\mathrm{GeV}$ at HERA lie in the transition region between the soft and hard sectors and, therefore, non-perturbative effects obscure the obtention of clearcut experimental signatures.

2.2 RHIC results

The expectation of enhanced parton saturation effects in the *nuclear* wave functions accelerated at ultra-relativistic energies, eq. (3), has been one of the primary physics motivations for the heavy-ion programme at RHIC[3] [12]. Furthermore, the properties of the high-density matter produced in the final state of AA interactions cannot be properly interpreted without having isolated first the influence of *initial state* modifications of the nuclear PDFs. After five years of operation, two main experimental results at RHIC have been found consistent with CGC predictions: i) the modest hadron multiplicities measured in AuAu reactions, and ii) the suppressed hadron yield observed at forward rapidities in dAu collisions. (In addition, a recent analysis of the existing nuclear DIS F_2 data also confirms the existence of "geometrical scaling" for $x < 0.017$ [29].)

---

[3] At $y = 0$, the saturation scale for an Au nucleus at RHIC ($Q_s^2 \sim 2\,\mathrm{GeV}$) is larger than that of protons probed at HERA ($Q_s^2 \sim 0.5\,\mathrm{GeV}$).

The bulk hadron multiplicities measured at mid-rapidity in central AuAu at $\sqrt{s_{NN}} = 200\,\text{GeV}$ are $dN_{ch}/d\eta|_{\eta=0} \approx 700$, comparatively lower than the $dN_{ch}/d\eta|_{\eta=0} \approx 1000$ expectations of "minijet" dominated scenarios [30], soft Regge models [31] (without accounting for strong shadowing effects [32]), or extrapolations from an incoherent sum of proton-proton collisions [33]. On the other hand, CGC approaches [11,29] which effectively take into account a reduced number of scattering centers in the nuclear PDFs, $f_{a/A}(x, Q^2) < A \cdot f_{a/N}(x, Q^2)$ reproduce well not only the measured hadron multiplicities but —based on the general expression (3)— also the centrality and c.m. energy dependences of the bulk AA hadron production (fig. 8).

The second manifestation of saturation-like effects in the RHIC data is the BRAHMS observation [34] of suppressed yields of moderately high-p_T hadrons ($p_T \approx 2$–$4\,\text{GeV}/c$) in dAu relative to pp collisions at $\eta \approx 3.2$ (fig. 9). Hadron production at such small angles is sensitive to partons in the Au nucleus with $x_2 \approx \mathcal{O}(10^{-3})$. The observed nuclear modification factor, $R_{dAu} \approx 0.8$, cannot be reproduced by p QCD calculations that include standard *leading-twist* shadowing of the nuclear PDFs [18, 19] but can be described by CGC approaches [36,37] that parametrize the Au nucleus as a saturated gluon wave function. As in the HERA case, it is worth noting however that at RHIC energies the saturation scale is in the transition between the soft and hard regimes ($Q_s^2 \approx 2\,\text{GeV}^2$) and the results consistent with the CGC predictions are in a kinematic range with relatively low momentum scales ($\langle p_T \rangle \sim 0.5\,\text{GeV}$ for the hadron multiplicities and $\langle p_T \rangle \sim 2.5\,\text{GeV}$ for forward inclusive hadron production) where non-perturbative effects can blur a simple interpretation based on partonic degrees of freedom alone.

3 Low-x QCD at the LHC

The Large Hadron Collider (LHC) at CERN will provide pp, pA and AA collisions at $\sqrt{s_{NN}} = 14$, 8.8 and 5.5 TeV, respectively with luminosities $\mathcal{L} \sim 10^{34}$, 10^{29} and $5 \cdot 10^{26}\,\text{cm}^{-2}\,\text{s}^{-1}$. Such large c.m. energies and luminosities will allow detailed QCD studies at unprecedented low x values thanks to the copious production of hard probes (jets, quarkonia, heavy quarks, prompt γ, Drell-Yan pairs, etc.). The advance in the study of low-x QCD phenomena will be specially substantial for nuclear systems since the saturation momentum, eq. (3), $Q_s^2 \approx 5$–$10\,\text{GeV}^2$, will be in the perturbative range [11], and the relevant x values, eq. (6), will be 30–70 times lower than AA and pA reactions at RHIC: $x \approx 10^{-3}(10^{-5})$ at central (forward) rapidities for processes with $Q^2 \sim 10\,\text{GeV}^2$ (fig. 10).

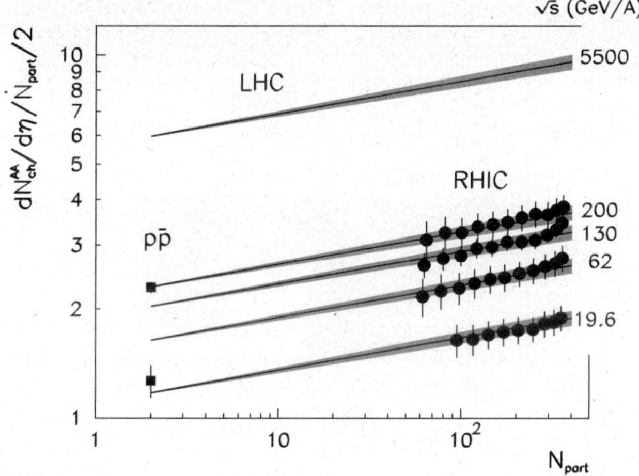

Fig. 8. Dependences on c.m. energy and centrality (given in terms of the number of nucleons participating in the collision, N_{part}) of $dN_{ch}/d\eta|_{\eta=0}$ (normalized by N_{part}): PHOBOS AuAu data [35] *vs.* the predictions of the saturation approach [29].

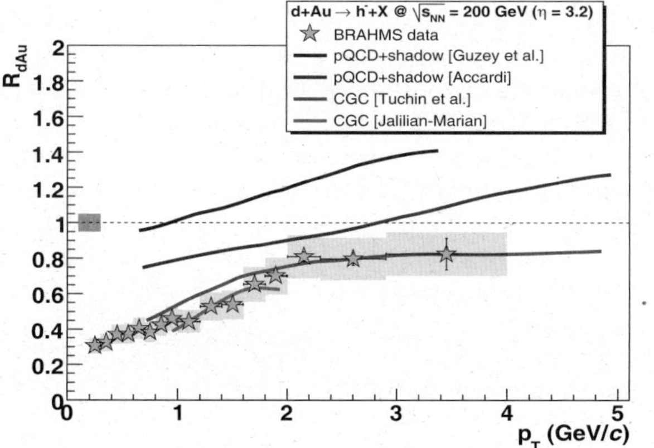

Fig. 9. Nuclear modification factor $R_{dAu}(p_T)$ for negative hadrons at $\eta = 3.2$ in dAu at $\sqrt{s_{NN}} = 200\,\text{GeV}$: BRAHMS data [34] compared to leading-twist shadowing p QCD [18,19] and CGC [36,37] predictions.

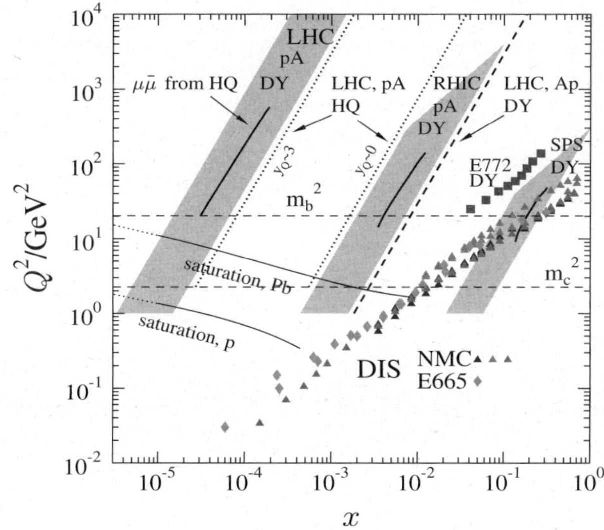

Fig. 10. Kinematical (x, Q^2) range probed at various rapidities y and c.m. energies in $\sqrt{s_{NN}} = 8.8\,\text{TeV}$ pA collisions at the LHC [23].

3.1 The LHC experiments

The four LHC experiments —*i.e.* the two large general-purpose and high-luminosity ATLAS and CMS detector systems as well as the heavy-ion–dedicated ALICE and the heavy-flavour–oriented LHCb experiments— have all detection capabilities in the forward direction very well adapted for the study of low-x QCD phenomena with hard processes in collisions with proton and ion beams:

i) Both CMS and ATLAS feature hadronic calorimeters in the range $3 < |\eta| < 5$ which allow them to measure jet cross-sections at very forward rapidities. Both experiments feature also zero-degree calorimeters (ZDC, $|\eta| \gtrsim 8.5$ for neutrals), which are a basic tool for neutron-tagging "ultra-peripheral" PbPb photoproduction interactions. CMS has an additional electromagnetic/hadronic calorimeter (CASTOR, $5.3 < |\eta| < 6.7$) and shares the interaction point with the TOTEM experiment providing two extra trackers at very forward rapidities (T1, $3.1 < |\eta| < 4.7$, and T2, $5.5 < |\eta| < 6.6$) well-suited for DY measurements.

ii) The ALICE forward muon spectrometer at $2.5 < \eta < 4$, can reconstruct J/ψ and Υ (as well as Z) in the dimuon channel, as well as statistically measure single inclusive heavy-quark production via semileptonic (muon) decays. ALICE counts also on ZDCs in both sides of the interaction point (IP) for forward neutron triggering of PbPb photoproduction processes.

iii) LHCb is a single-arm spectrometer covering rapidities $1.8 < \eta < 4.9$, with very good particle identification capabilities designed to accurately reconstruct b and c mesons. The detector is also well-suited to measure jets, $Q\bar{Q}$ and $Z \to \mu\mu$ production in the forward hemisphere.

3.2 Low-x QCD measurements at LHC

Measurement at the LHC forward rapidities of any of the processes shown in fig. 4 provides an excellent means to look for signatures of high-gluon-density phenomena at low x. Four representative measurements are discussed in the last section of the paper.

– Case study I: forward (di)jets (pp, pA, AA)

The jet measurements in $\bar{p}p$ collisions at Tevatron energies have provide valuable information on the proton PDFs (see fig. 5, top). According to eq. (6), the measurement of jets with $p_T \approx 20$–$200\,\mathrm{GeV}/c$ in pp collisions at $14\,\mathrm{TeV}$ in the ATLAS or CMS forward calorimeters ($3 < |\eta| < 5$) allows one to probe the PDFs at x values as low as $x_2 \approx 10^{-4}$–10^{-5}. Figure 11 shows the actual $\log(x_{1,2})$ distribution of two partons scattering at LHC and producing at least one forward jet as obtained with PYTHIA 6.403 [38]. As expected in forward scattering, the collision is very asymmetric with x_2 (x_1) peaked at $\sim 10^{-4}$ ($\sim 10^{-1}$) and thus provides direct information

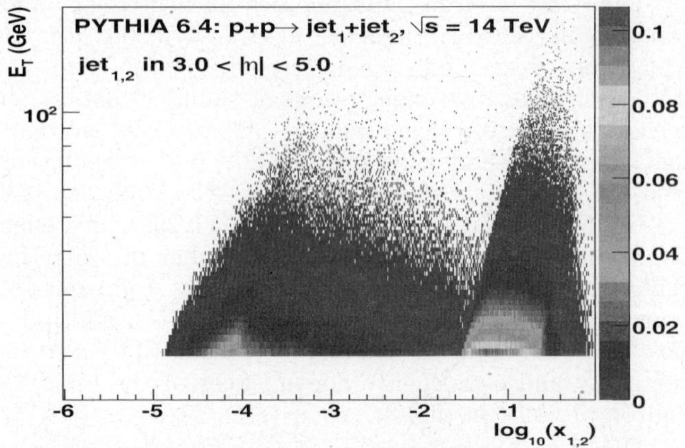

Fig. 11. $\log(x_{1,2})$ distribution of two partons colliding in pp collisions at $\sqrt{s} = 14\,\mathrm{TeV}$ and producing at least one jet within ATLAS/CMS forward calorimeters acceptances as determined with PYTHIA.

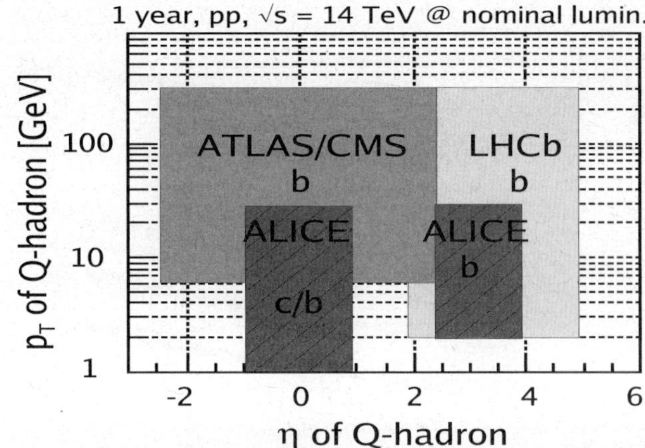

Fig. 12. Acceptances in (η, p_T) for open charm and bottom hadrons in the four LHC experiments for 1-year nominal luminosity [15].

on the low-x parton densities. Not only the single inclusive cross-section but the forward-backward dijet production, "Müller-Navelet jets", is a particularly sensitive measure of BFKL [39] as well as non-linear [6] parton evolutions. In the presence of low-x saturation effects, the Müller-Navelet cross-section for two jets separated by $\Delta\eta \sim 9$ (and, thus, measurable in each one of the forward calorimeters) is expected to be suppressed by a factor of ~ 2 compared to BFKL predictions [6]. A study is underway to determine the feasibility of both forward (di)jet measurements in CMS [40].

– Case study II: forward heavy quarks (pp, pA, AA)

Studies of small-x effects on heavy-flavour production at the LHC in two different approaches, based on collinear and k_T factorization, including non-linear terms in the parton evolution, lead to two different predictions (enhancement *vs.* suppression) for the measured c and b

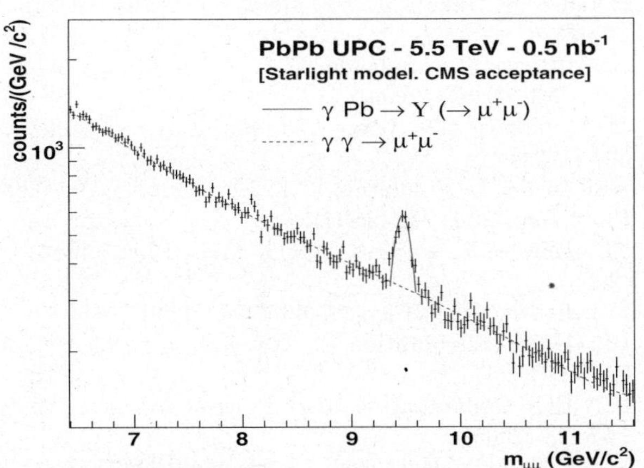

Fig. 13. Expected $\mu^+\mu^-$ invariant mass from $\gamma\,\mathrm{Pb} \to \Upsilon\,\mathrm{Pb}^\star \to \mu^+\mu^-\,\mathrm{Pb}^\star$ and $\gamma\gamma \to \mu^+\mu^-$ processes predicted by *Starlight* [44] for UPC PbPb collisions at $\sqrt{s_{NN}} = 5.5\,\mathrm{TeV}$ in the CMS acceptance.

cross-sections [15]. The possibility of ALICE and LHCb to reconstruct D and B mesons in a large rapidity range (fig. 12) will put stringent constraints on the gluon structure and evolution at low-x. In the case of ALICE, the heavy-Q pp studies will have a natural extension in AA and pA collisions [41], providing a precise probe of non-linear effects in the nuclear wave function.

– Case study III: $Q\bar{Q}$ photoproduction (electromagnetic AA)

High-energy diffractive photoproduction of heavy vector mesons $(J/\psi, \Upsilon)$ proceeds through colourless two-gluon exchange (which subsequently couples to $\gamma \to Q\bar{Q}$) and is thus a sensitive probe of the gluon densities at small x, see eq. (4). Ultra-peripheral interactions (UPCs) of high-energy heavy ions generate strong electromagnetic fields which help constrain the low-x behaviour of xG via quarkonia [42], or other hard probes [43], produced in γ-nucleus collisions. Lead beams at 2.75 TeV have Lorentz factors $\gamma = 2930$ leading to maximum (equivalent) photon energies $\omega_{max} \approx \gamma/R \sim 100\,\mathrm{GeV}$, and c.m. energies $W_{\gamma\gamma}^{max} \approx 160\,\mathrm{GeV}$ and $W_{\gamma A}^{max} \approx 1\,\mathrm{TeV}$. From eq. (5), the x values probed in $\gamma A \to J/\psi\,A$ processes at $y = 2$ can be as low as $x \sim 10^{-5}$. The ALICE, ATLAS and CMS experiment can measure $J/\psi, \Upsilon \to e^+e^-, \mu^+\mu^-$ produced in electromagnetic PbPb collisions tagged with neutrons detected in the ZDCs (as done at RHIC [42]). Figure 13 shows the expected dimuon invariant mass distributions around the Υ mass predicted by *Starlight* [44] within the CMS central acceptance ($\eta < 2.5$) for an integrated PbPb luminosity of $0.5\,\mathrm{nb}^{-1}$ [45].

– Case study IV: forward Drell-Yan pairs (pp, pA, AA)

High-mass Drell-Yan pair production at the very forward rapidities covered by LHCb and by the CMS CASTOR and TOTEM T2 detectors can probe the parton densities

down to $x \sim 10^{-6}$ at much higher virtualities M^2 than those accessible in other measurements discussed here. A study is currently underway in CMS [40] to combine the CASTOR electromagnetic energy measurement together with the good position resolution of T2 for charged tracks, to trigger on and reconstruct the e^+e^- invariant mass in pp collisions at 14 TeV, and scrutinize xg in the M^2 and x plane.

4 Conclusion

We have reviewed the physics of non-linear QCD and high gluon densities at small fractional momenta x, with emphasis on the existing data at HERA (proton) and RHIC (nucleus) which support the existence of a parton saturation regime as described, *e.g.*, in the framework of the Colour Glass Condensate effective field theory. The future perspectives at the LHC have been presented, including the promising capabilities of the forward detectors of the ALICE, ATLAS, CMS and LHCb experiments to study the parton densities down to $x \sim 10^{-6}$ with various hard probes (jets, quarkonia, heavy quarks, Drell-Yan). The programme of investigating the dynamics of low-x QCD is not only appealing in its own right but it is an essential prerequisite for predicting a large variety of hadron, photon and neutrino scattering cross-sections at very high energies.

The author thanks A. de Roeck and H. Jung as well as A. Dobado and F.J. Llanes Estrada for their invitation to present this overview talk at the *2nd HERA-LHC Workshop* and at the *QNP'06 International Conference*. This work is supported by the 6th EU Framework Programme contract MEIF-CT-2005-025073.

References

1. H1 Collaboration (C. Adloff *et al.*), Phys. Lett. B **520**, 183 (2001).
2. V.N. Gribov, L.N. Lipatov, Sov. J. Nucl. Phys. **15**, 438 (1972); G. Altarelli, G. Parisi, Nucl. Phys. B **126**, 298 (1977); Yu.L. Dokshitzer, Sov. Phys. JETP **46**, 641 (1977).
3. M. Dittmar *et al.*, in *Proceedings of HERA-LHC Workshop*, hep-ph/0511119.
4. L.N. Lipatov, Sov. J. Nucl. Phys. **23**, 338 (1976); E.A. Kuraev, L.N. Lipatov, V.S. Fadin, Zh. Eksp. Teor. Fiz **72**, 3 (1977); Ya.Ya. Balitsky, L.N. Lipatov, Sov. J. Nucl. Phys. **28**, 822 (1978).
5. ZEUS Collaboration (S. Chekanov *et al.*), Phys. Lett. B **632**, 13 (2006); H1 Collaboration (A. Aktas *et al.*), Eur. Phys. J. C **46**, 27 (2006).
6. C. Marquet, C. Royon, Nucl. Phys. B **739**, 131 (2006).
7. S. Catani, F. Fiorani, G. Marchesini, Nucl. Phys. B **336**, 18 (1990); S. Catani, M. Ciafaloni, F. Hautmann, Nucl. Phys. B **366**, 135 (1991).
8. E. Iancu, C. Marquet, G. Soyez, hep-ph/0605174.
9. L. Gribov, E.M. Levin, M.G. Ryskin, Phys. Rep. **100**, 1 (1983).

10. A.H. Mueller, J.-w. Qiu, Nucl. Phys. B **268**, 427 (1986).
11. D. Kharzeev, M. Nardi, Phys. Lett. B **507**, 121 (2001); D. Kharzeev, E. Levin, M. Nardi, Nucl. Phys. A **747**, 609 (2005).
12. See, *e.g.*, E. Iancu, R. Venugopalan, in *QGP*, edited by R.C. Hwa *et al.*, Vol. **3** (World Scientific, Singapore, 2003) hep-ph/0303204; J. Jalilian-Marian, Y.V. Kovchegov, Prog. Part. Nucl. Phys. **56**, 104 (2006) and references therein.
13. J. Jalilian-Marian, A. Kovner, A. Leonidov, H. Weigert, Nucl. Phys. B **504**, 415 (1997); Phys. Rev. D **59**, 014014 (1999); E. Iancu, A. Leonidov, L. McLerran, Nucl. Phys. A **692**, 583 (2001).
14. I. Balitsky, Nucl. Phys. B **463**, 99 (1996); Yu.V. Kovchegov, Phys. Rev. D **61**, 074018 (2000).
15. J. Baines *et al.*, in *Proceedings of HERA-LHC Workshop*, hep-ph/0601164.
16. M.G. Ryskin *et al.*, Z. Phys. C **76**, 231 (1997).
17. T. Teubner, *Proceedings of DIS'05*, AIP Conf. Proc. **792**, 416 (2006).
18. V. Guzey, M. Strikman, W. Vogelsang, Phys. Lett. B **603**, 173 (2004).
19. A. Accardi, Acta Phys. Hung. A **22**, 289 (2005).
20. A. Dumitru, A. Hayashigaki, J. Jalilian-Marian, Nucl. Phys. A **765**, 464 (2006).
21. P. Newman, Int. J. Mod. Phys. A **19**, 1061 (2004).
22. D. d'Enterria, J. Phys. G **30**, S767 (2004).
23. A. Accardi *et al.*, in *Hard probes in Heavy Ion collisions at the LHC*, CERN Yellow report, hep-ph/0308248.
24. N. Armesto, J. Phys. G **32**, R367 (2006).
25. K. Golec-Biernat, M. Wüsthoff, Phys. Rev. D **59**, 014017 (1999); **60**, 114023 (1999).
26. J.R. Forshaw, G. Shaw, JHEP **0412**, 052 (2004).
27. C. Marquet, L. Schoeffel, hep-ph/0606079.
28. E. Iancu, K. Itakura, L. McLerran, Nucl. Phys. A **708**, 327 (2002).
29. N. Armesto, C.A. Salgado, U.A. Wiedemann, Phys. Rev. Lett. **94**, 022002 (2005).
30. M. Gyulassy, X.N. Wang, Comput. Phys. Commun. **83**, 307 (1994).
31. A. Capella, U. Sukhatme, C.I. Tan, Van J. Tran Thanh, Phys. Rep. **236**, 225 (1994).
32. N. Armesto, C. Pajares, Int. J. Mod. Phys. A **15**, 2019 (2000).
33. K.J. Eskola, Nucl. Phys. A **698**, 78 (2002).
34. BRAHMS Collaboration (I. Arsene *et al.*), Nucl. Phys. A **757**, 1 (2005).
35. PHOBOS Collaboration (B.B. Back *et al.*), Nucl. Phys. A **757**, 28 (2005).
36. D. Kharzeev, Y. Kovchegov, K. Tuchin, Phys. Lett. B **599**, 23 (2004).
37. J. Jalilian-Marian, Nucl. Phys. A **748**, 664 (2005).
38. T. Sjostrand, S. Mrenna, P. Skands, JHEP **0605**, 026 (2006).
39. A.H. Mueller, H. Navelet, Nucl. Phys. B **282**, 727 (1987).
40. CMS and TOTEM Collaborations, *Prospects for Diffractive and Forward Physics at LHC*, CERN-LHCC-2006-039/G-124.
41. A. Dainese, *Proceedings of Hard Probes'06*, nucl-ex/0609042.
42. D. d'Enterria, *Proceedings of Quark Matter'05*, nucl-ex/0601001.
43. M. Strikman, R. Vogt, S. White, Phys. Rev. Lett. **96**, 082001 (2006).
44. S.R. Klein, J. Nystrand, Phys. Rev. C **60**, 014903 (1999); A. Baltz, S. Klein, J. Nystrand, Phys. Rev. Lett. **89**, 012301 (2002).
45. CMS Collaboration (D. d'Enterria), Eur. Phys. J. C **49**, 155 (2007).

Eur. Phys. J. A **31**, 824–827 (2007)

DOI 10.1140/epja/i2006-10260-0

Special Article – QNP 2006

Phase diagram of neutron star quark matter in nonlocal chiral models

D. Gómez Dumm[1,2,a], D.B. Blaschke[3,4,5], A.G. Grunfeld[1,6], T. Klähn[4,7], and N.N. Scoccola[1,6,8]

[1] CONICET, Rivadavia 1917, 1033 Buenos Aires, Argentina
[2] Dpto. de Física, Universidad Nacional de La Plata, C.C. 67, 1900 La Plata, Argentina
[3] Institut für Physik, Universität Rostock, Universitätsplatz 3, D-18051 Rostock, Germany
[4] Instytut Fizyki Teoretycznej, Uniwersytet Wrocławski, pl. M. Borna 9, 50-204 Wrocław, Poland
[5] Bogoliubov Laboratory of Theoretical Physics, JINR Dubna, Joliot-Curie Street 6, 141980 Dubna, Russia
[6] Physics Department, Comisión Nacional de Energía Atómica, Av. Libertador 8250, 1429 Buenos Aires, Argentina
[7] Gesellschaft für Schwerionenforschung mbH, Planckstr. 1, D-64291 Darmstadt, Germany
[8] Universidad Favaloro, Solís 453, 1078 Buenos Aires, Argentina

Received: 23 November 2006
Published online: 15 March 2007 – © Società Italiana di Fisica / Springer-Verlag 2007

Abstract. We analyze the phase diagram of two-flavor quark matter under neutron star constraints for a nonlocal covariant quark model within the mean-field approximation. Applications to cold compact stars are discussed.

PACS. 12.38.Mh Quark-gluon plasma – 24.85.+p Quarks, gluons, and QCD in nuclei and nuclear processes – 26.60.+c Nuclear matter aspects of neutron stars – 97.60.-s Late stages of stellar evolution (including black holes)

1 Introduction

The characteristics of the QCD phase diagram is presently an important open subject of research in particle physics. In particular, the behavior of strongly interacting matter in the region of low temperatures and large baryon densities could be tested against observational constraints from neutron stars [1]. Since in this thermodynamical region one finds strong difficulties when trying to perform lattice QCD calculations, the pictures emerging from different effective models of strong interactions deserve great interest from the theoretical point of view. Here we focus on a two-flavor chiral quark model which includes covariant nonlocal four-fermion interactions, motivated by an effective one-gluon exchange (OGE) picture. Nonlocality arises naturally in the context of several successful approaches to low-energy quark dynamics, such as the instanton liquid model [2] and the Schwinger-Dyson resummation techniques [3], and it is also a well-known feature of lattice QCD [4].

2 Formalism

The Euclidean action for the nonlocal model considered here, in the case of two light flavors and anti-triplet di-

quark interactions, is given by

$$S_E = \int \mathrm{d}^4x \left\{ \bar{\psi}(x) \left(-i\slashed{\partial} + m\right) \psi(x) \right.$$
$$\left. -\frac{G}{2} j_M^f(x) j_M^f(x) - \frac{H}{2} \left[j_D^a(x)\right]^\dagger j_D^a(x) \right\}. \quad (1)$$

Here m is the current quark mass, which is assumed to be equal for u and d quarks, and $j_{M,D}$ are mesonic and diquark nonlocal currents. The nonlocality is introduced here in a covariant way, through a separable interaction arising from an effective OGE picture. In this way, we have

$$j_M^f(x) = \int \mathrm{d}^4z \, g(z) \, \bar{\psi}\left(x + \frac{z}{2}\right) \, \Gamma_f \, \psi\left(x - \frac{z}{2}\right),$$
$$j_D^a(x) = \int \mathrm{d}^4z \, g(z) \, \bar{\psi}_c\left(x + \frac{z}{2}\right) \, i\gamma_5\tau_2\lambda_a \, \psi\left(x - \frac{z}{2}\right), \quad (2)$$

where $\psi_c(x) = \gamma_2\gamma_4 \bar{\psi}^T(x)$, $\Gamma_f = (\mathbf{1}, i\gamma_5\boldsymbol{\tau})$, and $\boldsymbol{\tau}$ and λ_a, with $a = 2, 5, 7$, stand for Pauli and Gell-Mann matrices acting on flavor and color spaces, respectively. The function $g(z)$ is a form factor that characterizes the nonlocal interaction.

The effective action in eq. (1) might arise via Fierz rearrangement from some underlying more fundamental interactions, and is understood to be used —at the mean-field level— in the Hartree approximation. In general, the

a e-mail: dumm@fisica.unlp.edu.ar

ratio of coupling constants H/G would be determined by these microscopic couplings; for example, OGE interactions lead to $H/G = 0.75$. Since the precise derivation of the effective couplings from QCD is not known, here we will leave H/G as a free parameter.

Standard bosonization of the theory leads, in the mean-field approximation, to a thermodynamical potential per unit volume given by

$$\Omega^{MFA} = \frac{\bar{\sigma}^2}{2G} + \frac{|\bar{\Delta}|^2}{2H} - \frac{T}{2} \sum_{n=-\infty}^{\infty} \int \frac{\mathrm{d}^3\boldsymbol{p}}{(2\pi)^3} \ln \det \left[\frac{S^{-1}}{T} \right], \quad (3)$$

where the inverse propagator $S^{-1}(\bar{\sigma}, \bar{\Delta})$ is a 48×48 matrix in Dirac, flavor, color and Nambu-Gorkov spaces (its explicit form is given in ref. [5]). Here $\bar{\sigma}$ and $\bar{\Delta}$ stand for the mean-field values of scalar meson and diquark fields. Owing to the nonlocality, they come together with momentum-dependent form factors. The values of $\bar{\sigma}$ and $\bar{\Delta}$ can be obtained from the coupled gap equations

$$\frac{\mathrm{d}\Omega^{MFA}}{\mathrm{d}\bar{\Delta}} = 0, \qquad \frac{\mathrm{d}\Omega^{MFA}}{\mathrm{d}\bar{\sigma}} = 0. \quad (4)$$

In general one has to consider a different chemical potential μ_{fc} for each quark flavor f and color c. However, when the system is in chemical equilibrium, not all μ_{fc} are independent. In our case, it can be seen [5] that they can be written in terms of only three quantities: the baryon chemical potential μ_B, the quark electric chemical potential μ_Q and the color chemical potential μ_8. Defining $\mu = \mu_B/3$, the corresponding relations read

$$\mu_{qr} = \mu_{qg} = \mu + Q_q \mu_Q + \mu_8/3,$$
$$\mu_{qb} = \mu + Q_q \mu_Q - 2\mu_8/3, \quad (5)$$

where, $q = u, d$, and Q_q are quark electric charges.

In the core of neutron stars, in addition to quark matter we have electrons. The latter can be thermodynamically treated as a free Fermi gas, and their contribution has to be added to the grand canonical thermodynamical potential. Moreover, quark matter and electrons have to be in β- equilibrium. Thus, assuming that antineutrinos escape from the stellar core, we must have

$$\mu_{dc} - \mu_{uc} = -\mu_Q = \mu_e. \quad (6)$$

If we now require the system to be electric and color charge neutral, the number of independent chemical potentials reduces further: μ_e and μ_8 are fixed by the conditions of vanishing electric and color densities. In this way, for each value of T and μ we should find the values of $\bar{\Delta}$, $\bar{\sigma}$, μ_e and μ_8 that solve eqs. (4), supplemented by the β-equilibrium and electric and charge neutrality conditions.

3 Numerical results

According to previous analyses carried out within nonlocal scenarios [6], the results are not expected to show a strong qualitative dependence on the shape of the form

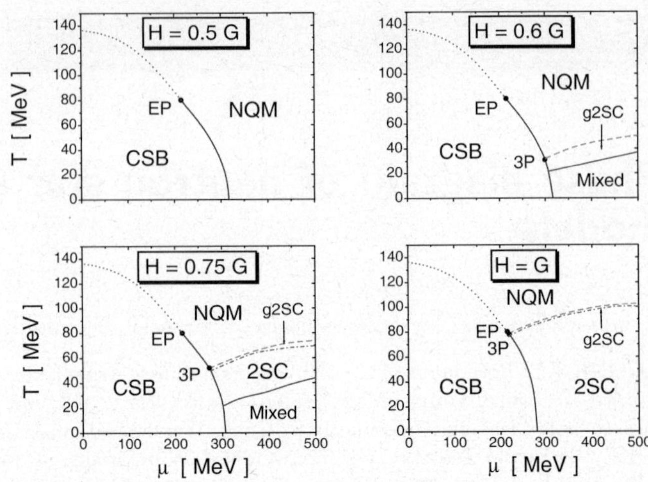

Fig. 1. Phase diagrams for the nonlocal OGE-based model, for different values of the ratio H/G, under neutron star constraints.

factor. Thus we will consider here a simple, well-behaved Gaussian function

$$g(p^2) = \exp(-p^2/\Lambda^2), \quad (7)$$

where Λ is a free parameter, playing the rôle of an ultraviolet cut-off.

For definiteness, for the input parameters we choose $m = 5.12\,\mathrm{MeV}$, $\Lambda = 827\,\mathrm{MeV}$ and $G\Lambda^2 = 18.78$. These values are fixed so as to reproduce the empirical values for the pion mass m_π and decay constant f_π, and lead to a phenomenologically reasonable value for the chiral condensate [7] at vanishing T and μ_B, namely $\langle 0|\bar{q}q|0\rangle^{1/3} = -250\,\mathrm{MeV}$. The only remaining free parameter is the coupling strength H in the scalar diquark channel. We choose here values for H/G in the range from 0.5 to 1, *i.e.* around the Fierz value $H/G = 0.75$ discussed above.

Our results for the phase diagrams are shown in fig. 1, where we plot the phase transition curves on T-μ diagrams for different ratios H/G. In the graphs we show the regions corresponding to different phases, as well as the position of triple points (3P) and end points (EP). Besides the region of low T and μ, in which the chiral symmetry is broken (CSB), one finds normal quark matter (NQM) and two-flavor superconducting (2SC) phases. Between CSB and NQM phases one has first order and crossover transitions —represented by solid and dotted lines, respectively—, whereas between NQM and 2SC regions, in all cases, we find a second order phase transition —dashed lines in the diagrams of fig. 1. Close to this NQM/2SC phase border, the dashed-dotted lines in the graphs delimit a band that corresponds to the so-called gapless 2SC (g2SC) phase. In this region, in addition to the two gapless modes corresponding to the unpaired blue quarks, the presence of flavor asymmetric chemical potentials gives rise to another two gapless fermionic quasiparticles. For the range of parameters considered here, however, the g2SC region is too narrow to lead to sizeable effects. Finally, for intermediate values of H/G we find a region in which a certain volume fraction of the quark matter undergoes a transition to the 2SC phase coexisting with the remaining NQM phase. This is a 2SC-NQM mixed phase in which the system realizes the constraint of electric neutrality globally: the coexisting phases have opposite electric charges which neutralize each other, at a common equilibrium pressure.

It can be seen that the 2SC phase region becomes larger when the ratio H/G is increased. This is not surprising, since H is the effective coupling governing the quark-quark interaction that gives rise to the pairing. As a general conclusion, it can be stated that even under neutron star constraints, provided the ratio H/G is not too low, the nonlocal scheme favors the existence of color superconducting phases at low temperatures and moderate chemical potentials (we do not find 2SC only for $H/G = 0.5$). This is in contrast to the situation in the NJL model [8], where the existence of a 2SC phase turns out to be rather dependent on the input parameters. Our results are also qualitatively different from those obtained in the case of noncovariant nonlocal models [9], where above the chiral phase transition the NQM phase is preferable for values of the coupling ratio $H/G \lesssim 0.75$, and a color superconducting phase can be found only for $H/G \approx 1$. It is now desirable to extend the present studies to other pairing channels relevant for neutron star cooling, such as spin-1 pairing, where at present only the NJL [10] and noncovariant nonlocal models [11] have been considered.

4 Application to cold compact stars

One of the most important present applications of the microscopic approaches to quark matter is to study whether such a state of matter can exist in the interior of cold compact stars. Let us briefly discuss our ongoing investigations on this issue. Unfortunately, a consistent relativistic approach to the quark-hadron phase transition, where hadrons appear as bound states of quarks, has not been developed up to now. Thus, we apply a so-called two-phase description, in which the nuclear matter phase is described within the relativistic Dirac-Brueckner-Hartree-Fock (DBHF) approach (see, *e.g.*, ref. [12]) and the transition to a quark matter phase is obtained by a Maxwell construction. From the curves presented in the previous section, it can be seen that in our OGE-inspired nonlocal model one finds a relatively low value of the critical density at $T = 0$, hence some extra repulsion is needed in order to obtain a more realistic value. We have found that this can be achieved by including some small additional interaction in the omega vector meson channel, which does not affect in general the qualitative features of the phase diagrams discussed in the previous section. Given the corresponding equation of state, the mass and structure of spherical, nonrotating stars is obtained by solving the Tolman-Oppenheimer-Volkov equations. Our preliminary results confirm the findings of ref. [13], *i.e.* that compact stars with quark matter cores are consistent with modern observations. Moreover, for the nonlocal quark models described here, this is found to be possible for values of H/G closer to the standard, OGE motivated ratio $H/G = 0.75$.

5 Conclusions

We have studied the phase diagram of two-flavor quark matter under neutron star constraints for a nonlocal, covariant quark model within the mean field approximation. The form of the nonlocal coupling has been motivated by a separable approximation of the OGE interaction. We have considered a nonlocal form factor of a Gaussian shape, and the model parameters (current quark mass m, coupling strength G, UV cutoff Λ) have been fixed so as to obtain adequate values for the pion mass, the pion decay constant and the chiral condensate at vanishing T and μ_B.

After the numerical evaluation of the gap equations at finite temperature and chemical potential, considering different values for the coupling strength in the scalar diquark channel, we have found that different low-temperature quark matter phases can occur at intermediate densities: normal quark matter (NQM), pure superconducting (2SC) quark matter and mixed 2SC-NQM phases. A band of gapless 2SC phase appears at the border of the superconducting region, but this occurs in general at nonzero temperatures and should not represent a robust feature for compact star applications. Finally, in the context of the nonlocal theory discussed here, we have obtained preliminary results showing that compact stars with quark matter cores turn out to be consistent with modern observations.

This work has been supported in part by CONICET and ANPCyT (Argentina), under grants PIP 6009, PIP 6084, PICT02-03-10718 and PICT04-03-25374, and by a scientist exchange program between Germany and Argentina funded jointly by DAAD and ANTORCHAS under grants No. DE/04/27956 and 4248-6, respectively. TK acknowledges support by the Virtual Institute VH-VI-041 of the Helmholtz Association and by the GSI Darmstadt.

References

1. D. Blaschke, D. Sedrakian (Editors), *Superdense QCD matter and compact stars*, NATO Sci. Ser. **II/197** (Springer, 2006).
2. T. Schäfer, E.V. Shuryak, Rev. Mod. Phys. **70**, 323 (1998).
3. C.D. Roberts, A.G. Williams, Prog. Part. Nucl. Phys. **33**, 477 (1994); C.D. Roberts, S.M. Schmidt, Prog. Part. Nucl. Phys. **45**, S1 (2000).
4. J. Skullerud, D.B. Leinweber, A.G. Williams, Phys. Rev. D **64**, 074508 (2001).
5. D. Gomez Dumm, D.B. Blaschke, A.G. Grunfeld, N.N. Scoccola, Phys. Rev. D **73**, 114019 (2006).
6. R.S. Duhau, A.G. Grunfeld, N.N. Scoccola, Phys. Rev. D **70**, 074026 (2004).
7. D. Gómez Dumm, A.G. Grunfeld, N.N. Scoccola, Phys. Rev. D **74**, 054026 (2006); D. Gómez Dumm, N.N. Scoccola, Phys. Rev. C **72**, 014909 (2005).

8. M. Buballa, Phys. Rep. **407**, 205 (2005).
9. D.N. Aguilera, D. Blaschke, H. Grigorian, Nucl. Phys. A **757**, 527 (2005).
10. D.N. Aguilera, D. Blaschke, M. Buballa, V.L. Yudichev, Phys. Rev. D **72**, 034008 (2005).
11. D.N. Aguilera, D. Blaschke, H. Grigorian, N.N. Scoccola, Phys. Rev. D **74**, 114005 (2006).
12. T. Klähn *et al.*, Phys. Rev. C **74**, 035802 (2006).
13. T. Klähn *et al.*, arXiv:nucl-th/0609067.

Eur. Phys. J. A **31**, 828–830 (2007)
DOI 10.1140/epja/i2006-10288-0

THE EUROPEAN
PHYSICAL JOURNAL A

Special Article – QNP 2006

Neutral color-spin-locking phase in neutron stars

D.N. Aguilera[a]

Department of Applied Physics, University of Alicante, Apartado Correos 99, 03080 Alicante, Spain

Received: 18 December 2006
Published online: 26 March 2007 – © Società Italiana di Fisica / Springer-Verlag 2007

Abstract. We present results for the spin-1 color-spin-locking (CSL) phase using a NJL-type model in two-flavor quark matter for compact stars applications. The CSL condensate is flavor symmetric and therefore charge and color neutrality can easily be satisfied. We find small energy gaps $\simeq 1\,\mathrm{MeV}$, which make the CSL matter composition and the EoS not very different from the normal quark matter phase. We keep finite quark masses in our calculations and obtain no gapless modes that could have strong consequences in the late cooling of neutron stars. Finally, we show that the region of the phase diagram relevant for neutron star cores, when asymmetric flavor pairing is suppressed, could be covered by the CSL phase.

PACS. 24.85.+p Quarks, gluons, and QCD in nuclei and nuclear processes – 26.60.+c Nuclear matter aspects of neutron stars

1 Introduction

The investigation of cold dense quark matter has received special attention due to the possible consequences for compact stars [1]. In particular, color superconducting quark matter phases enforcing color and charge neutrality has been widely studied [2]. Model calculations have shown that the intermediate density region of the neutral QCD phase diagram, where the quark chemical potential is not sufficiently large to have the strange quark deconfined, might be dominated by u, d quarks [3]. If this is the case, two-flavor quark matter phases may occupy a large volume in the core of compact stars [4].

On the other hand, local charge neutrality disfavors the occurrence of phases with large gaps where quarks with different flavor pair in a spin-0 condensate, such as the 2SC phase [5], in which quarks pair in e.g. $(u_r d_g)$ and $(d_r u_g)$ diquarks leaving the u_b, d_b unpaired. Therefore, while the occurrence of neutral 2SC pure phase is rather model dependent and might be unlikely for moderate coupling constants [6], other phases such e.g. with spin-1 pairings [7], could be relevant for neutron star phenomenology. Specially, because these condensates with small energy gaps $(\Delta \simeq 1\,\mathrm{MeV})$ do not influence the equation of state (EoS) (it is not distinguishable from the normal-quark (NQ) matter EoS) but they strongly affect the transport and thermal properties of quark matter [8] and consequently the neutron star cooling. The unpaired quarks in the core lead to rapid cooling via the direct Urca process, incom-

patible with the observations. Phases that present no gapless modes prevent the direct Urca to work uncontrolled suppressing the neutrino emissivities and could explain the observed data [9].

We consider in this work the color-antitriplet single-flavor spin-1 pairing in the color-spin-locking (CSL) phase [7] and compare it with the 2SC phase. Our results for the CSL phase are obtained using the Nambu–Jona-Lasinio (NJL) model [10] keeping finite quark masses and thus obtaining no gapless modes. Since these condensates are color neutral and single-flavor, neutral beta-equilibrated CSL quark matter is obtained easily. We present also the thermal behavior of the neutral CSL phase showing the phase diagram. Finally, we stress important features of the CSL phase that could give a consistent picture of a compact star with a superconducting quark matter core.

2 Flavor symmetric (CSL) vs. flavor asymmetric (2SC) pairing

We consider two-flavor $(f = u, d)$ quark matter, assuming that the strange-quark mass is large enough to appear only at higher densities. In the NJL-type[1] models the quarks

---

[1] We consider also nonlocal extensions of the NJL model: the quark interactions act over a certain range introducing momentum-dependent form factors in the the current-current interaction terms. The inclusion of high-momenta states beyond the NJL cutoff reduces the diquark condensates and lowers the density for the chiral phase transition (for a discussion see [11]).

---

[a] e-mail: deborah.aguilera@ua.es; supported by VESF-Fellowships EGO-DIR-112/2005.

Table 1. Two-flavor quark matter phases: flavor asymmetric spin-0 2SC and flavor symmetric spin-1 CSL. For $[S \ldots A]$ the following pairs of indices apply: $S, A = 3, 2; 1, 7; 2, 5$.

Phase	Condensate Δ	Diquarks	Free
2SC (spin-0)	$2G_1 \langle\, \psi^T\, C\, \gamma_5\, \tau_2\, \lambda_2\, \psi\, \rangle$	$u_r d_g,\ d_r u_g$	$u_b,\ d_b$
CSL (spin-1)	$4H_v \langle\, \psi^T\, C\, \gamma_{[S}\, \lambda_{A]}\, \psi\, \rangle$	$u_r u_g,\ u_g u_b,$ $u_b u_r,\ (u \to d)$	—

interact locally by a 4-point vertex effective force. The NJL Lagrangian $\mathcal{L}_{\text{eff}} = \mathcal{L}_0 + \mathcal{L}_{q\bar{q}} + \mathcal{L}_{qq}$ contains a free part $\mathcal{L}_0$, a quark-antiquark channel $\mathcal{L}_{q\bar{q}}$ that causes spontaneous chiral symmetry breaking with condensates $\sigma = \langle \bar{q}q \rangle$ and a diquark channel $\mathcal{L}_{qq}$ that describes color superconductivity with condensate Δ. The constituent-quark mass is defined as $M = m - 4G\sigma$ and the energy gaps Δ are listed in table 1 for the phases: spin-0 2SC and spin-1 CSL.

Linearizing $\mathcal{L}_{\text{eff}}$ in the presence of the condensates the thermodynamical potential $\Omega(T, \{\mu_f\})$ can be derived. For the quark sector, in the 2SC case we obtain

$$\Omega_q = 4G\sigma^2 + \frac{|\Delta|^2}{4G_1} - 2 \sum_{\pm, j=1}^{3} \int \left[E_j^{\pm} + 2T \ln\left(1 + e^{-E_j^{\pm}/T}\right) \right]$$

and for CSL, since the flavors decouple, $\Omega_q = \sum_f \Omega_f(T, \mu_f)$

$$\Omega_f = 4G\sigma^2 + \frac{3|\Delta_f|^2}{8H_v} - \sum_{k=1}^{6} \int \left[E_k^f + 2T \ln\left(1 + e^{-E_k^f/T}\right) \right],$$

where $E_j^{\pm}$ and E_k^f are the corresponding dispersion relations and the integration is in the momentum space.

The lepton sector is modeled as an ideal electron gas with chemical potential μ_e. At the mean-field level, the stationary points $\delta\Omega/\delta\Delta = \delta\Omega/\delta M = 0$ define a set of gap equations for Δ and M. The stable solution is the one which corresponds to the absolute minimum of Ω.

The parameters (quark mass $m = 5\,\text{MeV}$, the coupling $G = 4.66\,\text{GeV}^{-2}$ and the cutoff $\Lambda = 664\,\text{MeV}$) are chosen to fit the pion mass and the decay constant to a vacuum constituent-quark mass equals $300\,\text{MeV}$.

3 Results and discussion

We solve self-consistently the gap equations for the dynamical mass M and the energy gap Δ for beta-equilibrated neutral matter: our solutions satisfy that the total quark electric charge $\mu_Q = -\mu_e$ and that the total charge density $n_Q - n_e = \frac{2}{3}n_u - \frac{1}{3}n_d - n_e$ vanishes. In fig. 1 we show M and Δ as a function of the baryon chemical potential μ_B for the flavor asymmetric spin-0 2SC phase on the left and for the flavor symmetric spin-1 CSL phase on the right. While, for a fixed μ_B, M presents a similar magnitude for both phases, Δ differs by order of magnitudes: $\simeq 100\,\text{MeV}$ for 2SC, $\simeq 1\,\text{MeV}$ for CSL. Moreover, the strength of the coupling constant determines whether

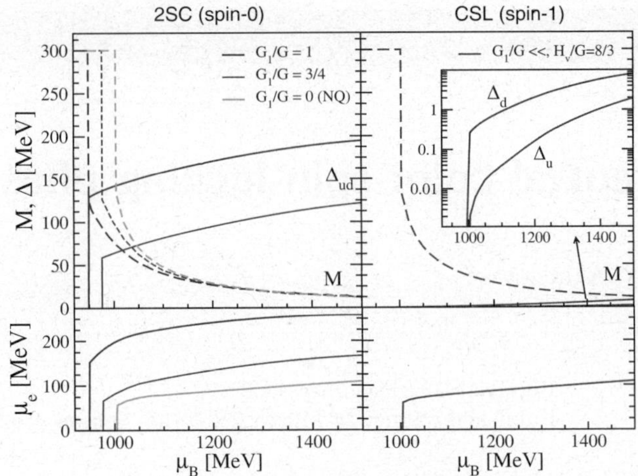

Fig. 1. Gap equations solutions for neutral matter in the flavor asymmetric spin-0 2SC phase (on the left) and in the flavor symmetric spin-1 CSL phase (on the right) at $T = 0$. The inset shows the small CSL gaps $\simeq 0.01$–$1\,\text{MeV}$. Upper panel: dynamical mass M and energy gap Δ as a function of the baryon chemical potential μ_B. Lower panel: electron chemical potential μ_e vs. μ_B. Different 2SC couplings are considered from $G_1/G = 1$ (dominant 2SC, see fig. 3) down to $G_1/G = 0$ ($\Delta \equiv 0$, NQ matter). The CSL coupling is taken as $H_v/G = 8/3$.

the 2SC phase occurs: for *strong coupling* $G_1/G = 1$, it presents large gaps $\Delta \simeq 150$–$200\,\text{MeV}$ that decrease as soon as the coupling does ($\Delta \simeq 50$–$100\,\text{MeV}$ for the usual Fierz value $G_1/G = 3/4$) and vanishing for values lower than a critical one. The crucial point is that the coupling should be large enough to pair quarks of different flavors overcoming a large Fermi sea mismatch ($\simeq 60$–$80\,\text{MeV}$). As a consequence, asymmetric flavor pairing might not be favorable unless the coupling is very strong (at least larger than $G_1/G = 3/4$)[2]. Therefore, flavor symmetric pairing becomes important when G_1/G is not large enough to have 2SC superconductivity.

On the other hand, the occurrence of the flavor symmetric CSL pairing is not affected by the charge neutrality constraint. The CSL condensates, having small energy gaps in comparison to the free energy of the system, do not modify the thermodynamic properties respect to the normal phase. In fig. 2 we show the number density for quarks, n_u, n_d, and for electrons, n_e, as a function of μ_B for the 2SC (on the left) and for the CSL phase (on the right). Different values of the coupling for the 2SC phase from $G_1/G = 1$ down to $G_1/G = 0$ (NQ) are considered. We clearly see that $n_i^{2SC, G_1/G=0} = n_i^{NQ} \simeq n_i^{CSL}$ for each particle specie i, so the two phases, CSL and NQ, are indistinguishable from the matter composition. This conclusion holds also for other thermodynamic quantities like pressure or energy density and therefore for the EoS.

---

[2] Actually, the critical value of G_1/G at which the 2SC condensate breaks down is model and parameterization dependent. NJL calculations have obtained no pure 2SC at intermediate densities ($\mu_B = 1200$ MeV) below $G_1/G = 3/4$, see [3,6].

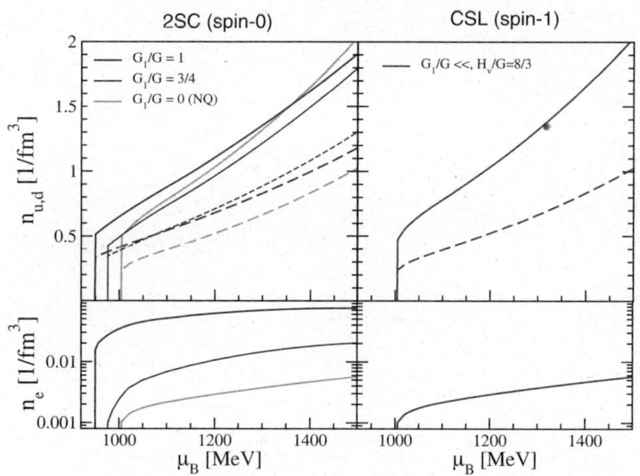

Fig. 2. Quark (d solid line, u dashed line) and electron number densities for 2SC (on the left) and CSL (on the right) as a function of μ_B for neutral matter at $T = 0$.

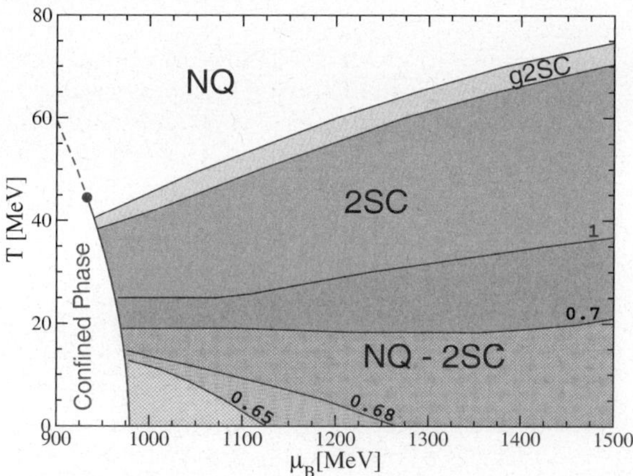

Fig. 3. Phase diagram for the intermediate density region relevant for neutron star cores. For *strong coupling* ($G_1/G \simeq 1$), flavor asymmetric spin-0 2SC phase is dominant. The mixed phase NQ-2SC assures global charge neutrality. The volume fraction of the 2SC sub-phase is indicated by numbers over the corresponding lines (Gaussian form factor [6]). As T increases, the volume fraction of the 2SC increases up to pure 2SC. Gapless 2SC is found before the transition to NQ occurs.

Finally, we found that while the phase diagram for *strong coupling* is dominated by the 2SC phase (fig. 3) for *intermediate or weak coupling*, the CSL phase is favorable (fig. 4). Note the low CSL critical temperatures $T_c^{CSL} \simeq 5\,\mathrm{MeV}$ in contrast to the 2SC case, for which $T_c^{2SC} \simeq 50\,\mathrm{MeV}$. Thus, we expect that in the cooling of a neutron star, when the temperature has fallen below the MeV scale, a CSL superconducting quark core could develop. Stable hybrid-star configurations have been obtained with a relatively large NQ matter core [4], therefore, hybrid stars with a CSL superconducting core will be stable as well. Finally, a qualitative study of the interaction of the magnetic field with the CSL phase shows that a CSL core is consistent with recent observations and models of magnetized neutron stars [12].

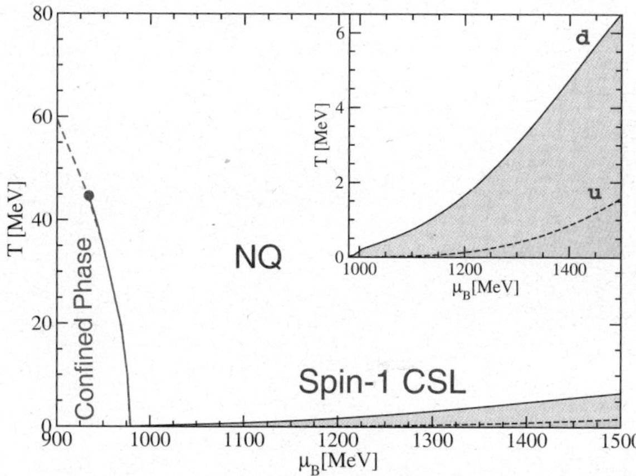

Fig. 4. Same as fig. 3 for *intermediate or weak coupling* ($G_1/G \leq 3/4$), for which flavor asymmetric pairing is no longer favorable. The volume fraction of the 2SC phase becomes very small ($< 10\%$) and the structure of fig. 3 disappears. Thus, the matter could be in the normal state (NQ) or in a phase with flavor symmetric pairing such as the spin-1 CSL phase.

D.N.A. thanks J.A. Pons, D. Blaschke, N.N. Scoccola and J.A. Miralles for fruitful discussions.

References

1. M. Alford, Lect. Notes Phys. **583**, 81 (2002); R. Rapp, T. Schäfer, E.V. Shuryak, M. Velkovsky, Ann. Phys. **280**, 35 (2000); K. Rajagopal, F. Wilczek, hep-ph/0011333.
2. M.G. Alford, J.A. Bowers, K. Rajagopal, J. Phys. G **27**, 541 (2001); A.W. Steiner, S. Reddy, M. Prakash, Phys. Rev. D **66**, 094007 (2002).
3. S.B. Rüster *et al.*, Phys. Rev. D **72**, 034004 (2005); D. Blaschke *et al.*, Phys. Rev. D **72**, 065020 (2005).
4. H. Grigorian, D. Blaschke, D.N. Aguilera, Phys. Rev. C **69**, 065802 (2004); I. Shovkovy, M. Hanauske, M. Huang, Phys. Rev. D **67**, 103004 (2003).
5. M. Alford, K. Rajagopal, JHEP **0206**, 031 (2002).
6. D.N. Aguilera, D. Blaschke, H. Grigorian, Nucl. Phys. A **757**, 527 (2005); D. Gomez Dumm *et al.*, Phys. Rev. D **73**, 114019 (2006).
7. T. Schäfer, Phys. Rev. D **62**, 094007 (2000); M.G. Alford *et al.*, Phys. Rev. D **67**, 054018 (2003); A. Schmitt, nucl-th/0405076.
8. A. Schmitt, I.A. Shovkovy, Q. Wang, Phys. Rev. D **73**, 034012 (2006).
9. D. Page *et al.*, Phys. Rev. Lett. **85**, 2048 (2000); D. Page *et al.*, Astrophys. J. Suppl. Ser. **155**, Issue 2, 623 (2004); D.G. Yakovlev, C.J. Pethick, Annu. Rev. Astron. Astrophys. **42**, 169 (2004).
10. D.N. Aguilera *et al.*, Phys. Rev. D **72**, 034008 (2005).
11. D.N. Aguilera, D. Blaschke, H. Grigorian, N.N. Scoccola, hep-ph/0604196.
12. D.N. Aguilera, to be published in Astrophys. Space Sci. J., *Conference Proceedings of Isolated Neutron Stars: from the Interior to the Surface, London, April 2006*, hep-ph/0608041.

Eur. Phys. J. A **31**, 831–835 (2007)
DOI 10.1140/epja/i2006-10253-y

Special Article – QNP 2006

Dilepton production in pp and CC collisions with HADES

I. Fröhlich[10,a], G. Agakishiev[11], C. Agodi[1], A. Balanda[5], G. Bellia[1,2], D. Belver[19], A. Belyaev[9], A. Blanco[3],
M. Böhmer[15], J.L. Boyard[17], P. Braun-Munzinger[6], P. Cabanelas[19], E. Castro[19], S. Chernenko[9], T. Christ[15],
M. Destefanis[11], J. Díaz[20], F. Dohrmann[7], T. Eberl[15], L. Fabbietti[15], O. Fateev[9], P. Finocchiaro[1], P.J.R. Fonte[3,4],
J. Friese[15], T. Galatyuk[6], J.A. Garzón[19], R. Gernhäuser[15], C. Gilardi[11], M. Golubeva[14], D. González-Díaz[6],
E. Grosse[7,8], F. Guber[14], Ch. Hadjivasiliou[16], M. Heilmann[10], T. Hennino[17], R. Holzmann[6], A. Ierusalimov[9],
I. Iori[12,13], A. Ivashkin[14], M. Jurkovic[15], B. Kämpfer[7], K. Kanaki[7], T. Karavicheva[14], D. Kirschner[11], I. Koenig[6],
W. Koenig[6], B.W. Kolb[6], R. Kotte[7], A. Kozuch[5], F. Krizek[18], R. Krücken[15], A. Kugler[18], W. Kühn[11], A. Kurepin[14],
J. Lamas-Valverde[19], S. Lang[6], S. Lange[6], L. Lopes[3], A. Mangiarotti[3], J. Marín[19], J. Markert[10], V. Metag[11],
B. Michalska[5], D. Mishra[11], E. Moriniere[17], J. Mousa[16], C. Müntz[10], L. Naumann[7], R. Novotny[11], J. Otwinowski[5],
Y.C. Pachmayer[10], M. Palka[5], V. Pechenov[11], O. Pechenova[11], T. Pérez Cavalcanti[11], J. Pietraszko[6], W. Przygoda[5],
B. Ramstein[17], A. Reshetin[14], M. Roy-Stephan[17], A. Rustamov[6], A. Sadovsky[7], B. Sailer[15], P. Salabura[5],
A. Schmah[6], R. Simon[6], S. Spataro[1], B. Spruck[11], H. Ströbele[10], J. Stroth[10,6], C. Sturm[6], M. Sudol[10,6], K. Teilab[10],
P. Tlusty[18], M. Traxler[6], R. Trebacz[5], H. Tsertos[16], I. Veretenkin[14], V. Wagner[18], H. Wen[11], M. Wisniowski[5],
T. Wojcik[5], J. Wüstenfeld[7], Y. Zanevsky[9], and P. Zumbruch[6]

1 Istituto Nazionale di Fisica Nucleare - Laboratori Nazionali del Sud, 95125 Catania, Italy
2 Dipartimento di Fisica e Astronomia, Università di Catania, 95125, Catania, Italy
3 LIP, Departamento de Física da Universidade de Coimbra, 3004-516 Coimbra, Portugal
4 ISEC Coimbra, Portugal
5 Smoluchowski Institute of Physics, Jagiellonian University of Cracow, 30059 Cracow, Poland
6 Gesellschaft für Schwerionenforschung mbH, 64291 Darmstadt, Germany
7 Institut für Kern- und Hadronenphysik, Forschungszentrum Rossendorf, PF 510119, 01314 Dresden, Germany
8 Technische Universität Dresden, 01062 Dresden, Germany
9 Joint Institute of Nuclear Research, 141980 Dubna, Russia
10 Institut für Kernphysik, Johann Wolfgang Goethe-Universität, 60486 Frankfurt, Germany
11 II.Physikalisches Institut, Justus Liebig Universität Giessen, 35392 Giessen, Germany
12 Istituto Nazionale di Fisica Nucleare, Sezione di Milano, 20133 Milano, Italy
13 Dipartimento di Fisica, Università di Milano, 20133 Milano, Italy
14 Institute for Nuclear Research, Russian Academy of Science, 117312 Moscow, Russia
15 Physik Department E12, Technische Universität München, 85748 Garching, Germany
16 Department of Physics, University of Cyprus, 1678 Nicosia, Cyprus
17 Institut de Physique Nucléaire d'Orsay, CNRS/IN2P3, 91406 Orsay Cedex, France
18 Nuclear Physics Institute, Academy of Sciences of Czech Republic, 25068 Rez, Czech Republic
19 Departamento de Física de Partículas. University of Santiago de Compostela. 15782 Santiago de Compostela, Spain
20 Instituto de Física Corpuscular, Universidad de Valencia-CSIC, 46971 Valencia, Spain

Received: 23 November 2006
Published online: 15 March 2007 – © Società Italiana di Fisica / Springer-Verlag 2007

Abstract. e^+e^- production was studied using the High Acceptance DiElectron Spectrometer (HADES). In pp collisions at 2.2 GeV kinetic beam energy, the exclusive η production and the Dalitz decay $\eta \to \gamma e^+e^-$ have been reconstructed. The electromagnetic form factor of the latter decay was found to be in good agreement with the existing theoretical predictions. In addition, an inclusive e^+e^- invariant-mass spectrum from the $^{12}C + ^{12}C$ reaction at 2 AGeV is presented and compared with a simplified thermal model.

PACS. 13.25.Jx Decays of other mesons – 13.40.Hq Electromagnetic decays – 14.40.Aq π, K, and η mesons – 25.40.Ve Other reactions above meson production thresholds (energies > 400 MeV)

1 Introduction

One of the main open questions in QCD is the origin of the hadron masses. Beside the so-called Goldstone bosons

such as π and η, the typical mass scale of hadrons in the vacuum is in the order of 1 GeV, whereas the current quark masses m_u, m_d are within 5–15 MeV. Based on this fundamental question it has been proposed that the mass of the hadrons is related to the spontaneous breaking of chiral

---

a e-mail: Froehlich@physik.uni-frankfurt.de

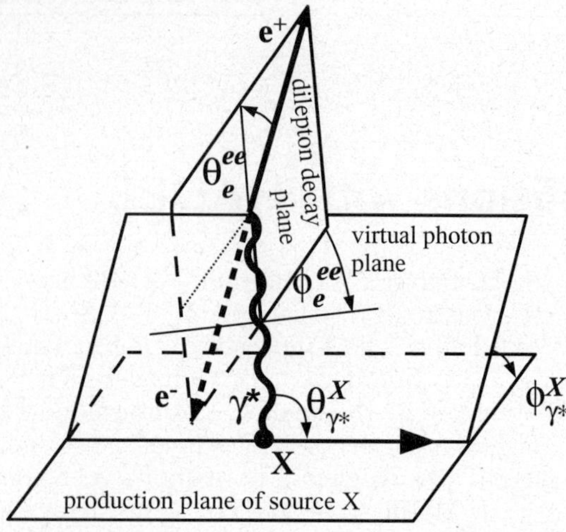

Fig. 1. Overview of the dilepton properties. In addition to the invariant mass $M^{inv}_{\gamma*}$ and the momentum $P^X_{\gamma*}$, 4 angles have to be taken into account, depending on the production plane and momentum of the source X.

symmetry, which is expected to be partially restored at finite baryon and energy densities. In connection to this question, one of the topics in modern hadron physics is how hadrons behave in strongly interacting matter. This means that their properties —like mass and width— have to be measured inside a cold, dense or hot nuclear environment. In this context, vector mesons have been proposed as an ideal probe for such studies, since they decay via an intermediate virtual photon γ^* into dileptons (e^+e^- or $\mu^+\mu^-$) which do not undergo strong interaction. Recently, the NA60 Collaboration [1] has extracted the ρ spectral function using the dimuon channel in In+In collisions at $158\,\mathrm{AGeV}$, which would correspond to a hot environment. This measurement indicates broadening of the line shape rather than a dropping of the mass.

However, for a complete understanding of the hadronic properties it is important to measure not only the hot, but also the dense region of the phase space. At beam energies of 1–$2\,\mathrm{AGeV}$, which correspond to moderate densities (2–$3\,\rho_0$), the production of mesons is dominated by multi-step excitation of a limited number of resonances and their subsequent decays, like $\Delta^{+,0} \to N\pi^0 \to N\gamma e^+e^-$ (π-Dalitz), $\Delta \to Ne^+e^-$ (Δ-Dalitz), $N^*(1535) \to N\eta \to N\gamma e^+e^-$ (η-Dalitz) and the decay of virtual resonances in $N(\omega,\rho)$. Here, most of the production mechanisms are at or even below threshold, which means that mesons are more likely produced in the dense phase of the fireball evolution.

The result of such experiments is usually a dilepton invariant-mass spectrum containing all these sources (dilepton cocktail). Before a conclusion on the properties of ρ and ω can be drawn, the contribution of Dalitz decays has to be evaluated within the detector acceptance and subtracted. In this context, it should be pointed out that a virtual photon (decaying into 2 stable particles) has 6 degrees of freedom, which are outlined in fig. 1: beside the invariant mass $M^{inv}_{\gamma*}$ these are the momentum $P^X_{\gamma*}$,

the polar $\theta^X_{\gamma*}$ and the azimuthal emitting angle $\phi^X_{\gamma*}$ of the virtual photon in the rest frame of the source X. In addition, the 2 decay angles of the photon into dilepton pairs, which are usually described with the helicity angle θ^{ee}_e, and the Treiman-Yang angle ϕ^{ee}_e.

2 The virtual-photon decay properties and the dilepton cocktail

For any interpretation of dilepton invariant-mass spectrum, the decay features of the virtual photons emitted by different sources into the dilepton pair have to be discussed. For the pseudoscalar mesons, which are spin-less, no alignment information can be carried from the production mechanism to the decay, so $\theta^X_{\gamma*}$, $\phi^X_{\gamma*}$ and ϕ^{ee}_e are isotropic. The helicity angle distribution is proposed to be $1 + \cos^2\theta^{ee}_e$ [2]. For a given mass of γ^* its momentum is fixed by the mass of the meson. The only degree of freedom is the mass spectrum, which is based on the electromagnetic form factor [3].

This is very different for the Δ Dalitz decay, because for particles carrying spin, production and decay do not factorize. In addition, the form factor as well as the branching ratio are based on calculations [4] and the helicity angle has quite some uncertainties [2]. For the direct decay of vector mesons, the helicity angle has to be isotropic, but the Treiman-Yang angle could contain higher-order contributions. Moreover, the $\omega \to \pi^0 e^+e^-$ transition form factor cannot be consistently described within the Vector Meson Dominance (VMD) model. On the Dalitz decays of the N^* resonances no information is available at all. Therefore, one of the goals of the HADES detector system (installed at GSI, Darmstadt), is to measure the dilepton properties in heavy-ion collisions as well as in elementary reactions. Consequently, the HADES program spans from $p+p$, $\pi+p$ to $\pi+A$ and $A+A$ collisions. First data has been taken in $C+C$ collisions at 1 and $2\,\mathrm{AGeV}$, $Ar+KCl$ collisions at $1.78\,\mathrm{AGeV}$, and $p+p$ at 1.25 and $2.2\,\mathrm{GeV}$. More details on the detector system can be found in [5,6]. Shortly, HADES is a magnetic spectrometer, consisting of up to 4 planes of Mini Drift Chambers (MDC) with a toroidal magnetic field. Particle identification is based on momentum and time-of-flight measurements. In addition, a Ring Imaging Cherenkov detector (RICH) and an electromagnetic Pre-Shower detector (PS) provide lepton identification capabilities.

3 The pp → ppη reaction at 2.2 GeV

In January 2004 the first pp run has been carried out at a kinetic beam energy of $2.2\,\mathrm{GeV}$, in order to study the performance of the detector using the exclusive reaction $pp \to pp\eta$ which has been measured by the DISTO Collaboration previously [7]. Since the decays $\eta \to \pi^+\pi^-\pi^0$ [8] and $\eta \to \gamma e^+e^-$ [3] are known, they served as a calibration measurement. Moreover, the performance of the HADES setup in extracting electromagnetic form factors has been

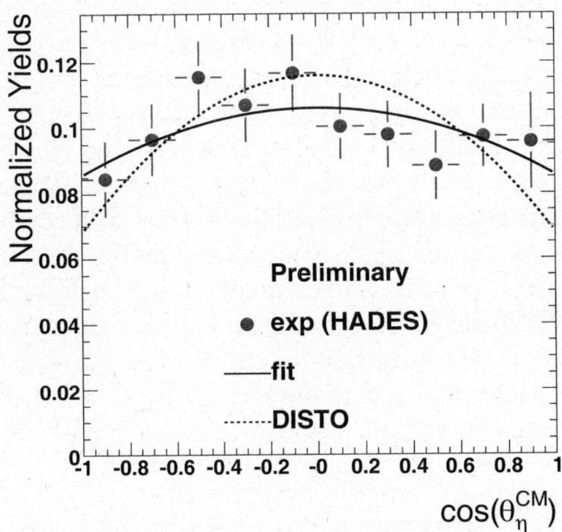

Fig. 2. Distribution of the η polar production angle in the CM frame (statistical errors only, corrected for efficiency and acceptance). The solid line represents a fit to the data, whereas the dashed line shows a parameterization for the existing data [7].

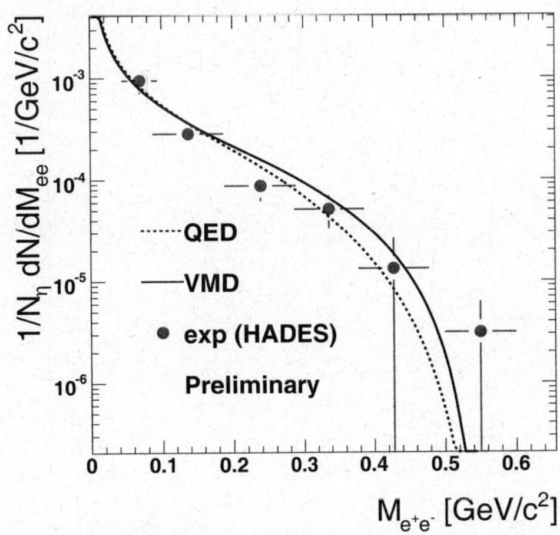

Fig. 3. Invariant-mass distribution M_{ee}^{inv} for the decay $\eta \to \gamma e^+ e^-$ [11] (statistical errors only). The dashed line is showing the prediction for a simple QED form factor, while the solid one represents the full VMD calculation [3].

checked. The analysis techniques are described in detail elsewhere [9]. Basically, for both reaction types kinematical constraints were used to identify the missing particle mass of π^0 or γ, respectively. In addition, a kinematic refit reduced the background.

3.1 The hadronic decay $\eta \to \pi^+\pi^-\pi^0$

For the evaluation of the η production, the polar-angle distribution of the η-meson emission has been analyzed. To subtract the background, first a selection on $\cos(\theta_\eta^{CM})$ and a fit on the corresponding *pp* missing-mass spectrum has been done, which resulted in the number of measured η-mesons in intervals of $\cos(\theta_\eta^{CM})$. The extrapolation to the full solid angle has been done using a full Monte Carlo simulation: Events of the type $pp \to pM \to pp\eta$ (with the shape $M \to p\eta$ taken from [7]) have been generated using the Pluto package [10] and processed through the full analysis chain. To evaluate the acceptance and efficiency for each angular interval, the number of reconstructed events has been divided by the number of generated events. Finally, each data point has been corrected by the value obtained by this method.

The result is shown in fig. 2 together with the anisotropy obtained by DISTO with a 2nd-order Legendre coefficient of $c_2 = -0.32 \pm 0.10$. A fit to the data gives $c_2 = -0.14 \pm 0.09$, which tends more to an isotropic η production, but is still consistent within errors with the previous result. This implies that the hadron efficiency is understood as a function of phase space.

3.2 The Dalitz decay $\eta \to \gamma e^+ e^-$

Similar methods have been used for the η Dalitz decay. In addition, selecting only events with an opening angle larger than 4°, the contribution of $\eta \to \gamma\gamma$ followed by

conversion in the detector material is strongly suppressed. Again, by fitting the *pp* missing-mass spectrum for each invariant-mass slice, the yield has been extracted. For the acceptance correction in this case the measured production angle (as described in the previous section) and the helicity angle (as outlined in sect. 2) have been fixed in the Monto Carlo simulation.

Figure 3 shows the corrected invariant-mass spectrum [11] together with functions using only the form factor from QED as well as the VMD correction [3]. Both curves have been normalized to the total number of measured η's. Since there is a strong dependence on the invariant mass, each data point has been corrected in mass position according to the given statistical mean in each corresponding interval. It can be seen that HADES is not sensitive to distinguish between these 2 models, but the result agrees with both predictions within the errors. This demonstrates that the efficiency is understood as a function of the invariant mass, which is the main important observable in the heavy-ion data. In addition, the ratio between the 2 η decay channels $\left(R = \frac{N_{\eta \to \pi^+\pi^-\pi^0}}{N_{\eta \to \gamma e^+ e^-}} \right)$ has been calculated for a full-cocktail simulation and the data. The result is $R_{exp} = 15.3 \pm 1.8_{stat}$ and $R_{sim} = 15.6 \pm 0.9_{stat}$, a nice agreement confirming that the lepton response of the HADES spectrometer is under control [9].

4 Results on C + C at 2 AGeV

The first result on dilepton production in C + C collisions at a kinetic beam energy of 2 AGeV has been published recently [12]. Details of the basic analysis steps can also be found in [13]. Basically, dileptons are identified via the RICH detector and reconstructed using the MDCs. A selection on pairs with opening angles larger then 9° was made in order to suppress the background from γ conversion. The combinatorial background (*i.e.*,

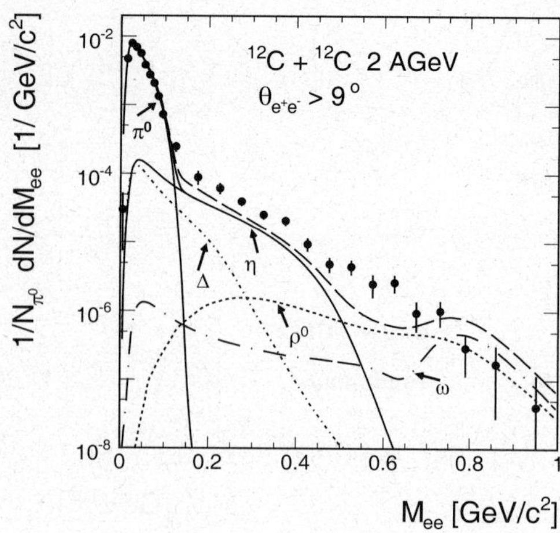

Fig. 4. Invariant-mass distribution M_{ee}^{inv} for CC at 2 AGeV [13]. The data points contain statistical error only. The lines are a thermal-model simulation as described in the text.

pairs mixed from different sources) has been removed by means of the like-sign method, where the number $N_{CB} = 2\sqrt{N_{++}N_{--}}$ has been obtained by forming pairs with equal charge in the same event. For events with masses ($M_{ee}^{inv} > 0.5\,\mathrm{GeV}/c^2$) the combinatorial background has been evaluated using the event-mixing technique. Here, opposite-sign pairs have been formed from different events but the same target segment.

In contrast to exclusive reactions, heavy-ion reactions depict only the final dilepton cocktail, without any selection on the different sources. As explained above, the virtual-photon properties might vary, thus the extrapolation to 4π is difficult, since 6 independent variables should be taken into account. While the integration over these variables cannot be done, a different method has been chosen. To allow for a model-independent comparison to our data, a single-track efficiency as a function of $p_e^{lab}, \theta_e^{lab}, \phi_e^{lab}$ has been extracted by simulation and applied to each lepton track, and finally combined to a dilepton efficiency. No attempt has been made to correct the resulting dilepton spectra for the geometrical acceptance. This was taken into account by a matrix using the same parameters per track for acceptance filtering of the models [14]. Figure 4 shows the invariant-mass spectrum after all analysis steps. Absolute normalization has been done by means of the measured charged pions, whose average value is the expected number of π^0 in an isospin-symmetric reaction.

In order to account for the contribution from the known Dalitz decays of the pseudoscalar mesons, a simulation has been made based on a simple thermal model [10] and filtered with the detector acceptance matrix as described above. The resulting curves are shown in fig. 4 as well. The solid line contains the π and η Dalitz decays with the decay properties as already mentioned and the known production cross-sections and distribution from TAPS [15]. It can be seen, that the π^0 region is well described, however, already in the η region model and data disagree. It is clear that additional sources are needed.

For a more complete view, Δ production has been added by π^0 scaling and the vector meson production by m_T scaling [16]. Still a factor of around 2 remains unaccounted for in the mass region of $0.2\,\mathrm{GeV}/c^2 < M_{ee}^{inv} < 0.7\,\mathrm{GeV}/c^2$, where the Δ decay is part of the contribution, as well as the low-mass tail of the ρ-meson. It has to be clarified if this enhancement is due to a different Δ yield and/or decay properties, or modified vector meson shapes in the dense medium. This means that the HADES data has to be compared to advanced model calculations [17–19], which, on the other hand, need more constraints extracted by elementary collisions. In this context, the decay properties of the virtual photon play an important role.

5 Summary and outlook

In summary, results for dilepton production obtained with HADES in elementary as well as in heavy-ion collisions have been presented. While the analysis of data sets, briefly described in this work, is almost finished, the CC collision at 1 AGeV is under analysis. In addition, a *pp* run at 1.25 GeV (below the η threshold) dedicated to Δ production has allowed us to collect promising statistics in spring 2006. Systematic studies using elementary reactions will continue, which are of particular importance for the interpretation of the inclusive invariant-mass spectra in heavy-ion collisions.

The HADES Collaboration gratefully acknowledges the support by BMBF grants 06TM970I, 06GI146I, 06F-140, and 06DR120 (Germany), by GSI (TM-FR1,GI/ME3,OF/STR), by grants GA CR 202/00/1668 and GA AS CR IAA1048304 (Czech Republic), by grant KBN 1P03B05629 (Poland), by INFN (Italy), by CNRS/IN2P3 (France), by grants MCYT FPA2000-2041-C02-02 and XUGA PGID T02PXIC20605PN (Spain), by grant UCY-10.3.11.12 (Cyprus), by INTAS grant 03-51-3208 and EU contract RII3-CT-2004-506078.

References

1. R. Arnaldi *et al.*, Phys. Rev. Lett. **96**, 162302 (2006).
2. E.L. Bratkovskaya *et al.*, Phys. Lett. B **348**, 283 (1995).
3. L.G. Landsberg, Phys. Rep. **128**, 301 (1985).
4. C. Ernst *et al.*, Phys. Rev. C **58**, 447 (1998).
5. R. Schicker *et al.*, Nucl. Instrum. Methods A **380**, 586 (1996).
6. A. Agakichiev *et al.*, to be published in Nucl. Instrum. Methods.
7. F. Balestra *et al.*, Phys. Rev. C **69**, 064003 (2004).
8. C. Amsler *et al.*, Phys. Lett. B **346**, 203 (1995).
9. S. Spataro (for the HADES Collaboration), *Proceedings of the Meson 2006*, Int. J. Mod. Phys. A **22**, 533 (2007).
10. M.A. Kagarlis, GSI Report 200-03 (2000) unpublished.
11. B. Spruck, PhD Thesis, University of Gießen.
12. A. Agakichiev *et al.*, Phys. Rev. Lett. **98**, 052302 (2007).
13. Th. Eberl (for the HADES Collaboration), Eur. Phys. J. C **49**, 261 (2007).
14. R. Holzmann, unpublished (r.holzmann@gsi.de).

15. R. Averbeck *et al.*, Z. Phys. A **359**, 65 (1997).
16. E.L. Bratkovskaya, W. Cassing, U. Mosel, Phys. Lett. B **424**, 244 (1998).
17. W. Cassing, E.L. Bratkovskaya, Phys. Rep. **308**, 65 (1999).
18. K. Shekhter *et al.*, Phys. Rev. C **68**, 014904 (2003); M.D. Cozma, C. Fuchs, E. Santini, A. Faessler, Phys. Lett. B **640**, 150 (2006).
19. D. Schumacher, S. Vogel, M. Bleicher, nucl-th/0608041.

Eur. Phys. J. A **31**, 836–841 (2007)
DOI 10.1140/epja/i2006-10223-5

THE EUROPEAN
PHYSICAL JOURNAL A

Special Article – QNP 2006

Low-mass dielectrons from the PHENIX experiment at RHIC

A. Kozlov[a]

For the PHENIX Collaboration
Weizmann Institute of Science, Rehovot 76100, Israel

Received: 8 November 2006
Published online: 6 March 2007 – © Società Italiana di Fisica / Springer-Verlag 2007

Abstract. The production of the low-mass dielectrons is considered to be a powerful tool to study the properties of the hot and dense matter created in the ultra-relativistic heavy-ion collisions. We present the preliminary results on the first measurements of the low-mass dielectron continuum in Au + Au collisions and the ϕ-meson production measured in Au + Au and d + Au collisions at $\sqrt{s_{NN}} = 200$ GeV performed by the PHENIX experiment.

PACS. 25.75.-q Relativistic heavy-ion collisions – 12.38.Mh Quark-gluon plasma

1 Introduction

Among the many diagnostic tools for the hot and dense matter produced in high-energy heavy-ion collisions the production of low-mass lepton pairs[1] plays an important role. Dileptons are emitted during the entire lifetime of the collision and interacting only electromagnetically, escape the interaction region almost freely carrying information directly to the detector. This makes dileptons an excellent tool to study the thermal radiation emitted by the dense medium and possible in-medium modifications of the light vector mesons properties (mass and/or width). This modifications are considered important signals of the restoration of chiral symmetry.

The production of low-mass electron pairs was extensively explored by the CERES Collaboration at the SPS [1]. CERES discovered an excess of the dielectron yield in the mass range 0.2–0.6 GeV/c^2 as compared to the one expected from the hadronic sources. A similar excess was observed for different colliding energies and species by DLS at the BEVALAC [2], E325 at KEK [3] and recently by the second-generation experiments NA60 at the SPS [4] and HADES at GSI [5]. For a summary of the most recent experimental results see [6].

The absolute yield and shape of the low-mass dielectron spectra measured by CERES are described satisfactorily by various theoretical models involving thermal radiation from the $\pi\bar{\pi}$ annihilation and in-medium modifications of the ρ, ω and ϕ vector mesons spectral functions [7].

The ϕ-meson is considered to be a sensitive probe of the possible restoration of chiral symmetry in relativistic

____
[a] e-mail: alex.kozlov@weizmann.ac.il

[1] Low-mass dilepton pairs include the light vector mesons and the dielectron continuum with mass $m_{e^+e^-} \lesssim 1$ GeV/c^2.

heavy-ion collisions which could manifest itself in the modification of its spectral properties (peak position and/or width) and in the changes of the relative yield as measured through the e^+e^- _vs._ the K^+K^- decay channels [8,9]. Since m_ϕ is slightly larger than $2m_K$, even small changes in the spectral properties of the ϕ or K can induce significant changes in the branching ratio of the $\phi \to K^+K^-$ decay.

The lifetime of the ϕ-meson is long compared to the lifetime of the fireball at RHIC energies ($\tau_\phi \approx 46$ fm/c _vs._ $\tau_{fireball} \approx 10$ fm/c [10]). In this case only a small fraction of the ϕ-mesons will decay inside the fireball, which could lead to a distortion of the line-shape if the ϕ-mesons spectral function is affected by the medium. For example, a low-mass tail should develop if the ϕ-meson mass decreases in the medium.

The PHENIX detector has the potential to measure accurately e^+e^- pairs in the low-mass region and the light vector meson properties. The observation of a ϕ-meson shape distortion is difficult but could be possible with the excellent mass resolution, of about 1%, of the PHENIX detector in the e^+e^- decay channel provided that a good signal-to-background, S/B, ratio can be achieved. In addition to that, the PHENIX detector has the unique capability of to measure simultaneously the ϕ-meson production through the e^+e^- and K^+K^- decay channels.

During RHIC run 4 in 2004, PHENIX has collected about $241 \mu b^{-1}$ integrated luminosity which allowed us to perform a first measurement of the low-mass dielectron continuum and the ϕ-meson production via e^+e^- and K^+K^- decay channels in Au + Au collisions at $\sqrt{s_{NN}} = 200$ GeV.

In this paper the low-mass dielectron mass spectra measured in Au + Au collisions at $\sqrt{s_{NN}} = 200$ GeV [11] and the results of the ϕ-meson production in d + Au [12] and Au + Au [13] collisions at $\sqrt{s_{NN}} = 200\,\text{GeV}$ as determined by the e^+e^- and K^+K^- channels are presented.

The measurement of low-mass e^+e^- pairs with the present PHENIX configuration is challenging due to huge combinatorial background originating from unrecognized conversions and π^0 Dalitz decays. Indeed, the S/B ratio measured in Au + Au collisions at $\sqrt{s_{NN}} = 200\,\text{GeV}$ is of the order of $\sim 1/150$ and $\sim 1/60$ at $m_{e^+e^-} \approx 400\,\text{MeV}/c^2$ [11] and ϕ mass [13], respectively.

A PHENIX upgrade with a Hadron Blind Detector foreseen in 2006 will significantly reduce the combinatorial background, improving the capability of PHENIX to measure low-mass dielectron pairs in heavy-ion collisions. The details of the HBD project are presented below.

2 The PHENIX experiment

2.1 Experimental apparatus

PHENIX is one of four experiments at the Relativistic Heavy Ion Collider (RHIC) at Brookhaven National Laboratory. Among them, PHENIX is the only one which has the capability to measure low-mass electron pairs. The PHENIX spectrometer [14] consists of two central arms which cover a pseudorapidity range $|\eta| < 0.35$ and $2 \times 90°$ in azimuthal angle. The momentum and charge of the particles are determined using a Drift Chamber (DC) and a Pad Chamber (PC1). Electrons are identified by a Ring Imaging Cherenkov (RICH) detector and by requiring the energy in a Electromagnetic Calorimeter (EMCal) to match the measured momentum of the tracks. Kaons are identified using the timing information from a Time-Of-Flight (TOF) detector and EMCal which have very good π/K separation in the momentum range $0.3 < p < 2.5\,\text{GeV}/c$ and $0.3 < p < 1.0\,\text{GeV}/c$, respectively. Valid DC-PC1 tracks are confirmed by the matching of the associated hit information to the RICH and EMCal in the case of electrons and to the TOF or EMCal in the case of kaons.

The beam-beam counters (BBC) are used to determine the z-coordinate of the collision vertex (z_{vtx}) and in combination with the zero-degree calorimeters (ZDC) provide the trigger and determine the event centrality.

In order to benefit from the high luminosity of RHIC and provide efficient detection of the rare electron and dielectron events PHENIX successfully implemented an electron trigger (ERT) which requires spatial matching between EMCal and RICH and an energy above a certain threshold in the EMCal.

2.2 Electron pair analysis with PHENIX spectrometer

The electron pair analysis is performed with the sample of identified electrons using a statistical procedure in which all particles in a given event are combined into pairs to generate unlike- and like-sign invariant-mass spectra. By construction the unlike-sign spectrum contains both the signal and an inherent combinatorial background of uncorrelated pairs. The size and shape of the combinatorial background are determined using the event mixing technique in which the particles from one event are combined with the particles from different events provided that all events belong to the same centrality and vertex classes.

The unlike-sign mixed event integral yield is normalized to the measured $2 \times \sqrt{N^{++}N^{--}}$ yield. All yields are calculated above $m_{e^+e^-} = 200\,\text{MeV}/c^2$ to exclude the correlated e^+e^+ and e^-e^- pairs from double π^0 Dalitz decays and double conversions. PHENIX has refined the mixed-event technique to a very high precision of less than $\pm 0.1\%$ which is confirmed by comparing the mixed event like-sign invariant-mass spectra to the measured one. The normalization procedure has been tested in four different approaches and found to be stable within 0.5% [11]. Finally, the signal mass distribution is derived by subtracting the normalized mixed event spectrum from the measured one.

3 ϕ-meson production

The ϕ-meson yield ($\text{d}N/\text{d}y$) and temperature (T) were derived from the invariant m_T-distribution:

$$\frac{1}{2\pi m_T}\frac{\text{d}^2 N}{\text{d}m_T \text{d}y} = \frac{N^\phi_{raw}(m_T) \cdot CF(m_T) \cdot \epsilon_{trigger}(m_T)}{2\pi m_T \cdot N_{events} \cdot \epsilon_{emb} \cdot \epsilon_{rbr} \cdot BR \cdot \Delta m_T},$$
(1)

where $N^\phi_{raw}(m_T)$ is the raw ϕ yield, $CF(m_T)$ is the correction factor to account for acceptance and pair reconstruction efficiency, $\epsilon_{trigger}(m_T)$ is the electron trigger efficiency which is equal to one for the analysis of minimum bias events, N_{events} is the number of analyzed events, ϵ_{emb} is the pair embedding efficiency which accounts for the reconstruction efficiency losses due to detector occupancy, ϵ_{rbr} is an efficiency due to run-by-run variations of the detector performance. BR is the branching ratio for $\phi \to e^+e^-$ or K^+K^- and Δm_T is the bin size.

The raw ϕ yield, N^ϕ_{raw}, is derived by summing the content of the spectrum over a mass interval of $\pm 3\sigma_{tot}$ where σ_{tot} is the total width calculated from the quadrature sum of the experimental mass resolution and natural width of the ϕ-meson.

The correction factor $CF(m_T)$ is determined using a Monte Carlo simulation. Single ϕ-mesons were generated with an exponential transverse momentum distribution. The ϕ's are decayed, propagated through an emulator of the PHENIX detector and the resulting output is passed through the whole analysis chain. For each m_T bin the ratio of the generated yield to the reconstructed one gives the correction factor $CF(m_T)$.

Finally, the $\text{d}N/\text{d}y$ and T are extracted from the corrected invariant m_T spectra fitted with the following exponential function, having $\text{d}N/\text{d}y$ and T as parameters:

$$\frac{1}{2\pi m_T}\frac{\text{d}^2 N}{\text{d}m_T \text{d}y} = \frac{\text{d}N/\text{d}y}{2\pi T(T + M_\phi)} \exp\frac{-(m_T - M_\phi)}{T},$$
(2)

where M_ϕ is the PDG value of the ϕ-meson mass.

Table 1. dN/dy and T for the $\phi \to e^+e^-$ and $\phi \to K^+K^-$ analysis in d + Au collisions.

Decay channel	dN/dy	T (MeV)
$\phi \to e^+e^-$	$0.056 \pm 0.015(\text{stat}) \pm 0.028(\text{syst})$	$326 \pm 94(\text{stat}) \pm 118(\text{syst})$
$\phi \to K^+K^-$	$0.047 \pm 0.009(\text{stat}) \pm 0.010(\text{syst})$	$414 \pm 13(\text{stat}) \pm 23(\text{syst})$

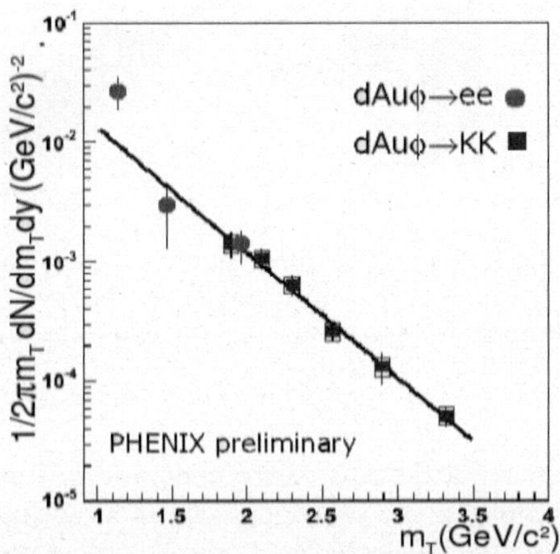

Fig. 1. (Color online) m_T spectra of the ϕ-mesons from the $\phi \to e^+e^-$ and $\phi \to K^+K^-$ analyses in d + Au collisions. Statistical and systematic errors are shown by vertical bars and open rectangles, respectively. The line represents the common fit to the exponential function, eq. (2).

3.1 d + Au collisions

The measurements in d + Au collisions establish the baseline information for possible nuclear matter effects and provide an essential reference for the comparison to the measurements in Au + Au collisions.

The presented results are based on about 31 million triggered d + Au events with a threshold of 600 MeV (see sect. 2.1) and about 62 million minimum-bias d + Au events for the $\phi \to e^+e^-$ and $\phi \to K^+K^-$ analyses, respectively.

The ϕ-meson invariant m_T spectra from both e^+e^- and K^+K^- analyses are shown in fig. 1 by the filled circles and squares, respectively [12]. The line represents the fit to all the data points with the exponential function given by eq. (2).

dN/dy and temperatures extracted from the $\phi \to e^+e^-$ and $\phi \to K^+K^-$ analyses in d + Au collisions are listed in table 1.

3.2 Au + Au collisions

In this section we present the m_T spectra, invariant yield and temperature results for the different centralities derived from the measurements of the ϕ-meson through the e^+e^- and K^+K^- decay channels in Au + Au collisions at $\sqrt{s_{NN}} = 200$ GeV. The $\phi \to e^+e^-$ analysis uses 903 million minimum-bias events. Kaons are combined in pairs using different combinations of the TOF and EMCal used

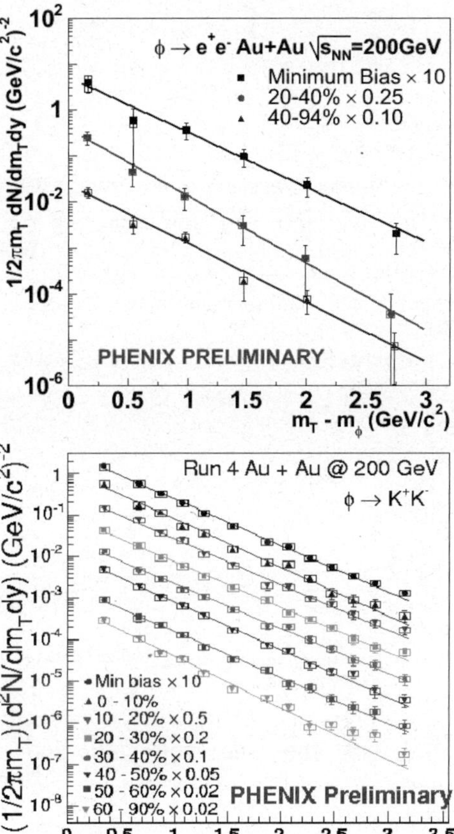

Fig. 2. (Color online) Invariant m_T spectra of the $\phi \to e^+e^-$ (top) and $\phi \to K^+K^-$ (bottom) for minimum bias and several centrality bins in Au + Au collisions. Statistical and systematic errors are shown by vertical bars and open rectangles, respectively. Each line represents the fit to the exponential function, eq. (2).

for the kaon identification (see sect. 2.1). The analysis performed with 409×10^6 minimum-bias events for TOF-TOF and 170×10^6 minimum-bias events for TOF-EMCal and EMCal-EMCal detector combinations.

The invariant m_T spectra for minimum bias and several centrality bins, fitted with the exponential function eq. (2), are shown in the top and bottom panels of fig. 2 for $\phi \to e^+e^-$ and $\phi \to K^+K^-$ analyses, respectively.

Figure 3 shows the centrality dependence of the ϕ-meson yield per participant pair extracted from the m_T distributions in both analyses. The highest centrality bin shown by the triangle in the top panel of fig. 3 has limited statistics and is derived from an independent analysis. In this analysis the invariant yield of the ϕ-meson is ob-

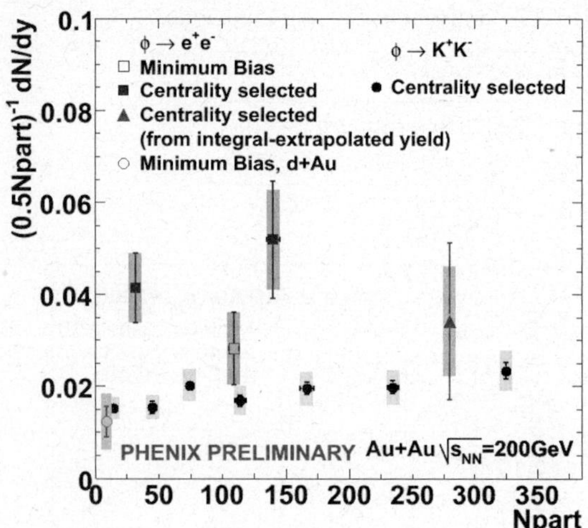

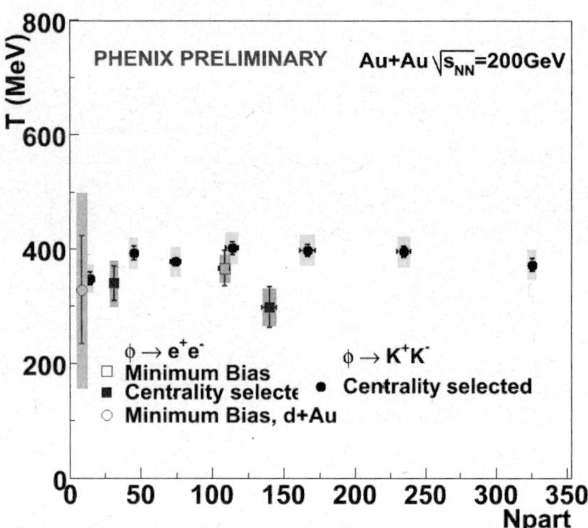

Fig. 3. (Color online) Multiplicity dependence of the ϕ-meson yield normalized to the number of participant pairs, $(0.5 \cdot N_{part})^{-1} \, dN/dy$ (top) and temperature, T (bottom) for e^+e^- and K^+K^- decay channels. Open and filled squares represent minimum bias and centrality selected events, respectively. The triangle shows the ϕ yield derived from an independent analysis. The open circle represents the reference measurements of dN/dy in d + Au collisions. Statistical and systematic errors are shown by vertical bars and shaded bands, respectively.

tained using the integral yield and the correction factor integrated over m_T assuming $T = 366 \,\mathrm{MeV}$.

The yield per participant pair shows a $\sim 50\%$ increase in the K^+K^- decay channel from peripheral to central collisions. The present statistical and systematic uncertainties do not allow us to infer the centrality dependence of the yield measured in the e^+e^- decay channel (top panel of fig. 3).

The comparison of the ϕ-meson production measured via the e^+e^- and K^+K^- decay channels shown in fig. 3 may indicate a possible increase of the yield in the dielectron channel compared to the kaon one. However, the statistical and systematic errors in the dielectron channel

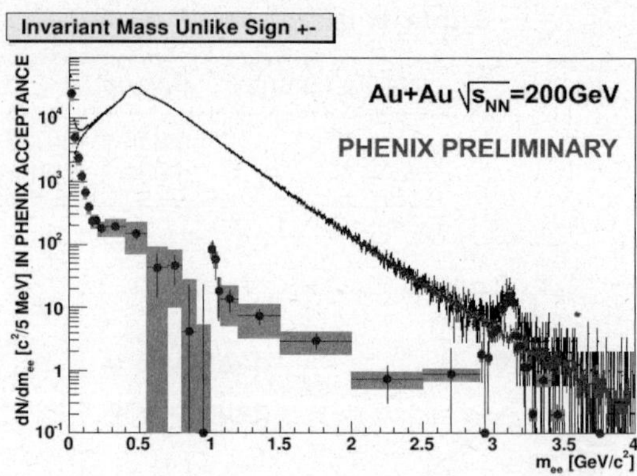

Fig. 4. (Color online) The foreground (black), background (red) and subtracted e^+e^- invariant-mass spectra. Statistical and systematic errors are shown by vertical bars and shaded bands, respectively.

are too large for a definite statement and within the error bars the yield in the two decay channels are consistent. The temperatures measured in Au + Au collisions through the e^+e^- and K^+K^- decay channels (see bottom panel of fig. 3) are centrality independent and agree within the statistical and systematic uncertainties.

The open circles in figs. 3 represent the reference measurements of the yield and T in d + Au collisions and both are found to be in a good agreement with dN/dy and T measured in the peripheral Au + Au collisions via the K^+K^- decay channel. The temperatures measured via the e^+e^- decay channel in d + Au and Au + Au collisions are consistent within the error bars. The yield per participant pairs measured in d + Au collisions is about two times smaller than the one measured in peripheral Au + Au collisions in the e^+e^- channel but the large statistical and systematic uncertainties do not allow for a conclusive statement.

4 Low-mass dielectron continuum

The results presented in this section are based on the analysis of 800 million minimum-bias events collected in Au + Au collisions at $\sqrt{s_{NN}} = 200\,\mathrm{GeV}$.

Figure 4 shows the foreground, background, and subtracted invariant-mass spectra of the dielectron pairs. The background distribution was obtained with an event mixing procedure with remarkable precision (see sect. 2.2). One can see that the signal-to-background ratio, S/B, at masses $m_{e^+e^-} \approx 400\,\mathrm{MeV}/c^2$ is of the order of $\sim 1/150$. With such a small S/B ratio the systematic errors, shown in fig. 4 by shaded bands, are dominated by the normalization of the background spectrum.

Figure 5 shows the comparison of the measured e^+e^- invariant-mass spectrum to the expected cocktail of hadron decays and to the theoretical predictions from the $\pi\pi$ annihilation channel without (dashed line) and with (dotted and dash-dotted lines) modification of the ρ-meson spectral function [15,16]. The results shown in

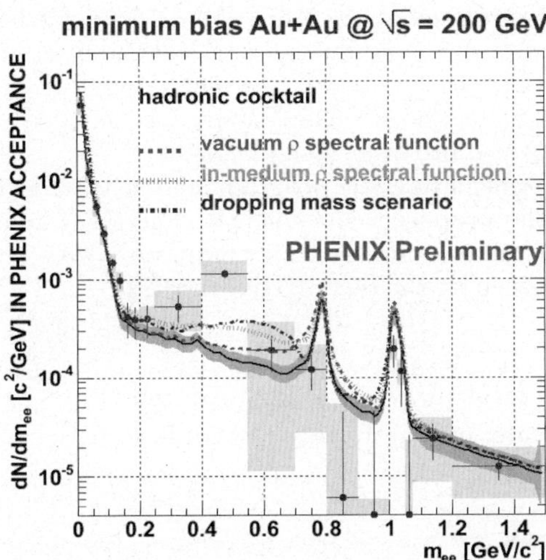

Fig. 5. (Color online) The measured e^+e^- invariant-mass spectrum compared to the hadronic cocktail (solid line) and calculations assuming the vacuum ρ spectral function (dashed line), in-medium broadening of the ρ width (dotted line) or ρ dropping mass (dash-dotted line).

fig. 5 may indicate an enhancement of the dielectron yield in the mass range between 0.2 to 0.6 GeV/c^2 over the expected hadronic cocktail and even the calculations involving in-medium modification of the ρ-meson, although the present uncertainties do not allow us to draw a strong conclusion.

Figure 6 shows the measured e^+e^- spectra for the central (0–20%) and semi-peripheral (20–92%) centrality classes compared to the hadronic cocktail. The possible excess of e^+e^- pairs within mass region 0.2–0.6 GeV/c^2 is seen in the central collisions whereas no excess is seen in the semi-peripheral ones.

The in-medium modification of the dielectron continuum can be studied also by looking at the ratio of the e^+e^- yield in different mass regions with respect to the π^0 Dalitz region ($m_{e^+e^-} < 100\,\text{MeV}/c^2$) which is independent of the number of participants. The top panel of fig. 7 shows the ratio of the yield in the mass region between 150–450 MeV/c^2 overlaid with the pion yield per participant pair shown by the crosses [17]. The pion yield scales with N_{part} while the ratio could have an indication of nonlinearity. The ratio for the mass region 1.1–2.9 GeV/c^2 (bottom panel) is within the uncertainties independent of N_{part} although it is expected to increase following the scaling with the number of participants of the charm production [18] which is the main source of the dielectrons in this mass region. The statistical significance of these results, however, is limited due to the low S/B ratio and does not allow us to draw a conclusion.

5 The Hadron Blind Detector

The capability of the PHENIX detector to measure low-mass dielectron pairs will be greatly improved with an upgrade that will add a Hadron Blind Detector (HBD) [19].

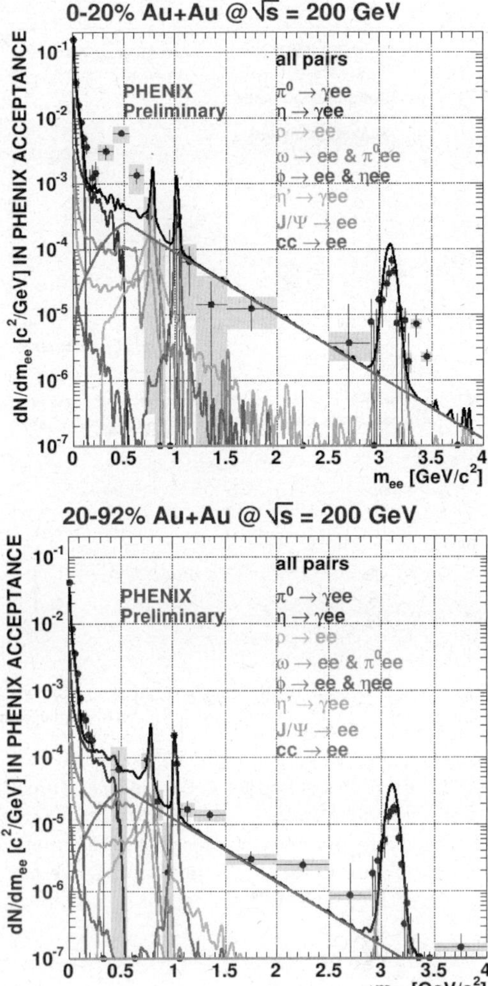

Fig. 6. (Color online) The comparison of the data to a hadronic cocktail and charm contribution for centrality classes 0–20% (top) and 20–92% (bottom). Statistical and systematic errors are shown by vertical bars and shaded bands, respectively.

The detector is a conceptually new Čerenkov detector operated with pure CF_4 in a proximity focus configuration. It is coupled directly to a triple Gas Electron Multiplier (GEM) [20] detector with a CsI photocathode layer evaporated on the top face of the first GEM foil. The detector has a pad readout scheme.

The HBD is located in the field free region extending up to $r \leq 60\,\text{cm}$ of the inner part of the PHENIX detector (fig. 8), which is realized by running the inner coil, recently installed in PHENIX for this purpose, with opposite current to compensate for the field from the outer coil. Čerenkov photons from an electron passing through the radiator are directly collected on the CsI photocathode. In this configuration photons form a circular blob, not a ring as in a RICH detector. The photoelectrons emitted from the photocathode are amplified by the triple GEM detector. Finally, the electron avalanche is read out in a pad plane located at the bottom of the detector element.

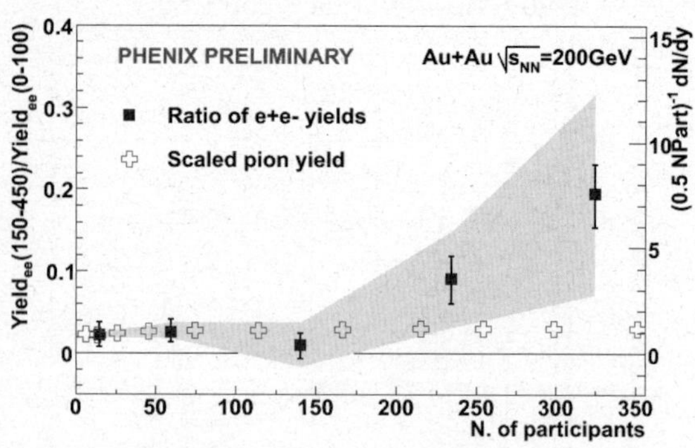

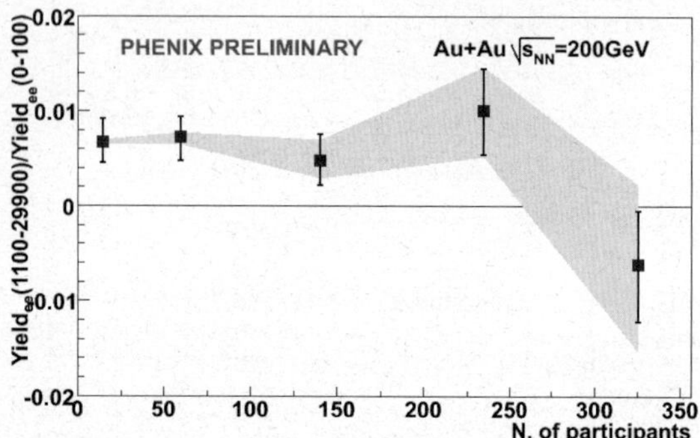

Fig. 7. Ratio of the dielectron yield measured in the mass region 150–450 MeV/c^2 (top) and 1.1–2.9 GeV/c^2 (bottom) with respect to the π^0 yield ($m_{e+e-} < 100$ MeV/c^2). Crosses represent the pion yield per participant pair as a function of the number of participants [17]. Statistical and systematic errors are shown by vertical bars and shaded bands, respectively.

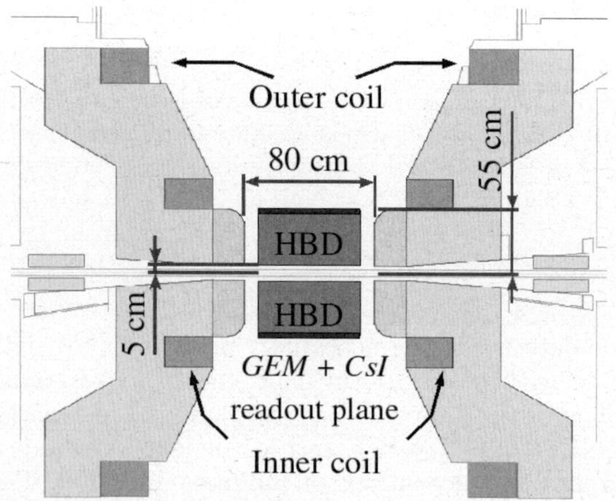

Fig. 8. Layout of the inner part of the PHENIX detector.

The HBD exploits the fact that the opening angle of electron pairs from γ conversions and π^0 Dalitz decays is small compared to that of pairs from the light vector mesons. In the field-free region this angle is preserved and by applying an opening angle cut one can reject more than 90% of the conversions and π^0 Dalitz decays, while keeping most of the signal.

The rejection ability of the HBD is confirmed by realistic Monte Carlo simulations which demonstrated that with the HBD the combinatorial background originating from conversions and π^0 Dalitz decays is reduced by approximately two orders of magnitude.

A comprehensive R&D program demonstrated the feasibility of the proposed concept [21]. The HBD construction and installation are nearing completion with first operation foreseen at the next RHIC run [22].

The author acknowledges support by the Israel Science Foundation, the MINERVA Foundation and the Nella and Leon Benoziyo Center of High Energy Physics Research.

References

1. G. Agakichev *et al.*, Eur. Phys. J. C **41**, 475 (2005); D. Adamova *et al.*, Phys. Rev. Lett. **91**, 042301 (2003).
2. R.J. Porter *et al.*, Phys. Rev. Lett. **79**, 1229 (1997).
3. K. Ozawa *et al.*, Phys. Rev. Lett. **86**, 5019 (2001).
4. R. Arnaldi *et al.*, Phys. Rev. Lett. **96**, 162302 (2006).
5. G. Agakichev *et al.*, nucl-ex/0608031.
6. I. Tserruya, Nucl. Phys. A **774**, 415 (2006).
7. R. Rapp, J. Wambach, Adv. Nucl. Phys. **25**, 1 (2000); H. van Hees, R. Rapp, Phys. Rev. Lett. **97**, 102301 (2006).
8. D. Lissauer, E.V. Shuryak, Phys. Lett. B **253**, 15 (1991).
9. S. Pal, C.M. Ko, Z. Lin, Nucl. Phys. A **707**, 525 (2002).
10. J.W. Harris, B. Muller, Annu. Rev. Nucl. Part. Sci. **46**, 71 (1996).
11. PHENIX Collaboration (A. Toia *et al.*), Eur. Phys. J. C **49**, 243 (2007).
12. PHENIX Collaboration (R. Seto *et al.*), J. Phys. G **30**, S1017 (2004).
13. PHENIX Collaboration (A. Kozlov *et al.*), Nucl. Phys. A **774**, 739 (2006).
14. PHENIX Collaboration (K. Adcox *et al.*), Nucl. Instrum. Methods A **499**, 469 (2003).
15. R. Rapp, *Proceedings of the 18th Winter Workshop on Nuclear Dynamics, 2002*, edited by R. Bellwied, J. Harris, W. Bauer, to be published in Heavy Ion Phys., nucl-th/0204003.
16. R. Rapp, Phys. Rev. C **63**, 054907 (2001).
17. PHENIX Collaboration (S.S. Adler *et al.*), Phys. Rev. C **69**, 034909 (2004).
18. PHENIX Collaboration (S.S. Adler *et al.*), Phys. Rev. Lett. **94**, 082301 (2005).
19. The PHENIX Collaboration (Z. Fraenkel, B. Khachaturov, A. Kozlov, A. Milov, D. Mukhopadhyay, D. Pal, I. Ravinovich, I. Tserruya, S. Zhou), Technical Note 391, http://www.phenix.bnl.gov/phenix/WWW/publish/ tserruya/1-06-HBD-proposal.ps.
20. F. Sauli, Nucl. Instrum. Methods A **386**, 531 (1997).
21. A. Kozlov *et al.*, Nucl. Instrum. Methods A **523**, 345 (2004); Z. Fraenkel *et al.*, Nucl. Instrum. Methods A **546**, 466 (2005); I. Ravinovich *et al.*, Nucl. Phys. A **774**, 903 (2006).
22. I. Tserruya, Nucl. Instrum. Methods A **563**, 333 (2006).

Eur. Phys. J. A **31**, 842–844 (2007)

DOI 10.1140/epja/i2006-10257-7

THE EUROPEAN
PHYSICAL JOURNAL A

Special Article – QNP 2006

Phase transitions in quark matter and behaviour of physical observables in the vicinity of the critical end point

P. Costa[1,2,a], C.A. de Sousa[1], M.C. Ruivo[1], and Yu.L. Kalinovsky[2,3]

[1] Departamento de Física, Universidade de Coimbra, P-3004-516 Coimbra, Portugal
[2] Laboratory of Information Technologies, Joint Institute for Nuclear Research, Dubna, Russia
[3] Université de Liège, Départment de Physique B5, Sart Tilman, B-4000, Liège 1, Belgium

Received: 23 November 2006
Published online: 15 March 2007 – © Società Italiana di Fisica / Springer-Verlag 2007

Abstract. We study the chiral phase transition at finite T and μ_B within the framework of the $SU(3)$ Nambu–Jona-Lasinio (NJL) model. The QCD critical end point (CEP) and the critical line at finite temperature and baryonic chemical potential are investigated: the study of physical quantities, such as the baryon number susceptibility near the CEP, will provide complementary information concerning the order of the phase transition. We also analyze the information provided by the study of the critical exponents around the CEP.

PACS. 11.30.Rd Chiral symmetries – 25.75.-q Relativistic heavy-ion collisions – 11.10.Wx Finite-temperature field theory

Understanding the behaviour of matter under extreme conditions is one of the most challenging issues in the physics of strong interactions. In the last years, major theoretical and experimental efforts have been dedicated to the physics of relativistic heavy-ion collisions, looking for signatures of the quark gluon plasma (QGP). A region of particular interest, that has been so far explored, is the $T - \mu_B$ phase boundary in order to try to explore different regions of the QCD phase diagram. The existence of the CEP in QCD was suggested in the end of the eighties, and its properties have been studied since then (for a review see ref. [1]). The most recent lattice results for the study of dynamical QCD with $N_f = 2 + 1$ staggered quarks of physical masses indicate the location of the CEP at $T_E = 162 \pm 2\,\mathrm{MeV}$, $\mu_E = 360 \pm 40\,\mathrm{MeV}$ [2], however its exact location is not yet known (the location of the CEP depends strongly of the mass of the strange quark). At the CEP the phase transition is of second order, belonging to the three-dimensional Ising universality class, and this kind of phase transitions are characterized by long-wavelength fluctuations of the order parameter.

As pointed out in [3], the critical region around the CEP is not pointlike but has a very rich structure. This critical region is defined as the region where the mean-field theory of phase transitions breaks down and nontrivial critical exponents emerge. The size of the critical region is important for future searches for the CEP in heavy ion-collisions. Some studies have been done in the $SU(2)$

sector [3,4] but less attention has been given to the effects of the strange quark. In this paper we aim to investigate the chiral phase transition and the CEP in quark matter with strange quarks.

We perform our calculations in the framework of the three-flavor NJL model, including the determinantal 't Hooft interaction that breaks the $U_A(1)$ symmetry, which has the following Lagrangian:

$$\mathcal{L} = \bar{q}\,(i\partial \cdot \gamma - \hat{m})\,q + \frac{g_S}{2} \sum_{a=0}^{8} \left[(\bar{q}\lambda^a q)^2 + (\bar{q}(i\gamma_5)\lambda^a q)^2 \right]$$
$$+ g_D \left[\det\left[\bar{q}(1 + \gamma_5)q\right] + \det\left[\bar{q}(1 - \gamma_5)q\right] \right]. \quad (1)$$

Here $q = (u, d, s)$ is the quark field with three flavors, $N_f = 3$, and three colors, $N_c = 3$, $\hat{m} = \mathrm{diag}(m_u, m_d, m_s)$ is the current quark mass matrix and λ^a are the Gell-Mann matrices, $a = 0, 1, \ldots, 8$, $\lambda^0 = \sqrt{\frac{2}{3}}\,\mathbf{I}$. The model is fixed by the coupling constants g_S, g_D in the Lagrangian (1), the cutoff parameter Λ, which regularizes the divergent integrals, and the current quark masses m_i. For our numerical calculations we use the parameter set: $m_u = m_d = 5.5\,\mathrm{MeV}$, $m_s = 140.7\,\mathrm{MeV}$, $g_S\Lambda^2 = 3.67$, $g_D\Lambda^5 = -12.36$ and $\Lambda = 602.3\,\mathrm{MeV}$, and we follow the methodology presented in detail in refs. [5,6].

The baryonic thermodynamic potential has the following form:

$$\Omega(\mu_i, T) = E - TS - \sum_{i=u,d,s} \mu_i N_i, \quad (2)$$

[a] e-mail: pcosta@teor.fis.uc.pt

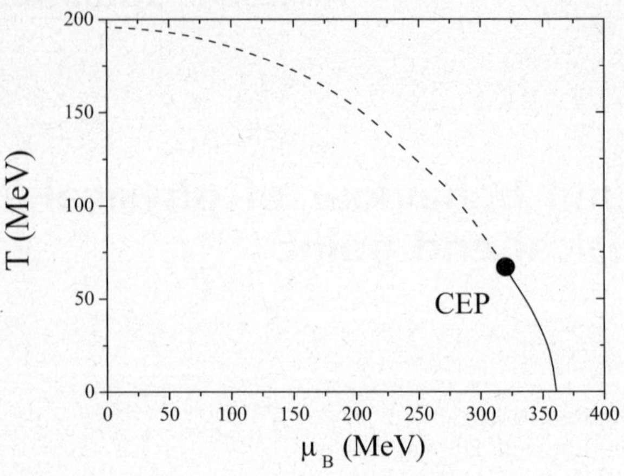
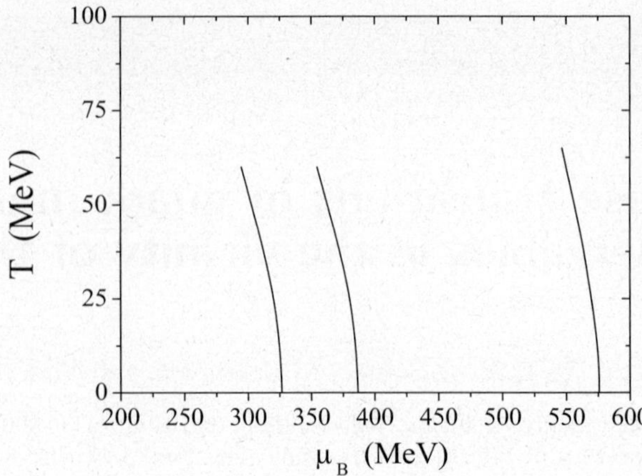

Fig. 1. Phase diagrams in the (T, μ_B)-plane. Left panel: $g_D = -12.36$. The solid line corresponds to first-order phase boundary and the dashed line corresponds to the crossover region. Right panel: $g_D = 0$. The lines correspond to first-order phase boundaries of the u, d and s sectors (from left). We have used the accepted values $\mu_I = 10\,\mathrm{MeV}$ and $\mu_Y = 12.5\,\mathrm{MeV}$.

where E, S and N_i are, respectively, the internal energy, the entropy and the number of particles of the i-th quark (the expressions are given in sect. IV of ref. [5]). The quark density for the i quark is determined by the relation

$$\rho_i = N_i/V = \frac{N_c}{\pi^2} \int p^2 \, \mathrm{d}p \, (n_i(\mu_i, T) - \bar{n}_i(\mu_i, T)), \quad (3)$$

where n_i and $\bar{n}_i$ are the Fermi distribution functions for quarks and antiquarks. Along this work we impose the condition $\mu_e = 0$ so the chemical equilibrium condition is $\mu_u = \mu_d = \mu_s = \mu_B$ (with $\mu_B = (\mu_u + \mu_d + \mu_s)/3$).

In general, the baryon number susceptibility is the response of the baryon number density $\rho_B(T, \mu_i)$ to an infinitesimal variation of the quark chemical potential μ_i:

$$\chi_B(\mu_B, T) = \frac{1}{3} \sum_{i=u,d,s} \left(\frac{\partial \rho_i}{\partial \mu_i} \right)_T. \quad (4)$$

This observable is one of the most relevant quantities in the context of possible signatures for chiral symmetry restoration in the hadron-quark transition and for transition from hadronic matter to the QGP [7].

The nature of the chiral phase transition in NJL type models at finite T and/ or μ_B has been discussed by different authors [5,8]. At zero density and finite temperature there is a smooth crossover, at nonzero densities, different situations can occur (for details see ref. [5]).

The phase diagram for the NJL model is presented in fig. 1, left panel. For zero temperature the transition is of first order. As the temperature increases, the first-order transition line persists up to a critical endpoint. In the CEP the chiral transition becomes of second order.

We will start our study by investigating the influence of the 't Hooft determinant in the (T, μ_B)-plane in the $SU(3)$ NJL model. Once for heavy-ion collisions the isospin chemical potential (μ_I) is supposed to be non-zero, the possible existence of more than one CEP in the phase diagram was proposed in [9]. On the other hand, to study whether a kaon condensate is formed, the non-zero hypercharge chemical potential (μ_Y) is also important. Having

these facts in mind, we start by taking $g_D = 0$ and, if we fix nonzero values of μ_I and μ_Y, there are three first-order phase transitions at low T and high μ_B and thus three CEP as we can see in fig. 1, right panel. However, in our specific case, where $g_D = -12.36$, there is only one first-order transition line and, consequently, only one CEP. This is a consequence of the flavor-mixing effects that cannot be neglected in the discussion of the phase diagram [10].

A bound to the size of the critical region around the CEP can be found by calculating the baryon number susceptibility (χ_B) and its critical behaviour. If the critical region of the CEP is small, it is expected that most of the fluctuations associated with the CEP will come from the mean-field region around the critical region [3].

In the left panel of fig. 2 the baryon number density for three different temperatures around the CEP is plotted. For temperatures below T^{CEP} we have a first-order phase transition and χ_B has a discontinuity (right panel of fig. 2). For $T = T^{CEP}$ the slope of the baryon number density tends to infinity at $\mu_B = \mu_B^{CEP}$ which implies a diverging susceptibility (this behaviour was found in [3,4] using different models in the $SU(2)$ sector). For temperatures above T^{CEP}, in the crossover region, the discontinuity of χ_B vanishes at the transition line, and the density changes gradually in a continuous way as we can see in the right panel of fig. 2.

We will focus our attention on the critical behaviour of χ_B in the vicinity of the CEP. At the CEP the baryon number susceptibility diverges with a certain critical exponent. As pointed out in [11], the form of the divergence depends on the route by which one approaches the critical point. If the path chosen is asymptotically parallel to the first-order transition line, the divergence of the baryon number susceptibility scales with an exponent γ_q. In the mean-field approximation, it is expected $\gamma_q = 1$ for this path. For any other path not parallel to the first-order line, the divergence scales with the exponent $\epsilon = 1 - 1/\delta$. Once in the mean-field approximation $\delta = 3$, we will have

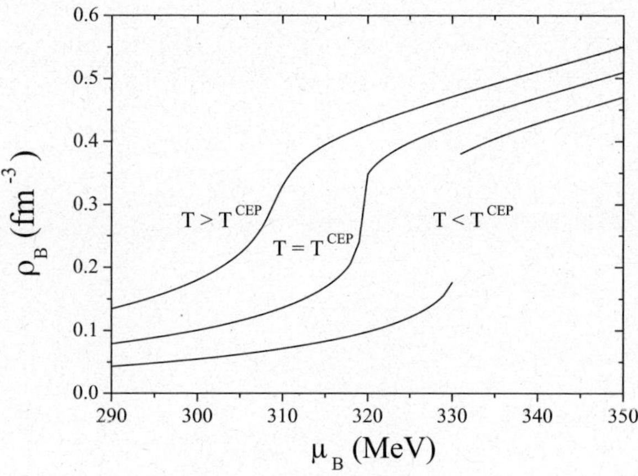

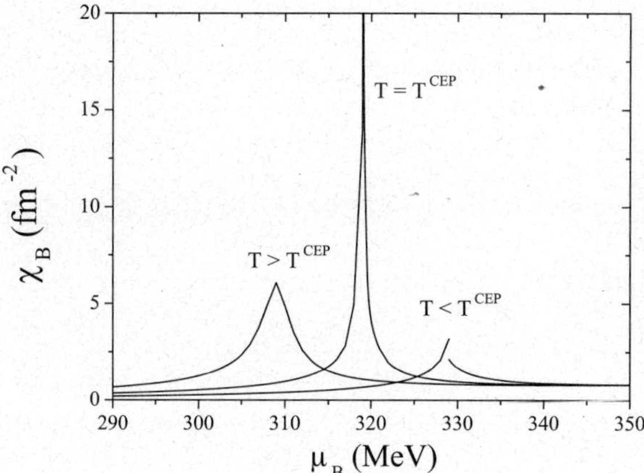

Fig. 2. Baryonic density (left panel) and baryon number susceptibility (right panel) as a function of μ_B for different temperatures around the CEP: $T^{CEP} = 67.7\,\text{MeV}$ and $T = T^{CEP} \pm 10\,\text{MeV}$.

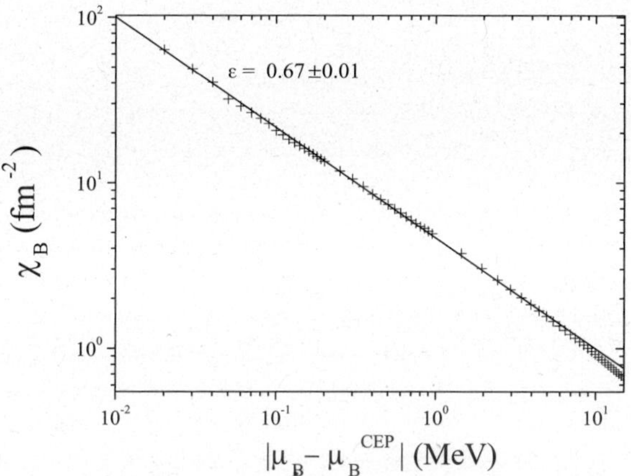

Fig. 3. The baryon number susceptibility χ_B as a function of $|\mu_B - \mu_B^{CEP}|$ at fixed temperature $T^{CEP} = 67.7\,\text{MeV}$.

$\epsilon = 2/3$ and $\gamma_q > \epsilon$ is verified. The last condition is responsible for the elongation of the critical region, χ_B being enhanced in the direction parallel to the first-order transition line; we verified that this is also true in the $SU(3)$ NJL model [12].

For the baryon number susceptibility we will use a path parallel to the μ_B-axis in the (T, μ_B)-plane from lower μ_B towards the critical $\mu_B^{CEP} = 318.5\,\text{MeV}$ at fixed temperature $T^{CEP} = 67.7\,\text{MeV}$. In fig. 3 we plot χ_B as a function of $|\mu_B - \mu_B^{CEP}|$ close to the CEP. To calculate the critical exponent ϵ we will use a linear logarithmic fit

$$\ln \chi_B = -\epsilon \ln |\mu_B - \mu_B^{CEP}| + \text{const}, \qquad (5)$$

where the term const is independent of μ_B. The result that we obtain is $\epsilon = 0.67 \pm 0.01$, which is consistent with the mean-field theory prediction: $\epsilon = 2/3$.

We have analyzed the phase diagram in the $SU(3)$ NJL which reproduces the essential features of QCD, such as a first-order phase transition for low temperatures and the existence of the CEP. We verified that, in the chiral limit and differently from the $SU(2)$ model, there is no TCP, which agrees with what it is expected: the chiral phase transition is of second order for $N_f = 2$ and first order for $N_f \geq 3$ [13]. A richer scenario, in $SU(3)$, is found when only the nonstrange quark masses are zero (the location of the CEP and the TCP depend strongly on the strange-quark mass) [12]. Around the CEP we have studied the baryon number susceptibility which is related with event-by-event fluctuations of μ in heavy-ion collisions. We also conclude that in our model the critical exponent ϵ obtained is consistent with the mean-field value $\epsilon = 2/3$.

Work supported by grant RFBR 06-01-00228 (Yu. Kalinovsky) and by grant SFRH/BPD/23252/2005 from FCT (P. Costa).

References

1. M.A. Stephanov, Prog. Theor. Phys. Suppl. **153**, 139 (2004); Int. J. Mod. Phys. A **20**, 4387 (2005).
2. Z. Fodor, S.D. Katz, JHEP **0204**, 050 (2004).
3. Y. Hatta, T. Ikeda, Phys. Rev. D **67**, 014028 (2003).
4. B.-J. Schaefer, J. Wambach, hep-ph/0603256.
5. P. Costa, M.C. Ruivo, Y.L. Kalinovsky, C.A. de Sousa, Phys. Rev. C **70**, 025204 (2004).
6. P. Costa, M.C. Ruivo, Yu.L. Kalinovsky, Phys. Lett. B **560**, 171 (2003).
7. M. Stephanov, K. Rajagopal, E. Shuryak, Phys. Rev. Lett. **81**, 4816 (1998).
8. M. Buballa, Phys. Rep. **407**, 205 (2005).
9. D. Toublan, J.B. Kogut, Phys. Lett. B **564**, 212 (2003).
10. M. Frank, M. Buballa, M. Oertel, Phys. Lett. B **562**, 221 (2003).
11. R.B. Griffiths, J. Wheeler, Phys. Rev. A **2**, 1047 (1970).
12. P. Costa, C.A. de Sousa, M.C. Ruivo, Y.L. Kalinovsky, hep-ph/0701135, to be published in Phys. Lett. B.
13. R.D. Pisarski, F. Wilczek, Phys. Rev. D **29**, 338 (1984).

Eur. Phys. J. A **31**, 845–847 (2007)
DOI 10.1140/epja/i2006-10204-8

THE EUROPEAN
PHYSICAL JOURNAL A

Special Article – QNP 2006

Fluctuations and correlations in the string clustering approach

L. Cunqueiro[a], E.G. Ferreiro, and C. Pajares

Instituto Galego de Física de Altas Enerxías and Departamento de Física de Partículas, Universidade de Santiago de Compostela, 15782 Santiago de Compostela, Spain

Received: 25 October 2006
Published online: 26 February 2007 – © Società Italiana di Fisica / Springer-Verlag 2007

Abstract. The fluctuations due to the clustering of color sources can explain the behaviour of the scaled multiplicity variance and transverse momentum fluctuations with centrality. They also predict a nonmonotonic behaviour with centrality for the multiplicity associated to high-p_T events. The clustering of color sources gives rise to an increase in the long-range correlations with centrality as well as to a supression at high centrality with respect to superposition models.

PACS. 25.75.Nq Quark deconfinement, quark-gluon plasma production, and phase transitions – 12.38.Mh Quark-gluon plasma – 24.85.+p Quarks, gluons, and QCD in nuclei and nuclear processes

The NA49 Collaboration has presented its data on multiplicity fluctuations as a function of centrality for Pb-Pb collisions [1]. A nonmonotonic centrality (system size) dependence of the multiplicty scaled variance was found. Concretely, the nonstatistical normalized fluctuations grow as the centrality increases, with a maximum around $N_{part} \simeq 100$, followed by a decrease at large centralities. This behaviour is similar to the one obtained for the $\Phi(p_T)$ and for the $F_{P_T}(p_T)$ measure, used to quantify the p_T fluctuations by the NA49 Collaboration [2,3] and by PHENIX Collaboration [4,5] respectively. This suggests that multiplicity and transverse momentum fluctuations are related [6]. Moreover, the $\Phi(p_T)$ variation with impact parameter implies that the collision is not a simple superposition of nucleon-nucleon collisions but rather collective effects are at work. In the framework of string clustering [7] such a behaviour is naturally explained [8,9]. The fluctuations in the size of the color clusters govern the behaviour of both multiplicity and transverse momentum fluctuations.

In a nucleus-nucleus collision strings are streched between partons from the projectile and the target [10–12]. With the increase of energy and/or the the atomic number of the colliding nuclei, the number of strings grows and so does its density in the nuclear overlap area: $\eta = N_S \frac{S_1}{S_A}$, being $S_A = \pi R_A^2$ for central collisions and S_1 the area of a single string. As the density grows, strings begin to overlap forming clusters. At a certain critical density a macroscopic cluster appears marking the percolation phase transition. A cluster consisting of n strings will have an area S_n and and a color charge which corresponds to

the vectorial sum of the charges of each individual string: $\langle Q_n \rangle = \sqrt{\frac{nS_n}{S_1}} Q_1$. Clusters fragment via the Schwinger mechanism. The fragmentation is controlled by the cluster tension that is related to the composed color field of the cluster. According to the Schwinger mechanism, we obtain the average multiplicity and p_T of a cluster composed by n strings: $\langle \mu \rangle_n = \sqrt{\frac{nS_n}{S_1}} \langle \mu \rangle_1$ and $\langle p_T^2 \rangle_n = (\frac{nS_1}{S_n})^{\frac{1}{2}} \langle p_T^2 \rangle_1$, being $\langle \mu \rangle_1$ and $\langle p_T \rangle_1$ the mean multiplicity and transverse momentum of a single string. In [9] we derive the analytical expression for the scaled multiplicity variance:

$$\frac{Var(\mu)}{\langle \mu \rangle} = 1 + \frac{\left\langle \left(\sum_j \langle \mu \rangle_n j \right)^2 \right\rangle - \left\langle \sum_j \langle \mu \rangle_n j \right\rangle^2}{\left\langle \sum_j \langle \mu \rangle_n j \right\rangle} \quad (1)$$

that is essentially a measure of the fluctuations in $\sum_j \sqrt{\frac{n_j S_{nj}}{S_1}}$. The sum over j goes over all individual clusters j, each one formed by n_j strings and occupying an area S_{nj}.

We use a Monte Carlo code based on the QSM [11] to obtain the different cluster configurations at a fixed number of participants. Once we know the number of clusters, the area occupied by each of them and the number of strings they contain, we apply the analytical formula in eq. (1). Our results for the scaled variance for negative particles are presented in fig. 1. The agreement with experimental data is acceptable. However, there is a discrepancy at the most central region where our results are well above the data. The reason for this are the simplifications coming from the use of analytical expressions.

The fluctuations on the number of target participants at a fixed number of proyectile participants have been pointed out to be an important contribution to the scaled

---

[a] e-mail: `leticia@fpaxp1.usc.es`

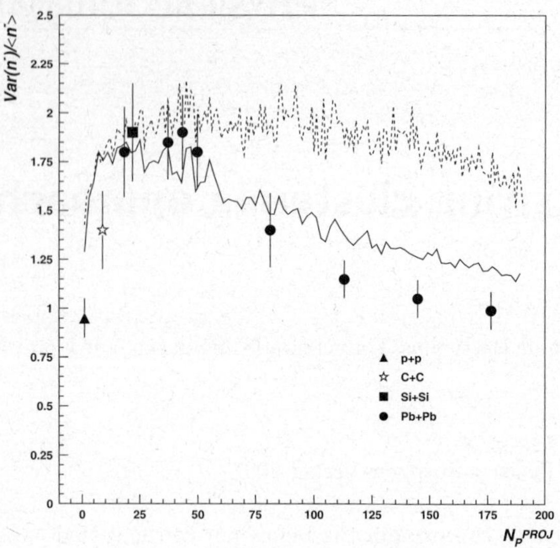

Fig. 1. Our results for the scaled variance of negatively charged particles in Pb-Pb collisions at SPS energies compared to NA49 data. Solid line: clustering of colour sources. Dashed line: independent strings.

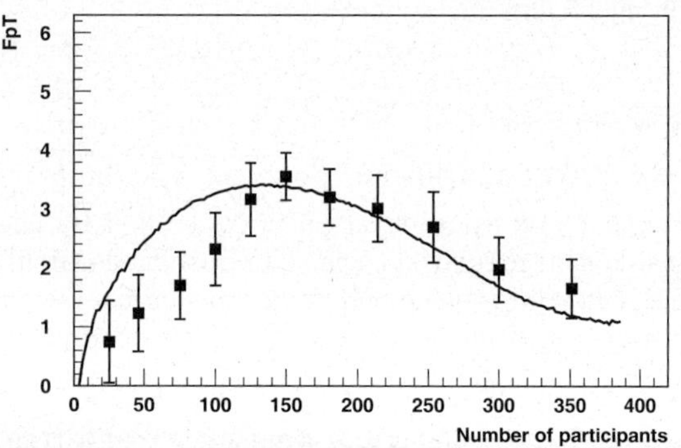

Fig. 2. F_{P_T} *versus* the number of participants. $F_{P_T} = \frac{w_{data} - w_{random}}{w_{random}}$ and $w = \frac{\sqrt{\langle p_T^2 \rangle - \langle p_T \rangle^2}}{\langle p_T \rangle}$. Experimental data from PHENIX at $\sqrt{s} = 200\,\text{GeV}$ are compared to our results (solid line).

multiplicity variance [13]. These fluctuations are included in our approach because most of our strings are formed between partons from the projectile and the target, conecting both hemispheres.

The PHENIX Collaboration [4,5] has measured the centrality dependence of the transverse momentum fluctuations using the observable F_{P_T} that quantifies the deviation of the observed fluctuations from statistically independent particle emission. The comparison of our results for the dependence of F_{P_T} on the number of participants N_p to the PHENIX data is shown in fig. 2. In fig. 3 our results for the observable Φ_{P_T} of charged particles in Pb-Pb central collisions at $158\,\text{AGeV}$ are compared to the data of NA49 [2]. A good agreement is obtained.

The behaviour of the transverse momentum fluctuations as well as the behaviour of the multiplicity fluctuations can be understood as follows: at low density, most

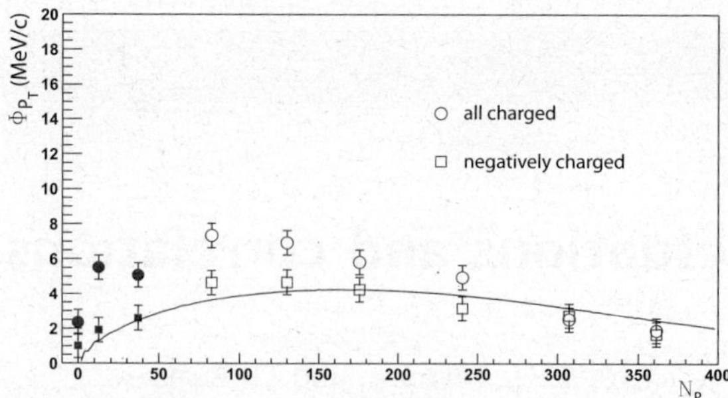

Fig. 3. Φ_{P_T} *versus* the number of participants. $\Phi_{P_T} = \sqrt{\frac{\left\langle \left(\sum_{i=1}^{\mu} (p_{T_i} - \overline{p_T}) \right)^2 \right\rangle}{\langle \mu \rangle}} - \sqrt{\overline{(p_T - \overline{p_T})^2}}$. The overline means averaging over a single-particle inclusive distribution and $\langle \ \rangle$ means an average over all events. Experimental data from NA49 at $\sqrt{s} = 158\,\text{GeV}$ are compared to our results (solid line).

of the particles are produced by individual strings with the same $\langle p_T \rangle$ and $\langle \mu \rangle$, so the fluctuations are small. Similarly, at large density above the percolation critical point there is essentially only one cluster formed by most of the strings created in the collision and therefore fluctuations are not expected either. There is a density in between these two limits for which the cluster configuration is the most diverse and thus the fluctuations are the highest.

1 Multiplicity associated to high-p_T events and multiplicity fluctuations

Assuming nuclear collisions to be a superposition of elementary interactions, a universal relation between the multiplicity distribution associated to the production of rare and self-shadowed events [14] and the total multiplicity distribution is obtained [15]:

$$P_C(n) \simeq \frac{nP(n)}{\langle n \rangle}. \qquad (2)$$

In [16] we argue that, although eq. (2) has been obtained in a superposition scheme, it remains approximately valid in a more general frame. From eq. (2) we derive [16] the following relation between the average multiplicity associated to rare events, the average total multiplicity and the scaled multiplicity variance:

$$\langle n \rangle_C - \langle n \rangle = \frac{\langle n^2 \rangle - \langle n \rangle^2}{\langle n \rangle}. \qquad (3)$$

High-p_T events are rare and self-shadowed. This means that their associated multiplicity distribution will satisfy eq. (2) and that their mean associated multiplicity will also satisfy eq. (3). The average multiplicity associated to high-p_T events will show a nonmonotonic behaviour with centrality due to the nonmonotonic centrality dependence of the scaled multiplicity variance at the right-hand side of eq. (3) This can be checked experimentally.

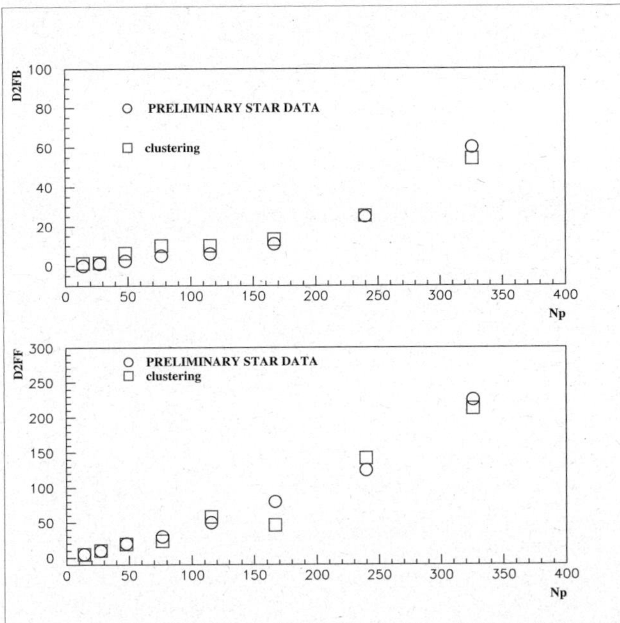

Fig. 4. DFB^2 and DFF^2 *versus* the number of participants compared to PRELIMINARY STAR data.

2 Forward-backward long-range correlations

In any model based on a superposition of elementary and statistically independent collisions, the squared forward-backward dispersion is proportional to the squared dispersion of the number of elementary collisions [17]. In fact, we have

$$D_{FB}^2 \equiv \langle n_F n_B \rangle - \langle n_F \rangle \langle n_B \rangle =$$
$$\langle N \rangle (\langle n_{0F} n_{0B} \rangle - \langle n_{0F} \rangle \langle n_{0B} \rangle)$$
$$+ (\langle N^2 \rangle - \langle N \rangle^2) \langle n_{0F} \rangle \langle n_{0B} \rangle, \qquad (4)$$

where N stands for the number of elementary interactions, n_{0F} (n_{0B}) for the number of forward (backward) produced particles in an elementary interaction and n_F (n_B) for the total number of forward (backward) particles. The first term of eq. (4) is the correlation between particles produced in the same elementary interaction. Assuming these correlations to have short range in rapidity, if one takes a rapidity gap between the forward and backward rapidity intervals large enough (1–1.5 units) this first term vanishes. In this way, one is left with the last term in eq. (4). This long-range rapidity correlation is due to the fluctuation in the number of elementary interactions, controlled by unitarity. It increases with the number of elementary interactions, therefore we expect it to increase with energy and the size of the nucleus in the collision. However, if there are interactions among strings, the number of independent elementary interactions translates approximately into the number of clusters of strings. Therefore a clear supression

of long-range correlations relative to the expected in a superposition picture is predicted [18,19]. Preliminary STAR data show that there is in fact a strong supression of long-range correlations. In fig. 4 we compare the preliminary data, obtained with a rapidity gap of 1.6 units in the central rapidity region, and a forward and backward intervals of 0.2 units, to our results [20] of percolation of strings. A good agreements is obtained.

We thank the organizers for such a nice meeting. This work was done under the contract FPA2005-01963 of CICYT of SPAIN.

References

1. NA49 Collaboration (M. Gazdzicki *et al.*), J. Phys. G **30**, 5701 (2004); NA49 Collaboration (M. Rybczynski *et al.*), J. Phys. Conf. Serv. **5**, 74 (2005).
2. NA49 Collaboration (H. Appelshauser *et al.*), Phys. Lett. B **459**, 679 (1999).
3. NA49 Collaboration (T. Anticic *et al.*), Phys. Rev. C **70**, 034902 (2004).
4. PHENIX Collaboration (K. Adox *et al.*), Phys. Rev. C **66**, 024901 (2002).
5. PHENIX Collaboration (S.S. Adler *et al.*), Phys. Rev. Lett. **93**, 092301 (2004).
6. St. Mrowczynski, M. Rybczynski, Z. Wlodorczyk, Phys. Rev. C **70**, 054906 (2004); G. Wilk, Z. Wlodorczyk Phys. Rev. Lett. **84**, 2770 (2000).
7. N. Armesto, M.A. Braun, E.G. Ferreiro, C. Pajares, Phys. Rev. Lett. **77**, 3736 (1996).
8. E.G. Ferreiro, F. del Moral, C. Pajares, Phys. C **69**, 034901 (2004).
9. L. Cunqueiro, E.G. Ferreiro, F. del Moral, C. Pajares, Phys. Rev. C **72**, 024907 (2005).
10. A. Capella, U.P. Shukatme, C.I. Tan, J. Tran Thanh Van, Phys. Rep. **236**, 225 (1994); A. Capella, C. Pajares, A.V. Ramallo, Nucl. Phys. B **241**, 75 (1984).
11. N. Armesto, C. Pajares, D. Sousa, Phys. Lett. B **527**, 92 (2002).
12. M.A. Braun, C. Pajares, Phys. Rev. Lett. **85**, 4864 (2001).
13. V.P. Konchakovski, S. Haussler, M.I. Gorenstein, E.L. Bratkovskaya, M. Bleicher, H. Stocker, Phys. Rev. C **73**, 034902 (2006).
14. R. Blankenbecler, A. Capella, J. Tran Thanh Van, C. Pajares, A.V. Ramallo, Phys. Lett. B **107**, 106 (1981); A. Kaidalov, Nucl. Phys. A **525**, 39 (1991).
15. J. Dias de Deus, C. Pajares, C.A. Salgado, Phys. Lett. B **407**, 335 (1997).
16. L. Cunqueiro, J. Dias de Deus, C. Pajares, Phys. Rev. C **74**, 034901 (2006).
17. A. Capella, A. Krzywicki, Phys. Rev. D **18**, 4120 (1978).
18. N.S. Amelin, N. Armesto, M.A. Braun, E.G. Ferreiro, C. Pajares, Phys. Rev. Lett. **73**, 2813 (1994).
19. M.A. Braun, R.S. Kolevatov, C. Pajares, V.V. Vechernin, Eur. Phys. J. C **32**, 535 (2004).
20. N. Armesto, L. Cunqueiro, E.G. Ferreiro, C. Pajares, in preparation.

Eur. Phys. J. A **31**, 848–850 (2007)
DOI 10.1140/epja/i2006-10194-5

THE EUROPEAN
PHYSICAL JOURNAL A

Special Article – QNP 2006

Transport coefficients in Chiral Perturbation Theory

D. Fernández-Fraile[a] and A. Gómez Nicola[b]

Departamentos de Física Teórica I y II, Universidad Complutense, 28040 Madrid, Spain

Received: 8 October 2006
Published online: 22 February 2007 – © Società Italiana di Fisica / Springer-Verlag 2007

Abstract. We present recent results on the calculation of transport coefficients for a pion gas at zero chemical potential in Chiral Perturbation Theory (ChPT) using the Linear Response Theory (LRT). More precisely, we show the behavior of DC conductivity and shear viscosity at low temperatures. To compute transport coefficients, the standard power counting of ChPT has to be modified. The effects derived from imposing unitarity are also analyzed. As physical applications in relativistic heavy-ion collisions, we show the relation of the DC conductivity to soft-photon production and phenomenological effects related to a non-zero shear viscosity. In addition, our values for the shear viscosity to entropy ratio satisfy the KSS bound.

PACS. 11.10.Wx Finite-temperature field theory – 12.39.Fe Chiral Lagrangians – 25.75.-q Relativistic heavy-ion collisions

1 Introduction

We are interested in studying transport properties in a pion gas with zero baryon density. Pions are the lightest hadronic degrees of freedom and their dynamics is well described for low-enough energies and temperatures by the Chiral Perturbation Theory (ChPT) [1]. ChPT is a low-energy expansion performed in terms of p^2 (p represents a momentum, a mass or a temperature) against some scale $\Lambda_\chi^2 \sim (1\,\mathrm{GeV})^2$. However, as we shall see, the computation of transport coefficients in ChPT is intrinsically non-perturbative, leading to a modification of the standard ChPT power-counting scheme to take into account collisions in the plasma (*i.e.*, a non-zero pion width).

Considering small external perturbations which put the system slightly out of equilibrium, we can employ the Linear Response Theory (LRT) [2–4], in which a transport coefficient t is given by:

$$\mathsf{t} = C \lim_{q^0 \to 0^+} \lim_{|\boldsymbol{q}| \to 0^+} \frac{\partial \rho(q^0, \boldsymbol{q})}{\partial q^0}, \qquad (1)$$

with C a constant and ρ the spectral density of a current-current correlator.

In the calculation of transport coefficients, there is a dominant contribution coming from products of the kind $G_A G_R \propto 1/\Gamma$ (G_A and G_R are the advanced and retarded propagators, respectively), called *pinching poles* [3, 4], where Γ is the particle width (which is perturbatively small: $\Gamma = \mathcal{O}(p^5)$ in ChPT [5]). From Kinetic Theory (KT), a behavior $\sim 1/\Gamma$ for transport coefficients is also

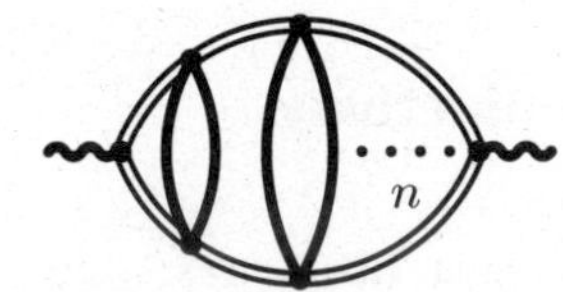

Fig. 1. Generic ladder diagram with n rungs ($n \geq 1$). Double lines represent particles with $\Gamma \neq 0$.

Fig. 2. Generic bubble diagram with n bubbles ($n \geq 1$). Double lines represent particles with $\Gamma \neq 0$.

expected, since $1/\Gamma$ represents the mean time between two collisions of the particles in the plasma. Each of these pinching poles comes from a pair of lines in a Feynman diagram contributing to the spectral density that share the same four-momentum when the limit in (1) is taken. Accordingly, there are some classes of diagrams potentially non-perturbative, being the most harmful the type called *ladder diagrams* (see fig. 1). Another type, the so-called *bubble diagrams* (see fig. 2), which in principle would be the most harmful, for $n \geq 2$ turn out to give a negligible contribution to the electrical conductivity and the shear viscosity. Thus, the standard power-counting scheme from ChPT has to be modified to compute transport coefficients [6]. In the case of a ladder diagram with n rungs, its contribution is in principle of order $\mathcal{O}(p^{2n}Y^{n+1})$, where Y is the contribution from a single-bubble diagram. For $T \ll M_\pi \simeq 140\,\mathrm{MeV}$, a detailed analysis of the spec-

a e-mail: danfer@fis.ucm.es
b e-mail: gomez@fis.ucm.es

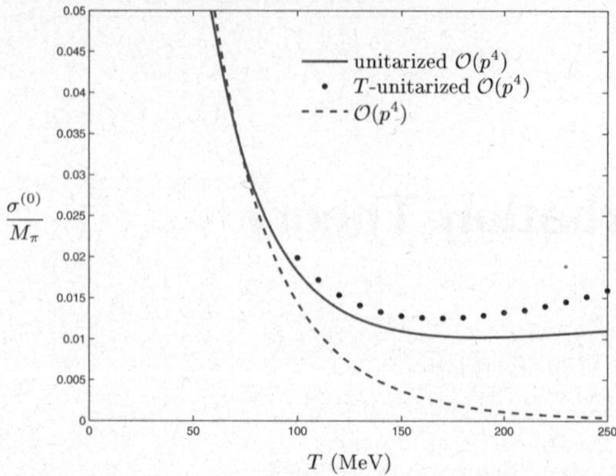

Fig. 3. Contribution to the electrical conductivity from the single-bubble diagram as a function of temperature. Three curves for the pion width are shown: non-unitarized $\mathcal{O}(p^4)$ (dashed line), $T = 0$ unitarized and finite-T unitarized.

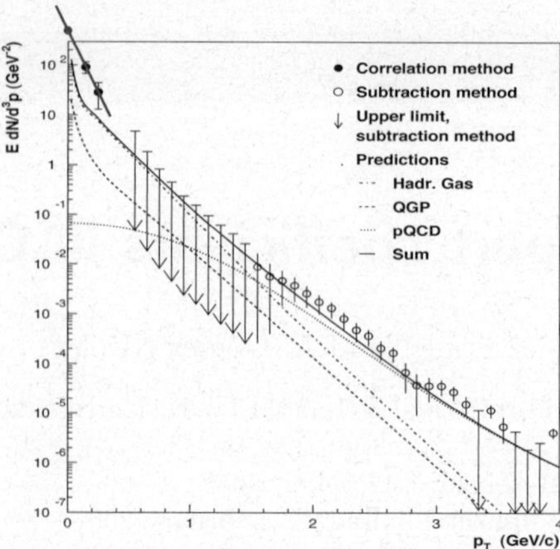

Fig. 4. Photon spectrum obtained by the experiment WA98 [10]. We show the linear extrapolation from the two lowest-energy points, which reaches a point at the origin compatible with our estimate. Note that theoretical predictions tend to underestimate the lowest-energy points [11].

tral function [6] shows that it contributes $\mathcal{O}(p^{2n}Y)$. For $T \gtrsim M_\pi$, Y becomes of order 1 or larger (due to unitarity effects) so that diagrams with one or more rungs could be non-negligible in this regime. Also note that, for high enough temperatures where typical momenta $p \sim T$, diagrams with derivative vertices become potentially dominant against those with constant vertices.

2 Electrical conductivity

We are interested in the DC electrical conductivity, which represents the response of the gas under the action of a constant electric field. The expression for this transport coefficient in LRT is

$$\sigma = - \lim_{q^0 \to 0^+} \lim_{|\boldsymbol{q}| \to 0^+} \frac{\rho_\sigma(q^0, \boldsymbol{q})}{6q^0},$$

with

$$\rho_\sigma(q^0, \boldsymbol{q}) = 2\,\mathrm{Im}\,\mathrm{i} \int \mathrm{d}^4 x\; \mathrm{e}^{\mathrm{i}q\cdot x} \theta(t) \langle [J_i(x), J^i(0)] \rangle,$$

and J^i is the electric current density in the gas. For this transport coefficient, at $T \ll M_\pi$, $Y \simeq \sqrt{M_\pi/T}$. So the contribution from a single-bubble diagram to the conductivity at these temperature is $\sigma^{(0)} \simeq e^2 M_\pi \sqrt{M_\pi/T}$, where e is the charge of the electron. The pion width is calculated in ChPT [5] from the pion-pion scattering amplitudes. To consider the unitarity effects on the transport coefficients, we impose unitarity on these amplitudes by means of the Inverse Amplitude Method (IAM) [7]. This unitarized scattering amplitudes have also been calculated at finite temperature [8]. We have implemented unitarization in the *dilute gas approximation*, where we neglect $\mathcal{O}(n_B^2)$ terms, with n_B the Bose-Einstein distribution evaluated at a pion energy. This is approximately valid for $T \leq M_\pi$. In fig. 3 we represent the behavior with temperature of the $\sigma^{(0)}$ contribution to conductivity. We see that unitarity makes conductivity grow near $T = 170\,\mathrm{MeV}$, as is expected in the QGP phase [9]. The contribution from

ladder diagrams with derivative vertices has to be taken into account for temperatures $T \gtrsim M_\pi$.

The DC conductivity is related to the zero-energy (soft) photon spectrum emitted by the pion gas. Considering a pion gas produced after a relativistic heavy-ion collision which expands cylindrically (Bjorken's hydrodynamical model), we obtain [6] the estimate $E\mathrm{d}N_\gamma/\mathrm{d}^3\boldsymbol{p}(p_T \to 0^+) \simeq 5.6 \times 10^2\,\mathrm{GeV}^{-2}$ for the photon spectrum at zero transverse momentum. In fig. 4 we see that this estimate is compatible with the experimental results obtained by the experiment WA98 [10] and other theoretical analysis [11]. This is reasonable since it is expected that the hadronic contribution dominates at low p_T.

3 Shear viscosity

The expression for shear viscosity in LRT is

$$\eta = \lim_{q^0 \to 0^+} \lim_{|\boldsymbol{q}| \to 0^+} \frac{\rho_\eta(q^0, \boldsymbol{q})}{20q^0},$$

where

$$\rho_\eta(q^0, \boldsymbol{q}) = 2\,\mathrm{Im}\,\mathrm{i} \int \mathrm{d}^4 x\; \mathrm{e}^{\mathrm{i}q\cdot x} \theta(t) \langle [\pi_{ij}(x), \pi^{ij}(0)] \rangle,$$

and π_{ij} is the traceless part of the spatial energy momentum tensor, *i.e.*

$$\pi_{ij} \equiv T_{ij} - \frac{1}{3} g_{ij} T^l{}_l.$$

The shear viscosity has been previously calculated for a meson gas in the KT framework [12–14]. For instance, in the work [12] (referred as DLE), a treatment based on a relativistic version of the Boltzmann equation is followed, using unitarized amplitudes from ChPT. In fig. 5 we show our leading-order result for η, where we see that unitarity

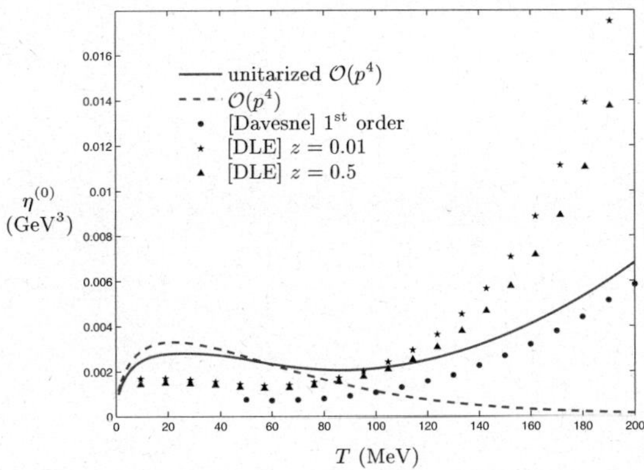

Fig. 5. Behavior of the unitarized shear viscosity with temperature and comparison with the results obtained in [12,13].

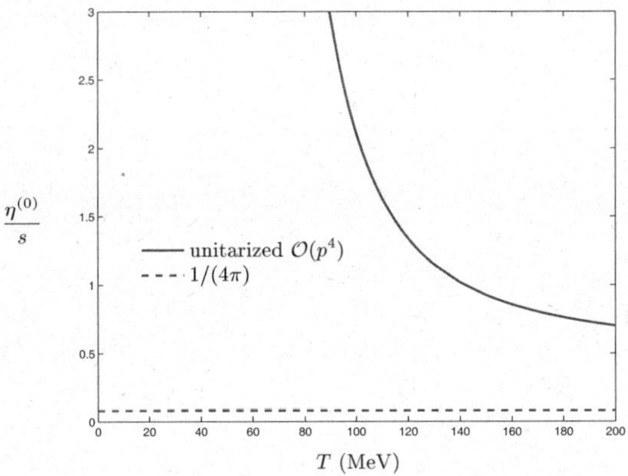

Fig. 6. Dependence of the η/s ratio with temperature in ChPT. The entropy density s is calculated to $\mathcal{O}(p^5)$ [19].

also makes the shear viscosity change its behavior with T. Following the same steps as in [6], we readily get the behavior for $T \ll M_\pi$ as $\eta \simeq 37F_\pi^4\sqrt{T}/M_\pi^{3/2}$ ($F_\pi \simeq 93\,\mathrm{MeV}$ is the pion decay constant). This behavior is consistent with non-relativistic KT [12].

In fig. 5 we compare our leading-order result with [12, 13]. The analysis in [12] is performed for fixed fugacity $z \equiv e^{\beta(\mu_\pi - M_\pi)}$. We work at $\mu_\pi = 0$, so that for the two values of z shown in fig. 5, we must compare at temperatures $T \simeq 30\,\mathrm{MeV}$ (for $z = 0.01$) and $T \simeq 200\,\mathrm{MeV}$ (for $z = 0.5$). We get a reasonable agreement with [13] and a lower η than [12] for $T > M_\pi$. The results in [14], where kaons are also included in the gas, also agree with ours for low and moderate temperatures. Nevertheless, our results should be taken with care, since for the shear viscosity, ladder diagrams with derivative vertices could become important already at $T \simeq 50\,\mathrm{MeV}$.

An interesting quantity is the ratio η/s, where s is the entropy density. It has been conjectured [15] that the value $1/(4\pi)$ is a universal lower bound for this quantity in any physical system. On the other hand, the sound attenuation length $\Gamma_s \equiv 4\eta/(3sT)$ (neglecting bulk viscosity) enters directly in phenomenological effects observed in relativistic heavy-ion collisions, such as elliptic flow or HBT radii [16]. In fig. 6 we plot this ratio for the pion gas. We see that in our analysis η/s decreases monotonously, but it respects the KSS bound for temperatures up to the chiral phase transition $T_c \simeq 180$–$200\,\mathrm{MeV}$. On general grounds, one expects η/s to decrease below T_c and increase logarithmically for very high temperatures [17]. The numerical values we get for $\Gamma_s (\simeq 1.1\,\mathrm{fm}$ at $T = 180\,\mathrm{MeV})$ are in remarkable agreement with those used in [16] and at high T our curve is not far from recent lattice and model estimates giving $\eta/s \sim 0.4$–0.5 above T_c [18].

In conclusion, we have presented a systematic method for evaluating transport coefficients in ChPT. The standard ChPT has to be modified, and as we go to higher temperatures, more diagrams have to be taken into account and eventually resummed. Even though the limitations of our approach are important, we get a reasonable agreement for physical quantities such as the very low-p_T photon spectrum or the viscosity to entropy ratio. We are currently analyzing the high-temperature diagrams, as well as the inclusion of kaons and other transport coefficients such as bulk viscosity and thermal conductivity. Detailed results will be reported elsewhere.

We are grateful to A. Dobado and F. Llanes-Estrada for their useful comments. We also acknowledge financial support from the Spanish research projects FPA2004-02602, PR27/05-13955-BSCH, FPA2005-02327 and FPI fellowship BES-2005-6726.

References

1. J. Gasser, H. Leutwyler, Ann. Phys. (N.Y.) **158**, 142 (1984).
2. M. Le Bellac, *Thermal Field Theory* (Cambridge University Press, 2000).
3. S. Jeon, Phys. Rev. D **52**, 3591 (1995).
4. M.A. Valle Basagoiti, Phys. Rev. D **66**, 045005 (2002).
5. J.L. Goity, H. Leutwyler, Phys. Lett. B **228**, 517 (2002).
6. D. Fernández-Fraile, A. Gómez Nicola, Phys. Rev. D **73**, 045025 (2006).
7. A. Gómez Nicola, J.R. Pelaez, Phys. Rev. D **65**, 054009 (2002).
8. A. Dobado, A. Gómez Nicola, F.J. Llanes-Estrada, J.R. Pelaez, Phys. Rev. C **66**, 055201 (2002).
9. P. Arnold, G.D. Moore, L.G. Yaffe, JHEP **0011**, 001 (2000).
10. WA98 Collaboration (M.M. Aggarwal *et al.*), Phys. Rev. Lett. **93**, 022301 (2004).
11. S. Turbide, R. Rapp, C. Gale, Phys. Rev. C **69**, 014903 (2004); W. Liu, R. Rapp, nucl-th/0604031.
12. A. Dobado, F.J. Llanes-Estrada, Phys. Rev. D **69**, 116004 (2004).
13. D. Davesne, Phys. Rev. C **53**, 3069 (1996).
14. M. Prakash, M. Prakash, R. Venugopalan, G.M. Welke, Phys. Rev. Lett. **70**, 1228 (1993).
15. P. Kovtun, D.T. Son, A.O. Starinets, Phys. Rev. Lett. **94**, 111601 (2005).
16. D. Teaney, Phys. Rev. D **68**, 034913 (2003).
17. L.P. Csernai, J.I. Kapusta, L.D. McLerran, Phys. Rev. Lett. **97**, 152303 (2006).
18. A. Nakamura, S. Sakai, Phys. Rev. Lett. **94**, 72305 (2005); B.A. Gelman, E.V. Shuryak, I. Zahed, Phys. Rev. C. **74**, 044908 (2006).
19. P. Gerber, H. Leutwyler, Nucl. Phys. B **321**, 387 (1989).

Eur. Phys. J. A **31**, 851–853 (2007)
DOI 10.1140/epja/i2006-10259-5

Special Article – QNP 2006

Scaling properties of fluctuation results from the PHENIX experiment at RHIC

J.T. Mitchell[a]

For the PHENIX Collaboration
Brookhaven National Laboratory, P.O. Box 5000, Building 510C, Upton, NY 11973-5000, USA

Received: 23 November 2006
Published online: 22 March 2007 – © Società Italiana di Fisica / Springer-Verlag 2007

Abstract. The PHENIX experiment at the Relativistic Heavy Ion Collider (RHIC) has made measurements of event-by-event fluctuations in the charged-particle multiplicity as a function of collision energy, centrality, collision species, and transverse momentum in several heavy-ion collision systems. It is observed that the fluctuations in terms of σ^2/μ^2 exhibit a universal power law scaling as a function of $N_{participants}$ that is independent of the transverse momentum range of the measurement.

PACS. 25.75.-q Relativistic heavy-ion collisions – 24.60.Ky Fluctuation phenomena

1 Introduction

The topic of event-by-event fluctuations of the inclusive charged particle multiplicity in relativistic heavy-ion collisions has been revived by the observation of non-monotonic behavior in the scaled variance as a function of system size at SPS energies [1]. The scaled variance is defined as σ^2/μ, where σ^2 represents the variance of the multiplicity distribution in a given centrality bin, and μ is the mean of the distribution. For reference, the scaled variance of a Poisson distribution is 1.0, independent of N. PHENIX has studied the behavior of inclusive charged-particle multiplicity fluctuations as a function of centrality and transverse momentum in $\sqrt{s_{NN}} = 62\,\text{GeV}$ and $200\,\text{GeV}$ Au+Au and Cu+Cu collisions in order to investigate whether the non-monotonic behavior persists at RHIC energies.

Details about the PHENIX experimental configuration can be found elsewhere [2]. All of the measurements described here utilized the PHENIX central-arm detectors. The maximum PHENIX acceptance of $|\eta| < 0.35$ in pseudorapidity and 180° in azimuthal angle is considered small for event-by-event measurements. However, the event-by-event multiplicities are high enough in RHIC heavy-ion collisions that PHENIX has a competitive sensitivity for the detection of many fluctuation signals. For example, a detailed examination of the PHENIX sensitivity to temperature fluctuations derived from the measurement of event-by-event mean p_T fluctuations is described in [3].

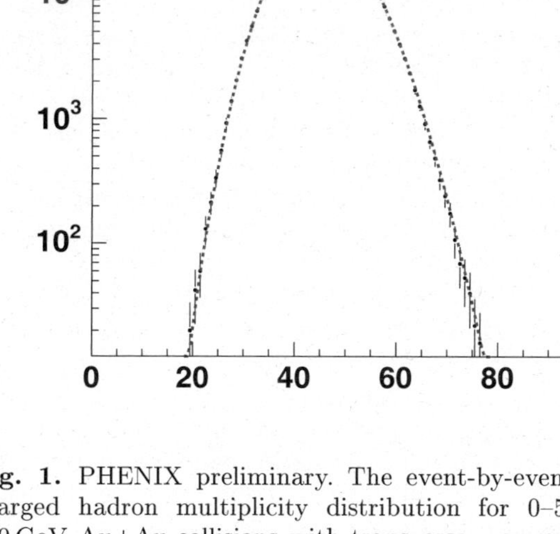

Fig. 1. PHENIX preliminary. The event-by-event inclusive charged hadron multiplicity distribution for 0–5% central 200 GeV Au+Au collisions with transverse momentum in the range $0.2 < p_T < 2.0\,\text{GeV}/c$. The dashed line is a Negative Binomial Distribution fit to the data.

2 Data analysis

It has been demonstrated that charged-particle multiplicity fluctuation distributions in elementary and heavy-ion collisions are well described by negative binomial distributions (NBD) [4]. The NBD of an integer m is defined by

$$P(m) = \frac{(m + k - 1)!}{m!(k - 1)!} \frac{(\mu/k)^m}{(1 + \mu/k)^{m+k}}, \qquad (1)$$

---
[a] e-mail: `mitchell@bnl.gov`

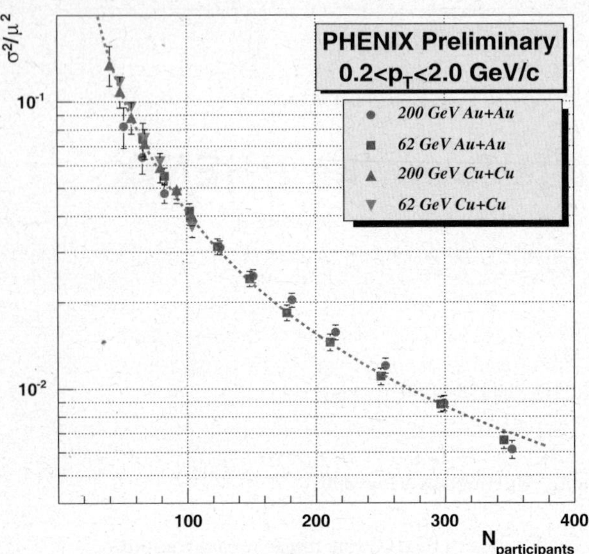

Fig. 2. Multiplicity fluctuations for inclusive charged hadrons in the transverse momentum range $0.2 < p_T < 2.0\,\text{GeV}/c$ in terms of σ^2/μ^2 as a function of $N_{participants}$ for 200 GeV Au+Au, 62 GeV Au+Au, 200 GeV Cu+Cu, and 62 GeV Cu+Cu collisions. The data have been scaled by factors of 1.0, 0.75, 1.35, and 0.75, respectively, in order to emphasize the universal scaling of all species. The dashed curve is a power law fit as described in the text.

where $P(m)$ is normalized for $0 \le m \le \infty$, $\mu \equiv \langle m \rangle$. The NBD contains an additional parameter, k, when compared to a Poisson distribution. The NBD becomes a Poisson distribution in the limit $k \to \infty$. The variance and the mean of the NBD is related to k by $1/k = \sigma^2/\mu^2 - 1/\mu$. The PHENIX multiplicity distributions are well described by NBD fits for all species, centralities, and transverse momentum ranges. The data presented here are results of NBD fits of the multiplicity distributions, an example of which is shown in fig. 1.

Each 5% wide centrality bin selects a range of impact parameters. This introduces a component to the multiplicity fluctuations that can be attributed to fluctuations in the geometry of the collision. It is desireable to estimate and remove this known source of fluctuations so that only fluctuations due to the dynamics of the collision remain. The contribution of geometrical fluctuations is estimated using the HIJING event generator [5], which well reproduces the mean multiplicity of RHIC collisions [6]. The estimate is performed by comparing fluctuations from events with a fixed impact parameter to fluctuations from events with a range of impact parameters covering the width of each centrality bin. The HIJING estimates are confirmed by comparing the HIJING fixed/ranged fluctuation ratios to measured 1%/5% bin width fluctuations. A 15% systematic error for this estimate is included in the errors shown.

By measuring the scaled variance in successively wider azimuthal ranges, a linear dependence on azimuthal acceptance is observed. In order to facilitate direct compar-

isons with other experiments, the multiplicity fluctuations quoted here have been linearly extrapolated to 2π acceptance by fitting the azimuthal dependance within the detector acceptance. Systematic errors due to the extrapolation have been included in the total errors shown.

3 Results

In the Grand Canonical Ensemble (GCE), the scaled variance of the particle number normalized by the mean can be directly related to the compressibility, $\sigma^2/\mu^2 = k_B(T/V)k_T$, where k_B is Boltzmann's constant, T is the system temperature, and V is its volume [7]. The multiplicity fluctuations in terms of σ^2/μ^2 are shown in fig. 2 as a function of centrality for 200 and 62 GeV Au+Au and Cu+Cu collisions. In order to demonstrate their scaling properties as a function of centrality, each curve has been scaled uniformly as a function of centrality to best correspond to the 200 GeV Au+Au curve. The 62 GeV Au+Au, 200 GeV Cu+Cu, and 62 GeV Cu+Cu data have been scaled by factors of 0.75, 1.35, and 0.75, respectively. All four datasets exhibit identical scaling as a function of $N_{participants}$.

It is expected that the compressibility diverges as one approaches the critical point, and the rate of divergence is described by a power law, $k_T = A((T - T_C)/T_C)^{-\gamma}$, where T_C is the value of the temperature at the critical point, and γ is the critical exponent for isothermal compressibility [7]. For illustration, the dashed curve on fig. 2 is a fit to the critical exponent power law function with

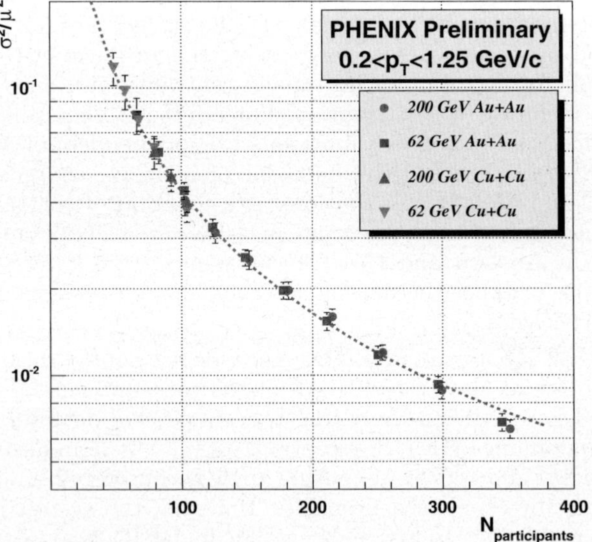

Fig. 3. Multiplicity fluctuations for inclusive charged hadrons in the transverse momentum range $0.2 < p_T < 1.25\,\text{GeV}/c$ in terms of σ^2/μ^2 as a function of $N_{participants}$ for 200 GeV Au+Au, 62 GeV Au+Au, 200 GeV Cu+Cu, and 62 GeV Cu+Cu collisions. The data have been scaled by factors of 1.0, 0.75, 1.35, and 0.75, respectively, in order to emphasize the universal scaling of all species. The dashed curve is a power law fit as described in the text.

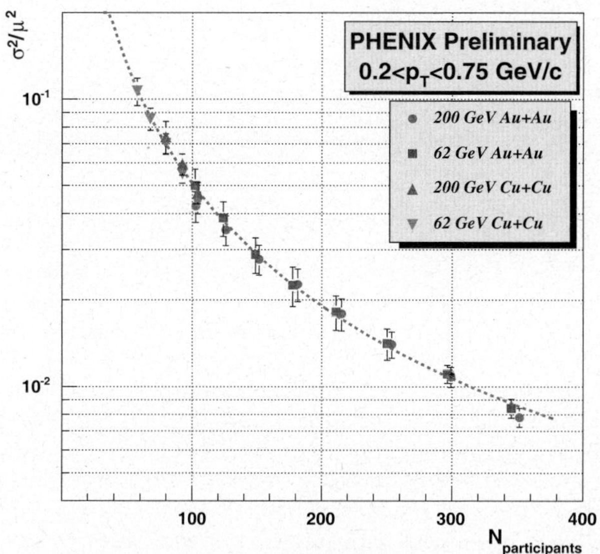

Fig. 4. Multiplicity fluctuations for inclusive charged hadrons in the transverse-momentum range $0.2 < p_T < 0.75\,\mathrm{GeV}/c$ in terms of σ^2/μ^2 as a function of $N_{participants}$ for $200\,\mathrm{GeV}$ Au+Au, $62\,\mathrm{GeV}$ Au+Au, $200\,\mathrm{GeV}$ Cu+Cu, and $62\,\mathrm{GeV}$ Cu+Cu collisions. The data have been scaled by factors of 1.0, 0.75, 1.35, and 0.75, respectively, in order to emphasize the universal scaling of all species. The dashed curve is a power law fit as described in the text.

$V \propto N_{participants}$, $T \propto N_{part}^{1/3}$, and $N_{part,c} \to 0$. With these assumptions, the critical exponent $\gamma = 1.09 \pm 0.06$. If QCD behaves like the 3D Ising model, as suggested in [8], one expects $\gamma = 1.0$. Note that the onset of critical behavior is only one possible explanation for the observed scaling properties in the multiplicity distributions, and this interpretation is contingent on the assumption that the system temperature varies as a function of centrality.

Figures 3 and 4 show the multiplicity fluctuations as a function of centrality for different p_T ranges. In each case, the scaling properties remain unchanged. Hence the influence of p_T-dependent processes at higher p_T, such as hard scattering have little effect on the scaling properties. An analysis of $22\,\mathrm{GeV}$ Cu+Cu collisions in the PHENIX detector is currently underway in order to determine if the scaling properties persist at lower collision energy.

References

1. NA49 Collaboration (M. Rybczynski *et al.*), Preprint, nucl-ex/0409009 (2004).
2. PHENIX Collaboration (K. Adcox *et al.*), Nucl. Instrum. Methods A **499**, 469 (2003).
3. PHENIX Collaboration (K. Adcox *et al.*), Phys. Rev. C **66**, 024901 (2002).
4. E-802 Collaboration (T. Abbott *et al.*), Phys. Rev. C **52**, 2663 (1995).
5. X. Wang, M. Gyulassy, Phys. Rev. D **44**, 3501 (1991).
6. PHENIX Collaboration (S. Adler *et al.*), Phys. Rev. C **71**, 034908 (2005).
7. H. Stanley, *Introduction to Phase Transitions and Critical Phenomena* (Oxford University Press, Inc., 1971).
8. M. Stephanov *et al.*, Phys. Rev. D **60**, 114028 (1999).

Eur. Phys. J. A **31**, 854–857 (2007)

DOI 10.1140/epja/i2006-10280-8

THE EUROPEAN
PHYSICAL JOURNAL A

Special Article – QNP 2006

Freeze-out of the expanding system

V.K. Magas[1,a], L.P. Csernai[2,3], and E. Molnár[2]

[1] Departament d'Estructura i Constituents de la Matéria, Universitat de Barcelona, Diagonal 647, 08028 Barcelona, Spain
[2] Section for Theoretical and Computational Physics, University of Bergen, Allegaten 55, 5007 Bergen, Norway
[3] MTA-KFKI, Research Institute of Particle and Nuclear Physics, H-1525 Budapest 114, P.O. Box 49, Hungary

Received: 18 December 2006
Published online: 20 March 2007 – © Società Italiana di Fisica / Springer-Verlag 2007

Abstract. The freeze-out (FO) of the expanding systems, created in relativistic heavy-ion collisions, is discussed. We start with kinetic FO model, which realizes complete physical FO in a layer of given thickness, and then combine our gradual FO equations with Bjorken-type system expansion into a unified model. We shall see that the basic FO features, pointed out in the earlier works, are not smeared out by the expansion.

PACS. 51.10.+y Kinetic and transport theory of gases – 24.10.Nz Hydrodynamic models – 25.75.-q Relativistic heavy-ion collisions

At the highest energies available nowadays in relativistic heavy-ion collisions at RHIC the total number of the produced particles exceeds 6000, therefore one can expect that the produced system behaves as a "matter" and generates collective effects. Indeed strong collective flow patterns have been measured at RHIC, which suggests that the hydrodynamical models are well justified during the intermediate stages of the reaction: from the time when local equilibrium is reached until the freeze-out (FO), when the hydrodynamical description breaks down. During this FO stage, the matter becomes so dilute and cold that particles stop interacting and stream towards the detectors freely, their momentum distribution freezes out. The FO stage is essentially the last part of a collision process and the main source for observables.

Nowadays, FO is usually simulated in two extreme ways: A) FO on a hypersurface with zero thickness, B) FO described by a volume emission model or hadron cascade, which in principle requires an infinite time and space for a complete FO. At first glance it seems that one can avoid troubles with FO modeling using a hydro+cascade two-module model [1], since in hadron cascades gradual FO is realized automatically. However, in a such a scenario there is an uncertain point, actually uncertain hypersurface, where one switchs from hydrodynamical to kinetic modeling. First of all it is not clear how to determine such a hypersurface. This hypersurface in general may have both time-like and space-like parts. Mathematically this problem is very similar to hydro-to-FO phase transition on the infinitely narrow FO hypersurface, therefore for example all the problems discussed for FO on the hypersurface

with space-like normal vectors will take place here. Another complication is that while for the post FO domain we have a mixture of non-interacting ideal gases, now for the hadron cascade we should generate distributions for the interacting hadronic gas of all possible species, as a starting point for the further cascade evolution. The volume emission models are based on the kinetic equations [2, 3] defining the evolution of the distribution functions, and therefore these also require to generate initial distribution functions for the interacting hadronic species on some hypersurface.

In this work we present a FO model which allows us to study FO in a layer of any thichness, L, from 0 to ∞, and which connects the pre-FO hydrodynamical quantities, like energy density, e, baryon density, n, with the post-FO distribution function in a relatively simple way. Many building blocks of the model are Lorentz invariant and can be applied to both time-like and space-like FO layers. In this work we are going to include a Bjorken-like expansion in our FO model, in contrast to the older versions [3–7]. In this latter case the FO layer is a domain restricted by two hypersurfaces $\tau = \tau_1$ and $\tau = \tau_1 + L$ (τ is the proper time).

We are going to review briefly all the steps done to derive our model. We will skip all the detailed derivations, refering to the corresponding publication, and will show and discuss only a small part of results, due to the limited space.

Starting from the Boltzmann Transport Equation, introducing two components of the distribution function, f, the interacting, f^i, and the frozen-out, f^f, ones, ($f = f^i + f^f$), and assuming that FO is a directed process (i.e. neglecting the gradients of the distribution functions in

<hr>

[a] e-mail: vladimir@ecm.ub.es

the directions perpendicular to the FO direction with respect to that in the FO direction) we can obtain the following system of equations [6,8]:

$$\frac{df^i}{ds} = -\frac{P_{esc}}{\tau_{FO}}f^i + \frac{f_{eq}(s) - f^i}{\tau_{th}}, \quad \frac{df^f}{ds} = \frac{P_{esc}}{\tau_{FO}}f^i. \quad (1)$$

The FO direction is defined by the unit vector $d\sigma_\mu$. FO happens in a layer of given thickness L with two parallel boundary hypersurfaces perpendicular to $d\sigma_\mu$, and $s = d\sigma_\mu x^\mu$ is a variable in the FO direction. We work in the reference frame of the front, where $d\sigma_\mu$ is either $(1,0,0,0)$ for the time-like FO, or $(0,1,0,0)$ for the space-like FO. The τ_{FO} is some characteristic length scale, like mean free path or mean collision time for time-like FO. The rethermalization of the interacting component is taken into account via the relaxation time approximation, where f_i approaches the equilibrated Jüttner distribution, $f_{eq}(s)$, with a relaxation length, τ_{th}. The system (1) can be solved semianalytically in the fast rethermalization limit [6].

According to the above references, the basis of the model, *i.e.* the invariant escape probability within the FO layer of thickness L, for both time-like and space-like normal vectors is given as [6,7,9]

$$P_{esc} = \left(\frac{L}{L - x^\mu d\sigma_\mu}\right)\left(\frac{p^\mu d\sigma_\mu}{p^\mu u_\mu}\right)\Theta(p^\mu d\sigma_\mu), \quad (2)$$

where p^μ is the particle four-momentum, u^μ is the flow velocity. In fact the model based on the escape rate (2) is a generalization of simple kinetic models studied in refs. [3–5], which can be restored in the $L \to \infty$ limit. Here we will concentrate on the time-like case only, where the above Θ-function is unity.

Simple semianalytically solvable FO models studied in [3–7] are missing an important ingredient —the expansion of the freezing-out system. The open question is whether the features of the FO, found in those papers, will survive if the system expasion is included. In this work we present a model which includes both gradual FO and Bjorken-like expansion of the system. And we will see that the answer is "yes" —the basic features of the post FO distribuitions will not be smeared out by the expansion.

First, let us remind the reader the basics of the famous Bjorken model. Bjorken model is one-dimensional in the same sense as discussed before eq. (1) —only the proper time, $\tau = \sqrt{t^2 - x^2}$, gradients are considered. Here the reference frame of the front, $d\sigma^\mu = (1,0,0,0)$, is the same as the local rest frame, $u^\mu = (1,0,0,0)$. The evolution of the energy density and baryon density is given by the following equations:

$$\frac{de}{d\tau} = -\frac{e + P}{\tau}, \quad \frac{dn}{d\tau} = -\frac{n}{\tau}, \quad (3)$$

where P is the pressure. The initial conditions are given at some $\tau = \tau_0$.

Applying our FO model to such a system, we obtain

$$df^i(\tau') = -\frac{d\tau'}{\tau_{FO}}\frac{L}{L - \tau'}f^i(\tau') + \frac{d\tau'}{\tau_{th}}\left[f_{eq}(\tau') - f^i(\tau')\right], \quad (4)$$

$$df^f(\tau') = +\frac{d\tau'}{\tau_{FO}}\frac{L}{L - \tau'}f^i(\tau'), \quad (5)$$

where FO begins at $\tau = \tau_1$ and $\tau' = \tau - \tau_1$. Taking the fast rethermalization limit, similarly to what is done in [5], we can obtain simplified equations for f^i, which is a thermal distribution $f^i(\tau) = f_{eq}(\tau)$, f^f as well as for e^i, n^i and e^f, n^f:

$$\frac{de^i}{d\tau'} = -\frac{e^i}{\tau_{FO}}\frac{L}{L - \tau'}, \quad \frac{dn^i}{d\tau'} = -\frac{n^i}{\tau_{FO}}\frac{L}{L - \tau'}, \quad (6)$$

$$\frac{de^f}{d\tau'} = +\frac{e^i}{\tau_{FO}}\frac{L}{L - \tau'}, \quad \frac{dn^f}{d\tau'} = +\frac{n^i}{\tau_{FO}}\frac{L}{L - \tau'}. \quad (7)$$

Now the idea is to create a system of equations which could describe a fireball which simultaneously expands and freezes out. Let us put our two components ($e = e^i + e^f$) into the first equation of (3) and do some simple algebra:

$$\frac{de^i}{d\tau} + \frac{de^f}{d\tau} = -\frac{e^i + P^i}{\tau} - \frac{e^f}{\tau} - \frac{e^i}{\tau_{FO}}\frac{L}{L - \tau'} + \frac{e^i}{\tau_{FO}}\frac{L}{L - \tau'}, \quad (8)$$

where last two terms add up to zero; the free component, of course, has no pressure. So far our eq. (8) is completely identical to the first equation of (3). Our assumption is that our system evolves in such a way that eq. (8) is satisfied as a system of two separate equations for interacting and free components [10]:

$$\frac{de^i}{d\tau} = -\frac{e^i + P^i}{\tau} - \frac{e^i}{\tau_{FO}}\frac{L}{L + \tau_1 - \tau}, \quad (9)$$

$$\frac{de^f}{d\tau} = -\frac{e^f}{\tau} + \frac{e^i}{\tau_{FO}}\frac{L}{L + \tau_1 - \tau}. \quad (10)$$

Similarly we can obtain equations for baryon density [10]:

$$\frac{dn^i}{d\tau} = -\frac{n^i}{\tau} - \frac{n^i}{\tau_{FO}}\frac{L}{L + \tau_1 - \tau}, \quad (11)$$

$$\frac{dn^f}{d\tau} = -\frac{n^f}{\tau} + \frac{n^i}{\tau_{FO}}\frac{L}{L + \tau_1 - \tau}. \quad (12)$$

Thus, finally, we have the following simple model of fireball created in relativistic heavy-ion collision.

Initial state, $\tau = \tau_0$: $e(\tau_0) = e_0$, $n(\tau_0) = n_0$.

Phase I, pure Bjorken hydrodynamics, $\tau_0 \leq \tau \leq \tau_1$,

$$e(\tau) = e_0\left(\frac{\tau_0}{\tau}\right)^{1+c_o^2}, \quad n(\tau) = n_0\left(\frac{\tau_0}{\tau}\right), \quad (13)$$

where $P = c_o^2 e$ —equation of state (EoS) in general form.

Phase II, Bjorken expansion and gradual FO, $\tau_1 \leq \tau \leq \tau_1 + L$,

$$e^i(\tau) = e_0\left(\frac{\tau_0}{\tau}\right)^{1+c_o^2}\left(\frac{L + \tau_1 - \tau}{L}\right)^{L/\tau_{FO}}, \quad (14)$$

$$n^i(\tau) = n_0\left(\frac{\tau_0}{\tau}\right)\left(\frac{L + \tau_1 - \tau}{L}\right)^{L/\tau_{FO}}. \quad (15)$$

With these last equations we have completely determined the evolution of the interacting component [10]. Knowing

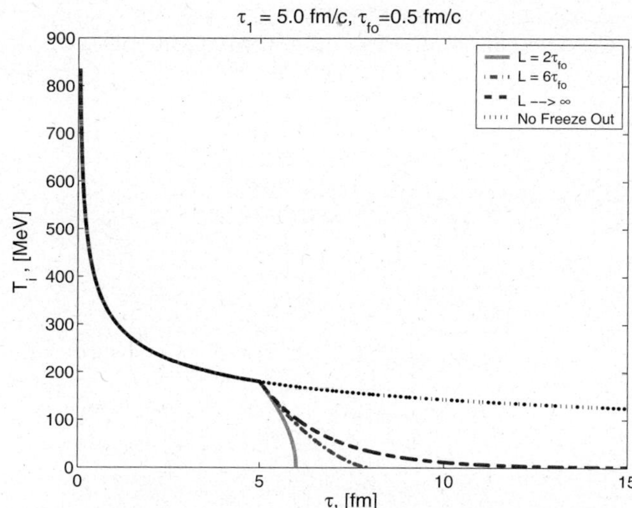

Fig. 1. Evolution of the temperature of the interacting matter for different FO layers. $T_i(\tau_0 = 0.05\,\text{fm}) = 835\,\text{MeV}$, $T_{FO} = 180\,\text{MeV}$. "No Freeze-Out" means that we used standard Bjorken hydrodynamics even in phase II.

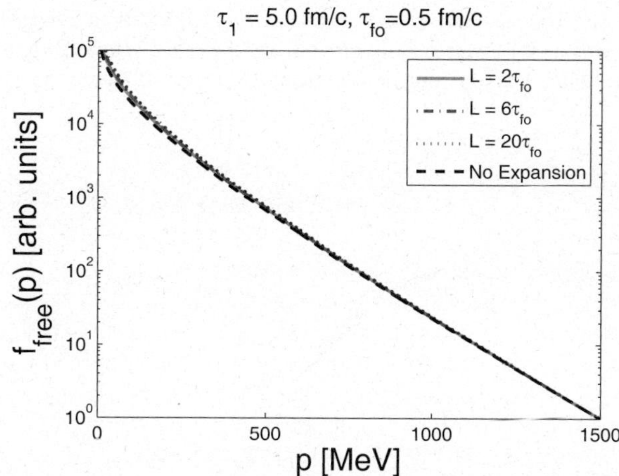

Fig. 2. Final post-FO distribution for different FO layers as a function of the momentum in the FO direction, $p = p^x$ in our case ($p^y = p^z = 0$). The initial conditions are specified in the text. The "No Expansion" curve is given by the analytical expression [5]: $f^f(p) = -\frac{4}{(2\pi)^3} Ei(-\frac{p}{T_{FO}})$.

$e^i(\tau)$ and EoS, we can find the temperature, $T_i(\tau)$. Due to the symmetry of the system $u_i^\mu(\tau) = u^\mu(\tau_0) = (1,0,0,0)$. Finally, $f^i(\tau)$ is a thermal distribution with given $T_i(\tau)$, $n^i(\tau)$, $u_i^\mu(\tau)$.

However, the most interesting for us is the free component, which is the source of the observables. Equations (9), (10) give us the evolution of the e_f and n_f, and one can easily check that these two equations are equivalent to the following equation for the distribution function:

$$\frac{\mathrm{d}f^f}{\mathrm{d}\tau} = -\frac{f^f}{\tau} + \frac{f^i}{\tau_{FO}} \frac{L}{L + \tau_1 - \tau}. \qquad (16)$$

The measured post-FO spectra are given by $f^f(L + \tau_1)$.

Aiming for a qualitative illustration of the FO process, we show below the results for the massless ideal gas without conserved charges with Jüttner equilibrated distribution ($P^i = e^i/3$, $e^i = \frac{3}{\pi^2}T_i^4$, $f^i(\tau, |\boldsymbol{p}|) = \frac{1}{(2\pi)^3}e^{-|\boldsymbol{p}|/T_i(\tau)}$). We have taken the following values of the parameters: $\tau_0 = 0.05\,\text{fm}$, $T_i(\tau_0) = 835\,\text{MeV}$; $\tau_1 = 5\,\text{fm}$, $T_i(\tau_1) = T_{FO} = 180\,\text{MeV}$; $\tau_{FO} = 0.5\,\text{fm}$ and we present results for different values of FO time L.

Figure 1 shows the evolution of the temperature of the interacting matter. As was already shown in [5,7] the final post-FO particle distributions are non-equilibrated distributions, which deviate from thermal ones particularly in the low-momentum region. By introducing and varying the thickness of the FO layer, L, we are strongly affecting the evolution of the interacting component, see fig. 1, but we again see the universality of the final post-FO distribution: for $L > 2\tau_{FO}$ it already looks very close to that for an infinitely long FO calculations: see fig. 2. The inclusion of the expansion into our consideration does not smear out this very important feature of FO. More results can be found in [10].

In our opinion these results may justify the use of FO hypersurface in hydrodynamical models for heavy-ion collisions, but with a proper non-thermal post FO distributions. If the FO layer is thick enough, say $L > 2\tau_{FO}$, then it does not matter how thick the FO layer was; we do not need to model the FO dynamics in details. Once we have a good parameterization of the post-FO spectrum (still asymmetric, non-thermal), for example the analytical post-FO distribution obtained in ref. [5] (see the caption to fig. 2), then the parameters of this distribution can be found from the conservation laws, as is usually done for sharp FO, with some volume scaling factor to effectively account for the expansion during FO. It is important to always check the non-decreasing entropy condition [10,11] to see whether such a process is physically possible.

References

1. D. Teaney *et al.*, Phys. Rev. Lett. **83**, 4951 (1999); S.A. Bass, A. Dumitru, Phys. Rev. C **61**, 064909 (2000); C. Nonaka, S.A. Bass, Nucl. Phys. A **774**, 873 (2006).
2. F. Grassi *et al.*, Phys. Lett. B **355**, 9 (1995); Z. Phys. C **73**, 153 (1996); Yu.M. Sinyukov *et al.*, Phys. Rev. Lett. **89**, 052301 (2002).
3. V.K. Magas *et al.*, Heavy Ion Phys. **9**, 193 (1999).
4. Cs. Anderlik *et al.*, Phys. Rev. C **59**, 388 (1999); Phys. Lett. B **459**, 33 (1999); V.K. Magas *et al.*, Nucl. Phys. A **661**, 596 (99).
5. V.K. Magas *et al.*, Eur. Phys. J. C **30**, 255 (2003).
6. E. Molnár *et al.*, Phys. Rev. C **70**, 024907 (2006); V.K. Magas *et al.*, nucl-th/0510066.
7. E. Molnár *et al.*, nucl-th/0503048.
8. V.K. Magas *et al.*, Nucl. Phys. A **749**, 202 (2005); L.P. Csernai *et al.*, hep-ph/0406082; Eur. Phys. J. A **25**, 65 (2005).
9. L.P. Csernai *et al.*, hep-ph/0401005; E. Molnár *et al.*, nucl-th/0510062.

10. V.K. Magas, talk at the *International Workshop Critical Point and Onset of Deconfinement, Florence, Italy, July 3-9, 2006*, `http://hep.fi.infn.it/becattini/deco.html`; E. Molnár *et al.*, nucl-th/0702083.

11. Cs. Anderlik *et al.*, Phys. Rev. C **59**, 3309 (1999).

Eur. Phys. J. A **31**, 858–861 (2007)
DOI 10.1140/epja/i2006-10237-y

Special Article – QNP 2006

$Q\bar{Q}$ modes in the Quark-Gluon Plasma

D. Cabrera[a] and R. Rapp

Cyclotron Institute and Physics Department, Texas A&M University, College Station, TX 77843-3366, USA

Received: 8 November 2006
Published online: 15 March 2007 – © Società Italiana di Fisica / Springer-Verlag 2007

Abstract. We study the evolution of heavy quarkonium states with temperature in a Quark-Gluon Plasma (QGP) by evaluating an in-medium $Q\bar{Q}$ T-matrix within a reduced Bethe-Salpeter equation in S- and P-wave channels. The interaction kernel is extracted from finite-temperature QCD lattice calculations of the singlet free energy of a $Q\bar{Q}$ pair. Quarkonium bound states are found to gradually move across the $Q\bar{Q}$ threshold after which they rapidly dissolve in the hot system. We calculate Euclidean-time correlation functions and compare to results from lattice QCD. We also study finite-width effects in the heavy-quark propagators.

PACS. 25.75.Dw Particle and resonance production – 12.38.Gc Lattice QCD calculations – 24.85.+p Quarks, gluons, and QCD in nuclei and nuclear processes – 25.75.Nq Quark deconfinement, quark-gluon plasma production, and phase transitions

1 Introduction

Bound states of heavy (charm and bottom) quarks ($Q = b, c$) are valuable spectroscopic objects in Quantum Chromodynamics (QCD) [1]. When embedded into hot and/or dense matter a large class of medium modifications can be studied, including (Debye-) color-screening of the $Q\bar{Q}$ interaction, dissociation reactions induced by constituents of the medium, and the change in thresholds caused by mass (or width) modifications of open heavy-flavor states. Lattice QCD (lQCD) calculations have made substantial progress in characterizing in-medium quarkonium properties from first principles. In particular, it has been found that ground-state charmonia [2–4] and bottomonia [5] do not dissolve until significantly above the critical temperature, T_c, which has been supported in model calculations based on potentials extracted from lQCD, using either a Schrödinger equation for the bound-state problem [6–9], or a T-matrix approach which additionally accounts for scattering states [10]. More reliable comparisons to lQCD can be performed using (space-like) Euclidean-time correlation functions [11,12], which are readily evaluated in lQCD. The conversion of (time-like) model spectral functions requires a description not only of the bound-state part of the spectrum but also its continuum and threshold properties.

In the present work we evaluate Euclidean correlation functions for charmonium and bottomonium in a T-matrix approach. The basic input consists of in-medium $Q\bar{Q}$ potentials extracted from lQCD, inserted in a scattering

equation to calculate the in-medium $Q\bar{Q}$ T-matrix [10]. This incorporates bound and scattering states on an equal footing (based on the same interaction), and additionally enables a straightforward implementation of in-medium single-particle (quark) properties via self-energy insertions in the two-particle Green's function.

2 Scattering equation and bound states

The T-matrix equation for $Q\bar{Q}$ scattering in the center-of-mass frame and in partial-wave basis reads [10]

$$T_l(E; q', q) = V_l(q', q)$$
$$-\frac{2}{\pi} \int_0^\infty \mathrm{d}k\, k^2\, V_l(q', k)\, G_{\bar{Q}Q}(E; k)\, T_l(E; k, q)\,, \quad (1)$$

which follows from a standard 3-dimensional reduction of the Bethe-Salpeter equation [13]. $G_{\bar{Q}Q}(E; k)$ denotes the intermediate two-particle propagator including quark self-energies (Σ) and Pauli blocking. The T-matrix equation (1) is solved with the Haftel-Tabakin algorithm [14], in which the integral equation is solved by discretizing 3-momentum, $\sum_{k=1}^{N} \mathcal{F}(E)_{ik}\, T(E)_{kj} = V_{ij}$, and subsequent matrix inversion. The zeroes of $\det \mathcal{F}(E)$ for $E < E_{th}$ determine heavy quark-antiquark bound states.

The quark self-energy, Σ, encodes the interactions with (light) quarks and gluons from the heat bath [10]. In the present work we consider a fixed heavy-quark mass m_Q (*i.e.*, $\mathrm{Re}\,\Sigma = 0$) together with a small imaginary part, $\mathrm{Im}\,\Sigma = -0.01\,\mathrm{GeV}$, for numerical purposes. Effects of

[a] e-mail: dcabrera@comp.tamu.edu

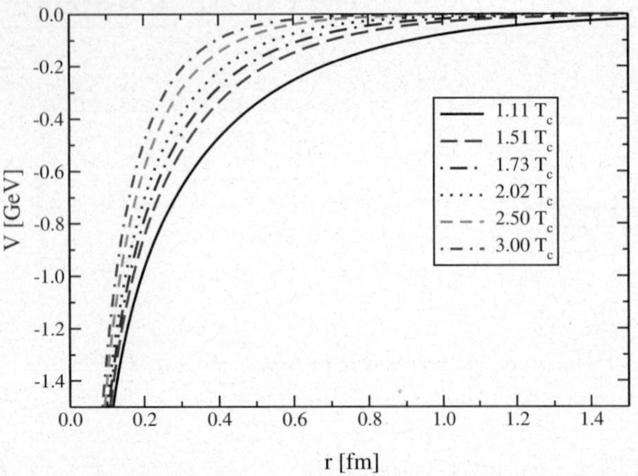

Fig. 1. $Q\bar{Q}$ potential for several temperatures above T_c based on the color-singlet internal energy.

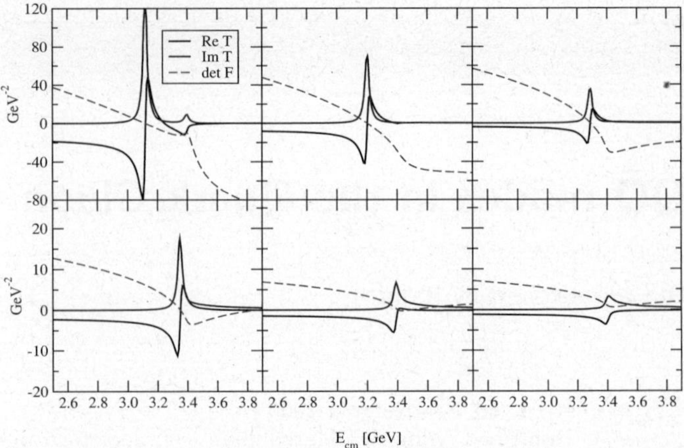

Fig. 2. T-matrix for S-wave $c\bar{c}$ scattering based on the potential in fig. 1. Also shown is $\det\mathcal{F}$ (dashed line, arbitrary units). From left to right (up-down) the temperatures are $(1.1, 1.5, 2.0, 2.5, 3.0, 3.3)\,T_c$.

temperature-dependent heavy-quark masses and widths are investigated in ref. [15].

The kernel of the scattering equation, V, can be estimated from the lQCD heavy-quark free energies, F_1, even though there is still an ongoing discussion on how to properly do it. Here we identify the heavy-quark potential with the color-singlet internal energy, $U_1 = F_1 - T\frac{dF_1}{dT}$, which reproduces ground-state charmonium dissociation temperatures as found in lattice analysis of spectral functions [7–10]. In fig. 1 we show the $Q\bar{Q}$ potential, $V(r,T) = U_1(r,T) - U_1(r \to \infty, T)$, as obtained from a fit to the lQCD color-singlet free-energy data [16], previously employed in [10]. The potential evolves smoothly with temperature, decreasing both in magnitude and range. Different parameterizations of the lattice data imply sizable uncertainties in the potential through the thermal derivative of the free energy. These uncertainties are studied in ref. [15] by alternatively obtaining the heavy-quark potential from a direct fit to the lQCD internal energy data of ref. [17]. $V_l(q',q)$ follows from a Fourier transform and partial-wave expansion.

3 Quarkonium T-matrices in the QGP

We start the calculation of the in-medium $Q\bar{Q}$ T-matrices by fixing the heavy-quark masses so that the corresponding quarkonium ground states approximately agree with their vacuum masses for the lowest considered temperature ($T = 1.1T_c$), i.e., $m_c = 1.7\,\text{GeV}$ and $m_b = 5.1\,\text{GeV}$. Figure 2 summarizes the on-shell S-wave $c\bar{c}$ scattering amplitude as a function of the CM energy, for several temperatures. We do not include the hyperfine (spin-spin) interactions and therefore η_c (η_b) and J/ψ (Υ) states are degenerate. At the lowest temperature, we recover the charmonium ground state at $E \approx 3.10\,\text{GeV}$, whereas the first-excited state (ψ') has just about melted. As the temperature increases, the $J/\psi(1S)$ gradually moves toward threshold and the T-matrix is appreciably reduced. The

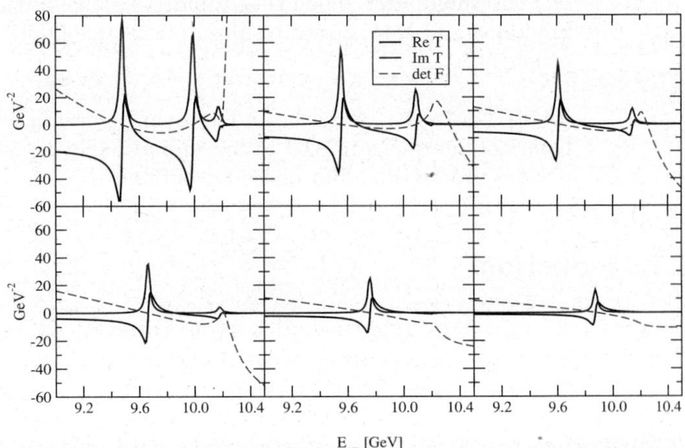

Fig. 3. Same as in fig. 2 but for S-wave $b\bar{b}$ scattering. From left to right and up to down the temperatures are $(1.1, 1.5, 1.8, 2.1, 2.7, 3.5)\,T_c$.

bound state survives up to $\sim 3\,T_c$, where it crosses the $c\bar{c}$ threshold and rapidly melts in the hot system.

The S-wave $b\bar{b}$ T-matrix exhibits two bound states at the lowest temperature ($E \approx 9.45, 9.95\,\text{GeV}$ for $\Upsilon(1S)$, η_b and $\Upsilon(2S)$, η_b', respectively) and the remnant of a third one, which is (almost) melted in the medium, cf. fig. 3. The $\Upsilon(2S)$ moves across the $b\bar{b}$ threshold at $T \approx 2.1T_c$, whereas the $1S$ state survives in the QGP until much higher temperatures, beyond $T \approx 3.5T_c$.

We only find one P-wave bound state for the charm system at the lowest temperature, at $E \approx 3.4\,\text{GeV}$, which we associate with the χ_c. The P-wave $b\bar{b}$ system exhibits two bound states at $1.1T_c$, with energies $E = 9.90, 10.15\,\text{GeV}$, in good agreement with the nominal values for $\chi_b(1P)$ and $\chi_b(2P)$ in the vacuum. The latter moves beyond threshold at $T \approx 1.3T_c$, and the $(1P)$ state at $T \approx 2.3T_c$. Masses and binding energies, $(E_B = E_{th} - M)$, of the P-wave states are summarized in table 1 for several temperatures.

Table 1. Masses and binding energies (in GeV) for P-wave quarkonia as obtained from $\det \mathcal{F}(E) = 0$.

T/T_c	1.1	1.3	1.5	2	2.3
$M[\chi_c(1P)]$	3.4	–	–	–	–
$E_B[\chi_c(1P)]$	≈ 0	–	–	–	–
$M[\chi_b(1P)]$	9.90	9.98	10.04	10.14	10.20
$E_B[\chi_b(1P)]$	0.30	0.22	0.16	0.06	≈ 0
$M[\chi_b(2P)]$	10.15	10.20	–	–	–
$E_B[\chi_b(2P)]$	0.05	≈ 0	–	–	–

4 Spectral functions and Euclidean correlators

The Euclidean-time correlation function is defined as the thermal two-point mesonic correlation function in a mixed Euclidean-time momentum representation (here $\boldsymbol{p} = \boldsymbol{0}$),

$$G(\tau, T) = \int_0^\infty \mathrm{d}\omega\, \sigma(\omega, T)\, \frac{\cosh[\omega(\tau - \beta/2)]}{\sinh(\omega\beta/2)}, \quad (2)$$

where σ is the spectral function obtained from closing external legs of the in-medium T-matrix, schematically as

$$G(E) = \int G_{\bar{Q}Q} + \int G_{\bar{Q}Q} T G_{\bar{Q}Q}, \quad \sigma(E) \propto \mathrm{Im}\, G(E). \quad (3)$$

As expected, the S-wave charmonium spectral function (top panel of fig. 4) reflects the bound states of the

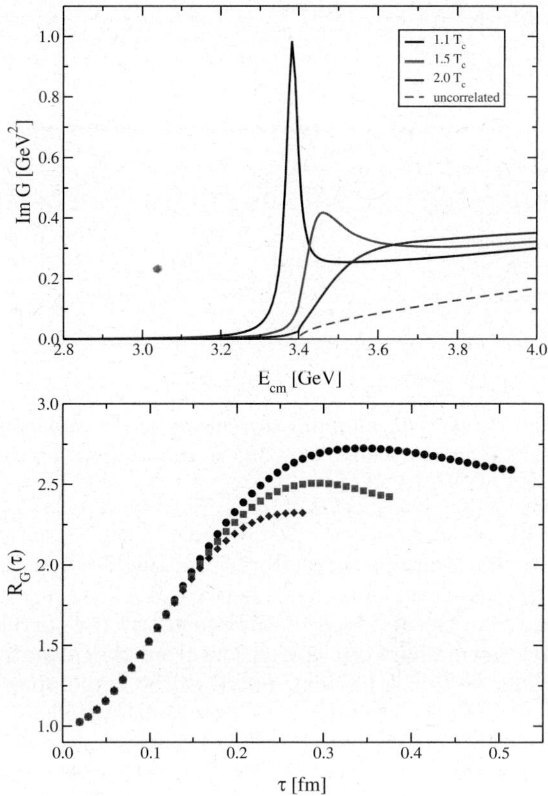

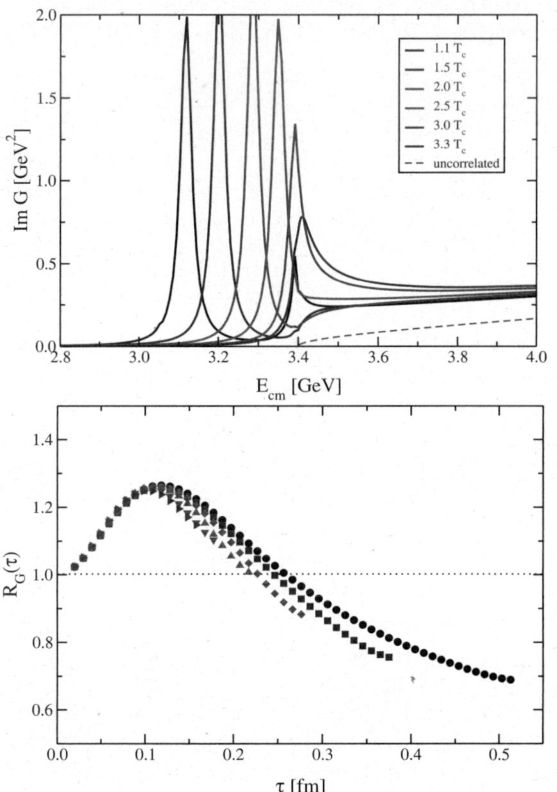

Fig. 4. Top: $c\bar{c}$ spectral function for S-wave scattering at several temperatures. Bottom: normalized correlation function at several temperatures.

Fig. 5. Same as in fig. 4 for $c\bar{c}$ P-wave scattering.

T-matrix, but the (non-perturbative) $c\bar{c}$ rescattering also generates appreciable strength above threshold, where the remnant of the first-excited state ($\psi(2S)$) can be still inferred at $T = 1.1 T_c$. The corresponding correlation function (bottom panel of fig. 4) is normalized to a "reconstructed" correlator represented by a zero-temperature spectral function consisting of a δ-function–like bound-state spectrum and perturbative continuum with onset at $E_{cont} = 4.5\,\mathrm{GeV}$ [9]. The temperature evolution of the correlator is a combined result of a decrease in binding energy of the bound states and the contribution of the non-perturbative continuum. The sizable drop at large τ is in qualitative agreement with lQCD [3]. The latter exhibits somewhat less reduction, leaving room for the effects of a threshold reduction with temperature in our T-matrix.

The P-wave charmonium spectral function (fig. 5, top) exhibits one bound state (χ_c) just below the threshold, which rapidly melts into the continuum accompanied by a sizable threshold enhancement. The normalized correlator steeply rises in the low-τ regime, due to: i) non-perturbative rescattering and ii) a larger threshold in the "reconstructed" correlator. While this is qualitatively consistent with lQCD, the evolution with temperature is opposite, which could improve by a shift of strength to lower energies due to a decreasing heavy-quark mass with temperature. The bottomonium correlation functions follow a similar pattern as for the charmonium system.

It turns out that the τ-dependence of the (normalized) Euclidean correlators is rather sensitive to the "reconstructed" correlator used for normalization. However,

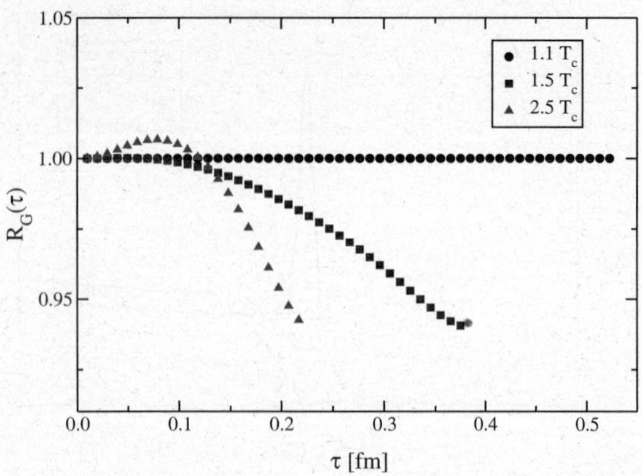

Fig. 6. S-wave charmonium correlators when replacing the "reconstructed" correlator (used for normalization) by the calculated result at $1.1\,T_c$.

S-wave charmonium correlators from lQCD show rather little temperature dependence below $\sim 1.5\,T_c$. Thus, to reduce ambiguities induced by the reconstructed correlator, we have normalized our correlators from the T-matrix to the result at $T = 1.1\,T_c$ calculated in the same approach, cf. fig. 6. The τ-dependence is now substantially weaker than in the bottom panel of fig. 4 (limited to less than 10%), and the agreement with lQCD [3] is much improved.

We have also studied finite width effects for quarkonia by implementing c-quark widths of $0.05\,\mathrm{GeV}$ [18], inducing $\Gamma_\Psi \approx 0.1\,\mathrm{GeV}$. The correlators show variations of the order of a few percent, indicating a small sensitivity to phenomenological (sizable) values of quarkonium widths.

We thank M. Mannarelli and F. Zantow for providing their (fits to) lattice QCD results, and M. Mannarelli and H. van Hees for useful discussions. This work is supported in part by Ministerio de Educación y Ciencia (Spain) via a postdoctoral fellowship and by a U.S. National Science Foundation CAREER Award under Grant No. PHY0449489.

References

1. N. Brambilla *et al.*, arXiv:hep-ph/0412158.
2. M. Asakawa, T. Hatsuda, Phys. Rev. Lett. **92**, 012001 (2004).
3. S. Datta, F. Karsch, P. Petreczky, I. Wetzorke, Phys. Rev. D **69**, 094507 (2004).
4. T. Umeda, K. Nomura, H. Matsufuru, Eur. Phys. J. C **39**, s1, 9 (2005).
5. K. Petrov, A. Jakovac, P. Petreczky, A. Velytsky, PoS (LAT2005) 153 (2006).
6. E.V. Shuryak, I. Zahed, Phys. Rev. C **70**, 021901 (2004).
7. C.Y. Wong, Phys. Rev. C **72**, 034906 (2005).
8. W.M. Alberico, A. Beraudo, A. De Pace, A. Molinari, Phys. Rev. D **72**, 114011 (2005).
9. A. Mocsy, P. Petreczky, Phys. Rev. D **73**, 074007 (2006).
10. M. Mannarelli, R. Rapp, Phys. Rev. C **72**, 064905 (2005).
11. R. Rapp, Eur. Phys. J. A **18**, 459 (2003).
12. A. Mocsy, P. Petreczky, Eur. Phys. J. C **43**, 77 (2005).
13. R. Blankenbecler, R. Sugar, Phys. Rev. **142**, 1051 (1966).
14. M.I. Haftel, F. Tabakin, Nucl. Phys. A **158**, 1 (1970).
15. D. Cabrera, R. Rapp, arXiv:hep-ph/0611134.
16. P. Petreczky, private communcation (2004).
17. O. Kaczmarek, F. Zantow, arXiv:hep-lat/0506019.
18. H. van Hees, R. Rapp, Phys. Rev. C **71**, 034907 (2005).

Eur. Phys. J. A **31**, 862–867 (2007)
DOI 10.1140/epja/i2006-10233-3

Special Article – QNP 2006

Fluctuations and correlations in STAR

B.K. Srivastava[a]

For the STAR Collaboration
Department of Physics, Purdue University, West Lafayette, IN 47907, USA

Received: 8 November 2006
Published online: 21 March 2007 – © Società Italiana di Fisica / Springer-Verlag 2007

Abstract. The study of correlations and fluctuations can provide evidence for the production of the quark-gluon plasma (QGP) in relativistic heavy-ion collisions. Various theories predict that the production of a QGP phase in relativistic heavy-ion collisions could produce significant event-by-event correlations and fluctuations in transverse momentum, multiplicity, etc. Some of the recent results using STAR at RHIC will be presented along with results from other experiments at RHIC. The focus is on forward-backward multiplicity correlations, balance function, charge and transverse-momentum fluctuations, and correlations.

PACS. 25.75.-q Relativistic heavy-ion collisions – 25.75.Gz Particle correlations

1 Introduction

The investigation of high-energy nucleus-nucleus collisions provides a unique tool to study the properties of hot and dense matter. The motivation is drawn from lattice QCD calculations, which predicts a phase transition from hadronic matter to a system of deconfined quarks and gluons (QGP) at high temperature [1]. The study of event-by-event fluctuations provides a novel probe to explore such transition in the search for the QGP. It is now possible with large-acceptance experiments at SPS and RHIC.

2 Forward-backward multiplicity correlations

The measurement of particle correlations has been suggested as a method to search for the existence of a phase transition in ultra-relativistic heavy-ion collisions [2–4]. If the quark-gluon plasma (QGP) is formed in these collisions, the existence or absence of particle correlations could lead to a determination of the presence of partonic degrees of freedom. A linear relationship has been found in high-energy colliding hadron experiments between the multiplicity in a forward η region (N_f) and average multiplicity in a backward η region (N_b) [5,6]:

$$\langle N_b(N_f)\rangle = a + bN_f. \tag{1}$$

The coefficient b is referred to as the correlation coefficient and it can be expressed in terms of the expectation value [5]:

$$b = \frac{\langle N_f N_b\rangle - \langle N_f\rangle\langle N_b\rangle}{\langle N_f^2\rangle - \langle N_f\rangle^2} = \frac{D_{bf}^2}{D_{ff}^2}, \tag{2}$$

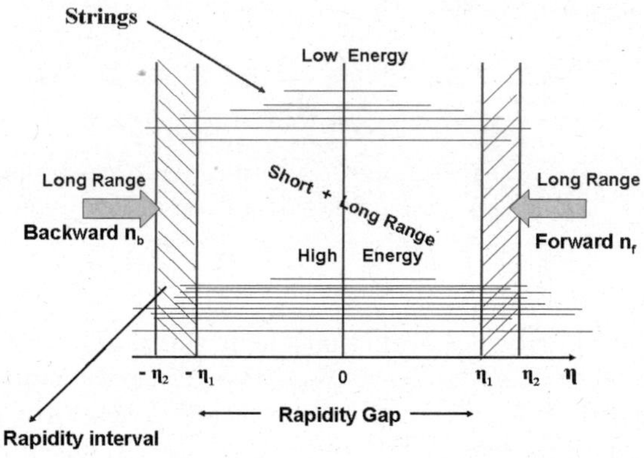

Fig. 1. Pictorial representation of forward-backward correlations in pseudorapidity.

where D_{bf}^2 and D_{ff}^2 are the backward-forward and forward-forward dispersions, respectively. This result is exact and model independent [5]. Short- and long-range (in rapidity) multiplicity correlations are predicted as a signature of string fusion [7,8]. The short-range correlations are confined to midrapidity, while long-range correlations are extended more than 2 units in rapidity. When strings fuse, a reduction in the long-range forward-backward correlation is expected. The existence of long-range multiplicity correlations may indicate the presence of multiple partonic inelastic collisions. These correlations arise from the superposition of a fluctuating number of strings, such that [5]

$$\langle N_f N_b\rangle - \langle N_f\rangle\langle N_b\rangle \propto \left[(\langle n^2\rangle - \langle n\rangle^2)\right]\langle N_{0f}\rangle\langle N_{0b}\rangle \tag{3}$$

[a] e-mail: brijesh@physics.purdue.edu

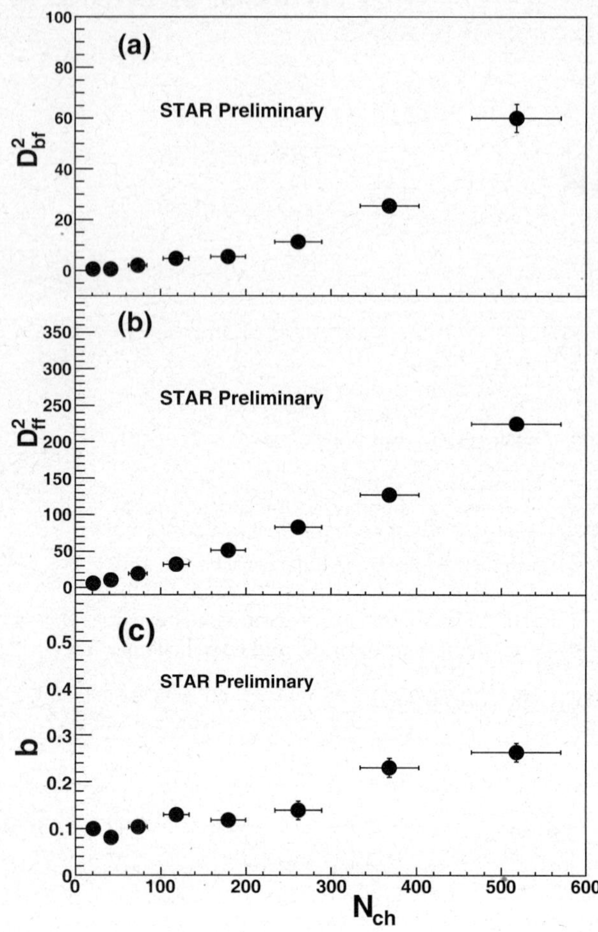

Fig. 2. (a) D^2_{bf}, (b) D^2_{ff}, and (c) b from Au + Au collisions at $\sqrt{s_{NN}} = 200\,\text{GeV}$, as a function of the STAR reference multiplicity, N_{ch}.

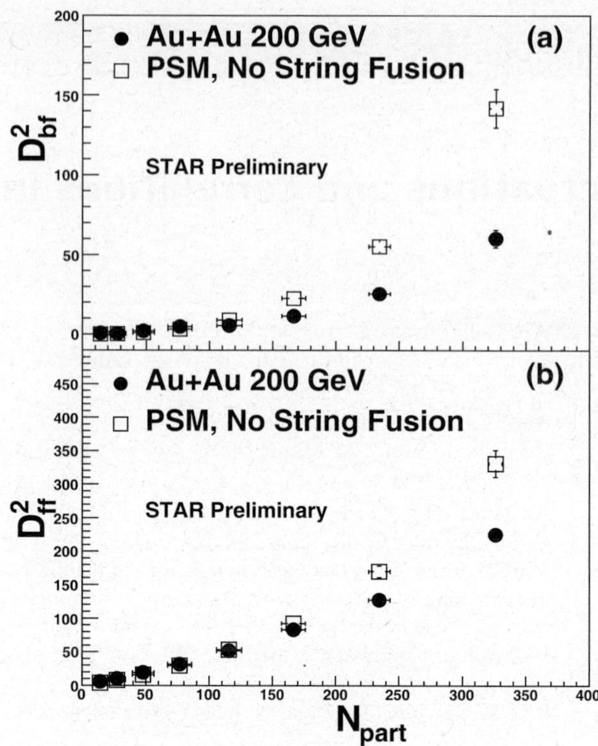

Fig. 3. (a) D^2_{bf} and (b) D^2_{ff} as a function of the N_{part} in 200 GeV Au + Au collisions, compared to the PSM with no string fusion (independent strings).

with $(\langle n^2 \rangle - \langle n \rangle^2)$ the fluctuation in the number of inelastic collisions and $\langle N_{0f} \rangle$, $\langle N_{0b} \rangle$ the average multiplicity produced from a single inelastic collision. Therefore, D^2_{bf} should be sensitive to the presence of long-range multiplicity correlations.

We discuss the result on forward-backward multiplicity correlations from 200 GeV Au + Au collisions for all charged particles with p_T in the range from 0.1 to 1.2 GeV/c, to ensure a sampling of soft particles only. To eliminate short-range correlations, a gap in pseudorapidity (η) of 1.6 units is considered [4]. The forward pseudorapidity interval was $0.8 < \eta < 1.0$ and the backward one was $-1.0 < \eta < -0.8$ [9]. Figure 1 depicts a schematic representation of the forward-backward correlations and their measurements.

Figure 2 shows the results for D^2_{bf}, D^2_{ff}, and the correlation coefficient, b, as a function of N_{ch} for the 8 centrality bins. The presence of long-range multiplicity correlations is evident from D^2_{bf} and b. The growth of D^2_{bf} as a function of N_{ch} is consistent with an increasing long-range correlation from peripheral to central heavy-ion collisions, corresponding to a greater number of fluctuating strings.

The comparison of D^2_{bf} and D^2_{ff} as calculated from the Au + Au data to that from the Parton String Model (PSM) [10,11] with the string fusion on or off is presented in figs. 3 and 4, respectively. The PSM (with two-string fusion) minbias multiplicity distribution closely matches that of the corrected Au + Au data. It also describes the $\langle p_T \rangle$ enhancement, particle ratios, strangeness production, etc.; seen in Au + Au collisions at RHIC [10]. As such, the centrality cuts in the PSM correspond to the same percentage of the minimum bias cross-section as in the data and are plotted as a function of the number of participant nucleons in the collision (N_{part}), calculated from Monte Carlo Glauber model [12]. The statistical and systematic uncertainties in figs. 3 and 4 are the same as those described for fig. 2. Figures 3 and 4 show good agreement in peripheral collisions between data and the independent (no string fusion) or collective (with string fusion) model. In central Au + Au collisions, D^2_{ff} with the independent string description deviates from the data, but shows good agreement for fusion of two soft strings. This confirms the agreement in multiplicity between the PSM (with two-string fusion) and data. However, there is a large discrepancy in D^2_{bf} for central collisions for both the independent and collective PSM, compared to the data. This discrepancy is greater for the case of independent strings (fig. 3). This suggests an additional, dynamical reduction in the number of particle sources in central Au + Au collisions, greater than that provided by the fusion of two soft strings (fig. 4).

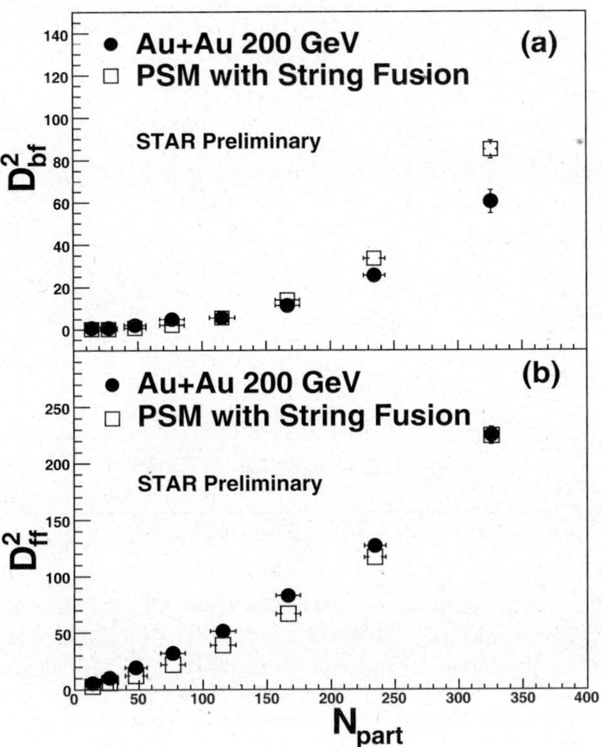

Fig. 4. (a) D^2_{bf} and (b) D^2_{ff} as a function of the N_{part} in 200 GeV Au + Au collisions, compared to the PSM with string fusion (collective strings).

3 Transverse momentum distributions and string percolation

It is postulated that in the collision of two nuclei at high energy, color strings are formed between projectile and target partons. These color strings decay into additional strings via $q - \bar{q}$ production, and ultimately hadronize to produce the observed hadron yields [13]. In the collision process, partons from different nucleons begin to overlap and form clusters in transverse space. The color strings are of radius $r_0 = 0.20$–0.25 fm [13]. The fusion of strings to form clusters is an evolution of the Dual Parton Model (DPM) [6], which utilizes independent strings as particle emitters, to the Parton String Model (PSM), which implements interactions (fusion) between strings [14]. At some point, a cluster will form which spans the entire system. This is referred to as the maximal cluster and marks the onset of the percolation threshold. An overview of percolation theory can be found in the following reference [15]. The quantity ρ, the percolation density parameter, can be used to describe overall cluster density. It can be expressed as

$$\rho = \frac{N \pi r_0^2}{S} \tag{4}$$

with N the number of strings, S the total nuclear overlap area, and πr_0^2 the transverse disc area. At some critical value of $\rho = \rho_c$, the percolation threshold is reached. ρ_c is referred to as the critical percolation density parame-

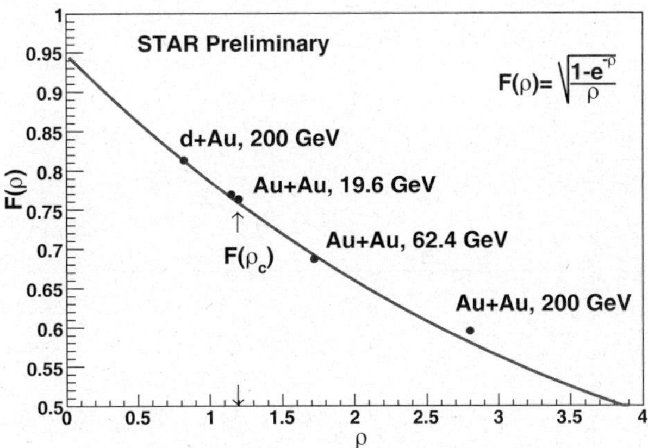

Fig. 5. Multiplicity suppression factor, $F(\rho)$ *versus* the percolation density parameter, ρ. The line is the function $F(\rho)$ and is drawn to guide the eye, not as a fit to the points. The estimated critical percolation density for 2D overlapping discs in the continuum limit, ρ_c, is shown.

ter. In two dimensions, for uniform string density, in the continuum limit, $\rho_c = 1.175$ [16].

To calculate the percolation parameter, ρ, a parameterization of pp events at 200 GeV is used to compute the p_T distribution

$$\frac{dN}{dp_T^2} = \frac{a}{(p_0 + p_T)^n} \tag{5}$$

where a, p_0, and n are parameters fit to the data. This parameterization can be used for nucleus-nucleus collisions if one takes into account the percolation of strings by [13]

$$p_0 \longrightarrow p_0 \left(\frac{\left\langle \frac{nS_1}{S_n} \right\rangle_{\text{Au-Au}}}{\left\langle \frac{nS_1}{S_n} \right\rangle_{pp}} \right)^{\frac{1}{4}}, \tag{6}$$

where S_1 and S_n are the transverse overlap area produced by a single and N number of strings, respectively. In pp collisions at 200 GeV, the quantity $\left\langle \frac{nS_1}{S_n} \right\rangle_{pp} = 1.0 \pm 0.1$, due to low string overlap probability in pp collisions. Once the p_T distribution for nucleus-nucleus collisions is determined, the multiplicity damping factor can be defined in the thermodynamic limit as [17]

$$F(\rho) = \sqrt{\frac{1 - e^{-\rho}}{\rho}} \tag{7}$$

which accounts for the overlapping of discs, with $1 - e^{-\rho}$ corresponding to the fractional area covered by discs.

The percolation density parameter, ρ, has been determined for several collision systems and energies. These results have been compared to the predicted value of the critical percolation density, ρ_c. If ρ_c is exceeded it is expected that the percolation threshold has been reached, indicating the formation of a maximal cluster that spans the system under study. Figure 5 is a plot of the quantity $F(\rho)$ *versus* the percolation density parameter, (ρ),

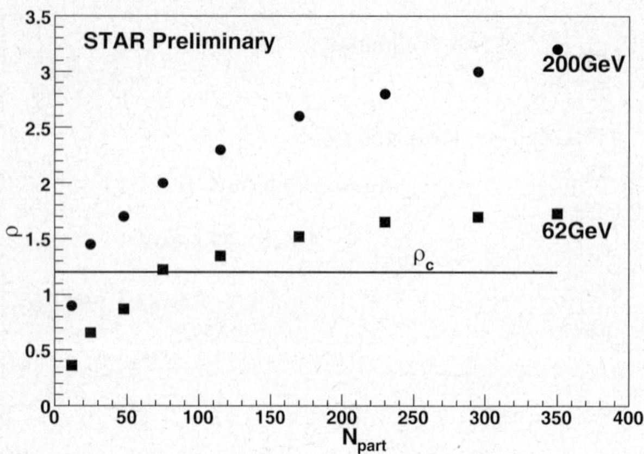

Fig. 6. The percolation density parameter, ρ, as a function of collision centrality (N_{part}) in 62.4 and 200 GeV Au + Au collisions.

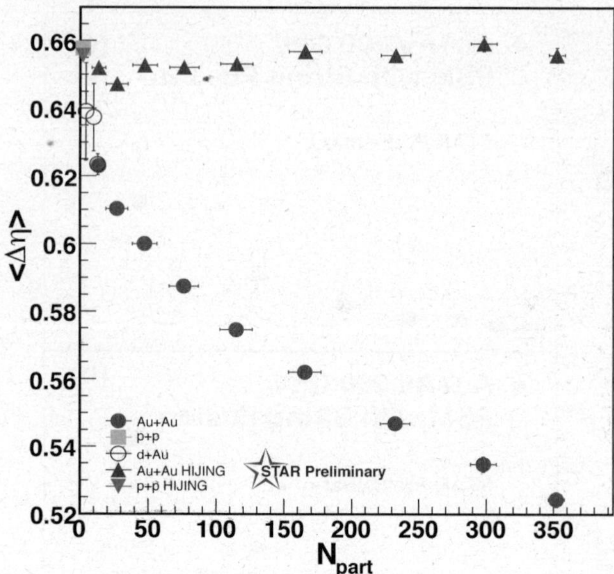

Fig. 7. The balance function widths for $p + p$, $d + Au$ and Au + Au collisions at $\sqrt{(s_{NN})} = 200$ GeV as a function of the number of participating nucleons along with HIJING calculations.

for central collisions. One can also consider the percolation density as a function of centrality in Au + Au collisions. The centrality is expressed in terms of the number of participating nucleons (N_{part}) as found from Monte Carlo Glauber calculations [9]. More central collisions correspond to greater values of N_{part}. Figure 6 shows ρ as a function of the number of participant nucleons in Au + Au collisions at 200 and 62.4 GeV [18]. For almost all collision centralities, 200 GeV Au + Au exceeds the critical percolation threshold, ρ_c. In 62.4 GeV Au + Au, all except the three most peripheral bins exceed ρ_c.

4 Balance function

The balance function is based on the principle that charge is locally conserved when particles are produced in pairs [19, 20]:

$$B(\Delta\eta) = \frac{1}{2}\left\{ \frac{N_{+-}(\Delta\eta) - N_{++}(\Delta\eta)}{N_+} + \frac{N_{-+}(\Delta\eta) - N_{--}(\Delta\eta)}{N_-} \right\}, \quad (8)$$

where N_{+-} is the number of charged pairs in a given pseudorapidity range, similarly for N_{++}, N_{-+} and N_{--}. N_+ (N_-) is the number of positive (negative) charged-particles sum over all the events. $\Delta\eta = |\eta_2 - \eta_1|$ is the width of the balance function. If the system exists in a deconfined phase for an extended time and the charged pairs are produced at hadronization, the pairs will retain more of their correlation in rapidity. The balance function may be sensitive to whether the transition to a hadronic phase was delayed, as expected if the quark-gluon phase were to exist for a longer time. In fig. 7 the balance function widths for $p + p$, $d + Au$ and Au + Au collisions at $\sqrt{(s_{NN})} = 200$ GeV are shown as a function of the number of participants [21]. The balance function widths scale with N_{part}. The widths predicted using HIJING calculations are also shown in fig. 7, along with data, and show

little dependence on centrality and are similar to those measured for pp. The measured $B(\Delta\eta)$ in central Au + Au collisions is consistent with trends of the model incorporating late hadronization.

5 Charge fluctuations

Combined analysis of fluctuations in, *e.g.*, total and net charge for positively and negatively charged particles, as well as their ratios can reveal interesting physics. The quantity related to fluctuations is the net charge fluctuation and is the difference of the number of produced positively and negatively charged particles measured in a fixed rapidity range, defined as [22, 23]

$$\nu_{+-} = \left\langle \left(\frac{N_+}{\langle N_+ \rangle} + \frac{N_-}{\langle N_- \rangle} \right)^2 \right\rangle, \quad (9)$$

where N_+ and N_- are multiplicities of positive and negative particles. The magnitude of the variance, ν_{+-}, is determined both by statistical and dynamical fluctuations, $\nu_{+-} = \nu_{+-,stat} + \nu_{+-,dyn}$. Details of net charge fluctuations analysis can be found in ref. [24]. In fig. 8 $\nu_{+-,dyn}$ is shown for Au + Au collisions at 20, 130, and 200 GeV along with the pp data for all charged particles with $|\eta| \leq 0.5$ and $0.1 \leq p_T \leq 5$ GeV/c as a function of number of participants [25]. One can see that $\nu_{+-,dyn}$ shows very little dependence on incident energy. One must however correct measured $\nu_{+-,dyn}$ to account for charge conservation [25]. Figure 9 shows the plot of $\nu_{+-,dyn} = \nu_{+-,dyn} + 4/N_{ch}$ as a function of beam energy. The total charged-particle multiplicity is given by N_{ch}. Figure 9 also shows the results from other experiments

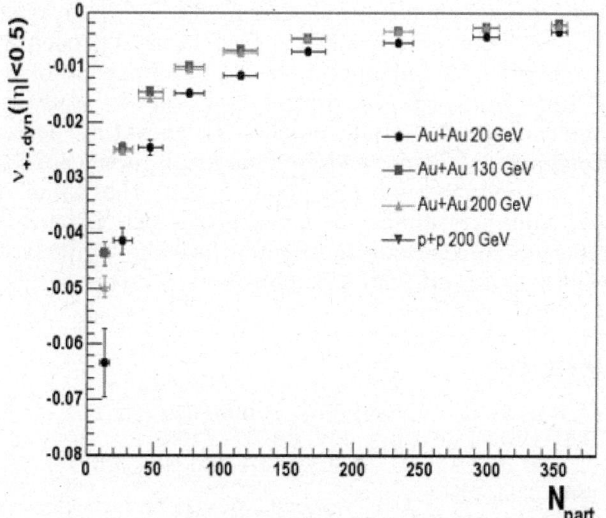

Fig. 8. $\nu_{+-,dyn}$ for all charged particles with $|\eta| < 0.5$ from Au + Au collisions at 20, 130, and 200 GeV and pp at 200 GeV as a function of the number of participating nucleons.

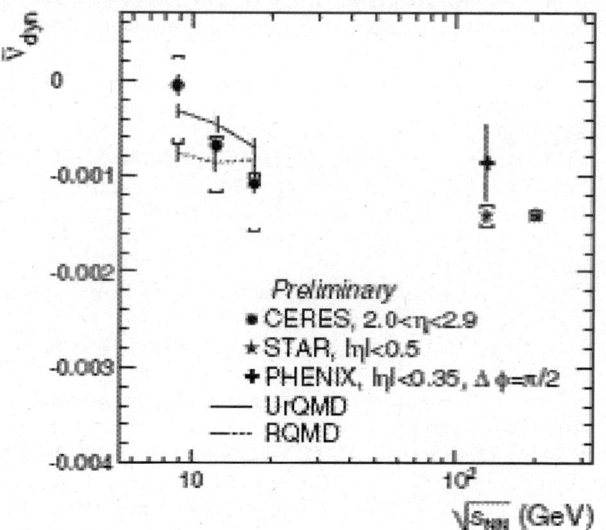

Fig. 9. $\nu_{+-,dyn}$ corrected for charge conservation as a function of collision energy.

for comparison. The net charge fluctuations are smaller than expected based on predictions from a resonance gas or a quark gluon gas, which undergoes fast hadronization.

6 p_T fluctuations

Event-by-event fluctuations of mean p_T have been proposed as a possible signature to search for the phase transition [26–28]. Fluctuations involve a statistical component arising from the stochastic nature of particle production, as well as a dynamic component determined by correlations arising in various particle production processes. There arc several measures to evaluate mean-

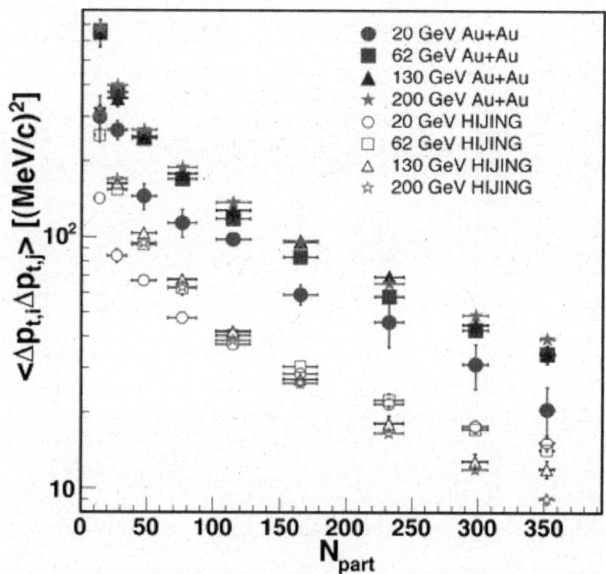

Fig. 10. $\langle \Delta p_{T,i} \Delta p_{T,j} \rangle$ as a function of centrality and incident energy for Au + Au collisions compared with HIJING results.

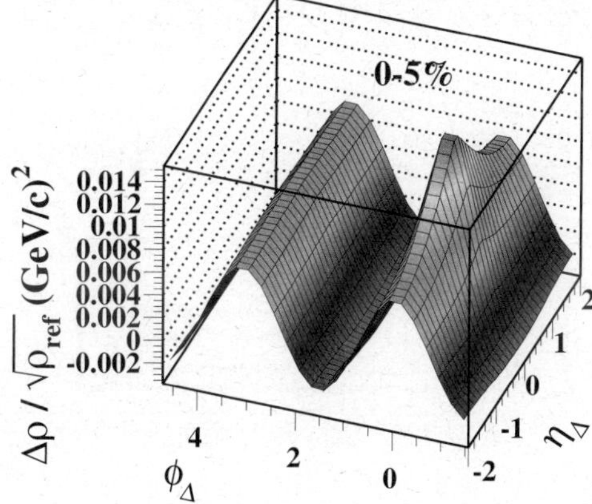

Fig. 11. Autocorrelations, $\Delta\rho/\sqrt{\rho_{ref}}$ on difference variables $(\eta\Delta, \phi\Delta)$ for 0–5% central Au + Au collisions.

p_T fluctuations. STAR has used $\langle \Delta p_{T,i} \Delta p_{T,j} \rangle$ [29] and $\Delta\sigma_{p_T}$ [30] as the measure of the dynamical fluctuations. $\sum_{p_T}$ gives the normalized fluctuation in CERES experiment [31]. Another measure, F_{p_T}, is defined as a deviation from 1 of the ratio of the r.m.s. of the event-by-event mean-p_T distribution in real events to that in mixed events (PHENIX) [32]. To characterize transverse-momentum correlation, the quantity $\langle \Delta p_{T,i} \Delta p_{T,j} \rangle$ was calculated. Figure 10 shows $\langle \Delta p_{T,i} \Delta p_{T,j} \rangle$ for Au + Au collisions at $\sqrt{s_{NN}} = 20$, 62, 130 and 200 GeV. One observes that $\langle \Delta p_{T,i} \Delta p_{T,j} \rangle$ decreases with centrality. These results are also compared with HIJING calculations as shown in fig. 10. The values for $\langle \Delta p_{T,i} \Delta p_{T,j} \rangle$ predicted by HIJING are always smaller than the data [29].

The pseudorapidity and azimuth (η, ϕ) bin size dependence of event-wise mean-transverse-momentum fluctuations has been measured in terms of $\Delta\sigma_{p_T}$ [30,33]. To access underlying dynamics, the corresponding autocorrelations $(\Delta\rho/\sqrt{\rho_{ref}})$ were extracted and are shown in

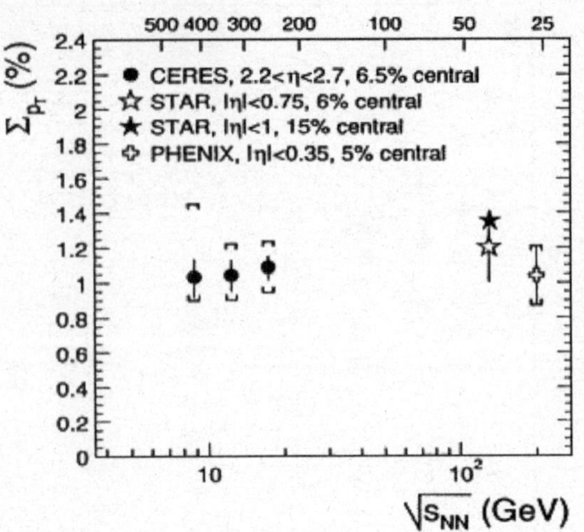

Fig. 12. $\sum_{p_T}$ as a function of $\sqrt{s_{NN}}$ in central events.

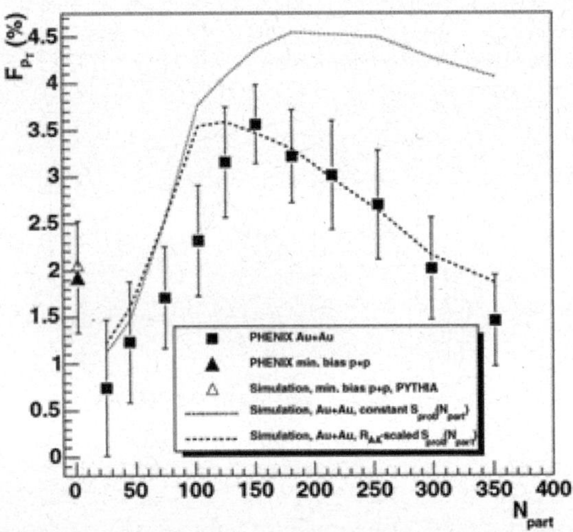

Fig. 13. F_{p_T} as a function of centrality. The result from PYTHIA simulation is also shown.

fig. 11 for 0–5% central Au + Au collisions. The general form of the autocorrelations suggests that the basic correlation mechanism is parton fragmentation [33].

$\sum_{p_T}$ also measures the dynamical fluctuations and is proportional to mean covariance of all charged-particle pairs per event. Figure 12 shows $\sum_{p_T}$ as a function of the nucleon-nucleon center-of-mass energy from SPS to RHIC energies. The observed fluctuations at SPS and at RHIC are about 1%. F_{p_T} is approximately proportional to $\langle N \rangle \sum_{p_T}^2$, where $\langle N \rangle$ is the mean charged-particle multiplicity. Figure 13 shows the magnitude of F_{p_T} as a function of centrality for Au + Au collisions with $p_T^{max} = 2.0\,\mathrm{GeV}/c$. A significant non-random fluctuation is seen to peak in mid central collisions [32].

7 Summary

Some results from fluctuation and correlation measurements in STAR have been presented. The forward-backward multiplicity correlation in Au + Au shows collective behavior and a fusion/percolation approach has been explored to understand this. The narrowing of balance function width in central collisions is consistent with trends predicted by models incorporating delayed hadronization. The net charge dynamical fluctuations are found to be negative and smaller than the value expected from a resonance or quark-gluon gas. Transverse-momentum correlations and fluctuations show interesting physics which need further exploration.

References

1. F. Karsch *et al.*, Nucl. Phys. B **605**, 579 (2001).
2. R.C. Hwa, Eur. Phys. J. C **38**, 277 (1988).
3. J. Dias de Deus, C. Pajares, C.A. Salgado, Phys. Lett. B **407**, 335 (1997).
4. M.A. Braun *et al.*, Eur. Phys. J. C **32**, 535 (2004).
5. A. Capella, A. Krzywicki, Phys. Rev. D **184**, 120 (1978).
6. A. Capella *et al.*, Phys. Rep. **236**, 225 (1994).
7. N.S. Amelin *et al.*, Phys. Rev. Lett. **73**, 2813 (1994).
8. N. Armesto *et al.*, Z. Phys. C **67**, 489 (1995).
9. STAR Collaboration (T.J. Tarnowsky), nucl-ex/0606018.
10. N.S. Amelin *et al.*, Eur. Phys. J. C **22**, 149 (2001).
11. N. Armesto, C. Pajares, D. Sousa, Phys. Lett. B **527**, 92 (2002).
12. STAR Collaboration (J. Adams *et al.*), nucl-ex/0311017.
13. M.A. Braun, F. del Moral, C. Pajares, Nucl. Phys. A **715**, 791 (2003).
14. M.A. Braun, C. Pajares, Nucl. Phys. B **390**, 542 (1993).
15. M. Isichenko, Rev. Mod. Phys. **64**, 861 (1992).
16. H. Satz, Rep. Prog. Phys. **63**, 1511 (2000).
17. M.A. Braun, F. del Moral, C. Pajares, Phys. Rev. C **65**, 024907 (2002).
18. STAR Collaboration (B.K. Srivastava, R.P. Scharenberg, T.J. Tarnowsky), nucl-ex/0606019.
19. S.A. Bass, P. Danielewicz, S. Pratt, Phys. Rev. Lett. **85**, 2689 (2000).
20. STAR Collaboration (J. Adams *et al.*), Phys. Rev. Lett. **90**, 172301 (2003).
21. STAR Collaboration (G.D. Westfall), J. Phys. G **30**, S1389 (2004).
22. C. Pruneau, S. Gavin, S. Voloshin, Phys. Rev. C **66**, 044904 (2002).
23. V. Koch *et al.*, Nucl. Phys. A **698**, 261 (2002).
24. STAR Collaboration (J. Adams *et al.*), Phys. Rev. C **68**, 044905 (2003).
25. STAR Collaboration (C. Pruneau), Nucl. Phys. A **774**, 651 (2006).
26. M. Stephanov, K. Rajagopal, E. Shuryak, Phys. Rev. Lett. **81**, 4816 (1998).
27. S.A. Voloshin, V. Koch, H.G. Ritter, Phys. Rev. C **60**, 024901 (1999).
28. H. Heiselberg, Phys. Rep. **351**, 161 (2001).
29. STAR Collaboration (J. Adams *et al.*), Phys. Rev. C **72**, 044902 (2005).
30. STAR Collaboration (J. Adams *et al.*), Phys. Rev. C **71**, 064906 (2005).
31. CERES Collaboration (H. Sako, H. Appelshauser), J. Phys. G **30**, S1371 (2004).
32. PHENIX Collaboration (S.S. Adler *et al.*), Phys. Rev. Lett. **93**, 092301 (2004).
33. STAR Collaboration (J. Adams *et al.*), J. Phys. G **32**, L37 (2006).

Eur. Phys. J. A **31**, 868–874 (2007)
DOI 10.1140/epja/i2006-10240-4

Special Article – QNP 2006

Latest results on the hot dense partonic matter at RHIC

M.J. Leitch[a]

Los Alamos National Laboratory, Los Alamos, NM 87545, USA

Received: 8 November 2006
Published online: 9 March 2007 – © Società Italiana di Fisica / Springer-Verlag 2007

Abstract. At the Relativistic Heavy-Ion Collider (RHIC) collisions of heavy ions at nucleon-nucleon energies of 200 GeV appear to have created a new form of matter thought to be a deconfined state of the partons that ordinarily are bound in nucleons. We discuss the evidence that a thermalized partonic medium, usually called a Quark Gluon Plasma (QGP), has been produced. Then, we discuss the effect of this high-density medium on the production of jets and their pair correlations. Next, we look at direct photons as a clean electro-magnetic probe to constrain the initial hard scatterings. Finally, we review the developing picture for the effect of this medium on the production of open heavy quarks and on the screening by the QGP of heavy-quark bound states.

PACS. 25.75.-q Relativistic heavy-ion collisions – 24.85.+p Quarks, gluons, and QCD in nuclei and nuclear processes

1 Introduction

The collisions of high-energy heavy ions are thought to form very high temperatures and densities similar to those that occurred in the earliest stages of our Universe. These collisions are thought to create matter with energy densities up to 30 times normal nuclear-matter density in a small region ($\sim 10^{-14}$ m) and for a short time ($\sim 10^{-23}$ s). Different experimental probes examine different stages of the matter as it expands and evolves back to normal matter. Hard probes created in the initial collisions before thermalization of the medium probe the medium through their final-state interactions. Soft probes come from the medium itself and provide a picture of its thermalization and spatial evolution.

Here we will review the most significant results from the RHIC heavy-ion program and discuss their implications in terms of the nature of the hot-dense matter that is created in these collisions. I thank many colleagues at RHIC for helping me prepare this review, especially Akiba, Constantin, d'Enterria, Granier de Cassagnac, Jacak, Nagle, Seto, and Zajc.

2 RHIC and its detectors

The Relativistic Heavy-Ion Collider (RHIC) collides Au ions at center-of-mass energies per nucleon pair of up to 200 GeV, substantially higher in energy than the 17 GeV

at the CERN SPS, but lower than the 5.5 TeV anticipated at the LHC. At RHIC four experiments, two large (PHENIX and STAR) and two small (BRAHMS and PHOBOS) experiments observe the collisions and together have advanced our understanding considerably since the first collisions at RHIC. Each experiment has different strengths with BRAHMS having two classic dipole spectrometers, PHOBOS focusing on extensive silicon-strip detectors, PHENIX being a complex many-subsystem detector optimized for rare probes, and STAR centered around a large solid-angle TPC.

In heavy-ion collisions, an important aspect one studies for all observables is their dependence on the centrality of the collision, *i.e.* whether a particular collision is "head-on" (central or, *e.g.*, 0–10% centrality) or peripheral (*e.g.*, 80–90%) with only the edges of the two colliding nuclei passing through each other. An approximate centrality measurement for each collision is obtained with a combination of the measured charged-particle multiplicity and the yield of spectator neutrons that are observed in zero-degree calorimeters (ZDC). These measurements are related to the actual centrality using a simple Glauber model of the collision.

3 Thermalization

One of the key issues in high-energy heavy-ion collisions is the time scale and the degree to which the available energy is thermalized into a medium. Several of the soft-sector observables give us important information on this aspect of the collisions.

[a] e-mail: `leitch@bnl.gov`

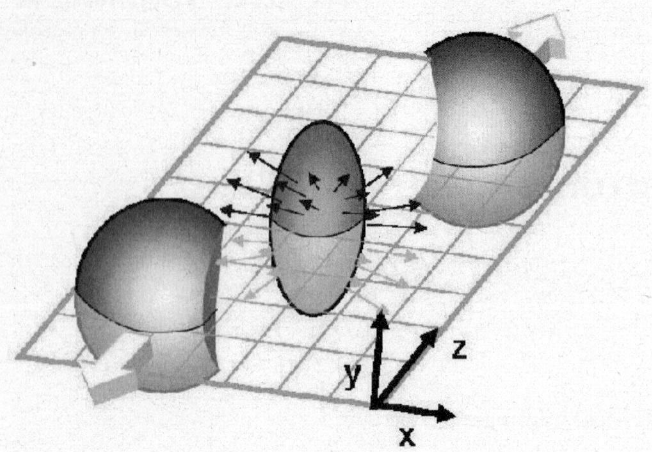

Fig. 1. Cartoon showing the initial geometrical asymmetry in a non-central heavy-ion collision.

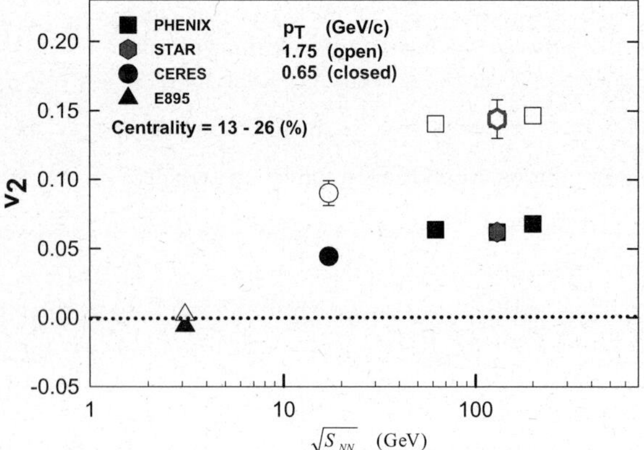

Fig. 2. Dependence of the flow or v_2 of hadrons on center-of-mass energy, $\sqrt{s_{NN}}$, for two different transverse momentum (p_T) values [1].

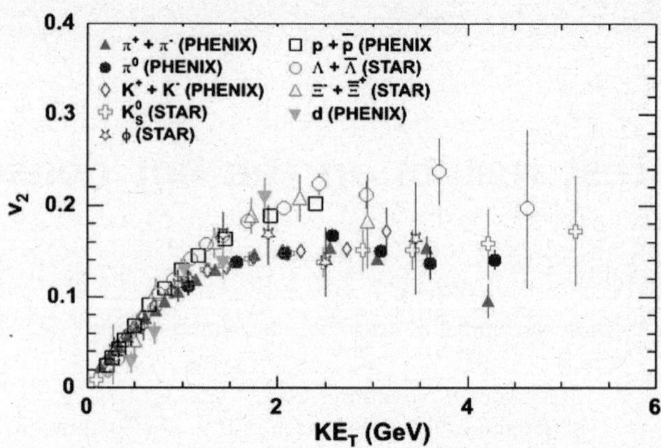

Fig. 3. Flow for hadrons *vs.* transverse energy, $KE_T = m_T - \text{mass}$.

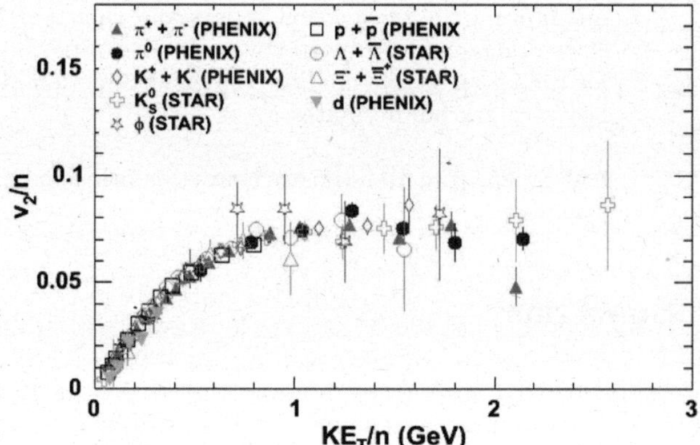

Fig. 4. Flow per quark *vs.* transverse energy per quark, KE_T/n.

The first is the extent to which the observed lower-momentum produced particles exhibit "hydrodynamic flow" as would be expected from a thermalized medium. For not fully head-on collisions, as shown in fig. 1, an asymmetric or almond-shaped collision region is produced. This initial anisotropy is converted into a corresponding momentum anisotropy which results in an azimuthal asymmetry relative to the reaction plane in the yield of produced particles at a given momentum. The efficiency of this conversion depends on the properties of the medium, with large asymmetries corresponding to early thermalization. This asymmetry is usually represented by "v_2" which is the 2nd Fourier coefficient of the momentum anisotropy,

$$\mathrm{d}n/\mathrm{d}\phi \sim 1 + 2v_2(p_T)\cos(2\phi) + \dots \qquad (1)$$

The observed flow tends to saturate near RHIC energies [1], as shown in fig. 2, suggesting that early thermalization has been achieved. Also supporting early thermalization are comparisons with hydrodynamics calculations which indicate that the flow observed at RHIC is near maximal flow predicted by the models. However, ongoing

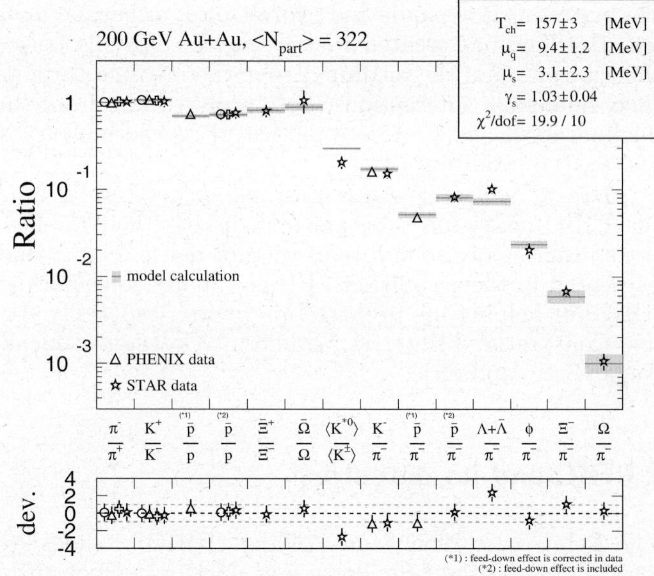

Fig. 5. Ratios of hadron yields are well reproduced by statistical combination models based on quark degrees of freedom.

improvements in the hydrodynamic models may shed new light on what the true limits are.

Another important aspect of the flow measurements is that when comparisons of the flow for different particles are made there are substantial differences observed, see fig. 3. However, if one plots the flow normalized to the number of valence quarks in the observed hadron *vs.* the transverse kinetic energy per quark, fig. 4, a universal behavior emerges. This supports a picture where the relevant degrees of freedom of the thermalized medium are quarks, *i.e.* the thermalization occurs when the matter is not hadrons but consists of deconfined quarks.

Finally, statistical models that reproduce ratios of hadron yields [2], such as those in fig. 5, also support a picture where the degrees of freedom are quarks up to the "freezeout" time when the observed hadrons form.

4 Jets and energy loss

Arguably, the most dramatic effect seen at RHIC so far, is the strong suppression of high-p_T hadrons in the most central Au + Au collisions compared to p + p collisions. Although it has not been possible in the high-multiplicity environment of Au + Au collisions at RHIC to fully reconstruct jets, these high-p_T particles should be good surrogates for the jets. As shown in fig. 6, π^0's are suppressed by more than a factor of two in central Au + Au collisions, but have little or no modification for the cold–nuclear-matter control measurement in d + Au collisions. This is interpreted as evidence for large energy losses of the jets in the final state where they traverse the hot high-density matter created in the Au + Au collisions. Models infer from this that matter densities of ~ 15 times normal nuclear-matter density are involved in the earliest stages, before expansion, of the created medium.

In addition to studying the yield of hadrons *vs.* p_T, the correlation of hadron pairs has also been studied. In this case a "tag" is provided by one high-p_T hadron, and then the distribution of other hadrons in the same event

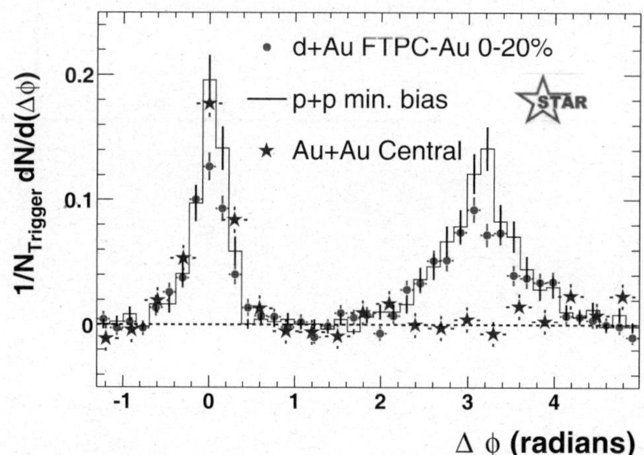

Fig. 7. (Colour on-line) Distribution in azimuthal angle (ϕ), with respect to a tagging high-p_T ($p_T = 4$–$6\,\mathrm{GeV}/c$) particle, of other particles ($p_T > 2\,\mathrm{GeV}/c$) seen in the same event for central Au + Au collisions (blue stars), compared to d + Au collisions (red dots) and p + p collisions (black histogram and vertical bars) [4].

is observed. Figure 7 shows these correlations for central Au + Au, central d + Au and p + p collisions. In d + Au collisions the correlation is essentially unaltered from that for p + p collisions, while for central Au + Au collisions the "away-side" correlation ($\Delta\phi \sim \pi$) disappears, with the "near-side" peak (other particles associated with the jet that the tagging particle is from) remains like that of p + p and d + Au. This result is usually interpreted as a tagging jet coming from near the surface and a partner "away-side" jet that is degraded by the thick high-density medium it has to pass through on the other side. Further studies looking at lower momenta for the "away-side" hadrons appear to find remnants of this jet in broad distributions of low-momentum particles. In some momentum windows these "away-side" particles even show a very interesting split distribution with a depression in their yield for exactly back-to-back angles, and side maxima. This phenomena has been interpreted by some as evidence for "mach-cone" effects in the medium [5].

5 Direct photons and the initial state

Direct photons, although difficult to isolate experimentally, because of their weak electromagnetic interaction with the medium in the final state, connect directly to the initial state where the hard interactions take place —before the thermalization of the medium. In fig. 8 the direct photons in central Au + Au collisions show that the initial state is not modified from that of p + p collisions, while the neutral mesons (as discussed above) are strongly suppressed due to final-state effects in the hot dense medium. This comparison affirms the idea discussed above, that the observed hadron jet modifications do come from the final state.

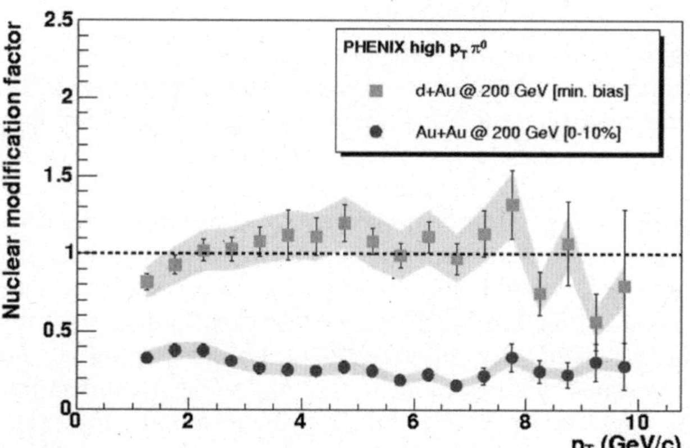

Fig. 6. (Colour on-line) Large suppression of the π^0 yield per binary collision in Au + Au collisions (red circles) contrasted with lack of suppression for d + Au collisions (green squares) [3].

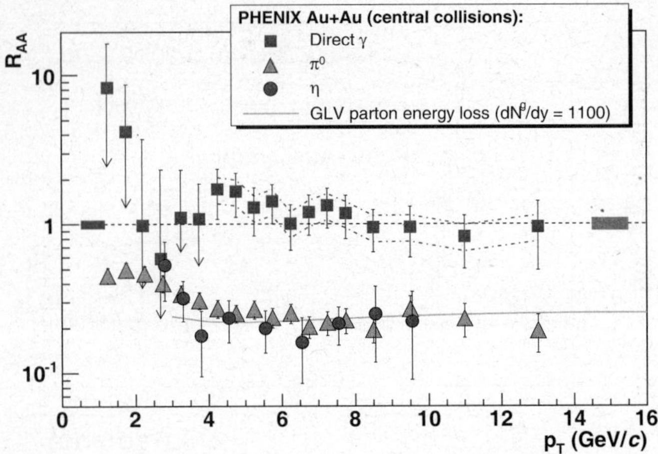

Fig. 8. (Colour on-line) Direct photons (purple squares) are not suppressed in central Au + Au collisions as compared to the strong suppression seen for π^0's (yellow triangles) and η's (red circles) [6]. The solid curve is a parton energy loss calculation [7].

6 Heavy quarks

So far we have discussed light hadrons, including the large energy loss observed in the final state when they pass through the high-density matter created in central Au + Au collisions. Heavy quarks are expected to suffer different effects in this medium, but they are more difficult to measure both because of their smaller production cross-sections and due to the difficulty of separating them from backgrounds. Despite these difficulties, initial measurements primarily through measurements of leptons from the semi-leptonic decay of the heavy (charm and

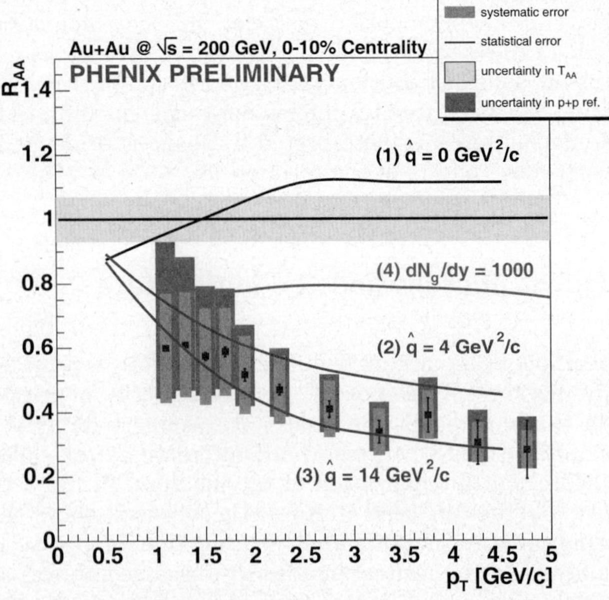

Fig. 9. Nuclear modification factor, R_{AA}, for heavy quarks at central rapidity observed using non-photonic electrons in central Au + Au collisions at 200 GeV [8].

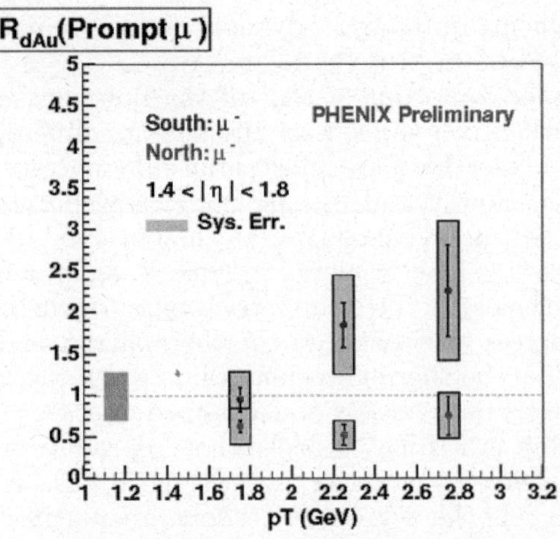

Fig. 10. Nuclear suppression factor, R_{dAu}, at forward and backward rapidity for heavy quarks observed using prompt muons (heavy quarks) in d + Au collisions at 200 GeV [9].

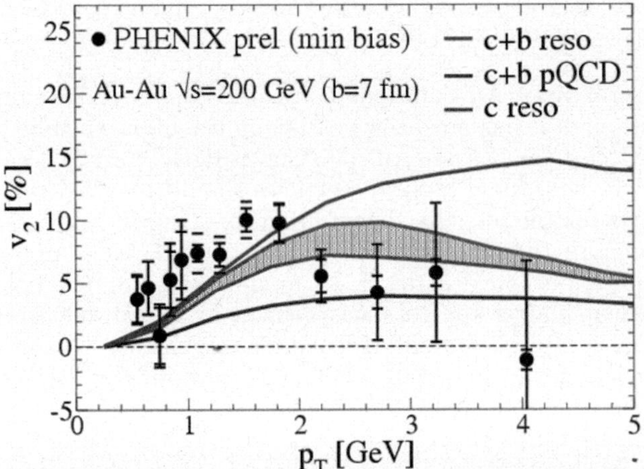

Fig. 11. (Colour on-line) Flow of heavy quarks at mid-rapidity for 200 GeV minimum-bias Au + Au collisions, as obtained from non-photonic electrons [8]. Data is compared to theoretical calculations [10] that have no flow (blue curve), only charm flow (purple curve), or both charm and beauty flow (red curve).

beauty) mesons are now available. In fig. 9 the suppression of non-photonic electrons at mid-rapidity from charm and beauty decays is shown. The large suppression seen, like that for light hadrons, is interpreted as energy loss in the high-density medium created in these central Au + Au collisions. Similar measurements at forward rapidity are being worked on using decays to muons, but so far results are only available for d + Au and p + p collisions. The latter are shown in fig. 10 where one sees that d + Au heavy-quark yields are suppressed at forward rapidity (small-x or shadowing region in the Au nucleus) and enhanced at backward rapidity (larger-x or anti-shadowing region). The suppression in the forward or small-x region is usually attributed to nuclear shadowing of the gluon distributions [11] or to gluon saturation models [12]; but could also involve other cold nuclear matter effects such as initial-state gluon energy loss.

The electrons from heavy-quark decays have also been analyzed in terms of their elliptic flow, analogous to what was discussed above for hadrons. Surprisingly, at least for small p_T values, these heavy quarks also seem to exhibit flow (non-zero v_2), as shown in fig. 11. Although the data uncertainties remain large due to the low rates of heavy-quark production and the large systematics background subtraction, the usual interpretation of these results is that for the lowest transverse momentum the charm quarks do flow with the light quarks, but as the momentum goes up they punch through the thermalized medium and the flow disappears. However, more accurate measurements will be needed to firm up the true characteristics of heavy-quark flow (sect. 8).

7 J/ψ's and color screening in the medium

The J/ψ and other heavy quarkonia are thought to provide a key signature for a deconfined medium. Early predictions were that the two heavy quarks that would form the bound state would be screened from each other in the high-density deconfined medium [13]. This effect would depend on the size or binding energy of the specific state and so different states (J/Ψ, ψ', χ_C, Υ_{1S}, Υ_{2S}, Υ_{3S}) would "melt" at different energy densities. However, more recently, lattice QCD calculations have suggested that, in fact, the J/ψ would not be screened unless effective densities of the hot plasma created in these heavy-ion collisions exceeded twice the critical temperature [14].

The J/ψ suppression at RHIC was predicted to be larger than that observed at the SPS by most of the theoretical models that were successful in describing the SPS data [15,16]. This reflects the expectation that the matter created at RHIC would be hotter and longer-lived than that for the lower-energy SPS measurements. Contrary to this expectation, the measurements at RHIC show suppression very similar to that at lower energy. In fig. 12 are shown preliminary J/ψ measurements for Au + Au collisions from PHENIX [17], along with earlier measurements of the cold–nuclear-matter effects as seen in d + Au collisions [18] at the same energy. First, one can see that simple models that describe the poor statistics d + Au measurements for a range of absorption cross-sections give a large uncertainty in the expected cold–nuclear-matter effects for Au + Au collisions (blue on-line bands). Clearly, better d + Au data is needed to more accurately constrain the cold–nuclear-matter effects and to allow a more quantitative understanding of how much of the Au + Au suppression does not come from these effects, and how much may come from additional effects of the hot dense matter. Nevertheless, for the most central collisions, the Au + Au data clearly show a stronger suppression than one would expect from cold–nuclear-matter effects alone.

At present, we are left with two theoretical pictures which both provide a plausible explanation of the level of suppression seen in the RHIC data. One of these, shown in fig. 13, was actually a prediction from before the experimental results were obtained [16]. This model includes

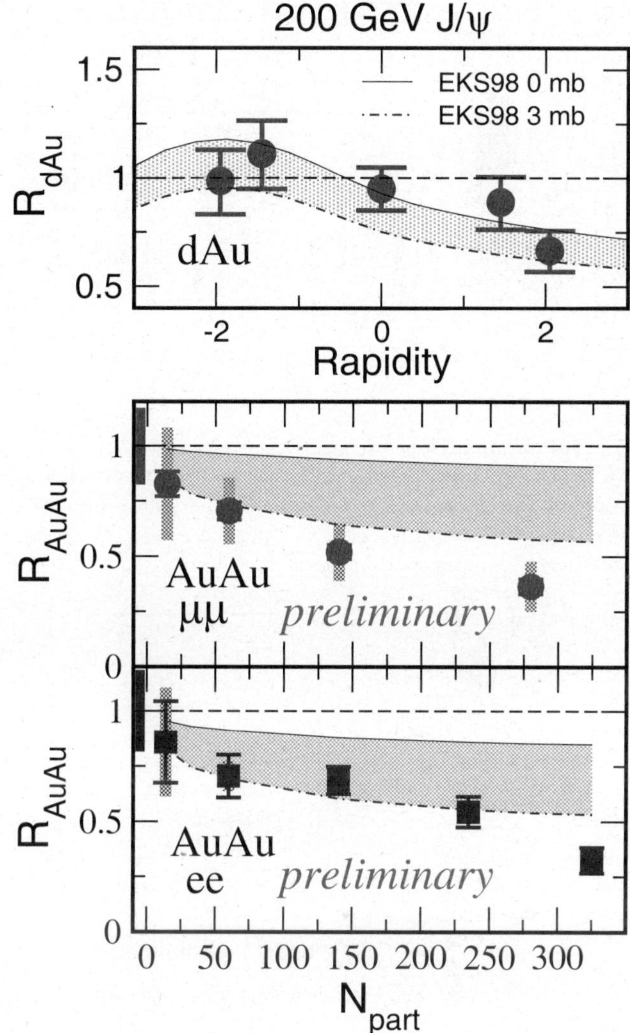

Fig. 12. (Colour on-line) Cold–nuclear-matter effects on J/Ψ production are constrained by the limited accuracy of our present d + Au data [18] (blue band in top panel). Corresponding calculations for Au + Au collisions give bands on the bottom panels for forward rapidity (middle panel) and central rapidity (bottom panel) data [17]. N_{part} is the number of participants, and is a measure of the centrality of the Au + Au collision.

strong dissociation of the charm pairs from the large gluon density created in the collisions (analogous to screening), but also includes regeneration of bound charm pairs (J/ψ's) in the later stages of the expansion driven by the large production and high density of independently produced charm quarks. So in this model, there is a stronger "screening" effect at RHIC than at the SPS but it is compensated for by the regeneration, resulting in a net suppression very similar to that seen at the SPS.

The second picture, sequential screening [19], asserts that (as suggested by recent lattice calculations) the J/ψ is not screened in central Au + Au collisions at RHIC or at SPS energies. Rather the observed suppression comes only from screening of the higher-mass states, χ_C and ψ',

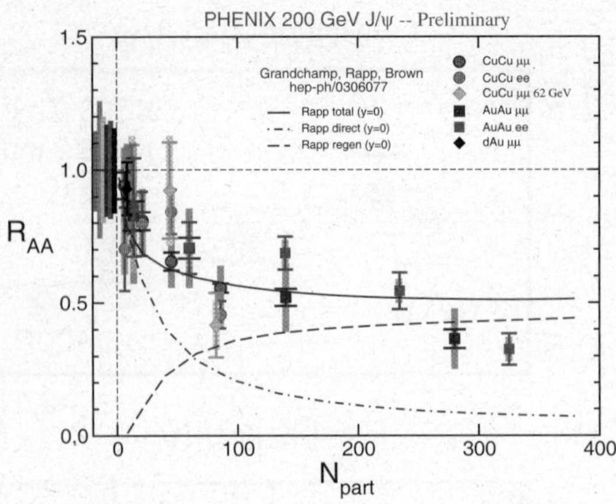

Fig. 13. Comparison *vs.* centrality of the Au + Au J/ψ data [17] to theoretical calculations that include both dissociation of the $c\bar{c}$ by a large gluon density and regeneration effects [16].

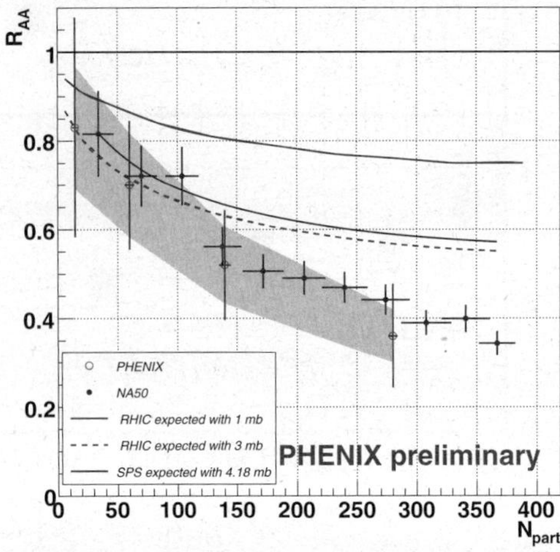

Fig. 14. (Colour on-line) 200 GeV RHIC J/ψ data [17] and 17.4 GeV NA50 [20] data plotted together *vs.* N_{part}, showing the universal dependence on number of participants. The curves are calculations for RHIC with shadowing and absorption cross-sections of 1 mb (red solid line) and 3 mb (red dashed line) [21]; and for NA50 with an absorption cross-section of 4.18 mb (blue solid line).

that through feed down normally provide about $\sim 40\%$ of the J/ψ production. This picture provides a simple explanation for the nearly identical suppression observed at RHIC and the SPS seen in fig. 14. The J/ψ itself would be screened only for higher-energy collisions at the Large Hadron Collider (LHC).

8 Future - RHIC-II and detector upgrades

The luminosities obtained at RHIC are just beginning to reach levels where the more rare probes such as the Υ can be studied. In the future this will be improved further when RHIC-II, with its electron cooling, allows even higher luminosities. In addition upgrades of the RHIC detectors are underway, most notably adding high-resolution

silicon vertex tracking in the inner regions of the large detectors (PHENIX and STAR) in order to make identification of heavy quarks more explicit via the separation of the primary and secondary vertices associated with the creation, propagation and subsequent decay of the heavy mesons. These detached vertex measurements of heavy quarks, along with the increasing luminosity will dramatically reduce systematic and statistical uncertainties in these measurements and will enable exclusive measurements of heavy quarks such as $B \to J/\psi X$.

9 Summary

In summary, evidence is mounting at RHIC for the creation of a new form of matter that is 1) dense and gives large energy losses and modifications of jet correlations for hadrons, 2) appears to be thermalized very early and to exhibit maximal flow as predicted by hydrodynamics models, with quark degrees of freedom, 3) also causes large energy loss and flow for heavy quarks, and 4) causes strong suppression, beyond that expected from cold–nuclear-matter effects, for the J/ψ. Future luminosity and detector upgrades will enable firming up and extending our understanding of these phenomena.

References

1. PHENIX Collaboration (S.S. Adler *et al.*), Phys. Rev. Lett. **94**, 232302 (2005).
2. M. Kaneta, N. Xu, nucl-th/0405068; P. Braun-Munzinger, K. Redlich, J. Stachel, nucl-th/0304013.
3. PHENIX Collaboration (S.S. Adler *et al.*), Phys. Rev. Lett. **91**, 072303 (2003).
4. STAR Collaboration (C. Adler *et al.*), Phys. Rev. Lett. **91**, 072304 (2003).
5. E. Shuryak, Nucl. Phys. A **783**, 31 (2007).
6. PHENIX Collaboration (S.S. Adler *et al.*), Phys. Rev. Lett. **96**, 202301 (2006).
7. I. Vitev, M. Gyulassy, Phys. Rev. lett. **89**, 252301 (2002); I. Vitev, J. Phys. G **30**, S791 (2004).
8. PHENIX Collaboration (S. Butsyk *et al.*), Nucl. Phys. A **774**, 669 (2006).
9. PHENIX Collaboration (X. Wang *et al.*), J. Phys. G **32**, S511 (2006).
10. H. van Hees, V. Greco, R. Rapp, Phys. Rev. C **73**, 034913 (2006); nucl-th/0608033.
11. K.J. Eskola, V.J. Kolhinen, R. Vogt, Nucl. Phys. A **696**, 729 (2001); L. Frankfurt, M. Strikman, Eur. Phys. J A **5**, 293 (1999).
12. L. McLerran, R. Venugopalan, Phys. Rev. D **49**, 2233; 3352 (1994).
13. T. Matsui, H. Satz, Phys. Lett. B **178**, 416 (1986).
14. F. Datta *et al.*, hep-lat/0409147.
15. S. Digal, S. Fortunato, H. Satz, Eur. Phys. J. C **32**, 547 (2004); hep-ph/0310354; A. Capella, E. Ferreiro, Eur. Phys. J. C **42**, 419 (2005); hep-ph/0505032.
16. L. Grandchamp, R. Rapp, G.E. Brown, Phys. Rev. Lett. **92**, 212301 (2004); hep-ph/0306077.

17. PHENIX Collaboration (H. Pereira da Costa *et al.*), Nucl. Phys. A **774**, 747 (2006).
18. PHENIX Collaboration (S.S. Adler *et al.*), Phys. Rev. Lett. **96**, 012304 (2006).
19. F. Karsch, D. Kharzeev, H. Satz, Phys. Rev. Lett. B **637**, 75 (2006); hep-ph/0512239.
20. NA50 Collaboration (B. Alessandro *et al.*), Eur. Phys. J. C **39**, 335 (2005).
21. R. Vogt, nucl-th/0507027 (2005) and private communication.

Eur. Phys. J. A **31**, 875–882 (2007)
DOI 10.1140/epja/i2006-10284-4

THE EUROPEAN
PHYSICAL JOURNAL A

Special Article – QNP 2006

Scenario of instabilities driven equilibration of the quark-gluon plasma

St. Mrówczyński[a]

Institute of Physics, Świętokrzyska Academy, ul. Świętokrzyska 15, PL - 25-406 Kielce, Poland and
Sołtan Institute for Nuclear Studies, ul. Hoża 69, PL - 00-681 Warsaw, Poland

Received: 18 December 2006
Published online: 20 March 2007 – © Società Italiana di Fisica / Springer-Verlag 2007

Abstract. Due to anisotropic momentum distribution the parton system produced at the early stage of relativistic heavy-ion collisions is unstable with respect to the magnetic plasma modes. The instabilities isotropize the system and thus speed up the process of its equilibration. The scenario of instabilities-driven isotropization is reviewed.

PACS. 12.38.Mh Quark-gluon plasma – 25.75.-q Relativistic heavy-ion collisions

1 Introduction

The matter created in relativistic heavy-ion collisions manifests a strongly collective hydrodynamic behaviour [1] which is particularly evident in studies of the so-called elliptic flow [2]. Hydrodynamic description requires, strictly speaking, a local thermal equilibrium and experimental data on the particle spectra and elliptic flow suggest, when analysed within the hydrodynamic model, that an equilibration time of the parton[1] system is as short as $0.6\,\mathrm{fm}/c$ [1]. Such a fast equilibration can be explained assuming that the quark-gluon plasma (QGP) is strongly coupled [3]. However, it is far not excluded that due to the high-energy density at the early stage of the collision, when the elliptic flow is generated, the plasma is weakly coupled because of the asymptotic freedom. Thus, the question arises whether the weakly interacting plasma can be equilibrated within $1\,\mathrm{fm}/c$.

Calculations, which assume that the parton-parton collisions are responsible for the equilibration of the weakly interacting QGP, provide an equilibration time of at least $2.6\,\mathrm{fm}/c$ [4]. To thermalize the system one needs either a few hard collisions of the momentum transfer of order of the characteristic parton momentum[2], which is denoted here as T (as the temperature of equilibrium system), or many collisions of smaller transfer. As discussed in *e.g.* [5], the inverse time scale of the collisional equilibration is of order $g^4 \ln(1/g)\,T$, where g is the QCD cou-

pling constant. However, the equilibration is speeded up by instabilities generated in an anisotropic QGP [6–8], as growth of the unstable modes is associated with the system's isotropization. The characteristic inverse time of instability development is roughly of order gT for a sufficiently anisotropic momentum distribution [6,9–11]. Thus, the instabilities are much "faster" than the collisions in the weak-coupling regime. Recent numerical simulation [12] shows that the instabilities-driven isotropization is indeed very efficient.

The isotropization should be clearly distinguished from the equilibration. The instabilities-driven isotropization is a mean-field reversible phenomenon which is *not* accompanied with the entropy production [6,12]. Therefore, the collisions, which are responsible for the dissipation, are needed to reach the equilibrium state of maximal entropy. The instabilities contribute to the equilibration indirectly, shaping the parton momenta distribution.

A large variety of instabilities of the electron-ion plasma are known [13]. Those caused by coordinate space inhomogeneities, in particular by the system's boundaries, are usually called *hydrodynamic* instabilities while those due to non-equilibrium momentum distribution of plasma particles are called *kinetic* instabilities. Hardly anything is known about QGP hydrodynamic instabilities, and I will not speculate about them. The kinetic instabilities are initiated either by the charge or current fluctuations. In the first case, the electric field ($\mathbf{E}$) is longitudinal ($\mathbf{E} \parallel \mathbf{k}$, where $\mathbf{k}$ is the wave vector), while in the second case the field is transverse ($\mathbf{E} \perp \mathbf{k}$). For this reason, the kinetic instabilities caused by the charge fluctuations are usually called *longitudinal* while those caused by the current fluctuations are called *transverse*. Since the electric field plays a crucial role in the longitudinal mode generation, the

---

[a] e-mail: mrow@fuw.edu.pl

[1] The term "parton" is used to denote quark or gluon.

[2] Although an anisotropic system is considered, the characteristic momentum in all directions is assumed to be of the same order.

longitudinal instabilities are also called *electric* while the transverse ones are called *magnetic*. The electric instabilities, which occur in systems with multi-bump momentum distributions [13], are rather irrelevant for QGP produced in relativistic heavy-ion collisions. For this reason the longitudinal modes are not discussed here. The magnetic mode known as the filamentation or Weibel instability appears to be relevant because a momentum anisotropy is a sufficient condition for its existence. In the following sections a whole scenario of the instabilities-driven isotropization is discussed. A more extensive account of the scenario is given in [14] where, in particular, other approaches to the problem of QGP thermalization are also presented. Here, however, very recent developments are briefly discussed.

2 Seeds of the filamentation and its mechanism

Let me consider a non-equilibrium parton system which is homogeneous but the parton momentum distribution is anisotropic. The system is on average locally colourless but colour fluctuations are possible. Therefore, $\langle j_a^\mu(x) \rangle = 0$, where $j_a^\mu(x)$ is a local colour four-current in the adjoint representation of $SU(N_c)$ gauge group with $\mu = 0, 1, 2, 3$ and $a = 1, 2, \ldots, N_c^2 - 1$ being the Lorentz and colour index, respectively; $x = (t, \mathbf{x})$ denotes a four-position in the coordinate space. As discussed in [15], the current correlator for a classical system of free partons is

$$M_{ab}^{\mu\nu}(t, \mathbf{x}) \stackrel{\text{def}}{=} \langle j_a^\mu(t_1, \mathbf{x}_1) j_b^\nu(t_2, \mathbf{x}_2) \rangle =$$

$$\frac{1}{8} g^2 \delta^{ab} \int \frac{d^3 p}{(2\pi)^3} \frac{p^\mu p^\nu}{E_p^2} f(\mathbf{p}) \, \delta^{(3)}(\mathbf{x} - \mathbf{v}t), \quad (1)$$

where $(t, \mathbf{x}) \equiv (t_2 - t_1, \mathbf{x}_2 - \mathbf{x}_1)$, $\mathbf{v} \equiv \mathbf{p}/E_p$ and the effective parton distribution function $f(\mathbf{p})$ equals $n(\mathbf{p}) + \bar{n}(\mathbf{p}) + 2N_c n_g(\mathbf{p})$ with $n(\mathbf{p})$, $\bar{n}(\mathbf{p})$ and $n_g(\mathbf{p})$ giving the average colourless distribution function of quarks $Q^{ij}(x, \mathbf{p}) = \delta^{ij} n(\mathbf{p})$, of antiquarks $\bar{Q}^{ij}(x, \mathbf{p}) = \delta^{ij} \bar{n}(\mathbf{p})$, and of gluons $G^{ab}(x, \mathbf{p}) = \delta^{ab} n_g(\mathbf{p})$. We note that the distribution function of (anti)quarks belongs to the fundamental representation of the $SU(N_c)$ group while that of gluons to the adjoint representation. Therefore, $i, j = 1, 2, \ldots, N_c$ and $a, b = 1, 2, \ldots, N_c^2 - 1$.

Due to the average space-time homogeneity, the correlation tensor (1) depends only on the difference $(t_2 - t_1, \mathbf{x}_2 - \mathbf{x}_1)$. The space-time points $(t_1, \mathbf{x}_1)$ and $(t_2, \mathbf{x}_2)$ are correlated in the system of non-interacting particles if a particle travels from $(t_1, \mathbf{x}_1)$ to $(t_2, \mathbf{x}_2)$. For this reason the delta $\delta^{(3)}(\mathbf{x} - \mathbf{v}t)$ is present in the formula (1). The momentum integral of the distribution function simply represents the sum over particles. The fluctuation spectrum is found as a Fourier transform of the tensor (1),

$$M_{ab}^{\mu\nu}(\omega, \mathbf{k}) = \frac{1}{8} g^2 \, \delta^{ab} \int \frac{d^3 p}{(2\pi)^3} \frac{p^\mu p^\nu}{E_p^2} f(\mathbf{p}) \, 2\pi \delta(\omega - \mathbf{kv}).$$

$$(2)$$

To compute the fluctuation spectrum, the parton momentum distribution has to be specified. Such calculations with two forms of the anisotropic momentum distribution are presented in [15]. Here I only qualitatively discuss eqs. (1), (2). I assume that the momentum distribution is elongated in, say, the z-direction. Then, eqs. (1), (2) clearly show that the correlator M^{zz} is larger than M^{xx} or M^{yy}. It is also clear that M^{zz} is the largest when the wave vector $\mathbf{k}$ is along the direction of the momentum deficit. Then, the delta function $\delta(\omega - \mathbf{kv})$ does not much constrain the integral in eq. (2). Since the momentum distribution is elongated in the z-direction, the current fluctuations are the largest when the wave vector $\mathbf{k}$ is the xy-plane. Thus, I conclude that some fluctuations in the anisotropic system are large, much larger than in the isotropic one. An anisotropic system has a natural tendency to split into the current filaments parallel to the direction of the momentum surplus. These currents are seeds of the filamentation instability.

Let me now explain in terms of elementary physics why the fluctuating currents, which flow in the direction of the momentum surplus, can grow in time. To simplify the discussion, which follows [15], I consider an electromagnetic anisotropic system. The form of the fluctuating current is chosen to be

$$\mathbf{j}(x) = j \, \hat{\mathbf{e}}_z \, \cos(k_x x), \quad (3)$$

where $\hat{\mathbf{e}}_z$ is the unit vector in the z-direction. We have the current filaments of thickness $\pi/|k_x|$ with the current flowing in the opposite directions in the neighbouring filaments. The magnetic field $\mathbf{B}$ generated by the current (3) is oriented along the axis y and the Lorentz force acting on the partons, which fly in the z-direction, is

$$\mathbf{F}(x) = q \, \mathbf{v} \times \mathbf{B}(x) = -q \, v_z \, \frac{j}{k_x} \, \hat{\mathbf{e}}_x \, \sin(k_x x),$$

where q is the electric charge. One observes, see fig. 1, that the force distributes the partons in such a way that those, which positively contribute to the current in a given filament, are focused in the filament centre while those, which negatively contribute, are moved to the neighbouring one. Thus, the initial current grows. For a somewhat different explanation see [11].

3 Dispersion equation

The equation of motion of the Fourier-transformed chromodynamic field $A^\mu(k)$ is

$$\left[k^2 g^{\mu\nu} - k^\mu k^\nu - \Pi^{\mu\nu}(k) \right] A_\nu(k) = 0, \quad (4)$$

where $\Pi^{\mu\nu}(k)$ is the polarization tensor or gluon self-energy which is discussed later on. A general plasmon dispersion equation is of the form

$$\det \left[k^2 g^{\mu\nu} - k^\mu k^\nu - \Pi^{\mu\nu}(k) \right] = 0. \quad (5)$$

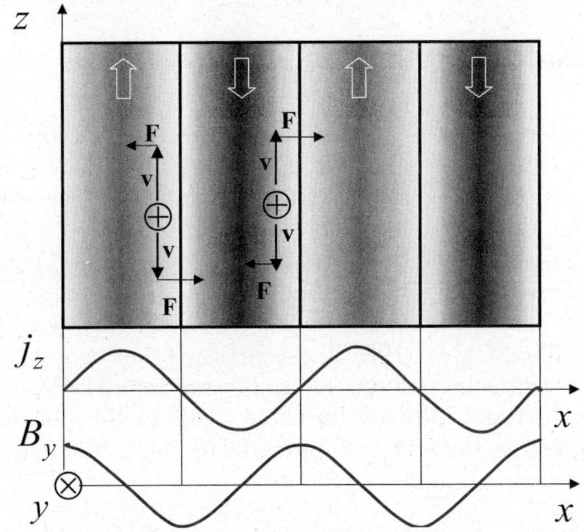

Fig. 1. The mechanism of filamentation instability, see text for the description.

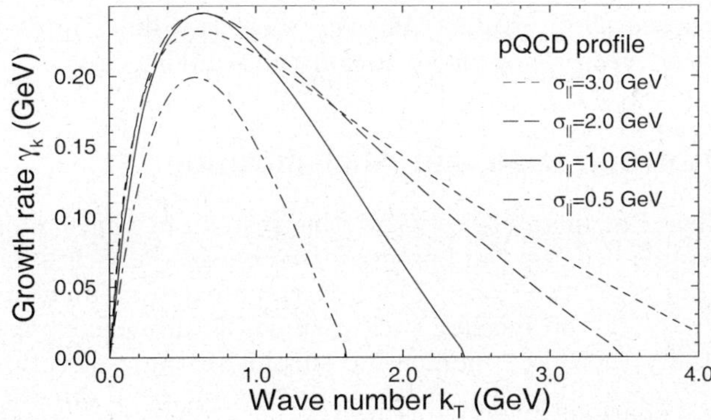

Fig. 2. The growth rate of the unstable mode as a function of the wave vector $\mathbf{k} = (k_\perp, 0, 0)$ for $\sigma_\perp = 0.3\,\mathrm{GeV}$ and 4 values of the parameter $\sigma_\parallel$ which controls the system's anisotropy. The figure is taken from [9].

The polarization tensor, which can be derived in the linear response approximation either within the transport theory or diagrammatically [16], is

$$\Pi^{\mu\nu}(k) = \frac{g^2}{2} \int \frac{\mathrm{d}^3 p}{(2\pi)^3} \frac{f(\mathbf{p})}{|\mathbf{p}|} \left[\frac{(p \cdot k)(k^\mu p^\nu + p^\mu k^\nu)}{(p \cdot k)^2} - \frac{k^2 p^\mu p^\nu + (p \cdot k)^2 g^{\mu\nu}}{(p \cdot k)^2} \right], \tag{6}$$

where $f(\mathbf{p})$ is the already defined effective parton distribution function; the spin and flavour are treated as parton internal degrees of freedom. The quarks and gluons are assumed to be massless. Since $\Pi^{\mu\nu}(k)$ is the unit matrix in the colour space, the colour indices are suppressed.

Due to the transversality of $\Pi^{\mu\nu}(k)$ $(k_\mu \Pi^{\mu\nu}(k) = k_\nu \Pi^{\mu\nu}(k) = 0)$ not all components of $\Pi^{\mu\nu}(k)$ are independent of each other, and consequently the dispersion equation (5), which involves a determinant of 4×4 matrix, can be simplified to the determinant of 3×3 matrix. For this purpose, I introduce the chromoelectric permittivity tensor $\epsilon^{lm}(k)$, where the indices $l, m, n = 1, 2, 3$ label three-vector and tensor components. Because $\epsilon^{lm}(k)E^l(k)E^m(k) = \Pi^{\mu\nu}(k)A_\mu(k)A_\nu(k)$, where $\mathbf{E}$ is the chromoelectric vector, the permittivity can be expressed through the polarization tensor as $\epsilon^{lm}(k) = \delta^{lm} + \Pi^{lm}(k)/\omega^2$. Then, the dispersion equation gets the form

$$\det\left[\mathbf{k}^2 \delta^{lm} - k^l k^m - \omega^2 \epsilon^{lm}(k)\right] = 0. \tag{7}$$

Substituting the permittivity $\epsilon^{lm}(k)$ into eq. (7), one fully specifies the dispersion equation (7) which provides a spectrum of quasi-particle bosonic excitations. A solution $\omega(\mathbf{k})$ of eq. (7) is called *stable* when $\mathrm{Im}\,\omega \leq 0$ and it is called *unstable* when $\mathrm{Im}\,\omega > 0$. In the first case the amplitude is constant or it exponentially decreases in time while in the second one there is an exponential growth of the amplitude. In practice, it appears difficult to find solutions of eq. (7) because of the rather complicated structure of the dielectric tensor. A quite general analysis of

the dispersion equation of the anisotropic system is given in [10]. The problem simplifies as we are interested in specific modes which are expected to be unstable. Namely, we look for solutions corresponding to the fluctuating current in the direction of the momentum surplus and the wave vector perpendicular to it.

As previously, the momentum distribution is assumed to be elongated in the z-direction, and consequently the fluctuating current also flows in this direction. The magnetic field has a non-vanishing component along the y-direction and the electric filed is in the z-direction. Finally, the wave vector is parallel to the axis x, see fig. 1. It is also assumed that the momentum distribution obeys the mirror symmetry $f(-\mathbf{p}) = f(\mathbf{p})$, and then the permittivity tensor has only non-vanishing diagonal components. Taking into account all these conditions, one simplifies the dispersion equation (7) to the form

$$H(\omega) \equiv k_x^2 - \omega^2 \epsilon^{zz}(\omega, k_x) = 0. \tag{8}$$

The existence of unstable solutions of eq. (8) can be proved without solving it. The so-called Penrose criterion [13], which follows from analytic properties of the permittivity as a function of ω, states that *the dispersion equation $H(\omega) = 0$ has unstable solutions if $H(\omega = 0) < 0$*. The Penrose criterion was applied to eq. (8) in [6]. A more general discussion of the instability condition is presented in [11]. Without entering into details, there exist unstable modes if the momentum distribution averaged (with a proper weight) over momentum length is anisotropic.

To solve the dispersion equation (8), the parton momentum distribution has to be specified. Several analytic (usually approximate) solutions of the dispersion equation can be found in [6,10,11]. A typical example of the numerical solution, which gives the unstable mode frequency in the full range of wave vectors is shown in fig. 2 taken from [9]. The mode is pure imaginary and $\gamma_k \equiv \mathrm{Im}\,\omega(k_\perp)$. The parameters $\sigma_\parallel$ and $\sigma_\perp$ control the widths of longitudinal (z) and transverse momentum distributions; the coupling is $\alpha_s \equiv g^2/4\pi = 0.3$, and the effective parton density is chosen to be $6\,\mathrm{fm}^{-3}$. As seen, there is a finite interval of wave vectors for which the unstable modes exist.

4 Isotropization and Abelianization

When the instabilities grow the system becomes more isotropic because the Lorentz force changes particle's momenta and the growing fields carry an extra momentum. To explain the mechanism I assume, as previously, that initially there is a momentum surplus in the z-direction. The fluctuating current tends to flow in the z-direction with the wave vector pointing in the x-direction. Since the magnetic field has a y component, the Lorentz force, which acts on partons flying along the z-axis, pushes the partons in the x-direction where there is a momentum deficit. Numerical simulations discussed in the next section show the efficiency of the mechanism.

The system isotropizes not only due to the effect of the Lorentz force but also due to the momentum carried by the growing field. When the magnetic and electric fields are oriented along the y and z axes, respectively, the Poynting vector points in the direction x that is along the wave vector. Thus, the momentum carried by the fields is oriented in the direction of the momentum deficit of particles.

One wonders whether non-Abelian non-linearities do not stabilize the unstable modes. An elegant argument [17] suggests that this is not the case as the system spontaneously chooses an Abelian configuration in the course of instability development. Let me explain the idea.

In the Coulomb gauge the effective potential of the unstable configuration has the form

$$V_{\mathrm{eff}}[\mathbf{A}^a] = -\mu^2 \mathbf{A}^a \cdot \mathbf{A}^a + \frac{1}{4}g^2 f^{abc} f^{ade}(\mathbf{A}^b \mathbf{A}^d)(\mathbf{A}^c \mathbf{A}^e),$$

which is shown in fig. 3 taken from [17]. The first term (with $\mu^2 > 0$) is responsible for the very existence of the instability. The second term, which comes from the Yang-Mills Lagrangian, is of pure non-Abelian nature. The term is positive and thus it counteracts the instability growth. However, the non-Abelian term vanishes when the potential $\mathbf{A}^a$ is effectively Abelian, and consequently, such a configuration corresponds to the steepest decrease of the effective potential. Thus, the system spontaneously abelianizes in the course of instability growth. The abelianization is further discussed in sect. 7.

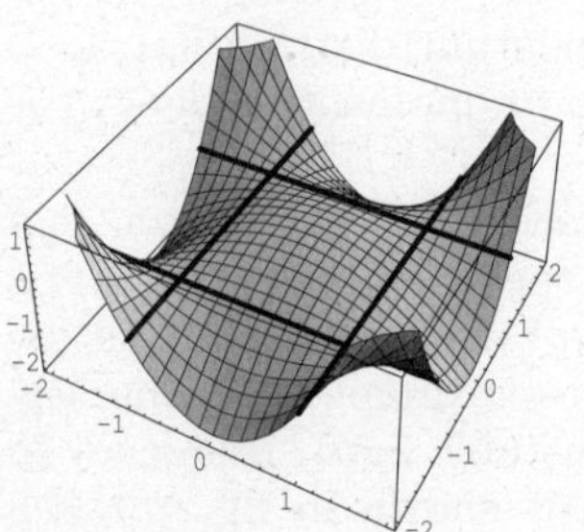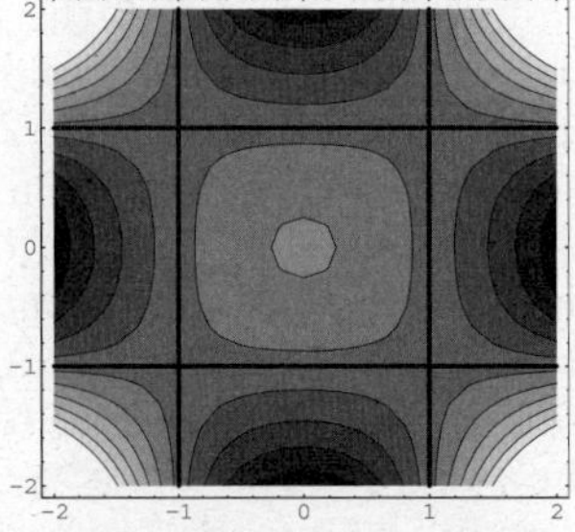

Fig. 3. The effective potential of the unstable magnetic mode as a function of magnitude of two colour components of $\mathbf{A}^a$ belonging to the $SU(2)$ gauge group. The figure is taken from [17].

5 Hard-loop effective action

Knowledge of the gluon polarization tensor or, equivalently, the chromoelectric permittivity tensor is sufficient to discuss the system's stability and the dispersion relations of unstable modes. For more detailed dynamical studies the effective action of anisotropic QGP is needed. Such an action for a system, which is on average locally colour neutral, stationary and homogeneous, was derived in [18], see also [19]. The starting point was the effective action which describes an interaction of classical fields with currents induced by these fields in the plasma. The Lagrangian density is quadratic in the gluon and quark fields and it equals

$$\mathcal{L}_2(x) = -\int \mathrm{d}^4 y \left(\frac{1}{2} A_\mu^a(x) \Pi_{ab}^{\mu\nu}(x-y) A_\nu^b(y) \right.$$
$$\left. + \bar{\Psi}(x) \Sigma(x-y) \Psi(y) \right); \tag{9}$$

the Fourier-transformed gluon polarization tensor $\Pi_{ab}^{\mu\nu}(k)$ is given by eq. (6) while the quark self-energy $\Sigma(k)$ reads [16,20]

$$\Sigma(k) = g^2 \frac{N_c^2 - 1}{8N_c} \int \frac{\mathrm{d}^3 p}{(2\pi)^3} \frac{\tilde{f}(\mathbf{p})}{|\mathbf{p}|} \frac{p \cdot \gamma}{p \cdot k}, \tag{10}$$

where $\tilde{f}(\mathbf{p}) \equiv n(\mathbf{p}) + \bar{n}(\mathbf{p}) + 2n_g(\mathbf{p})$. The action (9) holds under the assumption that the field amplitude is much smaller than T/g, where T denotes the characteristic momentum of (hard) partons.

Following Braaten and Pisarski [21], the Lagrangian (9) was modified to comply with the requirement of gauge invariance. The final result, which is non-local but manifestly gauge invariant, is [18]

$$\mathcal{L}_{\mathrm{HL}}(x) = \frac{g^2}{2} \int \frac{\mathrm{d}^3 p}{(2\pi)^3} \left[f(\mathbf{p}) F_{\mu\nu}^a(x) \left(\frac{p^\nu p^\rho}{(p \cdot D)^2} \right)_{ab} F_\rho^{\ b\mu}(x) \right.$$
$$\left. + i \frac{N_c^2 - 1}{4N_c} \tilde{f}(\mathbf{p}) \bar{\Psi}(x) \frac{p \cdot \gamma}{p \cdot D} \Psi(x) \right], \tag{11}$$

where $F_a^{\mu\nu}$ is the strength tensor and D denotes the covariant derivative. The effective action (11) generates n-point functions which obey the Ward-Takahashi identities. For the equilibrium plasma the action (11) is equivalent to that derived in [22] and in the explicitly gauge-invariant form in [21]. The equilibrium hard-loop action was also found within the semiclassical kinetic theory [23,24].

6 Equations of motion

Transport theory provides a natural framework to study temporal evolution of non-equilibrium systems and it has been applied to QGP for a long time. The distribution functions of quarks (Q), antiquarks ($\bar{Q}$), and gluons (G), which are the $N_c \times N_c$ and $(N_c^2 - 1) \times (N_c^2 - 1)$ matrices, respectively, satisfy the transport equations of the form [25,26]:

$$p^\mu D_\mu Q(\mathbf{p}, x) + \frac{g}{2} p^\mu \left\{ F_{\mu\nu}(x), \frac{\partial Q(\mathbf{p}, x)}{\partial p_\nu} \right\} = 0,$$

$$p^\mu D_\mu \bar{Q}(\mathbf{p}, x) - \frac{g}{2} p^\mu \left\{ F_{\mu\nu}(x), \frac{\partial \bar{Q}(\mathbf{p}, x)}{\partial p_\nu} \right\} = 0, \qquad (12)$$

$$p^\mu \mathcal{D}_\mu G(\mathbf{p}, x) + \frac{g}{2} p^\mu \left\{ \mathcal{F}_{\mu\nu}(x), \frac{\partial G(\mathbf{p}, x)}{\partial p_\nu} \right\} = 0,$$

where $\{\ldots, \ldots\}$ denotes the anticommutator; the transport equation of (anti)quarks is written down in the fundamental representation while that of gluons in the adjoint one. Since the instabilities of interest are very fast, much faster than the inter-parton collisions, the collision terms are neglected in eqs. (12). The gauge field, which enters the transport equations (12), is generated self-consistently by the quarks and gluons. Thus, the transport equations (12) should be supplemented by the Yang-Mills equation

$$D_\mu F^{\mu\nu}(x) = j^\nu(x), \qquad (13)$$

where the colour current is given as

$$j^\mu(x) = -g \int \frac{\mathrm{d}^3 p}{(2\pi)^3} \frac{p^\mu}{|\mathbf{p}|} \tau_a \, \mathrm{Tr} \left[\tau_a (Q - \bar{Q}) + T_a G \right], \qquad (14)$$

with τ_a and T_a being the $SU(N_c)$ group generators in the fundamental and adjoint representation, respectively. There is a version of eqs. (12), (13) where colour charges of partons are treated as a classical variable [27]. Then, the distribution functions depend not only on x and $\mathbf{p}$ but on the colour variable as well.

When eqs. (12), (13) are linearized around the state, which is stationary, homogeneous and locally colourless, the equations provide the hard-loop dynamics encoded in the effective action (11). The equations are of particularly simple and elegant form when the quark $\delta Q(\mathbf{p}, x)$, antiquark $\delta \bar{Q}(\mathbf{p}, x)$ and gluon $\delta G(\mathbf{p}, x)$ deviations from the stationary state described by $Q_0^{ij}(\mathbf{p}) = \delta^{ij} n(\mathbf{p})$, $\bar{Q}_0^{ij}(\mathbf{p}) = \delta^{ij} \bar{n}(\mathbf{p})$, and $G_0^{ab}(\mathbf{p}) = \delta^{ab} n_g(\mathbf{p})$ are parameterised by the field $W^\mu(\mathbf{v}, x)$ through the relations

$$\delta Q(\mathbf{p}, x) = g \frac{\partial n(\mathbf{p})}{\partial p^\mu} W^\mu(\mathbf{v}, x),$$

$$\delta \bar{Q}(\mathbf{p}, x) = -g \frac{\partial \bar{n}(\mathbf{p})}{\partial p^\mu} W^\mu(\mathbf{v}, x),$$

$$\delta G(\mathbf{p}, x) = g \frac{\partial n_g(\mathbf{p})}{\partial p^\mu} T_a \, \mathrm{Tr} \left[\tau_a W^\mu(\mathbf{v}, x) \right],$$

where $\mathbf{v} \equiv \mathbf{p}/|\mathbf{p}|$. Then, instead of the three transport equations (12) one has one equation

$$v_\mu D^\mu W^\nu(\mathbf{v}, x) = -v_\rho F^{\rho\nu}(x) \qquad (15)$$

while the Yang-Mills equation (13) reads

$$D_\mu F^{\mu\nu}(x) = -g^2 \int \frac{\mathrm{d}^3 p}{(2\pi)^3} \frac{p^\nu}{|\mathbf{p}|} \frac{\partial f(\mathbf{p})}{\partial p^\rho} W^\rho(\mathbf{v}, x), \qquad (16)$$

where $v^\mu = (1, \mathbf{v})$. In contrast to the effective action (11), eqs. (15), (16) are local in coordinate space. Therefore, the transport equation (15) combined with eq. (16) is often called the local representation of the hard-loop dynamics. Equations (15), (16), which for the isotropic equilibrium plasma were first given in [28], are used in the numerical simulations [7,29–32] discussed in the next section.

Recently the fluid equations, which are applicable to short-time scale colour phenomena in QGP, have been derived [33] from the kinetic equations (12). The quantities, which enter the equations, like the hydrodynamic velocity or pressure are gauge dependent matrices in the colour space. The chromo-hydrodynamic approach is designed for numerical studies of the dynamics of the unstable QGP.

7 Numerical simulations

Temporal evolution of the anisotropic QGP was studied by means of numerical simulations [7,12,29–31,34,32, 35]. The dynamics governed by a complete hard-loop action [18], was simulated in [7,29–32]. These simulations provide fully reliable information on the field dynamics but particles are included as a stationary (anisotropic) background. The simulations [12,34,35] treat the quark-gluon system completely classically: partons, which carry classical colour charges, interact with a self-consistently generated classical chromodynamic field. The simulations [7,12] were effectively performed in $1+1$ dimensions as the chromodynamic potentials depend on time and one space variable. The calculations [29,30] represent full $1+3$ dimensional dynamics. In most cases the $SU(2)$ gauge group was studied but some $SU(3)$ results, which are qualitatively very similar to $SU(2)$ ones, are given in [30]. The techniques of discretization used in [7,12,29,30] are rather different while the initial conditions are quite similar. The initial field amplitudes are distributed according to the Gaussian noise and the momentum distribution of partons is strongly anisotropic.

In fig. 4, taken from [7], the results of the hard-loop simulation performed in $1+1$ dimensions are shown. One observes exponential growth of the field energy density which is dominated, as expected, by the magnetic field which is transverse to the direction of the momentum deficit. The growth rate appears to be equal to that of the fastest unstable mode (γ^*). Figure 5, taken from [12], shows results of the classical simulation on the $(1 + 1)$-dimensional lattice of physical size $L = 40 \, \mathrm{fm}$. As in fig. 4, the amount of field energy grows exponentially and the magnetic contribution dominates.

The Abelian $(U(1))$ and non-Abelian $(SU(2))$ results of the $(1 + 1)$-dimensional simulation presented in fig. 5 are remarkably similar to each other. The abelianization appears to be very efficient in $1+1$ dimensions, as shown in

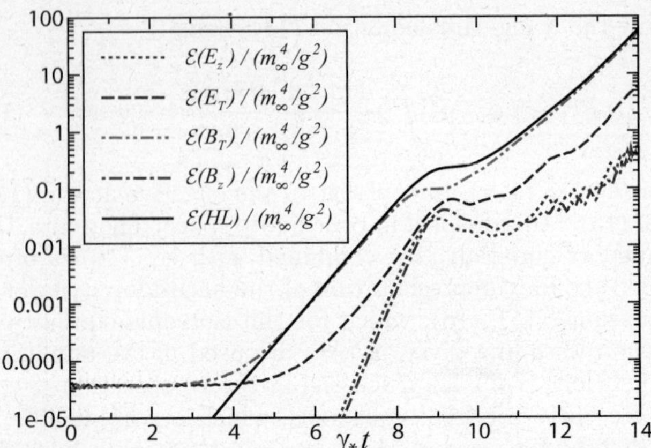

Fig. 4. Time evolution of the (scaled) energy density (split into various electric and magnetic components) which is carried by the chromodynamic field. The simulation is $(1+1)$-dimensional and the gauge group is $SU(2)$. The parton momentum distribution is squeezed along the z-axis. The solid line corresponds to the total energy transferred from the particles. The figure is taken from [7].

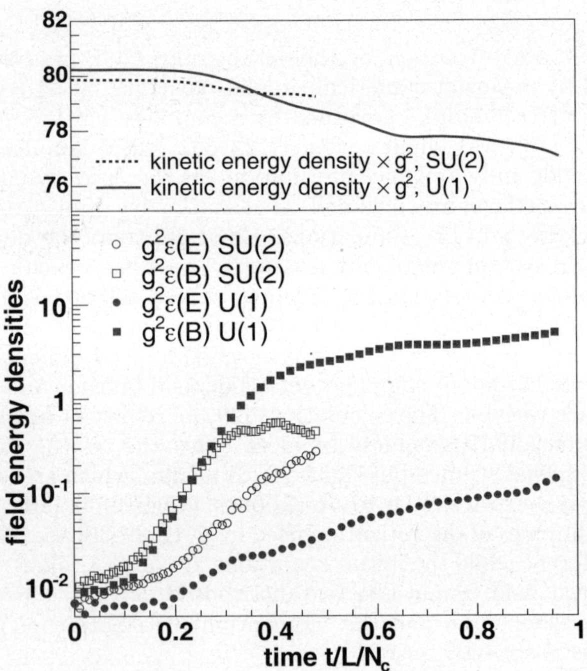

Fig. 5. Time evolution of the kinetic energy of particles (upper panel) and of the energy of electric and magnetic fields (lower panel). The figure is taken from [12].

figs. 6, 7, taken from [12] and [7], respectively. The authors of [12] analysed the functionals:

$$\phi_{\mathrm{rms}} \equiv \sqrt{\int_0^L \frac{\mathrm{d}x}{L}\left(A_y^a A_y^a + A_z^a A_z^a\right)},$$

$$\bar{C} \equiv \int_0^L \frac{\mathrm{d}x}{L}\frac{\sqrt{\mathrm{Tr}[(i[A_y, A_z])^2]}}{\mathrm{Tr}[A_y^2 + A_z^2]}. \tag{17}$$

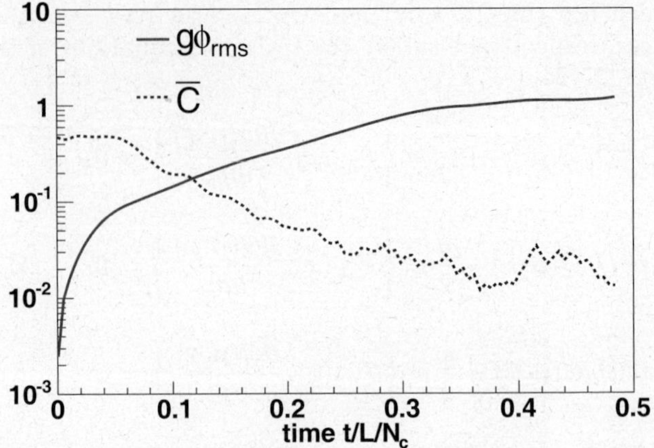

Fig. 6. Temporal evolution of the functionals $\bar{C}$ and ϕ_{rms} measured in GeV. The figure is taken from [12].

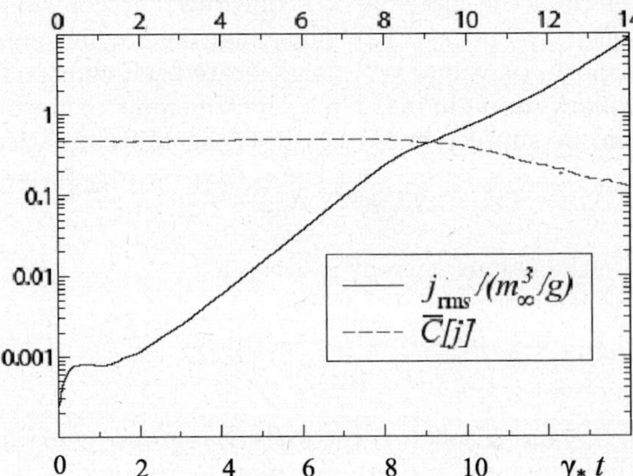

Fig. 7. Temporal evolution of the (scaled) functionals $\bar{C}$ and j_{rms}. The figure is taken from [7].

The quantities j_{rms} and $\bar{C}$, studied in [7] and shown in fig. 7, are fully analogous to ϕ_{rms} and $\bar{C}$ defined by eq. (17) but the components of the chromodynamic potential are replaced by the respective components of the colour current. As seen in figs. 6, 7, the field (current) commutator decreases in time although the magnitude of the field (current), as quantified by ϕ_{rms} (j_{rms}), grows.

The results of the $(1+3)$-dimensional simulations [29, 30] are qualitatively different from those of $1+1$ dimensions. As seen in figs. 8, 9, taken from [29,30], respectively, the growth of the field energy density is exponential only for some time, and then the growth becomes approximately linear. The regime changes when the field's amplitude is of order k/g, where k is the characteristic wave vector. Then, the non-Abelian effects start to be important. Figure 10 taken from [29] demonstrates that the abelianization is efficient in $1+3$ dimensions only for a finite interval of time. The commutator C shown in fig. 10 is a natural generalization of the $(1+1)$-dimensional commutator defined by eq. (17).

The regime of linear growth of the magnetic energy, shown in figs. 8, 9 was studied numerically in [31]. It was found that when the exponential growth of the magnetic

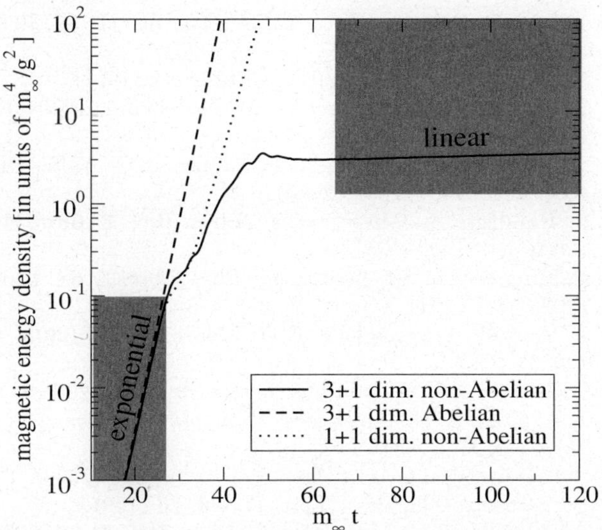

Fig. 8. Time evolution of the (scaled) chromomagnetic energy density in the $(1+3)$-dimensional simulation. The Abelian result and that of $1+1$ dimensions are also shown. The figure is taken from [29].

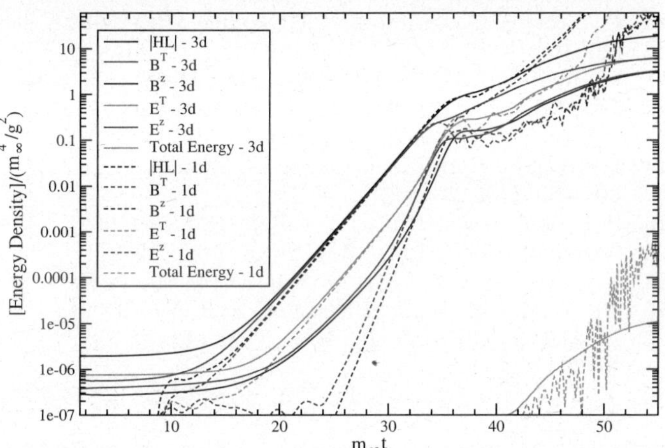

Fig. 9. Time evolution of the (scaled) energy density (split into various electric and magnetic components) of the chromodynamic field in the $1+1$ and $1+3$ simulations. "HL" denotes the total energy contributed by hard particles. The figure is taken from [30].

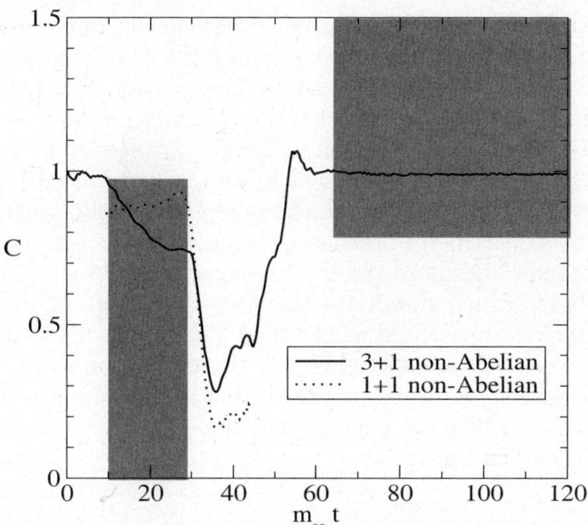

Fig. 10. Temporal evolution of the field commutator quantified by C. The figure is taken from [29].

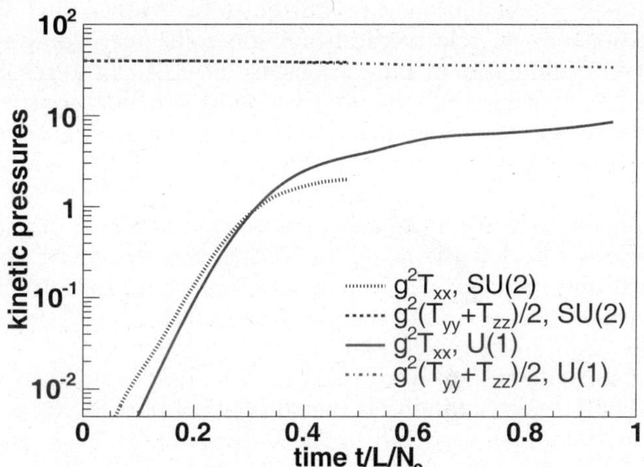

Fig. 11. Temporal evolution of the energy-momentum tensor components T^{xx} and $(T^{yy} + T^{zz})/2$. The Abelian and non-Abelian results are shown. The figure is taken from [12].

energy ends, the long-wavelength modes associated with the instability stop growing, but that they cascade energy towards the ultraviolet in the form of plasmon excitations and a quasi-stationary state with the power law distribution k^{-2} of the plasmon mode population appears. The phenomenon was argued [31] to be very similar to the Kolmogorov wave turbulence where the long-wavelength modes transfer their energy without dissipation to the shorter and shorter ones.

A different picture of the non-Abelian regime emerges from the classical simulation [35] where the system with strong momentum anisotropy was studied. When the field strength is high enough, the energy drained by the Weibel-like plasma instability from the particles does not build up exponentially in magnetic fields but instead returns

isotropically to the ultraviolet not via the quasi-stationary process, as argued in [31], but via a rapid avalanche.

The effect of isotropization of particle momentum distribution due to the action of the Lorentz force is nicely seen in the $(1+1)$-dimensional classical simulation [12]. In fig. 11 taken from [12] there are shown diagonal components of the energy-momentum tensor $T^{\mu\nu}$. The initial momentum distribution is such that $T^{xx} = 0$. As seen, T^{xx} exponentially grows.

The numerical studies discussed so far deal with the quark-gluon system of constant volume. A very elegant formulation of the hard-loop dynamics of the system, which experiences the boost invariant expansion in one direction, is given in [32]. In agreement with the earlier expectations [9,11], the expansion is shown both numerically and analytically [32] to slow down the growth of instabilities even when the initial state is highly anisotropic. The field amplitude does not grow exponentially with time but rather as the exponent of $\sqrt{t}$. The effect of expansion requires further quantitative analysis, as the instabilities might occur irrelevant for heavy-ion collisions, if they are not fast enough to cope with the system's expansion.

An attempt to study an unstable parton system in the conditions close to those, which are realized in relativistic heavy-ion collisions, was undertaken in [36–38]. The system was described in terms of the colour glass condensate approach [39] where small-x partons of large occupation numbers, which dominate the wave functions of incoming nuclei, are treated as classical Yang-Mills fields. Hard modes of the classical fields play the role of particles. The instabilities, identified as the Weibel modes, appear to be generated when the system of Yang-Mills fields expands into the vacuum.

8 Outlook

Although an impressive progress has been achieved, the numerical simulations are still quite far from a real situation met in relativistic heavy-ion collisions. Complete $(1 + 3)$-dimensional simulations are needed as the results of [29,30] show that the dimensionality crucially matters. The system expansion needs to be incorporated. The effect of back reaction of fields on the particles is fully included only in the classical simulations [12,35–38]. The effect is difficult to study in quantum field approaches as it goes beyond the hard-loop physics which has appeared very rich and complex [29,30]. An attempt to go beyond the hard-loop approximation was undertaken in [40] where the higher-order terms of the effective potential of the anisotropic system were found. Since these terms can be negative, the instability is then driven not only by the negative quadratic term but by the higher-order terms as well.

The coupling constant is assumed to be small in all studies of the unstable parton systems. This is certainly a severe limitation as the phenomenology of heavy-ion collisions suggests that QGP manifests very small viscosity characteristic for strongly coupled systems [3]. However, it has been recently argued [41,42] that an anomalously small viscosity of the quark-gluon system can arise from interactions with turbulent colour fields dynamically generated by the instabilities. Therefore, it might well be that the scenario of instabilities-driven equilibration does not only solve the problem of fast thermalization but other puzzling features of QGP.

References

1. U.W. Heinz, AIP Conf. Proc. **739**, 163 (2005).
2. F. Retiere, J. Phys. G **30**, S827 (2004).
3. E. Shuryak, J. Phys. G **30**, S1221 (2004).
4. R. Baier, A.H. Mueller, D. Schiff, D.T. Son, Phys. Lett. B **539**, 46 (2002).
5. P. Arnold, D.T. Son, L.G. Yaffe, Phys. Rev. D **59**, 105020 (1999).
6. St. Mrówczyński, Phys. Rev. C **49**, 2191 (1994).
7. A. Rebhan, P. Romatschke, M. Strickland, Phys. Rev. Lett. **94**, 102303 (2005).
8. P. Arnold, J. Lenaghan, G.D. Moore, L.G. Yaffe, Phys. Rev. Lett. **94**, 072302 (2005).
9. J. Randrup, St. Mrówczyński, Phys. Rev. C **68**, 034909 (2003).
10. P. Romatschke, M. Strickland, Phys. Rev. D **68**, 036004 (2003).
11. P. Arnold, J. Lenaghan, G.D. Moore, JHEP **0308**, 002 (2003).
12. A. Dumitru, Y. Nara, Phys. Lett. B **621**, 89 (2005).
13. N.A. Krall, A.W. Trivelpiece, *Principles of Plasma Physics* (McGraw-Hill, New York, 1973).
14. St. Mrówczyński, Acta Phys. Pol. B **37**, 427 (2006).
15. St. Mrówczyński, Phys. Lett. B **393**, 26 (1997).
16. St. Mrówczyński, M.H. Thoma, Phys. Rev. D **62**, 036011 (2000).
17. P. Arnold, J. Lenaghan, Phys. Rev. D **70**, 114007 (2004).
18. St. Mrówczyński, A. Rebhan, M. Strickland, Phys. Rev. D **70**, 025004 (2004).
19. R.D. Pisarski, arXiv:hep-ph/9710370.
20. P. Arnold, G.D. Moore, L.G. Yaffe, JHEP **0301**, 039 (2003).
21. E. Braaten, R.D. Pisarski, Phys. Rev. D **45**, 1827 (1992).
22. J.C. Taylor, S.M.H. Wong, Nucl. Phys. B **346**, 115 (1990).
23. J.P. Blaizot, E. Iancu, Nucl. Phys. B **417**, 608 (1994).
24. P.F. Kelly, Q. Liu, C. Lucchesi, C. Manuel, Phys. Rev. D **50**, 4209 (1994).
25. H.T. Elze, U.W. Heinz, Phys. Rep. **183**, 81 (1989).
26. St. Mrówczyński, Phys. Rev. D **39**, 1940 (1989).
27. U.W. Heinz, Ann. Phys. (N.Y.) **161**, 48 (1985).
28. J.P. Blaizot, E. Iancu, Phys. Rep. **359**, 355 (2002).
29. P. Arnold, G.D. Moore, L.G. Yaffe, Phys. Rev. D **72**, 054003 (2005).
30. A. Rebhan, P. Romatschke, M. Strickland, JHEP **0509**, 041 (2005).
31. P. Arnold, G.D. Moore, Phys. Rev. D **73**, 025006 (2006).
32. P. Romatschke, A. Rebhan, arXiv:hep-ph/0605064.
33. C. Manuel, St. Mrówczyński, arXiv:hep-ph/0606276.
34. A. Dumitru, Y. Nara, Eur. Phys. J. A **29**, 65 (2006).
35. A. Dumitru, Y. Nara, M. Strickland, arXiv:hep-ph/0604149.
36. P. Romatschke, R. Venugopalan, Phys. Rev. Lett. **96**, 062302 (2006).
37. P. Romatschke, R. Venugopalan, Eur. Phys. J. A **29**, 71 (2006).
38. P. Romatschke, R. Venugopalan, Phys. Rev. D **74**, 045011 (2006).
39. E. Iancu, R. Venugopalan, in *Quark-Gluon Plasma 3*, edited by R.C. Hwa, X.N. Wang (World Scientific, Singapore, 2004).
40. C. Manuel, St. Mrówczyński, Phys. Rev. D **72**, 034005 (2005).
41. M. Asakawa, S.A. Bass, B. Muller, Phys. Rev. Lett. **96**, 252301 (2006).
42. M. Asakawa, S.A. Bass, B. Muller, arXiv:hep-ph/0608270.

J/Ψ production at RHIC

Frédéric Fleuret

Laboratoire Leprince-Ringuet, École Polytechnique/IN2P3, Palaiseau 91128, France

Abstract. The study of J/Ψ production is expected to provide evidence for the production of the Quark-Gluon Plasma (QGP) in relativistic heavy ion collisions. Several models indicate that the production of a QGP will result in changes to the J/Ψ yield. At CERN, the NA50 experiment observed a strong suppression compared to expectations. At RHIC, the PHENIX experiment has measured J/Ψ production at 200 GeV, in pp and d+Au collisions to study cold nuclear effects, in Au+Au and Cu+Cu collisions to study hot and dense medium effects. Here the PHENIX results will be presented and compared to various theoretical models.

PACS.

The production of charmonium states in relativistic heavy ion collisions has been a subject of extreme interest along the past twenty years. As first predicted by Matsui and Satz [1], the suppression of charmonia production is expected to be an unambigous signature for the formation of a Quark Gluon Plasma (QGP). A first anomalous J/Ψ suppression, increasing with the centrality of the collision, has been observed by the NA50 experiment at CERN/SPS in Pb+Pb collisions at $\sqrt{s_{NN}} = 17.3$ GeV, leading to an intense theoretical work. Since 2001, the PHENIX experiment, one of the four experiments running at the Relativistic Heavy Ion Collider (RHIC) has measured the J/Ψ production in p+p, d+Au, Cu+Cu and Au+Au collisions at a center-of-mass energy per nucleon pair up to 200 GeV. The PHENIX experiment measures J/Ψ via its dilepton decay in two central arms (electron channel) covering the mid-rapidity region |y| < 0.35 and two forward arms (muon channel) covering forward and backward rapidity region 1.2< |y| <2.2. Here we will review the results obtained by the PHENIX experiment and compare them to the CERN/SPS results, as well as to theoretical expectations. In order to study "anomalous" suppression induced by Hot and Dense Matter (HDM) effects, we will, in a first part, cover "normal" suppression, the Cold Nuclear Matter (CNM) effects induced by the fact that J/Ψ's are produced in nuclei. These effects are studied with p+p and d+Au results. In a second part, we will review the results obtained at RHIC in Cu+Cu and Au+Au interactions and compare them with SPS results. Finally, we will compare these results with theoretical expectations.

1 Cold nuclear effects

The p+p measurement defines the baseline to study nuclear effects for J/Ψ production. It is used to compute the nuclear modification factor R_{AB}:

$$R_{AB} = \frac{dN_{AB}^{J\psi}}{\langle N_{coll} \rangle \times dN_{pp}^{J/\psi}} \tag{1}$$

where $dN_{AB}^{J/\psi}$ and $dN_{pp}^{J/\psi}$ are respectively the J/Ψ yield observed in A+B collisions and p+p collisions, and $\langle N_{coll} \rangle$ is the average number of nucleon-nucleon collisions (extracted from a Glauber model) occuring in a A+B collision. Any nuclear effect for J/Ψ production leads to a deviation of the R_{AB} ratio from unity. As already mentioned, nuclear effects can come either from Cold Nuclear Matter (CNM) or Hot and Dense Matter (HDM). While HDM effects are studied in A+B (Au+Au or Cu+Cu) collisions, CNM effects have been studied with d+Au collisions (where only CNM effects are expected to occur). Figure 1 shows R_{dAu} (the R_{AB} ratio for d+Au collisions) as a function of rapidity [2]. It exhibits modest CNM effects which can be fairly reproduced by models [4] incorporating weak gluon shadowing which induces a modification of the parton structure functions and weak nuclear absorption which comes from the fact that J/Ψ's are produced in nuclei and can thus interact with the surrounding nucleons [1]. Figure 2 shows the R_{dAu} ratio as a function of

---

[1] This last effect is well established. At SPS energies, the absorption cross section has been measured to be 4.18 ± 0.35 mb [3], while at RHIC, its value seems to stand between 1 (the favored one) and 3 mb (the maximum absorption).

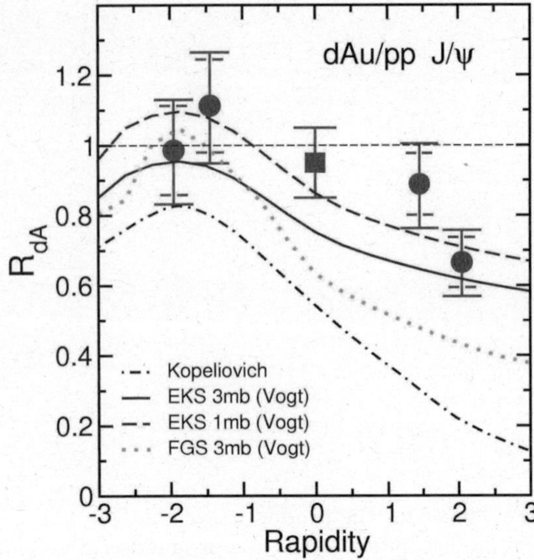

Fig. 1. Nuclear modification ratio for J/ψ production in d+Au collisions as function of rapidity. The lines correspond to several theoretical models [4][6]. Data are well reproduced by a model (EKS 1mb) involving weak gluon shadowing and small absorption cross-section (1 mb).

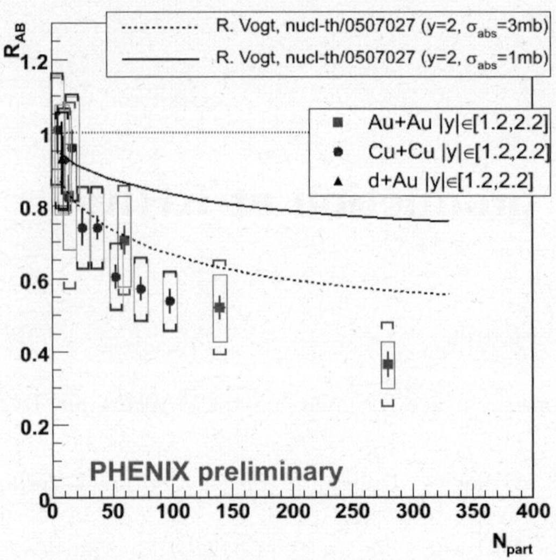

Fig. 3. Nuclear modification factor as a function of centrality (given here by the number of participants) for d+Au, Cu+Cu and Au+Au data at forward rapidity. The curves correspond to theoretical predictions [5] for Au+Au collisions, which include weak shadowing and small abosrption cross-section (1 mb, upper curve, and 3 mb, lower curve).

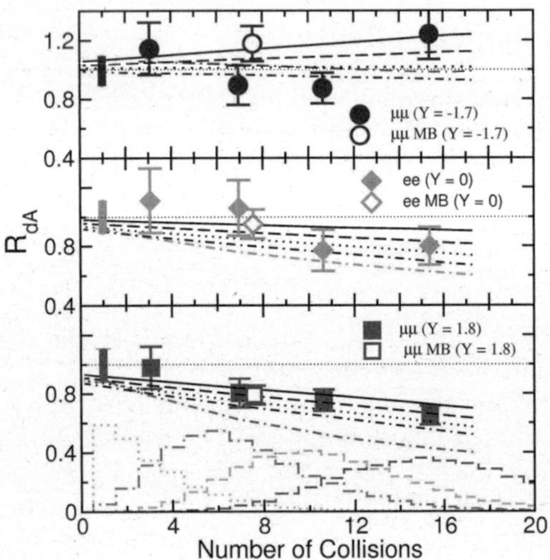

Fig. 2. Nuclear modification ratio for J/ψ production in d+Au collisions as a function of the centrality (given here by the number of collisions N_{coll}). Colored lines correspond to FGS parametrization [4]; black lines correspond to EKS parametrization [4] with several absorption cross-sections, from 0 (upper line) to 3 mb (lower line).

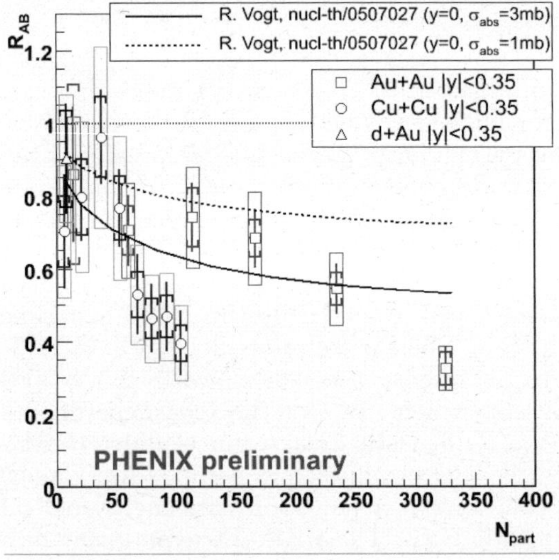

Fig. 4. Nuclear modification factor as a function of centrality (given here by the number of participants) for d+Au, Cu+Cu and Au+Au data at forward rapidity. The curves correspond to theoretical predictions [5] for Au+Au collisions, which include weak shadowing and small abosrption cross-section (upper curve : 1 mb, lower curve : 3 mb).

2 Cu+Cu and Au+Au results

the centrality. Again, weak gluon shadowing and weak absorption cross-section seem to be at play. It's important to note that CNM effects depend on the centrality of the collision. Consequently, one has to take these effects into account in order to estimate the Hot and Dense Matter effects.

In order to study HDM effects, we consider the results obtained with Au+Au collisions collected in 2004 (241 μb^{-1}) and Cu+Cu collisions collected in 2005 (3.06 nb^{-1}), both at $\sqrt{s_{NN}} = 200$ GeV. Figures 3 and 4 show the results obtained for d+Au, and the preliminary results for Cu+Cu and Au+Au data, both at forward and central rapidities [7]. The lines show theoretical predictions [5] in Au+Au

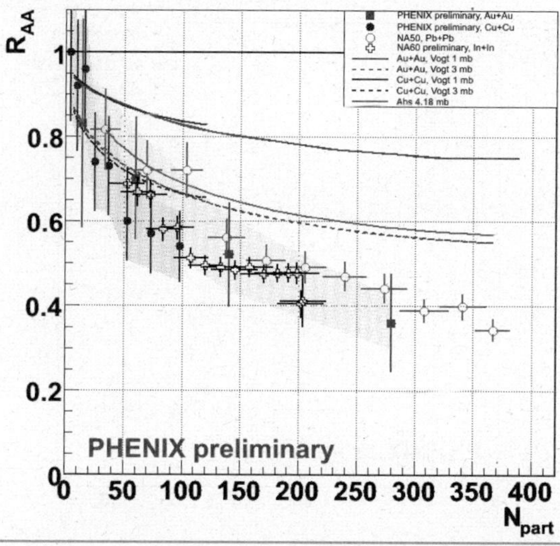

Fig. 5. Nuclear modification factor R_{AA} as a function of centrality (given here by the number of participants N_{part}) for NA50, NA60 and PHENIX data (at forward rapidity).

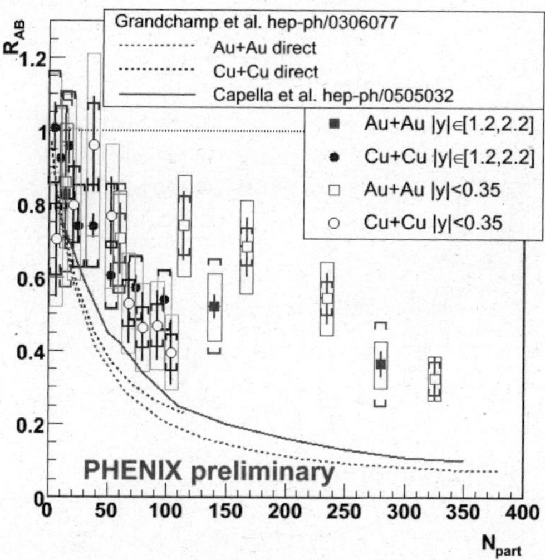

Fig. 6. Nuclear modification factor R_{AA} as a function of centrality (given here by the number of participants N_{part}) for Cu+Cu (blue circles) and Au+Au (red squares) PHENIX data at central (open symbols) and forward (closed symbols) rapidities. Lines are theoretical curves (see text).

collisions for cold nuclear effects. In both cases (at forward and backward rapidity), the J/Ψ suppression in most central Au+Au collisions goes significantly beyond the CNM effects. A factor of at least 2 is observed relative to expected CNM effects and a factor of 3 relative to binary scaled p+p results.

Figure 5 shows a comparison of the results obtained at $\sqrt{s_{NN}} = 200$ GeV at RHIC with the results obtained at lower energies ($\sqrt{s_{NN}} \sim 20$ GeV) at CERN by the NA50 [3] and NA60 [8] experiments. On the experimental point view, without taking into account any CNM effect, the results obtained at both energies are similar. At CERN, the CNM effects are well reproduced with a 4.18 mb absorption cross-section (without including any gluon shadowing effect). At RHIC, depending on the amount of nuclear absorption one considers, the suppression pattern due to CNM effect is similar (if taking 3 mb absorption cross-section) or smaller by a factor of ~ 1.3 (if taking 1 mb absorption cross-section), when comparing with SPS results. Depending of the amount of nuclear absorption one takes to reproduce RHIC data, one can conclude that the net additionnal suppression, coming from HDM effects, is either the same at SPS and RHIC, or larger at RHIC. This ambiguity points out the necessity to take additionnal d+Au data at RHIC to improve our knowledge of CNM effects observed at RHIC.

3 Comparison with models

Several models have been proposed to accomodate the results obtained both at SPS and RHIC. In the first subsection, we will cover the models which have been used to fit SPS data and their extensions to RHIC data. Then, we will show the regeneration models and finally we will present the expectations from a sequential suppression model.

3.1 Suppression models

Figure 6 shows Au+Au and Cu+Cu PHENIX results (at both rapidities) and their comparison with some theoretical curves which have been fitted on NA50 data and extrapolated to RHIC energy [9][10]. These models clearly overestimate the J/Ψ suppression observed at RHIC, suggesting that new mechanisms could be at work, J/Ψ regeneration, for instance.

3.2 Regeneration

The regeneration process is based on the idea that in a medium such as a Quark Gluon Plasma, if several c and $\bar{c}$ quarks are produced, then, a c quark from one initial $c\bar{c}$ pair can interact with a $\bar{c}$ quark from a different initial $c\bar{c}$ pair to form a J/Ψ, thus leading to an increase of the net J/Ψ production cross-section. The number of J/Ψ formed via such interactions is expected to be proportional to the number of $c\bar{c}$ combinations, which is roughly proportional to the square of the number of $c\bar{c}$ pair [9][11].

At CERN energies, this process doesn't occur since around 0.1 $c\bar{c}$ pairs are produced, on average, in each collision. On the other hand, at RHIC energies, the amount of $c\bar{c}$ pairs produced in a collision is larger than ten [12].

Figure 7 shows Au+Au and Cu+Cu PHENIX results (at both rapidities) and their comparison with several theoretical curves which include regeneration process. The lower decreasing dashed curves correspond to a direct suppression without regeneration (as in figure 6), the increasing dashed curve corresponds to the regeneration mechanism, and the full curves show the combination of direct suppression and regeneration. Even though the more central Au+Au points are not well reproduced by these curves, the agreement is much better than without any

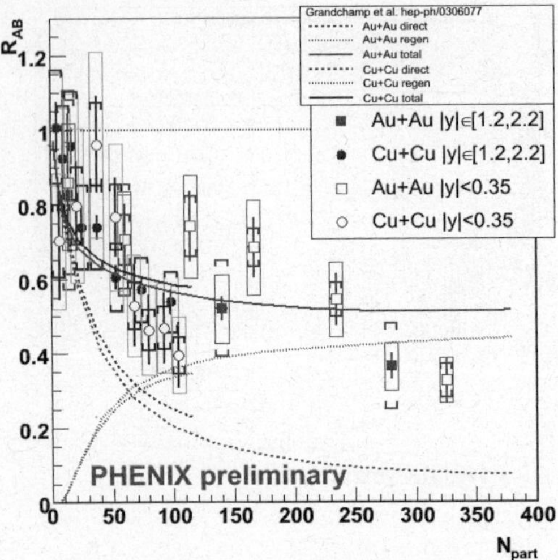

Fig. 7. Nuclear modification factor R_{AA} as a function of centrality (given here by the number of participants N_{part}) for Cu+Cu (blue circles) and Au+Au (red squares) PHENIX data at central (open symbols) and forward (closed symbols) rapidities. Lines are theoretical curves (see text).

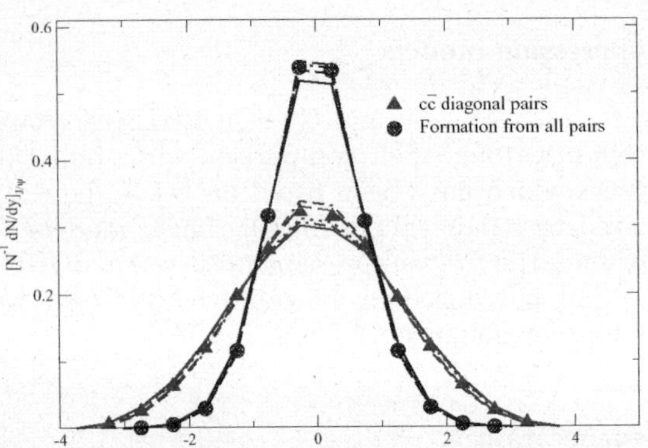

Fig. 8. Predicted rapidity spectra, within the regeneration framework [11], of J/Ψ in Au+Au reactions at 200 GeV.

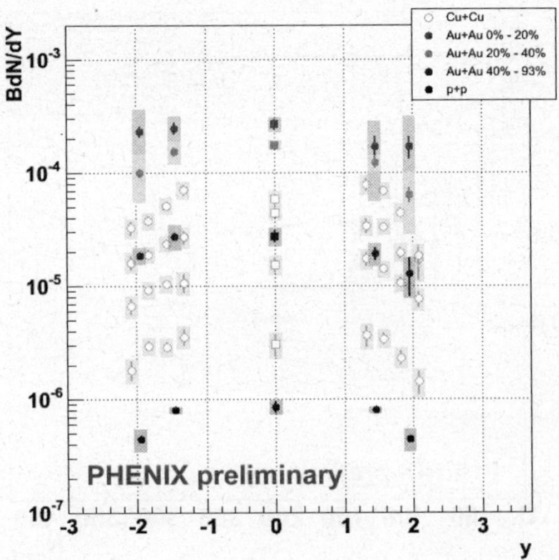

Fig. 9. Measured rapidity spectra for p+p (black points), Cu+Cu (gray points) and Au+Au (colored points). For Cu+Cu and Au+Au, data are displayed for several centrality bins (0% - 2% being the most central one).

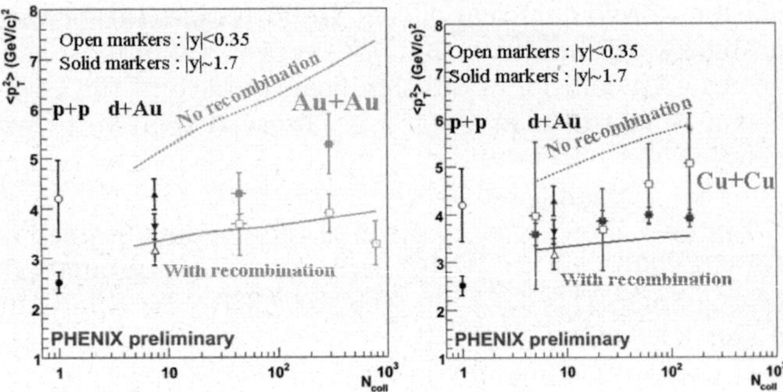

Fig. 10. Mean J/Ψ squared transverse momentum as a function of the number of binary collisions N_{coll}. Left plot shows p+p, d+Au and Au+Au data; right plot shows p+p, d+Au and Cu+Cu data. Lines correspond to theoretical curves with and without regeneration [11].

regeneration, advocating in favour of a regeneration mechanism interplay at RHIC energies.

Based on scenarios involving regeneration, predictions have been made, both for the J/Ψ rapidity and p_T distributions [11]. Figure 8 shows the predicted rapidity spectra of J/Ψ in Au+Au interactions at 200 GeV. The solid triangles correspond to initial J/Ψ production via diagonal $c\bar{c}$ pairs as it occurs in p+p interactions. Circles correspond to J/Ψ in-medium formation involving regeneration mechanism. As one can see, the spectrum involving regeneration is narrower than the one corresponding to a "standard" J/Ψ production.

Figure 9 shows the rapidity distributions measured by the PHENIX experiment for p+p, Cu+Cu and Au+Au interactions. Cu+Cu and Au+Au results are plotted for several bins of centrality (0% - 20% being the most central one). According to the data, the rapidity distributions are mostly flat for all the systems and all centrality bins. One can then draw the conclusion that no modification

in the rapidity shape is observed when comparing p+p to Cu+Cu and Au+Au data, which is in contradiction with the regeneration mechanism expectations.

Figure 10 shows $< p_T^2 >$, the mean J/Ψ squared transverse momentum, for various systems and centrality intervals (given here by the number of binary collisions N_{coll}). The observed broadening of the p_T spectra as a function of the number of collisions is usually attributed to the so-called Cronin effect, the initial state elastic scattering of the projectile parton in the nuclear target before the J/Ψ is produced. In Figure 10 the lines show the prediction made in the regeneration framework. Dashed lines correspond to the expected $< p_T^2 >$ distribution in case of no regeneration (usual Cronin effect); solid lines correspond to the expected $< p_T^2 >$ distribution in the scenario where full regeneration is at work. According to these plots, the "no recombination" (no regeneration) scenario doesn't fit the data. On the other hand the "with recombination" (regeneration) scenario exhibits a better agreement. Ac-

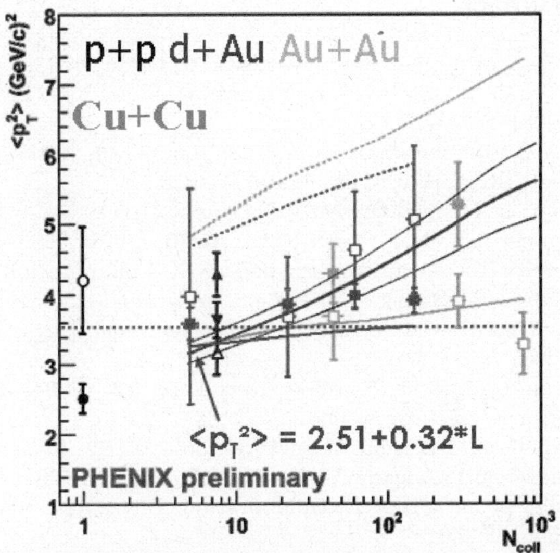

Fig. 11. Mean J/Ψ squared transverse momentum as a function of the number of binary collisions N_{coll} for p+p, d+Au, Cu+Cu and Au+Au data. Yellow and green lines are the same as in figure 10. Red line is a parametrization of Cronin effect derive from d+Au data [13]; the dotted lines on each side show the associated errors.

cording to this figure, the fit could well reproduce the data if one considers that both process are involved at the same time.

Within the PHENIX framework a recent phenomenological parametrization of the Cronin effect was done based on p+p and d+Au data [13]. The parametrization is made as a function of L the length of nuclear matter seen by the J/Ψ in the collision system :

$$\langle p_T^2 \rangle_{AA} = \langle p_T^2 \rangle_{pp} + \rho\sigma(\delta p_T^2) \times L = 2.51 + 0.32 \times L \quad (2)$$

where $\langle p_T^2 \rangle_{pp}$ is the mean squared transverse momentum measured in p+p interactions, ρ is the nuclear matter density, σ is the elastic parton-nucleon scattering cross-section, δp_T^2 is the average p_T quick accounting for one projectile parton scattering, and L is the lenght of nuclear matter seen by the J/Ψ [2]. The L values, corresponding to various N_{coll} values, are determined using a Glauber model. The value of $\rho\sigma(\delta p_T^2)$ given in equation 2 has been obtained by fitting p+p and d+Au data [3].

Figure 11 shows the results of this "Cronin" fit extrapolated to large N_{coll} values. One can see that, this phenomenological fit can fairly reproduce the Au+Au and Cu+Cu $\langle p_T^2 \rangle$ behavior at forward rapidity. At midrapidity, a better p+p baseline is needed to interpret the apparent flat $\langle p_T^2 \rangle$ pattern observed in Au+Au collisions.

As a summary of this section, one can conclude that the addition of the regeneration process in the J/Ψ pro-

---

[2] which is equivalent to the lenght of nuclear matter seen by the projectile parton since, on average, the J/Ψ is produced in the middle of the interaction region.

[3] Note that due to the very poor statistic of the p+p sample at midrapidity, the fit has been performed on forward rapidity data.

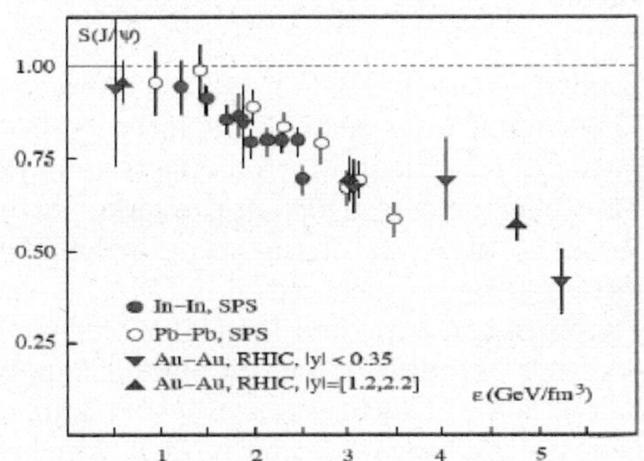

Fig. 12. J/Ψ suppression as a function of the energy density [14] for In+In (NA60), Pb+Pb (NA50) and Au+Au (PHENIX) data.

duction mechanisms is of good help to interpret the nuclear modification factor observed at RHIC. But its predictions for rapidity distribution behaviours are in contradiction with the data, and the predictions for $\langle p_T^2 \rangle$ distributions are not needed to interpret the $\langle p_T^2 \rangle$ behavior observed in the data.

3.3 Sequential suppression

Recently, a new approach has been proposed [14] to interpret the results observed at RHIC based on recent lattice QCD calculations. According to this new approach, the J/Ψ melting temperature in a QGP could be higher than initially predicted, leading to the fact that the suppression of direct[4] J/Ψ could be out of range at RHIC energies. Only the production of J/Ψ coming from χ_c and Ψ' decays could be suppressed (since their binding energy is smaller than the J/Ψ one). This "hierarchy" in the suppression mechanism should lead to a sequential suppression pattern. In [14], the authors define the J/Ψ survival probability $S_{J/\Psi}$ as follows :

$$S_{J/\Psi} = 0.6S_\Psi + 0.4S_x \quad (3)$$

where S_Ψ is the survival probability for direct J/Ψ and S_x is the survival probability of J/Ψ coming from χ_c [5] and Ψ' [6] decays. Since direct J/Ψ are expected to not be suppressed even at RHIC energies, one expects that at both SPS and RHIC energies, one should observe a decrease of $S_{J/\Psi}$ as a function of the energy density involved in the reaction, until the reach of a plateau at $S_{J/\Psi} = 0.6$.

Figure 12 shows $S_{J/\Psi}$ as a function of the energy density as described in [14] for both RHIC and SPS data. As expected by the authors, $S_{J/\Psi}$ is decreasing until an energy density of ~ 3 GeV/fm³. Wether data reach a plateau or not is debatable since the highest energy density point is below the 60% expected in this scenario. More data are needed, especially in Au+Au collisions at RHIC, to obtain a clearer picture.

---

[4] prompt J/Ψ which are not coming from χ_c and Ψ' decays.

[5] 30% of the total amount of produced J/Ψ.

[6] 10% of the total amount of produced J/Ψ.

4 Conclusion

The results obtained at RHIC on J/Ψ production by the PHENIX experiment have lead to an intense theoretical work on both Cold Nuclear Matter effects and Hot and Dense Matter effects. The J/Ψ suppression pattern observed in A+A collisions is of the same order of magnitude as the one already observed at CERN. It seems to exceed the suppression expected for Cold Nuclear Matter effects, suggesting an additionnal anomalous suppression. This anomalous suppression seems weaker than one expected in the past by extrapolating models which fit SPS data at lower energy. New models such as regeneration models or sequential suppression model better agree with the data, but several questions are still open and will need more data to be fully answered. At RHIC, new sets of p+p data have already been taken in 2005 and 2006 (providing more than 10 times more statistics than got in 2003) and are currently under analysis. In the near future, new Au+Au and d+Au data at $\sqrt{s_{NN}} = 200\,\mathrm{GeV}$ are expected to be stored at RHIC.

References

1. Matsui and Satz, Phys. Lett. **B178**, (1986) 416.
2. Adler et al., PHENIX collaboration, Phys. Rev. Lett. **96**, (2006) 012304.
3. NA50 Collaboration, Eur. Phys. Jour. **C39**, (2005) 335.
4. Vogt, Phys. Rev. **C71**, (2005) 054902.
5. Vogt, arXiv:nucl-th/0507027.
6. Kopeliovich et al., Nucl. Phys. **A696**, (2001) 669.
7. Pereira Da Costa, for the PHENIX Collaboration, Proceedings of Quark Matter 2005, arXiv:nucl-ex/0510051.
8. R. Arnaldi et al., NA60 collaboration, Nucl. Phys. **A774**, (2006) 719.
9. GrandChamp et al., Phys. Rev. Lett. **92**, (2004) 212301 and arXiv:hep-ph/0306077.
10. Capella and Ferreiro, Eur. Phys. Jour. **C42**, (2005) 419.
11. Thews and Mangano, arXiv:nucl-th/0505055 (2006).
12. Adler et al., PHENIX collaboration, Phys. Rev. Lett. **94**, (2005) 082301.
13. Tram, for the PHENIX collaboration, Proceedings of rencontres de Moriond 2006, arXiv:nucl-ex/0606017, and PhD thesis, http://pastel.paristech.org/bib/archive/00001917/
14. Karsh, Kharzeev and Satz, hep-ph/0512239.

Eur. Phys. J. A **31**, 883–885 (2007)
DOI 10.1140/epja/i2006-10254-x

THE EUROPEAN
PHYSICAL JOURNAL A

Special Article – QNP 2006

Out-of-equilibrium quantum field dynamics in external fields

F.J. Cao[a]

Departamento Física Atómica, Molecular y Nuclear, Universidad Complutense de Madrid, Avenida Complutense s/n, 28040 Madrid, Spain and
LERMA, Observatoire de Paris, Laboratoire Associé au CNRS UMR 8112, 61, Avenue de l'Observatoire, 75014 Paris, France

Received: 23 November 2006
Published online: 14 March 2007 – © Società Italiana di Fisica / Springer-Verlag 2007

Abstract. The quantum dynamics of the symmetry-broken $\lambda(\Phi^2)^2$ scalar-field theory in the presence of an homogeneous external field is investigated in the large-N limit. We consider an initial thermal state of temperature T for a constant external field $\mathcal{J}$. A subsequent sign flip of the external field, $\mathcal{J} \to -\mathcal{J}$, gives rise to an out-of-equilibrium nonperturbative quantum field dynamics. We review here the dynamics for the symmetry-broken $\lambda(\Phi^2)^2$ scalar N component field theory in the large-N limit, with particular stress in the comparison between the results when the initial temperature is zero and when it is finite. The presence of a finite temperature modifies the dynamical effective potential for the expectation value, and also makes that the transition between the two regimes of the early dynamics occurs for lower values of the external field. The two regimes are characterized by the presence or absence of a temporal trapping close to the metastable equilibrium position of the potential. In the cases when the trapping occurs it is shorter for larger initial temperatures.

PACS. 11.10.Wx Finite-temperature field theory – 11.15.Pg Expansions for large numbers of components (*e.g.*, $1/N_c$ expansions) – 11.30.Qc Spontaneous and radiative symmetry breaking

1 Introduction

Several important physical systems, as the ultrarelativistic heavy-ion collisions [1] and the early universe [2], present out-of-equilibrium dense concentrations of particles. The presence of these concentrations imply the need of out-of-equilibrium nonperturbative quantum field theory methods, as the large-N limit.

2 The model

We consider N scalar fields, $\boldsymbol{\Phi}$, with a $\lambda(\boldsymbol{\Phi}^2)^2$ self-interaction in the presence of an external field $\mathcal{J}$. The action and the Lagrangian density are given by

$$S = \int \mathrm{d}^4 x \mathcal{L}\,, \tag{1}$$

$$\mathcal{L} = \frac{1}{2}[\partial_\mu \boldsymbol{\Phi}(x)]^2 - \frac{1}{2}m^2 \boldsymbol{\Phi}^2 - \frac{\lambda}{8N}(\boldsymbol{\Phi}^2)^2 - \frac{m^4 N}{2\lambda} + \mathcal{J}\boldsymbol{\Phi}. \tag{2}$$

We restrict ourselves to the case where the symmetry is spontaneously broken, *i.e.*, $m^2 < 0$; and we mainly consider small coupling constants λ, because this slows the dynamics and allows a better study of its different parts.

---

[a] e-mail: `francao@fis.ucm.es`

We consider here the evolution of a initial thermal state of temperature T after a flip in the homogeneous external field $\mathcal{J} \to -\mathcal{J}$. We can choose the axes in the N-dimensional internal space such that

$$\mathcal{J} = \begin{cases} (\sqrt{N}J, 0, \ldots, 0) & \text{for } t \leq 0, \\ (-\sqrt{N}J, 0, \ldots, 0) & \text{for } t > 0. \end{cases} \tag{3}$$

For an initial thermal state we have an expectation value parallel to the external field. The following decomposition can be done:

$$\boldsymbol{\Phi}(x) = (\sigma(x), \boldsymbol{\pi}(x)) = \left(\sqrt{N}\phi(t) + \chi(x), \boldsymbol{\pi}(x)\right), \tag{4}$$

with $\sqrt{N}\phi(t) = \langle \sigma(x) \rangle$; thus, $\langle \chi(x) \rangle = 0$. While in the remaining $N - 1$ directions transversal to the expectation value, $\langle \boldsymbol{\pi}(x) \rangle = 0$.

3 Evolution equations in the large-N limit

As we have one direction parallel to the expectation value and $N - 1$ transversal, the fluctuations in the transverse directions dominate in the large-N limit, while those in the longitudinal direction only contribute to the evolution equations as corrections of order $1/N$.

The large-N limit provides explicit evolution equations for the expectation value and the modes of the quantum fluctuations. See refs. [3] and [4] for the detailed expressions. We will discuss here the results obtained with this evolution equations, in particular the effects of the initial temperature. In order to simplify the expression of the results, we introduce the following adimensional variables:

$$\tau \equiv |m|t\,, \quad \eta(\tau) \equiv \sqrt{\tfrac{\lambda}{2}}\,\frac{\phi(t)}{|m|}\,, \quad \jmath \equiv \sqrt{\tfrac{\lambda}{2N}}\,\frac{\mathcal{J}}{|m|^3}\,, \tag{5}$$

$$g \equiv \frac{\lambda}{8\pi^2}\,, \quad \beta \equiv \frac{\beta_d}{\cdot|m|}\,, \quad g\Sigma(\tau) \equiv \frac{\lambda}{2|m|^2}\frac{\langle\pi^2\rangle(t)}{N}\,. \tag{6}$$

The fact that the initial state is not the ground state but a thermal state increases the initial value of $g\Sigma$. It can be shown that $\Sigma(0)$ is approximately given by

$$\Sigma(0) = \Sigma^{T=0}(0) + \frac{\pi^2}{3}\beta^{-2}\,. \tag{7}$$

$\Sigma^{T=0}(0)$ is the zero-temperature value, that for $g \ll 1$ only depends on j (see ref. [3]). The second term on the right-hand side of eq. (7) is the thermal contribution computed in the hard thermal loop approximation [5]. This expression for $\Sigma(0)$ implies that it is greater for larger initial temperatures.

4 Effective dynamical potential for the expectation value

We can define a potential

$$V_{de;\tau>0}(\eta,\Sigma) \equiv \frac{\eta^4}{4} - \frac{\eta^2}{2} + \frac{1}{2}\eta^2 g\Sigma - \frac{g\Sigma}{2} + \frac{(g\Sigma)^2}{4} + \frac{1}{4} + j\eta\,. \tag{8}$$

that can be interpreted as a dynamical effective potential because the evolution equation for η (in the large-N limit) can be written as [3,4]

$$\ddot\eta(\tau) = -\frac{\partial}{\partial\eta}V_{de;\tau>0}(\eta,\Sigma)\,. \tag{9}$$

It must be stressed that $V_{de;\tau>0}$ is an effective potential *only* for η (and *not* for the modes).

5 Equilibrium states for the dynamical potential

The dynamical effective potential for $\tau \leq 0$ can be defined as $V_{de;\tau\leq0}(\eta,\Sigma) = V_{de;\tau>0}(\eta,\Sigma) - 2j\eta$. Therefore, the stationary states for the initial dynamics for times $\tau \leq 0$ (before the external field has been flipped) are the solutions of $V'_{de;\tau\leq0}(\eta) = 0$ (the prime means η derivative). Thus, they verify

$$\eta^3 + (-1 + g\Sigma)\eta - j = 0\,. \tag{10}$$

For small external fields,

$$j < j_d \equiv 2\sqrt{\frac{(1 - g\Sigma)}{27}}\,, \tag{11}$$

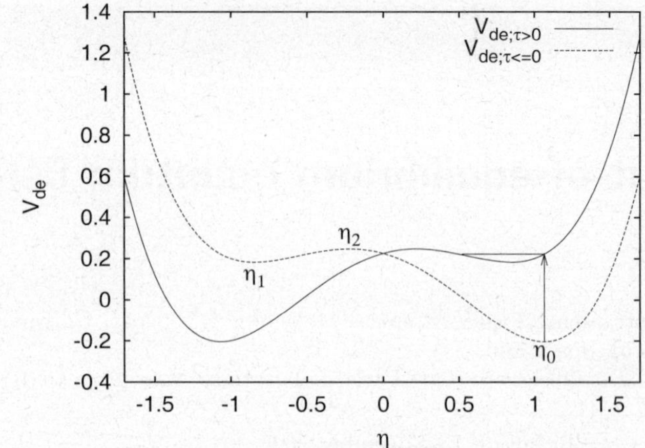

Fig. 1. Dynamical effective potential $V_{de;\tau\leq0}$ and $V_{de;\tau>0}$ as a function of η for $g\Sigma(0) = 0.05$. The value of the external field is $j = 0.2$. The positions of the global minimum η_0, the local minimum η_1, and the local maximum η_2 are shown. There is a potential barrier ($j < j_d$), and the systems gets temporally trapped ($j < j_c$).

we have three roots. There is a global minimum that corresponds to a stable equilibrium state, a local minimum that corresponds to a metastable equilibrium state, and a local maximum (unstable equilibrium) (see fig. 1). On the other hand, for larger external fields ($j > j_d$) there is a single extreme that corresponds to a global minimum (stable equilibrium).

As a larger initial temperature increases $g\Sigma(0)$, it contributes to restoring the symmetry and as consequence makes the value of j_d lower (see eq. (11)).

We consider here the more interesting case $j < j_d$, where a potential barrier is present.

6 Early-time dynamics and dynamical regimes

After the initial flip of the external field sign at $\tau = 0$ the positions of the absolute minimum and the relative minimum of the potential are interchanged, and the state of the system becomes a metastable state (fig. 1). The existence of a potential barrier gives rise to two different dynamical regimes. In the first one the system can directly overcome the barrier, and rapidly reaches the neighborhoods of the global minimum. While in the second regime the system cannot overcome the barrier directly, and it gets temporally trapped close to the metastable state. The external field critical value j_c that separates the two regimes, untrapped $|j| > j_c$, or trapped $|j| < j_c$, is

$$j_c(\beta) = \frac{j_c^{T=0}}{\left[1 - \left(\frac{\beta_c^{J=0}}{\beta}\right)^2\right]^{3/2}} \tag{12}$$

with $j_c^{T=0} = \sqrt{2\,\frac{(13^2+15\sqrt5)}{19^3}} = 0.243019\ldots$, and $\beta_c^{J=0} = \pi\sqrt{\frac{g}{3}}$. Therefore, we see that if the initial temperature is

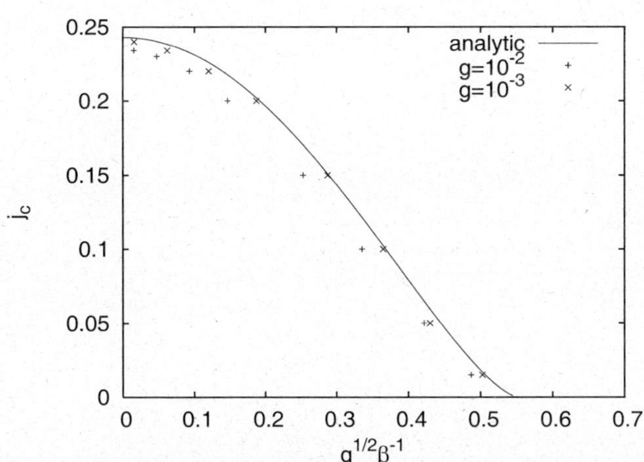

Fig. 2. Critical external field j_c as a function of β^{-1} for $g = 10^{-2}$ and $g = 10^{-3}$ obtained from numerical simulations (symbols) and from the analytical formula in eq. (12) (full line). There are two regions in the (j, β^{-1})-plane corresponding to the two dynamical regimes: for $j < j_c$ the system is temporally trapped in the metastable state; for $j > j_c$ the system rapidly reaches the neighborhood of the global minimum of the potential.

higher enough the system will not be trapped. In addition, eq. (12) states how the trapping can disappear due to a combination of the effects of both the initial temperature and the external field (see fig. 2).

When the trapping is present spinodal instabilities made the modes grow and finally their backreaction allows the system to go close to the stable state. A detailed analysis of the spinodal instability allows to obtain the trapping time, or spinodal time τ_s (see refs. [3,4]). At early times ($\tau < \tau_s$), the effective squared mass oscillates with a negative average, $-\mu^2 \simeq -j$, implying the growth of the modes due to spinodal instability. This implies a quasiexponential growth of $g\Sigma(\tau)$ for $\tau < \tau_s$

$$g\Sigma_s(\tau) \approx \begin{cases} \frac{2}{\beta\mu} g\Sigma_s^{T=0}(\tau) & \text{for } \beta^{-1} \gg \mu, \\ g\Sigma_s^{T=0}(\tau) & \text{for } \beta^{-1} \ll \mu, \end{cases} \quad (13)$$

with

$$g\Sigma_s^{T=0}(\tau) \approx \frac{g\sqrt{\pi}\mu e^{2\tau\mu}}{8\tau^{3/2}} \quad (14)$$

the value for zero temperature. After a certain time, the spinodal time τ_s, the quantum and thermal effects start to be important in the dynamics, $g\Sigma_s(\tau_s)$ compensates $-\mu^2$ and the exponential growth of the mode functions stops, then the mode functions start to have an oscillatory behavior. Thus, the spinodal time τ_s is defined as

$$g\Sigma_s(\tau_s) = \mu^2 . \quad (15)$$

A good approximate expression for the spinodal time is

$$\tau_s = \begin{cases} \frac{1}{2\mu}\left[\log\left(\frac{8}{g\sqrt{\pi}}\right) - \log\left(\frac{2}{\beta\mu}\right) + \frac{3}{2}\log(\mu\tau_s)\right] & \text{for } \beta^{-1} \gg \mu \\ \frac{1}{2\mu}\left[\log\left(\frac{8}{g\sqrt{\pi}}\right) + \frac{3}{2}\log(\mu\tau_s)\right] & \text{for } \beta^{-1} \ll \mu \end{cases}$$

$$(16)$$

For $\beta^{-1} \gg \mu$ we see that a higher initial temperature implies a shorter trapping period.

7 Intermediate-time dynamics

After the early dynamics described in the previous section, the system enters a quasiperiodic regime, both for $j < j_c$ and for $j > j_c$. The system at this intermediate times presents a clear separation between fast variables and slow variables. $\eta(\tau)$, $g\Sigma(\tau)$, and the effective squared mass oscillate fast, while the amplitude of their oscillations slowly decreases. This quasiperiodic regime in the large-N limit evolution equations is present both for zero and nonzero initial temperatures. (See refs. [3,4].)

8 Conclusions

One of the main features of the dynamics is the existence in some cases of a trapping stage in the dynamics that have been characterized. In particular, we have seen that the main influence of the initial temperature is to contribute to the restoration of the symmetry of the potential, and as a consequence a higher initial temperature shortens or even avoids the initial trapping. For intermediate times we have found a quasiperiodic regime. However, how much this dynamics will be modified by the inclusion of next-to-leading-order terms in the large-N approximation [6] is still an open question.

We thank H.J. de Vega and M. Feito for fruitful collaborations in this problem. We acknowledge financial support from the Ministerio de Educación y Ciencia through Research Projects Nos. BFM2003-02547/FISI and FIS2006-05895.

References

1. J.W. Harris, B. Muller, Annu. Rev. Nucl. Part. Sci. **46**, 71 (1996); K. Geiger, Phys. Rep. **258**, 237 (1995); X.N. Wang, Phys. Rep. **280**, 287 (1997); M.H. Thoma, in *Quark Gluon Plasma 2*, edited by R.C. Hwa (World Scientific, Singapore, 1995); Robert D. Pisarski, in the *Proceedings of the International School of Astrophysics D. Chalonge*, edited by N. Sánchez, A. Zichichi (Kluwer Academic Publishers, Dordrecht, 1998) p. 195.
2. F.J. Cao, H.J. de Vega, N.G. Sánchez, Phys. Rev. D **70**, 083528 (2004); D. Boyanovsky, F.J. Cao, H.J. de Vega, Nucl. Phys. B **632**, 121 (2002); D. Boyanovsky, D. Cormier, H.J. de Vega, R. Holman, S.P. Kumar, Phys. Rev. D **57**, 2166 (1998).
3. F.J. Cao, H.J. de Vega, Phys. Rev. D **65**, 045012 (2002).
4. F.J. Cao, M. Feito, Phys. Rev. D **73**, 045017 (2006).
5. M. Le Bellac, *Thermal Field Theory* (Cambridge University Press, New York, 2000).
6. J. Berges, Nucl. Phys. A **699**, 847 (2002); J. Berges, hep-ph/0409233; S. Borsányi, hep-ph/0512308.

Chiral condensate thermal evolution at finite baryon chemical potential within ChPT

R. García Martín and J. R. Peláez

Dpto. de Física Teórica II, UCM, 28040 Madrid, Spain

Abstract. We present a model independent study of the chiral condensate evolution in a hadronic gas, in terms of temperature and baryon chemical potential. The meson-meson interactions are described within Chiral Perturbation Theory and the pion-nucleon interaction by means of Heavy Baryon Chiral Perturbation Theory, both at one loop. Together with the virial expansion, this provides a model independent systematic expansion at low temperatures and chemical potentials, which includes the physical quark masses. This can serve as a guideline for further studies on the lattice. We also obtain estimates of the critical line of temperature and chemical potential where the chiral condensate melts, which systematically lie somewhat higher than recent lattice calculations but are consistent with several hadronic models.

PACS. 11.30.Rd – 11.30.Qc – 12.38.Aw – 12.39.Fe – 11.10.Wx

On a recent work [1] we approach the question of the phase diagram of QCD. In particular, we study the transition from the hadronic phase, in which the chiral symmetry is spontaneously broken, to a phase in which it is restored. This is done within Chiral Perturbation Theory (ChPT), the low energy effective theory of QCD. We also gave estimates for the melting curve of the quark condensate in presence of a baryon chemical potential.

Let us briefly describe ChPT. The spontaneous chiral symmetry breaking of QCD requires the existence of eight massless Goldstone Bosons (GB), that can be identified with the pions, kaons and eta. These are thus the most relevant degrees of freedom at low energies. In addition, there is an explicit symmetry breaking due to the non-vanishing quark masses that give rise to a small mass for the pions, kaons and eta, which are thus just pseudo-GB. ChPT is built as the most general derivative and mass expansion over the spontaneous symmetry breaking (SSB) scale $\Lambda_\chi \equiv 4\pi f_\pi \simeq 1.2$ GeV which respects the symmetry constraints of QCD. Baryons can be included in this formalism, although its treatment is more involved due to their large masses. Within Heavy Baryon ChPT (HBChPT) [3] this problem is overcome for the meson-baryon interaction by an additional expansion on inverse powers of the baryon mass. This approach has proven very successful, and works remarkably well within the meson sector, and with a somewhat slower convergence in the meson-baryon sector.

Next, the thermodynamics of the hadron gas is described by the virial expansion [4,5]. This is a low density series expansion for the grand canonical potential of a relativistic interacting multicomponent gas. We will assume that only the strong interactions are relevant, and that the baryon density defined as $n_B - n_{\bar{B}}$ is conserved. The thermodynamics of the gas will thus depend on the temperature T (usually written in terms of its inverse $\beta = 1/T$) and a baryon chemical potential μ_B. The virial expansion, written in terms of the pressure, reads:

$$\beta P = \sum_i B_i^{(1)}\,\xi_i + B_i^{(2)}\,\xi_i^2 + \sum_i \sum_{j \geq i} B_{ij}^{int}\,\xi_i \xi_j + \dots, \quad (1)$$

where $\xi_i = \exp\beta(\mu_i - M_i)$, with M_i the mass of the i-th species, $\mu_i = 0, \pm\mu_B$ for mesons/(anti)baryons respectively. The free virial coefficients

$$B_i^{(n)} = \frac{g_i \eta_i^{n+1}}{2\pi^2} \int_0^\infty dp\, p^2\, e^{-n\beta(\sqrt{p^2+M_i^2}-M_i)}, \quad (2)$$

, with $\eta_i = \pm 1$ for bosons/mesons and g_i the degeneracy of the i-th species, correspond simply to the series expansion of the pressure for a free gas, while the interactions are taken into account in the term:

$$B_{ij}^{int} = \frac{e^{\beta(M_i+M_j)}}{2\pi^3} \int_{M_i+M_j}^\infty dE\, E^2 K_1(\beta E)\Delta^{ij}(E), \quad (3)$$

where $K_1(x)$ is the modified Bessel function (of the second kind), and $\Delta^{ij} = \sum_{I,J,S}(2I+1)(2J+1)\delta^{ij}$ are the $ij \to ij$ elastic scattering phase shifts for a state ij with well defined isospin I, total angular momentum J and strangeness S, defined so that $\delta = 0$ at threshold.

The dependence of the chiral condensate with temperature and chemical potential is written in terms of the

"""

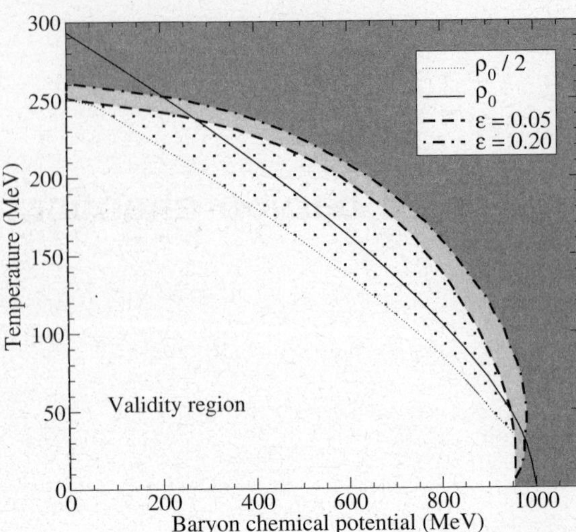

Fig. 1. Crude estimate of the region of validity of the virial expansion for a free gas. We have excluded the regions in which the virial expansion breaks, but also those regions in which nuclear density is large enough so that higher order effects are relevant.

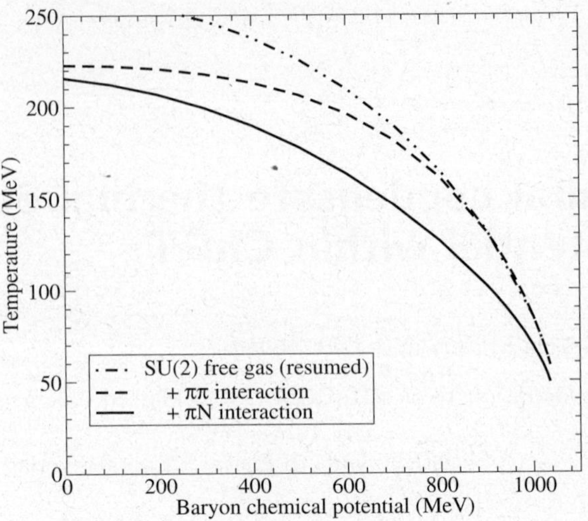

Fig. 2. Melting line in the (μ_B, T) plane of the chiral condensate for an SU(2) gas of hadrons. We show the result for the free gas, but also adding the $\pi\pi$ ChPT SU(2) interaction to one loop, and the line resulting from adding the πN interaction to third order in HBChPT.

pressure of the gas as follows:

$$\langle \bar{q}q \rangle = \langle 0|\bar{q}q|0 \rangle \left(1 + \sum_h \frac{c_h}{2 M_h F^2} \frac{\partial P}{\partial M_h} \right), \qquad (4)$$

where the constant F, which is the pion decay constant in the chiral limit, is introduced for further convenience, and the coefficients $c_h = -F^2 \frac{\partial M_h^2}{\partial \hat{m}} \langle 0|\bar{q}q|0 \rangle^{-1}$ encode the hadron mass dependence on the quark mass. Numerically, these coefficients amount to $c_\pi = 0.9^{+0.2}_{-0.4}$, $c_K = 0.5^{+0.4}_{-0.7}$, $c_\eta = 0.4^{+0.5}_{-0.7}$, for mesons [6], and $c_N = 3.6^{+1.5}_{-1.9}$ for nucleons [1].

First of all, we want to estimate the region of validity of our calculations. For this, we check the points for which the virial expansion breaks down, by computing the difference between the free condensate calculated exactly and with the virial expansion. In Figure 1 we show, as a dashed and a dashed-dotted lines, the points where the virial expansion is off the exact calculation by 5% and 20%, respectively. The expansion rapidly deteriorates after that. Also, as we want to neglect NN interactions, we have to restrict ourselves to the region where nucleon density is small enough. In ref. [7] it is shown that, for densities below $\rho_0/2$, with $\rho_0 \simeq 0.16 \, \text{fm}^{-3}$, and *as far as we are only concerned with the quark condensate*, the πN interaction dominates over the NN interaction. Thus, we restrict the validity of our calculation to the points below the $\rho_0/2$ line. This line is shown in the Figure as a dotted line, together with the ρ_0 line, which is shown for illustration as a continuous line. The white area in Fig.2 corresponds to our estimated "validity region".

In Figure 2 we show in the (μ_B, T) plane the condensate melting line of the free $SU(2)$ gas composed of pions and nucleons as a dash-dot-dotted line. We have then added the effects of the $\pi\pi$ interaction, which yields

the dashed line of the Figure. Finally, the πN interaction was included. We see that the effect of the πN interaction is small, but noticeable (~ 6 MeV at $\mu_B = 0$) at low baryon chemical potentials, and maximum at around $\mu_B \simeq 600 - 700$ MeV, where it produces a decrease in the melting temperature of about 40 MeV with respect to the gas without πN interactions. Up to a chemical potential of $\mu_B = 40 - 50$ MeV, the region of relevance for Relativistic Heavy Ion Collisions, the decrease in the melting temperature amounts roughly to 10 MeV when we include the πN interaction. It is important to note that, although the critical lines are actually extrapolations, at low temperature and chemical potential our calculation is model independent, and should be quite accurate. Still, we show the melting lines since they are useful to visualize and help quantifying the relative size of each contribution we add into the gas.

In a real gas, we should consider all hadrons. We will do this by including them as free particles. The only exception to this will be the kaons and etas, which are abundant enough up to 200 MeV to deserve a separate treatment and include their interaction with a pion. All other interactions are suppressed by Boltzmann (thermal) and c_h/M_h factors, and are therefore not included in our treatment. At very large μ_B the heavier nucleons may not be Boltzmann suppressed, and their interactions may become important, but this is outside the scope of this work. In Figure 3 we show the two lines corresponding to the $SU(2)$ gas with $\pi\pi$ interactions (dashed) and with $\pi\pi$ and πN interactions (continuous). We then plot the line corresponding to a gas in which also kaons, etas, and their interactions with a pion, now in $SU(3)$, are included (dotted line). Note that the effect is pronounced at $\mu_B = 0$, with a decrease of the melting temperature of around ~ 13 MeV, but becomes negligible at very high baryon chemical potential. Finally, we have added the effects of other heav-

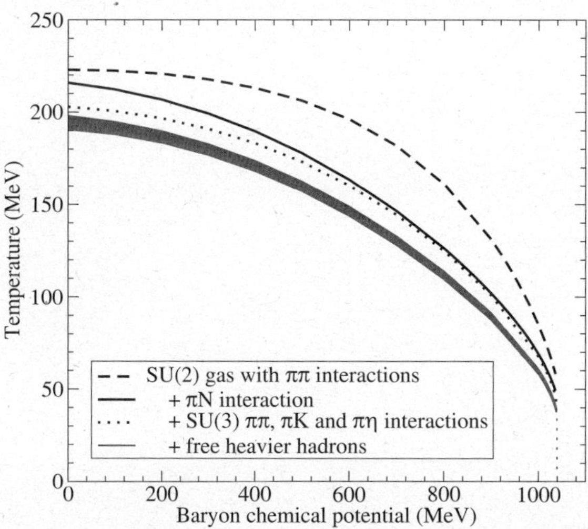

Fig. 3. Our final result for the chiral condensate melting line in the (μ_B, T) plane. Starting from the SU(2) gas with $\pi\pi$ interactions, we show the effect of πN interactions and of adding kaons and etas. The dark area covers the uncertainty due to heavier hadrons estimated as explained in the text.

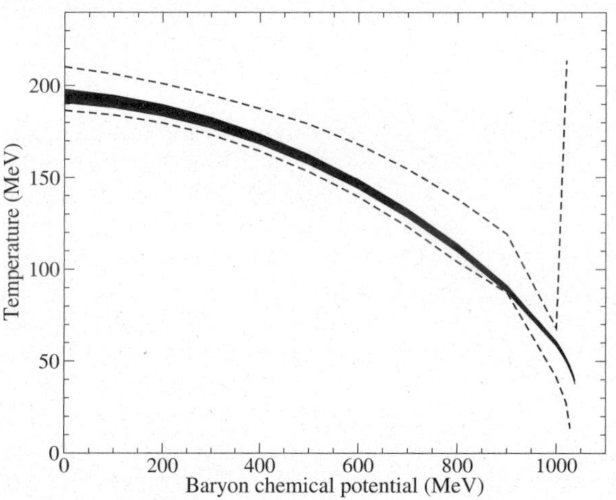

Fig. 4. Analysis of uncertainties. To the dark area, which corresponds to the case described in Fig. 3, we have added the uncertainties coming from the chiral parameters, shown as dashed lines (see [1] for discussion).

ier hadrons, included as free particles, since they suffer a larger Boltzmann suppression. We have estimated that, for heavier hadrons, $\partial M_h / \partial \hat{m} \simeq \alpha N^h_{u,d}$, where α is an adimensional constant and $N^h_{u,d}$ is the number of valence u or d quarks in the hadron. Based on the values of α that we have calculated for the pions, kaons, etas and nucleons, we have estimated that, for heavier hadrons, $\alpha \simeq 0.5 - 2.5$. The band in Figure 3 corresponds to this range of values for α. In order to properly analyse the uncertainties in our calculation, we also have to include the uncertainties that come from our imperfect knowledge of the chiral parameters. This was done by adding in quadrature the errors coming from independently varying each of the parameters. The final results are plotted in Figure 4, where

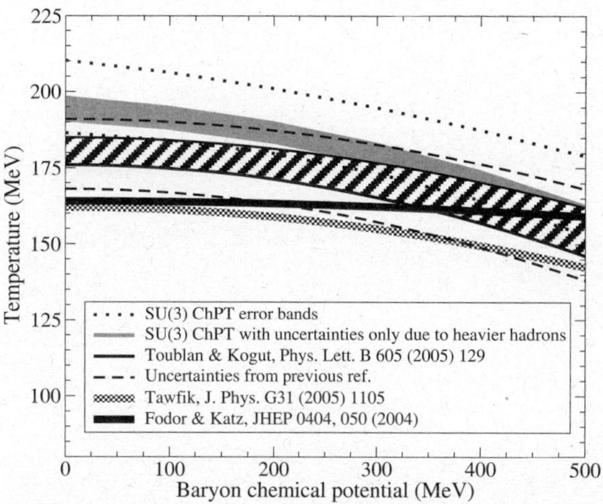

Fig. 5. Comparison between our results and others in the literature. Our melting temperatures are consistent within errors with ref. [8] but systematically higher than [9].

the dark band corresponds to the uncertainties coming from the α parameter in heavy hadrons and the dashed line includes also the uncertainty coming from the chiral parameters. Note that the error is highly asymmetric: even though every contribution lowers the melting temperature, the melting of the condensate accelerates near the melting point, and thus even if the contribution has a similar size to the previous one, its effect on the condensate melting seems smaller.

It is interesting to compare our results with those obtained by using lattice calculations. In Figure 5 we show our curves with their errors, plotted against other results recently appeared in the literature. Our approach agrees, within errors, with the hadron resonance gas model described in [8]. However, ChPT calculations [5,6], including ours [1], yield melting temperatures that are systematically above the results from the lattice calculations [9].

References

1. R. Garcia Martin and J.R. Pelaez, hep-ph/0608320.
2. S. Weinberg, Physica A96, (1979) 327. J. Gasser and H. Leutwyler, Ann. Phys. 158, (1984) 142; Nucl. Phys. B250 (1985) 465,517,539.
3. E. Jenkins and A. V. Manohar, Phys. Lett. B **255**, 558 (1991). V. Bernard, N. Kaiser and U. G. Meissner, Int. J. Mod. Phys. E **4**, 193 (1995)
4. R. Dashen, S. Ma, H.J. Bernstein,Phys. Rev. 187 (1969) 187. G.M. Welke, R. Venugopalan and M. Prakash,Phys. Lett. B245(1990) 137. V.L. Eletsky, J. I. Kapusta and R. Venugopalan,Phys. Rev. D48 (1993)4398. R. Venugopalan, M. Prakash, Nucl. Phys. A546 (1992)718.
5. P. Gerber and H. Leutwyler, Nucl. Phys. B321 (1989) 387.
6. J. R. Pelaez, Phys.Rev.D **66**(2002) 096007.
7. M. Lutz *et al.*, Phys. Lett. B **474**, 7 (2000).
8. D. Toublan and J. B. Kogut, Phys. Lett. B **605** (2005) 129
9. A. Tawfik, J. Phys. G **31** (2005) S1105; Z. Fodor and S. D. Katz, JHEP **0404**, 050 (2004)

Heat conductivity of a pion gas

Antonio Dobado, Felipe J. Llanes-Estrada and Juan M. Torres Rincón

Departamento de Física Teórica I, Universidad Complutense, 28040 Madrid, Spain

Abstract. We evaluate the heat conductivity of a dilute pion gas employing the Uehling-Uehlenbeck equation and experimental phase-shifts parameterized by means of the $SU(2)$ Inverse Amplitude Method. Our results are consistent with previous evaluations. For comparison we also give results for an (unphysical) hard sphere gas.

PACS. 05.20.Dd – 51.20.+d

1 Transport coefficients

Transport coefficients in heavy-ion collisions remain of interest as prospects for direct experimental measurement at RHIC improve. In particular, elliptic flow has been proposed [1] as a tell-tale for viscous effects. There have also recently been a number of papers emphasizing viscosity (for example [2], [3]) following recent insight from conformal field theory [4]. Follow-up studies are under preparation by several groups from the hadron and from the quark-gluon phases in RHIC-theory to see the effect of the phase transition on the viscosity.

Although we have presented a comprehensive study of the shear viscosity of a hadronic gas at low temperature [5], in this brief report we focus on another transport coefficient, the thermal conductivity, associated to energy conservation and flow in the hot gas.

We employ the same notation and conventions as in our earlier publication [5]. In particular we employ the $SU(2)$ Inverse Amplitude Method parametrization of the pion-pion scattering experimental phase shifts (as well as alternative parametrizations to check the sensitivity of the calculation presented). Note we also employ the Landau convention for the flows as opposed to the Eckart convention that has also been widely used [6].

To make this paper minimally self-contained, we collect here a few of the key results and assumptions.

We are working in a pion gas at temperatures well below any phase transition, in the dilute approximation near equilibrium. The equilibrium equation of state yields an enthalpy per unit volume $h = P + \rho$ as

$$h = \frac{g_\pi m_\pi^4}{4\pi^2} \int_0^\infty dx \frac{\sqrt{x}(1 + 4x/3)}{\sqrt{x+1}\left(z^{-1} e^{y(\sqrt{1+x}-1)} - 1\right)} \quad (1)$$

and the particle (pion) density is given by

$$n\left(y = \frac{m_\pi}{T}, z\right) = \frac{g_\pi m_\pi^3}{4\pi^2} \int_0^\infty dx \frac{x^{1/2}}{z^{-1} e^{y(\sqrt{1+x}-1)} - 1} \quad (2)$$

($z = e^{(\mu-m)/T}$ being the fugacity).

We take a constant perturbation around equilibrium proportional to the relativistic temperature-pressure gradient

$$\delta f(\mathbf{p}) = \frac{f_0}{T}\frac{\mathbf{p}}{|\mathbf{p}|} g(|\mathbf{p}|)\left[\boldsymbol{\nabla}T - \frac{T}{h}\boldsymbol{\nabla}P\right] \quad (3)$$

and change the variable to $x = (p/m_\pi)^2$. We perform a polynomial expansion of $g(x)$, and display results for the first order only, $g(x) = A_0$, with A_0 a constant. Convergence is known to be fast. Since pions are not exactly massless, no singularities appear that might require a more careful treatment [7] and the only reason to improve on this variational calculation would be to improve its accuracy beyond the 10% level. This program has been carried out at least for the non-relativistic pion gas viscosity in ref. [8]. We have also made minimum sensitivity checks based on a simple numeric Gram-Schmidt polynomial construction beyond A_0 but will report them elsewhere.

From the standard linearized (Boltzmann) Uehling-Uehlenbeck transport equation one can derive the following for this perturbation near equilibrium:

$$A_0 = \frac{I_1(y,z)}{I_2(y,z)} \quad (4)$$

$$I_1 = 2\pi m_\pi^3 y \int_0^\infty \frac{x dx}{\sqrt{1+x}} f(E(x))\left(\sqrt{1+x} - \frac{h}{m_\pi n}\right) \quad (5)$$

$$I_2 = \frac{1}{4A^2} \int [d\Phi] \left[e^{\beta(\omega - 2\mu)} f f' f_1 f_1'\right.$$

$$\Delta_i\left(\hat{\mathbf{p}}_i\left[1 - e^{-\beta(E_i - \mu)}\right]\right) \cdot \Delta_j\left(\hat{\mathbf{p}}_j\left[1 - e^{-\beta(E_j - \mu)}\right]\right)\right] \quad (6)$$

where ω denotes the total energy ($\omega = E + E_1 = E' + E_1'$), A is the inverse normalization constant of the distribution functions

$$A = \xi_\pi^{-1} = \frac{g_\pi}{(2\pi)^3}$$

($g_\pi = 3$ for isospin degeneracy) and the last term is a shorthand for the symmetrization

$$\Delta_i \left(\hat{\mathbf{p}}_\mathbf{i} \left[1 - e^{-\beta(E_i - \mu)} \right] \right) = \tag{7}$$

$$\left(\hat{\mathbf{p}}'_\mathbf{1} \left[1 - e^{-\beta(E'_1 - \mu)} \right] + \hat{\mathbf{p}}' \left[1 - e^{-\beta(E' - \mu)} \right] - \right.$$

$$\left. \hat{\mathbf{p}}_\mathbf{1} \left[1 - e^{-\beta(E_1 - \mu)} \right] - \hat{\mathbf{p}} \left[1 - e^{-\beta(E - \mu)} \right] \right) .$$

Once A_0 has been thus computed, the heat conductivity follows as

$$\kappa = -\frac{h m_\pi^3 2\pi A}{3nT} A_0 l_1(y, z) \tag{8}$$

with

$$l_1(y, z) = \int_0^\infty \frac{x\,dx}{\sqrt{1+x}\left(z^{-1} e^{y(\sqrt{1+x}-1)} - 1 \right)} \tag{9}$$

To evaluate the integral in I_2 in Eq. (4) we employ a Montecarlo computer program. Without loss of generality we can choose the total momentum directed along the OZ axis. The independent variables can be taken as P and ω (respectively the total momentum and energy in a binary collision), $p = |\mathbf{p}|$ and $p' = |\mathbf{p}'|$ (the incoming and outgoing pion momenta for one of the two pions, the other being fixed by momentum conservation). Finally the outgoing pion with momentum p' does not need to be in the same plane as $\mathbf{P}$ and $\mathbf{p}$, and therefore we need an azimuthal angle ϕ' for this momentum. The cosines of the polar angles associated to $\mathbf{p}$ and $\mathbf{p}'$ are fixed by the energy conservation relation

$$\delta(\omega - E - E_1) = \frac{E_1}{pP} \delta(x - x_0)$$

and the associated integrals are immediately performed, with

$$x_0 = \frac{P^2 + \omega(2E - \omega)}{2pP} \tag{10}$$

$$x'_0 = \frac{P^2 + \omega(2E' - \omega)}{2p'P} . \tag{11}$$

In addition, rigid rotations around $\mathbf{P}$ parametrized by ϕ are trivial and lead to a factor of 2π, and global rotations of the system (the angles associated to $\mathbf{P}$) are also trivial and yield another 4π.

Putting all together, the phase space integration weighted with the square scattering amplitude can be expressed as

$$\int [d\Phi] = \int v_{rel}\,d\sigma\,d\mathbf{p}\,d\mathbf{p}' = \tag{12}$$

$$\int dP dp dp' d\omega d\phi' |T|^2 \frac{4\pi 2\pi}{4\pi^2} \frac{1}{4EE_1} \frac{p'}{E'_1} \frac{p'}{E'} \frac{E_1}{pP} \frac{E'_1}{p'P} P^2 p^2$$

Our numerical results for the thermal conductivity are displayed in the figures.

In Fig. 1 we plot the heat conductivity as a function of temperature, at several μ's in the hard-sphere scattering

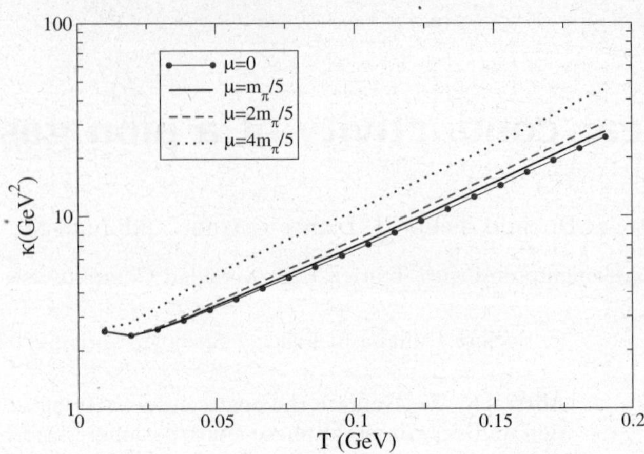

Fig. 1. We show the μ dependence of the conductivity for the hard-sphere gas appoximation

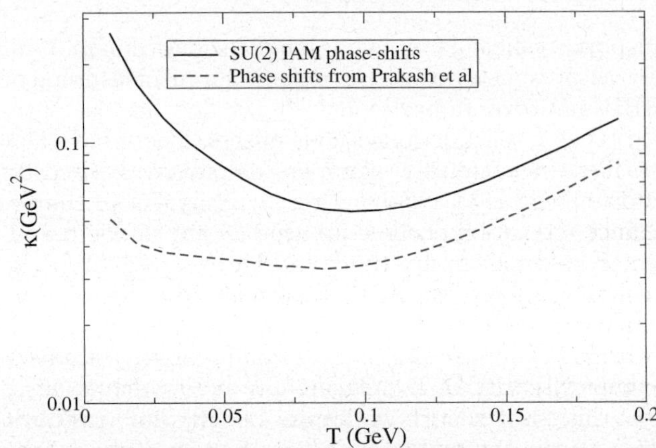

Fig. 2. Thermal conductivity κ from SU(2) chiral perturbation theory unitarized by means of the Inverse Amplitude Method. The pion chemical potential is taken to be $\mu_\pi = 0$.

case, that is, when we use a constant scattering amplitude based on Weinberg's low energy theorem:

$$|T|^2 = \frac{23}{3} \frac{m_\pi^4}{f_\pi^4} \tag{13}$$

For no μ does the heat conductivity diverge at low temperature to the reach of our Montecarlo. In the high temperature limit we expect on dimensional grounds, and numerically find, $\kappa = A \cdot T^2$, where the numerical constant is close to $A = 685$.

In Fig. 2, employing IAM phase shifts, that unitarize a higher order $O(p^4)$ in chiral perturbation theory [9], and therefore introduce a dependence with the energy for the pion-pion scattering amplitude, we now find that there is a minimum value near $T_c = 100\,\mathrm{MeV}$ and $\kappa \to \infty$ when $T \to 0$. The high T scaling is close to the dimensional analysis

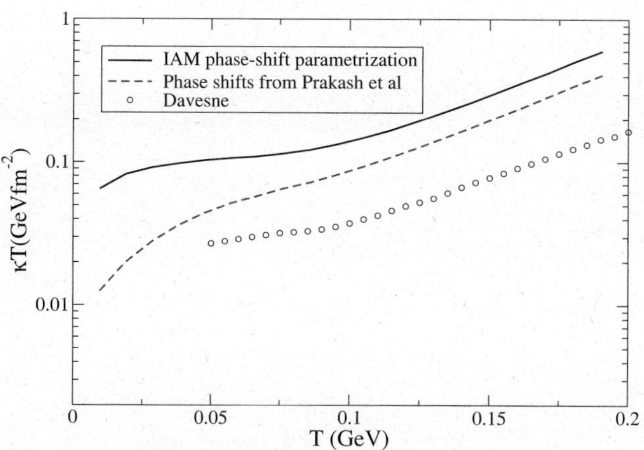

Fig. 3. Thermal conductivity times temperature $T\kappa$ from SU(2) chiral perturbation theory unitarized by means of the Inverse Amplitude Method. The pion chemical potential is taken to be $\mu_\pi = 0$. We also plot the computation with the phase-shifts of [10] and compare them to D. Davesne's evaluation.

even at the moderately large plotted temperatures,

$$\kappa \propto 3.4\,T^2 \ .$$

In the figure we also show our own Montecarlo-based calculation of the heat conductivity employing the simple resonance saturation parametrization for the isoscalar and isovector phase shifts by the Brookhaven group [10], for comparison and to give an idea of the sensitivity to the parametrization.

Next in Fig. 3 we show the comparison with the existing computation of D. Davesne [11] of both calculations.

Note also a recent calculation by [12] that employs chiral perturbation theory alone (without unitarization). This approach features an ever-increasing cross section, unphysical behavior that artificially shortens the mean free path and therefore lowers the transport coefficients. Therefore the validity of the results is limited to low temperatures. However a direct comparison with this approach is difficult since the authors directly include the effect of baryon resonances.

In conclusion, from published calculations and our own contribution we see that we have a fair theoretical idea on how the thermal conductivity of a pure pion gas should behave with temperature. We have further studied its behavior with chemical potential that is important because chemical freeze-out is expected to occur before thermal freeze out in Heavy-Ion Collisions, the reason being that the low energy hadron interactions are elastic scattering, largely mediated by resonances (automatically incorporated in our Inverse Amplitude Method Phase shifts).

This work has been supported by research grants FPA 2004-02602, 2005-02327, PR27/05-13955-BSCH and has been presented at the IVth International Conference on Quarks and Nuclear Physics, Madrid, Spain, June 5th-10th 2006. We thank A. Gómez Nicola and D. Fernández-Fraile for informing us that their own calculation along the lines of [13] in chiral perturbation theory also yields analogous results.

References

1. D. Teaney, Phys. Rev. C **68** (2003) 034913 [arXiv:nucl-th/0301099].
2. H. Liu, D. Hou and J. Li, arXiv:hep-ph/0609034.
3. E. Nakano, arXiv:hep-ph/0612255.
4. G. Policastro, D. T. Son and A. O. Starinets, Phys. Rev. Lett. **87** (2001) 081601 [arXiv:hep-th/0104066].
5. A. Dobado and F. J. Llanes-Estrada, Phys. Rev. D **69** (2004) 116004 [arXiv:hep-ph/0309324].
6. S. Gavin, Nucl. Phys. A **435** (1985) 826; W. A. van Leeuwen and S. R. de Groot, Physica **51** (1971) 1.
7. C. Manuel, A. Dobado and F. J. Llanes-Estrada, JHEP **0509** (2005) 076 [arXiv:hep-ph/0406058].
8. A. Dobado and S. N. Santalla, Phys. Rev. D **65** (2002) 096011 [arXiv:hep-ph/0112299].
9. A. Dobado, M. J. Herrero and T. N. Truong, Phys. Lett. B **235** (1990) 134.
10. M. Prakash, M. Prakash, R. Venugopalan and G. Welke, Phys. Rept. **227** (1993) 321.
11. D. Davesne, Phys. Rev. C **53** (1996) 3069.
12. S. Muroya and N. Sasaki, Prog. Theor. Phys. **113** (2005) 457 [arXiv:nucl-th/0408055].
13. D. Fernandez-Fraile and A. Gómez-Nicola, arXiv:hep-ph/0610197.

Finito di stampare nel mese
di maggio 2007 da
Compositori Industrie Grafiche